Satellite Communications

Frequency-Response Methods in Control Systems, *Edited by A.G.J. MacFarlane*
Programs for Digital Signal Processing, *Edited by the Digital Signal Processing Committee*
Automatic Speech & Speaker Recognition, *Edited by N. R. Dixon and T. B. Martin*
Speech Analysis, *Edited by R. W. Schafer and J. D. Markel*
The Engineer in Transition to Management, *I. Gray*
Multidimensional Systems: Theory & Applications, *Edited by N. K. Bose*
Analog Integrated Circuits, *Edited by A. B. Grebene*
Integrated-Circuit Operational Amplifiers, *Edited by R. G. Meyer*
Modern Spectrum Analysis, *Edited by D. G. Childers*
Digital Image Processing for Remote Sensing, *Edited by R. Bernstein*
Reflector Antennas, *Edited by A. W. Love*
Phase-Locked Loops & Their Application, *Edited by W. C. Lindsey and M. K. Simon*
Digital Signal Computers and Processors, *Edited by A. C. Salazar*
Systems Engineering: Methodology and Applications, *Edited by A. P. Sage*
Modern Crystal and Mechanical Filters, *Edited by D. F. Sheahan and R. A. Johnson*
Electrical Noise: Fundamentals and Sources, *Edited by M. S. Gupta*
Computer Methods in Image Analysis, *Edited by J. K. Aggarwal, R. O. Duda, and A. Rosenfeld*
Microprocessors: Fundamentals and Applications, *Edited by W. C. Lin*
Machine Recognition of Patterns, *Edited by A. K. Agrawala*
Turning Points in American Electrical History, *Edited by J. E. Brittain*
Charge-Coupled Devices: Technology and Applications, *Edited by R. Melen and D. Buss*
Spread Spectrum Techniques, *Edited by R. C. Dixon*
Electronic Switching: Central Office Systems of the World, *Edited by A. E. Joel, Jr.*
Electromagnetic Horn Antennas, *Edited by A. W. Love*
Waveform Quantization and Coding, *Edited by N. S. Jayant*
Communication Satellite Systems: An Overview of the Technology, *Edited by R. G. Gould and Y. F. Lum*
Literature Survey of Communication Satellite Systems and Technology, *Edited by J. H. W. Unger*
Solar Cells, *Edited by C. E. Backus*
Computer Networking, *Edited by R. P. Blanc and I. W. Cotton*
Communications Channels: Characterization and Behavior, *Edited by B. Goldberg*
Large-Scale Networks: Theory and Design, *Edited by F. T. Boesch*
Optical Fiber Technology, *Edited by D. Gloge*
Selected Papers in Digital Signal Processing, II, *Edited by the Digital Signal Processing Committee*
A Guide for Better Technical Presentations, *Edited by R. M. Woelfle*
Career Management: A Guide to Combating Obsolescence, *Edited by H. G. Kaufman*
Energy and Man: Technical and Social Aspects of Energy, *Edited by M. G. Morgan*
Magnetic Bubble Technology: Integrated-Circuit Magnetics for Digital Storage and Processing, *Edited by H. Chang*
Frequency Synthesis: Techniques and Applications, *Edited by J. Gorski-Popiel*
Literature in Digital Processing: Author and Permuted Title Index (Revised and Expanded Edition), *Edited by H. D. Helms,*
    J. F. Kaiser, and L. R. Rabiner
Data Communications via Fading Channels, *Edited by K. Brayer*
Nonlinear Networks: Theory and Analysis, *Edited by A. N. Willson, Jr.*
Computer Communications, *Edited by P. E. Green, Jr. and R. W. Lucky*
Stability of Large Electric Power Systems, *Edited by R. T. Byerly and E. W. Kimbark*
Automatic Test Equipment: Hardware, Software, and Management, *Edited by F. Liguori*
Key Papers in the Development of Coding Theory, *Edited by E. R. Berkekamp*
Technology and Social Institutions, *Edited by K. Chen*
Key Papers in the Development of Information Theory, *Edited by D. Slepian*
Computer-Aided Filter Design, *Edited by G. Szentirmai*
Laser Devices and Applications, *Edited by I. P. Kaminow and A. E. Siegman*
Integrated Optics, *Edited by D. Marcuse*
Laser Theory, *Edited by F. S. Barnes*
Digital Signal Processing, *Edited by L. R. Rabiner and C. M. Rader*
Minicomputers: Hardware, Software, and Applications, *Edited by J. D. Schoeffler and R. H. Temple*
Semiconductor Memories, *Edited by D. A. Hodges*
Power Semiconductor Applications, Volume II: Equipment and Systems, *Edited by J. D. Harnden, Jr. and F. B. Golden*
Power Semiconductor Applications, Volume I: General Considerations, *Edited by J. D. Harnden, Jr. and F. B. Golden*
A Practical Guide to Minicomputer Applications, *Edited by F. F. Coury*

Satellite Communications

Edited by

Harry L. Van Trees
**Principal Deputy Assistant Secretary for Communications,
Command, Control, and Intelligence
U.S. Department of Defense**

A volume in the IEEE PRESS Selected Reprint Series,
prepared under the sponsorship of the
IEEE Communications Society.

**IEEE
PRESS**

The Institute of Electrical and Electronics Engineers, Inc. New York

Contents

Preface

In the short period of time since the introduction of the first operational system in 1965, satellite communications has changed the entire nature of communications. In the international area, the most spectacular product is global television, with major events such as the Olympics and World Cup soccer being broadcast live to viewers in countries around the world. However, the traffic in the international system is dominated by voice, and the links to more than 100 countries have had a major impact on the communications patterns of both businesses and individuals. In the military area, satellite communications has become a vital part of strategic and tactical systems, and current plans project an even larger role in the future. In the mobile area, reliable communication to ships is now available on a worldwide basis. Domestic satellites have provided communications to remote Canadian villages as well as supplying voice trunking and wideband channels to many users.

While the accomplishments in the past have been dramatic, the prognosis for the future is even more exciting. With the advent of the space shuttle, the improvement of various technologies, and the increased demands for information transmission, satellite communications will continue to expand.

In spite of this widely recognized importance of satellite communications, a comprehensive book on satellite communications is not available. This reprint book is intended to be a partial step in filling this gap. The book is designed to fulfill four objectives. The first objective is to provide a general integrated framework of satellite communications and, within that framework, to identify a set of key reprint papers and supporting bibliography that will serve as a reference volume for both satellite communications engineers and researchers. The second objective is to provide a suitable reference volume for a short course (typically one week) in an advanced or continuing education program. The third objective is to provide a reference text for a one-semester graduate course in satellite communications. To help fulfill these latter two objectives, tutorial level introductions were written for many of the sections, and several tutorial articles are included. In addition, several papers are included which are of more interest to graduate students than practicing engineers. A fourth and somewhat subsidiary objective is to provide an overview of satellite communications for communications managers or users. To fulfill this objective, several overview papers are included.

The papers were selected to emphasize the communications aspects of satellite communications. In order to conserve space, many important questions in the overall satellite system design are omitted. For example, there is no discussion of launch vehicles, orbital dynamics, and very little discussion of the actual design of the spacecraft. I have tried to achieve a balance between papers dealing with current operational systems and technology and papers dealing with the techniques and technologies that will be useful in future systems.

In order to select the reprint papers, an overall framework for satellite communications was developed first. Then, papers which fit within the framework and contributed to one or more of the above objectives were selected. Unfortunately, due to stringent page limitations and the requirement for balance, I have had to omit many excellent papers. In addition, I have attempted to avoid an overlap with two readily available references, the Special Issue of the Proceedings of the IEEE on Satellite Communications, published in March 1977 and the AIAA Reprint Book on Satellite Communications (see the references for Part I). However, in some cases common papers are included in order to make this volume as self-contained as possible. The emphasis is on newer papers in the field, although classical papers are included in areas where they are still relevant. Papers from professional journals are used where possible. However, the satellite communications area appears to be somewhat different from some technical disciplines in that a number of interesting results are available in conference papers and commercial magazines. Papers from these sources are included where appropriate. In addition to the reprint papers, several invited papers were solicited. In some cases it was not possible to find an appropriate general paper, and I have tried to accommodate this by comments in the introduction. I have tried to achieve a balanced selection that will be useful to readers interested in various types of systems.

The book is divided into eight parts:

    I   General
    II Systems
    III The Communication System
    IV Spacecraft Design and Technology
    V   Propagation
    VI Earth Stations
    VII The Future
    VIII Regulatory, Political, Legal, and Economic Issues.

In each part there is an introduction to the material, including a list of references, and a set of reprint papers. In addition, there is a bibliography at the end of the book which is indexed to the parts. The larger parts of the book are divided into sections (and in some cases, subsections), each of which may contain an introduction and separate bibliography.

The first part of the book contains papers which provide general overviews of satellite communication and serves as an introduction to the remainder of the book.

The second part "Systems," discusses current systems and systems that will be implemented in the near future. As there are some 50 satellite systems in operation or in advanced planning stage, it was necessary to pick a representative collection. The first paper provides a synopsis of commercial satellite communications; the second discusses the SBS system, an advanced U.S. domestic system; the third paper describes a next-generation military satellite, the DSCS-III; and the last describes an innovative experimental satellite, the ATS-6.

The third part, "The Communications System," is the major element of the book. The introduction to this part sets the framework for the communication problem and outlines the various areas which are important. Then the various sections cover the areas in detail. The sections deal with source coding, analog modulation, digital modulation, channel coding, terrestrial interface, and multiple access techniques. The multiple access section is further subdivided to provide discussions of frequency division multiple access, time division multiple access, code division multiple access, demand access for voice and data, on-board processing, and intersatellite links.

The fourth part is "Spacecraft Design and Technology." A general paper gives an overview of the problems involved in designing communications satellites. The next three sections deal with the most important technologies in the communications system: the antenna, the receiver and transmitter, and the filters and multiplexers.

The fifth part, "Propagation," discusses the problems of rain attenuation and interference, depolarization, and ionospheric scintillation.

The sixth part discusses earth stations. Most early and many current satellite systems employ large earth stations, and the first paper discusses this technology. However, newer satellite systems will use medium-size/medium-cost earth stations or low-cost earth stations. Papers are included to outline this emerging technology.

The seventh part has a somewhat ambitious title, "The Future." Work in this area can be divided into three areas. The first area, "Advanced Systems," deals with definitive system planning for second- or third-generation systems. Examples of this are the planning for the INTELSAT VI satellite for the post-1985 area, the DSCS-III satellite system, or the second generation of the WESTAR satellite. In these cases, the user requirements can be projected, and reasonably definitive system studies can be carried out. The second area, "Advanced Concepts," discusses ideas and projections for future satellites which are more speculative or further in the future than in the Advanced Systems area. The third area, "Advanced Technology," focuses on technology for the post-1985 time frame. The fourth area concentrates on New Applications for satellite communications.

The eighth part discusses various regulatory, political, legal, and economic issues that are important in satellite communications. Topics such as spectrum management, orbital utilization, launch vehicle policies, and the Communications Act are discussed in this chapter.

It is a pleasure to acknowledge the contribution made by a number of colleagues who reviewed my original list of papers and suggested additions and deletions. In each case, the advice was considered but not always followed. These reviewers include J. Aein, N. Abramson, P. Bargellini, R. Binder, F. Bond, C. Cuccia, R. Fang, O. Gabbard, E. Hoversten, P. Jain, D. Jarrett, N. Jayant, A. Jefferis, L. Kleinrock, R. Kwan, S. Lam, G. LaVean, W. Lindsey, E. Matthews, C. Niessen, P. Nuspl, W. Osborne, L. Palmer, R. Pickholtz, W. Pritchard, L. Ricardi, W. Schmidt, T. Seay, T. Sekimoto, J. Spilker, R. Strauss, A. Viterbi, G. Welti, and W. Wu.

In addition, Fred Bond, Peter Nuspl, Jack Hammett, Bill Pritchard, and John McElroy supplied entries for the Bibliography. A particular word of thanks is due to my coauthors of "Communications Satellites: Looking to the 1980's," Estil Hoversten and Terry McGarty. Several of the introductory sections are based on sections of that document. The Comsat Library staff, Ms. Rosa Liu and Ms. Rita Carter were helpful in obtaining articles. Ms. Mary Lou Sweeney did an excellent job of typing all of the introductory material and the Bibliography.

HARRY L. VAN TREES

Part I
General

The papers reprinted in this part, and the references listed below, provide general discussions of satellite communications. The first paper, Reprint Paper 1.1, provides an overview of some of the problems and programs in satellite communications. It provides a background for many of the issues discussed later in the book. The second paper, Reprint Paper 2.1, provides a systems framework within which technology areas can be studied. Many of the technology areas discussed in this paper, such as multiple access, antennas, on-board processing, source coding, and intersatellite links, are the subject of sections presented later in the book.

The first group of references below are books, reprint volumes, reports, special journal issues, conference proceedings, and articles which discuss satellite communications. "Satellite Communications Engineering" [1] provides a detailed introduction to the INTELSAT system. "Digital Communications by Satellite" [2] discusses various aspects of digital transmission in satellites. References [3] and [4] are good sources of reference data. References [5]-[8] contain collections of papers in the area. References [9]-[12] are journals and conferences devoted wholly or partially to various aspects of satellite communications.

The last two references are articles dealing with historical aspects of satellite communications. Reference [13] is Arthur Clarke's frequently quoted article on satellite communications. Reference [14] is the paper by Pierce and Kompfner who derived similar ideas independently.

REFERENCES

[1] K. Miya, "Satellite communications engineering," Lattice Company, Tokyo, Japan, 1975.

[2] J. J. Spilker, *Digital Communications by Satellite,* Englewood Cliffs, NJ: Prentice-Hall, 1977.

[3] *Reference Data for Satellite Communications Earth Stations,* IT & T Space Communications, Inc., Ramsey, NJ, 1977.

[4] *Satellite Communications Reference Data Handbook,* AD 746-165, 1972.

[5] E. I. Podraczky, Ed., "Satellite communications," in *Proc. IEEE,* vol. 65, no. 3, Mar. 1977.

[6] W. E. Morrow, Jr., Ed., "Satellite communications," in *Proc. IEEE,* vol. 59, no. 2, Feb. 1971.

[7] I. Kadar, Ed., "Satellite communications systems," *AIAA selected Reprint Series,* vol. XVIII, Jan. 1976.

[8a] Y. F. Lum and R. G. Gould, Eds., "Communications satellite systems: An overview of the technology," in *IEEE Aerospace and Electronic Systems, IEEE Press.* New York, 1976.

[8b] J. H. W. Unger, "Literature survey of communication satellite technology," *IEEE Aerospace and Electronic Systems, IEEE Press.* New York, 1976.

[9] AIAA Progress in Astronautics and Aeronautic Series: These volumes contain selected papers from the biannual Satellite Conference of the American Institute of Aeronautics and Astronautics.
 a. Vol. 19: R. B. Marsten, "Communications satellite system technology," 1966.
 b. Vol. 25: N. E. Feldman and C. M. Kelly, Eds., "Communications satellites for the 70's—Technology," 1971.
 c. Vol. 26: N. E. Feldman and C. M. Kelly, Eds. "Communications satellites for the 70's—Systems," 1971.
 d. Vol. 32: P. L. Bargellini, Ed., "Communications satellite systems," 1974.
 e. Vol. 33: P. L. Bargellini, Ed., "Communications satellite technology," 1974.
 f. Vol. 41: G. E. LaVean and W. G. Schmidt, Eds., "Communications satellite developments: Systems," 1976.
 g. Vol. 42: G. E. LaVean and W. G. Schmidt, Eds., "Communications satellite developments: Technology," 1976.
 h. Vol. 54: D. Jarett, Ed., "Satellite communications: Future systems," 1977.
 i Vol. 55: D. Jarett, Ed., Satellite communications: Advanced technologies ," 1977.
 j. "A collection of technical papers," AIAA 7th Communications Satellite Systems Conf., San Diego, CA, Apr. 24-27, 1978.

[10] International Conferences on Digital Satellite Communications:
 a. First Conf., London, England, Nov. 25-27, 1969.
 b. Second Conf., Paris, France, Nov. 28-30, 1972.
 c. Third Conf., Kyoto, Japan, Nov. 11-13, 1975.
 d. Fourth Conf., Montreal, Canada, Oct. 23-25, 1978.

[11] *COMSAT Technical Review:* A journal published twice a year by the Communications Satellite Corporation (COMSAT) containing articles by COMSAT staff on communications satellites. Vol. 1, No. 1 was published in Fall 1971.

[12] These conferences are held annually, and there are usually 10-25 papers on satellite communications at each conference.
 a. *Proc. Int. Communications Conf.*
 b. *Proc. National Telecommunications Conf.*
 c. *Proc. IEEE Electronics and Aerospace Systems Conv.* (EASCON).
 d. Session papers at (WESCON).

[13] A. C. Clarke, "Extra-terrestrial relays," in *Wireless World,* vol. 51, no. 10, Oct. 1945, p. 305.

[14] J. R. Pierce and R. Kompfner, "Transoceanic communications by means of satellite," in *Proc. IRE,* vol. 47, Mar. 1959, p. 372.

Satellite Communication—An Overview of the Problems and Programs

WILBUR L. PRITCHARD, FELLOW, IEEE

Abstract—This paper introduces the subject of satellite communication in its broadest aspects, recounts its history and discusses the principal technical problems. It is primarily communications-oriented but relevant spacecraft and launch considerations are summarized. Tabular summaries of the world's satellite communication programs are given.

I. INTRODUCTION

COMMUNICATION satellites have several important characteristics. One is certainly the availability of bandwidths exceeding anything previously available for intercontinental communications. Although overland transmission of high-quality TV pictures by microwave radio relays or cable has been possible for some years, trans-Atlantic TV transmission took place for the first time only after the first active communication satellite had been put in orbit. Intercontinental relaying of TV programs, now commonplace, is done exclusively by satellites.

Another, perhaps the most important of all, is the unique ability to cover the globe. In the future, it is likely that cables will use much higher carrier frequencies, probably as high as the optical region of the spectrum. If so, a multitude of TV channels or their equivalent could be transmitted from one continent to another without satellites. However, a cable still has two fixed ends and there must be a connection between every pair of points to be in communication. Satellite systems offer, in this respect, a flexibility that cannot now be duplicated. Furthermore, this flexibility applies not only to fixed points on earth, but also to moving terminals, such as ships at sea, airplanes, and space vehicles.

With communication satellites, then, instant and reliable contact can be established rapidly between any points on earth, in addition to, and well beyond, the capabilities of available land lines, microwave line-of-sight relay systems, and other techniques. Satellites are the elements of a communications revolution analogous to that in transportation resulting from the airplane.

II. HISTORY

Although the origins of the whole idea of satellite communication are obscure, there is no question that the synchronous, or more accurately, geostationary, satellite was first proposed by Arthur C. Clarke in an article in *Wireless World* entitled, "Extraterrestrial Relays." He recognized the potential for rocket launches based on the German V2 work during the war and also the conspicuous advantage of the geostationary orbit. Prophetically, his proposal was for the use of these satellites for FM voice broadcast rather than for telephone service. Interestingly enough, Clarke also foresaw the use in space of

Manuscript received April, 1976; revised September 7, 1976.
The author is with Satellite Systems Engineering, Inc., Washington, DC 20014.

electric power generated by panels of solar cells. Implementation of his idea still had to wait for the Space Age (Sputnik— 1957) and solid-state technology.

Thirty-one years have passed since his prophecy, and there are now 22 satellite communication programs with satellites in orbit or under active construction. There are another score of programs with earth stations only using the satellites of others. A word of tribute to his exceptional vision is certainly in order.

A. The Early Years

Moon reflections for radar and communication purposes were repeatedly demonstrated in the late forties and early fifties. In July 1954, the first voice messages were transmitted by the US Navy over the earth-moon path. In 1956, a US Navy moon relay service was established between Washington, DC, and Hawaii. This circuit operated until 1962, offering reliable long-distance communication limited only by the "availability" of the moon at the transmitting and receiving sites. Power used was 100 kW, with 26-m diameter antennas at 430 MHz.

A metallized balloon of the correct size, launched by a rocket and placed in orbit can be used as a scatterer of electromagnetic waves generated by an earth transmitter. Part of the energy can be picked up by receiving stations at any point on earth from which the balloon is visible, thus obtaining a passive communications satellite system.

Through the joint action of Bell Telephone Labs, NASA, and JPL, the ECHO experiment was performed. Successful communications across the US were first established in early August 1960, between Goldstone, CA, and Holmdel, NJ, at frequencies of 960 MHz and 2290 MHz. The ECHO "balloon," in an inclined orbit at 1500-km altitude, was visible to the unaided human eye.

Later in the same month, the first trans-Atlantic transmission occurred between Holmdel, NJ, and a French receiving station [1]. This project alerted the entire world to be prospect of the new medium of communications although the specific method was never exploited commercially.

Although passive satellites have infinite capability for multiple-access communications, they are gravely handicapped by the inefficient use of transmitter power. In the ECHO experiment, for instance, only one part in 10^{18} of the transmitted power (10 kW) is returned to the receiving antenna. Since the signal has to compete with the noise coming from various sources, special low-noise receivers must be used. Luckily, the invention of the maser in 1954 and its successive development, permitted the construction of very low-noise receivers (with temperatures in the neighborhood of 10 K which, used with horn-reflector receiving antennas having an aperature of about 43 m^2, made possible the transmission of teletype, voice, and pictures).

The advantage of passive satellites is that they do not require sophisticated electronic equipment on board. A radio beacon transmitter might be required for tracking, but in general, neither elaborate electronics, nor, with spherical satellites, attitude stabilization is needed. Such simplicity, plus the lack of space-flyable electronics in the late fifties, made the passive system attractive in the early years of satellite communications.

As soon as space-flyable electronics became available, it was obvious that passive systems would be replaced by active satellites. The mathematics of the inverse square law for active satellites (versus the radar-like inverse fourth-power law applicable to passive satellites) are overwhelmingly in favor of the former.

The relative disadvantages of a passive system increase with orbital altitude and the on-board power availability of the active satellite. After the early experimental trials, all subsequent satellite communication experimental and operational systems have been of the active type, and there is nothing to indicate that the situation is likely to change.

It is interesting to note that the first active US communication satellite was a broadcast satellite. SCORE, launched on December 18, 1958, transmitted President Eisenhower's Christmas message to the world with a power of 8 W at a frequency of 122 MHz. SCORE was a delayed-repeater satellite receiving signals from earth stations at 150 MHz; the message was stored on tape and later retransmitted. The 68 kg payload was placed in rather low orbit (perigee 182 km, apogee 1048 km).

The communications equipment was battery powered and not intended to operate for a long time. After 12 days of operation, the batteries had fully discharged and transmission stopped.

B. The Experimental Years

Aside from early space probes like Sputnik, Explorer, and Vanguard, as well as the SCORE and Courier projects, which were early communication satellites of the record and retransmit type, the major experimental steps in active communication satellite technology were the Telstar, Relay, and Syncom projects.

Project Telstar is the best known of these probably because it was the first one capable of relaying TV programs across the Atlantic. This project was begun by AT&T and developed by the Bell Laboratories, which had acquired considerable knowledge from the early work of John R. Pierce and his associates, and from the work with the ECHO passive satellite. The first Telstar was launched from Cape Canaveral on July 10, 1962. It was a sphere of approximately 87 cm diameter, weighing 80 kg. The launch vehicle was a Thor-Delta rocket which placed the satellite into an elliptical orbit with an apogee of 5600 km, giving it a period of $2\frac{1}{2}$ h.

Telstar II was made more radiation resistant because of experience with Telstar I, but otherwise, it was identical to its predecessor. It was successfully launched on May 7, 1963.

The power of Telstars I and II was 2.25 W provided by a TWT, with an RF bandwidth of 50 MHz at 6 and 4 GHz. Both satellites were spin-stabilized. The overall communication capability was 600 voice telephone channels, or one TV channel. To overcome the low carrier-to-noise ratio available in the down-link, receivers at the earth stations used FM feedback in order to obtain an extended threshold. Even though the Telstar system was superbly engineered, it was designed as an experiment and was not intended for commer-

cial operation. Among other things, the orbit used made it only visible for brief periods. A project with similar objectives, Project Relay, was developed by the Radio Corporation of America under contract to NASA. It was similarly successful.

In early 1962, the President sent proposed legislation to Congress to start the commercial exploitation of these successes. After extensive hearings on the Bill, the US Congress passed the Communications Satellite Act of 1962, which led to the establishment of the Communications Satellite Corporation in 1963.

On August 20, 1964, a significant event occurred when agreements were signed by 11 sovereign nations which resulted in the establishment of a unique organization—the International Telecommunications Satellite Consortium, known as INTELSAT. This new organization was formed for the purpose of designing, developing, constructing, establishing, and maintaining the operation of the space segment of a global commercial communications satellite system.

C. The Commercial Era

Commercial communications by satellite began officially in April 1965, when the world's first commercial communication satellite, INTELSAT I (known as "Early Bird"), was launched from Cape Kennedy. It was decommissioned in January of 1969 when coverage of both the Atlantic and Pacific was accomplished by two series of satellites, INTELSAT's II and III. Interestingly enough, Early Bird was planned to operate for only 18 months. Instead, it lasted four years with 100 percent reliability.

The fully mature phase of satellite communications probably is best considered as having begun with the installation of the INTELSAT IV into the global system starting in 1971. These spacecraft weigh approximately 730 kg in orbit and provide not only earth coverage but also two "pencil" beams about $4°$ in diameter which can be used selectively to give spot coverage to Europe and North and South America. INTELSAT IV is a spinning satellite, as were its predecessors, but the entire antenna assembly, consisting of 13 different antennas, is stabilized to point continually toward the earth. Two large parabolic dishes form the two spot beams. Each satellite provides about 6000 voice circuits, or more, depending upon how the power in the satellite is split between the spot beams and the earth coverage beam. INTELSAT IV can carry 12 color TV channels at one time.

D. Military Satellites

The first military satellites, the DSCS-I, were launched by the US Air Force in June of 1966. These launches were interesting because 8 satellites were launched simultaneously. Finally, about 30 satellites of a very simple spinning type and without station-keeping were placed in near synchronous orbits. Some are still in operation today. The DSCS-II system was initiated several years ago and constitutes the present US military system although it has had both spacecraft and launch vehicle failures. DSCS-III is being planned.

III. CATEGORIES OF SYSTEMS

There are some 42 satellite communication systems in the world today, 22 of which include both satellite and terrestrial equipment (See Table I). By satellite system, we mean one which is in active operation or one for which the equipment is being built under funded contract. There are literally dozens

TABLE I
LIST OF PROGRAMS

Program	Class						Coverage		Status				First Operational Date -- Actual or Planning	Frequency Bands (MHz)
	Fixed	Mobile	Broadcast	Experimental	Military	Proposed	Regional	Domestic	Operational	Under Construction	In planning	Under Construction		
1 INTELSAT	X						X		X				1965	5925 - 6425 up 3700 - 4200 down (C Band)
2 USSR	X						X	X	X				1965	5725 - 6225 up 3400 - 3900, 800 - 1000 down
3 TELESAT - Canada	X							X	X				1973	5927 - 6403 3702 - 4178
4 Western Union	X							X	X				1974	5925 - 6425 3700 - 4200
5 RCA SATCOM	X							X	X	X			1974	Same as WESTAR
6 AMERICAN SATELLITE CORP.*	X							X	X				1974	WESTAR Transponder
7 COMSAT/ATT (COMSTAR)	X							X	X				1976	5925 - 6425 3700 - 4200
8 ESA (OTS/ECS)	X						X			X			1977	14152.5 - 14192.5 11490 - 11530 14242.5 - 14362.5 11580 - 11700 14455.0 - 14460 11792.5 - 11797.5
9 ALGERIA*	X							X	X				1975	INTELSAT Transponder
10 INDONESIA	X							X	X				1976	5925 - 6425 3700 - 4200
11 NORWAY/NORTH SEA	X						X			X			1975	INTELSAT Transponder
12 SBS	X							X			X		1978	14000 - 14500 11700 - 12200 Ku Band
13 PHILIPPINES*	X							X				X	1976	5925 - 6425 3700 - 4200
14 INDIA	X	X						X				X	?	C or Ku 2500 B'cast
15 ARAB SYSTEM	X						X					X	1980-90	C Band 2500 B'cast
16 COMSAT GENERAL (MARISAT)		X					X		X				1975	Satellite - Ship / Satellite - Shore 300 - 312 1638.5 - 1642.5 up 6174.5 - 6425 up 248 - 260 1537 - 1541 down 3945.4 - 4199.0 down
17 ESA/COMSAT GENERAL (AEROSAT)		X					X				X		1979	Satellite - Plane / Satellite - Ground 1645 - 1660 131.425 - 131.975 up 5000 - 5125 up 1543.5 - 1558.5 125.425 - 125.975 down 5125 - 5250 down
18 ESA (MAROTS)		X					X			X			1977	Satellite - Ship / Satellite - Shore 1641.5 - 1644.5 up 14490 - 14500 up 1540 - 1542.5 down 11690 - 11700 down
19 IMCO (INMARSAT)		X					X					X	1980+	Satellite - Ship / Satellite Shore L Band C or Ku Band
20 EUROPEAN BROADCASTING UNION			X				X					X	1980+	Ku Band
21 FEDERAL REPUBLIC OF GERMANY			X				X	X				X	~1985	Ku Band
22 SYMPHONIE				X			X		X				1975	5926 - 6424 up 3700 - 4200 down
23 ATS-6				X			X		X				1974	Wide variety of experiments in K, C, S, UHF, and VHF Bands
24 JAPAN - COMMUNICATION	X			X			X			X			1977	5925 - 6425 3700 - 4200 30 000 20 000
25 JAPAN - BROADCAST			X	X			X			X			1978	14 000 12 000
26 CTS				X			X		X				1975	14 000 12 000
27 SIRIO				X			X			X			1977	17009 - 17782 11331 - 11862
28 DSCS	X	X			X		X		X	X			1966	7900 - 8400 7250 - 7750
29 SKYNET	X	X			X		X		X				1969	7976 - 8005 7257 - 7286
30 NATO	X	X			X		X		X				1970	7975 - 8162 7250 - 7437
31 FLTSATCOM	X	X			X		X			X			1977	Satellite - Ship / Satellite - Shore 225 - 400 7900 - 8400
32 LES SERIES	X	X		X	X		X		X				1965	Wide variety of experiments at UHF and K Bands
33 ANDEAN NATIONS*	X							X	?	?		X	?	INTELSAT Transponder
34 ARGENTINA*	X							X		X		X	?	INTELSAT Transponder
35 AUSTRALIA	X							X		X		X	1980's	C Band, Ku, S Band
36 BRAZIL	X	?	?	X				X		X		X	1978	5925 - 6425 3700 - 4200
37 DENMARK*	X							X		X		X	?	INTELSAT Transponder
38 IRAN	X	?	?					X		X		X	?	--
39 MALAYSIA*	X							X		X		X	1975	INTELSAT Transponder
40 UNITED KINGDOM									-				-	--
41 SYSTEMS USING INTELSAT*	X	X						X	X	X		X	Various	INTELSAT Transponder
42 ASEAN NATIONS	X							X	?	X		X		

* Earth Stations Only

of other systems under study in various phases. They range, in likelihood of implementation, from remote possibilities to being on the verge of realization.

We can characterize and categorize satellite systems in a variety of manners, either by their technical characteristics or by their operational use. The latter seems more natural and, indeed, is instructive in the sense that after having categorized the satellites operationally, we can examine the diversity of technical methods used to achieve similar operational results.

The first category of system is certainly international civil telecommunications, of which we have only 2 examples today: the highly successful INTELSAT and the inchoate STATSIONAR of the Soviet Union.

The second category is that of regional and domestic satellites, principally for civil telecommunications but occasionally for the distribution of television programs. In this class are: Anik, Westar, Satcom, Comstar, Palapa (Indonesia), and Molniya. For military communications (either to fixed or mobile terminals), we have 4 systems: NATO, Skynet, DSCS, and FLTSATCOM.

There are as yet no operational direct broadcast systems although there are several experimental ones in that category. It is likely that even operational ones will be joined with telecommunications services.

And, finally, still in the operational rather than the experimental category, we have those systems planned specifically for communication with mobile terminals: MAROTS, MARISAT, and AEROSAT.

The experimental programs cover a wide range of purposes, from experimenting with operational problems and proving spacecraft technology to acquiring propagation data in frequency bands of potential interest, especially above 12 GHz.

In this category, we have: ATS-6, Canadian Technology Satellite, the Japanese Communication Satellite, Japanese Broadcast Satellite, OTS, Symphonie, SIRIO, and the Lincoln-Laboratory Series (LES). The gross technical features of these systems are summarized in Table II.

IV. Main Technical Considerations

In order to discuss the various ways in which the different systems have chosen to cope with the particular technical problems, we must examine, at least briefly, the technical and operational problems of satellite communications. These problems cover virtually the entire spectrum of physics and engineering. It is our intent here to look particularly at those that are special and even unique to satellite communication.

Clearly, a whole range of problems results from the necessity to insert the satellite into the desired orbit, normally geostationary, to control it remotely, and to maintain it in good operating condition. Even if satellite communications were to consist of passive reflectors, the later problems alone would be obviously not negligible. The wide bandwidths now easily available in the use of satellites in order to be exploited properly lead to another group of problems; many of which, of course, are familiar in connection with terrestrial microwave relay.

Most important of all, we have that category of problem unique to satellite communication that arises from the necessity and desirability of exploiting the geometric availability of a geostationary satellite to any point over almost a third of the earth's surface. Before this convenience can be realized, it is necessary to choose a system of multiple access. In a very real sense, we can call this *the* problem of satellite communications.

The problems in the three categories are not independent of each other. We will discuss, at least briefly, the most important interrelations among them.

A. The Communication Link

The exploitation of the wide bandwidth and the multiple access problem, and the related question of choosing appropriate modulation systems, are all best examined by starting with the elementary equation for the performance of a communication link.

The carrier power received at the earth station is given by

$$C = \frac{P_T G_T A_R}{4\pi R^2 L_i}, \qquad A_R = \frac{G_R \lambda^2}{4\pi}, \qquad \text{space loss} = \frac{4\pi R}{\lambda^2} \quad (1)$$

where

P_T satellite transmitter power
G_T transmitter antenna gain (over isotropic)
A_R earth station antenna effective area $= G_R \lambda^2 / 4\pi$
λ wavelength
R distance from satellite to earth station
L_i incidental loss.

The noise density N_0 is equal to kT_s where k is Boltzmann's constant and T_s is the equivalent system temperatures defined so as to include antenna noise and thermal noise generated at the receiver. Let us further define effective isotropic radiated power (EIRP) as equal to $P_T G_T$ and a dimensionless spaceloss equal to $(4\pi R/\lambda)^2$. It is routine to show

$$\frac{C}{N_0} = \frac{EIRP}{L_s L_i} \frac{G_R}{T_s} \frac{1}{k} \qquad \text{dBHz.} \quad (2)$$

Communications engineers usually make these calculations in dB by taking 10 times the $\log_{10}$ of both sides of (2). Care should be taken since some terms are dimensionless and some others are not. (e.g., (C/N_0) is in dBHz, L_i will be in dB, and EIRP in dBw.)

Even more interestingly, the transmitter antenna gain G_T is inversely proportional to the solid coverage angles which in turn depends on the terrestrial area to be covered A_{cov}

$$A_{\text{cov}} \left(\frac{C}{N_0}\right) \sim P_T \frac{A_R}{T_s} \frac{1}{L_i}. \quad (3)$$

Equation (3) has been written so as to highlight the most basic aspects of space communication. On the left side of the equation are the desired parameters C/N_0 and the area to be covered on the earth. On the right side, by using the reference simple relations, we have succeeded in eliminating all the terms other than the transmitted power in the satellite, the effective physical size of the receiving antenna, the dissipative losses, and the system temperature.

The carrier-to-noise density (C/N_0) can be written as $(C/N)B$ where B is the noise bandwidth and (C/N) is the desired carrier-to-noise ratio. (C/N) typically has a threshold value, the achievement of which will permit substantial improvements in demodulated signal-to-noise ratio. It will also be a function of bit-error rate in digital systems and in general, depends on the modulation used. It has values between 8 and 13 dB for most space communication systems. We now can write

$$A_{\text{cov}} B \sim P_T \frac{A_R}{T_s} \frac{1}{L_i} \left(\frac{C}{N}\right)^{-1} \quad (4)$$

Equations (3) and (4) are both fundamental to appreciating the essential problems of space communications. Although they have been written for a down-link, the same equations apply to the up-link from earth station to satellite if appropriate parameters are substituted.

Either one or both links may determine the overall performance. It is usually assumed that the noise from all sources (including intermodulation to be discussed later) is additive and it is then easy to show that

$$\left(\frac{C}{N_0}\right)_T^{-1} = \left(\frac{C}{N_0}\right)_U^{-1} + \left(\frac{C}{N_0}\right)_D^{-1} + \left(\frac{C}{N_0}\right)_I^{-1} \quad (5)$$

where the subscripts $U, D, I,$ and T apply to the carrier-to-noise density ratios as calculated for the up-link, down-link, equivalent intermodulation noise, and total, respectively.

Note that in both (3) and (4) certain terms do *not* appear, e.g., satellite altitude, carrier frequency, and earth station (G/T).

The carrier frequency has an important second-order effect since clearly the available transmitter power, receiver temperatures, etc., does depend on it's state of the art at this frequency, but the dependence is not a function of the communication performance equation. The commonly used figures of merit EIRP and G/T are not basic and should be used with caution. EIRP has an implied coverage and G/T an implied carrier frequency and when used as figures of merit, this must be remembered.

The implications of the term C/N_0 are fundamental and account for its frequent use in communications systems engineering. If one examines Shannon's expression for the information capacity of a channel and allows the bandwidth

TABLE II
CHARACTERISTICS

SYSTEM	SATELLITE MASS - STABILIZATION	LAUNCH VEHICLE	MODULATION AND MULTIPLE ACCESS	EARTH STATION ANTENNA DIAMETER	CAPACITY PER SATELLITE
1. INTELSAT IV INTELSAT IV-A	700 KG. after apogee motor firing 790 KG. after apogee motor firing S	ATLAS-CENTAUR "	FM/Video FDM/FM/FDMA "	29.5 M. for Std. A 13 M. for Std. B 29.5 M.	7500 channels + SPADE + TV 12,000 channels + SPADE + TV
2. U.S.S.R. - MOLNIA	~ 1,000 KG. B (elliptical orbit)	A-Z-ε and SL-12	FM	12 M. 25 M.	1 TV channel + unspecified telephones
3. TELESAT - Canada	272.2 KG. S	DELTA 1914	Single carrier FM Multi carrier FM Single channel/ carrier, Delta modulation PSK TDMA	Heavy Route - 30M. Network TV - 10.1 M. Northern 10.1 M. Tele-communications Remote TV 8.1M./ 4.7M. Thin Route 8.1M./ 4.7M.	12 transponders of 36 MHz bandwidth
4. WESTERN UNION - (WESTAR)	297 KG. S	DELTA 2914	SSB/FM/FDM Single & Multiple Access Video, SCPC: PSK/TDM/TDMA	15.5M.	12 Video channels (one way) or 14,400 FDM Voice channels (one way)
5. RCA SATCOM	461 KG. B	DELTA 3914	FDMA: FDM/FM & PCM/PSK for voice data, 4 PSK for digital data, FM for monochrome or color TV TDMA: PCM/PSK for voice/ data	10M.	24 Video channels w/34 MHz bandwidth 9000 channels/transponder
7. COMSAT/ATT- (COMSTAR)	750 KG. S	ATLAS-CENTAUR	FDM/FM Digital Transmission 4PSK	30M.	28,800 one-way telephony channels or >1,000 mb/s data
8. ESA (OTS/ECS)	324 KG. B	DELTA 3914	4-Phase PSK FM Video TDMA	Eurobeam A· 13M. + Spot Beam Eurobeam B 3M.	1-120 MHz Transponder 1-40 MHz Transponder 1-5 MHz Transponder
10. INDONESIA	300 KG. S	DELTA 2914	FDM/FM Multiple carriers per transponder SCPC/FM demand-assigned	9.8M. 7.3M. 4.0M.	
16. COMSAT GENERAL (MARISAT)	326.6 KG. on station S	DELTA 2914	Voice: SCPC-FM Data: 2-Phase coherent PSK	1.22M., mobile terminals 12.8M., shore terminals	9 voice channels (both ways) 110 teleprinter channels (both ways)
17. ESA/COMSAT GENERAL (AEROSAT)	470 KG. B	DELTA 3914	Voice: NBFM, PDM and/or VSDM Data: PSK/FSK	Low gain airborne antennas	5-80 kHz comm. ch. for ground-to-air and surveillance 15-40 kHz comm. ch. for air-to-ground 2-80 kHz comm. ch. for ground-to-ground 1-400 kHz or 10 MHz experimental channels
18. ESA (MAROTS)	466 KG. B	DELTA 3914	FDM TDM FDMA TDMA	About 1M.	Shore-to-ship: Up to 50 voice/high-speed data channels Ship-to-shore: Up to 60 voice/high-speed data channels Shore-to-shore: Up to 3 voice/high-speed data channels

to approach infinity as a limit, it turns out that the information that can be transmitted in a channel is proportional to C/N_0.

If the RF power budget, as determined from the above questions, yields a particular C/N_0 and if a particular bandwidth B is available from a frequency allocation point of view, then C/N_0 is quite simply divided by the bandwidth to determine the carrier-to-noise ratio (C/N) available for detectability with the particular modulation system. For a fixed (C/N) and modulation system, B is proportional to the number of channels. Thus the number of channels in a given coverage can depend only on the transmitter power, receiver antenna size, and system temperature as before. Multiple-beam antennas reduce the coverage and hence increase the number of channels for a constant total power. If the power is divided among the beams proportionally, then B remains constant for each beam and the total bandwidth available per satellite increases by the number of beams.

Increasing capacity by increasing B through frequency reuse so as to avoid the limitations of (4) costs considerably in

TABLE II (*con't*)

22.	SYMPHONIE	230 ± 5 KG.	B	DELTA 2914	Analog and Digital	16M. 12M.	1200 one-way telephony or 2 color TV channels

23.	ATS-6	1356 KG.	B	TITAN III-C	FM, Video	25.9M.
						3M. (Various)
							2 Video channels at 2.6 GHz
							or
							1 Video channel at UHF (860 MHz)
							C-band transponder has 40 MHz
							   bandwidth
							1.5 GHz transponder has
							   12 MHz bandwidth

26.	CTS	350 KG.	B	DELTA 2914	FM Video
					FM Sound
					   broadcast
					10 ch. (FDM) of
					   FM duplex voice
						10 - 0.91M.
						8 - 2.43M.
						2 - 3.05M.
						2 - 9.14M.
							1 TV Channel
							1 sound broadcast channel
							10 duplex voice channels

27.	SIRIO	188 KG.	S	THOR-DELTA	PCM-PSK, 2-phase for
					   voice (narrow band
					   communications)
					FM or Digital for TV
					   (wideband communi-
					   cations)
						14.5M. for stations
						   in Italy
						14M. for stations
						   in Finland
						12M. for stations
						   in USA
						Various smaller sizes
						   down to 1.2M. for
						   shipboard terminal
							12 - 100 kHz telephone channels,
							   total 1.5 MHz band-
							   width
							or
							1 - 4 MHz baseband for TV

28.	DSCS II	500 KG.	S	TITAN III-C	Stage 1a of program:
					   FDMA & CDMA
					Stage 1b of program:
					   FDMA & CDMA
					Stage 1c of program:
					   FDMA & CDMA, phasing
					   into TDM/PCM
					Stage 2 of program:
					   TDMA
						18.2M. for largest
						   terminals (Fixed)
						0.8M. for smallest
						   transportable
						   terminals (airborne)
							1300 duplex voice channels
							or
							100 Mb/s data
							Total of 410 MHz of
							   transponder bandwidth

29.	SKYNET	232 KG.	S	DELTA 2313 (SKYNET 2A & 2B)	CDMA in 20 MHz channel FDMA in 2 MHz channel	I 12.8M. II 12.2M. III 6.4M. IV 6.4M. V 1.8M. "SCOT" 1.1M.	1 - 20 MHz channel 1 - 2 MHz channel 24 (2400 bps) data channels or 280 voice channels
30.	NATO	340 KG.	S	DELTA 2914	FDMA/FDM (clear mode) CDMA (jamming mode)	12.8M.	
31.	FLTSATCOM	862 KG.	B	ATLAS-CENTAUR			9 UHF and 1 SHF uplink 10 UHF downlink Each UHF has 25 kHz bandwidth

32.	LES SERIES	450 KG.	B	TITAN III-C	DPSK downlink
					QPSK conferencing link
					8-ary FSK forward uplink
					8-ary MFSK, hopped at
					   200/sec.
						1.2M. for ABNCP
						   terminal (Lincoln
						   Labs)
						92 CM. for airborne
						   terminal, AN/ASC-22
						46 dM. for Navy term-
						   inal
							36-38 GHz:
							   10 kb/s, DPSK, to other
							      LES satellites
							   20 kb/s, DPSK, to ABNCP
							      terminal
							   8 ary FSK forward
							      uplink; QPSK
							      conferencing uplink
							   50 K-ary symbols/sec
							      from ABNCP terminal
							   75 b/s to Navy (shipboard)
							      terminal
							UHF:
							   50 8-ary symbol/sec to
							      aircraft
							   100 8-ary symbol/sec
							      from aircraft

spacecraft antenna size and complexity. A channel may be available only in a particular beam and must be switched if it is desired in another. The switching can be done slowly, mechanically, as is now done in INTELSAT IV and will be done in INTELSAT V. Increased capacity is obtained at the expense of increased transponder complexity—there will be hundreds of switches in INTELSAT V—and a loss of flexibility that is acceptable because of the traffic patterns for a world-wide fixed system. The switching can also be electronic and rapid, such as would be the case in a time-division switched satellite in which multiple access was achieved by burst transmissions from each earth station that would be switched in the satellite to the appropriate beam.

The multiple-beam configuration is suitable to high-traffic fixed systems, such as INTELSAT and large-area regional systems. It is less appropriate in domestic systems, such as Canada, Indonesia, and the United States.

When very high carrier frequencies are used because of increased crowding in the more desirable bands, the dissipative and scattering losses and lower available powers will probably force multiple-beam operation even where it might otherwise be undesirable.

Broadcast satellites, especially those for areas covering several time zones, will use them since the same programs do not necessarily have to go simultaneously to different areas. The European Broadcasting Union plans, when brought to fruition, envision multiple beams for Europe.

Satellite connections require up-links and down-links. Normally, since it is relatively easy to supply high transmitter powers and antenna gains at earth stations, the performance is determined by the down-link. The limitations of satellite power and the necessity for covering the appropriate terrestrial area limit the overall performance. The problem is complicated by the necessity for multiple access and the resulting possibilities for intermodulation noise in nonlinear transponders. In the case of small terminals, such as in mobile and data gathering systems, the limitation is often in the up-link.

As we see from (3), the ultimate limit in the down-link performance is the transmitted power in the satellite, and this limitation cannot be avoided for an assigned bandwidth and the requirement for terrestrial coverage.

Satellite power is directly translatable into weight in the spacecraft and, even more pointedly, into cost. Although the carrier frequency is not a first-order problem in satellite system planning, it has many very important second-order effects. External sources of noise, such as the galaxy and propagation through the ionosphere and atmosphere, are generally frequency dependent.

B. Multiple Access

To exploit the unique geometric properties of wide-area visibility and multiple connectivity that go with satellites, the various communications links using it must be separated from each other. This can be accomplished in several ways.

1) Space-Division Multiple Access (SDMA): One is to use different antenna beams and separate amplifiers within the satellite. This is SDMA. Flexibility is only possible at the expense of complications within the satellite, increased weight, and occasional operational difficulties.

2) Frequency-Division Multiple Access (FDMA): A second basic way, and the one in most common use, is that of using different carrier frequencies for each transmitting station. This is FDMA and permits many stations to use the same transponder amplifier until finally the overall noise level limits the capacity of that amplifier. Multiple carriers in any nonlinear amplifier produce intermodulation products which raises the apparent noise level. This requires a "back-off" of drive on the amplifier in order to reduce this intermodulation noise. The carrier level received is less and thus the effect of thermal noise generated in the earth station receiver is increased. This reduction in drive must thus be optimized. Even optimized, the effect is not trivial and the reduction in capability of a transponder over that it would have if all the available information was multiplexed on a single carrier frequency can be as much as 6 dB. Nevertheless, FDMA is the most popular technique for commercial communication satellites. It is efficient if one is not power limited, and it is the natural expansion of terrestrial communication methods.

FDMA can be implemented in two ways. One is to multiplex, in the conventional terrestrial manner, many channels on each carrier that is transmitted through the satellite. Another is to use a separate carrier frequency for each telephone or baseband channel within the satellite. If many carriers are used, the intermodulation problem is still more serious. On the other hand, it does approach, asymptotically, a limiting level that is usually acceptable. This single-channel-per-carrier approach has particular advantages in systems where there are many links to be made, each one having only a few circuits to be handled at any one time. Normal multiplexing is very convenient terrestrially but may be economical only if each carrier has traffic, for example, in a group of 12 channels or more.

Both systems are in extensive use today. INTELSAT uses both systems, the SPADE system being a single-channel-per-carrier multiple-access system. Canada, Indonesia, Algeria, to mention a few, use single-channel-per-carrier systems. The modulation for single-channel-per-carrier systems is a separate decision and in use today, we have PCM, Delta modulation, and narrow-band FM. The arguments as to which is best are rather complicated and are discussed elsewhere in this issue.

3) Time-Division Multiple Access (TDMA): The next basic method is TDMA, in which each earth station is assigned a time slot for its transmission, and all the earth stations use the same carrier frequency within a particular transponder. In terms of total satellite performance, this is the superior method because the intermodulation noise is eliminated and there is an increase in capacity. The required back-off is much less, just that required to achieve acceptable spectrum spreading. The price paid is a considerable increase in complexity of the ground equipment. It does seem as if the long-term trend will be toward more and more TDMA since it fits naturally with the digital communications systems that are so rapidly proliferating terrestrially, not only for data transmission but more and more for digitized voice.

Various experimental TDMA systems in the 6 Mbit/s to 60 Mbit/s range have been built and tested by INTELSAT and others. Their efficiency advantage over FDMA can be illustrated by comparing the approximate channel capacities of an INTELSAT IV global beam transponder operating with standard INTELSAT 30 m earth stations, using TDMA and FDMA, respectively. Assuming 10 accesses, the typical capacity using FM/FDMA is about 450 one-way voice channels [2]. With TDMA, using standard 64 kb/s voice frequency PCM encoding, the capacity of the same transponder is approximately 900 channels. If Digital Speech Interpolation (DSI) is used to process the PCM bit streams, the capacity is further increased to about 1800 channels.

A TDMA system went into commercial operation on Telesat, Canada's system, starting in May 1976. Numerous other TDMA systems are planned for regional and domestic satellite systems throughout the world.

This trend to digital systems both terrestrially and via satellite is reinforced by the ease with which the TDMA methods can be combined with SDMA by switching transmission bursts from one antenna beam to another depending on their ultimate destination. This notion of time-division switching, although not yet exploited in any satellite, seems inevitable for the reasons stated in connection with the discussion on the link equation. It is efficient in its exploitation of both the satellite power and the frequency spectrum, and both these resources are in short supply. The price paid is increasing complexity. That seems less and less of a price considering the awesome technology of large-scale integration and microcomputers.

Time-division switching will be a major factor in communication satellite technology. A satellite-switched TDMA system (SS-TDMA) using a microwave switch matrix of redundant design shows an increase of over 30 percent in available capacity over FDMA/TDMA [3] (separate frequency bands, each carrying TDMA). The satellite-switched TDMA concept uses a single 400 MHz channel, as distinguished from the FDMA/TDMA system, which uses 5 channels of 80 MHz. Its keying rate is 300 MBd/s, rather than 60 MBd/s. Four-phase PSK is used, as with FDMA/TDMA. A total channel capacity of 39 700 is achieved by SS-TDMA, compared with 29 870 for TDMA/TDMA. Note that this time-division switching must be done in nanoseconds so as to connect successive bursts to different spot beams. Diodes of the p-i-n type and similar solid-state switches will be necessary and are under development along with extensive ancillary logic circuitry.

4) Code-Division Multiple Access (CDMA): The final basic method of multiple access is that of CDMA, called occasionally "spread-spectrum multiple access." In either case, the idea is the same. The transmission from each earth station is combined with a pseudo-random code so as to cause the transmission to occupy the entire bandwidth of the transponder. The station for whom the transmission is intended has a duplicate of this pseudo-random code and by cross-correlating techniques can extract it from the "noise level" created by the simultaneous use of many other stations.

It has considerable advantage in military systems because the spread-spectrum technique must be used anyway to harden the satellite receiver against possible jamming and the pseudo-random sequences are necessary to provide cryptographic security.

The use of such crypto and anti-jam systems provides automatic multiple access. In a sense, it is free. The difficulty is that it is not nearly so efficient an exploitation of the resources of power and frequency spectrum as is even the FDMA system, not to mention TDMA. In addition, it requires extra equipment at both ends of the link.

Nevertheless, it is used and will continue to be used for military systems. The possibility of its limited use in commercial systems may appear as satellite users become increasingly concerned with the possibilities of both malicious interference and unauthorized listening. Users of satellite systems for commercial data transmission of the kind envisioned in domestic US systems may well be the first to consider at least the crypto secure aspects of these methods.

C. Multiplexing

Multiplexing is the process of combining a number of information-bearing signals into a single transmission band. This is either a terrestrial or satellite problem and is not to be confused with the related multiple-access question. Theoretically almost any sequence of terrestrial modulation—terrestrial multiplexing, carrier modulation to the satellite, multiple-access system—can be used. For instance, the standard INTELSAT, Telesat, DSCS-1, and Molniya systems use single-sideband AM and frequency-division multiplexing on the ground, FM to the satellite, and separate carrier frequencies for each earth station. In abbreviation, this system is SSB/FDM/FM/FDMA. The proposed TDMA referred to earlier would be described as PCM/TDM/QPSK/TDMA. The SPADE single-channel-per-carrier system is written as PCM/QPSK/FDMA. The most common terrestrial multiplex method in use is frequency-division multiple (FDM), which is used throughout the world. Frequency-division systems include:

a) single-sideband suppressed carrier (SSC or SSB);
b) single-sideband transmitted carrier (SSTC);
c) double-sideband suppressed carrier (DSSC);
d) double-sideband transmitted carrier (DSTC).

Most terrestrial and space systems use SSB, although some short-to-medium-haul systems use other techniques.

Time-division multiplexing (TDM) is becoming of increasing interest in satellite communications. Time-division systems can use many modulation systems, such as pulse-amplitude modulation (PAM) and pulse-duration modulation (PDM). By far the most important for satellite communication are pulse-code modulation (PCM) and delta modulation (DM). Within these headings there are variations, such as differential PCM and variable-slope delta modulation. The tradeoffs are complex.

Although FDM goes naturally with FDMA, and TDM with TDMA, nevertheless hybrid systems are entirely conceivable and will be used; e.g., a FDM-Master Group Codec (coder-decoder) has recently been designed for use in the Telesat TDMA system [4].

A low-loss multiplexer for satellite earth terminals has been developed to eliminate the broadband high-power transmitter and thereby improve satellite earth station reliability and efficiency [5]. The 5925- to 6425-MHz frequency band is divided into 12 contiguous channels, each 36 MHz wide. Each channel is amplified with a separate air-cooled TWTA. Channels can be added by using modular units consisting of two 3-dB quadrature filters. Time delay and amplitude responses are connected with waveguide equalizers placed before the TWT, thereby avoiding the equalizer loss in the high-power TWT output.

These units are expected to find wide application in small, unmanned earth terminals. Successful implementation of multiplexer and equalization circuits has demonstrated the practicality of the modular transmitter as an alternative to single, large, high-power transmitters currently used in satellite earth stations.

D. Demand Assignment (DA)

Earth stations having continuous traffic over a given number of channels use preassigned channels. However, many channel requirements, as in any communications plant, are of a short-term nature, so a channel and terminal equipment economy technique known as demand assignment is used.

Increased space segment efficiency in a fully variable DA network arises from the fact that all channels are pooled and may be used by any station, according to its instantaneous traffic load. This may be contrasted with a system using pre-assignment in which all channels are dedicated, i.e., both ends of the channel are fixed. With this system, when traffic to a particular destination is light, the utilization is poor. Also, for a given system traffic load, the blocking probability for a system employing preassignment is higher than for a system employing DA. This occurs because some number of channels are "locked in" to a particular link. In a system employing DA, unused channels may be made available to other users. Conversely, for a given blocking probability, the number of channels required to pass a given amount of traffic in a pre-assigned system is greater than in a DA system. The lighter the traffic per destination, the greater the advantage of the DA system.

DA offers two main advantages when compared to pre-assigned systems: 1) more efficient utilization of the space segment; 2) more efficient utilization of terrestrial interconnect facilities. Corollary advantages are more direct service (the need for "via" or transit routing is eliminated) and a consequent possible slight increase in communications quality on such links because of the elimination of tandem connections.

There are many possible forms which a DA communications system may take, and there are various ways of categorizing their makeup. If, for example, both ends of all channels in the system are undedicated so that any station may use any channel, then the system is termed "fully variable." On the other hand, if blocks of channels are reserved to an originating station or a destination station, but are still used only on demand, then it is a "semi variable" system.

Another way of categorizing DA systems is as follows: when carriers (in FDMA systems) or bursts (in TDMA systems) are assigned on demand, the approach may be termed DA multiple access (DAMA); when channels on existing carriers (FDMA) or time slots in existing bursts (TDMA) are assigned on demand, then the approach may be referred to as baseband DA (BDA).

Various combinations of these approaches may be created, the choice depending on traffic characteristics and on user requirements. For example, if a network has a multiplicity of users but only a few large earth stations, then a BDA approach may be most suitable. If there are a great many earth stations in a network, each with low traffic requirements, a fully variable DAMA approach would seem best. In an application where priority control of access is vital, the greater restriction of a semi-variable system may be advantageous, since it would enable a certain number of channels incoming to each station or each of several stations to be reserved for priority traffic.

It is also quite possible and perhaps desirable to mix approaches within the same system. The choice depends solely on the user's requirements.

DAMA systems, of which the INTELSAT SPADE system is an example, are characterized by a per-user access technique. In the SPADE system, for example, each user accesses the satellite with a single-channel-per-carrier (SCPC). Other possible approaches to DAMA are single-channel-per-burst (SCPB) TDMA and CDMA. With these approaches, the carrier (FDMA or CDMA) or burst (TDMA) is not transmitted until it is required. The arrival at the ground station via a terrestrial line of a call request results in the establishment of the carrier or the burst for the duration of the call. As stated earlier, these techniques are attractive in systems in which there are a large number of users whose individual traffic requirements are small.

The INTELSAT network uses a common TDM signalling channel on which all stations call each other and keep track of the available channels which are seized in turn on a first-come-first-served basis. This avoids the politically awkward problem of central controlled systems and the choice of country in which to locate such control—Canada and Algeria, for domestic service only, use central common control and avoid the expense of increased equipment complexity necessary to avoid it.

BDA may be implemented using digital speech interpolation (DSI) or speech-predictive encoding (SPEC). In the SPEC type system, each voice channel is monitored, and the present sample value for each channel is compared with the previous sample value for the same channel. If the samples differ by an amount which exceeds some threshold, (i.e., the present sample is not predictable from the previous sample) the present sample is transmitted. If the difference is less than or equal to the prediction threshold, the sample is not trans-

mitted. By making the prediction threshold adaptive, the system can be made able to respond to rapid changes in traffic loading.

In application, both SPEC and digital TASI (the digital counterpart of the well-known Bell system for intersyllabic channel sharing developed for trans-Atlantic cable use) exhibit approximately the same advantage but SPEC is superior in its freedom from speech clipping, simple algorithms, lower complexity, and lower cost.

Using this approach, a 2:1 reduction may be achieved in the bit rate required to transmit a group of voice channels. It is important to note that such economies may be achieved only when at least 30 voice channels are processed as a group. One of the advantages of SPEC is its graceful and slight degradation in the face of an overload condition.

E. On-Board Processing

Increases in circuit reliability that have accompanied advances in solid-state technology allow significant increases in the complexity of satellite on-board circuitry; consequently, designers now can give serious consideration to advantages of on-board processing previously considered too unreliable.

Although on-board processing is being done effectively in the case of earth-resources and weather satellites involved in large-scale data gathering missions, it has not been attempted with commercial communications satellites since their chief purpose has been the provision of a link between points on the ground for the unaltered transmission of voice, data, and television.

In military satellites, up-link signals accompanied by unwanted interference can be converted to baseband, processed for interference removal, and then remodulated on a down-link carrier. This at least prevents the repeating of the interference and the nonlinear "capture" effect of a strong signal in the transponder. The increased noise level must be dealt with by other methods.

Experiments are now being designed in which packets of information sent using TDMA will be sorted on-board the satellite and transmitted via one of the several spot beams. This is the previously discussed time-division switching. On-board DA, the "switchboard in the sky," also is expected to become important toward the end of the coming decade.

Since satellites historically have been power limited, while earth stations tend to be bandwidth limited, it is conceivable that a future system may be designed in which the up-link channels are transmitted using single sideband (amplitude modulation) to minimize bandwidth, while the down-link channels are transmitted using PCM/FM to minimize power required from the satellite. Other combinations of up-link and down-link modulation and multiple-access systems can also be conceived for various optimization plans. Such arrangements would also be accomplished with demodulation and remodulation on board the satellite.

F. Higher Frequencies

With the launch of CTS, Sirio, and other satellites in 1976, the 14/12 and 18/12 GHz bands will enter active use for satellite communication. The Japanese also plan use of these frequencies in their BS and CS satellites. Still higher frequencies are under active consideration, not only because of spectrum crowding in the lower frequency bands, but because of the desire for broader bandwidths to accommodate higher data rates than are now being sent commercially.

COMSAT Laboratories has used IMPATT amplifiers providing about one watt output at 19 GHz (16-dB gain, 700-MHz bandwidth) and at 28.5 GHz (21-dB gain, 1000-MHz bandwidth) for the AT&T Domestic Satellite Propagation Experiment [6].

Efforts to obtain a 1 to 2 Gbit/s data-transmission capability have lead to interest in the 60-GHz band for privacy and interference protection from the high oxygen absorption in the earth's atmospheric blanket; and in the 94-GHz band, which is the shortest wavelength atmospheric window beyond the infrared. Millimeter-wave travelling wave tubes can deliver kilowatts of power in the 50- to 100-GHz range, but they use very large solenoids or permanent magnets. For the space segment, tubes of lower power (e.g., 60 W) have been developed using periodic magnet focusing systems based on samarium cobalt magnet material.

Work has been done at Hughes [7, pp. 4-1–4-6] and the Air Force [7, pp. 4-7–4-12] at 10.6 μm, at which wavelength $N_2 HeCO_2$ lasers can be built with good efficiency. The 10.6 μm band is also being explored by AIL [8].

Work on coherent optical links is being done by TRW [9], by the Air Force [10], and by NASA. Common-carrier relay represents one possible use for optical links, but circuit reliabilities are marginal because of weather conditions. Another severe problem with optical links is the extremely narrow beamwidth, which would require mutual autotracking from the satellite and earth stations to keep a beam pointed properly. Such links may eventually be useful as supplements to saturated long-haul facilities.

Inter-Satellite Relays: Communications between earth stations that are not both visible to the same satellite require either the complexities and time delays of double-hop transmission or a link directly from one satellite to another. Because most of the paths between geostationary satellites would not involve transmission through the atmosphere, which would attenuate them, work on intersatellite relays has concentrated on the use of millimeter waves and optics, because of the small aperture requirements when using such wavelengths.

Wavelengths under consideration for such relays are 5 mm (60 GHz), which is highly attenuated by the oxygen absorption of the earth's atmospheric blanket but otherwise unaffected, and the optical wavelengths of 10.6 and 0.53 μm. At 10.6 μm, highly efficient $N_2 HeCO_2$ laser sources are available, while the 0.53-μm wavelength takes advantage of the simple detection properties of photomultipliers and the availability of energy from doubled Nd: YAG lasers.

A major difficulty for intersatellite laser links is the acquisition and tracking of the two widely separated space packages. Laboratory tests [7] by the Air Force have achieved pointing errors less than 1.2 μm rad peak-to-peak.

Apertures in the 1- to 2-m range and beamwidths of tenths of degrees are achieved in the millimeter (e.g., 60-GHz) systems, while apertures on the order of 25 cm are used for the optical systems. This 10:1 difference, despite a 10^4:1 wavelength difference, results from the facts that: 1) the noise levels at millimeter wavelengths are lower by more than two orders of magnitude; and 2) higher efficiency power generation can be used for millimeter waves at a level at least an order-of-magnitude higher than for lasers.

The principal issue with respect to millimeter-wave systems concerns their relative weight. Systems weighing on the order of 100 kg, drawing 300 W of prime power and having 2-m apertures, appear to be feasible.

Weights of laser transceivers are projected at less than 90 kg as a result of the relatively small apertures and higher laser efficiency which can be used effectively at this wavelength (10.6 μm).

The chief areas for research and development for intersatellite links, in addition to beam stabilization and system weight, are:

a) at 0.53 μm, electrically powered transmitter efficiency and reliability;

b) at 10.6 μm, the internal laser modulator and its driver electronics;

c) at 60 GHz, the reduction of receiver noise through passive cooling techniques.

The first test of an intersatellite relay will take place using LES-8 and LES-9 in the 36- to 38-GHz band.

G. Antennas

At geostationary altitude, the earth subtends an angle of approximately $18°$. This, plus the limited power available on board satellites, makes the concentration of RF output into narrow beams (e.g., $\leq 18°$) important. However, beamwidth is inversely proportional to antenna diameter, which is constrained by the space available within the fairing of the launch vehicle. Furthermore, attempts to obtain very small beamwidths (e.g., $\leq 1°$) may be thwarted by spacecraft attitude-control precision limitations (it is difficult and costly to point antennas to a high degree of accuracy) or by antenna reflector imperfections. One way of alleviating the problem of fairing size is by the use of an antenna that can be deployed in space, as was done on ATS-6, where a 9.1-m diameter antenna was contained in a torus of 2.0-m outside diameter prior to deployment.

Multiple antenna beams are increasing in importance because of the need to concentrate energy toward different parts of the world simultaneously. They are also attractive from the viewpoint of frequency reuse, i.e., transmitting different message groups on the same frequencies, but beaming the groups simultaneously in different directions toward different parts of the earth. A single antenna reflector can provide multiple beams by the use of feeds offset from the focal point. Separate reflectors, however, provide better efficiency and less crosstalk.

Omnidirectional antennas serve a useful purpose for telemetry and command during the launch and orbital injection phases of a spacecraft's life, but once the spacecraft's attitude becomes stabilized correctly, omni-antennas generally serve only for back-up purposes.

The polarization of an antenna's beam is governed by the polarization of its feeds. (Polarization refers to the orientation of the electric vector of the radiated field.) Polarization may be linear or circular. Two linear polarizations (vertical and horizontal) or two circular polarizations (left-hand and right-hand) can be used to achieve isolation of transmitted and received beams from one another, or for the transmission of two separate message groups in a given frequency band.

1) Polarization: Tests on frequency reuse via orthogonal polarization have been sufficiently successful that the COMSTAR satellites launched starting in 1976 have dual linear polarizations with a polarization isolation of 33 dB. The frequency plan calls for the transponder frequencies on orthogonal polarizations to be interleaved. The RCA Satcom is the first satellite to use dual polarization.

Following successful commercial operation of Comstar, as well as similar operation planned for INTELSAT IV-A, F-2, and F-3, it has been predicted [12] that the widespread use of dual polarization as a means of obtaining added channel capacity in the already crowded 6 and 4 GHz bands. For example, INTELSAT V will use both the present INTELSAT polarization and polarization orthogonal to it.

H. Orbits

To appreciate the various tradeoffs made in the satellite communications systems, it is necessary to look briefly at the various orbits in which communication satellites can be placed, how they get there, what the ensuing spacecraft problems are, and how they affect the possibilities for transponder design.

The period of an orbiting satellite is given by

$$T = 2\pi \sqrt{\frac{A^3}{\mu}} \tag{6}$$

where A is the semi-major axis of the eclipse and μ is the gravitational constant 3.99×10^5 km^3/s^2. For a circular orbit to have a period equal to that of the earth's rotation—a sidereal day 23 h, 56 min, 4.09 s—an altitude of 35803 km is required. In the equatorial plane, this satellite will remain fixed relative to any point on earth to be "geostationary." In other planes at this altitude, it will describe figure eights daily relative to the earth. The geostationary orbit is indeed delightful from many points of view. An earth station can work with a single satellite, or several with multiple antenna beams, without the need for frequent hand-over characteristics of nonstationary satellites.

Three satellite locations can be configured to permit covering almost the entire earth. Nevertheless, it does have some disadvantages. It is a difficult orbit to get into and it does not provide coverage of the polar regions. The civilized parts of the globe are overwhelmingly within the coverage area of geostationary satellites and the latter limitation has not been serious to date. Nevertheless, future marine and aeronautical systems may want to communicate to the far northern and southern latitudes. Certain other application such as data gathering and military communication may also have the same need. When one considers that orbits meeting this requirement, such as the medium-altitude polar, also permit the injection of much greater payloads into orbit, it may be that the future will see such orbits used for satellite communication.

Ten years ago there was concern that the combination of time delay and echo inevitably present on a synchronous altitude link with hybrid two-wire to four-wire transformers would impair intelligibility noticeably. This has simply not been a serious problem except when the required echo suppressors are defective. It is no longer a consideration by system planners if voice only is used. Data transmission with long time delay places special requirements on error-correction protocols. The automatic repeat-request (ARQ) error-detecting system that requires retransmission must have a block length chosen to optimize the throughput. This block length is sensitive to both the round-trip delay and the noise-bit-error rate, normally very low in a satellite link compared to terrestrial links. In tandem connections involving the bit-error rates of a mediocre terrestrial link with the time delay of a satellite connection, the overall throughput can be poor. We may expect that satellite links more and more will use forward error-correction codes that require no retransmission and thus the time delay again will be of little significance [11]–[15].

Another orbit of interest is that of the Soviet Union's Molniya used for their domestic communication satellites. It is uniquely tailored to the coverage requirements of the far northern latitudes while avoiding the payload handicaps of a launch site at these latitudes. A highly elliptical 12-h orbit with apogee over the northern hemisphere is used for far northern coverage.

Normally the major axis of any elliptical orbit, called the line of apsides, rotates slowly because of the nonspherical or "oblate" earth. There is one angle of orbit inclination in which the effects cancel and this angle is about 62°. A 12-h period orbit at this angle and with apogee of the ellipse over the northern hemisphere is reasonably convenient for northern coverage. It is also an easy orbit in which to launch payloads from sites at northern latitudes. The geometry of launches states that any orbit inclination less than the latitude of the launch site (for instance equatorial) requires a turn or "dog leg." The loss in useful payload can be quite noticeable for far northern sites. This undoubtedly contributed to the Soviet decision to use an inclined orbit from launch sites above 45°N and the French decision to locate its launch facilities at Kourou, French Guiana—almost on the equator. This inclined orbit system gets northern coverage and high payloads in orbit at the expense of multiple satellites and stringent backing and "hand-over" problems. It is not as convenient as a synchronous system for most applications.

I. Spacecraft

Several aspects of spacecraft design deserve discussion since they affect the communication performance in varying ways. They are attitude control, primary power sources, and propulsion.

Once a communications satellite is on station, its attitude must be held fixed so that its antenna beams are always directed as desired. Effects such as gravity gradient (the difference in gravitational attraction caused by the difference in distance to the earth's center of mass of different parts of the spacecraft), the earth's magnetic field, solar radiation pressure and uncompensated motion of internal motors, gear trains, and lever arms all constitute disturbing forces acting on the spacecraft. All but the internal torques are quite small but continuous, whereas the internal torques, although large, are of short duration.

The simplest form of stabilization is that of spinning the satellite in orbit at a rate of 30 to 100 rpm. This makes the satellite act as a gyro wheel with a high angular momentum. The satellite's angular-momentum vector provides attitude "stiffness." However, it requires that the antennas be "despun", i.e., located on a relatively low-inertia platform spinning in the opposite direction so that the net effect is a stationary antenna beam relative to the earth. A bearing and power transfer assembly then couples the spinning and despun portions of the spacecraft. Spin stabilization also means that a given solar cell is effectively illuminated by the sun only $1/\pi$ of the time, thus causing the primary power to be only $1/\pi$ of the value it would have been if the cells were not spinning.

Rather than spinning a substantial fraction of the satellite, angular momentum can be provided by using a fly wheel spinning about the pitch axis and mounted inside. In this case, the entire satellite is the "despun portion."

This trend in dual-spin designs is toward despinning a larger percentage of the satellite's mass. This trend will continue as multiple beam antennas become more common. Systems such as INTELSAT V and DSCS-III will have severe requirements of this kind.

The question arises of when the stabilization system is no longer to be classified "dual-spin" but rather "three-axis with spinning drum providing angular momentum." One possible definition of dual-spin stabilization is that the spinning portion of the satellite performs functions other than providing angular momentum.

As solar arrays and antennas become very large (10 m on a side, or in diameter), the problem of adequately balancing solar disturbing torques becomes difficult, and full three-axis stabilization becomes necessary. More and more satellite designs are of this type.

1) Attitude Control: A comparison of dual-spin versus three-axis stabilization is instructive. The following three points explore dual-spin advantages relative to three-axis stabilization.

a) Simpler attitude-sensing system: Scanning is provided by the spinner, and the spin momentum eliminates the need for direct measurements of yaw angle.

b) Minimum number of jet thrusters: The propulsion system obtains ullage control (i.e., the feeding of propellents to the nozzles) from the centrifugal force of the spinner; a minimum number of jet thrusters are required and the same relatively high thrust level can be used for station keeping as well as attitude control.

c) Attitude "Stiffness:" The spinning momentum creates attitude stiffness that reduces the effects of torques which are created within the spacecraft and also prevents a rapid accumulation of attitude error as a result of environmental torques. Ground command thus has enough time to provide compensation. This attitude "stiffness" also can be used for attitude control during an apogee motor burn (this also applies to a three-axis system, but to a lesser degree).

The following four points explore disadvantages of dual-spin relative to three-axis stabilization.

d) Vulnerability: A single catastrophic bearing-failure mode can cause a total telecommunications outage with dual-spin stabilization. Vulnerable slip-rings and brushes, and binding of the despin bearings can cut off communications, thus rendering the satellite useless. Furthermore, power losses associated with transferring RF signals increase with frequency, and redundant encoders/decoders may have to be used on both sides of the mechanical despin mechanism.

e) Spacecraft diameter limitations: A spinning body, to be stabilized about a desired axis, should have a higher stable shape than a pencil, for example. If the spacecraft diameter is limited by the launch vehicle fairing, then this constraint is very serious.

f) Nutational Instability: Mechanical damping is needed on the despun platform to compensate for nutational instability (i.e., "coning") that results from an unfavorable ratio of spin-to-lateral moments of inertia and by energy dissipation from fuel sloshing in the tanks in the spinning portion of the spacecraft.

g) Power: More solar cells are needed for a given power when mounted on a rotating drum, resulting in a weight and cost penalty. This factor is increasing in importance because of the need for more RF power output from any single antenna, the need for more channels, the use of higher frequencies with their lower efficiency transmitters, and more onboard data processing and automation.

Some general considerations are: reliability of the three-axis design is decreased by the more complex attitude-sensing system which it requires, but the sensing system can be made redundant; and dual-spin reliability is degraded by the platform despin system, which cannot easily be made redundant.

Spacecraft costs for the two design approaches appear to be comparable.

2) Primary Power

a) Solar Cells: Primary power for communication satellites mostly is obtained by the use of silicon solar cells. They may be fixed to the spacecraft body, or mounted so that they can be oriented continuously for maximum solar energy.

During the equinox seasons, a geostationary satellite will be eclipsed by the earth. This means that the satellite will be in the dark for up to 70 min per day, depending on the inclination of the orbit and the number of days before or after equinox. To maintain operation during such periods, communication satellites depend upon internal batteries, usually consisting of nickel-cadmium cells, although nickel hydrogen and other technologies are improving swiftly. The batteries represent a major tradeoff among weight, power, and performance.

To avoid the solar-cell battery limitations, consideration has been given to the use of nuclear cells for powering satellites. Either radio isotope thermoselectric generations (RTG) or nuclear reactor powered turbines can be used. A kg of U^{235} could supply 2.5 MWh of energy even at a 10 percent conversion efficiency. With a half-life of 10^8 years, it would outlast most spacecraft.

The advantage of the nuclear supply over solar power systems is that no solar orientation is required nor is any battery needed. However, heavy shielding is required to protect the payload from radiation. This disadvantage has caused solar cells to continue to be the preferred primary power source for communications satellites. Nuclear fuel handling continues to present safety problems both during manufacture and in the event of launch malfunctions. The safer fuels, such as Plutonium, Curium (Cm^{244}), etc., are very expensive. Strontium (Sr^{90}), although much cheaper and with a convenient half-life of 25 years, is very dangerous to handle.

b) Propulsion: After launch, one or two types of propulsion are required. Satellites launched by Thor-Delta or Atlas-Centaur launch vehicles inject into transfer orbit only and require the use of an apogee kick motor for injection of the satellite into geostationary orbit. The weight of this apogee motor and its propellant is typically equal to that of the weight of the rest of the spacecraft.

Spacecraft launched by Tital III-C "direct injection" launch vehicles do not require a separate apogee kick motor, the functions of orbit circulation and inclination removal being performed by the launch vehicle itself.

Because of anomalies in the earth's gravitational field and the perturbing effects of the sun and moon, all spacecraft require a small propulsion system for station keeping. Changes in longitudinal position may be desired from time to time and also require propulsion.

Hydrazine is very popular as a monopropellant because it has high density for storage, low molecular weight and high specific impulse. It is dense, storable, and catalytic; i.e., it needs no oxidizer but dissociates on its own.

The change in velocity of a spacecraft Δr that can be achieved (e.g., for station keeping or apogee-kick purposes) is

$$\Delta v = v_e \ln M_0/M_b \qquad (7)$$

where v_e = exhaust velocity, M_0 = mass of spacecraft plus hydrazine, and M_b = mass of spacecraft (all hydrazine burned).

The exhaust velocity is telated to I, the specific impulse, by the expression

$$v_e = gI \qquad (8)$$

where g, the acceleration due to gravity (at the earth's surface), is 9.8 m/s^2. I, specific impulse, is measured in seconds and is a property of the propellant.

By equating molecular kinetic energy to 1/2 kT per degree of freedom à la Boltzman, it is seen that velocity is proportional to the square root of the absolute temperature and inversely proportional to the square root of the molecular weight of the propellants; thus, the importance of high temperature and low molecular mass is readily seen. Equation (7) can be used for sizing apogee-kick engines and hydrazine station-keeping systems. Its important attribute is the "logarithm." This makes the increase in velocity changing ability of any propulsion system insensitive to increases in propellant weight carried. The efficient way to improve the capability is through high specific impulses, that is, higher escape velocities for the propellant molecules. It explains the great attractiveness for future work of ion engines where the particles are accelerated to high velocities electronically. Specific impulses of several thousand seconds are easily achieved. They are still experimental, but one can expect their use during the next decade. The correction of latitude in synchronous orbit, because of the perturbing effects of the gravity of the sun and moon, require a Δv capability of perhaps 100 m/s/yr over a long period—a high value and a natural for ion engines. Longitudinal corrections because of a noncircular earth are very much smaller—perhaps 5 m/s/yr—and would probably continue to be made by hydrazine engines. Even they will probably be improved by various techniques, the most promising of which seems to be heating the hydrazine thermally.

c) Engine Type and Propellant: The choice of engine type and propellant is another major tradeoff area. Accuracy in station keeping simplifies the earth station tracking problems, but at the expense of hundreds of kilograms of propellant in spacecraft of the INTELSAT class.

Ion engines [16], because of the high exhaust velocity provided by electronic acceleration, offer hope of large reductions in propellant requirements but their technology is still not mature enough to be acceptable to operational spacecraft designers.

J. Launch

The delivery of a communication satellite to its geostationary position takes place in four steps:

    a) ascent;
    b) parking orbit;
    c) transfer orbit;
    d) insertion into final orbit.

The spacecraft mass that can be placed into geostationary orbit is maximized by injecting the spacecraft into a transfer orbit at an equatorial crossing. This means that the spacecraft with its second and third states must coast in the parking orbit until the right time for the injection burn, which uses both the second and third stages and accelerates the spacecraft to 36 700 km/h.

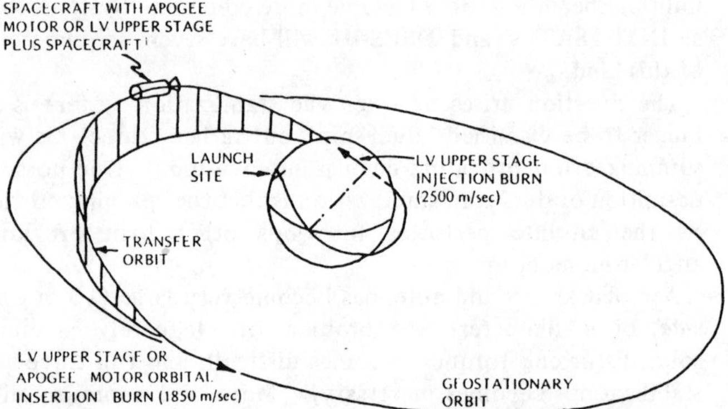

Fig. 1. Typical profile of a geostationary equatorial mission.

Fig. 1 shows the geometry and events for the transfer orbit and orbital insertion phases of a geostationary mission. It shows the transfer-injection burn occurring at the second equatorial crossing, where the launch vehicle places the spacecraft into a transfer orbit with its apogee equal to geosynchronous altitude.

Actually, to achieve geostationary orbit, another velocity impulse is required at the apogee of the transfer orbit to remove the orbital inclination caused by the launch site latitude and to make the final orbit circular. This last velocity increment can be obtained from the launch vehicle upper stage or from the spacecraft [17]. Current communication satellites launched from Cape Canaveral insert themselves into final orbit by use of a solid propellant moor. The Titan III-C, however, has an upper stage called the Transtage which performs both the transfer orbit and the final orbit injection.

Cape Canaveral in the US is used for launches in which use of the rotation of the earth is desired in order to increase the velocity of the vehicle, i.e., for eastward launches. Most communications satellite launches take place here. The Western Test Range (WTR) is used mainly for southernly launches into near-polar orbits.

The latitude of Cape Canaveral (nearly 29°N) places it at a disadvantage for launches into geostationary equatorial orbit compared with sites closer to the equator. Accordingly, the European Space Agency (ESA) is building an Ariane launch facility at its Kourou, French Guiana, launch site, which is at approximately 5°N latitude. Other near-equatorial sites are at Sriharikota and Thumba (Trivandrum) India and San Marco, an Italian mobile platform base off the coast of Kenya.

Launch vehicles available for satellite communication, especially to geostationary orbit, fall in several groups. The most important to date is that group putting spacecraft into synchronous transfer or low orbit only, e.g., the Thor-Delta in its many versions, Atlas-Centaur, and the Titan-Agena. The Titan III-C brings the spacecraft directly to synchronous orbit, without requiring the use of an apogee-kick motor. This is a very convenient method for the spacecraft designer since he does not have to design for the apogee kick and transfer orbit, but it is expensive.

On the horizon is a new vehicle being developed in France, the Ariane, which will be in the first class but with payload capability almost equal to Atlas Centaur. It will go from Kourou with all the advantages of an equatorial launch site.

Even more interesting will be the NASA shuttle. It will permit very large and complicated satellites to be placed in

200 km parking orbits, but it will be necessary to transfer them to the ultimate operational orbit, normally geostationary. Ultimately an auxiliary vehicle called the "tug" will be developed to do this transfer in a recoverable fashion.

Without the tug vehicle to do this, it will be necessary to provide both perigee and apogee stages on the satellite itself, and this will permit launching about one-quarter of the parking-orbit weight into the geostationary orbit.

The economic and operational tradeoffs are extraordinarily complicated. At this moment, it does, indeed, seem as if this may be an efficient and economic way of launching geostationary payloads although the final decision will depend on the total number of shuttle launches. It seems quite possible to design restartable liquid engines, or a combination of liquids and solids, that will transfer the satellites from parking to geostationary orbit efficiently.

Besides the ability to check a spacecraft before putting it into synchronous transfer and after it has experienced the worst in launch environment, the shuttle will have another feature of particular interest to communication satellite designers. It will permit the use not only of much heavier spacecraft but also a bigger spacecraft physically. Notably the diameter of spacecraft can go up to about 5 m. Current spinning satellite designs are seriously hindered by the limit of about 3 m on the diameter which forces large, high-capacity spacecraft to be long and slender. As mentioned previously, this makes them inherently unstable dynamically, and requires sophisticated damping in order to prevent catastrophic nutation.

An increase in diameter from 3 m to 5 m will increase the desired moment of inertia by almost 3 times and make the spacecraft a good deal more stable. In addition, recent developments of solar-cell technology also favor the continued use of spinning satellites because it again permits more primary power for a particular diameter.

V. Conclusions

In a sense this paper, as a survey of the satellite communication field, is its own conclusion. One need only glance at Table I, a list of the world's programs in all categories, to realize that as an industry, satellite communication has arrived. Table II lists more detail on those programs that include satellites. A complete listing of the characteristics of all those systems would occupy hundreds of pages [12] but we have tried to excerpt those characteristics that epitomize each system. The aggregate serves to make one appreciate the variety of programs already in existence and to make predictions of the future safe, in the sense that the magnitudes will clearly increase and risky in the sense that there are so many diverse possibilities.

With 94 nations participating and some 80 percent of the world's overseas traffic going by satellite, INTELSAT's role is clear. Yet this is only a small part—domestic traffic and services to mobile platforms will probably represent the greatest part of satellite traffic ten years from now. The satellites will continue to exploit the solid-state revolution so as to permit increasingly complicated spacecraft and communications services. One can expect on board message switching and processing in great quantities. The military organizations of the US and other countries and groups will expand their satellite activities so as to exploit the spectacular tactical possibilities. Digital technology will predominate in future development, but FM and FDMA will be around for a long time. Broadcast satellites, long possible technically but involved institutionally and sensitive politically, will slowly come into their own. Antenna techniques to restrict useful signal levels to within a national boundary will be developed. The ability of NASA's shuttle to launch large payloads—and of continuous rather than quantized sizes—will make the exploitation of all the techniques easier.

R&D programs of all kinds will continue—in orbit and on the ground so as to foster the continued development of a mature technologically oriented industry.

Any kind of extrapolation of the past ten years leads to a predicted activity for the next ten that is staggering.

Acknowledgment

The author is grateful to Horizon House-Microwave, Inc., of Dedham, MA, for permission to quote freely from their study entitled "Communications Satellite Systems Worldwide, 1975–1985." Much of the tabular material and some text has been extracted from this study.

References

[1] *Bell Syst. Tech. J.*, July 1961.
[2] D. G. Gabbard and P. Kaul, "Time-division multiple access," in *Eascon Rec.*, pp. 179–188, 1974.
[3] T. Muratami, "Satellite-switched time-domain multiple access," in *Eascon Rec.*, pp. 189–196, 1974.
[4] H. Kaneko, Y. Katagiri, and T. Okada, "The design of a PCM Master-Group Codec for the Telesat TDMA system," in *Conf. Rec., Int. Conf. Communications*, vol. 3, pp. 44-6 and 44-10, June 1975.
[5] R. W. Gruner and E. A. Williams, "A low-loss multiplexer for satellite earth terminals," *COMSAT Tech. Rev.*, vol. 5, no. 1, pp. 157–177, Spring 1975.
[6] R. Sicotte and M. Barett, "Centimeter-wave IMPATT amplifiers for space communications," in *Eascon Record*, p. 412, 1974.
[7] J. D. Barry, *et al.*, "1000 megabits per second intersatellite laser communications system technology," in *Conf. Rec., Inter. Conf. Communications*, vol. 1, pp. 4-1–4-6, 4-7–4-12, June 1975.
[8] B. J. Peyton and A. J. D. Nardo, "Systems measurements of a CO_2 laser communications link," in *Conf. Rec., Int. Conf. Communications*, vol. 3, pp. 4-13–4-17, June 1975.
[9] N. Killen, "Digital communications performance using both temporal and spatial modulation," in *Conf. Rec., Int. Conf. Communications*, vol. 3, pp. 4-18–4-22, June 1975.
[10] W. L. O'Hern, Jr., "Applications of optical communications in space," in *Conf. Rec. Int. Conf. Communications*, vol. 3, pp. 4-28–4-32, June 1975.
[11] Heller and Jacobs, "Viterbi Decoding for satellite and space communications," *IEEE Trans. Commun. Technol.*, vol. COM-19, pp. 835–848, Oct. 1971.
[12] W. L. Pritchard *et al.*, *Communications Satellite Systems Worldwide, 1975–1985*. Dedham, MA: Horizon-House Microwave, 1975.
[13] Cacciamani and Kim, "Circumventing the problem of propagation delay on satellite data channels," in *Data Communications*
[14] Cohen and Germano, "Gauging the effect of propagation delay and error rate on data transmission systems," *Telecommun. J.*, vol. 37, pp. 569–574, 1970.
[15] A. R. K. Sastry, "Performance of hybrid error control schemes on satellite channels," *IEEE Trans. Commun.*, vol. COM-23, pp. 689–694, July 1975.
[16] B. A. Free, "Chemical and electric propulsion tradeoffs for communication satellites," *COMSAT Tech. Rev.*, vol. 2, no. 1, p. 123, 1973.
[17] J. R. Mahon, "Launch vehicles for communication satellites," presented at the Nat. Electronics Conf., Chicago, IL, Oct. 1974.

Communications satellites: looking to the 1980s

It will be a decade of new and exciting applications, lower costs, and more reliable service

How will satellite communications evolve in the decade 1980–1990? In answer to that question, some prognostications appear reasonably certain of fulfillment:

• The INTELSAT global communications system will continue to grow, providing existing services at lower cost and introducing new services.

• Domestic and regional systems will continue to expand.

• The Space Transportation System will have a major impact.

• Technological advances will allow systems with complexity and capability probably an order of magnitude greater than present-day systems.

• Applications will expand from classical point-to-point telephony and broadcast into many new areas.

To provide a more detailed picture of the developments that can be expected, this article will examine some of the key technologies involved.

The simplified satellite communications system diagram in Fig. 1 emphasizes those subsystems that involve technology areas that appear likely to have a heavy impact on satellite communications applications and services in the 1980s. The areas of technology covered include: launch vehicles and satellite "buses"; antennas; on-board processing; intersatellite links; multiple access; error control; terminals, source coding, and multichannel processing; and terrestrial interface and interconnection.

Launch vehicles and satellite buses

The majority of the communications satellites of the 1980s will start their journey to geosynchronous orbit on the space shuttle instead of the expendable launch vehicles used today. The space shuttle will make bigger and heavier satellites possible and will reduce the cost per pound in orbit.

The satellite "bus" consists of the basic satellite structure and the equipment for positioning, orientation, power generation, thermal control, and telemetry, command, and control. The communications subsystems include receivers, transmitters, filters, and antennas. In designing a satellite, various tradeoffs are made among

H. L. Van Trees, E. V. Hoversten, T. P. McGarty
Communications Satellite Corporation

the elements of the bus and communications subsystem, and these tradeoffs are influenced by technological advances. For example, one of the current determinants of satellite life expectancy is the amount of fuel the satellite can carry for station-keeping. Ion engines and thermally augmented hydrazine thrusters may prove useful as propulsion devices for positioning, with a resultant increase in satellite life.

Although radioactive thermal generators have been flown on the LES-8 and -9 communications satellites, they are not expected to replace photovoltaic solar cells as the prime power source. For the future, higher-efficiency solar cells, such as "violet" cells (whose response is better matched to the sun's spectrum) and nonreflective cells, will increase the power per unit area. Satellites stabilized in three axes with larger deployable solar-cell arrays oriented toward the sun will be able to generate more power than present satellites. When a satellite is expected to function during eclipse, it must depend on batteries for power, and thus battery lifetime affects satellite lifetime. Longer-life batteries such as nickel-hydrogen units will increase satellite longevity.

As satellites become more complex, so will the requirements for telemetry, command, and control. For example, it will be necessary to monitor and control on-board processors and to check on the performance of multiple-beam antennas, solar arrays, batteries, and traveling-wave-tube amplifiers. Microprocessors on satellites will do much of the monitoring and control.

High-power traveling-wave-tube amplifiers for the 12–14-GHz and 20–30-GHz ranges will help to combat fading caused by rain. Linearized transponders containing multiple-collector tubes will permit modulation schemes that utilize bandwidth more effectively, such as combined amplitude and phase schemes. From the L-band through K_u-band frequency ranges, transistor amplifiers (both bipolar and GaAs FETs) will operate with higher powers and efficiencies and longer lifetimes.

Gallium arsenide field-effect transistors with low noise figures will improve receivers both in spacecraft and on the ground to provide increased capacity and lower-cost service. At higher frequencies (12 GHz and above), further development of parametric amplifiers, both cooled and uncooled, is needed. Passive radiation coolers may be used in satellite receivers.

Reflectors, lenses, and arrays

The antenna plays the role of the radio-frequency (RF) interface in the earth/space link. It provides directivity (gain), isolation, coverage, and connectivity. Prime concerns are mass, size, and cost; and these, in turn, impact

Reprinted from *IEEE Spectrum,* vol. 14, pp. 42–51, Dec. 1977.

on the viability of the systems that are designed.

A reflector antenna consists of a reflecting surface fed by a set of radiating elements at the focal plane, which are, in turn, fed by a distribution network. The reflecting surface can be parabolic, toroidal, or spherical in shape. Its size determines the antenna gain at a given frequency. The feeds may be horns, arrays, lenses, or other radiators. They may be movable so that the beams can be steered mechanically, or the steering may be done electronically. The distribution network contains phase shifters and provides power distribution to the radiating structures at the feed. Beam shaping and beam steering are performed in this network.

The reflector antenna is most commonly used in fixed earth stations. The INTELSAT system, for example, uses 30-meter reflectors for global telephony because of their high gain. For the new domestic U.S. systems, parabolic reflectors are being considered because of their light weight, low cost, and easy deployment. With the increase in the number of satellites that may provide service to a particular country or satellite communication application, there is a need for a multiple-beam earth-station antenna, and recent efforts in this area have led to the development of the toroidal reflector with a movable feed structure. Reflectors are also quite common as spacecraft antennas, and are used on some of the recent INTELSAT satellites and on a number of domestic satellites.

In contrast to the reflector antenna, the lens antenna is a transmission device that focuses electromagnetic energy from a feed structure. Metallic lenses consist of an array of waveguides of various lengths, and the differences in propagation time among the various waveguides shape the beam. Because the waveguides are hollow, the lens is light in weight, but such lenses have small fractional bandwidths (about 5 percent). Dielectric lenses are broadband but heavy, although lightweight dielectric materials are under development. Transverse electromagnetic (TEM) lenses (see lead illustration) consist of TEM lines that phase the electromagnetic field as required to shape the beam. These lenses are lighter than dielectric lenses and have broader bandwidth than metallic ones.

Lenses have been proposed for the military system satellites. For this application, the antenna must be able to change beam shape and direction and null out specific interference sources on earth. The lens can scan off-boresite easily and, with sufficiently complex feed structures, can null large signals from arbitrary locations. As a fixed earth-station antenna, the lens may have potential because of its beam-steering and nulling capa-

Role of the systems engineer

From a system point of view, the specification of the technology goals of the 1980s can be approached via the following methodology. Desired user applications determine a set of required communications services. Typically, several system concepts could provide these services, but their technical and economic viability will depend on the availability of adequate technology. It is then possible to select technology goals that, by providing either new capabilities or more cost-effective alternatives to existing capabilities, can significantly impact both the economic viability of serving various applications and the choice of the best system concepts.

Some of the new applications or services envisioned for the '80s will demand complex tradeoffs within the communications system as a whole, from end-user to end-user. The system must be viewed as a complete communications network, often one providing sophisticated services such as circuit and packet switching. The designers of such systems will have to apply additional disciplines such as networking and switching theory. They will have to consider new technologies such as optical communications and expanded roles for old technologies such as voice compression. And the designers will have to find ways to provide services directly to end-users and to provide a variety of services with a single system. Clearly, then, the system engineer will play a leading role in designing the satellite communications systems of the 1980s.

bilities, but for mobile earth antennas, it does not appear to be a viable alternative for the 1980s.

In an array antenna, the outputs of an array of individual radiating elements—individual dipoles, helical elements, open-ended waveguides, or any other kind of radiator—are combined to form the desired pattern. The elements may or may not be spaced uniformly. Phase shifters feed each element and are used to point and form the desired beams. A feed system distributes energy to the phase shifters and provides amplitude weighting as needed for beam shaping. The array antenna can form beams of arbitrary shapes, and can form multiple beams with little cross-beam interference. The shape and direction of the beams are controlled by digital or analog circuitry.

Although array antennas have been considered for use

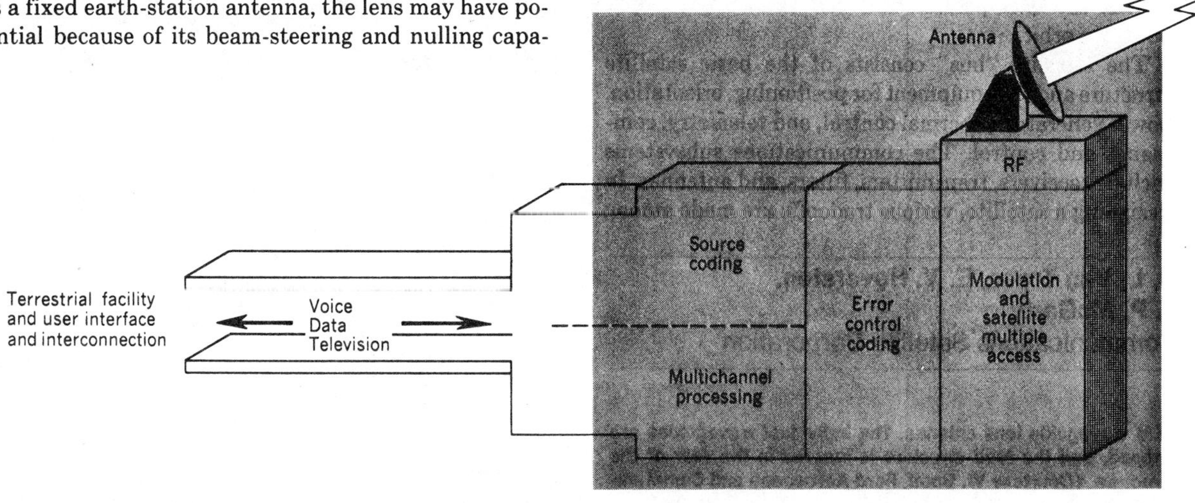

Earth segments

on satellites, they have been regarded as too big and inefficient—the feed and phase shifters are too lossy. The maritime-communication MARISAT satellites do use simple arrays, but the phase shift and feed are fixed. It is possible that the role of array antennas on satellites may be reassessed as satellite systems become more complex. For mobile earth-segment uses, the array shows a great deal of promise, particularly in aeronautical applications where it is necessary to minimize protuberances. Conformal arrays have been tested on aircraft at L and X band, and similar arrays can be developed for land and maritime mobile use. Mobile antennas usually must be small; however, even small arrays can steer and track and null out interference. Arrays are presently too costly for fixed earth stations, but as available orbit space fills up, their interference-nulling capability could become important.

Figure 2 presents a set of useful operating frequencies for various antenna technologies as a function of the specific implementation (space, fixed earth, mobile earth). Today, at most frequencies, only two alternatives are available, which makes the choice for the system designer fairly straightforward. In the Fig. 2 projection of the antenna technology ranges for 1990, the array range has been extended, based upon improved efficiencies, using strip-line and solid-state technologies. The lens range has been extended by developing lightweight dielectric materials, which, in turn, also increase bandwidth. The reflector range has been extended by improving surface tolerances and achieving increased feed efficiencies.

On-board processing

Satellites today are essentially repeaters. They provide a "transparent" channel to the users of the system. On-board processing represents a set of generic techniques, involving real- or near-real-time processing or switching, that can be used on the satellite to provide an improved performance level in the total communications system. With an on-board information processor, the satellite can become an active component of the system. It can dynamically reallocate its resources to maximize system efficiency, and it can decouple the uplinks and downlinks so that they perform independently. Complex integrated circuits will make it possible to introduce processors in satellites without adding significantly to weight or launch cost.

Connectivity and capacity are two system design characteristics that are significantly affected by the use of on-board processing. With a static satellite switch, the interconnectivity is fixed, but with an on-board processor, rapidly changing connectivity can be achieved. The usable capacity is increased with on-board processing by dynamically assigning the satellite's resources to meet demand.

An RF processor deals directly with the uplink RF signal and does its processing at RF. No translation of the signal to intermediate frequency (IF) or baseband is needed, although a simple frequency translation may be used so that operations can be performed at a convenient frequency. In its simplest form, an RF processor is a satellite switch matrix having n inputs and m outputs and can connect any input to any output.

A bit-stream processor, on the other hand, takes the RF signal, translates it to IF, demodulates the digital bit stream, and uses the bit stream to modulate another carrier. The remodulation operation would be done on a symbol-by-symbol basis; no use would be made of the signal structure to control the processing and there is essentially no storage capability. In one important advantage, the bit-stream processor can be used to suppress the uplink noise.

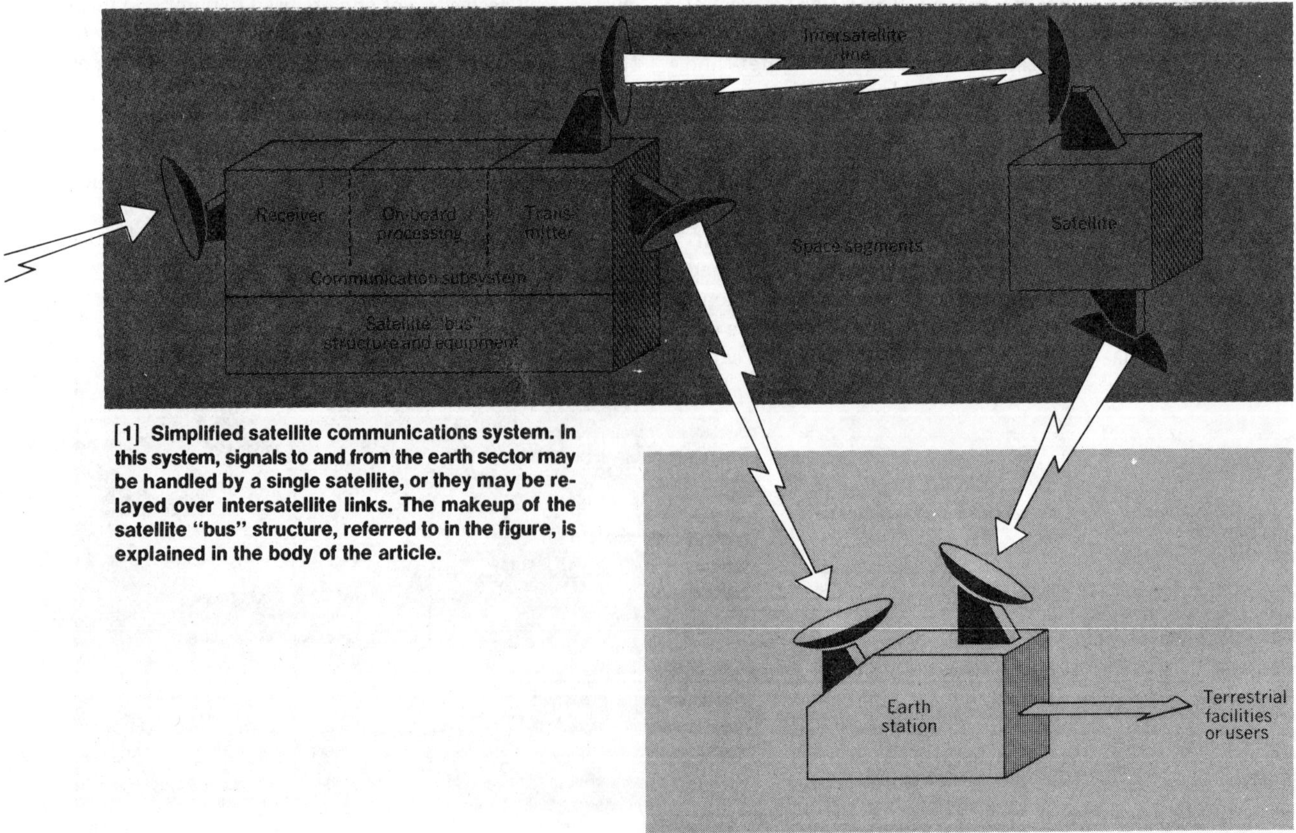

[1] **Simplified satellite communications system. In this system, signals to and from the earth sector may be handled by a single satellite, or they may be relayed over intersatellite links. The makeup of the satellite "bus" structure, referred to in the figure, is explained in the body of the article.**

The most complicated on-board processor is the full-baseband processor, in which a digital RF signal is demodulated, decoded, reencoded, and remodulated. Such a processor can use information in the decoded signal to route messages or otherwise process the signal.

A typical example of an RF on-board processor (developed but not yet in service) is a 16 × 16 RF switch for INTELSAT (Fig. 3). Advances in technology—in large-scale integration (LSI), microprocessors, surface acoustic wave devices, and charge-coupled devices—will make possible improved bit-stream and full-baseband processors.

Intersatellite links

An intersatellite link is a two-way communication channel between two or more satellites. Such links can

I. Current and future technology for implementing communications satellite technology

Application	Performance Goal	Current Status	Possible Future Technology (1980–90)
Intersatellite links			
High-data-rate digital data transfer system	1-Gb/s cross link of 10-degree separation	None	(Doubled) Nd; YAG (pump life-time improvements); CO_2 (modulator development)
Low-bandwidth dual ocean maritime system	10-MHz, 90-degree separation with C/N of 18 dB	Millimeter wave	GaAlAs; millimeter wave
Global telephony; applied to interregional interconnection	Large bandwidth (>500 MHz at 30-degree separation)	None	CO_2 (modulator development); GaAlAs (increased power output)
Global multisatellite; multi-ISL telephony system for regional coverage	100–200-MHz bandwidth at 3–9-degree synchronous spacing	HeNe; millimeter wave	GaAlAs (increased power); CO_2
Satellite multiple access			
Trunking types of applications in domestic, regional, or international use	Efficient, cost-effective multiple-access scheme to share single transponder among several accesses; variations in channel requirements assumed slow	Mostly FDMA; TDMA systems under development and test	Extensive use of TDMA. Typical TDMA systems will operat at rates of 100 Mb/s or higher. Costs will decrease greatly and the implementations will involve significant use of software and computer control
Applications requiring extensive frequency reuse or high EIRP	Efficient, cost-effective multiple-access scheme to use multibeam satellite	FDMA techniques could be used with significant filter weight penalty; SS-TDMA in the R&D phase; transponder hopping experiment on OTS	Fully developed SS-TDMA system; possible use of IF or baseband satellite processing for combining and switching uplink signals
Domestic systems providing thin-line telephony; provision of demand-assigned data services in domestic systems	Make available desired capacity and connectivity on an as-required basis	SPADE using SCPC techniques; demand-assigned TDMA under R&D; particularly for military	Continued application of SCPC with demand-assigned feature; full development and use of demand-assigned TDMA, with very flexible capacity assignments, as costs decrease
Satellite data networks; voice conferencing using packet voice	Efficient sharing of satellite channel among many users, characterized as low duty cycle and bursty; typically must satisfy a service delay constraint	Research and early experiments on packet-type multiple-access schemes	Development and use of packet-type schemes, particularly those with a reservation feature
Data collection	Minimal cost access technique for low-data-rate, low-duty-cycle users with minimal coordination requirements	Military CDMA systems exist, but expensive; some experimental use of Aloha scheme	Possible development of low-cost CDMA systems using SAW and CCD technology; use of satellite processing to convert FDMA-type uplinks to single downlink, e.g., TDM; use of random-access Aloha-type scheme
Error control			
Computer communications or electronic mail	Low-bit-error probability ($10^{-8} - 10^{-12}$)	Convolutional encoding with soft decision Viterbi decoding or hard decision sequential decoding.	High-data-rate soft decision sequential decoding. Viterbi decoding with larger constraint length codes. Concatenated codes with threshold or feedback decoding of inner code and Viterbi decoding of inner code
Mobile applications	Multipath immunity and error control	Mostly military applications, feedback decoding or interleaving with Viterbi decoding	Commercial application of current techniques with improved performance and reduced cost
Digital telephony	High rate, low cost, high bandwidth efficiency, with modest error improvement		Threshold decoding

connect users who access different satellites in the same or different orbits, without an intervening hop from satellite 1 to ground to satellite 2. This capability significantly influences the connectivity, capacity, coverage, orbit-utilization efficiency, and cost of the system.

The frequency ranges that appear most attractive for intersatellite links are the millimeter-wave and optical ranges. In the millimeter-wave band of 55 to 60 GHz al-located to intersatellite links, there are traveling-wave-tube amplifiers that can be used in the standard fashion. These tubes can put out 10 watts of peak power at 10–15 percent efficiency. For low-noise reception, a cooled parametric amplifier is available with a receiver noise temperature of 750K. At present, uncooled receiver noise temperatures are almost 2500K, but this value can be reduced with passive cooling. In the higher millimeter-

Applicaton	Performance Goal	Current Status	Possible Future Technology (1980–90)
Source coding and channel multiplication			
Telephony—particularly private-line and mobile applications	Cost-effective, low-data-rate digital voice terminals	64-kb/s PCM. Development activities on 16–48-kb/s systems. Experiments and research on less-than-4-kb/s systems	Extensive use of 16–32-kb/s terminals. Development of cost-effective high-voice-quality LPC terminals at <4 kb/s, with some use by late 1980s
Teleconferencing	Cost-effective, low-data-rate, full-motion video transmission	Intraframe coding: 22–33.6 Mb/s. Interframe coding: 6.3–22 Mb/s, but costly	Use of 22 Mb/s for network distribution with cost-effective terminals operating at rates as low as 6.3 Mb/s
Facsimile broadcast	Inexpensive, rapid hard-copy transmission	Digital facsimile terminals with 10:1 compression ratios and scan times per page of $\approx$ 1 minute	Digital facsimile terminals with greater compression ratios, improved image quality, higher transmission rates, and lower costs (scan times per page of a few seconds or less)
Telephony trunking	Efficient use of channel capacity for trunking applications	DSI techniques under development and test. Packet voice experiments	Extensive application of DSI, particularly as part of TDMA systems. Development of packet voice
Terrestrial interface and interconnection			
Telephony situations where it is desirable to use TDMA or digital modulation	Low-cost flexible digital interface with terrestrial transmission systems.	Most interface is with FDM signals and digital transmission requires A/D conversion and perhaps some demultiplexing	Digital terrestrial transmission will permit use of low-cost flexible TDM and switching equipment, perhaps microprocessor based
Interconnect of antenna and interconnect facility or other intrasite transmission	Interconnect with EMI immunity and ease of installation.	Cable or radio	Fiber optics
Interconnect of users to customer premises earth terminals	Low-cost, easily reconfigured earth station interconnect	Cable or radio	Packet radio using spread-spectrum techniques

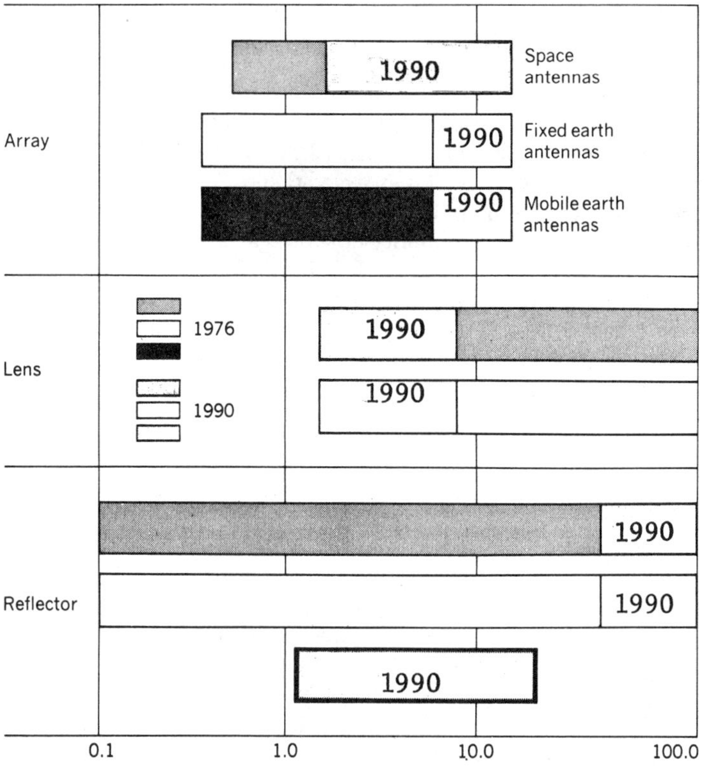

[2] Frequency capabilities of antenna types: 1976 and 1990. There are three areas of application for satellite communications antennas—spacecraft, fixed earth segment, and mobile earth segment. Many forms of antennas can be utilized for these application areas, but three generic types—reflectors, lenses, and arrays—are expected to dominate.

[3] This 16 × 16 satellite switch has been tested for use with a 4/6-GHz satellite and 120-Mb/s TDMA system.

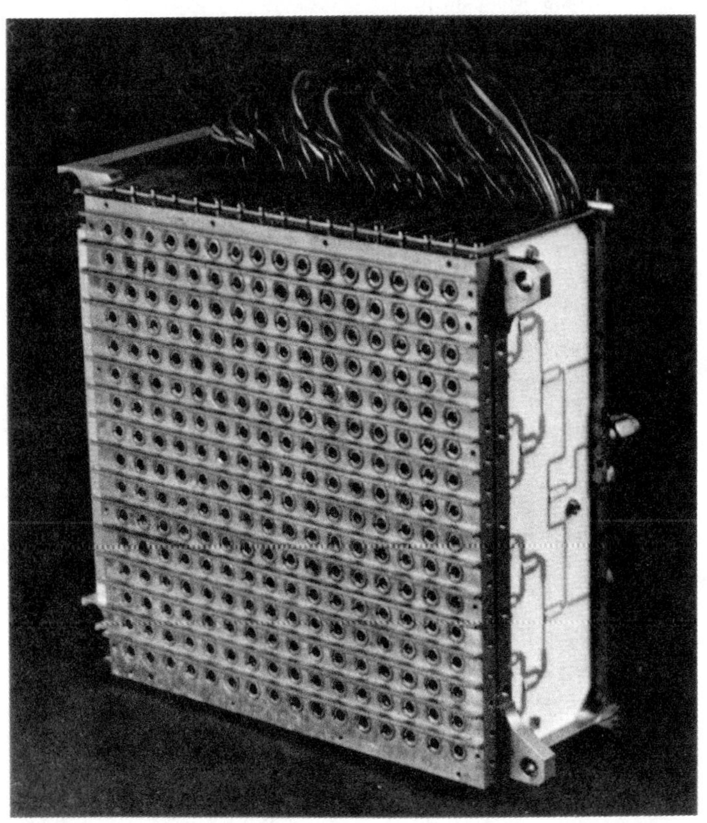

wave regions allocated for intersatellite links (above 100 GHz), no strong technology base exists.

Table I shows some performance goals, applications, and technologies for intersatellite links. It is assumed that significant advances will be made in the next decade in solid-state optical systems. Decreased reliance on millimeter-wave systems is predicted as optical modulators and detectors become efficient and optical systems demonstrate they can perform reliably in space.

Multiple access

A satellite must share its capacity among a number of earth terminals. This sharing is implemented by some form of multiple-access system: frequency-division multiple access (FDMA), time-division multiple access (TDMA), or code-division multiple access (CDMA, also referred to as spread spectrum). As shown in Fig. 4, the various schemes allocate portions of the time–frequency plane to various users. Although, in some instances, these allocations are preassigned, in many systems the satellite's power and bandwidth can be used more efficiently through dynamic demand assignment.

The multiple-access problem is fundamental to satellite communications because it is the means by which the wide geographic coverage capability and broadcast nature of the satellite channel are exploited. It affects all elements of the system, determines the system capacity and flexibility, and has a strong influence on costs.

At present, the most common scheme is FDMA, but the late 1970s and 1980s should see a trend toward TDMA for both commercial and military satellites, as a result of drastic cost reductions in digital logic and the continuing emphasis on digital terrestrial transmission. On the other hand, the simplicity and current investment in FDMA will justify its retention in some applications for the indefinite future.

Past TDMA development efforts have tended to emphasize designs and parameter values tailored to telephony applications. For example, the algorithms associated with demand-assigned TDMA systems have typically emphasized classical Erlang loading. With the increased interest in satellite data transmission, schemes suited to the various facets of this application are receiving attention. The appropriate system metrics are often different; e.g., the throughput–delay tradeoff in an interactive computer application.

Packet-switching schemes, in which individually addressed packets of data are transmitted, are expected to receive considerable attention for use in certain data network applications.

Table I indicates specific performance goals for multiple access that could pay off in the 1980s. Several of these goals can be achieved by further development of techniques now in the advanced development and prototype phase; notably, low-cost TDMA, satellite-switched TDMA (with an on-board processor), and demand-assigned TDMA (which rapidly reassigns time slots to accommodate changes in demand by earth stations). Device technology advances will not only play an important role in TDMA development, but could make the presently expensive CDMA techniques attractive for certain applications, such as data collection.

Error control

As the predominant transmission mode shifts from analog to digital, the measures of performance will change. In particular, the quality of a channel will be measured in terms of bit-error rate instead of signal-to-noise ratio. Various error-control techniques can be employed to obtain the desired error level.

On digital terrestrial communication links, the most common form of error control is the automatic repeat request (ARQ) procedure: When an error is detected, the sender is asked to repeat part of the message. The ARQ method is less attractive for satellite communication because the half-second round-trip propagation delay reduces the data throughput or makes large buffers necessary. Two specific techniques that tend to minimize the throughput penalty, at the expense of requiring large transmit buffers for high-speed circuits, are "go-back n" and "selective-repeat" ARQ. Selective-repeat ARQ also requires a large receiving buffer with a reordering capability.

Forward error correction (FEC) techniques provide a means, when properly integrated with the overall modulation design, of efficiently using the available signal energy at some price in bandwidth. Although block codes will continue to be used, it is expected that convolutional codes will predominate because of their ease of implementation and the availability of several attractive decoding schemes.

When the requirements demand a very low overall error probability and a return transmission path is available, hybrid schemes employing ARQ techniques in combination with FEC techniques are expected to find some use, but concatenated FEC techniques will also be

For further reading

General

- Van Trees, H. L. (Ed.), *Satellite Communications.* New York: IEEE Press, to be published.
- "Outlook for Space," NASA SP-386, Jan. 1976.
- Van Trees, H. L., Hoversten, E. V., and McGarty, T. P., "Communications Satellites—Projections for the 1980s," Advanced Systems Div. Report, Communications Satellite Corporation, Nov. 1976 (available from authors on request).
- Pritchard, W. L., "Satellite communications—an overview of the problems and programs," *Proc. IEEE,* vol. 65, Mar. 1977.
- Bargellini, P., and Edelson, B., "Progress and trends in commercial satellite communications," presented at IAF XXVI Congress, Lisbon, Portugal, Sept. 21–27, 1975.
- Bond, F. E., and Curry, W. H., Jr., "The evolution of the military satellite communications systems," *Signal,* pp. 39–44, Mar. 1976.
- Van Trees, H. L., "Future INTELSAT system (1986–1993) planning," *Progress in Astronautics and Astronautics,* AIAA, vol. 54, pp. 95–141.

Launch vehicles and satellite buses

- Bleviss, Z. O., "Expendable launch vehicles for synchronous communications satellites," Paper 76-274, AIAA/CAST Sixth Communications Satellite System Conf., Montreal, Que., Canada, Apr. 5–8, 1976.
- Kloves, T. R., "Space shuttle program—an overview," Paper 76-275, AIAA/CASI Sixth Communications Satellite System Conf., Montreal, Que., Canada, Apr. 5–8, 1976.
- Collette, R. C., and Berdan, B. L., "Design problems of spacecraft for communication missions," *Proc. IEEE,* vol. 65, pp. 342–356, Mar. 1977.

Reflectors, lenses, and arrays

- Matthews, E. W., Scott, W. G., and Han, C. C., "Advances in multibeam satellite antenna technology," *EASCON '76 Record,* pp. 122A–1220, Sept. 1976.
- Ricardi, L. J., "Communications satellite antennas," *Proc. IEEE,* vol. 65, pp. 356–369, Mar. 1977.
- Hellman, C., and Cuccia, C. L., "Small, low cost earth stations," *Conf. Record,* Internat'l Conf. Communications, June 14–16, 1976, pp. 21-13–21-18.

On-board processing

- Koga, K., Muraturi, T., and Ogawa, A., "On-board regenerative repeaters applied to digital satellite communications," *Proc. IEEE,* vol. 65, pp. 401–410, Mar. 1977.
- Rozee, X, and Assal, F., "Microwave switch matrix for communications satellites," *Conf. Record,* Internat'l Conf. Communications, June 14–16, 1976, pp. 35-13–35-17.
- Cuccia, C. L., "Modern transponder technology for baseline design of data transmission over space communication laser links," IAF XXVII Congress, Oct. 1976.

Intersatellite links

- Klein, P. I., and Soifer, R., "Intersatellite communications using the AMSAT-OSCAR 6 and AMSAT-OSCAR 7 radio amateur satellites," *Proc. IEEE,* vol. 63, pp. 1526–1527, Oct. 197.
- Van Trees, H. L., Palmer, L., McGarty, T., and Mohajeri, M., "Intersatellite links: Part I. Systems," to be published.
- Barry, J. D., "1000-Mbits/s intersatellite laser communications system technology," *IEEE Trans. Communications,* pp. 470–478, Apr. 1976.
- Sarles, F. W., Jr., Bowles, L. W., Farnsworth, L. P., and Waldron, P., "The Lincoln experimental satellites LES-8 and LES-9," *EASCON '77 Record,* pp. 21-1A–21-IV.

Multiple access

- Puente, J. G., Schmidt, W. G., and Werth, A. M., "Multiple access techniques for commercial satellites," *Proc. IEEE,* vol. 59, pp. 218–229, Feb. 1971.
- Lebow, I. L., Jordan, K. L., and Drouilhet, P. R., "Satellite communications to mobile platforms," *Proc. IEEE,* vol. 59, pp. 139–159, Feb. 1971.
- Jacobs, I. M., *et al.,* "C-PODA—A demand assignment protocol for SATNET," *Proc.,* Fifth Data Communications Symp., Snowbird, Utah, Sept. 27–29, 1977, pp. 2-5-2-9.

Error control

- Jacobs, I. M., "Practical applications of coding," *IEEE Trans. Infromation Theory,* pp. 305–310, May 1974.
- Heller, J. A., and Jacobs, I. M., "Viterbi decoding on satellite and space communications," *IEEE Trans. Communication Technology,* pp. 835–847, Oct. 1971.

Terminal devices, source coding, and channel multiplication

- Jayant, N. S., "Digital coding and speech waveforms: PCM, DPCM, and DM quantizers," *Proc. IEEE,* vol. 62, pp. 611–632, May 1974.
- Campanella, S. J., "Digital speech interpolation," *COMSAT Tech. Rev.,* pp. 127–158, Spring 1976.

Terrestrial interface and interconnect

- Chynoweth, A. G., "The fiber lightguide," *Physics Today,* pp. 28–37, May 1976.
- Conwell, E. M., "Integrated optics," *Physics Today,* pp. 48–59, May 1976.
- Kahn, R. E., "The organization of computer resources into a packet radio network," *Proc.,* 1975 National Computer Conf., June 1975.

competitive.

Coding can be effective in combating signal-dependent noise and multipath, certain modem implementation degradations, and interference. In the case of signal-dependent noise, the amount of degradation is related to the bit energy-to-noise ratio; coding reduces the value of bit energy-to-noise ratio required to achieve a given error probability and hence reduces the degradation. Thus, coding provides a signal-processing tradeoff against a requirement for greater multipath immunity through antenna design.

The great advances of recent years in solid-state electronics are expected to make low-cost, high-speed error control possible. An example of the impact of LSI on complex signal processing is a module in a 10-Mb/s Viterbi decoder, in which a single chip performs the add–compare–select operation that is central to the Viterbi algorithm. The resulting complete decoder has only 225 chips as compared with the 275 required in a corresponding 1.8-Mb/s decoder that does not use LSI circuits. Note that the more than fivefold increase in data rate is achieved with a decrease in the number of chips; due to the speed limitations of the components available for the implementation, such a rate increase requires parallel processing and would normally imply a more than fivefold increase in the number of chips.

Table I also indicates performance goals for error control in the 1980s. It predicts more extensive use of available techniques rather than new ideas.

Terminal devices

In the 1980s, digital terminal devices will increase data traffic and, probably, satellite data traffic as well. Such devices include optical character readers, terminals with magnetic tape, microfilm or microfiche input and output, digital facsimile terminals, and graphics terminals.

Digital terminal devices are improving in capability as their costs decrease, largely because of advances in solid-state memories and microprocessors. Video computer terminals now cost only about $1000. Digital facsimile terminals suitable for high-speed transmission—up to 1.5 Mb/s—are available. "Intelligent" terminals that can communicate with a central data base or a sophisticated computer should undergo rapid development in the next few years.

A number of digital voice-processing techniques have been developed or are under development. These techniques can reduce the bit rate per voice channel to anywhere from 40 to 2 kb/s; although, in the latter case, there is some loss of naturalness, quality, and ease of speaker identification.

The potential gain from voice source coding, in terms of the number of voice channels that can be handled with a given system capacity, is quite large. However, a number of factors must be considered before a particular method is applied, including sensitivity to errors, ability to "tandem" (that is, to undergo several digital-to-analog and analog-to digital conversions), ability to handle nonvoice analog signals such as modem signals, and cost.

Two basic digital channel multiplication schemes are included under the generic name of digital speech interpolation (DSI). One class of schemes covers various digital realizations of time-assigned speech interpolation (TASI). All of these systems exploit the fact that, on the average, a voice channel is being used only 40 percent of the time. For a large number of channels, it is probable that only 40 percent of the channels are active at any given time. Thus, if channels are transmitted only when they are active, more channels can be used for a given capacity—channels can be borrowed briefly for other calls when they are inactive.

Another speech interpolation scheme is speech predictive encoded communications (SPEC), which is based on the principle of sending a pulse-code-modulation (PCM) word for a channel only when that word differs significantly—that is, cannot be predicted—from the previous PCM word for that channel.

Table I lists goals and possible applications for source

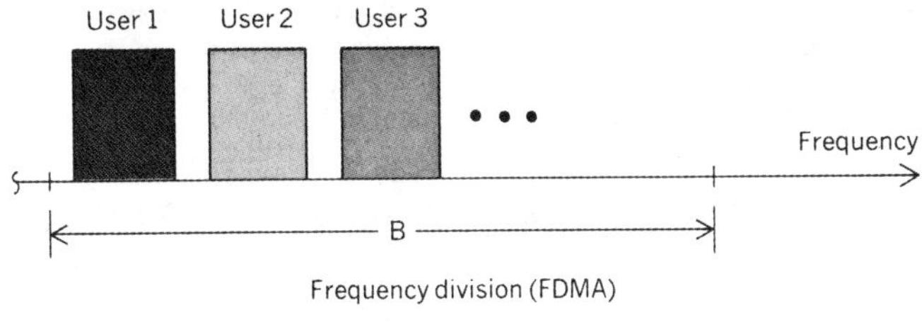

Frequency division (FDMA)

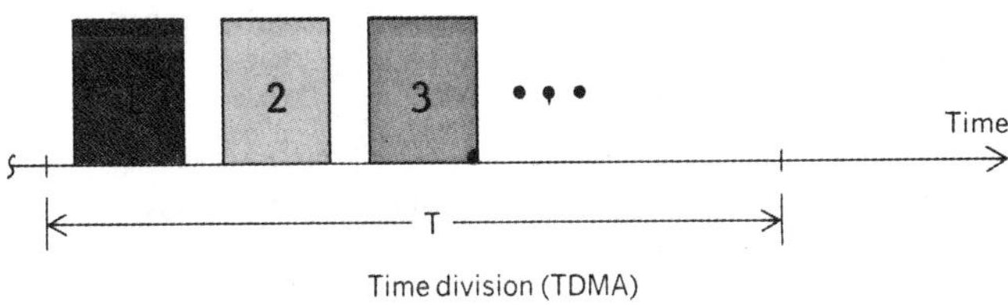

Time division (TDMA)

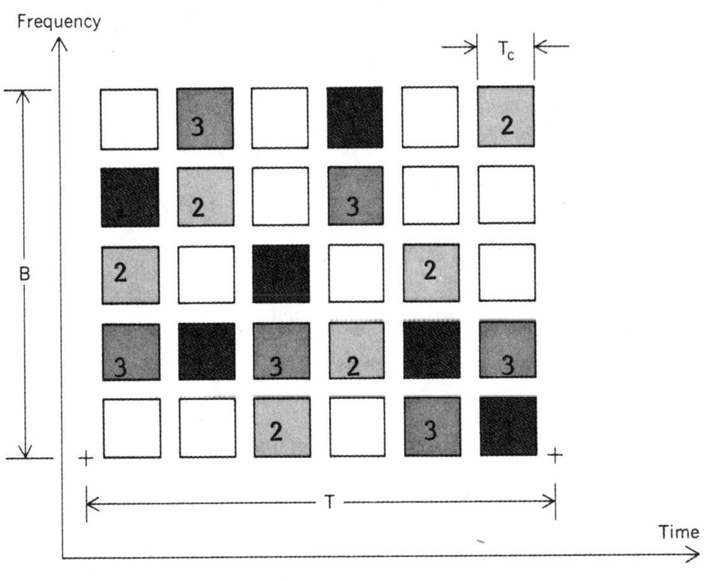

Code division (CDMA)

coding and channel multiplication in the 1980s. In the digital voice area, it is likely that PCM will dominate voice coding where "toll" quality is needed, and where there is an extensive interface with the established terrestrial network. Other coding techniques will be used in private networks and other applications where slightly less quality is needed. The more sophisticated coding methods, such as linear predictive coding, may become more widely used in the 1980s when terminal costs have dropped substantially.

Terrestrial interface and interconnect

The user of a satellite system must have a communication path to the earth terminal. Furthermore, various interface operations—formatting, signaling, multiplexing, buffering, etc.—must be performed by the user and the earth terminal equipment. At the very least, an appropriate communication protocol must be followed to support the flow of information between user and earth terminal.

In many cases (e.g., the INTELSAT system), the earth terminal does not interface with individual users but with a concentrator, such as a switch in the terrestrial communications system. In other cases, individual or small groups of users—for example, those connected to a PBX—will connect to the earth terminal through access lines obtained from the terrestrial communication plant. In still other cases, the physical distances will be small enough to justify direct user-owned access—e.g., via cable, microwave, or fiber optic link—to the earth terminal.

"Roof-top" or "customer premises" earth terminals are attractive because they free the satellite system from the costs and constraints of the terrestrial system. Such systems can provide highly flexible and economical service. This should also be true of less direct user–earth terminal interfaces as the cost of digital processing goes down, and it becomes possible to introduce a flexible processor that effectively isolates the terrestrial and satellite portions of an integrated communication system and allows these portions to be optimized separately. Indeed, an appropriate integration of terrestrial and satellite communications into a single system to provide various services efficiently and flexibly is one of the major challenges faced by the communications system designer.

For the interface of the future, fiber optics transmission is developing rapidly and will be cost-competitive with cable in the 1980s. This technology offers advantages in electromagnetic-interference immunity and flexibility of use, as well as large bandwidths (e.g., 500 MHz/km). Integrated optical processing is expected to flourish in the 1980s, raising the real possibility of integrated optical interface and connection systems. Optical fibers will also be used for signal transmission within the earth terminal.

Packet radio may connect users to earth terminals in a very flexible way without normal terrestrial communication facilities. The technique is being explored for both voice and data by the U.S. Defense Advanced Research Projects Agency. It uses radio repeaters organized so that information can be passed between various users even if they change location or are in motion. "Spread spectrum" transmission is employed so that packet radio can coexist with other uses in the same frequency band.

Table I gives the prognosis for terrestrial interfaces and interconnects in the 1980s.

In summary

Projections of communications satellite technology in the 1980s have been developed based on an applications and systems viewpoint. The specific areas selected were those that appear to have the potential of causing a major impact on the communications satellite system by improving one or more of the system design characteristics (capacity, connectivity, coverage, orbit or spectrum utilization, system cost, and flexibility) or one or more of the service characteristics (quality, reliability, and cost).

The pitfalls of forecasting the future are well documented. Political or regulatory decisions ranging from unexpected decisions at the 1979 World Administrative Radio Conference to new Government decisions could influence applications and system concepts, and hence the importance of various technologies. Genuine technological breakthroughs as contrasted to evolutionary developments could alter the technology projections significantly. However, the structure illustrated in Fig. 1 provides a basis for discussing the factors involved in future growth and a format for updating the projections. ◆

This article is based in part upon work performed at the Communications Satellite Corporation (COMSAT). The views expressed are those of the authors and do not necessarily represent the views of COMSAT.

Part II
Systems

There are currently a large number of communications satellites for international, domestic/regional, mobile, military, experimental, and other services.

The reprint papers discuss communication satellite systems that are currently in operation or scheduled for operation in the near term. Future or planned systems are discussed in Part VII. The systems are categorized by their usage using the hierarchy shown below.

Operational
  Commerical
    Fixed
      International
      Domestic/Regional
    Mobile
    Broadcast
  Military
    Fixed
    Mobile
  Experimental
    Civilian
    Military

Several systems with multiple functions have been categorized according to their primary function. The systems discussed in this part are shown in the following lists.

Operational Systems: Current and Near Term

Commerical
  Fixed
    International
      INTELSAT
      INTERSPUTNIK (USSR)
    Domestic/Regional
      TELESAT (Canada)
      Western Union (USA)
      RCA SATCOM
      COMSTAR
      SBS
      TDRSS/Advanced WESTAR
      ESA (OTS/ECS)
      Indonesia (Palapa)
      Arabsat
      Indian Satellite
  Mobile
    Marisat (Comsat General)
    MAROTS (ESA)
Military
  Fixed
    DSCS
    Skynet
    NATO
  Mobile
    FLTSATCOM

Experimental Systems: Current and Near Term

Civilian
  ATS-6
  Symphonie
  Canadian Technology Satellite
  Japanese Communication Satellite
  Japanese Broadcast Satellite
  SIRIO
  OTS
Military
  Lincoln Experimental Satellites (LES)

(Systems which obtain their space segment by lease from another system are not included.) The reference section contains one or more articles describing each of these systems. This introduction provides some general background.

INTELSAT

The INTELSAT system is presently composed of over 100 countries and provides global satellite communications services to three ocean regions.

At present, there are over 141 antennas, typically 30-m dishes, that operate in the 4–6 GHz band. There are three operational satellites in the Atlantic plus an operational spare. There is one satellite plus a spare in both the Indian and Pacific Ocean regions. There are a total of about 35 000 half-circuits in use worldwide (with approximately 22 000 in the Atlantic, 9 000 in the Indian, and 4 000 in the Pacific).

The INTELSAT IV satellite series, first launched in 1971, had a capacity of 8 000 half circuits (channels) plus one TV transponder and a SPADE transponder and was launched on an Atlas/Centaur. The INTELSAT IV-A provides an increased operational capacity of 12 000 channels plus TV and SPADE transponders (in the Atlantic primary) and was first launched in 1975. The INTELSAT V is scheduled for 1980 use in the Atlantic Ocean Region and will have a capacity of almost 25 000 channels. All of these capacities are relative to use in a frequency modulation-frequency division multiple access (FM/FDMA) mode. The increased capacity of INTELSAT V will be obtained through the use of dual polarization frequency reuse in the 4–6 GHz band plus beam isolation reuse in the same band. In addition, an 11–14 GHz capability has been included, with spot beam antenna coverages over North America and Europe in the Atlantic Ocean region. Reprint Paper 2.1 discusses the INTELSAT system in detail as part of a general synopsis of commercial communications satellites. References [1]–[3] discuss other aspects of the INTELSAT system.

DOMESTIC/REGIONAL

Many domestic satellite systems have become operational during the past several years. The first such geosynchronous system was initiated by TELESAT of Canada with the ANIK satellite launched in 1972. This system is composed of more than 80 earth stations and three satellites operating at 4–6 GHz, providing voice, video, and data communication services. The system also incorporates the first operational TDMA link at a transmission rate of 61 Mbit/s. Both heavy-route and thin-route services are provided, and a mix of earth segment equipment exists. Reprint Paper 2.1 and [4] discuss the Telesat system.

The U.S.S.R. has had, since the mid-1960's, an operational nonsynchronous system employing the Molniya satellites. These satellites are in elliptical orbit and use tracking earth-station antennas. In late 1975, the U.S.S.R. launched the first of the STATSIONAR series of synchronous satellites to be used for domestic service using the 4–6 GHz band. A total of ten STATSIONAR orbital slots are planned, seven of which would be used for global coverage (see [5]).

In the U.S. there are presently three operational systems—Western Union's WESTAR, RCA's SATCOM, and COMSAT General COMSTAR which is leased to AT&T and is used by both AT&T and GTE satellite earth stations. Both WESTAR and COMSTAR are spin-stabilized satellites, and all three operate in the 4–6 GHz band. The RCA SATCOM satellite is three-axis stabilized and was the first satellite to be launched on the Thor-Delta 3914 vehicle. In addition, the RCA and COMSTAR satellites use dual-linear polarization for frequency reuse, while WESTAR operates on a single polarization. Reprint Paper 2.1 and [6]–[8] discuss these systems.

Reprint Paper 2.2. describes the SBS system. This is an innovative system which will provide integrated voice, data, and image services using digital transmission at 12 and 14 GHz between 5- and 7-m customer premise earth stations. The use of a flexible TDMA allows efficient use of the satellite capacity.

Algeria, the Arab States, Brazil, Colombia, India, Indonesia, Iran, Japan, Malaysia, and Nigeria are also operating, developing, or seriously planning their own domestic or regional systems. The Algerian, Brazilian, Malaysian, and Nigerian systems currently use leased transponders on the INTELSAT satellites. Norway is also using leased INTELSAT capacity to provide service to the North Sea oil rigs. References [9]–[11] discuss some of these systems.

MOBILE

The MARISAT satellites, operated by COMSAT General, provide services to the U.S. Navy in the Atlantic, Pacific, and Indian Ocean regions and commercial voice and teletype communications in the Atlantic and Pacific Ocean regions [12]. The commercial portion of MARISAT is a C-band (4–6 GHz) earth-station satellite link crossed to an L-band (1.5–1.6 GHz) satellite/ship link. The ship terminals use 1.3-m tracking antennas. The military portion of the satellite uses frequencies in the lower UHF band.

The European Space Agency (ESA) is planning a maritime satellite called MAROTS [13]. An aeronautical satellite system, AEROSAT, had been planned for use over the Atlantic Ocean, but the program has been cancelled [14]. Several conference proceedings [15]–[16] are useful references on mobile communications.

MILITARY SYSTEMS

Military satellite communications systems have demonstrated the unique capabilities that they can provide in both the strategic and tactical environments and are currently an important part of the overall military communications system.

These systems are designed to satisfy the unique and vital requirements of the military as well as to ensure the existence of a communication capability in crisis situations. The military X-band (7–8 GHz) systems are primarily used for strategic communications including intelligence, trunking, extension and restoral of terrestrial-switched systems, and in support of strategic command and control requirements. They provide some communication to major Navy ships and potentially can provide communication to command aircraft. The UHF systems primarily provide tactical command and control communication for mobile platforms, including ships, aircraft, and ground vehicles.

The military systems must satisfy certain requirements which tend to make them more expensive and require technologically advanced designs.

The system must be able to provide secure communication to a large mix of earth terminal types with rapidly changing requirements, even in the advent of electronic warfare, or, in some cases, physical attack. Satisfying these requirements implies a need for such features as a secure and protected command and telemetry system, narrow-beam satellite antennas, satellite antennas with nulling capabilities, and an essentially real-time control capability.

26

By the early 1980's the U.S. strategic system, the Defense Satellite Communications System (DSCS), will be composed of DSCS II and DSCS III spacecraft. Reprint Paper 2.3 describes the DSCS-III satellite. References [17]-[21] discuss the DSCS system. The NATO system will be composed of NATO III spacecraft [22]-[23].

The U.S. UHF system consists of the leased UHF position of three MARISAT satellites and the FLTSATCOM spacecraft [24]. In the early 1980's, it will be augmented by a set of leased satellites.

The Lincoln Experimental Satellites (LES) have been used to develop and demonstrate many of the advanced technologies and system concepts. References [25]-[26] discuss the latest in the series, LES 8 and 9. Reprint Paper 3.6.7.1 discusses the intersatellite link on the LES 8/9.

EXPERIMENTAL

A number of satellites have been orbited in the last ten years for the purpose of communication satellite application and technology evaluation. Of particular interest are those systems which are pioneering the technology and operational experiments for broadcast satellite applications and systems which are major precursors to potential operational satellite-based telecommunication systems. The first class includes ATS-6, the Canadian Technology Satellite (CTS), the Japanese Broadcast Satellite Experiment (BSE), and the Russian Statsionar satellites. The second class includes the ESA Orbital Test Satellite (OTS), the French-German Symphonie satellite, the Italian SIRIO satellite and the Japanese Experimental Test Satellite (ETS-2), Experimental Communications Satellite (ECS), and medium capacity Communications Satellite (CS). Both classes of systems are not only important technology experiments, particularly in the areas of higher satellite EIRP and/or the use of higher frequency bands, but also involve significant applications oriented experiments and demonstrations.

Reprint Paper 2.4 discusses the significance of the ATS-6. References [27]-[28] describe the ATS-6 satellite. The CTS system is described in [29]-[30].

The Symphonie satellite is being used for numerous technical and operational experiments as described in [31]. OTS is a three-axis stabilized satellite developed by the European Space Agency [32]-[33] which operates in the 11-14 GHz band with polarization frequency reuse. It is designed as a precursor to the planned operational European Communications Satellite (ECS) system [34]-[35]. The ECS System is expected to provide intra-European international telephony service, exchange of TV programs among the members of the European Broadcasting Union, and new services such as high-speed data, TV broadcasting, teleconferencing, communications to North Sea oil rigs, and computer communications. The Italian SIRIO satellite which was launched in 1977 is being used for various propagation and communications experiments [36].

The Japanese ETS-2 satellite, which contains an *S*-band transponder, has the primary mission of enabling Japan's

National Space Development Agency (NASDA) to develop and test its ability to launch and control a spacecraft in synchronous orbit. It was launched using an *N*-rocket in 1977. It also contains a propagation experiment transmitter which can provide)*S*-band (1.7 GHz), *X*-band (11 GHz), and *Ka*-band (34 GHz) coherent signals.

The ECS satellite will be launched on an *N*-rocket. It will contain *C*-band and *K*-band (34.8-GHz uplink and 31.6-GHz downlink) transponders for digital data, wideband FM color television transmission (20-40 MHz), and *K*-band propagation experiments.

The CS satellite is the first commercial satellite operating in the 30-20 GHz region. The satellite contains six 200-MHz transponders in the 30-20 GHz band and two 200-MHz transponders in the 6-4 GHz band. It is being used with a variety of earth stations ranging in size from 3 to 13 m. References [37]-[41] describe the various Japanese satellite programs.

REFERENCES

[1] "A decade of Intelsat," *J. British Interplanetary Society,* vol. 28, no. 7, Jul. 1975.

[2] B. Edelson, H. William Wood, and C. J. Reber, "Cost effectiveness in global satellite communications," *IEEE Communications Society Magazine,* Jan. 1977.

[3] B. I. Edelson, R. Strauss, and P. L. Bargellini, "Intelsat system reliability," *Acta Astronautica,* vol. 2, nos. 7-8, Jul.-Aug. 1975.

[4] D. G. Thorpe, "Evolution of the Telesat Canada System into the mid-1980's," in *Satellite Communications: Future Systems,* from *Progress in Astronautics and Astronautics,* vol. 54, David Jarrett, Volume Ed., Martin Summerfield, Series Ed., New York: AIAA 1977.

[5] K. Johnsen, "Soviets plan seven-satellite global system," *Aviation Week & Space Technology,* Dec. 15, 1975.

[6] D. J. Lee, "System performance of America's first domestic communications satellite—WESTAR," *Proc. of EASCON 74,* pp. 386-393.

[7] J. Christopher and D. Greenspan, "RCA Satcom communication system," *EASCON '77,* Washington, DC, Sept. 26-28, 1977.

[8] M. Abutaleb, M. C. Kim, K. F. Manning, J. F. Phiel, Jr., and L. H. Westerlund, "The COMSTAR satellite system" *COMSAT Technical Review,* vol. 7, no. 1, 1977.

[9] J. S. Tengker, "Indonesian domestic satellite system," *EASCON 76,* Sept. 1976, pp. 11-A-11-U.

[10] M. Abdallah, "The Arab satellite," *Telecommunication J.,* vol. 44-IX, Sept. 1977, p. 422.

[11] A. Bairi and J. Leonhard, "A domestic satellite communications system for Algeria," *Int. Communications Conf.,* San Francisco, CA, June 1975.

[12] D. W. Lipke, D. W. Swearingen, J. F. Parker, E. E. Steinbrecher, T. O. Calvit, and H. Dodel, "MARISAT—A maritime satellite communication system," *COMSAT Technical Review,* vol. 7, no. 2, pp. 351-392, 1977.

[13] O. J. Haga and J. A. Vandenkerckhov, "MAROTS: A satellite for maritime communication," *IAF XXVIII Congress,* Prague, Sept. 25-Oct. 1, 1977.

[14] E. F. Binz, "A satellite concept for Aerosat," *AIAA Paper 76-259, AIAA/CASI 6th Communications Satellite Systems Conf.,* Apr. 1976.

[15] "Satellite systems for mobile communications," *IEEE Conf. Publication No. 95,* Mar. 1973.

[16] International Conference on Maritime and Aeronautical Satellite Communication and Navigation, *Conf. Proc.,* London, Mar. 1978.

[17] G. E. LaVean, "The defense satellite communications system," *AIAA Paper No. 74-457,* Los Angeles, Apr. 1974.

[18] F. E. Bond and W. H. Curry, Jr., "The evolution of military satellite communications systems, *Signal,* Mar. 1976, pp. 39-44.

[19] W. H. Curry, Jr., "The military satellite communications systems architecture," *Paper 76-268, 1976 AIAA Satellite Communications Conf.,* Montreal, Apr. 1976.

[20] J. H. Babcock, "Architecture and management of DOD satellite communications programs," *EASCON '77.*

[21] H. Wynne and D. E. Kendall, "Defense satellite communications system in the 1980's," *WESCON.*

[22] H. A. Kissinger, "NATO satellite communications: Past, present and future," *Signal,* Mar. 1976, pp. 53-57.

[23] E. T. Bobak and R. G. Clabaugh, "NATO phase III satellite design," *EASCON '77,* Washington, DC, Sept. 26-27, 1977.

[24a] N. L. Wardle, "U.S. Navy fleet satellite communications," *AIAA Paper, No. 74-458,* Los Angeles, Apr. 1974.

[24b] F. S. McCartney and E. K. Heist, "FLTSATCOM program review: Requirements, design, and performance," *WESCON,* Los Angeles, Sept. 1978, pp. 442-452.

[25] F. J. Solman, C. D. Bergland, R. W. Chick, and B. J. Clifton, "The K_a-band systems of the Lincoln Experimental Satellites LES-8 and LES-9," *Proc. 7th AIAA Communications Satellite Systems Conf.,* San Diego, Apr. 24-27, 1978, pp. 208-215.

[26] L. J. Collins, L. R. Jones, D. R. McElroy, D. A. Siegel, W. W. Ward, and D. K. Willim, "LES-8/9 communications system test results," *Proc. 7th AIAA Communications Satellite Systems Conf.,* San Diego, Apr. 24-27, 1978, pp. 471-478.

[27] D. B. Dobson, Ed., *Special Issue on IEEE Trans. on Aerospace and Electronic Systems on ATS-6 Satellite,* vol. AES-11, no. 6, Nov. 1975.

[28] W. M. Redisch, "ATS-6 Description," *Proc. of the IEEE International Conf. on Communications,* Jun. 16-18, 1975, pp. 18-1-18-5.

[29] C. A. Franklin and E. H. Davison, "A high-power communications technology satellite for the 12 and 14 GHz bands," *AIAA Fourth Communications Satellite Systems Conf.,* Washington, DC, Apr. 24-26, 1972.

[30] P. L. Donoughe, "United States societal experiments via the communications technology satellite," *Proc. of the IEEE Int. Conf. on Communications,* Jun. 14-16, 1976, pp. 1-17-1-22.

[31] G. Moesl and J. Muller, "In-orbit performance and experimental utilization of the symphonie satellites," *Paper 76-306, 1976 AIAA Satellite Communications Conf.,* Montreal, Apr. 1976.

[32] R. C. Collette and B. Stockwell, "The OTS Project, basis of future European space communications systems," *AIAA 5th Communications Satellite Systems Conf.,* Los Angeles, Apr. 22-24, 1974.

[33] S. Tirro and A. Bayliss, "The utilization programme of the orbital test satellite," *Paper 76-247, 1976 AIAA Satellite Communications Conf.,* Montreal, Apr. 1976.

[34] R. C. Collette, "The European communications satellite programme and the orbital test satellite," *Journal of the British Interplanetary Society,* vol. 29, 1976, pp. 335-353.

[35] P. Bartholome, "The European communications satellite system—A review of current and planned activities," *Paper 76-243, 1976 AIAA Satellite Communications Conference,* Montreal, Apr. 1976.

[36] F. Carassa, L. A. Ciavoli-Cortelli, and S. Tirro, "The SIRIO-SHF Experiments: First in-orbit measurements, results and possibilities of experimentation," *IAF XXVIII Congress,* Prague, Sept. 25-Oct. 1, 1977.

[37] T. Ishida, N. Fugono, J. Tabata, and M. Ohara, "Program of Experimental Communication Satellite (ECS) of Japan," *Paper 78-614, 1976 AIAA Satellite Communications Conf.,* Montreal, Apr. 1976.

[38] H. Kaneda, K. Tsukamoto, and H. Fuketa, "Experiments in the Japanese CS Program," *Paper 78-616, 1976 AIAA Satellite Communications Conf.,* Montreal, Apr. 1976.

[39] T. Ohtake, H. Reichert, and L. T. Seaman, "Japanese broadcast satellite," *Microwave Journal,* Sept. 1977.

[40] H. S. Braham, "Spacecraft for broadcast systems," *EASCON '77* Washington, DC, Sept. 26-28, 1977.

[41] H. Kaneda, K. Tsukamoto, and O. Ogawa, "Experiments with the Japanese medium-scale broadcast satellite," *Paper 78-573, 1976 AIAA Satellite Communications Conf.,* Montreal, Apr. 1976.

2.1

A SYNOPSIS OF COMMERCIAL SATELLITE COMMUNICATIONS SYSTEMS*

P. L. Bargellini, Senior Scientist
COMSAT Laboratories
Clarksburg, Maryland 20734

INTRODUCTION

With their impact upon people's lives and exchanges among nations, communications satellites, which constitute one of the outstanding fallouts of the space program, have totally altered the patterns of world communications. Communications by satellite, having evolved from science fiction through scientific and engineering developments to trials and operations, is now a thriving business. Satellite systems interface effectively with terrestrial networks; their second decade of operational service began three years ago, and substantial growth is foreseeable in the future.

Developments in the commercial field can be best projected by distinguishing three separate categories of satellite systems: the international INTELSAT system, domestic (U.S. and non-U.S.) systems, and mobile systems. With respect to the first category, the INTELSAT system today provides routine reliable worldwide communications. In the second category, satellite systems in Canada, the Soviet Union, Indonesia, and the United States provide domestic communications for telephone, television, facsimile, and data traffic. Finally, in the case of mobile services, communications via satellite have been extended to ships at sea with the MARISAT system.

This paper reviews the origins of satellite communications, surveys their technological and operational progress, and examines future trends.

HISTORICAL BACKGROUND

Satellite communications combines two distinct technologies, rocketry and microwave engineering, which had greatly advanced during World War II. Their combination, intended to provide new means of communications, was proposed shortly after the end of the War; early experiments occurred about 15 years later.

In 1945, Arthur C. Clarke conceived satellites in geosynchronous orbit providing global telecommunications. Clarke showed that three geosynchronous satellites, powered by solar energy converted into electricity by silicon cells, could provide worldwide communications "for all possible types of service with unrestricted use of a frequency band at least 1000 MHz wide (providing), with the use of beams, an almost unlimited number of channels."[1]

The translation of Clarke's concept into reality required adequate rockets for launching satellites and suitable onboard electronic devices. In the field of rocketry, the pioneering work of Constantin Tsiolkowsky, Robert Goddard, Herman Oberth, and Wernher Von Braun led through the V-1's and V-2's of World War II to the first Sputnik, launched by the U.S.S.R. in October 1957. In communications, electronic devices operating at ever increasing frequencies and a better understanding of radio wave propagation phenomena were major causes of progress. Curiously enough, however, until the late 1940's electrical communications systems had developed without a clear understanding of the commodity (information) which was being handled. In 1948, long after practical communications systems had been implemented, Statistical Communication Theory,

also known as Information Theory, was developed. This theory and its many extensions focus on the fundamental relationships among information transmission rate, bandwidth, signal and noise power, transmission impairments, and attenuation.[2]

Communications satellite systems benefited from all this knowledge, and as soon as rocket engineering was capable of safely injecting reliable electronic packages into orbit, success was ensured. Most importantly, a substantial increase in communications capacity and network flexibility with respect to previously available wire and wireless systems was achieved.

Arthur Clarke's concept went unnoticed by communications engineers for several years. In 1954, J. R. Pierce of Bell Telephone Laboratories independently studied the fundamentals of radio relaying via artificial satellites, and two years before Sputnik, prepared concrete technical proposals for satellite communications.[3] Passive satellites had the important advantages of simplicity and potentially unlimited multiple access. However, the inverse distance square law applicable to active satellites gives them a substantial advantage over passive satellites, to which the inverse distance fourth power law applies. Thus, although early experiments were performed with reflecting balloons, all operating systems have used active satellites.

The stage was set for satellite communications when the first commercial satellite, Early Bird, was orbited in 1965. With Early Bird, the status of transoceanic communications was drastically altered as 240 voice circuits became available, and high-quality transatlantic TV was possible for the first time.

SERVICES AND INSTITUTIONAL ARRANGEMENTS

The combined efforts of governments and industry were required to promote space exploration leading to satellite communications. In the United States, the Communications Satellite Act, enacted by the 87th Congress in August 1962, was signed by President John F. Kennedy on 31 August 1962. The Act declared it to be the policy of the United States "to establish, in conjunction and in cooperation with other countries, as expeditiously as practicable a commercial communications satellite system, as part of an improved global communications network, which will be responsive to public needs and national objectives, which will serve the communications needs of the United States and other countries, and which will contribute to world peace and understanding."

The Communications Satellite Corporation (COMSAT), incorporated in February 1963, led the development program for the global satellite system and its financing. In 1964, 11 participant nations entered into a unique international partnership originally known as the International Telecommunications Satellite Consortium (INTELSAT), and designated COMSAT as its manager.

Since its inception, INTELSAT has operated under two sets of agreements, referred to as interim and definitive. On 12 February 1973, INTELSAT, whose

*This paper is based on work performed at COMSAT Laboratories.

Manuscript received December 6, 1978.

membership had grown to 87, became the International Telecommunications Satellite Organization, with the definitive agreements becoming effective.

While the space segment, consisting of the satellites and the facilities required to support their operation, is owned and operated by INTELSAT, the earth segment, which comprises the earth stations in various countries, is owned by designated telecommunications entities in each country.

The expansion of the INTELSAT system, which continued with great vigor and has been documented in the literature,[4,5] will be illustrated in some detail in a later section.

Although in recent years domestic services have been provided by INTELSAT to numerous countries (at present 13 countries lease 12 transponders from INTELSAT and some 70 earth stations are in use for this purpose), it became clear in the late sixties that satellite communications for internal (domestic) services would be highly attractive to countries characterized by vast territory or geographical singularities which would make difficult and costly the deployment of earth-based communications systems.

The U.S.S.R. initiated domestic satellite communications in 1965 with nongeosynchronous Molniya satellites placed in highly elliptical orbits inclined by 65°. Their capabilities have been increased and the orbits have been replenished to accommodate the relatively short lifetime of these spacecraft. More recently the U.S.S.R. has also placed several communications and broadcast satellites in geosynchronous orbit (Statsionar, Raduga, Ekran, etc.). In November 1972, Canada launched the first geostationary satellite for domestic communications. The Canadian system is in its sixth year of operation; a second generation of spacecraft is under construction, and a third generation is planned.[6] In November 1975, Indonesia inaugurated its own domestic satellites.[7] Domestic satellites have permitted the establishment of new, highly reliable, high-capacity communications links where almost none existed previously.

The different situation in the U.S.,[8] due to existing ground-based communications networks, has led to the establishment of four separate, partly competing systems, and an additional innovative system is currently being developed.

Satellites are obviously ideal for establishing links with mobile terminals. In 1976, almost a decade of efforts directed toward the development of a satellite system to serve mobile maritime users led to the implementation of the MARISAT system.[9]

The extension of communications satellites to aeronautical services, TV distribution, educational and public services and also direct broadcasting to individual homes has been the object of numerous studies. All the above mentioned areas show considerable promise; yet difficult institutional and financial problems will have to be resolved before operational status is achieved.[10]

<u>CHOICE OF THE ORBIT</u>

Orbital height and inclination are fundamental parameters of satellite systems.[11] From Kepler's Laws, the orbital period is

$$P = \frac{2\pi\alpha^{3/2}}{\mu^{1/2}} \qquad (1)$$

where α is the semimajor axis of the ellipse, and μ is the gravitational constant x earth mass = 3.99×10^{14} m^3/s^2. As only modest payloads (~100 kg) could be placed in low- or medium-altitude orbits (<10,000 km) by early rockets, these orbits were used for experimental satellites. Satellites in low orbits pass rapidly overhead and require tracking. Continuous communications between points on the earth's surface would thus require traffic handover as a satellite sets beyond the horizon while another rises to take its place and many satellites to ensure continuous coverage. In spite of these difficulties, systems were proposed with as many as 50 spacecraft in medium-altitude orbits to serve the Atlantic Ocean region. Early experiments with single spacecraft (TELSTAR, 1962-1963, and RELAY, 1962-1964) proved the feasibility of active satellites.

Increasing the orbital altitude to 35,863 km, i.e., about six times the earth's radius, increases the period to a sidereal day (23 hr, 56 min). A synchronous satellite above a fixed spot, subtends an angle of about 18°, and can provide coverage of about four-tenths of the earth's surface. Additional advantages of the "geostationary" orbit are zero Doppler, infrequent thermal stress cycles, moderate energy storage requirements for eclipse operation, mild radiation environment above the Van Allen belts, and minor perturbations by the earth's magnetic field.

Yet the difficulties of achieving operation in geostationary orbit were considerable. The Thor-Delta vehicle, which had been used for launching the TELSTAR (AT&T, 1962-1963) and RELAY (RCA-NASA, 1962-1964) satellites, was inadequate to inject a payload of around 80 kg directly into geostationary orbit from Cape Kennedy (28° north latitude). A brilliant solution, consisting of first injecting the payload in a highly elliptical orbit with apogee at synchronous altitude, was eventually adopted.[12] An added rocket, weighing about one-half of the payload in transfer orbit and fired at apogee, allowed circularization of the orbit. Auxiliary thrusters were used to change the orbital plane, thus achieving zero inclination. Such a mission was regarded as highly complicated, given the state-of-the-art of space technology at the time it was first proposed. The first experimental geosynchronous satellite, SYNCOM I (February 1963), failed to attain its orbit; the second and third launches of SYNCOM II (July 1963) and SYNCOM III (August 1964) were successful. These satellites were developed for NASA by the Hughes Aircraft Co.

A geostationary satellite must be kept in position as solar and lunar gravitational forces and the non-sphericity of the earth cause orbital perturbations. Solar radiation pressure is an additional nongravitational perturbation. Stationkeeping is required to counteract these forces, corrections being obtained by activating onboard thrusters upon command from the earth. Ideally, longitudinal drift and orbit inclination can be controlled independently, but in practice some interaction is encountered. Usually drift is allowed to build up to some set value beyond which corrective maneuvers are effected. Information from earth, sun, and star sensors provide the inputs for correction maneuvers.

In addition to orbital positioning, attitude control is necessary to point the satellite's directional antennas toward the earth. A common method of attitude control consists of spinning the satellite body for gyroscopic stiffness. Despinning the communications antenna permits beam pointing accuracy of the order of about one-tenth of a degree. This simple, reliable method eases the problem of thermal control of the structure, but also results in only partial utilization of the total area of the solar cell panels, limiting the amount of electrical power available.

Early Bird (INTELSAT I) was an "experimental satellite with operational capabilities" which provided answers to questions related to the applicability of geostationary satellites within the existing global terrestrial telephone network.[13] Early Bird remained in operational commercial service for more than four years.

THE COMMUNICATIONS PROBLEMS

Satellite communications involve at least two cascaded links: the up-link from an earth station to the active repeater in orbit, and the down-link from the satellite to another earth station. The two channels in opposite directions constitute a full circuit. A satellite in geosynchronous orbit which sees about four-tenths of the earth's surface can, in principle, link any pair of stations separated by great circle distances up to 17,000 km. It is this capability of multiple access which makes satellites a truly unique tool for communications.

For m stations "visible" from a satellite, the number of potentially available communications circuits is

$$n = \frac{m(m-1)}{2} \qquad (2)$$

Clearly, the n-port network topology of the satellite system offers conspicuous advantages vis-a-vis the inflexibility of 2-port networks such as cable and land circuits.

For a single link, the rate of information transmission relates to certain fundamental parameters through the equation

$$R = \frac{BP_t G_t A_r}{4\pi r^2 \beta N} = \frac{B}{\beta}\frac{C}{N} \qquad (3)$$

where

R = information rate (bit/s)
r = link distance (m)
B = bandwidth (Hz)
P_t = transmitter power (W)
G_t = transmitting antenna gain
A_r = receiving antenna area (m^2)
β = communications efficiency = $\dfrac{\text{energy per bit}}{\text{noise power density}}$
C = carrier power (W)
N = noise power (W)

The upper bound for R is the Shannon channel capacity:

$$R \leq B \log_2\left(1 + \frac{S}{N}\right) \qquad (4)$$

When B is allowed to go to infinity, the signal-to-noise ratio goes to zero and the communications efficiency, β, assumes the asymptotic minimum value $\log_e 2 = 0.693$. The actual value of β greater than the above-mentioned minimum depends upon the modulation-demodulation scheme. Equation (3) indicates that communications capacity is proportional to bandwidth and power when other quantities are fixed. Optimum operating frequencies and bandwidth defined, in principle, on the basis of minimum noise power density and favorable propagation conditions depend in practice on spectrum availability in terms of international and regional agreements. Power, easily available on the ground, is limited by spacecraft mass and hence launch vehicle capability.

Since onboard transponders have, until now, been operated as frequency translators between the up- and down-links, two cascaded links are considered. The system transmission rate is obtained by introducing into equation (3) the overall carrier-to-noise ratio, which is equal in analog systems to the inverse of the sum of the inverses of the carrier-to-noise ratios in the up- and down-links [$(C/N)_u$ and $(C/N)_d$, respectively], i.e.,

$$R = \frac{B}{\beta}\frac{(C/N)_u (C/N)_d}{(C/N)_u + (C/N)_d} \qquad (5)$$

Among the modulation schemes capable of trading signal-to-noise (S/N) ratio for bandwidth, frequency modulation (FM) has prevailed because of its effectiveness, simplicity, and adaptability to the interfacing of satellite and ground communications systems. Frequency modulation, combined with frequency-division multiplexing (FDM) of voice channels at baseband and frequency-domain multiple access (FDMA) for several RF transmissions through the satellite, constitutes the transmission scheme known as FDM/FM/FDMA.*

Unfortunately, the situation is complicated by nonlinearities, especially in traveling wave tube amplifiers (TWTAs), which cause undesirable modulation conversion effects (AM/AM and AM/PM) in both the satellite repeaters and the transmitting earth stations. Ultimately, the maximum information transmission rate—in practice, the number of telephone circuits a satellite can handle—is dictated by the overall carrier-to-noise ratio, C/N_{tot}, whose inverse equals the sum of the inverse carrier-to-noise ratios arising in the up-link, down-link, and in the intermodulation processes:

$$\frac{1}{C/N_{tot}} = \frac{1}{C/N_{up\text{-}link}} + \frac{1}{C/N_{down\text{-}link}} + \frac{1}{C/N_{im}} \qquad (6)$$

Other impairments arise from earth RF out-of-band emission, cochannel interference due to imperfect beam isolation, transponder group-delay and dual path distortion, and adjacent transponder interference. Finally, in the case of frequency reuse through orthogonal polarizations, another kind of interference appears in the form of cross coupling due to rain depolarization effects (amplitude and phase). Clearly, transmission system planning requires careful analysis of several interactive factors.

With FDM/FM/FDMA, the intermodulation noise contribution can be reduced by backing off the TWTAs until a maximum value for C/N_{tot} is achieved in equation (6). In general, the choice of a specific method of multiple access, or a combination of the methods of multiple access with appropriate forms of modulation and multiplexing, yields solutions which are ultimately evaluated in terms of the number of channels which can be provided.

With communications capacity proportional to transmitter power, and power in turn related to spacecraft mass and size, since the communications subsystem

---

*A notable exception is the phase shift keyed (PSK) FDMA single-channel-per-carrier (SCPC) access on demand transmission system (SPADE) introduced in INTELSAT in 1969 and presently used by more than 30 countries.

of a satellite is the useful payload, it is important to minimize the mass of all noncommunications subsystems. In the apportionment of mass among the different subsystems (communications, structure, power, positioning and orientation, and TT&C), communications satellites of different types have been investigated.[14] Trends have thus been identified which reveal that the use of advanced technologies would greatly increase communications capacity for a given total satellite mass. Although some of these concepts have been verified in experimental programs, the first decade of satellite communications has chiefly evolved around the gradual growth from the technology introduced in the SYNCOM and Early Bird satellites. In this sense, communications capacity has increased essentially in terms of larger and more powerful satellites. The variety of launch vehicle combinations and their capabilities are shown in Figure 1 for conventional disposable rockets.

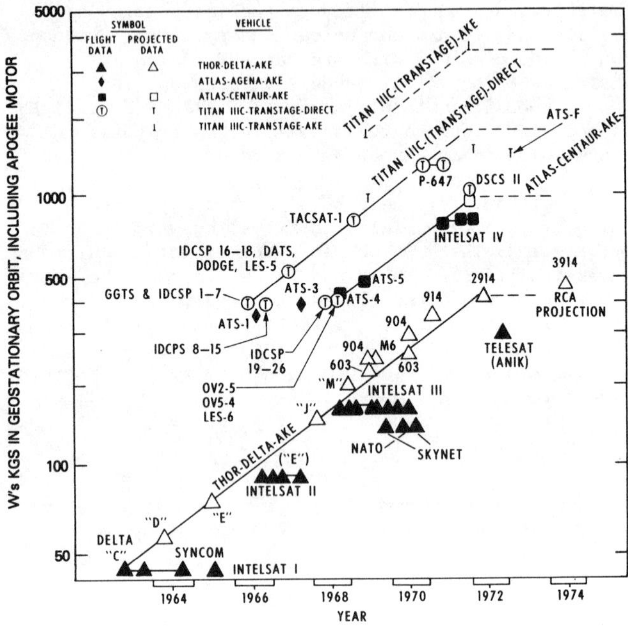

Figure 1. Launch Vehicle Evolution

In spite of the many potential launchers, from small rockets to the huge SATURN V vehicle, the choice of a practical means of placing communications satellites in geosynchronous orbit is limited, in the U.S.A., to the Thor, Atlas, and Titan rockets in combination with specific upper stages. Important factors are also the availability of launch facilities at suitable geographical locations, the life of a given program, its reliability record, and naturally, the cost of the vehicle.

In recent years NASA's efforts have been directed toward the construction of reusable vehicles which will take off as rockets and land as planes. The impact of this Space Transportation System, in combination with the upper stages required to inject payloads in geosynchronous orbit on future communications satellites, will certainly be conspicuous.[15-17]

THE INTELSAT SYSTEM

The INTELSAT global system, the first commercial system in operation (1965), is today by far the largest and most extensive. All INTELSAT satellites have operated in synchronous equatorial orbits 35,700 km above the equator. This choice allows continuous coverage, with tracking by earth stations necessary to maintain the antenna beams pointed at the satellite in the presence of very small drift angles. All of the satellites used so far are spin-stabilized; that is, they maintain their orientation in space through rotation of the satellite body along an axis parallel to the earth's axis. Silicon solar cells mounted on the spinning body convert solar energy into electrical energy and nickel-cadmium batteries provide energy during eclipses.

The satellites receive on frequencies in the 6-GHz band and transmit on frequencies in the 4-GHz band. Thus, the repeaters have all been frequency translators with wideband receivers and, usually, channelized transmitters. Tunnel diodes have been used in the front end and traveling wave tubes (TWTs) as final amplifiers delivering a few watts of RF power.

The evolution of the INTELSAT space segment is illustrated in Figures 2 and 3.

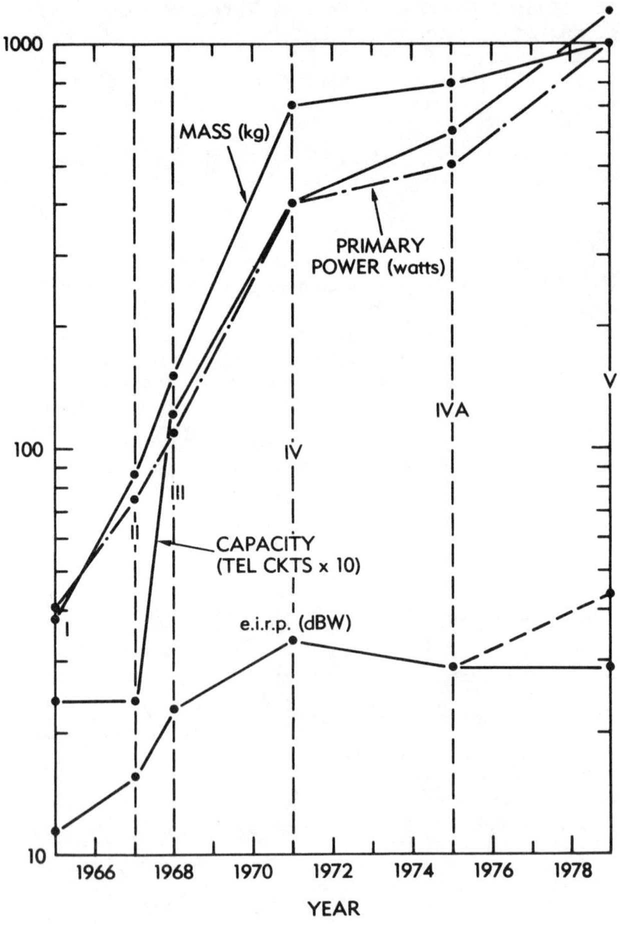

Figure 2. Evolution of INTELSAT Space Segment

As the effectiveness of a communications satellite depends upon its capacity, which in turn is related to radiated power and bandwidth, and upon lifetime in orbit, the technology developed during the 1965-75 decade by INTELSAT resulted in improvements in all three areas:

a. _Power_. More powerful launch vehicles allowed larger mass in orbit, thus providing a greater area for solar cells and more electrical power for radio transmission.

b. _Bandwidth_. Increased power and more sophisticated transponder design permitted more efficient use of the allocated bandwidth, and eventually its reuse.

c. _Lifetime_. Improved components, devices, and design, and more effective quality assurance techniques improved satellite reliability and operating lifetime.

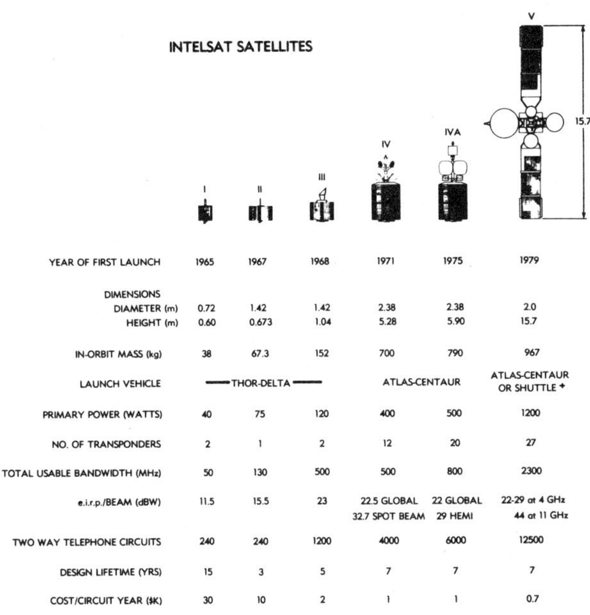

	I	II	III	IV	IVA	V
YEAR OF FIRST LAUNCH	1965	1967	1968	1971	1975	1979
DIMENSIONS						
DIAMETER (m)	0.72	1.42	1.42	2.38	2.38	2.0
HEIGHT (m)	0.60	0.673	1.04	5.28	5.90	15.7
IN-ORBIT MASS (kg)	38	67.3	152	700	790	967
LAUNCH VEHICLE	THOR-DELTA			ATLAS-CENTAUR		ATLAS-CENTAUR OR SHUTTLE +
PRIMARY POWER (WATTS)	40	75	120	400	500	1200
NO. OF TRANSPONDERS	2	1	2	12	20	27
TOTAL USABLE BANDWIDTH (MHz)	50	130	500	500	800	2300
e.i.r.p./BEAM (dBW)	11.5	15.5	23	22.5 GLOBAL 32.7 SPOT BEAM	22 GLOBAL 29 HEMI	22-29 at 4 GHz 44 at 11 GHz
TWO WAY TELEPHONE CIRCUITS	240	240	1200	4000	6000	12500
DESIGN LIFETIME (YRS)	15	3	5	7	7	7
COST/CIRCUIT YEAR ($K)	30	10	2	1	1	0.7

Figure 3. INTELSAT Spacecraft Characteristics

Early Bird, built by the Hughes Aircraft Company, was launched in April 1965. It weighed only 38 kg in orbit and had a total effective radiated power of 10 W in each of two transponders, using only 50 MHz of bandwidth. Its limited power output required that the spinning antenna radiation pattern be "squinted" to cover North America and Europe. Its potential capacity of 240 two-way telephone circuits allowed linking of only two earth stations at a time; i.e., multiple access was not available, nor was it possible to carry television and telephone simultaneously.

A second series of commercial satellites was developed to support the manned spaceflight operations of NASA. Reliable communications were urgently needed to connect a worldwide network of tracking stations for Project Apollo, some on islands and others on ships at sea. The Hughes-built INTELSAT II satellites were larger than Early Bird, having more power and bandwidth, and thus were able to provide coverage of a wider area of the earth. An important innovation in INTELSAT II was multiple-access capability; many pairs of earth stations could be connected through the satellites, each transponder carrying several radio frequency carriers simultaneously. The first INTELSAT II satellite entered service in January 1967.

The larger INTELSAT III series, built by TRW, was introduced in late 1968. By 1969, three of these satellites made possible the realization of a true global communications system. Each INTELSAT III had a nominal capacity of 1,200 telephone circuits. This increase was achieved by using a mechanically despun

antenna always pointed toward the earth, providing a so-called "global beam" which covered all of the earth visible from a given position of the synchronous orbit.

The first INTELSAT IV satellite was launched in January 1971. Built by Hughes, and weighing 720 kg in orbit, these satellites are still in use. The major advance of INTELSAT IV over its predecessors was the use of "spot-beam" transmit antennas covering only a small portion of the visible earth, in this case a beam angle of about 4.5°. The resulting concentration of radiated energy contributed to the increased capacity. The INTELSAT IV satellites are rated at about 4,000 circuits or greater, depending on the number of transponders connected to spot beams and the multiple-access system in use. The INTELSAT IV electrical power subsystem provides about 470 W generated by some 45,000 solar cells, and includes nickel-cadmium batteries used during solar eclipse.

INTELSAT IV was the first communications satellite to be bandwidth- rather than power-limited.[18] The communications subsystem is channelized into 12 transponders of 36-MHz bandwidth. All transponders receive from a global-beam antenna. Both global- and spot-beam transmit antennas provide respectively about 180 and 2,500 W of equivalent radiated power (22.5- and 34-dBW e.i.r.p.).

Improvements of the INTELSAT IV configuration resulted in the INTELSAT IV-A satellites, again built by the Hughes Aircraft Company. The first of these satellites entered service in 1975 in the Atlantic region to meet that area's high level of traffic requirements.

From 1976 to 1977, three other INTELSAT IV-A satellites entered service, two in the Atlantic and one in the Indian Ocean region. Frequency reuse techniques, through spatial separation, allow simultaneous use of the same portion of the spectrum in two separated areas; e.g., in the Atlantic region, one shaped beam will cover Europe and Africa, and the other will cover North and South America. A global beam is also available and 20 transponders, connected in various combinations, provide global, eastern, or western receive or transmit beams. With approximately the same weight and power as INTELSAT IV and using the same 6/4-GHz frequency band, the INTELSAT IV-A satellite has 50 percent greater capacity, i.e., about 6,000 circuits.

The INTELSAT V satellites,[19] the first of which is expected to become operational by late 1979 or early 1980, will differ from their predecessors as follows:

a. body stabilization,
b. fourfold frequency reuse by beam separation and cross polarization of the 500-MHz available bandwidth at 4/6 GHz, and
c. use of the 11/14-GHz band with twofold frequency reuse of the available 500-MHz bandwidth via beam separation.

Table 1 gives the major characteristics of this series of satellites. The configuration chosen for INTELSAT V, shown in detail in Figure 4, departs from that of the preceding INTELSAT satellites. Increased communications capacity will be achieved by the following means:

a. higher power (prime, RF, and e.i.r.p.);
b. more efficient bandwidth utilization;

c. addition of the 11/14-GHz bands; and

d. advances in the design of all subsystems, resulting in greater payload availability for communications.

The Ford Aerospace and Communications Corporation is the prime contractor for the INTELSAT V satellites.

Table 1. INTELSAT V Performance Specifications

Mass in Orbit	967 kg
Launch Vehicle	Atlas-Centaur or STS
Frequency Bands	6/4 and 14/11 GHz with interconnect capability and frequency reuse (separate beams at both 6/4 and 14/11 GHz, and orthogonal polarization at 6/4 GHz)
Transponders	27
6/4 GHz	16 with 80-MHz bandwidth and 5 with 40-MHz bandwidth
14/11 GHz	4 with 80-MHz bandwidth and 2 with 240-MHz bandwidth
Received Power Flux Density, dBW/m^2	
6 GHz	-60 to -69
14 GHz	-67 to -76.5
e.i.r.p.,* dBW	
4 GHz	22 to 29
11 GHz	44.1
Nominal Communications Capacity	12,000-14,500 voice circuits

*Specific coverage requirements differ according to the region of utilization.

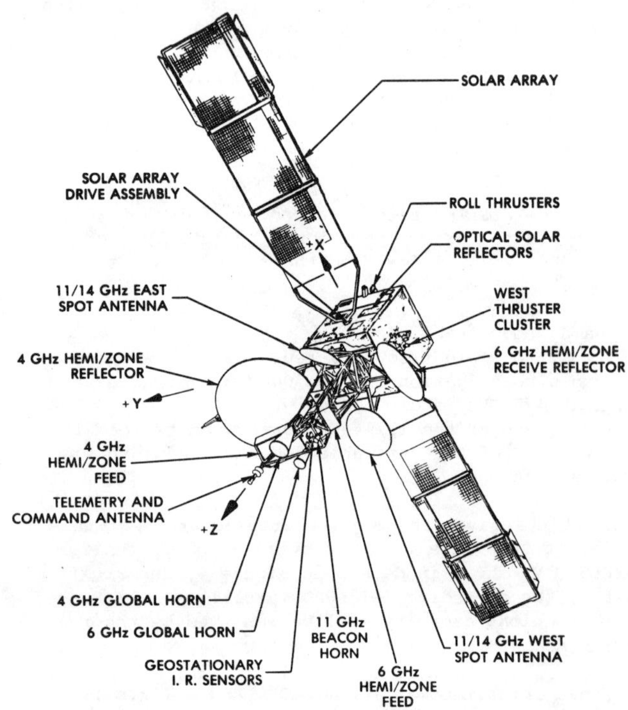

Figure 4. INTELSAT-V Configuration (Courtesy of Ford Aerospace Communications Corporation)

In conjunction with spacecraft developments, the earth segment has grown enormously, as shown in Figure 5. The INTELSAT system has grown from 5 earth stations in 1965 to 202 earth station antennas in 88 countries in 1978.

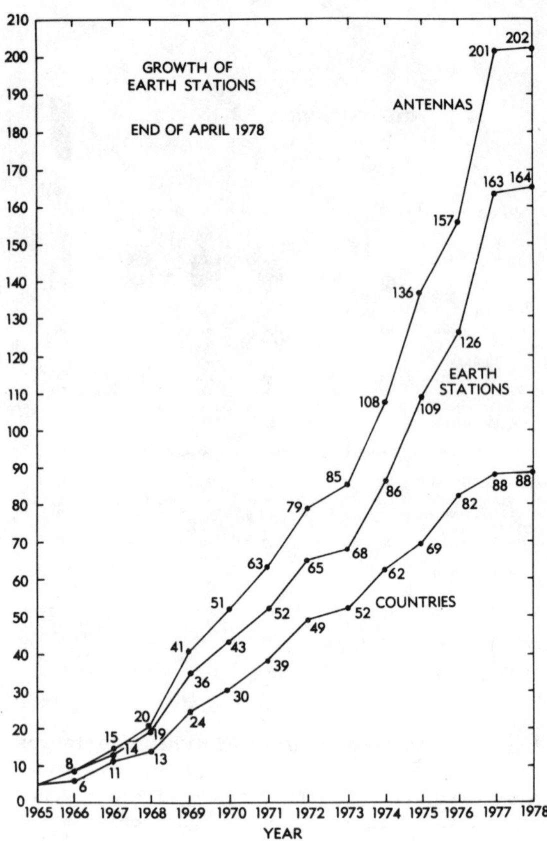

Figure 5. Earth Segment Growth in the INTELSAT System

After the early use of radome-covered horn antennas, a design previously developed for low-orbit satellites such as TELSTAR and RELAY, two kinds of earth stations have been adopted by INTELSAT. The Standard A antenna comprises a large reflector (about 30 m diameter) and liquid-helium-cooled parametric amplifiers, resulting in a G/T ratio of 40.7 dB/K. These antennas, fully steerable in azimuth and zenith, can track a geostationary satellite to within 0.02° and allow full versatility for all types of communications.

The Standard B antennas comprise a 12-m reflector with limited steerability, although the tracking requirements are identical to those of the Standard A antennas. The resultant G/T of 31.7 dB/K allows the convenient adoption of this antenna, which costs only a fraction of the previously mentioned Standard A, at terminals characterized by light traffic. A Standard C antenna will be introduced in the INTELSAT V era. This antenna will operate at 11/14 GHz and have a G/T ratio of 39.0 dB/K for 90 percent of the year.[20]

The INTELSAT system provides continuous global telephone service in addition to television, telegraph, and data transmission. Temporary service is also available on a short-term basis to accommodate special world events and special communications requirements. All such services meet or exceed international standards of quality, while maintaining an exceedingly high level of reliability.[21,22]

The primary operational mode of the system is multiple access, preassigned in the frequency domain. The previously mentioned SPADE, a demand-assignment multiple-access system, was introduced in 1973 and has been implemented in the Atlantic Ocean region, with 32 terminals in operation. Direct digital transmissions have been established between a number of points, and more are planned.

A transmission system known as single channel per carrier (SCPC), which provides preassigned circuits for voice and data, is available in all ocean regions. SCPC combined with pulse code modulation and phase shift keying is the standard transmission technique for Standard B earth stations.

The INTELSAT system provides restoration service for interrupted submarine cables on very short notice, and permits rerouting via satellite of cable circuits which would have remained suspended for a long time while repairs are made at sea.

All real-time transoceanic television is transmitted via satellite. TV service is characterized by large fluctuations, caused by singular world events such as international political events and sports competitions.

Finally, the extra capacity of the INTELSAT space segment provides domestic communications services to various countries. Communications capacity is leased in terms of multiples of one-quarter transponder utilizing various transmission techniques which include FDM/FM, SCPC/PCM/PSK, SCPC/companded FM, or delta modulation. This kind of service is very attractive in countries with geographical obstacles to conventional terrestrial communications. Plans for the post-1985 period envisage alternate concepts tied to advanced technologies. As the traffic projections by end of 1977 indicate the need for some 300,000 channels, new systems concepts and technologies will be required.[23]

THE CANADIAN DOMESTIC SYSTEM

Canada has been the first country in the Western hemisphere to establish a multipurpose communications satellite system for domestic service. The space segment comprises three satellites in orbit, the first launched in November 1972 and operational in January 1973 at 114° west longitude, the second launched in April 1973 at 109° west longitude, and the third launched in May 1975 at 104° west longitude. The satellites were manufactured by Hughes Aircraft Co. with Canadian firms as subcontractors for the onboard communications equipment (Northern Electric) and the main structure (Spar Aerospace). The launch vehicles (Thor-Delta) and facilities were provided through NASA.

Telephone, television, data, and facsimile transmission services are provided throughout Canada, increasing the capacity of the existing terrestrial systems and permitting their interconnection. The overall transmission performance is comparable to or better than that of terrestrial systems.

The earth segment consists of over 50 stations: 2 for heavy route traffic, 6 for network television,

18 for thin route traffic, 2 for northern telecommunications, 26 for remote television service, and finally, 1 TT&C station. The operating frequency bands are from 3.202 to 4.178 GHz in the down-link (horizontal polarization), and 5.927 to 6.403 GHz in the up-link (vertical polarization).

The first generation satellites built for TELESAT Canada by Hughes Aircraft were spinners with a diameter of 190.5 cm, height of 161 cm, and in-orbit mass of 600 kg. The platform, carrying 12 transponders, spins at 100 rpm with the outer drum, which supports 20,048 silicon cells, producing 300 W of electrical power at beginning of life. The despun upper structure carries a 152.4-cm-diameter lightweight parabolic reflector illuminated by a multiple horn offset feed. The radiation pattern (about 3° x 8.5°) is elliptically shaped to cover the territory of Canada. The e.i.r.p. per transponder is 33 dBW (minimum at the beam contour) obtained by a 5-W TWT in each transponder, and the transmit antenna gain is 27 dB. A TT&C antenna on top of the parabolic reflector brings the overall satellite height to 358.1 cm.

The repeater is a fixed gain, single-conversion, 12-transponder design with 36-MHz bandwidth allocated to each channel and a 4-MHz guard band between adjacent channels. A redundant (switchable) wideband receiver using tunnel diode amplifiers at 6 GHz is common to the 12 transponders. The receive G/T ratio is -7 dB/K. Each transponder can operate with the TWT at saturation in the single-access mode, providing a capacity of 960 voice circuits or one color TV and two 5-kHz audio circuits (earth station G/T = 37 dB/K). It can also operate at backoff in the multiple-access FDM/FM/FDMA mode with varying communications capacity, depending on the circumstances. During eclipses, only 10 of the 12 transponders can be fully operational.

The locations of the Anik system earth stations are shown in Figure 6; their variety is considerable. The TT&C facilities, located at the two heavy route stations on the east (Allan Park, Ontario) and west (Lake Cowichan, Vancouver Island) coasts, operate in conjunction with the Ottawa control center. During the transfer orbit phase of a mission, a third TT&C station operates on the island of Guam in the Pacific.

The two heavy route stations satisfy high-density traffic requirements as well as television services. The antennas are 29.87-m-diameter parabolas, with a gain of 63 dB in transmission and 59 dB in reception. The combination of the antenna and a parametric front end yields a G/T ratio of 37 dB/K. The e.i.r.p. per carrier is 84 dBW with 960 voice circuits, or one color TV plus two audio channels. Normally, five transmit and seven receive channels with hot standbys are provided. The stations operate continuously and have tracking capabilities.

The six network television stations located near major cities provide TV signal transmission and reception for use by the CBC network. The antennas are 10.5-m-diameter parabolas with gains of 52.5 and 50.5 dB, respectively, in transmission and reception. The antennas are fixed with manual steering over a limited range. Therefore, precise satellite station-keeping is required to remain within the antenna beamwidth. The G/T is 28 dB/K and the e.i.r.p. per carrier is 83 dBW. The communications capacity per carrier is one TV plus two audio channels. Some stations are permanently manned, while others require only part-time staffing.

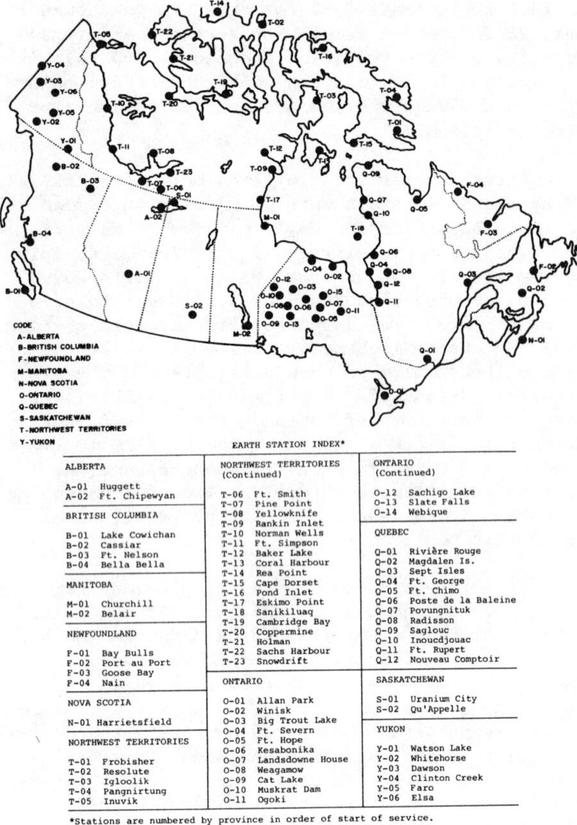

EARTH STATION INDEX*

ALBERTA		NORTHWEST TERRITORIES (Continued)		ONTARIO (Continued)	
A-01	Huggett	T-06	Ft. Smith	O-12	Sachigo Lake
A-02	Ft. Chipewyan	T-07	Pine Point	O-13	Slate Falls
		T-08	Yellowknife	O-14	Webique
BRITISH COLUMBIA		T-09	Rankin Inlet		
		T-10	Norman Wells	QUEBEC	
B-01	Lake Cowichan	T-11	Ft. Simpson		
B-02	Cassiar	T-12	Baker Lake	Q-01	Rivière Rouge
B-03	Ft. Nelson	T-13	Coral Harbour	Q-02	Magdalen Is.
B-04	Bella Bella	T-14	Rea Point	Q-03	Sept Isles
		T-15	Cape Dorset	Q-04	Ft. George
MANITOBA		T-16	Pond Inlet	Q-05	Ft. Chimo
		T-17	Eskimo Point	Q-06	Poste de la Baleine
M-01	Churchill	T-18	Sanikiluaq	Q-07	Povungnituk
M-02	Belair	T-19	Cambridge Bay	Q-08	Radisson
		T-20	Coppermine	Q-09	Saglouc
NEWFOUNDLAND		T-21	Holman	Q-10	Inoucdjouac
		T-22	Sachs Harbour	Q-11	Ft. Rupert
F-01	Bay Bulls	T-23	Snowdrift	Q-12	Nouveau Comptoir
F-02	Port au Port				
F-03	Goose Bay	ONTARIO		SASKATCHEWAN	
F-04	Nain				
		O-01	Allan Park	S-01	Uranium City
NOVA SCOTIA		O-02	Winisk	S-02	Qu'Appelle
		O-03	Big Trout Lake		
N-01	Harrietsfield	O-04	Ft. Severn	YUKON	
		O-05	Ft. Hope		
NORTHWEST TERRITORIES		O-06	Kesabonika	Y-01	Watson Lake
		O-07	Landsdowne House	Y-02	Whitehorse
T-01	Frobisher	O-08	Weagamow	Y-03	Dawson
T-02	Resolute	O-09	Cat Lake	Y-04	Clinton Creek
T-03	Igloolik	O-10	Muskrat Dam	Y-05	Faro
T-04	Pangnirtung	O-11	Ogoki	Y-06	Elsa
T-05	Inuvik				

*Stations are numbered by province in order of start of service.

Figure 6.　TELESAT Earth Stations (Courtesy)
of TELESAT Canada)

The northern telecommunications stations provide up to 132 voice channels at two far north locations. The stations are unattended and controlled from a nearby supervision and maintenance center. The antennas at these stations are 10.15-m parabolas (52.5-dB transmit and 50.5-dB receive gain), with a G/T of 28 dB/K and an e.i.r.p. of 73 dBW per carrier. Klystrons rated at 1.5 kW are used as final power amplifiers.

The thin route stations provide small communities with a limited number of telephone circuits (usually two to eight). SCPC techniques with delta modulation (32 kbit/s) and FDMA preassigned access are used, providing telephone service or combinations of voice, teletype, facsimile, and data. Increased capacity is achieved by voice activation and demand-assigned multiple-access techniques. Major characteristics of these stations are a G/T of 20 dB/K and an e.i.r.p. per voice carrier of 55-58 dBW.

The TELESAT system was the first to introduce TDMA, starting with early tests in 1971 and with commercial operation beginning in September 1975. Four hundred telephone circuits are handled at an overall transmission rate of 61.248 Mbit/s between a station at Harrietsfields, Nova Scotia (G/T = 31 dB/K), and another station at Allan Park, Ontario (G/T = 37.5 dB/K). This TDMA system replaced a 240-circuit FDMA system.

The outstanding characteristic of the TELESAT system is the variety of services provided. In general, as system organization for a given type of service does not necessarily coincide with optimization for other services, this conflict can be resolved by optimization of all services, matching different earth stations with satellite transponders of different bandwidth and power. However, this solution is not desirable because flexibility will be lost and, in addition, spacecraft mass and service costs will increase. Standardized transponder design, with compromises in communications capacity and quality of service, was the solution finally adopted.

The careful planning of the TELESAT system has contributed to its operational success since 1973. The advanced design of the spacecraft resulted in its adoption by other users. The Hughes Aircraft Co. HS-333 spacecraft has proved highly successful and was later adopted with minor changes by Western Union for its two WESTAR satellites (1974-1975) for U.S. domestic services, and by the Government of Indonesia for its PALAPA system (1976). Two HS-333 satellites constitute the space segment of the Indonesian system; fifty 10-m earth terminals located on the various islands of the Indonesian Archipelago constitute the earth segment.

U.S. DOMESTIC SATELLITE SYSTEMS

The existence of an extremely well-developed and efficient earth-based communications network in the U.S. delayed the introduction of satellites for domestic communications. In addition, as previously mentioned, complex politico-economical forces contributed to further delays.

In December 1972, the U.S. Federal Communications Commission, after almost seven years of arguments, ruled on the matter of domestic communications satellites, announcing the so-called "open-sky" policy whereby the arena was open to competing private enterprise. Approximately 12 companies filed applications for domestic systems. Three companies (American Satellite, RCA, and Western Union Telegraph) established ground stations and initiated services in 1973-1974. Initial service was established through leasing available transponder capacity from the Canadian TELESAT satellites.

Western Union launched the first U.S. domestic satellite, WESTAR I, in April 1974, followed by WESTAR II in June of the same year. Commercial service was established in August 1974 with five earth terminals in New York, Los Angeles, Chicago, Dallas, and Atlanta. These stations have 15-m parabolic antennas.

The WESTAR satellites have become an integral part of the Western Union network. Telex, TWX, hotline point-to-point, Central Telephone Bureau, telegrams, mailgrams, and Info-Master services are available in addition to point-to-point voice, data, and facsimile, and point-to-point or point-to-multipoint video.

Two satellites and five earth stations (Glenwood, New Jersey; Estill Fork, Alabama; Steele Valley, California; Cedar Hill, Texas; and Lake Geneva, Wisconsin) are now operational. Twenty satellite access cities and five television centers are connected via the Western Union surface microwave network.

The RCA domestic system began with leased transponders on the TELESAT Canada Anik II satellite. With the launches of the SATCOM spacecraft in December 1975 and March 1976, commercial services were provided to the 48 contiguous states and Alaska and Hawaii. The services include TV distribution in Alaska, toll messages and bush telephones in Alaska, private-line video, voice and data to government agencies, commercial TV and radio, and CATV program distribution to over 100 small receive-only stations.

In the case of COMSAT General Corporation, two COMSTAR satellites entirely leased to AT&T and GSAT have been integrated within the nation's telephone network. A third satellite was launched in June 1978 and is now in orbit. AT&T's earth stations are located at Three Peaks, California; Hanover, Illinois; Woodbury, Georgia; and Hawley, Pennsylvania. GSAT's earth stations are in Sunset, Hawaii; Triunfo Pass, California; and Homosassa, Florida.

The major characteristics of the three spacecraft used in the above mentioned systems are summarized in Table 2.

The RCA SATCOM spacecraft, the first operational body-stabilized communications satellite, was launched by a modified 3914 Thor-Delta vehicle. Its communications capacity is 14,400 voice circuits with 24 transponders and frequency reuse by means of orthogonal linear polarization.

The COMSTAR spacecraft, the largest of the three, has 24 transponders, and again dual polarization yields a total communications capacity of 18,000 voice circuits.

American Satellite Corporation, although it does not have satellites of its own, has installed 21 earth stations. SCPC techniques are used for digital transmission ranging from 56 Mbit/s to 1.344 Mbit/s. Modest antenna size (10-meter parabolas), unattended operation, installation at the user's premises, high reliability, and low cost have been key elements in the success of the system. American Satellite Corporation has also been carrying commercial traffic for a number of users such as Dow Jones (_Wall Street Journal_), Sperry, and Boeing.

Certain characteristics of the four domestic systems now operational in the U.S. are indicated in Table 3. Currently, there are six operational spacecraft with 96 available transponders. The total available communications capacity amounts to 57,600 voice circuits. Otherwise, each transponder can carry digital transmissions at the rate of 60 Mbit/s to yield a 5,760-Mbit/s total digital communications capability.

Notwithstanding the differences among the four operational U.S. domestic systems, the following goals have been achieved:

a. satisfactory integration with the nation's earth-based networks;
b. connectivity among widely dispersed points;
c. direct connection avoiding local interconnection problems;
d. high-speed, high-reliability data transmission permitting the efficient linking of computers; and
e. broadcast-mode TV distribution to receive-only stations.

Table 2. Major Characteristics of U.S. Domestic Satellites

Quantity	HS332	SATCOM	COMSTAR
Mass in Orbit (kg)	297	465	810
Launch Vehicle	Delta 2914	Delta 3914	Atlas-Centaur
Primary Power (W)[a]	300	740	760
Frequencies (GHz)	4/6	4/6[b]	4/6[b]
Nominal Bandwidth (MHz)	500	1,000	1,000
No. of Transponders	12	24	24
e.i.r.p./ Transponder (dBW)	24.5–34	26–32	27–34
Communications Capacity (2-way voice circuits)	7,000	14,000	18,000

[a]Beginning of life.
[b]Frequency reuse with orthogonal polarizations.

Table 3. U.S. Operational Domestic Systems (1977)

Company	S/C Designation (No. of Transponders)	Launch Date	Orbital Position (°W)	Earth Station Dia (m)	No. of Earth Stations
Western Union	WESTAR[b] (12)	4/3/74 10/10/74	99 123.5	15.5	5
RCA American Comm.	SATCOM[c] (24)	12/13/75 3/26/76	119 135.8	10	5
COMSAT General Corp./ AT&T-GSAT	COMSTAR[c] (24)	15/13/76 7/22/76	128 94	30	5 + 3
American Satellite[d] Corp.	–	–	–	10 5	21

[a]All satellites use the 4/6-GHz frequency bands.
[b]HAC HS333 Type.
[c]Frequency reuse with orthogonal linear polarizations.
[d]Leases capacity from WESTAR satellites.

U.S. domestic satellite systems are expected to expand, particularly in those areas where earth-based systems are less developed (e.g., Alaska). The Public Broadcasting Service (PBS) will become the first nationwide TV broadcast network interconnected via satellite.[24] PBS stations, supported by the Federal government and by the contributions of viewers, carry no advertising. Terrestrial cable and microwave circuits currently leased from the common carriers at high cost will be replaced by satellites. The PBS satellite network will consist of a central program-originating station in Washington, D.C., five regional stations, and about 150 receive-only stations.

Other entities exploring satellite interconnections are the Corporation for Public Broadcasting and the Public Service Consortium. The Corporation for Public Broadcasting will seek FCC approval to provide satellite interconnections for over 200 radio stations, broadcasting audio monaural and stereophonic programs. The Public Service Satellite Consortium consists of consumer groups, educational institutions, medical organizations, and other similar entities. Plans to service communities, universities, and hospitals are being defined.

As a result of the growth in all three current areas of service, i.e., high-density trunking, thin routes, and distribution, the available communications capacity of the existing systems will eventually become fully utilized, although their saturation date will be delayed by more efficient transmission techniques. Recent estimates indicate that around the mid-1980's, utilization of higher frequency bands at 11/14 GHz will be necessary. By the late 1980's or early 1990's, the 20/30-GHz bands may be needed.[25]

The trend will also be toward an all-digital approach for the most diverse kinds of traffic. In this context, the system announced by Satellite Business Systems (SBS), a partnership of COMSAT General Corporation, International Business Machines Corporation, and Aetna Casualty and Surety Company, deserves special mention.[26] In December 1975, SBS filed with the FCC for approval to construct a domestic satellite system to provide private-line networks. On February 8, 1977, FCC approval was obtained; and since then, progress has been made toward the definition of the system. The system will include:

a. use of 11- and 14-GHz bands;
b. all-digital transmission with time domain on-demand multiple access;
c. integrated voice, data, and image services;
d. unattended earth stations;
e. 5- and 7-meter earth station antennas located at the customers' premises;
f. minimum dependence on terrestrial interconnections;
g. centralized system management facilities; and
h. means to enable customers to dynamically control their networks.

The space segment will consist of a primary operational satellite, a second operational spacecraft as an in-orbit backup, and a third satellite as a ground spare. Table 4 lists the major characteristics of the spacecraft. The Hughes Aircraft Company will construct these spacecraft, which will use a dual-spinner configuration. Spacecraft design will be compatible with both the STS or Delta 3910 launch vehicles. These launch vehicles will be employed to inject the satellite into a parking orbit, a perigee stage will be used for injection into transfer orbit, and an apogee motor incorporated in the satellite will be fired for injection into geostationary orbit.

Table 4. SBS Spacecraft Performance Specifications

Launch Vehicle	STS or Delta 3910
Launch Date	1980
Frequencies (GHz)	11/14
No. of Transponders	10
Transponder Bandwidth (MHz)	43
e.i.r.p./Transponder (dBW)	43, 39, 37
Nominal Communications Capacity (2-way voice circuits)	12,000-14,000

Each satellite will carry 10 transponders, with a nominal bandwidth of 49 MHz. Traveling wave tubes with 23-watt output power will be used as final amplifiers. Full solar eclipse capability will be maintained. The satellites will each carry one active and three spare wideband receivers and ten communications channels, including ten active and six spare TWTAs. Parabolic reflectors, one for receiving and two for transmitting, will be used with multiple feeds and a maximum gain of no less than 32 dB.

In the earth segment, earth stations interfacing with customer-provided PBXs, foreign exchange lines, data terminals, and other equipment will be designed for unattended operation. Five- and seven-meter antennas and 500-watt HPAs will be employed. Small earth stations at the user's premises will constitute a fully switched, wideband communications network with minimum need for terrestrial access facilities. Through the use of demand assignment, the satellite will enable low-density traffic nodes to be served as effectively as high-density nodes. The RF link is designed for a 99.5-percent availability (i.e., BER $\leq 1 \times 10^{-4}$). With a raw bit rate of 43 Mbit/s and 10 active transmission channels, the overall spacecraft capacity is 430 Mbit/s.

MOBILE SYSTEMS

In 1976 the MARISAT system inaugurated a new era in maritime communications, which previously suffered from the limited communications capacity and low reliability of conventional radio from VLF to HF. Satellites have made possible highly reliable, high-quality voice, data, facsimile, and teletype services to and from ships. Interconnection with domestic and international networks makes the system doubly attractive. Currently, 106 shipboard terminals are operational.

The MARISAT system is also unique because its spacecraft constitute the first example of multipurpose operational communication platforms, since UHF channels are available to the U.S. Navy while SHF channels are available to commercial users.

Figure 7 shows the deployment of the system, Figure 8 its coverage, and Figure 9 the spacecraft and its major characteristics. Three distinct transponders are available on the spacecraft. The first operates at 300/250 MHz, providing half-duplex transmission in three separate channels, one 480 kHz wide, and the other two 24 kHz wide used for U.S. Navy purposes. The second transponder is a 1.64/4.19-GHz wideband repeater used for ship-to-shore commercial traffic. The third is a 6.42/1.53 GHz wideband repeater used for shore-to-ship commercial traffic.

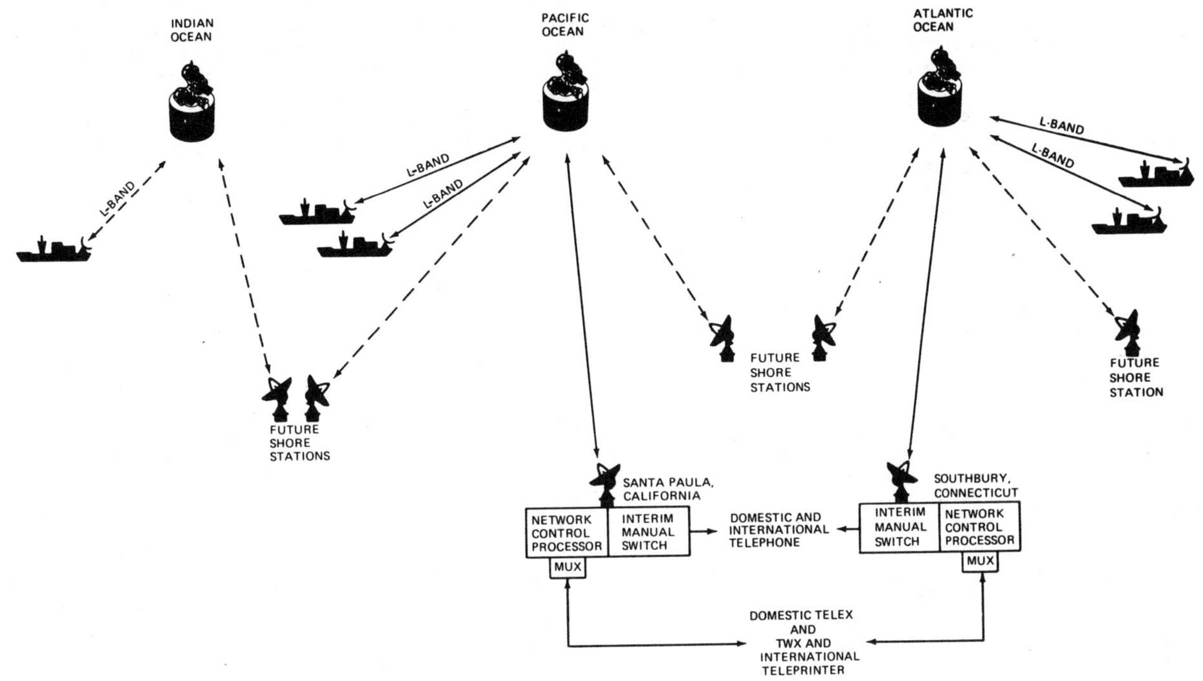

Figure 7. MARISAT System Deployment (from COMSAT Technical Review Vol. 7, No. 2, Fall 1977)

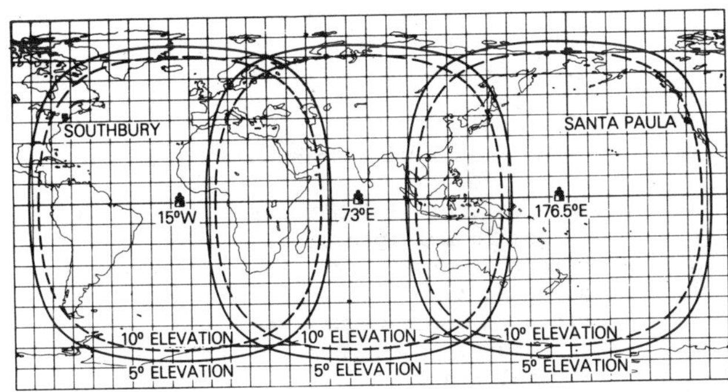

Figure 8. MARISAT Coverage (from COMSAT Technical Review, Vol. 7, No. 2, Fall 1977)

The commercial voice-grade channels are allocated on an SCPC basis using frequency modulation. The telegraph channels are time-division multiplexed (TDM) in the ship-to-ship direction and time-domain multiple accessed (TDMA) in the ship-to-shore direction with 22 channels sharing the same carrier. Channel assignment and signaling information is integrated into the telegraph TDM carrier. A single frequency is shared by the ship's terminals on an as-needed basis for channel requests. A voice-grade call is set up by selecting two frequencies at the shore station and assigning them to a specific mobile terminal.

MARISAT	
YEAR OF FIRST LAUNCH	1976
DIMENTIONS:	
DIAMETER (m)	2.16
HEIGHT (m)	1.63
IN-ORBIT MASS (kg)	325
LAUNCH VEHICLE	THOR DELTA 2914
PRIMARY POWER (W)	300

UHF REPEATER	
NOMINAL RECEIVE FREQUENCY (MHz)	300
G/T (dB/K)	-18
CHANNEL BANDWIDTH (kHz)	480, 24, 24
NOMINAL TRANSMIT FREQUENCY (MHz)	250
e.i.r.p. (dBW)	28, 23, 23

SHF REPEATER	SHORE/SHIP	SHIP/SHORE
RECEIVE BAND (GHz)	6.420 - 6.424	1.6385 - 1.6625
G/T (dB/K)	-25.4	-17
TRANSMIT BAND (GHz)	1.537 - 1.541	4.195 - 4.199
e.i.r.p. (dBW)	20, 26, 29.5	18.8

Figure 9. MARISAT Spacecraft Characteristics

For telegraph calls, the assignment of frequency pairs is different; the shore-to-ship TDM carrier is PSK-modulated by a 1,200-bit/s stream. The ship-to-shore TDMA channel is filled by gated bursts transmitted in a prearranged sequence by as many as 22 different ships. Each burst is PSK modulated at 4,800 bit/s and carries up to 12 message characters. As the time reference for the ship's transmissions is derived from the continuously received TDM signal, no synchronization is necessary. Adequate time between bursts accommodates different propagation delays within the coverage and voice area. Voice-grade circuits can handle speech, facsimile, and data up to 2,800 bit/s.

The communications capacity depends upon the operating mode of the L-band satellite transmitter. Distress and broadcast remote type messages are also possible. Reference 9 should be consulted for details.

Expansion of maritime communications via satellite is foreseen, and plans are being formulated for an international organization is this area. The Inter-Governmental Maritime Consultive Organization (IMCO) established in 1972 a panel of experts to study the technical, economic, and operational factors involved in future commercial maritime communications systems via satellite. During several important meetings the relationship between the existing operational MARISAT system and future possible systems such as MAROTS and MARECS have been investigated.

There are clearly other possibilities for satellite communications to mobile users. Plans for an aeronautical satellite system (AEROSAT) were actively pursued from 1975 to 1977 by the U.S., Canada, and various European countries. Although the project came to a standstill, mostly because of economical and institutional obstacles, the potential advantages of such a service are sufficiently well defined.

<u>FUTURE TRENDS, SPECTRUM AND ORBIT UTILIZATION</u>

Satellite communications systems will continue to expand because traffic growth forecasts indicate the need for greater communications capacity. For example, the yearly average growth rate in the INTELSAT system is expected to be at least 15 percent in the next decade. Satellites will therefore become more competitive vis-a-vis other systems—in terms of increasing communications capacity, reliability, flexibility, and cost effectiveness. In addition, satellites will permit many new communications services to be offered.

The two natural resources, the electromagnetic spectrum and the geosynchronous orbit, determine the bounds of satellite communications' ultimate growth, and will be briefly discussed in the following paragraphs.

As latecomers among the users of the electromagnetic spectrum, satellite services were not assigned "optimum" spectral regions. The assignment to fixed satellite services of 500 MHz at around 4 and 6 GHz for the down- and up-links, respectively, only approaches an optimum allocation since the noise power density spectrum exhibits a broad minimum between 1 and 2.5 GHz. In addition, power flux density limitations were imposed on satellite systems because the 4- and 6-GHz frequency bands must be shared with existing terrestrial microwave systems. In spite of these constraints, satellite systems have been extremely successful, the most serious limitation probably being that which derives from the constraints on the earth terminal site selection.

All commercial satellite systems at this time operate at 4 and 6 GHz. Among the frequency bands assigned to satellite communications at the World Administrative Radio Conference of 1971 two new pairs of bands are especially attractive: one at 11 and 14 GHz, and another at 19 and 29 GHz. At 11 and 14 GHz, a bandwidth of 500 MHz is available for the down- and up-links, respectively. These bands are shared with terrestrial services and are subject to power flux density limitations. There is 2,000 MHz of shared spectrum available between 17.7 and 19.7 for down-links, and another 2,000 MHz between 27.5 and 29.5 GHz for up-links. Finally, bandwidths of 1,500 MHz are exclusively available for down- and up-links, respectively, between 19.7 and 21.2 GHz and 29.5 and 31.0 GHz.

The WARC 1971 frequency assignment results in bandwidth availability eight times greater than that at 4 and 6 GHz. The decrease in antenna size and the reduction or elimination of the constraint related to sharing make the new frequency bands attractive. Future plans for INTELSAT and domestic or regional systems include the use of the 11/14-GHz bands. The 19/29-GHz bands will be used later as the art progresses and as the traffic needs increase. Attenuation resulting from rain will be counteracted by adequate power margins, diversity techniques, and special coding techniques combined with diversity.

Several years ago it was pointed out[27,28] that communications capacities of the order of 100,000 telephone circuits per satellite could be attained without excessive demands on satellite mass and power. Increases in capacity at least one order of magnitude above that of current satellites can be achieved by introducing advanced technologies such as:

a. 3-axis body stabilization,
b. higher efficiency solar cells (e.g., "Violet" and non-reflective cells),
c. higher efficiency energy storage devices (e.g., nickel-hydrogen batteries),
d. electrical propulsion for stationkeeping and positioning,
e. frequency reuse with orthogonal polarization and multiple-beam antennas,
f. onboard switching with combined time- and space-domain multiple-access,
g. onboard regeneration of digital signals,
h. transponder linearization,
i. hybrid modulation techniques,
j. source and baseband encoding, and
k. intersatellite links.

Table 5 summarizes the status and the potential of various technologies.

Antenna gain affects both e.i.r.p. and usable bandwidth because capacity is proportional to power and usable bandwidth, the latter being determined by the frequency reuse factor. Precise stationkeeping and positioning and orientation techniques are needed with narrow antenna beamwidths. The advantages obtained are clear, for example, a beamwidth of 1° yields a gain increase of 360 times with respect to a 19° beamwidth for global coverage, i.e., a 25.8-dB improvement.

Multibeam antennas yield higher e.i.r.p. toward given areas and allow frequency reuse, but require interconnection capability to provide multipoint services. Hence the combination of multibeam antennas and satellite-switched TDMA (SS-TDMA) techniques is especially promising.

Table 5. Status and Potential of Various Technologies

Item	Operational in Space	Experience in Space	Experience in Lab	Untried in Space	Increase in Communications Capacity	Other Advantages
Body Stabilization	√				30%-40%	Easier deployment of complex antennas
High-Efficiency Solar Cells	√				5%-7%	-
Ni-H$_2$ Batteries	√				5%-15%	Longer life
Ion Propulsion		√	√		30%-40%	Longer life, increased pointing accuracy
Multibeam Antennas	√					
Spectrum Reuse					Proportional to frequency reuse factor	-
- Via Orthogonal Polarization	√					
- Via Separate Beams	√					
Higher Frequencies		√			Proportional to available bandwidth	Smaller antennas
TDMA/(SS-TDMA)	√	√	(√)	(√)	Up to 100%	-
Source Encoding (DSI-SPEC-CODIT)	√				Up to 100%	-
Regenerative Repeaters			√	√	50% or more	Better error control
Intersatellite Links		√			N.A.	Increased connectivity

Figure 10 shows a possible system configuration combining TDMA with SS/TDMA. A spaceborne distribution center, consisting of a switching matrix and a control unit, provides the interconnection of different transponders. Information about traffic flow is stored in onboard memory circuits that control the switching matrix; command signals from the earth rearrange the connections in the matrix whenever necessary. Three important advantages result from SS/TDMA. With a single carrier present at any given time, the TWTs operate at saturation with maximum conversion efficiency, no intermodulation noise is produced, and the weight penalty of the multiplexing filters required in FDMA is eliminated.

SATELLITE SWITCHED-TDMA

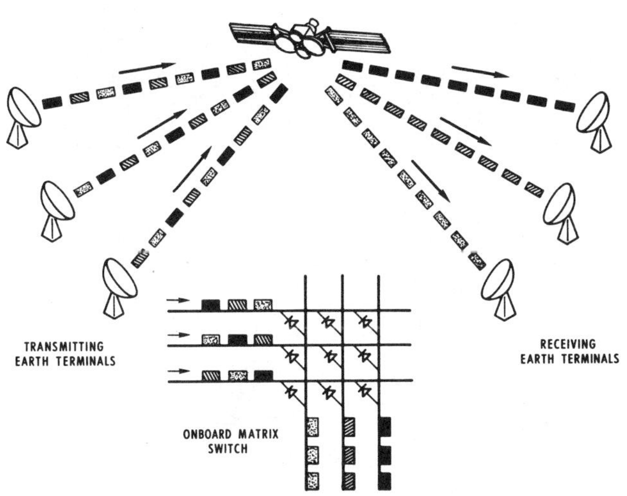

Figure 10. Satellite-Switched TDMA

The introduction of advanced technologies, plus the increased reliability and longer in-orbit lifetime of the spacecraft, will lead to a progressive reduction of the space segment cost.

The theoretical bound of communications capacity per unit of bandwidth and per unit of angular separation is determined by radiation spillover on the basis of the following assumptions.[29]

a. ideal modulation-demodulation processes;
b. neglect of thermal noise, i.e., no power limitations;
c. sharing of a given bandwidth by satellites uniformly spaced in equatorial synchronous orbit; and
d. earth station antennas with uniformly illuminated apertures.

The noise power is a function of a single geometric variable, the satellite spacing. When differences in slant range are neglected, an optimum space $\Delta\theta = \lambda/D$ yields a communications capacity per units of bandwidth and angle:

$$C = 2 \frac{D}{\lambda} \text{ bit/s/Hz/rad} \qquad (6)$$

where D is the earth station diameter and λ the operating wavelength. Consequently, the maximum transmission rate (bit/s) which can be handled by a satellite system using a bandwidth B and a segment of synchronous orbit spanning θ rad is

$$R = 2 \frac{D}{\lambda} B\theta \text{ bit/s} \qquad (7)$$

For instance, a bandwidth of 500 MHz and 30-m-diameter antennas at 4 GHz yield a theoretical global ($\theta = 2\pi$)

capacity of 2.51×10^{12} bit/s, i.e., approximately 40 million telephone channels.

In practice, results of this kind must be corrected to include possible mechanical problems resulting from tight orbital spacing, the lower efficiency of real modulation-demodulation processes, the unavoidable presence of thermal noise, intermodulation noise, and actual antenna radiation patterns. While the first four items lead to lower communications capacities, the last item can lead to an increase in communications capacity beyond the above mentioned theoretical bound.

Additional possibilities for augmenting communications capacity are intersatellite relaying, increase in the allowable interference ratio, channel interleaving, reversed use of frequencies, and pseudo-stationary satellites and 2-dimensional orbit space.[30]

Clearly, as the two problems of spectrum and orbit utilization are inseparable, tradeoffs are possible.[31] At 4/6 GHz the current value of the orbital spacing (around 4°) is primarily dictated by the earth station size within the current range of 10-30 meters. Increased satellite e.i.r.p., accompanied by reduction of the size (cost) of the earth station antennas, leads unavoidably to a less efficient utilization of the orbit because of the wider satellite spacing required. A reduction of the number of spacecraft in geosynchronous orbit, conceivable by combining services on multipurpose platforms (an approach which may be desirable for other reasons), would also lead to reduced communications capacity. Improvements in antenna sidelobe control techniques (including signal processing antennas) will ultimately be required to avoid wasteful use of the geosynchronous orbit.

In all cases, of course, the opening of higher frequency bands or the availability of additional bands leads to better orbit utilization and higher communications capacity.

CONCLUSIONS

Striking progress has been made in technology, operational use, institutional arrangements, and cost effectiveness of communications satellite systems in their first decade of deployment. Continued progress is expected through the expanded use of existing international, domestic/regional, and mobile systems. Further progress will occur as systems and services come into being.

Future systems will have greater capacity and flexibility to serve a variety of users. Increased reliability, longer lifetime, in-orbit servicing, and the introduction of new technologies will contribute to progressive cost reductions in both the space and earth segments.

The higher frequency bands will be occupied in addition to the continued use and improved exploitation of the lower frequency bands already occupied. Advanced spacecraft, signal processing, and microwave technologies, combined with earth terminal development slanted toward automated operations and greater reliability, will contribute to the progress of satellite communications systems well beyond the achievements recorded in the first decade of commercial operation.

ACKNOWLEDGMENT

Thanks are offered to the International Astronautical Federation (IAF) and in particular to Prof. L. G. Napolitano, Editor of the Proceedings of the XXVth IAF Congress, and Pergamon Press, the publishers, for their kind permission to allow parts of Reference 5 to be incorporated in this paper.

The encouragement of Dr. B. I. Edelson, Director of COMSAT Laboratories, is also gratefully acknowledged.

REFERENCES

1. A. C. Clarke, "Extraterrestrial Relays," _Wireless World_, October 1945, pp. 305-308.
2. C. E. Shannon, "The Mathematical Theory of Communication," _Bell System Technical Journal_, Vol. 27, No. 3, July 1948, pp. 379-428; Vol. 27, No. 4, October 1948, pp. 623-656.
3. J. R. Pierce, "Orbital Radio Relays," _Jet Propulsion_, Vol. 25, April 1955, pp. 153-157.
4. I. Goldstein, "INTELSAT and the Developing World," _IEEE Transactions on Communications_, COM-24, No. 7, July 1976, pp. 742-748.
5. P. L. Bargellini and B. I. Edelson, "Progress Trends in Commercial Satellite Communications--A Survey," _Proc. of the XXVI International Astronautical Congress_, Lisbon, Portugal, 1975, L. G. Napolitano, Pergamon Press, 1976, pp. 259-276.
6. D. G. Thorpe, "Evolution of the TELESAT Canada System into the Mid-1980's," _Satellite Communications: Future Systems_, D. Jarett, ed., New York: AIAA Press, 1977, pp. 39-55.
7. C. C. Sanderson and B. R. Elbert, "Communication System Design of the Indonesian Domestic Satellite System," WESCON 1976, Los Angeles, CA, Paper No. 9.2.
8. P. L. Bargellini, "Evolution of U.S. Domestic Satellite Communications," _Proc. of 3rd Jerusalem Conf. on Information Technology_, August 1978, J. Moneta, ed., North Holland Publishing Co.
9. D. W. Lipke et al., "MARISAT--A Maritime Satellite Communications System," _COMSAT Technical Review_, Vol. 7, No. 2, Fall 1977, pp. 351-391.
10. J. V. Charyk, "Communications Satellites," Von Karman Lectureship in Astronautics, AIAA 13th Annual Meeting, Washington, D.C., 1977, Paper No. 77-323.
11. W. L. Pritchard, "Satellite Communication--An Overview of the Problems and Programs," _Proc. IEEE_, Vol. 65, No. 3, March 1977, pp. 294-307.
12. H. Rosen, "Synchronous Communications Satellites," _Space Communications_, A.V. Balakrishnan, ed., New York: McGraw-Hill, 1963, Chapt. 17.
13. S. Metzger, "Foreword," _COMSAT Technical Review_, Vol. 2, No. 2, Fall 1972, pp. vii-xi.
14. J. Kiesling et al., "A Technique for Modeling Communications Satellites," _COMSAT Technical Review_, Vol. 2, No. 1, Spring 1972, pp. 73-104.
15. _New Space Transportation Systems--An AIAA Assessment_, J. Preston Layton and Jerry Grey, ed., New York: AIAA Press, 1973.
16. _Space Transportation Systems--User Handbook_, National Aeronautics and Space Administration, June 1977.
17. M. H. Kaplan, _Space Shuttle; America's Wings to the Future_, Fallbrook, CA: Aero Publishers, 1978.
18. "The INTELSAT-IV Communications System," P. L. Bargellini, ed., _COMSAT Technical Review_, Vol. 2, No. 2, Fall 1972, pp. 437-572.
19. R. J. Rusch et al., "INTELSAT V Spacecraft Design Summary," AIAA 7th Communications Satellite Systems Conference, San Diego, CA, April 24-27, 1978, _A Collection of Technical Papers_, pp. 8-20.
20. M. P. Brown, Jr., "INTELSAT V Standard C Earth Station Performance Objectives," EASCON '77, _Proceedings_, pp. 13-1A-13-1I.

21. S. Browne, "The INTELSAT Global Satellite Com-
    munications System," COMSAT Technical Review,
    Vol. 4, No. 2, Fall 1974, pp. 477-488.
22. B. I. Edelson et al., "INTELSAT System Reli-
    ability," International Astronautical Federation
    XXVth Congress, Amsterdam, The Netherlands,
    30 September-5 October 1974.
23. H. Van Trees et al., "Planning for the Post-1985
    INTELSAT System," AIAA 7th Satellite Systems
    Conference, San Diego, CA, April 24-27, 1978,
    A Collection of Technical Papers, pp. 43-54,
    Reprint Paper 7.1.
24. A. Washburn, "U.S. Domestic Communications Satel-
    lite Systems," International Astronautical
    Federation XXVIIIth Congress, Prague, Czecho-
    slovakia, 25 September-1 October 1977, Paper
    No. 77-01.
25. D. Jarett, "Meeting the Twin Challenges of Demand
    and Conservation of Spectrum and Orbit Through
    Technology," EASCON '77, Washington, D.C.,
    September 1977, Proceedings, pp. 25.1A-25.1H.
26. J. D. Barnla and F. R. Zitzmann, "The SBS Digital
    Communications Satellite System," EASCON '77,
    Washington, D.C., September 1977, Reprint
    Paper 2.2.3.
27. W. L. Pritchard and P. L. Bargellini, "Trends in
    Technology for Communications Satellites,"
    Astronautics and Aeronautics, Vol. 10, No. 4,
    April 1972, p. 26.
28. J. J. Knopow, "Next Generation Communications
    Satellites," Communications Satellite Systems
    (AIAA Progress in Astronautics and Aeronautics,
    Vol. 32), P. L. Bargellini, ed., Cambridge, MA:
    MIT Press, 1974, pp. 63-86.
29. W. E. Bradley, "Communications Strategy of Geo-
    stationary Orbit," Astronautics and Aeronautics,
    Vol. 6, No. 4, April 1968, p. 34.
30. P. L. Bargellini, "Extension of the Concept of
    Satellite Communication System Capacity to a Two-
    Dimensional Model," ICC '69, Boulder, CO,
    June 9-11, 1969, Conference Record,
    pp. 37-25-37-29.
31. R. C. Davis et al., "Future Trends in Communica-
    tions Satellite Systems," ACTA Astronautica,
    Vol. 5, No. 3-4, March-April 1978, pp. 275-298.

2.2

DIGITAL COMMUNICATIONS SATELLITE SYSTEM OF SBS

J.D. BARNLA and F.R. ZITZMANN

Satellite Business Systems, McLean, Virginia

Abstract

Satellite Business Systems will implement a domestic satellite system that will provide private-line switched networks for integrated voice, data and image services. The SBS System will employ all digital transmission in the 12 and 14 GHz bands. Each network will have dedicated full-period capacity for integrated services and on-demand (per call) satellite capacity for high-speed digital services. The System features five- and seven-meter antennas located on customer premises, demand assignment (DA), time-division multiple access (TDMA) and two in-orbit satellites designed for space shuttle launching.

Since December 22, 1975, when SBS applied with the Federal Communications Commission for its planned system, substantial company studies have been completed and requests for proposals have been issued for subsystem development. This paper provides an updated technical description of the SBS System, including principal system parameters and some of the factors that led to the selection of the current design.

Introduction

Satellite Business Systems (SBS) is proceeding with the development of its planned domestic satellite system. The objective of this paper is to provide a description of the system that reflects the latest refinements in the system design.

Background

SBS is a partnership sponsored by COMSAT General Corporation, International Business Machines Corporation and the Aetna Casualty and Surety Company. SBS succeeded to the interests of its predecessors (MCI Lockheed Satellite Corporation and CML Satellite Corporation). In December, 1975, SBS filed with the Federal Communications Commission (FCC) for approval to construct a domestic satellite system to provide private-line networks. On February 8, 1977, the FCC approved SBS's application and SBS has proceeded with further system studies that have led to improvements in transmission capacity and other system parameters.

The features of the SBS System concept reflect results from several recent studies and include:

- fully integrated voice, data and image services
- all digital transmission
- RF operation in the 12 & 14 GHz bands
- 5- and 7-meter earth station antennas
- customer premises earth stations
- unattended operation of earth stations
- minimum dependence on terrestrial interconnecting facilities
- centralized system management facilities
- Time Division Multiple Access (TDMA) of satellite communications channels
- Demand Assignment (DA) of satellite capacity to earth stations and networks
- Voice Activity Compression (VAC) to provide efficient utilization of channel capacity (similar methods are planned for digital data channels)
- facilities to enable customers to dynamically control and manage the use of their networks

In May and June of 1977, SBS issued requests for proposals for development of the satellite, RF Terminal and TDMA Burst Modem. It is the objective of SBS to award contracts in these areas by the end of 1977.

Functional Description

Figure 1 shows the principal elements of the SBS System.

Earth stations will interface with interconnecting facilities to customer-provided PBX's, foreign exchange lines, data terminals and other equipment. Customer locations and equipment that are co-located on the same customer premise as the earth station may be connected via in-plant cabling. Locations which are remote from earth stations will be interconnected via conventional terrestrial facilities such as those that are available from the terrestrial common carriers. The earth station will provide the necessary interface adapters to ensure electrical and supervisory compatibility with the interconnecting facilities. Where needed, service between on-net and off-net customer locations can be provided as well.

Reprinted from *IEEE Electron. & Aerosp. Syst. Conv.*, Sept. 26–28, 1977, pp. 7-2A-7-21.

Voice and other analog signals will be converted
to a digital format at the originating earth station and converted back to analog signals at each
terminating earth station. Although not all
design decisions have been made concerning the
digital codec to be used, a possible implementation as of this writing is a 32 kbps, syllabically companded delta modulation implementation
that has excellent performance characteristics
and leads to efficient utilization of satellite
capacity.

Digital data service will be provided at a variety
of data rates from 2.4 kbps to 6.312 Mbps.

The user of network facilities will have available
an assortment of advanced features including:

  . flexible voice and data conference
    arrangements
  . multipoint distribution of digital data,
    which can include document distribution
  . teleconferencing, including multipoint video
    conferences
  . hot-line and other priority connecting
    features
  . network access control features

The exploded view of an earth station on Figure 1
shows the principal components of the station:

  . Satellite Communications Controller (SCC)
    which is a highly integrated hardware/

software device comprised of processors,
storage units and control programs. It performs the essential TDMA, DA, switching and
other control and processing functions, as
well as analog/digital conversion for voice-
grade signals.

  . TDMA Burst Modem which performs the modula-
    tion, demodulation and associated functions
    that enable bursts of digital information to
    be transmitted through each satellite commu-
    nications channel on a time-shared basis.

  . RF Terminal which provides the radio fre-
    quency transmit and receive functions and
    provides the frequency translations between
    the 14 and 12 GHz transmit and receive fre-
    quencies and the 70 MHz interface with the
    TDMA Burst Modem.

  . Monitor and Command loop which enables the
    remotely-located SBS System Management Facil-
    ity (SMF) to determine the status and health
    of earth station equipment and to issue
    diagnostic and corrective commands.

The space segment of the SBS System includes the
satellites; Telemetry, Tracking and Command (TT&C)
earth stations; and the Satellite Control Facility
(SCF). When system operations are initiated in
early 1981, there will be two satellites in orbit
and a third satellite available on the ground as
a backup. The TT&C earth stations receive

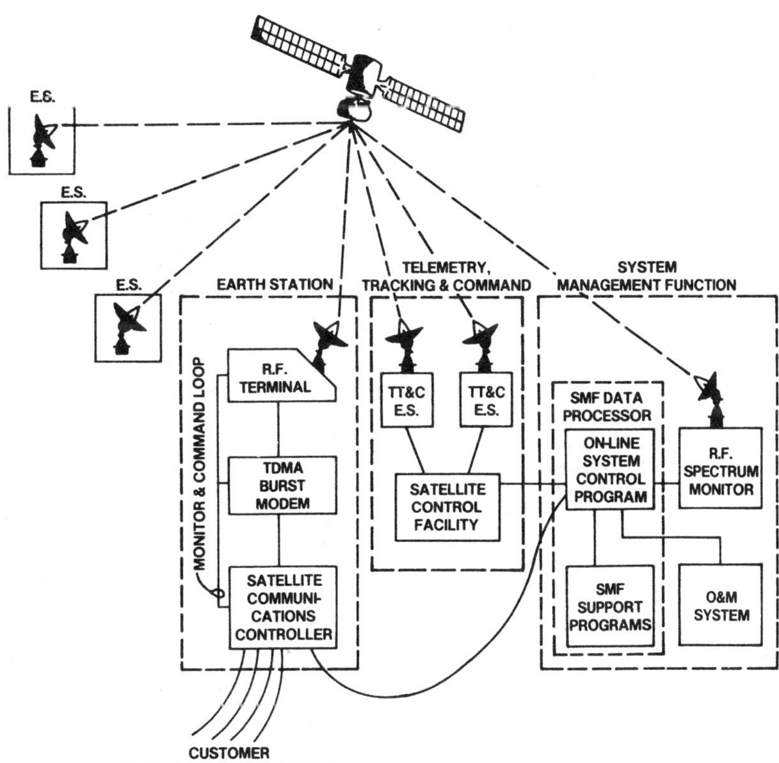

Figure 1. SBS System Block Diagram

telemetry data from sensors and from satellite
components, accurately track the position of each
satellite, and transmit commands - to correct orbital positions or to switch to redundant satellite components, as examples. The SCF is the
central facility for processing and analyzing data
received from the satellite, for determining when
to transmit commands to the satellite and serves
as the principal control center for all actions
involving the satellite.

The System Management Function includes the System
Management Facility (SMF), the RF Spectrum Monitor
and a network of distributed Maintenance Centers.
The SMF is a large-scale data processing facility
that maintains cognizance of the status of all
earth stations and other system components, collects traffic data and billing information, provides support to the maintenance organization via
automated system diagnostics, controls changes to
the system configuration and assignment of
resources, and furnishes a variety of administrative support functions. The RF Spectrum Monitor
is a sensitive and highly instrumented earth station that monitors the communications signals
being transmitted by each communications channel
in the satellite to detect any problems in the
frequencies being transmitted or in the timing of
information bursts. Maintenance Centers are distributed geographically in accordance with the
location of earth stations. Each Center will have
responsibility for repairs and preventive maintenance for ten to twenty earth stations.

The following sections summarize the principal
parameters of the SBS System and elaborate on the
design concepts for each of the system elements.

System Characteristics

The SBS System is designed to serve the full spectrum of communications needs of a large community
of business and government organizations. The
greatest benefits will be derived from the SBS
System by those customers that generate communications traffic at a number of locations geographically dispersed throughout the 48 contiguous
states and whose requirements include the need for
high-speed transmission services.

By utilizing small earth stations operating at 12
& 14 GHz and located on the customer's premises in
close proximity to the customer's major traffic
nodes, the SBS System will permit direct access to
a fully switched, wideband communications network
with minimum dependence on extensive terrestrial
access facilities.

A large nationwide enterprise may require twenty
or more earth stations to support its communications requirements. Through the use of demand
assignment, the satellite acts as a central concentrator that enables a large ensemble of relatively low intensity traffic nodes (earth stations)
to achieve the trunk usage normally associated
with high traffic intensity nodes. The net effect
is that all earth station locations can benefit
from the traffic engineering advantages normally
achieved in the design of high density trunk groups.

By permitting continuous real-time reassignment of
the satellite capacity to meet the varying needs
of the user, the SBS System will offer an attractive alternative to traditional point-to-point
terrestrial communications systems. Unused communications capacity on the West Coast in a
terrestrial system has little or no utility in
meeting the traffic demand in the Boston-
Washington corridor. In contrast, the SBS satellite networks can permit the total satellite
transmission capacity dedicated to the user to be
applied to the instantaneous traffic demands of
that user without respect to location.

SBS will offer service to its customers via networks that are tailored to meet their telecommunications needs with respect to geographical
distribution, traffic mix, traffic intensity,
quality of service, and grade of service. These
networks will employ space segment transmission
capacity on a full-period basis. In this mode,
a customer may obtain a fixed fraction of the
TDMA frame adequate to meet his traffic loading
and grade of service objectives.

The RF channel performance objectives of the SBS
System design are indicated in the table below.

RF Channel Performance Objectives

Maximum Channel BER*	Availability
1×10^{-6}	95.0%
1×10^{-4}	99.5%
1×10^{-2}	99.8%

*(Bit Error Rate)

Essentially, the system breakpoint is the 1×10^{-2}
BER point, since the TDMA Burst Modem has been
specified to maintain synchronism at least at
that error rate.

An option will be available for the customer to
obtain even higher performance for digital data
transmission, if desired. The SCC will apply
forward error correction (FEC) coding to the
digital data information of ports for which the
customer has exercised the option. The effect of
the application of FEC will be to provide a BER
of 1×10^{-7} or better, 99.5% of the time, using
a coding rate of approximately two-thirds.

The RF link has been designed for a 99.5% availability (i.e., BER $\leq 1 \times 10^{-4}$). At that design
point, excellent quality is expected for voice
conversations. The link will readily support a
raw bit rate of 43 Mbps at that design objective.
Since each satellite will have ten active transmission channels, the capacity of the satellite
can be expressed as 430 Mbps, in gross terms.
SBS plans to initiate service at the link bit
rate of 43 Mbps as a conservative design measure.
As experience with link margins is gained, the
bit rate may be increased to 48 Mbps; the TDMA
Burst Modem has been specified to facilitate that
increase in rate. The capacity of the satellite
expressed in terms of voice circuits, for example,

is a complex function of the bit rate per voice channel, the effectiveness of the voice activity compression (VAC) algorithm, the overhead of the TDMA frame format, the number and sizes of earth stations served by each satellite channel, the traffic characteristics of each customer network, and other factors. For one range of assumptions that present studies indicate will be representative of actual traffic, it was estimated that the satellite will provide the equivalent useful capacity of 12,000 voice circuits at 43 Mbps (or 14,000 voice circuits at 48 Mbps).

<u>Space Segment</u>

Three satellites will be procured initially, consisting of a primary operational satellite, and a secondary operational satellite which will also serve as an in-orbit backup. The third satellite will serve as a ground spare to be launched to meet additional traffic demands or in the event of failure or unacceptable degradation of one of the orbiting satellites. The estimated date for the first satellite launch is July 1980 with the second satellite launch to occur three months later.

Spacecraft design will be compatible with both the STS or Delta launch vehicles. Although SBS is enthusiastic about the launch cost savings achievable by the use of the shuttle, the Delta backup is considered necessary for a commercial venture planning to use the first operational shuttle launch.

Station keeping limits of 0.1 degrees are standard for communication satellite systems at the present time and SBS's early planning was based on this value. Recently, however, in reviewing the earth station step track consideration, it has become apparent that tighter station keeping limits are desirable. We are at the present time investigating the possibility of achieving .03 degrees as a station keeping tolerance. Early results of this investigation show promise that this may be achievable without a substantial increase in fuel utilization. Further development of programs to define the fine grain structure of solar and lunar pertubations will be required before new station-keeping limits can be adopted.

Telemetry, tracking, and command (TT&C) for the launching and transfer orbit maneuvers of the satellites will operate in the 4 & 6 GHz frequency bands to provide compatibility with existing TT&C ground facilities.

After the satellites have been positioned in synchronous orbit, the TT&C function will be accomplished via ground facilities operating in the 12 & 14 GHz frequency bands.

The communications subsystem in the satellite will include ten active communications channels. Each channel will occupy a nominal bandwidth (channel spacing) of 49 MHz and a usable bandwidth of 43 MHz. Each channel will be capable of generating 23 watts at the output of the final traveling-wave tube amplifier (TWTA). A summary of space segment parameters is given in Table 1.

During solar eclipses there will be no capacity reductions. The satellite will have sufficient battery capacity to support all ten active transponder channels during these eclipse periods.

The STS of Delta 3910 launch vehicle will be employed to inject the satellite into parking

TABLE 1 SPACE SEGMENT PARAMETERS

Satellite Characteristics	Type or Value
Launch Vehicle	STS or Delta 3910
Launch Date	1980
Satellite Design Life	7 years
North-South Stationkeeping	$\pm$ 0.03 degree goal; $\pm$ .05 degree maximum
East-West Stationkeeping	$\pm$ 0.03 degree goal; $\pm$ .05 degree maximum
Eclipse Capability	100%
Redundancy	100% all elements, except 60% for TWTA's
Stabilization	Spin or body stabilized
Total Communications RF Output Power	230 Watts
Communications Channelization	Ten Operational 43-MHz Transporder Channels
Communications EIRP per Transponder	CONUS weighted Region 1: $\geq$ 43 dBW $\geq$ 39 dBW $\geq$ 37 dBW
Communications Receiver G/T	Region 1 $\geq$ + 1.5 dB/^{o}K $\geq$ - 2.5 dB/^{o}K $\geq$ - 4.5 dB/^{o}K
12 & 14 GHz Frequency Band	Transmit: 11.7 to 12.2 GHz Receive: 14.0 to 14.5 GHz
Communications Polarization	Transmit: linear horizontal Receive: linear vertical

orbit, and a perigee stage will inject the satellite into transfer orbit. Final injection into a nominally geostationary orbit will be accomplished by means of an apogee kick motor incorporated in the satellite.

The Delta 3910 performance capability will limit the payload mass at lift-off to 1090 kilograms or less, including the perigee stage/satellite adapter, separation devices and third stage telemetry. The performance of applicable apogee kick motors will permit an initial mass of the satellite after injection into near geostationary orbit of approximately 555 kilograms.

A preferred stabilization technique for the satellite has not been selected at this writing. Proposals have been invited from qualified satellite manufacturers for both spin and body-stabilized designs, and it is the opinion of SBS that either type can be considered for this application.

Each satellite will contain one active and three spare wideband receivers and ten communications channels, including ten active and six spare TWTA's. The ten center frequencies are listed below.

| Active | Frequency, GHz | |
Channel	Uplink	Downlink
1	14.025	11.725
2	14.074	11.774
3	14.123	11.823
4	14.172	11.872
5	14.221	11.921
6	14.270	11.970
7	14.319	12.019
8	14.368	12.068
9	14.417	12.117
10	14.466	12.166

The satellite antennas will provide shaped beams to cover the contiguous 48 states. In the baseline antenna configuration the receive antenna will employ a parabolic reflector. There will be either one or two transmit antennas that employ a parabolic reflector. Final characteristics of the satellite antennas are subject to variation depending on the contractor to be selected. Each antenna of the baseline design will have multiple feeds with the appropriate input/output ports connected to a directional coupler. In the direction of maximum radiation the antenna gain is estimated as greater than 32 dB at 12 GHz and at 14.3 GHz.

Figure 2 represents the required coverage of the satellite antenna. The respective EIRP's and G/T are indicated for the three regions. Figure 3 is a representative coverage pattern as determined by SBS-funded studies. The final pattern will depend on the results of the satellite development contract, but it will meet the minimum requirements of Figure 2. Tables 2 and 3 list additional budgeted parameters of the satellite for each of the three regions in the coverage area.

Earth Stations

By incorporating high quality components and redundancy, SBS earth stations will provide highly reliable operation on an unattended basis.

The RF Terminal equipment will include a parabolic antenna and feed assembly, a high-power amplifier (HPA), a low-noise amplifier (LNA), and up- and down-converters that will be interconnected with the TDMA Burst Modem via an interfacility link (IFL) that will carry signals at the 70 MHz intermediate frequency.

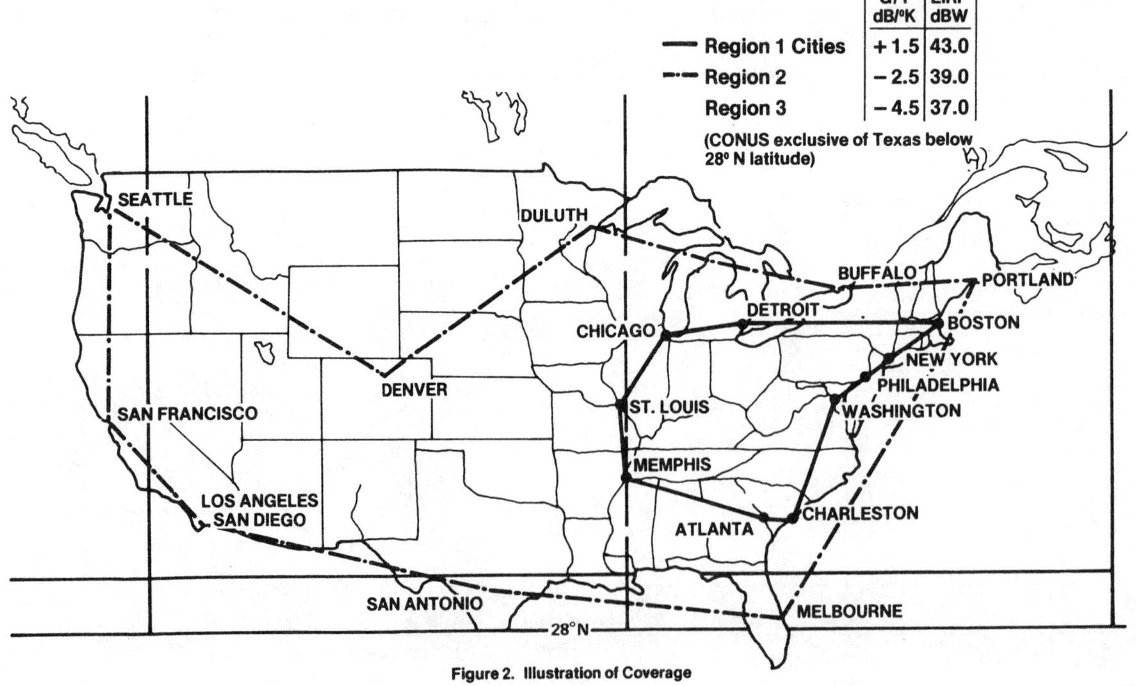

Figure 2. Illustration of Coverage

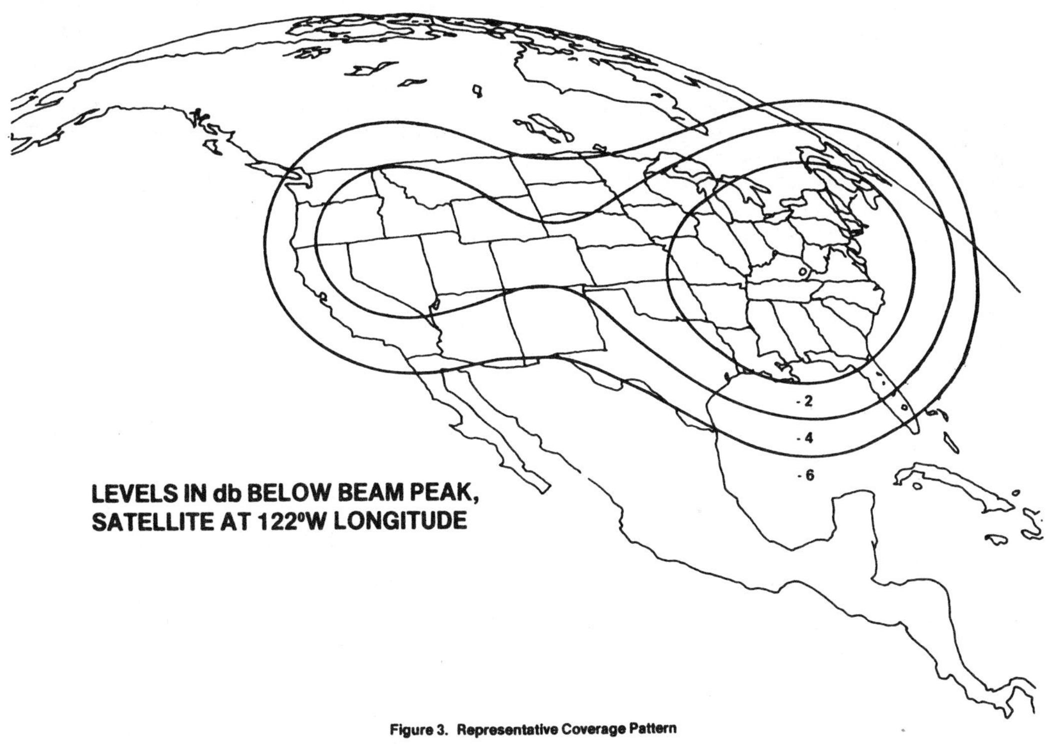

Figure 3. Representative Coverage Pattern

TABLE 2   SATELLITE G/T BUDGET

Region	1	2	3
Minimum Antenna Gain over Coverage Region (Including Pointing Error Losses, etc.)	31.5 dB	27.5 dB	25.5 dB
Transmission Losses Between Antenna and Preamp	1.0 dB	1.0 dB	1.0 dB
System Noise Temperature at Preamp Input	800°K	800°K	800°K
Minimum G/T	+1.5 dB/$^{\circ}$K	−2.5 dB/$^{\circ}$K	−4.5 dB/$^{\circ}$K

TABLE 3   SATELLITE EIRP BUDGET

Region	1	2	3
TWTA Output	+13.6 dBW	+13.6 dBW	+13.6 dBW
Losses Between TWTA and Antenna	1.7 dB	− 1.7 dB	− 1.7 dB
Antenna Gain at Edge of Coverage (Includes Pointing Error Losses, etc.)	31.5 dBi	27.5 dBi	25.5 dBi
Minimum E.I.R.P.	43 dBW	39 dBW	37 dBW

Generally, 5-meter antennas will be employed in Region 1 of the satellite coverage area and 7-meter antennas in Region 2. In parts of Region 3 and in areas where rain attenuation is unusually severe, system performance objectives will be met through the use of larger than 7-meter antennas or by higher performance HPA's or LNA's to increase up-link EIRP and downlink G/T, respectively. The nominal HPA size will be 500 watts, but stations can be operated at lower transmit levels so that the RF bursts from all earth stations will arrive at the satellite receiver at the same level, under clear sky conditions. The antennas will have limited motion capability to permit repointing to any satellite position within an orbital arc from 100 to 130 degrees West Longitude.

Table 4 summarizes the parameters of the earth station antennas.

TABLE 4   ANTENNA CHARACTERISTICS (COMMUNICATIONS)

CHARACTERISTIC	REGION 1	REGION 2
Type	Parabolic	Parabolic
Size	5m (16 ft)	7m (23 ft)
Gain (mid-band)		
Receive	53.9 dB	56.8 dB
Transmit	55.5 dB	58.4 dB
Beamwidth		
Receive	0.37^{o}	0.43^{o}
Transmit	0.31^{o}	0.67^{o}
Sidelobes		
Receive: first	0.60^{o}	0.43^{o}
second	0.93^{o}	0.67^{o}
Transmit: first	0.50^{o}	0.36^{o}
second	0.77^{o}	0.55^{o}
Polarization		
Receive	Horizontal	Horizontal
Transmit	Vertical	Vertical
Tracking Mode	Fixed	Fixed, with option for step track
$G/T^{(1)}$	30.4 dB/^{o}K	33.3 dB/^{o}K
EIRP (mid-band)		
Main Beam	$86.0^{(2)}$ dBW 80.4 dBW*	$88.9^{(2)}$ dBW 83.3 dBW*
Horizontal Plane	$30.0^{(3)}$ dBW	$27.6^{(3)}$ dBW

(1) Calculated for receiver system temperature of 225°K and at a center frequency of 11.950 GHz.
(2) Calculated for 14.250 GHz, with maximum HPA output of 2000 watts.
  * Added entry: Values for SBS nominal system with 500 watt HPA.
(3) Calculated for smallest elevation angle, with maximum HPA output.

The TDMA Burst Modem will employ quaternary phase-
shift-keying (QPSK) or its equivalent, such as
OS-QPSK, MSK, or FFSK, with two bits per symbol.
The modulation approach will be determined during
the design and development phase of the procurement. SBS plans to procure prototypes for test
and evaluation purposes that will operate at 43 or
48 Mbps, depending on the timing modules installed
in the basic modem configuration. The basic configuration will be capable of operating over a
data rate range from 35 to 55 Mbps, with appropriate timing modules installed.

The functions of the Burst Modem are illustrated
in Figure 4. In summary, they include:

. Modulation and demodulation
. Generation of the preamble for each burst
. Overlay and removal, upon demodulation, of an
  energy dispersion sequence to reduce adjacent
  channel interference levels
. Codeword detection
. Acquisition and synchronization with each
  burst

The Satellite Communications Controller (SCC)
functions as a time division switch, information
processor and controller. It is an integrated
hardware and software device which itself includes
a small data processor. The SCC performs the
following principal functions:

. Analog/digital voice conversion
. Data signal forward error correction coding/
  decoding
. Voice activity compression (VAC)
. Signalling
. Call processing
. Circuit switching

. Echo suppression
. Formatting, framing, synchronizing
. Multiplexing
. Multiple access control
. Demand assignment (DA) control

System Maintenance and Management

Earth Stations in the SBS System will be operated
unattended. Maintenance (both routine and un-
scheduled) will be accomplished from Operations
and Maintenance (O&M) Centers. Each O&M Center
will provide support to earth stations in the
same general geographic area. Centers will be
supported by higher echelon maintenance groups and
the SMF, as illustrated in Figure 5.

An essential role of the SMF is to monitor and
control the system. Under normal conditions, the
SCC's in customer networks will operate independently of the SMF. The SMF will intervene when
malfunctions or performance degradations occur
and when additional services or networks are to be
added. The SMF obtains status and health information for each earth station by polling each operating SCC via multidrop satellite circuits. In
effect, the SCC's in each satellite communications
channel function as remote communications controllers on a 9600 bps multidrop satellite circuit
that is polled by the SMF host processor. Each
SCC, in turn, obtains status information on the
RF Terminal, Burst Modem and other earth station
equipment via the Monitor and Command (M&C) loop.
This 1200 bps loop is driven synchronously by the
SCC and collects periodic status and alarm information from loop/equipment interface terminals;
it also distributes commands to the equipment.
Diagnostic commands can be sent over the loop by
the SCC, using stored diagnostic routines, or by
remote command from the SMF.

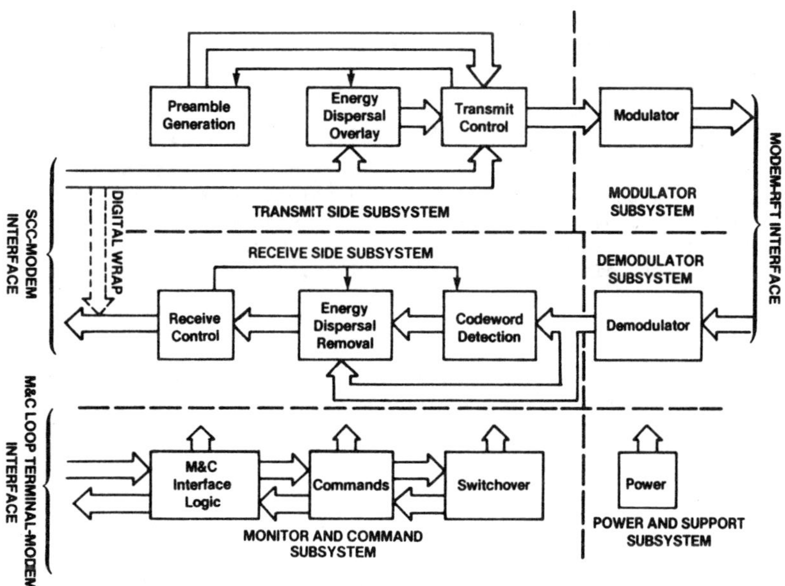

Figure 4. TDMA Burst Modem Functional Block Diagram

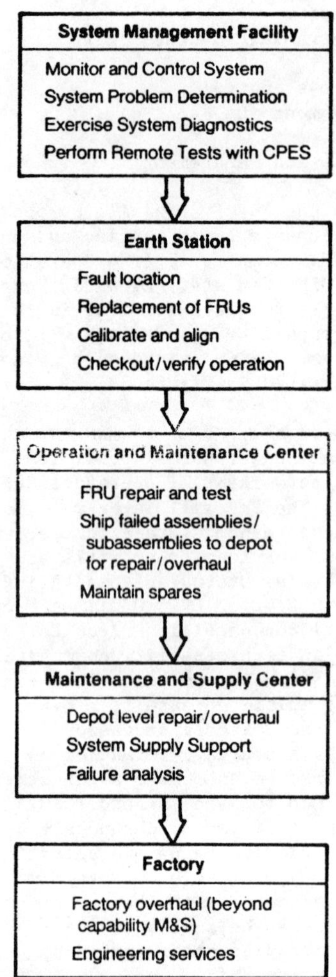

Figure 5. System Functional and Organization Relationships

A high degree of redundancy is planned for earth station equipment, yielding an earth station availability (due to equipment failures) of .9991 or better. Repairs will usually be made on the redundant equipment within a 24-hour period after a unit fails. In the small number of cases where a failure of a unit also causes an earth station outage, a service technician will be dispatched immediately by direction of the SMF. O&M Centers will be distributed to permit the technician to reach a site in less than 1½ hours. Repairs at the earth station will be made by substituting a Field Replaceable Unit (FRU). More complex repairs will be accomplished at higher echelon locations, as indicated in Figure 5.

The recently distributed specifications for development of SBS earth stations have established MTBF objectives of 15,000 hours for the Burst Modem and 30,000 hours for the RF Terminal. These objectives assume the application of redundancy.

Conclusions

Since approval of the SBS System application by the FCC in February, 1977, SBS has completed system studies and has taken steps to develop the equipment for its system. These efforts have led to improvements and refinements in planned system technical parameters. It is the intention of SBS to award development contracts by the end of 1977 and to begin serving customers with its operational 12 & 14 GHz system in early 1981.

References

1. _Applications of Satellite Business Systems for a Domestic Communications Satellite System_, Volume II, "System Description", December 1975 (Filed with Federal Communications Commission, Washington, D.C.)

2. _An Amendment to Operational System Application_, April 16, 1976, Satellite Business Systems filing with the Federal Communications Commission, Washington, D.C.

3. Kittiver, C. and Zitzmann, F.R., "The SBS System – An Innovative Domestic Satellite System for Private-Line Networks", Presented at AIAA/CASI Sixth Communication Satellite Systems Conference, Montreal, Canada, April 5, 1976.

4. Kittiver, C. and Westwood, D.H., "The Composite Margin Plane -- A Tool to Compute Overall Satellite Communications Link Availability", Presented at NTC '76, National Telecommunications Conference, Dallas, Texas, December 1, 1976.

2.3

THE DSCS III SATELLITE
A DEFENSE COMMUNICATION SYSTEM FOR THE 80's

I. S. Haas*
The General Electric Company
Valley Forge, Pennsylvania

A. T. Finney**
The Aerospace Corporation
El Segundo, California

Abstract

Critical needs of the government have now dictated development of another generation of military communication satellite to operate at X-band frequencies. This system, designated as DSCS III, will be designed to accommodate long haul, point-to-point communications, and also tactical users. Several major advances in technology are embodied in the design of the X-band communications payload. In particular, a set of multibeam antennas will permit transmission and reception via antenna patterns which will vary between single beams of 1.5^O diameter to total earth coverage. The X-band communication transponder will contain six separate channels two of which will have separate 40 watt traveling wave tube amplifiers and four of which will contain separate ten watt traveling wave tube amplifiers. Vehicle commanding will be done at a selected frequency in the communications band while telemetry will be superimposed on an X-band beacon which is also used for timing and doppler tracking by certain users. In addition to the X-band command capability, the vehicle will have a redundant S-band tracking and telemetry system which will operate via the U.S. Air Force Satellite Control Facility. In order to achieve the performance permitted by the multibeam antenna, a precise attitude control system will be required. This function will be performed by a three axis system which maintains virtually zero momentum about vehicle axes.

Introduction

Two generations of satellites have carried military communications traffic since the early 1960's when a unique and vital role for satellite communications was first identified in the Defense Communication System. The first satellite system to be implemented in response to military communications requirements was the Initial Defense Communications Satellite System, a constellation of small drifting satellites which was first established in 1966. These satellites were of a modest design, each containing a single 2.5 watt X-band transponder. Their simplicity was, in part, responsible for their high reliability; the satellites far exceeded reliability design goals and continued to provide dependable service through the early 70's.

A second generation of military satellites was developed in response to the increasing require-

ments of the Defense Communication System. This is the DSCS II satellite, which will provide service to government users through the early 1980's. These satellites represent a substantial increase in capacity over the IDCSS. Each spacecraft has a complement of two straight through 20 watt X-band channels together with an array of earth coverage and narrow coverage antennas for achievement of high EIRP's.

Current projections for the 1980's indicate a rapid increase in the number and variety of military users of X-band communications. At present confined almost totally to long haul, point-to-point communications, it is apparent that the community of users will soon include shipboard terminals as well as small tactical terminals. In response to this requirement for increased diversity, a third generation of DSCS satellites is under development.

When operational, this new system will be comprised of four synchronous satellites providing essentially world wide coverage. Each three-axis stabilized satellite will contain a Super High Frequency (SHF) communications payload consisting of multiple-beam antennas and a six-channel transponder. The communications system will be designed for both Frequency Division Multiple Access (FDMA) and Time Division Multiple Access (TDMA). An artist's concept is contained in Fig. 1. As illustrated, the antenna complement will include a 61-beam receive antenna with selective coverage and uplink anti-jam capability and two 19-beam transmit antennas with selective coverage to maximize EIRP allocation. Control of the system will be exercised by a two-channel Tracking, Telemetry and Command (TT&C) subsystem. One channel of the TT&C system will operate at S-band and will be used for housekeeping and backup payload control. The second system will operate at X-band through one of the communications channels for primary payload control and backup housekeeping. The DSCS III satellite is being developed by the Space Division of The General Electric Company under contract from the Space and Missile Systems Organization (SAMSO), U.S. Air Force.

In the following paragraphs satellite design features are described together with some aspects of the operational concept.

Communications Payload

The DSCS III satellite carries two communication payload packages. The primary payload is an SHF antenna/transponder which will satisfy the space link requirements of standard military communications networks. The secondary payload

---

* Program General Manager for DSCS III Program

**DSCS III Program Manager

is an Air Force Satellite Communications Single Channel Transponder which will transmit messages between Military Command Posts and Force Elements (B52's, FB 111, submarines, missile sites, etc.).

DSCS III uses electronic beam control of multibeam lens antennas allowing arbitrary shaped coverage and jammer discrimination. The channelization provided by a 6-channel transponder enables compatible operation with many user groups with EIRP varying from 72 to 97 dBW. The transponder is designed for operation in both TDMA and FDMA networks. The DSCS III satellite therefore represents a considerable advance in satisfying user needs which vary in geographic location, traffic demand and terminal size.

The SHF payload connects a series of receive antennas to 6 channels having bandwidths as shown in the frequency plan of Figure 2. (A communications payload block diagram is presented in Figure 3.) The receive band is 7900 to 8400 MHz with Channels 1 through 5 translated 725 MHz and Channel 6 translated 200 MHz. Channels 1 through 4 may be switched to receive on either the multibeam antenna or the earth coverage antenna in any combination. This provides the simultaneous capability to provide world wide receive coverage on some channels while concentrating high gain across a specified area for other channels.

On transmit, Channels 1 and 2 are switchable to the multibeam antennas or the high gain narrow coverage gimballed reflector. Channel 3 has access to either a multibeam antenna (MBA) or an earth coverage horn. Channel 4 can be switched between the MBA, the earth coverage horn, or the narrow beam reflector. Channels 5 and 6 are always connected to the earth coverage transmit horns. This flexible connectivity allows the operators in near real-time to configure the satellite coverage to satisfy narrow coverage high gain users, multiple shaped area coverages and earth coverage users simultaneously. Ground software provides the capability to optimize the channelization and antenna coverage patterns to suit the changing military communications scenario.

The top level communications performance is summarized in Table 1. The narrow coverage MBA receive performance is defined across a 0.5° radius including all spacecraft pointing errors to be better than -1 dB G/T. The MBA can create shaped patterns from this narrow coverage beam out to earth coverage where the G/T is -16 dB. The EIRP figures differ for channels based on their antenna configuration and due to the different traveling wave tube amplifiers (TWTA's) used. Channels 1 and 2 have 40-watt TWTA's while 3 through 6 use 10-watt TWTA's. The narrow coverage provided by the 33-inch reflector is measured at 1.5° radius with all pointing error degradation factors included.

Corresponding link budgets have been developed for a wide range of ground terminals. The traffic capability varies from 6 PCM channels operating through an airborne command post to over 1500 in earth coverage taking advantage of convolutional encoding for transmission. Table 2 shows some sample link calculations for different types of users.

The communications payload block diagram of Figure 3 illustrates the flexible interconnectivity between the transponder and the multibeam antennas to exploit the selective coverage and uplink jammer discrimination available in the multibeam antennas. The uplink receiver front end contains a preselector and a low-noise GaAs FET preamplifier which achieves an overall noise figure at the receiver input terminals of less than 4.9 dB. The preamplified signals are separated according to frequency by 8 section Chebychev filters in the demultiplexer. Channels 1, 2, 3 and 4 may be connected individually or in any combination to the receive MBA or the earth coverage horns by commanding switches respectively as shown in Figure 3. Channels 5 and 6 are dedicated to earth coverage.

The six channels of the transponder are identical in configuration except for power level and redundancy. Phase compensation is provided with a 5-section equalizer (E). The tunnel diode amplifier-limiter (TL) provides four gain steps in increments of 6 dB for a total commandable gain range of 24 dB. The driver amplifiers employ GaAs FET devices and are similar in design to the preamplifiers. Channels 1 and 2 operate with a 40-watt TWTA and are fully redundant as shown in Figure 3. Channels 3, 4, 5 and 6 operate with 10-watt TWTA's and have a single tube backing up two channels as shown in the figure.

The payload configuration has an additional intrinsic redundancy in that Channels 1, 3 and 5 are nearly identical to 2, 4 and 6 except for frequency assignment. This interconnective flexibility allows the system operator to shift traffic among the channels to accommodate wide variations in user requirements. This flexibility in antenna coverage and interconnectivity allows the system operator to respond rapidly to wide variations in the deployment of forces precipitated by changing world events.

<u>Multibeam Antenna (MBA) Design and Performance</u>

Flexible geographic coverage is afforded by the utilization of multibeam antennas for receive and transmit. These antennas will provide shaped patterns to arbitrary selected coverage areas and similarly, will create nulls for jammer discrimination.

Figure 4 shows the 61 individual beam patterns which can be overlayed on the earth's disk by the receive antenna. Figure 5 shows the elements of the receive antenna. The amplitude and phase of each of the beams is controlled electronically in a Beam Forming Network (BFN).

The receive MBA shown in Figure 6 was designed to provide maximum shaped coverage consistent with theoretical limitations of the 45-inch aperture and 61 beams. The relative amplitude and phase allocated to each of the 61 beams is derived from beam synthesis algorithms that optimize coverage and is implemented with an all electronic beam forming network made up of ferrite variable power dividers. The amplitude accuracy can be set to within 0.4 dB; the phase accuracy can be set to within $\pm$ 3° (typically $\pm$ 2^O).

Ground control of the MBA is accomplished through commanding the Beam Forming Network (BFN), a schematic of which is shown in Figure 7. The receive BFN is assembled from 60 variable power dividers and 61 phase shifters developed by Electromagnetic Sciences Incorporated. The 60 VPD's form a binary ladder network to collect and weigh the signals from the 61 receive ports. Port to port phase differential across all 61 ports is maintained to within 10°. Both the VPD phase shifters and the 360° phase shifters are driven with an 8-bit command, and are individually linearized prior to assembly to produce a phase shift which is proportional to the drive voltage. A 45° phase shifter is included at the output to prevent phase transients from occurring during BFN reconfiguration. During reconfiguration, the 45° phase shifter is switched as each level of the BFN is reset. This insures that the BFN total insertion phase changes less than 15° during reconfiguration. The BFN insertion loss of 2.6 dB is constant to within 1.0 dB over the entire receive frequency band and and over a temperature range of 25°C to 40°C. The BFN is an all electronic device with no moving parts switchable in a fraction of a second for minimum transient time in the transmission channel.

The 19-beam transmit MBA is a waveguide lens 28 inches in diameter with an integral 19 port BFN optimized for a gain and EIRP. Full amplitude control for creating desired selective coverage exists in all 19 beams. The measured narrow coverage gain is 27.0 dB with greater than 55% lens antenna efficiency. The transmit BFN is similar in design to the receive BFN except that no phase setting capability is provided. The BFN assembled during the first phase of the program had measured fifth order intermodulation products below -115 dBm when operated with the 40-watt TWTA in Channel 1 and a 10-watt TWTA in Channel 3.

The DSCS III MBA selective coverage feature allows the DCA operation center to allocate antenna gain and consequently EIRP to ground terminal locations according to terminal size and traffic flow requirements. A selective coverage algorithm for generating desired coverage contours which vary from the narrow coverage cone to irregular shapes to full earth coverage has been developed and verified by range test. The measured contours shown in Figures 8 and 9 demonstrate the flexibility and operational feasibility of this MBA selective coverage algorithm. The antenna pattern formed for any desired coverage is the result of superposition of the 19 individual beam patterns weighted in amplitude by the BFN. The amplitude weights are computed by an algorithm which solves up to 7 simultaneous equations. Figure 9 illustrates a narrow coverage pattern (0.75^O radius) located arbitrarily on the earth disk and having nearly perfect circular symmetry with all sidelobes below 18 dB. The selective coverage capability is illustrated in Figure 9 with a narrow coverage cone located in the upper left hand portion of the figure, an elliptical coverage shape in the upper right hand corner and a boot shaped coverage pattern in the bottom portion of the figure. The shapes possible are constrained only by the 28-inch aperture size which determines the resolution at which the pattern may vary across the field-of-view.

Channel Performance

A breadboard of the 6-channel transponder for DSCS III has been extensively tested. The amplitude and phase response of Channels 1 and 3 are shown in Figure 10. The amplitude response is flat to within 0.1 dB across the band. The channel bandwidth is 60 MHz and 85 MHz for channels 1 and 3 respectively, with an effective noise bandwidth of 1.2 times the channel bandwidth. The adjacent channel rejection is greater than 45 dB. The departure from linear phase is less than $\pm$ 5.0^O across the channels.

With the transponder channel operating in hard limiting, these characteristics combine to produce an excellent (QPSK) bit error rate performance as shown in Figure 11a. Figure 11b shows the same bit error rate data with the transponder operating in a 3 dB power backoff. Both sets of data were taken with QPSK having a 0.5 bit offset. Curve A in both figures is the theoretical curve; curve B is the bit error rate measured with the test set connected back-to-back (without the transponder). Curve C is the bit error rate versus E_b/N_o measured with the transponder connected into the test set. With the transponder operating in saturation, the degradation is 2 dB at a bit error rate of 10^{-6}. With the transponder operating in the linear mode with 3 dB power backoff, the transponder test data is 1.6 dB poorer than theoretical at a bit error rate of 10^{-6}. The test set contains a data detector having a butterworth low pass filter with a sampled output. The transponder gain of the 40-watt channels is a maximum of 135 dB with a minimum gain of 111 dB. The gain of the 10-watt channels, 3-6 is a maximum of 129 dB. When connected to the earth coverage horn the overall subsystem gain is a minimum of 145 dB and maximum of 169 dB.

Support Subsystems

The satellite has been designed to provide an accurately pointing, highly reliable long life platform for support of the communications mission. Table 3 highlights the key features of the satellite bus.

The DSCS III satellite utilizes a three-axis stabilization system to provide a fixed antenna platform and a sun oriented solar array for efficient power generation. The structurally integrated design is completely modular with all communications components mounted only on the north panel and housekeeping on the south panel. This results in reduced cable weight, short waveguide runs, takes advantage of structural shielding, and provides an equivalent Faraday cage around the component electronics for spacecraft charging protection. Electrical integration is achieved through standardized telemetry and command interfaces, a regulated central power controller, and a common spacecraft ground.

The following sections describe the Attitude Control Subsystem (ACS), the TT&C subsystem, and the Electric Power and Distribution (EP&D) subsystem.

Attitude Control Subsystem

The ACS design shown in Figure 12 is based upon a "zero momentum" control approach which has been breadboarded and tested. The ACS orients the satellite with four skewed momentum wheels controlled by processed signals from the earth and sun sensors.

The earth sensor detects pitch and roll errors and the sun sensors detect yaw errors. The sun sensors are mounted on the solar array yokes to eliminate shadowing, minimize the number of detectors, and simplify geometry related signal processing. Derived rate computed by the electronics eliminates the need for rate gyros to maintain yaw control during eclipse periods. When the wheel momentum storage capability is reached, the ACS automatically reduces the internal wheel momentum level through operation of the appropriate propulsion subsystem thrusters.

The ACS uses the same control laws and sensors for initial stabilization, normal pointing, momentum unloading and stationkeeping. Time shared digital electronics reduces the complexity, weight and cost of the subsystem. A gyro is provided for use in the acquisition mode which ensures acquisition even at tipoff rates greater than those expected. The acquisition sequence is independent of sun/earth relative geometry and allows rapid orientation of the array to the sun regardless of array orientation with respect to the main structure.

The subsystem design is capable of growing to control a spacecraft of larger size with minimum impact. The ACS is also adaptable to controlling a spinning orbit transfer state (which is a potential growth mode using a Shuttle/SSUS launch system).

Telemetry, Tracking, & Command Subsystem (TT&C)

The TT&C Subsystem will offer a completely dual capability for performing all satellite command and control functions. An S-band system will be available for operation via the Air Force Satellite Control Facility and an X-band system will permit commanding via communication ground terminals. Its performance meets the DSCS III requirements of varied command and telemetry interfaces (SCF, DCS terminals, and Shuttle); rapid MBA reconfiguration; incorporation of S-band and SHF COMSEC. A block diagram of the subsystem is provided in Figure 13.

SHF commanding is accomplished on the communication subsystem, Channels 1 or 5. These signals are doubly downconverted and level adjusted by the SHF/S and S/IF converters to match the input requirements of the KI-24 COMSEC. The KI-24 decrypts the data and transfers it to the command decoder for processing. The KI-24 also encrypts plain text telemetry data which modulates a sub-carrier and is combined with the spacecraft generated PN code in the SHF beacon. This com-

posite signal modulates the SHF beacon at 7.6 GHz and 7.605 GHz carriers which are transmitted at 0.7 watts (minimum) to a ground station via the communication subsystem earth coverage antennas.

The subsystem provides telemetry, tracking and command capabilities via both S-band and SHF links. The S-band link employs a block redundant transponder to receive encrypted commands, transmit encrypted and plaintext telemetry, and to transpond ranging signals.

The S-band received signal is decrypted by the KIR-23A COMSEC and processed by redundant command decoders. Coherent S-band doppler tracking is achieved by providing the carrier tracking loop reference to the transmitter which, after frequency multiplication is used as the downlink carrier. This carrier is phase modulated by the composite telemetry and received ranging code signal. The telemetry signal is formed by redundant master and remote telemetry units which multiplex telemetry data into a serial data stream for encryption by the KGX-28A COMSEC. The modulated carrier is transmitted at 2 watts (minimum) to the SCF ground station via the wide beam S-band antennas. An encrypter bypass is provided to transmit clear text telemetry data in the event of an encrypter failure.

The data handling systems are tailored to the DSCS III requirements by employing a combination of centralized and distributed features. Each BFN, the ACS and the SCT contain specialized circuits and registers for remote decoding of high volume, short time serial messages and for remote encoding of similar telemetry data. Otherwise, a centralized decoder and telemetry unit approach keeps down weight and cost below that of full distributed systems. A remote telemetry unit, located in the north panel, is dedicated to transponder telemetry.

Electrical Power Subsystem

The Direct Energy Transfer approach provides a high efficiency regulated bus voltage and is directly derived from current state-of-the-art subsystems on Broadcast Satellite and Global Positioning System.

The DSCS III implementation provides 827 watts of daylight power and 950 watts of eclipse power after 10 years of on-orbit life which allows a minimum margin in excess of 115 watts.

The approach provides a regulated bus voltage of +28 Vdc $\pm$ 1% for distribution to all subsystem equipment. This tightly controlled bus voltage approach was selected over an unregulated voltage system principally because of the higher overall energy transfer efficiency which is reflected in terms of lower solar array area. The regulated system also results in lower costs for other user sybsystem components due both to the reduced design, testing and integration effort associated with a regulated input voltage, and to the elimination of voltage regulation circuitry in the user equipment.

The solar array provides 126 ft^2 of total panel area on two hinged panels on each side of the satellite which compactly fold in the launch configuration. A yoke structure eliminates shadowing and supports shunt dissipator circuits on each of two wings. The solar array provides 1188 watts (or 9.4 watt/ft^2) under worst case design conditions at beginning-of-mission at the autumnal equinox. The folded array provides sufficient power to the spacecraft during transfer orbit.

A block redundant drive assembly for each solar array wing orients the panels to the sun. A flight proven slip ring assembly transfers power and signals across the rotating drive shaft.

Batteries provide power during prelaunch, launch and ascent, for on-orbit eclipse operations, and provide fuse flow current to clear faults. Three 32 ampere-hour sealed nickel cadmium 16-cell batteries satisfy the total energy storage requirements with conservative depths of discharge to provide 10-year design life.

The redundant electronic power conditioning circuits are on-line for utilization without relay or contactor transfer, and the transition from battery discharge to battery charge to shunt occurs without a disturbance to the regulated bus. Each battery has dedicated charge and reconditioning circuitry. A reconditioning cycle can be initiated by ground command with self-terminating discharge, or the discharge can proceed to zero volts.

Three power controllers control distribution of all power by ground command, provide load protection, enable thermal control, and switch communication subsystem elements. The subsystem provides safing, arming and actuation functions for all electro-explosive device initiators to separate a second satellite and release the folded solar array.

Operational Concept

The DSCS III Space Segment orbital configuration consists of four operational satellites and two on-orbit spares in synchronous orbit located such that there is substantial overlap in earth coverage. The on-orbit spares provide rapid and reliable replacement for any degraded operational satellite. The system operates at SHF (X-band) serving the Department of Defense World-Wide Military Command and Control Communications System (WWMCCS), the ground mobile forces, the Defense Communications System (DCS), Navy Fleet Communications, the White House Communications Agency and Allied Communications Networks such as NATO.

The DSCS III will initially be launched on a Titan IIIC-Transtage in July 1979 in combination with a DSCS II satellite. A second launch is planned in 1980, again with a DSCS II. After 1981 DSCS III launches will take place using the Space Transportation System (Shuttle) - IUS combination to initialize and replenish the operational system. The versatile satellite bus has been designed to be compatible with all of these booster combinations.

After being placed on station, the communications payload will be commanded from one of several Satellite Configuration Control Element (SCCE) Stations under the direction of the DCA. These stations will communicate with the satellite at SHF to optimize operational flexibility. The satellite will operate as shown in Figure 2 with large or small terminals, and with either FDMA or TDMA. Its six independent channels can provide resources on command, grouping users by their operational needs or geographical situation, and allocating receiver sensitivity or transmitter power among them for maximum efficiency.

The satellite housekeeping subsystems are controlled from the Air Force Satellite Control Facility (SCF) and its remote tracking station (RTS) network. The satellite itself is autonomous, but telemetry status is monitored by the SCF on a daily basis to ensure the health of the spacecraft. The SCF also provides backup communications configuration control functions. The satellites have also been hardened against the defined nuclear environment.

Conclusion

The DSCS III satellite represents a significant advance in satisfying user needs which vary dramatically in traffic demand and terminal size. The channelization afforded by the six channel transponder enables compatible operation with many user groups whose ground terminals vary in size from 3 ft. antennas to 60 ft. antennas with EIRP varying from 72 dBw to 97 dBw. The operational flexibility of the six channel transponder is enhanced by flexible interconnectivity between multibeam antennas and the earth coverage horns which allows the operator to shift traffic among channels and antennas. The transmitters in the six channel transponder are designed for operation in both FDMA and TDMA networks. The amplitude response, phase linearity and intermodulation performance verify the feasibility of FDMA operation. The measured QPSK bit error rate data illustrates performance to within 2 dB of theoretical at a bit error rate of 10^{-6}. Another major improvement which will be provided is the dual commanding capability at both X and S-band. All other vehicle support systems are designed to represent a minimum risk, state-of-the-art capability.

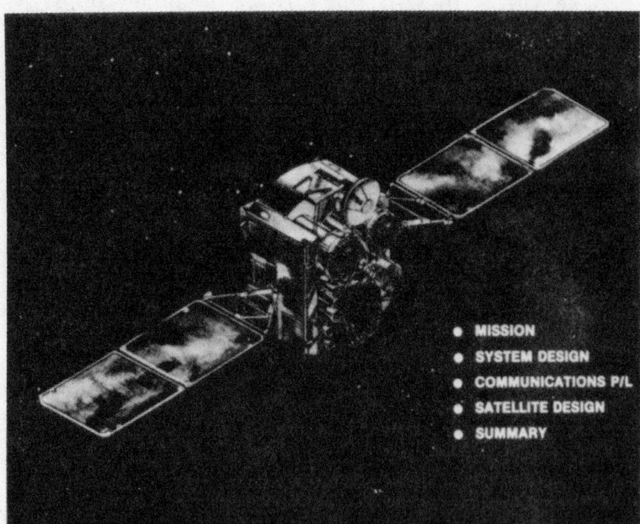

Figure 1. Defense Satellite Communications System Phase III

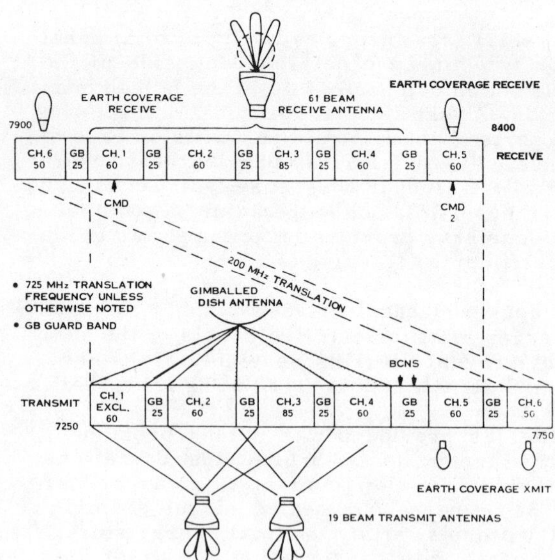

Figure 2. Frequency Plan and Antenna Interconnectivity

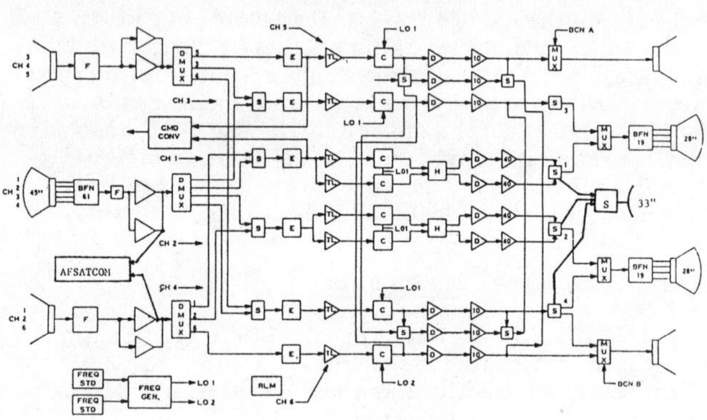

Figure 3. Communication Payload Block Diagram

Table 1. Communications Performance Summary

G/T*	
MULTIBEAM (NARROW COVERAGE)	−1.0 dB
MULTIBEAM (EARTH COVERAGE)	−16.0 dB
EARTH COVERAGE HORN	−15.0 dB
EIRP*	
CHANNEL 1, 2 – MBA – NARROW COVERAGE	40 dBw
– MBA – EARTH COVERAGE	29 dBw
– REFLECTOR	44 dBw
CHANNEL 3, 4 – MBA – NARROW COVERAGE	34 dBw
– MBA – EARTH COVERAGE	23 dBw
– REFLECTOR (CHANNEL 4)	37.5 dBw
– EC HORN	25 dBw
CHANNEL 5, 6 – EC HORN	25 dBw

*PERFORMANCE INCLUDES DEGRADATION FACTORS FOR ALL POINTING ERRORS.

Table 2. Communications Capacity

SYSTEM PARAMETER	WSC 2 4' (8')	ASC–18 AABNCP	TSC–86 (LT)	MSC–61 (MT)	FSC–78(HT) (MSC–60)
SATELLITE ANTENNA	MBA-NC	MBA-NC	MBA-NC	EC-HORN	EC-HORN
MAXIMUM UPLINK EIRP (dBW)	71.0 (77.0)	72.0	71.7	87	97
PATH LOSSES +3dB MARGIN (dB) (EARTH EDGE)	207	207	207	207	207
SATELLITE G/T (dB/$^\circ$K)	−1	−1	−1	−15	−15
BOLTZMANN'S CONSTANT (dBW/$^\circ$K)	−228.6	−228.6	−228.6	−228.6	−228–6
UPLINK $C/\kappa T_s$ (dB–Hz)	91.7 (97.7)	92.6	92.3	93.7	103.7
MULTIPLE ACCESS MODE	TDMA	TDMA	TDMA	TDMA	TDMA
SATELLITE TRANSMIT ANTENNA	MBA-NC	MBA-NC	MBA-NC	EC-HORN	EC-HORN
SATELLITE CHANNEL	4	1	2	3	5
SATELLITE EIRP (dBW)	34	40	40	25	25
PATH LOSSES +3 dB MARGIN (dB) (EARTH EDGE)	206.2	206.2	206.2	206.2	206.2
EARTH TERMINAL G/T (dB/$^\circ$K)	12 (18)	6	18	27	39
BOLTZMANN'S CONSTANT (dBW/$^\circ$K)	228.6	228.6	228.6	228.6	228.6
DOWNLINK C/kT_s (dB–Hz)	68.4 (74.4)	68.4	80.4	74.4	86.3
COMBINED C/kT_s (dB–Hz) QPSK DATA RATE (MBITS/SEC) BER – 1 X 10^{-6}	68.4 (74.4)	68.4	80.2	74.3	86.3
• UNCODED	0.35 (1.4)	0.35	5.25	1.35	21.4
• CONVOLUTIONAL ENCODING	1.4 (5.6)	1.4	20.9	5.4	85.2
6 BIT PCM CHANNELS					
• UNCODED	7 (29)	7	109	28	425
• CONVOLUTIONAL ENCODING	29 (116)	29	435	113	1774

UNCODED E_b/N_o = 13 dB, RATE 1/2 CODE, E_b/N_o = 7.0 dB BER = 10^{-6}
6 BIT PCM = 48 kB/SEC/VOICE CHANNEL

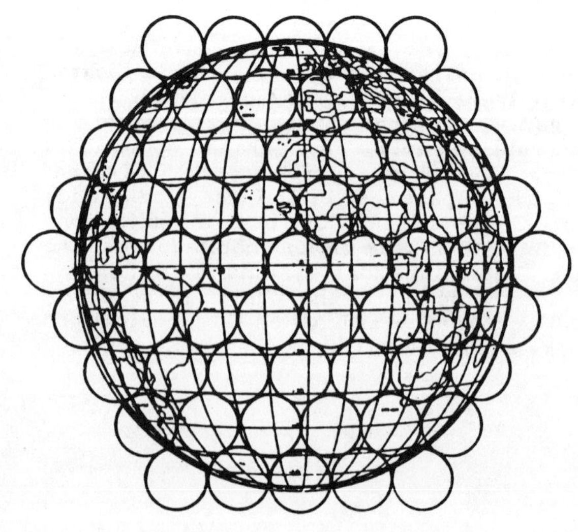

Figure 4. 61-Beam Receive Antenna

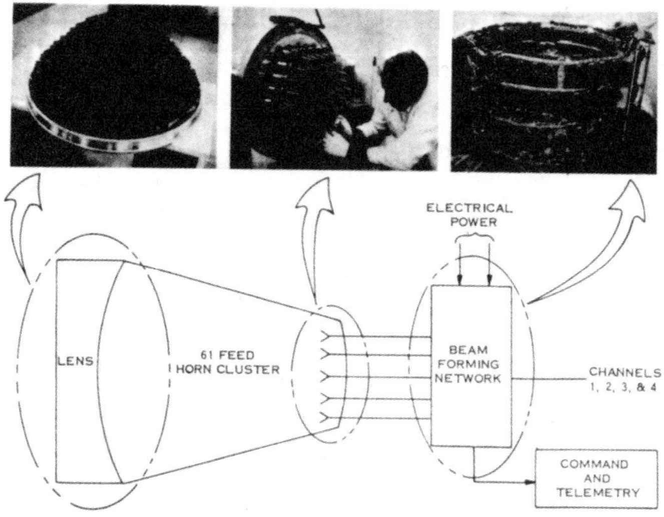

Figure 5. Elements of 61 Beam Receive MBA

Figure 6. Receive MBA

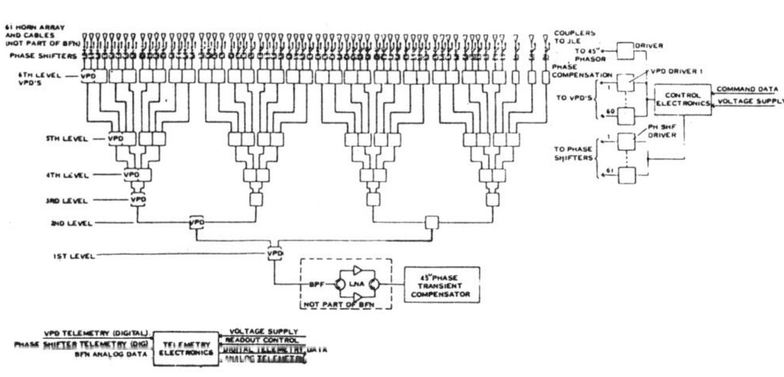

Figure 7. Electronic Beamforming Network

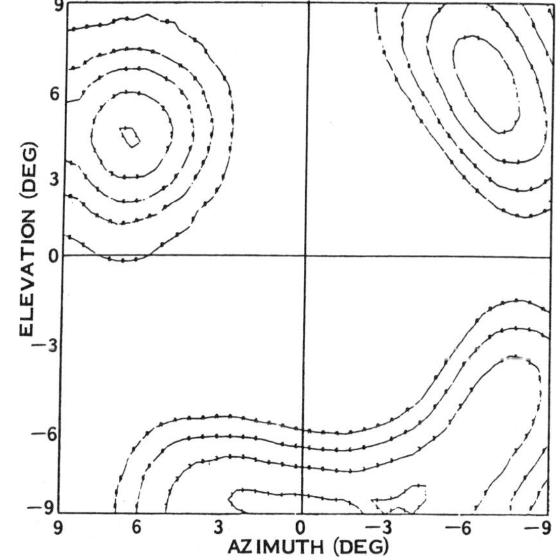

Figure 8. Transmit Selective Coverage

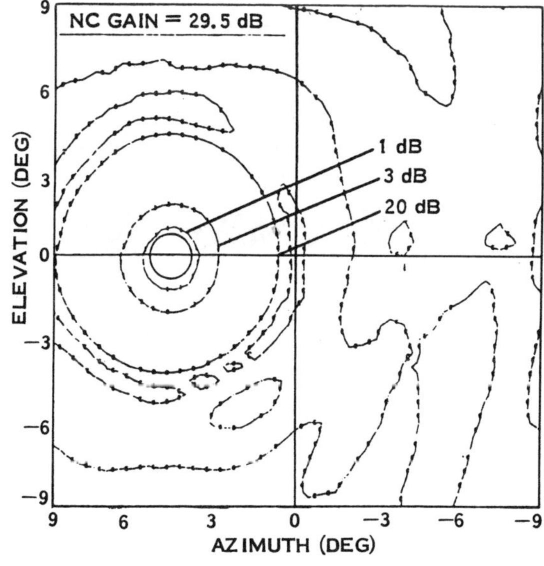

Figure 9. Transmit Narrow Coverage Beam

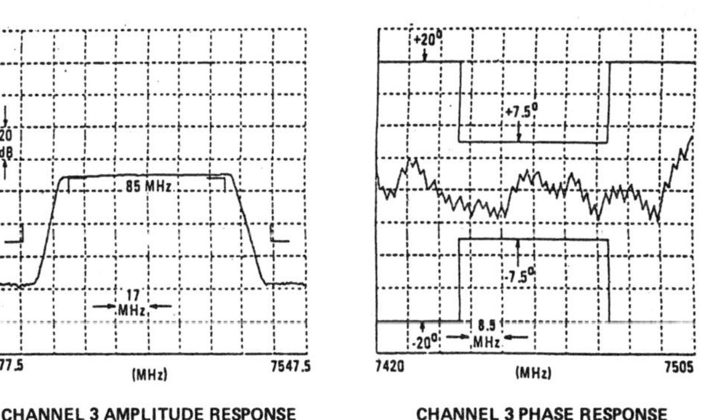

Figure 10. Wideband Channel Response

Figure 11. Mode Performance

Table 3. Key Features

ITEM	KEY FEATURES	ITEM	KEY FEATURES
SPACECRAFT SUMMARY	• 3 AXIS STABILIZED • FIXED PAYLOAD ANTENNA PLATFORM • ORIENTED SOLAR ARRAY FOR EFFICIENT POWER OPERATION • LARGE NORTH/SOUTH VIEWING PANELS FOR PASSIVE HEAT REJECTION • LOW LIFE CYCLE COST • BROADBAND NULLING • HARDENED DESIGN • FULL 61 BEAM CONTROL • ON BOARD JAMMING SENSING • 10-YEAR LIFE • NORTH/SOUTH STATIONKEEPING • GIMBALLED ANTENNA FOR SPOT COVERAGE • RELIABILITY EXCEEDS .7 AT 7 YEARS	ELECTRICAL POWER SUBSYSTEM	• REDUNDANT SOLAR ARRAY DRIVE • FULLY AUTONOMOUSLY REDUNDANT • RAPID RESPONSE TO LOAD CHANGES • LOAD FAULT ISOLATION • TRANSIENT PROTECTION • DUAL REDUNDANT ORDNANCE CONTROL • BATTERY RECONDITIONING TO ZERO VOLTS • SYSTEM OPERATION IN FAILED BATTERIES MODE
		PROPULSION SUBSYSTEM	• HYDRAZINE PROPULSION SYSTEM WITH REDUNDANT THRUSTERS AND TANKS • MODULARIZED • 600 LB. CAPACITY • FLTSATCOM TANKS VIKING P 95 • LMSC 1.0 LB. THRUSTER P 50 MODIFIED
TT&C SUBSYSTEM	• PERFORMANCE MEETS THE UNIQUE DSCS III REQUIREMENTS OF VARIED COMMAND AND TELEMETRY INTERFACES (SCF, DCS TERMINALS, AND SHUTTLE) • RAPID MBA RECONFIGURATION • INCORPORATION OF S-BAND AND SHF COMSEC EQUIPMENT	THERMAL SUBSYSTEM	• PASSIVE • FLIGHT PROVEN COATINGS AND COMPONENTS • NORTH/SOUTH RADIATOR PANELS WITH INCREASED THICKNESS LOCALLY • IMPOSES NO SATELLITE OPERATIONAL RESTRICTIONS • COMPATIBLE WITH TITAN IIIC OR SHUTTLE • SURVIVE FAILURE MODES INCLUDING ATTITUDE LOSS AND TOTAL BATTERY FAILURE
ATTITUDE CONTROL SUBSYSTEM	• AUTONOMOUS • TIME SHARED CENTRAL DIGITAL PROCESSOR FOR ALL CONTROL MODES • PERFORMANCE .08" R, .08" P, 0.8" Y • FOUR SKEWED REACTION WHEELS • ACQUISITION AUTONOMOUS WITH GROUND BACKUP • GROWTH FLEXIBILITY • USES EARTH & SUN SENSORS FOR ATTITUDE SENSING	STRUCTURE SUBSYSTEM	• ACCESSIBILITY/MODULARITY • PARALLEL SUBSYSTEM ASSEMBLY AND TEST • NORTH/SOUTH PANEL CONCEPT • THROUGH SOLAR ARRAY DRIVE SHAFT • INDEPENDENT PROPULSION MODULE • VIBRATION DAMPED EQUIPMENT PANELS • LIGHTWEIGHT/STIFF/DIMENSIONALLY STABLE • GROWTH AND OPTION FLEXIBILITY

ELECTRICAL POWER	• HIGH EFFICIENCY
	• REGULATED BUS
	  REDUCED USER SUBSYSTEM COST
	  NO SINGLE POINT FAILURE
	  ACCEPT FAILURE WITHOUT CHANGE IN SPECIFIED PERFORMANCE
	  LOWER SOLAR ARRAY AREA
		SURVIVABILITY	• OVERALL HARDENING APPROACH IS BASED UPON HARD COMPONENT INTERFACES

		LAUNCH VEHICLE	• TITAN IIIC • SHUTTLE • TITAN 34D IUS

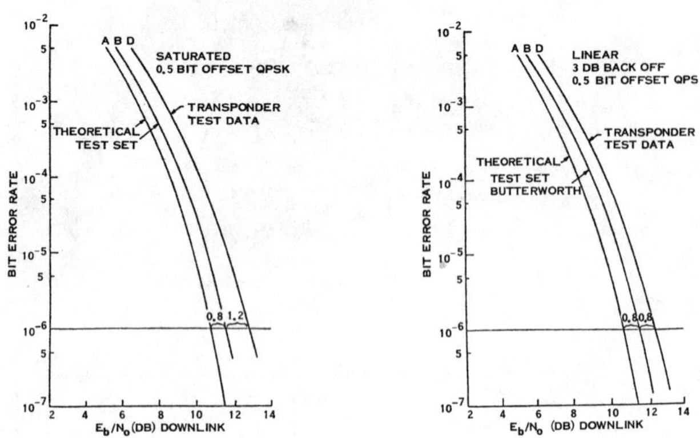
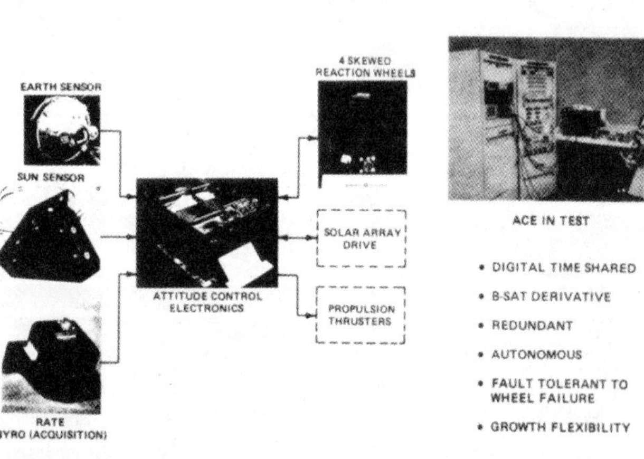

Figure 12. Attitude Control Subsystem

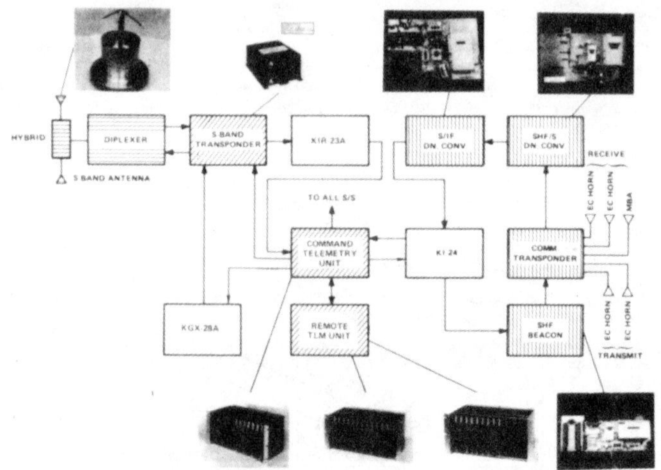

Figure 13. Telemetry, Tracking and Command Subsystem

2.4

ATS-6 Significance

RICHARD B. MARSTEN, Fellow, IEEE
The City College, City University of New York
New York, N. Y. 10031

Abstract

The Applications Technology Satellite (ATS-6), the most powerful, most sophisticated, most versatile communications satellite flown to date, is the last of NASA's experimental satellites intended to demonstrate major advances in communications and spacecraft technology. It is a multipurpose, multidisciplinary spacecraft whose principal objectives were to demonstrate a large, unfurlable antenna structure and precise pointing and attitude control in the synchronous orbit. The spacecraft carries 27 different experiments, 3 of which demonstrate users' applications of satellite communications. Significant advances in antenna technology, precise attitude control, materials technology, spacecraft structures, and thermal control have been successfully demonstrated. The most significant accomplishments of the ATS-6 mission are the demonstration of the practicality of satellite broadcasting to small, simple, inexpensive ground stations and the uses of this potential service in the solution of social problems involving education and health care. The success of these initial demonstrations has led ATS-6 experimenters and potential users to incorporate a Public Service Satellite Consortium dedicated to the provision of satellite broadcasting services for educational and health-care applications.

Manuscript received August 1, 1975.

Applications Technology Satellite (ATS-6) is the last of the National Aeronautics and Space Administration's (NASA's) line of advanced, experimental, applications satellites intended for operation in the geosynchronous, [1] equatorial orbit. This most sophisticated satellite of the ATS series was launched successfully from the NASA Kennedy Space Center, Cape Canaveral, Fla., at 9:00 A.M. EDT on May 30, 1974. Launch was accomplished with a U.S. Air Force Titan III-C rocket. Actual launch, parking orbit, and transfer orbit parameters were nominal, well within three-σ linits. Injection into geosynchronous orbit occurred on schedule: 6 h, 28 min, and 39 s after liftoff. Actual orbital parameters compare to nominal, as shown in Table I. The

TABLE I
ATS-6 Orbital Parameters

Parameter	Nominal	Actual
Apogee radius (km)	42 168	42 195
Perigee radius (km)	42 157	42 139
Inclination (deg)	1.82	1.78
Longitude (deg W)	94.56	94.41

nearly perfect orbit provided by the Air Force launch virtually eliminated the need to use the spacecraft propulsion system to remove injection errors. As a result, the 8.6 kg of fuel budgeted for that purpose became available to extend mission life beyond the 2-year design requirement, to cope with any operating contingencies that might develop.

ATS-6 is a multipurpose, multidisciplinary satellite whose configuration was determined by the objective of providing good-quality signals from synchronous orbit to small, simple, inexpensive Earth or mobile terminals. All told, it carries 27 different technological applications, and scientific experiments, many of them international in scope. Consequently, spacecraft and experiment designs represent practical compromises to attain maximum returns from each experiment rather than an optimization for any particular one.

Primary objectives of the ATS-6 mission were:

1) to inject the spacecraft into a near-geostationary orbit;
2) to erect a large antenna structure (nominally 9 m in diameter) in orbit which, with its associated communications system, is capable of providing a good-quality television signal to small, inexpensive ground receivers, and to measure and evaluate the performance of such a system;
3) to stabilize the spacecraft using a three-axis control system, with a slewing capability in roll and pitch, permitting antenna pointing with an accuracy commensurate with the antenna characteristics;

[1] The geosynchronous orbit has a period of 24 h. This occurs at an altitude of 35 680 km (22 300 mi) above the Earth's surface.

Reprinted from *IEEE Trans. Aerosp. & Electron. Syst.*, vol. AES-11, pp. 984–993, Nov. 1975.

4) to support and demonstrate user-oriented applications experiments utilizing the spacecraft capabilities.

Mission success would be judged on the basis of these objectives.

In addition, there were numerous secondary objectives:

1) to demonstrate new technology in space appropriate to:
   a) aeronautical and maritime traffic control;
   b) infrared Earth observation;
2) to acquire useful systems data for communications applications in space by:
   c) demonstrating a two-axis, steerable antenna beam;
   d) measuring radio-frequency interference in shared frequency bands;
   e) measuring propagation characteristics of millimeter waves;
   f) performing spacecraft-to-spacecraft communications and tracking experiments;
3) to test and obtain useful data on the following spacecraft technologies:
   a) three-axis stabilization with a precision pointing goal of 0.1° in all three axes and a slewing capability in the roll and pitch axis,
   b) microwave interferometer with precision compatible with the 0.1° pointing objective;
4) to make particle and radiation measurements in the geosynchronous environment.

To meet these objectives, the spacecraft was designed with zero-momentum, three-axis attitude stabilization to produce a pointing precision of better than 0.1° and a rate stability of better than 0.001°-s. These attitude control capabilities, together with deployment and demonstration of performance of the large antenna structure at several frequencies, constituted the spacecraft's experimental basis for achieving mission success.

For the first year of its 2-year design life, ATS-6 was stationed at 94° W longitude. On this station it was planned to perform all the experiments listed in Table III, except for the\Satellite Instructional Television Experiment (SITE). On May 20, 1975, it was planned to move the satellite toward a station at 35° E longitude, where it would be used to support the Indian Government's year-long SITE project. Continuation of the millimeter-waves propagation and communication and PLACE experiment, with Indian and European participation, was also planned, although SITE would be the principal objective of the second year.

The move toward 35° was planned for about 45 days. Completion on June 29 was followed by a period of checkout with the SITE ground complex in India. NASA control of the spacecraft shifted from the ATS ground station site at Rosman, N.C., to the ATS transportable ground station at NASA's site in Madrid, Spain. From July 15 to 26, 1975, ATS-6 provided tracking, data relay, and video transmission support to the US/USSR Apollo-Soyuz Test Project. SITE would begin August 1, 1975, and end July 31, 1976, after

which ATS-6 would be returned to the United States, on a permanent station at 105° W longitude. Plans for a third year of experimentation are presently under way.

II. Spacecraft and Experiments

The 1360-kg spacecraft consists of an Earth-viewing module (EVM) connected by a tubular support truss of graphite-fiber-reinforced plastic (GFRP) to an unfurlable, parabolic reflector 9 m in diameter. Two erectable aluminum structural booms supporting a solar array are mounted on a GFRP truss above the aluminum reflector hub. Fig. 1 shows the orbital configuration for the spacecraft. A weight summary is given in Table II.

The EVM is constructed of three subsections. The bottom one, the experiments module, contains most of the experiments and all of the Earth-viewing sensors. The central subsection, the service module, contains components of the electrical subsystem, the telemetry and command subsystem, and the attitude control subsystem. The top subsection, the communications module, contains the transponder and three experiments closely associated with the communications subsystem. The prime-focus feed of the high-gain antenna system is mounted on the top surface of the EVM.

The 9-m parabolic reflector of the antenna system consists of 48 flexible aluminum radial ribs supporting a copper-coated Dacron mesh. During launch the ribs were folded 90° at their hinges on the hub and wrapped tangentially around the hub. The stowed, furled antenna was restrained by 24 hinged doors held closed by a cable. The cable was cut after launch and the ribs unfurled to form the parabolic reflector. A self-portrait taken in orbit is shown in Fig. 2. The central hole within the hub is covered by a flat mesh screen to shield the environmental measurements experiments from direct illumination by the prime-focus feed. A more detailed description is given in [6].

Separate equipments are provided for all experiments other than those of the basic spacecraft and the users. Experiments, in order of priority, are listed in Table III. Detailed descriptions will be found elsewhere in this issue.

The HET experiment, sponsored by the Department of Health, Education, and Welfare, was intended to demonstrate applications of educational broadcasting and health-care delivery and teleconferencing. It contained six experimental components, each exploring different aspects of the delivery of educational and health services to citizens in remote regions of the country. Operating on *S*-band transponders in the 2.5-2.69 GHz broadcast band, ATS-6 provided video with up to four voice channels to simple, small, low-cost terminals in the Appalachian region, the Federation of Rocky Mountain States, and the States of Washington and Alaska. ATS-6 television coverage was supplemented by interactive voice links from small terminals back to the control center via ATS-1 and ATS-3 spacecraft at 136 MHz.

PLACE, operating in the radio location and communication band at 1535-1660 MHz, was designed to demonstrate the practicality of real-time position location and communication with aircraft and ships. The U.S. Coast Guard, the

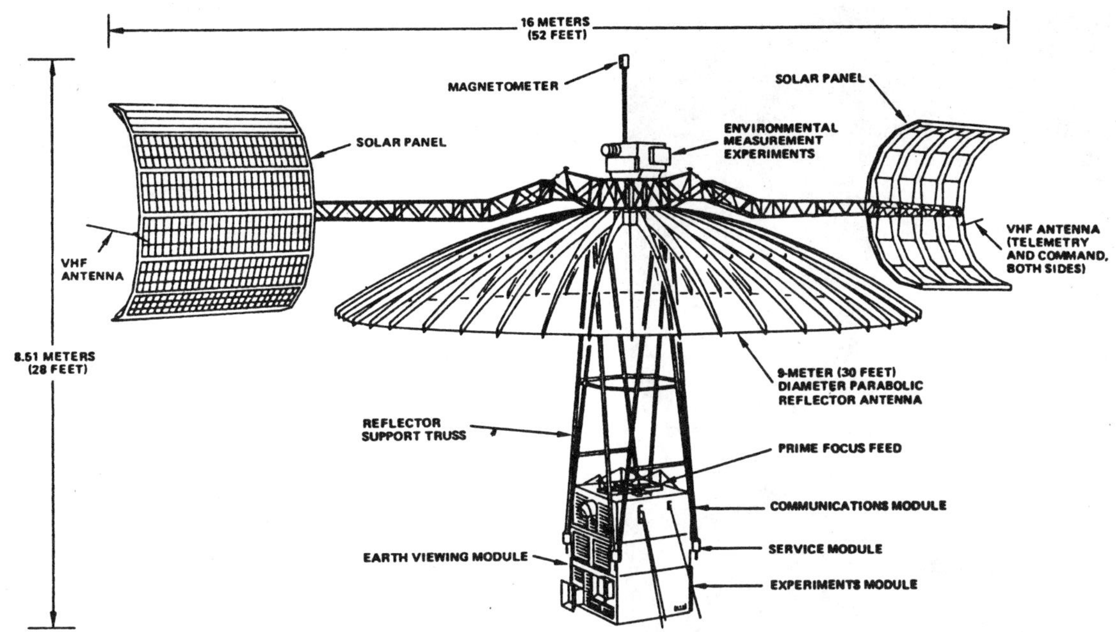

Fig. 1. Orbital Configuration for ATS-6

TABLE II

ATS-6 Weight Summary

Subsystem		Weight (kg)
Propulsion/attitude control		185.7
Telemetry and command		42.6
Communications		234.5
parabolic reflector	87	
antenna feed	26.8	
transponder	120.6	
Electrical (power/utilities)		116.8
power generation	100	
utility electronics	16.5	
Structure		366.2
primary structure	192.7	
solar array and deployment mechanism	144	
mounting provisions —subsystems	6.9	
mounting provisions—experiments	12.7	
Thermal control		64
Experiments		216
Harness		140
Total Spacecraft		1360
Titan III-C Adapter		46.7

Maritime Administration of the Department of Commerce, the Federal Aviation Administration, the Canadian Department of Communications, and the European Space Research Organization were active participants.

On May 20, 1975, ATS-6 was moved eastward to a location over Lake Victoria, Africa, at 35° E longitude. There, it is being used by the Indian Government to support SITE, an experiment in educational broadcasting to 5000 Indian villages, 2500 of which will have direct-receiving terminals. Ground transmitters and receivers, programming, broadcasting, and all operations of the ground complex are the responsibility of the Indian Government. Broadcasts will emphasize family planning, basic literacy, basic hygiene, and improved agricultural techniques.

The broadcasting experiments included demonstrations of television reception using small terminals. The small terminal for the HET experiment (Fig. 3) consists of outdoor and indoor units. The outdoor unit is a fixed-position, 3-m (10-ft) receiving antenna whose feed is integrated with a front end of 4.2-dB noise figure that converts the incoming S-band signal to intermediate frequency (IF). The indoor unit is a modulation converter that transforms the frequency-modulated, IF signal to a standard National Television System Committee (NTSC) signal that can drive the video portion of any commercial television set. The only control on the small terminal is an ON-OFF switch, accompanied by a pilot light to show power on. A field-strength meter indicates the presence of a received signal. These stations, emplaced, boresighted, and checked out, cost $3700 (1973 dollars) each in an experimental lot quantity of 130.

As listed in Table III, experimental data on propagation, atmospheric losses, and effects of station diversity were conducted in new bands allocated but not yet in use for satellite communication applications. Experimental data will be used to support U.S. views on frequency allocations at World Administrative Radio Conferences.

Experiments were performed with new technologies associated with precise attitude measurement and pointing; ground-based, adaptive, precise stabilization and attitude control; electric propulsion for spacecraft station keeping; measurement of contamination from propulsion-system exhaust; solar-cell degradation in orbit; and day/night cloud-cover imaging for meteorology.

Two-dimensional, electronic steering of satellite antenna beams was demonstrated, and ATS-6 was used to track and

Fig. 2. ATS-6 Self-Portrait, Deployed in Orbit.

TABLE III

ATS-6 Experiments

Category	Description
Basic spacecraft	unfurlable antenna deployment 0.1° stabilization and pointing; television reception using small terminals
User experiments	Health/Education Telecommunications (HET) Experiment; Position Location and Communicatior Experiment (PLACE); Satellite Instructional Television Experiment (SITE)
Satellite communications	millimeter wave propagation and communication at 13, 18, 20, and 30 GHz; station diversity at 13 and 18 GHz; RF interference at C-band
Space technology	very high resolution, day and night observation of cloud cover and motion; RF interferometer for precise angle measurement; advanced thermal control; spacecraft-attitude precision pointing and slewing adaptive control; solar cell degradation; cesium bombardment ion engine; quartz crystal microbalance
NASA support	Tracking and Data Relay Experiment; radio beacon for electron density measurement
Space science	five experiments to measure charged particles; solar cosmic ray measurements; magnetic field measurement

relay data from three low-altitude orbiting spacecraft: Geos-3, Nimbus-6, and Apollo-Soyuz. Results from these short-duration experiments are being used to develop NASA requirements for operational tracking and data relay satellite systems.

And finally, to provide continuing support to an ongoing scientific activity, several experiments are being performed to determine electron density, cosmic ray activity, and magnetic field at the spacecraft locations in the synchronous orbit.

III. How It Performed

The auspicious launch performance of the Titan III-C that gave an 8.6-kg fuel reserve to ATS-6 for extension of its orbital life began a year of unprecedented success in orbit. Successful deployment of booms, antenna, and solar array (Fig. 2) were achieved, and the spacecraft was locked on the Earth and stabilized by 11:00 p.m. on the day of the launch. Functional control had been established, and the initial checkout period of 30 days began early on June 1, 1974. The 30 days ended with the following results.

Performance of the attitude control system exceeded specifications in every mode (Table IV). The ability to point and stabilize the large, flexible antenna with both accuracy and stability well in excess of the specified 0.1° was demonstrated. Precise pointing and stabilization of large structures in the synchronous orbit, using inertia wheels driven by monopulse sensors or an array of star and Earth sensors, was consistently achieved. Demonstration of the capability of precisely tracking and communicating with low-altitude spacecraft was especially significant. Functionally redundant sensors all operated well, and the very high precision expected of the interferometer was demonstrated. Because of the very high precision, propellant consumption was significantly less than calculated, and it remained so for the duration of the first year in orbit (1.6 kg spent versus nearly 2.7 kg calculated).

Antenna pattern measurements, taken at all spacecraft frequencies, were obtained by precision slewing of the spacecraft boresight across calibrating ground stations. Observed gain figures at the higher frequencies verified the accuracy of the reflector contour.

Table V indicates in-orbit results, obtained during the initial 30 day checkout period, showing communications transponder performance very close to ground test results. Performance of redundant elements was also satisfactorily

TABLE IV

ACS Summary Performance

MODE	PARAMETER	ACTUAL*	SPEC
ABC SUN ACQUISITION	TIME TO ACQUIRE	10 MIN	30 MIN
	POINTING ACCURACY	2°	4.5^o
ABC EARTH ACQUISITION	TIME TO ACQUIRE	50 MIN	80 MIN
	POINTING ACCURACY	0.35^o	1.0^o
ABC LOCAL VERTICAL	POINTING ACCURACY	0.35^o	1.0^o
DOC VHF MONOPULSE	POINTING STABILITY	0.5^o	1.0^o
DOC S-BAND MONOPULSE	POINTING STABILITY	0.01^o	0.3^o
DOC C-BAND MONOPULSE	POINTING STABILITY	0.002^o	0.1^o
DOC OFFSET POINT GROUND	POINTING ACCURACY	0.04^o	0.1^o
	POINTING STABILITY	0.01^o	0.1^o
DOC LOW JITTER	POINTING ACCURACY	0.01^o	0.5^o
	POINTING STABILITY	0.005^o	0.01^o
	RATE STABILITY (LOW FREQ)	$0.0003^o/sec$	$0.001^o/sec$
DOC SATELLITE TRACK	POINTING ACCURACY	$< 0.2^o$	0.5^o
DOC LOCAL VERTICAL	POINTING ACCURACY	$< 0.1^o$	0.1^o
DOC OFFSET POINT SLEW	RATE	$1.2^o/sec$	$0.5^o/sec$
	SETTLING TIME	< 3 MIN	10 MIN
*DOC OPERATIONAL MODES	YAW ACCURACY	$< 0.1^o$	0.1^o

demonstrated. This performance remained satisfactory throughout the year

A minor anomaly, attributed to outgassing at the *C*-band transmitter output, appeared at the start of the mission but abated as time passed. *C*-band signals at the ground receivers indicated power drops of 1 to 2 dB, initially occurring at 1-min intervals and lasting about 80 ms. At the end of the 30-day checkout, the period had lengthened to 2 min and the severity of power drops had decreased to under 1 dB. An attenuator in the driver circuit of the traveling-wave tube (TWT) eliminated the anomaly without introducing any operating constraint.

All supporting systems operated nominally and without significant anomalies. Experiments generally operated satisfactorily although one of the two 20-GHz TWT's appeared to have failed during launch. The end-to-end performance of the HET link, specified to produce video signals of 47 dB peak-to-peak signal to rms noise ratio, was measured at 52 dB during a test run. This is equivalent to International Radio Consultative Committee (CCIR) relay quality, far in excess of the performance needed for successful user experiments.

By September 1974 all experiments had been successfully demonstrated and user experiments were run on a regular schedule in accordance with prelaunch planning. One of the two cesium-bombardment ion thrusters had been fired successfully, but because of faulty nozzle design the orifice blocked with solidified cesium as it cooled after shutoff. The dual-wavelength, very high resolution radiometer, having taken over 700 excellent images in both visible and infrared bands, experienced a failure in the chopper drive motor during October. But enough day/night cloud-cover images were obtained for data reduction and analysis to assure that the experimenter's objectives would be met in spite of the failure. The preliminary assessment indicates that the resolution of 4.8 km for visible-band data and 9.6 km for infrared has been met.

Successful, controllable, repeatable operation of spacecraft and experiments was well established by September. The practicality of wide-area broadcasting of high-quality television signals to small, simple, inexpensive ground terminals was consistently demonstrated, and the spacecraft was in routine use for support and demonstration of user applications experiments. Having met or exceeded its primary mission objectives, ATS-6 was declared a mission success on September 26, 1974. It continued to serve its experimenters until May 20, 1975, when it was started on its way toward 35° E longitude. The move was completed satisfactorily on June 29.

On June 26, a stabilization anomaly was observed in the roll axis, with deviations from stable position of $\pm6^\circ$. Attitude control was switched from wheels to jets, and the spacecraft was stabilized with expenditure of 0.2 kg of fuel. Analysis suggested a failure of one side of the two-phase, reversible-phase drive circuit for the roll wheel. Simulation showed that mixed-mode stabilization using pitch and yaw wheels with roll-axis thrusters was feasible and could be achieved with a minimum additional expenditure of fuel. One of the two digital operational controllers (DOC's) was reprogrammed for this mode, and test runs were made using ATS-6 to track the Geos-3 and Nimbus-6 spacecraft live. Results were compared with those from a run on Nimbus-6 using jet thrusters on all three axes. It was found that the mixed mode would require under 2.7 kg of fuel to support all 133 orbits of Apollo-Soyuz, compared with 12.2 kg for the "all-thrusters" mose and compared with 4.5 kg estimated from simulation. The conclusion was that the mixed mode of attitude control is not only practical, but would not impact either Apollo-Soyuz or the SITE mission and would still allow the spacecraft to return home with adequate fuel reserve for a third year of operation. As this is written, procedures for mixed-mose control have been implemented and ATS-6 is ready for Apollo-Soyuz and SITE. The entire process described above was completed in a week.

Technically, and in its versatility, reliability, and responsiveness, ATS-6 has performed beyond all expectations. System reliability in the HET experiment, once the initial learning period was completed, exceeded 99 percent. Users agreed that, in both signal quality and reliability of transmission, technical operation permitted them to concentrate on their user applications objectives 96 percent of the time. An overall user assessment stated that the ATS-6 performance in HET exceeded specifications and enabled the users to underrun costs [1]. Further assessments credit the HET experiments themselves with great advances in delivery of educational and health-care services [2],[3]. A series of detailed papers on these was given at a symposium in Denver in July 1975 [4]. We now await only the results of SITE, the last ATS-6 experiment of the initially planned 2-year mission.

IV. The Satellite Communications Environment

Contracting for design, development, and construction of ATS-6 was settled early in 1971. At that time there was one commercial satellite communications system operating: the global system of the International Telecommunications

TABLE V

Communications Subsystems Performance Characteristics

Band/ Mode	Frequencies Bandwidth		Antenna Field of View		Transmit Performance, EIRP over FOV				Receive Performance, G/T over FOV			
	Transmit	Receive	Transmit	Receive	P_O dBw	Gant, dB	Expected EIRP	Actual Δ	T_s, dB-°K	Gant dB	Expected G/T	Actual Δ
VHF	136 MHz/ ± 100 kHz	152 MHz + 4 MHz	15.0*, on axis		0.3	9.9	10.2	+0.9 dB	36.4	10.9	-25.5	**
UHF	800 MHz ± 20 MHz		2.8°, on axis	---------	19.6/18.6	30.2	49.8/48.8	+1.0 dB	---------			
L-Band, Fan	1550 MHz ± 15 MHz	1650 MHz + 20 MHz	1.0° X 7.5°, off axis		15.3	26.3	41.6	-1.8 dB*	31.1	24.8	-6.2	**
L-Band, Pencil	(Same)	(Same)	1.3°, off axis		15.3	34.8	50.1		31.1	35.4	4.3	
S-Band, Scan	2075 MHz ± 20 MHz	2250 MHz + 20 MHz	9.8° X 9.8°, using switched cross array		11.4	35.6	47.0		30.2	35.5	5.3	
S-Band Mono	(Same)	(Same)	1.0°, on axis		11.4	36.6	48.0	+1.4 dB	30.2	36.2	6.0	**
HET, Channel 1	2570 MHz ± 20 MHz		S_1: 0.85°, offset 0.7° S_2: 0.85°, offset 1.6°		11.1	37.6	48.7		---------			
HET, Channel 2	2670 MHz ± 15 MHz		(same as HET, 2570 MHz given above)		11.6	37.3	48.9	+0.1 dB	---------			
C-Band, PFF	3750, 3950 & 4150 MHz ± 20 MHz Each	5950, 6150 & 6350 MHz + 20 MHz Each	0.6°, on axis	0.4°, on axis	7.3	36.2	43.5		32.9	42.0	9.1	**
C-Band, ECH	(Same)	(Same)	13° X 20°, on axis	20° X 20°, on axis	7.8	16.3	24.1	+2.0 dB	32.9	13.5	-19.4	**

*Not Pointed At Rosman
**Indirect Indications That The G/T Is As Expected

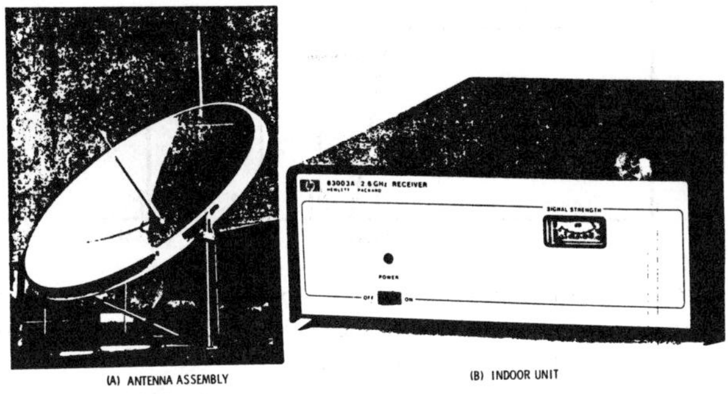

Fig. 3. Simple, small terminal for HET experiment.

Fig. 4. Intelsat IV: typical of large, spin-stabilized communications satellite.

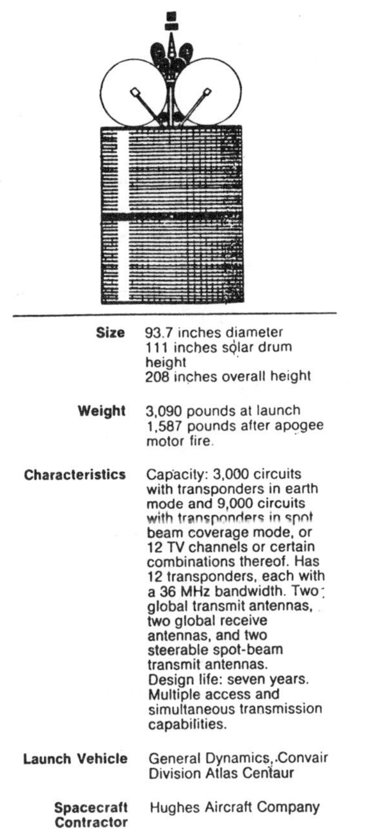

Size	93.7 inches diameter 111 inches solar drum height 208 inches overall height
Weight	3,090 pounds at launch 1,587 pounds after apogee motor fire.
Characteristics	Capacity: 3,000 circuits with transponders in earth mode and 9,000 circuits with transponders in spot beam coverage mode, or 12 TV channels or certain combinations thereof. Has 12 transponders, each with a 36 MHz bandwidth. Two global transmit antennas, two global receive antennas, and two steerable spot-beam transmit antennas. Design life: seven years. Multiple access and simultaneous transmission capabilities.
Launch Vehicle	General Dynamics, Convair Division Atlas Centaur
Spacecraft Contractor	Hughes Aircraft Company

Satellite Organization (INTELSAT). The first of the fourth-generation Intelsat-IV satellites had been launched and placed on station over the Atlantic Ocean. Satellites for the Canadian domestic commercial system of Telesat (Canada) were under construction. In the United States, the Federal Communications Commission was preparing to invite applications for competing commercial domestic systems. And NASA was operating ATS-1, ATS-3, and ATS-5 for the benefit of users interested in experimenting with satellite communications for their own needs.

In this environment the extant technology was that of two-body, spinning spacecraft with low-power transmitters and relatively small, rigid antennas. A spinning, cylindrical shell with solar cells rigidly fixed to the outer, cylindrical surface was connected to a despun platform on which the antenna assembly was mounted. Controls, electronics, and spacecraft propulsion subsystems were all contained within the shell. This technology was coupled with that of large ground antennas, frequently provided with tracking mounts, and low-temperature, cooled receivers.

Intelsat-IV, the most advanced of this type of satellite currently in orbit, is characterized as shown in Fig. 4. Each of the 12 transponders has a 6-W output. The global antennas have an 18° beamwidth and a gain of about 16 dB. Stabilization capability in the spacecraft is ±0.5°, roughly ±0.1 beamwidt

Such spacecraft work typically with Earth station antennas of diameter greater than 9 m, with beamwidths under 0.4°. An Earth station sited near the 3-dB contour of the spot beam would have to be equipped with antenna tracking, and it is not uncommon to have tracking capability associated with Earth station antennas of diameters exceeding 10 m. The antennas are conventially coupled to uncooled parametric amplifiers having temperatures of about 55 K.

Today there are seven Intelsat-IV spacecraft in orbit, in full operational service or functioning as in-orbit spares. The domestic system of Telesat (Canada) has three, somewhat smaller, spinning spacecraft in orbit. Western Union's WESTAR system, serving the United States, has three 12-transponder spacecraft, also spinners, in orbit. And the domestic system of AT&T and Comsat is approaching the launch of its first 24-transponder spinning satellite. This last spacecraft, like the modification to Intelsat-IV to double the number of transponders, is probably as large a spinner as can be provided practically, as power and beam requirements on spacecraft increase.

Power available from the Intelsat-IV body-mounted array is nearly 500 W; from the Intelsat IV A modification for 24 transponders, about 600 W. More beams are provided in these latest model spinners to maximize the channel capacity of the 4-GHz downlink band by maximizing reuse through spatial diversity. Their weight at launch is 1480 kg, compared with 1387 kg for the Intelsat-IV. Both the Intelsat IV-A and the first AT&T-Comsat domestic satellite are planned for launch in the last quarter of 1975.

All of these communication satellites are designed primarily for telephone (message and data) traffic: high-density, multiple-message traffic between large terminals connected to multiplexing, switching, and distribution centers. Whole transponders are available for television transmission; but with these, spacecraft still must work into relatively large and sophisticated Earth stations. For this sort of point-to-point service, the use of many transponders of low power with relatively few, large, complex Earth stations has proven economically sound. The objective is to maximize channel capacity at minimum cost.

Satellite channel capacity is maximized by increasing the number of beams and transponders, and this approach results in a corresponding reduction of power [5]. Such changes in design require much narrower beams and a corresponding increase in pointing accuracy, and as traffic continues to increase this could lead to performance requirements that spinners may not be capable of meeting. Coupled with the capability of large Earth stations (26 through 29-m antenna diameter) to support greater traffic, the result may be expected to lead to more power per unit weight in the satellite. For a fixed weight, this can be translated into greater communications channel capacity. Alternatively, a fixed capacity may be designed for a weight reduction sufficient to permit launch on a much less costly rocket. It has long been recognized that the use of inertial stabilization makes this possible: such recognition was one of the driving considerations behind the ATS-6 program. That same approach was later taken by RCA in its own domestic satellite communications system application. The RCA design uses a momentum-bias, inertial stabilization system for a 24-transponder spacecraft whose launch weight is about 900 kg. Since this spacecraft is also intended for the high-density, telephone-traffic service, other characteristics are comparable to those of the spinners in the systems described above. The first of these lightweight, inertially stabilized domestic satellites is also scheduled for launch in the last quarter of 1975. Launch will be on a Thor-Delta booster at a cost of about $9 million. This compares with launch of the 24-transponder, spinner class of satellite on an Atlas-Centaur, at a cost of about $25 million.

Two applications of the types of systems described herein have recently been receiving attention: service to remote areas, where the traffic is light, and television broadcast networking. The first of these is restricted by the need for small, low-cost stations that are economical in the context of relatively many widely distributed, isolated locations. The state of Alaska could be an example, with well over 150 such places scattered throughout its territory and a traffic load equivalent to 4 voice circuits (8 channels) as a probable maximum for any location. Typical designs for small, low-cost ground stations to work with the domestic communication satellites consist of a 4.6-m antenna, an uncooled receiver of about 1000°K noise temperature, and single-carrier-per-channel equipment that can support up to 4 voice circuits. The two-way station would cost from $54 000 to $62 000, depending upon the number of voice circuits (1 or 4).

The second application is currently being pursued by the Public Broadcasting Service in its program of networking 150 of its stations by satellite. Here, the ground station must be able to receive television signals of a quality sufficient to allow television rebroadcasting: 53-dB signal-to-noise ratio and a noise temperature of about 80 K is required. To work with the capability available from the domestic communications of today, such stations could be built for approximately $80 000 each. Neither this type of station nor the thin-route small terminal is economically attractive for the kinds of services demonstrated experimentally by ATS-6: in-service educational and instructional broadcasting; multipoint professional teleconferencing; and various health-care services. These new classes of services cannot be provided by present commercial satellite communication systems. They require capabilities that make possible the widespread use of ground terminals like those associated with the HET experiment on ATS-6.

The HET ground terminal of Fig. 3 has the capability of

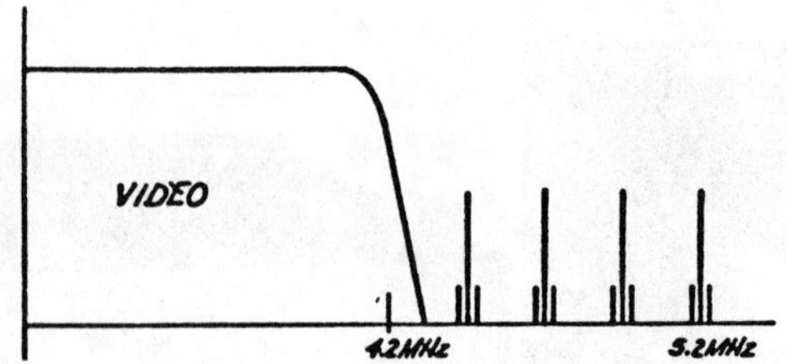

F ig. 5. HET broadcast signal format. Composite baseband: 1 color video plus 4 aural channels; aural subcarrier RF bandwidth: 130kHz; aural information bandwidth: 10 kHz.

Fig. 6. RF spectrum of composite signal (in megahertz). F_i: video carrier center frequency; master station uplink frequencies: one of three F_i in 6-GHz bank; remote video transmitter uplink frequency: 2247.5 MHz; downlink center frequencies: 2569.2 MHz, 2670 MHz and one of three F_i in 4-GHz band; remote audio uplink frequencies: 2.25-GHz band.

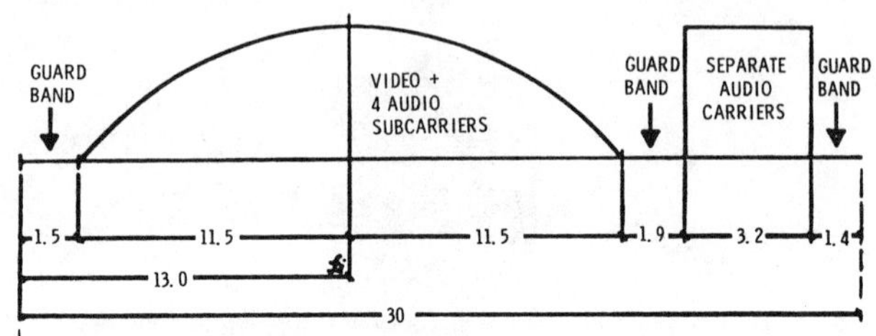

receiving a full-bandwidth color signal with 4 audio channels (Fig. 5). As stated earlier, the cost for this receive-only capability, in a single-quantity run of 130, was $3700 (1973 dollars) installed, checked out, and operational. If used in a direct-viewing mode at, say, a school or a hospital, the cost of a conventional television receiver must be added. Originally, ATS-6 had been planned to accommodate audio talkback, according to the spectrum plan of Fig. 6. Costs for audio transmission, added to the HET receive-only capability, would increase the total terminal cost by $8000. Although the audio interaction capability was provided through the VHF transponders of ATS-1 and ATS-3, the costs held valid. In Alaska, terminals having two-way video capability were provided; here, the cost over that for receive-only was under $30 000 each for a total "run" of only three. These figures compare very favorably with those for the 4-GHz stations described for thin-route and television networking services. They appear attractive and potentially even economical for some of the services mentioned in Section III.

These features of the HET ground stations were made possible by the basic capabilities of ATS-6: the large unfurlable antenna, with its gain of 37.6 dB over the field of view at 2.57 GHz, and the precision pointing accuracy of 0.04° or better obtained with inertial stabilization. High levels of effective radiated power, coupled with coverage only of the area in which service was desired and the concomitant ability to point the spacecraft beam precisely to cover that area, were recognized in 1968 as the major technical advances required beyond already projected improvements in the extant technology of that time. High ERP could be provided by a combination of higher antenna gain and higher transmitter power. The choice of a 9-m antenna with an 80-W transmitter was brought about by cooperative systems studies with the Department of Atomic Energy in India, leading to a choice of UHF at about 860 MHz for what is now SITE: the combination of power and antenna size provides an actual ERP of about 50 dBW (specification value 49) over a field of view of about 2.8°, covering essentially the country of India. The HET experiment, coming in mid-1971 well after the spacecraft design had been frozen and special accommodations would have to be made for the S-band experiment, resulted in fields of view of only 0.85° for each beam. For SITE, therefore, demonstration of the technology of instructional broadcasting includes effective, nationwide coverage as well as a satellite capability compatible with widely distributed, small, simple, inexpensive ground receiving terminals. For HET, the restricted experimental coverage of 0.85° per beam has been viewed by some as inadequate for the applications of educational broadcasting in the United States.

V. The ATS-6 Promise

To pass from the limitations of an experiment, especially one as successful as HET has been [1], [2], [3], to the design of an optimized spacecraft for an operational service is not difficult when the experimental spacecraft has demonstrated the technological advances required and has exhibited two aspects of the design tradeoffs needed for optimization. The promise of ATS-6 is that this passage is technically in our hands. The technology of large, unfurlable antennas has been developed and demonstrated: materials, mechanics, energy storage and release, holding of contour to tolerances as high as X band, and thermal stability. The technology of lightweight, fiber-reinforced, plastic structures having near-zero temperature coefficient and great strength has been demonstrated, in this case proving that antenna f/D ratios can be precisely maintained in space over a very wide range of temperatures and after severe vibration stresses. The technology of precise pointing, attitude measurement and control, and precise station keeping in an inertially stabilized spacecraft in the synchronous orbit has been demonstrated. They have been applied, in a spacecraft that has integrated these technologies successfully, in two different gain-power combinations in different bands, to demonstrate EIRP and coverage capabilities with small, simple, inexpensive ground terminals producing video signals of TASO-1 grade.

Additionally, the ability to track on low-orbiting spacecraft, and the ability to reprogram the attitude control system or to operate it in a ground-controlled, adaptive mode with selected attitude sensors have been demonstrated. Performance of operational tracking and data relay satellite systems may be predicated on these results, and several versatile, low-weight approaches to precise attitude control may be derived from them.

The promise of applying these major technological advances to demonstrating new applications of space for the public benefit has been met. The promise of the technologies themselves will depend on their effective transfer to operational use to bring about new services and to enable space-craft of greater capability to be produced: economically, one hopes. In that regard, it bears repeating that ATS-6, a multipurpose, multidisciplinary, experimental spacecraft, was expensive because of its multipurpose nature. Application of its technologies to a spacecraft optimized for one intended service would have a high probability of being economical.

VI. Some Outcomes of the ATS-6 Mission

With the promise come limitations, largely because of the multipurpose nature of ATS-6. These are primarily concerned with time. Because ATS-6 was available for only a year before being moved toward India, first-year experimenters were tightly scheduled and very restricted in their time on the spacecraft. Their experiments, accordingly limited in scope, could not completely satisfy the full range of objectives desired. This was particularly constraining to the HET, L-band, and communications experiments of Table III.

The L-band experiments operated up to termination of first-year operations and hope to resume at 35° E longitude. Resolution of position for ships was assessed at between 0.3 and 5.1 nautical miles in latitude, 0.7 to 5.2 in longitude. For aircraft, resolution was assessed at ±5 km. Prelaunch specification was ±1 km for both. Restricted time on the satellite did not permit full assessment of computer programs, which are currently being reviewed. And because of this fundamental problem, experiments in traffic-control and communication procedures directed toward determining satisfactory operational practices could not be run.

In somewhat comparable circumstances, limited and controlled time on the satellite prevented the communications experimenters from acquiring all the results they would like to have had in bad weather. These results are especially important to the establishment of sound technical criteria for the use of newly allocated bands at 12, 14, 18, 20 and 30 GHz, where data is very meager. Some work is anticipated with these experiments at 35° E, and it may well be that work in the United States will resume when ATS-6 returns home after SITE.

The HET experimenters had been planning a long-range series of activities in educational and health-care broadcasting aimed at defining characteristics and traffic for operational services. Without ATS-6, or the continuity of a direct-line follow-on, they have turned their attention to the 12-GHz broadcasting experiment being jointly executed by NASA and Canada's Department of Communications. This Communications Technology Satellite, planned for launch in December 1975, will provide an *ERP* of about 52 dBW over a 2.5° field of view, thus continuing the capability of working with small, relatively inexpensive ground terminals into the 12-GHz band allocated to satellite broadcasting. But the success of the HET experiments, and their promise, has already been great enough, in the users' views [1]-[4], for them to have gathered together a membership of current and potential users into a Public Service Satellite Consortium incorporated in Delaware in March 1975. The users—states, hospitals, universities, regional school systems, clinics, medical groups, and library systems—are engaged now in defining the uses they will make of satellite broadcasting, and in determining their financially supportable traffic load. Preliminary indications are that a satellite broadcasting service for educational and health-care applications may be economically viable, and that a market can be aggregated. The technology, thanks to ATS-6, is already here.

VII. Conclusions

ATS-6 is the first satellite whose capabilities were intentionally marshalled to apply space technology to the solution of social problems while the satellite was still in development. The success of its very advanced technological ventures may affect satellite communications for a long time to come. But its greatest significance is in its demonstrating, with the enthusiastic participation of large groups of users whose interest was in their own disciplines and not in technology, the practicality of service as well as technology and the potential of aggregating a market for satellite broadcasting to help solve problems in education and health care. High-*ERP* satellite services incorporating small, simple, inexpensive ground terminals may form the next major change in characterizing domestic satellite communications systems.

References

[1] C.W. Weinberger, Honors Banquet Address, AIAA, Washington, D.C., February 26, 1975

[2] ______, Remarks presented at the NASA Earth Resources Symp., Houston, Tex., June 12, 1975.

[3] H.C. Cannon, Statement presented to the Committee on Science and Technology, U.S. House of Representatives, Washington, D.C.; February 5, 1975.

[4] AIAA Conf. on Communication Satellites for Health/Education Applications, Denver, Colo., July 21–23, 1975. Nine papers on HET (Sessions II and III: The Year on ATS-6):
a) M.R. Schwarz and M. Johnson, (School of Medicine, Univ. of Washington), "The role of satellite broadcasting in regionalized medical education."
b) G. Odland and E.A. Einser, (School of Medicine, Univ. of Washington), "The feasibility of dermatological consultations to remote areas via two-way, color satellite transmissions."
c) C. Dohner, E. Zinzer, and T. Collins (School of Medicine, Univ. of Washington), "Evaluating satellite broadcasting in basic science and clinical teaching in medicine."
d) M.R. Wilson and C. Brady (Indian Health Service), "Health care in Alaska via satellite."
e) W.M. Domm (Dep. of Medicine and Surgery, Veterans Administration), "Veterans Administration satellite-transmitted experiments in biomedical communications."
f) A.A. Whalen (NASA Goddard Space Flight Center) and W.A. Johnston (Fairchild Industries), "ATS-6: A satellite for human needs."
g) C.M. Northrip (Office of the Governor of Alaska), "Building history: Communications technology as a cultural goal."
h) L.A. Bransford (Federation of Rocky Mountain States), "Humanizing satellite service."
i) H.E. Morse (Satellite Project, Appalachian Regional Commission), "Institutional change through the use of satellites."

[5] S. Metzger, "Communications satellite systems in the 70's," presented at the Symp. on Long Term Prospects for Satellite Communications, Genoa, Italy, June 1971.

[6] W.N. Redisch, "ATS-6: Description and performance," IEEE Trans. Aerospace and Electronic Systems, 994-1003

Part III
The Communications System

A general model of a satellite communications system is shown in Fig. 1. The signal is generated by a user and enters the terrestrial system. In some systems, this terrestrial system is simply a dedicated link to the earth station, while in other cases it is a switched telephone network. At the earth station, the baseband signal is processed and then transmitted at an RF frequency to the satellite where it is processed and retransmitted to the receiving earth station. The receiving earth station processes the signal down to a baseband signal which is sent through the terrestrial network to the user. A simplified functional diagram of the earth station is shown in Fig. 2. The various functional elements will be described later in this chapter.

The first communication system model of interest is the IF-to-IF model shown in Fig. 3. This model assumes that there is a single beam from the satellite antenna which illuminates all of the earth stations in the system. The features of interest in the model which will be discussed are

  (1) the multiple access technique,
  (2) the earth station power and antenna gain,
  (3) the noise and interference,
  (4) the uplink channel,
  (5) the satellite receiving antenna gain,
  (6) the satellite receiver,
  (7) the satellite transponder,
  (8) the satellite power and antenna gain,
  (9) the downlink channel,
 (10) the earth station antenna and receiver.

First, consider the transmission path from a single earth station through the satellite to another earth station. The initial step is to derive the link equations. The model of the uplink is shown in Fig. 4. The flux density at the satellite is given by

$$\Omega_u = \frac{P_T G_T}{4\pi R_u^2} L_u \quad \text{W/m}^2, \tag{1}$$

where

P_T transmitted power (watts),
G_T transmit antenna gain,
R_u uplink range (meters),
L_u additional uplink loss factor.

The signal power received at the satellite is

$$P_u - \Omega_u A_{su} - \Omega_u \frac{G_{su} \lambda_u^2}{4\pi} \quad \text{W,} \tag{2}$$

where

A_{su} effective area of the satellite antenna,
G_{su} satellite antenna gain,
λ_u wavelength on uplink.

The noise consists of thermal noise in the receiver, rain-induced noise, and the earth background noise. The noise is assumed to have a flat spectral density over the receiver bandwidth of N_{0u} W/Hz. Thus, the signal power to noise density ratio is

$$\frac{C_u}{N_{0u}} = \frac{P_u}{N_{0u}} \tag{3}$$

Frequently N_{0u} is expressed in terms of an effective noise temperature,

$$N_{0u} = kT_s \tag{4}$$

where

k Boltzman's constant, 1.38×10^{-23} J/K,
T_s effective receiver input noise temperature (K).

Substituting (1), (2), and (4) into (3) gives

$$\frac{C_u}{N_{0u}} = (P_T G_T)\left(\frac{\lambda_u}{4\pi R_u}\right)^2 \left(\frac{G_{su}}{T_s}\right)(L_u)\left(\frac{1}{k}\right). \tag{5}$$

Expressing in decibels gives

$$\left.\frac{C_u}{N_{0u}}\right)_{dB} = \underbrace{10 \log P_T G_T}_{\substack{\text{earth station} \\ \text{EIRP}}} - \underbrace{20 \log\left(\frac{4\pi R_u}{\lambda_u}\right)}_{\substack{\text{free space} \\ \text{loss}}} + \underbrace{10 \log\left(\frac{G_{su}}{T_s}\right)}_{\substack{\text{satellite} \\ \text{G/T}}}$$

$$\underbrace{+ \; 10 \log L_u}_{\substack{\text{additional} \\ \text{uplink} \\ \text{losses}}} - \; 10 \log k.$$

This is the basic uplink equation with its various components identified. Several subsidiary relations are also useful. The antenna gain for a parabolic antenna is

$$G = \eta \left(\frac{\pi D}{\lambda}\right)^2, \tag{7}$$

where

η antenna efficiency (typically 0.55),
D antenna diameter,
λ wavelength,

or in decibels,

$$G_{dB} = 20 \log f + 20 \log D - 52.6 \tag{8}$$

(with D in ft and f in MHz). The antenna beamwidth (between 3 dB points) is

$$\theta = 70 \frac{\lambda}{D} \simeq \frac{70\,000}{fD} \quad \text{deg.} \tag{9}$$

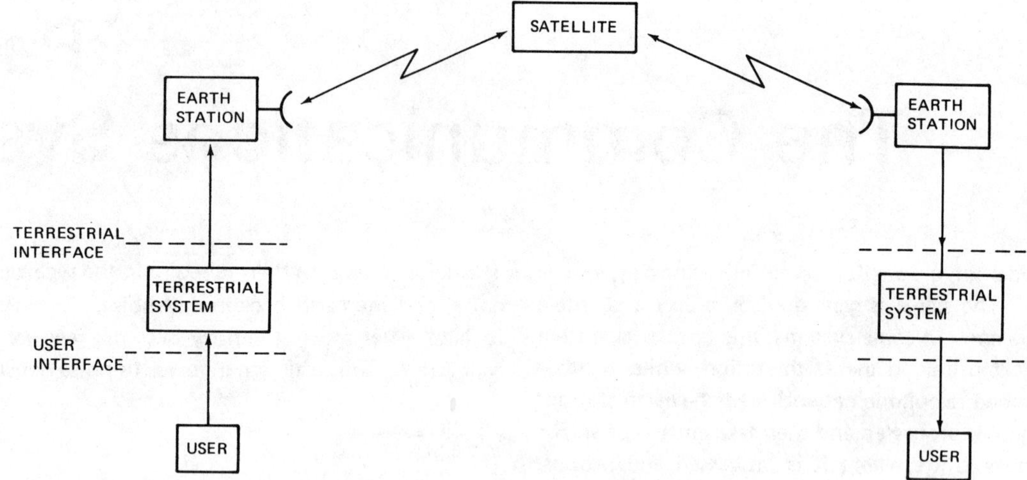

Fig. 1. Satellite communication systems.

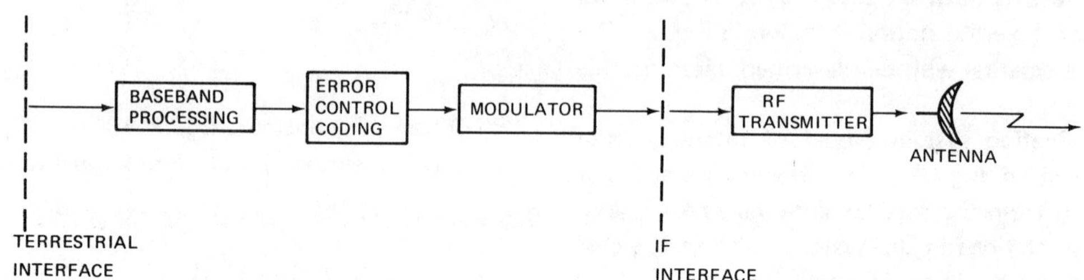

Fig. 2. Earth station.

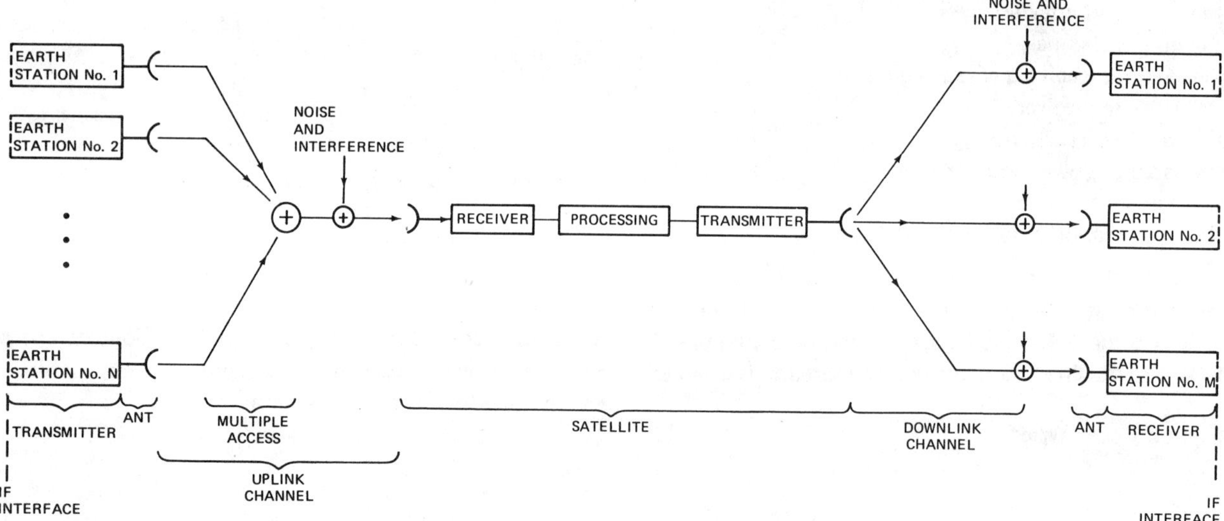

Fig. 3. Communication system model (IF-to-IF).

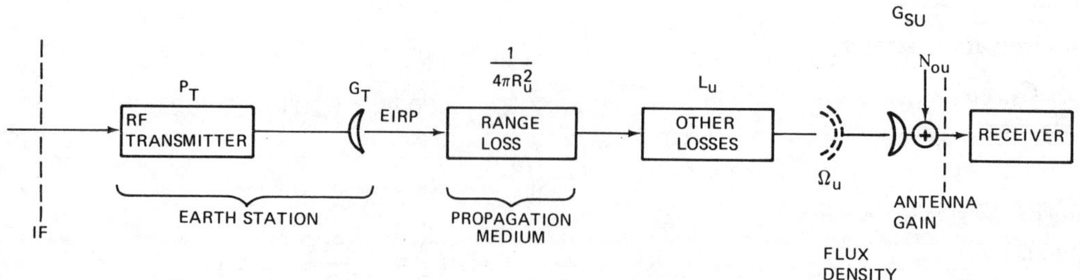

Fig. 4. Uplink model.

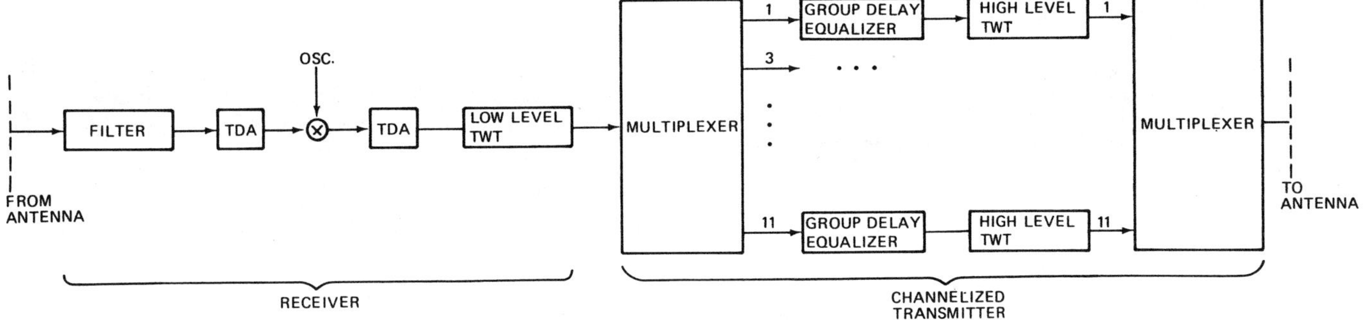

Fig. 5. Satellite model.

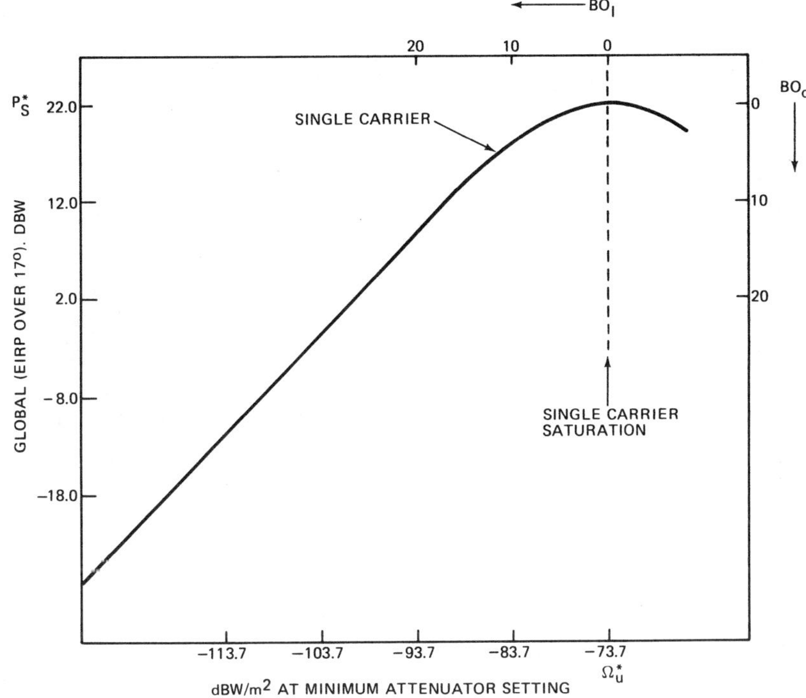

Fig. 6. TWTA characteristic.

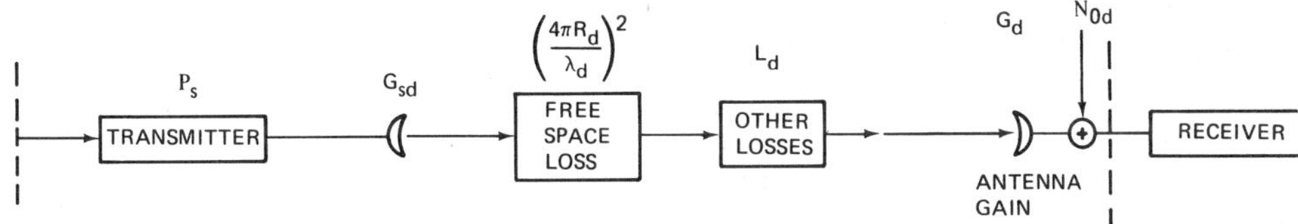

Fig. 7. Downlink model.

The free space (FS) loss can be written in decibels as

$$L_{FS} = 36.6 + 20 \log f + 20 \log R \tag{10}$$

(with R in statute miles and f in MHz).

The satellite transponder is shown in Fig. 5. The model will be developed in more detail in Section 3.2 and 3.3. For the present discussion, only the nonlinearity of TWTA is of interest. A typical characteristic is shown in Fig. 6. The flux density necessary to saturate the TWTA is denoted by Ω_u^* and the corresponding output power is P_s^*. Frequently, the satellite is operated in a backed-off from saturation mode to avoid nonlinear distortions. The input and output back-offs are denoted by BO$_I$ and BO$_O$, respectively.

The downlink channel is shown in Fig. 7. Using the same procedure as in the uplink channel gives,

$$\left.\frac{C_d}{N_{0d}}\right)_{dB} = \underbrace{10 \log (P_s G_{sd})}_{\substack{\text{satellite} \\ \text{EIRP}}} - \underbrace{20 \log \left(\frac{4\pi R_d}{\lambda_d}\right)}_{\substack{\text{free space} \\ \text{loss}}} + \underbrace{10 \log \left(\frac{G_d}{T_d}\right)}_{\substack{\text{earth station} \\ G/T}}$$

$$+ \underbrace{10 \log L_D}_{\substack{\text{additional} \\ \text{downlink} \\ \text{losses}}} - 10 \log k. \tag{11}$$

The satellite EIRP can also be expressed as

$$P_s G_{sd} = P_s^* \, BO_O \, G_{sd}. \tag{12}$$

The model in Fig. 5 is a classical frequency-translating receiver. In this case, the signal power-to-noise density ratio for the overall link is given by the relation,

$$\frac{C}{N_0}\bigg)_T = \left(\left(\frac{C_u}{N_{0u}}\right)^{-1} + \left(\frac{C_d}{N_{0d}}\right)^{-1}\right)^{-1}. \tag{13}$$

This relationship will be modified to include other factors as a more detailed model is developed. In Section 3.6.6, satellites with on-board signal processing are discussed. In this case, the uplink and downlink are uncoupled and (13) is no longer valid.

A simple example will illustrate numerical values for a typical system.

Example: Typical values for the INTELSAT IV satellite global beam and a large earth station are

$$P_T G_T = \quad 90.1 \text{ dBW} \tag{14a}$$

$$G_s/T_s = -18.6 \text{ dB/K} \tag{14b}$$

$$P_s G_{sd} = \quad 22.5 \text{ dBW} \tag{15a}$$

$$G_d/T_d = \quad 40.7 \text{ dB/K} \tag{15b}$$

which results in a total C/N_0 of

$$(C/N_0)_T = 94.6 \quad \text{dB-Hz}. \tag{16}$$

The $(C/N_0)_T$ relationship in (13) is the basis for the analysis of modulation schemes for satellite communications.

The next step in the development of the communication model is to investigate suitable modulation and coding schemes for transmitting information over the satellite channel. A prerequisite to the modulation discussion is a brief discussion of the user input signals.

Section 3.1 discusses the characteristics of the various signals that appear at the user interface to the satellite system interface. These include single-voice channels, groups (12 channels) and supergroups (60 channels) of voice channels that have been frequency division multiplexed, analog television signals, and digital bit streams. Effective methods of encoding these sources into digital bit streams are also discussed in that section.

Section 3.2 discusses analog modulation. For the purpose of illustrating the key features of analog modulation, an FDM signal containing n_v voice channels is an adequate example. The most common modulation scheme is frequency modulation (FM).

The performance of an FM system is given by the classical FM equation and Carson's bandwidth expression:

$$\left(\frac{S}{N}\right) = \left(\frac{C}{N_0}\right)\frac{1}{b}\left(\frac{f_r}{f_m}\right)^2 p \cdot w, \tag{17}$$

or alternatively,

$$\left(\frac{S}{N}\right) = \left(\frac{C}{N}\right)\frac{B}{b}\left(\frac{f_r}{f_m}\right)^2 p \cdot w \tag{18}$$

where

(S/N) weighted signal-to-noise ratio at 1-mW test-tone level = 51.2 dB for 7500-pWp channel noise,

(C/N) carrier-to-noise ratio over the Carson's rule bandwidth B,

b channel bandwidth = 3.1 kHz,

f_r rms test-tone deviation,

f_m maximum baseband frequency $\simeq 4.2 \times n_v$, in kHz,

n_v number of telephone channels,

p psophometric weighting factor = 1.78 ($P \triangleq 10 \log p = 2.5$ dB),

w preemphasis weighting factor = 2.5 ($W \triangleq 10 \log w = 4.0$ dB),

The bandwidth B is given by Carson's rule:

$$B = 2(f_p + f_m) \tag{19}$$

where

B the required predetection bandwidth,

f_p peak deviation.

Since the signal is Gaussian, the peak deviation is not defined. Generally, a peak factor, the ratio of the peak to rms deviation of $g = 3.16$ (10 dB) to 8.5 (18.6 dB) is assumed. The lower value is applicable for large n_v.

In addition, rms multichannel deviation is related to rms test-tone deviation by load factor g:

$$20 \log l = L = \begin{cases} -15 + 10 \log n_v & n \leqslant 240 \\ -1 + 4 \log n_v & 12 \leqslant n \leqslant 240. \end{cases} \tag{20}$$

Now, (19) becomes

$$B = 2(g \cdot l \cdot f_r + f_m) \tag{21a}$$

$$= 2(3.16l \cdot f_r + f_m). \tag{21b}$$

Equation (17) assumes the demodulator is above threshold. The threshold level is

$$\frac{C}{N} = \frac{C}{N_0 B} \geqslant a_0. \tag{22}$$

The value of a_0 depends on the demodulator used in the system. For an FM limiter-discriminator,

$$a_0 = 13 \text{ dB} \tag{23a}$$

while with a phase-lock loop demodulator,

$$a_0 = 10 \text{ dB}. \tag{23b}$$

In this introduction, only the simple case of a single carrier per transponder is considered. In this case, the transponder bandwidth B_T and the available C/N_0 are given. To maximize the number of channels, f_r is increased until the signal occupies the entire bandwidth, subject to the threshold constraint in (23). The result for the INTELSAT IV transponder is shown in Fig. 8.

In many applications, there are multiple FM carriers in a transponder. This problem differs from the previous single-carrier case in two ways. The first is that the output power

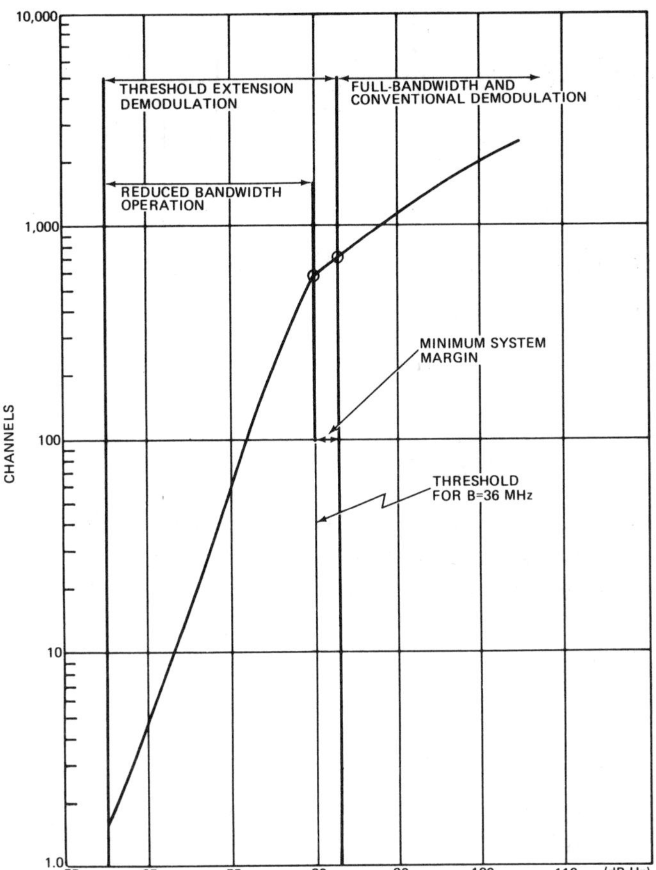

Fig. 8. Channel capacity as a function of $(C/N_0)T$ in dB-Hz. Single carrier per transponder using FDM/FM.

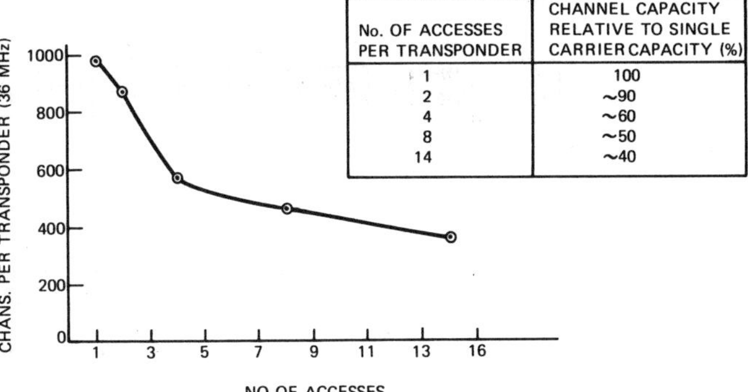

No. OF ACCESSES PER TRANSPONDER	CHANNEL CAPACITY RELATIVE TO SINGLE CARRIER CAPACITY (%)
1	100
2	~90
4	~60
8	~50
14	~40

Fig. 9. Number of telephone channels as a function of the number of accesses.

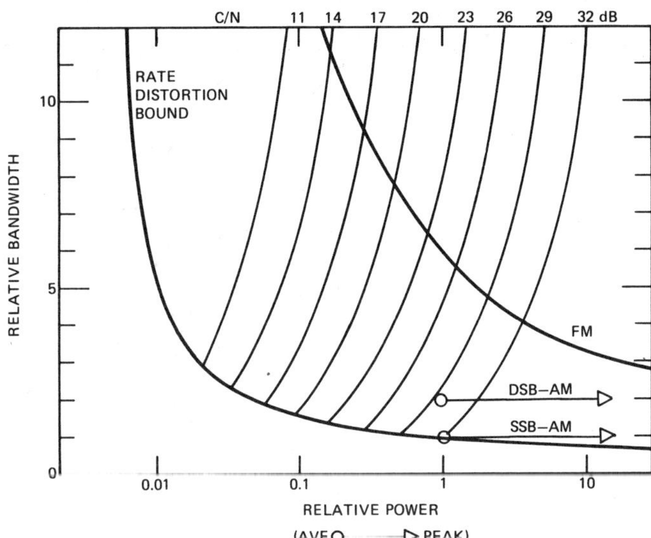

Fig. 10. Power-bandwidth relationship for analog modulation schemes.

is divided among the various input signals in a ratio proportional to their input powers. The second is that the nonlinear characteristic of the traveling-wave tube amplifier causes intermodulation products to appear at the output. These intermodulation products degrade the system in two ways: by using part of the output power and by acting as interference on the downlink. To reduce these effects, the traveling-wave tube is normally operated in a backed-off mode. This problem is discussed in detail in Section 3.2. The key result is shown by the typical curve shown in Fig. 9. As the number of carriers increases, there is a dramatic reduction in the capacity of a transponder. This capacity reduction is due to two factors: the reduction in $(C/N_0)_T$ and the change in the load factor due to a smaller number of channels per carrier.

The generic relation for modulation schemes to transmit analog signals is of the form

$$\left(\frac{S}{N}\right) = f_1 \left(M_0, \left(\frac{C}{N_0}\right)_T, B\right). \tag{24}$$

In other words, the *fidelity* of the signal supplied to the user, as measured by (S/N) is a function of the modulation and multiplex scheme (denoted by M_0), the (C/N_0) available in the channel, and the bandwidth available. Using Shannon's rate distortion theory, a bound on the performance of any modulation scheme can be obtained. A typical result is shown in Fig. 10.

For a given fidelity, the number of channels that can be transmitted is

$$n_v = f_2 \left(M_0, \left(\frac{C}{N_0}\right)_T, B, \frac{S}{N}\right). \tag{25}$$

Section 3.3 develops digital modulation in detail. Only a few key ideas will be summarized here. A simplified model for a digital modulation system is shown in Fig. 11. Typical signaling schemes include quadrature phase shift keying (QPSK), frequency shift keying (FSK), and differential phase shift keying (DPSK). In any digital communication system, the modulation scheme is chosen so as to effectively use the available power and bandwidth. In addition, in a digital satellite system the modulation scheme must be robust in the presence of channel nonlinearities, channel filters, and interference. The probability of error for this simple model is a function of the modulation scheme and of E_b/N_0, where E_b is the energy per bit.

For example, with QPSK,

$$P_e = \text{erfc}_* \sqrt{\frac{2E_b}{N_0}} \tag{26}$$

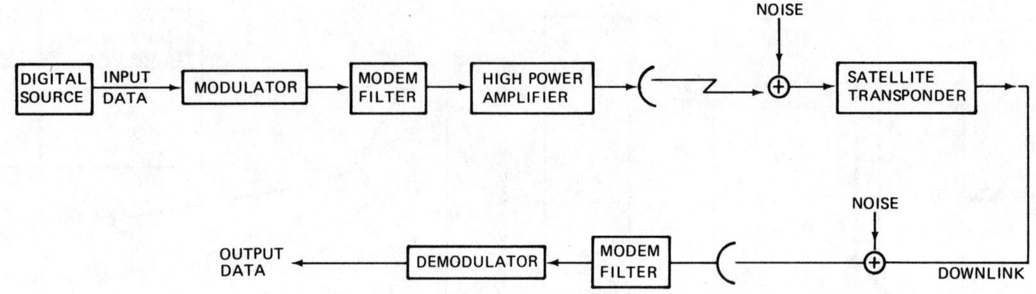

Fig. 11. Digital modulation system.

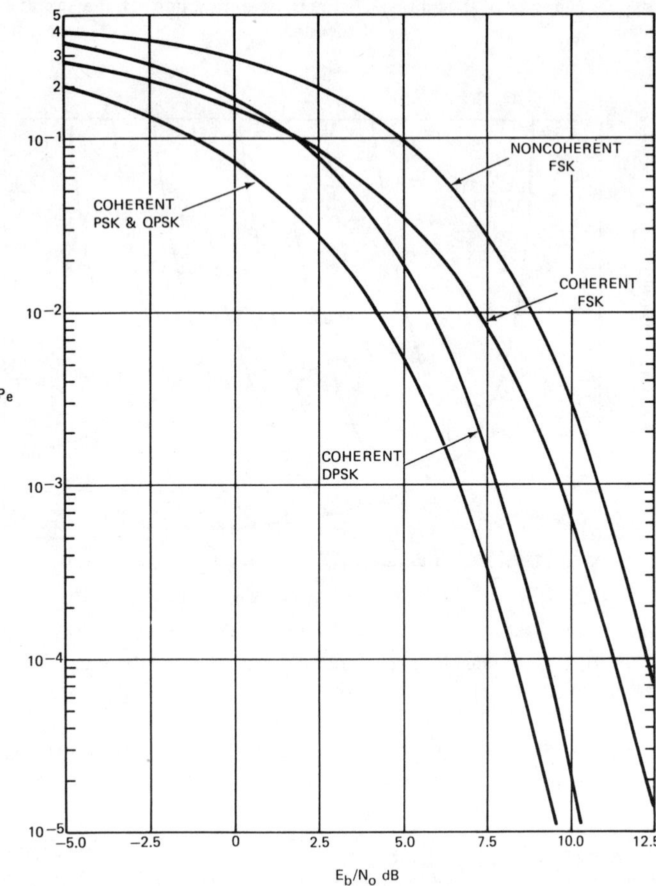

Fig. 12. P_e versus E_b/N_0.

where erfc$_*$ is the complementary error function. The generic form is

$$P_e = g_1 (M_0, E_b/N_0). \qquad (27)$$

P_e curves for typical signaling schemes are shown in Fig. 12. This result can be written in terms of the channel C/N_0 by using the relation

$$\frac{E_b}{N_0} = \frac{C}{N_0 R} \qquad (28)$$

where R is the rate in bits/s.

Thus,

$$P_e = g_1 \left(M_0, \frac{C}{N_0 R} \right) \qquad (29)$$

or the allowable rate is

$$R_P = g_2 \left(M_0, \frac{C}{N_0}, P_e \right) \qquad (30)$$

where the subscript P denotes a power-limited rate. The actual form of the $g_i(\cdot)$ functions depends on M_0.

A second constraint that must be imposed is a bandwidth constraint:

$$B = g_3 (M_0, R), \qquad (31)$$

or

$$R_B = g_4 (M_0, B) \qquad (32)$$

where the subscript B denotes a band-limited rate. Combining (30) and (32),

$$R = \min [R_P, R_B] = g_5 \left(M_0, \frac{C}{N_0}, B, P_e \right). \qquad (33)$$

Thus, the rate is a function of the available C/N_0, the bandwidth, and the desired error rate. Fig. 13 shows the ratio of the information rate to the channel bandwidth as a function of E_b/N_0 for various modulation schemes.

A bound on the transmission rate for error free communication is given by Shannon's capacity formula

$$R \leqslant C_B = B \log_2 (1 + C/N_0 B). \qquad (34)$$

One measure of bandwidth efficiency is R/B. Using (34) gives a lower bound on E_b/N_0 as a function of the bandwidth efficiency,

$$E_b/N_0 \geqslant \frac{\exp \left(0.69 \frac{R}{B} \right) - 1}{(R/B)}. \qquad (35)$$

Figure 14 shows this bound and the performance of several modulation systems. (Note that this is the same type of plot as Fig. 13, but it emphasizes the lower E_b/N_0 range. The coding curve will be discussed below.

The discussion up to this point has provided a framework for the digital modulation problem and has introduced the ideas of power and bandwidth efficiency, and the concept of a bound on the performance of any modulation technique. It has not, however, focused on one of the central problems in digital satellite communications. This problem is the effect of the nonlinearities in the satellite and earth station, the effect of the filters in the transmission path, and the effect of interferences. Section 3.3 concentrates on this aspect of the problem:

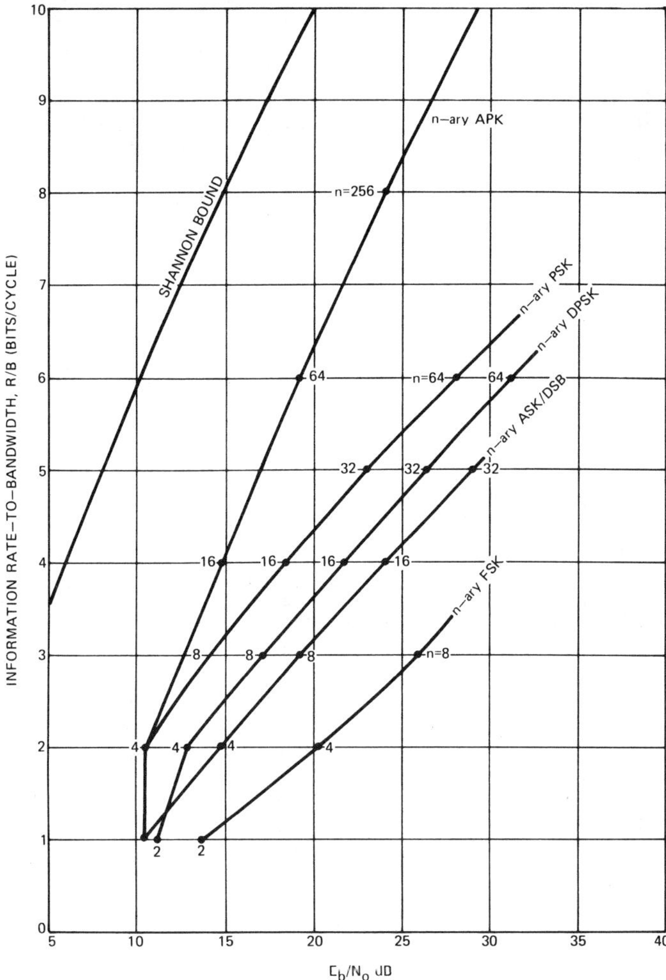

Fig. 13. Comparison of digital modulation schemes.

analyzing the effects and then discussing schemes to improve the performance.

In many satellite systems, the communication system is power limited so that it is impossible to achieve the desired error probability P_e and transmission rate R using one of the conventional modulation schemes such as QPSK. In these cases, error control coding may be able to achieve the desired results. Section 3.4 discusses error control techniques with emphasis on convolutional codes and Viterbi decoding, techniques that are particularly useful in the satellite environment. The basic idea in an error correcting code is to introduce redundancy into the bit stream (which increases the rate and required bandwidth) and then exploit this redundancy at the receiver to reduce the probability of error. Fig. 15 shows the probability of error for several coding schemes (a description of these techniques is given in Section 3.4). Significant reductions in the required E_b/N_0 can be achieved by the use of coding (e.g., 4.5 dB at $P_e = 10^{-5}$ with Viterbi decoding). Fig. 14 shows the bandwidth-power performance. Note that coding gives the system designer an ability to choose an operating point near the bound.

The last element in the model in Fig. 1 is the terrestrial system and, in particular, the terrestrial interface which is discussed in Section 3.5. The characteristics of this interface will vary widely in different systems. Some of the issues discussed in Section 3.5 are echo control, FDM/TDM transmultiplexers, multiplexing, and synchronization.

Up to this point, the communication system model assumed that a single beam illuminated all of the earth stations in the system. There are two disadvantages to this type of system. The first is that the capacity is limited by the available bandwidth, and the second is that the antenna gain is inversely proportional to the coverage area. To illustrate the first limitation, consider a satellite operating at 6/4 GHz with 12 transponders of 40-MHz bandwidth. With guard bands, this completely utilizes the allocated 500-MHz bandwidth. If all of the transponders were used for voice traffic in an FDM/FM with multiple accesses, then, from Fig. 9, the total capacity of the satellite would be approximately

$$n_T = 500 \times 12 = 6000 \text{ channels} \qquad (36)$$

In general, the satellite capacity is the product of the bandwidth efficiency of the modulation scheme in channels/MHz times total satellite bandwidth:

$$n_T = \gamma_v B_{sat} \qquad (37)$$

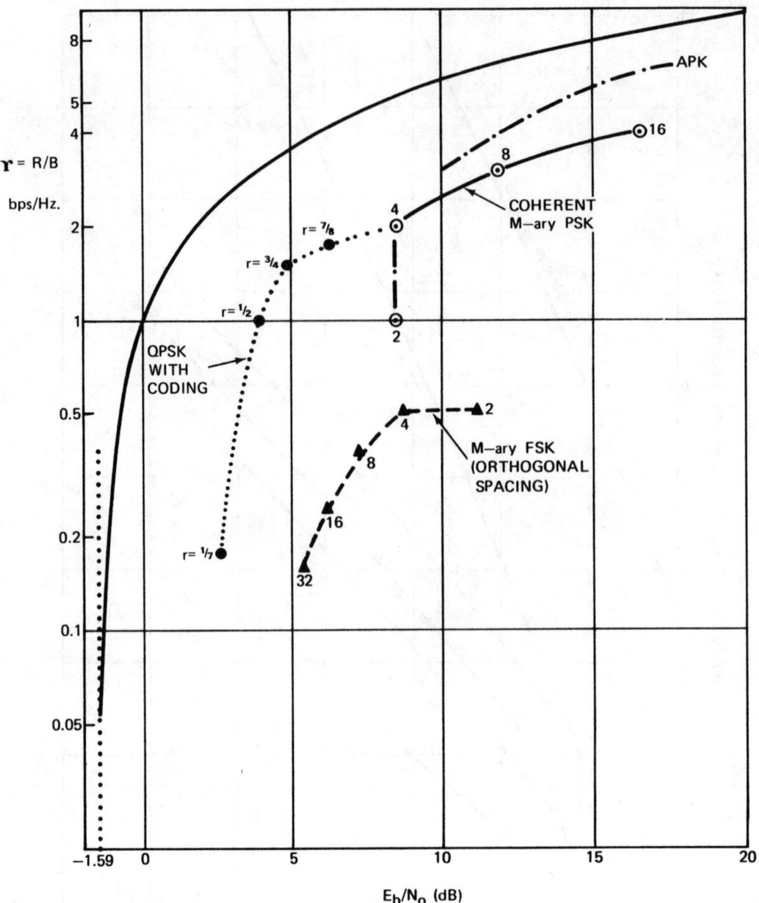

Fig. 14. Bandwidth efficiency of various modulation schemes.

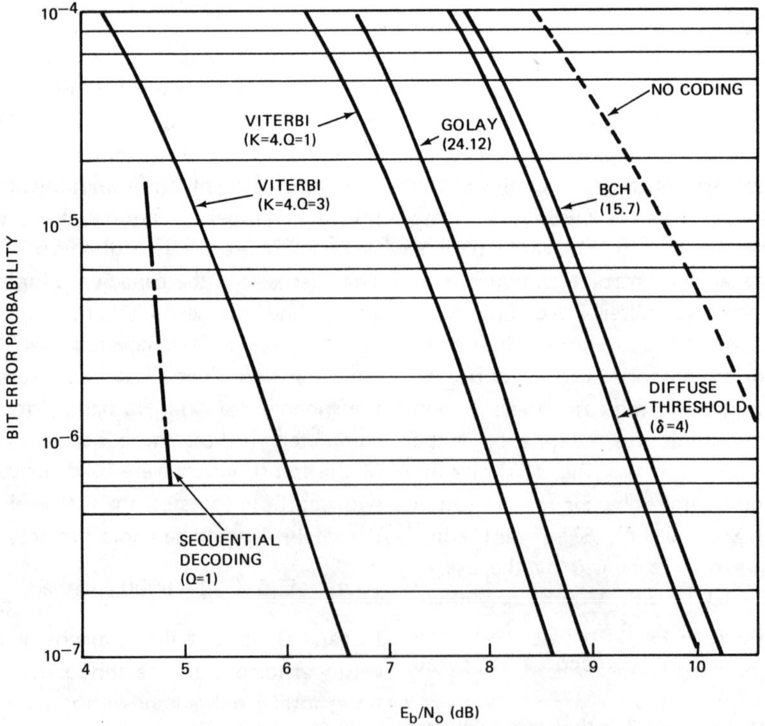

Fig. 15. P_e versus E_b/N_0 for various decoders.

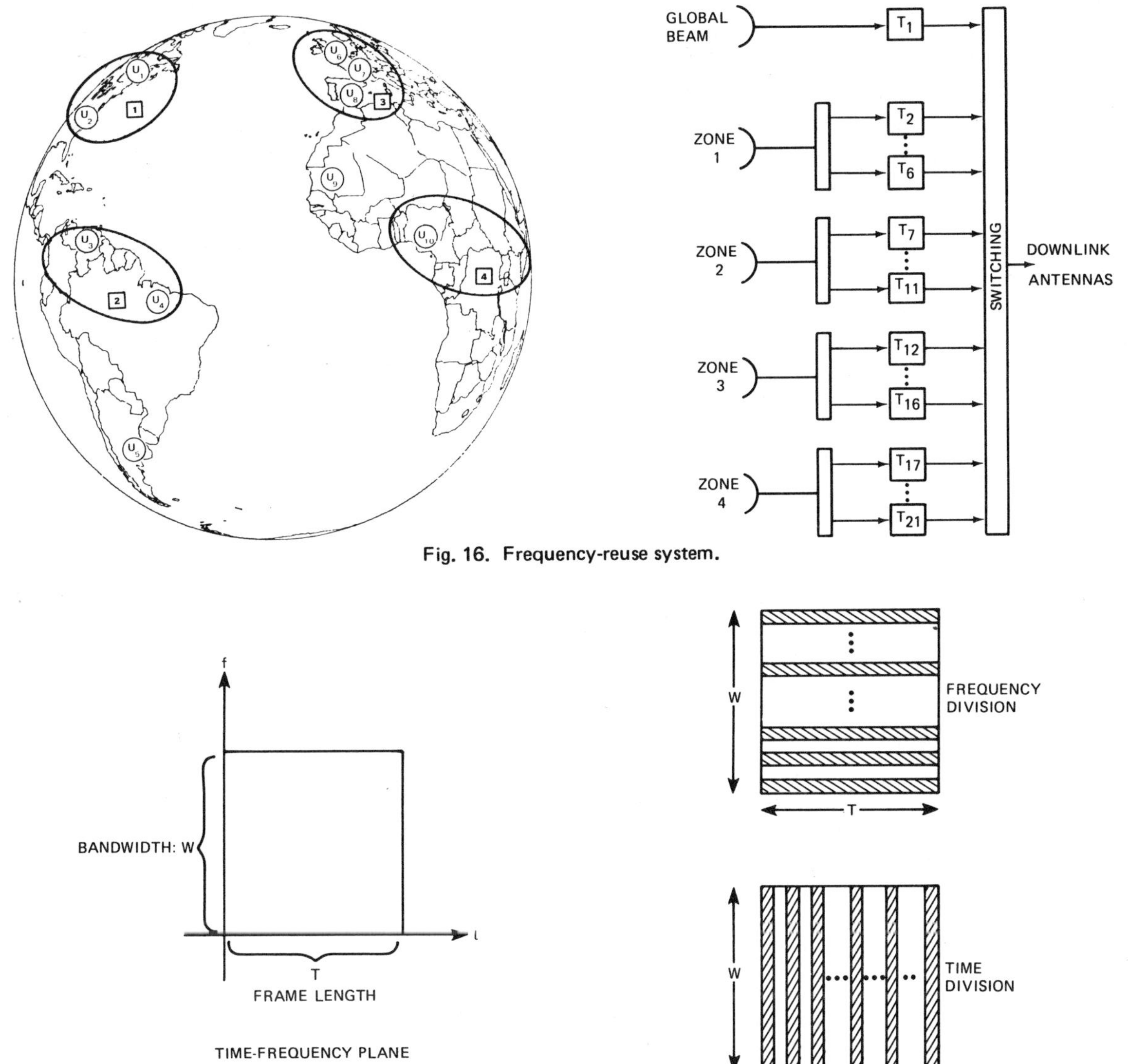

Fig. 16. Frequency-reuse system.

Fig. 17. Division of time-frequency space.

Sections 3.1 through 3.4 discuss techniques for increasing γ_v. There are two ways to increase B_{sat}:

1) use higher frequency bands,
2) reuse the 500-MHz available at 6/4 GHz.

As discussed in Section 2, civilian experimental satellite systems have operated at 14/11 GHz and 30/20 GHz. From the standpoint of the communications model, the basic difference is the propagation characteristics of the channel. These characteristics are discussed in Section 5. From the standpoint of practical implementation, the problem is the availability and reliability of devices. Device technology is discussed in Section 4.3.

Frequency reuse can be accomplished through dual polarization or multiple antenna beams. In the dual polarization case, there is one complete transmission channel operating at one polarization (e.g. vertical polarization or right-hand circular polarization) and a second channel operating at the opposite polarization (e.g. horizontal polarization or left-hand circular polarization). This technique doubles the bandwidth available and has been used in several operational satellite systems. The key factor in the system is maintaining the orthogonality of the polarizations. In practice, isolations of 30 dB or greater are achievable. Papers in Sections 3.4, 3.5, and 3.6 discuss various aspects of the problem.

A simple example of multiple beam system providing increased bandwidth is shown in Fig. 16. There is one global beam and four zone beams. The global beam utilizes one 80-MHz transponder, and *each* zone beam has five 80-MHz transponders. Thus, the total bandwidth available is

$$B_{sat} = 1 \times 80 + 4 \times 5 \times 80 = 1680 \quad \text{MHz} \tag{38}$$

In addition, the gain of the zone beam antennas will be higher so that the capacity per transponder will be higher. Operational systems have employed moderate frequency reuse, and systems with extensive reuse have been proposed. The two central problems in frequency reuse systems are

1) antenna isolation in order to achieve satisfactory cochannel interference levels, and

2) connectivity between the different beams.
The multiple-beam antenna design problem is discussed in Section 4.2. The effect of the cochannel interference on the communications system performance is covered in Section 3.3. The connectivity problem is discussed in Section 3.6.

The final area of interest in the communication system is that of multiple access. It is the multiple access and broadcast characteristics of satellite systems that give them unique capabilities. The frequency reuse systems just discussed can be thought of as space-division multiple-access systems (SDMA). The other two dimensions available are time and frequency. The two basic methods of dividing the time-frequency plane are frequency division multiple access (FDMA) and time division multiple access (TDMA) as shown in Fig. 17. These two methods are developed in Section 3.6. A third method, code-division multiple access (CDMA), which is less obvious, as well as various hybrid schemes are also developed. A major, but not the only element in the choice of the multiple-access scheme is the trade-off between the system capacity and the cost complexity, and reliability of the system. These issues are discussed in detail in Section 3.6.

This introduction has developed the framework for the overall satellite communications problem. The introduction to each section will further develop this framework in order to place the papers reprinted and those listed in the bibliography in context.

Section 3.1
Source Coding

Typical sources of signals in a satellite communications systems include voice, video, images, facsimile, and digital data. In most current operational systems, the traffic is predominantly voice, and it is transmitted in analog form. For satellite systems using digital transmission, it is necessary to transform the analog signal into a digital bit stream. At the receiving end, the digital signal is converted back to an analog signal. The quality of the received signal will be a function of the A/D conversion algorithm and the quality of the overall communications link between the users. Source encoding exploits the characteristics of the source in order to achieve acceptable quality while reducing the bit rate. The acceptability of a particular quality requires a subjective evaluation and depends on the system application.

In the first portion of this section, source encoding for single channels (i.e., a single voice or single TV signal) is discussed. This discussion is divided into three parts: voice, image compression, and facsimile. It is important to note that in some cases, the source encoding will be accomplished at the user interface with the terrestrial communication system. In these cases, it affects the satellite system by specifying the required bit rate and probability of error. In other cases, the encoding will occur in the earth terminal and will be part of the overall communications satellite system design.

In the second portion of the section, multiple channel techniques are discussed. These techniques are applicable to sets of channels, e.g., a group or supergroup of analog voice channels. These techniques exploit the voice activity statistics in the voice channel group by detecting voice activity and only transmitting the active channels. The earliest of these systems was time-assigned speech interpolation (TASI) for use on cable circuits. This class of technique is generally referred to as channel multiplication or baseband demand assignment. Analogous schemes for digital signals are usually referred to as statistical multiplexing and is discussed in Section 3.5.

Source encoding for voice has received a significant amount of attention because of the volume of voice traffic in communications sytems. Reprint Paper 3.1.1 provides a good summary of the current state of the art. A number of digital voice processing techniques are either developed or under development. The voice source coding techniques fall into two broad categories, waveform coding and analysis-synthesis techniques. The first category includes linear and companded PCM, differential PCM (DPCM), adaptive DPCM using either adaptive quantization, adaptive prediction, or both, and linear and adaptive delta modulation (ADM). Companded 64 Kb/s pulse code modulation (PCM) establishes the toll quality standard for commercial communications systems. The other techniques can be used to reduce the required bit rate per voice channel to the 10–40 Kb/s range with varying amounts of degradation (e.g. Reprint Paper 3.1.1).

Two of the most promising techniques are ADPCM in the 24–32 Kb/s range and continuously variable slope delta (CVSD) modulation, in the 16–32 Kb/s region. ADPCM provides the better quality, but CVSD is very rugged in terms of error performance and inexpensive to implement. ADPCM is expected to find utility in some mobile radio applications, and CVSD will be extensively used by the U.S. military. Reprint Paper 3.1.2 discusses PCM, DPCM, and CM encoding techniques for speech and video signals.

Analysis-synthesis techniques, which include the various well-known vocoder techniques, have recently been an area of intense research, spurred on in part by military requirements. The result has been a set of modern vocoder-like strategies which rely on linear and adaptive prediction techniques. (See [4] and [5].) With these techniques, the bit rates can be reduced to the 2–8 Kb/s range with reasonable quality.

A summary of bit rates achievable with various techniques and their relative complexity is given in Reprint Paper 3.1.1. The potential gains achievable with voice source coding are quite large. A number of factors such as their sensitivity to errors, their ability to tandem (i.e., undergo several D/A and A/D conversions), their ability to accommodate nonvoice analog signals (e.g., modem) must be considered before selecting a technique for a particular system. The books by Flanagan [1] and Jayant [2] provide thorough discussions of voice and video encoding areas. Supplementary references are included in the Bibliography at the end of this volume.

Digital television compression techniques have received significant theoretical and experimental attention.

Systems which use intraframe techniques (e.g., COMSAT's DITEC system and Hadamard transform systems) have achieved network quality with transmission rates in the 22 to 33.6 Mb/s range [6], [7]. Interframe techniques have been used by Nippon Electric Corporation in its NETEC system to achieve rates of 6.3, 16, and 22 MB/s [8]. For teleconferencing and similar applications, other techniques are useful. The AT&T Picturephone® transmission rate is 6.3 Mb/s [9]. Current research efforts are aimed at reducing the videophone rate to 1.544 Mb/s.

Facsimile compression techniques have advanced significantly in recent years. Compression ratios of 10:1 or more black and while documents are within the present state of the art. Typical techniques employ forms of run length coding, with the more sophisticated systems processing several scan lines at a time, so as to achieve data reduction by using the redundancy in both the vertical and horizontal directions [10], [11]. More advanced systems use such techniques as area-based or contour coding, which is based on sending only boundary information. Compression techniques can also be applied to gray-tone documents and colored documents.

The image compression area has also seen significant progress in recent years. DPCM and transform coding are the most advanced in development. The recent Special Issue on Image Compression of the IEEE TRANSACTIONS ON COMMUNICATIONS (November 1977, [12]) and [13] provide good overviews of the image compression and video encoding area.

The second major area of interest is the multiple channel coding area. There are two basic digital channel multiplication schemes which go under the generic name of digital speech interpolation (DSI).

One class of schemes includes various digital realizations of time-assigned speech interpolation (TASI). Systems of this type, which are under development or field test, include Japan's "ITT," France's "CELTIC," and Italy's "ATIC." INTELSAT is conducting field trials with this type of system in conjunction with its TDMA system. These systems exploit the fact that, on the average, a voice channel is being used only 40 percent of the time; and, if the number of channels is large enough, this implies that, with high probability, only 40 percent of the channels will be active. Thus, by detecting and only sending the channels which have voice activity, DSI gains—defined as the ratio of the number of channels provided service to the required number of physical channels—can be two or larger, as long as the number of input channels exceeds 40 or 50. The DSI gain is essentially 1.5 for 15 input channels and grows roughly as the number of channels to the one-sixth power until it saturates at approximately 2.2. These gains, of course, depend on the quality criterion used.

The second DSI scheme is speech predictive encoded communications (SPEC). It is based upon the principle of only sending a PCM word for a given channel when that work differs significantly (i.e., cannot be predicted) from the previous PCM word associated with that channel. The criterion used to set the allowed prediction error is a function of load. Again, gains on the order of two can be realized, although a somewhat smaller minimum number of channels is required ($\cong 20$) to realize a gain of 2 or greater. Reprint Paper 3.1.3 compares the two classes of digital speech interpolation schemes.

REFERENCES

[1] J. L. Flanagan, *Speech Analysis, Synthesis, and Perception,* 2nd ed. New York: Springer-Verlag, 1972.

[2] N. S. Jayant, "Waveform Quantization and Coding" *IEEE Press Reprint Book,* 1976.

[3] P. Noll, "A comparative study of various quantization schemes for speech encoding," *Bell Syst. Tech. J.,* pp. 1597–1614, Nov. 1975.

[4] J. L. Flanagan, "Focal points in speech communication research," *IEEE Trans. Commun. Tech.,* pp. 1006–1015, Dec. 1971.

[5] J. N. Birch, "Low-bit voice processors," *Proc. EASCON,* pp. 59–63, 1974.

[6] L. S. Golding and R. K. Garlow, "Frequency interleaved sampling of a color television signal," *IEEE Trans. Commun. Tech.,* pp. 972–979, Dec. 1971.

[7] T. Ohira *et al.,* "Comparison of Hadamard transform coding and DPCM based on the subjective evaluation of difference error signals from original pictures," *Proc. IEEE Int. Conf. Communications,* pp. 26-24–26-29, June 16–18, 1975.

[8] K. Iinuma *et al.,* "Interframe coding for 4-MHz color television signals," *IEEE Trans. Commun.,* pp. 1461–1466, Dec. 1975.

[9] I. Dorros, "The Picturephone® system: The network," *Bell Syst. Tech. J.,* pp. 221–233, Feb. 1971.

[10] S. W. Golomb, "Run length coding," *IEEE Trans. Inform. Theory,* vol. IT-12, pp. 399–401, July 1966.

[11] H. Meyr, H. C. Roadolsky, and T. S. Huang, "Optimum run length codes," *IEEE Trans. Commun.,* vol. COM-22, pp. 826–835, June 1974.

[12] A. Habibi, Ed., "Special issue on bandwidth compression," *IEEE Trans. Commun.,* vol. COM-25, no. 11, Nov. 1977.

[13] T. S. Huang and O. J. Tretiak, Eds., *Picture Bandwidth Compression,* New York: Gordon and Breach, 1972.

3.1.1

OPPORTUNITIES AND ISSUES IN DIGITIZED VOICE

J. L. FLANAGAN

Acoustics Research Department
Bell Laboratories
Murray Hill, New Jersey   07974

Speech Coding Rates

The figure shows the "spectrum" (in terms of transmission bit rate) of speech coding techniques currently of interest. A vertical dividing line is drawn above the bit-rate axis between 7.2 and 4.8k bits/sec. The higher transmission rates (to the left) are achieved by coders which typically strive to reproduce the signal waveform in a way that is acceptable to a human listener. The liberties that can be taken in the waveform representation, and the efficiency of the coding, are therefore dictated by auditory criteria. For this reason, acoustic signals other than speech can be transmitted satisfactorily.

The lower transmission rates (to the right) depend upon source coding, where acoustic properties and physical constraints of speech generation are exploited in the coder designs. These coders are generically referred to as vocoders. They depend upon approximating the speech signal by a "source-system" model in which the sound generation (source) and the intelligence-modulation (system) are linearly separable. The coding depends upon perceptually-acceptable parametric descriptions of both (typically in terms of voice pitch, voiced/unvoiced switching, and spectrum envelope). At best, these quan-tities (especially the source parameters) are fragile and delicate, and are susceptible to input noise, transmission errors and talker characteristics.

Digital Speech Quality

Our present understanding permits coder designs that provide the speech qualities indicated just beneath the bit-rate axis. We presently know how to make digital coders that achieve telephone toll quality† at coding rates of 16k bits/sec and above. Circuit complexity typically increases as coding rate decreases. At the present time we know of no way, with any amount of complexity, to achieve toll quality at rates much below 16k bits/sec.

At rates below 16k bits/sec, and specifically the data-speed range 9.6k to 7.2k bits/sec, we know how to make waveform coders that provide communications quality

---

†We use the term "toll quality" somewhat loosely, but typically to imply quality comparable to an analog signal having approximately the properties:
Bandwith ≥ 200-3200Hz; Signal-to-noise ratio ≥ 30db; Harmonic distortion ≤ 2-3%.

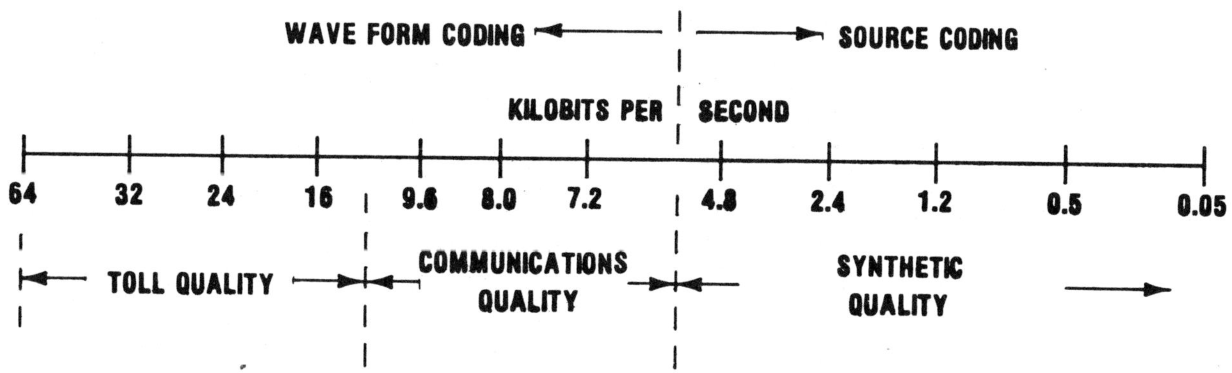

Reprinted from the *Record of the IEEE Electron. & Aerosp. Syst. Conv., (EASCON),* Sept. 25-27, 1978, pp. 709-712.

speech. The signal is highly intelligible but has noticeable quality reduction, some detectable distortion, and perhaps lessened talker recognition. Again, circuit complexity is a function of transmission rate. Typical bit error rates tolerable in waveform coders are of the order of 10^{-3}. Special design and some reduction in quality can push the tolerable rate toward 10^{-2}.

Coders in the source-coding range (vocoders), 4.8k bits/sec and below, provide synthetic quality where the signal usually has lost substantial naturalness, typically sounds "reedy" and sometimes automaton-like. Acceptable transmission error rates are in the range 10^{-3} to 10^{-2}. Talker recognition often is substantially degraded, and is talker dependent. This characteristic, along with the susceptibility of the parameters to errors in extraction, seem to be a consequence of the source-system model of the signal. So long as this first-order model is adhered to, it appears that no amount of processing or expanded bit-rate can improve quality or the fragility of the coding.

Coding Techniques and Circuit Complexity

We have devised, studied and quantified performance for a number of waveform coders and vocoders. Representative coders, and their approximate complexity (essentially a gate count) relative to a simple adaptive delta modulator, are:

Relative Complexity	Coder	
1	ΔM:	adaptive delta modulator
1	ADPCM:	adaptive differential PCM
5	SUB-BAND:	sub-band coder
5	P-P ADPCM:	pitch-predictive ADPCM
50	APC:	adaptive predictive coder
50	ATC:	adaptive transform coder
50	ΦV:	phase vocoder
50	VEV:	voice-excited vocoder
100	LPC:	linear-predictive coefficient (vocoder)
100	CV:	channel vocoder
200	ORTHOG:	LPC vocoder with orthogonalized coefficients
500	FORMANT:	formant vocoder
1000	ARTICULATORY:	vocal-tract synthesizer; synthesis from printed English text.

These coders and their performance data have been described in the technical literature.

Our experience to date suggests that toll-quality coding can be obtained with the following coders running at, or above, the indicated transmission bit rates:

Toll-Quality Transmission

Coder	Bits/Sec.
ΔM	32k
SUB-BAND	24k
P-P ADPCM	24k
APC,ATC,ϕV,VEV	16k

Similarly, our experience suggests that communications quality can be achieved by the following combinations of coders and minimal bit rates:

Communications-Quality Transmission

Coder	Bits/Sec.
ADPCM	16k
SUB-BAND	9.6k
APC,ATC,ϕV,VEV	7.2k

Crossing down into the synthetic-quality range, consistent with source-coding techniques (vocoders), characterizations become even more difficult, but performance is typified by:

Synthetic-Quality Transmission

Coder	Bits/Sec.
CV,LPC	2.4k
ORTHOG	1.2k
FORMANT	500
ARTICULATORY	50

Trades Among Quality, Complexity and Coding Rate

Comprehensive data on the possible trades among quality, complexity and bit-rate - across a broad spectrum of coders - is not available, but some speculation and hypothesizing can be done on the basis of experience. In particular, some comparisons can be drawn between waveform coders and vocoders.

The quality of waveform coders diminishes significantly as bit rate is reduced below 16k. Their quality tends to diminish precipitously below about 8k. Vocoders achieve maximum quality at about 4.8k and cannot rise much further, no matter how much the bit-rate is increased. Reduction of bit rate to the order of 2.4k is accompanied by relatively small reduction from the initial (synthetic) quality. Additional processing to orthogonalize the parametric description of the speech signal

results in similar quality. Although
theoretically unquantified, but practically
founded, further bit rate reduction, to
the order of 500 and below, imposes significant quality reductions.

Present limitations of the source-system
model of the speech signal seem to impose
a distinct ceiling on the quality of vocoders. A more sophisticated signal model,
that better duplicates the short-time signal spectrum and the natural interactions
between source and system (such as the
acoustic interaction between vocal cords
and vocal tract), might elevate the
ceiling. Presently, the quality crossover
point between waveform coders and vocoders
possibly lies in the vicinity of 7.2k,
with a difference in coder complexity of
the order of 2.

Some comparisons can also be drawn, quite
approximately, between word intelligibility (typically from so-called Diagnostic Rhyme Tests) and subjective user-
acceptance ratings. Vocoders (LPC and CV)
at 2.4k typically achieve 75-85% DRT and a
user acceptance of 0.5-0.7. Sophisticated
waveform coders at 9.6k do appreciably
better, and at 16k are comparable to toll-
quality log PCM, which is in the high 90's
for DRT and approaches 1.0 in user acceptance.

Another important factor (and one subject
to substantial modification with time),
is the cost of various speech coders
versus their complexity. The advantages
that LSI can provide are obviously greater
in the more complex circuits. For example,
a CV or LPC vocoder of complexity 100,
running at 2.4k, might be expected to decrease in cost by an order-of-magnitude
in perhaps several years. ATC and APC
would profit commensurately, and would
provide improved quality at higher transmission rates. The least improvement in
cost, not surprisingly, can be expected
in coders of lower complexity, such as ΔM.
Their cost advantages, from single-chip
fabrication, have already been realized.

Where the Action Is

In view of these factors, where should
research and development effort be placed?

Clearly, toll quality digital transmission
can be achieved with simple coders at 32k
bits/sec (ADPCM) and 24k bits/sec (Sub-
Band). Mobile radio telephone quality at
24k with the same relatively-simple coders
also seems feasible. With increased complexity, toll quality at 16k bits/sec is
possible. The necessary complexity for
this is becoming much less frightening -
with circuit complexities and densities
comparable to complete Fast Fourier Trans
forms being planned for single chip fabrication.

New designs for digital transmission and

for customer systems might benefit substantially from these possibilities. The
understanding is essentially in hand. Development is needed.

Research efforts might press in two directions. Look again at the figure. The
boundary between toll quality and communications quality probably is a "soft" one,
and susceptible to new fundamental understanding. This boundary is, in fact, the
wall that we're "leaning" on at present,
with new studies in Adaptive Predictive
Coding and Adaptive Transform Coding.
Pushing this boundary down to the region
of 14k-9.6k would constitute a valuable
advance. Why? Private lines, adaptively
equalized, can transmit 9.6k bits/sec with
10^{-5} or better bit-error-rate (ber). At
16k bits/sec, however, current expectations
are closer to 10^{-2} ber, not good enough for
toll quality speech transmission. Toll
quality speech transmission at bit rates
of 10k to 14k might therefore be interesting both for digital radio and for wire
transmission with modems which could
achieve 10^{-3} ber, or better.

By the same token, conventional, dialed-up
terrestrial lines can be expected to
support 7.2 to 9.6k bits/sec at 10^{-3} to
10^{-2} ber. These rates for speech coding
and bit error presently correspond to
less-than-desirable communications quality.
Again, finding sophisticated ways to elevate speech quality at data coding speeds
and to moderate the effects of transmission
error on conventional channels is of substantial interest. Even relatively small
advances here might make the difference
between achieving good useful systems and
unacceptable performance.

In one direction, then, speech research in
progress aims to push the toll-quality
bit-rate boundary down toward the neighborhood of 9.6k, while at the same time
exploring possibilities for error detection and partial correction that would
allow the coder to perform in a more
severe error environment. In another
direction, reducing the bit error rate on
the 9.6k channel by clever, sophisticated
processing is a research challenge that
properly belongs to data transmission.

Myriad other issues have not been touched
on here. Not the least among these are:
tandeming of digital codes, direct digital
conversion of codes, digital conferencing,
variable-quality coding, packet transmission of speech, and privacy and encryption
techniques. We have given elsewhere an
outline of some of these issues.

Demonstration Tape

An important facet, alluded to earlier, is
the auditory assessment of various coder
designs and transmission bit rates. Tape-
recorded samples are included in the paper
to illustrate a variety of digital speech
coders.

3.1.2

Digital Coding of Speech Waveforms: PCM, DPCM, and DM Quantizers

NUGGEHALLY S. JAYANT, MEMBER, IEEE

Invited Paper

Abstract—A study is presented on the digital coding of speech by means of a straightforward approximation of the time waveform. In particular, the closely related discrete-time discrete-amplitude signal representations that are rather well known as pulse-code modulation (PCM), differential pulse-code modulation (DPCM), and delta modulation (DM) are discussed.

Speech is recognized as a nonstationary signal, and emphasis is therefore placed on "companding" and "adaptive" strategies for waveform quantization and prediction. With signal-to-quantization-error ratio SNR as a performance measure, techniques are suggested which are most likely to be appropriate for given specifications of information rate.

It is pointed out that error waveforms in speech quantization cannot be regarded as additive white noise, in general. This means that for finer assessments of speech coders, either relative or absolute, one needs to supplement SNR-based observations with corrections for subjective and perceptual factors. The latter seem to defy quantification as a rule. Invaluable, therefore, are explicit preference tests for direct comparisons of coders from a perceptual standpoint, and notions such as isopreference and multidimensional scaling are naturally appropriate in interpreting the results of such tests.

Final points of concern are communication questions such as multiple encodings of speech by tandem coder–decoder pairs; conversions among different digital code formats; and the effects of additive and multiplicative noise in the communication channel, as manifest in the erroneous reception of speech-carrying bits. Information on these topics tends to be heterogeneous and nontheoretical, and the present digression into the subject is cursory by intent.

This invited paper is one of a series planned on topics of general interest— The Editor.

Manuscript received September 10, 1973; revised January 28, 1974. The author is with Bell Laboratories, Murray Hill, N. J. 07974.

I. INTRODUCTION

THE ADVANTAGES of coding a signal digitally are well known and are widely discussed in the literature. Briefly, digital representation offers ruggedness, efficient signal regeneration, easy encryption, the possibility of combining transmission and switching functions, and the advantage of a uniform format for different types of signals. The price paid for these benefits is the need for increased bandwidth. Futuristic carriers, as well as some recently utilized channels, are characterized by rapidly increasing bandwidth availability. This, together with the emergence of the requisite device technology, is the reason why digital techniques today are so promising. Papers by Flanagan [1] and Schindler [2] include illustrative sections on digital coding as applied to speech transmission, synthesis, and storage.

The treatment in the present paper will be confined to the encoding of speech by means of a straightforward reconstruction of the acoustic time waveform: specifically, by means of the closely related discrete-time, discrete-amplitude representations known as pulse-code modulation (PCM), differential pulse-code modulation (DPCM), and delta modulation (DM). We are interested in straightforward approximations of the time waveform in the sense that we will not consider questions on the reproduction of speech in terms of features such as excitation descriptions, vocal-tract resonances, or articulatory parameters. Many dramatic results in speech compression have indeed depended on the utilization of these fundamental features [1]. However, straightforward waveform coding still proves to be an important subject on two counts. First, from the point of view of coder complexity, waveform-approximating techniques are the most likely candidates for wide-scale applications of digital speech coding; this has been especially true of systems that are required to reproduce speech with a quality sufficient for commercial telephony. Second, waveform quantization is, after all, the most generally applicable approach to signal coding. Thus, with the exception of some coders that are especially tailored for the speech waveform, the PCM, DPCM, and DM coders that we discuss are appropriate to the communication of any band-limited time function: for example, the intensity waveform in a picture signal, the channel waveform corresponding to a sequence of data bits, or even the variation with time of a specific resonance in the acoustic waveform.

In the treatment of waveform coding, we shall keep implementation details as well as mathematical digressions to a minimum. The purpose will be to explain the basic quantizing techniques, to describe the relatively recent adaptive versions thereof (APCM, ADPCM, and ADM), and to include performance data and demonstrations that enable a comparative evaluation of alternative coding strategies.

Talking of performance data, the signal-to-quantization-error ratio SNR will be ubiquitously employed as a performance criterion in this paper, and although toward the end of the paper we discuss subjective and perceptual effects in coder assessment, the terms "best" and "optimal" will be used in reference to techniques or designs that maximize the SNR. Wherever possible, the SNR will be a measurement on a real speech signal. At times, however, we will also be invoking results derived for Gaussian and sinusoidal signals. These references should maximize our use of available quantizer literature and, at the same time, not minimize our performance with real speech signals!

Sections II, III, and IV will be devoted to individual discussions of PCM, DPCM, and DM, respectively. The results of these sections will be used for comparative evaluations in Section V. This section will also discuss perceptual and subjective factors in coder assessment and will include brief comments on transmission problems such as multiple encodings, code conversions, and the effects of channel errors on coder performance.

In Section II, we recapitulate a basic SNR formula for PCM coding. The formula says that the maximum attainable SNR increases exponentially with information (bit) rate. We discuss how the realization of maximal SNR necessitates the matching of quantizer step size to the signal dynamic range. For a nonstationary signal such as speech, such matching is achieved either by nonuniform quantization or—as in some recent proposals—by the use of adaptive time-varying quantizers. An additional topic included in Section II is the use of dithering in speech quantization.

Section III discusses how differential coding is naturally appropriate in speech quantization. The simplest differential PCM (DPCM) coder essentially encodes speech in terms of the first differences in the sampled waveform, while more sophisticated versions of DPCM represent a speech sample in terms of the error in a suitable prediction thereof. In any case, assuming a fixed sampling frequency (typically the Nyquist frequency for the band-limited speech), differential PCM provides a straightforward SNR gain over PCM. The gain is a function of the correlations present in the samples of speech waveform, and of how well these are exploited in the differential strategies, but is independent, in large measure, of the

number of bits used to code an input sample. Finally, practical versions of differential coding include adaptive techniques, and our discussion will include adaptive predictors in addition to time-varying quantizers.

The subject of Section IV is delta modulation (DM), a differential encoding procedure based on the use of a 1-bit (2-level) quantizer. The technique usually depends on signal-oversampling for satisfactory speech reproduction. In fact, the SNR in DM is seen to increase approximately as the cube of sampling frequency. Adaptive quantizing techniques for DM generally fall into two classes: those based on fast, instantaneous step-size changes, and those that employ slower "syllabic" companding. The latter class is especially suited for relatively lower sampling rates and/or relatively higher error rates in bit transmission.

Section V compares typical adaptive versions of PCM, DPCM, and DM from an SNR viewpoint. ADPCM, for example, is seen to exhibit a constant SNR gain over companded PCM. In the ADM–PCM comparison, on the other hand, there is a crossover point in bit rate below which ADM (which was originally conceived for its simplicity) can in fact outperform PCM. Finally, ADPCM seems to be superior, SNR-wise, to ADM—except, possibly, at bit rates that are too small to be of more-than-academic significance.

The insufficiency of SNR as an adequate measure of performance is next discussed. This inadequacy is related to the observation that error waveforms in speech quantization cannot be modeled by additive white noise, in general. Thus for finer assessments of speech coders, either relative or absolute, one needs to supplement SNR-based observations with corrections for so-called subjective and perceptual factors. The latter seem to defy quantification, as a rule. Invaluable, therefore, are explicit preference tests for direct comparisons of coders from a perceptual standpoint, and we briefly discuss the notions of isopreference and multidimensional scaling in this context. These notions are naturally appropriate in interpreting the results of preference tests.

Finally, we are concerned briefly with what may be termed communication questions: multiple encodings of speech by tandem coder–decoder pairs; conversions among different digital code formats; and the effects of additive and multiplicative noise in the communication channel, as manifest in the erroneous reception of speech-carrying bits.

While SNR is still the single most informative measure of quantizer performance, the choice of a specific speech quantizer for a given application should, wherever possible, follow a consideration of how well a coder can reproduce speech perceptually and an evaluation of how well the coder can stand up to the imperfections of a real communication channel.

The Appendix describes the contents of an accompanying gramophone record.* The recordings are intended to provide the reader with a flavor of the different kinds of speech degradation that waveform coders and real channels cause.

It will be appropriate, at this point, to summarize the notation employed in this paper.

List of Symbols	*Explanation*
X_r	rth sample of quantizer input.
Y_r	rth sample of quantizer output.
$E_r \triangleq Y_r - X_r$	rth sample of quantization error.
$L_r \triangleq X_r - Y_{r-1}$	rth sample of quantizer lag.
$m_r \triangleq Y_r - Y_{r-1}$	rth step in quantizer output.

*Not included in this reprint volume.

Z_r — rth sample of filtered quantizer output.

Δ_r — Quantizer step size at sampling instant r.

Δ — Constant step size in nonadaptive quantization.

$\Delta_{\min}, \Delta_{\max}$ — Minimum and maximum values of step size.

M, P, Q — Step-size multipliers.

H_r — A dimensionless quantity, proportional to Y_r, in a uniform quantizer.

N_r — rth sample of additive random noise.

b_r — Binary code element at sampling instant r in DM.

B — Number of bits used to code a sample in PCM, DPCM ($B=1$ for DM).

W — Highest frequency in input signal = bandwidth of a low-pass-filtered signal.

f_0 — Sampling frequency.

$F \triangleq f_0/2W$ — Oversampling factor in DM ($F=1$ in PCM, DPCM).

f_s — Frequency of a sinusoidal input signal.

$S(f)$ — Power density spectrum.

λ — Dimensionless quantity in the expression for an integrated power spectrum.

$T_0 = 1/f_0$ — Sampling duration.

T_{AD} — Adaptation time.

α — Dummy variable.

$p(\alpha)$ — Probability density function of α.

$\langle \alpha \rangle$ — Mean value of α.

$\langle [\alpha - \langle \alpha \rangle]^2 \rangle$ — Variance of α

$\langle X \rangle, \langle Y \rangle, \langle E \rangle$ — Assumed to be zero.

$X_{\mathrm{rms}} \triangleq \langle X_r^2 \rangle^{1/2}$ — Root-mean-squared value (square root of the variance) of X.

$\mathrm{SNR} \triangleq \langle X_r^2 \rangle / \langle E_r^2 \rangle$ — Signal-to-noise (quantization-error) ratio.

$\mathrm{SNR(dB)}$ — $10 \log_{10} \mathrm{SNR}$.

G — SNR gain in DPCM (over PCM).

$\tilde{\alpha}, \alpha_Q$ — Alternative symbols for quantized value of α.

$\hat{\alpha}$ — Predicted value of α.

$e_r \triangleq \alpha_r - \hat{\alpha}_r$ — rth sample of prediction error.

n — Order of predictor (number of predictor coefficients).

a_i — ith predictor coefficient.

δa_i — Perturbation of a_i in adaptive prediction

$$D_r(a_1, a_2, \cdots, a_i) \triangleq X_r - \sum_{s=1}^{i} a_s X_{r-s}.$$

A — Vector of predictor coefficients.

$C_i \triangleq \langle X_r X_{r-i} \rangle / \langle X_r^2 \rangle$ — Correlation between samples of X separated by a duration iT_0 (i samples).

Γ, Σ — Correlation matrices.

γ, t — Coefficients of an adaptive predictor.

V — Overload level of speech compressor in nonuniform quantization.

v — Output of speech compressor.

μ — Dimensionless quantity in expression for the logarithmic compression function.

θ — Dimensionless constant in SNR expression for DM.

K — Dimensionless constant in formula for step size of a uniform quantizer.

β_1, β_2 — Dimensionless constants in formulas for adaptation of predictor coefficients.

Coder Acronyms

PCM	Pulse-code modulation.
DPCM	Differential PCM.
APCM	Adaptive PCM.
ADPCM	Adaptive DPCM.
DM	Delta modulation.
LDM	Linear (nonadaptive) DM.
ADM	Adaptive DM.
CDM	Continuous DM.
DCDM	Digitally controlled DM.

II. Pulse-Code Modulation

Waveform coding by PCM [3]–[6] involves the following steps.

a) The (band-limited) waveform is sampled at a rate of at least $2W$ Hz (the Nyquist frequency), where W is the highest frequency contained in the waveform. Such sampling insures perfect reconstruction of the analog signal by an appropriate, subsequent desampler. Also, the minimum sampling rate can be less than $2W$ if the lowest signal frequency is nonzero [5].

b) The amplitude of each signal sample is quantized into one of 2^B levels. This implies an information of B bits per sample, and an overall information rate of $2WB$ bits per second (bits/s) for a low-pass-filtered signal.

c) The discrete amplitude levels are represented by distinct binary words of length B. For example, with $B=2$, one can represent 4 distinct levels using the code words 00, 01, 10, and 11.

d) For decoding, the binary words are mapped back into amplitude levels, and the amplitude-time pulse sequence is low-pass filtered with a filter whose cutoff frequency is W.

We shall now assume that steps a), c), and d) can be carried out perfectly in representative implementations, and concentrate on coder performance as determined by the quantization errors introduced in step b). Fig. 1 illustrates the simple case of a uniform quantizer and displays waveforms of signal and quantization error.

Let the quantizer step size be denoted by Δ. If the number of quantizer levels is large, one assumes that the quantization error has the following uniform distribution:

$$p(E) = \frac{1}{\Delta}, \qquad -\frac{\Delta}{2} \leq E < \frac{\Delta}{2}. \tag{1}$$

This will not be true if the signal overloads the quantizer. For example, in an implementation of the quantizer in Fig. 1, the quantizer output might saturate at 5 for inputs exceeding that number, and the quantization error during such overload would be a linearly increasing function of the input.

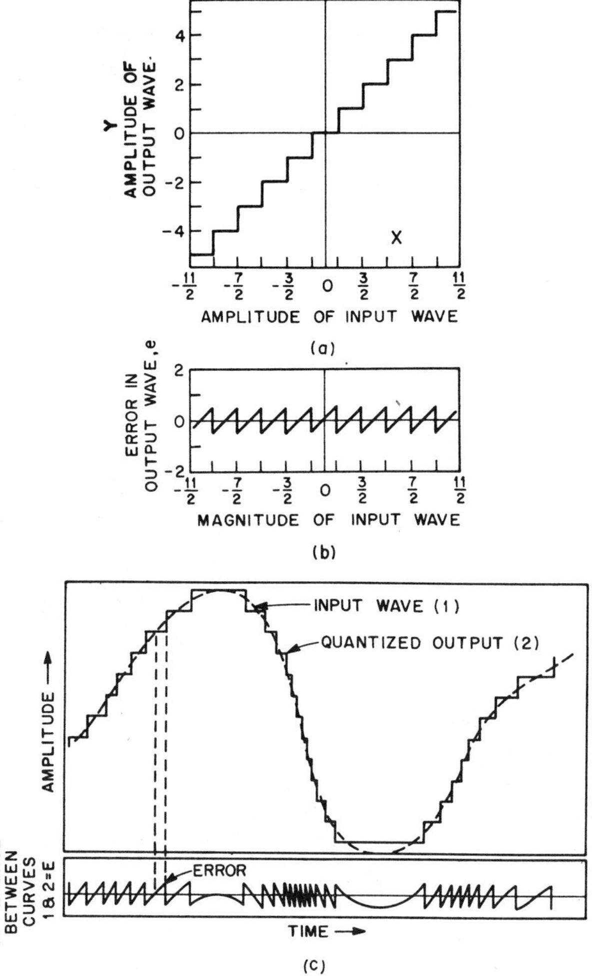

Fig. 1. The quantizing principle. (After Black [5].) (a) Quantizing characteristic. (b) Characteristic of errors in quantizing. (c) A quantized signal wave and the corresponding error wave.

We shall come back to this problem very soon. For the present, let us assume that the quantizer is not overloaded. Then, by virtue of (1), the mean-square value of the quantization error is

$$\int_{-\Delta/2}^{\Delta/2} E^2 p(E) dE = \frac{\Delta^2}{12}. \qquad (2)$$

If the rms value of input X is X_{rms}, the signal-to-error ratio is

$$\text{SNR} = X_{rms}^2 / [\Delta^2/12]. \qquad (3)$$

For example, let the quantizer include the amplitude range $-4X_{rms}$ to $+4X_{rms}$. (Although conventional in theoretical work, the quantizer width of $8X_{rms}$ has no sanctity, except that for most specifications of input signal density, the width of $8X_{rms}$ is more than sufficient to encompass a significant range of the input. For example, if the signal PDF $p(x)$ is modeled by a zero-mean Gaussian function, signal samples will fall outside the $8X_{rms}$ quantizer with a probability (of overload) that is less than 1 in 10 000.) The step size of the uniform quantizer can now be expressed as the ratio of amplitude range to number of steps (output levels):

$$\Delta = 8X_{rms}/2^B. \qquad (4)$$

From (3) and (4),

$$\text{SNR (dB)} = 10 \log_{10} \text{SNR} = 6B - 7.2. \qquad (5)$$

The formula provides a good description of PCM performance under the following conditions.

1) The system operates with a clean channel (or bit-storage medium), so that performance is limited by quantization error (and not by the corruption of code-word letters (bits) by channel noise).

2) The input signal is complex enough to prevent obvious temporal structure in the error waveform, so that the statistical description of the error in (1) is meaningful.

3) The quantization is fine enough (say, $B > 6$) to prevent signal-correlated patterns in the error waveform, so that the effect of errors can be measured in terms of a noise power or the error variance (2).

4) The quantizer is indeed aligned with the amplitude range $(-4X_{rms}, +4X_{rms})$.

A. Use of Dithering in Speech Quantization

Conditions 1)–3) are generally true of toll-quality speech links. When demands on speech quality are not severe, crude quantizers ($B < 6$) are of interest. Speech degradation in such cases is reflected not only by a lower SNR, but also by an undesirable presence of signal-dependent patterns in the waveform of quantization error. The situation can be alleviated by dithering [7], [8], as follows: a pseudorandom noise sequence is added to the speech that is to be quantized; subsequent subtraction of the pseudorandom sequence from the quantizer output provides a white-noise-like error sequence (which is found to be less objectionable, perceptually, than a signal-dependent error waveform) and also preserves the original value of SNR. The technique is most useful in the narrow but practically significant range $4 \leq B \leq 6$ [9].

The use of a pseudorandom generator is demonstrated in Fig. 2. In Fig. 3 are shown waveforms of speech and quantization error with and without dithering. It is clear that dithering is unwarranted when B is large (say, $B \geq 8$). This is because there are no (obvious) signal-dependent patterns in the error waveform, even without dithering, when the quantization is sufficiently fine.

We shall now concentrate on the fact that condition 4) is very rarely met in practice. This is because of the nonstationarity of the speech signal, leading to quantizer-input mismatches, especially when several speakers or circuits are to be handled by a single encoder–decoder system. Unless somehow corrected, mismatches between quantizer range and signal power entail obvious deteriorations of quantizer performance, reflected by values of SNR that are lower, by definition, than what is predicted by (5). We now consider two (not mutually exclusive) solutions to the mismatch problem.

B. Nonuniform Quantization

Consider a nonuniform quantizer with the feature that the step size increases as we go away from the center of the (zero-mean) quantizer. The advantage of such a quantizer is that, without increasing the total number of quantization levels (and hence the needed bit rate), one can allow large end steps in the quantizer to take care of possible excursions of the speech signal into the (relatively infrequent) large amplitude

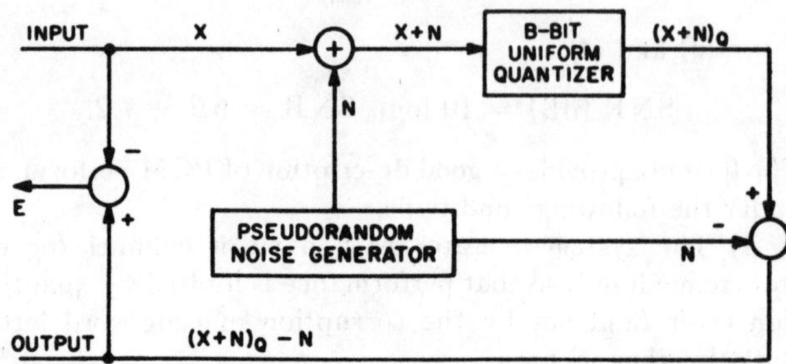

Fig. 2. Schematic of the use of dither in quantization.

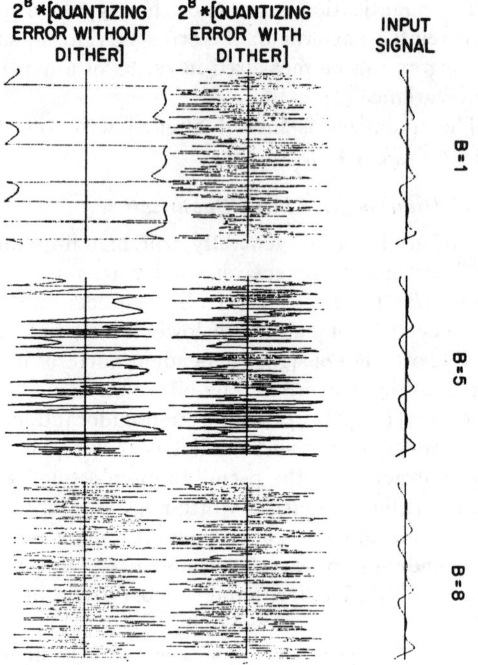

Fig. 3. Quantization error waveforms with and without dither. (After Jayant and Rabiner [8].)

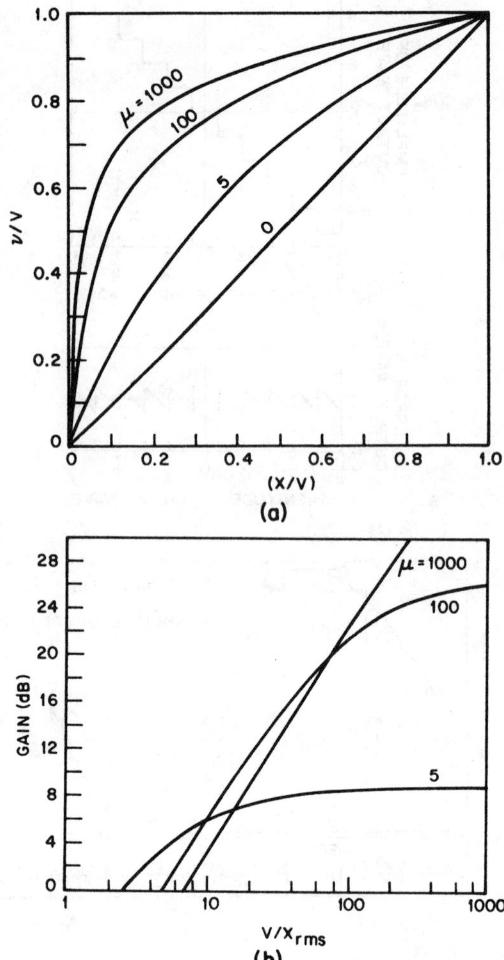

Fig. 4. Logarithmic quantizer. (a) Companding characteristics. (b) Performance curves. (After Smith [10].)

ranges. Equivalently, for a given quality of encoding, over a specified dynamic range of the signal, a nonlinear quantizer permits a reduction of bit rate. For example, the logarithmic quantizer of Smith [10] can typically be designed to provide toll-quality speech quantization with 128 levels or 7 bits per sample, while a uniform quantizer (Fig. 1) needs about 11 bits for a similar performance. [Note that, according to (5), if one is always dealing with a stationary signal of known variance, an 11-bit quantizer can be designed to provide a 24-dB improvement in SNR over a 7-bit quantizer!]

The use of a nonuniform quantizer is equivalent to the presentation of a compressed signal to a uniform quantizer, and a subsequent expansion of the output. The compression law used in the Smith quantizer is given by

$$|v| = V \log\left(1 + \frac{\mu |X|}{V}\right) \Big/ \log(1 + \mu), \quad \mu > 0. \quad (6)$$

Fig. 4(a) illustrates this compression characteristic, and shows that linear (uniform) quantization corresponds to $\mu = 0$. Smith also calculates the SNR gain over uniform quantization as a function of the ratio of the compressor overload level

V to X_{rms}, the square root of speech variance, and uses the SNR results in conjunction with statistical observations on the dynamic range of the speech signal to conclude that desirable values of μ are in the order of 100. The point is illustrated in Fig. 4(b) where the $\mu = 100$ curve includes the greatest of three areas over the abscissa. These areas are measures of how well quantizer performance holds up when X_{rms} changes. The specific value of $\mu = 255$ is frequently used in current implementations.

Finally, it is helpful in compander implementations to approximate the logarithmic curve (6) for small signals by a linear segment. This procedure is referred to as A-law companding. The reciprocal of A is defined as the value of the abscissa, in Fig. 4(a), after which the curve transitions smoothly to a true logarithmic form.

Panter and Dite [6] show how the advantage of nonuniform quantization increases with the crest factor (ratio of peak to rms value) of the signal, and indicate that the choice of μ should also be governed by the value of B. Larger values of B call for larger values of μ, or greater nonuniformity of the quantizer. More recently, Paez and Glisson [11] utilize the gamma distribution as a good model of speech amplitudes and derive a corresponding optimum nonuniform quantizer. The term "optimum quantizer" (whether uniform or nonuniform) refers to an SNR-maximizing quantizer. It is as-

N	2		4		8		16		32	
i	x_i	y_i	x_i	y_i	x_i	y_i	x_i	y_i	x_i	y_i
1	∞	0.577	1.205	0.302	0.504	0.149	0.229	0.072	0.101	0.033
2			∞	2.108	1.401	0.859	0.588	0.386	0.252	0.169
3					2.872	1.944	1.045	0.791	0.429	0.334
4					∞	3.799	1.623	1.300	0.630	0.523
5							2.372	1.945	0.857	0.737
6							3.407	3.798	1.111	0.976
7							5.050	4.015	1.397	1.245
8								6.085	1.720	1.548
9									2.089	1.892
10									2.517	2.287
11									3.022	2.747
12									3.633	3.296
13									4.404	3.970
14									5.444	4.838
15									7.046	6.050
16									∞	8.043
MSE	0.6680		0.2326		0.0712		0.0196		0.0052	
SNR dB	1.77		6.33		11.47		17.07		22.83	

Within the table: $p(x) = \dfrac{\sqrt{k}}{2\sqrt{\pi}} \dfrac{e^{-k|x|}}{\sqrt{|x|}}$; $k = 0.866$

$\text{VARIANCE} = 0.75/k^2 = 1$

Fig. 5. Optimum quantizers for signals with gamma density; mean $\mu = 0$, variance $\sigma^2 = 1$. (After Paez and Glisson [11].)

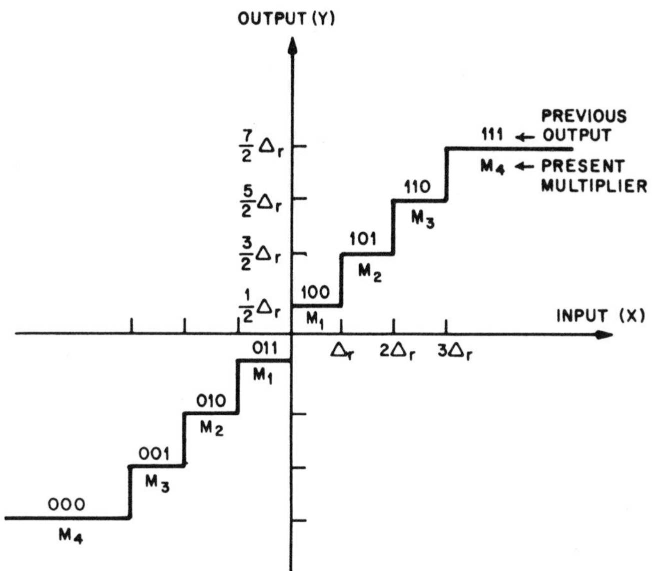

Fig. 6. Adaptive quantization with a one-word memory.

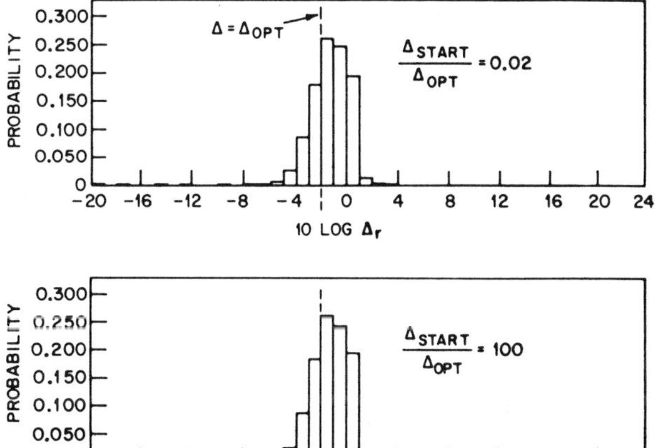

Fig. 7. Histogram of step sizes in adaptive quantization of a Gauss–Markov input; $B = 3$, $C_1 = 0.5$. (After Jayant [13].)

sumed that the SNR computation includes the effect of overload errors, which were neglected in the simple exposition preceding (5). The Paez–Glisson quantizer is summarized in Fig. 5. The x values define the ends of quantizer input ranges, and the y values are corresponding outputs. For example, in the 3-bit quantizer, inputs between 0.504 and 1.401 are quantized as 0.859, and all inputs exceeding 2.872 are quantized into 3.799. The derivation of the quantizer is based on an iterative determination of input ranges and output levels, as suggested by a general theory due to Max [12]. Results in [11] and [12] also specify the best uniform quantizers for signals with gamma, Laplacian [11], and Gaussian [12] density functions. For example, Max derives the following SNR-maximizing step sizes for uniform 1-, 2-, 3-, and 4-bit quantizers with Gaussian inputs:

$$\Delta_{\text{opt}} = K(B) \cdot X_{\text{rms}}$$
$$K(1) = 1.596$$
$$K(2) = 0.996$$
$$K(3) = 0.586$$
$$K(4) = 0.335. \tag{7}$$

C. Adaptive Quantization

A more recently studied, and presumably more flexible, means of matching quantizer step size to signal variance is the use of step-size adaptation based on quantizer memory. The idea is to work with a basic quantizer that is very simple (uniform, if necessary), but to modify its step size (for every new input sample, in general) by a factor depending on the knowledge of which quantizer slots were occupied by the previous samples. In its simplest form, the scheme operates with a one-word memory—a suggestion due to Flanagan, developed at length by Cummiskey and Jayant [13]. Let the output of a B-bit (uniform) quantizer be

$$Y_r = H_r \frac{\Delta_r}{2}$$
$$\pm H_r = 1, 3, 5, \cdots, 2^B - 1 \quad (\Delta_r > 0, B \geq 2). \tag{8}$$

The step size Δ_{r+1} is now chosen to be the previous step size multiplied by a time-invariant function of the code word magnitude $|H_r|$:

$$\Delta_{r+1} = \Delta_r \cdot M(|H_r|). \tag{9}$$

When the multiplier function is properly designed, the adaptation logic (9) serves to match the step size, at every sample, to an updated estimate of signal variance. Fig. 6 explains a 3-bit adaptive quantizer. Shown in the figure are the eight possible values of (the latest) coder output (000 through 111) and the corresponding step-size multipliers M. Note that the value of M depends on the magnitude of (the latest) coder output—equivalently, on the magnitude $|H_r|$ in (9)—and not on the sign of the output. (This strategy is a simple consequence of the observation that the input probability density function $p(X)$ is expected to be symmetric about a mean value of zero.)

Fig. 7 shows histograms of step sizes encountered in the simulation of a 4-bit adaptive quantizer with a Gaussian signal at the input. The step-size multipliers were selected to maximize an appropriately defined SNR [13], and Fig. 7

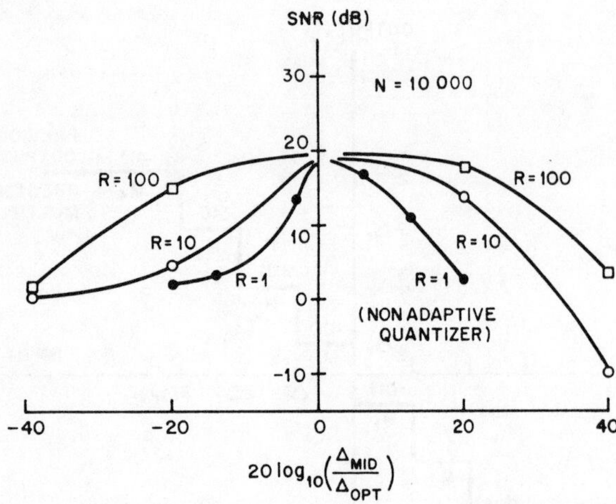

Fig. 8. Companding characteristics of adaptive quantizer; $B=4$, $C_1=0.5$, $\Delta_{\mathrm{mid}}=[\Delta_{\min}\cdot\Delta_{\max}]^{1/2}$, $R=\Delta_{\max}/\Delta_{\min}$.

CODER	PCM			DPCM		
B	2	3	4	2	3	4
M_1	0.60	0.85	0.80	0.80	0.90	0.90
M_2	2.20	1.00	0.80	1.60	0.90	0.90
M_3		1.00	0.85		1.25	0.90
M_4		1.50	0.80		1.75	0.90
M_5			1.20			1.20
M_6			1.60			1.60
M_7			2.00			2.00
M_8			2.40			2.40

Fig. 9. Step-size multipliers for $B=2$, 3, and 4. (After Jayant [13].)

shows how these multipliers indeed maintain the variable Δ, in a region centered on Δ_{opt}, the optimal (constant) step size for a nonadaptive quantizer (even when the starting step size is severely suboptimal). Fig. 8 describes the companding characteristics of an adaptive quantizer for the example of $B=4$. The quantities $\Delta_{\max}$ and $\Delta_{\min}$ represent practical constraints on the adaptation logic (9), and these constraints determine, through the ratio R, the dynamic range of the adaptive quantizer. The dynamic range is the range of input variance which the quantizer can handle—equivalently, the range of Δ_{mid} which a given input signal can tolerate for a specified minimum performance (SNR). Note that for the example of Fig. 8, the advantage of adaptation consists in increasing the dynamic range rather than in increasing the peak value of SNR. This observation is fairly general and carries over to speech quantization. Step-size adaptations provide considerable gains in peak SNR when the correlation C_1 between adjacent samples approaches unity, or when the nonadaptive design is based—as it often is—on the long-term statistics of a nonstationary signal.

Fig. 9 lists step-size multipliers found to be optimal for adaptive PCM (APCM) coding of a low-pass-filtered speech sample. The recommended step-size multipliers do not in general constitute overly critical target values. It is possible

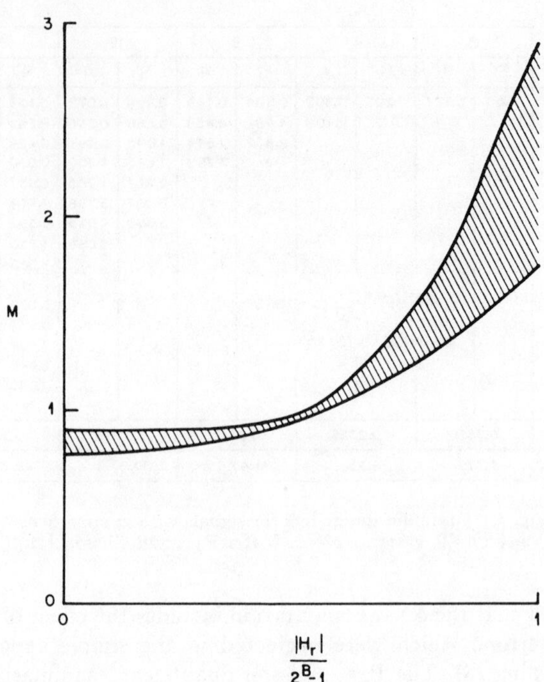

Fig. 10. General shape of optimal multiplier function in speech quantization; $B>2$. (After Jayant [13].)

to get away, for instance, with trivial values of unity for some of the central multipliers. For example,

$$M_1 < 1, \quad M_2 = M_3 = 1, \quad M_4 > 1; \quad B = 3. \quad (10)$$

What is critical, however, is the observation that step-size increases should, in general, be more rapid than step-size decreases. This has to do with the following comparison of two basic types of quantization error: "overload" errors that occur when Δ_r is too small, and a signal sample falls outside the quantizer range; and "granular" errors that are inherent in quantization even when the input falls within a quantizer slot or step. Overload errors have a density function that has a theoretically infinite support, while granular errors can be modeled by uniform distribution (1) with a finite support equal to the step size itself. As a result, granular errors tend to be less harmful to SNR than overload errors. To mitigate the contribution of the latter to the total noise power, one seeks to correct for the occurrence of overload errors more expeditiously. The end result is an optimal multiplier function that has the general form depicted in Fig. 10. The figure shows how the M function of (9) steadily increases with $|H_r|$ during step-size increases ($M>1$; to correct for overload). On the other hand, during step-size decreases ($M<1$), the optimal M function is remarkably close to unity. The shaded region in Fig. 10 expresses variations in the multiplier function when there are changes in B and the input signal statistics [13]. The exclusion of $B=2$ from the adaptation rule in the figure has to do with the fact that when the quantization gets to be crude enough, the distinction between expected magnitudes of granular and overload errors diminishes—and so does the disparity between desired rates of step-size decrease and increase.

Schlink describes a slightly different approach to adaptive quantization [14]. Here adaptations consist of switches between two invariant quantizers. The switches are made, once again, by the use of quantizer memory. The idea of utilizing such memory for the APCM coding of speech has been fairly recent [13], [14], and so far the most widely used compander for speech coding using PCM has been the (memoryless) logarithmic compander due to Smith [10]. Adaptive quantization with memory has also been applied, however, to differential encoding, or ADPCM [15], [16]. Here the combination of differential and adaptive strategies has proved to be extremely fruitful from an application viewpoint.

Before going into differential PCM, we should also refer to the so-called syllabic techniques for companding or adaptation, as against the instantaneously adaptive techniques discussed in Subsections II-B and C. In these schemes the step size Δ_r is adapted with a time constant of about 5–10 ms, rather than for every sample. Wilkinson has described a syllabic-companded PCM that provides very useful speech reproduction at low bit rates [17]. The most extensive applications of syllabic adaptation, however, have been in the field of delta modulation (Section IV). Here, as in PCM, a principal advantage of syllabic companding is that such a strategy makes an adaptive system more tolerant to channel errors. Instantaneous adaptation, on the other hand, may be simpler to implement when time division multiplexing of several speech channels is involved.

III. DIFFERENTIAL PULSE-CODE MODULATION

Speech that is sampled at the Nyquist rate exhibits a very significant correlation C_1 between successive samples. One consequence of this correlation is that the variance of the first difference

$$D_r(1) = X_r - X_{r-1} \tag{11}$$

is smaller than the variance of the speech signal itself. With basic statistical notation,

$$\begin{aligned}\langle D_r^2(1)\rangle &= \langle (X_r - X_{r-1})^2\rangle \\ &= \langle X_r^2\rangle + \langle X_{r-1}^2\rangle - 2\langle X_r X_{r-1}\rangle \\ &= \langle X_r^2\rangle[2(1 - C_1)] \end{aligned} \tag{12}$$

where C_1 is the correlation between adjacent samples. Obviously, if C_1 is greater than 0.5, $D(1)$ has a smaller variance than X. As a result, it is advantageous to quantize $D(1)$ instead of X, and use an integrator to reconstruct X from the quantized values of $D(1)$. This is because, for a given fineness of quantization B, the quantization error power is proportional to the variance of the signal present at the quantizer input [as in (2) and (4)]. The use of $D(1)$ as quantizer input therefore permits a smaller error variance and a better SNR. Equivalently, for a given value of SNR, the use of the differential technique permits the employment of a smaller value of B and hence leads to straightforward bandwidth reduction. It is important, of course, that in integrating the quantizer values of $D(1)$, we do not accumulate quantization error. This is insured by a simple feedback procedure which inputs to the quantizer the quantity $D_r(1) + E_{r-1}$, rather than $D_r(1)$, where E_{r-1} is the quantization error for the previous sample.

As a generalization of (12), consider

$$\langle D_r^2(a_1)\rangle = \langle (X_r - a_1 X_{r-1})^2\rangle = \langle X_r^2\rangle(1 + a^2 - 2a_1 C_1). \tag{13}$$

It is readily seen that the preceding variance has a minimum value when $a_1 = C_1$:

$$\langle (X_r - a_1 X_{r-1})^2\rangle_{\min}$$

$$= \langle (X_r - C_1 X_{r-1})^2\rangle = \langle X_r^2\rangle(1 - C_1^2). \tag{14}$$

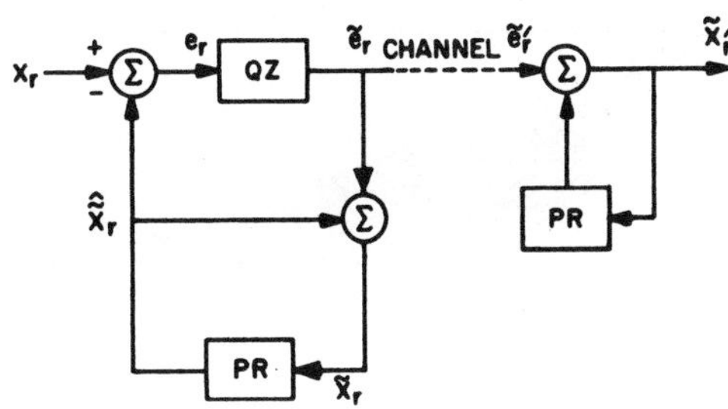

Fig. 11. Differential PCM. QZ: quantizer; PR: predictor: $\hat{X}_r = \sum_{i=1}^n a_i \tilde{X}_{r-j}$. If channel is error-free, $\tilde{X}_r' = \tilde{X}_r =$ reconstructed signal. (After McDonald [20].)

Note that the variance reduction in (14) is positive for all C_1, while that in (12) is positive only if $C_1 > 0.5$.

A further generalization is possible if we use as quantizer input a difference of the form

$$D_r(a_1, a_2, a_3 \cdots a_n) = X_r - \sum_{i=1}^n a_i X_{r-i} = X_r - \hat{X}_r \tag{15}$$

where $\hat{X}_r$ denotes a (linear) prediction of X_r. This form of differential encoding [18], [19] is formalized in the (coder portion of the) differential pulse-code modulation (DPCM) circuit [20], [21] of Fig. 11.

Note that the feedback around the quantizer insures that the error in the reconstructed (quantized) signal $\tilde{X}_r$ is precisely the quantization error for the sample e_r and not an accumulation of previous quantization errors:

$$E_r = e_r - \tilde{e}_r = (X_r - \hat{X}_r - \tilde{e}_r) = X_r - \tilde{X}_r. \tag{16}$$

The signal-to-noise ratio

$$\text{SNR} = \langle X_r^2\rangle / \langle E_r^2\rangle \tag{17}$$

shows an SNR gain G over PCM, given by the ratio of the variance of X_r to that of e_r:

$$G = \frac{\langle X_r^2\rangle}{\langle e_r^2\rangle} = \frac{\langle X_r^2\rangle}{\langle (X_r - \hat{X}_r)^2\rangle} \tag{18}$$

where $\hat{X}_r$ is the prediction of X_r based on n quantized past values:

$$\hat{X}_r = \sum_{i=1}^n a_i \tilde{X}_{r-i} = \sum_{i=1}^n a_i (X_{r-i} + E_{r-i}). \tag{19}$$

To maximize G, the predictor coefficients are chosen to minimize the denominator of (18):

$$\min_{a_i} \left[\langle e_r^2\rangle \simeq \left\langle \left(X_r - \sum_{i=1}^n a_i X_{r-i}\right)^2\right\rangle + \langle E_r^2\rangle \sum_{i=1}^n a_i^2 \right]. \tag{20}$$

Error-signal correlation terms of the form $\langle EX\rangle$ are assumed to be negligible in arriving at (20). If we can also assume that the coder is good enough to ensure that

$$\langle E_r^2\rangle \ll \langle X_r^2\rangle \tag{21}$$

the predictor which minimizes (20) is given by

$$A_{\text{opt}} = \Gamma^{-1}\Sigma \tag{22}$$

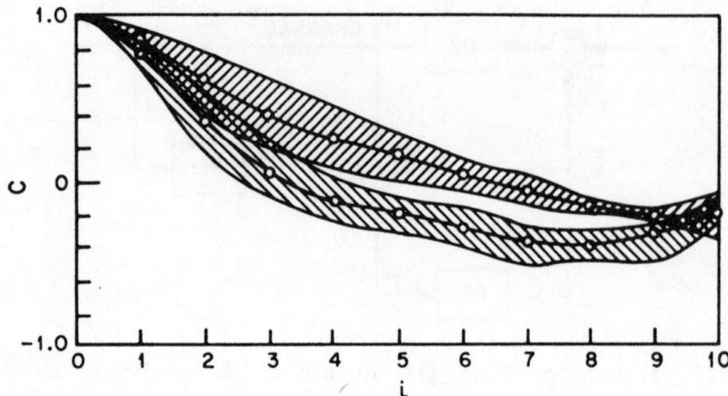

Fig. 12. Autocorrelation functions of speech signals. Upper curves: low-pass speech; lower curves: bandpass speech. (After Noll [22].)

where

$$\Gamma = \begin{bmatrix} C_0 & C_1 & \cdot & \cdot & C_{n-1} \\ C_1 & \cdot & \cdot & \cdot & C_{n-2} \\ \cdot & \cdot & \cdot & \cdot & \cdot \\ \cdot & \cdot & \cdot & \cdot & \cdot \\ \cdot & \cdot & \cdot & \cdot & \cdot \\ C_{n-1} & \cdot & \cdot & \cdot & C_0 \end{bmatrix},$$

$$\Sigma = \begin{bmatrix} C_1 \\ C_2 \\ \cdot \\ \cdot \\ \cdot \\ C_n \end{bmatrix}, \quad A = \begin{bmatrix} a_1 \\ a_2 \\ \cdot \\ \cdot \\ \cdot \\ a_n \end{bmatrix}. \tag{23}$$

The resulting gain over PCM is

$$G_{\text{opt}} = \left[1 - \sum_{i=1}^{n} a_i C_i \right]^{-1}. \tag{24}$$

Note that in Γ the diagonal entry $C_0 = 1$. Notice also that the SNR gain is independent of the number of bits B.

For the special case of $n = 1$, the predictor (22) is simply a leaky integrator with $a_1 = C_1$, while if $a_1 = 1$, the feedback network provides perfect integration. We now reproduce illustrative results on DPCM from a recent paper by Noll [22]. The results are based on measurements on a 55-s speech sample that is either low-pass-filtered (LPF: 0 to 3400 Hz) or bandpass-filtered (BPF: 300 to 3400 Hz). The sampling rate in either case is 8 kHz. Fig. 12 shows plots of C_i versus i. The upper set of curves refers to LPF speech, and the lower set to BPF speech. In each set, the upper and lower limits represent maxima and minima, respectively, over four speakers (two male and two female), and the middle curve gives mean values (averages over four speakers). Note that in no case is $C_1 < 0.5$. This means that the use of even the simplest DPCM logic [as in (11) and (12)] should provide an SNR gain over PCM. The gains in an optimal DPCM logic (22) are, of course, even greater; these G values are plotted in Fig. 13(a) (LPF speech) and Fig. 13(b) (BPF speech) as a function of n, the number of predictor coefficients. As before, each set of curves includes a maximum, a minimum, and an average, over the population of four speakers. It is interesting that the G function typically saturates for all practical purposes at $n = 2$. Notice also that the asymptotic G value is greater for LPF speech than for BPF speech. This is expected because LPF speech has greater low-frequency energy, and hence greater adjacent sample correlations. This, in turn, reflects the possibility of greater redundancy removal by

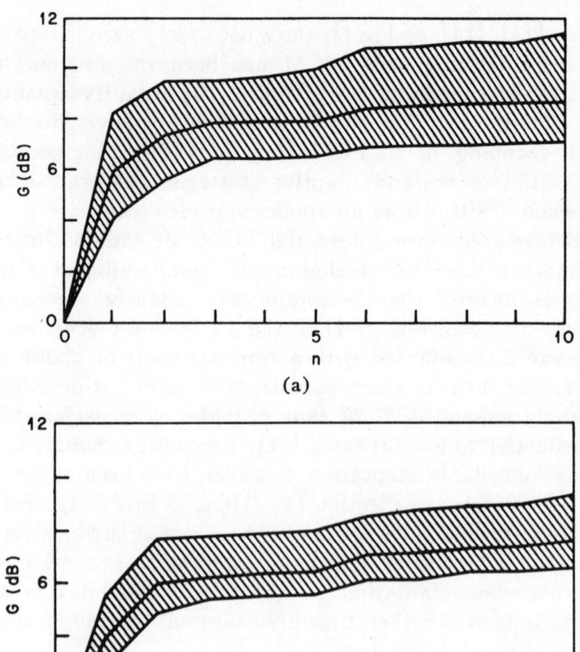

(a)

(b)

Fig. 13. Optimum SNR gain G versus number of predictor coefficients. (a) LPF speech. (b) BPF speech. (After Noll [22].)

differential encoding; equivalently, a greater asymptotic value for G. It is significant, however, that even for LPF speech, G never reaches a value of 12 dB [which would signify a 2-bit advantage over PCM by virtue of (5)].

Finally, note the undesirably strong speaker dependence of G in Fig. 13. An even more undesirable property, not demonstrated in the figures, is that G is very sensitive to the matching of the predictor vector A to the correlation matrices Γ and Σ. In other words, changing the input speech material degrades, in general, the performance of a DPCM system operating with a predictor designed on the basis of average speech statistics. By the same token, it is suboptimal to use a single predictor for encoding voiced and voiceless segments in the speech waveform. Voiced segments, for example, are generally characterized by $C_1 > 0.5$, while noise-like fricatives can be expected to exhibit a near-zero correlation, $C_1 \rightarrow 0$. (Overall speech quality, however, is controlled mainly by how well the voiced sections are coded, because they predominate in probability of occurrence; furthermore, fricatives are more tolerant to the addition of granular quantization noise, from a perceptual viewpoint.)

The reader should expect, by now, that practical versions of differential PCM are likely to employ adaptive quantizers and/or adaptive predictors—the former to follow changes in signal power, and the latter to respond to changes in the short-term spectrum of speech. Coders with these adaptive features have been referred to in the literature as adaptive DPCM (ADPCM) coders.

A. Adaptive Quantizers

The use of adaptive quantization in ADCPM [15], [16] has the same motivation as the use of an adaptive quantizer in APCM. The variance of the predictor error, which is the quantizer input in DPCM (Fig. 1), is proportional to the variance of the input signal X, and the input signal variance is either unknown or highly variable in a quasi-stationary signal such as speech. The variance can, however, be usefully estimated by the use of quantizer memory when the input is locally stationary. Stroh describes a straightforward estimation procedure for Gaussian signals [16]. The technique of Cummiskey, Jayant, and Flanagan [15] makes a more explicit use of "overload" and "underload" cues to provide what is believed to be a very efficient adaptation logic. The technique uses a minimal one-word memory, and it has been studied for both Gaussian and real speech inputs. The reader is referred back to Fig. 6 for a recapitulation of the technique, and to Fig. 9 for numerical data on step-size adaptation.

The results on DPCM in Fig. 9 refer to the case of a simple first-order predictor (11). Notice that optimum (SNR-maximizing) step-size multipliers for ADPCM are slightly different from those for APCM. This has to do with the fact that the correlation between adjacent signal samples X is quite high at the input of a DPCM quantizer [being equal to the correlation between adjacent Nyquist-sampled speech samples [20]; the input samples of a DPCM quantizer are typically much less correlated because of the differentiating (or high-passing) process involved (11)]. In fact, the step-size multipliers recommended for adaptive quantization of speech in DPCM are extremely close to those theoretically derived to be optimum for a white ($C_1 = 0$) Gaussian signal [13]:

$$M_r{}^{\text{opt}} = \left[\frac{1}{2} + \frac{K^2(B) H_{r-1}{}^2}{8}\right]^{1/2} + \delta^2(|H_{r-1}|)$$

$$\delta^2(|H_{r-1}|) = 0, \qquad \text{if } H_{r-1} \neq 2^{B-1} \qquad (25)$$

where $K(B)$ is the step-size-to-X_{rms} ratio whose values for $B = 1$ to 4 are given in (1). Recall that H_{r-1} defines the most recent quantizer output (8), and note that δ^2 is an unspecified positive correction, significant only for the last slot of the B-bit quantizer.

B. Adaptive Predictors

The desirability of adapting predictor coefficients to the changing spectral properties of speech signals has been mentioned earlier. A simple way to do this is to store finite sections of speech, calculate the autocorrelation function for this section, and then determine an optimum predictor vector using (22). The predictor is periodically updated at time intervals given by the length T_{AD} of the stored section, and the predictor coefficients transmitted to the receiver. (Such transmission does not consume excessive channel capacity because the coefficients tolerate coarse quantization and slow updating.) Adaptive prediction of this type makes the G function saturate at a larger value, and for a larger n, as compared with a nonadaptive predictor. An illustration taken from Noll [22] is provided in Fig. 14. Note that with a predictor-update time of $T_{\text{AD}} = 4$ ms (32 Nyquist samples), the gain over PCM is 13 dB. This represents a more-than-2-bit advantage. If the DPCM quantizer is not too coarse ($B > 2$), transmission of predictor coefficients can be avoided, with a small sacrifice of SNR gain, by determining the coefficients automatically (at both the transmitter and the receiver).

ENTRIES ARE SNR GAINS OVER PCM, IN dB	n= 1	3	5	10
NONADAPTIVE DPCM	5.4	8.4	8.6	9.0
ADAPTIVE DPCM; $T_{\text{AD}} = 4$ ms	5.6	10.0	11.5	13.0
ADAPTIVE DPCM; $T_{\text{AD}} = 32$ ms	5.6	9.8	11.1	12.6

Fig. 14. Comparison of nonadaptive and adaptive predictors. (After Noll [22].)

This has been done by means of a steepest descent gradient search [23]. We shall now review this method very briefly.

Neglecting the effect of quantization errors, consider a positive definite function of predictor error such as

$$|e_r| = e_r \operatorname{sgn} e_r \qquad (26)$$

$$e_r = X_r - \hat{X}_r = X_r - \sum_{i=1}^{n} a_i X_{r-i}. \qquad (27)$$

The gradient of this is now taken with respect to the predictor vector, and the coefficients are adapted in a direction opposite the gradient. The result is a predictor perturbation of the type

$$\delta a_j = \beta_1 \cdot \operatorname{sgn}(e_r) \cdot X_{r-j} \bigg/ \sum_{k=1}^{n} |X_{r-k}| \qquad (28)$$

where β_1 is an adaptation speed that can be suitably optimized (to a noncritical value). Employing the square function

$$e_r{}^2 = (X_r - \hat{X}_r)^2 \qquad (29)$$

instead of (26) results in a predictor adaptation algorithm

$$\delta a_j = \beta_2 e_r X_{r-j} \bigg/ \sum_{k=1}^{n} X_{r-k}{}^2. \qquad (30)$$

The use of quantized values $\tilde{X}$ instead of X in (28) and (30) can provide converging prediction algorithms if $B \geq 2$ [16], [23]. This means that predictor coefficients need not be transmitted to a DPCM receiver. Also, note that the adaptations given are at every Nyquist sample. Values of $n = 8$, $\beta_1 = 0.09$, and $G = 9$ dB are typical for BPF speech [23]. [This value, although less than some figures cited earlier for nonadaptive DPCM, is considered to be significant. This is because prepredictor-speech mismatches can lead to asymptotic G values that are significantly lower than 9 dB—lowermost curve of Fig. 13(b)—if the predictor is nonadaptive.] This figure of 9 dB is from a simulation that included an adaptive quantizer. The step-size information was coded separately and transmitted to the DPCM receiver. An ADPCM system where both step-size information and predictor coefficients are generated automatically by the bit stream is described in [16] by Stroh. It is shown here that the benefits of adaptive quantization and adaptive prediction in ADPCM are nearly complementary, and that only second-order interactions exist between the quantizer and the predictor.

Finally, it may be relevant to mention that as the predictor order n is increased, the quantizer input tends to have a Gaussian density function, and that small perturbations of predictor coefficients from their optimal values do not alter the Gaussian nature of the error PDF.

One of the most sophisticated adaptive predictors designed

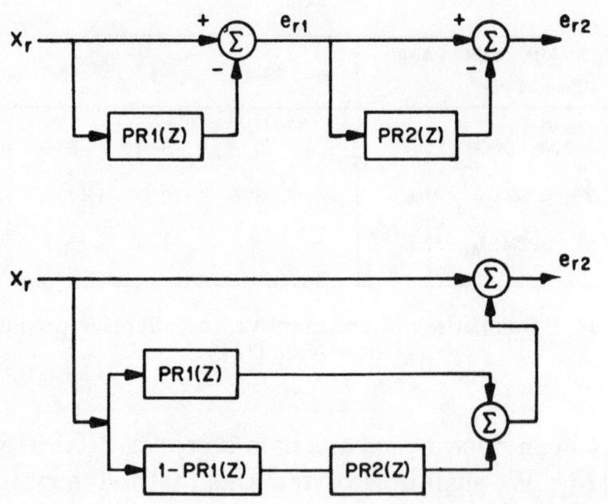

Fig. 15. Two-stage predictor for adaptive predictive coding.
(After Atal and Schroeder [24].)

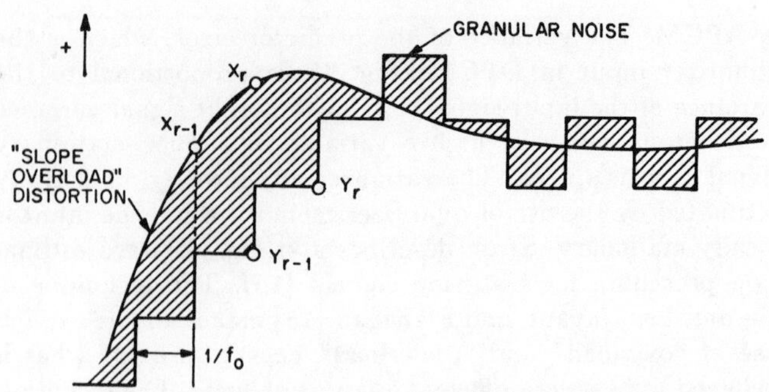

Fig. 16. Linear delta modulation (LDM); $Y_r - Y_{r-1} = \Delta \cdot b_r$, sgn $(X_r - Y_{r-1}) = b_r$, and $b_r =$ transmitted "channel symbol."

for speech is the one due to Atal and Schroeder [24]. Their scheme exploits the quasi-periodic nature of the speech signal to obtain more complete signal prediction than that provided by the short-term ($n \leq 10$, say) linear predictors discussed so far. In other words, the signal redundancy is now removed in two stages: first, by a predictor that removes the quasi-periodic nature of the signal and, second, by a predictor that removes formant information from the spectral envelope. The first predictor is simply a gain-and-delay adjustment, and the second is a linear combination of past values of the first predictor output. The equivalent operations are shown in Fig. 15 which, in z-transform notation, can be summarized as follows:

$$PR1 (z) = \gamma z^{-t}$$

$$PR2 (z) = \sum_{i=1}^{8} a_i z^{-i}$$

$$PR (z) = \{PR1 (z) + PR2 (z)[1 - PR1 (z)]\}. \quad (31)$$

The coefficients for predictor PR2 (z) are calculated as described previously in this section. Those for PR1 (z), i.e., γ and t, are obtained as follows. First, t is selected to maximize the normalized correlation

$$C_t = \frac{\sum_{r=1}^{n} X_r X_{r-t}}{\left[\sum_{r=1}^{n} X_r^2 \sum_{r=1}^{n} X_{r-t}^2 \right]^{1/2}}. \quad (32)$$

The optimum value t_{opt} is found by a search of computed and tabulated values of C_t. Having determined t_{opt}, γ is set in the manner of (14) to the value

$$\gamma_{\text{opt}} = \frac{\sum_{r=1}^{n} X_r X_{r-t_{\text{opt}}}}{\sum_{r=1}^{n} X^2_{r-t_{\text{opt}}}}. \quad (33)$$

Typically, with $n = 75$, $t_{\max} = 150$, and reasonable quantization (and transmission) of updated predictor coefficients, the scheme can achieve SNR gains over PCM which are of the order of $G = 20$ dB. However, due to the large amount of computation required to determine pitch period, adaptive

predictors of this type are presently too complex for application to most real-time communication systems. On the other hand, simpleminded approaches toward exploiting the quasi-periodicity of speech (such as techniques based on one or more replications of an entire pitch-period-long segment of speech waveform [25], [26]) prove to be too crude, and not particularly simple, to enable any kind of general application, although they can provide useful speech reproduction for special low-bit-rate applications [26]. The limiting factors in the subject are clearly the existence, and the perceptual significance, of very small variations of pitch period in voiced speech. Due to these variations, the exploitation of periodicity in speech is rendered extremely nontrivial. Speech researchers in this context cannot help envying their counterparts in picture coding, where the obvious time invariance of the frame period has permitted far more remarkable strides in differential encoding, such as the technique of conditional frame-replenishment [27].

We shall now shift our attention to the other (least complex) end of the DPCM spectrum, and examine what can be achieved with an extremely simple first-order predictor and a 2-level quantizer ($B = 1$); we make up for these demands, however, by agreeing to oversample ($f_0 > 2W$) the band-limited acoustic waveform.

IV. Delta Modulation (DM)

The exploitation of signal correlations in DPCM suggests the further possibility of oversampling a signal to increase the adjacent sample correlation C_1, and thus to permit the use of a simple quantizing strategy. Delta modulation (DM), the 1-bit version of DPCM, is precisely such a scheme. In its original form, the DM coder [28] operates on the basis of approximating an input time function by a series of linear segments of constant slope. Such a coder is therefore referred to as a linear delta modulator [29], [30]. We shall use the term "nonadaptive" synonymously with "linear," in anticipation of delta modulators where the slope of the approximating function is variable or "adaptive."

A. Linear (Nonadaptive) Delta Modulation (LDM)

The principle of LDM is illustrated in Fig. 16. The band-limited input signal (bandwidth W) is sampled at a rate f_0 which is much higher than the Nyquist frequency, and a staircase approximation to the input is constructed as follows. One notes, at every sample, the sign b_r of the difference between the input sample X_r and the latest staircase approximation to it, Y_{r-1}, and increments Y by a step Δ in the direction

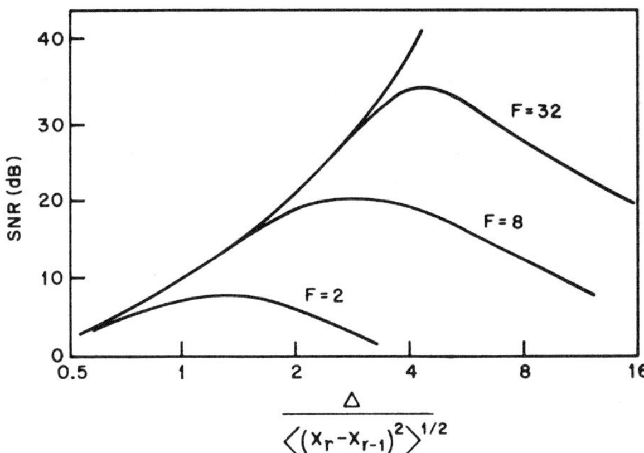

Fig. 17.   Performance curves for LDM. (After Abate [29].)

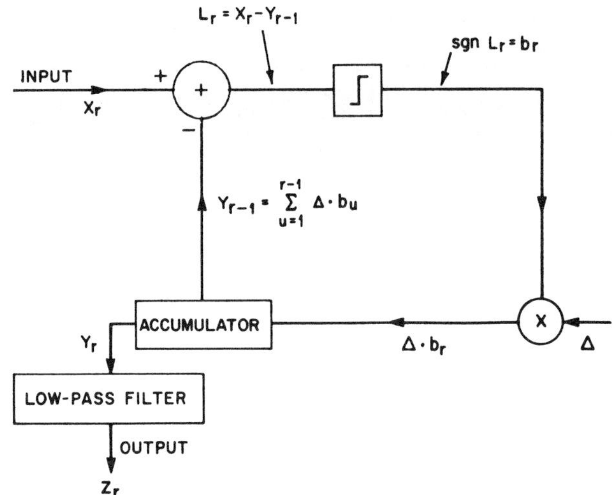

Fig. 18.   Block diagram of LDM.

of b_r. Finally, the high-frequency waveform Y is low-pass-filtered to the original signal bandwidth.

The basic DM principle can therefore be formalized in these equations:

$$b_r = \text{sgn}\,(X_r - Y_{r-1}) \tag{34}$$

$$Y_r = Y_{r-1} + \Delta \cdot b_r. \tag{35}$$

For each signal sample, the transmitted "channel" symbol is the single bit b_r, and the information rate is simply the sampling rate f_0.

Finally, Fig. 16 illustrates two types of quantizing error in DM—slope-overload distortion and granular noise. Slope overload is said to occur when the step size Δ is too small to follow a steep segment of the input waveform. Granularity, on the other hand, refers to a situation where the staircase function Y_r hunts around a relatively flat segment of the input function, with a step size that is too large relative to the local slope characteristics of the input. For given statistics of input signal slope, it is therefore clear that relatively small values of Δ accentuate slope overload, while relatively large values of Δ increase granularity. It should therefore be possible to tailor the (time-invariant) step size Δ to provide a minimum total error power. This optimum (SNR-maximizing) step size Δ_{opt} can be related to the rms value of the first differences in the input signal through a rule-of-thumb formula due to Abate [29]:

$$\Delta_{opt} = \langle (X_r - X_{r-1})^2 \rangle^{1/2} \cdot \ln\,(2F) \tag{36}$$

$$F = f_0/2W. \tag{37}$$

The oversampling index F is generally much greater than 1 for useful DM. Fig. 17 shows the SNR in LDM as a function of step size, with F as a parameter. The results are from a simulation with Gaussian signals with a uniform power spectrum. Notice that for each sampling frequency, the SNR curve peaks (at a corresponding Δ_{opt}). To the left of this maximum, the SNR is mainly controlled by slope-overload distortion, while to the right of the maximum, the contribution of granular noise predominates in the total quantization error variance. Finally, it can be seen from Fig. 17 that the maximum SNR increases approximately as the cube of sampling frequency (resulting in an SNR gain of about 9 dB per octave of frequency increase). This result has also been noted for sinusoidal inputs [28].

As mentioned earlier, the results of Fig. 17 are for a band-limited white Gaussian noise input. Slightly larger SNR values are obtained (because of a greater adjacent-sample correlation for a given F) when one delta modulates a signal with an integrated power spectrum:

$$S(f) = \frac{\lambda}{\arc\tan \lambda} \cdot \frac{1}{W} \cdot \frac{\lambda}{1+(\lambda^2 f^2/W^2)}, \quad \text{for } f < W, \lambda > 1$$

$$= 0, \qquad\qquad\qquad\qquad\qquad \text{for } f > W. \tag{38}$$

The spectrum (38), obtained by passing white noise through an RC integrating network, is a useful model for the long-term power spectrum of speech, if $\lambda = 4$ [29].

For mathematically oriented work on LDM quantization noise, the reader is referred to papers in [31]–[36]. Earlier approaches were based on separate treatments of granular noise (see Van De Weg [31] and Goodman [33]) and slope-overload distortion (see Protanotarios [32] and Greenstein [34]). Simple estimates of total noise power result when one simply adds the variances of granular and overload errors, as explained by O'Neal [35]. But the approach is known to give inexact results and is mathematically nonrigorous. A recent paper by Slepian [36] eschews the overload–granularity distinction and gives exact curves of SNR and Δ_{opt} for a Gaussian signal with a (non-band-limited) rational power spectrum. Finally, one notes that some of the earliest (and rather elegant) results in the subject are the SNR formulas as given by DeJager [28], and Schouten, DeJager, and Greefkes [30] for sinusoidal inputs. It is clear that Gaussian signals (with an integrated power spectrum), as well as 800-Hz sine waves, are tempting models for speech in some instances, and these models have been extensively invoked by DM researchers. Experience has shown, however, that complete assessments of DM coders for speech must necessarily come from simulations and experimentation with real speech inputs, and we feel that this has not been sufficiently emphasized in the literature.

A block diagram of LDM appears in Fig. 18. Note its basic similarity with the DPCM network in Fig. 11. Important differences are the use of a 2-level (1-bit) quantizer in DM and the replacement of a general predictor network with a simple integrator [which, of course, can be regarded as a first-order linear predictor: $a_1 = 1$ in (19) corresponds to per-

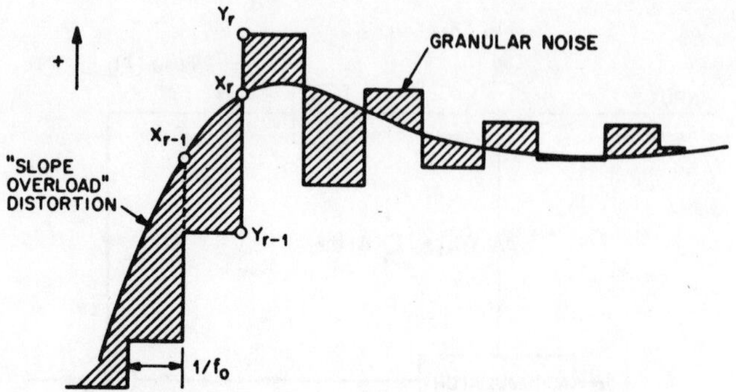

Fig. 19. Adaptive delta modulation (ADM); $Y_r - Y_{r-1} = \Delta_r \cdot b_r$, sgn $[X_r - Y_{r-1}] = b_r$, and $b_r =$ transmitted "channel symbol."

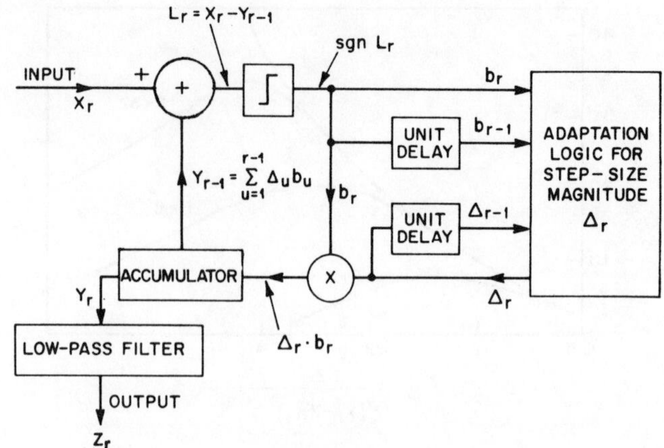

Fig. 20. Block diagram of ADM.

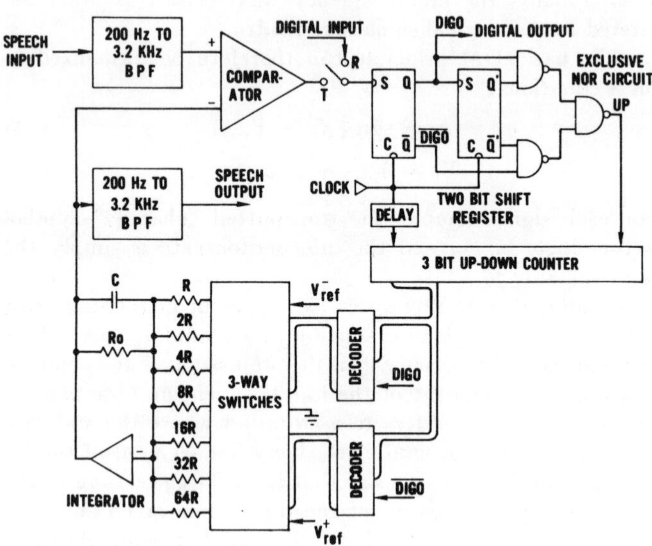

Fig. 21. Realization of ADM with integrated-circuit hardware. (After Cummiskey [23].)

fect integration, and $a_1 < 1$ to sloppy (leaky) integration]. Finally, while desampling filters are employed in DPCM as well, the low-pass filter in the DM circuit of Fig. 18 does the critical job of rejecting the out-of-band quantization noise in the high-frequency staircase function Y, and this has no parallel in DPCM or PCM with Nyquist sampling.

B. Adaptive Delta Modulation (ADM)

As in PCM and DPCM, it is possible to improve the dynamic range of the DM coder by the use of adaptive (variable step-size) techniques [29], [37], [38]. The idea is illustrated in Fig. 19. Here the variable step size Δ_r increases during a steep segment of the input and decreases when the delta modulator is quantizing a slowly varying segment of X waveform (compare this with Fig. 16). The problem in ADM, of course, is to specify suitable rules for step-size variation. Note that the PCM–DPCM adaptation rule [(8) and (9)] does not apply for $B = 1$. This is because, with a 2-level quantizer, the observation of a single sample of quantizer output does not provide any indication of overload or underload (granularity). One needs, on the other hand, to inspect sequences (of length ≥ 2) of quantizer outputs for meaningful step-size adaptations in DM.

In a typical, and conceptually least complex, realization, described by Jayant [38], successive bits b_r and b_{r-1} are compared to detect probable slope overload ($b_r = b_{r-1}$) or probable granularity ($b_r \neq b_{r-1}$). Fig. 19 illustrates how slope overload and granularity tend to correspond, respectively, to occurrences of like and unlike (successive) bits. The specific adaptation rule in [38] is the following:

$$\Delta_r = \Delta_{r-1} \cdot P^{b_r b_{r-1}}, \qquad P \geq 1. \tag{39}$$

The rate of step-size increase (or decrease) is given by a single factor P. Notice that $P = 1$ represents (nonadaptive) LDM. Typically, a value of $P_{opt} = 1.5$ minimizes quantization error power for speech encoding and it can be shown from statistical considerations that the constraint [38]

$$1 < P_{opt} < 2 \tag{40}$$

is important for a wider class of input signals. The value of P_{opt} is not too critical, however, and the adaptation logic obeying (40) provides, in general, a dynamic range much greater than that obtained with LDM. (Recall the sharply tuned maxima in Fig. 17. For ADM, dynamic ranges of 30–40 dB are typical in speech coding [23], [29].) In an illustrative computer simulation with real speech, with $W = 3.3$ kHz and

$f_0 = 60$ kHz, ADM ($P = 1.5$) showed an SNR advantage of more than 10 dB over LDM [38].

Fig. 20 is a block diagram of the exponentially adaptive DM with a one-bit memory (39), and Fig. 21 depicts an integrated-circuit realization of the coder. (The EXCLUSIVE NOR compares b_r and b_{r-1}, and a resulting state of the UP–DOWN counter leads to selection of one of eight values of Δ_r, as determined by the dictionary R, $2R$, $4R$, $\cdots$, $128R$. The polarity of the rth step is of course b_r, as determined by digital output (DIGO).)

Fig. 22 illustrates waveform reconstruction with ADM, and Fig. 23 compares unfiltered ADM and LDM waveforms. The sampling frequency in each case is three times the Nyquist rate, and the signals are time waveforms of formants (time-varying frequencies of resonance in voiced speech). The results are from an application of DM to the formant synthesis of voiced speech at bit rates as low as 500 bits/s [39]. The subject of low-bit-rate speech synthesis, as such, is beyond the scope of this paper. Nevertheless, the waveforms in Figs. 22 and 23 were included because they are quite representative of waveform construction using LDM and ADM. Note again that the quality of reconstruction is critically

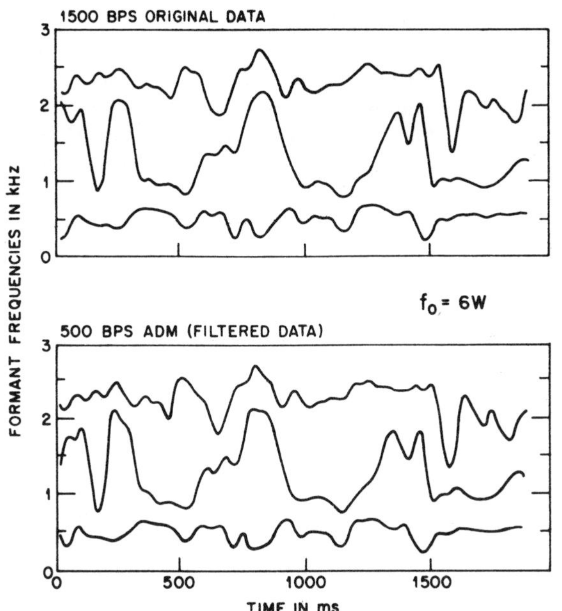

Fig. 22. Waveform coding with ADM; $F = 3$. (After Jayant [38].)

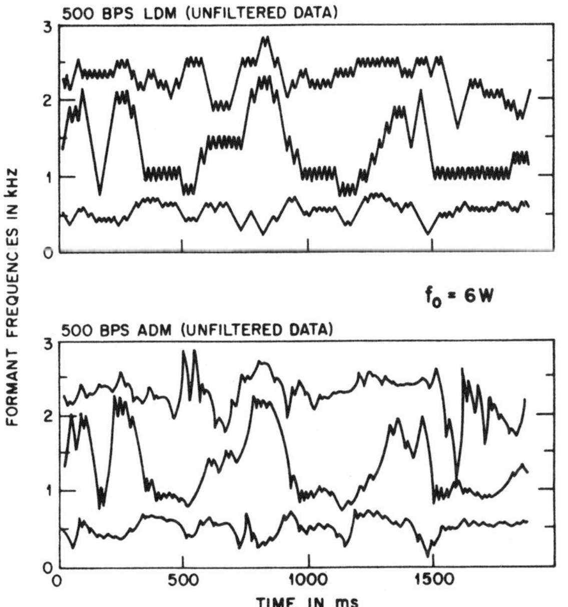

Fig. 23. Comparison of unfiltered waveforms in
LDM and ADM; $F = 3$. (After Jayant [38].)

determined by the oversampling index F. In fact, the value of $F = 3$ used in these figures is often inadequate for satisfactory DM's of the acoustic waveform (Section V).

C. An SNR Formula

The following upper bound has been proposed [40] for the SNR of delta modulators which include a single integrator in the feedback loop (Figs. 18 and 20):

$$ \mathrm{SNR}^U = \theta \frac{F}{1 - C_1(F)}, \qquad \text{for } F \gg 1 \qquad (41) $$

where θ is of the order of unity.

The bound is heuristically derived with the idea that the signal being quantized is basically the first difference $X_r - X_{r-1}$ whose variance is related to that of the speech input X_r [see (12)] via the denominator in (41). The adjacent sample correlation C_1 is expressed as an explicit function of sampling frequecy to emphasize the oversampling of the signal in DM. Further, the quantization noise spectrum is assumed to be flat from zero to one-half the Nyquist rate. The in-band noise ac e)ted by a low-pass filter therefore has a variance which is smaller than the total noise power by a factor equal to the oversampling index. Hence the term F in the numerator of (41).

For several important characterizations of input power spectrum, including the integrated power spectrum in (38), the denominator in (41) is inversely related to F^2 (if F is much greater than one). This implies a cubic law for the SNR bound

$$ \mathrm{SNR}^U \propto F^3, \qquad F \gg 1. \qquad (42) $$

Note that since the bit rate in DM is numerically equal to F, (42) implies that the SNR in DM increases as the cube of the bit rate. This should be contrasted with an exponential performance improvement (5) in PCM and DPCM, where the effect of increased bit rate is reflected in a greater number of quantizer levels, rather than an increased signal-sampling frequency.

The 9-dB-per-octave gain predicted by (42) has been consistently noted in existing DM literature. It is a characteristic of delta modulators with a first-order predictor (single integrator) in the feedback loop. For example, we recapitulate below a well-known peak SNR formula [28] for the DM of a sine wave whose frequency is f_s Hz:

$$ \mathrm{SNR\ (dB)}\Big|_{\mathrm{sine}}^{\mathrm{SI}} = 10 \log_{10}\left[\frac{f_0{}^3}{f_s{}^2 W}\right] - 14 \qquad (43) $$

where SI stands for "single integration," and W is the upper cutoff frequency of the low-pass filter at the coder output.

D. Double Integration (DI)

When the feedback network incorporates double integration, the peak SNR for a sine wave input is derived to be [28]

$$ \mathrm{SNR\ (dB)}\Big|_{\mathrm{sine}}^{\mathrm{DI}} = 10 \log_{10}\left[\frac{f_0{}^5}{f_s{}^2 f_m{}^2 W}\right] - 32 \qquad (44) $$

where DI stands for "double integration," and f_m is the frequency above which DI (second-order prediction) takes place in the feedback loop. Other symbols have the same connotation as in (43). The idea of DI is merely to provide better signal prediction. Double-integrating networks do this by facilitating, in effect, more storage (or input signal memory) for the prediction of the higher frequency components in the signal being coded.

It is not claimed that SNR improvements with DI are proportional to $f_0{}^5$ when the input signal is speech. Nevertheless, the use of DI in the DM of speech can provide a nearly constant overload probability for all components of the average speech spectrum. For this, the frequency above which DI is allowed is typically about 2 kHz, and the design of the appropriate feedback network is described in [41]. The higher order prediction involved in DI raises stability issues [23], [42]. Thus, while DI can be straightforwardly incorporated in linear and syllabically companded delta modulators, it can cause instability when used with a very strong instan-

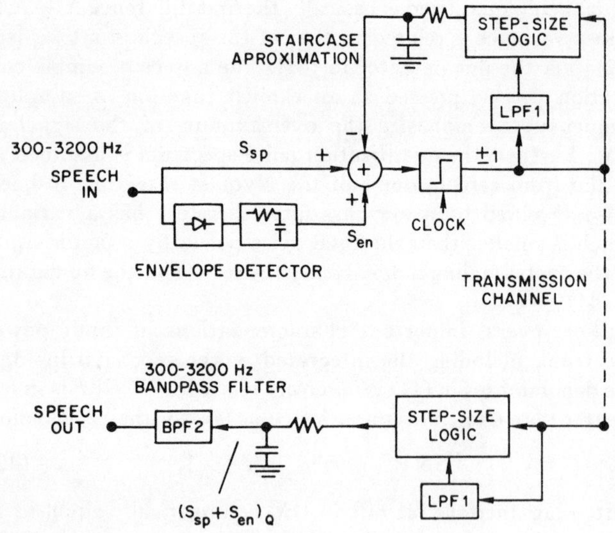

Fig. 24.   Continuous delta modulation (CDM).
(After Greefkes and DeJager [46].)

taneously adaptive logic, such as the one-bit memory scheme in (39). In such cases, the use of DI necessitates the toning down of fast adaptations, either by the use of the adaptation logic in conjunction with delayed encoding, where the coder is allowed to "look ahead" into the input signal [43], or by increasing the length of quantizer memory used for step-size adaptations. The latter technique has been widely used in the ADM of speech [41], [44].

E. Syllabic Adaptation

As mentioned in the context of APCM, syllabic companding [45], [46] provides a very useful adaptation strategy from the point of view of resistance to bit errors. Continuous delta modulation (CDM) was among the first DM techniques that incorporated syllabic companding for step-size adaptations [47]. Unlike the case of an instantaneous compandor (Figs. 20 and 21), the step size is now adapted, more smoothly in time, with a time constant that is in the order of 5-10 ms. A block diagram of the coder appears in Fig. 24. Step-size control is derived (both at the transmitter and at the receiver) from the mean number of ones in the bit stream. This number, in turn, reflects the input signal level through the feed-forward control provided by the low-frequency envelope signal S_{en}. The companding is syllabic because of the low-pass filter LPF1 (typically a 100-Hz filter). The slow adaptation that characterizes syllabic companding (in contradistinction with instantaneous companding) has the effect of decreasing granular noise in the output speech, at the cost of a significant increase of slope-overload distortion. This leads to a certain loss of "crispiness" in syllabically companded speech. The slope-overloading strategy ensures, on the other hand, a "clean-sounding" signal at relatively low bit rates (say, below 25 kbits/s). At such rates, the performance of most instantaneous compandors is seriously impaired by excessive granular noise. Syllabic adaptation also offers, as mentioned earlier, the additional advantage of increased resistance to bit errors.

Digitally controlled delta modulation (DCDM), as proposed more recently by Greefkes and Riemens [48], [41], also incorporates syllabic companding. However, it eschews the need for speech envelope detection at the delta coder, and derives step-size information directly from the bit sequence. In one of the latest versions of DCDM, for example, step-size increases follow the detection of four consecutive bits of the same polarity. At sampling rates below about 16 kHz, the correlation between adjacent signal samples decreases to a point where observations on the bit stream seem to be less useful, for step-size control, than the direct monitoring of the speech envelope. For this reason, CDM is reported to provide better speech coding than DCDM at bit rates below 16 kbits/s.

In concluding our discussion of DM, we must point out that the extreme simplicity of the coding technique has inspired an extent of theoretical and application-oriented work which is too wide to be exhaustively mentioned in this paper. Contributions that come to mind are: a study of the functional type of staircase (Y) function which offers best matches to the acoustic waveform [49]; a DM refinement incorporating slope prediction [50] that waives the basic DM equation (34); a very general theoretical approach to one-bit coding in [51]; a technique for pitch-period detection using delta-modulator bit sequences [26]; and, finally, useful review papers on the subject [52], [2].

V. Comparison of Techniques

A. SNR Data

So far in this paper, the signal-to-quantization-error ratio SNR has been ubiquitously employed as a measure of quantizer performance. We shall now plot SNR as a function of bit rate for three illustrative coders: logarithmic PCM, (6); ADPCM [with adaptive quantizer and a simple first-order predictor, (8) and Figs. 5-9]; and ADM [with a one-bit memory, (39)]. The results are from computer simulations [23], [38], [53] which used, as coder input, bandpass-filtered male utterances of "A lathe is a big tool." Two bandwidths were considered: 200-3200 Hz and 200-2400 Hz. For PCM and ADPCM, the respective sampling frequencies f_0 were 8 kHz and 6.6 kHz. The number of bits per sample B was a variable, and the product Bf_0 determined the bit rate. For ADM, f_0 was a variable and numerically equal to the bit rate. Results are displayed in Fig. 25(a)—200-3200 Hz—and in Fig. 25(b)—200-2400 Hz. The following conclusions emerge from the SNR plots.

a) Notice in both plots that ADPCM has a constant SNR gain over PCM [12 dB in Fig. 25(a) and 8 dB in Fig. 25(b)]. The gain is due, essentially, to the differential encoding advantage in (18) and, as such, is not a function of B.

b) The SNR comparison between ADM and PCM, on the other hand, is dependent on the bit rate. This is because SNR increases as the cube of the bit rate for ADM (42), while the increase is exponential (5) for PCM. The interesting consequence is a crossover point in bit rate [50 kbits/s in Fig. 25(a) and 30 kbits/s in Fig. 25(b)] below which ADM, which was originally conceived for its simplicity, can in fact outperform logarithmic PCM, which is the best established digital coding technique at this time.

c) Finally, ADPCM is always seen to outperform ADM on the basis of SNR; however, the SNR advantage is fairly small at low bit rates, and it should be weighed against the added complexity of a multibit quantizer.

The reason for using a logarithmic companded PCM (rather than adaptively quantized APCM) in our SNR com-

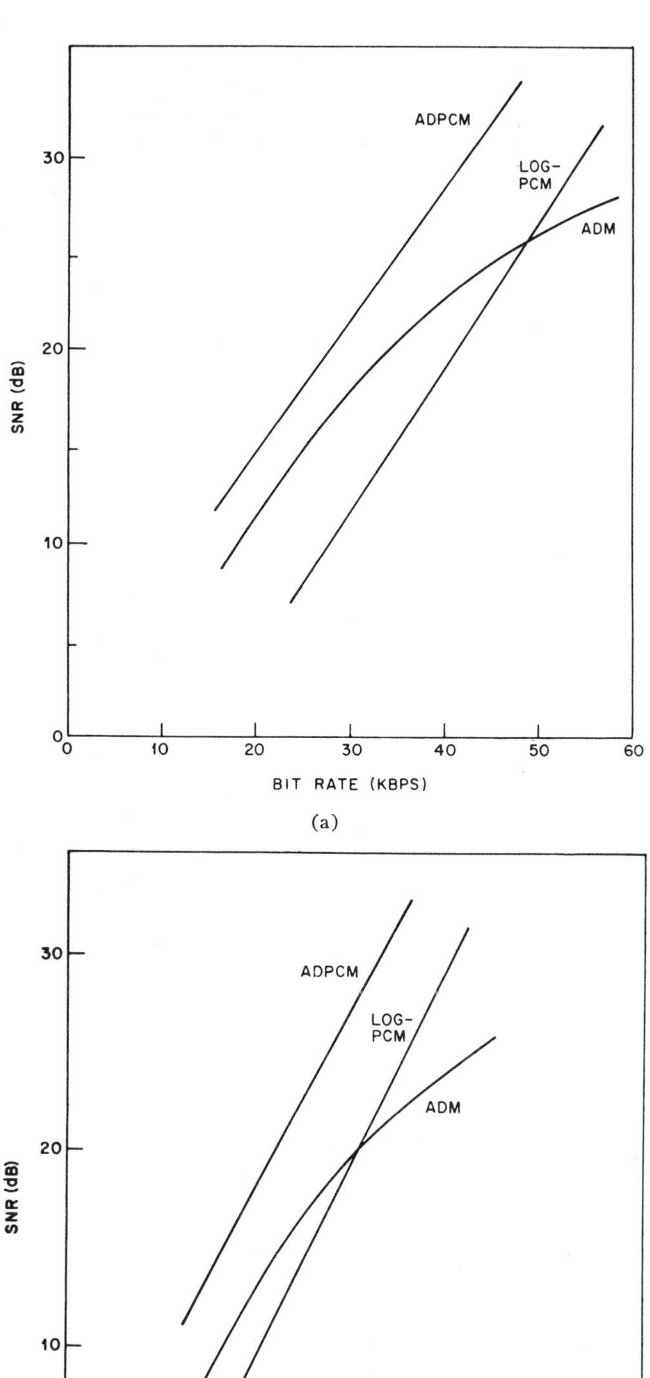

Fig. 25. Comparison of SNR versus bit rate functions. (a) Bandwidth: 200 to 3200 Hz. (b) Bandwidth: 200 to 2400 Hz. (After Rosenberg [53].)

parisons is the fact that log-PCM is very well established and understood, and hence serves as a standard reference in speech quality comparisons. It may be mentioned, however, that our simulations showed APCM to have a very clear SNR ad-

vantage over log-PCM, at least for values of B up to 5. (The APCM simulations did not include values of $B > 5$.) For example, the advantage was 6 dB for $B = 3$.

The plots in Fig. 25 suggest, from an information rate point of view, coder classes which are best for a given specification of SNR. The speech bandwidth in Fig. 25(a) is representative of telephone quality speech, while the lower bandwidth of Fig. 25(b) typifies the type of speech that may be used in a mobile radio or military application.

We must emphasize that the results of Fig. 25, while being representative, should be used only as guidelines of performance for the following reasons.

1) The results are based on the use of a very short and specific speech sample.

2) The coders employed in our simulations are really members of fairly wide classes of differential and nondifferential coders; our coders can therefore provide only estimates of average performance.

3) The coder assessments in specific applications will depend not only on SNR, but on subjective and perceptual factors, on coder tolerance to channel errors, etc. These will be detailed subsequently; however, it may be mentioned at this point that perceptual considerations seem to indicate that the advantage of differential coding (over nondifferential PCM) is even more than what is suggested by the SNR comparisons of Fig. 25. There is a strong suggestion, therefore, that for toll-quality applications (SNR > 30 dB, say), as well as for sub-toll-quality encoding, both ADCPM and ADM offer serious challenges to time-honored log-PCM. The merit of ADPCM is its greater efficiency (consistent SNR gain over PCM), while the strong feature of ADM is its very simple implementation.

A word about nonadaptive techniques. For the sake of illustration, at $f_0 = 56$ kHz, the ADM in our simulation had a 10-dB advantage over LDM, and with $B = 3$, ADPCM had a 3-dB advantage over nonadaptive DPCM; and it is accepted generally that, in PCM, logarithmic companding provides a 4- to 5-bit advantage over uncompanded PCM (on a long-term basis), when encoding speech with toll quality. We have not included any nonadaptive techniques in Fig. 25 because the performances of these techniques, understandably, are strong functions of the specific speech material used as coder input, and the only fair way to assess the three basic coders relatively appeared to be to compare illustrative adaptive versions. The interested reader, however, is strongly recommended to study the nonadaptive coder comparisons in Zetterberg [54] and O'Neal [55]. O'Neal's results are for a Gaussian signal; as such, they are often at variance with results that we have noted for speech. For example, O'Neal's LDM outperforms DPCM when the bit rate is sufficiently reduced. This is not indicated in Fig. 25. Again, O'Neal's DPCM is inferior to an information-theoretically derived optimal coder by a mere 4 dB [55]. For speech coding, the suboptimality of simple DPCM would be much greater; this is because the quasi-periodic nature of the speech signal is not exploited in simple DPCM with short-term predictions ($n \leq 10$, say).

B. Perceptual and Subjective Effects

Inadequacies of SNR as a performance measure have been continually recognized in speech coding literature. We will illustrate the point with examples from PCM, ADPCM, and ADM. In each case, the insufficiency of the SNR measure has to do with the fact that the quantization error sequence has signal-dependent (or signal-correlated) components, and signal-dependent noise (or distortion), with a variance of $\langle E^2 \rangle$, does not have the same annoyance value as indpendent additive noise of equal variance. Consequently, the perceptual quality provided by a quantizer with signal-correlated errors cannot be precisely, or completely, described by the ratio of signal power $\langle X^2 \rangle$ to $\langle E^2 \rangle$.

1) PCM: In discussing the use of dither in PCM, we noted that, if $B \leq 6$, the quantization error waveform has noticeable signal-dependent components which are perceptually annoying, and we described the advantages of redistributing a given error variance into signal-independent patterns by dithering. It is implied, in this case, that SNR overstates the performance of PCM speech in some sense, at low bit rates ($B < 6$).

Signal-dependent errors become significant at high bit rates as well, if nonuniform quantization is employed. Here the step size is proportional to the instantaneous input amplitude, and, since the quantization error E_r has a uniform distribution with a support equal to the step size (1), the magnitude and the variance of E_r are obviously correlated with the input amplitude.

Richards [56] has made perceptual studies of the relative annoyance values of uncorrelated (additive) and amplitude-correlated noise, and reports the following equivalence between SNR_A, the speech-to-additive noise ratio, and SNR_C, the speech-to-amplitude-correlated noise ratio:

$$\mathrm{SNR}_A \text{ (dB)} = 0.062[\mathrm{SNR}_C \text{ (dB)}]^2$$
$$+ 0.285[\mathrm{SNR}_C \text{ (dB)}] + 1.07. \quad (45)$$

Padula [57], in a recent unpublished work, proposes a similar equivalence curve:

$$\mathrm{SNR}_A \text{ (dB)} = 0.027[\mathrm{SNR}_C \text{ (dB)}]^2$$
$$+ 1.26[\mathrm{SNR}_C \text{ (dB)}] - 5.08. \quad (46)$$

Note the following specific equivalences, as determined from (46):

$$\mathrm{SNR}_C = \ 0 \text{ dB} \leftrightarrow \mathrm{SNR}_A = -5 \text{ dB} \quad (47)$$

$$\mathrm{SNR}_C = 20 \text{ dB} \leftrightarrow \mathrm{SNR}_A = 31 \text{ dB}. \quad (48)$$

The first of these results says that at low values of SNR (at low values of B), amplitude-correlated noise is more objectionable than additive noise of the same variance. This tends to support the perceptual effects mentioned in the dither problem. Notice, however, that undithered quantization noise, although signal dependent, is not strictly amplitude correlated in the sense that the noise in nonuniform quantization is. Also, our argument unfortunately falls apart if one prefers to follow (45) in lieu of (46)!

At higher values of SNR (at higher values of B), amplitude-correlated noise is less annoying than additive noise of equal variance, according to (48), as well as according to (45). This implies the observation that amplitude-correlated noise is less easily perceived than additive noise when the noise levels under consideration are below a threshold given by setting $\mathrm{SNR}_A = \mathrm{SNR}_C$ in an equation like (46) or (45).

2) DPCM: It has been noted [15] that the quality of speech output in an ADPCM coder, relative to log-PCM quality, is better than what is predicted by the SNR curves in Fig. 25. This is demonstrated in Fig. 26 which compares the results of a preliminary perceptual experiment with con-

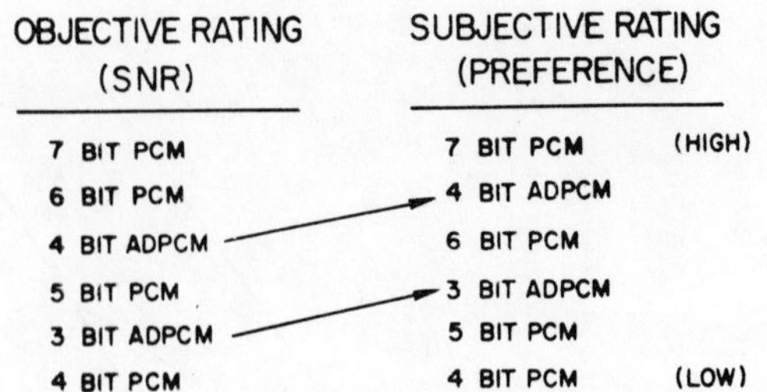

Fig. 26. Comparison of objective and subjective performance of ADPCM and log-PCM. (After Cummiskey [23].)

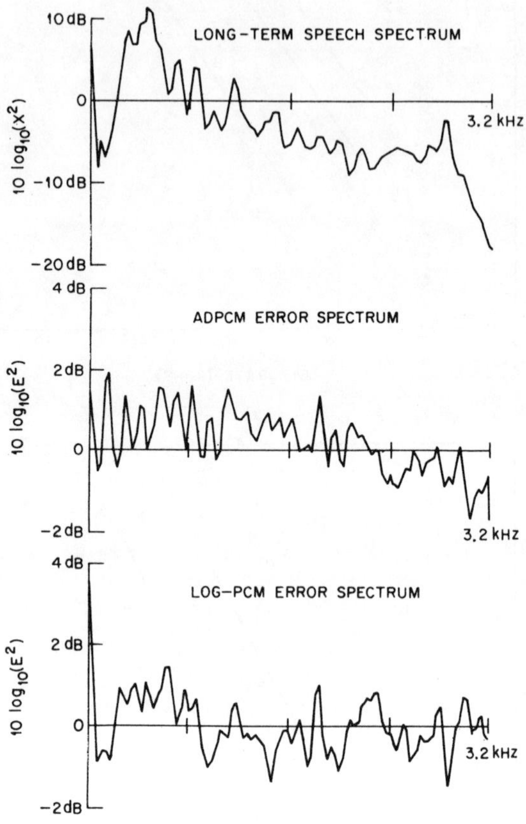

Fig. 27. Long-term error spectra in ADPCM and log-PCM. (After Cummiskey [23].)

clusions drawn from SNR data. It is seen, for example, that 4-bit ADPCM is subjectively assessed to be better than 6-bit log-PCM, although the SNR for the latter is actually about 1.5-dB higher, according to Fig. 25(a). Similarly, perceptual factors seem to promote 3-bit ADPCM to a rank higher than that of 5-bit PCM.

The difference between objective (SNR-based) and subjective assessments is attributed to the following perceptual effects.

1) The quantization error (power) spectrum in ADPCM is matched to the long-term spectrum of speech itself, in that it drops at the high-frequency end. The error spectrum in log-PCM, on the other hand, is relatively white. This comparison is demonstrated in Fig. 27. The perceptual effect involved is that ADPCM errors have a greater proportion of low-frequency (slope-overload) distortion and, therefore, are more correlated with speech than with PCM noise.

2) There is a certain pitch-related buzziness in ADPCM noise. Notice the vertical striations in the ADPCM error

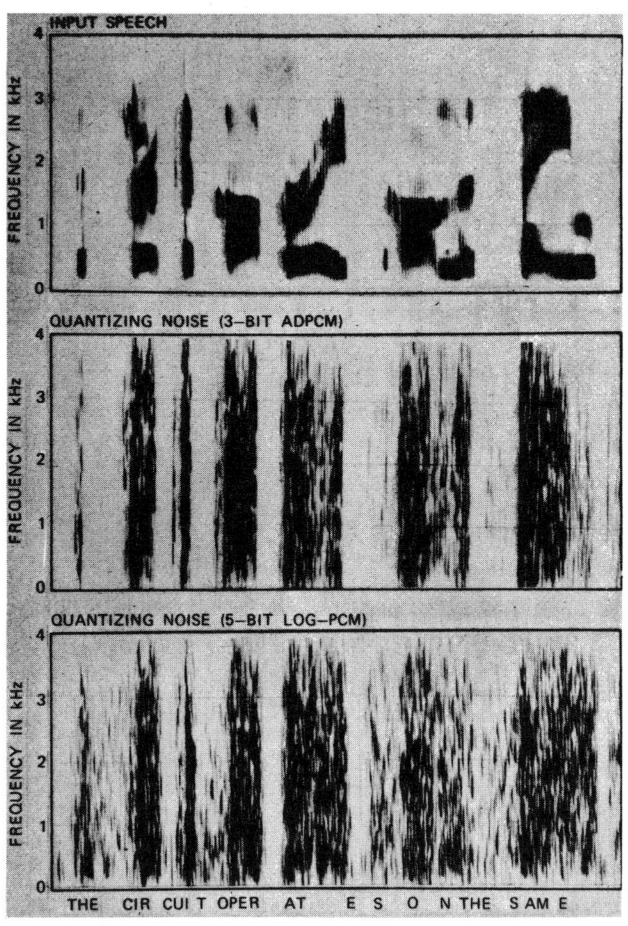

Fig. 28. Spectrograms of speech and quantization error in PCM and log-PCM. (After Cummiskey [23].)

spectrogram in Fig. 28. Pitch striations in the PCM error spectrogram are not quite as well marked.

3) The suppression of quantizing noise during the silent intervals of speech is better in ADPCM than in log-PCM. This is also demonstrated in the spectrograms of Fig. 28.

It is not clear how much of effects 1) and 2) are peculiar to the specific ADPCM coder used, but it is apparent that the better idle-channel noise suppression in ADPCM [effect 3)] is directly attributable to the use of an adaptive differential quantizer. (Although the maximum-to-minimum step-size ratio of 100 in the ADPCM matched the value $\mu = 100$ of the logarithmic PCM, the fact that the quantizer in ADPCM has to handle the first differences in speech, rather than the speech amplitudes themselves, means that the ADPCM can quantize the zero amplitudes in silent intervals with much smaller nonzero amplitudes than the PCM quantizer can.)

The point of such observations is that even if an ADPCM quantizer had the same SNR as a log-PCM quantizer, the errors in the former have a better, signal-correlated, distribution in time; this makes the same error variance less objectionable in ADPCM.

3) DM: In delta modulation, signal-correlated errors take the specific form of temporal bursts of slope-overload distortion. In ADM studies at 20 and 40 kHz, it has been noted that although slope-overload distortion constitutes the major component of total error power, the perceptual quality of ADM is largely controlled by granular errors [58]. For example, consider the use of $P = 1.2$ instead of the SNR-maximiz-

ing $P = 1.5$ in the step-size adaptation for the ADM with a one-bit memory at 20 kHz (39). The result is a 40-percent increase in overload noise power and a 1-dB decrease in SNR. However, the granular noise, which constitutes less than 2 percent of the total noise power, decreases by 30 percent when $P = 1.2$, and this makes the speech perceptually more pleasing! In other words, slope-overload distortion is much less annoying at the bit rate in question than signal-uncorrelated granular noise; by the same token, the overload distortion is less annoying than additive noise of equal variance. It is tempting to contrast this trend with an earlier observation (48) regarding the lesser objection to correlated noise. Note, however, that the latter observation had been made on the basis of studies with amplitude-correlated, rather than slope-correlated, noise.

In any case, the point of this section is that, at least at bit rates above 20 kbits/s, SNR results are believed to provide a conservative estimate of DM performance. The relative acceptability of slope-overload distortion has been exploited in delta-modulator design in many ways. One of them is the recommendation of "slow" adaptation strategies (for example $P = 1.2$ instead of 1.5 in recent discussion). Another is the recommendation of error power spectra that roll off at higher frequencies (signifying greater acceptance of low-frequency distortion). Finally, the perceptibility of slope-overload distortion, as such, has been the subject of a very interesting formal investigation [59].

Having discussed the insufficiency of SNR as a performance measure for PCM, DPCM, and DM quantizers, one is faced with the problem of developing a single uniform performance criterion that would measure all quantizers in these three classes. We are not suggesting at all that the problem has a simple solution. After all, quantizing errors take on special forms in each case, and perhaps the best that can be done by way of unifying performance results would be the specification of as many (distinct) equivalent SNR equations, like (45) and (46), as there are (distinct) types of signal-correlated errors: amplitude-correlated errors, slope-correlated errors, and envelope-correlated errors (in some syllabically companded systems), to mention a relevant few. Such an approach would evidently be exceedingly impractical when one is involved with large populations of different kinds of coders.

Besides, it is doubtful if such approaches can provide reliable indications over different levels of perceptual distortion: for example, distortion that is just perceptible, distortion that is annoying, and distortion that reduces intelligibility. The second of these levels is subject to substantial individual differences [see Section V-D] which are evidently unreflected in an SNR-equivalence curve. All in all, there seems to be no genuine alternative at this point to the brute-force technique of actually testing out a given set of coders in elaborate perceptual experiments (involving direct comparisons of all possible coder pairs). One such perceptual experiment is being carried out, with the coders of Fig. 25 (at different bit rates) as stimuli [53].

At least two formal techniques have been discussed in the literature for interpreting the results from perceptual tests involving results of paired ($A-B$) comparisons. We shall now discuss these procedures very briefly.

C. Isopreference Curves

Fig. 29 illustrates the results from a perceptual test involving DPCM stimuli [60]. The input speech was low-pass-

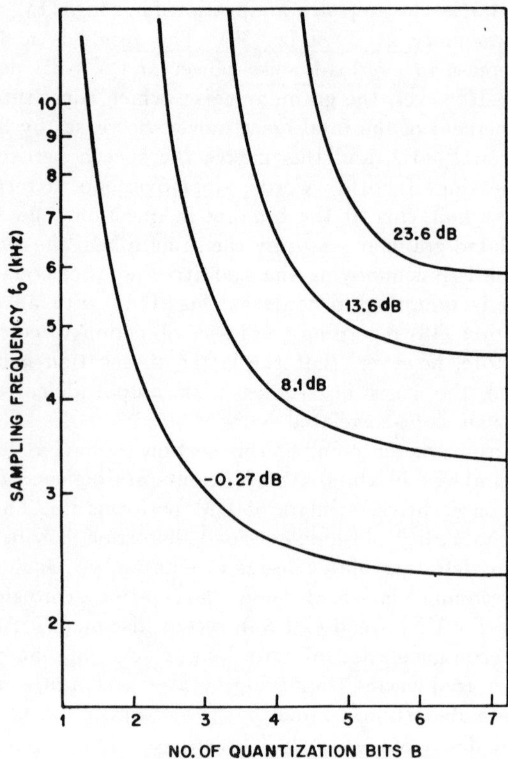

Fig. 29. DPCM Isopreference contours.
(After Chan and Donaldson [60].)

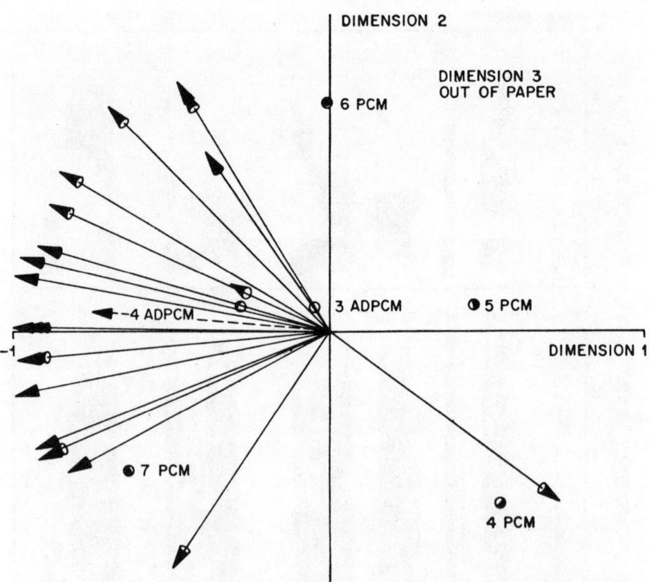

Fig. 30. Subjective preference judgments of various ADPCM and log-PCM codings. Dimensions 1 and 2 account for most of intersubject variance. Increasing preference is in the $-x$ direction. Individual subject vectors are plotted, and projection of coding conditions onto a subject's vector indicates how that individual rank-ordered the coding system. (After Cummiskey, Jayant, and Flanagan [15].)

filtered to 4 kHz, the quantizer used was logarithmic, and the results were from preference judgments made by a pool of 17 listeners. The variable coder parameters were the sampling rate f_0 and the number of bits B used to code each sample. Fig. 29 illustrates the effects of f_0 and B on the subjective quality of DPCM speech by means of isopreference contours that connect points of equal subjective quality. Two coders are isopreferent if in a (statistically representative) listener population, one-half of the listeners prefer one code to the other. Thus the construction of isopreference curves involves forced-choice $A-B$ comparisons of test and reference stimuli, followed by determination of 50-percent probability (of preference) points on appropriate psychometric functions [60], [61]. The isopreference curves in Fig. 29 have also been calibrated (in terms of the SNR of a white-noise-degraded speech signal, judged equal in quality to the reference signal associated with a given isopreference curve). Finally, the dotted line in Fig. 29 is a locus of points of minimum channel capacity (bit rate). It is seen that Nyquist sampling is not too inefficient for high-quality DPCM codes if one operates at a point (in a $[f_0, B]$ plane) that is not far from the minimum channel capacity locus.

A weakness of the isopreference method is that individual differences in judgment are not displayed in the end result. There are situations where one desires to preserve such differences and to distinguish between "perceptual" and "subjective" factors, so to speak. (In fact, the same individual may change his preference judgment while evaluating a coder for different applications: for example, a transatlantic personal call versus local time-information service.) Another subtlety that is inherently excluded in the isopreference display is the possibility of "nontransitive" preference patterns: (stimulus) A is preferred to B, B is preferred to C; but A is not preferred to C. We shall now illustrate a procedure which eschews the transitivity assumption and, furthermore, preserves individual differences in the final preference display.

D. Multidimensional Scaling

We shall demonstrate the technique with the example of a perceptual test mentioned earlier. The experiment involved 3- and 4-bit ADPCM stimuli and 4-, 5-, 6-, and 7-bit log-PCM stimuli [13]. The total number of cross comparisons possible was 16 (2 stimuli $\times$ 4 stimuli $\times$ 2 orders of presentation). Twenty-two listeners participated in the tests and made preference judgments of signal quality for each of the 16 $A-B$ comparisons. The judgments were submitted to a multidimensional scaling program [62], and the results were plotted in terms of two subjective dimensions which accounted for most of the perceived differences. One dimension, in particular, accounted for 75 percent of the variance in the preference data.

The end results are shown in Fig. 30. Individual subject vectors are displayed (solid lines), and projection of the coding conditions onto a subject's vector reveals how that individual rank-ordered the signal qualities. (One subject, the vector in quadrant IV, apparently misunderstood the test instructions and gave essentially complementary preference judgments.) A resultant of the subject vectors is also shown (dashed), and projections onto this resultant indicate subject consensus in ranking the qualities. This averaged subjective ranking corresponds, indeed, to the subjective ranking listed earlier in Fig. 26.

E. Communication Questions

Hitherto we have discussed quantizers primarily as source encoders. While transmission questions are not the primary issue of this paper, we would like to mention some basic considerations. Some of these can be quite relevant to the problem of selecting a source encoder (PCM, DPCM, or DM). In fact, integrated speech communication systems of the future are likely to use all of the three basic quantizers discussed in this paper, in both nonadaptive and adaptive formats. Therefore, there has been considerable interest in questions such as direct digital conversions among such formats [63]–[65], and repeated encoding of a speech signal by identical [66] or nonidentical [64] quantizers in tandem.

For example, in a simulation of identical encoder–decoder pairs in tandem, the following relations were noted for the degradation of SNR with the number I of encoder–decoder pairs [66]:

$$\text{SNR}\,(I)\big|_{\text{PCM}} = \text{SNR}\,(1)\big|_{\text{PCM}} - 10\log I \qquad (49)$$

$$\text{SNR}\,(I)\big|_{\text{ADM}} = \text{SNR}\,(1)\big|_{\text{ADM}}$$
$$- 10\log\,(0.9 + 3.5I + 1.4I^2) \qquad (50)$$

$$\text{SNR}\,(I)\big|_{\text{LDM}} = \text{SNR}\,(1)\big|_{\text{LDM}}$$
$$- 10\log\,(3.15 + 0.35I^2). \qquad (51)$$

The delta modulations were performed at 60 kbits/s.

It should be mentioned at this point that although adaptive techniques have been emphasized throughout this paper for efficient speech coding, nonadaptive quantizers are far from being only academic in importance. A very good case in point is LDM. In situations where the need is for extreme simplicity, and the sampling frequency is not seriously limited (for example, in communication terminals, rather than in transmission channels), LDM provides a very attractive means of analog-to-digital conversion [67].

Direct conversion between LDM and ADM formats has been considered in at least one simulation [64]. Here speech was encoded twice by a high-bit-rate LDM encoder with an intermediate stage of ADM encoding. The LDM-to-ADM and ADM-to-LDM conversions were performed on the respective high-frequency staircase functions, and the speech was not detected (low-pass-filtered) until after the second LDM. It was found that with an ADM bit rate of about 40 kbits/s, the LDM sampling frequency had to be in the order of 1 MHz in order not to degrade speech beyond a quality that 40-kHz ADM could provide.

F. Effects of Random Channel Errors

Many toll-quality speech links maintain bit-error rates which are too small (say, less than 1 in 10^5) to affect quantizer performance. If the error probability should be much higher, SNR degradations are implied, and the extent of these degradations is clearly quantizer-dependent.

DPCM quantizers, for example, are affected differently by bit errors than a PCM quantizer. This is because the feedback action in DPCM leads to a certain propagation of errors in the coder output [68]. Notwithstanding such error propagation, DPCM has been reported to be more tolerant to (randomly occurring) bit errors than PCM, from a perceptual viewpoint [69], [70]. This is because error spikes caused in the reconstruction of a PCM waveform (due to a wrongly received bit) can have maximum amplitudes which are in the order of $|X_r|_{\max}$, the peak value of the input signal, while corresponding spike magnitudes in DPCM decoding are really related to $|X_r - X_{r-1}|_{\max}$, the peak value of the first differences in the input. The consequent greater magnitude of a typical PCM error spike makes it more annoying, in spite of the fact that it does not propagate in time.

For the same reason, delta modulators can be designed for greater channel error resistance than PCM (from a perceptual viewpoint), in spite of an error propagation effect present in DM [41]. The higher sampling rate of DM also contributes

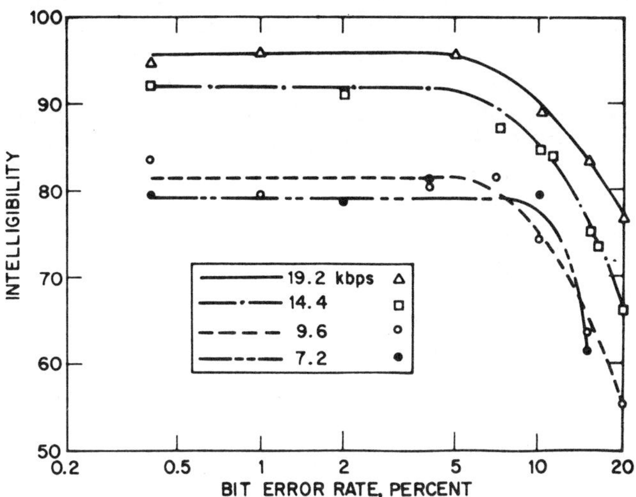

Fig. 31. Intelligibility performance of variable slope delta modulation. (After Melnick [71].)

to this perceptual advantage (when comparing equal bit-rate DM and PCM with a given percentage of digital errors), and the error propagation problem in DM is also alleviated in large measure by the use of an imperfect integrator [38]. Finally, within the class of DM coders, instantaneously adapting delta modulators are more vulnerable to channel noise than slowly companding (or linear) coders. Therefore, although instantaneously adapting delta modulators are very simple and efficient (SNR-wise) in relatively noise-protected environments (say, bit-error probability $<10^{-4}$), a number of syllabically companded delta modulators have been designed for use over noisy channels, and these coders are particularly attractive for low-bit-rate (<20 kbits/s) applications. We hope to summarize the performance of such delta modulators by reproducing, in Fig. 31, the recently published intelligibility curves due to Melnick [71]. The intelligibility scores refer to word intelligibility, as measured in modified rhyme tests (MRT) [72]. The MRT is a multiple-choice test in which the listener selects the word he feels is nearest to one which he has heard from a closed set of six possible choices (differing in either the initial or the final consonant sound). In interpreting word-intelligibility scores, it should be remembered that, in many situations, perfect sentence intelligibility does not require 100-percent word intelligibility. For example, acceptable MRT scores for air traffic control purposes can be as low as 75 percent [71]. Notice, in Fig. 31, that such a requirement is satisfied by a 7.2-kbits/s (syllabic) coder even when the bit error rate is as high as 10 percent. Comparable performance figures are claimed by the recently described "two-channel delta modulation" due to Greefkes [73]. The system has an auxiliary delta modulator to transmit a slope-envelope signal that provides a useful basis for step-size control. (As mentioned in the context of CDM and DCDM, short-term properties of the DM bit sequence are not expected to be a very useful basis for step-size variation at sampling frequencies as low as 7.2 kHz. At these sampling rates, the correlation between adjacent speech samples is too small to regard snapshots of short bit sequences as indicators of immediately following slope patterns.)

G. Clustered Bit Errors

Our foregoing observations, especially the quantitative ones, refer to independently (randomly) occurring bit errors.

Such errors are characteristic of signal channels corrupted by signal-independent additive noise. Many channels fit this description and the independent-error assumption has accordingly been made in most of the quantizer literature. A notable exception is the analytical treatment by Wolf [74]. This includes a study of errors occurring in bursts, as generated by a first-order Markov process. A typical real channel with clustered errors is a mobile radio link. The slow signal fading (colored multiplicative noise) over such a channel causes bit-error patterns in which temporal correlations are usually very obvious.

The effect of correlated errors on digital speech codes is being currently investigated. One of the earlier indications of these studies has been that, in spite of the strong bunching of the errors, differential codes such as ADM can be designed to provide a surprising amount of impulse-noise suppression. Unfortunately, however, such designs invariably seem to necessitate the operation of the delta modulator in a slope-overloading mode [75].

Appendix*

In the following description of the gramophone record, dots and dashes refer to tone signals used as file markers.

Band 1: Linear (uncompanded) PCM (*FS*: sampling frequency in kHz; *B*: bits of quantization per sample) [· · · · · · · ·] [9 versions of a 4-kHz-bandwidth utterance, corresponding to the following combinations of (*FS, B*): (10, 12), (10, 9), (10, 4), (10, 2), (10, 1), and (10, 12), (5, 12), (2.5, 12), (1.25, 12)]

Band 2: DPCM (*I*: information rate in kbits/s; *N*: number of predictor coefficients) [· · · ·] [$I = 38.4$, $N = 1$, $FS = 9.6$, $B = 4$] [—] [$I = 25.6$, $N = 1$, $FS = 6.4$, $B = 4$] [—] [$I = 19.2$, $N \gg 1$ (pitch adaptive), $FS = 8$]

Band 3: DM (LDM: linear DM; DCDM: digitally controlled DM; ADM: adaptive DM with a 1-bit memory. Numbers in parentheses are the sampling frequency *FS* in kHz, number of integrators *NI* in the feedback network, signal level *SL* (relative to most desirable value) in dB, in the following order: (*FS, NI, SL*)) [— ·] [LDM(40, 1, 0)] [· ·] [LDM(40, 1, −20)] [· · ·] [LDM(40, 2, −20)] [—] [DCDM (40, 2, −20)] [— — — — ·] [ADM(60, 1)] [ADM(40, 1)] [ADM(20, 1)]

Band 4: Coder comparisons (ADPCM uses adaptive quantization with a 1-word memory. Companding for PCM is logarithmic, $\mu = 100$. Information rate *I* is in kbits/s). [— ·] [input] [log-PCM ($I = 48$)] [ADPCM($I = 48$)] [ADM ($I = 48$)] [— · ·] [log-PCM($I = 32$)] [ADPCM($I = 32$)] [ADM($I = 32$)] [DCDM($I = 32$)] [— · · ·] [input] [ADPCM ($I = 16$)] [ADM($I = 16$)] [DCDM($I = 16$)]

Band 5: Perceptual factors (*E*: quantization error signal in ADM; *B*: bits/sample in linear PCM; $D = 1$ for dithered PCM; $D = 0$ for undithered PCM) [— — · ·] [$E(I = 56)$] [$E(I = 40)$] [$E(I = 24)$] [$E(I = 8)$] [input to ADM] [· · ·] [PCM($B = 4$, $D = 0$)] [PCM($B = 4$, $D = 1$)] [PCM($B = 3$, $D = 0$)] [PCM($B = 3$, $D = 1$)]

Band 6: Effects of random bit errors (numbers in parentheses are information rate *I* in kbits/s, and bit-error probability PER in following format: (*I*, PER)) [— — — — ·] [piecewise linear PCM(56, 0.001)] [· ·] [DCDM(56, 0.001)] [· · ·] [PCM(56, 0.01)] [· · · ·] [DCDM (56, 0.01)] [— · ·] [DCDM(32, 0)] [·] [DCDM(32, 0.01)] [·] [DCDM(32, 0.1)] [—] [2-channel DM(7, 0.01)] [· — · — ·]

Band 7: Effects of signal fading in a typical mobile radio link using: [24-kHz frequency modulation, signal-to-channel-noise ratio $= 9$ dB], [24 kbits/s ADM, average bit-error probability $= 0.03$] [· — · —]

Acknowledgment

The author wishes to thank Dr. L. Rabiner for prompting him to write this paper after the author's lecture in an in-hours course program at Bell Laboratories. The paper owes its present form to helpful comments from Dr. Rabiner, Prof. H. Levitt of the City University of New York, and an unknown reviewer. It is a privilege to acknowledge the cooperation of Prof. J. A. Greefkes and Dr. J. A. M. deBrouwer (Technische Hogeschool, Eindhoven, The Netherlands), Dr. K. Riemens (Philips Research Laboratories), and Dr. J. W. Bayless (Defense Communications Agency, Virginia) in the preparation of the demonstration record. Finally, a great deal of the author's research effort reported in this paper was made possible by invaluable encouragement from Dr. J. Flanagan and the excellent programming assistance of Mrs. Kathie Shipley.

References

[1] J. L. Flanagan, "Focal points in speech communication research," *IEEE Trans. Commun. Tech.*, vol. COM-19, pp. 1006–1015, Dec. 1971.

[2] H. R. Schindler, "Delta modulation," *IEEE Spectrum*, vol. 7, pp. 69–78, Oct. 1970.

[3] A. H. Reeves, French Patent 852 183, Oct. 3, 1938.

[4] B. M. Oliver, J. R. Pierce, and C. E. Shannon, "The philosophy of PCM," *Proc. IRE*, vol. 36, pp. 1324–1331, Oct. 1948.

[5] H. S. Black, *Modulation Theory.* Princeton, N. J.: Van Nostrand, 1953.

[6] P. F. Panter and W. Dite, "Quantization distortion in pulse-count modulation with non-uniform spacing of levels," *Proc. IRE*, vol. 39, pp. 44–48, Jan. 1951.

[7] L. G. Roberts, "Picture coding using pseudo-random noise," *IRE Trans. Inform. Theory*, vol. IT-8, pp. 145–154, Feb. 1962.

[8] N. S. Jayant and L. R. Rabiner, "The application of dither to the quantization of speech signals," *Bell Syst. Tech. J.*, pp. 1293–1304, July–Aug. 1972.

[9] L. R. Rabiner and J. A. Johnson, "Perceptual evaluations of the effects of dither on low bit-rate PCM systems," *Bell Syst. Tech. J.*, pp. 1487–1494, Sept. 1972.

[10] B. Smith, "Instantaneous companding of quantized signals," *Bell Syst. Tech. J.*, pp. 653–709, 1957.

[11] M. D. Paez and T. H. Glisson, "Minimum mean squared-error quantization in speech, PCM and DPCM systems," *IEEE Trans. Commun.*, vol. COM-20, pp. 225–230, Apr. 1972.

[12] J. Max, "Quantizing for minimum distortion," *IRE Trans. Inform. Theory*, vol. IT-6, pp. 7–12, Mar. 1960.

[13] N. S. Jayant, "Adaptive quantization with a one-word memory," *Bell Syst. Tech. J.*, pp. 1119–1144, Sept. 1973.

[14] W. Schlink, "A redundancy reducing PCM system for speech signals," in *Proc. Int. Zurich Seminar Integrated Systems for Speech, Video and Data Communications*, Mar. 1972, pp. F41-1–4.

[15] P. Cummiskey, N. S. Jayant, and J. L. Flanagan, "Adaptive quantization in differential PCM coding of speech," *Bell Syst. Tech. J.*, pp. 1105–1118, Sept. 1973.

[16] R. W. Stroh, "Optimum and Adaptive Differential PCM," Ph.D. dissertation, Polytech. Inst. Brooklyn, Farmingdale, N. Y., 1970.

[17] R. M. Wilkinson, "An adaptive pulse code modulator for speech," in *Proc. IEEE Int. Conf. Communications* (Montreal, Canada, June 1971), pp. 1-11–1-15.

[18] C. C. Cutler, "Differential quantization of communications," U. S. Patent 2 605 361, July 29, 1952.

[19] P. Elias, "Predictive coding," *IRE Trans. Inform. Theory*, vol. IT-1, pp. 16–33, Mar. 1955.

[20] R. A. McDonald, "Signal-to-noise and idle channel performance of DPCM systems—Particular application to voice signals," *Bell Syst. Tech. J.*, pp. 1123–1151, 1966.

[21] K. Nitadori, "Statistical analysis of DPCM," *J. Inst. Electron. Commun. Eng. Jap.*, pp. 4–6, 1965.

[22] P. Noll, "Non-adaptive and adaptive DPCM of speech signals," *Overdruk uit Polytech. Tijdschr. Ed. Elektrotech./Eletron.* (The Netherlands), no. 19, 1972.

[23] P. Cummiskey, "Adaptive differential pulse-code modulation for

*The record described is not included in this reprint volume but is available on request from N. S. Jayant, Bell Laboratories, Murray Hill, NJ 07974.

speech processing," Ph.D. dissertation, Newark College of Engineering, Newark, N. J., 1973.

[24] B. S. Atal and M. R. Schroeder, "Adaptive predictive coding of speech signals," *Bell Syst. Tech. J.*, pp. 1973–1986, 1970.

[25] E. E. David and H. S. McDonald, "Note on pitch-synchronous processing of speech," *J. Acoust. Soc. Amer.*, pp. 1261–1266, Nov. 1956.

[26] A. H. Frei, H. R. Schindler, P. Vettiger, and E. Von Felten, "Adaptive predictive speech coding based on pitch-controlled interruption/reiteration techniques," in *Proc. IEEE Int. Conf. Communications* (Seattle, Wash., June 1973), pp. 46-12–46-16.

[27] F. W. Mounts, "A video encoding system with conditional picture-element replenishment," *Bell Syst. Tech. J.*, vol. 4A, pp. 2545–2554, Sept. 1969.

[28] F. DeJager, "Delta modulation, a method of PCM transmission using a 1-unit code," *Philips. Res. Rep.*, pp. 442–466, Dec. 1952.

[29] J. E. Abate, "Linear and adaptive delta modulation," *Proc. IEEE*, vol. 55, pp. 298–308, Mar. 1967.

[30] J. S. Schouten, F. DeJager, and J. A. Greefkes, "Delta modulation, a new modulation system for telecommunications," *Philips Tech. Rep.*, pp. 237–245, Mar. 1952.

[31] H. Van De Weg, "Quantization noise of a single integration delta modulation system with an N-digit code," *Philips Res. Rep.*, pp. 367–385, Oct. 1953.

[32] E. N. Protanotarios, "Slope-overload noise in differential PCM systems," *Bell Syst. Tech. J.*, pp. 2118–2161, 1967.

[33] D. J. Goodman, "Delta modulation granular quantization noise," *Bell Syst. Tech. J.*, pp. 1197–1218, May–June 1969.

[34] L. J. Greenstein, "Slope overload noise in linear delta modulators with Gaussian inputs," *Bell Syst. Tech. J.*, pp. 387–422, Mar. 1973.

[35] J. B. O'Neal, Jr., "Delta modulation quantizing noise: Analytical and computer simulation results for Gaussian and television input signals," *Bell Syst. Tech. J.*, pp. 117–142, Jan. 1966.

[36] D. Slepian, "On delta modulation," *Bell Syst. Tech. J.*, pp. 2101–2137, Dec. 1972.

[37] M. R. Winkler, "High information delta modulation," in *IEEE Int. Conv. Rec.*, pt. 8, pp. 260–265, Mar. 1963.

[38] N. S. Jayant, "Adaptive delta modulation with a one-bit memory," *Bell Syst. Tech. J.*, pp. 321–342, Mar. 1970.

[39] ——, "Delta modulation of pitch, formant and amplitude signals for the synthesis of voiced speech," *IEEE Trans. Audio Electroacoust.*, vol. AU-21, pp. 135–140, June 1973.

[40] ——, "Characteristics of a delta modulator," *Proc. IEEE* (Lett.), vol. 59, pp. 428–429, Mar. 1971; also unpublished work.

[41] J. A. Greefkes and K. Riemens, "Code modulation with digitally controlled companding for speech transmission," *Philips Tech. Rev.*, vol. 31, no. 11/12, pp. 335–353, 1970.

[42] P. T. Nielsen, "On the stability of a double integration delta modulator," *IEEE Trans. Commun. Technol.*, vol. COM-19, pp. 364–366, June 1971.

[43] C. C. Cutler, "Delayed-encoding: Stabilizer for adaptive coders," *IEEE Trans. Commun. Technol.*, vol. COM-19, pp. 898–907, Dec. 1971.

[44] I. M. McNair, Jr., "Subscriber-loop multiplexer—A high pair gain system for upgrading and growth in rural areas," *IEEE Trans. Commun. Technol.*, vol. COM-19, pp. 523–527, Aug. 1967.

[45] S. J. Brolin and J. M. Brown, "Companded delta modulator for telephony," *IEEE Trans. Commun. Technol.*, vol. COM-16, pp. 157–162, Feb. 1968.

[46] A. Tomozowa and H. Kaneko, "Companded delta modulator for telephone transmission," *IEEE Trans. Commun. Technol.*, vol. COM-16, pp. 149–157, Feb. 1968.

[47] J. A. Greefkes and F. DeJager, "Continuous delta modulation," *Philips Res. Rep.*, vol. 23/2, pp. 233–246, 1968.

[48] J. A. Greefkes, "A digitally companded delta modulation modem for speech transmission," in *Proc. IEEE Int. Conf. Communications*, June 1970, pp. 7-33–7-48.

[49] S. Hosokawa, S. Tazaki, and K. Yamashira, "Companded delta modulation coders of the 1/2-power and 2/3-power types," *Electron. Commun. Jap.*, vol. 50, no. 8, pp. 99–108, 1967.

[50] C. M. B. Newton, "Delta modulation with slope-overload prediction," *Electron. Lett.*, vol. 6, pp. 272–274, Mar. 1970.

[51] P. A. Bello, R. N. Lincoln, and H. Gish, "Statistical delta modulation," *Proc. IEEE*, vol. 55, pp. 308–319, Mar. 1967.

[52] J. W. Bayless, S. J. Campanella, and A. J. Goldberg, "Voice signals: Bit-by-bit," *IEEE Spectrum*, vol. 10, pp. 28–39, Oct. 1973.

[53] A. Rosenberg, Bell Labs., private communication.

[54] L. H. Zetterberg, "A comparison between delta and pulse-code-modulation," *Ericsson Tech.*, vol. 11, pp. 95–154, 1955.

[55] J. B. O'Neal, Jr., "An upper bound on the SNR of digital encoding systems," *Proc. IEEE*, vol. 55, pp. 287–292, Mar. 1967.

[56] D. L. Richards and R. W. Berry, "Impairment of telephone speech by quantizing," in *4th Int. Symp. Human Factors in Telephony* (Band Wiesse, Germany, Sept. 1969), pp. 29–46. (Copyright 1970 by VDE Verlag GmbH, Berlin, Germany.)

[57] L. M. Padula, Bell Labs., unpublished work.

[58] N. S. Jayant and A. E. Rosenberg, "The preference of slope-overload to granularity in the delta modulation of speech," *Bell Syst. Tech. J.*, pp. 3117–3125, Dec. 1971.

[59] H. Levitt, C. A. McGonegal, and L. L. Cherry, "Perception of slope-overload distortion in delta modulated speech signals," *IEEE Trans. Audio Electroacoust.*, vol. AU-18, pp. 240–247, Sept. 1970.

[60] D. Chan and R. W. Donaldson, "Subjective evaluation of pre- and postfiltering in PAM, PCM, and DPCM voice communication systems," *IEEE Trans. Commun. Technol.*, vol. COM-19, pp. 601–612, Oct. 1971.

[61] W. A. Munson and J. E. Karlin, "Isopreference method for evaluating speech transmission circuits," *J. Acoust. Soc. Amer.*, pp. 762–774, June 1962.

[62] J. D. Caroll, "Individual differences and multidimensional scaling," in *Multidimensional Scaling: Theory and Applications in the Behavioral Sciences*, vol. I, R. N. Shepard, A. K. Romney, and S. Nerlove, Eds. New York: Seminar Press, 1972, pp. 105–155.

[63] D. J. Goodman and L. J. Greenstein, "Quantizing noise of ΔM/PCM encoders," *Bell Syst. Tech. J.*, pp. 183–204, Feb. 1973.

[64] D. J. Goodman and J. L. Flanagan, "Direct digital conversion between linear and adaptive delta modulation formats," in *Proc. IEEE Int. Conf. Communications* (Montreal, Canada, June 1971), pp. 1-33-1-36.

[65] P. A. Deschenes and M. Villeret, "Numerical converters from PCM to DM and vice versa," in *Proc. IEEE Int. Conf. Communications* (Montreal, Canada, June 1971), pp. 1-22-1-26.

[66] N. S. Jayant and K. Shipley, "Multiple delta modulation of a speech signal," *Proc. IEEE* (Lett.), vol. 59, p. 1382, Sept. 1971.

[67] D. J. Goodman, "The application of delta modulation to analog-to-PCM encoding," *Bell Syst. Tech. J.*, pp. 321–343, Feb. 1969.

[68] J. B. O'Neal, "Predictive quantizing systems for the transmission of television signals," *Bell Syst. Tech. J.*, pp. 689–719, May–June 1966.

[69] K. Y. Chang and R. W. Donaldson, "Analysis, optimization, and sensitivity study of PCM systems operating on noisy communication channels," *IEEE Trans. Commun.*, vol. COM-20, pp. 338–350, June 1972.

[70] J. Yan and R. W. Donaldson, "Subjective effects of channel transmission errors on PCM and DPCM voice communications systems," *IEEE Trans. Commun.*, vol. COM-20, pp. 281–290, June 1972.

[71] M. Melnick, "Intelligibility performance of a variable slope delta modulation," in *Proc. IEEE Int. Conf. Communications* (Seattle, Wash., June 1973), pp. 46-5–46-7.

[72] A. S. House, C. E. Williams, M. H. L. Hecker, and D. Kryter, "Articulation-testing methods: Consonantal differentiation with a closed-response set," *J. Acoust. Soc. Amer.*, vol. 37, pp. 158–166, Jan. 1967.

[73] J. A. Greefkes, "Code modulation system for voice signals using bit rates below 8 KBPS," *Proc. IEEE Int. Conf. Communications* (Seattle, Wash., June 1973), pp. 46-8–46-11.

[74] J. K. Wolf, "Effects of channel errors on delta modulation," *IEEE Trans. Commun. Technol.*, vol. COM-14, pp. 2–7, Feb. 1966.

[75] N. S. Jayant and R. W. Schafer, Bell Labs., Murray Hill, N. J., unpublished work.

Index: SPEC (speech predictive encoding communications) system, digital speech interpolation, telephone channels, time-assigned speech interpolation.

3.1.3
Digital speech interpolation

S. J. CAMPANELLA

(Manuscript received August 29, 1975)

Abstract

This paper discusses two different methods of digital speech interpolation (DSI). One uses a digitized version of time-assigned speech interpolation (TASI) and the other a technique known as SPEC (speech predictive encoded communications). The digitized TASI system employs the low speech activity on incoming telephone channels to interpolate speech spurts onto a smaller number of transmission channels. A technique known as bit reduction makes it possible to generate additional channels for TASI transmission during overload conditions so that the probability of disruptive speech clips can be significantly reduced. The SPEC method incorporates PCM sample prediction to achieve a substantial reduction in the number of samples needed for transmission and inherently avoids the initial clips of speech spurts which are encountered in TASI.

The operation of each method is analyzed to evaluate its performance as a function of the number of channels processed. TASI is considered in terms of degradation due to initial clips of speech spurts whose duration is sufficient to damage the intelligibility of initial consonants, while SPEC is evaluated in terms

---

This paper is based upon work performed in COMSAT Laboratories under the sponsorship of the International Telecommunications Satellite Organization (INTELSAT). Views expressed in this paper are not necessarily those of INTELSAT.

A condensed version of this paper was presented at the INTELSAT/IECE/ITE 3rd International Conference on Digital Satellite Communications in Kyoto, Japan, November 11–13, 1975. The SPEC system is a COMSAT corporate development.

of degradation due to added prediction distortion. In each case the ratio of the number of incoming channels to the number of transmission channels (the interpolation advantage) as a function of the total number of transmission channels is calculated for amounts of degradation that are considered subjectively insignificant. Both methods are competitive in terms of interpolation advantage, although SPEC achieves slightly better performance, especially when the number of channels is small.

Introduction

Since it is customary during 2-way conversation for one talker to pause while the other speaks, active speech signals are present on a transmission channel for only a fraction of the total conversation time. In addition, even when only one talker is speaking, pauses occur between utterances and there are times when the circuit is idle. On the average, speech is present for less than 50 percent of the time; in fact, actual measurements show that speech is present on a typical telephone channel approximately 40 percent of the time [1]. This paper discusses digital time-assigned speech interpolation (TASI) [2]–[5] and speech predictive encoded communications (SPEC) [6]–[8] techniques which, by exploiting these activity properties, lead to a reduction of the total information rate needed to handle a multiplicity of telephone channels.

Time-assigned speech interpolation

General considerations

TASI [9]–[11] is a technique in which the idle time between calls and the conversation pauses during calls are used to accommodate additional calls. With a sufficiently large number of channels, most of the idle time on the transmission link can be filled, giving an enhancement in transmission capacity greater than two.

Observations of transatlantic cable circuits have indicated that speech is actively present on a busy channel about 40 percent of the time. Each period of time occupied by a caller's speech is called a speech spurt. The low activity is largely due to the facts that a subscriber speaks less than half the time during a conversation and that each subscriber's speech is carried on a separate channel. If all connections supplied to a TASI terminal are busy, the speech spurt activity will average 40 percent. If, on the other hand, all circuits are not busy, the average speech activity experienced by the TASI terminal will be further decreased. The percentage of busy circuits is called the incoming channel activity and the percentage of time that speech spurts occupy a channel is the speech spurt activity, or simply speech activity. Thus, if the incoming channel activity is 85 percent, the speech spurt activity will be reduced to 34 percent.

Quality aspects

COMPETITIVE CLIPPING AND FREEZE-OUT

TASI exploits the low speech spurt activity by assigning transmission channels only when a speech spurt is present. It is evident that the process becomes more efficient as the number of channels increases. An attempt to interpolate two independent conversations on a single transmission channel will cause a large percentage of the speech to be lost due to competition for simultaneous channel occupancy. Until it is terminated, the conversation occupying the transmission channel will "freeze out" any attempt by the other conversations to occupy the channel.

When a large number of independent conversations compete for some smaller number of transmission channels, the same type of competition prevails and there is always a finite probability that the number of conversations demanding service will exceed the number of available transmission channels. This competition manifests itself in the form of clipping of the initial portion of a speech spurt, or competitive clipping. The percentage of time that speech is lost due to such competition is called the percent freeze-out or freeze-out fraction. Provided that the population of incoming channels processed is sufficiently large and the ratio of the number of incoming channels to the number of channels available for transmission (the interpolation advantage, or in a digital implementation, the DSI advantage) is sufficiently small, the fraction of speech lost to freeze-out can be rendered acceptably small. TASI systems have been designed so that the freeze-out fraction is 0.5 percent.

The freeze-outs consist principally of short clips of the initial portions of speech spurts ranging from zero to a few hundred milliseconds. Speech clips longer than 50 ms cause perceptible mutilation of initial plosive, stop, fricative, and nasal consonants [12]. It is important to minimize the frequency of occurrence of speech clips longer than 50 ms if high quality is to be maintained. In this paper the performance of TASI is analyzed in terms of the criterion that the percentage of clips longer than 50 ms is less than 2 percent. This criterion is considered to be preferable to the constant freeze-out fraction criterion traditionally used by designers of TASI systems [6] because it relates more directly to the cause of degradation, viz, clipping of initial consonants.

CONNECT CLIPPING

At each TASI terminal the presence of speech on a telephone channel is sensed by a speech detector which initiates a request for a transmission channel. A channel assignment processor assigns an idle transmission channel to the incoming channel in response to the request and also sends a connect signal to the TASI terminal at the far end specifying the outgoing channel to which the transmission channel is to be connected. During the time required to make the channel assignment and to connect the listener and talker, the speech can be clipped. This phenomenon is called a connect clip and is to be distinguished from the competitive clip discussed previously.

The connect clip is constrained to a very short duration to minimize subjective degradation consistent with reliable connect signaling. The statistics of this type of clip are controlled by the competition of connect messages for space on the assignment channel. During periods of light loading on the incoming channels, transmission channels may remain assigned even during idle intervals so that the impact of connect clipping is absent. During periods of heavy loading on incoming channels, the connect clip may occur on every speech spurt. Thus, the degradation caused by connect clipping will vary with trunk activity.

SPEECH DETECTOR CLIPPING

In addition to competitive and connect clipping, some time is required to accomplish speech detection and a detector clip can occur. Since it is possible to virtually eliminate detector clipping by appropriate voice detector design, detector clipping can for all practical purposes be ignored.

Digital TASI implementation

A functional implementation of a digital TASI system is shown in Figure 1. The incoming telephone channels are digitized into conventional 8-bit-per-sample, 8,000-sample-per-second, pulse code modulated (PCM) time-division-multiplexed (TDM) format as part of the existing telephone plant or specifically to interface with the DSI system. The digitized signals are then processed by the transmit assignment processor, which is sequentially shared among all the incoming telephone channels.

The transmission channels consist of time slots arrayed consecutively in a TDM frame. The PCM samples selected for transmission are assigned to slots which become available after the assignment processor demands a transmission channel to service a particular incoming channel. The control channel needed to carry the channel assignment information to the far end is included in the same TDM frame.

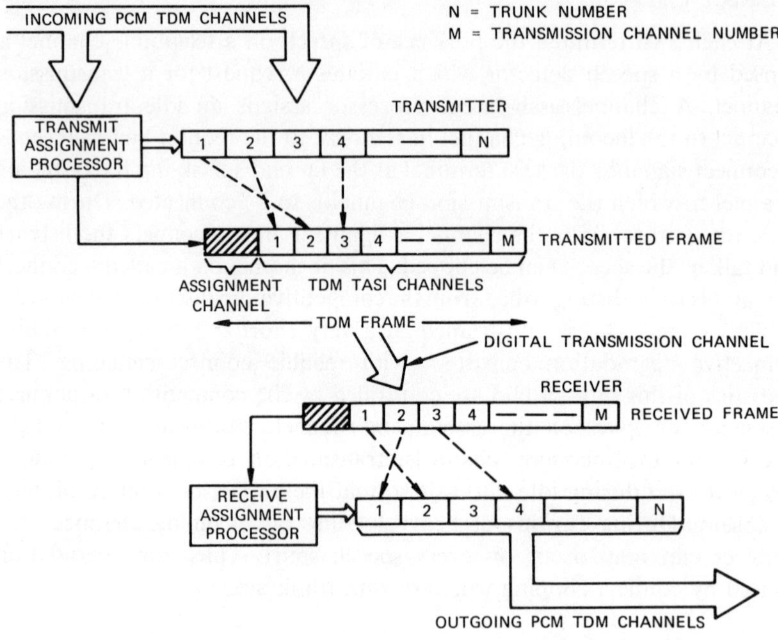

Figure 1. *TASI-Type DSI System*

Digital TASI has a number of advantages over analog TASI. For example, digital voice detectors perform better than their analog counterparts.* In addition, more precise and efficient switching of digital speech samples among channel slots of the TDM time frame is inherent to digital techniques. An all-digital approach is also compatible with the use of a digital channel assignment processor for recording the channel assignments at any instant and communicating this information to a companion digital channel assignment processor at the receiver.

---

*In a DSI system, the digital voice detector is part of a central assignment processor that is time shared among all the telephone channels. Therefore, a more sophisticated design can be incorporated. An equivalent analog design may be much more costly since it must essentially be repeated in every detail for each incoming telephone channel.

An additional advantage of the digital implementation of TASI is the ability to expand the number of transmission channels to avoid freeze-out during instants of overload by reappropriating the least significant bits of the digital transmission time slots. This technique, known as channel augmentation by bit reduction, can be invoked during overload conditions to avoid excessive competitive speech clipping. Although reducing the number of quantizing levels per slot from 256 (8 bits) to 128 (7 bits) produces a 6-dB increase in quantization noise, the fraction of time that bit reduction is required is very low. Hence, its presence is not apparent.

Analysis of digital TASI performance

STATISTICS

As mentioned previously, a principal factor governing TASI performance from the subjective point of view is the probability of occurrence of voice spurt clips of 50 ms or more. A 2-percent probability of occurrence of clips with durations equal to or greater than 50 ms is used as a threshold of acceptability in the following analysis. Hence, on the basis of an average voice spurt duration of 1.5 seconds and an activity (average percentage of time during which the speech is present on a channel) of 40 percent, a clip of this type will occur once every 1.5 minutes in a telephone conversation to one of the subscribers.

At any particular instant, the probability that the number of simultaneous talkers on N incoming channels with activity α will equal or exceed c (where c is the number of transmission channels) is given by the binomial distribution

$$B_{c,N,\alpha} = \sum_{x=c}^{N} \frac{N!}{x!\,(n-x)\,!}\, \alpha^x (1-p)^{N-x} \tag{1}$$

If it is further assumed that the voice spurts have durations that are exponentially distributed with mean L, then the probability that a spurt is frozen out (i.e., clipped) for longer than time t is given by $B_{c,n,\theta}$ [9], where

$$\theta = \alpha \epsilon^{-t/L} \tag{2}$$

From these expressions, the probability of occurrence of clips with $t > 50$ ms and $L = 1.5$ seconds as a function of the number of transmission channels, c, for $N = 15, 30, 60, 120, 180,$ and 240 incoming channels has

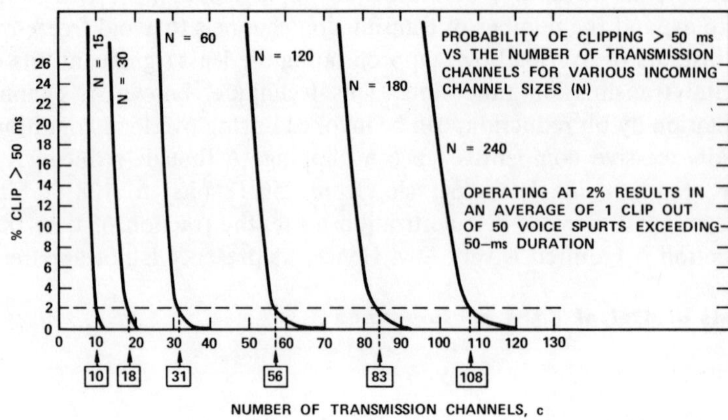

Figure 2. *TASI Performance in Terms of Competitive Clipping*

been calculated. The results are shown in Figure 2, which also indicates the 2-percent boundary and the number of transmission channels needed to accommodate a given number of incoming channels, N. The ratio of N/c, or the DSI advantage, is plotted in Figure 3 as a function of N. It

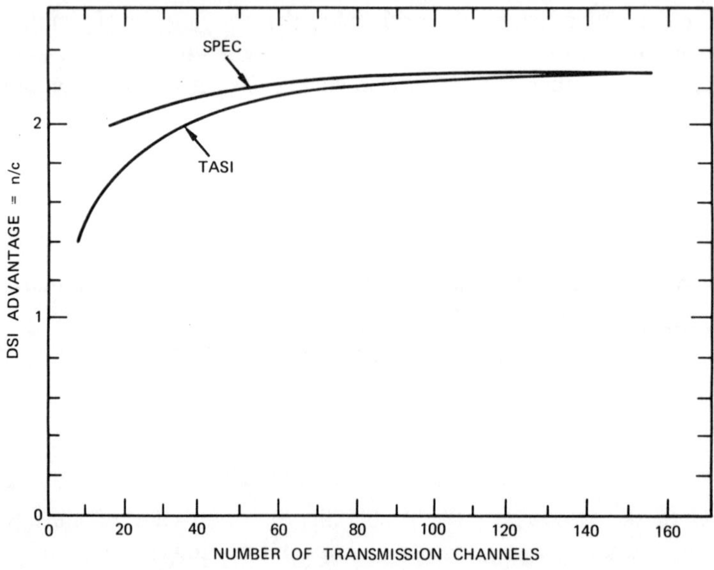

Figure 3. *DSI Advantage for TASI and SPEC*

can be seen that the TASI advantage, as defined here, reaches a value of 2.2 for 240 incoming channels and exceeds 2 for all cases in which the number of available transmission channels exceeds 38. For comparison, the advantage of SPEC for the same number of transmission channels is also shown in Figure 3. The curve for SPEC is not based on clipping occurrence, since SPEC totally avoids the clipping problem. Instead, the SPEC performance is based on the probability that degradation due to distortion produced by the predictor exceeds 0.5 dB for 25 percent of the time, a point which will be discussed in detail later.

The distribution of competitive clip durations is a function of the number of incoming channels, as shown in Figure 4, which indicates the distribu-

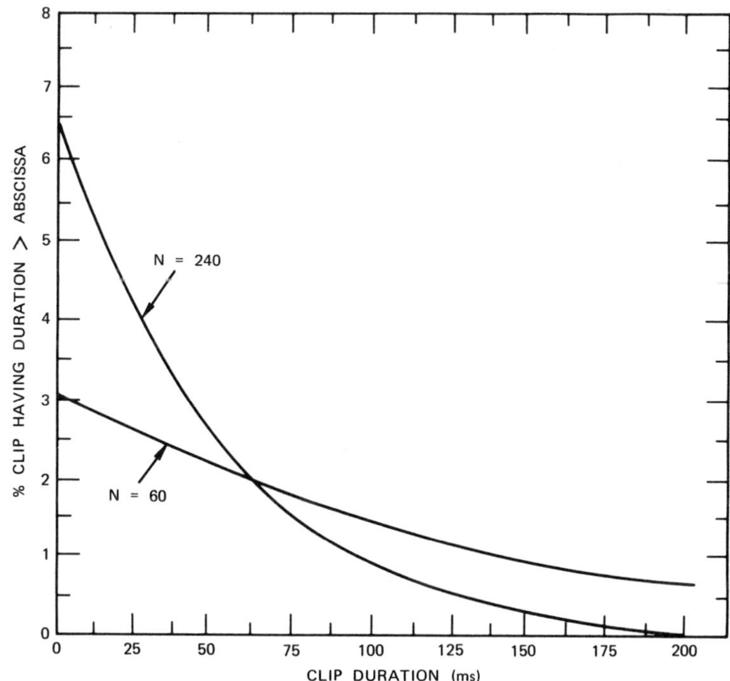

Figure 4. *Distribution of Competitive Clip Duration for N = 60 and 240 Channels*

tions of competitive clip duration for $N = 60$ and 240. Both distributions have about the same probability of competitive clip duration greater than 50 ms, but longer clips are more likely to occur for $N = 60$ than for $N = 240$.

INFLUENCE OF BIT REDUCTION

As mentioned previously, bit reduction can be used to absorb overloads that would otherwise result in clipping. Bit reduction is required when the number of telephone channels demanding service exceeds the number of normal transmission channels. For a TASI system, if 240 channels are interpolated for the DSI gain shown in Figure 3, bit reduction will be required 5 percent of the time.

Additional 7-bit PCM words are generated without increasing the number of bits allocated to PCM samples in the DSI subburst by reappropriating the least significant bit of all 8-bit PCM words and hence reducing their resolution to 7 bits. Thus, for a DSI subburst composed of c 8-bit PCM words, bit reduction will augment the number of transmission PCM words to yield

$$c_{BR} = c + \left\lfloor \frac{c}{7} \right\rfloor \tag{3}$$

where $\lfloor \ \rfloor$ indicates the next lower integer. When bit reduction is used, the probability of occurrence of clips longer than 50 ms drops significantly. This is shown in Table 1, where the probability that the clip duration exceeds 50 ms for 40-percent speech spurt activity is compared for systems with and without bit reduction.

TABLE 1. PROBABILITY OF CLIP DURATION >50 ms
WITH AND WITHOUT BIT REDUCTION ($\alpha = 40\%$)

Number of Incoming Channels, N	Probability of Clip Duration >50 ms (%)	
	With Bit Reduction	Without Bit Reduction
60	0.13	2
120	0.07	2
180	0.01	2
240	0.004	2

Table 1 indicates that bit reduction can effectively reduce the clipping problem encountered in TASI. Although the introduction of bit reduction causes a 6-dB increase in quantization noise on all channels affected, as long as the system is not heavily overloaded, the subjective impact of this increased quantization noise is negligible.

ASSIGNMENT MESSAGE FREEZE-OUT

In a version of DSI being considered for TDMA use, assignment messages are transmitted once every 750 μs as part of the DSI subburst. Each assignment message consists of 24 bits of assignment information which is rate one-half coded for transmission error protection and can accommodate one connect or disconnect operation. Use of the bit reduction method requires that both a connect and a disconnect message be sent to accommodate each voice spurt.

The probability that the assignment messages will be frozen out can be analyzed in terms of the binomial probability expression $B_{c,N,q}$ introduced in equation (1). Since there is one assignment transmission channel, c is equal to one. This channel must serve the randomly occurring assignment messages arriving on the N parallel incoming channels. The activity, q, is the ratio τ/l, where τ is the length of the TDMA frame and l the average time interval between assignment messages. If both connect and disconnect messages are needed, then

$$l = \frac{L}{2\alpha} \tag{4}$$

If only connect messages are needed, then l is twice this value.

For the TDMA equipment, in which both connect and disconnect messages are needed, $l = 1.875$ seconds and $q = 4 \times 10^{-4}$ for $L = 1.5$ seconds and $\alpha = 40$ percent. In this case, the freeze-out probability of an assignment message for $N = 240$ channels, where $c = 1$ and $q = 4 \times 10^{-4}$, is $B_{c,N,q} = 10^{-1}$. This is in effect the probability of having to wait more than one TDMA frame period, i.e., 750 μs. This situation will occur once every five voice spurts or once every 18.75 seconds in each direction of conversation. If statistical independence is assumed, the probability of having to wait s TDMA frames for assignment is simply $(B_{c,N,q})^s$. For $s = 4$, the probability of having to wait longer than 3 ms is 10^{-4} and the time between occurrences is 5.2 hours. Hence, it is obvious that speech clips caused by assignment message channel congestion are of negligible consequence for the case of interest here.

Speech predictive encoded communications

General considerations

SPEC is a form of digital speech interpolation which differs significantly from digital TASI. One of its principal merits is total avoidance of the competitive clip problem experienced by TASI. Its operation protocol does not require recordkeeping for connections from incoming channel to transmission channel to outgoing channel since all channel assignment information is contained in each frame. Hence, channel assignment memory and assignment message channel implementation is unnecessary. Its adaptive processing method exhibits only a slight increase in quantizing noise when confronted with overload. This event occurs so infrequently at the channel augmentation ratios shown in Figure 3 that it is of negligible subjective consequence.

Quality aspects

PREDICTOR DISTORTION

SPEC requires the signal to pass a speech detector before samples are admitted to the predictor. The signals passed by the speech detector exhibit an activity similar to that experienced with TASI. The predictor algorithm reduces this activity by eliminating unnecessary samples in the instantaneous speech waveform or in the short intersyllabic pauses not sensed by the voice detector. Under average load conditions the SPEC predictor removes more than 25 percent of the PCM samples during voice spurts 25 percent of the time. This results in a signal-to-distortion (S/D) ratio decrease of only 0.5 dB. The achievable DSI advantage ratios are shown in Figure 3 as functions of the number of incoming channels.

COMPETITIVE CLIPPING

SPEC cannot cause clips such as those encountered in TASI since samples would have to be frozen out for a succession of 80 SPEC frames to produce a 10-ms clip. The probabiiity of occurrence of such an event is very small.

SPEECH DETECTOR CLIPPING

SPEC incorporates a speech detector and, as in the case of digital TASI, the inclusion of adaptive threshold features in the design results in negligible clipping.

CONNECT CLIPPING

Since there is one bit in the sample assignment word (SAW) for each incoming channel, the SPEC SAW allows fully flexible connectivity for all incoming channels. Hence, there cannot be any connect clipping caused by waiting for channel assignment.

Implementation

In the SPEC system, shown functionally in Figure 5, the speech information contained on N incoming 8-bit-per-sample PCM telephone channels is transmitted in the space of $(N/8) + c$ 8-bit-per-sample transmission channels, where c is the number of PCM sample channels used for transmission and is typically equal to slightly more than $N/3$. SPEC processes all incoming trunks and transmits a frame once per period of the PCM Nyquist rate, which corresponds to a 125-μs sampling interval in typical commercial telephone usage.

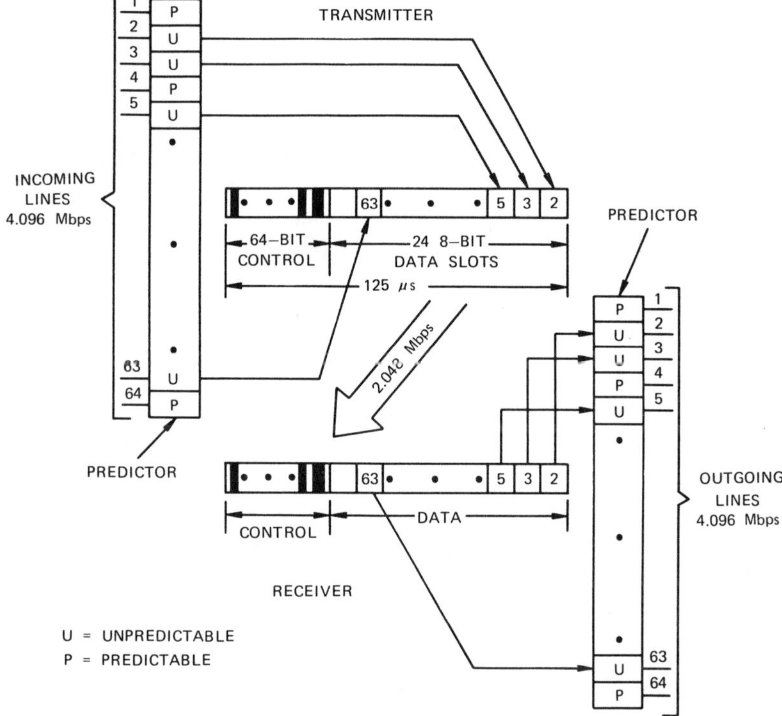

Figure 5. *SPEC-Type DSI System*

SPEC operation is conveniently described by using an actual example of its implementation* for accommodating 64 incoming telephone channels (corresponding to a bit rate of 4.096 Mbps) in the transmission space

---

*A system such as that described has been built and tested in demonstration service between the U.S. mainland and Hawaii on the INTELSAT IV satellite [8].

normally allotted to 32 PCM channels (corresponding to a bit rate of 2.048 Mbps). The PCM samples derived during each sample period from the 64 incoming channels are compared with the samples previously sent to the receiver and stored in a memory at the transmitter. Any that differ by an amount equal to or less than some given number of quantizing steps, called the aperture, are discarded and not sent to the receiver. These are referred to as "predictable" samples. The remaining "unpredictable" samples are transmitted to the receiver and replace the values formerly stored in memories at both the transmitter and receiver. The aperture is automatically adjusted as a function of activity observed over the 64 incoming channels on each frame so that the number of samples transmitted is nearly constant.

The SPEC transmission frame is composed of an initial SAW followed by a number of 8-bit time slots that carry the individual PCM samples judged unpredictable by the transmitter's prediction algorithm. The SAW contains one bit for each of the incoming telephone channels. Thus, for the 64-terrestrial-channel system, it contains 64 bits. The bit corresponding to a given channel is a "1" if the frame contains a sample for that particular channel and a "0" if it does not. Thus, the SAW contains all of the information needed to distribute the samples among the 64 outgoing channels at the receive end.

At the receiver the unpredictable samples received in the SPEC frame replace previously stored samples in the receiver's 64-channel memory as directed by the SAW. The samples in the memory, in the form of a conventional PCM/TDM frame, are strobed into the outgoing channels at the proper rate. The most recent frame thus contains new samples on the channels that have been updated by the most recent SPEC frame and repetitions of the samples that have not been updated.

Analysis of SPEC performance

ACTIVITY REDUCTION BY PREDICTION

The PCM samples transmitted by SPEC are determined by selecting that set in which the differences between the samples previously sent and those currently being sent are equal to or greater than some number of quantizing steps called the aperture, a. The value of the aperture is adjusted so that the number of samples transmitted does not exceed the number of available transmission channel slots.

If the incoming channel activity for a given frame is such that all new samples can be transmitted with zero aperture, i.e., if the number of samples to be transmitted with differences greater than zero is equal to or less than the number of transmission slots, no error is made in the transmission. If the incoming channel activity for a given frame is such that the aperture $a = 1$, samples differing from the previously transmitted samples by a number of quantizing steps greater than 1 are transmitted. Hence, the number of samples requiring transmission is reduced by culling out those with differences less than or equal to 1. In this case errors equal to one quantizing step will be made for those samples with differences of zero. The activity resulting from this action is

$$\alpha_1 = \alpha[1 - P(\Delta = 0) - P(\Delta = 1)] \tag{5}$$

where
$$\alpha = \text{activity experienced on the ensemble of incoming terrestrial channels with no prediction}$$
$$\alpha_1 = \text{activity after culling out samples with differences less than or equal to 1}$$
$$P(\Delta = k) = \text{probability of samples differing from those previously sampled by } k \text{ quantizing steps.}$$

In this case, errors of the magnitude of one quantizing step will be made for a fraction of the samples equal to $P(\Delta = 0)$.

In the general case, if a set of SPEC processed values is transmitted with an aperture of value a, then the resulting activity for that set is

$$\alpha_a = \alpha \left[1 - \sum_{k=0}^{a} P(\Delta = k) \right] = \alpha(1 - P_a) \quad . \tag{6}$$

The amount given by the summation and designated as P_a is known as sample predictability for aperture a. On the average, the value of incoming channel activity, α, is approximately 40 percent.

SPEC is designed to operate at an average transmit channel activity (given by the ratio of the number of transmission channels available for samples, c, to the number of incoming channels, N) of $c/N = 0.3$. The aperture must be such that

$$\alpha_a \leqq \frac{c}{N} \tag{7}$$

and the predictability must be such that

$$P_a \geqq 1 - \frac{c}{N\alpha} \quad . \tag{8}$$

These rules establish the average aperture at which SPEC operates.

DISTRIBUTION OF PREDICTION

For a given incoming channel activity, α, the number of incoming channels, n, requiring transmission due to the presence of voice spurts is a random variable. The probability that the number of incoming channels requiring service at a given instant will exceed the number of transmission channels is given by the binomial distribution

$$P(n > c) = B_{c,n,\alpha} = \sum_{x=c}^{n} \frac{n!}{x! \, (n-x)!} \, \alpha^x (1-\alpha)^{n-x} \quad . \tag{9}$$

The system is constrained so that the number of samples needing transmission never exceeds c. This is accomplished by using prediction to reduce the activity in the entire set of incoming channels to a new activity, α_a, described previously. The average number of channels requiring service under these circumstances is then

$$n_a = \alpha_a N \quad . \tag{10}$$

Substituting the new value of activity into the binomial distribution yields

$$P(n > c) = B_{c,n,\alpha_a} \tag{11}$$

where $\alpha_a = \alpha(1 - P_a)$, which expresses the modification in activity resulting from the introduction of the predictor. Since the average number of transmission channels used with prediction is

$$c = N\alpha(1 - P_a) \tag{12}$$

the binomial distribution can be redesignated as

$$P[n > N\alpha(1 - P_a)] = B_{c,n,\alpha_a} \tag{13}$$

which can be further modified as follows:

$$P\left[\alpha(1 - P_a) < \frac{n}{N}\right] = P\left[\alpha < \frac{n}{N(1 - P_a)}\right]$$
$$= P\left[P_a > 1 - \frac{n}{N\alpha}\right]$$
$$= P\left[P_a > \left(1 - \frac{\alpha_a}{\alpha}\right)\right]$$
$$= B_{c,n,\alpha_a} \quad . \tag{14}$$

Equation (14), which gives the probability that the prediction will exceed the value $(1 - \alpha_a)/\alpha$, is very useful in assessing the level of degradation introduced by the predictor.

Figure 6 shows the performance of SPEC in terms of the predictability, $P_a = 0.25$, calculated from equation (14) for $\alpha = 0.40$ and $N = 30, 60, 120, 180,$ and 240 incoming channels plotted against the number of transmission channels, c. The ordinate is the probability that the indicated value of P_a is exceeded for the number of transmission channels on the abscissa, which includes those channels needed to accommodate the SAW. Each transmission channel is assumed to be 8 bits wide. A predictability of $P_a = 0.25$ means that, on the average, one sample out of four is predicted during a voice spurt. In the case of $N = 120$, it can be seen that, for $c = 54$ transmission channels, the probability that the predictability will exceed $P_a = 0.25$ is 25 percent. This can also be interpreted to mean that for 75 percent of the time the predictability will be such that less than one in four samples will be predicted during a voice spurt when 120 incoming channels are carried in the space of 54 transmission channels. In this case, the SPEC DSI advantage is $120/54 = 2.22$.

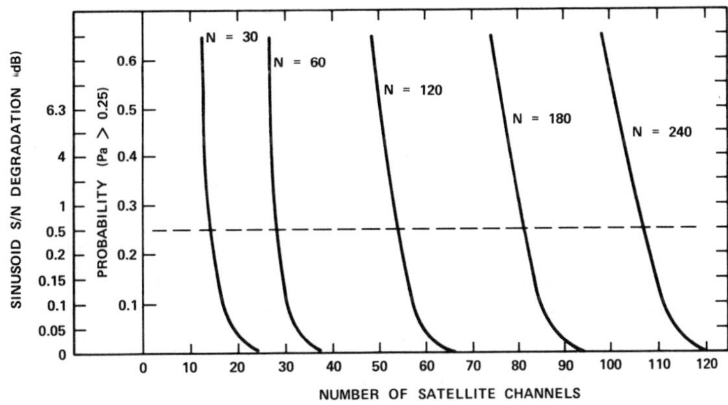

Figure 6. *SPEC Predictor Performance*

SIGNAL-TO-DISTORTION RATIO

It is possible to relate the probability that $P_a > 0.25$ to degradation in the S/D ratio over SPEC channels. This is done by using experimental data regarding the S/D ratio experienced by a 0-dBm0 sinusoidal signal passed through one channel of an experimental SPEC terminal built to accommodate 64 incoming channels in 32 PCM transmission channels plus 64 SAW bits [8]. The experimental data, given in Figure 7, show the degradation of the sinusoidal S/D ratio as the average activity on the 64 incoming

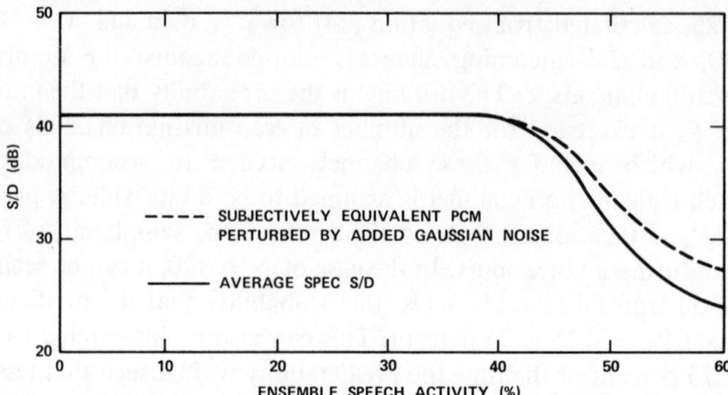

Figure 7. *Measured and Subjectively Equivalent S/D as Functions of Ensemble Speech Activity*

channels varies over a wide range of values. The overall degradation is due to a combination of PCM quantization noise and prediction distortion. As the incoming channel activity increases, the predictor must handle a greater number of samples by increasing its aperture, thus causing the total distortion to increase. Figure 7 also shows the signal-to-Gaussian-noise ratio (S/N) for a 0-dBm0 sinusoidal signal on a Gaussian-noise-perturbed PCM channel with speech quality equivalent to that experienced with prediction distortion. This curve indicates that a given level of prediction distortion corresponds to a lower level of Gaussian noise.

The data presented in Figure 7 can be directly related to predictability by using equation (14). The results are found in Figure 8, which shows the dB degradation in sinusoidal S/D ratio incurred by prediction as a function of the probability that $P_a > 0.25$.

Also shown in Figure 8 is a curve giving the equivalent performance of the PCM channel perturbed by additive Gaussian noise. This curve indicates that, for conditions such that $P(P_a > 0.25) = 25$ percent, the dB degradation in the sinusoidal S/D is only 0.75 dB. The dB degradation in the sinusoidal S/N on a subjectively equivalent channel perturbed by Gaussian noise is only 0.5 dB. Thus, it appears that $P(P_a > 0.25) = 25$ percent is an acceptable threshold for determining the advantage offered by the SPEC system. The ordinate of Figure 6 also contains a scale giving the dB deg-

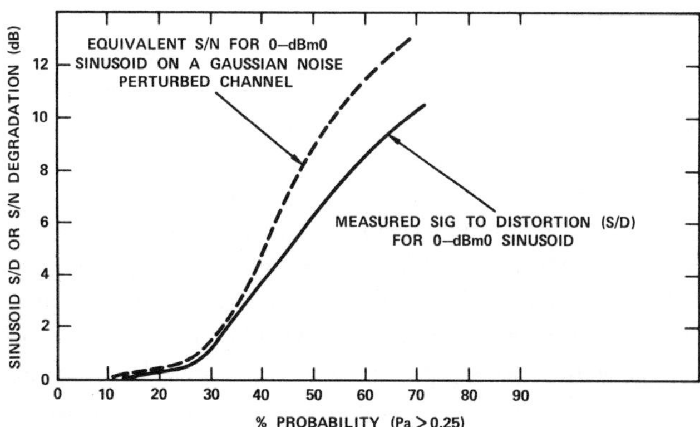

Figure 8. *Sinusoid S/D or Equivalent S/N versus* $P(P_a > 0.25)$

radation in the subjectively equivalent S/N obtained by using the data provided in Figure 8.

The number of transmission channels needed to accommodate each number of incoming channels is determined from the intersection of each curve shown in Figure 6 with the 0.5-dB degradation boundary. The plot of SPEC advantage versus the number of transmission channels (including SAW bits) shown in Figure 3 has been obtained from this result.

Digital speech interpolation in multidestinational satellite use

Digital TASI

In multidestinational satellite DSI, voice spurts originating on incoming terrestrial lines must be connected to appropriate outgoing terrestrial lines at the desired destinations. For digital TASI, these connections are accomplished in two steps, as shown by the connectivity maps of Figure 9. First, voice spurts occurring on the incoming terrestrial lines must be connected to satellite channel slots in the DSI burst. Then the DSI burst channels must be connected to the appropriate outgoing terrestrial lines at the designated receive terminal.

As shown in Figure 9, there are two maps at transmit terminal A. Map I indicates the connections between the incoming terrestrial lines on which voice spurts are occurring and the individual satellite transmission channel numbers that appear in the DSI burst. The connections at all receive terminals with which A communicates are indicated in map II, which shows that terminal A has distributed its satellite channels among destination terminals B, C, and D. Three satellite channels are assigned to destination B, two to destination C, and three to destination D. The assignments made at terminal A are communicated to the various destinations by the assignment message portion of the DSI burst.

Consider next the outgoing connectivity map at a typical receive terminal such as the 3-section map at terminal B. One section corresponds to channels in the DSI burst from terminal A, and the other two correspond to channels in bursts received from terminals C and D. In that portion of the map corresponding to reception from A, it can be seen that satellite channel slots 0, 2, and 3 are connected to outgoing terrestrial lines 1, 2, and 4. This is of course a duplicate of the map which has been set up at terminal A. The other sections of terminal B's map duplicate the assignments at terminals C and D and direct the connections at B to receive the voice spurts from terminals C and D, respectively.

To make the appropriate destination line assignments in multidestinational operation, a transmit terminal must be familiar with the connectivity maps at each destination with which it communicates. Thus, at terminal A there are replicas of the connectivity maps which designate the assignments of satellite channels to outgoing terrestrial lines for each destination with which terminal A is to communicate. If it is established in advance that terminal A has exclusive use of a preassigned set of outgoing lines at each destination, then the maps which are part of map II can be established at terminal A without knowledge of assignments on outgoing lines at the destination terminal made by other terminals communicating with the same destination. If, however, the outgoing lines at each of the destinations with which A communicates are to be shared among all other terminals communicating with the same destinations, then all terminals must possess the complete connectivity maps for all destinations and these must be updated to show the most recent connectivities. This will result in a fully variable demand-assignment DSI communications network. In either case, regardless of destination, the interpolation process occurs over the entire pool of incoming lines since satellite channels are fully variable in terms of both source and destination assignment.

SPEC

The method of channel assignment used in the SPEC system is different from that used in a digital TASI system. The SPEC frame, which is transmitted

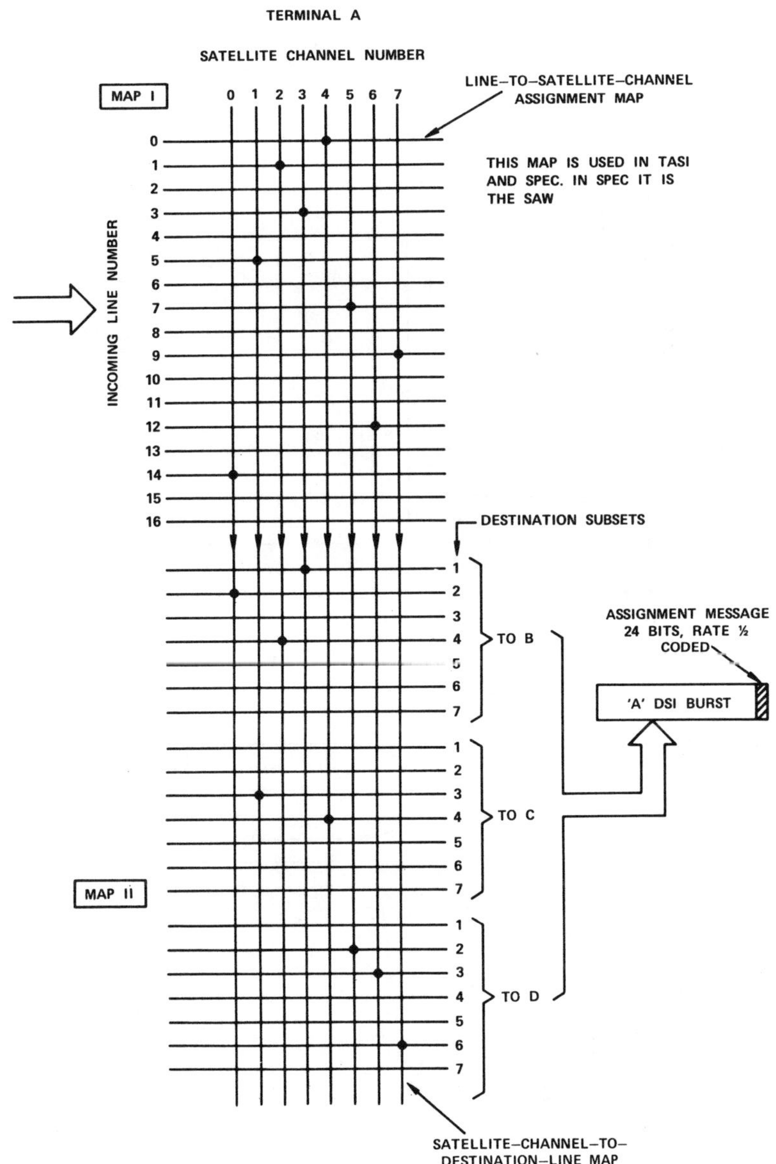

Figure 9a. *TDMA/DSI Connectivity Maps from Terminal A to Terminals B, C, and D*

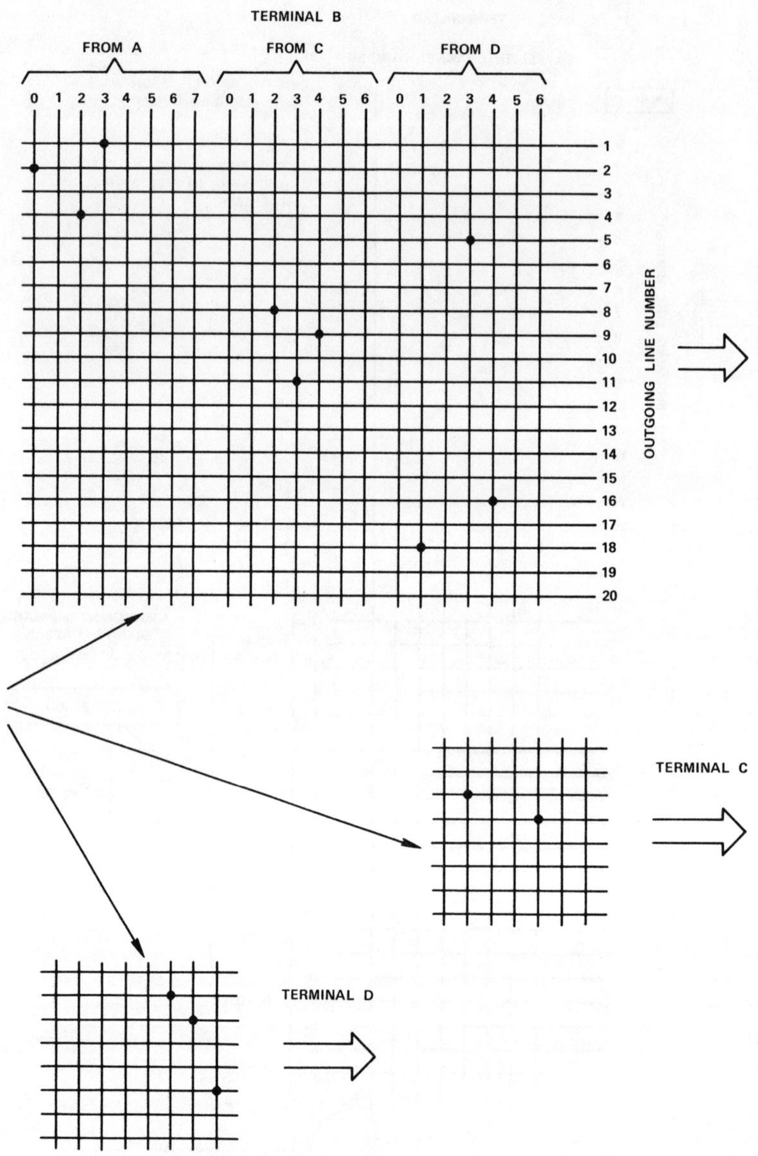

Figure 9b. *TDMA/DSI Connectivity Maps from Terminal A to Terminals B, C, and D*

once every 125 μs, consists of a SAW followed by unpredictable PCM samples of the speech channels which are to be transmitted to a set of destinations. The number of bits in the SAW corresponds to the number of destination lines to which the burst is directed. If a channel is to receive information, a "1" is transmitted in the SAW in the position corresponding to that channel; otherwise a "0" is transmitted. The samples that are transmitted in the PCM sample part of the SPEC frame correspond one-for-one to the "1's" in the SAW part of the frame. Thus, if the first terrestrial channel to which a sample is to be transmitted is number 3, then a "1" appearing in the third bit of the SAW directs the first sample of the sample portion of the SPEC frame to terrestrial channel number 3. If the second channel to which a sample is to be transmitted is channel number 7, then the next "1" appearing in the seventh bit of the SAW directs the second sample in the PCM part of the frame to terrestrial channel number 7. The number of samples transmitted in the PCM part of the frame will correspond to the number of "1's" present in the SAW part of the frame. The SAW may thus be considered to be a sample activity mask.

The SAW is a replica of map I in Figure 9. If the SPEC frame is to be communicated to one destination, then all of the channel numbers appearing in the SAW part of the frame correspond to channel numbers at that destination. If SPEC is to be operated in a multidestinational mode, then the channel numbers appearing in the SAW part of the frame are grouped according to destination. Thus, channel numbers 0 through 30 may go to destination A, 31 to 60 to destination B, 61 to 90 destination C, and so on. This constitutes a preassignment multidestinational mode of operation.

At the receiver, each burst received from a given source is stored in two parts, the SAW part and the PCM sample part. Only samples designated for a given destination are transferred from the SPEC frame into the appropriate outgoing lines as dictated by the SAW part of the frame. SPEC transmits a new connection map every frame, thus permitting highly flexible reassignment of the transmitted PCM samples among the outgoing lines at the destinations as a function of the instantaneous activity on all incoming lines at each transmit terminal. Hence, the SPEC system interpolates over the entire pool of channels going to all destinations.

It is also possible to reassign the traffic capacities among various destinations in the SPEC system by redesignating the transmission channels as traffic patterns change. This process, which can be accomplished by a low-data-rate channel transmitted as part of the SPEC frame, makes it possible to adapt to variations in traffic among destinations.

DSI implementation for TDMA

Methods for implementing the TASI and SPEC DSI systems for TDMA operation will now be described. Although actual equipment designs may differ in specific details, the structures presented are basic and should illustrate the functions needed to realize operational equipment.

TASI/TDMA implementation

TRANSMIT TERMINAL

The transmit terminal of a digital TASI system for possible use in TDMA equipment is shown in Figure 10. The terminal is configured to accommo-

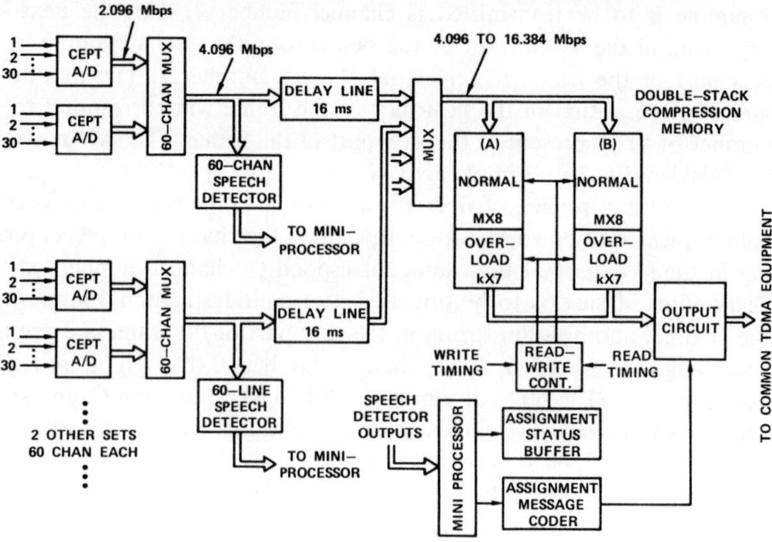

Figure 10. *TASI/TDMA Transmit Terminal*

date 240 terrestrial channels. Analog terrestrial input lines in groups of 30 are supplied to CEPT-32 PCM analog-to-digital (A/D) converters. Each of these converters delivers a PCM/TDM digital stream at a bit rate of 2.048 Mbps. These outputs are then multiplexed by a 60-channel multiplexer into a single stream at a bit rate of 4.096 Mbps. Modular grouping into 60 channels is used as the basic building block of the system. The system can handle additional channels in groups of 60 by adding more building blocks of this type in parallel.

The digital output of each 60-channel multiplexer is next fed to a digital delay line having a delay of 16 ms to offset any delay in the voice spurt detection circuits. The same digital stream is also fed to a voice detector, which simultaneously processes 60 channels in digital form. The voice detector decisions are supplied as input to the miniprocessor section of the DSI equipment. The miniprocessor then assigns active voice spurts to appropriate satellite channels as they become available in accordance with a satellite channel map stored in its memory. It also assigns the relationship, in the form of a destination map, between the satellite channel and an outgoing terrestrial channel at the desired destination and formulates assignment messages to inform the destination of the connection for each new voice spurt or the disconnection for each terminated voice spurt. [Disconnects are needed in overload (bit reduction) operation only.] Finally, the miniprocessor updates the transmit assignment status buffer so that selected TDM time slots coming from the multiplexer are assigned to their appropriate channel slot positions in the TDMA compression buffer.

The TDMA compression buffer is a double-stack design in which one stack is filled while the other is emptied to prevent conflict due to read/ write function overlapping. It stores as many PCM samples as there are time slots available in the DSI burst. Each DSI burst contains six Nyquist frame sections, and each section can hold up to 108 eight-bit samples, which is the number required for 240 input terrestrial channels. In addition, each section has an overload memory to accomplish bit reduction. The bit reduction strategy essentially increases the number of time slots available in the DSI burst by a factor of 8/7 rounded to the next lower integer. Thus, if the system is operating with 108 eight-bit time slots, the capacity is increased to 123 seven-bit time slots when bit reduction is used. The miniprocessor controls the onset of the bit reduction mode and establishes the necessary signaling to inform the receiver that bit reduction is being used.

RECEIVE TERMINAL

The receive side of the DSI subsystem, shown in Figure 11, must be capable of accepting bursts from multiple sources and appropriately distributing the information contained in these bursts to the terrestrial channels. Assignment messages from all sources are stored in the miniprocessor in the form of maps which associate channel slots in the various received DSI subbursts with outgoing terrestrial channels. The expansion buffer is a dual-stack structure. Each stack is capable of storing the contents of all DSI frames destined to the receive terminal and stores as many

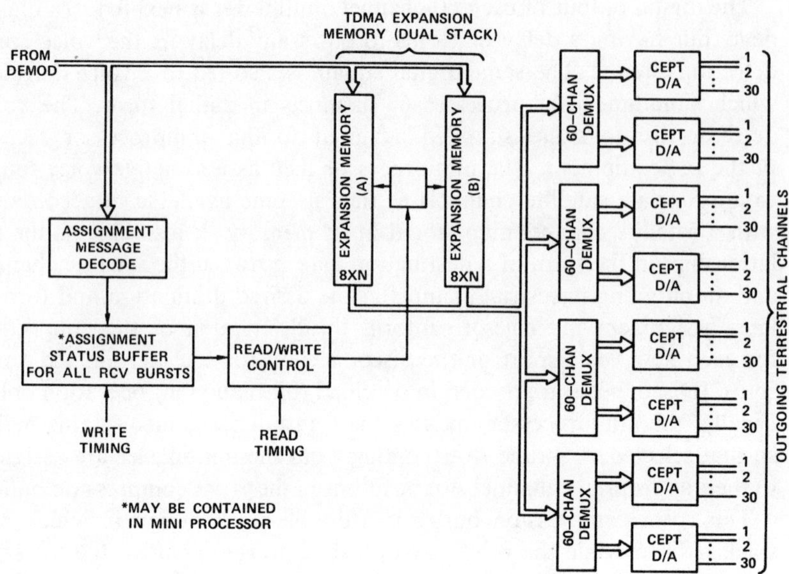

Figure 11. *TASI/TDMA Receive Terminal*

PCM samples as there are outgoing terrestrial channels. Under control of the assignment status buffer at the receive terminal, the samples stored in the expansion buffer are read out to the appropriate terrestrial channels.

SPEC/TDMA implementation

TRANSMIT TERMINAL

Figure 12 is a block diagram of a SPEC transmit terminal configured for 240 incoming terrestrial channels handled in modules of 60 channels each. Each group of 60 channels is processed through a pair of CEPT-32 A/D converters. The outputs of all CEPT units are multiplexed in a single multiplexer to a bit rate of 16.384 Mbps. The bit stream is then supplied to two devices: a 5-ms digital delay line and the 240-channel speech detector. The speech detector detects the presence of voice spurts on each of the 240 incoming terrestrial channels. Whenever a voice spurt is present it permits the PCM samples to pass to the intermediate frame memory (IFM) and the zero-order predictor (ZOP).

The IFM is capable of storing 240 eight-bit PCM samples. Depending on decisions made by the ZOP, the samples stored in the IFM will be transferred to the predictor frame memory (PFM). Specifically, the ZOP calculates the difference between values stored in the PFM and the most recent set of values which has been supplied to the IFM. For those PCM values whose difference is greater than the aperture, the values in the IFM are transferred to the PFM to replace the old values and are stored in the transmit frame memory (TFM) for transmission to the appropriate set of destinations.

Within the predictor there are a number of 240-bit-long storage units in which SAWs are stored for several aperture values. The SAW used in a given frame which controls the transfer of values from the IFM to PFM and TFM is the one corresponding to an aperture such that the number of samples transmitted is just less than the number of sample slots available in the TFM. For a 240-channel system the number of sample slots available in the TFM is 90. With low activity on the incoming channels, it will be found that all samples transmitted are the result of applying a small aperture value to yield a low quantization noise. If, on the other hand, the

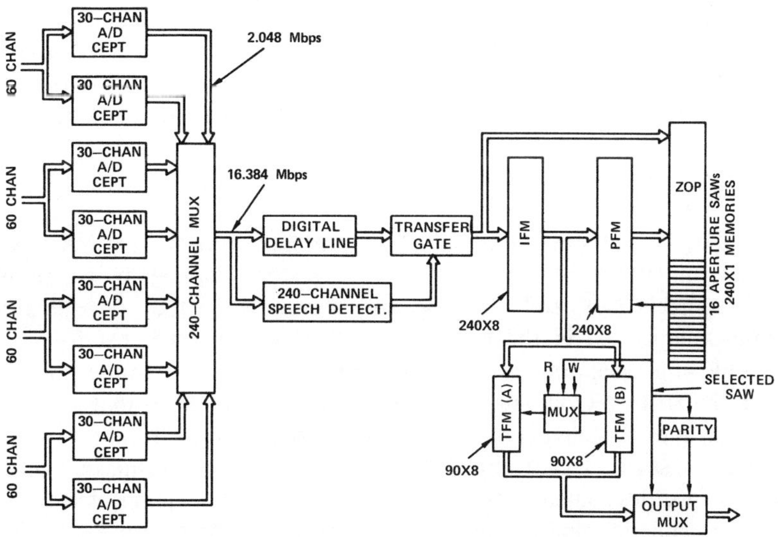

Figure 12. *SPEC Transmit Terminal*

activity on the incoming terrestrial channels is high, the aperture needed to reduce the population to a number below 90 will be greater and the quantization noise will be correspondingly greater. Thus, this implementation of the SPEC system permits the aperture occurring from each SPEC burst to vary over a large range to accommodate wide fluctuations in activity.

All values which are updated in the PFM each time the ZOP executes its function are called unpredictable values and are also transferred to the TFM where they await transmission to their destinations. The TFM is a dual-stack buffer configuration consisting of two 90×8 sections, each able to store an entire TFM. While one half is being filled the other half is being emptied, thus permitting a continuous flow of signal information to the outgoing communications link. An output multiplexer combines the samples of the TFM with the appropriate SAW to constitute the transmission burst. The frame duration upon which the SPEC transmitter operates is 125 μs and is synchronized to the Nyquist sampling frame used in the A/D PCM converters at the input.

For the SPEC system implementation to correspond to the 750-μs frame specified by INTELSAT, the frames generated by the equipment shown in Figure 12 must be accumulated in a TDMA compression buffer. This compression buffer consists of a dual-stack configuration which stores six SPEC frames and outputs them at the burst rate of the TDMA terminal equipment.

RECEIVE TERMINAL

A SPEC receive terminal must select from each received DSI burst only those samples destined to its outgoing terrestrial channels by examining each source's SAW and selecting and storing only the corresponding samples contained in the PCM sample part of each SPEC DSI frame. A possible implementation is shown in Figure 13. Each SPEC frame obtained from the TDMA demodulator is supplied to a demultiplexer that separates the SAW and PCM sample parts. The SAW is supplied to a PARITY CHECK and a SAW SUBSET SELECT unit. If the parity check is successful, that subset of SAW bits [designated as $(\text{SAW})_x$] which identifies the samples destined to the particular terminal of interest is passed on to the SAW memory. The SAW SUBSET SELECTOR then admits to the TFM only the appropriate subset of PCM samples, designated as $(\text{TFM})_x$. For any DSI subburst on which the SAW parity check fails, none of the received samples are stored in the TFM and all samples for that subburst are treated as predictable. The $(\text{SAW})_x$ and corresponding $(\text{TFM})_x$ samples are passed into the PFM, which reconstitutes the conventional PCM/TDM format with all predicted samples appropriately filled in. The conventional PCM/TDM stream is then supplied to the digital-to-analog (D/A) sections of CEPT-21 PCM units and hence to the outgoing terrestrial lines.

Comparison of TASI and SPEC implementations

The TASI and SPEC DSI implementation schemes which have been described herein have certain common features. In particular, each implementation requires the same kinds of A/D and D/A conversion interfaces between terrestrial links and satellite channels. Each also requires speech detectors and digital delay lines to appropriately implement these detectors. There are essentially no differences in the voice detector performance requirements of the two systems.

There are differences in the manner in which the samples are prepared for transmission, however. In digital TASI, incoming terrestrial channels on which speech spurts occur are connected to and disconnected from satellite channels and the satellite channels are connected to and disconnected from destination terrestrial channels on the basis of voice

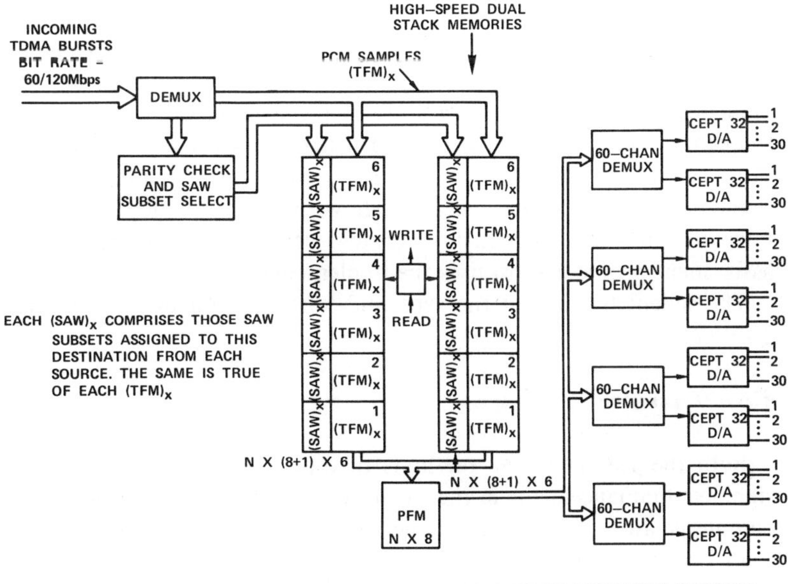

Figure 13. *SPEC Receive Terminal for TDMA Operation*

spurt demand. PCM samples on active channels are carried in TDM form. TASI uses a separate assignment message channel to make the appropriate connects and disconnects at various destinations.

By comparison, SPEC PCM samples occurring at the Nyquist rate on each terrestrial channel are compared in a ZOP and those determined to be unpredictable are sent to appropriate destinations on a sample-by-sample demand basis. Similarly to digital TASI, the SPEC samples are carried in TDM form. Unlike the TASI system, however, the SPEC system does not use a separate assignment message channel for connect and disconnect information, but instead incorporates a set of bits called a SAW, which is included as part of each SPEC frame, to direct individual samples to their destination. TASI hardware also uses a miniprocessor which retains maps giving all of the connectivities needed to steer speech spurts to satellite channels and satellite channels to all destinations. Storage of these maps is not required in SPEC, since all the destination information needed to distribute the PCM samples for each frame is transmitted in the SAW.

SPEC and TASI also differ with respect to operation during high peaks of activity. In SPEC, because of its operating characteristics, the predictor works harder, removing a greater fraction of the incoming speech samples by increasing its aperture value. Although this action does increase the quantization noise, this increase is not subjectively significant under operational conditions. In TASI, unless special precautions are taken, high peaks of activity will produce perceptible ($>$50-ms) clips with a frequency that may be unacceptable when operating at a DSI advantage ratio of two. This deficiency of TASI is overcome by adopting the bit reduction strategy, which has been shown to be very effective. However, introduction of the bit reduction strategy complicates voice spurt channel assignment protocol and raises the implementation cost.

The salient features of the SPEC and TDMA techniques are compared in Table 2.

Conclusions

Both the TASI and SPEC methods of digital speech interpolation offer significant enhancements in the capacity of digital transmission for speech communications. Both methods achieve interpolation advantages greater than two, with SPEC attaining slightly higher values than TASI when the number of channels processed is small.

The principal cause of degradation in the TASI method is initial clips of speech spurts. The frequency of occurrence of destructive initial clips can

TABLE 2. COMPARISON OF SPEC AND TDMA TECHNIQUES

	SPEC	TASI
DSI Advantage for 240 Incoming Channels*	2.2	2.2
DSI Advantage for 60 Incoming Channels*	2.1	2.0
Susceptible to Speech Spurt Clipping without Special Precautions	No	Yes
Requires Changes in Channel Assignment Protocol During Overload	No	Yes
Uses Miniprocessor for Channel Assignment Control Connectivity Map	No	Yes
Susceptible to Connect Message Channel Overload	No	Yes
Relative Hardware Complexity	Lower	Higher
Relative Cost of Transmit Terminal	Lower	Higher
Relative Cost of Receive Terminal	Same	Same
Efficient Application in Multidestinational Service	Same	Same

*Based on an assumed voice spurt activity of 40 percent.

be kept low enough to produce little degradation by properly adjusting the interpolation advantage. The threshold of acceptability is based on the condition that the probability of initial clips longer than 50 ms must be no greater than 2 percent. Under this condition, clips longer than 50 ms will occur once every 1.5 minutes. The technique of bit reduction permits additional transmission channels to be made available during occasions of overload, which reduces the probability of occurrence of initial clips. If the least significant bit of each 8-bit PCM sample in the transmit frame is used to generate 7-bit PCM sample slots, the probability of occurrence of initial clips is reduced by more than an order of magnitude.

The principal cause of degradation in SPEC is the production of prediction distortion. The amount of distortion varies as the fraction of samples predicted varies in response to changes in the ensemble average of voice spurt activity. The fraction of samples predicted and hence the amount of prediction noise produced is controlled by the interpolation advantage, as expected. The design objective is to adjust the interpolation advantage so that the probability that more than 25 percent of the samples are predicted during speech spurts is 0.25. It is shown that on the average this results in approximately 0.5-dB degradation in the subjectively assessed speech power-to-quantization-noise power ratio. SPEC is an adaptive system that yields to occasions of higher than average voice spurt activity by automatically increasing the fraction of samples predicted. This is an inherent feature of this technique and at no time does SPEC produce damaging initial clips.

Both TASI and SPEC methods are suited to transmission systems using digital modulation with each carrier capable of operating in preassignment multidestinational service. In terms of implementation complexity SPEC is simpler because it does not require an assignment map memory nor does it need the bit reduction option suggested for use in TASI. Each method can also be adapted to demand-assignment multidestinational service.

Acknowledgment

The author would like to acknowledge the valuable contributions made by Messrs. H. Suyderhoud and R. Ridings in providing information needed to quantify the SPEC system performance.

References

[1] P. T. Brady, "A Statistical Analysis of On-Off Patterns in 16 Conversations," *Bell System Technical Journal*, Vol. 47, No. 1, January 1968, pp. 73–91.

[2] F. D. Daynonnet, A. Jowsset, and A. Profit, "LeCELTIC: Concentrateur Exploitant les Temps d'Inactivete des Circuits," *L'Onde Electrique*, Vol. 42, No. 426, September 1962.

[3] M. Hashimoto et al., "An Application of the Digital Speech Interpolation Technique to a PCM-TDMA Demand Assignment System," International Conference on Space and Communications, Paris, France, March 1971.

[4] E. Lyghounis, "Il Sistema A.T.I.C.," *Telecommunicazioni*, No. 26, March 1968, pp. 21–30.

[5] I. Poretti, G. Monty, and A. Bagnoli, "Speech Interpolation Systems and Their Applications in TDM/TDMA Systems," International Conference on Digital Satellite Communications, Paris, France, 1972, p. 348.

[6] S. J. Campanella and J. A. Sciulli, "Speech Predictive Encoded Communications," International Conference on Digital Satellite Communications, Paris, France, 1972, p. 342.

[7] J. Sciulli and S. J. Campanella, "A Speech Predictive Encoding Communication System for Multichannel Telephony," *IEEE Transactions on Communications*, COM-21, No. 7, July 1973.

[8] H. Suyderhoud, J. A. Jankowski, and R. P. Ridings, "Results and Analysis of the Speech Predictive Encoding Communications System Field Trial," *COMSAT Technical Review*, Vol. 4, No. 2, Fall 1974, pp. 371–393.

[9] K. Bullington and J. M. Fraser, "Engineering Aspects of TASI," *Bell System Technical Journal*, Vol. 38, No. 2, March 1959.

[10] J. M. Fraser, D. B. Bullock, and H. G. Long, "Overall Characteristics of a TASI System," *Bell System Technical Journal*, Vol. 41, No. 4, July 1962.

[11] H. Midema and M. G. Schachtman, "TASI Quality-Effect of Speech Detectors and Interpolation," *Bell System Technical Journal*, Vol. 41, No. 4, July 1962.

[12] R. Ahmend and R. Fatechand, "Effect of Sample Duration on the Articulation of Sounds in Normal and Clipped Speech," *Journal of the Acoustical Society of America*, Vol. 31, No. 7, July 1959, p. 1022.

Section 3.2
Analog Modulation

The dominant modulation technique in current satellite communications systems is frequency modulation (FM). FM is widely used in broadcasting and terrestrial radio links so that the theory was well understood and equipment was available when satellite communications systems began operation. There are two new issues that arise in the application of FM in satellite communications:

1) In order to use the available satellite power efficiently, it is desirable to operate the satellite TWTA near saturation.

2) In most cases, there will be several earth stations accessing the transponder at the same time using different carriers. The transmission of multiple carriers through a nonlinear amplifier generates intermodulation products.

The papers in this Section address these issues as well as other impairments encountered in the communications channel.

The signals received at the earth stations are of four types:

1) groups of voice-band analog signals,
2) single-channel signals,
3) analog video signals,
4) wideband (56 kbit/s or 1.544 Mbit/s) digital signals.

In a trunking system such as INTELSAT or COMSTAR, the telephony signals are in groups or supergroups consisting of 12 or 60 voice channels which have been frequency division multiplexed (FDM) onto a carrier. Each voice channel is a nominal 4-kHz channel which occupies a 300–3400 Hz frequency band. The standard group occupies the 60–108 kHz band. Note that these channels will generally contain voice signals but may also carry multiplexed teletype or data signals. If there is more than one carrier accessing the transponder at the same time, they are separated in frequency, and the system is referred to as a frequency division multiple access (FDMA) system.[1] Thus, the first problem of interest is an FDM/FM/FDMA system. Reprint Paper 3.2.1 discusses transmission planning for the INTELSAT IV system and demonstrates the analysis techniques commonly used for FDM/FM/FDMA systems (see [1] also).

Paper 3.2.2 contains a detailed discussion of the modeling necessary to adequately characterize and analyze the communication system. The importance of utilizing analysis, simulation, and measurement in an integrated manner is illustrated. Although both of these reprint papers discuss the INTELSAT IV satellite, the techniques apply to most satellite systems.

The second type of signal received at the earth station is a single voice-band (i.e., 4-kHz) channel signal. These signals occur in thin route systems (e.g., the Telesat system discussed in Reprint Paper 2.1) or on the small traffic routes in larger systems. One method of transmitting these signals is to modulate each voice channel onto a separate carrier using FM or companded FM. These systems are referred to FM/SCPC (single-channel-per-carrier) or CFM/SCPC systems and are discussed in Section 3.6.2.

The third type of signal is an analog video signal. The FM transmission of television signals is discussed in Reprint Paper 3.2.1. The fourth type of signal is a wideband data signal. These signals will be transmitted using digital modulation which is discussed in Section 3.3.

The final reprint paper in this section, Paper 3.2.3, is a theoretical analysis of the effects of intermodulation and AM/PM conversion in multicarrier TWT systems. References [2]–[7] discuss various interface results for FM systems.

REFERENCES

[1] R. J. Westcott, "Investigation of multiple FM/FDM carriers through a satellite TWT operating near to saturation," *Proc. IEE,* vol. 114, no. 6, pp. 726–740, June 1967.

[2] N. K. M. Chitre and J. C. Fuenzalida, "Baseband distortion caused by intermodulation in multicarrier FM systems," *COMSAT Tech. Rev.,* vol. 2, no. 1, Spring 1972.

[3] R. C. Chapman, Jr. and J. B. Millard, "Intelligible crosstalk between frequency modulated carriers through AM-PM conversion," *IEEE Trans. Commun. Tech.,* vol. COM-12, no. 2, pp. 160–166, June 1964.

[4] A. L. Berman and C. E. Mahle, "Non-linear phase shift in traveling wave tubes as applied to multiple-access communications satellites," *IEEE Trans. Commun. Tech.,* vol. COM-18, no. 1, pp. 37–48, Feb. 1970.

[5] R. G. Medhurst, "FM interfering carrier distortion: General formula," *Proc. IEE* (UK), vol. 1098, no. 44, pp. 149–150, Mar. 1962.

[6] M. Wachs, "Analysis of adjacent channel interference in a multicarrier FM communications system," *COMSAT Tech. Rev.,* vol. 1, no. 1, Fall 1971.

[7] B. A. Pontano, J. C. Fuenzalida, and N. K. M. Chitre, "Interference into angle-modulated systems carrying multichannel telephony," *IEEE Trans. Commun.,* June 1973.

[1] FDMA will be discussed in detail in the multiple access section (3.6.2). However, it is necessary to introduce multiple carriers at this point in order to meaningfully discuss analog modulation.

3.2.1
Systems planning

J. L. DICKS, P. H. SCHULTZE, AND C. H. SCHMITT

Transmission system planning

After the essential satellite characteristics of INTELSAT IV had been determined and the satellite specification developed, it was possible to prepare detailed transmission plans and standards for earth stations to work with INTELSAT IV. The objective was to achieve the maximum channel capacity and specified channel performance within the constraints imposed by satellite and earth station characteristics. For the initial planning stage, only conventional multiple-access FDM/FM was considered. This limitation was necessary to provide a gradual transition, in terms of operational considerations and equipment changes, from INTELSAT III to INTELSAT IV satellites. Transmission characteristics for the more advanced techniques such as TDMA and a SPADE-type system were developed later.

Telephony Performance Objectives. The INTELSAT IV system has been designed to meet all C.C.I.R. and C.C.I.T.T. recommendations applicable to satellite communications. For example, the C.C.I.R. recommendation on channel noise performance (Rec. 353–2 [1]) states that the noise power in a telephone channel should not exceed the following values:

a. 10,000-pW psophometrically weighted mean power in any hour,
b. 10,000-pW psophometrically weighted 1-minute mean power for more than 20 percent of any month,
c. 50,000-pW psophometrically weighted 1-minute mean power for more than 0.3 percent of any month, and
d. 1,000,000-pW unweighted (with an integrating time of 5 ms) for more than 0.03 percent of any month.

This recommendation obviously had a direct impact on the design of the transmission parameters. In line with this recommendation, the following noise breakdown was adopted for the design of the INTELSAT IV satellite communications link:

thermal noise at the satellite (up-link) + thermal noise at the earth station (down-link) + satellite intermodulation noise:	7,500 pWp
earth station out-of-band intermodulation:	500 pWp
interference noise:	1,000 pWp
earth station equipment noise:	1,000 pWp
total	10,000 pWp

The appropriate recommendations for intelligible crosstalk and impulse noise were considered in the design of carrier sizes and their arrangement in the transponder, as well as in the selection of earth station IF filters. Other factors which had a direct impact on the design of the transmission parameters were group-delay distortion, multipath transmission, and adjacent transponder intermodulation.

Typical INTELSAT IV Transponder Operating Characteristics. Transmission characteristics and FDM/FM channel capacities of an INTELSAT IV transponder are largely a function of its mode of operation. For example, the use of a single carrier to access the transponder permits full utilization of the transponder output power and hence maximum channel capacity. However, if the transponders are accessed by multiple FM carriers, traveling wave tube (TWT) nonlinearities will produce intermodulation noise. To limit the intermodulation noise at which the transponder must be operated, an appropriate input backoff, resulting in a less efficient utilization of transponder output power, is employed. Table 4 has been prepared for both spot- and global-beam operation to demonstrate the impact of single- and multiple-carrier operation on transponder operating parameters and achievable channel capacities.

TABLE 4. TYPICAL INTELSAT IV TRANSPONDER OPERATING CHARACTERISTICS

Operating Characteristics	Global Beam		Spot Beam	
	Single Carrier	Multiple Carriers	Single Carrier	Multiple Carriers
Satellite e.i.r.p. at Beam Edge (dBW)				
Specified	22.0	22.0	33.7	33.7
Expected Average	22.5	22.5	34.2	34.2
Satellite Receive G/T (dB/°K)				
Specified	−18.6	−18.6	−18.6	−18.6
Expected Average	−17.6	−17.6	−17.6	−17.6
Selected Gain Setting No.	2	4	4	7
Saturation Flux Density at Beam Edge (dBW/m^2)	−73.0	−67.0	−66.0	−57.0
Optimum TWT Operating Point (dB)	0	−11.0	0	−16.0
Total Carrier-to-Noise Density Ratio, $(C/N_o)_T$ (dB-Hz)	94.6	88.1	105.1	96.0
Total Carrier-to-Noise Ratio, $(C/N)_T$ (dB)	19.0	13.0	29.5	20.9
Transponder Capacity (channels)	960	500	1,800	900

In Table 4, it should be noted that an increase in satellite e.i.r.p. of approximately 12.0 dB when switching from the global-beam mode to the spot-beam mode results in a capacity increase of only 3 dB. This indicates that the achievable transponder capacity is not merely a function of the available satellite e.i.r.p., but that it is also a function of the achievable total carrier-to-noise density ratio and the available transponder bandwidth.

Optimization of Carrier-to-Noise Density Ratio for Multicarrier FDMA. The impairments suffered by FDM/FM carriers passing through a transponder operating in the multiple-access mode are caused by a number of factors. The nonlinear characteristic of the satellite TWT will create an intermodulation spectrum which causes interference noise in the wanted carriers. The amount of intermodulation noise encountered in a telephony channel is primarily a function of the TWT operating point, the level and composition of the RF spectrum, and the relationship of the RF spectrum to the wanted carrier spectrum. Extensive studies have been performed at COMSAT Laboratories to accurately calculate and predict the intermodulation noise in a telephony channel. From these studies, a computer model has been developed which has been successfully verified in a laboratory simulation test program.

In systems calculations, it is useful to express the channel intermodulation noise as an equivalent carrier-to-intermodulation noise density, (C/N_o). Hence, the equivalent average intermodulation noise density in the transponder and the appropriate carrier power level adjustment for equalization of the channel noise performance can be determined. Figure 9 shows the RF intermodulation spectrum for a typical frequency plan, the computed equivalent carrier-to-noise density ratio for each carrier, and the

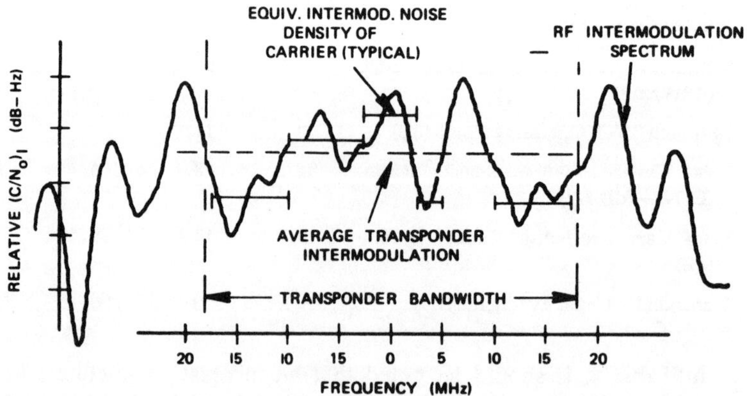

Figure 9. *Intermodulation Spectrum of Typical Frequency Plan*

corresponding average transponder intermodulation. Figure 10 shows the typical variation of the intermodulation as a function of the TWT operating point. The total intermodulation carrier-to-noise density ratio is a composite of intermodulation noise caused by amplitude and phase nonlinearities of the satellite TWT.

The total carrier-to-noise density ratio, $(C/N_o)_T$, as a function of the TWT operating point is the sum of thermal up-link noise, thermal down-link noise, and satellite intermodulation noise:

$$\frac{1}{\left(\dfrac{C}{N_o}\right)_T} = \frac{1}{\left(\dfrac{C}{N_o}\right)_U} + \frac{1}{\left(\dfrac{C}{N_o}\right)_D} + \frac{1}{\left(\dfrac{C}{N_o}\right)_I} \, . \tag{1}$$

The carrier-to-noise density ratio in the up-link is

$$\left(\frac{C}{N_o}\right)_U = W_S + \left(\frac{G}{T}\right)_S - 10 \log \frac{4\pi}{\lambda^2} - 10 \log k - BO_i \tag{2}$$

where W_S = saturation flux density at beam center, in dBW/m²

    $\left(\dfrac{G}{T}\right)_S$ = gain-to-noise-temperature ratio of the satellite at beam center, in dB/°K

    $10 \log \dfrac{4\pi}{\lambda^2}$ = gain of 1-m² aperture at the transponder center frequency, in dB

    $10 \log k$ = Boltzmann's constant (-228.6 dBW/°K Hz)

    BO_i = TWT input backoff relative to single-carrier saturation, in dB.

The carrier-to-noise density ratio in the down-link is

$$\left(\frac{C}{N_o}\right)_D = e.i.r.p._{sat} - PL_{4\,GHz} + \left(\frac{G}{T}\right)_E - \Delta D - 10 \log k - BO_o \tag{3}$$

where $e.i.r.p._{sat}$ = satellite e.i.r.p. at beam center and single-carrier saturation, in dBW

    $PL_{4\,GHz}$ = 195.6-dB path loss from the satellite to the sub-satellite point at 4 GHz

    $\left(\dfrac{G}{T}\right)_E$ = 40.7-dB/°K earth station gain-to-noise temperature ratio at 4 GHz

    ΔD = average down-link adjustment factor for path loss and antenna gain in dB

    BO_o = TWT output backoff relative to single-carrier saturation in dB.

Since path loss and earth station antenna gain have the same frequency dependence, equation (3) can be applied to frequencies other than 4 GHz.

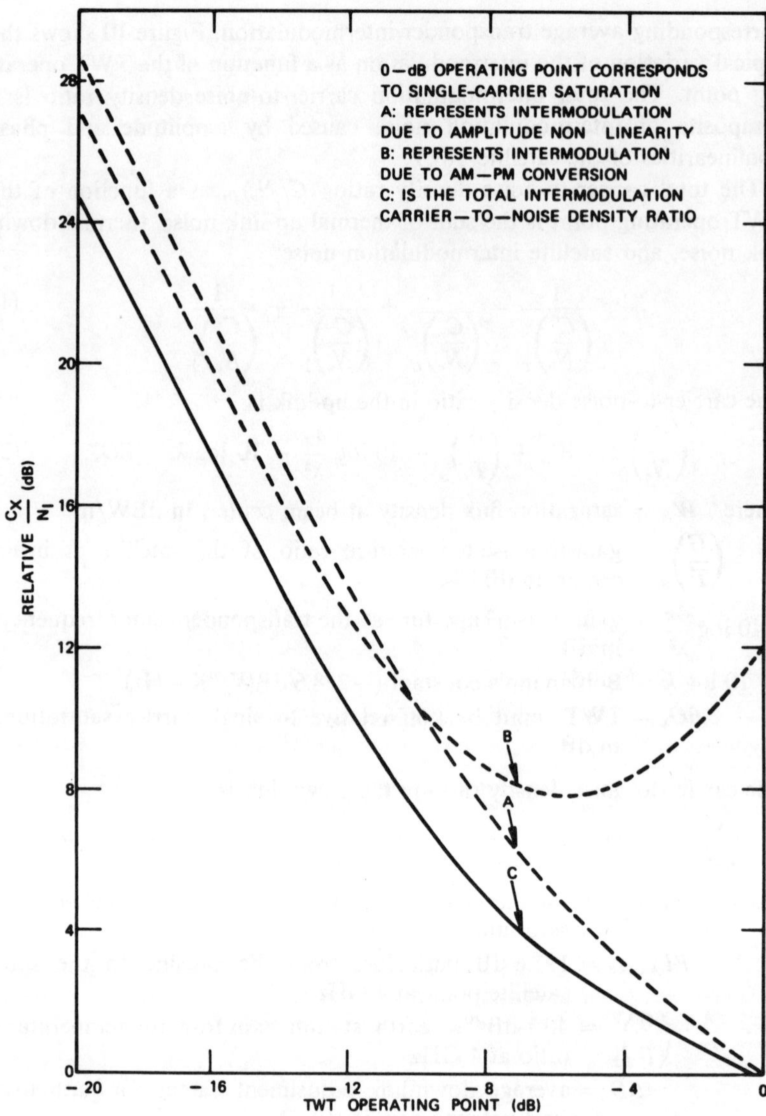

Figure 10. *Relative Carrier-to-Intermodulation Noise Ratio vs TWT Operating Point*

Figure 11 is a graphical presentation of the resulting $(C/N_o)_T$ as a function of the TWT operating point. [The optimum TWT operating point is defined as the point where $(C/N_o)_T$ reaches its maximum.]

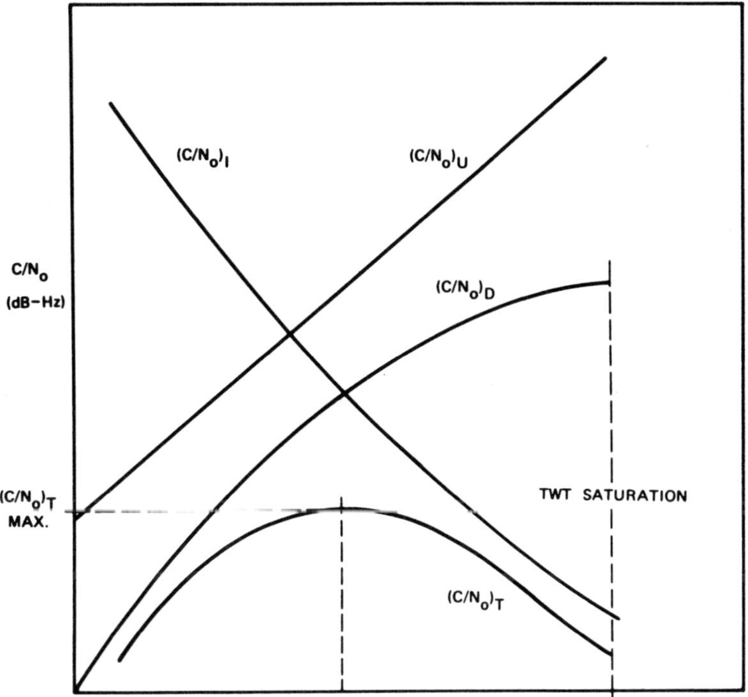

Figure 11. *Optimum TWT Operating Point*

Transponder Gain Settings. Switchable attenuator settings in the INTELSAT IV satellite make it possible to vary the saturation flux density in discrete steps. If the gain of the satellite is reduced, the earth station transmit power must be increased, resulting in a better up-link carrier-to-noise density ratio. An attenuation range of 24 dB makes it possible to adjust the up-link noise for various modes of operation so that it does not become the noise-limiting factor.

Figure 12 shows a typical gain setting tradeoff for global-beam multi-carrier operation. It can be seen that reducing the gain from gain setting no. 1 to gain setting no. 4 results in a capacity increase of 25 percent, whereas a further gain reduction will have no significant impact on the transponder capacity. Once thermal down-link and satellite intermodulation noise become the limiting factors, a further reduction of thermal up-link noise becomes relatively inconsequential.

The required earth station power per channel was used as a criterion for selecting the appropriate gain settings for a particular mode of operation. Power requirements of approximately 1 to 2 watts/channel yield a

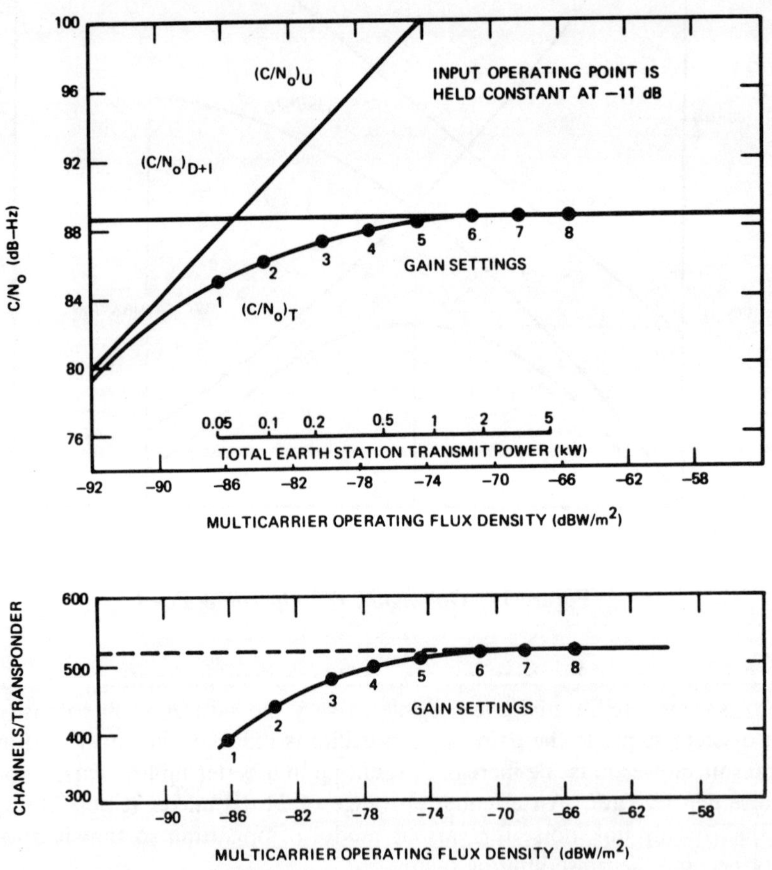

Figure 12. *Carrier-to-Noise Density Ratio and Channel Capacity as a Function of Flux Density (Global Beam)*

reasonable tradeoff between satellite capacity and earth station power requirements.

Selection of Carrier Sizes and Parameters. Because of the traffic requirements in the global network, a wide range of carrier capacities had to be accommodated by the satellite. Hence, it was necessary to introduce a large number of carrier sizes in reasonable increments so that a reasonable ratio between available and actual used capacity, or fill factor, could be achieved. Bandwidth units of multiples of 2.5 MHz were selected to allow a flexible interchange of carriers within a given frequency range without requiring frequency changes among adjacent carriers. Furthermore, it was desirable to obtain the carrier sizes from multiples of channel groups and supergroups.

The transmission parameters for each carrier size were calculated by using the standard FM equation and Carson's Rule of bandwidth,

$$\left[\frac{S}{N}\right] = \left[\frac{C}{N}\right]\frac{B}{b}\left[\frac{f_r}{f_m}\right]^2 P \cdot W$$

$$B = 2[3.16\, gf_r + f_m]$$

where

$\left[\dfrac{S}{N}\right]$ = weighted signal-to-noise ratio at the 1-mW test-tone level = 51.2 dB for 7500-pWp channel noise

$\left[\dfrac{C}{N}\right]$ = carrier-to-noise ratio over the Carson's Rule bandwidth, B

b = channel bandwidth = 3.1 kHz

f_r = rms test-tone deviation

f_m = maximum baseband frequency $\simeq 4.2 \times n$, in kHz

n = number of telephone channels

P = psophometric weighting factor = 2.5 dB

W = pre-emphasis weighting factor = 4 dB

g = antilog $\left[\dfrac{L}{20}\right]$

$L = -15 + 10 \log n,\ n \geqslant 240$ channels,

$\quad = -1 + 4 \log n,\quad n < 240$ channels.

Once the number of channels per carrier and the desired signal-to-noise ratio, S/N, are fixed, the carrier-to-noise temperature ratio, C/T, can be traded for RF bandwidth, B. The only constraint in this tradeoff is that the carrier-to-noise ratio, C/N, should not fall below a minimum of 13 dB to avoid operation in the threshold region of the demodulator.

Figure 13 shows the number of channels per carrier as a function of RF bandwidth for spot- and global-beam operation, a noise allocation of 7,500 pWp, and guardbands of 10 and 20 percent. These curves and the points discussed previously were used to select the INTELSAT IV carrier sizes. Tables 5 and 6 show the transmission parameters for all spot- and global-beam carrier sizes selected for use on INTELSAT IV satellites.

Transmission Impairments. As shown in Table 7, transmission impairments in a communications satellite link can be classified according to the sources of degradation. These impairments can occur in the satellite or at the earth stations, or they can be caused by linear or nonlinear devices in the communications link. In addition, there can be dual-path transmissions, in which signal transmitted through two independent paths, produces interference noise. More specifically, the various sources of transmission impairments which can be encountered in a communications satellite link are

a. thermal up-link and down-link noise,
b. group-delay noise,
c. intelligible crosstalk,
d. dual-path transmission,
e. impulse noise, and
f. RF out-of-band intermodulation.

Thermal up-link noise, thermal down-link noise, and satellite intermodulation noise have already been considered in the carrier-to-noise density optimization procedure and will therefore be excluded from this discussion.

Group-delay noise caused by the nonlinear phase characteristics of the transmission path can occur at the earth stations as well as in the satellite. The use of group-delay equalization at the transmitting earth station makes it possible to keep the group-delay noise in a link within the 500-pWp allowance. (See Fig. 25.)

Intelligible crosstalk is caused by nonlinearities in the gain-frequency response of a transmission path which includes an active device such as an earth station high-power amplifier (HPA) or a satellite TWT with inherent AM/PM conversion characteristics. Intelligible crosstalk can be kept at an acceptable level by equalizing the gain-frequency response or, in the case of nonlinearities in the satellite, by predistorting the gain-frequency response at the transmitting earth station.

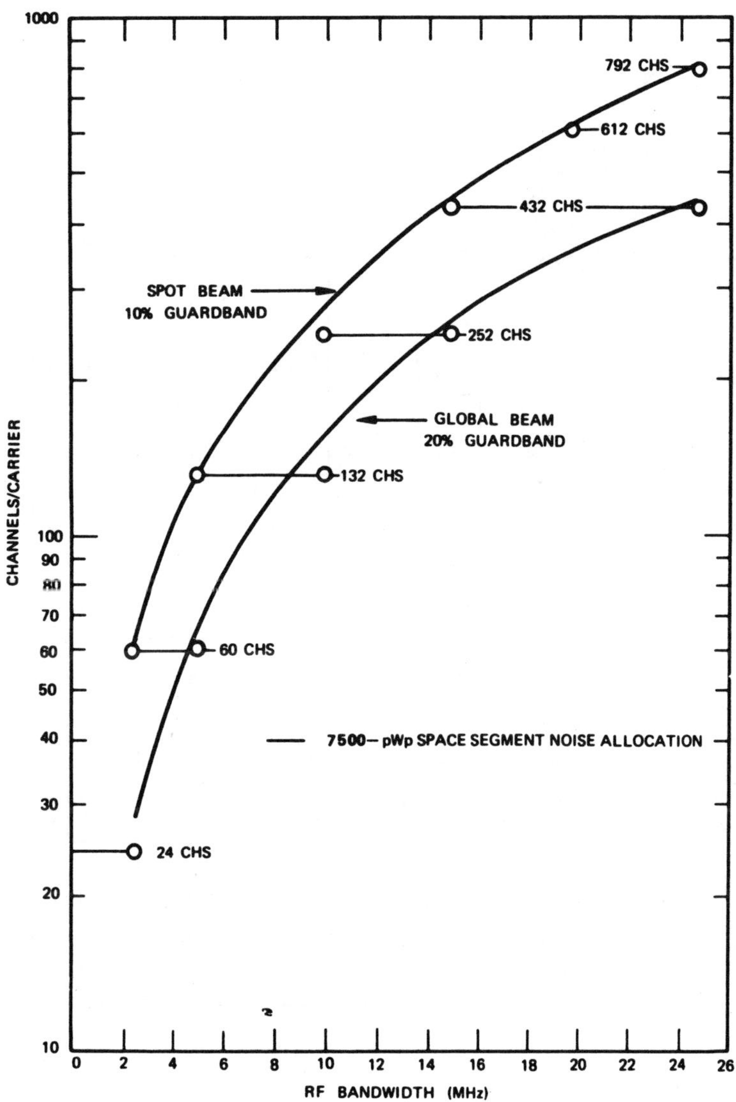

Figure 13. *Channel Capacity per Carrier as a Function of RF Bandwidth*

TABLE 5. GLOBAL-BEAM INTELSAT IV TRANSMISSION PARAMETERS (8,000 pWp) FOR STANDARD AND EXPANDED CARRIERS

Channels/Carrier, n	Top Baseband Frequency, f_m (kHz)	Allocated Satellite BW Unit (MHz)	Occupied BW ($b)_{IF}$ at Earth Station) (MHz)	rms Multicarrier Deviation, f_{rms} (kHz)	Carrier-To-Total Noise Temp. Ratio, $(C/T)_T$ (dBW/°K)	Carrier-To-Noise Ratio in Occupied BW, C/N (dB)	Satellite e.i.r.p. at Beam Edge (dBW)	Earth Station e.i.r.p. at 10° Elevation (dBW)
24	108.0	2.5	2.00	275.0	−153.0	12.7	5.8	74.7
36	156.0	2.5	2.25	307.0	−150.0	15.1	8.8	77.7
60	252.0	2.5	2.25	276.0	−144.0	21.1	14.8	83.7
60	252.0	5.0	4.00	546.0	−149.9	12.7	8.9	77.8
72	300.0	5.0	4.50	616.0	−149.1	13.0	9.7	78.6
96	408.0	5.0	4.50	584.0	−145.5	16.6	13.3	82.2
132	552.0	5.0	4.40	529.0	−141.4	20.7	17.4	86.3
96	408.0	7.5	5.90	799.0	−148.2	12.7	10.6	79.5
132	552.0	7.5	6.75	891.0	−145.9	14.4	12.9	81.8
192	804.0	7.5	6.40	758.0	−140.6	19.9	18.2	87.1
132	552.0	10.0	7.50	1020.0	−147.1	12.7	11.7	80.6
192	804.0	10.0	9.00	1167.0	−144.4	14.7	14.4	83.3
252	1052.0	10.0	8.50	1009.0	−139.9	19.4	18.9	87.8
252	1052.0	15.0	12.40	1627.0	−144.1	13.6	13.9	82.8
312	1300.0	15.0	13.50	1716.0	−141.7	15.6	16.3	85.2
432	1796.0	17.5	15.75	1919.0	−138.5	18.2	18.0	88.0
432	1796.0	20.0	18.00	2276.0	−139.9	16.1	17.7	86.6
432	1796.0	25.0	20.70	2688.0	−141.4	14.1	16.2	85.1
972	4028.0	36.0	36.00	4417.0	−135.2	17.8	22.4	90.1
1092	4892.0	36.0	36.00	4118.0	−132.4	20.7	23.7	93.6

TABLE 6. SPOT-BEAM INTELSAT IV TRANSMISSION PARAMETERS (8,000 pWp) FOR STANDARD AND EXPANDED CARRIERS

Channels/Carrier, n	Top Baseband Frequency, f_m (kHz)	Allocated Satellite BW Unit (MHz)	Occupied BW (b_{IF}) at Earth Station (MHz)	rms Multicarrier Deviation, f_{mc} (kHz)	Carrier-To-Total Noise Temp. Ratio, $(C/T)_T$ (dBW/K)	Carrier-To-Noise Ratio in Occupied BW, C/N (dB)	Satellite e.i.r.p. Beam Edge (dBW)	Earth Station e.i.r.p. at 10° Elevation (dBW)
60	252.0	2.5	2.25	276.0	−144.0	21.1	15.2	81.4
72	300.0	2.5	2.25	261.0	−141.7	23.4	17.5	83.7
132	552.0	5.0	4.40	529.0	−141.4	20.7	17.7	83.9
192	804.0	5.0	4.50	459.0	−136.3	25.8	22.8	89.0
192	804.0	7.5	6,40	758.0	−140.6	19.9	18.5	84.7
252	1052.0	7.5	6.75	733.0	−137.1	23.2	22.0	88.2
252	1052.0	10.0	8.50	1009.0	−139.9	19.4	19.2	85.4
312	1300.0	10.0	9.00	1005.0	−137.1	22.0	22.0	88.2
432	1796.0	15.0	13.00	1479.0	−136.2	21.2	22.2	88.4
612	2540.0	20.0	17.80	1996.0	−134.2	21.9	23.9	90.1
792	3284.0	20.0	18.00	1784.0	−129.9	26.2	28.2	94.4
792	3284.0	25.0	22.40	2494.0	−132.8	22.3	25.3	91.5
972	4028.0	25.0	22.50	2274.0	−129.4	25.7	28.7	94.9
1872	8120.0	36.0	36.00	3181.0	−123.5	29.5	34.2	98.6

TABLE 7. CLASSIFICATION OF TRANSMISSION IMPAIRMENTS

| Transmission Path | Satellite | | Earth Station |
	Single Transponder	Two Transponders	
Linear	Thermal Noise (up-link)	Dual-Path Group-Delay Distortion	Thermal Noise (down-link)
	Group-Delay Distortion		Group-Delay Distortion (envelope distortion)
			Amplitude/Frequency Distortion
			Impulse Noise
Nonlinear	Intermodulation	Adjacent Transponder Intermodulation	Out-of-Band Emission due to
	Intelligible In-Band Crosstalk	Intelligible Out-of-Band Crosstalk	HPA multicarrier operation

Dual-path transmission occurs when a signal near the band edge of a transponder is also transmitted through the adjacent transponder at a reduced level. This impairment source will introduce phase nonlinearities in the wanted signal path and can be treated as group-delay noise. The amount of noise encountered is primarily a function of the filter response of the adjacent transponder, the gain difference between the two transponders, and the mode of operation of the two transponders (spot or global beam). The dual-path transmission effect was one of the major inputs to the INTELSAT IV filter specification and it is one of the factors which must be considered in the development of frequency plans.

Impulse noise is formed in the demodulator when the instantaneous noise or interference power level equals or exceeds the carrier power level. This impairment may be caused by the threshold effect at low carrier-to-noise ratios (not expected to be encountered on INTELSAT IV operation), bandlimiting within a carrier selection filter, or adjacent carrier interference. Thus, the major factors controlling impulse noise are carrier arrangements, relative power levels resulting from different carrier sizes and geographical advantages, frequency deviations, and filter performance. These factors and their variations must be considered in the development of frequency plans. The transmit and receive IF filter specifications for INTELSAT IV carriers were primarily based on impulse noise considerations.

RF out-of-band intermodulation in the satellite and earth station occurs when a number of carriers are transmitted through the same nonlinear amplifier. The RF out-of-band emission from the earth station is caused by the transmission of two or more carriers through the same HPA.

A noise allowance of 500 pWp for RF out-of-band emission restricts the earth station HPA drive to a level which will produce a power flux density of 26 dBW/4 kHz at a 10° elevation angle. At higher angles, the flux density should be reduced by the empirical factor of 0.06 $(\alpha - 10)$. RF out-of-band emission in the satellite is a function of the frequency plan, the TWT operating point, the gain-frequency response of the satellite output multiplexer, and finally the relationship between the wanted and interfering spectra, which determines the amount of noise in a baseband channel.

In developing the satellite filter specifications it was found that for typical frequency plans the RF out-of-band intermodulation noise could be neglected. Except in cases in which two large carriers are transmitted through a transponder, the interference noise into the adjacent transponder must be carefully evaluated.

Before the INTELSAT IV carriers were introduced into the operational system, a laboratory transmission simulation was performed through INTELSAT IV transponders. The purpose of this system simulation program was to ensure that all transmission impairments had been properly considered in the design of the system and that the overall systems performance would be as expected. Mathematical models were developed for individual transmission effects and then verified by measurements. The results of the simulation program indicated that all transmission impairments were within the expected limits.

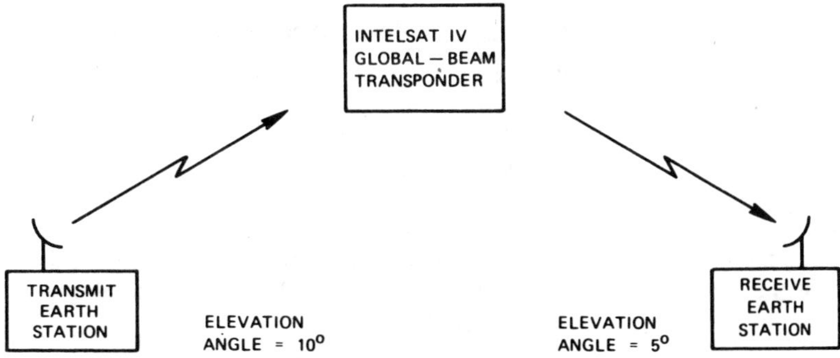

Figure 14. INTELSAT IV *Television Reference Link*

Television Performance. The transmission parameters and performance objectives for the INTELSAT IV television channel are described in Ref. 2. It is convenient to apply these parameters and objectives to a hypothetical earth station-to-earth station link, which will be called the INTELSAT IV television reference link. Such a link, which is shown in Figure 14, is consistent with the transmit earth station elevation angle of 10° and the reference receive station elevation angle of 5° for television.

The television signal-to-noise ratio can be written as

$$\frac{S}{N} = \frac{3}{2} d^2 \frac{C}{T} \frac{1}{f_{v_n}} \frac{1}{k} \tag{4}$$

where S = sinusoidal signal average power

N = noise power in the video noise bandwidth.

$d = f_d/f_v$ = peak FM deviation index

C = received carrier power, in W, over the video noise bandwidth

T = system noise temperature, in °K

$k = 1.38 \times 10^{-23}$, in J/°K

f_v = highest modulating frequency, in Hz

f_d = peak frequency deviation, in Hz

f_{v_n} = video noise bandwidth, in Hz.

The C.C.I.R. recommends that the signal-to-noise ratio for TV signals be expressed in terms of the ratio:

$$\frac{\text{peak-to-peak luminance signal}}{\text{weighted rms noise}}. \tag{5}$$

Since for a sine-wave the (power) ratio of the peak-to-peak and rms values is eight, and since the peak-to-peak value of the luminance component is 0.707 of the peak-to-peak value of the composite video signal, the above-mentioned (power) ratio must be halved.

Substitution into equation (4) yields

$$\frac{\text{peak-to-peak luminance signal}}{\text{average noise power}} = \frac{S_{p-p}}{N} = 6d^2 \frac{C}{T} \frac{1}{f_{v_n}} \frac{1}{k} \cdot \tag{6}$$

If pre-emphasis of the video signal is not employed, the noise can be assumed to be triangular in shape and a weighting improvement, w,* can be included in equation (6). Thus,

$$\frac{\text{peak-to-peak luminance signal power}}{\text{weighted average noise power}} = \frac{S_{p-p}}{N_w} = 6d^2 \frac{C}{T} \frac{1}{f_{v_n}} \frac{1}{k} w \cdot \tag{7}$$

If pre-emphasis of the video signal is employed, the complementary de-emphasis network, d, will modify the noise spectrum and $q(w,d)$ will be the weighting improvement after de-emphasis. Therefore,

$$\frac{\text{peak-to-peak luminance signal power}}{\text{weighted average noise power}} = \frac{S_{p-p}}{N_q} = 6d^2 \frac{C}{T} \frac{1}{f_{v_n}} \frac{1}{k} q \cdot \tag{8}$$

When equation (8) is expressed in dB,

$$\left[\frac{S_{p-p}}{N_q}\right]_{dB} = 236.4 + 20 \log d + \left[\frac{C}{T}\right]_T - 10 \log f_{v_n} + Q \cdot \tag{9}$$

In equation (9), it was assumed that the bandwidth allocation of a TV carrier would have the value listed in Table 8. The signal-to-noise ratios, S_{p-p}/N_q, for the various systems are plotted in Figure 15 as a function of $(C/T)_T$.

TABLE 8. BANDWIDTH AND FREQUENCY ALLOCATIONS FOR
ONE VIDEO CARRIER AND TWO AUDIO CARRIERS

Carrier	Bandwidth (MHz)	Frequency Allocation for Transponder No. 12 (MHz)
Video Carrier	30	$f_o = 4178 \pm 15$
Audio Carrier No. 1	2.5	$f_c = 4158.25 \pm 1.25$
Audio Carrier No. 2	2.5	$f_c = 4160.75 \pm 1.25$

* Weighting networks as per C.C.I.R. rec. 421-1 and 451 [3], [4]

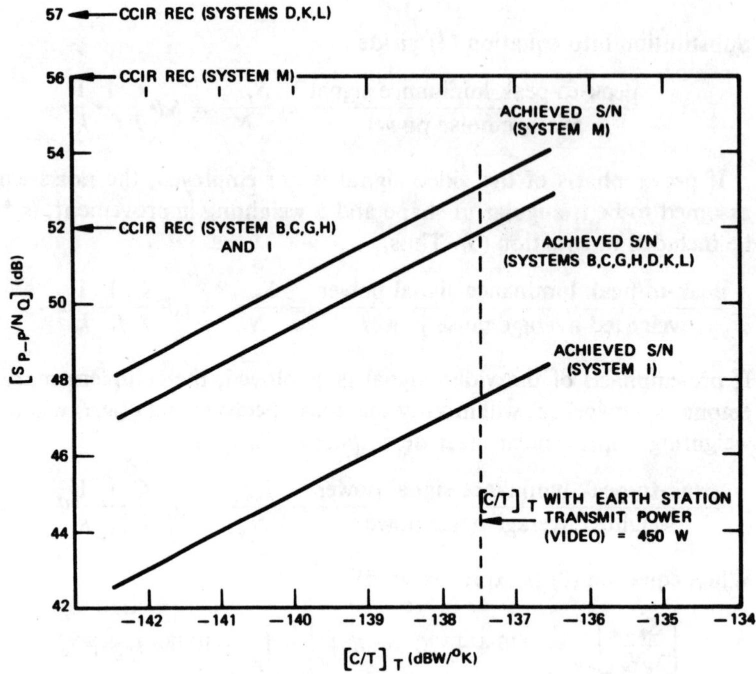

Figure 15. *Video Channel S/N as a Function of* $[C/T]_T$

Transmission modeling

A. L. BERMAN, C. MAHLE, AND M. R. WACHS

Typical signal path

Figure 26 will be used to model the INTELSAT IV communications system. Initially, one carrier will be traced from the modulator, through the spacecraft transponder, to the demodulator to define the distortion effects encountered in this case. Subsequently, the additional distortion effects caused by multicarrier operation will be presented, and finally, the detailed modeling of each of these effects will be described.

Single-carrier operation

At baseband, a stack of FDM telephone channels is represented by the equivalent white noise load from f_a to f_b (see Figure 27). After pre-emphasis, this composite signal band frequency modulates an IF carrier. The process is essentially distortion-free up to this point. A transmit filter following the modulator limits the out-of-band spectrum of the modulated carrier. Out-of-band attenuation, which is dictated by adjacent channel

---

A. L. Berman is Manager, Transponders Department, Technology Division, COMSAT Laboratories.

C. Mahle is Manager, Transponders Design, Technology Division, COMSAT Laboratories.

M. R. Wachs is a Technical Staff Member, COMSAT Laboratories.

Reprinted with permission from *COMSAT Tech. Rev.,* vol. 2, pp. 489–527, Fall 1972.

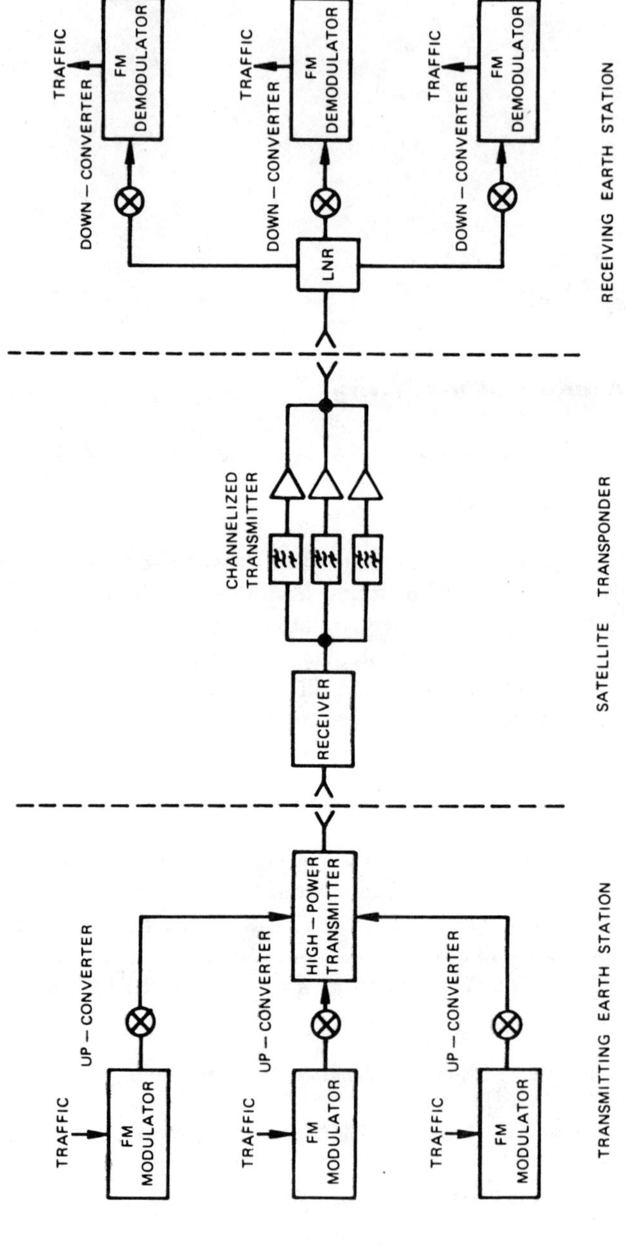

Figure 26. *Block Diagram of* INTELSAT IV *Communications System*

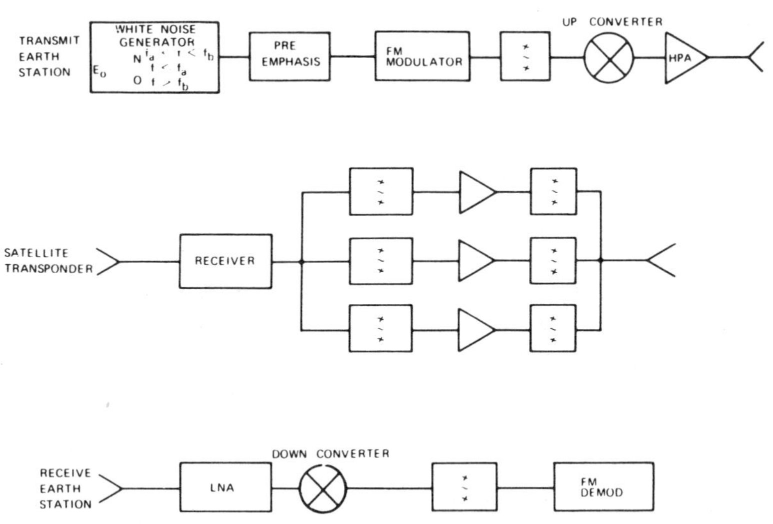

Figure 27. *Single-Carrier Path*

considerations, will be described further in the discussion of dual-carrier operation.

The ideal in-band characteristics of the filter are flat amplitude and group delay over the largest bandwidth which will satisfy the attenuation requirements in the adjacent band. With a realizable filter, several distortion components of the signal will exist at the output of an "ideal" frequency demodulator:

    a. intermodulation noise at baseband caused by finite residual group-delay variation across the Carson's Rule bandwidth [10], [11],*

    b. truncation noise at baseband caused by the abrupt rise of either group delay or attenuation at the filter band edge [12],

    c. an increase in the onset of the impulse threshold caused by momentary suppressions of the modulated carrier when detection is performed in the presence of noise, and

    d. amplitude modulation (AM) components in the carrier envelope that are coherent with the original baseband modulation because of their finite amplitude slope. These typically low-level AM compo-

---

* To compute the intermodulation distortions across the baseband, it is sufficient to resolve the group-delay variation into linear, parabolic, and ripple components and to utilize the formulas given in References 10 and 11.

nents are of interest primarily because they produce crosstalk onto other carriers in subsequent common path amplifiers exhibiting AM/PM conversion effects. (These effects will be described in the discussion of dual-carrier operation.)

The signal path through the up-converter, high-power transmitter, and satellite receiver can be made essentially distortion free for single-carrier operation. [The illumination level and satellite G/T determine the up-link carrier-to-noise temperature ratio (C/T), and requirements for multiple-carrier operation determine the other major performance characteristics.] At the input to the channelized satellite transmitter, two signal paths are possible:

a. a path through the intended input channel filter, TWT, output combining filter, and antenna to the receiving earth station at approximately zero relative attenuation, or

b. a path through the adjacent channel filters, TWT, and antenna to the receiving earth station at a relative attenuation determined by the skirt selectivity of the adjacent channel filters across the desired channel bandwidth.

The sum of the microwave carrier vectors (i.e., the vector of the intended path plus the vector of the adjacent channel path at a reduced level and different phase) results in the well-known multipath ripple, with overall effect of amplitude and group-delay distortion. The dual-path effect is most severe at the band edge and drops to negligible levels toward midband as the attenuation of the adjacent channel increases. This effect cannot be equalized because of the rapid variations of the undesired channel path length with illumination in that channel, filter tuning variations with temperature, and pointing variations of the spacecraft.

At the input to the earth station receiver, the distortion levels previously described for the single carrier are increased as a result of dual-path and unequalized residual group-delay contributions of the channelizing filters. At this point, the down-link C/T is determined by the received illumination and the earth station G/T. Again, the signal path through the earth station receiver and down-converter may be made essentially distortion free for a single carrier. The function of the receive filter (in conjunction with the transmit filter of the adjacent channel) is primarily that of reducing adjacent channel interference. In the absence of adjacent channel considerations, the receive filter will define a bandwidth just wide enough to be

noise presented to the demodulator. This will provide a composite (carrier plus noise) signal that can be used to derive the automatic gain control at the detector input.*

The distortion components at the output of the receive filter are identical to those described previously for the transmit filter.

The state of the art in demodulator design [13] permits an essentially linear detection of all signals with intermodulation contributions of less than 100 pWp.

Multicarrier operation

In addition to the baseband distortion components of each single carrier (primarily in-band linear, parabolic, and ripple group-delay coefficients), three other components have a significant effect on multicarrier operation of the INTELSAT IV satellite: adjacent channel interference, RF intermodulation distortion (caused by nonlinear P_{out} vs P_{in} and $\Delta\phi_{out}$ vs P_{in} characteristics), and coherent crosstalk.

Adjacent Channel Interference. To achieve the maximum channel capacity in a bandwidth-limited mode such as that encountered by the INTELSAT IV satellite, the number of channels within each bandwidth unit is maximized, and the guardband between adjacent channels is reduced until the resulting distortions rise to a level of several hundred picowatts.† In the case of FDM/FM telephony transmissions, two interference effects occur as the guardband is reduced:

    a. Spectrum overlap giving rise to "convolution" noise in the baseband (measured in pWp). For purposes of computation, this noise component, calculated from the convolution of two spectra, may be assumed to have a Gaussian amplitude distribution in the baseband.

    b. Impulse noise caused when the resultant RF vector at the demodulator input undergoes an abrupt rotation of 2π radians as the

---

* In the case of phase-lock demodulators, a constant level into the phase detector is required to achieve a constant loop gain. For conventional detection above threshold with a limiter discriminator, the filter bandwidth will determine the C/N ratio.

† This would provide a reasonable margin (3–6 dB) against accidental overdeviation.

adjacent carrier momentarily deviates into the desired channel bandwidth. This effect produces impulses at baseband, similar to those observed at the onset of the FM threshold, which can be measured as average noise in a 4-kHz voice channel and on a counter to determine the number of impulses above the -22-dBm0 reference level in one minute.

For FDM/FM telephony, the multiplexed baseband has a Gaussian amplitude statistic (i.e., links are tested with bandlimited white noise); therefore, large deviations for small periods of time are inherent. In this case, receive filters alone are insufficient to control the interference effects described in the preceding paragraph. As noted previously, a combination of a transmit filter on the interfering channel (to limit its instantaneous frequency excursions and spectrum) and a receive filter on the "desired" channel (to limit the spectrum presented to the demodulator) are required.

For INTELSAT IV, representative adjacent channel cases were selected. (A 10-MHz carrier positioned next to a 2.5-MHz carrier was assumed to be a practical worst case before frequency plan readjustment.) Each of the transmit/receive filter pairs was then narrowed until the effects of in-band distortion were balanced against adjacent channel interference. In this tradeoff, the ultimate limit of in-band noise caused by the truncation effect was used [14] with the understanding that the group-delay response of the filters would be equalized [15] so that truncation noise would be the dominant contributor.

RF Intermodulation Caused by Multicarrier Operation. Analysis and experimental confirmation of RF intermodulation levels produced by amplitude and phase nonlinearities typical of traveling wave tubes have been previously published [16-19]. A brief summary of the base model used for computation, the series representations of the nonlinearities, and the logic algorithm used to sort the products will be given in a subsequent subsection.

Crosstalk Caused by FM/AM/PM Conversion. The modeling technique presented here was originally developed to determine the extent to which the transponder channelizing filters could be permitted to have rounded corners caused by finite Qs and still meet the multicarrier crosstalk specifications. To evaluate this effect for more than two carriers, a statistical approach was used to compute the AM component, and the 2-carrier AM/PM conversion coefficient was replaced by a direct computation of the crosstalk term of interest from the series representation of $\Delta\phi_{out}$ vs P_{in}.

Description of the individual distortion effects

For convenience, the distortion components in this subsection will be grouped into several relatively independent effects:

a. multipath considerations (satellite transmitter channelizing filter selection),
b. TWT nonlinearities resulting in RF intermodulation products,
c. crosstalk considerations, and
d. adjacent channel interference (balancing of in-band and adjacent channel effects).

Multipath. The dual path under consideration is shown in Figure 28. The basic procedure for evaluating the distortion consists of computing the desired (intended path) vector and adding it to the unwanted (adjacent channel) vector. The resultant vector across the band of interest gives the channel amplitude response. Differentiating* the phase response with respect to frequency makes it possible to obtain the group delay. In this case, with the unwanted path length independent of the desired path length, the computation has been performed for initial mid-band phase differences of 0° to 360° in increments of 30°.

To optimize the INTELSAT IV satellite, the following constraints were assumed at the beginning of the study:

a. no more than 21 microwave cavities per channel should be used for the input filter, output filter, and input filter equalizer because of weight budget limitations;

b. the center-to-center spacing between channels should be 40 MHz, with full communications performance over the center 36 MHz for an arbitrary mixture of RF carriers; and

c. the loss in the output combining network (multiplexer) should be no more than 0.5 dB.

An examination of microwave filters potentially capable of meeting the gain slope requirements (typically 0.01-dB Tchebycheff filters) leads to the following initial configurations:

---

* High-resolution differentiation may be performed efficiently on a digital computer by computing a set of phase values typically every 100 kHz, determining a corresponding second set of phase values displaced by 1 kHz from the first set, and then forming $\bar{\tau}(\omega) = \Delta\phi/\Delta\omega$.

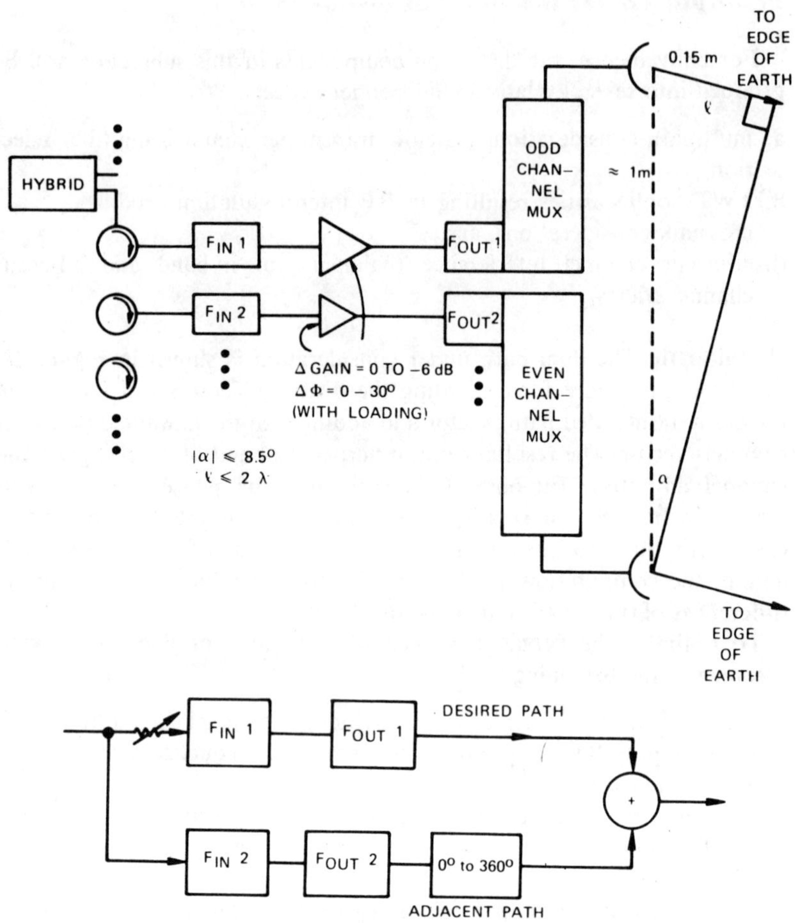

Figure 28. *Multipath Diagram*

   a. For realizable unloaded Qs of 7,000 to 10,000, the loss in the
output section can be met with six sections or less. Two output multiplexers and antennas (one set for even channels and one for odd channels) are required to successfully combine the required output with the
selectivity available from 6-pole filters. Initial computations of required intermodulation product suppression in the adjacent channel
indicated that the output filter should have six sections.

b. Again, based on unloaded Qs of 7,000 to 10,000, the rate of attenuation at the band edge of the adjacent channel reaches the point of diminishing returns for filters with 9 to 11 sections.

c. Since elliptic function filters were not realizable at microwave frequencies in the early stages of the project, only Tchebycheff filters were considered.

d. A minimum of five equalizer poles is required to achieve equalization over most of the 36-MHz band.

On the basis of these assumptions, a set of multipath calculations was performed for the two most probable filter configurations:

a. a 10-pole input filter, a 5-pole equalizer, and a 6-pole output filter; and

b. a 9-pole input filter, a 6-pole equalizer, and a 6-pole output filter.

For both input filter configurations, the equiripple bandwidths were set at 35, 36, 37, 38, and 40 MHz. In addition, for each of these cases, the output filter bandwidth was initially 2 MHz wider than that of the input filter* (to limit the number of cases studied), and the equalizer poles were set to give the best equalization over the center 36 MHz.

The results for the nominal case of a 38-MHz input filter and a 40-MHz output filter are presented in Figures 29–32. The other cases, which are summarized in Tables 11 and 12, show that, for each of the filter sets, the use of a 40-MHz equiripple bandwidth yields a very low inherent group delay (caused only by the filters) across the center 36 MHz. However, the multipath contribution (indicated by the spreading of the curves as the initial phase of the adjacent channel is varied) severely limits the channel

---

* As previously mentioned, because of initial loss and design complexity considerations, there is one output multiplexer for the even channels, and another one for the odd channels. Hence, the diplexer at the output is no longer required to provide total reflection at the edge of the adjacent channel. Instead, it must provide "total reflection" of other signals at least one channel bandwidth (40 MHz) from band edge.

The major factor in determining the output filter bandwidth was the tradeoff between the rapid rise of group delay over the last 10 percent of its equiripple bandwidth (leading to widening of the filter to facilitate equalization) and the requirement for attenuating those intermodulation products falling into the adjacent channel. For the input filter selected, further computations showed a broad maximum of system performance, varying slowly with the choice of output filter equiripple bandwidth.

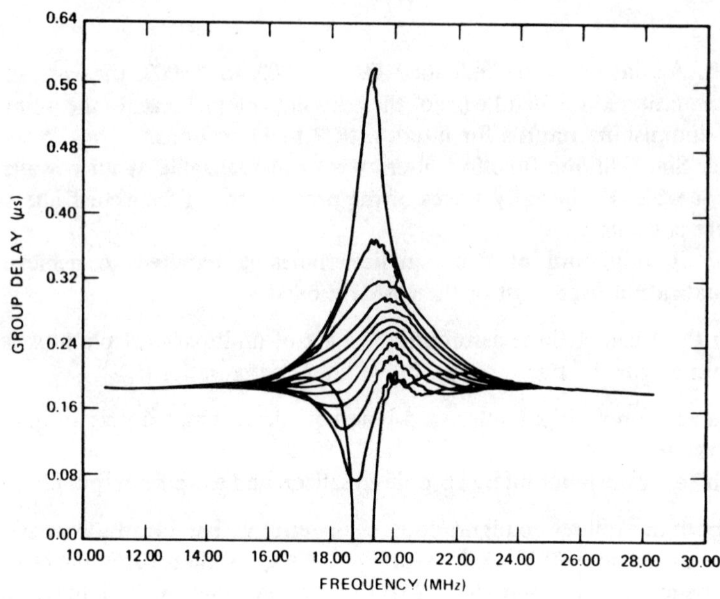

Figure 29. *Sample Output, Group Delay (9-section input filter)*

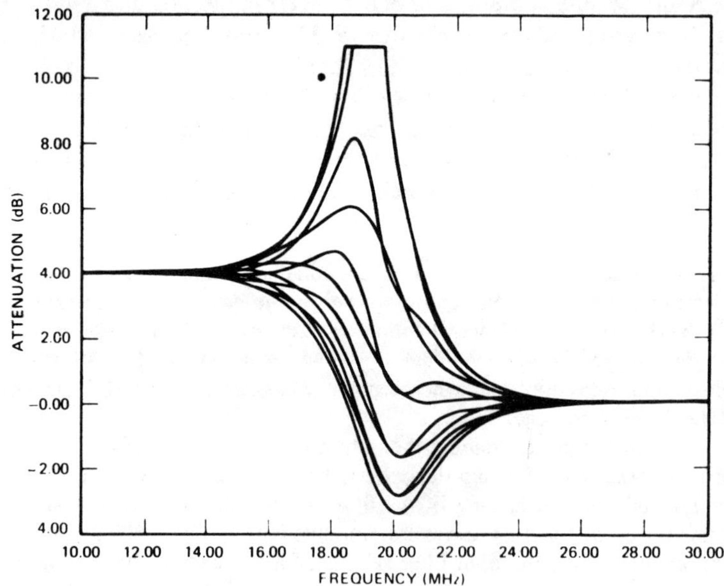

Figure 30. *Sample Output, Attenuation (9-section input filter)*

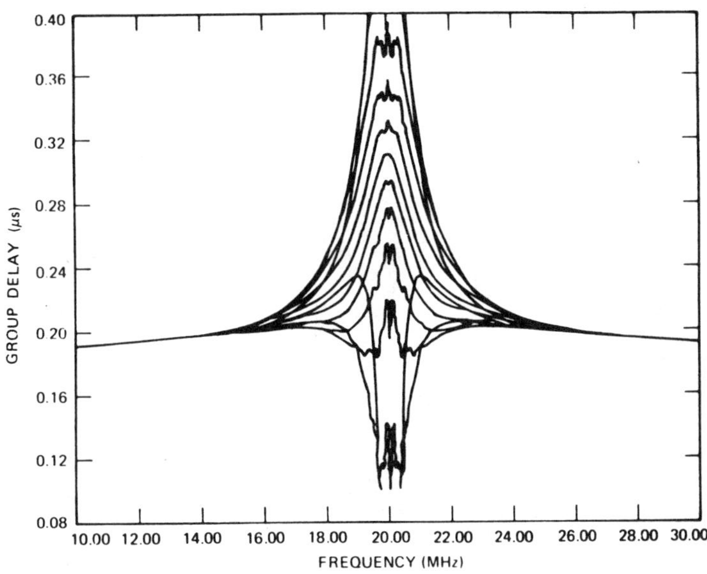

Figure 31. *Sample Output, Group Delay (10-section input filter)*

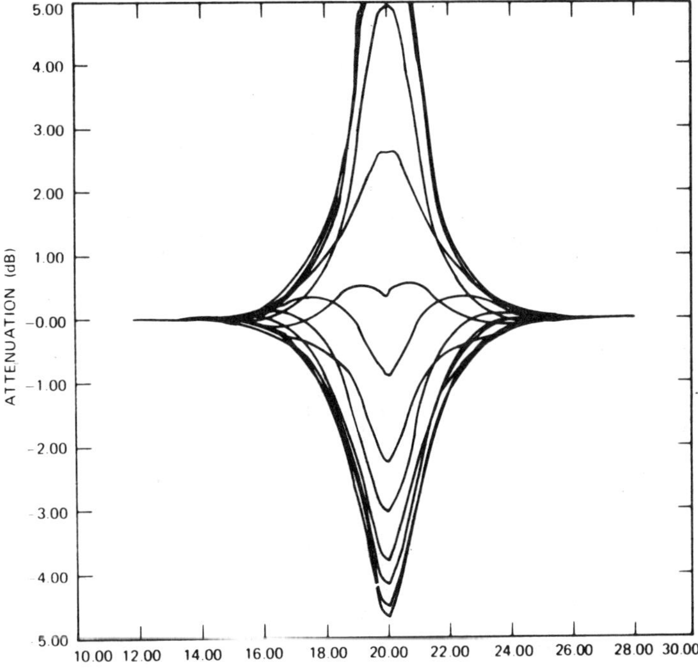

TABLE 11. MULTIPATH RESULTS FOR A 9-SECTION TCHEBYCHEFF INPUT FILTER AND A 6-SECTION TCHEBYCHEFF OUTPUT FILTER (6-SECTION NETWORK DELAY EQUALIZER)

Filter Bandwidths (MHz)	Percent of 36-MHz Bandwidth	Additional Group Delay (ns)		Gain Slope (dB/MHz)	
		Maximum	Minimum	Maximum	Minimum
Input = 38.9, Output = 41	70	0.6	−0.6	0.0354	−0.0302
	80	2.68	−2.69	0.14	−0.154
	90	11.1	−11.7	0.59	−0.697
	100	22.4	−55.6	1.9	−4.17
Input = 37, Output = 39	70	0.209	−0.209	0.932×10^{-2}	−0.0135
	80	0.947	−0.947	0.0518	−0.0517
	90	4.56	−4.6	0.236	−0.267
	100	18.1	−21.5	1.03	−1.33
Input = 35, Output = 36.9	70	0.0667	−0.0667	$−0.161 \times 10^{-2}$	$−0.89 \times 10^{-2}$
	80	0.292	−0.292	0.0198	−0.0121
	90	1.48	−1.48	0.074	−0.883
	100	8.48	−8.66	−0.314	−1.25
Input = 33.1, Output = 34.8	70	0.0215	−0.0215	$−0.114 \times 10^{-2}$	$−0.349 \times 10^{-2}$
	80	0.09	−0.09	$−0.176 \times 10^{-2}$	$−0.116 \times 10^{-1}$
	90	0.443	−0.443	0.046	$−0.242 \times 10^{-2}$
	100	5.82	−5.84	−5.25	−5.89

TABLE 12. MULTIPATH RESULTS FOR A 10-SECTION TCHEBYCHEFF INPUT FILTER AND A 6-SECTION TCHEBYCHEFF OUTPUT FILTER (5-SECTION NETWORK DELAY EQUALIZER)

Filter Bandwidths (MHz)	Percent of 36-MHz Bandwidth	Additional Group Delay (ns)		Gain Slope (dB/MHz)	
		Maximum	Minimum	Maximum	Minimum
Input = 38.9, Output = 41	70	0.266	−0.266	0.025	−0.416 $\times$ 10^{-2}
	80	1.33	−1.33	0.0658	−0.0794
	90	6.57	−6.68	0.351	−0.379
	100	26.4	−39.2	1.63	−2.66
Input = 37, Output = 39	70	0.0863	−0.0863	0.0127	−0.33 $\times$ 10^{-2}
	80	0.0432	−0.0432	0.014	−0.0333
	90	2.36	−2.36	0.13	−0.128
	100	12.4	−13	0.685	−0.74
Input = 35, Output = 36.9	70	0.0256	−0.0256	−0.224 $\times$ 10^{-3}	−0.302 $\times$ 10^{-2}
	80	0.123	−0.123	0.663 $\times$ 10^{-2}	−0.684 $\times$ 10^{-2}
	90	0.696	−0.696	0.025	−0.0511
	100	4.91	−4.94	−1.06	−1.6
Input = 33.1, Output = 34.8	70	0.769 $\times$ 10^{-2}	−0.769 $\times$ 10^{-2}	−0.0104	−0.0113
	80	0.0351	−0.0351	0.643 $\times$ 10^{-2}	0.26 $\times$ 10^{-2}
	90	0.191	−0.191	0.0428	0.022
	100	4.15	−4.15	−8.13	−8.58

capacity. As the equiripple bandwidth is narrowed, the multipath effect is reduced, and it becomes more difficult to equalize across the center 36 MHz. For both filter sets, the optimum bandwidth choice (where inherent group-delay and multipath effects are approximately equal and their sum is a minimum) is a bandwidth between 37.5 and 38.5 MHz, and the 10-cavity filter yields slightly better results.

Amplitude and phase nonlinearities

For the purpose of discussion, the traveling wave tubes (earth station HPA, satellite receiver driver, and satellite transmitter) will be assumed to dominate the nonlinear amplitude and phase behavior of the INTELSAT IV system.*

In general, a traveling wave tube may be represented by cascaded filters and two nonlinear effects, as shown in Figure 33.† Since the nonlinear

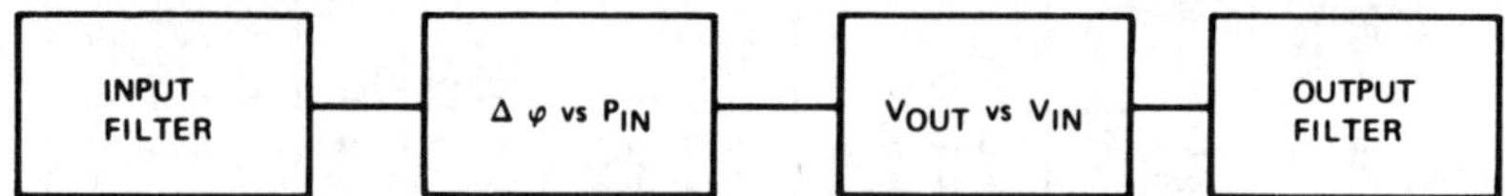

Figure 33. *Simplified Model of Helix TWT*

element (the beam interacting with the helix) has a bandwidth which is much larger than the possible rates of change of the envelope, the nonlinearities are assumed to be memoryless and are represented by a series expansion. To choose a series expansion, several factors were considered. First, the series approximation of measured single-carrier behavior should provide high accuracy over the range of output drive from zero to the drive required for saturation, reasonable accuracy for inputs up to twice the drive required for saturation, and bounded values (for which the magnitude of the computed value should not exceed the saturated value) for inputs from twice to at least four times the drive required for saturation.

---

* Detailed measurements of individual circuits have shown that C/I contributions of the 6- and 4-GHz TDAs in the satellite receiver may not be neglected. However, the bandwidth of the nonlinear element (the diode) is large enough so that the instantaneous (memoryless) modeling developed for the TWTs may also be used to predict their low-level nonlinear performance.

† The present state of the art in traveling wave tubes makes it possible to omit the filters in the case of the satellite tubes, and to reduce them to input and output linear slopes for a majority of the HPAs.

Second, the series selected should be suited to efficient solution by computer and should typically yield closed-form algorithm solutions.

Amplitude Nonlinearity. As shown in Figure 34 and Reference 16, the amplitude nonlinearity is characterized by

$$e_{out} = c_1 \sin (\alpha e_{in}) + c_2 \sin (2\alpha e_{in})$$
$$+ c_3 \sin (3\alpha e_{in}) + c_5 \sin (5\alpha e_{in})$$

where the coefficients are determined graphically from the measured single-carrier data.

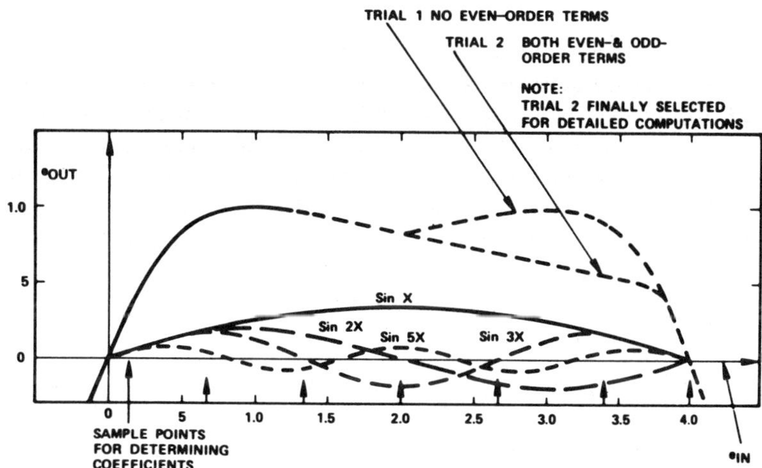

Figure 34. *Instantaneous Voltage Transfer Characteristic*

For the general case of an arbitrary set of inputs,

$$e_{in} = \sum_{i=1}^{n} A_i \sin \left[\omega_i t + \int \psi_i \, dt \right]$$

where A_i = amplitude of the ith carrier
ω_i = center frequency of the ith carrier
ψ_i = rms value of the frequency deviation when the RF carrier is modulated with a noise-like baseband signal.

The output, e_{out}, may be written as

$$e_{out} = c_1 \sin \left[\alpha \sum_{i=1}^{n} A_i \sin \omega_i t + \int \psi_i \, dt \right] + \ldots .$$

The output of any terms may be obtained by appropriately manipulating the subscripts; i.e.,

$$e_{out} = k_1 \sum_a \sum_b \sum_c \dots \sum_d J_a(\alpha A_1) \cdot J_b(\alpha A_2)$$

$$\cdot J_c(\alpha A_3) \dots J_d(\alpha A_n)$$

$$\cdot \cos\left[(a\omega_1 + b\omega_2 + c\omega_3 + \dots d\omega_n)t\right.$$

$$\left. + \int \psi_r \, dt\right] + \dots$$

where $\psi_r = \sqrt{(a\psi_1)^2 + (b\psi_2)^2 + (c\psi_3)^2 + \dots (d\psi_n)^2}$

in the case of high-index noise modulation. If the ith subscript is assumed to be 1 (and all others are assumed to be 0), the useful power output of the ith carrier may be computed. Similarly, if the remaining subscripts are appropriately evaluated, all of the output products can be computed. For certain applications with small numbers of carriers, the useful power output and product heights are utilized directly. However, in the general multicarrier case, it is desirable to use an appropriate algorithm to further process the products before the computer prints out the final values.

For the INTELSAT IV multiple-carrier computations, the following sorting modes have been provided:

a. classifying product types, with options to compute or neglect (see Table 13),
b. sort/store by frequency,
c. product spreading caused by modulation, and
d. coherence between products [19].

TABLE 13. CLASSIFYING PRODUCT TYPE

Subscript Forms	Type
1,0,0,0, . . ., 0	1st-Order Desired Carrier Output
2,1,0,0,0, . . ., 0	3rd-Order A^2B Intermodulation Product
1,1,1,0,0,0, . . ., 0	3rd-Order ABC Intermodulation Product
1,1,1,1,1,00, . . ., 0	5th-Order ABCDE Intermodulation Product
2,2,1,00, . . ., 0	5th-Order A^2B^2C Intermodulation Product
3,1,100, . . ., 0	5th-Order A^3BC Intermodulation Product
3,2,00, . . ., 0	5th-Order A^3B^2 Intermodulation Product
4,1,00, . . ., 0	5th-Order A^4B Intermodulation Product

In the sort/store by frequency mode, the 500-MHz communications band is subdivided into 5,000 cells. For each product type, an algorithm gives the frequencies of the product (i.e., an $A_i^2 A_j$ product falls at $|2\omega_i - \omega_j|$), enters the product into the appropriate cell, and keeps a running power total for each of the cells. A plot of the total in the cells yields the intermodulation density.

In the case of FDM/FM telephony (sorting mode c), the spectrum of an RF carrier is approximately Gaussian, with an rms value equal to the rms deviation of the baseband noise load.* For any product of the form $A_i^m \cdot A_k^p$ (rigorously obtained from the multiple convolution of the carriers comprising it), the spectral shape will be Gaussian with an rms index given by

$$\sigma^2 = (m\sigma_i)^2 + (n\sigma_j)^2 \ldots (p\sigma_k)^2.$$

The spectral density of each product is computed, and the portion falling into each frequency cell is entered into the total of that cell (sorting mode b).

It should be noted that, for other forms of modulation, the spectral shape must be computed only once for each product type; it can then be stored as a "library" function.

If a significant number of carrier pairs have instantaneous difference frequencies that are smaller than the baseband widths of other large carriers within the multiple-access community, the fourth sorting mode should be used to properly interpret the intermodulation noise density when deriving the resulting baseband (detected) noise spectrum.

For FM carriers, products having the form† $\omega_i \pm (\omega_i - \omega_j) = \omega_i \pm \Delta\omega$ will appear as an amplitude modulation about ω_i. If $\Delta\omega$ is less than the Carson's Rule bandwidth of ω_i, the product pair is discarded. On the other hand, if $\Delta\omega$ is greater than the Carson's Rule bandwidth of ω_i, then each of the individual terms will appear as an unrelated intermodulation product about some other carrier; each term must then be retained and added (in power) to the appropriate cell (sorting mode b).

---

* To compute the spectral density of the intermodulation products, the Gaussian approximation is sufficiently accurate for the range of rms indexes encountered by the INTELSAT IV system.

† For most cases of interest, only ABC products are significant; fifth-order products may be neglected entirely.

Phase Nonlinearity. The single-carrier phase nonlinearity curve is shown in Figure 35. The curve is approximated by the series [17]

$$\phi = k_1[1 - \exp(-k_2 P_{in})] + k_3 P_{in}$$

where k_1, k_2, k_3 = tube constants
P_{in} = normalized input power
ϕ = relative phase shift.

With an arbitrary set of input carriers,

$$e_{in} = \sum_{i=1}^{p} A_i \cos \omega_i t$$

The squared envelope is given by

$$\overline{[e_{in}(t)]^2} = \sum_{i=1}^{p} A_i^2 + 2 \sum_{l=1}^{p-1} \sum_{k=l+1}^{p} A_l A_k \cos (\omega_l - \omega_k)t$$

leading to output terms of the form

$$e_{out} = \sum_{i=1}^{p} A_i \sum_{l=1}^{p-1} \sum_{k=l+1}^{p} \prod_{m=1}^{\infty} \sum_{n_{mlk}=-\infty}^{+\infty} J_{n_{mlk}}(\alpha_{mlk})$$
$$\cdot \cos [\omega_i + mn_{mlk}(\omega_l - \omega_k)]t$$

where
$$\alpha_{mlk} = 2k_1 \exp \left(-k_2 \sum_{i=1}^{p} A_i^2 \right)$$
$$\cdot I_m(2k_2 A_l A_k) + \begin{cases} 2k_3 A_l A_k, & m = 1 \\ 0, & m \neq 1 \end{cases} .$$

As in the case of the AM nonlinearities, sorting modes a through c, listed previously, are used.

For phase nonlinearity, sorting mode d is altered as follows. Products of the form $\omega_i \pm \Delta\omega$ produce coherent phase modulation about ω_i. Since the output of the computations is a spectral density which will be added to other noise densities during the course of system computations, the product value is doubled (in power) before it is added to the totals in the cells if $\Delta\omega \leq$ Carson's Rule bandwidth of ω_i. Then, the amplitude intermodulation density and the thermal noise density are added in power.

Subsequent work in this area has shown that the effects of both phase and amplitude nonlinearities can be derived from a single nonlinear

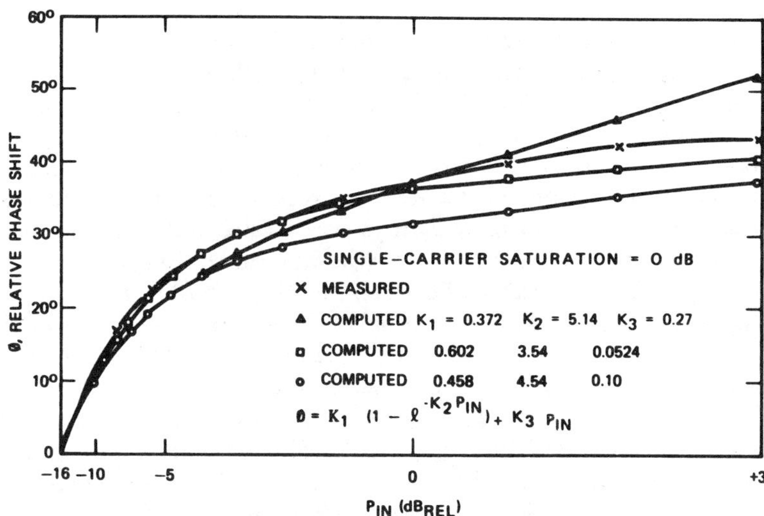

Figure 35. *Measured and Computed Phase Shift Transfer Function*

transfer function expanded as a series with complex coefficients. This characterization has resulted in a unified formulation which allows convenient sorting for coherence relationships, and direct computation of baseband noise for a variety of modulation types.*

Crosstalk computations

In multicarrier FM systems, crosstalk results from a sequence of two phenomena: first, an amplitude response which varies with frequency, producing amplitude modulation coherent with the original frequency modulation of an RF carrier (FM/AM transfer); and secondly, a coherent amplitude modulation, which phase modulates all carriers in a subsequent traveling wave tube amplifier exhibiting AM/PM conversion.

For those cases in which the gain slope is small and may be characterized by a linear coefficient (in dB/MHz), and the phase shift vs drive may be represented by an average conversion coefficient (in deg/dB), accurate results may be obtained from Reference 20. However, several cases of in-

---

* This work will be reported in a forthcoming paper.

terest in the INTELSAT IV system required further refinement. For example, in the satellite transmitter (and again in the earth station transmitter), signals intended for channel II were slope detected on the skirt of channel I, producing coherent AM components and crosstalk in channel I. In this case, the use of an unmodified gain slope (in dB/MHz) resulted in grossly inaccurate predictions of the AM produced in channel I. As a second example, when two or more carriers were located within one input filter bandwidth, the carrier near band edge experienced a high-order unsymmetrical gain slope because of the filter's finite unloaded Q (and consequent rounded corners). In this case, the use of either an instantaneous gain slope or an average chord would yield predictions of the coherent AM component that may be in error by as much as a factor of 2 to 5.

In addition, when there is a wide variety of carrier combinations, the use of an average AM/PM conversion coefficient results in a poor correlation between predicted and measured results. The latter problem is readily overcome (once the AM component is known) by using the phase non-linearity series described previously* to directly compute the one term of interest.

To compute the resulting AM component, a test tone inserted into the top baseband telephone channel is represented by a pair of low-index sidebands about the carrier. The remainder of the multiplex signal (having an average frequency lower than that of the top baseband signal) is assumed to be a low-frequency slewing signal with a Gaussian probability density. To provide computational assistance, a program has been written to slew the carrier and sideband pair across a given filter response (one of the program inputs) in small increments. For each position, the imbalance in sideband heights is resolved into an AM component and multiplied by the probability that the carrier is in that position (determined from the Gaussian statistic of the noise load). The probabilistic weighted sum over all positions gives the average AM component that would cause crosstalk into another 4-kHz voice channel.

Experimental verification of this model, detailed in the following, indicates good agreement with theory over a large range of rms indexes (≥ 0.5), gain slopes (≥ 0), and locations on the filter curve.

---

* To simplify the computations, the AM component is represented at the input to the phase shift program as a small carrier next to the AM producing carrier; the spacing is set to yield the desired rate, and the height of the small carrier is set to yield the desired AM index.

Verification of the crosstalk model

Theoretical Model. Figure 36 is a 2-block model used to represent the satellite repeater. The first block of Figure 36 contains the amplitude and group-delay frequency response of the communications path up to the input of a TWTA. The second block models the AM/PM conversion exhibited by a TWTA or other active device.

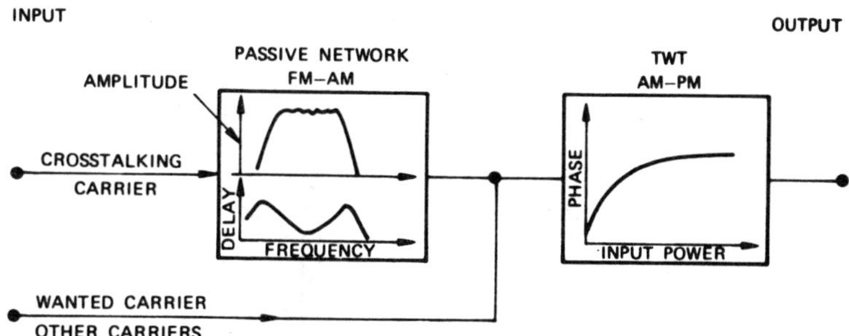

Figure 36. *Model of a Satellite Repeater*

In this analysis, the crosstalking carrier bears a multicarrier frequency modulation, the wanted carrier is an unmodulated carrier, and other carriers (with arbitrary angle modulation) are inserted to achieve the actual operating conditions for the TWT characteristic. The intelligible crosstalk originates from a baseband channel on the crosstalking carrier and is transferred onto the wanted carrier. In this case, only the crosstalking carrier must pass through the FM/AM conversion block of the model. The wanted carrier and the other carriers are injected at the input of the AM/PM converter so that any frequency response in their path may be neglected.

For this analysis, intelligible crosstalk is defined as a performance requirement specified in terms of the ratio between wanted and unwanted levels of a sine wave test tone inserted into the baseband channel of interest. The actual level used to represent an average talker is not significant in the computations; however, this level must be taken into account when comparing experimental results measured with an NPR test set with theoretical predictions referenced to a 0-dBm0 test tone.

It is well-known that the intelligible crosstalk increases proportionally with the baseband frequency. Therefore the worst case can be expected to occur at the high-frequency end of the baseband spectrum of a multichannel signal. For system performance evaluation, the worst case is usually of interest; consequently, the frequency of the test tone can be restricted to the high end of the baseband spectrum. This restriction leads to a significant simplification of the analysis; i.e., the test tone can be represented in the RF spectrum by a low index modulation, and the noise loaded baseband including the test tone can be represented by a high index modulation. Figure 37 shows the baseband and RF spectrum for this situation.

For clarity, the analysis has been divided into two parts. The first part follows the path through the model for the simple case of a single tone-

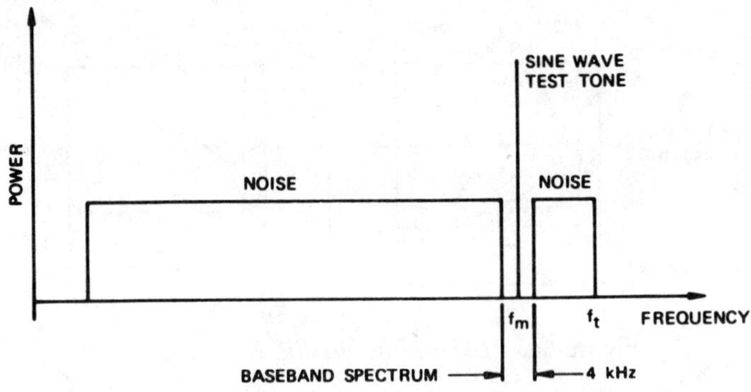

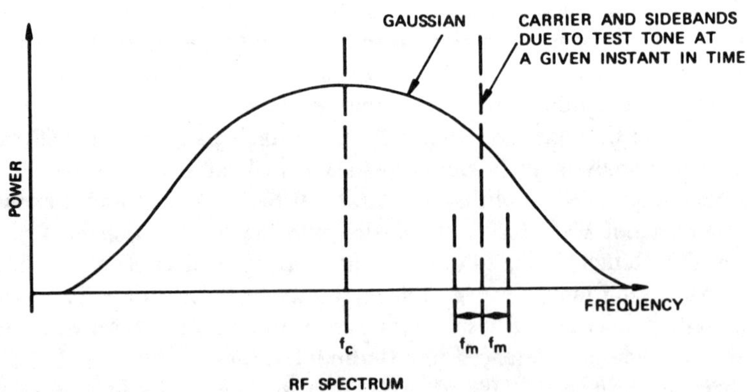

Figure 37. *Baseband and RF Spectrum Representation of Crosstalking Carriers*

modulated crosstalking carrier. The second part leads to an equation which allows computation of intelligible crosstalk for the noise loaded case.

Consider a crosstalking carrier modulated by a sine wave of frequency f_m, where β_c is the modulation index. In the time domain, this can be expressed as

$$A_c \sum_{n=-\infty}^{+\infty} J_n(\beta_c) \cdot \cos 2\pi(f_c + nf_m)\, t \qquad (10)$$

where A_c is the carrier amplitude and f_c the carrier frequency.

The carrier and sidebands described by equation (10) will be modified in accordance with the transfer characteristic of the passive network at the particular carrier and sideband frequencies. For the simple case of a carrier with only two sidebands and a passive network with amplitude response $G(f)$ and constant group delay, this process is shown in Figures 38 and 39. After passing through the amplitude response $G(f)$, the carrier and its sidebands will have new amplitudes:

$$\begin{aligned} \text{carrier amplitude} &= A_c \cdot G(f_c) \\ \text{upper sideband amplitude} &= A_c \cdot G(f_c + f_m) \cdot J_1(\beta_c) \\ \text{lower sideband amplitude} &= A_c \cdot G(f_c - f_m) \cdot J_1(\beta_c) \end{aligned} \qquad (11)$$

where it is assumed that $J_0(\beta_c) \approx 1$.

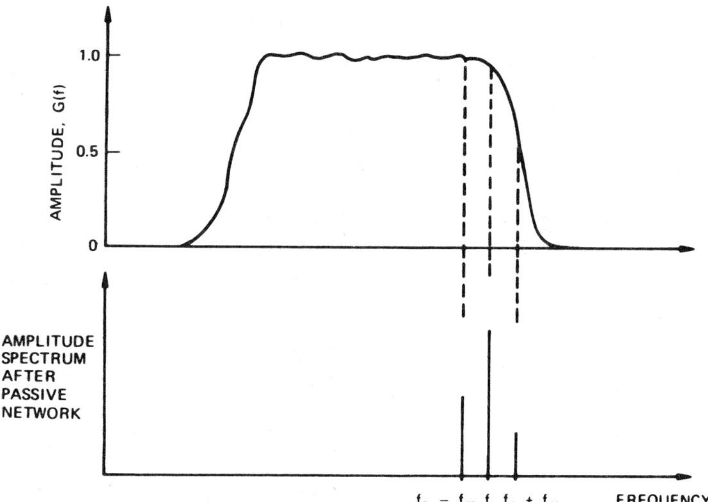

Figure 38. *Amplitude Response and Spectrum at Output of Passive Network*

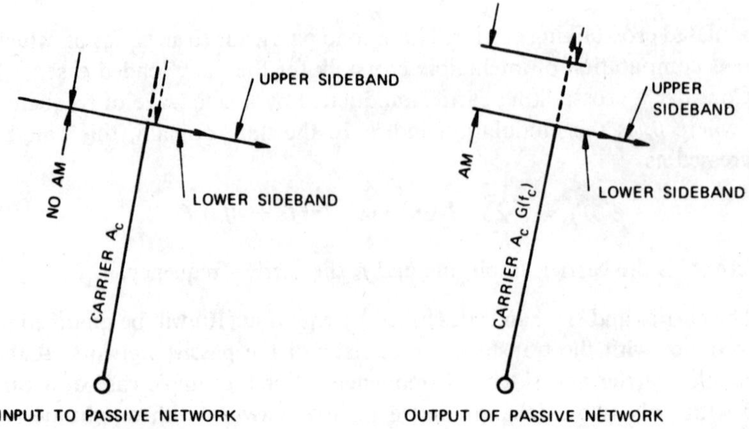

Figure 39. *Vector Diagram of the Crosstalking Carrier*

It is apparent that the sidebands become unbalanced so that AM is generated. In this simple case, the amount of AM generated is proportional to the amplitude difference of the two sidebands, which can be expressed as

$$A_c \cdot J_1(\beta_c) \cdot D \tag{12}$$

where $D = [|G(f_c + f_m) - G(f_c - f_m)|]$. The AM index can be obtained by dividing equation (12) by the actual carrier amplitude, i.e.,

$$m = \frac{J_1(\beta_c)}{G(f_c)} \cdot D. \tag{13}$$

On the crosstalking carrier in the subsequent AM/PM converter, amplitude modulation or envelope fluctuations coherent with f_m are the only components of interest. The crosstalking carrier at the output of the passive network can therefore be expressed as

$$A_c \cdot G(f_c)[1 + m \cos(2\pi f_m t)] \cdot \sin 2\pi f_c t. \tag{14}$$

This carrier, the wanted carrier, and other eventual carriers are fed to the AM/PM converter. The AM on the crosstalking carrier then gives rise to PM at the same rate on all other carriers passing through the AM/PM converter.

For clarity, the pertinent results of Reference 18 will be summarized in the following. Assume that there are three carriers, described by their amplitudes A_1, A_2, A_3, and their frequencies f_1, f_2, f_3, at the input of the AM/PM converter. For illustrative purposes, also assume that $f_1 < f_2$ and $f_2 \approx f_3$. After the AM/PM converter, carrier A_1 will bear sidebands at the rate of $|f_2 - f_3|$. Each sideband will consist of two components; i.e.,

$$
\begin{aligned}
SB = {} & A_1 \cdot J_1\{2k_1 e^{(-k_2 P_{in})} \cdot I_1(2k_2 A_2 A_3) + 2k_3 A_2 A_3\} \\
& \cdot \cos 2\pi[f_1 - (f_2 - f_3)]t \\
& + A_3 \cdot J_1\{2k_1 e^{(-k_2 P_{in})} \cdot I_1(2k_2 A_1 A_2) + 2k_3 A_1 A_2\} \\
& \cdot \cos 2\pi[f_3 + (f_1 - f_2)]t
\end{aligned}
\tag{15}
$$

where k_1, k_2, k_3 are constants characterizing the AM/PM converter, P_{in} is the input backoff (input power to the AM/PM converter normalized with respect to the saturation power), and $\Sigma_i A_i^2 = P_{in}$.

For this part of the analysis, the AM on the crosstalking carrier can be represented by a carrier whose amplitude $= A_c \cdot G(f_c)$ and one sideband whose amplitude $= A_c J_1(\beta_c) D$. If these amplitudes are substituted into equation (15), the amplitude of the PM sidebands generated on the wanted carrier is

$$
\begin{aligned}
SB = {} & A_w J_1\{2k_1 e^{(-k_2 P_{in})} \cdot I_1[2k_2 A_c^2 G(f_c) J_1(\beta_c) D] \\
& + 2k_3 A_c^2 G(f_c) J_1(\beta_c) D\} \\
& + A_c J_1(\beta_c) D \cdot J_1\{2k_1 e^{(-k_2 P_{in})} \cdot I_1[2k_2 A_c G(f_c) A_w] \\
& + 2k_3 A_c G(f_c) A_w\} .
\end{aligned}
\tag{16}
$$

This PM sideband amplitude will be much smaller than the wanted carrier amplitude, A_w. Therefore, the approximation $J_1(x) = x/2$ is valid and the modulation index of the interfering PM on the wanted carrier is

$$
\beta_i = \frac{2SB}{A_w} .
\tag{17}
$$

The crosstalk ratio for this simple case can now be calculated as

$$
\text{IXTR (single-tone FM)} = 20 \log \frac{\beta_i}{\beta_w}
\tag{18}
$$

where β_w is the modulation index of a similar test tone of frequency f_m on the wanted carrier. In satellite systems, β_c and β_w may or may not be the same. For example, for a 432-channel carrier crosstalking into a 252-channel carrier, the highest f_m of interest is obviously that of the 252-channel carrier, and depending on pre-emphasis, $\beta_c \neq \beta_w$.

For the general case of a noise loaded baseband, the same procedure will be followed. The baseband of the crosstalking carrier is as shown in Figure 37. If the quasi-stationary FM model is used and a Gaussian probability density is assumed, the RF spectrum can be written as

$$S(f) = \frac{1}{\sqrt{2\pi}\,f_{rms_c}} \cdot \exp - \left\{ \frac{(f - f_c)^2}{2 f^2_{rms_c}} \right\}$$
$$\cdot \sum_{n=-\infty}^{+\infty} A_c J_n(\beta_c) \cdot \cos 2\pi (f_c + n f_m) t \tag{19}$$

where f_{rms_c} is the rms deviation of the crosstalking carrier.

In a practical situation, the passive network given by the amplitude and group-delay response may contain sharp filters. Hence, if a polynomial representation is used, several higher order terms will be required.

The approach presented here uses directly measured responses. The envelope of the crosstalking carrier will contain a broad spectrum after passing through the passive network. For calculations of intelligible crosstalk, the envelope components at rates of f_m and integer multiples of f_m are of interest, and a Fourier series is an adequate representation. Since the energy of the crosstalking carrier is spread out over a bandwidth approximately equal to the Carson's Rule bandwidth, the coefficients of the Fourier series are functions of the instantaneous carrier frequency. This Fourier series can be pictured as a carrier amplitude modulated by a time-varying complex wave with a fundamental frequency f_m. All other envelope components may be neglected.

The amplitude modulated carrier, wanted carrier, and other carriers are fed to the AM/PM converter. At the output, all carriers will bear a similarly complex wave phase modulation, which, after demodulation, will result in a baseband complex periodic wave. The phase modulation sidebands on the wanted carrier can be calculated according to Reference 18, regardless of any angle modulation this carrier might bear before entering the model. The complex wave at baseband can be developed into another Fourier series, in which the coefficient of the fundamental represents the instantaneous amplitude of the crosstalking signal. A time average of this amplitude can then be calculated and compared with the level of a similar test tone on the wanted carrier, and the intelligible crosstalk ratio can be determined.

For engineering purposes, this process can be simplified considerably without appreciably degrading the accuracy of the results. First, the series describing the angle modulation of the test tone on the crosstalking carrier may be truncated so that only one pair of sidebands is considered, i.e., $n = \pm 1$ in equation (19). Hence, a Fourier analysis of the envelope of the crosstalking carrier after the passive network is no longer necessary. Although the AM of interest is still a complex wave, the amplitude of the first pair of AM sidebands can be calculated by using equation (11). This approximation is valid for $\beta_c \ll 1$. (For $\beta_c = 0.9$, the amplitude error is 1.5 dB; for $\beta_c = 0.6$, it is less than 0.7 dB.)

In addition, since experimental data indicate that, in practical networks such as microwave filters, group delay is only a minor worst-case contributor to the FM/AM conversion, it can usually be neglected. As a further simplification, the series describing the complex periodic AM wave may be truncated. Omitting the higher order ($2f_m$, $3f_m$, . . .) AM sidebands does not introduce any significant error, at least for all practical cases of interest. Since the transfer from AM to PM is small in practical communications systems, this approximation makes it possible to calculate the PM sidebands on the wanted carrier by using an expression similar to equation (16). A Fourier development of the wanted carrier's complex periodic wave at baseband is no longer necessary, and the time average can be taken directly on the PM sidebands. As a result of these three simplifications, the time-averaged PM sideband amplitude on the wanted carrier can be expressed as

$$
\begin{aligned}
\overline{SB} = \frac{1}{\sqrt{2\pi}\, f_{rms_c}} \int_{-\infty}^{+\infty} &\left[A_w J_1\{2k_1 e^{(-k_2 P_{in})} \cdot I_1[2k_2 A_c^2 G(f)\, J_1(\beta_c)\, D] \right. \\
&+ 2k_3 A_c^2 G(f)\, J_1(\beta_c)\, D\} + A_c J_1(\beta_c)\, D \\
&\left. \cdot J_1\{2k_1 e^{(-k_2 P_{in})} \cdot I_1[2k_2 A_c G(f)\, A_w] + 2k_3 A_c G(f)\, A_w\} \right] \\
&\cdot \exp - \left\{ \frac{((f - f_c))}{2f_{rms_c}^2} \right\} df .
\end{aligned}
\tag{20}
$$

The intelligible crosstalk ratio can now be calculated by using equations (17) and (18):

$$
\text{IXTR} = 20 \log_{10} \frac{2\overline{SB}}{A_w \beta_w} .
\tag{21}
$$

This equation, which has been slightly modified by using a finite sum of samples at equidistant frequeneics covering the Carson's Rule bandwidth, has been programmed on a digital computer. The spacing between samples is selected according to the frequency-response of the passive network.

Equation (21) describes the most common process involved in the generation of intelligible crosstalk in satellite repeaters. Actually, intelligible crosstalk can also be produced by more elaborate processes. An example that has been observed experimentally is as follows. Assume that a nonlinear amplifier passes two carriers at f_1 and f_2, generating a third-order intermodulation product at $2f_1 - f_2$. This product falls on a steep amplitude or group-delay slope of a passive network following the amplifier. It can be shown that, at the output of this passive network, the intermodulation product contains amplitude modulation coherent with the frequency modulation of both f_1 and f_2. Amplifying this intermodulation product and one of the two carriers in a subsequent AM/PM converter will result in intelligible crosstalk. This process can be calculated by modifying the model described here to achieve a 3-block model.

Comparison with Previous Results. The intelligible crosstalk ratio has previously been calculated by Chapman and Millard [21]; i.e.,

$$\text{IXTR} = 20 \log 2K_1 A_c^2 g f_m$$

where K_1 is a tube constant related to the phase shift and the input power of the TWT by their equation (56):

$$\phi = \frac{K_1 P_{in}}{2} \tag{22}$$

A_c^2 is the power of the crosstalking carrier referenced to single-carrier saturation, g is the gain slope defined by

$$G(f - f_1) = 1 + g(f - f_1) \tag{23}$$

and f_m is the baseband frequency under consideration.

It can be shown that Chapman and Millard's result is a special case of the foregoing analysis. If the operating range of the TWT is restricted to low input power, and small signal approximations are used, both results become identical.

For low input power, it can be assumed that $J_1(x) \approx x/2$, $I_1(x) \approx x/2$, and $\exp -(k_2 P_{in}) \approx 1$ in equation (20). Furthermore, according to Reference 18, the phase-input power characteristic of a TWT can be described as

$$\phi = k_1[1 + e^{(-k_2 P_{in})}] + k_3 P_{in} \tag{24}$$

which, for $P_{in} \ll 1$, reduces to

$$\phi = (k_1 k_2 + k_3)\, P_{in}\,. \tag{25}$$

A comparison of equations (22) and (25) indicates that

$$k_1 k_2 + k_3 = \frac{K_1}{2}\,. \tag{26}$$

These approximations can be used to simplify equation (21) as follows:

$$\text{IXTR} = 20 \log_{10} \frac{2 \int_{-\infty}^{+\infty} K_1 A_c^2 A_w G(f)\, J_1(\beta_c)\, D \cdot \exp - \left\{ \frac{(f-f_c)^2}{2 f_{rms_c}^2} \right\}\, df}{\sqrt{2\pi}\, A_w \beta_w f_{rms_c}}\,. \tag{27}$$

Assuming a linear gain slope as in equation (23), where $f - f_1 = f_m$ and $G(f) \approx 1$, makes it possible to simplify equation (27):

$$\text{IXTR} = 20 \log_{10} \frac{2 K_1 A_c^2 J(\beta_c)\, 2 g f_m}{\beta_w}$$

$$\cdot \frac{1}{\sqrt{2\pi}\, f_{rms_c}} \int_{-\infty}^{+\infty} \exp - \left\{ \frac{(f-f_c)^2}{2 f_{rms_c}^2} \right\}\, df. \tag{28}$$

The integral with the preceding factor $1/\sqrt{2\pi}\, f_{rms_c}$ is equal to 1. With further simplification, equation (28) can be expressed as

$$\text{IXTR} = 20 \log_{10} \frac{2 K_1 A_c^2 g f_m \beta_c}{\beta_w}\,. \tag{29}$$

For equal modulation parameters, equation (29) can now be reduced to yield the result of Chapman and Millard.

Experimental results

The objective of the experimental work was to verify specific parts of the model for which a correlation between experiment and analysis had not yet been established instead of using a baseband-to-baseband measurement of intelligible crosstalk to verify the overall model. Such an overall crosstalk measurement presents many possibilities for error that cannot be recognized as easily as in a part-by-part verification of the model.

Since the validity of the model for the AM/PM conversion mechanism had already been established, the experimental work concentrated on the first part of the model, the FM/AM conversion. The first objective consisted of matching measurements and computed predictions of AM at the output of a linear network. For convenience, the AM index was measured for three values of β. The test object was a filter characterized by the measured responses shown in Figure 40. A carrier was frequency modulated with a single tone at 100 kHz and fed through the filter; the AM index was then measured at the output of the filter. The results for three modulation indexes (Figures 41–43) show excellent agreement with computed data. These computations included up to seven pairs of FM sidebands, and both amplitude and group-delay responses. The AM index was

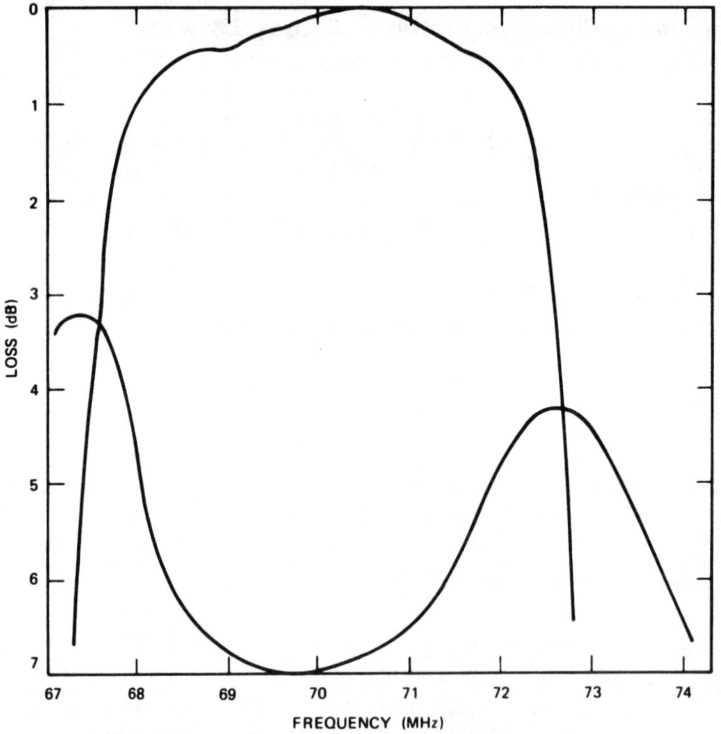

Figure 40. *Frequency Response for 5-Pole, 5-MHz, Elliptic Function Filter*

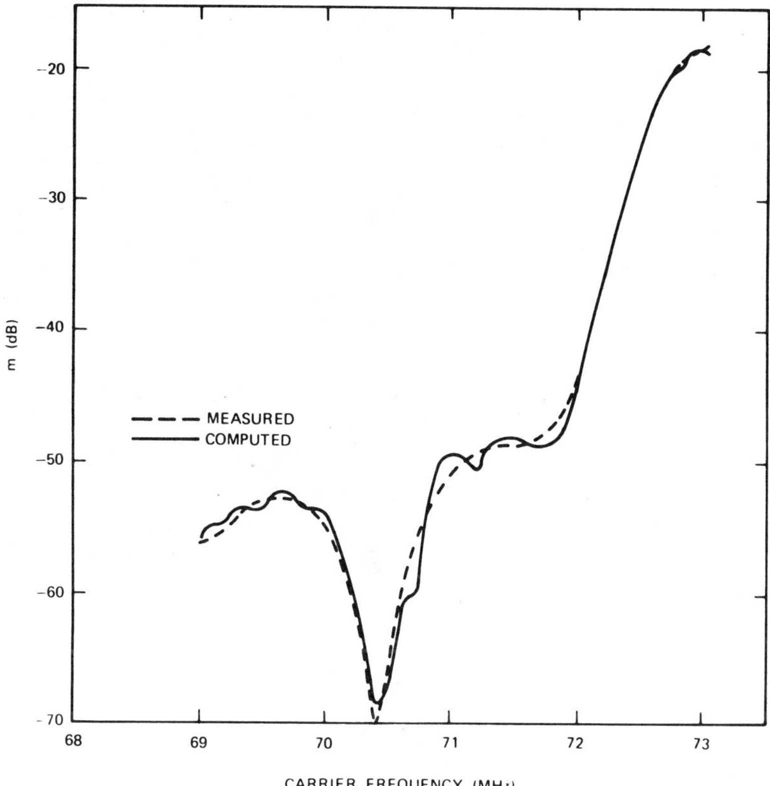

Figure 41. *AM Index for β = 0.5*

obtained from a Fourier evaluation of the envelope at the output of the filter.

Omitting the group delay in these computations affected only the AM close to the sharp dip occurring around 70.4 MHz; where the AM index peaks (for instance, at 69.7 MHz and above 71 MHz), no appreciable group-delay effect could be detected.

These single-tone measurements were used as a basis for two of the approximations made in the analysis. Another assumption used in the model, namely, the assumption that the test tone is represented by a low index modulation and the noise loading by a high index modulation, was verified as follows. A carrier was frequency modulated by a 1002-kHz sine wave with an index of 0.14, and a triangular wave with a peak deviation of

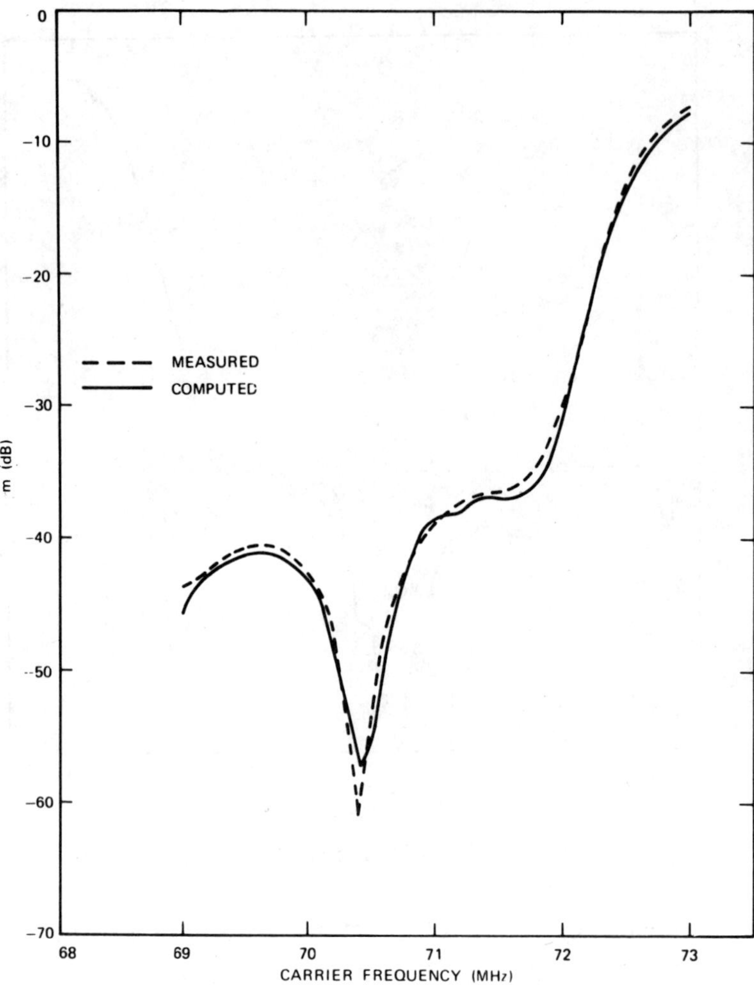

Figure 42. *AM Index for $\beta = 2.0$*

±3 MHz. (The triangular wave was chosen because it has a constant probability density so that the RF spectrum of the modulated carrier can be easily calculated.) This composite signal was fed through a 10-section Tchebycheff filter with the carrier placed close to the sharp dropoff in amplitude response. The triangular wave frequency could be varied from 20 to 100 kHz without appreciably changing the AM index measured at the output of the filter.

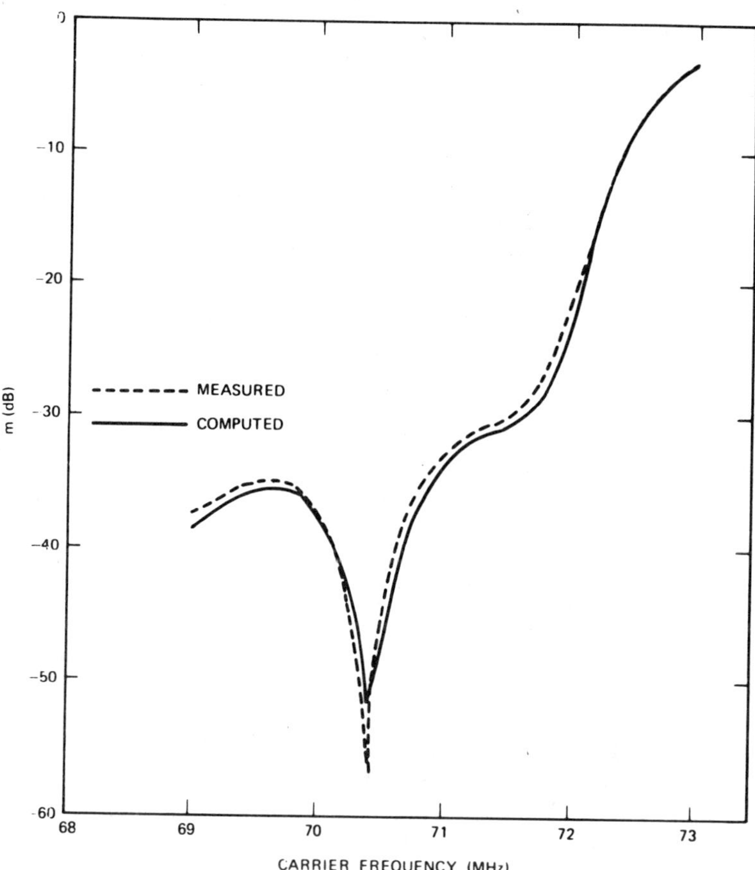

Figure 43. *AM Index for β = 4.0*

Next, noise loading was measured for a 252-channel carrier using standard parameters and a 1002-kHz test tone. The modulated carrier was fed through the same 10-section filter used previously, and a weighted AM index was measured at the output of the filter and compared with computed data (Figure 44). The computations omitted group delay and included only one pair of FM sidebands on the carrier. The weighted AM index was computed as follows:

$$M = \frac{J_1(\beta_c)}{\sqrt{2\pi}\, f_{rms\,c}} \int_{-\infty}^{+\infty} D \cdot \exp -\left\{\frac{(f-f_c)^2}{2f_{rms\,c}^2}\right\} df. \qquad (30)$$

To verify that the probability approach to the problem was correct, the noise baseband spectrum width was varied from about 252 kHz to above 2 MHz, while the rms deviation was kept constant. The modulation index, M, varied less than ± 0.5 dB over this range.

These measurements complete the verification of the first block of the model. Additional corroboration was obtained from a baseband-to-baseband measurement of a 132-channel carrier. The test setup, the measured filter response, and the computed intelligible crosstalk ratio are shown in Figure 45. Discrepancies, caused by an imperfect test setup, occur in the case of low crosstalk. In addition, the measurement accuracy for crosstalk ratios greater than 70 dB degrades very rapidly because of noise. Otherwise, the agreement between predictions and measurements is considered to be good.

Adjacent channel interference

Adjacent FDM/FM telephony channels are characterized as shown in Figure 46 [14]. At the output of the desired channel receive filter, two effects are noted. First, there is a spectral density caused by modulator 2 (modified by filters G_I and G_{RD1}) at the output of G_{RD}. It can be shown that, after demodulation, the resulting phase noise at baseband is given by the convolution of the desired spectrum with the residual adjacent channel spectrum (both referenced to the output of G_{RD}). The only restriction on this derivation is that the instantaneous complex amplitude of the desired carrier must always be greater than that of the interfering carrier.

Secondly, when viewed in the time domain (i.e., at output R_1) with a detector having a video bandwidth $\simeq 10$ times greater than the top baseband frequency of the interfering modulation, M_2, the instantaneous amplitude of the waveform caused by M_2 is greater than the amplitude of the waveform caused by the desired modulation, M_1, for short random time periods. This random impulse interference was the limiting factor in the guardband and filter requirements determined for the cases of interest in the INTELSAT IV system.

It can be shown [14] that the rate of impulses is

$$\text{rate} = \int_{-\infty}^{+\infty} |F_o - f'| \cdot p(f') \cdot p[A_D(f) < A_i(f')]\, df'$$

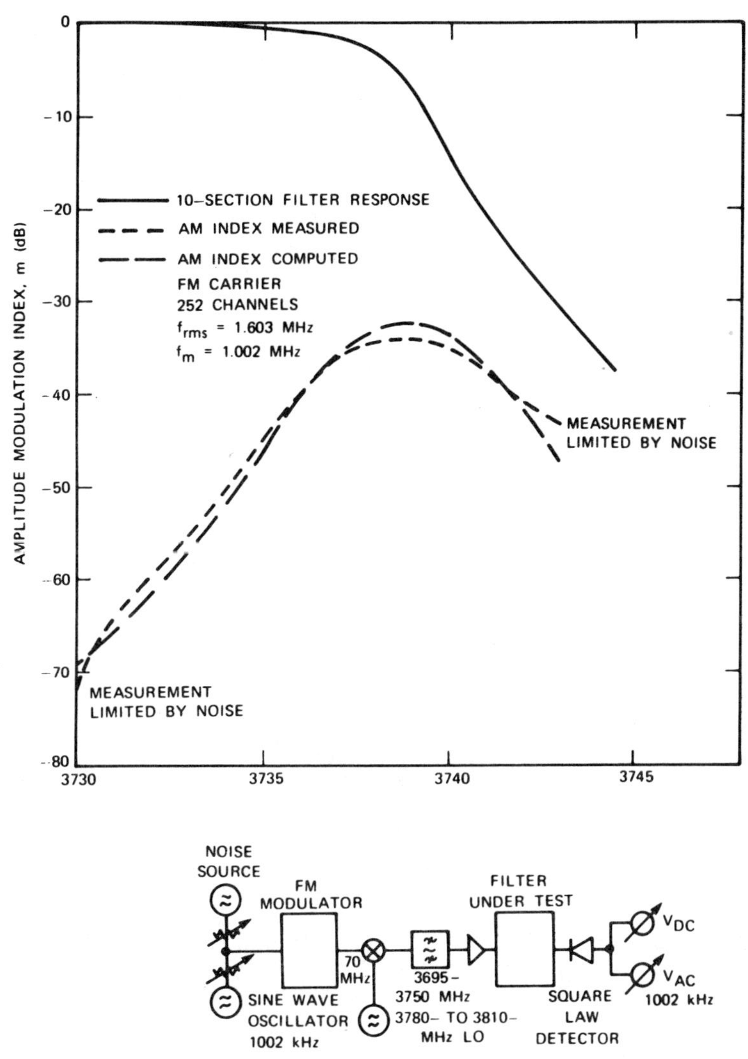

Figure 44. *AM Index Measurement with Noise Loading*

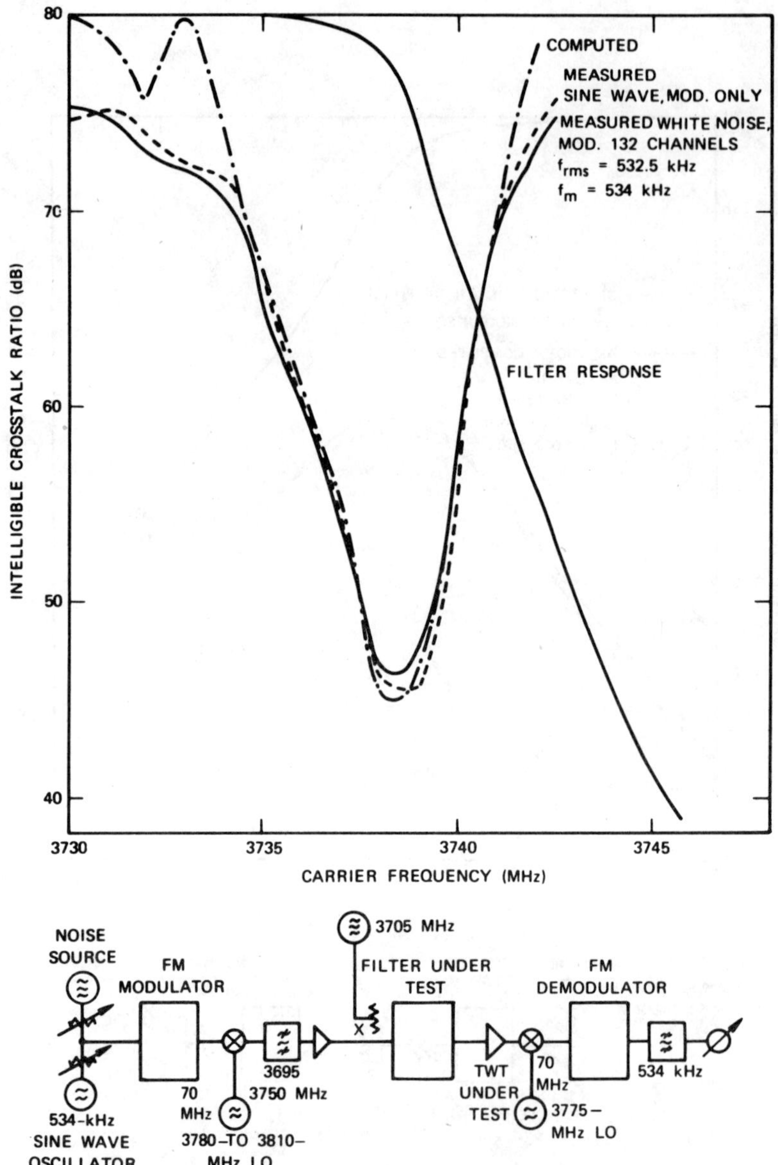

Figure 45. *Baseband-to-Baseband Measurements*

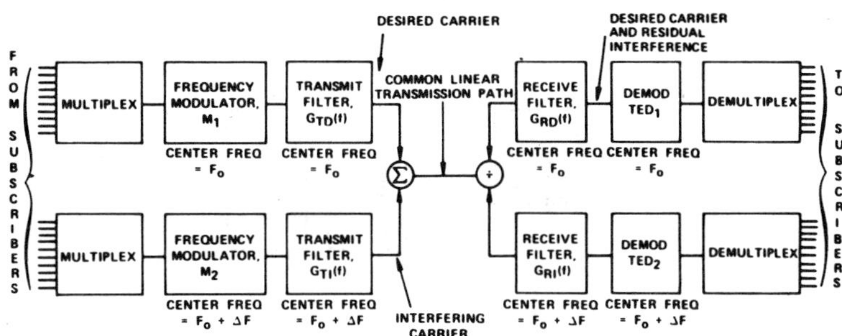

Figure 46. *Adjacent Channel Interference*

where

F_o = center frequency of the desired carrier

f' = frequency variable of integration

$P(f')$ = probability density for the interfering carrier at frequency f'

$P[A_D(f) < A_i(f')]$ = probability density when the amplitude of the desired carrier is less than that of the interfering carrier and the interfering carrier is at frequency f'.

In a bandwidth-limited mode, the system capacity increases in proportion to the useful bandwidth. To increase the useful bandwidth while maintaining the impulse interference at an acceptable level, all adjacent channels should have transmit filters with very high rates of attenuation for the initial 10 to 15 dB, and bandwidths as narrow as possible without distorting the modulation in the adjacent channel. Companion studies [15] have shown that elliptic function filters with up to seven poles may be used at intermediate frequencies (typically 70 MHz) before the achievable unloaded Q's begin to limit the achievable rate of attenuation. It has also been shown that these filters may be group-delay equalized over their full equiripple bandwidth.

For the case of interest, the residual ripple component of group-delay was set to contribute $\simeq 100$ pWp to the overall noise budget, and the performance limit was determined by truncation noise caused by the abrupt attenuation at the filter band edge. In the adjacent channel interference study, the filter characteristics (summarized in Table 10) were set to balance the in-band impulse effects caused by truncation noise against the impulse effects caused by the adjacent channel.

For the entire range of parameters in the INTELSAT IV system, the con-

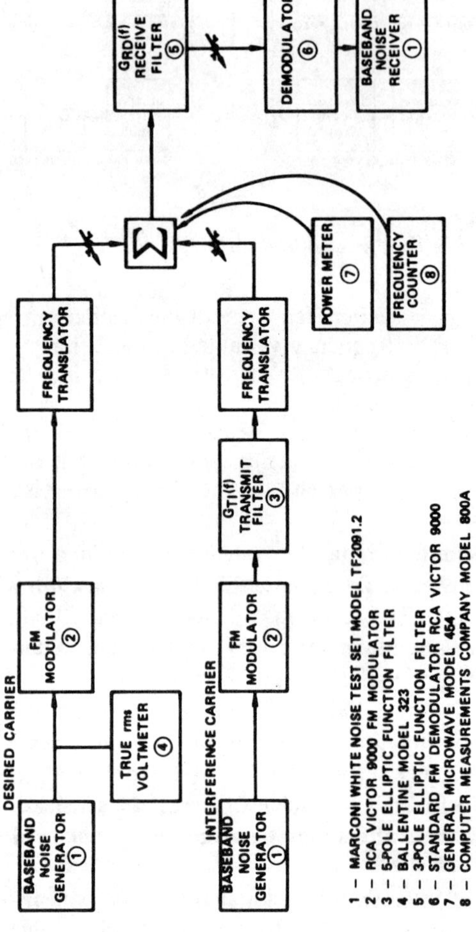

Figure 47. *Measurement Setup*

cept of instantaneous frequency was used to predict the number of impulses. In addition, a phase-lock loop was modeled as a single-pole tracking filter (whose bandwidth was set approximately equal to the closed-loop bandwidth of the receiver), and a statistical distribution of center frequency was determined from the modulation statistics. The validity of this modeling for engineering purposes was confirmed by the accuracy with which the modeling predicted the onset of impulses as the adjacent carrier and its transmit filter were translated in frequency toward the desired carrier and its receive filter, and the statistical improvement offered by the use of a threshold extending receiver that could be logically characterized as a "narrow" tracking filter (see Figures 47 and 48).

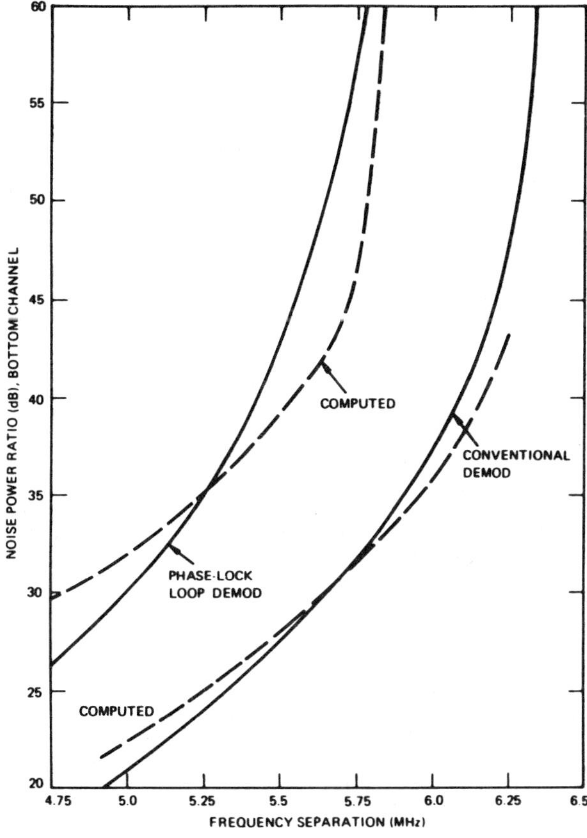

Figure 48. *Comparison of Measured and Computed Adjacent Channel Interference Performance*

Effects of Intermodulation, AM–PM Conversion, and Additive Noise in Multicarrier TWT Systems

OSAMU SHIMBO

Abstract—The intermodulation due to the traveling-wave tube (TWT) is analyzed for the case where arbitrarily modulated carriers and Gaussian noise are amplified through the TWT. Both AM–PM conversion effects and nonlinear amplification in the TWT are considered. The possibility of reducing intermodulation for system improvement is also discussed.

I. Introduction

IN SATELLITE communications, it is sometimes necessary that many modulated carriers be amplified through traveling-wave tube (TWT) amplifiers. In this paper, a formula to evaluate the intermodulation due to the nonlinear characteristic of TWT amplification and amplitude modulation–phase modulation (AM–PM) conversion when many arbitrarily modulated carriers and Gaussian noise are passed through a TWT is first established [1]. Secondly, using this formula, the possibility of reducing the intermodulation for system improvement is rigorously discussed. The nonlinear input–output amplitude characteristic of a TWT amplifier and the AM–PM conversion effects with Gaussian noise are considered simultaneously. The results will be useful for the following purposes: 1) design of TWT amplifiers for multicarrier systems including thermal noise, and 2) discussions of certain methods for reducing intermodulation.

In Section II a general formula for calculation of the intermodulation is analyzed under the condition that the input–output nonlinear characteristic of TWT amplification and the characteristic of AM–PM conversion are given for the case with thermal noise. In Section III a general formula for the case without thermal noise is analyzed. In Section IV a formula for the case where the noise is small is determined. Using the general formulas derived in these three sections, the cases where the nonlinear characteristics (including the AM–PM conversion) are given by special forms (for example, power series expansion of the input, Fourier series expansion, and a special type of function) are discussed in Section V. In Section VI a formula for the case of many carriers is derived. In Section VII the cases of two-modulated carriers and three-modulated carriers with noise are discussed. Based on the results in the previously mentioned sections, the possibility of reducing the intermodulations due to TWT for system improvement is discussed in Section VIII.

Manuscript received September 3, 1970. This paper is based upon work performed in COMSAT Laboratories under corporate sponsorship. The author is with COMSAT Laboratories, Clarksburg, Md. 20734.

II. General Formula for Intermodulations Due to Nonlinear Amplitude Characteristics and AM–PM Conversion with Noise

Represent the input signal of the TWT amplifier disturbed by noise by

$$e_i(t) = \sum_{k=1}^{n} A_k \cos\left[\omega_0 t + \omega_k t + \phi_k(t) + \lambda_k\right] + N_c(t) \cos \omega_0 t - N_s(t) \sin \omega_0 t \quad (1)$$

where $\omega_0 + \omega_k$, $\phi_k(t)$, λ_k, and A_k represent, respectively, the carrier frequency, the modulating signal, and the phase and amplitude of the kth carrier; $N_c(t)$ and $N_s(t)$ represent, respectively, the in-phase and quadrature components of the noise; ω_0 is conveniently taken as the midband angular frequency. Fig. 1 shows the carrier frequency arrangement.

Equation (1) can be rewritten as

$$e_i(t) = (x_1^2 + y_1^2)^{1/2} \cos\left(\omega_0 t + \tan^{-1} \frac{y_1}{x_1}\right) \quad (2)$$

where

$$x_1 = \sum_{k=1}^{n} A_k \cos\left[\omega_k t + \phi_k(t) + \lambda_k\right] + N_c(t)$$

$$y_1 = \sum_{k=1}^{n} A_k \sin\left[\omega_k t + \phi_k(t) + \lambda_k\right] + N_s(t). \quad (3)$$

It is assumed here that the output of the TWT amplifier can be represented by

$$e_0(t) = g[(x_1^2 + y_1^2)^{1/2}] \cos\left\{\omega_0 t + \tan^{-1}\left(\frac{y_1}{x_1}\right) + f[(x_1^2 + y_1^2)^{1/2}]\right\} \quad (4)$$

where $f[(x_1^2 + y_1^2)^{1/2}]$ represents the phase distortion due to the AM–PM conversion and $g[(x_1^2 + y_1^2)^{1/2}]$ represents the amplitude which is distorted due to the amplification of TWT. To obtain the power spectrum of the intermodulation, the autocorrelation function of the output $e_0(t)$ must be obtained.[1] The characteristic function method is used here to obtain it.

[1] The Fourier transformation of the autocorrelation function gives the power spectrum.

Reprinted from *Proc. IEEE*, vol. 59, pp. 230-238, Feb. 1971.

Fig. 1.   Frequency arrangement of input multicarriers.

Represent the characteristic function of x_1, y_1, $x_2 = x_1(t+\tau)$, and $y_2 = y_1(t+\tau)$ by

$$C(u_1, v_1, u_2, v_2)$$
$$= \text{avg}\{\exp(ju_1x_1 + jv_1y_1 + ju_2x_2 + jv_2y_2)\}. \quad (5)$$

Then the autocorrelation function of $e_0(t)$ is given by

$$\text{avg}\{e_0(t)e_0(t+\tau)\} = \tfrac{1}{2}\text{Re}\exp(-j\omega_0\tau)(2\pi)^{-4}$$

$$\int_{-\infty}^{\infty}\int_{-\infty}^{\infty}\int_{-\infty}^{\infty}\int_{-\infty}^{\infty}\int_{-\infty}^{\infty}\int_{-\infty}^{\infty}\int_{-\infty}^{\infty}\int_{-\infty}^{\infty}$$
$$\cdot C(u_1, v_1, u_2, v_2)g[(x_1^2+y_1^2)^{1/2}]$$
$$\cdot \exp[j[(x_1^2+y_1^2)^{1/2}]](x_1+jy_1)$$
$$\cdot (x_1^2+y_1^2)^{-1/2}g[(x_2^2+y_2^2)^{1/2}]$$
$$\cdot \exp[-j[(x_2^2+y_2^2)^{1/2}]]$$
$$\cdot (x_2-jy_2)(x_2^2+y_2^2)^{-1/2}$$
$$\cdot \exp(-ju_1x_1-jv_1y_1-ju_2x_2-jv_2y_2)$$
$$\cdot dx_1\,dy_1\,dx_2\,dy_2\,du_1\,dv_1\,du_2\,dv_2 \quad (6)$$

where Re denotes the real part of the function.

Averaging (5) with respect to $\lambda_1, \lambda_2, \cdots, \lambda_n, N_c(t), N_s(t), N_s(t+\tau), N_c(t+\tau)$, we have

$$C(u_1, v_1, u_2, v_2) = \exp\{-\tfrac{1}{2}[R(0)(u_1^2+v_1^2+u_2^2+v_2^2)$$
$$+ 2(u_1u_2+v_1v_2)R(\tau) + 2(u_1v_2-v_1u_2)I(\tau)]\}$$
$$\cdot \prod_{k=1}^{n} J_0\{A_k[u_1^2+v_1^2+u_2^2+v_2^2$$
$$+ 2(u_1u_2+v_1v_2)\cos(\omega_k\tau+\psi_k)$$
$$+ 2(u_1v_2-v_1u_2)\sin(\omega_k\tau+\psi_k)]^{1/2}\} \quad (7)$$

where $R(\tau)$ is the autocorrelation function of $N_c(t)$ or $N_s(t)$ [i.e., $R(\tau)$ is the cosine Fourier transformation of the power spectrum of $N_c(t)$ or $N_s(t)$] and $I(\tau)$ is the sine Fourier transformation of the power spectrum of $N_c(t)$ or $N_s(t)$, ψ_k is given by

$$\psi_k = \phi_k(t+\tau) - \phi_k(t). \quad (8)$$

To consider an arbitrary modulation, the average on ψ_k in (7) is not taken here; instead, it is done in the final stage by assuming some modulation [frequency division–frequency modulation (FDM–FM), digital FM, etc.].

Substituting (7) into (6), applying the polar coordinate transformations,

$$u_1 = r_1\cos\theta_1 \qquad u_2 = r_2\cos\theta_2$$
$$v_1 = r_1\sin\theta_1 \qquad v_2 = r_2\sin\theta_2$$
$$x_1 = \rho_1\cos\mu_1 \qquad x_2 = \rho_2\cos\mu_2$$
$$y_1 = \rho_1\sin\mu_1 \qquad y_2 = \rho_2\sin\mu_2, \quad (9)$$

and achieving the integrations on μ_1 and μ_2, we have

$$\text{avg}\{e_0(t)e_0(t+\tau)\}$$
$$= -\frac{1}{2}\text{Re}\exp(-j\omega_0\tau)(2\pi)^{-2}\int_0^{2\pi}d\theta_1\int_0^{2\pi}d\theta_2\int_0^{\infty}r_1\,dr_1\int_0^{\infty}r_2$$

$$\cdot \prod_{k=1}^{n} J_0\{A_k[r_1^2+r_2^2+2r_1r_2\cos(\omega_k\tau+\psi_k+\theta_1-\theta_2)]^{1/2}\}$$

$$\cdot \exp\left\{-R(0)\left(\frac{r_1^2+r_2^2}{2}\right) - U(\tau)r_1r_2\cos[\theta_1-\theta_2+\Lambda(\tau)]\right\}$$

$$\cdot \exp(j\theta_1-j\theta_2)\int_0^{\infty}\rho_1 g(\rho_1)\exp[jf(\rho_1)]J_1(r_1\rho_1)d\rho_1$$

$$\cdot \int_0^{\infty}\rho_2 g(\rho_2)\exp[-jf(\rho_2)]J_1(r_2\rho_2)d\rho_2\,dr_2 \quad (10)$$

where

$$U(\tau) = [R^2(\tau) + I^2(\tau)]^{1/2}$$

$$\Lambda(\tau) = \tan^{-1}\left[\frac{I(\tau)}{R(\tau)}\right]. \quad (11)$$

Using an expansion (see eq. (1), p. 358, in [2]) and achieving the integrations with respect to θ_1 and θ_2, (10) becomes

$$\frac{1}{2}\text{Re}\sum_{k_1,k_2,\cdots,k_n=-\infty}^{\infty}\exp\{j(\omega_0+k_1\omega_1+k_2\omega_2+\cdots+k_n\omega_n)\tau\}$$
$$\cdot \exp\{j(k_1\psi_1+k_2\psi_2+\cdots+k_n\psi_n)\}$$
$$\cdot \exp\{-j(k_1+k_2+\cdots+k_n-1)\Lambda(\tau)\}$$
$$\cdot \int_0^{\infty}\int_0^{\infty}I_{k_1+k_2+\cdots+k_n-1}[U(\tau)r_1r_2]\prod_{l=1}^{n}[J_{k_l}(A_lr_1)J_{k_l}(A_lr_2)]$$
$$\cdot \exp\left\{-R(0)\left(\frac{r_1^2+r_2^2}{2}\right)\right\}r_1r_2\,dr_1\,dr_2\int_0^{\infty}\rho_1 g(\rho_1)$$
$$\cdot \exp[jf(\rho_1)]d\rho_1\int_0^{\infty}\rho_2 g(\rho_2)\exp[-jf(\rho_2)]d\rho_2. \quad (12)$$

Applying a formula (eq. (1), p. 395, in [2]) to the modified Bessel I in (12), (12) can be rewritten as

$$\text{avg}\{e_0(t)e_0(t+\tau)\}$$
$$= \frac{1}{2}\text{Re}\sum_{k_1,k_2,\cdots,k_n=-\infty}^{\infty}\exp\{j(\omega_0+k_1\omega_1+k_2\omega_2+\cdots+k_n\omega_n)\}$$
$$\cdot \exp\{j(k_1\psi_1+k_2\psi_2+\cdots+k_n\psi_n)\}$$
$$\cdot \exp\{-(k_1+k_2+\cdots+k_n-1)j\Lambda(\tau)\}$$
$$\cdot \int_0^{\infty}t\exp\left\{-\frac{t^2}{2}\right\}\left|\int_0^{\infty}rJ_{k_1}(A_1r)J_{k_2}(A_2r)\cdots J_{k_n}(A_nr)\right.$$
$$\cdot J_{k_1+k_2+\cdots+k_n-1}\{[U(\tau)]^{1/2}rt\}$$
$$\cdot \exp\left\{-[R(0)-U(\tau)]\frac{r^2}{2}\right\}dr\int_0^{\infty}\rho g(\rho)$$
$$\left.\cdot \exp\{jf(\rho)\}J_1(r\rho)d\rho\right|^2 dt. \quad (13)$$

In (12) and (13) the summation is done over all $k_1, k_2 \cdots, k_n$, and k_l ($l=1, 2, \cdots, n$) can take the negative and positive integers and zero independently.

Equation (13) represents a basic formula for evaluating

the intermodulation in a multicarrier system due to TWT under Gaussian noise. From (13), the magnitude of the intermodulation whose frequency is $\omega_0 + k_1\omega_1 + k_2\omega_2 + \cdots + k_n\omega_n$ is given by

$$\int_0^\infty t \exp\left\{-\frac{t^2}{2}\right\}\left|\int_0^\infty r J_{k_1}(A_1 r) J_{k_2}(A_2 r) \cdots J_{k_n}(A_n r)\right.$$

$$\cdot J_{k_1 + k_2 + \cdots + k_n - 1}\{[U(\tau)]^{1/2} rt\}$$

$$\cdot \exp\left\{-[R(0) - U(\tau)]\frac{r^2}{2}\right\} dr \int_0^\infty g(\rho)$$

$$\left.\cdot \exp\{jf(\rho)\} J_1(r\rho)\, d\rho\right|^2 dt = \int_0^\infty t$$

$$\cdot \exp\left\{-\frac{t^2}{2}\right\} \left|M(t; k_1, k_2, \cdots, k_n)\right|^2 dt. \tag{14}$$

The power-spectrum spread of this intermodulation is given by the Fourier transformation of the following function:

$$\exp\{-j(k_1 + k_2 + \cdots + k_n - 1)\Lambda(\tau)\} \operatorname{avg}\{\exp[j(k_1\psi_1$$
$$+ k_2\psi_2 + \cdots + k_n\psi_n)]\} \tag{15}$$

where the average is taken over $\psi_k = \phi_k(t+\tau) - \phi_k(t)$ $(k = 1, 2, \cdots, n)$.

Note here that the modulation only affects (15), and that (14) is independent of the modulation. Also, the nonlinear characteristics of TWT $g(\rho)\exp\{jf(\rho)\}$ affect only (14). Such a separation between the modulation and nonlinear characteristics is convenient for the analysis and computation of the intermodulation.

The pure-signal component, whose angular frequency is $\omega_0 + \omega_l$ and whose modulating signal is $\phi_l(t)$, is given by putting $k_l = 1$, $k_1 = k_2 = \cdots = k_{l-1} = k_{l+1} = \cdots = k_n = 0$ and $U(\tau) = 0$. [Accordingly, $\Lambda(\tau) = 0$.]

The evaluation of the integral (14) is generally rather difficult. However, in some cases which are of practical importance, it becomes simple, as will be shown in the following section.

If the intermodulation is small, by taking only the first-order distortion, we can approximate it as

$$g(\rho)\exp\{jf(\rho)\} \simeq g(\rho) + jf(\rho)\rho \tag{16}$$

and (14) is reduced to

$$\int_0^\infty t \exp\left\{-\frac{t^2}{2}\right\}\left[\int_0^\infty r \exp\left\{-[R(0) - U(\tau)]\frac{r^2}{2}\right\}\right.$$

$$\cdot J_{k_1 + k_2 + \cdots + k_n - 1}\{[U(\tau)]^{1/2} rt\} \prod_{l=1}^n J_{k_l}(A_l r)\, dr$$

$$\cdot \int_0^\infty \rho g(\rho) J_1(r\rho)\, d\rho\Big]^2 dt + \int_0^\infty t \exp\left\{-\frac{t^2}{2}\right\}$$

$$\cdot \left[\int_0^\infty r \exp\left\{-[R(0) - U(\tau)]\frac{r^2}{2}\right\} J_{k_1 + k_2 + \cdots + k_n - 1}\right.$$

$$\left.\cdot \{[U(\tau)]^{1/2} rt\} \prod_{l=1}^n J_{k_l}(A_l r)\, dr \int_0^\infty \rho^2 f(\rho) J_1(r\rho)\, d\rho\right]^2 dt. \tag{17}$$

From (17), if the approximation of (16) is satisfied, there is no

cross term between the distortion due to TWT nonlinear amplification $g(\rho)$ and that due to AM–PM conversion $f(\rho)$; i.e., they can be evaluated independently.

III. Case Without Thermal Noise

In this case, we can put $U(\tau) = R(0) = 0$ in (13) to obtain

$$\operatorname{avg}\{e_0(t)e_0(t+\tau)\} = \frac{1}{2}\operatorname{Re} \sum_{\substack{k_1, k_2, \cdots, k_n = -\infty \\ (k_1 + k_2 + \cdots + k_n = 1)}}$$

$$\cdot \exp\{j(k_1\omega_1 + k_2\omega_2 + \cdots + k_n\omega_n + \omega_0)\tau\}$$

$$\cdot \exp\{j(k_1\psi_1 + k_2\psi_2 + \cdots + k_n\psi_n)\}$$

$$\cdot \left|\int_0^\infty r J_{k_1}(A_1 r) J_{k_2}(A_2 r) \cdots J_{k_n}(A_n r)\, dr\right.$$

$$\left.\cdot \int_0^\infty \rho g(\rho)\exp\{jf(\rho)\} J_1(r\rho)\, d\rho\right|^2 \tag{18}$$

where $k_1, k_2, \cdots, k_n$ can take the positive and negative integers and zero under the constraint that

$$k_1 + k_2 + \cdots + k_n = 1. \tag{19}$$

The magnitude of the intermodulation whose angular frequency is $k_1\omega_1 + k_2\omega_2 + \cdots + k_n\omega_n + \omega_0$ is given by the absolute value of the complex quantity

$$M(k_1, k_2, \cdots, k_n) = \int_0^\infty r \prod_{l=1}^n J_{k_l}(A_l r)\, dr$$

$$\cdot \int_0^\infty \rho g(\rho)\exp\{jf(\rho)\} J_1(r\rho)\, d\rho. \tag{20}$$

The power-spectrum spread of this component is given by the Fourier transformation of

$$\operatorname{avg}\{\exp[j(k_1\psi_1 + k_2\psi_2 + \cdots + k_n\psi_n)]\}. \tag{21}$$

If the intermodulation is small, the real part of (20) gives the intermodulation due to the amplitude characteristic $g(\rho)$ and the imaginary part gives the intermodulation due to the AM–PM conversion $f(\rho)$ as explained in (17). The pure-signal component, whose angular frequency is $\omega_0 + \omega_l$, is given by putting $k_l = 1$, $k_1 = k_2 = \cdots = k_{l-1} = k_{l+1} = \cdots = k_n = 0$ in (18).

IV. Small-Noise Case

In the small-noise region above the threshold, where the first-order noise approximation can be used, the following two terms appear in addition to (18):

$$\frac{1}{4}\operatorname{Re}[R(\tau) + jI(\tau)] \sum_{\substack{k_1, k_2, \cdots, k_n = -\infty \\ (k_1 + k_2 + \cdots + k_n = 2)}}^{\infty}$$

$$\cdot \exp\{j(k_1\omega_1 + k_2\omega_2 + \cdots + k_n\omega_n + \omega_0)\}$$

$$\cdot \exp\{jk_1\psi_1 + jk_2\psi_2 + \cdots + jk_n\psi_n\}|M(k_1, k_2, \cdots, k_n)|^2$$

$$+\frac{1}{4}\operatorname{Re}[R(\tau) - jI(\tau)] \sum_{\substack{k_1, k_2, \cdots, k_n = -\infty \\ (k_1 + k_2 + \cdots + k_n = 0)}}^{\infty}$$

$$\cdot \exp\{j(k_1\omega_1 + k_2\omega_2 + \cdots + k_n\omega_n)\}$$

$$\cdot \exp\{jk_1\psi_1 + jk_2\psi_2 + \cdots + jk_n\psi_n\}|M(k_1, k_2, \cdots, k_n)|^2 \tag{22}$$

where $k_1, k_2, \cdots, k_n$ can take the positive and negative integers and zero under the constraints that $k_1 + k_2 + \cdots + k_n = 2$ for the first term and $k_1 + k_2 + \cdots + k_n = 0$ for the second term.

V. Nonlinear Characteristic of Special Forms

In this section, to make numerical computation possible, special forms are assumed for $g(\rho) \exp\{jf(\rho)\}$ which represents both the characteristics of nonlinear TWT amplification and AM–PM conversion.

A. Case Where the Input–Output Characteristic Is Given by a Power Series

Since $g(\rho)$ is an odd function of ρ, and $f(\rho)$ is an even function, the following expansion is assumed:

$$g(\rho) \exp\{jf(\rho)\} = \sum_{m=0}^{\infty} \alpha_m \rho^{2m+1} \tag{23}$$

where α_m is generally a complex number. The integration with respect to ρ in (14) and (20) is now represented by

$$\int_0^{\infty} \rho g(\rho) \exp\{jf(\rho)\} J_1(\rho r) d\rho$$

$$= \sum_{m=0}^{\infty} \alpha_m \int_0^{\infty} \rho^{2(m+1)} J_1(\rho r) d\rho \tag{24}$$

where

$$\int_0^{\infty} \rho^{2(m+1)} J_1(r\rho) d\rho$$

is a distribution operator; i.e.,

$$\int_0^{\infty} G(r) dr \int_0^{\infty} \rho^{2(m+1)} J_1(r\rho) d\rho = (m+1)! \, 2^{m+1}$$

$$\cdot \left\{ \left(-\frac{1}{Y} \cdot \frac{\partial}{\partial Y} \right)^m \frac{G(Y)}{Y^2} \right\}_{Y=0} \tag{25}$$

where the existence of the right-hand term is assumed. If $G(r)$ is expanded by a power series of r as

$$G(r) = \sum_{l=1}^{\infty} e_l r^{2l} \tag{26}$$

(25) becomes

$$\int_0^{\infty} G(r) dr \int_0^{\infty} \rho^{2(m+1)} J_1(r\rho) d\rho$$

$$= (-1)^m m! \, (m+1)! \, 2^{2m+1} e_{m+1}. \tag{27}$$

Thus in this case, we have

$$\int_0^{\infty} G(r) dr \int_0^{\infty} \rho g(\rho) \exp\{jf(\rho)\} J_1(r\rho) d\rho$$

$$= \sum_{m=0}^{\infty} (-1)^m m! \, (m+1)! \, 2^{2m+1} \alpha_m e_{m+1}. \tag{28}$$

Therefore, if we take $G(r)$ as

$$G(r) = r J_{k_1}(A_1 r) J_{k_2}(A_2 r), \cdots, J_{k_n}(A_n r), \tag{29}$$

$M(k_1, k_2, \cdots, k_n)$ of (20) is given by (28). If we take $G(r)$ as

$$G(r) = r \exp\left\{ -[R(0) - U(\tau)] \frac{r^2}{2} \right\} J_{k_1 + k_2 + \cdots + k_n - 1}$$

$$\cdot \{[U(\tau)]^{1/2} rt\} \prod_{l=1}^{n} J_{k_l}(A_l r), \tag{30}$$

$M(t; k_1, k_2, \cdots, k_n)$ of (14) is given by (28), where e_m is now a function of t.

B. Case Where Input–Output Characteristic Is Expanded by a Fourier Series

Since $g(\rho) \exp\{jf(\rho)\}$ is an odd function of ρ, it is expanded here as

$$g(\rho) \exp\{jf(\rho)\} = \sum_{m=1}^{\infty} A_m \sin(mp\rho) \tag{31}$$

where A_m is generally a complex number. In this case, we have

$$\int_0^{\infty} G(r) dr \int_0^{\infty} \rho g(\rho) \exp\{jf(\rho)\} J_1(r\rho) d\rho$$

$$= \sum_{m=1}^{\infty} A_m \int_0^{\infty} G(r) dr \int_0^{\infty} \rho \sin(mp\rho) J_1(r\rho) d\rho$$

$$= \int_0^{\pi/2} \left\{ \sum_{m=1}^{\infty} A_m G'(mp \sin\theta) \right\} d\theta \tag{32}$$

where $G'(r)$ is the derivative of $G(r)$.

If $G(r)$ is given by (26), (32) can be expanded as

$$\sum_{m=1}^{\infty} A_m \sum_{l=0}^{\infty} \frac{l! \, (l+1)!}{(2l+1)!} 2^{2l+1} e_{l+1} (pm)^{2l+1} \tag{33}$$

where $G(r)$ is given by (29) for the noiseless case and by (30) for the noisy case.

C. Case Where AM–PM Conversion Takes a Special Form

As an example, a case where the AM–PM characteristic $f(\rho)$ takes a special form (eq. (4), p. 41, in [3]) is analyzed in this section; i.e.,

$$f(\rho) = C_1\{1 - \exp(-C_2\rho^2)\} + C_3\rho^2$$
$$g(\rho) = \rho. \tag{34}$$

In the region of normal operation where the intermodulation is not large, we can approximate $g(\rho) \exp\{jf(\rho)\}$ as

$$g(\rho) \exp\{jf(\rho)\}$$
$$\simeq \rho + j\rho[C_1\{1 - \exp(-C_2\rho^2)\} + C_3\rho^2]$$
$$= \{(1 + jC_1)\rho + jC_3\rho^3\} - jC_1\rho \exp\{-C_2\rho^2\}. \tag{35}$$

For the polynomial in the first term of (35), (28) can be applied to evaluate $M(t; k_1, k_2, \cdots, k_n)$ and $M(k_1, k_2, \cdots, k_n)$, with

$$\alpha_0 = 1 + jC_1$$
$$\alpha_1 = jC_3. \tag{36}$$

The term resulting from α_0 gives the pure-signal component and that resulting from α_1 gives the third-order intermodulation.

In the noiseless case, only the case where $k_l=1$ and $k_1=k_2=\cdots=k_{l-1}=k_{l+1}=\cdots=k_n=0$ $(l=1,2,\cdots,n)$, contributes the pure-signal component and the case where $k_l=k_m=1$, $k_p=-1$, and all the others are zero or the case where $k_l=2$, $k_m=-1$ and all the others are zero contributes the third-order intermodulation. The noisy case can be analyzed similarly from (30).

For the second term of (35), $M(t;k_1,k_2,\cdots,k_n)$ and $M(k_1,k_2,\cdots,k_n)$ are given by

$$M'(k_1,k_2,\cdots,k_n)$$
$$= -jC_1(2C_2)^{-1/2}\int_0^\infty J_{k_1}[(2C_2)^{1/2}A_1 r]J_{k_2}[(2C_2)^{1/2}A_2 r]\cdots$$
$$\cdot J_{k_n}[(2C_2)^{1/2}A_n r]\exp\left\{-\frac{r^2}{2}\right\}r^2 dr$$

$$M'(t;k_1,k_2,\cdots,k_n)$$
$$= -jC_1(2C_2)^{-1/2}\int_0^\infty J_{k_1}[(2C_2)^{1/2}A_1 r]\cdots$$
$$\cdot J_{k_n}[(2C_2)^{1/2}A_n r]J_{k_1+k_2+\cdots+k_n-1}$$
$$\cdot \{[2C_2 U(\tau)]^{1/2}rt\}\exp\left[-\{1+2C_2\right.$$
$$\cdot [R(0)-U(\tau)]\}\frac{r^2}{2}\right]dr. \tag{37}$$

In the normal operating region, where the intermodulation is not large, the following approximation is possible for (37):

$$J_{|k_l|}[(2C_2)^{1/2}A_l r]\simeq \frac{\{(C_2/2)^{1/2}(A_l r)\}^{|k_l|}}{|k_l|!}. \tag{38}$$

VI. Case of Many Carriers

As seen from (10), the autocorrelation function of the output of the TWT disturbed by noise can be obtained by an integration of the product of the 0th order Bessel functions. As is well known from the saddle-point approximation and the central limiting theorem, an approximation can be made as

$$\prod_{k=1}^n J_0(Z_k)\simeq \exp\left\{-\sum_{k=1}^n \frac{Z_k^2}{4}\right\}. \tag{39}$$

Using this approximation, the integral of (10) is reduced to

$$\text{avg}\{e_0(t)e_0(t+\tau)\}$$
$$\simeq \tfrac{1}{2}\text{Re}\{(B+R_\tau)-j(C+I_\tau)\}\exp\{-j\omega_0\tau\}$$
$$\cdot \{(B+R_\tau)^2+(C+I_\tau)^2\}^{-1/2}\int_0^\infty\int_0^\infty r_1 r_2$$
$$\cdot \exp\left\{-(A+R_0)\left(\frac{r_1^2+r_2^2}{2}\right)\right\}I_1[\{(B+R_\tau)^2$$
$$+(C+I_\tau)^2\}^{1/2}r_1 r_2]dr_1 dr_2$$
$$\cdot \int_0^\infty\int_0^\infty J_1(r_1\rho_1)J_1(r_2\rho_2)\rho_1\rho_2 g(\rho_1)g(\rho_2)$$
$$\cdot \exp\{jf(\rho_1)-jf(\rho_2)\}d\rho_1 d\rho_2 \tag{40}$$

where

$$A = \frac{1}{2}\sum_{k=1}^n A_k^2$$
$$B = \frac{1}{2}\sum_{k=1}^n A_k^2\cos(\omega_k\tau+\psi_k)$$
$$C = \frac{1}{2}\sum_{k=1}^n A_k^2\sin(\omega_k\tau+\psi_k) \tag{41}$$

and where $R(\tau)$ and $I(\tau)$ are abbreviated as R_τ and I_τ, respectively.

Applying the same formula which was used to obtain (13) and achieving the integration with respect to r, (40) is finally reduced to

$$\tfrac{1}{2}\text{Re}\exp\{-j\omega_0\tau\}\{(B+R_\tau)-j(C+I_\tau)\}\{(A+R_0)-\xi\}^{-2}\xi^{-2}$$
$$\cdot \int_0^\infty\exp\left\{-\left[\frac{1}{(A+R_0)-\xi}+\frac{2}{\xi}\right]\frac{t^2}{2}\right\}t\left|\int_0^\infty\rho g(\rho)\exp\{jf(\rho)\}\right.$$
$$\cdot \left. I_1\left(\frac{t}{\xi}\rho\right)\exp\left\{-\frac{\rho^2}{2\xi}\right\}d\rho\right|^2 dt \tag{42}$$

where

$$\xi = (A+R_0)-\{(B+R_\tau)^2+(C+I_\tau)^2\}^{1/2}. \tag{43}$$

As seen from (40) and (41), the approximation of (39) is equivalent to the assumption that the summation of many modulated carriers whose phases are random [as shown by the first terms in (3)] can be approximated as a Gaussian process.

A. Case of Power Series Expansion

If $g(\rho)\exp\{-jf(\rho)\}$ is expanded as a power series of (23), (42) can be evaluated as

$$\tfrac{1}{2}\text{Re}\exp\{-j\omega_0\tau\}\{(B+R_\tau)-j(C+I_\tau)\}\{(A+R_0)-\xi\}^{-2}$$
$$\cdot \int_0^\infty\exp\left\{-\xi(A+R_0-\xi)^{-1}\frac{t^2}{2}\right\}t^3\left|\sum_{m=0}^\infty \alpha_m(2)^m\xi^{m+1}\right.$$
$$\cdot \left.(m+1)!_1 F_1\left(-m;2;-\frac{t^2}{2}\right)\right|^2 dt \tag{44}$$

where $_1F_1$ is the confluent hypergeometric function and is given by

$$_1F_1\left(-m;2;-\frac{t^2}{2}\right)=1+\frac{1}{1!}\frac{m}{2}\frac{t^2}{2}+\frac{1}{2!}\frac{m(m-1)}{2\cdot 3}\left(\frac{t^2}{2}\right)^2$$
$$+\frac{1}{3!}\frac{m(m-1)(m-2)}{2\cdot 3\cdot 4}\left(\frac{t^2}{2}\right)^3+\cdots$$
$$+\frac{1}{m!}\frac{m!}{(m+1)!}\left(\frac{t^2}{2}\right)^m. \tag{45}$$

Therefore, we can expand as

$$\left|\sum_{m=0}^\infty \alpha_m 2^m\xi^{m+1}(m+1)!_1 F_1\left(-m;2;-\frac{t^2}{2}\right)\right|^2$$
$$= \sum_{m=0}^\infty \beta_m t^{2m} \tag{46}$$

where β_m is a function of ξ. Using this value of β_m, (44) is

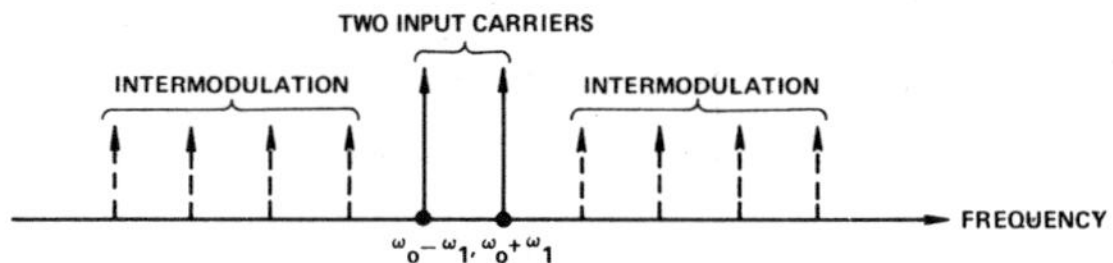

Fig. 2. Frequency arrangement of two input carriers
and intermodulations.

represented by

$$\mathrm{Re}\,\exp\{-j\omega_0\tau\}\{(B + R_\tau) - j(C + I_\tau)\}\xi^{-2}$$
$$\cdot \sum_{m=0}^{\infty} (m + 1)!\,2^m \beta_m \left(\frac{A + R_0}{\xi} - 1\right)^m. \quad (47)$$

B. Case Where AM–PM Conversion Takes a Special Form

If $g(\rho)\,\exp\{jf(\rho)\}$ is given by (35), (42) can be simply calculated as

$$\mathrm{Re}\,\exp\{-j\omega_0\tau\}[(B + R_\tau) - j(C + I_\tau)]$$
$$\cdot [C_1^2\{[1 + 2C_2(A + R_0)]^2 - (2C_2)^2 Z^2\}^{-2}$$
$$+ \{8C_3^2 + 16C_1 C_2 C_3 [1 + 2C_2(A + R_0)]^{-3}\}Z^2 + 1$$
$$+ \{C_1 + 4C_3(A + R_0)\}^2 - 2C_1[C_1 + 4C_3(A + R_0)]$$
$$\cdot [1 + 2C_2(A + R_0)]^{-2}] \quad (48)$$

where

$$Z = \{(B + R_\tau)^2 + (C + I_\tau)^2\}^{1/2}. \quad (49)$$

VII. Simple Examples

Since it is not possible to explain in detail the applications of the formulas shown previously because of page limitation, only the cases of two-carrier and three-carrier inputs will be analyzed in this section.

A. Two-Carrier Case

1) Noiseless Case: In this case, (18) can be used. Without losing any generality we can choose ω_0 as the center frequency between $\omega_0 + \omega_1$ and $\omega_0 + \omega_2$; i.e., $\omega_1 = -\omega_2$. Fig. 2 shows the carrier frequency arrangement.

Considering the constraint that $k_1 + k_2 = 1$, (18) becomes

$$\frac{1}{2} \sum_{k_1 = -\infty}^{\infty} \exp\left[j\{\omega_0 + (2k_1 - 1)\omega_1\}\tau\right]$$
$$\cdot \exp\{jk_1\psi_1 + j(1 - k_1)\psi_2\}|M(k_1, 1 - k_1)|^2 \quad (50)$$

where

$$M(k_1, 1 - k_1) = \int_0^\infty r J_{k_1}(A_1 r) J_{1-k_1}(A_2 r)\,dr \int_0^\infty \rho g(\rho)$$
$$\cdot \exp\{jf(\rho)\}J_1(r\rho)\,d\rho. \quad (51)$$

The two components given by $k_1 = 1$ and $k_1 = 0$, whose frequencies are $\omega_0 + \omega_1$ and $\omega_0 - \omega_1$, respectively, represent the pure-signal components. The other components whose frequencies are $\omega_0 \pm 3\omega_1$, $\omega_0 \pm 5\omega_1$, $\omega_0 \pm 7\omega_1$, $\cdots$, represent the intermodulations due to TWT. The absolute values of $M(k_1, 1 - k_1)$ represent the magnitudes of the intermodulations and the power-spectrum spreads of these components and are given by the Fourier transform of

$$\mathrm{avg}\,\big[\exp\{jk_1[\phi_1(t + \tau) - \phi_1(t)]$$
$$+ j(1 - k_1)[\phi_2(t + \tau) - \phi_2(t)]\}\big]. \quad (52)$$

If $g(\rho)\,\exp\{jf(\rho)\}$ is given by (23), we have

$$M(k_1, 1 - k_1) = \sum_{m=0}^{\infty} \alpha_m m!\,(m + 1)! \left\{ \sum_{\substack{l_1, l_2 \\ (l_1 + l_2 = 2m + 1)}}' \right.$$
$$\cdot \frac{A_1^{l_1}}{\left(\dfrac{l_1 + k_1}{2}\right)!\left(\dfrac{l_1 - k_1}{2}\right)!}$$
$$\left. \cdot \frac{A_2^{l_2}}{\left(\dfrac{l_2 + 1 - k_1}{2}\right)!\left(\dfrac{l_2 - 1 + k_1}{2}\right)!} \right\} \quad (53)$$

where the summation $\sum'$ is done over all the possible combinations under $l_1 + l_2 = 2m + 1$.

Using (32) for the expansion of (31), we have

$$M(k_1, 1 - k_1) = \int_0^{\pi/2} \left\{\sum_{m=0}^{\infty} A_m G'(mp \sin\theta)\right\} d\theta \quad (54)$$

where

$$G'(r) = \frac{d}{dr}\{r J_{k_1}(A_1 r) J_{1-k_1}(A_2 r)\}$$
$$= J_{k_1}(A_1 r) J_{1-k_1}(A_2 r) + \left(\frac{A_1 r}{2}\right)\{J_{k_1-1}(A_1 r) - J_{k_1+1}(A_1 r)\}$$
$$\cdot J_{1-k_1}(A_2 r) + \left(\frac{A_2 r}{2}\right)\{J_{-k_1}(A_2 r) - J_{2-k_1}(A_2 r)\}$$
$$\cdot J_{k_1}(A_1 r). \quad (55)$$

Equation (33) may also be used.

If (35) is used, it is possible to achieve the integral (37):

$$M'(k_1, 1 - k_1) = (-1)^{k_1} C_1 [A_2 I_{k_1}(2C_2 A_1 A_2)$$
$$- A_1 I_{1-k_1}(2C_2 A_1 A_2)]\exp\{-C_2(A_1^2 + A_2^2)\}. \quad (56)$$

In this case, we have pure-signal components:

$$\frac{1}{2}\,\mathrm{Re}\,\exp\{j(\omega_0 + \omega_1)\tau\}\exp\{j\psi_1\}$$
$$\cdot \left[A_1^2 + \left\{M'(1, 0) - C_1 A_1 - 2C_3 A_1\left(\frac{A_1^2}{2} + A_2^2\right)\right\}^2\right]$$

and

$$\frac{1}{2}\,\mathrm{Re}\,\exp\{j(\omega_0 - \omega_1)\tau\}\exp\{j\psi_2\}$$
$$\cdot \left[A_2^2 + \left\{M'(0, 1) - C_1 A_2 - 2C_3 A_2\left(\frac{A_2^2}{2} + A_1^2\right)\right\}^2\right], \quad (57)$$

third-order intermodulation:

$$\frac{1}{2}\,\mathrm{Re}\,\exp\{j(\omega_0 + 3\omega_1)\tau\}$$
$$\cdot \exp\{j2\psi_1 - j\psi_2\}\{M'(2, -1) - C_3 A_1^2$$

and

$$\frac{1}{2}\,\mathrm{Re}\,\exp\{j(\omega_0 - 3\omega_1)\tau\}$$
$$\cdot \exp\{j2\psi_2 - j\psi_1\}\{M'(-1, 2) - C_3 A_1 A_2^2\}^2, \quad (58)$$

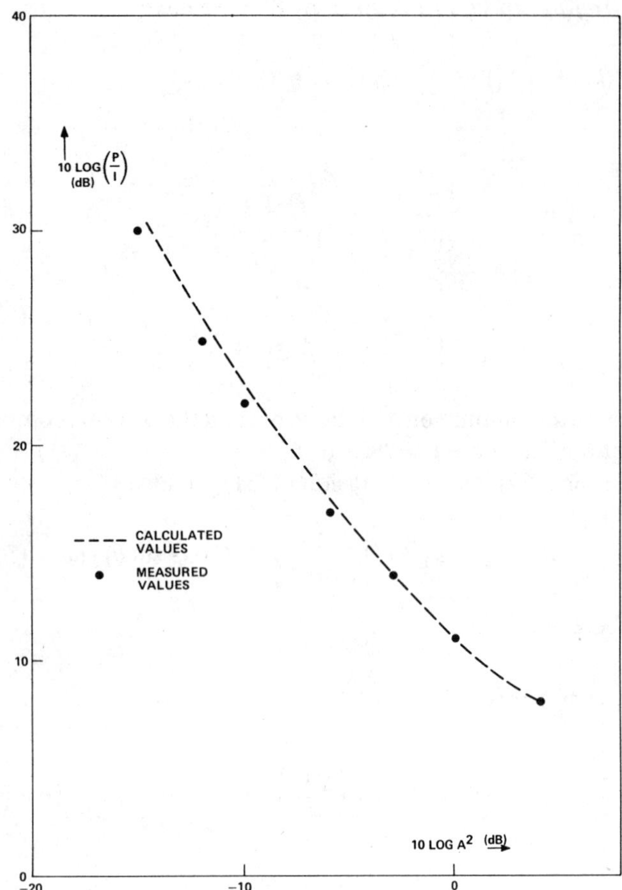

Fig. 3. Third-order intermodulation due to TWT (two-carrier case). A, amplitude of carrier ($A_1 = A_2 = A$); I, power of one of third-order intermodulations; P, power of one of pure signals in output.

higher order intermodulation:

$$\frac{1}{2} \operatorname{Re} \sum_{\substack{k_1 = -\infty \\ (k_1 \neq 0, \pm 1, 2)}}^{\infty} \exp\left[j\{\omega_0 + (2k_1 - 1)\omega_1\}\tau\right]$$

$$\cdot \exp\{jk_1\psi_1 + j(1 - k_1)\psi_2\}|M'(k_1, 1 - k_1)|^2 \quad (59)$$

where the first term in the last representation of (35) contributes only to the pure-signal and third-order intermodulation.

2) *Noisy Case:* In this case, (13) must be calculated for $n = 2$; i.e., taking $\omega_2 = -\omega_1$, we have

$$\frac{1}{2} \operatorname{Re} \sum_{k = -\infty}^{\infty} \exp\{j(\omega_0 + k\omega_1)\tau\} \exp\{-jk\psi_2 + jk\Lambda(\tau)\}$$

$$\cdot \left[\sum_{k_1 = -\infty}^{\infty} \exp\{jk_1(\psi_1 + \psi_2) - j(2k_1 - 1)\Lambda(\tau)\} \right.$$

$$\left. \cdot \int_0^\infty t \exp\left\{-\frac{t^2}{2}\right\}|M(t; k_1, k_1 - k)|^2 \, dt \right]. \quad (60)$$

If $g(\rho) \exp\{jf(\rho)\}$ takes the special forms given by (23), (31), and (34), the analysis can be done using (28), (33), and (37), as in Section VII-A1.

Fig. 3 shows a numerical example of (54) compared with an experimental result achieved in the COMSAT Laboratories [4].

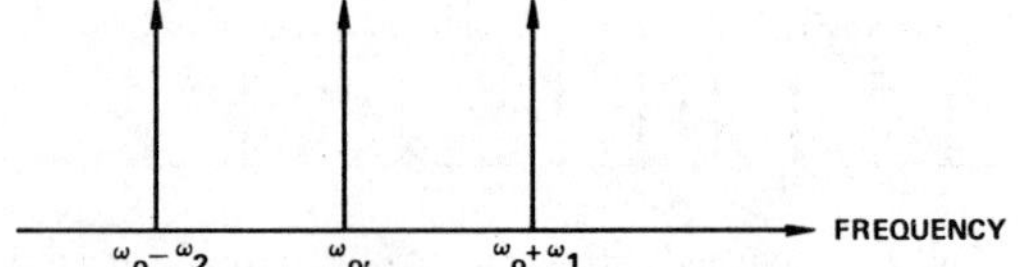

Fig. 4. Frequency arrangement of three input carriers.

B. Modulated Three-Carrier Case

The noiseless case is begun with (18), where $\omega_3 \to 0$, $\omega_2 \to -\omega_2$, and $\omega_1 \to \omega_1$ can be used without losing any generality, as shown in Fig. 4. Equation (18) becomes

$$\frac{1}{2} \operatorname{Re} \sum_{k_1 = -\infty}^{\infty} \sum_{k_2 = -\infty}^{\infty} \exp\{j(\omega_0 + k_1\omega_1 - k_2\omega_2)\tau\}$$

$$\cdot \exp\{jk_1\psi_1 + jk_2\psi_2 + j(1 - k_1 - k_2)\psi_3\}$$

$$\cdot |M(k_1, k_2, 1 - k_1 - k_2)|^2. \quad (61)$$

For the special forms of $g(\rho) \exp\{jf(\rho)\}$, (28), (32), and (37) can be used for the analysis.

Although in the two-carrier case the intermodulation carrier frequencies did not fall on the pure-signal carrier frequencies, in this case, some intermodulation frequencies fall on or near the pure-signal frequencies; for example, the cases of $k_1 = k_2$, $k_1 - k_2 = \pm 1$, $k_1 - k_2 = \pm 2, \cdots$.

For the noisy case, although the analysis is complex, it can be done by using formulas (28), (33), and (37).

VIII. METHOD FOR REDUCTION OF INTERMODULATIONS

In multicarrier satellite communications, it is very important to reduce the intermodulation due to the TWT so that the traveling-wave tubes on the satellite may be effectively utilized, since the operating point of the TWT is determined by the intermodulation levels.

As was seen from the analyses in the previous sections, the nonlinear amplitude characteristic of the TWT $g(\rho)$ and the AM–PM conversion characteristic $f(\rho)$ produce the intermodulation. The condition when there is no intermodulation but the pure-signal components is when $g(\rho) \cdot \exp\{jf(\rho)\}$ is a linear function of ρ. As far as the nonlinear amplitude characteristic is concerned, it is theoretically possible to equalize it after the TWT amplification; the distorted multicarrier signal (4) can be passed through the nonlinear amplitude characteristic $g_a[g(\rho)]$ such that

$$g_a[g(\rho)] = \rho. \quad (62)$$

Berman [4] discusses the characteristic $g(p)$ in practice and Sones and Standing [5] discuss the experimental result of the equalization. As for the AM–PM equalization, it is practically difficult to separate the distortion $f[(x_1^2 + y_1^2)^{1/2}]$ due to AM–PM conversion from the detected phase of (4). Therefore, detecting the amplitude $g[(x_1^2 + y_1^2)^{1/2}]$ and passing it through the amplitude nonlinear characteristic such that

$$g_p[g(\rho)] = f(\rho) \quad (63)$$

we obtain $f[(x_1^2 + y_1^2)^{1/2}]$. This change is then subtracted from the phase of the combined multicarrier signal (4). Two methods to achieve this equalization have been reported.

1) As shown by Fig. 5, at the input side of the TWT, the amplitude of the multicarrier signal (2) is detected and the

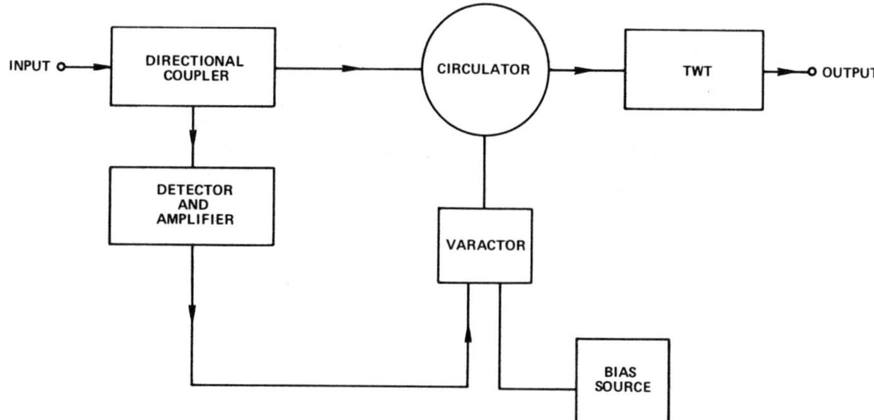

Fig. 5.   Equalizer for AM–PM conversion.

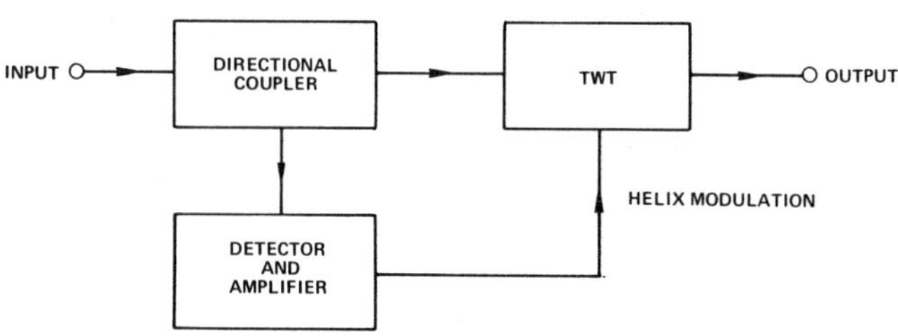

Fig. 6.   Equalizer by helix modulation for AM–PM conversion.

bias of the varactor is modulated by this detected amplitude. The change of the varactor bias changes the phase of the input signal to TWT (2) to cancel the AM–PM distortion $f[(x_1^2 + y_1^2)^{1/2}]$ [6].

2) At the input side of the TWT, the amplitude of the multicarrier signal (2) is detected and the helix voltage of the TWT is modulated by this detected amplitude. This helix voltage modulation can cancel the AM–PM distortion. Fig. 6 is a block diagram of this method [7].

However, for practical purposes, it is almost impossible to precisely construct the nonlinear devices g_a of (62) and g_p of (63). Since in the normal dynamic range of the TWT, the third-order intermodulation dominates all the other intermodulation products [8], if we can reduce the third-order intermodulation, there can be a large improvement in the intermodulation characteristic of the TWT. This can be achieved by subtracting $\alpha_1 \rho^3$ [in (23)] from $g(\rho) \exp \{jf(\rho)\}$.

It is seen from (25), (26), and (29) that the subtraction of this term only reduces the third-order intermodulation and that all the other higher order intermodulation products are invariant. For this purpose, $g_a(g)$ [in (62)], for the amplitude equalization, and $g_p(g)$ [in (63)], for the equalization of AM–PM conversion, can be approximated as[2]

$$g_a(g) = \frac{g}{g'(0)} - \frac{g'''(0)g^3}{3! \{g'(0)\}^3} \tag{64}$$

$$g_p(g) - \frac{f''(0)g^2}{2! \{g'(0)\}^2}. \tag{65}$$

---

[2] Although the third-order intermodulation is reduced greatly by the equalizations of (64) and (65), the higher order intermodulation is not invariant in this case since the equalization of (64) and (65) cannot subtract $\alpha_1 \beta^3$ only [in (23)]. However, if the intermodulation is not large, the change in the higher order intermodulation is negligible.

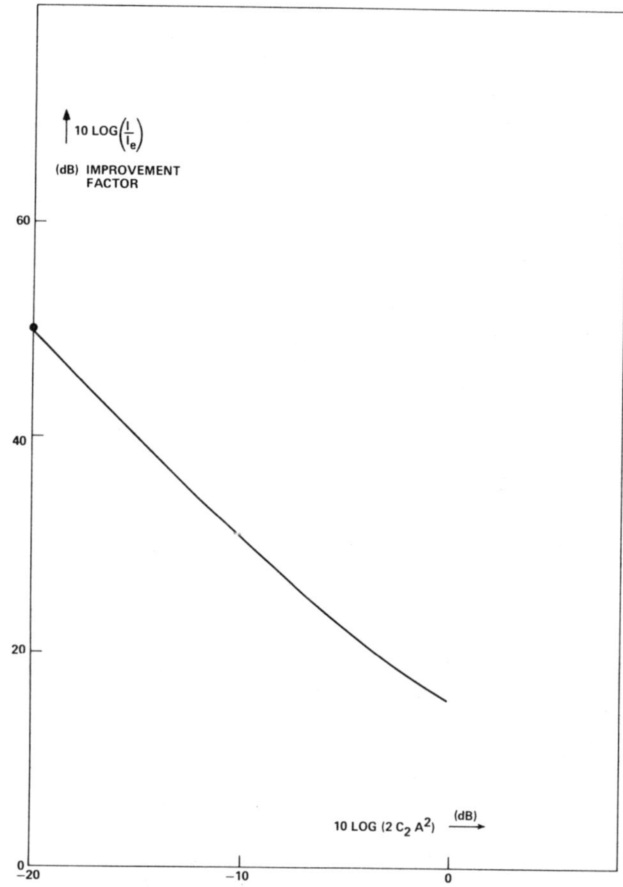

Fig. 7.   Reduction of third-order intermodulation in the two-carrier case. A, amplitude of one of input carriers ($A_1 = A_2 = A$); I, power of third-order intermodulation without equalizer; I_e, power of third-order intermodulation with equalizer.

Fig. 7 shows the use of this method to reduce the third-order intermodulation for the nonlinearity of (35) in the two-carrier case. It can be similarly shown that the reduction is also large in the case where the number of carriers is greater than two.

Since the thermal noise level is very small at the receiver of a satellite, (64) and (65) are at least theoretically possible on a satellite. However, at the receiver on earth, where the noise level is normally comparable with or dominant over the intermodulation level, the possibility of this equalization for the intermodulation due to the TWT on a satellite is small since the cross term between the intermodulation products and noise cannot be cancelled by the given equalizations.

IX. Conclusion

A general formula to calculate the intermodulation in multicarrier TWT systems, which includes the combined effects of nonlinear amplification, AM–PM conversion, and additive noise, was obtained in this paper. Several examples showing how to use this formula were also analyzed. Finally, on the basis of this formula, the possibility of reducing intermodulation was discussed.

Acknowledgment

The author wishes to thank Dr. M. I. Celebiler for his comments and encouragement.

References

[1] O. Shimbo, "Complete analysis of the combined effects of intermodulation, AM–PM conversion and additive noise in multi-carrier TWT systems," COMSAT Laboratories, Clarksburg, Md., Tech. Memo. CL-61-69, Dec. 3, 1969.

[2] G. N. Watson, *Theory of Bessel Functions.* London, England: Cambridge Univ. Press, 1958.

[3] A. L. Berman and C. E. Mahle, "Nonlinear phase shift in traveling-wave tubes as applied to multiple access communications satellites," *IEEE Trans. Commun. Technol.*, vol. 18, Feb. 1970, pp. 37–47.

[4] A. Berman, "Experimental determination of intermodulation distortion produced in a wideband communications repeater," Communications Satellite Corporation, Tech. Memo. ED-16-66, Aug. 26, 1966.

[5] W. K. Sones and A. F. Standing, "A theoretical evaluation of amplitude predistortion as a means of relaxing TWT backoff," COMSAT Laboratories, Clarksburg, Md., Tech. Memo. CL-57-69, Nov. 20, 1969.

[6] K. Yamaguchi, "Equalization of intermodulation due to TWT amplification," *Trans. Joint Conv. of the Four Elec. Inst. of Japan*, no. 1599, 1965.

[7] S. Owaku and D. Kuwagaki, "Common amplification of many signals by a traveling wave tube amplifier," *J. Inst. Electron. Commun. Eng. Jap.*, vol. 51-B, 1968, pp. 193–200.

[8] S. J. Campanella, "Analysis of PSK distortion spectra generated in satellite transponders," presented at INTELSAT/IEE International Conf. on Digital Satellite Commun., London, England, Nov. 25–27, 1969.

In this section, digital transmission over satellite channels is discussed. The general area of digital communication has received considerable attention, and there are a large number of papers and books available (e.g., [1]-[4]). Reprint Paper 3.3.1 provides an excellent tutorial discussion to digital communications. In addition to digital signaling, which is the topic of this section, it discusses the related aspects of source encoding and error control. Readers who are not familiar with digital communication should read this paper at this point. Many of the general results are directly applicable to satellite communications. However, the satellite channel has several unique characteristics which require special treatment. The introduction will develop a framework for the satellite communication problem and emphasize the particular characteristics of the satellite problem.

A model of the satellite communications channel is shown in Fig. 1. A similar model has been encountered in the discussion of analog modulation in Section 3.2. Just as in the analog modulation case, the TWTA in the satellite transponder is a key element in the digital communication system design. In order to efficiently use the available dc power on the satellite and to maximize the satellite power output it is desirable to operate the TWT at (or near) saturation. To minimize the interference due to AM/AM and AM/PM conversions, constant envelope modulation schemes such as phase-shift keying (PSK), differential phase-shift keying (DPSK) and continuous phase frequency-shift keying (CFSK) are normally used. When a single carrier is accessing the transponder at any given time (as in TDMA) it is possible to operate the TWT at or near saturation. As in the analog modulation case, operation in the FDMA mode requires that the transponder be backed-off. If the satellite transponder is operating in a linear region and the bandwidth of the transmitted signal is less than the various filters, then the model reduces to the classical additive white Gaussian noise channel. The error probability for various PSK and FSK schemes in an additive noise channel is shown in Fig. 2.

Most early satellite systems operated in power-limited mode. However, many current systems are bandwidth-limited. If the transponder is operating in a backed-off mode, then signaling schemes which utilize amplitude-shift, as well as phase-shift keying have the potential of improving the bandwidth utilization. These systems are referred to amplitude-phase keying (APK), quadrature amplitude modulation (QAM), or hybrid modulation (e.g., [5]-[8]).

One of the problems with BPSK and QPSK is that there are $180°$ phase reversals at the symbol times. When the signal is passed through a bandpass filter, envelope modulation nulls occur. This envelope variation causes increased adjacent channel interference and AM/PM conversion when the signal passes through the TWTA. Several phase-continuous schemes such as offset QPSK (OQPSK), minimum shift keying (MSK), and trellis phase modulation has been introduced to overcome this problem. Reprint paper 3.3.4 and [9]-[14] discuss some of these schemes.

In order to perform the coherent detection operation, carrier phase and symbol timing must be available. Phase-lock loops are used to recover the carrier phase. Several problems are of interest:

1) the phase error due to oscillator instabilities and noise,
2) the transient response of the PLL (this is of particular importance in the TDMA case),
3) the effect of phase and timing errors on the error probability.

There is an abundance of literature in this area. Reference [15] and the IEEE Reprint Book *Phase-Locked Loops and Their Applications* [16] give general discussions and excellent bibliographies. Other references include [17]-[19].

In an actual system the modulation and demodulation process is implemented by a modem. Some of the issues involved in modem implementation are discussed in [20]-[22].

FILTER DISTORTION AND INTERSYMBOL INTERFERENCE

The discussion to this point has ignored the effect of the various filters in the satellite communications link. The most commonly used filters are Chebyshev, dual-mode elliptic, and linear phase filters. Their amplitude and phase characteristics can have a significant effect on the system performance. This effect must be analyzed and, if possible, compensated for. A simplified baseband model for a linearized satellite channel using ASK or BPSK is shown in Fig. 3. Note that the uplink noise has been neglected. For BPSK, the a_n are a sequence of ± 1's. An excellent tutorial discussion for signaling in a linear channel is given in [23]. The kth sample is

$$y_k = a_k x_0 + \sum_{\substack{n \\ n \neq k}} a_n x_{k-n} + n_k, \tag{1}$$

where $x(t)$ is the impulse response corresponding to the cascade of filters. The first term is the desired signal, and the last term is the additive noise. The second term is the intersymbol interference term. The classic Nyquist problem is to choose the product $G_T(\omega)C(\omega)G_R(\omega)$ such that the intersymbol interference term is zero. If $C(\omega)$ is time-invariant, this can be done, and the model reduces to AWGN model (e.g., [23]-[25]).

In an actual channel, there will usually be some intersymbol interference because of filter tolerances and changes over time. The ISI term consists of the values of the x_n at the other sample points weighted with either +1 or -1 depending on the

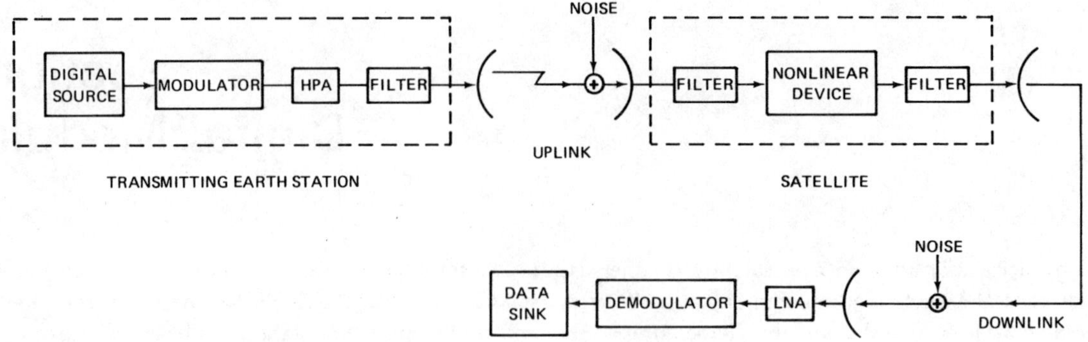

Fig. 1. Model of digital satellite communications channel.

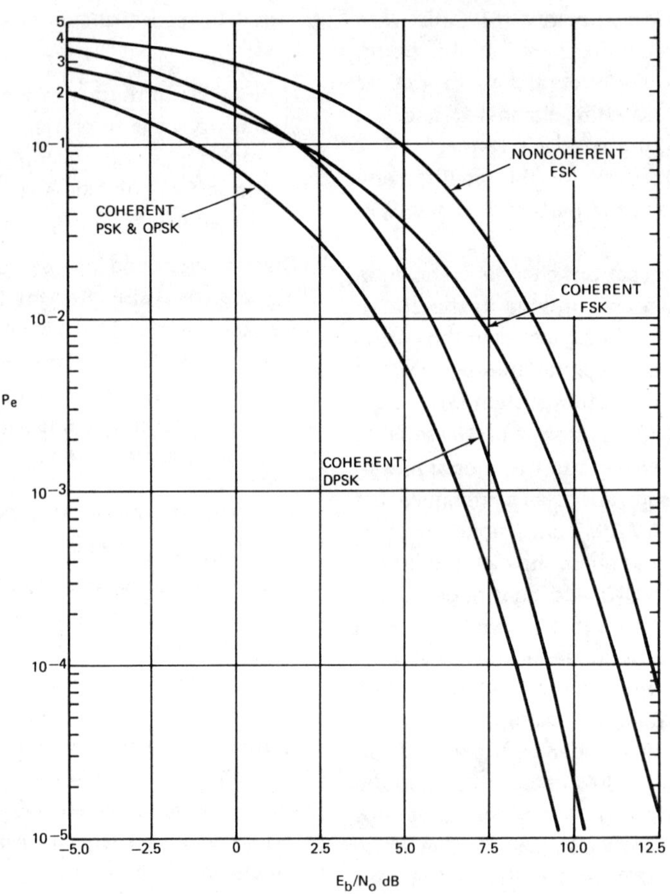

Fig. 2. P_e versus E_b/N_0.

a_n from the data sequence. It is convenient to study the effect of ISI through the use of an "eye patern." Experimentally, it is obtained from an oscilloscope display of $y(t)$ with a sweep rate of $1/T$. Eye patterns and their characteristics are discussed in [23].

There are three common approaches to evaluating the P_e in the presence of ISI:

1) exact expressions based on series expansions (e.g., [26]–[29]),

2) peak distortion bounds based on the worst possible data sequence (e.g., [23]),

3) various upper and lower error bounds (e.g., [30]–[32]).

Reprint Paper 3.3.3 includes ISI as a special case of their general result.

A typical result showing the performance degradation caused by a Chebyshev filter as a function of the BT product is shown in Fig. 4 (from [56]).

The performance can be improved if the receiver is optimized with the intersymbol interference included in the model. The first step in an optimum decision problem is to form a set of sufficient statistics (e.g., [1] or [2]). A key step in the ISI problem is the recognition that if $r(t)$ is passed through a matched filter and the output is sampled every T seconds, then the resulting $2N + 1$ samples form a set of sufficient statistics

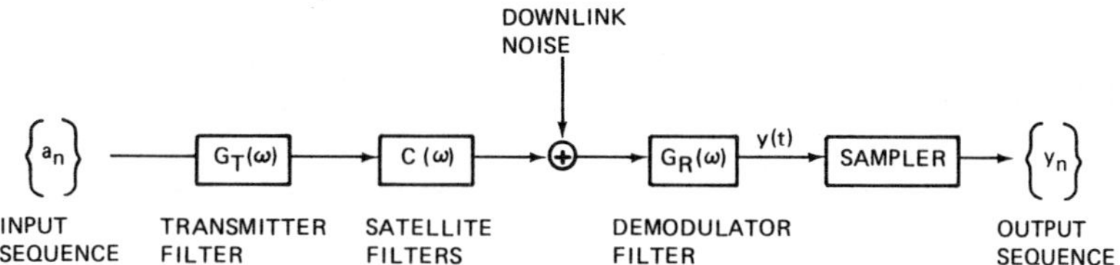

Fig. 3. Simplified intersymbol interference model.

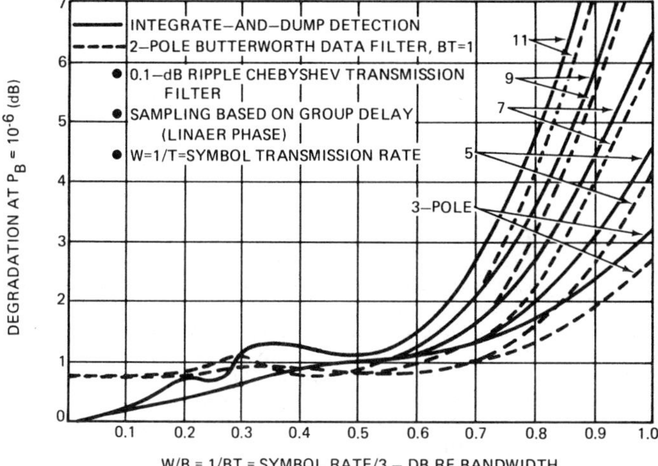

Fig. 4. Link SNR degradation of a quadraphase (QPSK) signal through a Chebyshev filter is shown for poles numbering from three to eleven, for two types of detection.

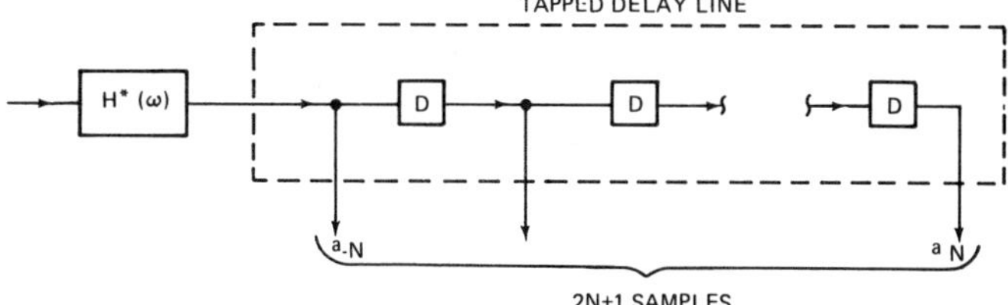

Fig. 5. Generation of sufficient statistics.

(e.g., [1, p. 160]). Thus, the first operation in the optimum receiver can be realized as shown in Fig. 5.

At this point one can find the optimum linear filter or the actual optimum receiver, which will be nonlinear. It can be shown that the optimum linear receiving filter is the cascade of the matched filter and a transversal filter. The tap gains can be chosen to satisfy various criterion (e.g., [33]). Finding the optimum nonlinear receiver is more challenging. The most significant result in this area was the recognition that the Viterbi algorithm provided the maximum likelihood estimator for this problem. Reference [34] develops this result for PAM signals. References [35] and [36] obtained related results independently. Subsequently, these results have been extended to MPSK signals. The application of these techniques to counter ISI in satellite communications systems is still in the research and development stage.

A second type of interference is also important in satellite systems. A typical system will have multiple carriers in the same transponder. The adjacent channels will interfere with each other. The magnitude of this interference is a function of: 1) the transmitter and receiver filters, 2) the frequency spacing, 3) the power levels in the adjacent channels, 4) the modulation schemes in the various channels. Several observations are useful. First, as the transmitter filter bandwidth is decreased, the adjacent channel interference (ACI) will decrease. However, the results in the previous paragraphs show that the ISI will increase. Thus, there is a trade-off between the ISI and ACI in order to minimize the error probability. Second, as the spacing between channels Δf is increased, the ACI will decrease. However, the number of carriers per transponder will decrease. Thus, there is a trade-off to maximize the total capacity. It is difficult to obtain a general solution to this problem. Most results deal with special cases of the problem, and frequently simulation is necessary.

Reprint Paper 3.3.3 includes ACI as a special case of the general analysis. Other papers in the Bibliography include bounds on error probability, analyses of the error probability, and simulation results.

A third type of interference is co-channel interference which arises in several satellite communications situations. Two important cases are those in which the system obtains frequency reuse by multiple beams (hemispheric, zone, or spot beams) or by dual polarization. In these cases, the interference is internal to the overall system, and various trade-offs are possible. Another case is the one in which the interference is caused by an adjacent satellite systems or terrestrial microwave systems. Here the interference is controlled by agreed-upon interference standards.

The co-channel interference problem is difficult to treat analytically, and most results have been obtained for special cases and/or by simulation. References [37]–[40] indicate the type of results that can be obtained. One result of interest is that treating the CCI as an equivalent Gaussian noise usually gives a pessimistic result. Reprint Paper 3.3.3 and [41]–[42] treat the combined effect of various interferences.

The discussion up to this point has alluded to the nonlinear nature of the satellite channel, but a model for the channel has not been developed. A typical system of interest is QPSK. In this case, the transmitted signal $s_i(t)$ is one of four waveforms,

$$s_i(t) = \sqrt{2P} \cos (\omega_c t + \theta_i), \theta_i = \frac{\Pi i}{2}, \quad i = 0, \cdots, 3. \quad (2)$$

Note that there is only one signal present at any given time as in the TDMA case.

The received signal at the satellite is

$$r_u(t) = \alpha s_i(t) + n_u(t)$$

$$\triangleq v_1(t) \cos (\omega_c t + \theta_i + n_1(t)), \quad (3)$$

where

$$v_1(t) = [(\alpha\sqrt{2P} + n_{uc}(t))^2 + n_{us}^2(t)]^{1/2}, \qquad (4)$$

and

$$n_1(t) = \tan^{-1}\left(\frac{n_{us}}{\alpha\sqrt{2P} + n_{uc}}\right). \qquad (5)$$

The nonlinear device is a TWTA which causes two types of distortions, an AM-to-AM conversion and an AM-to-PM conversion. The TWTA is followed by a zonal filter. The output of the nonlinearity and the zonal filter is

$$s_d(t) = f[v_1(t)] \cos(\omega_c t + \theta + g[v_1(t)] + n_1(t)) \qquad (6)$$

where

$f(\cdot) = $ AM-AM characteristic

and

$g(\cdot) = $ AM-PM characteristic.

The received waveform at the earth station is

$$r_d(t) = \beta s_d(t) + n_d(t).$$

In some cases (particularly military satellites) there is a hard limiter in the channel. Then $f(\cdot)$ is constant and $g(\cdot)$ is zero.

This model serves as the starting point for many studies and analyses. Several problems are of interest.

1) The first problem assumes the receiver is specified. Either the optimum linear receiver or a suboptimum configuration may be used. The object is to evaluate or bound the error probability. Simulation is generally used to augment or verify analyses. Reprint Paper 3.3.2 discusses the performance evaluation of a 4-phase PSK system using simulation. An analytic approach is in Reprint Papers 3.3.3 and 3.3.5, and [43]–[46].

2) A second category of problem retains the basic structure of the optimum linear receiver (filter, sampler, and independent symbol decisions) and attempts to optimize the transmitter or receiver filter or the shape of the nonlinearity through compensation). References dealing with this problem include [47]–[50].

3) A third category of problems investigates different types of signaling schemes to minimize the effects of the nonlinearity. One of the effects of the nonlinearity is called "spectrum spreading" (e.g., [51]). When a constant envelope scheme such as QPSK is filtered, the output has a nonconstant envelope. The AM/AM characteristic has the effect of providing a more constant envelope, and the AM/PM characteristic generates new sidebands. Two promising modulation techniques to reduce these effects are minimum shift keying (MSK) and offset QPSK (OQPSK). Reprint Paper 3.3.4 compares these two schemes for a nonlinear bandlimited channel. References dealing with various aspects of MSK and OQPSK, as well as other modulation schemes include [11] and [52]–[53].

4) The fourth category of problems is the design of the optimum receiver for the nonlinear bandlimited channel. Most of the recent work in this area is based on the Viterbi algorithm or various modifications of it. Two basic references are [54]–[55].

More complicated models include one or more of the following characteristics:

1) nonlinearity in the transmitter HPA,
2) co-channel interference,
3) timing and synchronization errors,
4) efforts of bursting in a TDMA system.

Many of these more complicated models can only be handled by simulation, while, in other cases, some analytic results are available.

The Biblography at the end of this volume contains numerous references to various aspects of the digital communications problem. As digital communications becomes more widely used in satellite systems, many of the important issues will develop further.

REFERENCES

[1] H. L. Van Trees, *Detection, Estimation, and Modulation Theory*, Part I. New York: Wiley, 1968.

[2] J. M. Wozencraft and I. M. Jacobs, *Principles of Communications Engineering*. New York: Wiley, 1965.

[3] W. C. Lindsey and M. K. Simon, *Telecommunications System Engineering*. Englewood Cliffs, NJ: Prentice-Hall, 1973.

[4] A. J. Viterbi, *Principles of Coherent Demodulation*. New York: McGraw-Hill, 1966.

[5] R. W. Lucky and J. C. Hancock, "On the optimum performance of *N*-ary systems having two degrees of freedom," *IRE Trans. Commun. Syst.*, vol. CS-10, pp. 185–192, June 1962.

[6] M. K. Simon and J. G. Smith, "Hexagonal multiple phase-and-amplitude-shift-keyed signal sets," *IEEE Trans. Comm.*, vol. COM-22, pp. 168–180.

[7] C. M. Thomas, M. Y. Weidner, and S. H. Durrani, "Digital amplitude-phase keying with *M*-ary alphabets," *IEEE Trans. Comm.*, vol. COM-22, pp. 168–180.

[8] G. R. Welti, "PCM/FDMA satellite telephony with 4-dimensionally coded quadrature amplitude modulation," *Comsat Tech. Rev.*, vol. 6, no. 2, pp. 323–338, Fall 1976.

[9] R. DeBuda, "Coherent demodulation of frequency-shift keying with low deviation ratio," *IEEE Trans. Comm. Tech.*, pp. 429–435, June 1972.

[10] M. L. Doelz and E. H. Heald, "Minimum-shift data communication system," *U.S. Patent No. 2,997,417,* (assigned to Collins Radio Company) March 28, 1961.

[11] H. R. Mathwich, J. F. Balcewicz, and M. Hecht, "The effect of tandem band and amplitude limiting on the E_b/N_0 performance of minimum (frequency) shift keying (MSK)," *IEEE Trans. Commun.*, vol. COM-22, pp. 1525–1540, Oct. 1974.

[12] W. P. Osborne and M. B. Lutz, "Coherent and noncoherent detection of CPFSK," *IEEE Trans. Comm.*, pp. 1023–1036, Aug. 1974.

[13] S. A. Rhodes, "Effect of noisy phase reference on coherent detection of offset QPSK signals," *IEEE Trans. Commun.*, vol. COM-22, pp. 1046–1055, Aug. 1974.

[14] J. B. Anderson and D. P. Taylor, "Trellis phase modulation coding: Minimum distance and spectral results," *EASCON 1977*, pp. 29-1A–29-1G.

[15] W. C. Lindsey, *Synchronization Systems in Communications*. Englewood Cliffs, NJ, Prentice-Hall, 1972.

[16] W. C. Lindsey and M. K. Simon, *Phase-Locked Loops and their Applications*, IEEE Press Reprint Book. New York: IEEE Press, 1978.

[17] L. C. Palmer, "Characterization of the effects of phase and symbol timing errors for QPSK transmission," *Int. Conf. Communications*, pp. 16.3-340 to 16.3-343, 1977.

[18] A. J. Viterbi, "Phase-locked loop dynamics in the presence of noise by Fokker–Planck techniques," *Proc. IEEE*, Dec. 1963.

[19] D. D. Falconer and J. Salz, "Optimal reception of digital data over the Gaussian channel with unknown delay and phase jitter," *IEEE Trans. Inform. Theory*, vol. IT-23, no. 1, Jan. 1977.

[20] A. M. Walker, "High data rate PSK modems for satellite communications," *Telecommunications*, July 1976.

[21] K. Nosaka, A. Ogawa, and T. Muratani, "PSK demodulator for the PCM-TDMA system," *IEEE Trans. Comm. Technol.*, vol. COM-18, pp. 427–434, Aug. 1970.

[22] C. Heegard, J. A. Heller, and A. J. Viterbi, "A microprocessor-based PSK modem for packet transmission over satellite channels," *IEEE Trans. Commun.*, vol. COM-26, no. 5, pp. 552–564, May 1978.

[23] R. W. Lucky, J. Salz, and E. J. Weddon, Jr., *Principles of Data Communication*. New York: McGraw-Hill, 1968.

[24] W. R. Bennett and J. R. Davey, *Data Transmission*. New York: McGraw-Hill, 1965.

[25] H. Nyquist, "Certain topics in telegraph transmission theory," *Trans. AIEE*, Apr. 1928.

[26] E. Y. Ho and Y. S. Yeh, "Error probability of multilevel digital system with intersymbol interference and Gaussian noise," *Bell System Tech. J.*, vol. 50, no. 3, pp. 1017–1023, Mar. 1971.

[27] O. Shimbo, M. I. Celebiler, and R. J. Fang, "Performance analysis of DPSK systems in both thermal noise and intersymbol interference," *IEEE Trans. Commun. Technol.*, vol. COM-19, no. 6, pp. 1170–1188, Dec. 1971.

[28] O. Shimbo, R. J. Fang, and M. Celebiler, "Performance of *M*-ary PSK systems in Gaussian noise and intersymbol interference," *IEEE Trans. Inform. Theory*, vol. IT-19, Jan. 1973.

[29] O. Shimbo and M. I. Celebiler, "The probability of error due to intersymbol interference and Gaussian noise in digital communication systems," *IEEE Trans. Commun. Technol.*, vol. COM-19, pp. 113–119, Apr. 1971.

[30] V. K. Prabhu, "Error probability performance of *M*-ary CPSK systems with intersymbol interference," *IEEE Trans. Commun.*, vol. COM-21, pp. 97–109, Feb. 1973.

[31] F. E. Glave, "An upper bound on the probability of error due to intersymbol interference for correlated digital signals," *IEEE Trans. Inform. Theory*, vol. IT-18, no. 3, pp. 356–363, May 1972.

[32] S. Benedetto, G. DeVincentiis, and A. Luvison, "Error probability in the presence of intersymbol interference and additive noise for multi-level digital signals," *IEEE Trans. Commun.*, vol. COM-21, pp. 181–190, Mar. 1973.

[33] D. A. George, "Matched filters for interfering signals," *IEEE Trans. Inform. Theory*, vol. IT-11, pp. 153–154, Jan. 1965.

[34] G. D. Forney, "Maximum-likelihood sequence estimation of digital sequences in the presence of intersymbol interference," *IEEE Trans. Inform. Theory*, vol. IT-18, no. 3, pp. 363–378, May 1972.

[35] J. K. Omura, "Optimal receiver design for convolutional codes and channels with memory via control theoretical concepts," *Information Science*, vol. 3, pp. 243–266, July 1971.

[36] H. Kobayashi, "Correlative level coding and maximum likelihood decoding," *IEEE Trans. Inform. Theory*, vol. IT-17, pp. 586–594, Sept. 1971.

[37] A. Rosenbaum, "Binary PSK error probabilities with multiple co-channel interferences," *IEEE Trans. Commun. Technol.*, vol. COM-18, no. 3, pp. 241–253, June 1970.

[38] J. M. Aein and R. Turner, "Effect of co-channel interference on CPSK carriers," *IEEE Trans. Commun.*, vol. COM-21, pp. 783–790, July 1973.

[39] V. K. Prabhu, "Error rate considerations for coherent phase-shift-keyed systems with co-channel interference," *Bell Syst. Tech. J.*, vol. 48, pp. 743–767, Mar. 1969.

[40] O. Shimbo and R. Fang, "Effects of co-channel interference and Gaussian noise in *M*-ary PSK systems," *COMSAT Tech. Rev.*, vol. 3, pp. 183–207, Spring 1973.

[41] M. Celebiler and G. Coupe, "Probability of error for coherent PSK in the presence of thermal noise and intersymbol, interchannel and co-channel interferences," *IEEE Int. Conf. Satellite Communication Systems Technology*, London, England, Apr. 7–10, 1975.

[42] S. Benedetto, E. Biglieri, and V. Castellani, "Combined effects of intersymbol, interchannel, and co-channel interferences in *M*-ary CPSK systems," *IEEE Trans. Commun.*, vol. COM-21, pp. 997–1008, Sept. 1973.

[43] P. C. Jain and N. M. Blachman, "Detection of a PSK signal transmitted through a hard-limited channel," *IEEE Trans. Inform. Theory*, vol. IT-19, no. 5, pp. 623–630, Sept. 1973.

[44] P. Hetrakul and D. P. Taylor, "The effects of transponder non-linearity on binary CPSK signal transmission," *IEEE Trans. Commun.*, vol. COM-24, no. 5, May 1976.

[45] R. G. Lyons, "The effect of a bandpass nonlinearity on signal detectability," *IEEE Trans. Commun.*, vol. COM-21, no. 1, Jan. 1973.

[46] J. M. Aein, "SNR and error rate calculations for coherent non-linear satellite channels," *Proc. INTELSAT/IEEE Conf. Digital Satellite Communications*, pp. 189–201, Nov. 1969.

[47] S. A. Fredricson, "Optimum receiver filters in digital quadrature phase-shift-keyed systems with a nonlinear repeater," *IEEE Trans. Commun.*, vol. COM-23, pp. 1389–1400, Dec. 1975.

[48] A. R. Kay, D. A. George, and M. J. Eric, "Analysis and compensation of bandpass nonlinearities for communications," *IEEE Trans. Commun. Technol.*, vol. COM-20, pp. 965–972, Oct. 1972.

[49] D. R. McCreath and P. J. McLane, "Filter parameter optimization for linear and nonlinear BPSK transmission," *Nat. Telecommunications Conf.*, pp. 26:3-1–26:3-7, 1977.

[50] S. H. R. Raghavan and R. E. Ziemer, "Improving QPSK transmission in band-limited channels with interchannel interference through equalization," *IEEE Trans. Commun.*, vol. COM-25, no. 10, pp. 1222–1226, Oct. 1977.

[51] G. Robinson *et al.*, "PSK signal power spectrum spread produced by memoryless nonlinear TWTs," *COMSAT Tech. Rev.*, pp. 227–256, Fall 1973.

[52] S. A. Rhodes, "Effects of hardlimiting on bandlimited transmissions with conventional and offset QPSK modulation," *Proc. Nat. Telecommunications Conf.*, pp. 20F/1–7, 1972.

[53] R. K. Kwan, "The effects of filtering and limiting a double-binary PSK signal," *IEEE Trans. Aerospace and Elect. Syst.*, vol. AES-5, pp. 589–594, July 1969.

[54] M. F. Mesiya, P. J. McLane, and L. L. Campbell, "Maximum likelihood sequence estimation of binary sequences transmitted over bandlimited nonlinear channels," *IEEE Trans. Commun.*, vol. COM-25, no. 7, July 1977.

[55] L. Biederman and J. K. Omura, "Performance of Viterbi demodulators for nonlinear satellite channels with intersymbol interference," *Nat. Telemetry Conf.*, 1977.

[56] C. L. Cuccia, "Bandwidth conservation is essential," *Microwave Systems News*, p. 67, Oct. 1978.

Alternatives in Digital Communications

MARLIN P. RISTENBATT

Invited Paper

Abstract—The following major interrelated aspects in digital communications are described: source encoding, channel encoding and transmission, and receiver methods. Methods for digitizing speech and video are summarized. Digitized voice can be transmitted at 10 kb/s, while phone video requires 1.544 Mb/s.

The major aspects of both forward-error control (FEC) and automatic request for repeats (ARQ) are considered. Power-limited forward channels are distinguished from bandwidth-limited ones. Signal distance concepts and maximum-likelihood detection principles provide the framework for describing both block-coded and convolutional-coded signaling. Convolutional coding with Viterbi decoding appears to achieve the most cost-effective forward-error control.

The improved FEC techniques benefit the transmission of real-time digitized voice and video and any communication over severely power-limited links. ARQ methods continue to be effective and practical for most cases where the delay associated with retransmission can be tolerated.

GLOSSARY OF SYMBOLS

ARQ   automatic request for repeats, using a feedback channel;

bauds   number of channel or modem signaling elements per second $=1/\tau$;

C channel capacity in bits per second;

chip   channel bit, as occurs when redundant bits are added to information bits;

E_b received signal energy per information bit;

E_s received signal energy per M-ary symbol;

FEC   forward-error control;

k number of information bits per M-ary channel symbol (or per m-ary source symbol);

m (alphabet size) $=$ number of source symbols;

M number of alternative channel symbols or signals;

n number of chips (channel bits) in an M-ary symbol or signal;

n_0 channel noise power per unit bandwidth;

P_B channel information-bit error rate;

P_E channel symbol error rate;

$P(n)$ probability that a block of n chips contains one or more errors;

R_I theoretical information rate of source in bits per second;

R_s actual source rate in bits per second;

R encoded source rate $=$ channel information bit rate in bits per second;

S average received signal power over interval T;

T time interval over which receiver makes final decisions about received symbol;

T_s time interval per source symbol;

τ length of channel signaling element;

W channel bandwidth.

This invited paper is one of a series planned on topics of general interest—The Editor.

Manuscript received November 27, 1972; revised March 12, 1973.

The author is with Cooley Electronics Laboratory, University of Michigan, Ann Arbor, Mich. 48105.

I. INTRODUCTION

DIGITAL COMMUNICATIONS has become increasingly attractive over the past decade both because of increased demand for data communication and because digital transmission offers options and flexibility not available with analog transmission. This paper describes the major current approaches and alternatives for implementing digital communications. The objective is twofold: 1) to provide a useful introduction and orientation to people in some other technical field; and 2) to provide an overview of current activity for those already versed in the general topic. Since the subject is wide in scope, details that can be found in the references are not given here. Also, no attempt is made to catalog the various contributions to this field historically, but instead the emphasis is on the techniques and their comparison.

Digital communications involves the transmission of discrete symbols over a potentially noisy channel. The discrete symbols may come from a discrete data source (Teletypewriter,® quantized telemetry, or computer symbols) or from a digitized analog source (voice, video, or continuous telemetry signal). Digital transmission implies that the transmitted signals are taken from a finite set of signals, corresponding to the discrete symbols, and that the receiver decides which of the finite set was received. Since transmission aspects depend also on source properties, the term digital *communications* is used to imply the larger area of digital transmission plus source considerations. Continuous analog sources must first be sampled and the samples then quantized for digital transmission.

Although the physical channel itself (wire, cable, RF channel) is analog, use of digital transmission has a number of advantages compared with the traditional continuous analog modulations (AM, FM, SSB) or the sampled analog modulations (PPM, PAM). In analog transmission a varying continuous voltage (or current) is transmitted to a receiver; the receiver attempts to track the original information waveform with as high a fidelity as possible. The effectiveness measure for analog signaling is the output signal-to-noise ratio, where the noise is the mean-square error between input and output waveform. For digital signaling, where one decides which of a finite set of signals was sent, the primary effectiveness measure is symbol probability of error.

Telephone, radio, and television, which extend the range of our analog ear and eye receptors, have traditionally used analog transmission because it is simpler and requires less bandwidth. Any written text is inherently digital, and it is interesting to note that Samuel Morse's digital telegraph (1832) preceded Bell's analog telephone system (1874). Digital transmission is increasingly being used on trunk portions of (otherwise) analog telephone networks, and new all-digital common-carrier facilities are being constructed [2]–[4].

One major advantage of the digital signaling is that *regener-*

ative repeaters can be used in the channel. *Analog* repeaters consist of amplifiers that amplify signal and any accumulated noise equally. Although the signal level with respect to the receiver front-end noise is raised, the signal level with respect to channel-induced distortions is not changed. The *digital* regenerative repeater, on the other hand, reconstructs the signal at each repeater. Assuming that a correct decision is made among the finite signal choices, a clean version of that signal is transmitted from the repeater. Thus, assuming no errors, channel-induced distortion or noise is removed at each repeater. So long as repeaters are placed at the necessary intervals, transmission may be made over an indefinitely long distance.

A second major advantage is the flexibility inherent in the "abstracted" nature of digital representation. Digital signals are handled identically in the channel, whether they represent a discrete data source or a digitized analog source. Also, digital (signaling) commands and logic information are handled identically to the subscriber or terminal information. Among other features this aspect facilitates implementing automatic switching and automatic tolling.

A third major advantage is that digital signaling tends to exploit the cost effectiveness of digital integrated circuits that has been achieved over the past decade. Also digital signals are easily encrypted to provide private or secure communication signals.

For an overview, the components and underlying factors in digital communications are shown in Fig. 1. The source symbols come either from a digital data source or a digitized analog source. Note that the source symbols are to be distinguished from channel symbols since one can *encode* source symbols into different channel symbols. The transmitter inputs channel symbols to the *forward* channel, and noise[1] in the channel may cause errors in the received symbols. If only forward transmission is used, the forward channel signal design or error-correcting code must meet the required error probability (P_E) for the given channel parameters. If an error-*detection* capability is added to the signal, and a feedback channel is available, the transmitter can request repeats of data segments found to contain errors (called automatic-repeat request—ARQ). Use of feedback ARQ can eliminate errors, but a time delay is experienced.

If hard-copy printed or stored information is to be transmitted, zero errors are usually required since the information is permanent in a record-keeping sense, and convenient data storage on paper or magnetic tape is feasible. Then usually time delay (in reception) is tolerable so that ARQ can be used. If the source is digitized voice or video, a more transient communication is effected. Now only a little delay is permitted, since the communication is real time, but a sufficiently small number of errors is more tolerable.

In the forward channel, received power and channel bandwidth are the two basic communication resources. The per-symbol transmission quality, measured by error rate, is (conceptually) entirely a function of the average received signal power, the noise level, the "distance" between alternative signals, and the symbol integrating time (reciprocal of symbol rate). *Power-limited* cases occur when it is difficult to increase transmitted power, reduce physical distance between trans-

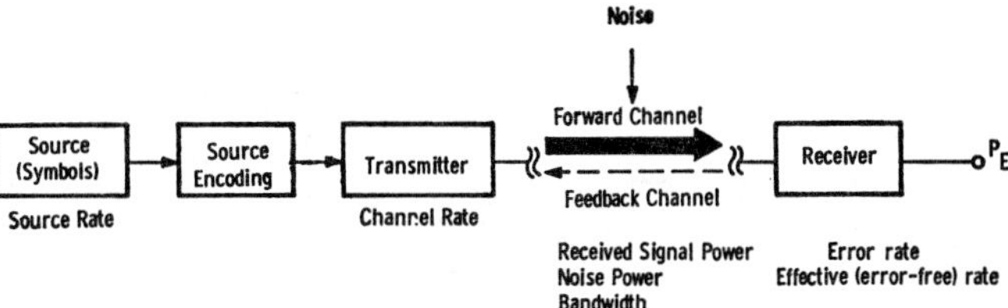

Fig. 1. Components in digital communication.

mitter and receiver, or improve the antenna. Space communication links are typically power limited.

If transmission quality must be improved in power-limited cases, one alternative is to allow the channel symbols to become longer and hence increase the final decision integrating time. This requires reducing the symbol rate, which is often not possible. A second alternative is to *encode* groups of source symbols and form channel signals which have a greater distance (in signal space) between the alternative signals. As will be seen, the data rate can be maintained if the bandwidth can be somewhat expanded.

Bandwidth-limited cases occur when adequate received power is more easily achieved than increased bandwidth. The radio spectrum channels may be bandwidth-limited due to competition among contending users. The switched network telephone system is bandwidth-limited for high-speed data and Picturephone® signals since it was engineered as a high signal-to-noise ratio voice-bandwidth facility. In these cases, one exploits the power adequacy by efficiently using the available bandwidth. One can achieve efficient use of bandwidth, given adequate power, by grouping source symbols and forming channel signals which are closer or more dense in signal space.

Since error detection with ARQ is potentially useful whenever the P_E requirement is stringent, it may be useful with either power-limited or bandwidth-limited forward channel cases.

The receiver (Fig. 1) decides which of the alternative (forward) signals was transmitted. If error control with forward-transmission-only (also called forward-error control, FEC) is used, the receiver design along with the signal design, power, and noise determines the error rate. If time delay is permitted and a feedback channel used for ARQ, the final performance measure is the average effective *rate* of the final correct symbols, and the undetected error rate.

In the ensuing material we first describe source encoding methods, where the objective is to minimize the channel symbol rate of any source by reducing any redundancy. Forward-transmission principles and methods then follow, and constitute a major part of this paper since forward techniques are used even in feedback situations. Dense signals for bandwidth-limited cases, and block-coded and convolutional-coded signals for power-limited cases are described. Power-limited cases are emphasized since these are the most difficult. Shift-register implementations are stressed since they are cost-effective parts of most constructions.

Next, feedback transmission methods and performance using block error-detecting codes are reviewed. Finally, channel-sharing concepts for digital signals are briefly reviewed.

II. Source Encoding

Source encoding refers to those techniques that alter the information source for efficient transmission over the channel. For digital transmission, it is usually desirable to minimize the source bit rate (measured in bits per second), while retaining

---

[1] The minimum possible noise is the front-end thermal noise of the first amplifier in the receiver. Other noise is contributed by interference from other users or from man-made sources (ignition noise, general radiation, or nearby equipment), internal disturbances, or atmospheric radiation.

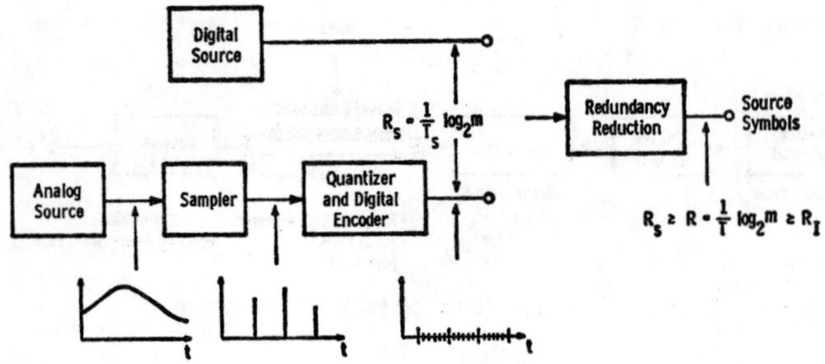

Fig. 2.   Elements in source encoding.

the desired information. For given channel power conditions, minimizing the source bit rate permits a lower P_E since one increases the energy per symbol.

The bit rate from any digital source (computer, Teletypewriter) can be reduced without losing information if the source symbols have *redundancy*. If an analog source, continuous in time and in value, is to be transmitted via digital transmission, an efficient analog-to-digital (A/D) conversion or digitization is required. Digitizing an analog source (voice, video, telemetry data) involves sampling the time-continuous waveform and then quantizing these samples, as shown in Fig. 2. Analog sources can be digitized in a way so as to minimize the source redundnacy and bit rate.

Let m denote the number (alphabet size) of different symbols available from either a natural digital source or a digitized analog source, occurring at a rate $1/T_s$. Since m symbols can be represented by $k = \log_2 m$ binary symbols, the source bit rate R_s is

$$R_s = \frac{1}{T_s} \log_2 m = \frac{k}{T_s} \; \frac{\text{source bits}}{\text{second}} \qquad (1)$$

where

R_s source rate in bits per second
T_s time per source symbol
k number of binary digits per word or symbol.

Note that the unit of measurement for the rate is bits per second (b/s) and is used whether or not binary representation (or digits) are actually involved. When binary representation is actually used, as in digital computers, one can either consider $m = 2$ with T_s equal to the bit interval or consider $m = 2^k$ with T_s equal to the k-bit word length.

A digital information source is said to possess *redundancy* if either: 1) the symbols are not equally likely, or 2) the symbols are not statistically independent. Source encoding can reduce the rate of source symbols if either of these conditions exists. To illustrate, it is well known that the frequency of English letters in written text is not equally likely; further, knowledge of a given letter (or a pair of letters) permits some prediction of the next letter. Digital symbols from computers and data processors, on the other hand, are usually (for all practical purposes) nonredundant.

If the source symbols are statistically independent the source is said to have *zero memory* since each symbol is selected without influence from any previous symbols. We have just seen that written language has memory although it is generally not feasible to exploit this redundancy. Another example of a digitized source with memory is the frame-to-frame content of either film or television frames. Since one is transmitting frames sequentially, one readily recognizes that much (video) information in a given frame remains the same in the next

frame, and hence the video source (outputting symbols from sampled and quantized video values, considered in a frame-to-frame fashion) has memory.

One major approach to reducing the bit rate with memory sources is to encode and transmit the *differential* between the present and the preceding value. This requires temporary electronic storage of the previous segment (an entire frame in the video case).

If the source has redundancy because of nonequally likely symbols, the average bit rate can be reduced by assigning short code words to the high-probability symbols and longer code words to the low-probability symbols. Two well-known procedures for doing this are the Shannon–Fano encoding and the Huffman encoding [5], [6]. In this way the average code-word length is reduced, which in turn reduces the source bit rate.

Conceptually, one can reduce the source bit rate down to the theoretical *information rate* or *entropy rate*. For a source with zero memory, it is known classically that the average amount of *information* bits per second from an alphabet $x_1, x_2, \cdots, x_m$ is given by

$$R_I = -\frac{1}{T_s} \sum_{i=1}^{m} P(x_i) \log_2 P(x_i)$$

$$= \frac{\text{symbols}}{\text{seconds}} \times \frac{\text{information bits}}{\text{symbol}} \qquad (2)$$

where

R_I information rate in information bits per second
T_s seconds per symbol
m alphabet size
$P(x_i)$ discrete occurrence probability of symbol x_i.

The summation term is the classical *entropy*, which evaluates the *information* bits per symbol. Note that the *source* bit of (1) is a binary representation or a unit of measure; the *information* bit of (2) is the theoretical information content.

For the source with memory, the equation corresponding to (2) is more complex, involving conditional probabilities, and will not be considered here. Usually the encoding procedures which exploit both nonequally likely and memory redundancy are *ad hoc* procedures. As an example, the Picture-phone signal uses frame-to-frame differential encoding for memory reduction and Shannon–Fano encoding of the resultant differential signals [7].

If a redundancy reduction technique is used, then the final source rate R should be smaller than R_s of (1) and greater than the theoretical R_I of (2). We define the final source rate in terms of m symbols per interval T:

$$R_s \geq R = \frac{1}{T} \log_2 m \geq R_I \quad \text{b/s} \qquad (3)$$

where $R =$ source-bit rate in bits per second.

The source rate R is equal to the information bit rate only if the m symbols are equally likely and statistically independent ($R \geq R_I$).

A. Digital Source Encoding

Morse code is one example of a digital source (alphabetic text) which minimizes the bit rate by reducing the nonequally likely redundancy. Most (Teletypewriter) printed text, however, uses a standard code-word length (7 b plus a parity-error bit, [46]) and makes no attempt to reduce redundancy.

Although redundancy reduction has not played a strong role in the common digital sources, it should be kept in mind as a possibility.

It is also of interest to note the possibility of *message formatting*. In communications, one typically uses the m-level alphabet to construct messages from a sequence of source symbols at will. The total number of messages possible with a sequence of k symbols is given by

$$\text{messages possible} = m^k \qquad (4)$$

where

m alphabet size or number of source symbols
k number of sequential source symbols.

In some communication situations it is desirable and permissible to designate the total number of messages at the outset. Here the total message itself is treated as the alternative among finite choices for the digital transmission, and this is termed *message formatting*. The entire transmission requires only k (m-ary) symbols. This can only be used if one can predict the total range of messages,[2] but message formatting results in very efficient (minimum symbol transmission) communication.

B. Analog Source Encoding

Redundancy reduction for analog sources usually involves reducing the memory redundancy among the sequential samples. The objective is to minimize the eventual bit rate while providing the necessary quality. Since it is difficult to evaluate accurately the true information rate of most analog sources, one usually searches for *ad hoc* ways to reduce the source bit rate while retaining necessary quality. One can exploit the particular properties of the source in doing this.

Rate Distortion Theory, first introduced by Shannon [1], is a theoretical framework for establishing the minimum bit rate for a given level of distortion in analog sources [56], [57]. It has proven quite difficult to derive the bounds (minimum bit rate) versus distortion for real signal sources such as speech and television. Some preliminary estimates are made in [58] but they are too detailed for inclusion here.

In the following we briefly review existing analog encoding methods and note the accompanying bit rate.

1) PCM: The classical analog digitization is *pulse-code modulation* (PCM)[3] ([8] and any communication text). In PCM the analog signal is first sampled at a rate exceeding twice the highest significant frequency present in the signal, and each sample is quantized to one of a finite number of (m) levels. Each such finite level is a digital symbol and is represented by a digital alphabet. Traditionally, PCM is associated with binary signaling because most past systems used binary. The digital signals are then sent sequentially, as indicated in Fig. 2. The reverse process is performed at the receiver: a quantized sample is reconstructed; these are smoothed or filtered to recover the analog waveform.

Theoretically, PCM is relatively inefficient for digitizing most analog sources; its symbol bit rate far exceeds the actual information (entropy) rate of the source. The source rate (R) would approximate the information rate (R_I) for PCM only if the analog source had: 1) uniform amplitude statistics, and 2) a flat (or white) power spectal density versus frequency. Voice, video, and most telemetry data signals are distinctly not of this form. Although PCM yields a highly redundant digitization for these sources, it is often useful and practical, especially when many sources can be multiplexed on a single channel (*trunk*). Digitization methods that improve the efficiency over PCM also increase the cost of the terminal.

The PCM format of grouping binary symbols to represent m levels is exactly the format needed for digital characters from computers (or alphabet symbols from Teletypewriters). Also, this grouping format directly fits time-division multiplexing (TDM) of multiple users (see Section V). For these reasons the common-carrier digital facilities (Bell System T1 and T2 digital carriers [2]) use the PCM format of grouping binary symbols into characters.

Before reviewing the efficient digitization techniques, we should note the role of quantization noise. Any quantizing of a continuous quantity produces an inevitable uncertainty (quantization noise) about the precise value within the *quantization interval* or step size. The signal's total amplitude span is divided into m levels or quanta. Increasing m reduces the quantization noise[4] but increases the symbol bit rate needed to represent the sample. Note that specifying m is tantamount to specifying or defining the tolerable distortion in the digitized source. Hence m is chosen in accordance with answering the question: What precision in amplitude level is required or desired by the user? Generally a 7-b representation ($m = 128$) for PCM yields excellent quality; 6 b is sometimes used when the quality required is less stringent. When varying signal volume (changing amplitude spans) is an issue, a fixed or a time-varying compression curve (nonlinear input–output amplitude relation) is used to make efficient use of the m levels.

While quantization noise usually represents a signal degradation imperceptible to the user, additional degradation occurs if (channel) errors are made during the digital transmission. Typically one chooses the m so that the quantization noise is barely imperceptible to the user and then uses a combination of signal power, encoding, and repeater spacing to keep channel errors negligible.[5]

The size of m directly affects the source bit rate but not the redundancy due to memory. In the following we will briefly review digitization techniques which produce less redundancy than PCM, and which result in lower source bit rates for any given m. First we should note the benchmark symbol rates for PCM. For a nominal 4-kHz voice signal, an m of 128 ($k = 7$) results in a source bit rate of 56 kb/s (48 kb/s for $m = 64$). The Bell System T carrier uses 64 kb/s ([7 information bits + 1 check bit] × 8 k samples/s) for one (3-kHz) digitized voice channel. If we consider a nominal 5-MHz bandwidth for commercial television (video) signals, a 70-

---

[2] One practical backup strategy with message formatting is to reserve one message for conveying change to another mode where longer messages are possible.

[3] The terminology PCM is justified as follows. The baseband digital signal is a pulse of some type. *Coding* here refers to expressing the m-level quantized samples via a sequence of binary symbols. The term *modulation* in PCM is somewhat a misnomer; the coded pulses can be sent over a channel using a wide variety of modulations. No particular channel modulation is implied by PCM.

[4] Quantization noise can be considered as a zero-mean added noise, where the noise power spectrum is uniform over the signal bandwidth. The quantization signal power-to-noise power ratio depends somewhat on the type of signal digitized, but is always proportional to m^2. If the sample is represented by k binary symbols ($k = \log_2 m$), then the power SNR is proportional to 2^{2k}.

[5] If encoding and repeater spacing are not available, then it may be of interest to compute the relation between quantization noise and noise due to channel errors. An optimum relation between m and channel power-to-noise can be found for this uncoded and unrepeated case [8], [9, pp. 256–258].

MHz bit rate results if $m = 128$ with PCM. With Picturephone the video requirement is relaxed somewhat; the analog bandwidth of 1 MHz would result in a bit rate of 14 Mb/s if $m = 128$ PCM were used. As seen, PCM requires a bandwidth more than 10 times as wide as the analog transmission would require. The resulting signal, however, is more rugged in noise and more flexible.

If one conveyed only the information content of speech, without voice quality and voice recognition, one could presumably transmit speech at a mere 60 b/s. Good quality speech does, of course, require a near-perfect reproduction of the specific voice signal. Speech digitized by PCM has redundancy primarily because the samples (sampled at slightly about twice the highest significant frequency) are quite nonindependent. This is due primarily to: 1) the quasiperiodicity of voiced speech, and 2) the structured (predictable) nature of the short-term spectral envelope. Consequently, most speech digitization attempts [10] pursuing minimum source bit rate try to predict the present sample value from past sample values and to transmit only the "error" signal. Thus whereas in PCM each sample is processed independently, predictive systems utilize stored past history to predict each present sample and then transmit only the difference between the predicted sample and the actual sample.

If video is digitized by straightforward PCM of the line-scanned signal, the resulting symbol stream is redundant primarily because many picture elements do not change from frame to frame. Secondarily, there are sample-to-sample and line-to-line correlations. Video digitization then [11] usually involves interframe predictive techniques, where one transmits only the change or differential from frame to frame. In addition, such interframe changes are themselves subjected to further line-to-line coding and run length coding to exploit the sample-to-sample correlation.

It should be emphasized that redundancy-reducing digitization techniques reduce the source bit rate, and hence the transmission cost, at the expense of increased complexity at the terminal.

We now review the major practical predictive methods that are available for using past samples to predict current samples. Many times the technique can be used for any sampled analog signal. Bayless *et al.* [10] give a convenient reference for current assessment of these alternate methods, especially for voice application.

2) Differential PCM (DPCM): In differential PCM (DPCM) the error between the current sample and a predicted value is PCM encoded and transmitted. The prediction computation is time invariant, and a basic tradeoff exists between the increasing use of sampling rates above twice the highest frequency and the m value of quantization levels used for the error signal. For speech, differential PCM gives "medium" quality at 20 kb/s and high quality at 40 kb/s. DPCM is a core technique in many video encodings [7], [11], [13]–[15]. It is applied in an interframe fashion, and from sample to sample. Often a variable length code (Shannon–Fano or Huffman) is applied to the DPCM values [13], [14]. This requires a buffer as previously noted. A range of video encoding techniques are surveyed in [17] and [18]. In Picturephone video, interframe differential PCM is used, along with a range of other encoding techniques (including Shannon–Fano) to effect a 1.544-Mb/s rate for the 1-MHz analog signal [7], [11].

3) Delta Modulation (DM): Delta modulation (DM) can be considered as the simplest form of differential PCM. Now the predictor simply integrates (with some variable gain) the past history and attempts to stay close to the actual waveform. Only a binary error signal is transmitted in DM, depending on whether the integrated prediction is higher or lower than the actual waveform at the sampling time, which can now be much larger than twice the highest frequency since there is only one binary symbol per sample. An adaptive delta modulation (ADM) is formed if the binary step size (commanded by the binary transmission) is varied from time to time, according to local waveform need. ADM can provide speech as low as 10 kb/s; a rate of 20 kb/s provides a good quality, while a 40-kb/s rate affords excellent quality. Although ADM could be used for video encoding [16], it is used almost solely for speech.

4) Adaptive Predictive Encoding (APD) and Linear Predictive Encoding (LPE): These encodings are aimed at speech and involve more elaborate prediction algorithms than DPCM or DM. While in the LPE the spectral envelope is predicted and the pitch is separately processed, the predictor in APD estimates both the pitch and the present spectral envelope. The error signal is transmitted in APD, and in LPE the error signal is characterized by sufficient parameters (pitch, voicing characteristics) so that it can be recreated at the receiver. In laboratory experiments [12] speech of high quality was obtained at 7.2 kb/s. At 4.8- and 2.4-kb/s rates the quality is somewhat degraded but better than any current compectitors. These devices are still in the research and development stage.

Finally, we note that *Vocoders* can be used to digitize voice signals. Although source rates as low as 1.2 kb/s and 0.6 kb/s [10] are possible, the quality is poor and the complexity remains substantial.

We see then that, relative to PCM, the voice bit rate can be reduced from approximately 50 kb/s to around 5 kb/s, and phone video can be reduced from about 14 Mb/s to 1.5 Mb/s. Although there is some inevitable quality reduction in these compressions, the degradation is barely discernible by the listener or viewer. Some of the basic source encoding methods surveyed here can be used on general telemetry data.

III. FORWARD-TRANSMISSION ALTERNATIVES AND PERFORMANCE

The fundamental issue in any communications is to convey messages *forward* accurately from the source to the receiver. As noted, a communications situation may use only forward transmission, or it may also use a *feedback* (receiver-to-source) channel. If forward transmission only is used, the desired accuracy must be achieved either by adequate (channel) signal-to-noise ratio and/or by an *error-coding* processing. With a feedback channel, the forward link need not consistently meet the final accuracy requirements. If means are provided for *error detection*, the receiver can automatically request a repeat (ARQ) of the segment found to contain errors by means of the feedback channel. With ARQ any incoming source symbols must be buffered or stored during a retransmission. The storage means are either electronic memories, paper tape, or magnetic tape. We noted before that the ARQ method is attractive whenever the accuracy requirement is severe, and some time delay is permissible, and especially when errors tend to be clustered.

It is interesting to note that written material and broadcast radio or television usually serve as forward-only links, while speakers or computers in conversation can effect an ARQ operation.

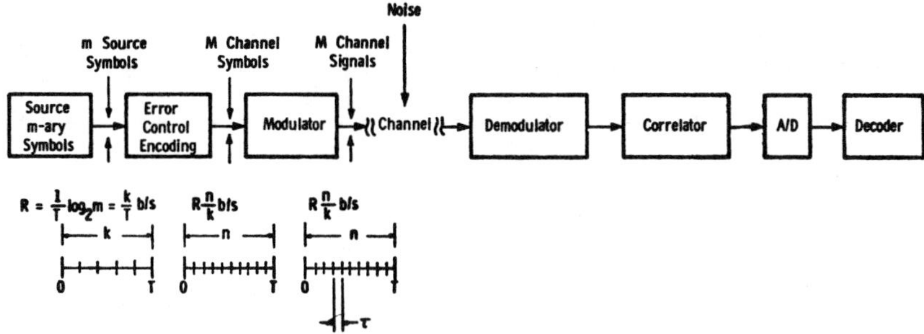

Fig. 3. Components of forward transmission.

We pursue the forward-link alternatives in this section and review ARQ in the ensuing subsection.

A. Concepts and Limitations

The goal in the forward channel is to transmit the source symbols at the rate (R) with an acceptable probability of information bit error (P_B) in the face of channel noise. We define a symbol decision interval T based on the time interval over which the receiver makes a final decision about the received symbol. The interval T contains one or more source symbols of length T_s. In any decision interval T the transmitter sends one of the M-alternative signals. Given the reception, perturbed by at least some noise, the receiver decides which symbol was sent.

If the transmission is asynchronous, the signaling intervals may occur at any time and are usually preceded by a special "start" signal and followed by a "stop" signal. Teleprinter signaling often uses the asynchronous mode, where the alphanumeric characters are represented by 5 or 7 elementary signaling intervals, with a start and stop signal added. In the *synchronous* mode the symbol decision intervals are contiguous in time, and synchronization of the receiver's quantized time axis to that of the transmitter is assumed. When using the voice grade telephone network, current practice uses asynchronous transmission for 0–1800 b/s and synchronous for 2000 b/s and up. The remainder of this paper assumes synchronous mode, but many aspects apply also to the asynchronous case.

The principal signaling components of a forward transmission link are shown in Fig. 3. The source produces one of m symbols at a rate R (bits per second). (Remember that we use "bits" as a fundamental unit of measure, even though the symbol may be nonbinary.) Also, the symbols emerging from the source can now be referred to as either source bits or information bits. The channel does not care whether the source symbols have redundancy, so it is no longer necessary to distinguish source bits from information bits.

Source symbols may be entered directly into the channel modulator, or they may first be *encoded*. In general, *coding* involves relating k source information bits so as to make efficient use of the communication (channel) resources: bandwidth and power. We shall see that the encoder can relate information bits in either a *block* fashion or a *convolutional* fashion. It is probably easiest to pursue the concepts by first assuming the block type (the type implied in Shannon's classical work, [1]). The convolutional type will be described later.

The block encoder output produces one of M *channel symbols* and the modulator converts the M-ary channel symbols into channel *signals*. If k source or information bits are combined in the encoder,[6] then

$$M = 2^k \qquad (5)$$

where

M number of channel symbols or signals
k number of information bits encoded.

The rate and type of the M channel symbols determine the required channel properties: received power (relative to noise) and bandwidth.

If the channel is bandwidth limited (with adequate power), the encoder groups k information bits to form M channel symbols. This increases the information bit rate per unit bandwidth (W) since now k information bits are sent in each symbol decision interval T.

If the channel is power limited, the encoder adds intentional redundancy to the k bits to effect a lower P_B at the expense of an increased channel bandwidth and receiver complexity. (If the source rate can be reduced, the channel bandwidth need not necessarily be expanded.) This redundancy coding may consist of signal design techniques or may involve the use of an algebraic *error-correcting* code.

The term *error-control encoding* refers to the encoding in either the bandwidth or the power-limited case. If feedback ARQ techniques are used, the encoding consists of an *error-detection* code, and may be used with either bandwidth or power-limited forward channels. In this section we consider forward-only techniques, and will include error-detection codes in Section IV.

The M symbols from the encoder are translated into signals suitable for transmission over the channel by the modulator. Although baseband pulse-type signals can be sent over cable, digital transmission usually involves digital modulation of a (sinewave) carrier. With amplitude-shift keying (ASK) one of M amplitudes is sent in the signaling interval. In M-phase-shift keying (MPSK) one transmits one of M phases of a sinewave segment (the term PSK alone implies binary $M = 2$). In frequency-shift keying (FSK) one of M frequencies is sent during the signaling interval. Sometimes combinations of these basic forms are used. The PSK channel signals require phase coherence at the receiver, while ASK and FSK are

---

[6] There is some arbitrariness about the term *coded signal*. Usually it implies combining *binary* source signals, to form new signals having more binary "chips" than original information bits. Thus k binary source symbols are combined: $m = 2$; $M = m^k = 2^k$. However, if m-ary source symbols are converted directly into M-ary channel signals $(m = M)$, then this may be considered a coded signal for $m > 2$ from the viewpoint that information bits are combined in the channel signal, or an uncoded signal since source symbols are not combined in any way.

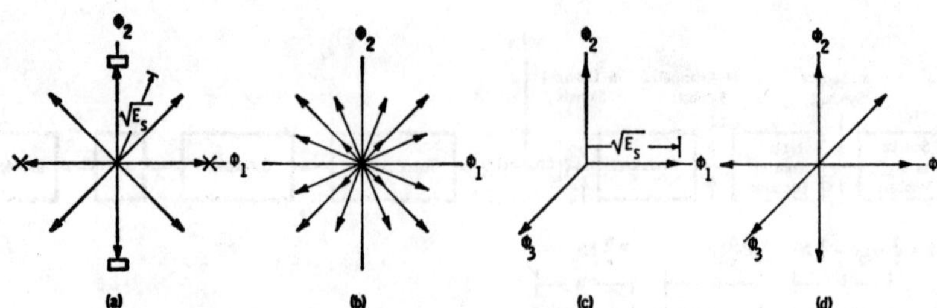

Fig. 4. Signals depicted in vector signal space. (a) 8-ary PSK. (b) $M=16$ combination ASK–PSK. (c) $N=M=3$, orthogonal. (d) $N=3$, $M=6$, biorthogonal.

usually operated with noncoherent detection. The terms "modem" (modulation/demodulation) or "data set" are widely used to refer to the equipment that interfaces the baseband signals with channel signals.

The digital signaling, consisting of encoding and modulating, can be described as follows. A set of M alternative signals is chosen, where the symbol decision interval is T and the channel bandwidth is W. If one permits indefinitely close spacing of signals (which are then more vulnerable to noise), one can let M become indefinitely large in a minimum TW space (i.e., MPSK or ASK signals). Such dense signals are desirable in bandwidth-limited cases, where the channel signal-to-noise ratio is adequate, since increasing M increases the information bits ($k = \log_2 M$) per interval T.

On the other hand, if the channel is power limited, but bandwidth is available, one should use signals having good "distance" properties. Signals that are orthogonal[7] to each other are one example of "distant" signals. Two related examples are: 1) the *simplex signals*, which have the maximum available negative correlation with respect to each other, and 2) the *biorthogonal signals*, where one adds the negative of each signal to the orthogonal set. These orthogonal-type signals (simplex, biorthogonal, and orthogonal) have similar performance if $M > 10$. In addition, the range of algebraic error-correcting codes provide distant signals.

For such distant signals, the number M is constrained by the dimension (N) of the signal space. In essence, the signal dimensions describe the number of independent descriptors of a signal T long and W in bandwidth. The descriptors may be independent samples (from the sampling theorem), Fourier coefficients, orthonormal coefficients, and others. The orthonormal description [19], [20] is universally used to provide a vector description of signal sets.

The theoretical dimensions (N) of an interval T having bandwidth W are limited to:

$$N \leq 2TW \qquad (6)$$

where

N dimensions (independent descriptors) in a time T and bandwidth W

W channel, and hence signal, bandwidth

T symbol decision interval.

In practice $N \approx TW$ due to the fact that most physical channels produce distortion, such as intersymbol interference.

Hence signals that are orthogonal when transmitted may not remain so at the receiver.

For orthogonal signals $N = M$; the dimensions must equal the total number in the signal set. Hence it can be seen that M can be increased with such signals only if one is able to directly increase W.

If one wishes to plot signal sets on a signal vector space, one must find the N orthonormal functions[8] for the given signal set [19], [20]. Fig. 4 shows such vector diagrams, for dimensions $N=2$ and $N=3$. On such a vector diagram the distance from the origin to the signal is $\sqrt{E_s}$:

$$E_s = \text{signal energy} = \int_0^T s^2(t)\,dt \qquad (7)$$

where $s(t)$ is the signal waveform during interval T.

In bandwidth-limited cases signals that are relatively dense are desired, since this increases the information bits per unit bandwidth. Fig. 4(a) shows a signal space vector diagram for an 8-ary PSK signal. With this signal three information bits would be grouped in the encoder and sent during every interval T, instead of one bit if (binary) PSK ($\times$) were used. Higher M dense signals use a combination of both ASK and PSK because they are more efficient. In addition, the phase jitter of practical channels would prevent use of high MPSK.

Fig. 4(b) shows an efficient $M=16$ dense signal configuration. These signals result from simultaneously transmitting two four-level double-sideband signals in quadrature phase (called 4-level quadrature amplitude modulation, QAM). The two signals combine to form a total of 12 phase values and 3 amplitude values. This signal, for example, can transmit 9600-b/s data over the 300–3000-Hz voice bandwidth analog telephone line [54].

The ($M=4$) quadraphase PSK signal (QPSK) is shown in Fig. 4(a) if one uses the $\times$ and $\square$ marked signals. This signal is efficient and widely used for channels that are both power and bandwidth limited, such as satellite-relay channels.

Signals perturbed by noise would be represented by adding noise vectors to the signal vectors shown in Fig. 4. It can be

---

[7] A signal set is orthogonal if:

$$\int_0^T s_i(t)s_j(t)\,dt = \begin{cases} E_s, & \text{if } i=j \\ 0, & \text{if } i \neq j. \end{cases}$$

[8] Expanding a signal in terms of orthonormal functions amounts to a generalization of a Fourier series expansion. The ϕ functions (kernel functions) are analogous to the sine or cosine terms in Fourier series. Each signal then is described by coefficients of ϕ functions, and the ϕ functions must fulfill:

$$\int_0^T \phi_i(t)\phi_j(t)\,dt = \begin{cases} 1, & \text{if } i = \\ 0, & \text{if } i \neq j. \end{cases}$$

Note that the vector axes in Fig. 4 correspond to the ϕ functions and represent actual time waveforms.

visualized that, for desired accuracy (P_E) and a given received signal energy, small noise permits dense packing of signals and the larger noise in power-limited cases requires distant signals. Any orthogonal (distant) signal set can be described by one signal per axis, since $N = M$, as depicted for $M = 3$ in Fig. 4(c). A biorthogonal signal set is formed by adding the negative of each of the orthogonal signals ($N = M/2$); Fig. 4(d) depicts the $M = 6$ biorthogonal case. The distance concepts of these vectors diagrams are general and apply for any N, even though it is difficult to visualize the signal space for $N > 3$. We may note that signals formed from algebraic codes can also be depicted on such a signal space diagram although many codes do not have a simple geometric property.

In power-limited cases, then, the encoder maps k information bits into M relatively distant signals. This entails an expansion of bandwidth, if the information rate is constant, since the dimensions grow directly with M (and $M = 2^k$). The actual channel signals can be formed with natural distant signals, such as the (orthogonal) MFSK, or one can form the distant M signals using a sequence of binary elements. When using binary elements, one represents the distant signals in the interval T by n-sequential channel bits ($n > k$), each τ long, as suggested in Fig. 3. These n *channel bits* (chips) must be distinguished from the k information bits. Thus $M = 2^k$ distant signals can be formed with n binary elements or chips. If the chip length is τ, then $T = n\tau$. Note that the chip sequences that are distant from each other are a small subset of the total 2^n possible chip sequences. The chips themselves are usually sent via PSK, if coherent detection is feasible, or FSK if noncoherent detection is used.

One can use these n chips to form orthogonal-type signals or one can form an algebraic block (error-correcting) code. In either case the objectives and techniques are grossly similar.[9] For the orthogonal-type signals $n = N$, and the following relations then hold: $n = M - 1$ for simplex; $n = M/2$ for biorthogonal; $n = M$ for orthogonal. From (6) we see that the channel bandwidth increases directly with increases in M for these distant signals.

For error-correcting codes, $n \leq M$ for all codes of interest. In general, for algebraic codes n grows linearly with $k = \log_2 M$.

With either type of signal, the power-limited cases require bandwidth expansion to improve accuracy since $n > k$.

The block-signaling situation for power-limited cases is included in Fig. 3. Assuming M is selected, the information-bit stream at rate R is blocked into $k = \log_2 M$ information-bit segments. The encoder then converts the k information bits into an assigned n-chip channel symbol or code word. If the rate R is kept constant, the chip rate out of the encoder must be $R \times n/k$ bits per second. (Note that we use the units bits per second for both information bits and channel bits.) This same rate holds for the signals from the modulator. One gains accuracy in this process because the n long code words are distant from each other while the k long information blocks are relatively close in signal space. We note the bandwidth expansion: if R bits per second enter the encoder, $R \times (n/k)$ chips per second emerge from the encoder.

Given that a signal set is chosen it is now classical knowledge that if the only significant channel noise is white Gaussian, the optimum receiver correlates the reception with each of the possible alternatives (in a synchronous fashion) and selects that signal whose correlator output, at the time T, is greatest. If the channel symbols are equally likely, such a receiver is called a maximum-likelihood receiver[10] since the receiver chooses that signal which, in light of the reception, was the most likely signal transmitted.

The correlations in the maximum-likelihood receiver can either be of the "multiply and integrate" correlation type or be a matched filter [19]. Either implementation provides the correlation values, and signaling-interval synchronization is assumed between transmitter and receiver.

Note that the ideal correlator is a linear analog device, operating over the interval T. For the dense signals such as MPSK, the correlators simply compare the incoming phase with the possible phases. For a natural (distant) signal such as MFSK the correlators consist of extremely narrow-band linear filters tuned to each permissible frequency, and are dumped at every interval T. These correlators may or may not use quantization in the choose-largest comparison.

If the signals are formed with binary elements, then the total correlation could conceivably be done in one step using a tapped analog delay line.[11] Although somewhat inferior, an alternative is to make decisions on each arriving τ long chip, via binary correlation, and then use digital computation (decoder) to compute the correlation comparison over the interval T. If one makes a binary decision on the arriving chips, it is termed a *hard decision*. A *soft decision* occurs on the chips if one quantizes the binary correlator output into more that two levels, to be used in the ensuing T interval computation. The receiver shown in Fig. 3 consists of a τ correlator followed by an A/D converter. The decoder completes the implementation of a correlation comparison over the T interval.

It is important now to note the relation between the (channel) symbol decision interval (T) and the actual channel signal interval (τ). Since different users will connect to a given (common-carrier) channel-plus-modem it is necessary to specify the decision rate of the demodulator-plus-τ-correlator. It is traditional to specify the channel signaling rate in terms of *bauds—the signal speed in bauds equals the number of signaling elements per second:*[12]

$$\text{bauds} = \frac{1}{\tau} = \frac{\text{signaling elements}}{\text{second}} \qquad (8)$$

where τ is the length of channel signaling element.

If a single M-ary symbol is sent in the channel signal interval τ, then $\tau = T$ and the information rate is

$$R = \frac{1}{\tau} \log_2 M = k \times \text{bauds} \quad \text{(b/s)}. \qquad (9)$$

---

[9] There is a somewhat different viewpoint between signal-design approaches and error-correcting code approaches. The signal-design case takes the viewpoint of minimizing the occurrence of information-bit errors while error-correcting codes take the viewpoint of *correcting* channel chip errors (which in turn minimizes information-bit errors). Also signal-design approaches model the physical channel as analog with added white Gaussian noise, while error coding often assumes a binary channel (a channel in which the analog reception is quantized to binary).

[10] We shall henceforth assume equally likely channel symbols. Even though English text letters are not equally likely, the binary representation of a string of letters is approximately balanced between ones and zeros.

If the symbols are not equally likely, the receiver should maximize the *a posteriori* probability, which is the product of the conditional probability (probability of reception given S_i) and the *a priori* probability [19], [21].

[11] The emerging surface-wave filter appears to be a cost-effective way to implement a tapped analog delay line for binary-encoded signals.

[12] Although baud is a signaling rate, one frequently sees the term baud used (incorrectly) to refer to a signaling interval itself.

Thus the *information* rate in bits per second is k times the channel signaling speed in bauds both for dense signals and for natural distant signals (MFSK). However, if an M-ary signal is sent via n chips, then $T = n\tau$ and

$$R = \frac{1}{n\tau} \log_2 M = \frac{k}{n} \times \text{bauds} \quad \text{(b/s)}. \tag{10}$$

Now the information rate equals the speed in bauds times k/n. Clearly the information rate of a given channel and modem depends upon how it is used.

When maximum-likelihood reception is used over the interval T, communication theory shows that the symbol error probability P_E depends only on the distance properties of the signals and the ratio of received energy per signal E_s to the noise power density n_0. Thus for a signal set:

$$P_E = f(E_s/n_0) \tag{11}$$

where

P_E	channel symbol error rate
E_s	received symbol signal energy $= S \times T$
S	average received signal power over interval T
T	symbol decision interval
n_0	noise power per unit bandwidth $= N/W$
N	mean-square noise power.

One way to depict forward transmission performance is to plot P_E versus E_s/n_0. However, since one wishes to plot such curves for varying k, it is necessary to normalize both the P_E and the E_s to their "per-information-bit" values. For this one can use:

$$P_B = \frac{1}{2} \frac{P_E}{1 - \frac{1}{2^k}}$$

$$E_b = \frac{E_s}{k} \tag{12}$$

where

P_B	information bit error rate
E_b	received energy per information bit.

The P_B relation in (12) holds strictly for orthogonal and regular-simplex codes and approximately for other cases.

One then plots the information-bit error rate P_B versus the bit-energy-to-noise-density ratio E_b/n_0 as shown, for example, in the signaling cases of Fig. 5. The rightmost curve describes P_B for the dense 8-ary PSK signal corresponding to the signal set in Fig. 4(a). As seen, such dense signals for bandwidth-limited cases require a higher E_b/n_0 than the distant signals for power-limited cases.

Shannon's pioneering work [1] and work done independently by Kotelnikov [20] showed that, when the channel is power limited, P_B can be reduced by increasing M while retaining distant signals. The basic reason for this is that one is increasing the distance in signal space between signals.

Consider a binary source ($M = 2$), and first assume binary transmission using orthogonal (FSK) or simplex signals (PSK). For a given received signal power S and rate R, the P_B will be specified. If this P_B is too large, one can either raise power or decrease the information bit rate (hence increasing the energy per bit by increasing the integration time). Another alterna-

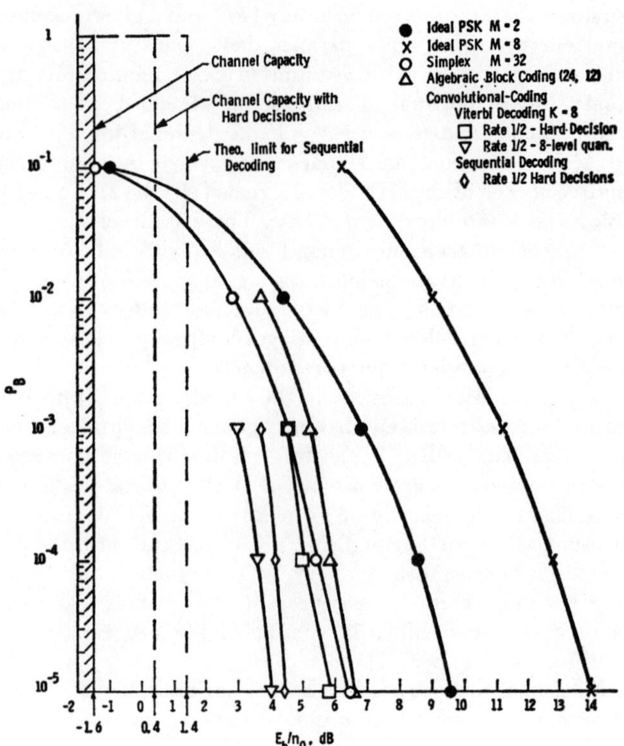

Fig. 5. P_B versus E_b/n_0 for forward-error control.

tive is to collect k source bits, encode them into $M = 2^k$ (orthogonal or algebraic code) signals, and effectively send k bits within a single receiver decision. Now we assume the power and the rate R remain the same.

One can compare the influence of M by plotting P_B versus E_b/n_0 for different M's, for an example distant signal. Fig. 5 includes the simplex signals for $M = 2$ (ideal PSK) and for $M = 32$ (plotted from [24, p. 129]). As seen, the P_B for a given E_b/n_0 decreases. Use of orthogonal-type signals with correlation detection is a costly solution, especially for large M because the required bandwidth grows directly with M, as previously explained. In addition one must in effect provide M parallel correlators in the receiver, or the equivalent.[13]

The algebraic error-correcting concepts, first introduced by Shannon and since pursued in many places, provide receiver techniques that are simpler and more practical than parallel correlators for large M. In error correcting one must first (usually) make *hard* decisions on the received chips. The block of n chips is then simultaneously processed to correct certain patterns of chip errors to estimate the k received information bits (to be explained later).

Shannon's work proved that the error-free capacity (the bit rate at which zero error can be achieved) for white Gaussian noise is a function of the average received signal power, the noise, and the bandwidth. The capacity relation is

$$\frac{C}{W} = \log_2 \left[1 + \frac{S}{n_0 C} \left(\frac{C}{W} \right) \right] = \log_2 \left[1 + \frac{E_b}{n_0} \left(\frac{C}{W} \right) \right] \tag{13}$$

where $C = \max \{R\}$ in bits per second, at which zero errors can be achieved.

---

[13] The correlations can be done simultaneously in a special-purpose computer, using the so-called Green machine, which operates roughly along the principles of the fast Fourier transform [25].

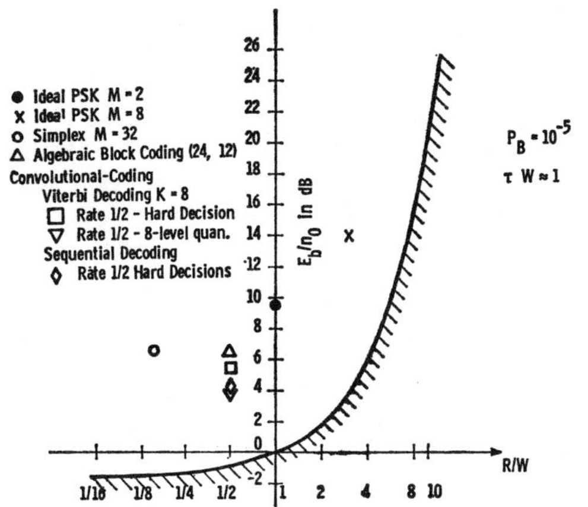

Fig. 6. E_b/n_0 versus R/W for alternative FEC signals for $P_B = 1 \times 10^{-5}$.

This relation settles, once and for all, the ultimate bounds on error-free communication. We see it is the *rate* that is limited by the SNR, not the error probability itself. From (13) we find that the minimum value of E_b/n_0 (for zero errors) is -1.6 dB, which occurs when the W approaches ∞ ($C/W \to 0$). Hence a zero-error transmission can in principle be accomplished if $E_b/n_0 > -1.6$ dB, when the noise is added white Gaussian. The P_B versus E_b/n_0 plot of Fig. 5 shows the capacity relation; it goes from a $P_B = 1$ to a $P_B = 0$ at $E_b/n_0 = -1.6$ dB.

Note that the channel capacity relation assumes full analog values for chip decisions. If hard decisions are used, the minimum (capacity) E_b/n_0 is $+0.4$ dB, as shown in Fig. 5. Hence the use of hard decisions results in performance loss which is equivalent to approximately a factor equal to $\pi/2$ (2 dB) in energy (this factor is asymptotically exact as $W \to \infty$).

Since the received signal energy and the channel bandwidth are the basic communication resources, it is sensible to also compare systems by plotting E_b/n_0 versus R/W for a given P_B. The capacity relation of (13) appears as the solid curve in Fig. 6 when E_b/n_0 in decibels is plotted versus R/W. Any actual signaling and receiver method will lie to the left and above this zero-error solid-curve barrier. Thus Fig. 5 shows the P_B versus E_b/n_0 behavior while Fig. 6 depicts the E_b/n_0 versus R/W for alternative signaling methods (for $P_B = 1 \times 10^{-5}$). These two depictions, considered together, form a complete description of the major performance aspects of forward-transmission methods. We will use these plots below to compare various forward-transmission methods. We confirm from Fig. 6 that if E_b/n_0 is adequate, the objective is to obtain a high bit rate per bandwidth (move to the right as much as possible) using dense signals; this is the bandwidth-limited case. On the other hand, if signal power (and hence energy) is inadequate for the desired P_B, then one must move to the left and downward as much as necessary (resulting in lower rates/bandwidth) in order to meet the P_B requirement. This is the power-limited case.

Traditionally, most FEC attention has emphasized the power-limited area since this seemed the more critical. Since error coding involves both collecting k information bits *and* using distant signals (orthogonal-type or algebraic), we see that coding is useful in the power-limited cases, but not in the bandwidth-limited cases. As already noted, in the bandwidth-limited area it is efficient to collect k information bits and use M-ary dense signals.

The channel bandwidth that a particular signaling method requires in practice is a function of the theoretical W treated here, along with the actual modulation type, and considerations of other users. The channel itself is not usually a baseband channel (zero to some cutoff frequency) but is rather a bandpass channel and hence requires modulation to a carrier frequency.

The modulation may be direct or may involve traditional modulation (AM, FM, SSB). If direct, the transmitter outputs M phases for MPSK or M frequencies via a frequency synthesizer for MFSK. If a traditional modulator is used, the bandwidth may be expanded due to the modulation (doubled for double-sideband AM). The bandwidths required in practice are also increased due to the need to provide a spectrum guard space between different users. Thus the bandwidth actually needed in practice has only a loose relation with the theoretical W discussed here. Nevertheless, the comparative relationships involving W are valid.

A common dense digital carrier is MPSK ($M \leq 8$). For binary-encoded distant signals, (binary) PSK is usually used to form the channel chips when coherent detection is feasible. Here the modulation can be considered AM:

$$s(t) = a(t) \cos (w_c t + \theta) \tag{14}$$

where

$$a(t) = \pm 1.$$

FSK is usually used when noncoherent detection is more practical. An important example of an error-coded system using noncoherent detection (with algebraic coding) is the TActical Transmission System (TATS) designed for tactical satellite-relay circuits by Lincoln Laboratory [48].

We can now summarize the forward-transmission (FEC) situation, using Fig. 3. If the M-ary dense signals are used, the encoder collects k information bits and outputs $M = 2^k$ channel symbols; the modulator produces M corresponding signals. The receiver now consists solely of correlators, over intervals $T = \tau$, for each alternate signal. A phase detector (or zero-crossing detector) is visualized for MPSK signals, and the reference phase is assumed known for the MPSK case. Since $N = 2$, one can in general implement dense signal detectors using only two correlators (in phase and quadrature) accompanied by an analog or digital weighting network.

For distant signals the encoder collects k information bits and one either adds redundant bits (to form n chips) or forms a natural M-ary signal (MFSK). The former is often preferred since this permits exploiting the natural binary electronic (integrated circuit) technology. If the signal consists of n chips, the receiver can either: 1) consist of $M = 2^k$ correlators integrating over the entire interval T to form a maximum-likelihood receiver, or 2) correlate only on the (binary) τ long chips, and then use a "decoding" step or computation to form the final decision (over T). Although the correlation receiver is optimum, one cannot let k become large because the receiving complexity grows as 2^k. Consequently, the decoding approach is more practical, especially if k must be made large to afford the necessary P_B. The decoding approach amounts to dividing the decision process into two parts: 1) a first tentative decision on the chip and 2) a final decision of the k information bits, considering the n chips.

We will pursue block and convolutional coded signals for power-limited situations later. First, however, it is useful to discuss shift-register implementations, since they are a central

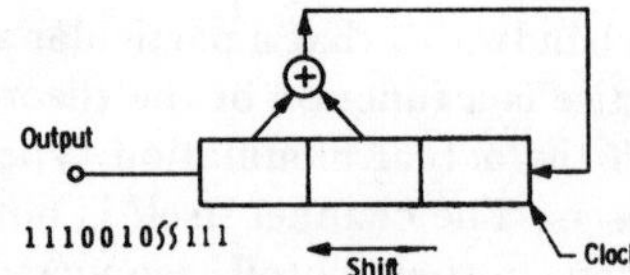

Fig. 7. Feedback shift-register generator.

part of the implementation of coded signals, both in generation and decoding.

B. Shift-Register Implementation of Coded Signals

Shift registers are a basic part of digital computers, digital filters, and general digital circuits. They form the core implementation for block-encoded and convolutional-encoded signals, as we shall see.

A shift register consists of cascaded (binary) flip-flops, where the binary content of all stages shifts to the neighboring (left)[14] stage when a clock pulse occurs. The basic function of temporary storage and shift capability, along with various external elements, permits a wide variety of digital manipulations. There is also provision for "resetting" simultaneously all the register stages. Modulo-2 addition[15] is the linear binary addition; a shift-register and modulo-two adders form a general *linear switching circuit*. A simple (unclocked or asynchronous) set–reset flip-flop can be used, as well as the clocked (synchronous) D type or J–K type.

Shift registers plus modulo-2 adders (plus other logic functions) are used both as generators of communication signal (sequences) and for decoding the received signal (sequences).

If a feedback loop is formed using modulo-two addition of certain stage outputs to determine the next input digit (Fig. 7), one can generate *maximal-length sequences*. These sequences are of especial interest because they serve as an illustration of orthogonal-type signals and are a useful introduction to algebraic block codes. They are also useful for general communication purposes (privacy coding, addressing).

A sequence is maximal length if, for k stages, the output sequence goes through each k-tuple except the all-zero-tuple before it repeats. Thus a maximal sequence has length (n):

$$n = 2^k - 1 \qquad (15)$$

where

n bit length of a maximal sequence
k number of register stages in the generator.

Maximal length sequences are also called *pseudonoise* sequences because their occurrence statistics resemble a sequence of truly random binary events (coin tosses). Thus there

is always just one more "one" than "zero" in a maximal sequence. Also, the periodic autocorrelation[16] is optimum: it has value 1 at $\tau = 0$, and has value $-1/n$ for all other τ values. It is of course periodic with n.

This autocorrelation property means that cyclic shifts of a given maximal sequence form a *simplex signal* set, which we discussed. The signals have a slight negative correlation with each other; for $N = M - 1 > 10$ this correlation is nearly equal to zero (orthogonal signals). However, the simplex signals here are especially easy to generate, as we have seen, and therefore may be useful as a block code.

C. Block-Coded Signals

As noted, the two major alternatives in power-limited cases are the block-coded signals of this section and the convolutional-coded signals of the next section. The block-coded signals are distant signals which can be constructed either via orthogonal-type or algebraic-coded signals. Sometimes the same signal can be looked at either way. The orthogonal-type signals usually imply M correlators for maximum-likelihood detection, while algebraic codes often use special (nonmaximum-likelihood) algebraic decoders. The major aim in algebraic decoding is to permit large k (and hence n) values for a feasible receiver complexity (simpler than 2^k correlators). A variety of algebraic decoding methods have been evolved.

All the distant signal methods essentially constitute a way of forming a coded signal in order to increase the distance between signals to combat noise at the receiver. As noted, Shannon's original concept was to improve the accuracy (for power-limited cases) by both grouping information bits and using distant signals.

1) Orthogonal-Type Signals: If the noise is additive white Gaussian, then the minimum P_E signal set (for any M), for energy constrained signals, is the *simplex set* [19, p. 260]. For a given energy, these signals have the maximum separation or distance between signals.[17] The value of M is chosen on cost-effective factors: the required E_b for a given accuracy goes down with M (Fig. 5), but the receiver complexity (hence cost) and bandwidth goes up directly with M.

Simplex signals can be formed with natural signals, such as the correct phases of FSK signals. However, usually orthogonal signals (approximately equal in performance for $M > 10$) are formed with natural signals (MFSK). We frequently prefer M-ary signals constructed from binary chips, to exploit digital electronic technology. As noted, maximal-length sequences form simplex signals, and these serve as a suitable illustration of orthogonal-type block-coded signals.

Assume that the register–generator in Fig. 7 is to be the encoder. First the information bits are *blocked* off in groups of $k = 3$. Then, for each transmitted signal during an interval T, the $k = 3$ information bits are first shifted into the register. Next the feedback link is closed, and the register is shifted

---

[14] We choose shifting to the left as a convention, so that the input binary digits can be read left to right in the usual manner.

[15] The table for modulo-2 addition is:

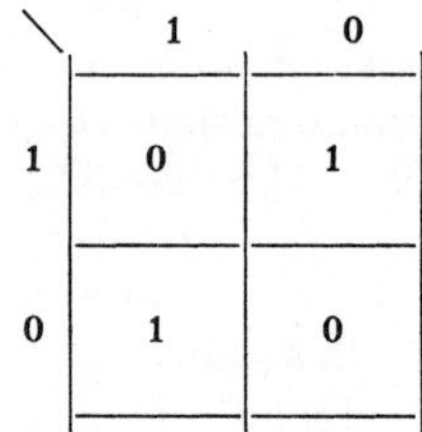

	1	0
1	0	1
0	1	0

This is also called an "exclusive-OR," and is identical to binary addition without the carry operation.

[16] The periodic autocorrelation refers to the correlation taken over a complete period, and where a periodic extension is assumed for the sequence that is "slid" past the other. The discrete correlation is given by:

$$\rho(\tau) = \frac{1}{n} \sum_{i=1}^{n} x_i x_{i+\tau}$$

where

$$x = +1, \ -1 (1 \rightarrow -1; \ 0 \rightarrow 1).$$

[17] The term simplex comes from mathematical geometry and refers to an N-dimensional hypercube centered on the origin.

TABLE I

BINARY-CODED $M = 8$ SIMPLEX SIGNALS

Info. Bits	Channel Bits (chips)
111	1110010
110	1100101
100	1001011
001	0010111
010	0101110
101	1011100
011	0111001
000	0000000

$2^k - 1 = n$ times. The resulting $M = 8$ signal set is given in Table I.

One can check and find that the correlation between any two such baseband signals is $-\frac{1}{7}$. In a similar manner, one could construct simplex signals for any M equal to powers of 2. Note that each code word (except the all-zero one) in Table I is an end-around shift of another code word. All maximal length sequences have this *cyclic* property. The feedback connections which, for a given k, produce the simplex (maximal length) signals can be found using the theory of irreducible polynomials[18] [22], [23].

Usually such baseband sequence signals are *modulated* to carrier frequency channels by having the (binary) phase shift key the carrier frequency using a balanced modulator.

One receiver possibility is (after heterodyning to baseband) to build M linear correlators correlating over the entire 7-chip-long interval T, using tapped analog delay lines. This would be fairly complex and expensive. Another possibility, using digital computation, is to correlate only over each chip, and use the quantized correlator output (from the soft decision) to form the quantized total correlation over T. Eight such computations would be required. Another possibility, explained in the next section, is to make hard decisions on each chip and perform the M correlations via shift-register logic. Some simplifications are possible since each signal is simply a shift of other code words.

One can also use orthogonal sequences in the same basic fashion. One way to form orthogonal sequences is to prefix all simplex sequences in a set with a common element (1). In general orthogonal sequences can be found using the mathematical technique of Hadamard matrices [24, p. 53]. Note that orthogonal sequences are not cyclic (shifts of a given word), and hence are less convenient both to generate and to decode.

The error performance of the (coherent) simplex signal is given in terms of the orthogonal signal performance [9, p. 243], since the latter is easier to calculate. The performance of noncoherent orthogonal signals nearly equals that of coherent simplex signals if M is large.

Two examples of simplex signals $M = 2$ (ideal PSK) and $M = 32$, [24], are shown plotted in Figs. 5 and 6. It can be seen (Fig. 6) that increasing M permits lower E_b/n_0 but provides less R/W. The plots in Fig. 6 have assumed that the nominal practical minimum τW equals 1, rather than the theoretical $\frac{1}{2}$. As noted, the actual channel bandwidth will differ somewhat due to practical factors.

2) Algebraic Error-Correcting Codes: Algebraic block codes can be used for either error *detecting* or error *correcting*. Here we are interested in error correcting; error detection is treated in Section IV.

Error-correcting codes use relatively distant signals and have the same objective as the orthogonal-type signals above. Error codes represent another set of alternatives for improving P_B in power-limited cases. In particular, the error codes are directed toward decoders which are simpler, and less expensive, than M parallel correlators. As a result, algebraic decoders can be used for large M values where maximum-likelihood decision would not be feasible.

An important feature of error codes is that the theory and structure of modulo-2 arithmetic[19] facilitates finding the error-correcting properties of the code and makes instrumentation of the codes practical.

The maximal length sequences previously used can also be viewed as algebraic error-correcting block codes. Now one views the code word as k information bits, and $n - k$ check bits. All such maximal length sequences are an example of an error code having the following important (code) properties:

Cyclic: Each code word is the end-around shift of another code word (except the all-zero one).

Systematic: The information bits are transmitted unchanged as part of the code word. (In Table I, the first 3 bits of each word are the information bits).

Parity check: Each check (parity) bit is a parity check on (the modulo-2 sum of) certain information bits. If one traces all the modulo-2 relations in the maximal length sequences, one finds that the check bits consist of all possible different parity checks on the information bits.

The cyclic codes, of which the maximal length sequences are an example, are the most important class of algebraic codes. Their advantage is, as we have seen, that simple feedback shift registers and modulo-2 adders can be used for both the encoding and the decoding operation.

First of all, if one uses algebraic decoding, one usually makes a hard decision on each received binary chip. One then uses the code relation between the received chips to decode or estimate the information bits. When using algebraic block codes and concepts, one can describe the performance in terms of "number of *channel-bit* (chip) errors correctable" in a given n-long block. Thus if the channel had burst errors of known maximum length, one could deterministically assure no errors by proper code selection (in practice this never happens). When the channel-bit errors are random, one can compute the *information-bit* error rate by computing the average number of times that channel-bit errors will exceed the correcting power of the code. Then we can plot P_B (of information bits) versus E_b/n_0 for the algebraic codes, as we previously did for various signals.

The correcting power of a code is determined by the distance between code words. The distance d (Hamming distance) between two binary code words is defined as the number of places in which they differ. If one adds two code words modulo-2, the distance equals the number of ones in the sum.

In general, an algebraic code can be guaranteed to correct a number of chip errors equal to the "greatest integer less than one-half the *distance* between two codewords." For the cyclic code (maximal length sequences) previously given, the distance is $M/2$. Thus one is guaranteed to correct $M/4 - 1$ errors

---

[18] The shift register with feedback taps can be described by a polynomial. Maximal sequences are associated with a subset of irreducible polynomials.

[19] Modulo-2 arithmetic is one example of an algebraic system. Algebraic systems are systems that satisfy certain rules or laws, similar to the laws that apply to our ordinary number system.

(one error in our $M = 8$ example). We will see below that $M/2$ errors can be *detected*. The information bit P_B for our $M = 8$ case would then be determined by first finding the P_E given by the number of times more than one chip error occurs in a 7-long block. This is then converted to P_B using the relation of (12).

Note that in the preceding section we could have spoken of correcting $M/4 - 1$ channel chips with the maximal length sequences if the correlators made hard decisions before comparing the reception to the alternatives. We recall that use of hard decisions results in about 2-dB loss in equivalent energy, compared to analog correlation. In effect, one would be choosing the correct sequence only if $(M/4 - 1)$ or less chip errors were made after the hard decisions. Such deterministic information about (information-bit) error preformance in terms of chip errors is useful if statistics of chip errors are known and reliable. In physical channels, however, one frequently cannot guarantee that the channel errors in a block will be limited to any given number.

One way to decode the cyclic maximal length sequences would be to search for some code word within $M/4 - 1$ of the reception. The $M = 8$ case of Table I is guaranteed to correct only one error; the $M = 16$ case (where $k = 4$) is guaranteed to correct 3 channel errors.

One decoding method, called *permutation decoding* [25] would simply enter the k information bits into a shift register and then use the feedback connection to generate the check bits. If the result agrees within $M/4 - 1$ chips, the k bits are assumed to be correct. If more than $M/4 - 1$ chip errors occur, another consecutive k received bits are treated as information bits and checked for agreement. This process will find the correct code word if the error criterion is not exceeded.

Another general decoding method (similar to the preceding) consists of using the *syndrome* with systematic codes. The syndrome is defined as the modulo-2 sum of the actual received check chips and the check chips generated at the receiver after entering the k information bits. If no chip errors occur, the syndrome will be all zero. If one isolated error occurs in the k information bits, a particular (designed) pattern of errors occur in the syndrome. If more than one error occurs (in the k bits), the syndrome output will be the (modulo-2) linear superposition of the syndrome patterns for each of the individual errors. Error-correcting coding is often described as associating information bit errors with unique syndrome patterns. In this way, logic circuits can be made to recognize the error patterns (up to $M/4 - 1$) and correct them. In general, error-correcting codes are most effective when the vast majority of error events result from a relatively few error patterns.

Another special class of codes whose syndromes permit simple logic correcting circuits are the codes that can be decoded with *majority logic*. Here the logic circuits need only determine the balance between ones and zeros in the sets of linear (modulo-2) combinations of received channel bits [26, p. 327], [27].

Although for a given code word (distance) selection, correlation detection would have a better P_B versus E_b/n_0 than algebraic decoding, it should be emphasized that algebraic decoding remains feasible for much longer code lengths (and hence k).

For illustration, the preformance of a (24, 12) block code with algebraic decoding [51] is shown in Figs. 5 and 6. The decoding here used an improved form of majority logic de-

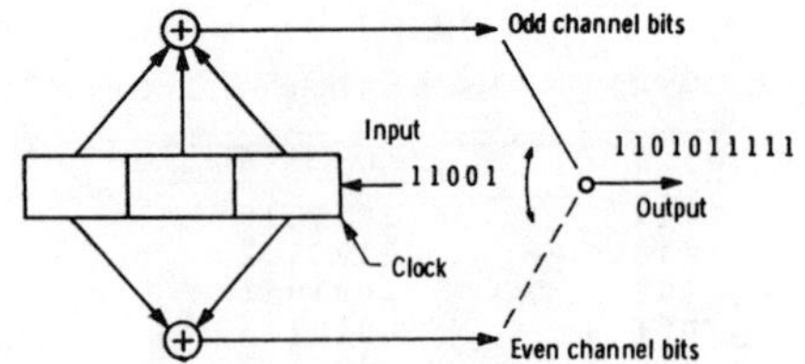

Fig. 8. Illustration of convolutional encoder ($K = 3$, $k = 1$, $v = 2$).

coding (called *a posteriori* probability decoding). A large range of performance can be achieved with block codes, depending on the code used and the complexity of the decoding.

Although a great variety of block-correcting codes [23] have been invented, along with decoding schemes, the algebraic block coding for error correction has not proven sufficiently attractive to be adopted in a large variety of applications. It appears that algebraic error-correcting codes have their main advantage where long codes are needed, since clearly for long codes the idea of implementing the maximum-likelihood comparison of all the code words is out of the question. In these cases the syndrome analysis, which is aimed at a particular circuit for each of the correctable error combinations, represents a shortcut. However, the difficulty is that on channels in which long codes are necessary to protect the errors the channels are not sufficiently well behaved to warrant the full-time use of such long codes. If one applies such a long code continuously, one suffers a constant inefficiency. On the other hand, when errors do occur they are relatively unpredictable as to their burst length and extent, so that in practice it is frequently not possible to exploit the particular advantages of the algebraic correcting code.

D. *Convolutional-Coded Signals*

A major alternative to the above *block* signaling is to construct channel symbols (and hence signals) by convolving the information bits with the impulse response of a shift-register encoder.[20] Convolutional encoding works on one (or k) information bits at a time and convolves the information bit stream with the impulse response of a shift register and attached modulo-2 adders. The convolutional process moves one (or k) bits at a time, whereas the previously mentioned block codes move n bits at a time.

Fig. 8 shows the basic arrangement for a convolutional encoder, which we will use as an illustrative example. The *span* or constraint length of the encoder is termed K, which specifies the number of k-bit shifts over which a given information bit can influence the encoder output. If k information bits are entered at a time, the number of shift-register stages required is kK, for a span of K. In Fig. 8 one information bit is shifted in at a time ($k = 1$). After each such information bit is entered, an output bit is taken sequentially from each of the v adders ($v = 2$ in Fig. 8). Thus one information bit (for $k = 1$) is placed into the register, and then $v = 2$ output channel bits (chips) are outputted to form a rate of one-half code. In general the codes are rate k/v; usually the codes are rate $\frac{1}{2}$ or rate $\frac{1}{3}$. Also, convolutional codes (similar to Fig. 8) are usually not systematic.

One can confirm that the output sequence is the convolu-

---

[20] The impulse response of a digital encoder is the sequence that occurs if one enters a single one, followed by all zeros. It is interesting to note here the analogy to a linear filter. The output of any linear filter is the convolution of the impulse response of that filter and the input waveform.

TABLE II
INPUT–OUTPUT SEQUENCE S_v FOR ENCODER OF FIG. 8

Input Information Bit	Sum of impulse responses
1	1 1 1 0 1 1 0 0 0 0 - - -
1	1 1 1 0 1 1 0 0 - - -
0	0 0 0 0 0 0 - - -
0	0 0 0 0 - - -
1	1 1 - - -
Output	1 1 0 1 0 1 1 1 1 1

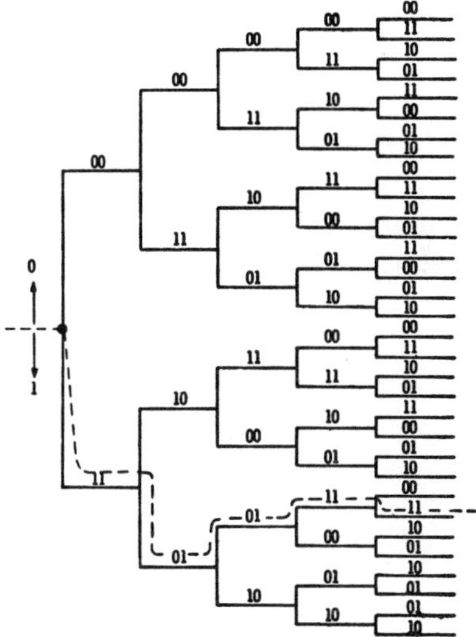

Fig. 9.   Code tree for convolutional coder of Fig. 8.

tion of the impulse response (sequence) and the input sequence by noting Table II, where an information bit sequence 1 1 0 0 1 . . . is assumed as input.

The action in a convolutional code can best be represented in basic form by showing the code tree. In fact, the convolutional code can be considered as a subclass of a more general tree code [23, p. 14]. Fig. 9 shows the tree diagram for the encoder shown in Fig. 8 and the path for the assumed input sequence. In any tree, the vertical members are called *nodes* and correspond to information bit time intervals. The horizontal members are *branches* and denote the output bits for that position in the tree. At each node (by convention) one goes down if the information bit is a one, and up if it is a zero. Then the total output chips generated are shown by successively writing the branch contents for a path determined by the input bits, as can be seen by tracing the dashed line in Fig. 9. Each unique code has a unique tree.

The receiver will receive a chip stream consisting of the generated stream perturbed by noise. It will be useful to describe the receiver action in terms of the generated code tree.

1) Viterbi Decoding Algorithm: Convolutional coding with the Viterbi decoding algorithm [28], [29] appears to be the most cost-effective forward error control technique for medium (P_B) accuracy requirements. It is relatively simple to implement and is effective because it essentially implements maximum-likelihood decoding.

We have noted that, in *block* coding, maximum likelihood concepts are applied independently from block to block (seg-

mented). With convolutional codes there is no block of chips that is independent of its neighbors. A maximum likelihood decoder then, ideally, would consider the entire received sequence and would compare it to all paths through the entire code tree. Ideally, then, the maximum likelihood decoder would choose that path in the tree whose code sequence "is the least distant" or most correlated with the received sequence. If the channel is binary (hard-decision) symmetric (binary symmetric channel, BSC)[21] the Hamming distance is the correct *metric*, and hence the "least distant" simply means the path having the minimum number of chips differing from the received chip stream. When a quantized (soft-decision) white Gaussian noise channel is assumed, the metric becomes the correlation (or inner product of signal vectors) of the tree path and the received chip stream.

Ideally one wishes to compute the metric for the *entire* information sequence and choose the path closest to the entire received chip sequence. However, since the encoder is only relating bits over a finite constraint length of K information bits (equivalent to Kv chips), it seems intuitive that one can sequentially compute the distance metric as one proceeds in decoding. Furthermore, the path length that must be considered simultaneously is dependent on: 1) the constraint length K and 2) the information bit path history prior to the current constraint length.

The Viterbi algorithm optimally exploits these factors and produces a decoding method which is essentially maximum likelihood and yet is relatively simple (and hence inexpensive). A major aspect is to fix the path memory at about four to five times the constraint length and then to compute the oldest bit on an arbitrary path each time it steps one level deeper into the tree.

As an illustration we will grossly pursue the steps in the metric computation for the encoder of Fig. 8. First, note that for $k = 1$, the encoder has $2^{K-1} = 4$ *states*. The state (or state of storage) of the encoder is the register contents of the two leftmost stages. (The state involves $K-1$ stages rather than the K stages since only the *past* digits must be stored; the modulo-2 adders can include the present incoming information without storage. However, it is convenient, from a clocking viewpoint, to use the K stages.) For the encoder of Fig. 8, then, there are four states and each of these states has two possible incoming paths.

The decoding operation can be summarized as follows [29]:

1) The distance or metric between the received signal and all paths entering state $i(i = 1, \cdots, 4)$ is calculated by adding the previous state metrics of predecessor states to the branch metrics of the upper and lower branch entering state i.

2) For each state, the largest of the two new path metrics is stored (called a survivor) as the new state metric for state i. Thus the new path history for state i is the path history of the state on the survivor path augmented (depending on which branch was largest) by a zero or one.

3) This add–compare–select operation is performed for all paths entering each of other states (three others in the case here).

4) The oldest bit on the path with the largest new path metric forms the decoder output at this state.

<hr>

[21] A BSC channel has symmetrical error probabilities: $P_{0,1} = P_{1,0} = p$, and $P_{1,1} = P_{0,0} = q$. $P_{x,y}$ is the discrete probability of deciding y when x is transmitted.

The length of the path calculated is usually four to five times the constraint length. It has theoretically and empirically been found [29, p. 839] that this provides negligible performance degradation from an optimum decoder having indefinite memory.

Fig. 5 shows the P_B versus E_b/n_0 behavior for a rate $\frac{1}{2}$ Viterbi decoding with $K=8$. The hard-decision case was taken from Fig. 7, and the 8-level quantization from [29, fig. 6]. The 8-level quantization provides almost 2 dB of improvement. As seen, the 8-level case results in an E_b/n_0 more than 5 dB lower than for uncoded PSK.

Since the storage and computation and complexity are proportional to $2K$ for any maximum length of the decoders, it becomes impractical to use Viterbi decoding for more than $K \geq 10$. For longer K sequential decoding becomes attractive. This is an algorithm that essentially searches the code tree in an attempt to find a path whose metric rises faster than some predetermined but variable threshold.

2) Sequential Decoding: A major alternative to Viterbi decoding for convolutional codes is sequential decoding. Here the decoder complexity is relatively insensitive to the code constraint length, and hence the constraint length can be made relatively large ($K=40$) to provide extremely low P_B.

The general approach with sequential decoding is to make temporary decisions, branch by branch, and calculate a metric between the received chip sequence and the sequentially evolving chosen path (in the code tree). Such a metric is computed over the constraint length, and tolerances on this metric are applied.

The decoder will proceed forward in step with incoming channel bits and hence will assume the current chosen path is correct, so long as the metric tolerance is not exceeded. If the metric tolerance is exceeded, the encoder is assumed to have departed from the actual code tree path, and it will: 1) temporarily send incoming bits into a buffer and 2) move backward one branch and then try an alternate forward path. If 2) fails to stay within the metric tolerance, the decoder will step back two branches and try alternate paths. In this way a forward and backward search is used to find the best path. We see that the computation required is a random variable, dependent on the current channel noise.

One major issue with sequential decoders is the buffer-overflow problem. During searches for alternate paths, the storage required is a function of the *computation rate* (speed of path searches). Hence a sequential decoder design is controlled by the buffer overflow probability, which is an inverse function of the buffer size and the machine speed.

The original sequential decoding algorithm, due to Wozencraft [30], was subjected to various modifications and generalizations [31, p. 157]. The Fano algorithm [19, p. 431], [31, p. 160] permitted simpler implementation and made the average computation rate insensitive to constraint length. Most of the applications are related to the Fano algorithm [32]–[34].

Although both sequential and Viterbi decoders can be operated at high speeds, the sequential type is usually limited to using hard decisions, since soft decisions would greatly increase the buffer length and computation cost. Figs. 5 and 6 show the simulated error performance for an example rate $\frac{1}{2}$ high-speed sequential decoder, when the input rate R is 9600 b/s [34, fig. 16]. This encoder can operate at 5 Mb/s, but then requires more E_b/n_0. Sequential decoders, like other decoders, can exhibit a large range of performance depending on the complexity used. As can be seen in Fig. 5, the sequential decoder typically has a steeper P_B curve (versus E_b/n_0) than the Viterbi case. This means that the advantage of sequential decoding increases for the very stringent P_B requirements.

On balance, it appears that most FEC applications will prefer Viterbi decoding. This results from the fact that it is usually better to perform only a cost-effective amount of FEC and then use feedback techniques (ARQ) with the modestly improved forward transmission (see Section IV) to obtain the very stringent P_B.

3) Feedback Decoding: Feedback decoding is a third type of decoding for convolutional codes. Although less capable than either Viterbi or sequential decoding, this decoding is the simplest to implement and was used exclusively in the first commercial convolutional decoders [25]. Frequently, systematic codes are used with the feedback decoder.

For certain codes having orthogonal properties, the syndrome is first formed by adding the received check bits to the receiver-generated check bits, as in the previous block decoding. Since the syndrome behavior reveals the particular errors present, errors can be corrected by altering the syndrome according to the diagnosed error pattern. In terms of the code tree, the feedback decoder makes a decision on the particular branch (information bit) in the code tree based on the received digits for a number of branches beyond this point.

Probably the simplest decision criterion to use in identifying error patterns in the syndrome is to use threshold decoding [36]. This uses a logical device whose output is active when, and only when, a weighted sum of binary variables exceeds some specific threshold. We should note that if all the weighting factors are equal, the threshold decoder is termed a *majority-logic decoder*, which we already encountered in block decoding.

In summary, feedback coders in general and threshold decoders in particular have extremely simple implementation techniques, but apply only to short codes correcting a few errors. They are uniquely useful, however, for channels where the errors are not random, but generally occur in bursts [28, p. 769].

This concludes the discussion of forward-error control methods.

IV. Feedback Transmission Alternatives

The major alternative to the forward-transmission-only techniques of the previous section is the additional use of a feedback channel (from data receiver to transmitter) to request repeats for sections of data found to contain an error. This ARQ operation makes use of forward-error *detection*, as opposed to the error correction of the previous section.

Although most of the communication literature for the last two decades has pursued development of forward-error coding, with the goal of approaching Shannon's channel capacity, most of the nonspace data communication systems use ARQ methods. FEC may of course be included in the forward transmission to improve the raw forward channel error rate. ARQ makes practical sense for permanent record sources where the data are already stored in convenient form. FEC, on the other hand, must be used for any real-time transmission such as is required by digitized video or conversational voice.

When ARQ is used, the stream of channel symbols is segmented into blocks and an error-detection code is applied. The algebraic error-detection codes are relatives of the block

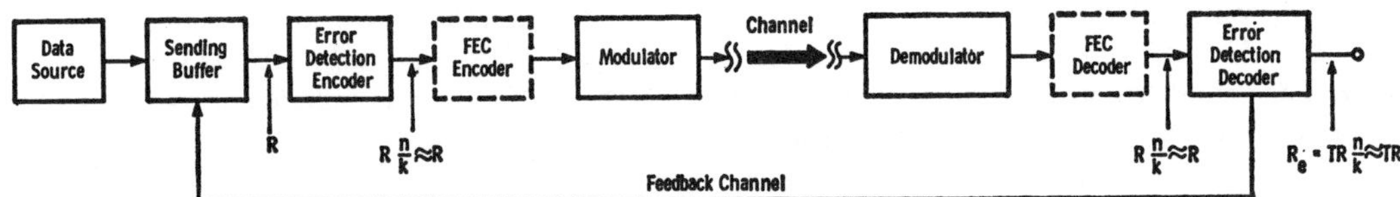

Fig. 10.   Elements of an ARQ system.

error-correction codes previously discussed. Since unpredictable retransmission will be required, the data at the transmitter must be temporarily stored in a buffer. Fig. 10 shows the major elements of an ARQ system. If an FEC code is used on the forward channel, it would appear in the position shown by the dashed boxes.

ARQ systems are either of the *stop-and-wait* type or the *continuous* type [37]. With stop-and-wait type the transmitter sends one block of encoded data at a time and waits for an acknowledge (ACK) signal from the receiver before proceeding. If a negative acknowledge (NACK) is received, the transmitter repeats the previous block. The advantage of stop-and-wait systems are: 1) they are easy to implement; 2) they require buffering of only one block; 3) a single channel can sometimes be used in a *turnaround* mode. In the turnaround mode a single channel serves alternately as a forward link and as a feedback link. Usually a settling time is involved in each turnaround, during which no symbols can be transmitted. Most of the past and existing ARQ systems use the stop-and-wait mode, for the above reasons.

In the continuous ARQ system the transmitter continuously sends the error-encoded blocked symbols. Here a simultaneous feedback channel is required. When an error is detected, the NACK signal received (at the transmitter) via the feedback channel causes the transmitter to repeat the block found in error and the subsequent blocks. The number of blocks retransmitted now depends on the channel round-trip delay, and is usually more than one. Although more efficient, this method is more complex to organize and requires more storage. In addition, a separate feedback channel is always needed. For channels with large delays (such as satellite circuits) the continuous mode is the only feasible ARQ approach [37].

We should note that in feedback parlance the ARQ techniques here use *decision feedback*. The receiver communicates a decision (about the accuracy) to the transmitter. *Information feedback* is another type of feedback; here the feedback channel is used to return information about how the signal was received. One simple method would be to have the feedback channel return the received signal, so that the transmitter can observe the received conditions; it would then adjust rate and power for optimized conditions. Another method [38], [39] being pursued theoretically uses an iterated procedure; for a given signal symbol the transmitter begins by transmitting the symbol (with low power). The receiver sends back its received value; the transmitter then sends an error signal. By iteration, the correct value can be sent using a minimum power. Note that information feedback attempts to send information correctly, similar to the preceding FEC methods, instead of taking a post-error point of view (as in ARQ). Ural and Haddad [40] describe a concept using both information and decision feedback, and review the previous work on feedback channels. Information feedback has thus far been pursued only theoretically, and we will not consider it further.

The algebraic error codes used for error detection with ARQ stem from the same basic ideas and theory as the error-correcting codes previously discussed, with some differences. Whereas the desirable correcting codes have large distance or separation between the n-dimensional code words, the desirable detecting-only codes have a high density in the n-dimensional code (signal) space. In addition, for a given code, it is far simpler to achieve error detection only than to correct the errors, and the detection error probablity is considerably lower than the correction probability [26, p. 360]. For this reason, it is possible to have large block (chip) lengths to achieve near-optimum conditions.

A variety of error-detecting codes have been used in past and existing systems. The simplest (and original) error-detecting code is odd- or even-parity code, where a single check bit is added to a block of information bits to cause an odd (or even) number of ones in the encoded block. This can only be applied to short blocks[22] since this approach fails if errors occur in pairs. The entire field of error codes can be considered a generalization of this basic code. Each code word of an (n, k) code has k information bits and $n-k$ check or redundancy bits. The efficiency of these codes is maximized when the number of errors detectable is maximized for a given $(n-k)$ number of check bits.

One variation is the *constant-ratio* code in which each code word has the same number of ones or zeros. A 7-long codeword set having 3 ones (3 out of 7) has long been used in ARQ radio telegraphy applications [52].

Another variation is to use two-dimensional block coding, where short blocks are arranged in a list. Then each short block is horizontally encoded, and the columns in the list of blocks are vertically encoded [55].

Various subclasses of the cyclic codes introduced before are more sophisticated error-detection codes. As noted, the cyclic codes are relatively easy to implement since identical feedback shift registers are used in both the encoder and decoder, and cyclic codes are especially suited for error-detection. For example, an $(n-k)$ BCH[23] code (a subset of cyclic codes) of arbitrary distance d is capable of detecting any combination of $d-1$, or fewer errors, or any burst length[24] $n-k$ or less [23, p. 230].

In the forward channel we noted that the important performance description is P_B versus E_b/n_0, and E_b/n_0 versus R/W. When ARQ is used, it is assumed that the final undetected errors are negligible for all practical purposes. Then the item of interest is the *effective rate* (R_e) after accounting for any acknowledge time and any retransmissions. The effective rate is lower than the information rate R out of the buffer (Fig. 10) due to: 1) the time spent in obtaining an acknowledge

---

[22] The alphanumeric characters, e.g., five information bits, are often encoded this way.

[23] Bose–Chaudhuri–Hocquenghem (BCH).

[24] An error burst of length x is a string of errors and nonerrors, the first and last of which are errors.

(for stop-and-wait systems); 2) the time spent in retransmission; and 3) the slight increase in chip rate over the information bit rate, n/k. The round-trip delay time is the major controlling factor. The effect of the n/k factor is slight, even though $n-k$ includes the check bits and bits for block synchronization and channel supervision.

The ratio of the R_e to the modem chip rate $(R \times (n/k))$ is an important figure of merit for ARQ systems, and is termed [41] the transmission efficiency or *throughput* (T):

$$T = \frac{R_e}{R\frac{n}{k}} = \frac{\text{average number of information bits accepted by receiver/unit time}}{\text{modem chip rate of forward channel}} . \tag{16}$$

To state the throughput in simple terms requires the simplifying assumptions that the feedback channel is error-free and that the error code detects all errors. Then the throughput [37] can be written:

$$T_{s-w} = \frac{k[1 - P(n)]}{n + D}$$

$$T_c = \frac{k[1 - P(n)]}{n + P(n) \cdot D} \tag{17}$$

where

T_{s-w} throughput for stop-and-wait

T_c throughput for continuous

$P(n)$ probability that a block of n chips contains one or more errors

n total number of channel bits per forward block

k number of information bits per block

D delay in chips during the time between the end of a block transmission and the end of reception of the corresponding acknowledgment message.

Since, for practical values of $P(n)$ in (17), the influence of $n > k$ is slight, we may note:

$$R_e = TR\frac{n}{k} \approx TR. \tag{18}$$

The effect of different block sizes for various delays is treated in [37] and [41]. For R's of 1200 to 4800 b/s, the throughputs for Bell System conditions range from 0.98 to 0.80.

In summary, ARQ techniques using feedback and instantaneous source rate control are practical and effective for a great many communications situations. They flexibly match the instantaneous channel conditions; when no errors occur in the continuous case the rate is approximately R; when errors occur, retransmission is used until success is achieved.

Burton and Sullivan [41] compare ARQ versus FEC for Bell System facilities. Since the actual error occurrence in the telephone network are neither random nor burstlike in nature, it has been found that the ARQ method remains preferable.

It is possible that certain applications will profit from using modest FEC on the forward channel and relying on ARQ to meet the stringent accuracy requirement. In any such arrangement, one must minimize any additional decoding delay required by the FEC since the ARQ throughput is sensitive to delay.

FEC methods are more competitive where transmission delay is unusually long (such as in satellite links) and where a feedback channel is difficult or expensive to achieve.

V. Channel-Sharing Alternatives

In concluding digital-communication alternatives, it seems desirable to summarize the alternatives for sharing a digital channel among multiple users. Such channel sharing is frequently desirable for economic reasons. Digital data sources (especially computers) often have conversational and delay parameters different from those of analog sources.

Circuit switching and *message switching* are the traditional methods for sharing the channels, in a total network consisting of many (dispersed) channels, in a common-carrier mode. In circuit switching one provides an actual (instantaneous) circuit or connection between the users. With message switching a data message is temporarily stored at the nodes between intervening channels to effect a *store-and-forward* operation.

The channel sharing alternatives summarized here refer to (nonnetwork) situations where a single channel is to be shared or *multiplexed* among cooperative users. There are two aspects to consider: 1) the physical basis for sharing or multiplexing the channel, and 2) the organization for gaining *access* to the shared facility. Here we will summarize the traditional linear filter, switched multiplexing, and cochannel sharing methods.

Three types of multiple access will be: 1) full-time access, 2) controlled access, and 3) random access. Full-time access refers to a permanent assignment of the channels to different users (subscribers or terminals). In controlled access the channels are instantaneously given to the "demanding" users, to form one type of demand–access capability. A requesting user may be temporarily blocked if all available channels are busy. Random access occurs when any subscriber can enter the channel at will without interfering with other subscribers.

A. Linear Filter Multiplexing (FDM)

Linear filter multiplexing refers to the conventional subdivision of a broad channel into subchannels via traditional frequency-division multiplexing (FDM). This method has often been used for analog-modulated signals, but is also used for digital signals [43]. The access for a given subscriber or terminal may be full-time or permanent, as for the entertainment broadcast stations. Satellite channels that use frequency-division multiple access (FDMA) are an example of controlled access. Here the available channels are assigned to subscribers or terminals on a demand basis. The access in FDM may also be user controlled (or contention), as in random time sharing or amateur- and citizens-band frequency channels.

B. Switching Multiplexing

Switching is frequently used to multiplex different digital sources onto a single channel. Message-switching multiplexing refers to devoting the entire channel to a single user for an entire sequence of information. Nonmessage-switching multiplexing arrangements regularly connect the subscriber for short intervals (for a single sample character, as in time-multiplexed PCM), and is called *time-division multiplexing (TDM)*.

1) Message-Switching Multiplexing: Message multiplexing requires that the message include address information of the destination (sometimes the source address is also attached). Computer-communication techniques have exploited message switching [42], [43] in the following ways:

a) Contention type: In a contention channel sharing any terminal user desiring to communicate waits until no traffic is being transmitted; he then attempts to sieze the line for his own use. This implements a user-controlled form of controlled access.

b) Polling type: With pooling, a channel master station sequentially queries (via terminal addresses) each "slave" terminal according to a predetermined sequence, asking if it has any messages to send. The master station authorizes only one terminal to transmit at a time, implementing controlled access.

c) Circuit switching: Whereas polling and contention permit the possibility of interference and eavesdropping by another user, this possibility is prevented in circuit switching by a connection, under control of a processor, between a user and the channel facility. Once connected, the user has full-time access.

2) Time-Division Multiplexing (TDM): In nonmessage switching, relatively short time intervals are repeatedly assigned to an individual user and this is universally called *time-division multiple access* (TDMA). The connection of users to time intervals may be controlled by supervisory techniques to form a switchable full-time access, or a demand-access control may be used. Various multiple-access methods for satellite links or trunks are compared in [44]. Although satellite links have heretofore used FDM, present plans call for TDMA.

C. Cochannel Sharing (Random Access)

Multiple users can also be multiplexed by permitting random access, where users can simultaneously (without control) use the channel. One possible random-access method is *spread spectrum multiple access* (SSMA), also called coding multiplexing. Here each user operates on the same nominal carrier frequency, but each signal has a different pseudorandom code (maximal length sequence). PSK is used to information modulate the pseudorandom code (this is termed direct sequence). Here the near orthogonality of the sequences can be used to separate the various users at the receivers, which receive the linear superposition of all the currently active subscribers. Each receiver consists of a correlation detector that synchronizes to its pseudonoise (or pseudorandom) code, demodulates the carrier, and recovers the digital information.

Experiments with SSMA for satellite links are reported in [45]. SSMA appears to utilize the required bandwidth expansion inefficiently [44].

The random-access discrete address (RADA) systems are another random-access implementation [53]. These systems are aimed at users with a low-duty factor, and use a frequency-hopping (FH) format rather than the direct sequence above. Each party, in a net situation, has a unique address or frequency-hop pattern. Under design conditions only the correct receiver will respond to a given address. The information modulation again uses PSK of the FH pattern.

A final cochannel sharing concept involves addressing types of information and sequentially repeating them on a common channel. Users can then selectively obtain desired information by matching the *function address*. This method is attractive when a relatively few messages are to be available to many users. A comprehensive motorist communication system using this organization has been proposed [47].

VI. Summary

Digital transmission is attractive both for digital sources and digitized analog sources. Source-encoding techniques permit transmission of reasonable quality speech at 10 kb/s and phone video at 1.544 Mb/s. The higher bit rate PCM will continue to be useful when many users are to be flexibly multiplexed on a switched network.

Improved FEC techniques can benefit the transmission of real-time digitized voice or video and data transmission over very long links. ARQ remains attractive for any data that can tolerate the necessary delay.

When considering the forward channel, it is desirable to distinguish power-limited from bandwidth-limited cases. Distance between signals, along with maximum-likelihood detection, are the central principles. While dense signals exploit the available power in bandwidth-limited cases, distant signals are needed in power-limited cases. Orthogonal-type signals and algebraic error-correcting codes are examples of block-coded distant signals. The distant convolutional-coded signals, along with Viterbi decoding, appear to be the most cost-effective FEC technique. Sequential decoding of convolutional codes should be considered in FEC cases with stringent P_B requirements. Further lowering of the required E_b/n_0 towards channel capacity may be unduly expensive.

The algebraic error-detecting codes used in ARQ form relatively dense forward-channel signals. Although convolutional coding has improved the performance of FEC, the inherently adaptive ARQ remains a very practical technique for most real terrestrial channels if the source can tolerate the required delay. There may be some cases where a modest FEC may be useful to improve the raw channel while retaining the ARQ operation.

FEC alone is attractive in severely power-limited cases, such as in the space communication links, and may be competitive with ARQ in very long (satellite-relay) links. Thus while the diligent research on FEC techniques since Shannon has probably brought us as close to channel capacity as we can afford to be, traditional ARQ techniques will remain a practical method of handling any data that can tolerate the retransmission format.

Finally, it seems safe to predict that the flexibility and many alternatives in digital communications, along with cost effectiveness of integrated-circuit technology, will lead to many new and useful communication services.

References

[1] C. E. Shannon, *The Mathematical Theory of Communication.* Urbana, Ill.: Univ. of Ill. Press, 1949.

[2] R. T. James and P. E. Muench, "AT&T facilities and services," *Proc. IEEE*, vol. 60, pp. 1342–1349, Nov. 1972.

[3] J. E. Cox, "Western Union digital services," *Proc. IEEE*, vol. 60, pp. 1350–1357, Nov. 1972.

[4] A. R. Worley, "The Datran system," *Proc. IEEE*, vol. 60, pp. 1357–1368, Nov. 1972.

[5] *Reference Data for Radio Engineers*, 5th ed., Howard W. Sams and Co., Inc., 1968, p. 38-6.

[6] R. M. Fano, *Transmission of Information.* Cambridge, Mass., and New York: MIT Press and Wiley, 1961.

[7] J. B. Millard, Bell Telephone Laboratories, Holmdel, N. J., private communication, Sept. 27, 1972.

[8] B. N. Oliver, J. R. Pierce, and C. E. Shannon, "The philosophy of PCM," *Proc. IRE*, vol. 36, pp. 1324–1331, Nov. 1948.

[9] A. J. Viterbi, *Principles of Coherent Communication.* New York: McGraw-Hill, 1966.

[10] J. W. Bayless, S. J. Campanella, and A. J. Goldberg, "A survey of

speech digitization techniques," in *Proc. 1972 Carnahan Conf. on Electronic Crime Countermeasures*, Bulletin UKY98, College of Engineering, University of Kentucky, Apr. 1972.

[11] J. B. Millard and H. I. Maunsell, "Digital encoding of the video signal," *Bell Syst. Tech. J.*, no. 2, p. 459, Feb. 1971.

[12] B. S. Atal and S. L. Hanauer, "Speech analysis and synthesis by linear prediction of the speech wave," *J. Acoust. Soc. Amer.*, vol. 50, pp. 637–655, 1971.

[13] M. C. Chow, "Variable length redundancy removal coders for differentially coded video telephone signals," *IEEE Trans. Commun. Technol.*, vol. COM-19, pp. 923–926, Dec. 1971.

[14] A. Habibi, "Comparison of nth order DPCM encoder with linear transformations and block quantization techniques," *IEEE Trans. Commun. Technol.*, vol. COM-19, pp. 948–956, Dec. 1971.

[15] J. B. O'Neal, Jr., "A bound on signal-to-quantizing noise ratios for digital encoding systems," *Proc. IEEE*, vol. 55, pp. 287–292, Mar. 1967.

[16] N. S. Jayant, "Adaptive delta modulation with a one-bit memory," *Bell Syst. Tech. J.*, no. 3, p. 321, Mar. 1970.

[17] D. J. Connor, R. C. Brainard, and J. O. Limb, "Intraframe coding for picture transmission," *Proc. IEEE*, vol. 60, pp. 779–791, July 1972.

[18] B. G. Haskell, F. W. Mounts, and J. C. Candy, "Interframe coding of videotelephone pictures," *Proc. IEEE*, vol. 60, pp. 792–800, July 1972.

[19] J. M. Wozencraft and I. M. Jacobs, *Principles of Communication Engineering*. New York: Wiley, 1965.

[20] V. A. Kotel'nikov, *The Theory of Optimum Noise Immunity* (translated by R. A. Silverman). New York: McGraw-Hill, 1959.

[21] M. P. Ristenbatt, "Notes in a digital communications systems," presented at Engineering Summer Conf. given at The University of Michigan, Ann Arbor, Mich., July 17–21, 1972.

[22] T. G. Birdsall and M. P. Ristenbatt, "Introduction to linear shift register generated sequences," Cooley Electronics Lab., Univ. of Michigan, Ann Arbor, Tech. Rep. 90, Oct. 1958.

[23] W. W. Peterson and E. J. Weldon, Jr., *Error Correcting Codes*, 2nd ed. Cambridge, Mass.: MIT Press, 1972.

[24] S. W. Golomb, *Digital Communications with Space Applications*. Englewood Cliffs, N. J.: Prentice-Hall, 1964.

[25] G. D. Forney, Jr., "Coding and its application in space communications," *IEEE Spectrum*, vol. 7, pp. 47–58, June 1970.

[26] R. W. Lucky, J. Salz, and E. J. Weldon, Jr., *Principles of Data Communication*. New York: McGraw-Hill, 1968.

[27] R. T. Chien, "Block-coding techniques for reliable data transmission," *IEEE Trans. Commun. Technol.*, vol. COM-19, pp. 743–751, Oct. 1971.

[28] A. J. Viterbi, "Convolutional codes and their performance in communication systems," *IEEE Trans. Commun. Technol.*, vol. COM-19, pp. 751–772, Oct. 1971.

[29] J. A. Heller and I. M. Jacobs, "Viterbi decoding for satellite and space communication," *IEEE Trans. Commun. Technol.*, vol. COM-19, pp. 835–848, Oct. 1971.

[30] J. M. Wozencraft and B. Reiffen, *Sequential Decoding*. Cambridge, Mass., and New York: MIT Press and Wiley, 1961.

[31] J. E. Savage, "Progress in sequential decoding," in *Advances in Communication Systems*. New York: Academic Press, 1968, p. 149.

[32] K. L. Jordan, Jr., "The performance of sequential decoding in conjunction with efficient modulation," *IEEE Trans. Commun. Technol.*, vol. COM-14, pp. 283–297, June 1966.

[33] J. W. Layland and W. A. Lushbaugh, "A flexible high-speed sequential decoder for deep space channels," *IEEE Trans. Commun. Technol.*, vol. COM-19, pp. 813–820, Oct. 1971.

[34] G. D. Forney, Jr., and E. K. Bower, "A high-speed sequential decoder: Prototype design and test," *IEEE Trans. Commun. Technol.*, vol. COM-19, pp. 821–855, Oct. 1971.

[35] J. L. Massey, "Advances in threshold decoding," in *Advances in Communication Systems*. New York: Academic Press, 1968, p. 91.

[36] J. L. Massey, *Threshold Decoding*. Cambridge, Mass.: MIT Press, 1963.

[37] W. J. McGruther, "Thruput of high speed data transmission systems using block retransmission error control schemes over voice bandwidth channels," in *Proc. Int. Conf. Communications* (Catalog No. 72-CHO-622-1-COM) (Philadelphia, Pa., June 19–21, 1972), pp. 15–19.

[38] T. M. Ebert, "Feedback in data transmission," in *Proc. 1970 Int. Telemetering Conf.*, vol. VI (Los Angeles, Calif., Oct. 1970), pp. 13–15.

[39] J. P. M. Schalwijk and T. Kailath, "A coding scheme for additive noise channels with feedback—Part I: No bandwidth constraint," *IEEE Trans. Inform. Theory*, vol. IT-12, pp. 172–182, Apr. 1966.

[40] A. T. Ural and A. H. Haddad, "A binary sequential communication scheme with information feedback," *IEEE Trans. Commun.* (Concise Papers), vol. COM-20, pt. 1, pp. 423–429, June 1972.

[41] H. O. Burton and D. D. Sullivan, "Errors and error control," *Proc. IEEE*, vol. 60, pp. 1293–1300, Nov. 1972.

[42] R. W. Sanders, "Multiplexing in the real world today," in *Proc. 1970 Int. Conf. on Communications* (San Francisco, Calif., June 8–10, 1970).

[43] D. R. Doll, "Multiplexing and concentration," *Proc. IEEE*, vol. 60, Nov. 1972, pp. 1313–1321.

[44] D. N. McGregor, M. E. Jones, and P. E. O'Neill, Jr., "Comparison of several demand assignment multiple access/modulation techniques for satellite communications," in *Proc. 1971 Int. Conf. on Communications* (Catalog No. 71C28-COM) (Montreal, Canada, June 14, 1971).

[45] N. G. Davies *et al.*, "A variable-rate capability in a modem for spread-spectrum multiple access," in *Proc. 1969 IEEE Int. Conf. on Communications* (Boulder, Colo., June 9–11, 1969), p. 38-1.

[46] P. Hersch, "Data communications," *IEEE Spectrum*, vol. 8, pp. 47–60, Feb. 1971.

[47] M. P. Ristenbatt, "A comprehensive motorist communication system," in *Proc. Int. Conf. on Communications* (Catalog No. 72-CHO-622-1-COM) (Philadelphia, Pa., June 19–21, 1972).

[48] S. L. Bernstein, "TATS—A case study in digital communication system design," in *Digital Communication Systems*, The University of Michigan Engineering Summer Conf., Ann Arbor, July 17–21, 1972.

[49] R. Gold, "Study of correlation properties of binary sequences," Magnavox Research Lab., Torrance, Calif., Tech. Rep. AFAL-TR-66-234, Aug. 1966.

[50] M. P. Ristenbatt, "Pseudo-random binary coded waveforms," in *Modern Radar Analysis*, R. S. Berkowitz, Ed. New York: Wiley, 1965, ch. 4.

[51] D. R. Lumb, "A study of codes for deep space telemetry," presented at 1969 Telemetry Conf., Los Angeles, Calif.

[52] M. Nesenbergs, "Comparison of the 3-out-of-7 ARQ with Bose-Chaudhuri-Hocquenghem coding systems," *IEEE Trans. Commun. Sys.*, vol. CS-11, pp. 202–212, June 1963.

[53] A. M. McCalmont, "Multiple access discrete address concepts for vehicular communications," *IEEE Trans. Veh. Commun.*, vol. VC-14, pp. 39–42, Mar. 1965.

[54] J. R. Davey, "Modems," *Proc. IEEE*, vol. 60, pp. 1284–1292, Nov. 1972.

[55] D. R. Doll, "Error control procedures in data communications," in *Digital Communication Systems*, The University of Michigan Engineering Summer Conf., Ann Arbor, July 17–21, 1972.

[56] T. Berger, *Rate Distortion Theory: A Mathematical Basis for Data Compression*. Englewood Cliffs, N. J.: Prentice-Hall, 1971.

[57] R. G. Gallager, "Source coding with a fidelity criterion," in *Information Theory and Reliable Communication*. New York: Wiley, 1968, ch. 9.

[58] J. B. O'Neal, "Bounds on subjective performance measures for source encoding systems," *IEEE Trans. Inform. Theory*, vol. IT-17, pp. 224–231, May 1971.

Unified Analysis of a Class of Digital Systems in Additive Noise and Interference

RUSSELL FANG AND OSAMU SHIMBO

Abstract—A unified approach is presented for evaluation of the error probabilities of a class of digital communications systems in additive noise and interference. This class of systems includes coherent systems such as M-ary amplitude-shift keying (ASK), M-ary phase-shift keying (PSK), and M-ary amplitude-and-phase keying (APK); it also includes differential coherent systems such as binary differential PSK (DPSK). The noise is not necessarily Gaussian. The interference can be intersymbol interference, co-channel interference, adjacent-channel interference, any of their linear combinations, or intermodulation products at the output of some nonlinear device.

This approach essentially expands the characteristic function of the interferences into a power series so that the desired error probability can be evaluated as the sum of terms representing perturbations around the error probability due to additive noise alone. Bounds on three kinds of truncation errors, which are simple and applicable to all aforementioned digital systems, are obtained. As a result, any desired accuracy in the evaluation of the error probabilities can be achieved with this approach. In the special case in which the noise is Gaussian, explicit bounds on truncation errors are also obtained.

Examples are given to illustrate how the unified analysis can be applied to evaluate the error probabilities of various digital systems. More specifically, the combined effects of Gaussian noise, intersymbol interference, and co-channel interference on the error performance of M-ary coherent PSK and APK (MCPSK and MCAPK) systems are computed. The probability of error of a binary DPSK (BDPSK) system in the presence of Gaussian noise and intersymbol interference is analyzed. The intermodulation products at the output of a hard-limiter are also determined.

I. INTRODUCTION

IN NEW digital communication systems design, a constant problem is that of determining the most effective way of using the available radiation power and channel bandwidth for a given data rate and an acceptable system performance, e.g., certain prescribed error probability. If the channel is Gaussian, there are many theoretical results that can serve as design guidelines. However, Gaussian channels rarely exist in practical applications, even in modern satellite communications. (Of course, because of severe power limitations on some older generation satellites, Gaussian noise was once a primary concern. Nevertheless, since high-power satellites are now readily available, the limitation on channel bandwidth due to hardware or legal constraints becomes increasingly important for future satellite communications. Consequently, other dis-

Paper approved by the Associate Editor for Communication Theory of the IEEE Communications Society for publication after presentation at the IEEE International Symposium on Information Theory, Pacific Grove, Calif., January 31–February 4, 1972. Manuscript received February 12, 1973; revised May 22, 1973. This work was performed at COMSAT Laboratories under the joint sponsorship of the Communications Satellite Corporation and the International Telecommunications Consortium (INTELSAT). Views expressed are not necessarily those of INTELSAT.

The authors are with COMSAT Laboratories, Clarksburg, Md. 20734.

turbances such as intersymbol interference and adjacent-channel interference are just as important as Gaussian noise.) Thus, all major disturbances as well as additive noise must be considered in the design. Since in many situations the system design margin is quite costly, an accurate performance analysis of various candidate digital communication systems to be operated over such non-Gaussian channels is necessary so that the most effective system can be chosen and the design margin can be minimized.

In this paper, symbol error probability is chosen as the criterion for evaluating system performance.

This paper is organized into two major parts: Sections II and III form the theoretical part while Sections IV–VII form the application part. Those readers interested only in the applications part may go through Sections II and III quickly, and directly get into the applications.

Section II poses the problem of calculating the probability of an n-dimensional random vector, defined as $z = S + N + \sum_{k=1}^{L} a_k$, falling into some region D in R^n,[1] i.e., $\Pr[z \in D]$, where S is an n-dimensional known signal vector, while N and a_k are n-dimensional random vectors representing noise and interference, respectively. This additive noise is not necessarily Gaussian, but it is assumed to have a probability density function in R^n satisfying certain minor regularity conditions. The interferences $\{a_k\}$ characterize disturbances such as those caused by intersymbol interference, co-channel interference, adjacent-channel interference, or any of their combinations, intermodulation products at the output of some nonlinear device. For simplicity, it is assumed that $N, a_1, a_2, \cdots,$ and a_L are statistically independent, and that $\{a_k\}$ are ordered according to their expected Euclidian norms, i.e., $E[\|a_{k+1}\|] \leqslant E[\|a_k\|]$ for $k = 1, 2, \cdots, L$. The number of interferences L may be infinite, for example, when $\{a_k\}$ represents the intersymbol interference due to all past symbols in the message sequence.

In general, since the probability density function of interference may not exist [1] or is just too difficult to obtain [2], [3], the probability density function of z is practically impossible to find. Therefore, the problem of evaluating $\Pr[z \in D]$ is by no means simple. However, since for almost all practical cases the characteristic functions of $\{a_k\}$ do exist and can be found without much difficulty, $\Pr[z \in D]$ can be expressed in terms of the characteristic functions of N and $\{a_k\}$. The characteristic function of the interference $\sum_{k=1}^{L} a_k$ is then expanded into a power series in the hope that $\Pr[z \in D]$ can be evaluated as the sum of terms representing perturba-

<hr>

[1] The meaning of the dimensionality n can be clearly seen from the examples in the applications part of this paper.

Reprinted from *IEEE Trans. Commun.*, vol. COM-21, pp. 1075–1091, Oct. 1973.

tions around $\Pr[z \in D \mid \Sigma_{k=1}^{L} a_k = 0]$. In what follows it is shown that this can be effectively done under some minor assumptions on $\{a_k\}$ and N.

In Section III, three kinds of truncation errors are present. The first kind is caused by integration over a finite region instead of over D, which is sometimes infinite. The second kind is caused by considering some finite number less than the total number of interferences. The third is caused by the termination of the power series. With these truncation errors it is shown that the power series expansion method presented here can achieve any desired accuracy in the evaluation of $\Pr[z \in D]$. Explicit bounds on these truncation errors are also obtained for the most usual case in which the additive noise is Gaussian.

Sections IV–VII then demonstrate how the results in the theoretical part, Sections II and III, can be applied to the problems of calculating the probabilities of error of a number of digital communications systems subject to both additive noise and interference. These systems may employ modulation techniques such as M-ary coherent amplitude-shift keying (MCASK), M-ary coherent phase-shift keying (MCPSK), and M-ary coherent amplitude-and-phase-keying (MCAPK) [7]–[9].

For systems subjected to Gaussian noise and intersymbol interference only, several methods have been previously proposed for calculating the error probabilities of digital systems such as MCASK [10]–[15], MCPSK [16]–[18], and binary differentially-coherent PSK (BDPSK) [19]–[21]. Methods for computing the probability of error for an MCPSK system with only Gaussian noise and co-channel interference have also been recently suggested [22]–[28]. Nevertheless, all of these methods are somewhat limited in their scope of applicability, whereas the method presented here is most general and very powerful, especially when numerical results are desired, as in most practical situations.

For illustrative purposes, the combined effects of Gaussian noise, intersymbol interference, and co-channel interference on the error performance of MCPSK and MCAPK systems will be analyzed in Section IV and Section V, respectively. The error probability of a binary differential PSK (DPSK) system in the presence of Gaussian noise and intersymbol interference will be determined in Section VI. Then, the intermodulation products due to hard-limiting multiple signals and noise are treated in Section VII. Finally, Section VIII gives the conclusion for this paper and includes some future research problems.

II. PROBLEM STATEMENT AND METHOD OF SOLUTION

Problem Statement

Let

$$z \equiv S + N + \sum_{k=1}^{L} a_k, \tag{1}$$

where S, N, and a_k are n-dimensional real vectors. S is deterministic, while N and a_k are random with associated characteristic functions denoted by ϕ_0 and ϕ_k, respectively. That is,

$$\phi_0(u) = \int \exp\left[jx \cdot u\right] dF_0(x) \tag{2}$$

$$\phi_k(u) = \int \exp\left[jx \cdot u\right] dF_k(x), \quad \text{for } k = 1, 2, \cdots, L, \tag{3}$$

where $x \cdot u$ denotes the scalar product of two vectors x and u, and F_0 and F_k are probability distribution functions of N and a_k, respectively. Assume that ϕ_0 is absolutely integrable over R^n so that the probability density function of N, denoted by p_N, exists and let

$$\phi_0(u) = \int \exp\left[jx \cdot u\right] p_N(x)\, dx. \tag{4}$$

Also assume the following.

Assumption 1) $N, a_1, a_2, \cdots,$ and a_L are zero-mean and statistically independent.

Assumption 2) For simplicity, ϕ_k for $k = 1, 2, \cdots, L$, are real; i.e., all odd moments of a_k are zero.

Assumption 3) $\Sigma_{k=1}^{L} E[\|a_k\|^2] = P < \infty$; when $L = \infty$, this condition guarantees that $\Pi_{k=1}^{\infty} \phi_k$ exists [4].

Assumption 4) $\sup_k \|a_k\| = A_0 < \infty$.

Assumption 5) $\int \|u\|^{2l} |\phi_0(u)|\, du < \infty$ for $l = 1, 2, \cdots$.

Then, the problem is to effectively evaluate $\Pr[z \in D]$ for some $D \subset R^n$ within any desired accuracy.

Method of Solution

Observe that the probability density function p_z of z exists, since the probability density function p_N of N and the probability distribution function G_L of $\Sigma_{k=1}^{L} a_k$ exist. [If $L = \infty$, then Assumption 3 implies the existence of G_∞]. Hence,

$$\Pr[z \in D] = E\left[\int_D p_z\left(x \,\middle|\, \sum_{k=1}^{L} a_k = t\right) dx\right]$$

$$= \iint_D p_N(x - S - t)\, dx\, dG_L(t)$$

$$= \iint_D p_N(x - S - t)\, dG_L(t)\, dx, \tag{5}$$

where $E[\cdot]$ denotes the ensemble average over $\Sigma_{k=1}^{L} a_k$.

Let the characteristic function of $\Sigma_{k=1}^{K} a_k$ be represented by

$$\Phi_K(u) = \int \exp\left(jt \cdot u\right) dG_K(t), \tag{6}$$

where G_K is the probability distribution function of $\Sigma_{k=1}^{K} a_k$. Then, because $\{a_k\}$ are statistically independent,

$$\Phi_K(u) = \prod_{k=1}^{K} \phi_k(u). \tag{7}$$

Now, application of Bochner's theorem [6] to (5) results in

$$\Pr[z \in D] = (2\pi)^{-n} \int_D dx \int_{R^n} \phi_0(u)\, \Phi_L(u)$$

$$\cdot \exp\left[-j(x - S) \cdot u\right] du. \tag{8}$$

Computation of (8) is extremely difficult, if not impossible, especially when $L = \infty$. A power series expansion method is proposed to evaluate this integral as follows. Construct an n-dimensional cube D_s of length a on each side, centered at the vector S, as illustrated in Fig. 1. Let

$$P(D \cap D_s; K) \equiv (2\pi)^{-n} \int_{D \cap D_s} dx \int_{R^n} du\, \phi_0(u)\, \Phi_K(u)$$
$$\cdot \exp\left[-j(x - S) \cdot u\right] \quad (9)$$

be the probability that the random variable $z_K \equiv S + N + \sum_{k=1}^{K} a_k$ falls into the set $D \cap D_s$. Also, define

$$P(D \cap D_s; K; R) \equiv (2\pi)^{-n} \int_{D \cap D_s} dx \int_{R^n} du\, \phi_0(u)$$
$$\cdot \sum_{l=0}^{R-1} \frac{(-1)^l}{(2l)!} \left[\left(\frac{\partial}{\partial S} \cdot \frac{\partial}{\partial t}\right)^{2l} \Phi_K(t)\right]_{t=0} \exp\left[-j(x - S) \cdot u\right], \quad (10)$$

where

$$\frac{\partial}{\partial y} \equiv \left(\frac{\partial}{\partial y_1}, \frac{\partial}{\partial y_2}, \cdots, \frac{\partial}{\partial y_n}\right)$$

and $\{y_k\}$ are the n components of the vector y. Then, by triangular inequality

$$|\Pr[z \in D] - P(D \cap D_s; K; R)| \leqslant |\Pr[z \in D] - P(D \cap D_s; K)|$$
$$+ |P(D \cap D_s; K) - P(D \cap D_s; K; R)|$$
$$\leqslant P(D \cap D_s^c; L) + |P(D \cap D_s; L) - P(D \cap D_s; K)|$$
$$+ |P(D \cap D_s; K) - P(D \cap D_s; K; R)|, \quad (11)$$

where D_s^c is the complement of D_s.

Since, as will be shown in (19), (26), and (33), respectively, for any $\epsilon > 0$

$$P(D \cap D_s^c; L) \leqslant P(D_s^c; L) \equiv E_0(a) < \epsilon/3 \quad (12a)$$

for sufficiently large a;

$$|P(D \cap D_s; L) - P(D \cap D_s; K)| < \epsilon/3 \quad (12b)$$

for sufficiently large K; and

$$|P(D \cap D_s; K) - P(D \cap D_s; K; R)| < \epsilon/3 \quad (12c)$$

for sufficiently large R, it follows by substituting (12a-c) into (11) that

$$|\Pr[z \in D] - P(D \cap D_s; K; R)| < \epsilon. \quad (13)$$

Thus, $P(D \cap D_s; K; R)$ can be used to approximate $\Pr[z \in D]$ within any desired accuracy. The question is simply how to compute $P(D \cap D_s; K; R)$ effectively. Before proceeding to the problem of computing $P(D \cap D_s; K; R)$, the following observations should be made.

The first term on the right-hand side (RHS) of (11) represents the truncation error caused by the confinement of D in a cube D_s centered at the vector S. The second term on the RHS of (11) represents the truncation error due to the approximation of $\Phi_L(u)$ by $\Phi_K(u)$ with $K \leqslant L$. This approximation is sometimes necessary, because $P(D \cap D_s; L)$ may

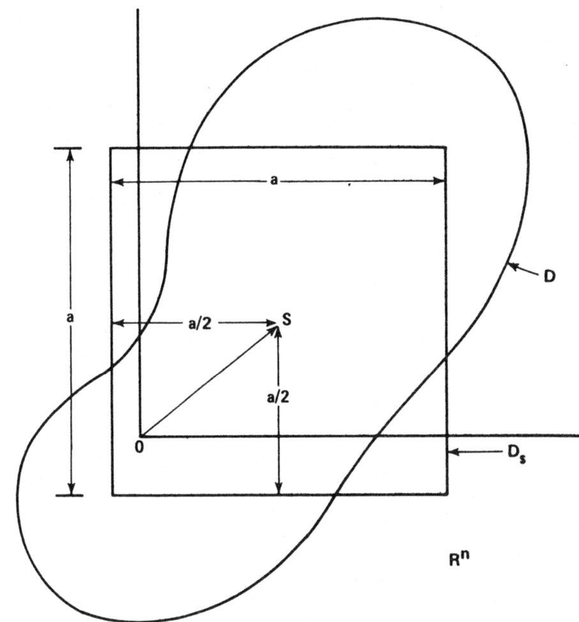

Fig. 1. Definition of D and D_S.

still be difficult to evaluate, especially if $L = \infty$. The third term on the RHS of (11) represents the error resultant from approximating $P(D \cap D_s; K)$ by $P(D \cap D_s; K; R)$. Since, from Lemma 1 of Appendix A, $\Phi_K(u)$ can be expanded into the power series

$$\Phi_K(u) = \sum_{l=0}^{\infty} \frac{(-1)^l}{(2l)!} \left[\left(u \cdot \frac{\partial}{\partial t}\right)^{2l} \Phi_K(t)\right]_{t=0}, \quad (14)$$

it follows by substituting (14) into (9) that

$$P(D \cap D_s; K) = (2\pi)^{-n} \int_{D \cap D_s} dx \int_{R^n} du\, \phi_0(u)$$
$$\cdot \sum_{l=0}^{\infty} \frac{(-1)^l}{(2l)!} \left[\left(\frac{\partial}{\partial S} \cdot \frac{\partial}{\partial t}\right)^{2l} \Phi_K(t)\right]_{t=0} \exp\left[-j(x - S) \cdot u\right]. \quad (15)$$

Consequently, a comparison of (10) and (15) reveals that the third term on the RHS of (11) is just the truncation error caused by terminating the power series in (14) at the $(R - 1)$th term.

Note that the individual integrals in (10) do exist under Assumptions 4 and 5; according to Lemma 2 in Appendix B the summation and the integration in (10) can be interchanged. Hence,

$$P(D \cap D_s; K; R) \equiv \sum_{l=0}^{R-1} \frac{(-1)^l}{(2l)!} (2\pi)^{-n} \int_{D \cap D_s} dx \int_{R^n} du\, \phi_0(u)$$
$$\cdot \left[\left(\frac{\partial}{\partial S} \cdot \frac{\partial}{\partial t}\right)^{2l} \Phi_K(t)\right]_{t=0} \exp\left[-j(x - S) \cdot u\right]$$
$$= \sum_{l=0}^{R-1} \frac{(-1)^l}{(2l)!} \left[\left(\frac{\partial}{\partial S} \cdot \frac{\partial}{\partial t}\right)^{2l} \Phi_K(t)\right]_{t=0}$$
$$\cdot \int_{D \cap D_s} p_N(x - S)\, dx, \quad (16)$$

where

$$\int_{D \cap D_s} p_N(x - S) \equiv (2\pi)^{-n} \int_{D \cap D_s} dx \int_{R^n} du\, \phi_0(u)$$
$$\cdot \exp\left[-j(x - S)\cdot u\right] \quad (17)$$

is the probability that $z \in D \cap D_s$ given $\Sigma_{k=1}^L a_k = 0$.

Therefore, if $\int_{D \cap D_s} p_N(x - S)\, dx$ and its derivatives with respect to S can be found by some means and if the coefficients of the power series expansion of $\Phi_K(u)$ can be found, e.g., through the recurrence method, then $\Pr[z \in D]$ can be evaluated within any desired accuracy by using (16), (17).

III. Truncation Error Bounds

There are three kinds of truncation errors. The first kind is due to the confinement of D in a cube D_s centered at the vector S. Since p_z is a probability density function that is by definition nonnegative and integrable over the whole R^n, it follows that the first kind of error bound as defined in (12a), $E_0(a)$, is a monotonically decreasing function of the volume of D_s and, consequently, a monotonically decreasing function of a. Let p_{z_i} be the ith marginal density of z. Then, from the definition of $P(D_s^c; L)$,

$$E_0(a) \equiv P(D_s^c; L)$$
$$= \int_{D_s^c} p_z(x - S)\, dx$$
$$\leqslant \sum_{i=1}^n \left(\int_{-\infty}^{-a/2} p_{z_i}(x_i)\, dx_i + \int_{a/2}^{\infty} p_{z_i}(x_i)\, dx_i \right). \quad (18)$$

Thus, for any $\epsilon > 0$,

$$E_0(a) < \epsilon/3 \quad (19)$$

for sufficiently large value of a.

The second kind of truncation error, which is caused by considering only K instead of L interference terms, can be bounded from above as follows:

$$|P(D \cap D_s; L) - P(D \cap D_s; K)|$$
$$= \left| (2\pi)^{-n} \int_{D \cap D_s} dx \int_{R^n} du\, \phi_0(u)\, \Phi_K(u) \right.$$
$$\left. \cdot \left[1 - \prod_{k=K+1}^{L} \phi_k(u)\right] \exp\left[-j(x - S)\cdot u\right] \right|$$
$$\leqslant A(2\pi)^{-n} \int_{R^n} |\phi_0(u)| \left| 1 - \prod_{k=K+1}^{L} \phi_k(u)\right| du, \quad (20)$$

where

$$A \equiv \int_{D \cap D_s} dx \leqslant \int_{D_s} dx = a^n < \infty. \quad (21)$$

However,

$$\prod_{k=K+1}^{L} \phi_k(u) \equiv E\left[\exp ju \cdot \left(\sum_{k=K+1}^{L} a_k\right)\right]. \quad (22)$$

Since the exponential term can be expanded into a Taylor series with an integral reminder [5, p. 309],

$$\exp\left(ju \cdot \sum_{k=1}^{L} a_k\right) = 1 + ju \cdot \sum_{k=K+1}^{L} a_k$$
$$- \frac{1}{2}\left(u \cdot \sum_{k=K+1}^{L} a_k\right)^2 \int_0^1 (1 - t) \exp\left(jtu \cdot \sum_{k=K+1}^{L} a_k\right) dt, \quad (23)$$

and since a_k has zero mean for all k, we can combine (22) and (23) to yield

$$\left| 1 - \prod_{k=K+1}^{L} \phi_k(u) \right|$$
$$= \frac{1}{2}\left| E\left[\int_0^1 \left(u \cdot \sum_{k=K+1}^{L} a_k\right)^2 \exp\left(jtu \cdot \sum_{k=K+1}^{L} a_k\right)(1 - t)\, dt\right] \right|$$
$$\leqslant \frac{1}{2} E\left[\left(u \cdot \sum_{k=K+1}^{L} a_k\right)^2\right] \int_0^1 (1 - t)\, dt$$
$$\leqslant \frac{1}{4} \|u\|^2 \sum_{k=K+1}^{L} E[\|a_k\|^2] < \infty. \quad (24)$$

Substituting (24) into (20) the following theorem is obtained.

Theorem 1: If p_N exists and if Assumptions 1–3 and 5 are satisfied, then

$$|P(D \cap D_s; L) - P(D \cap D_s; K)|$$
$$\leqslant \frac{A}{4(2\pi)^n} \int_{R^n} \|u\|^2\, |\phi_0(u)|\, du \sum_{k=K+1}^{L} E[\|a_k\|^2]$$
$$\equiv E_1(K) < \infty, \quad (25)$$

and $E_1(K)$ is obviously decreasing monotonically. Thus, for any $\epsilon > 0$, K can be selected large enough so that

$$E_1(K) < \epsilon/3. \quad (26)$$

It should be noted that this bound does not depend upon whether or not a power series expansion method is used, but it does depend upon K, the number of interference terms used.

The third kind of truncation error is caused by the termination of the power series expansion of $\Phi_K(u)$. In other words, it is due to the approximation of $P(D \cap D_s; K)$ by $P(D \cap D_s; K; R)$. According to Taylor's theorem [5, p. 245], the series in (14) can also be represented by only R terms plus a remainder; namely,

$$\Phi_K(u) = \sum_{l=0}^{R-1} \frac{1}{(2l)!} \left[\left(u \cdot \frac{\partial}{\partial t}\right)^{2l} \Phi_K(t)\right]_{t=0}$$
$$+ \frac{1}{(2R)!} \left[\left(u \cdot \frac{\partial}{\partial t}\right)^{2l} \Phi_K(t)\right]_{t=u_0}, \quad (27)$$

where u_0 is a vector on the line joining 0 and u. From (27), (15), and (10), it is evident that

$$|P(D \cap D_s; K) - P(D \cap D_s; K; R)|$$

$$= \left| (2\pi)^{-n} \int_{D \cap D_s} dx \int_{R^n} du \, \phi_0(u) \frac{1}{(2R)!} \right.$$

$$\left. \cdot \left[\left(u \cdot \frac{\partial}{\partial t} \right)^{2R} \Phi_K(t) \right]_{t=u_0} \exp\left[-j(x - S) \cdot u \right] \right|. \quad (28)$$

However,

$$\left| \left[\left(u \cdot \frac{\partial}{\partial t} \right)^{2R} \Phi_K(t) \right]_{t=u_0} \right|$$

$$= \left| \int \left(u \cdot \frac{\partial}{\partial t} \right)^{2R} \exp\left[jx \cdot t \right] dG_K(x) \Big|_{t=u_0} \right|$$

$$= \left| \int (u \cdot x)^{2R} \exp\left[jx \cdot t \right] dG_K(x) \Big|_{t=u_0} \right|$$

$$\leqslant \left| \int (u \cdot x)^{2R} dG_K(x) \Big|_{t=u_0} \right|$$

$$\leqslant \|u\|^{2R} \int \|x\|^{2R} dG_K(x) \quad \text{(Schwartz inequality)}$$

$$\equiv \|u\|^{2R} E\left[\left\| \sum_{k=1}^{K} a_k \right\|^{2R} \right], \quad (29)$$

which is obviously finite according to Assumption 4. Thus,

$$\left| \frac{1}{(2R)!} \left[\left(u \cdot \frac{\partial}{\partial t} \right)^{2R} \Phi_K(t) \right]_{t=u_0} \right|$$

$$\leqslant \|u\|^{2R} E\left[\left\| \sum_{k=1}^{K} a_k \right\|^{2R} \right] \Big/ (2R)!. \quad (30)$$

The RHS of (30) approaches zero as R increases to ∞ since all a_k are bounded.

Substituting (30) into (28) and applying Assumption 5 results in

$$|P(D \cap D_s; K) - P(D \cap D_s; K; R)|$$

$$\leqslant \frac{A}{(2\pi)^n (2R)!} E\left[\left\| \sum_{k=1}^{K} a_k \right\|^{2R} \right] \int_{R^n} |\phi_0(u)| \, \|u\|^{2R} \, du$$

$$\equiv E_2(R). \quad (31)$$

Whenever the condition

$$\frac{A}{(2\pi)^n (2R)!} E\left[\left\| \sum_{k=1}^{K} a_k \right\|^{2R} \right] \int_{R^n} |\phi_0(u)| \, \|u\|^{2R} \, du \searrow 0$$

$$\text{as } R \nearrow \infty \quad (32)$$

is satisfied, the truncation error bound $E_2(R)$ in (31) can be made arbitrarily small if R is made sufficiently large. Therefore, the following theorem can be stated.

Theorem 2: If Assumptions 1–5 and relation (32) are true, then, for any $\epsilon > 0$, R can be chosen sufficiently large so that

$$|P(D \cap D_s; K) - P(D \cap D_s; K; R)| \leqslant E_2(R) < \epsilon/3. \quad (33)$$

Gaussian Noise Case

For many practical applications, the special case in which the noise vector N is Gaussian is of primary interest. Without loss of generality, N can be assumed to have a zero mean and a covariance matrix $\sigma_n^2 I$, where I is an $n \times n$ identity matrix. That is,

$$\phi_0(u) = \exp\left(-\tfrac{1}{2} \sigma_n^2 \|u\|^2 \right). \quad (34)$$

Substituting (34) into (16) and (17) yields

$$P(D \cap D_s; K; R) = \sum_{l=0}^{R-1} \frac{(-1)^l}{(2l)!} \left[\left(\frac{\partial}{\partial S} \cdot \frac{\partial}{\partial t} \right)^{2l} \Phi_K(t) \right]_{t=0}$$

$$\cdot \int_{D \cap D_s} (2\pi\sigma_n^2)^{-n/2} \exp\left(-\frac{\|x - S\|^2}{2\sigma_n^2} \right) dx. \quad (35)$$

Now, represent $\Phi_K(u)$ by the power series

$$\Phi_K(u) \equiv \sum_{k_1, k_2, \cdots, k_n} b_{k_1, k_2, \cdots, k_n} u_1^{k_1} u_2^{k_2} \cdots u_n^{k_n}. \quad (36)$$

And, for $n = 0, 1, 2, \cdots$, define

$$\lambda_n(x) = (2\pi)^{-1/2} H_n(x) \exp\left(-\tfrac{1}{2} x^2 \right)$$

$$\lambda_n'(x) = -\lambda_{n+1}(x)$$

$$\lambda_{n+1}(x) = x \lambda_n(x) - n \lambda_{n-1}(x)$$

$$\lambda_{-1}(x) \equiv (2\pi)^{-1/2} \int_x^\infty \exp\left(-\frac{1}{2} t^2 \right) dt \equiv \frac{1}{2} \operatorname{erfc}(x/\sqrt{2}), \quad (37)$$

where H_n are the Hermite polynomials of degree n, and λ_n are the associated Hermite functions. Then (35) can be expressed as

$$P(D \cap D_s; K; R) = \sum_{k_1 + k_2 + \cdots + k_n = 0}^{R-1} b_{k_1, k_2, \cdots, k_n} (-j)^{\Sigma_i k_i}$$

$$\cdot \int_{D \cap D_s} \prod_{i=1}^{n} \lambda_{k_i} [(x_i - S_i)/\sigma_n] \, dx. \quad (38)$$

Since each term of (38) contains an integration of a product of n Hermite functions, the desired approximation $P(D \cap D_s; K; R)$ of $\Pr[z \in D]$ is simply a weighted sum of such integrals. The weighting factors $b_{k_1, k_2, \cdots, k_n}$ can be obtained by developing recurrence relations. For specific problems the recurrence relations may be developed differentially. This subject will be postponed until Sections IV–VII.

Here the attention will be restricted to investigation of the error bounds in (25) and (31) for the special case in which the noise is Gaussian. Clearly,

$$E_1(K) = \frac{A}{4(2\pi)^n} \int_{R^n} \|u\|^2 \exp\left(-\frac{1}{2} \sigma_n^2 \|u\|^2 \right) du$$

$$\cdot \sum_{k=K+1}^{L} E[\|a_k\|^2]$$

$$= \frac{a^n n \sigma_n^2}{4(2\pi\sigma_n^2)^{n/2}} \sum_{k=K+1}^{L} E[\|a_k\|^2], \quad (39)$$

which decreases monotonically with K. Substituting (34) into (31) yields

$$E_2(R) = \frac{a^n}{(2\pi)^n (2R)!} E\left[\left\|\sum_1^K a_k\right\|^{2R}\right]$$

$$\cdot \int_{R^n} \exp\left[-\frac{1}{2}\sigma_n^2 \|u\|^2\right] \|u\|^{2R} du$$

$$= \frac{a^n}{(2\pi)^{n/2} (2R)!} E\left[\left\|\sum_1^K a_k\right\|^{2R}\right]$$

$$\cdot \frac{1}{2^{n/2-1} \Gamma(n/2) \sigma_n^{2R+n}} \int_0^\infty r^{2R+n+1} \exp\left[-\frac{1}{2}r^2\right] dr$$

$$= \frac{\Gamma(R+n/2) 2^R}{\Gamma(n/2)(2R)!(2\pi)^{n/2}} \left(\frac{a}{\sigma_n}\right)^n E\left\{\left[\left\|\sum_{k=1}^K a_k\right\|^2/\sigma_n^2\right]^R\right\}.$$

$$(40)$$

The RHS of (40) is a monotonically decreasing function of $R(R \gg n/2)$ when R is sufficiently large. In the case that $E[\|\sum_{k=1}^K a_k\|^{2R}] < \sigma_n^{2R}$, the error bound $E_2(R)$ decreases faster than

$$E_2(R) \leqslant \frac{\Gamma(R+n/2) 2^R}{(2\pi)^{n/2} \Gamma(n/2)(2R)!} \left(\frac{a}{\sigma_n}\right)^n. \qquad (41)$$

Hence, when the noise is Gaussian, the truncation error bounds can be made more explicit, as illustrated in (39) and (41).

The applications part of this paper follows. It demonstrates how the above two theoretical sections can be applied to solve practical problems.

IV. Combined Effects of Gaussian Noise, Intersymbol Interference, and Co-Channel Interference on the Error Performance of MCPSK Systems

Let the M-ary coherent PSK system be depicted by Fig. 2, in which $W(t)$, a trapezoidal gating function, is defined as

$$W(t) = \begin{cases} 1, & |t| \leqslant (T-\tau_0)/2 \\ [(T+\tau_0)/2 - |t|]/\tau_0, & (T-\tau_0)/2 < |t| \leqslant (T+\tau_0)/2 \\ 0, & \text{elsewhere}, \end{cases}$$

$$(42)$$

where T is the inverse of symbol rate and τ_0 represents both the finite rise and delay time of the pulse. For convenience, M is assumed to be even. The transmitted symbol $\hat{\phi}_k$ is assumed to be selected with equal probability from the set $\{\pi/M + i\Omega, i = 0, 1, \cdots, (M-1)\}$ in which $\Omega \equiv 2\pi/M$. The receiver bandpass filter, represented by the complex envelope $q(t)$ of its impulse response, is usually matched to the transmitted signal

$$S(t) = \frac{1}{2} \sum_k \int_{-\infty}^\infty p(\tau) W(t-\tau-kT) d\tau \exp(j\hat{\phi}_k),$$

where $p(t)$ is the complex envelope of the impulse response of the transmit bandpass filter.

The receiver front-end noise is assumed to be zero-mean, white, stationary, and Gaussian with a two-sided power spectral density of $N_0/2$. This noise is represented by its complex envelope $N(t)$.

The lth co-channel interference can be represented by its complex envelope $A_l(t) \exp[j\theta_l(t) + j\xi_l]$, in which $A_l'(t)$, $\theta_l(t)$, and ξ_l are amplitude, phase modulation, and random carrier phase, respectively. For many practical applications, the random phase ξ_l may be assumed to have a uniform density of $[0, 2\pi]$.

The noise and co-channel interference terms at the input to the decision box can be represented by the two-dimensional vectors

$$N \equiv \begin{pmatrix} n_1 \\ n_2 \end{pmatrix} \equiv \begin{Bmatrix} \text{Re} \\ \text{Im} \end{Bmatrix} \int_{-\infty}^\infty q(t_0 - \tau) N(\tau) d\tau$$

$$\sum_{l=1}^{L_1} a_l \equiv \sum_{l=1}^{L_1} \begin{pmatrix} a_{l1} \\ a_{l2} \end{pmatrix} \equiv \begin{Bmatrix} \text{Re} \\ \text{Im} \end{Bmatrix} \int_{-\infty}^\infty \sum_{l=1}^{L_1} A_l(\tau) \exp[j\theta_l(\tau) + j\xi_l]$$

$$\cdot q(t_0 - \tau) d\tau, \quad (43)$$

respectively, where t_0 is the sampling time, and L_1 is the number of co-channel interferers at the input to the receiver. The first and second components of N and a_l denote the in-phase and quadrature components, respectively.

Let

$$\begin{pmatrix} R_k \\ I_k \end{pmatrix} \equiv \begin{pmatrix} R_k(t_0) \\ I_k(t_0) \end{pmatrix} \equiv \begin{Bmatrix} \text{Re} \\ \text{Im} \end{Bmatrix} \int_{-\infty}^\infty \int_{-\infty}^\infty p(t_0 - kT - \sigma - \tau)$$

$$\cdot q(\tau) W(\sigma) d\tau d\sigma \quad (44)$$

$$\alpha_k \equiv \frac{1}{2}(R_k \cos \hat{\phi}_k - I_k \sin \hat{\phi}_k)$$

$$\beta_k \equiv \frac{1}{2}(R_k \sin \hat{\phi}_k + I_k \cos \hat{\phi}_k). \qquad (45)$$

Then, the signal and intersymbol interference terms at the input to the decision box are

$$S \equiv \begin{pmatrix} \alpha_0 \\ \beta_0 \end{pmatrix} \qquad (46)$$

$$\sum_k' a_k^* \equiv \sum_k' \begin{pmatrix} \alpha_k \\ \beta_k \end{pmatrix}, \qquad (47)$$

respectively, where $\sum_k'$ denotes summation over all integers k except $k = 0$.

Without loss of generality, the order of summation in (47) can be rearranged so that the summation is taken over all positive integers greater than L_1 and the intersymbol interference vectors (α_k, β_k) are ordered in the descending order of their magnitudes. (Note that, from (45), $\alpha_k^2 + \beta_k^2 = \frac{1}{4}(R_k^2 + I_k^2)$, $\forall k$. This reordering of (α_k, β_k) can obviously be performed). Let $L = L_1 + L_2$, where L_2 is the total number of intersymbol interference terms and may be ∞. Then,

$$\sum_k' \begin{pmatrix} \alpha_k \\ \beta_k \end{pmatrix} \equiv \sum_{l=L_1+1}^L \begin{pmatrix} a_{l1} \\ a_{l2} \end{pmatrix} \equiv \sum_{l=L_1+1}^L a_l. \qquad (48)$$

Therefore, the following decision statistics can be formed at the decision box:

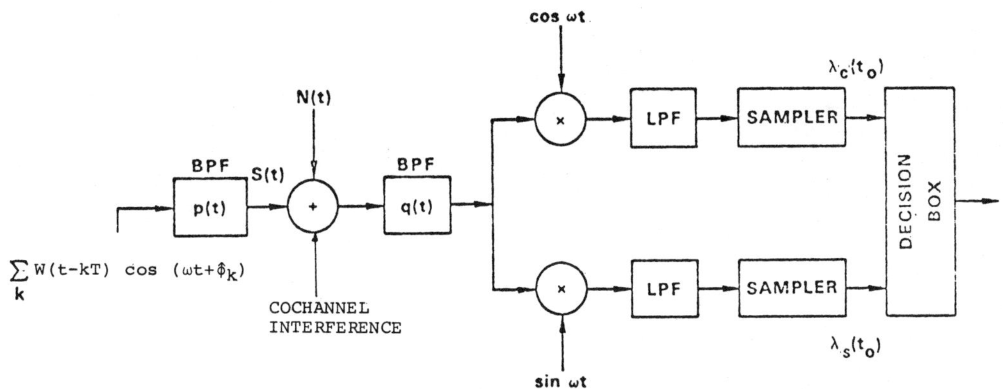

Fig. 2. Block diagram of MCPSK receiver.

$$\begin{pmatrix} \lambda_c \\ \lambda_s \end{pmatrix} \equiv z \equiv S + N + \sum_{k=1}^{L} a_k, \tag{49}$$

which is in the same form as (1) with $n = 2$.

The receiver decides that the ith symbol was sent if and only if

$$i\Omega \leqslant \underline{/(\lambda_c + j\lambda_s)} < (i + 1)\,\Omega. \tag{50}$$

For illustrative purposes, assume that the co-channel interferences are random carrier waves (CW's) or angle-modulated signals with bandwidths sufficiently narrow that the receiving filter $q(t)$ has negligible distortions on them.[2] Then, $A_l(\tau) = A_l$, $\forall l \leqslant L_1$, and the kth interference signal at the output of the receiving filter $q(t)$ has a uniform phase and an amplitude

$$|a_k| = A_k \left| \int_{-\infty}^{\infty} q(t_0 - \tau)\,d\tau \right| < \infty, \qquad \forall\, 1 \leqslant k \leqslant L_1. \tag{51}$$

It can easily be shown that the characteristic functions associated with a_k are

$$\phi_k(u) = J_0(a_k \sqrt{u_1^2 + u_2^2}), \qquad \text{for } 1 \leqslant k \leqslant L_1$$

$$= \frac{2}{M} \sum_{i=0}^{M/2-1} \cos\left[\frac{1}{2}(R_k u_1 + I_k u_2) \cos(\pi/M + i\Omega)\right.$$

$$\left. + \frac{1}{2}(R_k u_2 - I_k u_1) \sin(\pi/M + i\Omega)\right], \quad \text{for } L_1 < k \leqslant L, \tag{52}$$

where J_0 is the 0th order Bessel function of the first kind; M is even; and u is a column vector whose corresponding transpose is $u^+ = (u_1, u_2)$.

Since for almost all practical systems, the filter impulse responses are bounded and also have finite energy,

$$\sup_k \|a_k\| = \max\left\{ \sup_k \left(\frac{1}{2}\sqrt{R_k^2 + I_k^2}\right), \sup_l A_l \right\} < \infty$$

$$\sum_{k=1}^{L} E[\|a_k\|^2] = \frac{1}{2}\sum_{l=1}^{L_1} A_l^2 + \frac{1}{4}\sum_k{}' (R_k^2 + I_k^2) < \infty. \tag{53}$$

---

[2] For ASK or AM co-channel interference on M-ary coherent PSK systems, see [31, Appendix D].

Assumptions 2–4 in Section II are clearly satisfied. If the decision region D is identified as the wedge of angle Ω in Fig. 3, and if the characteristic functions ϕ_k are further equated to those in (52), then the correct decision probability can be obviously put into the form of (8); namely

$$\Pr[z \in D] = (2\pi)^{-2} \int_D dx \int_{R^2} \phi_0(u)\,\Phi_L(u)$$

$$\cdot \exp\,[-j(x - S)\cdot u]\,du, \tag{54}$$

where

$$\Phi_L(u) = \prod_{k=1}^{L} \phi_k(u)$$

$$\phi_0(u) = \exp\left(-\frac{1}{2}\sigma_n^2 \|u\|^2\right)$$

$$\sigma_n^2 = N_0/2 \int_{-\infty}^{\infty} |q(\tau)|^2\,d\tau.$$

To compute the error probability, or equivalently, the correct decision probability, it is sometimes sufficient to consider only L_1' (instead of L_1) of the more significant co-channel interference terms and only L_2' (instead of L_2) of the more significant intersymbol interference terms. Let $K \equiv L_1' + L_2'$, and

$$\sigma^2 \equiv \sigma_n^2 + \frac{\Delta_1}{2}\sum_{k=1}^{L_1'} A_k^2 + \frac{\Delta_2}{8}\sum_{l=1}^{L_2'} (R_l^2 + I_l^2)$$

$$\lambda^2 \equiv \frac{\Delta_1}{2\sigma^2}\sum_{k=1}^{L_1'} A_k^2 + \frac{\Delta_2}{8\sigma^2}\sum_{k=1}^{L_2'} (R_k^2 + I_k^2)$$

$$A_{k1} \equiv A_k/\sigma, \qquad R_{k1} \equiv R_k/\sigma, \qquad I_{k1} \equiv I_k/\sigma$$

$$\alpha' \equiv \alpha_0/\sigma, \qquad \beta' \equiv \beta_0/\sigma, \tag{55}$$

where $\Delta_i \in [0, 1]$ for $i = 1$ and 2 are weighting factors controlling the fraction of total interference power that can be effectively regarded as an equivalent Gaussian noise. (These factors Δ_i could be determined either by making educated guesses based upon specific problems, or by trial-and-error [16]). Moreover, construct a rectangle D_s centered at (α_0, β_0) as shown in Fig. 3. Then, the following approximation of (54)

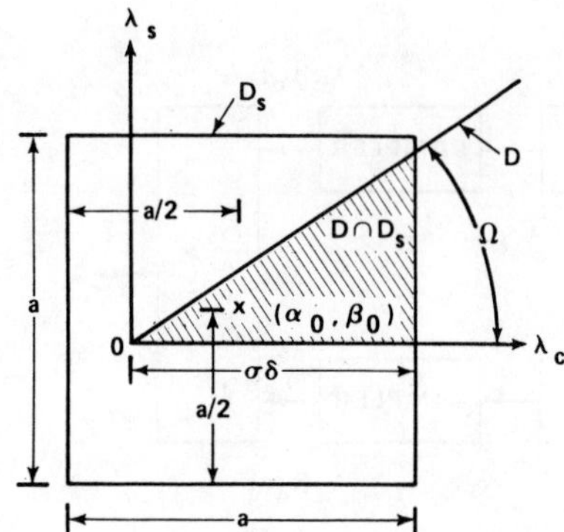

Fig. 3. Decision region for an MCPSK system.

is obtained:

$$
\begin{aligned}
P(D \cap D_s ; K) = (2\pi)^{-2} \int_{-\infty}^{\infty} \int_{(D \cap D_s)/\sigma^2} & \exp\left[-\frac{1}{2}(u_1^2 + u_2^2)\right] \\
& \cdot \Phi_{K_1}(u_1, u_2) \cdot \exp\left[\frac{1}{2}\lambda^2(u_1^2 + u_2^2)\right] \\
& \cdot \exp\left[-j(x_1 - \alpha')u_1 \right. \\
& \left. - j(x_2 - \beta')u_2\right] dx_2\, dx_1\, du_2\, du_1,
\end{aligned} \tag{56}
$$

where

$$
\Phi_{K_1}(u_1, u_2) \equiv \prod_{k=1}^{K} f^{(k)}(u_1, u_2), \tag{57}
$$

in which

$$
f^{(k)}(u_1, u_2) \equiv J_0(A_k \sqrt{u_1^2 + u_2^2}) \tag{58}
$$

for $k = 1, 2, \cdots, L_1$, and

$$
f^{(k)}(u_1, u_2) \equiv \frac{2}{M} \sum_{i=0}^{M/2-1} \cos(\xi_{ki} u_1 + \eta_{ki} u_2) \tag{59}
$$

$$
\xi_{ki} \equiv \frac{1}{2}[R_{(k-L_1')1} \cos(\pi/M + i\Omega)
$$

$$
- I_{(k-L_1')1} \sin(\pi/M + i\Omega)] \tag{60}
$$

$$
\eta_{ki} \equiv \frac{1}{2}[I_{(k-L_1')1} \cos(\pi/M + i\Omega)
$$

$$
+ R_{(k-L_1')1} \sin(\pi/M + i\Omega)] \tag{61}
$$

for $k = L_1' + 1, L_1' + 2, \cdots, K$.

Note that $\Phi_{K_1}(u_1, u_2) \exp\left[\frac{1}{2}\lambda^2(u_1^2 + u_2^2)\right]$ is unchanged if the signs of u_1 and u_2 are changed simultaneously. Thus, the product can be expressed by

$$
\Phi_{K_1}(u_1, u_2) \exp\left[\frac{1}{2}\lambda^2(u_1^2 + u_2^2)\right] = \sum_{k_1, k_2, 2} b_{k_1, k_2} u_1^{k_1} u_2^{k_2}, \tag{62}
$$

where $\Sigma_{k_1, k_2, 2}$ denotes the summation over all nonnegative k_1 and k_2 for $k_1 + k_2$ even. If $f^{(k)}(u_1, u_2)$ is represented by the power series

$$
f^{(k)}(u_1, u_2) = \sum_{m, n, 2} \tau_{m,n}^{(k)} u_1^m u_2^n, \tag{63}
$$

it is easy to verify

$$
\tau_{0,0}^{(k)} = 1 \ \forall k
$$

$$
\tau_{2m, 2n+1}^{(k)} = \tau_{2m+1, 2n}^{(k)} = 0 \ \forall k
$$

$$
\tau_{2m, 2n}^{(k)} = \frac{(-1)^{m+n} A_{k1}^{2(m+n)}}{[(m+n)!\, 2^{m+n}]^2} \frac{(m+n)!}{m!\, n!},
$$

$$
\text{for } k = 1, 2, \cdots, L_1'
$$

$$
\tau_{2m+1, 2n+1}^{(k)} = 0, \qquad \text{for } k = 1, 2, \cdots, L_1'
$$

$$
\tau_{m,n}^{(k)} = (-1)^{(m+n)/2} \frac{2}{M} \cdot \frac{1}{m!\, n!} \sum_{i=0}^{M/2-1} \xi_{ki}^m \eta_{ki}^n,
$$

$$
\text{for } k = L_1' + 1, L_1' + 2, \cdots, K. \tag{64}
$$

Now, by employing the results in [16, Section V] we can find b_{k_1, k_2} according to the following recurrence relations:

$$
\alpha_{0,0}^{(k)} = 1 \ \forall k \neq 0
$$

$$
\alpha_{m,n}^{(k)} = \sum_{p=0}^{m} \sum_{q=0}^{n} \tau_{p,q}^{(k)} \tau_{m-q, n-q}^{(k)},
$$

$$
\text{for even } m + n \tag{65}
$$

$$
\beta_{0,0}^{(k)} = 0 \ \forall k \neq 0
$$

$$
\beta_{m,0}^{(k)} = \tau_{m,0}^{(k)} - \sum_{p=1}^{m} (1 - p/m)\, \tau_{p,0}^{(k)} \beta_{m-p, 0}^{(k)},
$$

$$
\text{for even } m > 0 \tag{66}
$$

$$
\beta_{0,n}^{(k)} = \tau_{0,n}^{(k)} - \sum_{q=1}^{n} (1 - q/n)\, \tau_{0,q}^{(k)} \tau_{0, n-q}^{(k)},
$$

$$
\text{for even } n > 0 \tag{67}
$$

$$
\beta_{m,n}^{(k)} = \frac{1}{mn} \left[\lambda_{m,n}^{(k)} - \sum_{r=0}^{m} \sum_{t=0}^{n} \epsilon_{r,t}\, \alpha_{r,t}^{(k)} \beta_{m-r, n-t}^{(k)} (m-r)(n-t) \right],
$$

$$
\text{for } m > 0,\, n > 0 \text{ and even } m + n > 0, \tag{68}
$$

where $\epsilon_{r,t} = 0$ if $r = t = 0$ and $\epsilon_{r,t} = 1$ if otherwise; and

$$
\lambda_{m,n}^{(k)} = \sum_{p=0}^{m} \sum_{q=0}^{n} (2p - m)\, q\, \tau_{p,q}^{(k)} \tau_{m-p, n-q}^{(k)} \tag{69}
$$

$$
\sigma_{2,0} = \frac{1}{2}\lambda^2 + \sum_{k=1}^{K} \beta_{2,0}^{(k)}
$$

$$
\sigma_{0,2} = \frac{1}{2}\lambda^2 + \sum_{k=1}^{K} \beta_{0,2}^{(k)}
$$

$$
\sigma_{m,n} = \sum_{k=0}^{K} \beta_{m,n}^{(k)}, \qquad \text{for } (m,n) \neq (2,0) \text{ or } (m,n) \neq (0,2) \tag{70}
$$

$$b_{0,0} = 1$$

$$b_{k_1,0} = \sum_{m=0}^{k_1-1} \left(1 - \frac{m}{k_1}\right) \sigma_{k_1-m,0}\, b_{m,0}, \quad \text{for even } k_1 > 0 \tag{71}$$

$$b_{0,k_2} = \sum_{n=0}^{k_2-1} \left(1 - \frac{n}{k_2}\right) \sigma_{0,k_2-n}\, b_{0,n}, \quad \text{for even } k_2 > 0 \tag{72}$$

$$b_{k_1,k_2} = \frac{1}{k_1} \sum_{m=0}^{k_1} \sum_{n=0}^{k_2} \frac{1}{2}[1 + (-1)^{m+n}]\, m\, b_{k_1-m,k_2-n}\, \sigma_{m,n}. \tag{73}$$

Now, the truncated version $P(D \cap D_s; K; R)$ of $P(D \cap D_s; K)$ can be evaluated as follows.

$$P(D \cap D_s; K; R) = \sum_{k_1,k_2,2}^{R-1} b_{k_1,k_2} (-1)^{(k_1+k_2)/2}$$

$$\cdot \frac{1}{2\pi} \iint_{(D \cap D_s)/\sigma^2} H_{k_1}(x_1 - \alpha')\, H_{k_2}(x_2 - \beta')$$

$$\cdot \exp\left[-\frac{1}{2}\{(x_1 - \alpha')^2 + (x_2 - \beta')^2\}\right] dx_2\, dx_1$$

$$= \sum_{k_1,k_2,2}^{R-1} (-1)^{(k_1+k_2)/2} b_{k_1,k_2} \int_0^\delta dx_1$$

$$\cdot \lambda_{k_1}(x_1 - \alpha') \int_0^{\Gamma x_1} dx_2\, \lambda_{k_2}(x_2 - \beta'),$$

where

$$\Gamma \equiv \tan \Omega, \qquad \delta \equiv \alpha' + a/2\sigma, \tag{74}$$

and λ_n are Hermite functions. Let

$$I_{k_1,k_2}(\alpha', \beta', \rho, \delta) \equiv \int_0^\delta \lambda_{k_1}(x - \alpha')\, \lambda_{k_2}(\rho x - \beta')\, dx. \tag{75}$$

Then,

$$P(D \cap D_s; K; R) = \int_0^\delta \int_0^{\Gamma x_1} \lambda_0(x_1 - \alpha')\, \lambda_0(x_2 - \beta')\, dx_2\, dx_1$$

$$+ \sum_{k_1,k_2,2}^{R-1}{}' (-1)^{(k_1+k_2)/2}\, b_{k_1,k_2}$$

$$\cdot [I_{k_1,k_2-1}(\alpha', \beta', 0, \delta)$$

$$- I_{k_1,k_2-1}(\alpha', \beta', \Gamma, \delta)]. \tag{76}$$

The prime after the summation sign in (76) denotes the exclusion of the term with $k_1 = k_2 = 0$.

The first term of (76) can be evaluated by direct integration, i.e.,

$$\int_0^\delta \int_0^{\Gamma x_1} \lambda_0(x_1 - \alpha')\, \lambda_0(x_2 - \beta')\, dx_2\, dx_1$$

$$= (2\pi)^{-1/2} \int_0^\delta \exp\left[-\frac{1}{2}(x_1 - \alpha')^2\right] \frac{1}{2}\{\mathrm{erfc}\,(-\beta'/\sqrt{2})$$

$$- \mathrm{erfc}\,[(\Gamma x_1 - \beta')/\sqrt{2}]\}\, dx_1. \tag{77}$$

Since, as mentioned before, the coefficients b_{k_1,k_2} can be computed by using the results in [16], the summation terms in (76) can thus be evaluated if $I_{k_1,k_2}(\alpha', \beta', \rho, \delta)$ are known.

$$I_{k,l}(\alpha', \beta', \rho, \delta) = \lambda_{k-1}(-\alpha')\, \lambda_l(-\beta') - \lambda_{k-1}(\delta - \alpha')\, \lambda_l(\rho\delta - \beta')$$

$$- \Gamma I_{k-1,l+1}(\alpha', \beta', \rho, \delta). \tag{78}$$

Hence, if $I_{0,l}(\alpha', \beta', \rho, \delta)$ are known for $l = 1, 2, \cdots, R-1$, then all $I_{k,l}$ for $0 \leqslant k, l \leqslant R-1$ can be computed from the recurrence relationships in (78).

Since

$$I_{0,0}(\alpha', \beta', \rho, \delta) \equiv \int_0^\delta \lambda_0(x - \alpha')\, \lambda_0(\rho x - \beta')\, dx$$

$$= (2\pi)^{-1} \int_0^\delta \exp\left[-\frac{1}{2}(x - \alpha')^2 - \frac{1}{2}(\rho x - \beta')^2\right] dx$$

$$= (2\pi)^{-1/2}\,(1 + \rho^2)^{-1/2} \exp\left[-(\alpha'\rho - \beta')^2/2(1 + m^2)\right]$$

$$\cdot \{\lambda_{-1}[-(\alpha' + \beta'\rho)/\sqrt{1 + \rho^2}]$$

$$- \lambda_{-1}[\delta\sqrt{1 + \rho^2} - (\alpha' + \beta'\rho)/\sqrt{1 + \rho^2}]\}, \tag{79}$$

it follows, by differentiating (79) l times with respect to β',

$$I_{0,l}(\alpha', \beta', \rho, \delta) \equiv \int_0^\delta \lambda_0(x - \alpha')\, \lambda_l(\rho x - \beta)\, dx$$

$$= \sum_{p=0}^{l} \binom{l}{p} \lambda_p\left(\frac{\alpha'\rho - \beta'}{\sqrt{1 + \rho^2}}\right) \left\{\lambda_{l-p-1}\left(-\frac{\alpha' + \beta'\rho}{\sqrt{1 + \rho^2}}\right)\right.$$

$$\left. - \lambda_{l-p-1}\left(\delta\sqrt{1 + \rho^2} - \frac{\alpha' + \beta'\rho}{\sqrt{1 + \rho^2}}\right)\right\} \frac{m^{l-p}}{(1 + \rho^2)^{(l+1)/2}}. \tag{80}$$

Indeed, $I_{0,l}(\alpha', \beta', \rho, \delta)$, for $l = 0, 1, \cdots, R-1$, can all be computed by (79) and (80) and, consequently, $I_{k,l}(\alpha', \beta', \rho, \delta)$ can be obtained from (78).

Therefore, $P(D \cap D_s; K; R)$ can be evaluated from (76), which in turn gives the desired error probability

$$P_e = 1 - \Pr(z \in D) \doteq 1 - P(D \cap D_s; K; R). \tag{81}$$

It seems appropriate to note that, when $L_1 = 0$, namely, when there is no co-channel interference at all, the results in this section obviously reduce to those derived for Gaussian noise and intersymbol interference only [16]. On the other hand, when there is no intersymbol interference (i.e., $L_2 = 0$), the results in this section can be simplified to those results obtained for Gaussian noise and co-channel interference only [22]. In general, the combined effects of Gaussian noise, intersymbol interference, and co-channel interference on the error probability of an MCPSK system are quite different from their separate effects, e.g., measured in terms of the degradation in signal-to-noise ratio needed to achieve a certain error performance.

It also seems appropriate to make the following remarks. It has been shown in [16] that, if the channel is symmetric, the error probability of a four-phase coherent PSK system can be derived from that of a binary coherent PSK system over the same channel, even though there is intersymbol interference in addition to Gaussian noise. The reason is simple. When $M = 4$ and when the channel is symmetric, the characteristic function $\Phi_L(u_1, u_2)$, corresponding to Gaussian noise and all intersymbol interferences can be factorized into a product of two functions. One is a function of u_1 only and another is a function of u_2 only. However, whenever there is random CW co-channel interference, the error probability of a four-phase PSK system can no longer be obtained from that of a two-phase PSK system because, in the presence of such co-channel interference, the characteristic function $\Phi_L(u_1, u_2)$ contains such factors as $J_0(A_l \sqrt{u_1^2 + u_1^2})$ for $l = 1, 2, \cdots, L_1$, which obviously cannot be factorized into products of two single-variable functions of u_1 and u_2. Hence it can be asserted that whenever there is random CW or narrow-band angle-modulated interference, the in-phase and the quadrature-phase components of a four-phase coherent PSK system are always correlated even though the channel is symmetric.

Next is a numerical example that is computed by employing the recurrence relationships given in (55), (60), (61), (64)–(74), (76)–(81).

Example 1. Four-Phase PSK/Butterworth Channel

The channel is assumed to be characterized by a three-pole Butterworth RF filter. The complex envelope of the impulse response of such a filter can be expressed by

$$h_3(t) = 4\pi B \, [\exp(-2\pi Bt)$$

$$+ \, 3^{-1/2} \exp(-\pi Bt) \sin(\sqrt{3}\,\pi Bt - \pi/3)], \quad \text{for } t \geq 0$$

$$= 0, \qquad\qquad\qquad\qquad \text{for } t < 0.$$

The symbol rates used in this example for the four-phase PSK signals are $2B/1.25$ and $2B$ symbols/s, respectively. Two cases of co-channel interferences are investigated: one has only a single interfering random CW$(L_1 = 1)$, and the other has four equal-amplitude interfering random CW's$(L_1 = 4; A_l = A; \forall l = 1, 2, 3, 4)$.

For five values of the signal-to-total co-channel interference-power ratios, $S/I(= 10, 15, 20, 25,$ and 40 dB, respectively), the results in this section have been used to evaluate the symbol error probabilities as functions of SNR's at both signaling rates.[3] In this example the following parameters have been chosen: $K \equiv L_1' + L_2' = L_1 + L_2 = 4 + 6 = 10$, $R = 7$, $\Delta_2 = 4\Delta_1 = \Delta = 0.5$, and $a = 2\sqrt{2}$. The computed results are shown in Fig. 4(a) and (b). It can be observed from Fig. 4 that, for same S/I, four equal-amplitude

---

[3]S/I is defined as $(R_0^2 + I_0^2)/4(\Sigma_k A_k^2)$ and SNR is defined as $(R_0^2 + I_0^2)/8\sigma_n^2$.

random CW's degrade the system performance more than a single random CW, in agreement with the conclusions of [22]. Also, note that at high S/I (for which the co-channel interference is negligible), the error probability curves agree well with the results in [16] as expected.

V. Combined Effects of Gaussian Noise, Intersymbol Interference, and Co-Channel Interference on the Error Performance of MCAPK Systems

The block diagram of the MCAPK system is identical to Fig. 2, which is for an MCPSK system. However, the M message symbols, instead of being placed on a unit circle in the in-phase and quadrature-phase plane, are now located anywhere in this plane. That is, they are now represented by the M points $(\hat{a}_k, \hat{b}_k)$ instead of by the points $(\cos \hat{\phi}_k, \sin \hat{\phi}_k)$. Of course, if $\hat{a}_k = \cos \hat{\phi}_k$, and $\hat{b}_k = \sin \hat{\phi}_k$, MCPSK can be treated as a special case of MCAPK systems. Fig. 5(a) shows the "rectangular" design of a 32-ary APK signal set, whereas Fig. 5(b), (c), and (d) illustrate the rectangular, the (4, 4) circular, and the triangular design of the 8-ary APK signal sets, respectively. Obviously, if a_k(or $\hat{b}_k$) is set equal to zero for all k and $\hat{b}_k$ (or $\hat{a}_k$) is set equal to any of a prescribed M levels, the M-ary coherent ASK [or pulse-amplitude modulation (PAM)] signal set can be treated as a special case of the MCAPK signal set. Thus, the MCAPK signals, in theory, are both amplitude-and-phase keyed (APK).

The receiver forms the same kind of decision statistics as in (49), but the $(\hat{a}_k, \hat{b}_k)$ are no longer $(\cos \hat{\phi}_k, \sin \hat{\phi}_k)$; instead they are determined by the particular APK signal set of interest. The decision regions for high SNR's generally have linear decision boundaries such as those shown in Fig. 5(a)–(d). The receiver's decision rule is simply to decide that the ith symbol was sent if and only if z falls into the ith decision region, Ω_i. Thus, a symbol error occurs if z falls outside the decision region corresponding to the correct symbol.

The average symbol error probability can be written as

$$P_e = \frac{1}{M} \sum_{i=1}^{M} (1 - P_r[\Omega_i]), \qquad (82)$$

where the MCAPK message symbols are assumed to be equally likely, and $P_r[\Omega_i]$ denotes the correct decision probability for the ith symbol given that the ith symbol was sent.

Assume that there are L_1 interfering random CW's. Obviously, the characteristic function, $\psi_c(u_1, u_2)$, corresponding to these CW's is

$$\psi_c(u_1, u_2) \equiv \prod_{k=1}^{L_1} \phi_k(u_1, u_2) = \prod_{l=1}^{L_1} J_0(A_l \sqrt{u_1^2 + u_2^2}), \quad (83)$$

where $\phi_k(u_1, u_2)$ is the characteristic function of the kth co-channel interference.

For illustrative purposes, consider the 32-ary APK system with the rectangular design shown in Fig. 5(a). Since the in-phase and quadrature components of the reordered lth intersymbol interference term are

$$\alpha_l \equiv \tfrac{1}{2}(\hat{a}_l R_l - \hat{b}_l I_l)$$

$$\beta_l \equiv \tfrac{1}{2}(\hat{a}_l I_l + \hat{b}_l R_l), \qquad (84)$$

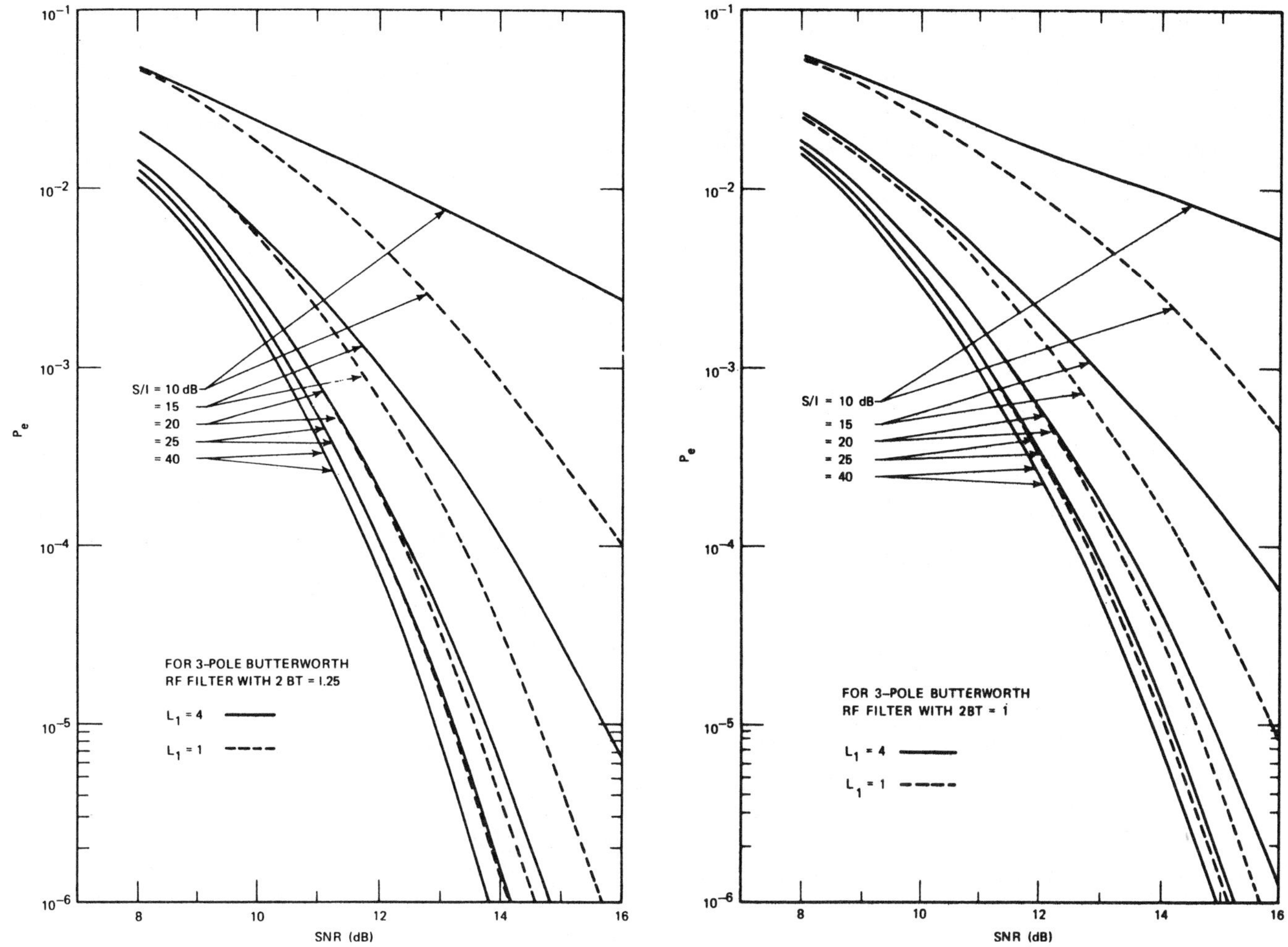

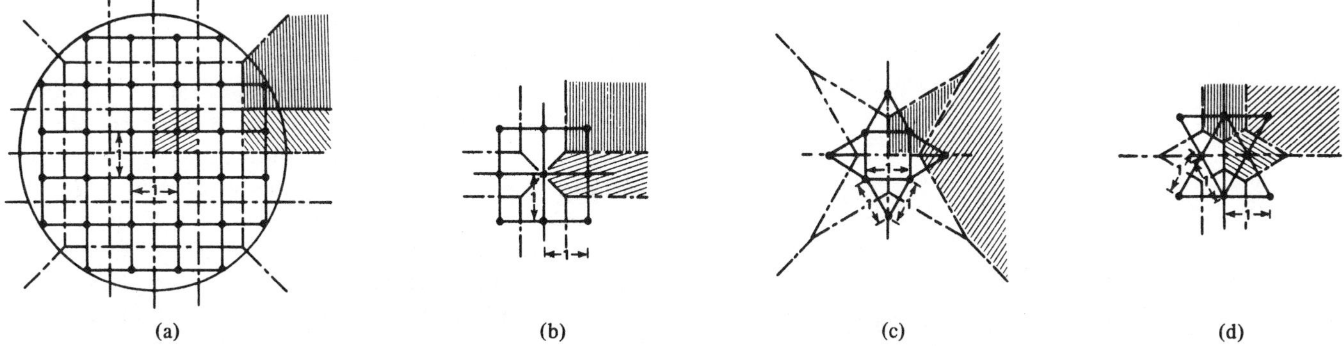

Fig. 4. (a) Probability of symbol error versus SNR, 2 BT = 1.25. (b) Probability of symbol error versus SNR, 2 BT = 1.

Fig. 5. APK signal sets. (a) 32-ary rectangular design, $\overline{A} = \sqrt{5}$. (b) 8-ary rectangular design, $\overline{A} = \sqrt{3}/2$. (c) 8-ary (4, 4) circular design, $\overline{A} = (7/8 + \sqrt{3}/4)^{1/2}$. (d) 8-ary triangular design, $\overline{A} = \sqrt{7}/2$.

respectively, it follows

$$E[\alpha_l^2 + \beta_l^2] = (R_l^2 + I_l^2)\overline{A}^2/4 \qquad (85)$$

where

$$\overline{A} = \{E[\hat{a}_l^2 + \hat{b}_l^2]\}^{1/2} = 5^{1/2} \qquad (86)$$

is the rms amplitude of the signals. It can be easily verified that the characteristic function $\psi_I(u_1, u_2)$ corresponding to the intersymbol interference is

$$\psi_I(u_1, u_2) = \prod_{k=L_1+1}^{L_1+L_2} \phi_k(u_1, u_2), \qquad (87)$$

where, for $l = 1, 2, \cdots, L_2$,

$$\phi_{L_1+l}(u_1, u_2) \equiv \frac{1}{16} \sum_{i=0}^{15} \cos\left[u_1(\pm X_i R_l - Y_i I_l)\right.$$

$$\left. + u_2(\pm X_i I_l + Y_i R_l)\right] \qquad (88)$$

is the characteristic function of the lth intersymbol interference. In (88), the summation is taken over the 16 pairs of $(\pm X_i, Y_i)$: $\{(\pm 1, 1), (\pm 1, 3), (\pm 3, 1), (\pm 3, 3), (\pm 1, 5), (\pm 5, 1), (\pm 3, 5), (\pm 5, 3)\}/4\overline{A}$.

Redefine α^2, λ^2, α', β', and $\tau_{m,n}^{(k)}$ as follows instead of as in (55) and (64):

$$\sigma^2 \equiv \sigma_n^2 + \frac{\Delta_1}{2} \sum_{k=1}^{L_1'} A_k^2 + \frac{\Delta_2}{8} \overline{A}^2 \sum_{k=1}^{L_1'} (R_k^2 + I_k^2) \quad (89)$$

$$\lambda^2 \equiv \frac{\Delta_1}{2\sigma^2} \sum_{k=1}^{L_1'} A_k^2 + \frac{\Delta_2}{8\sigma^2} \overline{A}^2 \sum_{l=1}^{L_2'} (R_l^2 + I_l^2) \quad (90)$$

$$\alpha' \equiv (\hat{a}_0 R_0 - \hat{b}_0 I_0)/2\sigma\overline{A} \quad (91)$$

$$\beta' \equiv (\hat{b}_0 R_0 + \hat{a}_0 I_0)/2\sigma\overline{A} \quad (92)$$

$$\tau_{m,n}^{(k)} \equiv (-1)^{(m+n)/2} \frac{1}{16} \frac{1}{m!\,n!} \sum_{i=0}^{15} \xi_{ki}^m \eta_{ki}^n, \quad (93)$$

for $k = L_1' + 1, L_1' + 2, \cdots, K$, where the 16 pairs of (ξ_{ki}, η_{ki}) for every k are defined by

$$(\xi_{ki}, \eta_{ki}) \equiv \{(\pm X_i R_{(k-L_1')1} - Y_i I_{(k-L_1')1}), (\pm X_i I_{(k-L_1')1}$$
$$+ Y_i R_{(k-L_1')1})\} \quad \text{for } i = 0, 1, \cdots, 15. \quad (94)$$

Then the coefficients b_{k_1, k_2} can be computed with the same recurrence relationships as in (55), (60), (61), (64)–(74), (76)–(81).

Since the decision regions are assumed to have linear boundaries, they must be polygons, which of course can be partitioned into a direct sum of disjoint triangles. Through a proper choice of coordinates, the decision polygons can all be partitioned into a direct sum of triangles defined by the lines $y = \Gamma x$, $y = 0$, and $x = \delta$ in the x-y plane. The probabilities that z falls into such triangles given that the signal is at (α', β') can then be evaluated by using recurrence relationships (74) and (77)–(80). From these probabilities, the desired error probability can therefore be obtained according to (82).

Example 2. 32-ary APK System with Rectangular Signal Set

Because of the speed of logic elements in the hardware implementation, a contemplated 33.3-Msymbol/s 32-ary APK system has the following window function corresponding to a 10 percent rise time and a 10 percent delay time:

$$W(t) = \begin{cases} 1, & \text{for } |t|/T \leqslant 9/20 \\ 11/2 - 10\,|t|/T, & \text{for } 9/20 < |t|/T \leqslant 11/20 \\ 0, & \text{otherwise.} \end{cases} \quad (95)$$

The amplitude and group-delay characteristics of the transmit and receive filters are given in Fig. 6(a) and (b). The computed average symbol error probabilities are shown in Fig. 7(a) and (b), respectively, for one-co-channel interference and for four equal-amplitude co-channel interference terms. From these curves, it can be seen that, for a given signal-to-total co-channel interference power ratio, S/I,[4] four co-channel interference

---

[4] S/I is defined as $\overline{A}^2 (R_0^2 + I_0^2)/4(\sum_k A_k^2)$ and SNR is defined as $\overline{A}^2 (R_0^2 + I_0^2)/8\sigma_n^2$.

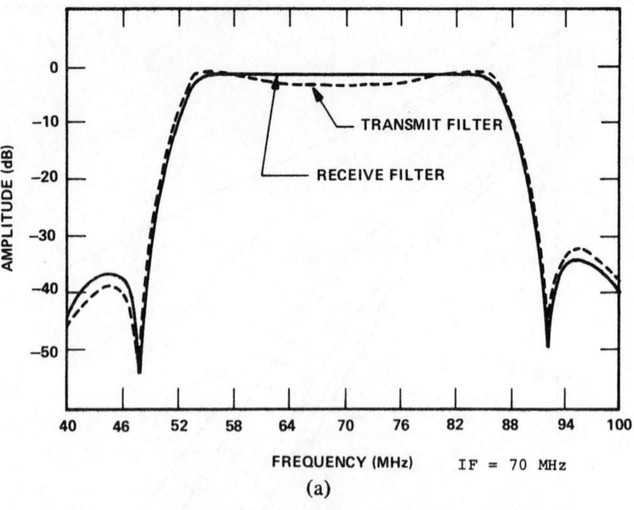

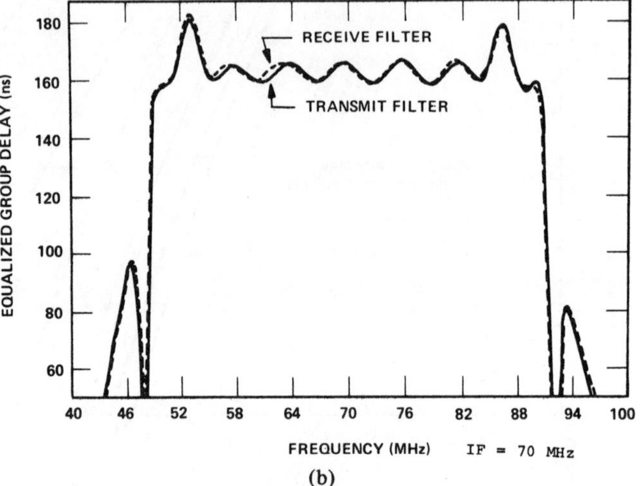

Fig. 6. (a) Amplitude characteristics of modem (elliptic) filters (adjusted to 33.3 Mbd). (b) Modem-filter group delay response (equalized).

terms cause more degradation in the error probabilities then a single co-channel interference term.

It is an easy matter to obtain the error probabilities as functions of SNR for this APK system in the absence of intersymbol interferences by simply setting R_{k1} and I_{k1} in (95) equal to zero for all $k \neq 0$ and by setting $R_0 = 1$ and $I_0 = 0$ in (91) and (92). On the other hand, if only the effects of Gaussian noise and intersymbol interference are desired, then the co-channel interference A_k can be simply set equal to zero for all k in the computation of the coefficients b_{k_1, k_2}. These matters are trivial, however. What is important is that the error performance of APK systems in the presence of Gaussian noise, intersymbol interference, and co-channel interference can be calculated accurately by using the general approach presented here.

If the ϕ_k in (83) are replaced by the ϕ_k in [31, D-8] and the definitions given in [31, D-10–D-12] are used for σ^2, $f^{(k)}$, and $\tau_{2m, 2n}^{(k)}$, respectively, then it is also possible to compute the error probabilities of APK systems in the presence of Gaussian noise, intersymbol interference, and AM co-channel interference. Of course, other extensions are just as possible, but the key is that the characteristic function of the co-channel interference terms at the sampler outputs must be obtainable.

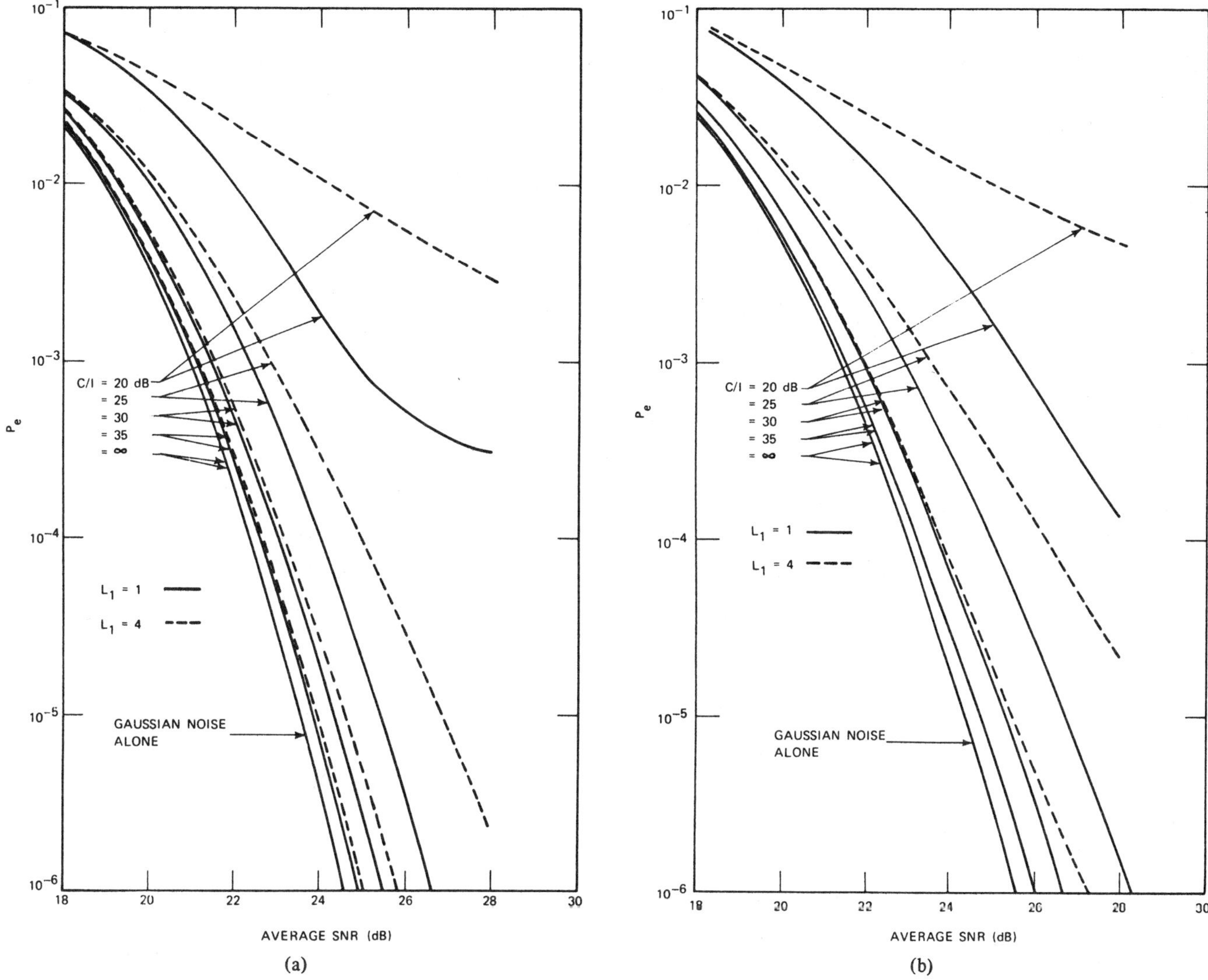

Fig. 7. (a) Symbol error probability versus average SNR for a 32-ary rectangular APK signal without intersymbol interference but with cochannel interference. (b) Symbol error probability versus average SNR for a 32-ary APK signal with intersymbol interference and cochannel interference.

VI. ERROR PROBABILITY OF A BINARY DPSK SYSTEM IN THE PRESENCE OF GAUSSIAN NOISE AND INTERSYMBOL INTERFERENCES

From [19, eq. (36)], the error probability of a binary DPSK system in the presence of Gaussian noise and intersymbol interference is

$$P_e = (1 - P_2)/2, \tag{96}$$

where

$$P_2 \equiv \frac{1}{\pi^2} \int_{-\infty}^{\infty} \int_{-\infty}^{\infty} \exp\left[-\frac{1}{2}(u_1^2 + u_2^2)\right] \{\hat{K}_1(u_1 u_2)$$

$$- \rho \hat{K}_0(u_1 u_2)\} \sin(h_0 u_1 + h_1 u_2) \sin(h_{-1} u_1 + h_0 u_2)$$

$$\cdot \prod_{k \neq 0, 1} \cos(h_k u_1 + h_{k+1} u_2) \, du_1 \, du_2. \tag{97}$$

In (97), $\hat{K}_0(x) \equiv K_0(|x|)$ and $\hat{K}_1(x) \equiv K_1(x)$ if $x \geq 0$, and $\hat{K}_1(x) \equiv -K_1(|x|)$ if $x < 0$; K_0 and K_1 are the zero-order and first-order modified Bessel functions of the second kind, respectively; ρ is the correlation of the circularly symmetric narrow-band, zero-mean Gaussian noise at the front end of the

receiver; $h_k \equiv h_k(t_0)/\sigma$, where t_0 is the sampling time, σ is the standard deviation of the Gaussian noise; and

$$h_k(t) \equiv \int_{-\infty}^{\infty} W(\tau - kT) h(t - \tau) \, d\tau. \tag{98}$$

In (98) W is the window function defined in (42) with $\tau_0 = 0$, and h is half of the complex envelope of the impulse response of the symmetric bandpass channel filter.

With $\sin \alpha \sin \beta = \frac{1}{2}[\cos(\alpha - \beta) - \cos(\alpha + \beta)]$, (97) can be rewritten as

$$P_2 = \frac{1}{2\pi^2} \sum_{l=0}^{1} (-1)^l \int_{-\infty}^{\infty} \int_{-\infty}^{\infty} \exp\left[-(u_1^2 + u_2^2)/2\right] \{\hat{K}_1(u_1 u_2)$$

$$- \rho \hat{K}_0(u_1 u_2)\} \prod_{k \neq 0, 1} \cos(h_k u_1 + h_{k+1} u_2)$$

$$\cdot [\cos x_l u_1 \cos y_l u_2 - \sin x_l u_1 \sin x_l u_2] \, du_1 \, du_2, \tag{99}$$

where

$$x_0 \equiv h_0 - h_{-1'} \qquad y_0 \equiv -h_0 + h_1$$

$$x_1 \equiv h_0 + h_{-1'} \qquad y_0 \equiv h_0 + h_1.$$

Suppose that it is possible to expand

$$\prod_{k \neq 0, 1} \cos (h_k u_1 + h_{k+1} u_2) = \sum_{k_1=0}^{\infty} \sum_{k_2=0}^{\infty} b_{k_1, k_2} u_1^{k_1} u_2^{k_2},$$

$$(100)$$

as will be performed later. Then, since $\hat{K}_0$ is even and $\hat{K}_1$ is odd, from (99) and (100),

$$P_2 = \frac{2}{\pi^2} \sum_{p=0}^{1} (-1)^p \left\{ \sum_{k_1} \sum_{k_2} b_{2k_1, 2k_2} \int_0^{\infty} \int_0^{\infty} u_1^{2k_1} u_2^{2k_2} \right.$$

$$\cdot \exp \left[-(u_1^2 + u_2^2)/2 \right] \left[-K_1(uv) \sin x_p u_1 \sin y_p u_2 \right.$$

$$\left. - \rho K_0(u_1 u_2) \cos x_p u_1 \cos y_p u_2 \right] du_1 du_2$$

$$+ \sum_{k_1} \sum_{k_2} b_{2k_1+1, 2k_2+1} \int_0^{\infty} \int_0^{\infty} u_1^{2k+1} u_2^{2l+1}$$

$$\cdot \exp \left[-(u_1^2 + u_2^2)/2 \right] \left[K_1(u_1 u_2) \cos x_p u_1 \cos u_p u_2 \right.$$

$$\left. \left. + \rho K_0(u_1 u_2) \sin x_p u_1 \sin u_p u_2 \right] du_1 du_2 \right\}, \quad (101)$$

where term-wise integration can be justified by showing that the integrated series is actually absolutely convergent.

Let

$$\alpha_{0, 0}(x, y) \equiv \int_0^{\infty} \int_0^{\infty} K_0(u_1 u_2) \exp \left[-(u_1^2 + u_2^2)/2 \right]$$

$$\cdot \cos x u_1 \cos y u_2 \, du_1 \, du_2 \quad (102)$$

$$\beta_{0, 0}(x, y) \equiv \int_0^{\infty} \int_0^{\infty} K_1(u_1 u_2) \exp \left[-(u_1^2 + u_2^2)/2 \right]$$

$$\cdot \sin x u_1 \sin y u_2 \, du_1 \, du_2 \quad (103)$$

$$\alpha_{k_1, k_2}(x, y) \equiv \frac{\partial^{k_1 + k_2}}{\partial_x^{k_1} \partial_y^{k_2}} \alpha_{0, 0}(x, y) \quad (104)$$

$$\beta_{k_1, k_2}(x, y) \equiv \frac{\partial^{k_1 + k_2}}{\partial_x^{k_1} \partial_y^{k_2}} \beta_{0, 0}(x, y). \quad (105)$$

Then,

$$P_2 = \frac{2}{\pi^2} \sum_{p=0}^{1} (-1)^{p+1} \sum_{k_1, k_2, 2} (-1)^{(k_1 + k_2)/2} b_{k_1, k_2}$$

$$\cdot \left[\beta_{k_1, k_2}(x_p, y_p) + \rho \alpha_{k_1, k_2}(x_p, y_p) \right], \quad (106)$$

in which α_{k_1, k_2} and β_{k_1, k_2} can be computed by using (56), (60), and [19, eq. (61)]. Hence, the question is how to compute b_{k_1, k_2}.

Without loss of generality, the product in (100) can be reordered so that the left-hand side (LHS) of (100) is a product of

$$f^{(k)}(u_1, u_2) \equiv \cos (\xi_k u_1 + \eta_k u_2)$$

$$\equiv \sum_{m, n, 2} \tau_{m, n}^{(k)} u_1^m u_2^n, \quad \text{for } k = 0, 1, 2, \cdots.$$

$$(107)$$

The reordering may be performed according to the sequence $k = -1, 2, -2, 3, -3, 4, -4, \cdots$, in (100). Obviously, then

$$\tau_{0, 0}^{(k)} = 1 \; \forall k$$

$$\tau_{2m, 2n+1}^{(k)} = \tau_{2m+1, 2n}^{(k)} = 0 \; \forall k$$

$$\tau_{m, n}^{(k)} = (-1)^{(m+n)/2} \frac{1}{m! \, n!} \xi_k^m \eta_k^n, \quad \text{for all even } m + n > 0.$$

$$(108)$$

The recurrence relationships in (65)–(73) can thus be used to obtain the coefficients b_{k_1, k_2}.

Therefore, the error probability of a binary DPSK system in the presence of Gaussian noise and intersymbol interference can be computed.

VII. Intermodulations due to Hard-Limiting of L Signals and Noise

Let the hard-limiter be represented by

$$y = \frac{2}{\pi} \int_0^{\infty} \frac{\sin xu}{u} \, du = \begin{cases} 1, & \text{if } x > 0 \\ -1, & \text{if } x \leqslant 0. \end{cases} \quad (109)$$

The input x to this hard-limiter is a sum of multi-carriers plus noise:

$$x(t) = S_i(t) + n(t) \quad (110)$$

where

$$S_i(t) = \sum_{k=0}^{L} a_k(t) \cos \left[(\omega_c + \omega_k) t + \tilde{\phi}_k(t) + \theta_k \right] \quad (111)$$

$$n(t) = r(t) \cos (\omega_c t + \phi(t))$$

$$\equiv n_c(t) \cos \omega_c t - n_s(t) \sin \omega_c t. \quad (112)$$

In the above equations, a_k, $\tilde{\phi}_k$, and $\omega_c + \omega_k$ are the amplitude modulation, the phase modulation, and the carrier frequency of the kth carrier, respectively. The noise $n(t)$ is a narrow-band stationary Gaussian random process with mean zero and variance σ_n^2. That is, for any t, $r(t)$ and $\phi(t)$ have a Rayleigh and a uniform density, respectively. The phase angles θ_k, for $k = 0, 1, 2, \cdots, L$, are assumed to be independent identically distributed random variables, each uniformly distributed on $[0, 2\pi]$.

Jain has shown [29] that the $s \times s$ term at the hard-limiter output is

$$S_0(t) = \frac{1}{2} \sum_{k_0, k_1, k_2, \cdots, k_L = -\infty}^{\infty} A_{k_0, k_1, \cdots, k_L}$$

$$\cdot \sin \left[\sum_{l=0}^{L} k_l \left\{ (\omega_c + \omega_l) t + \tilde{\phi}_l(t) + \theta_l + \frac{\pi}{2} \right\} \right], \quad (113)$$

where

$$A_{k_0, k_1, \cdots, k_L} \equiv \frac{4}{\pi} \int_0^{\infty} \frac{1}{u} \prod_{l=0}^{L} J_{k_l}(u a_l) \exp (-u^2 \sigma_n^2/2) \, du \quad (114)$$

are the amplitudes of the output signal and intermodulation components.

In (113), only those terms that fall into the channel bandwidth of the hard-limiter output filter centered at ω_c are usually of interest. Let the ideally filtered output be $S(t)$. Then,

$$S(t) = \frac{1}{2} \sum_{k_0, k_1, \cdots, k_L = -\infty}^{\infty} {}^* A_{k_0, k_1, \cdots, k_L}$$

$$\cdot \cos\left[\omega_c t + \sum_{l=0}^{L} k_l(\omega_l t + \tilde{\phi}_l(t) + \theta_l)\right], \quad (115)$$

where the asterisk * after the summation sign denotes that the summation is taken under the condition $\sum_{l=1}^{L} k_l = 1$.

Suppose that $a_0(t) \cos[(\omega_c + \omega_0)t + \tilde{\phi}_0(t) + \theta_0]$ is the desired signal at the input of the limiter. Then

$$I(t) \equiv \sum_{k=1}^{L} a_k(t) \cos[(\omega_c + \omega_k)t + \tilde{\phi}_k(t) + \theta_k] + n(t) \quad (116)$$

is simply the undesired signal. As a result of hard-limiting, the term $A_{1,0,0,\cdots,0}$ represents the distortion on the pure signal $a_0(t) \cos[(\omega_c + \omega_0)t + \tilde{\phi}_0(t) + \theta_0]$ due to the presence of the undesired signal "interference."

$$A_{1,0,\cdots,0} = \frac{4}{\pi} \int_0^\infty \frac{1}{u} J_1(a_0 u) \left[\prod_{l=1}^{L} J_0(a_l u)\right] \exp(-u^2 \sigma_n^2/2)\, du. \quad (117)$$

However, the factor $[\prod_{l=1}^{L} J_0(a_l u)] \exp(-u^2 \sigma_n^2/2)$ in the integral of (117) is just the characteristic function of the undesired signal $I(t)$ in (116). What it amounts to is that the hard-limiter simply "transposes" the undesired carriers $\sum_{k=1}^{L} a_k(t) \cos[(\omega_c + \omega_k)t + \tilde{\phi}_k(t) + \theta_k]$ onto the desired carrier, and effectively results in co-channel interference. Therefore, it should not be surprising to see that the integral in (117) also appears in the error rate analysis of the CPSK system in the presence of Gaussian noise and co-channel interference.

Consider the evaluation of (114) or the evaluation of (117) in particular. If L is a small integer (e.g., ≤ 3) then after some manipulations with Bessel functions these integrals may be computed by numerical means. If L is a moderate or a large number, however, those integrals become extremely difficult to evaluate even with high-speed digital computers. Although, for example, the integral in (117) may be expressed as a generalized hyper-geometric function of L variables, this hyper-geometric function has no tabulation yet and its series expansion involves $(L + 1)$-fold summation, which is of course impractical for numerical computation. On the other hand, accurate bounds on (117) are not easy to obtain. Hence, power-series expansion method presented here will be used to integrate (117).

$$\sigma^2 = \sigma_n^2 + \frac{\Delta}{2} \sum_{l=1}^{L} a_l^2$$

$$a_l' \equiv a_l/\sigma \ \forall l$$

$$\lambda^2 \equiv \frac{\Delta}{2} \sum_{l=1}^{L} a_l'^2 < 1. \quad (118)$$

Then,

$$A_{1,0,\cdots,0} = \frac{4}{\pi} \int_0^\infty \frac{1}{v} J_1(a_0' v) \left[\prod_{l=1}^{L} J_0(a_l' v)\right]$$

$$\cdot \exp(\lambda^2 v^2/2) \exp(-v^2/2)\, dv. \quad (119)$$

Since $[\prod_{l=1}^{L} J_0(a_l' v)] \exp(\lambda^2 v^2/2)$ is even in v, it can be expanded into a power series:

$$\Psi(v) \equiv \exp(\lambda^2 v^2/2) \prod_{l=1}^{L} J_0(a_l' v) = \sum_{m=0}^{\infty} b_{2m} v^{2m}. \quad (120)$$

Substitute (120) into (119) and note that [30]

$$\int_0^\infty x^\mu \exp(-\alpha x^2) J_\nu(\beta x)\, dx = \frac{\Gamma[(\mu + \nu + 1)/2]}{\beta \alpha^{\mu/2}\, \Gamma(\nu + 1)} \exp(-\beta^2/8\alpha)$$

$$\cdot M_{\mu/2, \nu/2}(\beta^2/4\alpha), \quad \text{for Re } \alpha > 0 \text{ and Re } (\mu + \nu) > -1, \quad (121)$$

in which $M_{m,n}$ is the Whittaker function, yields

$$A_{100\cdots 0} = \sqrt{2/\pi}\, a_0' \exp(-a_0'^2/4)\, [I_0(a_0'^2/4) + I_1(a_0'^2/4)]$$

$$+ \sum_{m=1}^{\infty} b_{2m} \pi^{-1}\, 2^{m-3/2}\, a_0'^{-1}\, \Gamma(m + 1/2)$$

$$\cdot \exp(-a_0'^2/4)\, M_{m-1/2,\, 1/2}(a_0'^2/2). \quad (122)$$

The first term in (121) represents the amplitude distortion on the pure signal component when the L intermodulation or "interference" signals in (116) are regarded as an equivalent Gaussian noise with power equal to $\Delta/2 \sum_{l=1}^{L} a_l^2$. The rest of (122) thus represents the corrections that are necessary to obtain an accurate $A_{100\cdots 0}$. Hence, it is only necessary to find an effective way of computing b_{2m}.

Parallel to the derivation of the recurrence relations for $b_{k,l}$ in [16], it can be shown [31] that b_{2n} admits the following one-dimensional version of similar recurrence relations:

$$b_{2n} = \sum_{m=0}^{n-1} (1 - m/n)\, b_{2m}\, \sigma_{2(n-m)}, \quad (123)$$

where

$$\sigma_2 = \lambda^2/2 + \sum_{l=1}^{L} \beta_2^{(l)}$$

$$\sigma_{2m} = \sum_{l=1}^{L} \beta_{2m}^{(l)}, \quad \text{for all } m \neq 1$$

$$\beta_{2m}^{(l)} = \left[\tau_{2m}^{(l)} - \sum_{n-0}^{m-1} n\, \tau_{2(m-n)}^{(l)}\, \beta_{2n}\right]\Big/ \tau_0^{(l)}$$

$$\tau_{2n}^{(l)} \equiv \frac{(-1)^n\, a_l'^{2n}}{(n!\, 2^n)^2}.$$

Similarly, to compute other intermodulation products in (114), the relationship in (123) can be used. Therefore, the problem of evaluating intermodulation products after hard-limiting is nothing but some form of interference and of course can be handled by our approach.

VIII. CONCLUSION

In this paper it has been shown that the wide range of interference or intermodulation problems in digital communications systems can be formulated on a general setting, and that a power series expansion method can be effectively applied to solve these problems. This is the basis of our unified analysis.

Truncation error bounds have been found. Although they are not claimed to be tight, they do show that the method presented here is theoretically rigorous. Of course, to find tighter truncation error bounds may be a significant future research subject.

The power of the unified approach has been demonstrated with examples, for which there has been no other method to provide effective solutions. More specifically, the combined effects of Gaussian noise, intersymbol interference, and co-channel interference on the error performance of MCPSK and MCAPK systems which includes the MCASK (or PAM) systems, have been evaluated. The error probability of a BDPSK system in the presence of Gaussian noise and intersymbol interference has been computed. Also, the intermodulation products at the output of a hard-limiter have been analyzed.

Finally, it should be pointed out that the Chernoff-type bound may be found for the general problem formulated in this paper. However, a really tight or "optimized" Chernoff bound may be extremely difficult to find. The effort required to find an optimized Chernoff bound may be even greater than required to implement the power series expansion approach. After all, the approach presented here provides accurate answers for almost all practical applications, as can be seen from Sections IV–VII of this paper.

APPENDIX A. EXISTENCE OF TAYLOR EXPANSION FOR $\Pi_{k=1}^{K} \phi_k(u)$

Lemma 1

If Assumptions 1–4 (see Section II of the text) are true, then the Taylor series of $\Pi_{k=1}^{K} \phi_k(u)$ exists for any finite $K \leqslant L$; i.e.,

$$\Phi_K(u) \equiv \prod_{k=1}^{K} \phi_k(u)$$

$$= \sum_{l=0}^{\infty} \frac{(-1)^l}{(2l)!} E\left[\left(u \cdot \sum_{k=1}^{K} a_k\right)^{2l}\right]$$

$$= \sum_{l=0}^{\infty} \frac{(-1)^l}{(2l)!} \left(u \cdot \frac{\partial}{\partial t}\right)^{2l} \Phi_K(t)\bigg|_{t=0} . \quad \text{(A-1)}$$

Proof: See [31, Appendix B].

APPENDIX B

Lemma 2

The integral

$$\Lambda_l \equiv \frac{1}{(2\pi)^n} \int_{D \cap D_s} dx \int_{R^n} du \, \phi_0(u) \left[\left(\frac{\partial}{\partial S} \cdot \frac{\partial}{\partial t}\right)^{2l} \Phi_K(t)\right]_{t=0}$$

$$\cdot \exp\left[-j(x - S) \cdot u\right] \quad \text{(B-1)}$$

exists for $l = 0, 1, 2, \cdots, R - 1$, and

$$\Lambda_l = \left[\left(\frac{\partial}{\partial S} \cdot \frac{\partial}{\partial t}\right)^{2l} \Phi_K(t)\right]_{t=0} \int_{D \cap D_s} p_N(x - S) \, dx, \quad \text{(B-2)}$$

where the integral in (B-2) is as defined in (17).

Proof: See [31, Appendix C].

ACKNOWLEDGMENT

The authors would like to thank Dr. K. Bhatnagar for his computer support.

REFERENCES

[1] O. Shimbo and M. I. Celebiler, "The probability of error due to intersymbol interference and Gaussian noise in digital communication systems," *IEEE Trans. Commun. Technol.*, vol. COM-19, pp. 113–119, Apr. 1971.

[2] B. R. Saltzberg, "Intersymbol interference error bounds with application to ideal bandlimited signaling," *IEEE Trans. Inform. Theory*, vol. IT-14, pp. 563–568, July 1968.

[3] R. W. Lucky, J. Salz, and E. J. Weldon, Jr., *Principle of Data Communication*. New York: McGraw-Hill, 1968, p. 65.

[4] W. Feller, *An Introduction to Probability Theory and Its Applications*, vol. II. New York: Wiley, 1966, p. 259.

[5] R. G. Bartle, *The Elements of Real Analysis*. New York: Wiley, 1964, pp. 245, 309.

[6] S. Bochner, *Lectures on Fourier Integrals*. Princeton, N.J.: Princeton University Press, 1959, p. 318.

[7] C. R. Cahn, "Combined digital phase and amplitude modulation communication systems," *IRE Trans. Commun. Syst.*, vol. CS-8, pp. 150–155, Sept. 1960.

[8] R. W. Lucky and J. C. Hancock, "On the optimum performance of N-ary systems having two degrees of freedom," *IRE Trans. Commun. Syst.*, vol. CS-10, pp. 185–192, June 1962.

[9] C. N. Campopiano and B. G. Glazer, "A coherent digital amplitude and phase modulation scheme," *IRE Trans. Commun. Syst.*, vol. CS-10, pp. 90–95, Mar. 1962.

[10] E. Y. Ho and Y. S. Yeh, "Error probability of a multilevel digital system with intersymbol interference and Gaussian noise," *Bell Syst. Tech. J.*, vol. 50, pp. 1017–1023, Mar. 1971.

[11] ——, "A new approach for evaluating the error probability in the presence of intersymbol interference and additive Gaussian noise," *Bell Syst. Tech. J.*, vol. 49, pp. 2249–2265, Nov. 1970.

[12] M. R. Aaron and D. W. Tufts, "Intersymbol interference and error probability," *IEEE Trans. Inform. Theory*, vol. IT-12, pp. 26–34, Jan. 1966.

[13] R. Lugannani, "Intersymbol interference and probability of error in digital systems," *IEEE Trans. Inform. Theory*, vol. IT-15, pp. 682–688, Nov. 1969.

[14] J. M. Aein and J. C. Hancock, "Reducing the effects of intersymbol interference with correlation receivers," *IEEE Trans. Inform. Theory*, vol. IT-9, pp. 167–175, July 1963.

[15] F. E. Glave, "An upper bound on the probability of error due to intersymbol interference for correlated digital signals," *IEEE Trans. Inform. Theory*, vol. IT-18, pp. 356–363, May 1972.

[16] O. Shimbo, R. J. Fang, and M. I. Celebiler, "Performance of M-ary PSK systems in Gaussian noise and intersymbol interference," *IEEE Trans. Inform. Theory*, vol. IT-19, pp. 44–58, Jan. 1973.
—— , "Correction to 'Performance of M-ary PSK systems in Gaussian noise and intersymbol interference,' " *IEEE Trans. Inform. Theory* (Corresp.), vol. IT-19, p. 365, May 1973.

[17] V. K. Prabhu, "Performance of coherent phase-shift-keyed systems with intersymbol interference," *IEEE Trans. Inform. Theory*, vol. IT-17, pp. 418–431, July 1971.

[18] ——, "Some consideration of error bounds in digital systems," *Bell Syst. Tech. J.*, vol. 50, pp. 3127–3151, Dec. 1971.

[19] O. Shimbo, M. I. Celebiler, and R. Fang, "Performance analysis of DPSK systems in both thermal noise and intersymbol interference," *IEEE Trans. Commun. Technol.*, vol. COM-19, pp. 1179–1188, Dec. 1971.

[20] W. M. Hubbard, "The effect of intersymbol interference on error rate in binary differentially coherent phase-shift-keyed system," *Bell Syst. Tech. J.*, vol. 46, pp. 1149–1172, July–Aug. 1967.

[21] J. J. Bussgang and M. Leiter, "Error performance of differential phase-shift transmission over a telephone line," *IEEE Trans.*

Commun. Technol., vol. COM-16, pp. 411–419, June 1968.

[22] O. Shimbo and R. Fang, "Effects of cochannel interference and Gaussian noise in *M*-ary PSK systems," *COMSAT Tech. Rev.*, vol. 3, pp. 183–207, Spring 1973.

[23] A. S. Rosenbaum, "PSK error performance with Gaussian noise and interference," *Bell Syst. Tech. J.*, vol. 48, pp. 413–442, Feb. 1969.

[24] ——, "Binary PSK error probabilities with multiple cochannel interferences," *IEEE Trans. Commun. Technol.*, vol. COM-18, pp. 241–253, June 1970.

[25] V. K. Prabhu, "Error rate consideration for coherent phase-shift keyed systems with cochannel interference," *Bell Syst. Tech. J.*, vol. 48, pp. 743–767, Mar. 1969.

[26] ——, "Error-probability upper bound for coherently detected PSK signals with cochannel interference," *Electron. Lett.*, vol. 5, Aug. 7, 1969.

[27] J. Goldman, "Multiple error performance of PSK systems with cochannel interference and noise," *IEEE Trans. Commun. Technol.*, vol. COM-19, pp. 420–430, Aug. 1971.

[28] J. M. Aein, "On the effects of undesired signal interference to a coherent digital carrier," Institute for Defense Analysis, Paper p-812, IDA Log HQ71-13729, Feb. 1972.

[29] P. C. Jain, "Limiting of signals in random noise," *IEEE Trans. Inform. Theory*, vol. IT-18, pp. 332–340, May 1972.

[30] I. S. Gradshteyn and I. M. Ryzhik, *Table of Integrals, Series and Products* (4th ed. prepared by Yu. V. Geronimus and M. Yu. Tseytlin). New York: Academic, 1965.

[31] R. Fang and O. Shimbo, "Unified analysis of a class of digital systems in additive noise and interferences," Comsat Tech. Memorandum, CL-57-72, Dec. 18, 1972.

COMPUTATIONAL METHOD FOR THE PERFORMANCE EVALUATION OF 4-PHASE PSK TRANSMISSION THROUGH LINEAR AND NONLINEAR CHANNELS*

Francois Assal and Khodadad Betaharon
COMSAT Laboratories
Clarksburg, Maryland 20734

ABSTRACT

This paper describes a computer simulation program to evaluate the performance of 4-phase PSK-PCM transmission through linear and nonlinear communications satellite channels. Measured and computed performance characteristics are presented for linear and nonlinear channels which have been purposefully distorted to demonstrate the validity and accuracy of the computer program.

INTRODUCTION

The 4-phase PSK-PCM satellite communications channel under consideration is illustrated in Figure 1. Two random bit streams having theoretically infinite bandwidth modulate two orthogonal carriers. An earth station transmitter filter restricts the bandwidth of the combined signal components to protect adjacent channels in the common transmission path. In a typical communications satellite with channelized transponders, e.g., INTELSAT IV,[1] TWT input and output filter multiplexers separate the received signals before and recombine them after power amplification, respectively. The band-edge attenuation roll-offs of the input and output filters are selected to minimize time-varying dual path amplification[2] through adjacent transponders. Since helix traveling wave tube amplifiers (TWTAs) with the typical nonlinearities shown in Figure 2 generally spread the 4-phase PSK-PCM signal spectrum,[3] an output filter protects adjacent channels by restricting the bandwidth of the earthbound signals. The earth station receiver filter suppresses adjacent channel interference and maintains the up- and down-link noise before demodulation and detection.

In linear channels, the definition of the filter characteristics is a compromise between adjacent channel interference, multipath, and in-band distortions while attempting to maximize the communications capacity of the system. When instantaneous sampling is utilized as the detection method, filter characteristics are derived, designed, and realized at IF for optimum data transmission.[4] To evaluate the performance of single-carrier transmission through linear channels, O. Shimbo and M. J. Celebiler[5] used the superposition principle to develop a rapidly converging technique for the computation of error probability due to intersymbol interference and Gaussian noise.

In the nonlinear satellite communications channel of Figure 1, where severe restrictions of the signal bandwidth produce large envelope variations at the TWTA input, superposition cannot be used for channel performance evaluation and channel filter optimization. Therefore, Shimbo and Celebiler's computational techniques do not apply and computer simulation programs[6-9] have been developed to evaluate the performance of the channels, including the phase and amplitude nonlinearities of the TWT, by averaging the error probability for several detected samples. Furthermore, as a guide to performance evaluation and critical parameter identification of channels and modems, probability of error characteristics have been computed and presented[8,10] as functions of TWT input backoffs, channel characteristics and modem bit timing, carrier phase recovery, and sampler set time.

To verify the validity and accuracy of the computer simulation program, measured and computed bit error rate characteristics will be compared in this paper for several linear and nonlinear channels which were modified to yield attenuation and group delay ripples. The simulation program will be described and the engineering definition for convergence used throughout this work will be proposed for reasonably well behaved channels.

SIMULATION PROGRAM

Figure 3 is a block diagram of the computer simulation program. Since the modulation of Figure 1 corresponds to bit stream frequency translations by orthogonal carriers, the carrier frequency, f_o, is reassigned as the origin of a low-pass system.

The input signals $X_c(t)$ and $X_s(t)$ consist of pulse waveforms $p(t)$ that are delayed by $k\tau$, where τ is the mid-bit separation between the adjacent pulses, and multiplied by independent random variables C_k, $S_k = \pm 1$ which have a probability 1/2. Typically $p(t)$ is a nearly rectangular waveform of duration τ.

Using a fast Fourier transform subroutine (FFT), the filter computations are performed in the frequency domain. The filter characteristics may be obtained from measured data, analytic expressions, or their pole and zero locations in the complex plane. To simplify bit-timing recovery in the computer program, the filter characteristics are advanced in time so that their mid-band group delay is equal to zero.

The helix TWTA used in a communications satellite responds only to the envelope of the RF signal and may be modeled as a phase nonlinearity followed by an amplitude nonlinearity.[11] These computations are performed in the time domain as stated in the equations of Figure 3. The phase and amplitude characteristics of the TWT may be obtained from measured data or from analytic expressions. If measured data are used, the points should be sufficiently close together since a linear interpolation has been used to estimate the output waveforms.

To obtain optimum bit timing, the periodic instants of time that simultaneously produce the largest eye-pattern[12,13] opening and the smallest scatter between sampled bits of the demodulated 4-phase PSK-PCM signals are selected by a computational process which "averages" over the whole 2N pulse sequence of a particular run of the simulation program. Therefore, minimizing the following expression should produce optimum bit timing, t_s:

$$\frac{\overline{(a_s - \overline{a}_s)^2} + \overline{(b_s - \overline{b}_s)^2}}{a_s^2 + b_s^2}$$

---

*This paper is based upon work performed in COMSAT Laboratories under the sponsorship of the International Telecommunications Satellite Organization (INTELSAT). Views expressed in this paper are not necessarily those of INTELSAT.

Manuscript received December 1, 1978.

where
$$t_s \leq \tau$$

$$a_s = C_k a(kt_s)$$

$$\bar{a}_s = \frac{\displaystyle\sum_{k=1}^{N} C_k a(kt_s)}{N}$$

$$b_s = S_k b(kt_s)$$

$$\bar{b}_s = \frac{\displaystyle\sum_{k=1}^{N} S_k b(kt_s)}{N}$$

and $a(t) - jb(t)$ is the earth station receiver filter output and $2N$ is the total number of consecutive pulses taken in the averaging process.

After the optimum sampling instant is obtained, the carrier phase is recovered by rotating the average sample vector $\bar{a}_s - j\bar{b}_s$ such that the new vector components are equal. To simplify the probability of error computations, the detected symbols are rotated to the first quadrant of the complex phase plane.

In this paper, the carrier-to-noise ratio is computed, as shown in Figure 3, as the ratio between the satellite transmitted average power and the filtered noise power at the sampler input.

Once a sample is detected and measured, the probability that it is in error when Gaussian noise is added in the down-link is easily computed by using the statistics of bandwidth restricted noise. Then, if all the possible combinations within the memory of the filters were considered, the error probability of the system could be computed accurately by averaging. Unfortunately, an extremely large number of pulses would be required for the filter bandwidths and attenuation selectivities that are normally used for efficient INTELSAT satellite transmission, i.e., when the excess bandwidth[10] is approximately 15 percent. For example, if the bit/pulse transient response of the 4-phase PSK channel filters were contained within a 16-bit duration, then each detected pulse would have been perturbed by 15 adjacent and 16 quadrature bits. Then, approximately 10^9 bits would have to be included in the simulation program to precisely compute the probability of error in the nonlinear channel.

For sufficiently well behaved channel filters and nonlinearities such as the communications satellite systems under consideration, it is possible, as indicated below, to take relatively few bits for each computer execution, thereby minimizing the computer resources allocated to the program, and to run the program until convergence is established. The minimum number of samples defining a bit in $\Delta t = \tau *$ and the minimum number of pseudo-random bits defining the length of the bit stream sequence must be selected to avoid aliasing before and after the nonlinearities in both the time and frequency domains, since fast Fourier transform subroutines compute periodic structures in these domains.

If it is assumed that errors are never caused by channel imperfections, the probability of error characteristics of noisy channels are essentially

controlled by those detected samples which are close to the origin and are hence most likely to produce errors. Therefore, practical convergence for a specific nonlinear channel is determined as follows:

a. the transient behavior of the channel should be clearly understood; (see Sections 2.9 and 4.1 of Reference 8)

b. computations of histograms and/or distribution functions of the detected samples, as shown in Figure 4, should be analyzed to identify and explain anomalies such as isolated samples specifically near the origin; and

c. computations of carrier-to-noise ratio degradations for a constant probability of error such as 10^{-4} versus the number of program runs, as shown in Figure 5, should indicate that an average carrier-to-noise ratio is reached within some acceptable engineering bounds (e.g., 0.1 dB peak-to-peak).

Results obtained by the simulation and the Shimbo and Celebiler programs were compared for typical linear channels. The resulting probability of error characteristics in the range of 10^{-2} to 10^{-9} were well within 0.05 dB after the fifth run of the 100-bit/run simulation program.

Once the number of bits per run is selected, the total number of runs required for an accurate simulation must be reviewed for each case, since it is highly dependent on the filters and since filtering after the TWT nonlinearities produces an increased scatter of the detected sample voltages as a function of input backoff from saturation. To highlight the change in the distribution of detected samples, Figure 6 presents the 1000-bit histograms for a nonlinear channel driven to 6-dB input backoff and to saturation, respectively. Since the distribution of detected samples for the TWT at saturation is broader, displays some irregularities for sample voltages ranging from 0.37 to 0.60, and has samples that are closer to the origin, additional computer runs may be required at this drive level to ensure convergence. The convergence test for this specific case has been performed, and is displayed in Figure 5.

<u>EXPERIMENTAL VERIFICATION</u>

Bit error rate characteristics were measured for several linear and nonlinear channels, i.e., with group delay equalized filter and with controlled mismatches that produce both attenuation and group delay ripples simultaneously. The measured transfer characteristics were used in the simulation program to compute the bit error rate characteristics for each case. A comparison of the results provides experimental verification of this time and frequency domain simulation program.

Since the test setup used for the nonlinear measurements requires up- and down-conversion to access the TWT at RF, the resulting transfer characteristics are different for the linear and nonlinear cases. Therefore, the test setups and the performance characteristics are presented separately.

<u>The Linear Channel</u>

The experimental test setup used for the measurements is shown in Figure 7. The inputs to the 4-phase PSK modulator are P and Q generated pseudo-random bit streams that are clocked in at 67.2 MHz. Low-pass-filtered bit streams modulate orthogonal 70-MHz carriers. The resulting signals are combined and bandpass filtered before transmission through a noisy channel. The received noisy signal is bandwidth limited before

---

*150-200 data points are taken in τ.

coherent demodulation with a hardwired carrier and
detected with a recovered clock. The detected P' and
Q' bit streams and the transmitted P and Q bit streams
are compared to produce a bit error signal. Measured
bit error rates as a function of the carrier-to-noise
ratios measured after the receiver bandpass filter are
used to verify the computed characteristics.

A 67.2-Mbit/s 4-phase PSK modem (modulator-demodu-
lator) was modified for the measurements since the computer model does not simulate exactly the clock and
carrier recovery loops of the circuits. The clock recovery loop was not bypassed since it caused negligible
degradation after lengths ℓ_p and ℓ_Q were properly adjusted (see Figure 7). The carrier was hardwired to
eliminate phase ambiguity. Since the carrier frequency
(70 MHz) is very low compared with the (sin x/x) spectrum of the 33-Mbaud digital signal composed of
rectangular waveforms, the performance of the experimental test setup was improved by adding four 25-MHz
elliptic function low-pass filters (Figure 8) with
transmission nulls at 70 MHz. Further improvements
were obtained by adjusting the modulator and the
demodulator of the test setup to ensure orthogonality
between, and equal power in, the P and Q channel waveforms.

After the modem was optimized as stated above for
4-phase PSK transmission, the transfer characteristics
of the input and output filters were measured and are
presented in Figures 9 and 10, respectively. These
characteristics which include the overall frequency
response from the modulator output to the transmit
filter output and from the transmit filter output to
the demodulator input, respectively were used in the
simulation program. The computed and measured performance characteristics of the linear channel which
are presented in Figure 11, provided a first milestone
in establishing both the necessary match between the
simulation program and the test setup, and the accuracy of the computer program.

Attenuation and group delay ripples were produced
simultaneously, as shown in Figure 7, by introducing a
resistive mismatch followed by cable lengths in a
T-junction to control both the magnitude and frequency
of the ripples. For example, Figure 12 presents the
transfer characteristics measured between points E and
G of Figure 7, when the attenuator was set at 9 dB.
Enough cables were introduced to produce 9.5 ripples in
the passband with 0.8-dB attenuation ripples and
28-ns peak-to-peak group delay ripples. With 9.5
ripples in the passband, Figure 13 compares measured
and computed performance characteristics for five
ripple sizes.

Figure 14 presents one set of transfer characteristics with approximately 3.4 ripples. Figure 15
compares measured and computed performance characteristics for three different ripple magnitudes. Finally,
Figure 16 compares the measured and computed relative
degradation caused by bit timing when the error probability is set at 10^{-4}.

The Nonlinear (TWT) Channel

The experimental test setup for the nonlinear
channel is obtained by adding the components shown in
Figure 17 between points D and D' of the linear test
setup of Figure 7. The additional components consist
of up- and down-converters using the same local oscillator source, a low-level TWT with a bandpass filter
at its input, a high-level TWT, and an attenuator
between the two TWTs to control the drive levels into
the nonlinear high-level TWT. This arrangement ensures
linear operation of the low-level TWT, since a constant

25-dB input backoff from saturation can be maintained
for all the measurements.

Configured intentionally to simulate up-link noise,
the low-level TWT produces a broadband thermal noise
that is subsequently amplified by the nonlinear high-
level TWT, causing further distortion of 4-phase PSK-
PCM signals. Therefore, broadband up-link noise has
been included in the simulation program, as indicated
in Figure 3, to approximate the experimental conditions
of the aforementioned test setup. This noise is added
in the time domain using a Gaussian random generator.
The resulting spectrum is subsequently bandpass filtered to approximate the measured conditions that were
observed on a spectrum analyzer at the input to the
high-level TWT.

In general, accurate predictions of P(BER) < 10^{-3}
may require an extremely large number of $\sim$100-bit computer runs. However, the up-link signal-to-noise
density ratio used in all the measurements is always
equal to approximately 22 dB, the largest down-link
carrier-to-noise ratios considered in this paper are
always less than or equal to 15 dB, and the high-level
TWT output is always bandwidth limited with filter
cutoff frequencies less than 1.3 times the Nyquist
frequency. Therefore, it was observed through numerous
computations that 10 runs or approximately 1,000 bits
provide a practical computational compromise that is
sufficiently accurate.

As the test setup was being prepared, the error
probability characteristics were measured and computed
for a number of cases as outlined in Figure 18. Before
the high-level TWT was added, the performance characteristics were computed, first with the low-level TWT
noise added statistically in the down-link, and then
with the up-link contribution added in the time domain
as described above. The results were identical.
Figure 19 presents calculated and measured performance
characteristics for three ripple cases. Finally,
Figure 20 compares the measured and computed relative
degradation caused by bit timing when the error probability is set at 10^{-4} and the high level TWT is at
saturation.

Conclusions

This paper has described a computational approach
that can be used to evaluate the performance of 4-phase
PSK transmission through linear and nonlinear channels.
Computed characteristics have been experimentally
verified for several channels having group delay and
attenuation ripples. In the nonlinear case, a broad
band of noise generated by the low-level TWT was also
included in the time domain to simulate the actual
conditions of the test setup.

For all the channel characteristics presented
above, the computer predictions, using at least ten
100-bit runs, produced results which are well within
0.2 dB of the measurements.

ACKNOWLEDGMENTS

The authors would like to thank Mr. A. Berman and
Dr. C. Mahle for technical discussions and guidance,
and Dr. H. Hung for support during the experimental
phase of this program.

REFERENCES

1. "The INTELSAT IV Communications System,"
   P. L. Bargellini, Editor, COMSAT Technical Review,
   Vol. 2, No. 2, Fall 1972, pp. 427-572.

2. A. L. Berman, C. Mahle, and M. R. Wachs, "Transmission Modeling," COMSAT Technical Review, Vol. 2, No. 2, Fall 1972, pp. 489-528.

3. G. Robinson, O. Shimbo, and R. Fang, "PSK Signal Power Spectrum Spread Produced by Memoryless Nonlinear TWTs," COMSAT Technical Review, Vol. 3, No. 2, Fall 1973, pp. 227-256.

4. F. T. Assal, "Approach to a Near-optimum Transmitter-Receiver Filter Design for Data Transmission Pulse-shaping Networks," COMSAT Technical Review, Vol. 3, No. 2, Fall 1973, pp. 301-322.

5. O. Shimbo and M. J. Celebiler, "The Probability of Error Due to Intersymbol Interference and Noise in Digital Communication Systems," IEEE Transactions on Communication Technology, Vol. COM-19, No. 2, April 1971, pp. 113-120.

6. T. Muratani, I. Matsushita, Y. Tsuji, and T. Hara, "Simulation of a PSK Transmission System Including the Nonlinear Satellite Channel," Second International Conference on Digital Satellite Communications, Paris, November 1972, Proc., pp. 182-192.

7. D. L. Hedderly and L. Lundquist, "Computer-simulation of a Digital Satellite Communications Link," IEEE Transactions on Communication Technology, Vol. COM-21, April 1973, pp. 321-325.

8. F. Assal and K. Betaharon, "Performance Degradation Mechanisms and Signal Performance Evaluation for Single-Carrier 4-Phase PSK-PCM Satellite Communications Channels," COMSAT Laboratories Technical Memorandum CL-29-74, September 1974.

9. H. P. Chan, D. P. Taylor, and S. S. Haykin, "A Comparative Study of Digital Modulation Techniques for Single Carrier Satellite Communications," Third International Conference on Digital Satellite Communications, Kyoto, Japan, November 1975, Proc., pp. 56-64.

10. F. Assal and K. Betaharon, "In-Band Degradations for Single-Carrier Transmission in Nonlinear Satellite Channels," NATO Advanced Study Institute Series, Series E: Applied Science, No. 12, Noordhoff-Leyden, 1975, pp. 347-358.

11. A. L. Berman and C. E. Mahle, "Nonlinear Phase Shift in Traveling Wave Tubes as Applied to Multiple Access Communications Satellites," IEEE Transactions on Communication Technology, Vol. COM-18, No. 1, February 1970, pp. 37-48.

12. W. R. Bennet and J. R. Davey, Data Transmission, New York: McGraw-Hill Co., 1965, pp. 61-63, 99-110.

13. R. W. Lucky, J. Salz, and E. J. Weldon, Jr., Principles of Data Communication, New York: McGraw-Hill Co., 1968, pp. 45-65.

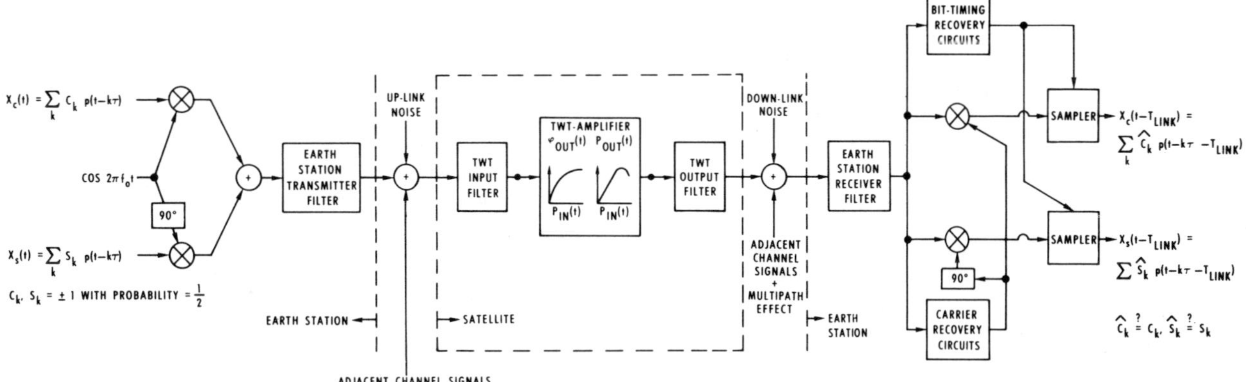

Figure 1.   4-Phase PSK-PCM Satellite Communications Channel

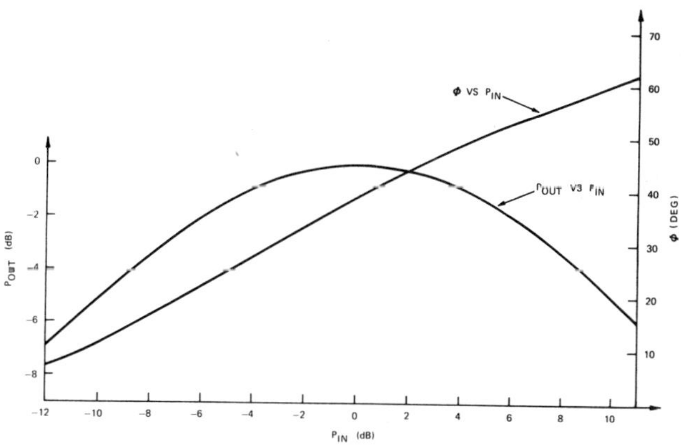

Figure 2.   Typical TWT Characteristics

251

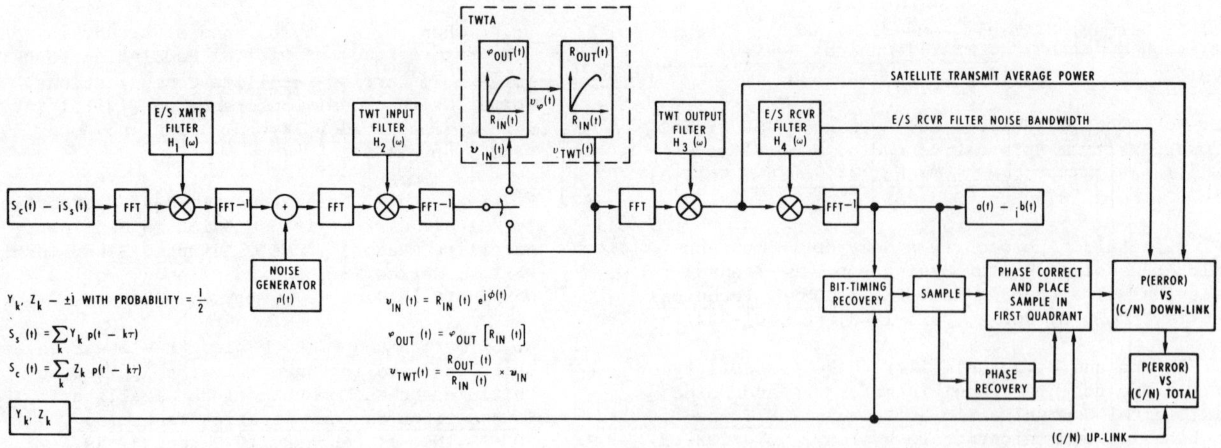

$Y_k,\ Z_k = \pm i$ WITH PROBABILITY $= \frac{1}{2}$

$S_s(t) = \sum_k Y_k\, p(t - k\tau)$

$S_c(t) = \sum_k Z_k\, p(t - k\tau)$

$\nu_{IN}(t) = R_{IN}(t)\, e^{i\phi(t)}$

$\varphi_{OUT}(t) = \varphi_{OUT}\left[R_{IN}(t)\right]$

$\nu_{TWT}(t) = \dfrac{R_{OUT}(t)}{R_{IN}(t)} \times \nu_{IN}$

Figure 3. Computer Simulation Block Diagram (FFT and FFT^{-1} are time-to-frequency and frequency-to-time fast Fourier transform subroutines.)

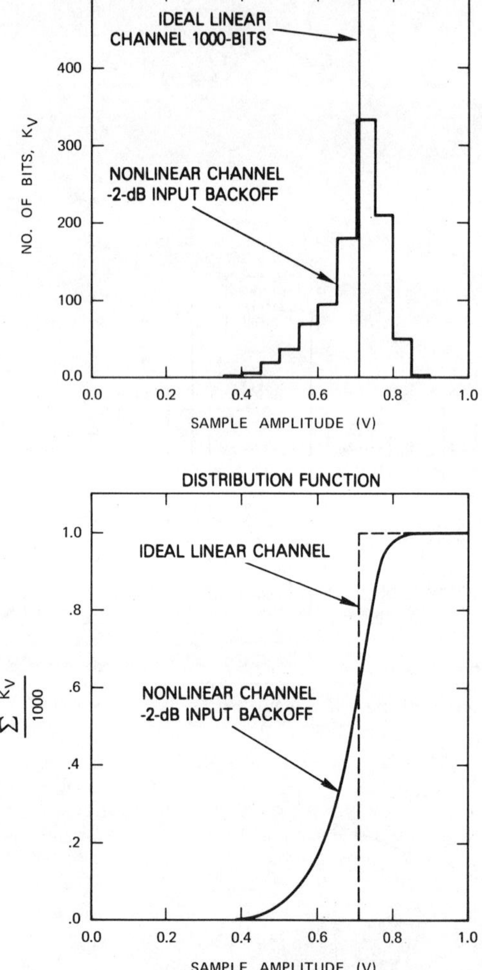

Figure 4. 1000-Bit Histogram and Distribution Functions of Detected Samples (filter characteristics optimized as indicated in Ref. 4, to avoid intersymbol interference)

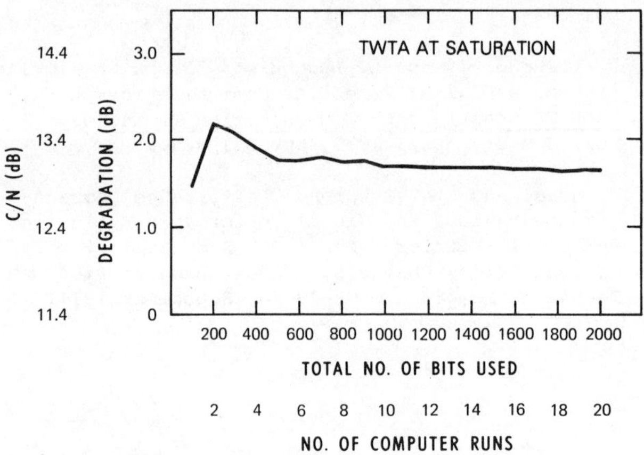

Figure 5. Convergence Test for Average Performance Degradation at P (BER) = 10^{-4} vs Number of Computed 100-Bit Simulation Runs

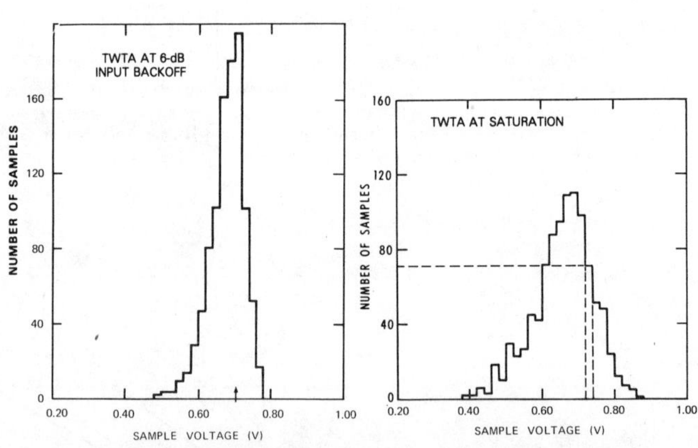

Figure 6. 1000-Bit Histograms for Satellite TWTA Amplifier at 6-dB Input Backoff and at Saturation, Respectively

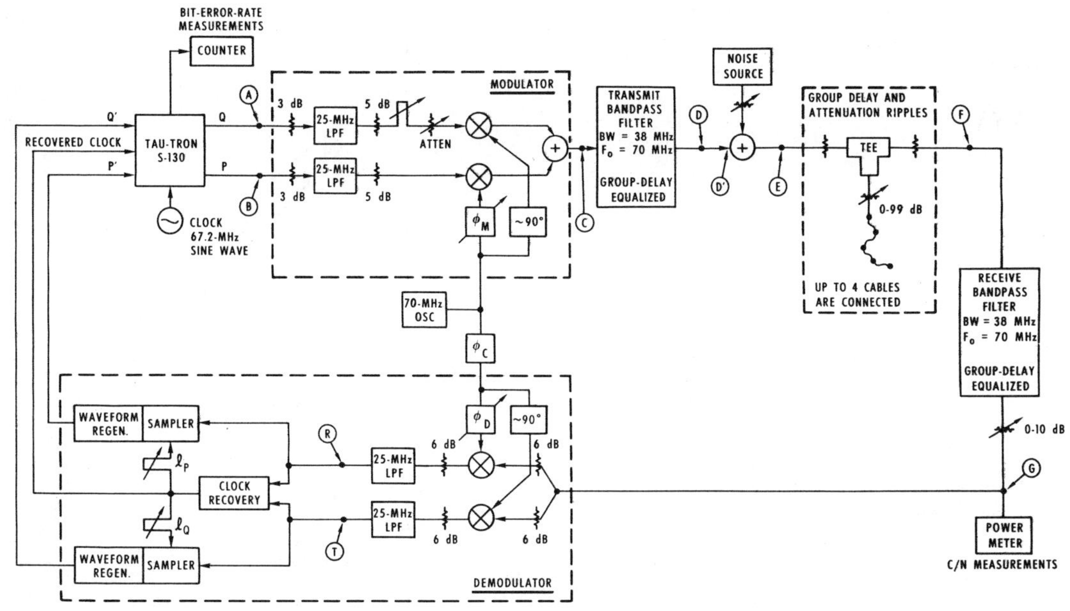

Figure 7.   Block Diagram of the Experimental Test Setup (linear case)

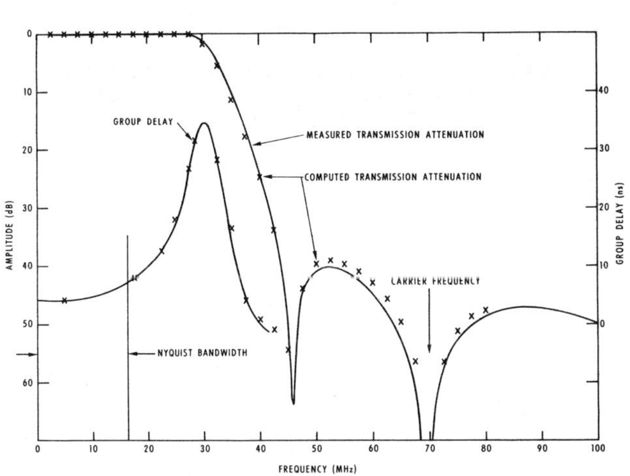

Figure 8.   Measured and Computed Characteristic
for One of the Low-pass Elliptic Function
Filters Used in the Modem at Baseband

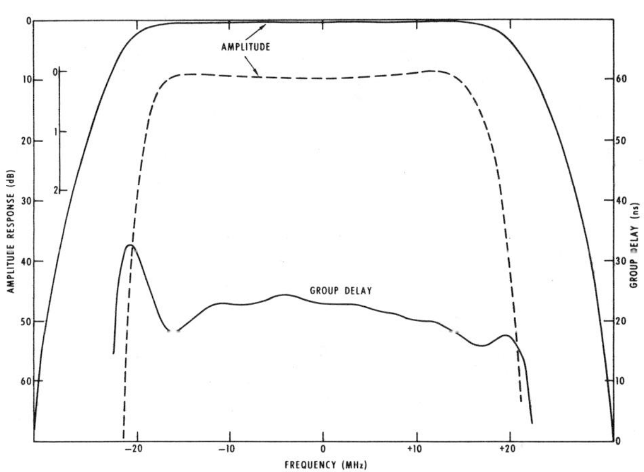

Figure 9.   Transmission Characteristics
of the Transmit IF Filter

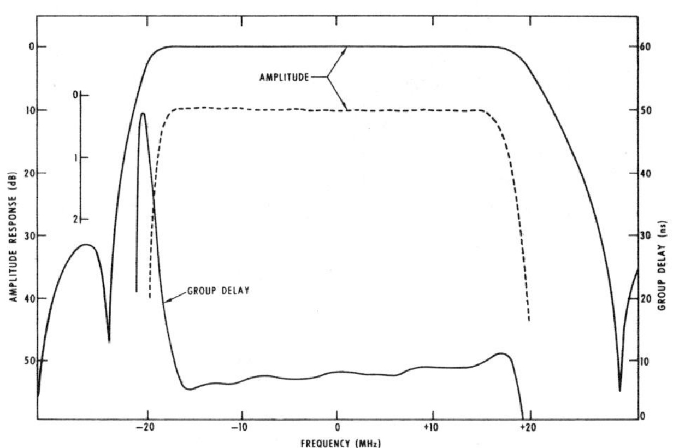

Figure 10.   Measured Transmission Attenuation
Characteristics of the Receive IF Filter

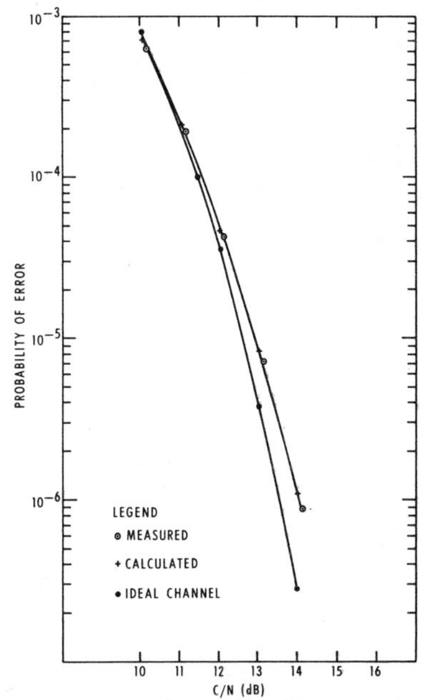

Figure 11.   Computed and Measured Performance
Characteristics of the Linear
Channel (back-to-back modem)

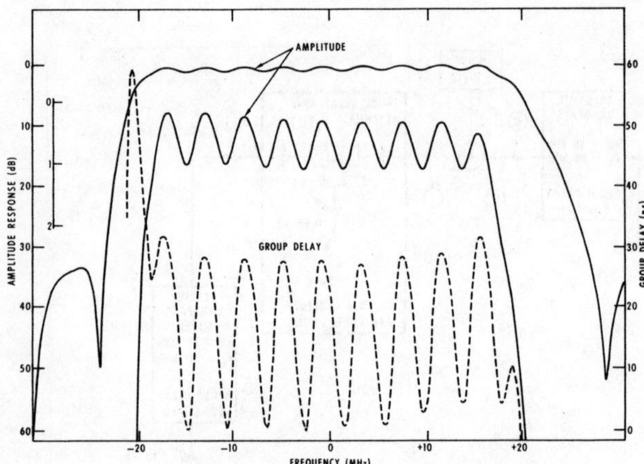

Figure 12.   9.5 Ripple Cycles Introduced in the
Amplitude and Group Delay of the
Received IF Filter

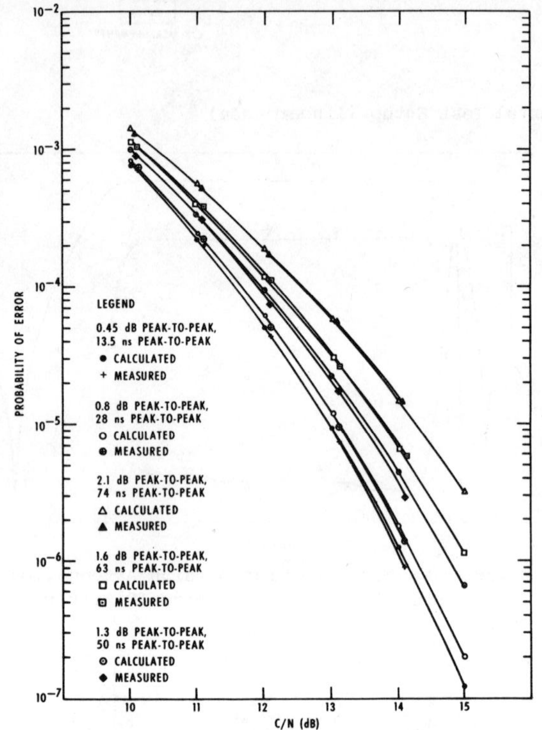

Figure 13.   Probability of Error vs
Carrier-to-Noise Ratio for 9.5 Amplitude
and Group Delay Ripples (linear case)

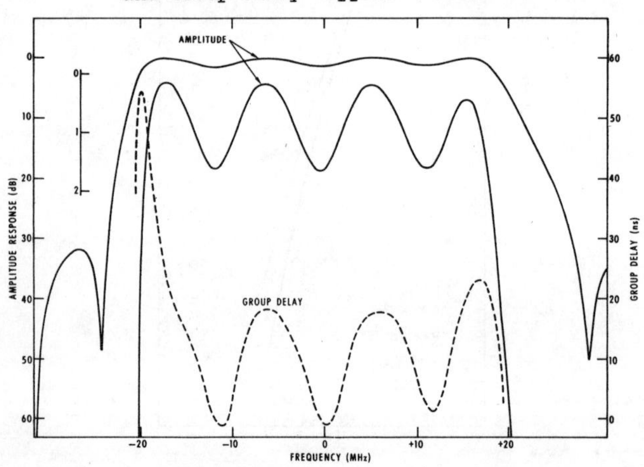

Figure 14.   3.4 Ripple Cycles Introduced in the
Amplitude and Group Delay of the
Received IF Filter

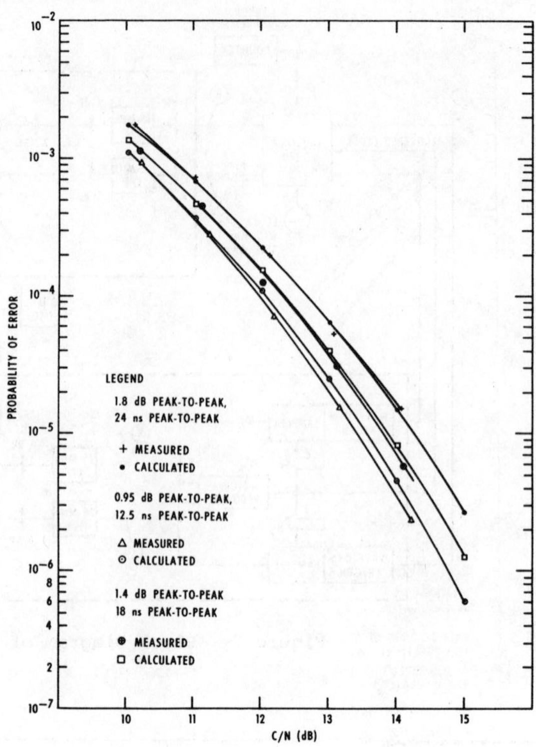

Figure 15.   Comparison Between Measurements
and Calculations for 3.4 Amplitude and
Group Delay Ripples

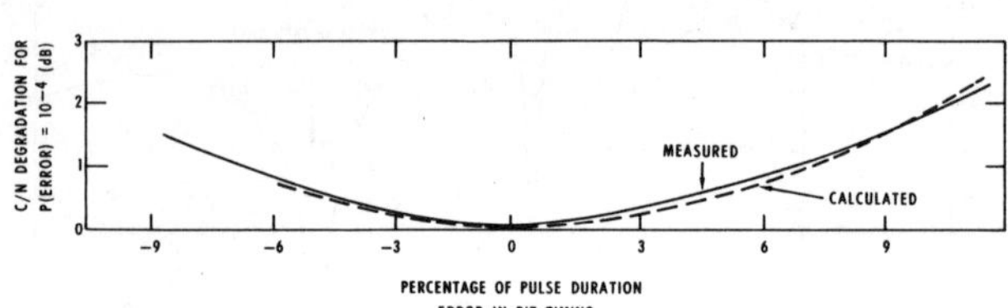

Figure 16.   Bit Timing Effect on Probability
of Error (linear case)

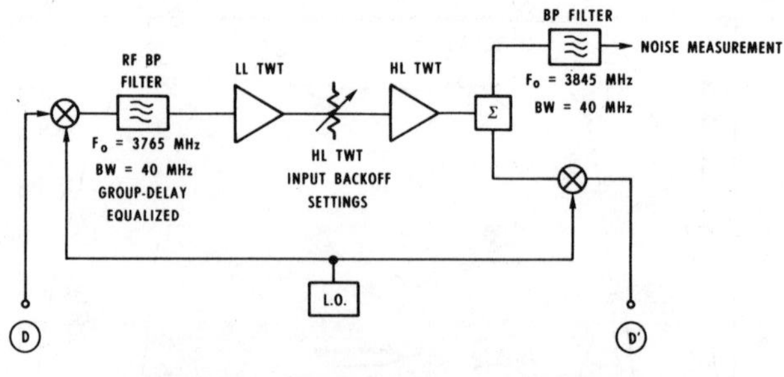

Figure 17.   Block Diagram of Nonlinear
Experimental Test Setup Inserted in
Points D and D' of Figure 20

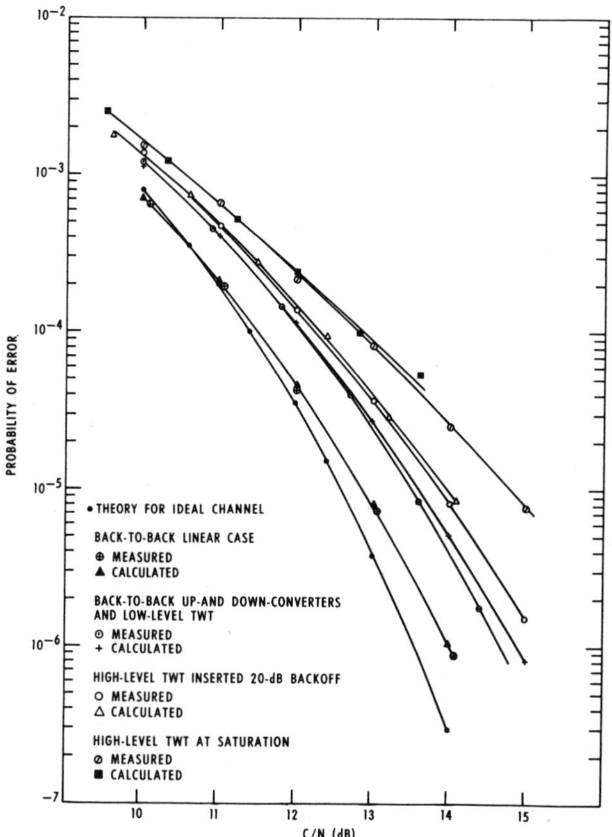

Figure 18. Comparison of Computed and Measured
Performance of Linear (IF only), Linear
(IF and RF only), and Nonlinear Cases

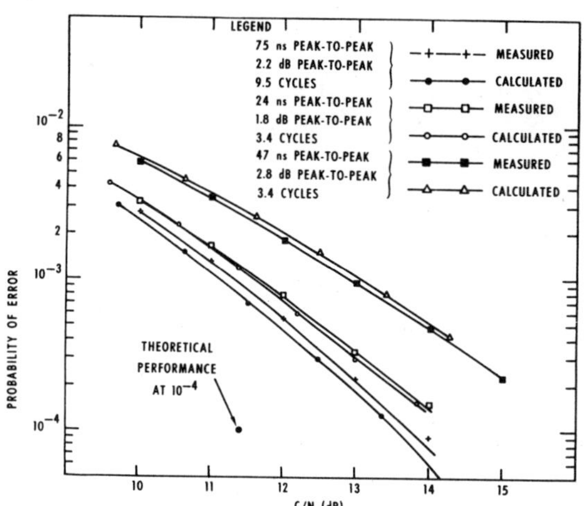

Figure 19. Comparison of Computed and Measured
Performance of Nonlinear Case (HL TWT at
saturation) When Amplitude and Group
Delay Ripples are Introduced

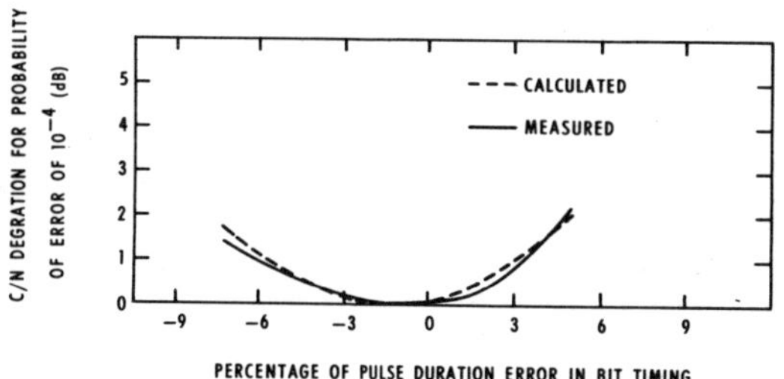

Figure 20. Bit Timing Error Effect for the
Nonlinear Case (at saturation)

3.3.4
MSK and Offset QPSK Modulation

STEVEN A. GRONEMEYER, MEMBER, IEEE, AND ALAN L. McBRIDE, MEMBER, IEEE

Abstract—Minimum shift keying (MSK) and offset keyed quadrature phase shift keying (OK-QPSK) modulation techniques are often proposed for use on nonlinear, severely band-limited communication channels because both techniques retain low sidelobe levels on such channels, while allowing efficient detection performance. A more detailed performance comparison of the two techniques on such channels is, therefore, of interest. In this paper a Markov process representation is developed which is applicable to either the MSK or OK-QPSK waveform. This representation is employed to illustrate the similarity between the modulation processes and to obtain the autocorrelations and power spectral densities of the two waveforms. This Markov process representation may be similarly employed with other modulation waveforms of the same class. The autocorrelations and power spectral densities of MSK and offset QPSK provide initial insight to expected performance on band-limited channels.

Paper approved by the Editor for Communication Theory of the IEEE Communications Society for publication after presentation at the National Telecommunications Conference, New Orleans, LA, December 1975. Manuscript received September 20, 1975; revised February 9, 1976.

The authors are with the Collins Radio Group, Rockwell International Corporation, Dallas, TX 75207.

The results of a digital computer simulation are presented. The simulation compares the bit error rates (BER's) of MSK and offset QPSK on nonlinear, band-limited double-hop links such as encountered in satellite communications. The simulation results are presented as E_b/N_0 degradation with respect to ideal detection versus channel noise bandwidth. The error probability was used as a performance metric, and equal adjacent channel interference as a constraint. For the channels simulated, MSK is found to provide superior performance when the channel noise bandwidth exceeds about 1.1 times the binary data rate. For narrower bandwidths, offset QPSK provides superior performance.

I. INTRODUCTION

BOTH minimum shift keying (MSK) [1] and offset keyed quadrature phase shift keying (OK-QPSK) modulation have been considered for use on band-limited, nonlinear channels as an alternative to conventional (nonoffset) QPSK for several reasons. If either an MSK or offset QPSK waveform is band-limited and then hard-limited, the degree of regeneration of the filtered sidelobes is less than is the case for conventional QPSK [2]. Both techniques achieve the matched filter coherent detection bit error rate (BER) performance of antipodal PSK on linear, infinite bandwidth, white Gaussian noise (WGN), perfect reference channels. Either technique has carrier reference recovery features providing an advantage with respect to conventional QPSK [3]-[5]. A feature of MSK which is often useful is that it can be noncoherently detected by a discriminator [6], whereas QPSK systems require either a fully coherent or differentially coherent detection system. This noncoherent detection property of MSK permits inexpensive demodulation when the received signal-to-noise ratio is adequate, yet allows for coherent detection with efficiency identical to coherent QPSK in limited signal-to-noise ratio situations. Investigations of the coherent detection performance of MSK and conventional QPSK on band-limited or combined band-limited and nonlinear channels have previously been reported [7], [8]. This paper develops the theory and connections between MSK and offset QPSK and presents a performance comparison between the two techniques.

MSK can be viewed as either a special case of continuous phase frequency shift keying (CPFSK), or a special case of offset QPSK with sinusoidal symbol weighting. In this paper offset QPSK, referenced without further qualification, will refer to "square-pulse" offset QPSK. Special cases of offset QPSK will be qualified by the appropriate symbol weighting used, such as the half-cycle sinusoidal weighting of MSK. The relationship between the two views of MSK is developed by constructing a first-order Markov process representation which has a transition probability matrix which is common to both MSK and offset QPSK. The Markov representation is then used to derive the autocorrelation and power spectral density properties of MSK and offset QPSK which provide insight to the BER performance of the two modulation techniques.

The results of a digital computer simulation of the two techniques over nonlinear, band-limited channels are presented. A definitive comparison of the two techniques on such channels is difficult, since the details of channel nonlinearities and passband characteristics strongly influence system performance. The approach used in comparing the two techniques

is to simulate the BER performance of each over identical channels, with post-simulation adjustment to account for an adjacent channel interference constraint. While the receiver implementations are not optimum over the channels considered, the simulation results provide relative performance measures of the MSK and offset QPSK modulation techniques as a function of system parameters.

The results of the simulation show that both MSK and offset QPSK require a certain critical bandwidth and, for narrower channels, performance rapidly degrades due to intersymbol interference and pulse distortion. For the type of filter used in the simulation, the BER performance of MSK is found to be superior to that of offset QPSK only when the channel bandwidth exceeds about 1.1 times the binary data rate. This relationship holds when both modulation techniques are subjected to the same channel filtering, or when a constant adjacent channel interference constraint is applied in a multiple carrier situation.

II. CHARACTERIZATION OF MSK

When viewed as CPFSK, the MSK waveform can be expressed as [9]

$$y_{\text{MSK}}(t) = \cos\left(\omega_c t + \frac{\pi u_k}{2T} t + x_k\right), \qquad kT \leqslant t \leqslant (k+1)T$$

$$(1)$$

where ω_c is the carrier, or center, radian frequency, $u_k = \pm 1$ is bipolar data being transmitted at a rate $R = 1/T$, and x_k is a phase constant which is valid over the kth binary data interval $kT \leqslant t \leqslant (k+1)T$. Fig. 1(a) illustrates the FSK nature of the MSK waveform, with a radian frequency $\omega_c + \pi/2T$ being transmitted for $u_k = 1$ and radian frequency $\omega_c - \pi/2T$ being transmitted for $u_k = -1$. The tone spacing in MSK is one-half that employed in conventional orthogonal FSK modulation, giving rise to the name "minimum" shift keying. During each T second data interval, the value of x_k is a constant determined by the requirement that the phase of the waveform be continuous at the bit transition instants $t = kT$. Applying this requirement to the argument of (1) results in the recursive phase constraint

$$x_k = x_{k-1} + (u_{k-1} - u_k)\frac{\pi k}{2}. \qquad (2)$$

Fig. 1(b) illustrates the continuous phase, constant amplitude MSK waveform. For coherent detection, a reference value of x_k, say x_0, can be set to zero without loss of generality. This assumption will be used in all that follows, with the result, using (2), that $x_k = 0$ or π, modulo 2π.

Define $\theta(t)$ as

$$\theta(t) = x_k + \left(\frac{\pi u_k}{2T}\right) t. \qquad (3)$$

$\theta(t)$ is a piecewise-linear phase function of the MSK waveform

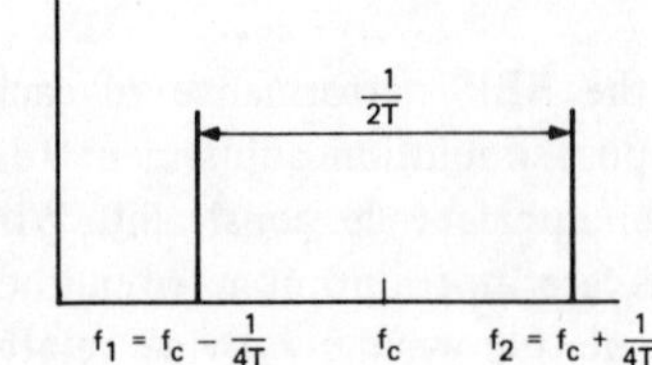

(A) TONE SPACING IN MSK. DATA RATE IS 1 BIT PER T SECONDS. CARRIER FREQUENCY IS $\omega_c = 2\pi f_c$ RAD/SEC

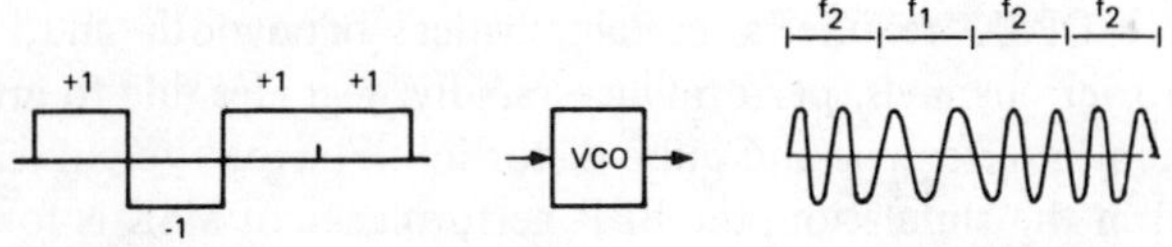

(B) CONTINUOUS PHASE NATURE OF MSK

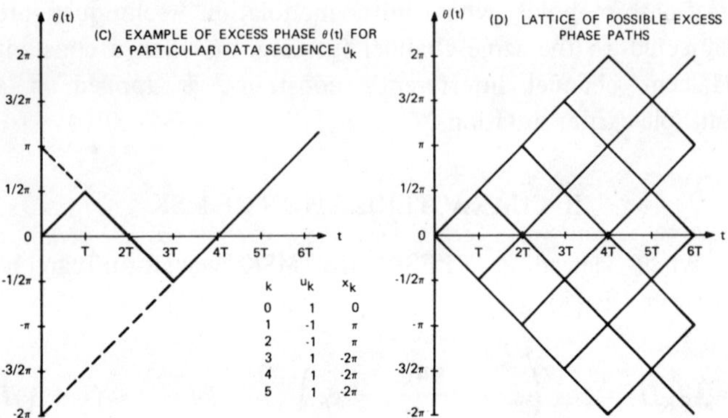

Fig. 1.   Characteristics of MSK. (a) MSK tone spacing. (b) Continuous phase MSK waveform. (c) Example of excess phase function. (d) MSK excess phase trellis.

in excess of the carrier term's linearly increasing phase. Using the recursive phase constraint, $\theta(t)$ is plotted in Fig. 1(c) for a particular data sequence u_k. The phase constant x_k is the phase axis intercept and $\pi u_k/2T$ is the slope of the linear phase function over each T second interval. The phase function in Fig. 1(c) is a particular path along the phase trellis of possible paths shown in Fig. 1(d). Fig. 1(d) illustrates that over each T second interval, the phase of the MSK waveform is advanced or retarded precisely $90°$ with respect to carrier phase, depending upon whether the data for that interval is +1 or −1, respectively.

Using trigonometric identities and the property that $x_k = 0, \pi$ modulo 2π, the MSK waveform representation of (1) can be rewritten as

$$y_{\text{MSK}}(t) = \cos(x_k)C(t)\cos(\omega_c t) - u_k \cos(x_k)S(t)\sin(\omega_c t)$$

$$kt \leqslant t \leqslant (k+1)T \qquad (4)$$

where

$$C(t) = \cos(\pi t/2T)$$

$$S(t) = \sin(\pi t/2T).$$

This representation of the MSK waveform can be viewed as being composed of two quadrature data channels. The in-phase channel, or I channel, is identified as $\cos(x_k)C(t)\cos(\omega_c t)$, where $\cos(\omega_c t)$ is the carrier, $C(t)$ is the sinusoidal symbol

weighting, and $\cos(x_k)$ is the data-dependent term. Similarly, the quadrature-phase channel, or Q channel, is identified as $u_k \cos(x_k)S(t)\sin(\omega_c t)$, where $\sin(\omega_c t)$ is the quadrature carrier term, $S(t)$ is the sinusoidal symbol weighting, and $u_k \cos(x_k)$ is the data-dependent term. Since the data, u_k, can change every T seconds, it might appear that the data terms $\cos(x_k)$ and $u_k \cos(x_k)$ in (4) can also change every T seconds. To the contrary, it is shown in Appendix A that as a result of the continuous phase constraint the term $\cos(x_k)$ can only change value at the zero crossings of $C(t)$ and the term $u_k \cos(x_k)$ can only change value at the zero crossings of $S(t)$. Thus, the symbol weighting in either the I or Q channel is a half-cycle sinusoidal pulse of duration $2T$ seconds and alternating sign. The I and Q channel pulses are skewed T seconds with respect to one another. Finally, the data are conveyed at a rate of one bit per $2T$ seconds in each the I and Q channels by weighting the I and Q channel pulses by $\cos(x_k)$ and $u_k \cos(x_k)$, respectively. Recall that for coherent detection, $x_k = 0, \pi$ modulo 2π, so that both $\cos(x_k)$ and $u_k \cos(x_k)$ take on only the values ±1.

In the discussion above, the MSK waveform was initially viewed as a CPFSK waveform. As a result, the quadrature signaling waveform, derived as (4), contained data terms in which the binary data u_k appeared encoded as $\cos(x_k)$ and $u_k \cos(x_k)$ and in addition, the symbol weighting pulses in either the I- or Q-channel alternated in sign. However, for bit-to-bit independent data u_k, the signs of successive I- or Q-channel pulses are also random from one $2T$ second pulse interval to the next. Thus, when viewed as a quadrature signaling waveform, (4) can be rewritten with a more straight-forward data "encoding" as

$$y_{\text{MSK}}(t) = \begin{cases} u_{2k-1}C[t - 2kT]\cos(\omega_c t) \\ \quad - u_{2k-2}S[t - (2k-2)T]\sin(\omega_c t), \\ \quad (2k-1)T \leqslant t \leqslant 2kT \\ u_{2k-1}C[t - 2kT]\cos(\omega_c t) \\ \quad - u_{2k}S(t - 2kT)\sin(\omega_c t), \\ \quad 2kT \leqslant t \leqslant (2k+1)T. \end{cases} \qquad (5)$$

The "encoding" used in (5) is to demultiplex the data stream u_k into odd and even data streams which are used to determine the symbol pulse signs of the I and Q channels during odd intervals, $(2k-1)T \leqslant t \leqslant (2k+1)T$, and even intervals, $2kT \leqslant t \leqslant (2k+2)T$, respectively. This process is illustrated in Fig. 2.

III. COMPARISON AND MARKOV PROCESS REPRESENTATION OF MSK AND OFFSET QPSK

Equation (5) is one case of a class of modulation waveforms which can be defined by replacing the symbol weighting shapes $C(t)$ and $S(t)$ by a variety of other pulse shapes. In this paper we give particular attention to (rectangular pulse) offset QPSK, defined as

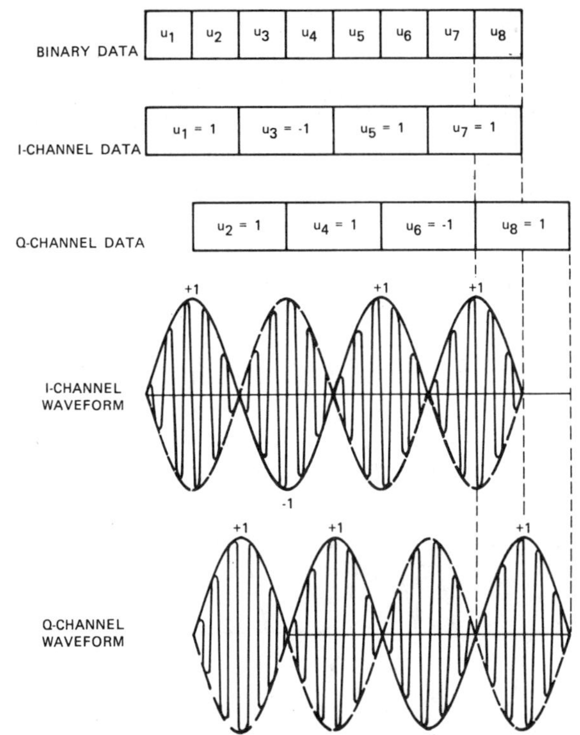

Fig. 2.   Data and symbol timing relationships in MSK waveform.

$$y_{\text{O-QPSK}}(t) = \begin{cases} u_{2k-1}S_{qc}(t-2kt)\cos(\omega_c t) \\ \quad -u_{2k-2}S_{qs}[t-(2k-2)T]\sin(\omega_c t), \\ \quad (2k-1)T \leqslant t \leqslant 2kT \\ u_{2k-1}S_{qs}(t-2kT)\cos(\omega_c t) \\ \quad -u_{2k}S_{qs}(t-2kT)\sin(\omega_c t), \\ \quad 2kT \leqslant t \leqslant (2k+1)T \end{cases} \tag{6}$$

where

$$S_{qs}(t) = \begin{cases} 1/\sqrt{2}, & 0 \leqslant t \leqslant 2T \\ 0, & \text{elsewhere} \end{cases}$$

$$S_{qc}(t) = \begin{cases} 1/\sqrt{2}, & -T \leqslant t \leqslant T \\ 0, & \text{elsewhere.} \end{cases}$$

The only difference between the MSK waveform (5) and the offset QPSK waveform (6) is that the former employs a half-cycle sinusoidal symbol pulse and the latter employs a rectangular pulse.

Fig. 3 illustrates typical waveforms for MSK, offset QPSK, and conventional QPSK waveforms. The binary data rate in each case is assumed to be $1/T$, so that the phase transitions in the MSK and offset QPSK waveforms occur every T seconds, but the conventional QPSK waveform has transitions only every $2T$ seconds. Typical phase transitions for each waveform are shown. The phase transitions shown for MSK are referenced to carrier phase, and accumulate linearly over T seconds. Offset and conventional QPSK have abrupt phase

transitions. In offset QPSK, the transitions can only be $+90°$ or $-90°$, while in conventional QPSK, $180°$ phase reversals are also possible. As will be shown later, the continuous phase nature of MSK is responsible for the more rapid spectral density fall-off for this waveform as compared to offset or conventional QPSK.

The class of quadrature modulation techniques employing offset, time-skewed symbol pulses in the quadrature channels, can be represented by a first-order Markov process. This process is completely described by the state transition probability matrix, initial state probabilities, and the waveforms associated with each state. Fig. 4 illustrates the state transition diagram of the Markov process of interest. The process remains in a given state for T seconds, during which time the corresponding state waveform is generated. Tables I and II show the state waveforms for MSK and offset QPSK, respectively. For convenience, low-pass equivalent notation [10] is used, wherein the state waveform for state i is represented by

$$s_i(t) = s_{ic}(t) + js_{is}(t) \tag{7}$$

where $s_{ic}(t)$ is the I-channel component and $s_{is}(t)$ is the Q-channel component and j is $\sqrt{-1}$. The actual signal is given by

$$y(t) = \text{Re}\,\{s_i(t)e^{-j\omega_c t}\}. \tag{8}$$

Each T seconds the state of the process is modified according to the next data value u_k, as illustrated in Fig. 4. In the figure transition arrows indicate allowable state transitions. Each arrow is labeled to indicate the data value, ± 1, which causes this transition, and the corresponding probability of occurrance of this data value. It will be assumed in subsequent analysis that the data values are equally likely, as indicated in the transition diagram. The transition probability matrix P corresponding to the transition diagram is given by

$$P = \begin{bmatrix} 0 & 0 & 0 & 0 & ½ & 0 & ½ & 0 \\ 0 & 0 & 0 & 0 & 0 & ½ & 0 & ½ \\ 0 & 0 & 0 & 0 & ½ & 0 & ½ & 0 \\ 0 & 0 & 0 & 0 & 0 & ½ & 0 & ½ \\ ½ & ½ & 0 & 0 & 0 & 0 & 0 & 0 \\ ½ & ½ & 0 & 0 & 0 & 0 & 0 & 0 \\ 0 & 0 & ½ & ½ & 0 & 0 & 0 & 0 \\ 0 & 0 & ½ & ½ & 0 & 0 & 0 & 0 \end{bmatrix}$$

The (i, j)th element of P, denoted by $p(j/i)$, is defined as the probability of transition from state i to state j at the transition instant, given that the process is currently in state i.

The Markov representation described above can be used to compute the autocorrelation functions of MSK, offset QPSK, or any of the other modulation processes of the same class. Using techniques similar to those discussed in [11], [12], it can be shown that the low-pass equivalent autocorrelation function is given by

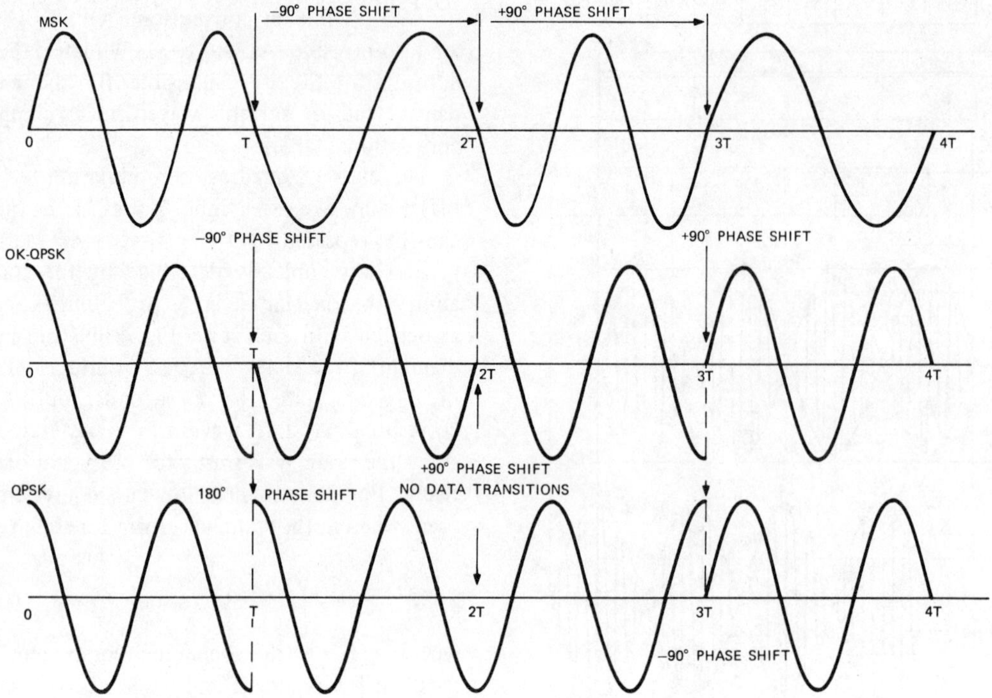

Fig. 3.   Typical waveforms for MSK, offset QPSK, and conventional QPSK.

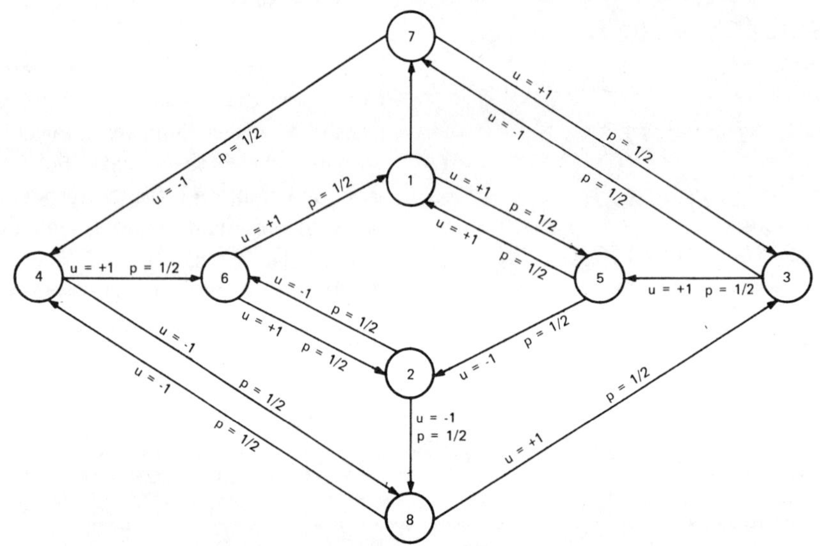

Fig. 4.   Markov process state transition diagram.

$$R_y(\tau) = R_y(\tau' + mT)$$

$$= \sum_{i=1}^{a}\sum_{j=1}^{a} p(i)\,\{p(j\,|\,i,m)s_{ij}(T-\tau')$$

$$+ p(j\,|\,i,m+1)s_{ji}*(\tau')\} \qquad (9)$$

for

$$\tau = \tau' + mT \geqslant 0, \qquad m = \text{integer} \geqslant 0, \qquad 0 \leqslant \tau' \leqslant T.$$

$$R_y(\tau) = R_y*(-\tau) \qquad \text{for } \tau < 0$$

where

$$s_{ij}(x) = \frac{1}{T}\int_0^x s_i(u)s_j * (u + T - x)\,du, \qquad 0 \leqslant x \leqslant T$$

$s_{ij}*(x)$ complex conjugate of $s_{ji}(x)$
$s_i(t)$ waveform of state i
a number of states = 8
$p(i)$ probability of state i on $0 \leqslant t \leqslant T = 1/8$
$p(j/i, m)$ probability of state j on $mT \leqslant t \leqslant (m+1)T$
    given state i on $0 \leqslant t \leqslant T$
    (i, j)th element of the matrix P^m.

Evaluting the autocorrelation expression (9) for MSK and offset QPSK results in the autocorrelation functions in Fig. 5.

TABLE I
MSK LOW-PASS EQUIVALENT STATE WAVEFORMS
$s_i(t) = s_{ic}(t) + js_{is}(t)$

STATE	I-CHANNEL $s_{ic}(t)$	Q-CHANNEL $s_{is}(t)$
1	$\cos\left[\frac{\pi}{2T}(t-nT)\right]$	$\sin\left[\frac{\pi}{2T}(t-nT)\right]$
2	$\cos\left[\frac{\pi}{2T}(t-nT)\right]$	$-\sin\left[\frac{\pi}{2T}(t-nT)\right]$
3	$-\cos\left[\frac{\pi}{2T}(t-nT)\right]$	$\sin\left[\frac{\pi}{2T}(t-nT)\right]$
4	$-\cos\left[\frac{\pi}{2T}(t-nT)\right]$	$-\sin\left[\frac{\pi}{2T}(t-nT)\right]$
5	$\sin\left[\frac{\pi}{2T}(t-nT)\right]$	$\cos\left[\frac{\pi}{2T}(t-nT)\right]$
6	$\sin\left[\frac{\pi}{2T}(t-nT)\right]$	$-\cos\left[\frac{\pi}{2T}(t-nT)\right]$
7	$-\sin\left[\frac{\pi}{2T}(t-nT)\right]$	$\cos\left[\frac{\pi}{2T}(t-nT)\right]$
8	$-\sin\left[\frac{\pi}{2T}(t-nT)\right]$	$-\cos\left[\frac{\pi}{2T}(t-nT)\right]$

TABLE II
OFFSET QPSK LOW-PASS EQUIVALENT STATE WAVEFORMS
$s_i(t) = s_{ic}(t) + js_{is}(t)$

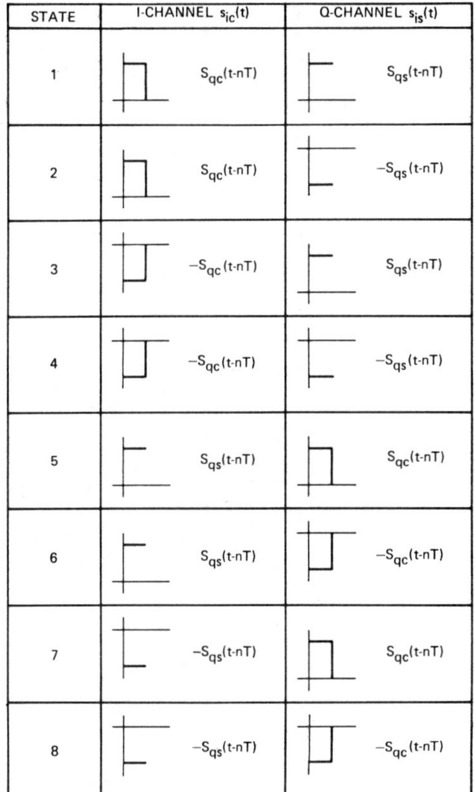

STATE	I-CHANNEL $s_{ic}(t)$	Q-CHANNEL $s_{is}(t)$
1	$S_{qc}(t-nT)$	$S_{qs}(t-nT)$
2	$S_{qc}(t-nT)$	$-S_{qs}(t-nT)$
3	$-S_{qc}(t-nT)$	$S_{qs}(t-nT)$
4	$-S_{qc}(t-nT)$	$-S_{qs}(t-nT)$
5	$S_{qs}(t-nT)$	$S_{qc}(t-nT)$
6	$S_{qs}(t-nT)$	$-S_{qc}(t-nT)$
7	$-S_{qs}(t-nT)$	$S_{qc}(t-nT)$
8	$-S_{qs}(t-nT)$	$-S_{qc}(t-nT)$

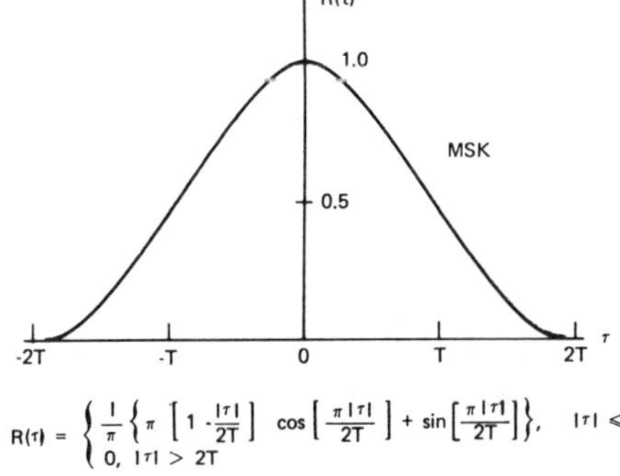

$$R(\tau) = \begin{cases} \frac{1}{\pi}\left\{\pi\left[1-\frac{|\tau|}{2T}\right]\cos\left[\frac{\pi|\tau|}{2T}\right]+\sin\left[\frac{\pi|\tau|}{2T}\right]\right\}, & |\tau| \leq 2T \\ 0, & |\tau| > 2T \end{cases}$$

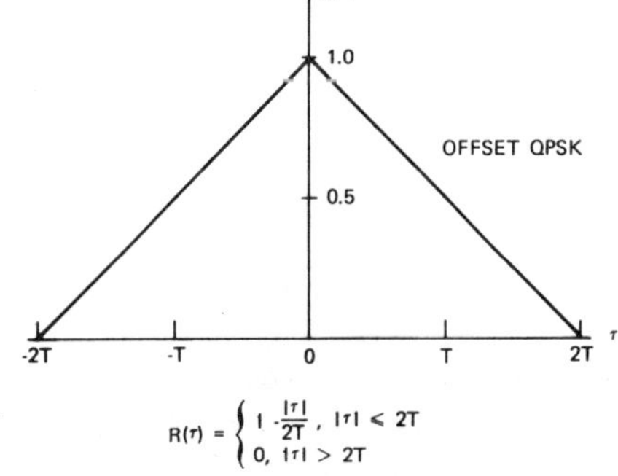

$$R(\tau) = \begin{cases} 1-\frac{|\tau|}{2T}, & |\tau| \leq 2T \\ 0, & |\tau| > 2T \end{cases}$$

Fig. 5. Normalized autocorrelation functions for MSK and offset QPSK.

Note that both the autocorrelation functions are zero for time lags exceeding two bit times. If either MSK or offset QPSK is transmitted through an infinite bandwidth, linear, additive WGN channel, no crosstalk will be produced between the in-phase and quadrature-phase channels at matched filter sample times. Thus, both systems can be viewed as two orthogonal binary channels, where antipodal symbols of length $2T$ seconds are used in each channnel, and a half-symbol timing skew of T seconds exists between the orthogonal channels. For independent binary data bits, the signs of the symbols in each binary channel are independent from one $2T$ interval to the next, and the autocorrelation function would be expected to go to zero for lags exceeding $2T$.

For later comparison, the BER performance of MSK and offset QPSK on an ideal channel is of interest. An ideal channel is defined as a linear, infinite bandwidth channel, corrupted only by additive WGN. In addition, perfect carrier and timing references are assumed available at the receiver. With these assumptions, and viewing both MSK and offset QPSK as orthogonal binary channels with antipodal signaling, the binary error probability is known to be [13]

$$P_e = Q(\sqrt{2E_b/N_0}) = \int_\lambda^\infty \frac{1}{\sqrt{2\pi}}\, e^{-y^2/2}\, dy \qquad (10)$$

where

$$\lambda = \sqrt{2E_b/N_0}$$

E_b = signal energy per bit

$N_0/2$ = two-sided spectral density of WGN.

Note that the detection of each bit takes place after observing

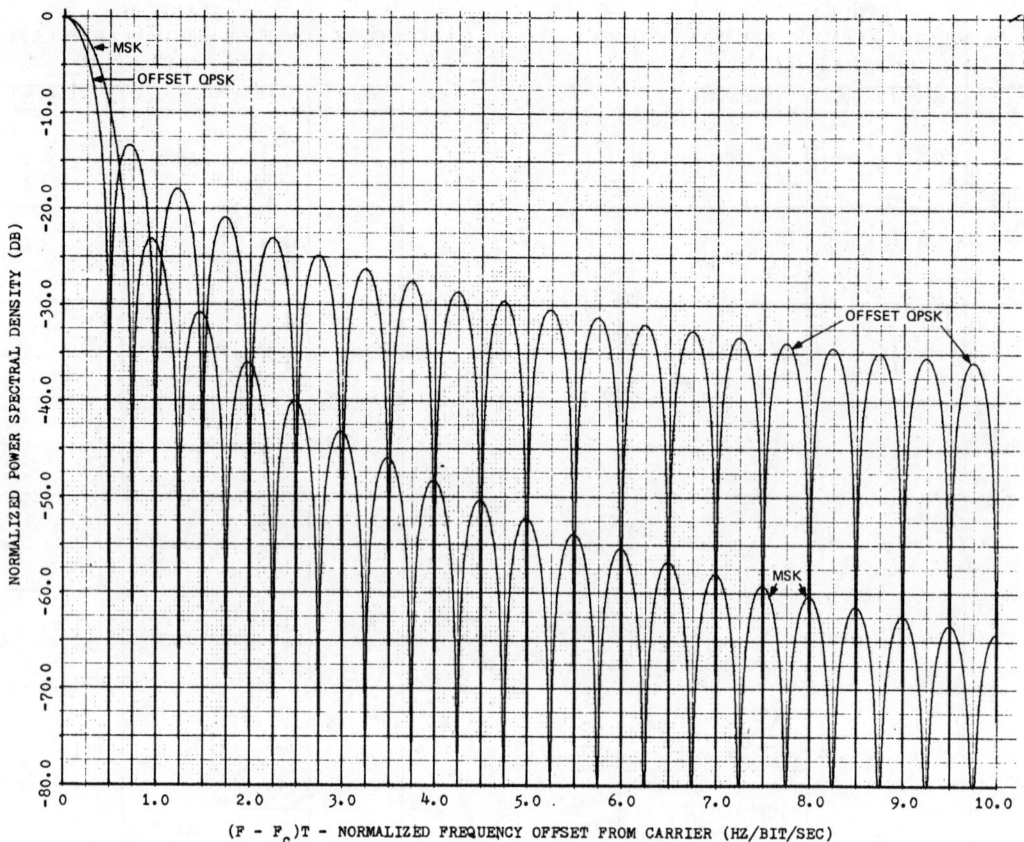

Fig. 6.   Power spectral densities for MSK and offset QPSK. ($T = 1$ s, $P_c = 1$ W.)

one of two antipodal symbols in noise for $2T$ seconds. We observe that the optimum performance for coherently detected orthogonal FSK (tone spacing $1/T$) is

$$P_e = Q(\sqrt{E_b/N_0}). \tag{11}$$

Thus, detecting MSK as two orthogonal binary channels provides a 3-dB advantage over detection of orthogonal FSK.

All of the foregoing discussion has centered on ideal channels having infinite bandwidth, additive WGN, and perfectly recovered carrier and timing references. Section IV will consider the performance of MSK and offset QPSK on band-limited, nonlinear channels. As a preliminary discussion to this topic, it is useful to consider the power spectral densities of the MSK and offset QPSK modulation processes. The power spectral density of each process is the Fourier transform of the corresponding autocorrelation function, and are given by the following expressions:

$$G_{\mathrm{MSK}}(f) = \frac{8P_c T(1 + \cos 4\pi f T)}{\pi^2(1 - 16T^2 f^2)^2} \tag{12}$$

$$G_{\mathrm{O\text{-}QPSK}}(f) = 2P_c T\left[\frac{\sin 2\pi f T}{2\pi f T}\right]^2 \tag{13}$$

where

f frequency offset from carrier
P_c power in modulated waveform.

Equations (12) and (13) are plotted in Fig. 6 as a function of f

normalized to the binary data rate $R = 1/T$. The MSK spectrum falls off at a rate proportional to $(f/R)^{-4}$ for large values of f/R. In contrast, the offset QPSK spectrum falls off at a rate proportional to only $(f/R)^{-2}$. The main lobe of the MSK spectrum, however, is wider than that of the offset QPSK spectrum, the first nulls falling at $f/R = 0.75$ and $f/R = 0.5$, respectively.

A measure of the compactness of a modulation waveform's spectrum is provided by the fractional out-of-band power, P_{ob}, defined as

$$P_{ob} = 1 - \left\{\int_{-B}^{B} G(f)\,df \bigg/ \int_{\infty}^{\infty} G(f)\,df\right\}. \tag{14}$$

Equation (14) has been evaluated for the MSK and offset QPSK power spectral density expressions (12) and (13) as a function of bandwidth B normalized to the binary data rate, and the results plotted in Fig. 7.

Figs. 6 and 7 suggest that for system bandwidths exceeding about $1.5\,R$, MSK should provide lower error rate performance than offset QPSK for equal transmitter power. However, as system bandwidth is decreased from $1.5\,R$ to $1.0\,R$, a point should be reached at which the performance of offset QPSK will be superior to that of MSK. As system bandwidth is increased, the performance of the two systems converges to the infinite bandwidth case (10). The precise boundaries of the regions of superior performance for each technique are difficult to determine in practical situations, since the detailed characteristics of the channel must be considered in evaluating

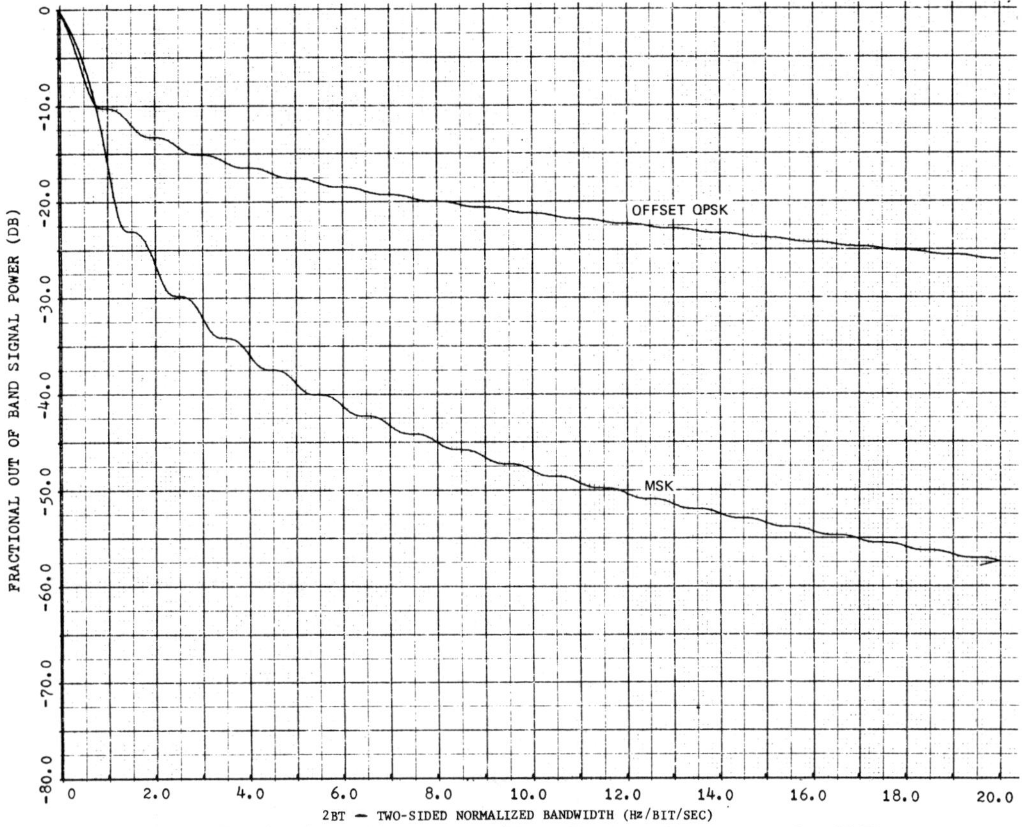

Fig. 7. Fractional out-of-band power. (Normalized two-sided bandwidth = $2BT$.)

intersymbol interference effects, intermodulation distortion, etc. One approach to determining the error-rate performance of a modulation technique on a complex channel is through the use of digital computer simulation techniques.

IV. COMPUTER SIMULATION RESULTS FOR SATELLITE LINK

A computer simulation was performed to examine the BER performance of MSK and offset QPSK on band-limited, nonlinear channels, such as those encountered in satellite communications systems. Fig. 8 is a block diagram of the simulated system. The simulation is carried out at baseband to minimize computer run time. The elements in the model correspond to the elements in a satellite communication link as follows. Filter number 1 simulates band-limiting elements at the transmitting earth station. Noise added at the satellite receive system is represented by WGN source number 1. The bandpass hard-limiter which simulates the satellite output power amplifier obeys the input–output relationship

$$a_0(t) + jb_0(t) = \frac{a_i(t) + jb_i(t)}{\sqrt{a_i{}^2(t) + b_i{}^2(t)}} . \qquad (15)$$

Downlink noise which is added at the receiving earth station is represented by WGN source number 2. A second filter is used to simulate band-limiting elements in the receive system. The matched filter is matched to the transmitter pulse shape. Both MSK and offset QPSK were simulated with this system model, simply by switching between rectangular and half-cycle sinusoid shapes in the "pulse shaping" and "matched filter"

system elements. Power is normalized prior to the points at which the uplink (WGN source 1) and downlink (WGN source 2) noise is added to facilitate setting the desired up and downlink values of E_b/N_0 (energy per bit to additive noise spectral density). This normalization also results in equal uplink transmitted power at the output of filter number 1 for either modulation technique which simulates fixed EIRP earth stations. In running the simulation, the noise samples of WGN source 2 are not actually added to the simulation sample stream. Rather, an estimation technique similar to that employed by Lebowitz and Palmer [8] is employed. As long as the system additive noise is dominated by the down-link component, this technique provides simulation run time savings relative to an error counting BER estimation technique.

In the simulations reported here, the two system filters were chosen to be identical. The channel bandwidth B is defined as the noise bandwidth of the two identical filters in cascade. The simulated filters were 7 pole, 0.1-dB ripple Chebyshev low-pass equivalent filters. A high filter Q-factor was chosen so that a high degree of filter symmetry was achieved. This resulted in the primary mechanism of crosstalk between the I and Q channels being the bandpass hard-limiter, via (15). Since many practical satellite links are limited by downlink thermal noise, the uplink E_b/N_0 was maintained at a high level for the results reported here. Under the above restrictions, several MSK and offset QPSK system simulation runs were performed for varying values of downlink E_b/N_0 and channel bandwidth B. The resulting data were used to derive the curves shown in Fig. 9. Each curve applies to either MSK or offset QPSK, as labeled. The curves show the variation

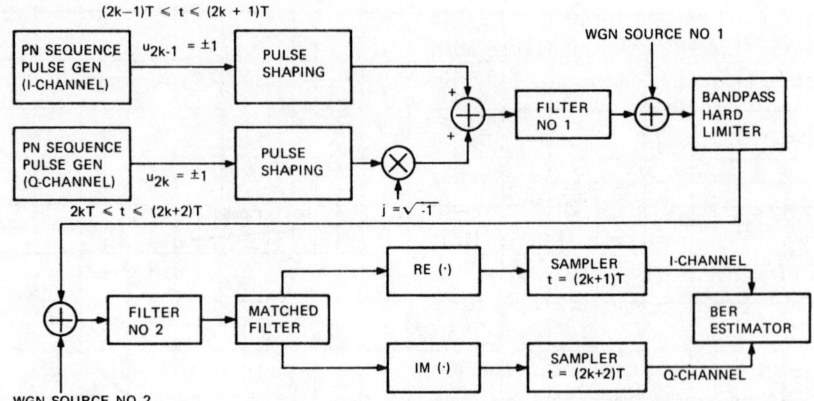

Fig. 8.   Simulation block diagram.

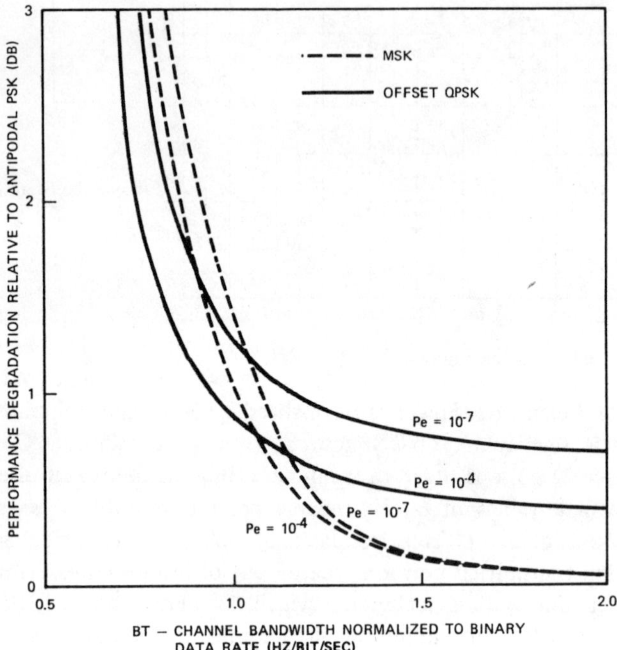

Fig. 9.   MSK and offset QPSK E_b/N_0 performance degradation with respect to ideal antipodal PSK as a function of cascaded filter noise bandwidth with error probability as a parameter.

of performance degradation as a function of channel bandwidth for the indicated constant value of bit-error probability.

Performance degradation is defined as the ratio of the total link E_b/N_0 required on the simulated link to achieve the desired BER to the E_b/N_0 required on an infinite bandwidth, WGN channel to achieve the same BER. The ideal link performance is given by (10). Timing in the simulated system is near ideal. Several samples of the matched filter near the expected optimum sample time are inspected to determine the optimum sample for use in the BER estimator. The sample time is not, however, bit-by-bit adaptive, but is constant for the entire simulated bit sequence. As conjectured in Section III, the performance curves show a crossover point, such that for channel bandwidths exceeding approximately $1.1\,R$, where R is the binary data rate, the performance of MSK is superior to that of offset QPSK. Conversely, for $B < 1.1\,R$, offset QPSK has an advantage over MSK. The reason for the performance crossover is basically the narrower main lobe of the spectral density of offset QPSK as compared to MSK. Both

techniques begin to suffer severe degradations as their main spectral lobes encounter the channel band edge. Since the main lobe of MSK is wider than that of offset QPSK, MSK begins to degrade sooner than offset-QPSK as channel bandwidth is decreased.

The above discussion is based upon a parametric variation of filter bandwidth only. Other system elements can influence the effect of this parametric variation. For example, Poza and Berger [14] briefly touch on a simulated comparison of offset QPSK[1] and MSK on a channel model containing an AM/PM system element rather than a hard-limiter. Their results indicate that the E_b/N_0 advantage of MSK over offset QPSK on such a channel extends to slightly narrower channel bandwidths than the $1.1\,R$ crossover bandwidth reported in the present paper.

In satellite communication systems, and indeed in most radio systems, adjacent channel interference is a critical design factor. The question arises as to whether MSK, since its unfiltered spectrum falls off faster than that of offset QPSK, requires less filtering than offset QPSK for practical values of interchannel spacing and adjacent channel interference. If so, the performance crossover points of Fig. 9 must be re-examined. Define a carrier to adjacent channel interference ratio as

$$C/A = \int_{\infty}^{\infty} G(f)\,|\,H(f)\,|^4\,df \bigg/ \int_{\infty}^{\infty} G(f)$$

$$\cdot\,|\,H(f)\,|^2\,|\,H(f+\Delta f)\,|^2\,df \qquad (16)$$

where $G(f)$ is the power spectral density of the MSK or offset QPSK waveform [(12) and (13)], and $H(f)$ is the common transfer function of the two system filters. The factor $|\,H(f)\,|^4$ is the magnitude-squared transfer function of two identical filters in cascade, with a common center frequency and cascaded noise bandwidth B. The factor $|\,H(f)\,|^2\,|\,H(f+\Delta f)\,|^2$ represents a cascade of the same two filters with the exception that one of the filters is now offset from the other by the channel spacing variable Δf. Equation (16) was calculated for a range of filter bandwidths and channel spacings, and the

---

[1] While referred to as QPSK in [14], it was pointed out during the presentation of the paper at ICC '75 that the modulation technique actually considered was offset QPSK.

results plotted as curves of channel spacing normalized to data rate versus channel bandwidth normalized to data range with C/A as a constant parameter. Two such curves are shown in Fig. 10 for MSK and for offset QPSK. These curves can be used to adjust the performance degradation curves of Fig. 9 to conform to a constant adjacent channel interference criterion. As an example[2] of the use of Figs. 9 and 10, consider Table III. The table presents a comparison of BER degradations for MSK and offset QPSK as a function of allowable channel spacing with BER and C/A specified to by 10^{-4} and 25 dB, respectively. The classifications of the various bandwidth restrictions are somewhat arbitrary, and are meant to apply to the 7-pole Chebyshev filters used in the simulation. For either conventional or offset QPSK, narrower bandwidths can be used without significant degradation if delay equalization is used [15]. It remains for future work to determine the allowable bandwidth for MSK when equalization is employed. Some initial work in this area is reported in [16]. For the simulation reported here, Table III indicates that offset QPSK is most advantageous in bandwidth limited systems, while MSK is superior in power-limited systems with adequate bandwidth for the wider main spectral lobe of MSK.

V. CONCLUSIONS

MSK and offset QPSK have been compared as modulation techniques and their performance on band-limited, nonlinear channels has been determined by means of digital simulation. The generation of both can be represented as two orthogonal, antipodal binary systems with the symbol timing in the two channels offset by one-half of a symbol duration. MSK uses half-cycle sinusoid pulse shapes while offset QPSK uses rectangular pulse shapes. Because of the sinusoidal pulse shaping in MSK, it was shown that MSK can be viewed as continuous phase FSK with one-half the tone separation of conventional orthogonal symbol FSK, or coherent, sine-weighted, symbol offset, QPSK where, in the latter view, the detection efficiency of antipodal PSK on an ideal channel is achieved. Incoherent detection of MSK, e.g., discriminator detection, could provide a low-cost flexibility feature in some systems. A Markov process representation of the two modulation processes was developed and used to derive autocorrelation and power spectral density properties for the processes.

Computer simulation of MSK and offset QPSK on channels such as those encountered in satellite communication systems resulted in the definition of ranges of channel bandwidth for which one or the other technique was superior with respect to signal-to-noise ratio required to achieve a given BER. For the Chebyshev filters used, MSK was found to have a performance advantage when the allowed channel bandwidth exceeded about 1.1 times the binary data rate. At lower bandwidths, offset QPSK was superior. The effect of imposing an adjacent channel interference constraint was illustrated. For the example system simulated, the added constraint did not strongly influence the crossover point in the choice between

²Suggested by an anonymous reviewer.

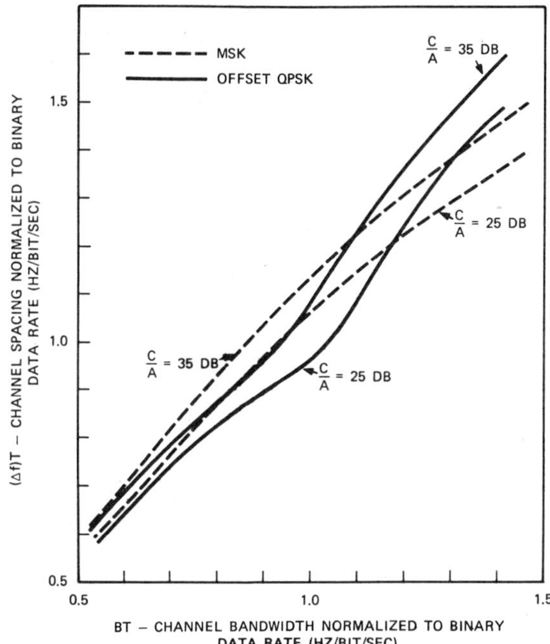

Fig. 10. Permissible channel spacing as a function of channel bandwidth for constant carrier to adjacent channel interference. ($R = 1/T = $ binary data rate.)

MSK and offset QPSK. The main thrust of the simulation was to compare the performance of MSK and offset QPSK based on a BER criterion. The performance of either system could be improved by using a noise whitening matched filter, transversal equalizer, or phase equalization. In certain satellite applications it may be necessary to include an additional filter in the satellite subsequent to the bandpass hard-limiter. In these cases MSK enjoys an advantage since it would suffer less satellite output power loss due to the filtering process. The advantage could be significant in a power-limited satellite system. The results of the analysis and simulation indicate the advantages of one modulation technique over another are not always clear cut, and the nature of the application and communication channel characteristics are critical to choosing a technique for implementation.

APPENDIX A

We show the following relationships among the data terms of (4):

$$\cos(x_{2K-1}) = \cos(x_{2k}) \tag{A1}$$

$$u_{2k}\cos(x_{2k}) = u_{2k+1}\cos(x_{2k+1}) \tag{A2}$$

$$\cos(x_{2k+1}) = \begin{cases} \cos(x_{2k}), & u_{2k+1} = u_{2k} \\ -\cos(x_{2k}), & u_{2k+1} = -u_{2k} \end{cases} \tag{A3}$$

$$u_{2k}\cos(x_{2k}) = \begin{cases} u_{2k-1}\cos(x_{2k-1}), \\ \qquad\qquad u_{2k} = u_{2k-1} \\ -u_{2k-1}\cos(x_{2k-1}), \\ \qquad\qquad u_{2k} = u_{2k-1}. \end{cases} \tag{A4}$$

TABLE III
PERFORMANCE DEGRADATION AS A FUNCTION OF NORMALIZED CHANNEL SPACING
FOR 10^{-4} BER AND 25 dB C/A

Channel Frequency Spacing	Classification of Bandwidth Restriction	Required Bandwidth for 25 dB C/A		Performance Degradation (dB) at 10^{-4}	
		MSK	O-QPSK	MSK	O-QPSK
1.4 R	Very Slight	1.45 R[a]	1.30 R	0.15	0.50
1.2 R	Slight	1.15 R	1.15 R	0.50	0.65
1.1 R	Moderate	1.05 R	1.10 R	0.75	0.75
1.0 R	Tight[b]	0.93 R	1.02 R	1.30	0.85
0.8 R	Very Tight	0.82 R	0.90 R	2.40	1.10

[a] Note that the filter bandwidth of $B = 1.45 R$ for each of two adjacent channels with a frequency separation of $1.4 R$ between band centers results in some overlap in passbands for adjacent channels.

[b] Tight filtering for sharp-cutoff case without delay equalization. With equalization, a bandwidth of $0.625 R$ and comparable spacing is allowable for offset QPSK.

The proof of the above relationships is based on the recursive phase relationship (2), the assumption that $x_k = 0, \pi$ modulo 2π, and trigonometric identities. First, substitute (2) into $\cos x_k$ and expand by trigonometric identities

$$\cos x_k = \cos \left[x_{k-1} + (u_{k-1} - u_k)(\pi k/2) \right]$$

$$= \cos (x_{k-1}) \cos \left[(u_{k-1} - u_k)(\pi k/2) \right]$$

$$-\sin (x_{k-1}) \sin \left[(u_{k-1} - u_k)(\pi k/2) \right].$$

We note the following:

1) $\sin (x_{k-1}) = 0$ for $x_{k-1} = 0, \pi$ modulo 2π

2) $u_{k-1} - u_k = -2, 0, +2$ for $u_k = \pm 1$

3) $\sin \left[(u_{k-1} - u_k)(\pi k/2) \right] = 0$ for k any integer

4) $\cos \left[(u_{k-1} - u_k)(\pi k/2) \right]$

$$= \begin{cases} 1 & \text{for } k \text{ even or for } k \text{ odd and } u_k = u_{k-1} \\ -1 & \text{for } k \text{ odd and } u_k \neq u_{k-1}. \end{cases}$$

Therefore,

$$\cos (x_k) = \cos (x_{k-1}) \text{ for } k \text{ even or for } k \text{ odd and } u_k = u_{k-1}$$

$$\cos (x_k) = -\cos (x_{k-1}) \text{ for } k \text{ odd and } u_k \neq u_{k-1}.$$

Equivalently, replacing k evey by $2k$ and k odd by $2k + 1$, we have

$$\cos (x_{2k}) = \cos (x_{2k-1})$$

$$\cos (x_{2k+1}) = \cos (x_{2k+1-1}) = \cos (x_{2k}) \text{ for } u_{2k+1} = u_{2k}$$

$$\cos (x_{2k+1}) = -\cos (x_{2k}) \text{ for } u_{2k+1} = -u_{2k}.$$

This established (A1) and (A3). The relationships (A2) and (A4) follow in a similar manner.

REFERENCES

[1] M. L. Doelz and E. H. Heald, "Minimum-shift data communication system," U.S. Patent No. 2,977,417, Mar. 28, 1961 (assigned to Collins Radio Company).

[2] S. A. Rhodes, "Effects of hardlimiting on bandlimited transmissions with conventional and offset QPSK modulation," in *Proc. Nat. Telecommunications Conf.*, 1972, pp. 20F/1–7.

[3] M. K. Simon and J. G. Smith, "Offset quadrature communications with decision feedback synchronization," *IEEE Trans. Commun.*, vol. COM-22, pp. 1576–1584, Oct. 1974.

[4] R. DeBuda, "Coherent demodulation of frequency shift keying with low deviation ratio," *IEEE Trans. Commun.*, vol. COM-20, pp. 429–435, June 1972.

[5] S. A. Rhodes, "Effect of noisy phase reference on coherent detection of offset QPSK signals," *IEEE Trans. Commun.*, vol. COM-22, pp. 1046–1055, Aug. 1974.

[6] W. R. Bennett and J. Salz, "Binary data transmission by FM over a real channel," *Bell Syst. Tech. J.*, vol. 42, pp. 2387–2427, Sept. 1963.

[7] H. R. Mathwich, J. F. Balcewicz, and M. Hecht, "The effect of tandem band and amplitude limiting on the E_b/N_0 performance of minimum (frequency) shift keying (MSK)," *IEEE Trans. Commun.*, vol. COM-22, pp. 1525–1540, Oct. 1974.

[8] S. Lebowitz and L. Palmer, "Simulation results for intersymbol interference loss; QPSK transmission through several filter types," in *Proc. Nat. Telecommunications Conf.*, 1973, pp. 32D/1–7.

[9] G. D. Forney, "The Viterbi algorithm," *Proc. IEEE*, vol. 61, pp. 268–278, Mar. 1973.

[10] S. Stein and J. J. Jones, *Modern Communication Principles with Application to Digital Signaling.* New York: McGraw-Hill, 1967, ch. 3.

[11] M. Hecht and A. Guida, "Delay modulation," *Proc. IEEE* (Lett.), vol. 57, pp. 1314–1316, July 1969.

[12] R. C. Titsworth and C. R. Welch, "Power spectra of signals modulated by random and pseudorandom sequences," Jet Propulsion Laboratory, Tech. Rep. 32-140, Oct. 10, 1961.

[13] J. M. Wozencraft and I. M. Jacobs, *Principles of Communication Engineering.* New York: Wiley, 1965.

[14] H. B. Poza and H. L. Berger, "Performance characterization of advanced wideband data links," in *Proc. IEEE Int. Conf. Communications*, 1975, pp. 4/23–27.

[15] F. Assal, "Approach to a near-optimum transmitter-receiver filter designator data transmission pulse-shaping networks," *Comsat Tech. Rev.*, vol. 3, pp. 301–322, Fall 1973.

[16] V. Z. Viskanta, "Effect of filtering on MSK signals," in *Proc. Nat. Telecommunications Conf.*, 1975, pp. 30/1–5.

DETECTION OF MPSK SIGNALS TRANSMITTED THROUGH *
A NONLINEAR SATELLITE REPEATER

Pravin Jain
Defense Communication Agency
Washington, D. C.

T. C. Huang, K. T. Woo, and J. K. Omura
LinCom Corporation, W. C. Lindsey,
University of Southern California

ABSTRACT

Error probabilities and channel transition probabilities are derived and evaluated for the detection of MPSK signals transmitted through a bandlimited nonlinear satellite communication channel. Two general formulas characterizing the effect of AM-AM and AM-PM distortions exhibited in high power amplifiers such as traveling-wave tubes (TWTs) are obtained. Numerical evaluation of the two approaches is carried out for BPSK, QPSK and 8-ary signaling and a comparison is made in terms of the accuracy and speed of numerical computation. The theory is required in performing channel throughput (bits/sec/Hz) and modulation/coding tradeoff studies.

I. Introduction

A common criterion for evaluating the reliability of a digital communication system is the error probability of signal detection at the receiver. For a coded system the channel throughput parameter, R_{comp}, should be added to measure the effectiveness in utilizing available channels. Transition probabilities, required in evaluating R_{comp}, are generally difficult to compute, especially for nonlinear channels. Thus the transiton probabilities as well as R_{comp} are quite often not available even though the error probability has been estensively studied for various types of communication channels.[1]

In this paper, the evaluation of both error probabilities and transition probabilities of MPSK signaling through bandlimited nonlinear channels, such as a nonlinear satellite communication repeater, is formulated in two ways. The nonlinearity appeared in amplifiers, such as hard limiters, traveling wave tubes (TWT), may exhibit two kinds of distortion, called AM-to-AM and AM-to-PM conversions. The effect of these nonlinearities on the system performance in terms of error probability or R_{comp}, can also be evaluated through the two general formulations.

One of the formulas given here involves one integration and one infinite summation over special functions, viz., hypergeometric functions and modified Bessel functions $I_n(x)$. The other formula is double integrals of error functions, using Gaussian quadrature rules for integrals, have been compared in terms of computation speed and accuracy.

As numerical examples, the error rates for the bandlimited, hard or soft limiter in cascade with a TWT channel are plotted for QPSK and 8PSK signaling, respectively.

II. Analytical Model of the Signal and System

Denote the transmitted signal by

$$s_i(t) = A \cos(\omega_c t + \theta_i) \qquad (1)$$

where $\omega_c/2\pi$ is the carrier frequency, $A^2/2$ be the up-link signal power and θ is one of the phases $2i\pi/M$, $i=0,1,\ldots,M-1$. At the input of the satellite repeater,

*This work was done at LinCom Corporation for the Military Satellite Communications Systems Branch, Defense Communications Agency, Arlington, VA, under contract No. DCA 100-76-C-0062.

the received signal is

$$x(t) = S(t) + n_u(t)$$
$$= v_1(t)\cos(\omega_c t + \theta + n_1(t)) \qquad (2)$$

where

$$v_1(t) = [[A+n_{uc}]^2 + n_{us}^2]^{1/2}$$

$$n_1(t) = \tan^{-1} \frac{n_{us}^2}{A+n_{uc}}$$

$$n_u(t) = n_{us}\cos(\omega_c t+\theta) - n_{uc}\sin(\omega_c t+\theta)$$

$$\left.\begin{array}{c} n_{us} \\ n_{uc} \end{array}\right\} = \text{In-phase and quadrature component of up-link noise process}$$

The output of the nonlinearity and its zonal filter can be expressed as follows

$$y(t) = f(v_1(t))\cos(\omega_c t+\theta+g(v_1(t))+n_1(t)) \qquad (3)$$

where

$$f(\cdot) = \text{AM-AM conversion of the nonlinearity} \qquad (4a)$$

$$g(\cdot) = \text{AM-PM conversion of the nonlinearity} \qquad (4b)$$

These two functions $f(\cdot)$ and $g(\cdot)$ are assumed to be known. For the hard limiter case $f(v) = $ constant and $g(v) = 0$. Typical $f(v)$ and $g(v)$ characteristics are shown in Figure 2.
The
The output signal $y(t)$ is transmitted over a down-link channel and received with a downlink channel noise

$$r(t) = y(t) + n_d(t)$$
$$= v_2(t)\cos(\omega_c t+n_2(t)) \qquad (5)$$

At the receiver it is frequency-down-converted into baseband signals and sampled for signal detection. The decision variables are

$$z_c = v_2 \cos n_2 \qquad (6)$$

$$z_s = v_2 \sin n_2 \qquad (7)$$

where v_2 and n_2 are envelope and phase of signals sampled. Without loss of generality, we assume the transmitted phase θ to be zero. The transition probabilities can be denoted as

$$P_M(j) = \Pr\{j\frac{2\pi}{M} - \frac{\pi}{M} \le n_2 \le j\frac{2\pi}{M} + \frac{\pi}{M}|\theta=0\} \qquad (8)$$

$$j = 0,1,2,\ldots,M-1$$

and the symbol error probability becomes

$$P_E(M) = 1 - P_M(0) \qquad (9)$$

In order to characterize the above probabilities, one has to obtain the probability density functions of random variables involved. Based on the assumption

Reprinted from *National Telecommun. Conf.*, vol. 2, Dec. 5–7, 1977, pp. 26.2-1–26.2-4.

pertaining to the channel noises, the probability density function of ν_1 and η_1 or ν_2 and η_2 can be written as

$$p(\nu_1,\eta_1) = \frac{\nu_1}{2\pi\sigma_1^2} \exp\left[-\frac{\nu_1^2+A^2-2\nu_1 A \cos(\eta_1-\theta)}{2\sigma_1^2}\right] \quad (10)$$

and

$$p(\nu_2,\eta_2|\nu_1,\eta_1) = \frac{\nu_2}{2\pi\sigma_2^2} \exp\left[-\frac{1}{2\sigma_2^2}(\nu_2^2 + f^2(\nu_1)\right.$$
$$\left. - 2\nu_2 f(\nu_1)\cos(\eta_2-\eta_1-g(\nu_1))\right] \quad (11)$$

respectively.

Now for given ν_1 and η_1, we can express the conditional transition probabilities as

$$P_M(j|\nu_1,\eta_1) = \int_{\frac{\pi}{M}(2j-1)}^{\frac{\pi}{M}(2j+1)} p(\nu_2,\eta_2|\nu_1,\eta_2)d\nu_2\, d\eta_2 \quad (12)$$

Then the unconditional transition probabilities $P_M(j)$ are obtained by averaging $P_M(j|\nu_1,\eta_1)$ over ν_1 and η_1 as follows

$$P_M(j) = \int_0^\infty \int_0^{2\pi} P(\nu_1,\eta_1)P_M(j|\nu_1,\eta_1)d\eta_1\, d\nu_1 \quad (13)$$

Eqs. (12) and (13) are the basic formulations used for the development of two different results for the $P_M(j)$ one is in terms of hypergeometric functions and modified Bessel functions while the other is in terms of error functions. Both results can be evaluated using methods of numerical integrations.

III. Confluent Hypergeometric Expansion

Integrating (12) with respect to ν_2, then using the result

$$\int_{-\frac{\pi}{M}}^{+\frac{\pi}{M}} \cos n(\eta_2-\phi)d\eta_2 = \begin{cases} \frac{2\pi}{M}; & n = 0 \\[6pt] \frac{2}{n}\sin\frac{n\pi}{M}\cos n\phi; & n\neq 0 \end{cases} \quad (14)$$

one can show that the conditional transition probabilities become

$$P_M(j|\nu_1,\eta_1) = \frac{1}{M} + \frac{2}{\pi}\sum_{n=1}^\infty \frac{\sin\frac{n\pi}{M}}{n}$$

$$\frac{\Gamma(\frac{n}{2}+1)}{\Gamma(n+1)} \rho_2^n\, {}_1F_1(\tfrac{n}{2};n+1;-\rho_2^2)\cos n[j\tfrac{2\pi}{M}-\eta_1-g(\nu_1)] \quad (15)$$

$$j = 0,1,2,\ldots,M-1$$

where

$$\rho_2 = \frac{f(\nu_1)}{\sqrt{2}\sigma_2} \quad (16)$$

$$\Gamma(x) = \text{Gamma function}$$

Substituting (15) into (13) and integrating it with respect to η_1 one has a general expression for

transition probability for MPSK channels, viz.,

$$P_M(j) = \frac{1}{M} + \int_0^\infty \frac{2}{\pi}\sum_{n=1}^\infty \frac{\sin\frac{n\pi}{M}}{n} \frac{\Gamma(\frac{n}{2}+1)}{\Gamma(n+1)} \exp(-\rho_1^2)$$

$$\cdot \left(\frac{\nu_1}{\sigma_1^2}\right)\exp\left(-\frac{\nu_1^2}{2\sigma_1^2}\right) I_n\left(\frac{\nu_1 A}{\sigma_1^2}\right)\left\{\left(\frac{f(\nu_1)}{\sqrt{2}\sigma_2}\right)^n\right.$$

$$\cdot\, {}_1F_1\left[\frac{n}{2},n+1;-\left(\frac{f(\nu_1)}{\sqrt{2}\sigma_2}\right)^2\right]\cos n(j\tfrac{2\pi}{M}-g(\nu_1))\bigg\}d\nu_1 \quad (17)$$

$$j = 0,1,2,\ldots,M-1$$

where

$$I_n(x) = \text{modified Bessel function of } n^{th} \text{ order}$$

$$\rho_1^2 = \frac{A^2}{2\sigma_1^2} \quad \text{up-link SNR}$$

The symbol error probability is then found from (9). Some observations on the results are interestingly noted. For a hard limiter case, denote $f(\nu_1)/\sqrt{2}\sigma_2 = \rho_2$ and $g(\nu_1) = 0$. For this case, one can reduce (9) and (17) to

$$P_E(M) = \frac{M-1}{M} - \frac{2}{\pi}\sum_{n=1}^\infty \frac{\sin\frac{n\pi}{M}}{n}\left[\frac{\Gamma(\frac{n}{2}+1)}{\Gamma(n+1)}\right]^2 \rho_1^n \rho_2^n$$

$$\cdot\, {}_1F_1(\tfrac{n}{2},n+1;-\rho_1^2)\, {}_1F_1(\tfrac{n}{2},n+1-\rho_2^2) \quad (18)$$

which was obtained by Mizuno et al.[1]. For the special case of BPSK, (18) further reduces to a result obtained earlier by Jain and Balchman[2].

IV. Error Function Expansion

Integrating (12) with respect to η_1, one can show that

$$P_M(j|\nu_1,\eta_1) = \frac{1}{M} + \frac{1}{4}\,\text{erf}(A_+) + \frac{1}{4}\,\text{erf}(A_-)$$

$$+ \frac{1}{4}\int_0^{A_+} \frac{2}{\sqrt{\pi}} e^{-\lambda_1^2} \text{erf}\left(\lambda_1\cot\frac{\pi}{M}+\frac{Q\cos\frac{2\pi}{M}}{\sqrt{2}\sin\frac{\pi}{M}}\right)d\lambda_1$$

$$+ \frac{1}{4}\int_0^{A_-} \frac{2}{\sqrt{\pi}} e^{-\lambda_2^2} \text{erf}\left(\lambda_2\cot\frac{\pi}{M}+\frac{Q\cos\frac{2\pi}{M}}{\sqrt{2}\sin\frac{\pi}{M}}\right)d\lambda_2$$

$$(19)$$

where

$$A_\pm = \frac{f(\nu_1)}{\sqrt{2}\sigma_2}\sin(\tfrac{\pi}{M}\pm(j\tfrac{2\pi}{M}+g(\nu_1)+\eta_1)) \quad (20)$$

The expression (19) can be further simplified for the case of BPSK and QPSK signalings. In particular, for BPSK signaling

$$P_2(j|\nu_1,\eta_1) = \frac{1}{2} + \text{erf}\left(\frac{f(\nu_1)\cos(\eta_1+g(\nu_1))}{\sqrt{2}\sigma_2}\right) \quad (21)$$

for $j=0,1$. For QPSK (19) reduces to

$$P_4(j|\nu_1,n_2) = \frac{1}{4} + \frac{1}{4}\,\mathrm{erf}(A_+) + \frac{1}{4}\,\mathrm{erf}(A_-)$$

$$+ \frac{1}{4}\,\mathrm{erf}(A_+)\mathrm{erf}(A_-) \qquad (22)$$

where

$$A_\pm = \frac{f(\nu_1)}{\sqrt{2}\sigma_2}\,\sin(\frac{\pi}{M} \pm (j\,\frac{2\pi}{M} + g(\nu_1) + n_1)) \qquad (23)$$

and j = 0,1,2,3. Substituting these results into (13) one may numerically evaluate $P_M(j)$ through standard integration routines.

V. Numerical Examples

The general expressions of symbol error probability and transition probabilities, though they are in integral forms, can be conveniently computed using standard numerical integration techniques. Our experience shows that the convergence of the sum of hypergeometric functions is rather slow when the SNR ρ_1 is high (>15 dB) and that the evaluation of double integrals of the error function expansion for BPSK and QPSK can be easily done with a reasonable computer time, which may be shorter than that for the hypergeometric function expansions.

The symbol error probability for 8PSK hard limited channel is given in Fig. 3, in terms of down-link SNR with the up-link SNR as a parameter. A channel with soft limiter, cascaded with a TWT, is also taken as an example for the evaluation of symbol erorr probability of BPSK and QPSK signalings, as shown in Figs. 4 and 5. The operating point of the soft-limiter is set at near saturation when the input signal is at 20 dB while the TWT is operated at the saturation. The dashed lines in Figs. 3 to 5 are the symbol error probability for a linear channel.

References

1. Mizuno, T., Morinaga, N., and Nmakawa, T., "Transmission Characteristics on an M-ary Coherent PSK Signal Via Cascade of N Bandpass Hard Limiters," *IEEE Trans. on Comm.*, Vol. COM-24, No. 5, 1976.

2. Jain, P., and Blachman, N., "Detection of a PSK Signal Transmitted Through a Hard-Limited Channel," *IEEE Trans. Inf. Theory*, Vol. IT-19, Sept. 1973.

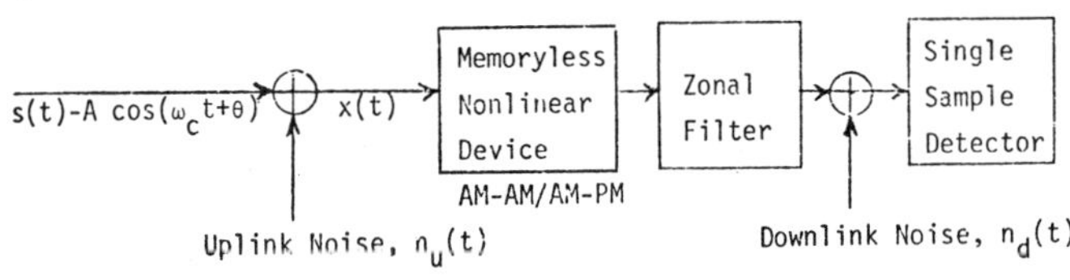

Figure 1. MPSK Satellite Communication Link.

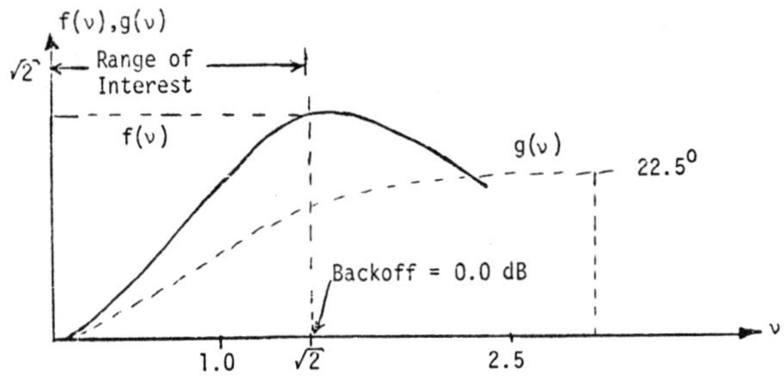

Figure 2. The TWT Model.

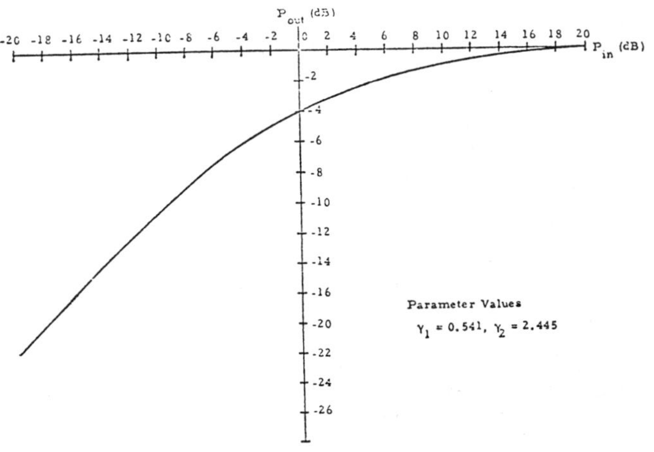

Figure 3. Power Transfer Characteristics of the Soft-Limiter Model.

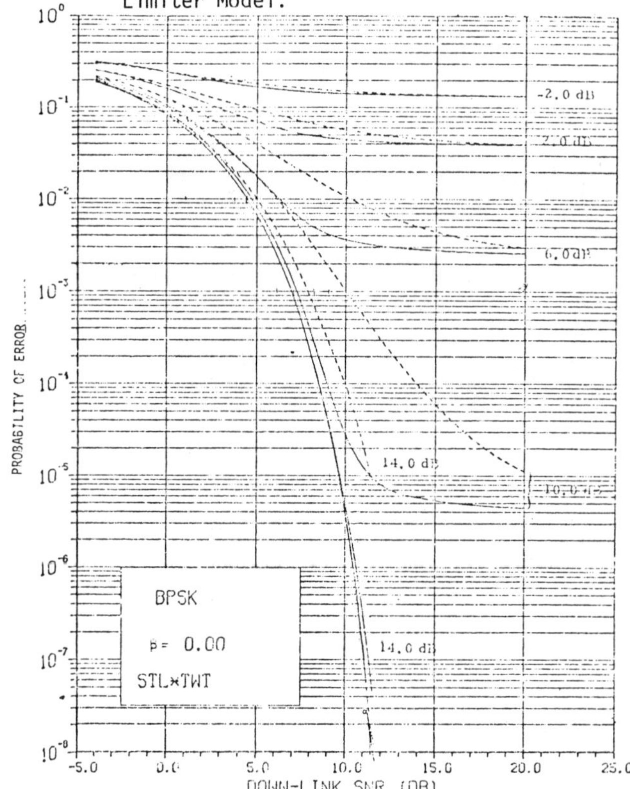

Figure 4. Symbol Error Probability vs Downlink SNR with Uplink SNR as a Parameter.

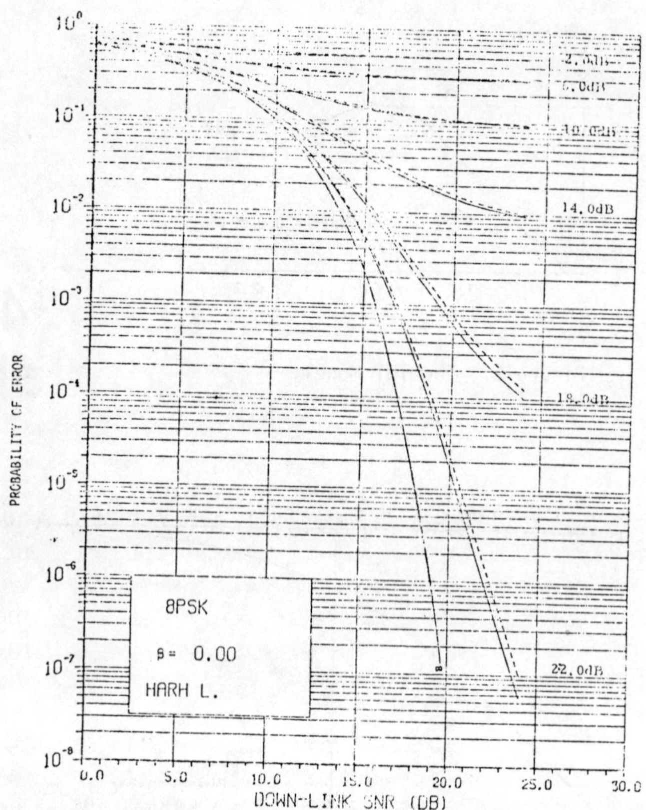

Figure 5. Symbol Error Probability vs Downlink SNR with Uplink SNR as a Parameter.

Figure 6. Symbol Error Probability vs Downlink SNR with Uplink SNR as a Parameter.

Section 3.4
Error Control

As digital transmission becomes more predominant, the importance of error control techniques will increase. Reprint Paper 3.3.1 introduced the basic ideas of error control as part of the general discussion of digital communication. The two classes of techniques are error detection and error correction. In an error detection system, the system performance is achieved by encoding the signal at the transmitter, detecting the errors at the receiver, and requesting retransmission of the information when an error is detected. In an error correction system, only the forward channel is used, and the receiver corrects errors by exploiting redundancy introduced in the encoding process. References [1] and [2] provide good introductions to coding theory.

The error control techniques most used on digital terrestrial links involve detection with an automatic repeat request (ARQ) procedure. These techniques are less attractive for satellite communication because of the long round-trip propagation delays, which lead to a significant reduction in throughput or large buffer requirements. Two specific ARQ techniques which tend to minimize the throughput penalty, at the expense of requiring large transmit buffers for high-speed circuits, are GO-BACK N and SELECTIVE REPEAT ARQ (e.g., [3] or [4]).

The two alternatives for forward error correction (FEC) codes are block codes and convolutional codes. These codes are designed to use the available signal energy in an efficient manner. The cost of this improved performance is an increase in the required channel bandwidth. Block codes are discussed in [5]. Block codes will be used in some cases, but it is expected that convolutional codes will predominate in satellite communications systems because of their ease of implementation and the availability of several attractive decoding schemes. The classic

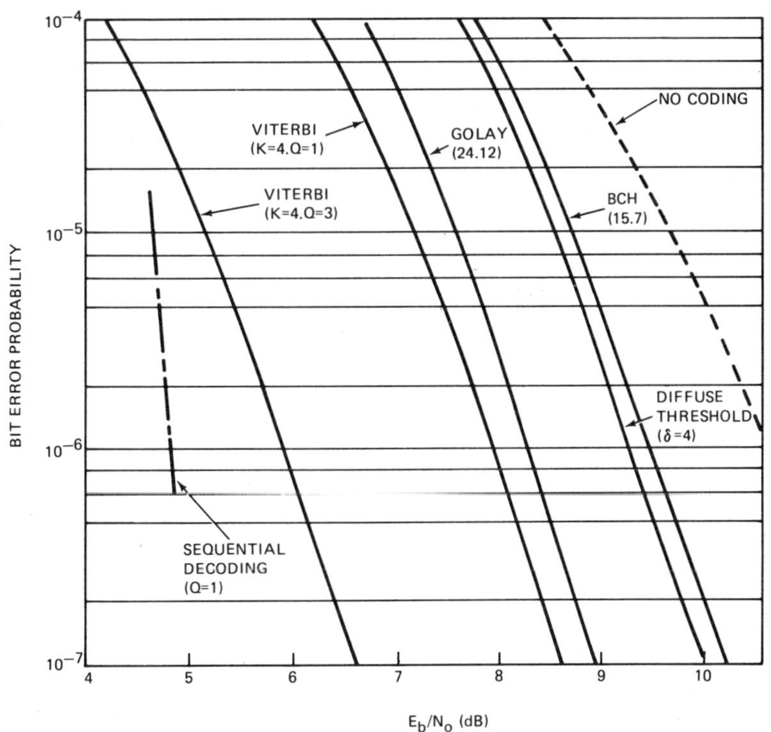

Fig. 1. P_e versus E_b/N_0 for various decoders.

paper by Viterbi [6] provides an excellent discussion of convolutional codes. The Viterbi decoding algorithm is discussed in Reprint Paper 3.4.1. The theoretical error performance as a function of E_b/N_0 is shown in Fig. 1. In practice, at a bit error probability of 10^{-5}, a rate one-half code and an existing Viterbi decoders using soft decisions can reduce, in comparison with the uncoded case, the required signal energy per bit by 5.1 dB at the cost of bandwidth expansions of 2.

Sequential decoders can achieve even greater savings and are potentially useful in situations with very stringent error requirements ($<10^{-8}$). Threshold decoding of convolutional codes does not provide as dramatic energy savings, but does provide the potential for improving the error performance of a reasonably good channel with modest equipment and data rate (bandwidth expansion) costs.

The Viterbi algorithm was first proposed as a method of decoding convolutional codes. Subsequently, it has been applied to a wide variety of digital estimation problems. Reprint Paper 3.4.2. develops some of these applications.

REFERENCES

[1] J. K. Wolf, "A survey of coding theory," *IEEE Trans. Inform. Theory,* pp. 381–388, July 1973.

[2] I. M. Jacobs, "Practical applications of coding theory," *IEEE Trans. Inform. Theory,* vol. IT-20, no. 3, pp. 305–310, May 1974.

[3] A. R. K. Sastry, "Error control techniques for satellite communications: An overview," *Conf. Record Int. Conf. Communications,* pp. 6A-1–6A-5, June 1974.

[4] K. Brayer, "ARQ and hybrid FEC-ARQ system design to meet tight performance constraints," *Proc. National Telecommunications Conf.,* 1976.

[5] G. D. Forney, "Generalized minimum distance decoding," *IEEE Trans. Inform. Theory,* Apr. 1966.

[6] A. J. Viterbi, "Convolutional codes and their performance in communication systems," *IEEE Trans. Commun. Technol.,* vol. COM-19, pp. 751–772, Oct. 1971.

Viterbi Decoding for Satellite and Space Communication

JERROLD A. HELLER, MEMBER, IEEE, AND IRWIN MARK JACOBS, MEMBER, IEEE

Abstract—Convolutional coding and Viterbi decoding, along with binary phase-shift keyed modulation, is presented as an efficient system for reliable communication on power limited satellite and space channels. Performance results, obtained theoretically and through computer simulation, are given for optimum short constraint length codes for a range of code constraint lengths and code rates. System efficiency is compared for hard receiver quantization and 4 and 8 level soft quantization. The effects on performance of varying of certain parameters relevant to decoder complexity and cost is examined. Quantitative performance degradation due to imperfect carrier phase coherence is evaluated and compared to that of an uncoded system. As an example of decoder performance versus complexity, a recently implemented 2-Mbit/s constraint length 7 Viterbi decoder is discussed. Finally a comparison is made between Viterbi and sequential decoding in terms of suitability to various system requirements.

Paper approved by the Communication Theory Committee of the IEEE Communication Technology Group for publication without oral presentation. This work was supported in part by NASA Ames Research Center under Contract NAS2-6024. Manuscript received June 10, 1971.

The authors are with the Linkabit Corporation, San Diego, Calif.

I. INTRODUCTION

THE SATELLITE and space communication channels are likely candidates for the cost-effective use of coding to improve communication efficiency. The primary additive disturbance on these channels can usually be accurately modeled by Gaussian noise which is "white" enough to be essentially independent from one bit time interval to the next, and, particularly on the space channel but also in many instances on satellite channels, sufficient bandwidth is available to permit moderate bandwidth expansion. Two effective decoding algorithms for independent noise (memoryless) channels have been developed and refined, namely sequential and Viterbi decoding of convolutional codes. These theoretical accomplishments, combined with real communication

Reprinted from *IEEE Trans. Commun. Technol.*, vol. COM-19, pp. 835–848, Oct. 1971.

needs and the availability of low-cost complex digital integrated circuits, make possible practical and powerful high-speed decoders for satellite and space communication.

Communication from a distant and isolated object in space to a ground-based station presents certain system problems which are not nearly as critical in earth-based communication systems. The most obvious among these is the high cost of space-platform power. It is desirable to design a system which is as efficient as practical in order to minimize the spacecraft weight necessary to generate power.

The modulated signal power at a ground station receiver front end P depends upon the transmitted power, the transmitting and receiving antenna gains, and propagation path losses. Primarily due to thermal activity at the receiver front end, wideband noise is added to the received signal, resulting in a received signal power-to-noise ratio (P/N_0), where N_0 is the single-sided noise spectral density. The noise is usually accurately modeled as being both white and Gaussian. Other perturbations caused by uncertainty in carrier phase at the demodulator and inaccuracies in receiver AGC are treated in Sections IV and V.

The efficiency of a communication system is usefully measured by the received energy per bit to noise ratio (E_b/N_0) required to achieve a specified system bit error rate. The E_b/N_0 is expressable in terms of the modulating signal power by the relationship

$$\frac{E_b}{N_0} = \frac{P}{N_0} \cdot \frac{1}{R} \tag{1}$$

where R is the information rate in bits per second. Alternatively, (1) can be written as

$$R = \frac{P/N_0}{E_b/N_0}. \tag{2}$$

The payoff for using modulation and/or coding techniques which reduce the E_b/N_0 required for a given bit error probability is an increase in allowable data rate and/or a decrease in necessary received P/N_0.

As a point of reference, it is traditional to compare the efficiency of modulation-coding schemes with that of a hypothetical system operating at channel capacity. Channel capacity for an infinite bandwidth white Gaussian noise channel with average power P is [1]

$$C_\infty = \frac{P}{N_0 \ln 2} \text{ bit/s.} \tag{3}$$

From (1), when $R = C_\infty$,

$$E_b/N_0 = \ln 2 \triangleq \frac{E_{b,\,min}}{N_0}. \tag{4}$$

Thus, the lower bound on achievable E_b/N_0 is about -1.6 dB.

Without coding, required E_b/N_0 can be minimized by selecting an efficient modulation technique. For example,

$180°$ binary phase-shift keying (BPSK) is more efficient than binary frequency shift keying (BFSK). For a desired bit error rate of 10^{-5}, an E_b/N_0 of 9.6 dB is required using BPSK (antipodal) modulation, whereas, 12.6 dB is required with BFSK (orthogonal) modulation. Quadraphase-shift keying (QPSK) is often used to conserve bandwith. Under the assumption of perfect phase coherence, QPSK has the same performance as BPSK.

In designing a communication system to operate at a specified data rate, the improvement in efficiency to be realized using coding must be weighed against the relative costs. Potential alternatives include increasing the transmitted power, increasing the transmitting antenna gain, and/or the receiving antenna area, and accepting a higher probability of bit error. In many applications, a minimum P_e is required and the incremental cost per decibel increase in P/N_0 is now greater (often much greater) than the cost of reducing the needed E_b/N_0 through coding. Soft decision Viterbi and hard decision sequential decoding can provide a relatively inexpensive 4–6-dB improvement in required E_b/N_0 (at a 10^{-5} bit error rate), even at multimegabit data rates. Sequential decoding is extensively discussed in [5]. In Section VII, we compare these techniques. Sections II and III examine various aspects of Viterbi decoding and present curves permitting system tradeoffs. In Section VI, a particular implementation of a Viterbi decoder is discussed to provide one benchmark for cost-complexity discussions.

In the discussion that follows, we assume that the channel is power limited rather than bandwidth limited. This assumption is realistic for many present day and future systems; however, the trend, especially in satellite repeaters, is to larger P/N_0 without a proportional increase in available bandwidth. For this reason, we will limit consideration to codes which involve a "bandwidth expansion" of 3 or less; that is, we assume that from 1 to 3 binary symbols can be transmitted over the channel for each bit of information communicated without appreciable intersymbol interference.

II. System

A. Convolutional Encoder

Fig. 1 shows a general binary-input binary-output convolutional coder. The encoder consists of a kK stage binary shift register and v mod-2 adders. Each of the mod-2 adders is connected to certain of the shift register stages. The pattern of connections specifies the code. Information bits are shifted into the encoder shift register k bits at a time. After each k bit shift, the outputs of the mod-2 adders are sampled sequentially yielding the code symbols. These code symbols are then used by the modulator to specify the waveforms to be sent over the channel. Since v code symbols are generated for each set of k information bits, the code rate R_N is k/v information bits per code symbol, where $k < v$. The constraint length of the code is K, since that is the number of k bit shifts over which a single information bit can

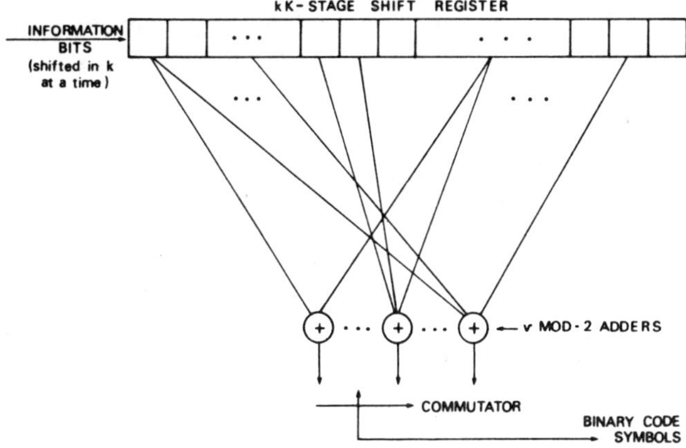

Fig. 1.   Rate k/n convolutional encoder.

influence the encoder output. The state of the convolutional encoder is the contents of the first $k(K-1)$ shift register stages. The encoder state together with the next k input bits uniquely specify the v output symbols.

As an example, a $K = 3$, $k = 1$, $v = 2$ encoder is shown in Fig. 2(a). The first two coder stages specify the state of the encoder; thus, there are 4 possible states. The code words, or sequences of code symbols, generated by the encoder for various input information bit sequences is shown in the code "trellis" [2] of Fig. 2(b). The code trellis is really just a state diagram for the encoder of Fig. 2(a). The four states are represented by circled binary numbers corresponding to the contents of the first two stages of the encoder. The lines or "branches" joining states indicate state transitions due to the input of single information bits. Dashed and solid lines correspond to "1" and "0" input information bits, respectively. The trellis is drawn under the assumption that the encoder is in state 00 at time 0. If the first information bit were a 1, the encoder would go to state 10 and would output the code symbols 11. Code symbols generated are shown adjacent to the trellis branches. As an example, the input data sequence $101\cdots$ generates the code symbol sequence $111000\cdots$. Further interpretations of the encoder state diagram and a discussion of "good" convolutional codes is presented in [3].

B. Modulation

The binary symbols output by the encoder are used to modulate an RF carrier sinusoid. Here we restrict our attention to the case of 180° BPSK modulation. Each code symbol results in the transmission of a pulse of carrier at either of two 180° separated phases. A squence of code symbols produces a uniformly spaced sequence of biphase pulses. The signal component of the received waveform thus has the form

$$s(t) = \sum_i \sqrt{2E_s}\, p(t - iT_s) \cos\left(2\pi f_c t + x_i\pi/2 + \theta\right)$$

$$= \sqrt{2E_s} \cos\left(2\pi f_c t + \theta\right) \sum_i \cdot x_i p(t - iT_s). \tag{5}$$

Here x_i is ± 1 depending on whether the ith code symbol

is 0 or 1. The function $p(t)$ is a convenient unit energy low-pass pulse waveform, f_c is the carrier frequency, E_s is the energy per pulse, and T_s is the time between successive code symbols. E_s and T_s are defined by the relationships

$$E_s = R_N E_b = kE_b/v \tag{6}$$

and

$$T_s = R_N/R. \tag{7}$$

There are several reasons for restricting attention to BPSK modulation. Three important ones are as follows.

1) BPSK signals are convenient to generate and amplify. Traveling wave tube amplifiers operate most efficiently at or near saturation. This nonlinear amplification would degrade performance with multilevel amplitude modulated waveforms.

2) It can be shown that antipodal (BPSK) modulation results in little increase in required E_b/N_0 compared to optimum signaling when E_s/N_0 is low [4].

3) BPSK modulation of quadrature carriers is equivalent to quadraphase (QPSK) modulation of one carrier. Thus, QPSK need not be separately treated except for synchronization and phase error requirements.

C. Demodulation and Quantization

At the receiver, the signal $s(t)$ of (5), is observed added to white Gaussian noise. When the carrier phase θ is known, the optimum demodulator consists of an integrate and dump filter matched to $p(t) \cos(2\pi f_c t + \theta)$. At time jT_s, the demodulator outputs data r_i relevant to the jth code symbol. Normalizing the matched filter output by dividing by $\sqrt{N_0/2}$ yields

$$r_i = x_i \sqrt{2E_s/N_0} + n_i \tag{8}$$

when n_j is a zero-mean unit variance Gaussian random variable. Each n_j is independent of all others.

To facilitate digital processing by the decoder, the continuous r_j must be quantized. The simplest quantization is a hard decision with 0 output if r_j is greater than zero and 1 output otherwise. Here, the received data are represented by only one bit per code symbol. Without coding, the matched filter sampler hard quantizer is an optimum receiver.

When coding is used, hard quantization of the received data usually entails a loss of about 2 dB in E_b/N_0 compared with infinitely fine quantization [4], [5]. Much of this loss can be recouped by quantizing r_j to 4 or 8 levels instead of merely 2. Adding additional levels of quantization necessitates a 2- or 3-bit representation of each r_j. Fig. 3(a) and (b) shows two quantization schemes with 4 and 8 levels, respectively. Here the quantization level thresholds are spaced evenly. The spacing is 1.0 for 4 levels and 0.5 for 8 levels. Uniform quantization threshold spacings of 1.0 and 0.5 can be shown by analytical means and through simulation to be very close to optimum for 4- and 8-level quantiza-

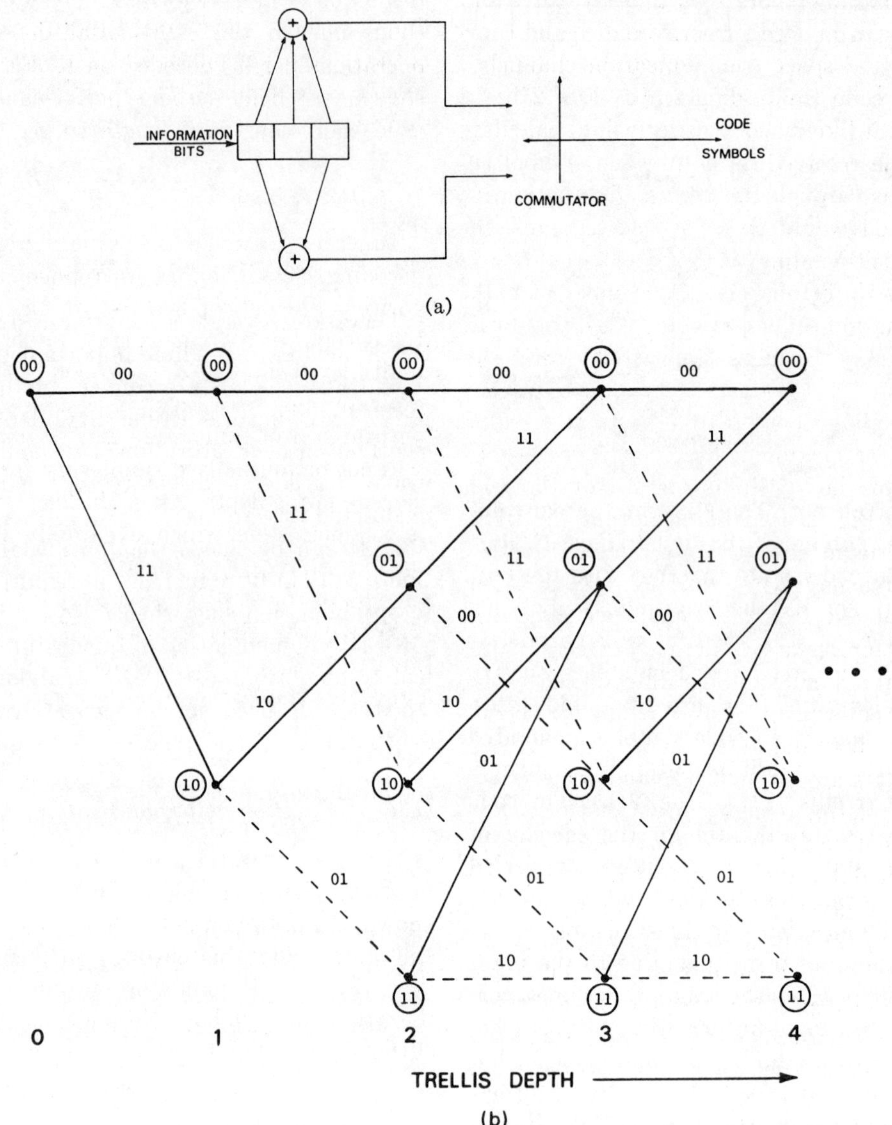

Fig. 2. (a) $K = 3$, $R_N = 1/2$ convolutional encoder. (b) Code trellis diagram.

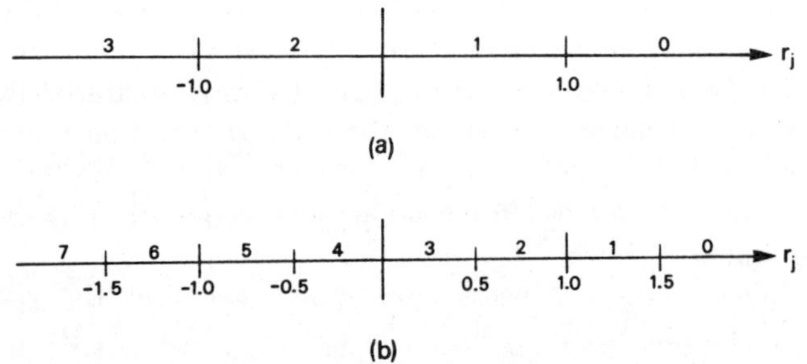

Fig. 3. Receiver quantization thresholds and intervals for (a) 4-level and (b) 8-level quantization.

tion. Furthermore, 8-level quantization results in a loss of less than 0.25 dB compared to infinitely fine quantization; therefore, quantization to more than 8 levels can yield little performance improvement. We confine our attention to hard decision quantization and the 4 and 8 level schemes shown in Fig. 3.

Receiver quantization converts the modulator, Gaussian channel, and demodulator into a discrete channel with two inputs (0 or 1, the code symbols) and 2, 4, or 8 outputs. The channel transition probabilities are a function only of the symbol signal-to-noise ratio E_s/N_0. For example, with 8-level quantization, the probability of receiving interval 6 given that a 0-code symbol is sent is the probability that a unit-variance Gaussian random variable with mean $\sqrt{2E_s/N_0}$ lies between -1.0 and -1.5 in value.

III. Viterbi Decoding

A. Basic Algorithm

The maximum likelihood or Viterbi decoding algorithm was discovered and analyzed by Viterbi [6] in 1967. Viterbi decoding was first shown to be an efficient and practical decoding technique for short constraint length codes by Heller [7], [8]. Forney [2] and Omura [12] demonstrated that the algorithm was in fact maximum likelihood.

A thorough discussion of the Viterbi decoding algorithm is presented by Viterbi [3]. Here, it will suffice to briefly review the algorithm and elaborate on those features and parameters which bear on decoder performance and complexity on satellite and space communication channels.

Referring to the code trellis diagram of Fig. 2(b), a brute-force maximum likelihood decoder would calculate the likelihood of the received data for code symbol sequences on all paths through the trellis. The path with the largest likelihood would then be selected, and the information bits corresponding to that path would form the decoder output. Unfortunately, the number of paths for an L bit information sequence is 2^L; thus, this brute force decoding quickly becomes impractical as L increases.

With Viterbi decoding, it is possible to greatly reduce the effort required for maximum likelihood decoding by taking advantage of the special structure of the code trellis. Referring to Fig. 2(b), it is clear that the trellis assumes a fixed periodic structure after trellis depth 3 (in general, K) is reached. After this point, each of the 4 states can be entered from either of two preceding states. At depth 3, for instance, there are 8 code paths, 2 entering each state. For example, state 00 at level 3 has the two paths entering it corresponding to the information sequences 000 and 100. These paths are said to have diverged at state 00, depth 0 and remerged at state 00, depth 3. Paths remerge after 2 [in general $k(K - 1)$] consecutive identical information bits. A Viterbi decoder calculates the likelihood of each of the 2^k paths entering a given state and eliminates from further consideration all but the most likely path that leads to that state. This is done for each of the $2^{k(K-1)}$ states at a given trellis depth; after each decoding operation only one path remains leading to each state. The decoder then proceeds one level deeper into the trellis and repeats the process.

For the $K = 3$ code trellis of Fig. 2(b), there are 8 paths at depth 3. Decoding at depth 3 eliminates 1 path entering each state. The result is that 4 paths are left. Going on to depth 4, the decoder is again faced with 8 paths. Decoding again eliminates 4 of these paths, and so on. Note that in eliminating the less likely paths entering each state, the Viterbi decoder will not reject any path which would have been selected by the brute force maximum likelihood decoder.

The decoder as described thus far never actually decides upon one most likely path. It always retains a set of $2^{k(K-1)}$ paths after each decoding step. Each retained path is the most likely path to have entered a given encoder state. One way of selecting a single most likely path is to periodically force the encoder into a prearranged state by inputting a $K - k$ bit fixed information sequence to the encoder after each set of L information bits. The decoder can then select that path leading to the known encoder state as its (1 bit) output.

The great advantage of the Viterbi maximum likelihood decoder is that the number of decoder operations performed in decoding L bits is only $L2^{k(K-1)}$, which is linear in L. Of course, Viterbi decoding as a practical technique is limited to relatively short constraint length codes due to the exponential dependence of decoder operations per bit decoded on K. Fortunately, as will be shown, excellent decoder performance is possible with good short constraint length codes.

B. Path Memory

In order to make the Viterbi algorithm a practical decoding technique, certain refinements on the basic algorithm are desirable. First of all, periodically forcing the encoder into a known state by using preset sequences multiplexed into the data stream is neither operationally desirable nor necessary. It can be shown [2], [9] that with high probability, the $2^{k(K-1)}$ decoder selected paths will not be mutually disjoint very far back from the present decoding depth. All of the $2^{k(K-1)}$ paths tend to have a common stem which eventually branches off to the various states. This suggests that if the decoder stores enough of the past information bit history of each of the $2^{k(K-1)}$ paths, then the oldest bits on all paths will be identical. If a fixed amount of path history storage is provided, the decoder can output the oldest bit on an arbitrary path each time it steps one level deeper into the trellis. The amount of path storage required u is equal to the number of states, $2^{k(K-1)}$ multiplied by the length of the information bit path history per state h,

$$u = h2^{k(K-1)}. \tag{9}$$

Since the path memory represents a significant portion of the total cost of a Viterbi decoder, it is desirable to minimize the required path history length h. One refinement which allows for a smaller value of h is to use the oldest bit on the most likely of the $2^{k(K-1)}$ paths as the decoder output, rather than the oldest bit on an arbitrary path. It has been demonstrated theoretically [2] and through simulation [9] that a value of h of 4 or 5 times the code constraint length is sufficient for negligible degradation from optimum decoder performance. Simulation results showing performance degradation incurred with smaller path history lengths are presented and discussed in Section IV.

C. State and Branch Metric Quantization

The path comparisons made for paths entering each state require the calculation of the likelihood of each path involved for the particular received information. Since the channel is memoryless, the path likelihood function is the product of the likelihoods of the individual code symbols [3]

$$P(\mathbf{r}^*/\mathbf{x}^l) = \prod_i P(r_i^*/x_i^l) \tag{10}$$

where $\mathbf{r}^* = (r_1^*, r_2^*, \cdots, r_i^*, \cdots)$ is the vector of quantized receiver outputs and $\mathbf{x}^l = (x_1^l, x_2^l, \cdots, x_i^l, \cdots)$ is the code symbol vector for the lth trellis path. In order to avoid multiplication, the logarithm of the likelihood is a preferable path metric

$$M_l = \log p(\mathbf{r}^*/\mathbf{x}^l)$$

$$= \sum_i \log p(r_i^*/x_i^{\,l}) \triangleq \sum_i m_i^{\,l} \qquad (11)$$

where M_l is the metric of the lth path and $m_j^{\,l}$ is the metric of the jth code symbol on the lth path. With this type of additive metric, when a path is extended by one branch, the metric of the new path is the sum of the new branch symbol metrics and the old path metric. To facilitate this calculation, the path metric for the best path leading to each state must be stored by the decoder as a state metric. This is an addition to the path information bit history storage required.

Viterbi decoder operation can then be summarized as follows, taking the $K = 3$ case of Fig. 2 as an example.

1) The metric for the 2 paths entering state 00 are calculated by adding the previous state metrics of states 00 and 01 to the branch metrics of the upper and lower branches entering state 00, respectively.

2) The largest of the two new path metrics is stored as the new state metric for state 00. The new path history for state 00 is the path history of the state on the winning path augmented by a 0 or 1 depending on whether state 00 or 01 was on the winning path.

3) This add–compare–select (ACS) operation is performed for the paths entering each of the other 3 states.

4) The oldest bit on the path with the largest new path metric forms the decoder output.

Since the code symbol metrics must be represented in digital form in the decoder, the effects of metric quantization come into question. Simulation has shown that decoder performance is quite insensitive to symbol metric quantization. In fact, use of the integers as symbol metrics instead of log likelihoods results in a negligible performance degradation with 2-, 4-, or 8-level receiver quantization [7], [9]. Fig. 4 shows such a set of metrics for the 8-level quantized channel. Use of these symbol metrics implies that symbol metrics as well as the received symbols themselves may be represented by 1, 2, or 3 bits for 2-, 4-, and 8-level receiver quantization, respectively.

D. Unknown Starting State

It has been assumed thus far that a Viterbi decoder has knowledge of the encoder starting state before decoding begins. Thus, in Fig. 2(b), the starting state is assumed to be 00. A known starting state may be operationally undesirable since it requires that the decoder know when transmission commences. In reality, it has been found through simulation that a Viterbi decoder may start decoding at any arbitrary point in a transmission, if all state metrics are initially reset to zero. The first 3–4 constraint lengths worth of data output by the decoder will be more or less unreliable because of the unknown encoder starting state. However, after about 4 constraint lengths, the state metrics with high probability have values independent of the starting values and steady-state reliable operation results.

QUANTIZATION LEVEL

CODE SYMBOL	0	1	2	3	4	5	6	7
0	7	6	5	4	3	2	1	0
1	0	1	2	3	4	5	6	7

Fig. 4. Integer code symbol metrics for 8-level receiver quantization.

IV. Simulation and Error Probability Bound Results

A. Tradeoffs Between Bit Error Probability and E_b/N_0 for Rate 1/2 Codes

Viterbi has derived tight upper bounds to bit error probability for Viterbi decoding based on the convolutional code transfer function [3]. These bounds are particularly tight for the white Gaussian noise channel for error probabilities less than about 10^{-4}. This bound has been numerically evaluated over a range of E_b/N_0 for a variety of codes. The upper bound is presented along with some of the 8-level receiver quantized simulation results for comparison. The upper bound also provides performance data at very-low bit error rates, where simulation results are not available due to excessive computer time required. In comparing the upper bounds to the simulation results, it is important to keep in mind that the upper bound was derived for an infinitely finely quantized receiver output.

The convolutional codes used in the simulations were found through exhaustive computer search [9], [10]. The search criterion was maximization of the minimum free distance for a given code constraint length [3]. Where two codes had the same minimum free distance, the number of codewords at that distance and the higher order free distances were used for code selection. Simulations have consistently shown that the free distance criterion yields codes with the minimum error probability. The principal results of the simulations and code transfer function bounds are shown in Figs. 5, 6, and 7. All of these figures show bit error rate versus E_b/N_0 for Viterbi decoders using optimum rate 1/2 convolutional codes. In all cases, the decoder path history length was 32 bits. In all simulation runs, at least 25 error events contributed to the compiled statistics.

B. Performance Depending on Quantization, Path History, and Receiver Automatic Gain Control

The simulation results in Figs. 5 and 6 are for soft (8-level) receiver quantization. Equally spaced demodulation thresholds are used as shown in Fig. 3(b). This choice of 8-level quantizer thresholds is within a broad range of near optimum values, as will be shown presently. The transfer function bound is for infinitely finely quantized received data, although tight bounds for any degree of quantization can be obtained. Allowing for the 0.20–0.25 dB loss usually associated with 8-level receiver quantization compared with infinite quantization, the transfer function bound curves are in excellent

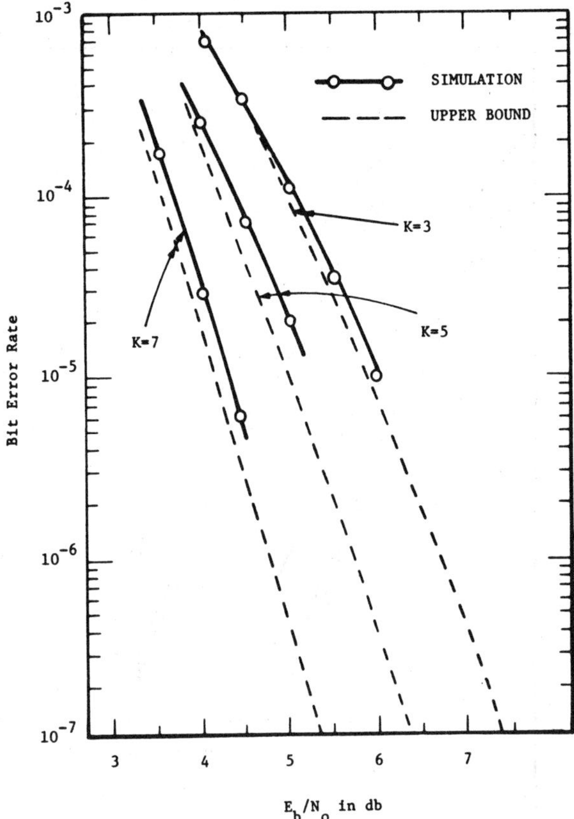

Fig. 5. Bit error rate versus E_b/N_0 for rate 1/2 Viterbi decoding. 8-level quantized simulations with 32-bit paths, and infinitely finely quantized transfer function bound, $K = 3, 5, 7$.

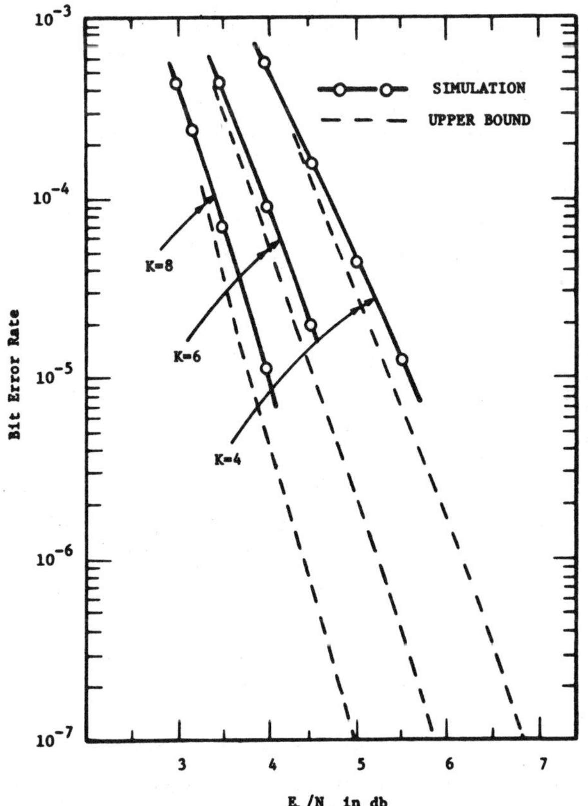

Fig. 6. Bit error rate versus E_b/N_0 for rate 1/2 Viterbi decoding 8-level quantized simulations with 32-bit paths, and infinitely finely quantized transfer function bound, $K = 4, 6, 8$.

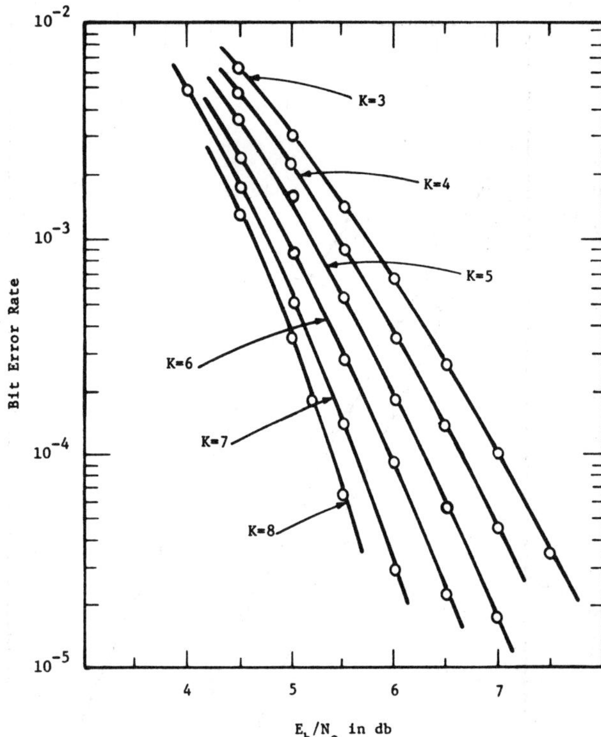

Fig. 7. Bit error rate versus E_b/N_0 for rate 1/2 Viterbi decoding. Hard quantized received data with 32-bit paths; $K = 3$ through 8.

agreement with simulation results in the 10^{-4} to 10^{-5} bit error rate range.

Since the accuracy of the transfer function bound increases with E_b/N_0, decoder performance can be ascertained accurately in the 10^{-5} to 10^{-8} region even in the absence of simulation. The symbol metrics used in the simulation were the equally spaced integers as shown in Fig. 4.

Fig. 7 gives the simulation results for Viterbi decoding with hard receiver quantization. The same optimum rate 1/2, $K = 3$ through $K = 8$ codes were used here as in the 8-level quantized simulations.

The following points are obvious from the performance curves.

1) 2-level quantization is everywhere close to 2-dB inferior to 8-level quantization.

2) Each increment in K provides an improvement in efficiency of something less than 0.5 dB at a bit error rate of 10^{-5}.

3) Performance improvement versus K increases with decreasing bit error rate.

To observe the effects of varying receiver quantization more closely, simulation performance data are presented in Fig. 8 for the $K = 5$, rate 1/2 code, with 2-, 4-, and 8-level receiver quantization. The $Q = 8$- and $Q = 4$-level thresholds are those of Fig. 3.

Fig. 9 shows bit error rate performance versus E_b/N_0 for three values of path history length (8, 16, and 32) using the rate 1/2, $K = 5$ code, for both 2- and 8-level received data quantization. (The length 32 path curve is identical to the $K = 5$ curve in Fig. 5.) Performance with

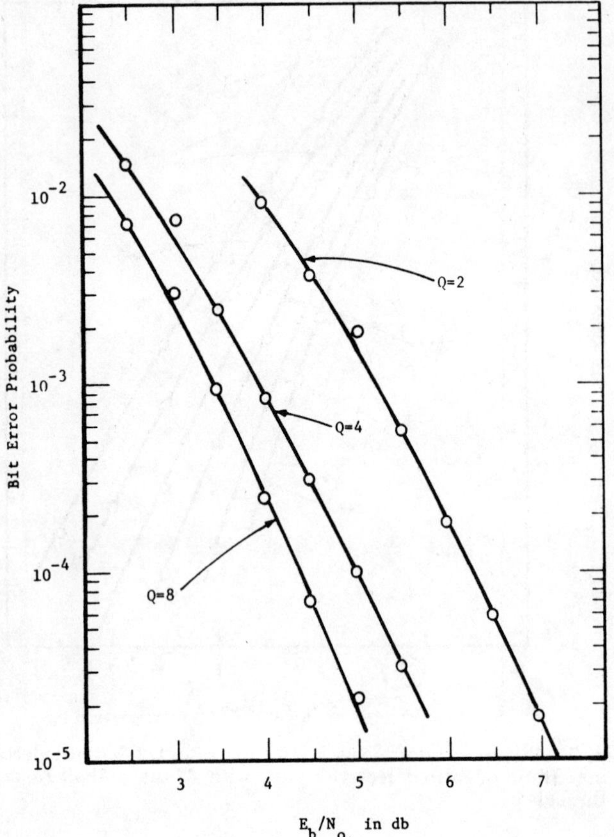

Fig. 8. Performance comparison of Viterbi decoding using rate 1/2, $K = 5$ code with 2-, 4-, and 8-level quantization. Path length = 32 bits.

length 32 paths is essentially identical to that of an infinite path decoder. Even for a path length of only 16, there is only a small degradation in performance. As previously mentioned, other simulations have shown that a path length of 4–5 constraint lengths is sufficient for other constraint lengths as well.

Coded systems that make use of receiver outputs quantized to more than two levels require an analog-to-digital converter at the modem matched filter output, with thresholds that depend on correct measurement of the noise variance. Since the level settings are effectively controlled by the automatic gain control (AGC) circuitry in the modem, it is of interest to investigate the sensitivity of decoder performance to an inaccurate or drifting AGC signal. Fig. 10 shows the decoder performance variation as a function of A–D converter level threshold spacing. In all cases, the thresholds are uniformly spaced. These simulations use the $K = 5$ rate 1/2 code with $E_b/N_0 = 3.5$ dB. It is evident that Viterbi decoding performance is quite insensitive to wide variations in AGC gain. In fact, performance is essentially constant over a range of spacing from 0.5 to 0.7. This allows for a variation in AGC gain of better than ± 20 percent with no significant performance degradation.

C. Performance of Codes of Other Rates

The preceding simulation results have concentrated on Viterbi decoding of rate 1/2 convolutional codes. The

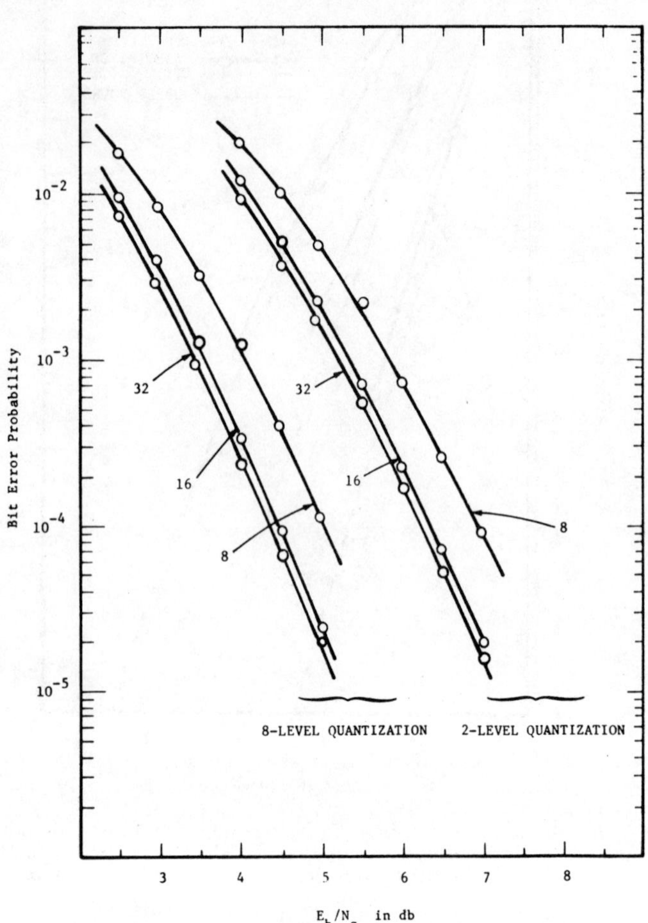

Fig. 9. Performance comparison of Viterbi decoding using rate 1/2, $K = 5$ code with 8-, 16-, and 32-bit path lengths and 2- and 8-level quantization.

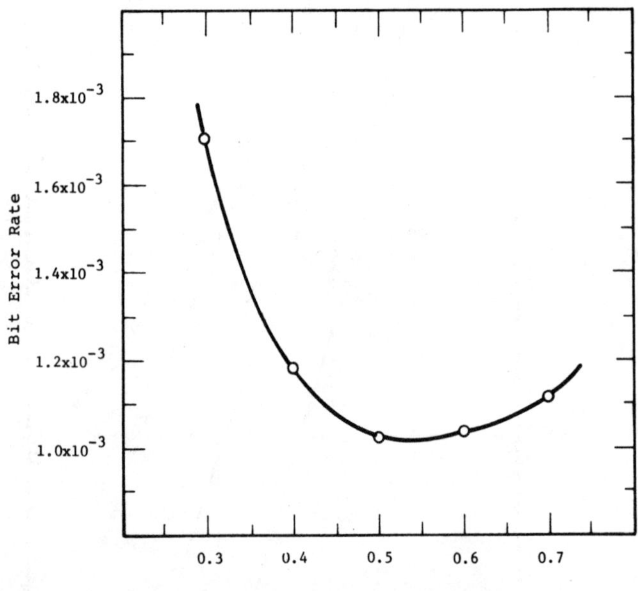

Fig. 10. Viterbi decoder bit error rate performance as function of quantizer threshold level spacing; $K = 5$, rate 1/2, $E_b/N_0 = 3.5$ dB, 8-level quantization with equally spaced thresholds.

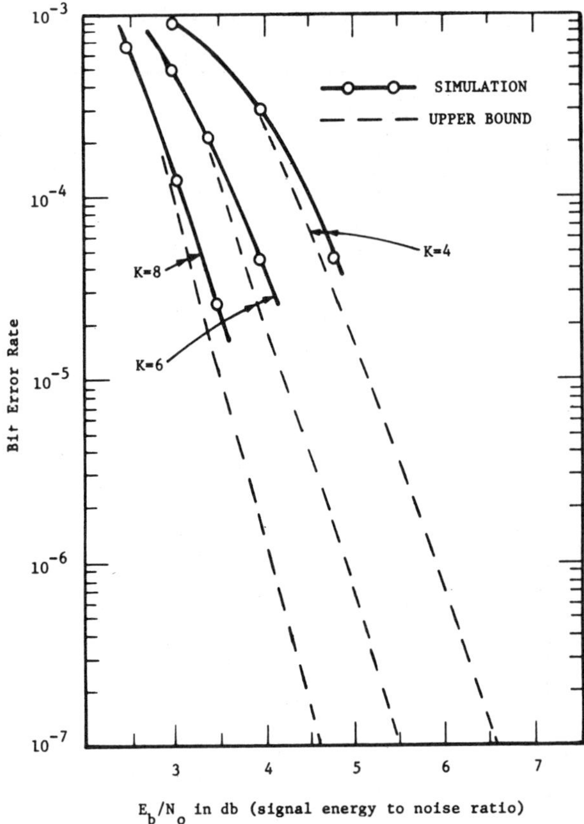

Fig. 11. Performance of rate 1/3, $K = 4$, 6, and 8 codes with Viterbi decoding.

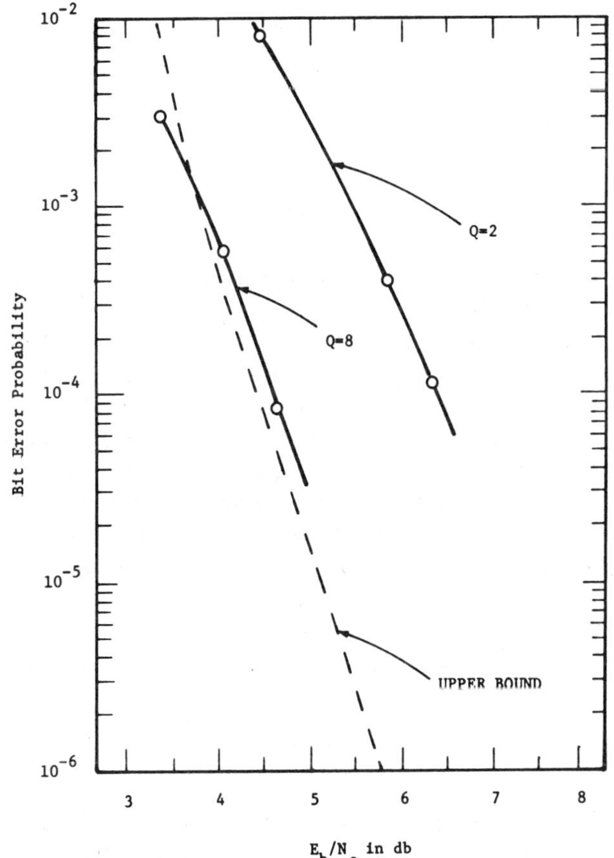

Fig. 12. Performance of rate 2/3 $K = 3$ code with Viterbi decoding. Numerical bound and simulation results.

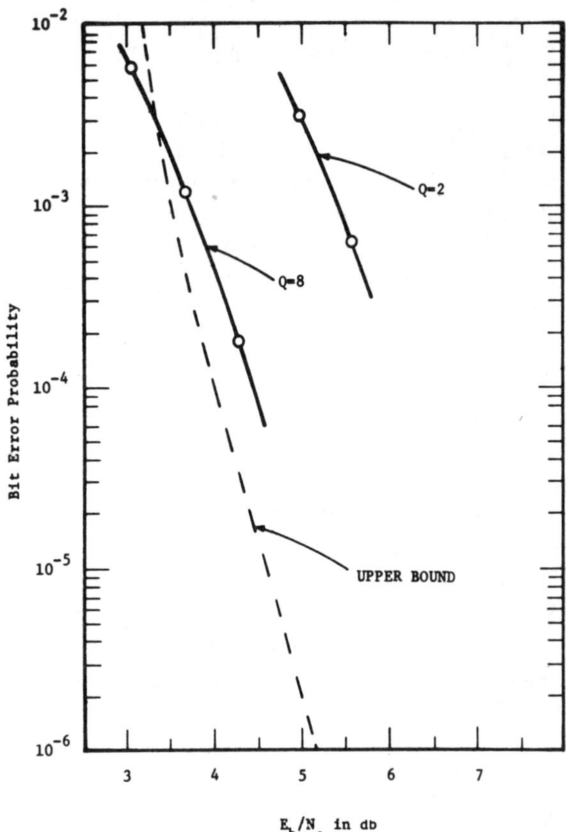

Fig. 13. Performance of rate 2/3 $K = 4$ code with Viterbi decoding. Numerical bound and simulation results.

results on performance fluctuation due to decoder parameter variation carry over to other code rates with minor changes.

Code rates less than 1/2 buy improved performance at the expense of increased bandwith expansion and more difficult symbol tracking due to decreased symbol energy-to-noise ratios. Rates above 1/2 conserve bandwidth but are less efficient in energy.

Fig. 11 shows bit error rate versus E_b/N_0 performance obtained from simulations of Viterbi decoding with optimum rate 1/3, $K = 4$, 6, and 8 codes, and 8-level quantization. Figs. 12 and 13 show numerical bound and simulation performance results for rate 2/3 $K = 3$ and $K = 4$ codes, respectively. Simulation curves are for 2- and 8-level quantization, while the numerical bound curves are for infinitely fine receiver quantization.

Comparing the performance data obtained through simulations of Viterbi decoders with rate 1/2 (Figs. 5, 6, and 7), and rate 1/3 codes, it is apparent that the latter offers a 0.3-to-0.5-dB improvement over the former for fixed K, in the range reported. This is close to the improvement in efficiency of a channel with capacity 1/3 compared with one of capacity 1/2, and is therefore expected.

Comparison of the higher rate codes with the rate 1/2 codes may also be made over the range spanned by the simulation and analytical data. The fairest comparison is probably between decoders with similar number of states, and hence similar decoder complexity. Thus, the $K = 3$ rate 2/3 data should be compared with the $K = 5$, rate 1/2 data.

Fig. 14 shows the union bounds on performances for the rate 2/3, $K = 3$, and rate 1/2, $K = 5$ codes. Both encoders have 16 states. The free distance d_f equals 7 for the rate 1/2 code and 5 for the rate 2/3 codes. At very high E_b/N_0, the rate 1/2 must be superior. This is because asymptotically, at high E_b/N_0, the error probability varies as

$$P_e \sim n_e \exp(-d_f E_s/N_0) = n_e \exp(-d_f R_N E_b/N_0)$$

where n_e is the number of bit errors contributed by codewords at distance d_f. This gives the rate 1/2 code an advantage of about 0.2 dB in the limit.

In Fig. 14, the difference between the two curves is about 0.1 dB in the error probability range of 10^{-6} to 10^{-9}. This small difference is due to the fact that the rate 2/3 code used happens to be a particularly good code; the value of n_e is smaller for it than for the rate 1/2 code and this difference is significant even for P_e as small as 10^{-9}.

V. Imperfect Carrier Phase Coherence

Thus far it has been assumd that carrier phase is known exactly at the receiver. In real systems this is usually not the case. Oscillator instabilities and uncompensated doppler shifts necessitate closed loop carrier phase tracking at the receiver. Since the carrier loop tracks a noisy received signal, the phase reference it provides for demodulation will not be perfect.

An inaccurate carrier phase reference at the demodulator will degrade system performance. In particular a constant error ϕ in the demodulator phase will cause the signal component of the matched filter output to be suppressed by the factor $\cos \phi$ (see [4, ch. 7]).

$$r_i = \pm\sqrt{\frac{2E_s}{N_0}} \cos \phi + n_i. \tag{12}$$

The effect of an imperfect carrier phase reference on performance is always worse for coded than uncoded systems. This is because coded systems are characterized by steeper error probability versus E_b/N_0 curves than uncoded systems. An imperfect carrier phase reference causes an apparent loss in received energy-to-noise ratio. Since the coded curve is steeper, the loss in E_b/N_0 degrades error probability to a greater extent. Furthermore, unless care is taken in the design of the phase-tracking loop, the phase error might be higher for the coded system than for an uncoded system, since loop performance may depend upon E_s/N_0, which is significantly smaller for coded than uncoded systems.

For convolutional coding with phase coherent demodulation and Viterbi decoding, exact analytical expressions for bit error rate P_e versus E_b/N_0 are not attainable. The simulation results of the preceding section, however, define a relationship between P_e and E_b/N_0 that can be

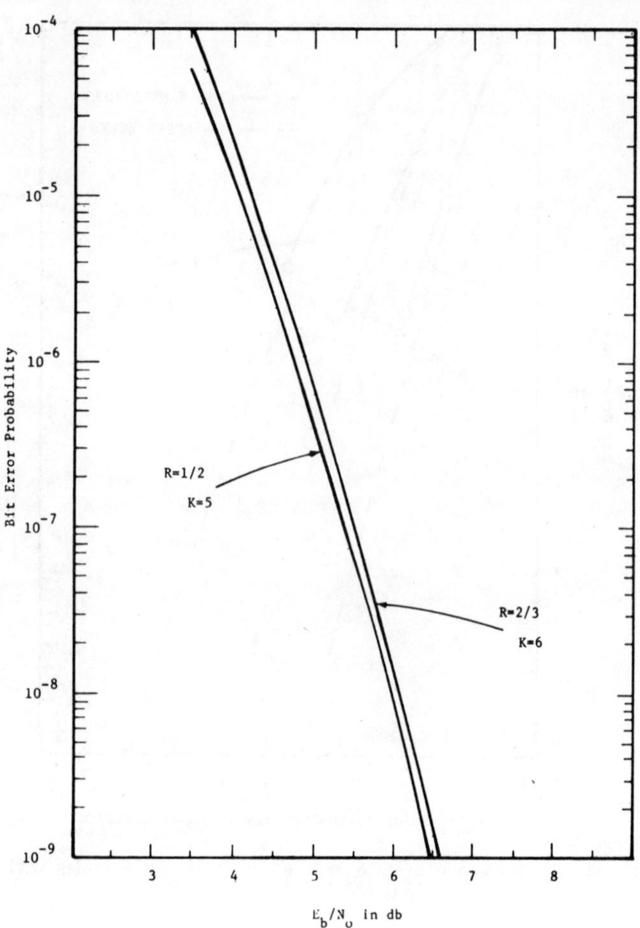

Fig. 14. Bit error probability bound for rate 1/2, $K = 5$, and rate 2/3, $K = 3$ code.

written formally as

$$P_e = f\left(\frac{E_b}{N_0}\right) \tag{13}$$

for a given code, receiver quantization, and Viterbi decoder. Since the carrier phase is being tracked in the presence of noise the phase error ϕ will vary with time. To simplify analysis, assume that the data rate is large compared to the carrier loop bandwidth so that the phase error does not vary significantly during perhaps 20–30 information bit times. Viterbi decoder output errors are typically several bits in length and are very rarely longer than 10–20 bits when the overall decoder bit error probability is less than 10^{-3}. Therefore, the phase error is assumed to be constant over the length of almost any decoder error. This being the case, the bit error probability for a constant phase error ϕ, can be written as

$$P_e(\phi) = f\left(\frac{E_b}{N_0} \cos^2 \phi\right) \tag{14}$$

from (12) and (13). This result uses the fact that received signal energy is degraded by $\cos^2 \phi$. If ϕ is a random variable with distribution $p(\phi)$, the resulting error probability averaged on ϕ is

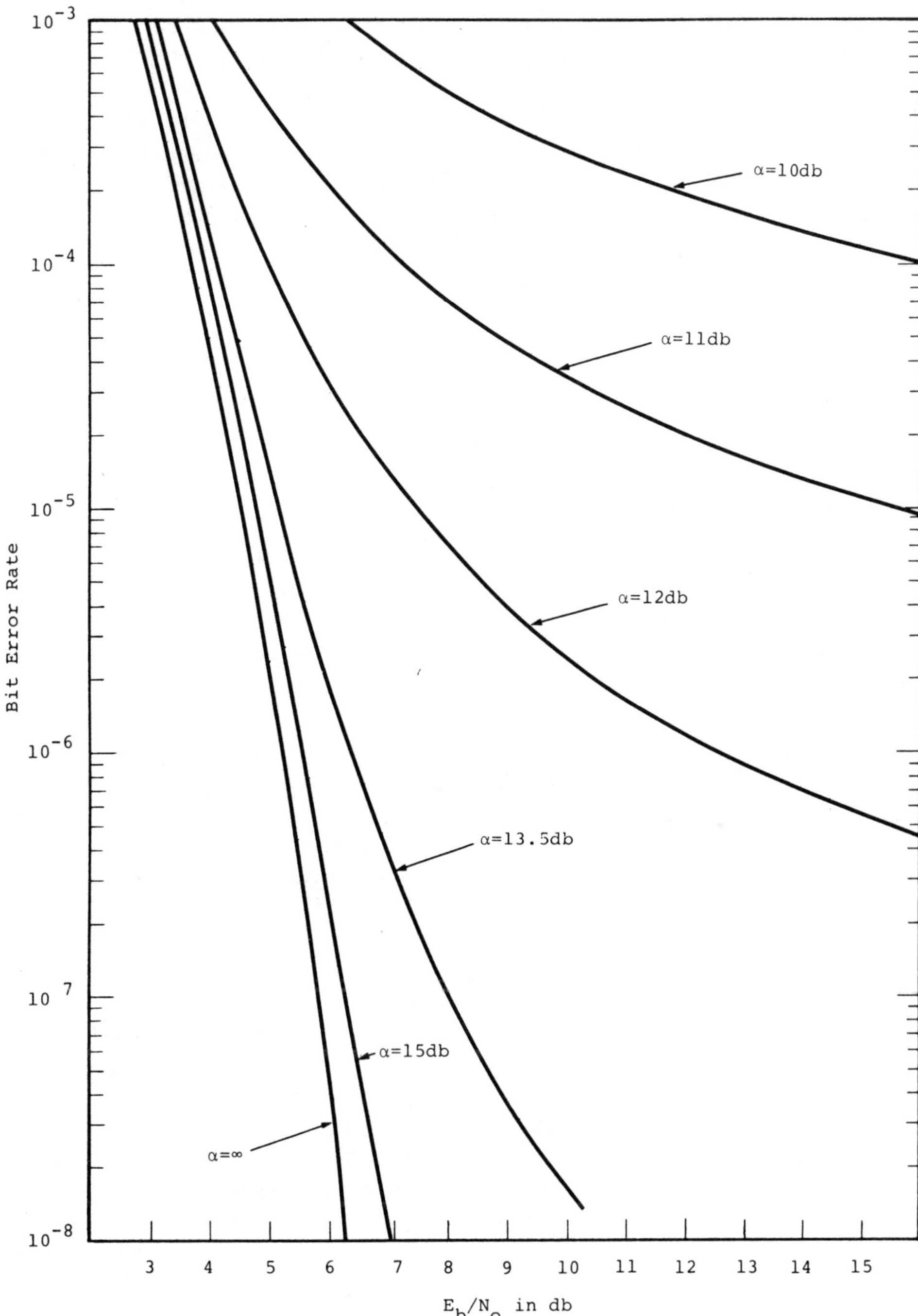

Fig. 15. Performance curves for rate 1/2; $K = 7$ Viterbi decoder with 8-level quantization as a function of carrier phase tracking loop signal-to-noise ratio α.

$$P_e' = \int_{-\pi}^{\pi} p(\phi) P_e(\phi) \, d\phi. \tag{15}$$

For the second-order phase-locked loop

$$p(\phi) = \frac{e^{\alpha \cos \phi}}{2\pi I_0(\alpha)}, \qquad \alpha \gg 1 \tag{16}$$

where $I_0(\cdot)$ is the zeroth order modified Bessel function and α is the loop signal-to-noise ratio [11]. Using this distribution and the P_e versus E_b/N_0 curve for the $K = 7$, rate 1/2 code of Fig. 5, the P_e integral of (15) has been evaluated for several values of α. The results are shown in Fig. 15 as curves of P_e' versus E_b/N_0 with α as

a parameter (the $K = 7$, rate 1/2 simulation curve of Fig. 5 was extrapolated to get the high E_b/N_0 results shown in this figure). These curves exhibit the same general shape as those for uncoded binary PSK modulation with phase coherence provided by a carrier tracking loop. As expected, the losses due to imperfect coherence are somewhat greater with than without coding. Fig. 16 shows the additional E_b/N_0 required to maintain a 10^{-5} bit error rate as a function of loop signal to noise ratio α. Curves are shown for the case of uncoded BPSK and rate 1/2, $K = 7$ convolutional encoding—Viterbi decoding.

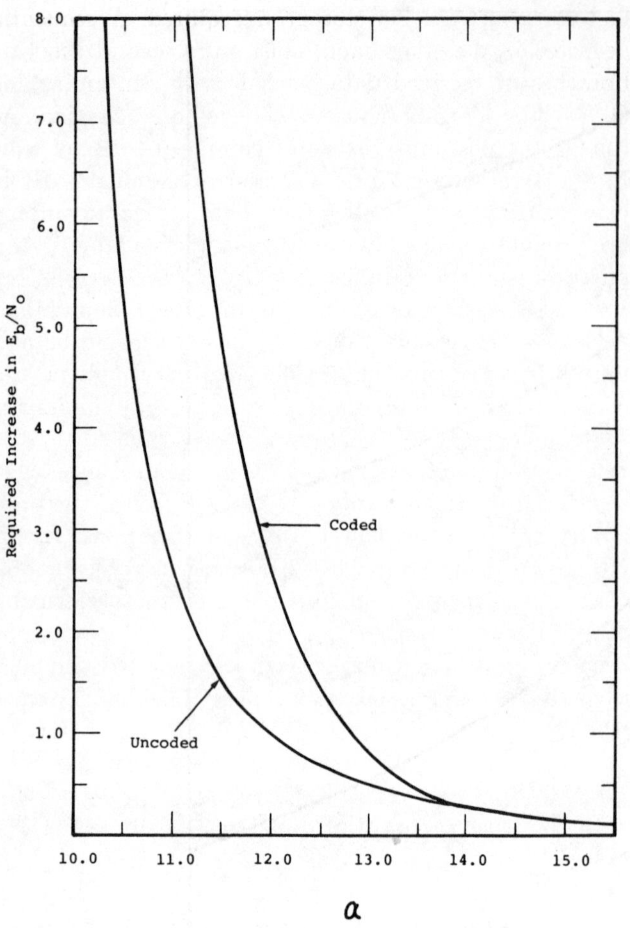

Fig. 16. Comparison to increase in E_b/N_0 due to imperfect phase coherence necessary to maintain 10^{-5} bit error rate for uncoded BPSK and $K = 7$; rate 1/2, $Q = 8$ Viterbi decoding.

VI. Implementation of a Viterbi Decoder

It is convenient to break the basic Viterbi decoder into five functional units; an input or branch metric calculation section, an ACS arithmetic section, and a path memory and output section. Information can be thought of as passing successively from one section to the next.

The branch metric calculation section accepts the input data and calculates (or looks up) the metric for each distinct branch. For a rate 1/2 code, four branches are possible corresponding to transmission of 00, 01, 10, and 11. For a rate 1/3 or rate 2/3 code, eight distinct branch metrics are possible. Note that this is the only section of the decoder that is directly concerned with the number of bits of quantization of the received data, and hence, the only section whose complexity is directly dependent on quantization. (The complexity of the ACS section also depends on quantization indirectly, in that the number of bits required for storing state metrics increases with the number of bits of quantization.) The input section is generally not critical in terms of either complexity or speed limitations. Its complexity does double, however, for each increase of the denominator of the rate R_N by one.

The ACS sections perform the basic arithmetic calcula-

tions of the decoder. For a rate $1/v$ decoder, an ACS is used to add the state metrics for two states to the appropriate branch metrics, to compare the resulting two sums, and to select the larger. The decision is transmitted to the path memory section and the larger of the two sums becomes a new state metric. One ACS function must be performed for each of the 2^{K-1} states. In a fully parallel very-high-speed decoder, 2^{K-1} ACS units are required. In general, the speed of the ACS unit places an upper bound on the speed of the decoder. For slower decoders, e.g., R less than several megabits per second for T²L logic, ACS units may be time shared, decreasing decoder cost significantly. Complexity of the ACS unit is strongly dependent upon required decoder speed. It should be noted that implementation of Viterbi decoders is greatly simplified by the fact that all ACS units perform identical functions and can be realized by a set of identical circuits.

The path memory section must store about a 4 constraint length history of decisions for each state. The memory requirements are thus nontrivial. Considerable advantage can be taken of new integrated-circuits memories to keep the equipment cost small. However, the complexity of the path memory and the ACS units both increase by a factor slightly larger than 2 for each increase in constraint length of 1. Thus, an increase in system performance of about 0.4 dB at a bit error rate of 10^{-5}, which can be achieved by increasing K by 1, comes at a cost of slightly more than doubling decoder complexity.

A complete decoder also must include interface circuits, synchronization circuits, timing circuits, and generally an encoder. A recent implementation[1] of a $K = 7$, rate 1/2 self-synchronized Viterbi decoder capable of operating at up to $R = 2$ Mbit/s with 2-, 4-, or 8-level quantized data required a total of 356 TTL integrated circuits for all functions. As noted in Fig. 5, this relatively simple decoder provides over 5-dB E_b/N_0 advantage over an uncoded BPSK system at $P_e = 10^{-5}$, and 6-dB advantage at $P^e = 10^{-8}$, when soft quantization is used.

VII. Comparison of Sequential and Viterbi Decoding

Both sequential and Viterbi decoding offer practical alternatives to a communications engineer designing a high-performance efficient communication system. The two decoders have significant differences which are noted below. Both are capable of very-high-speed operation.[2]

---

[1] The Linkabit LV7026 decoder is designed for use with differentially encoded BPSK or QPSK systems. It automatically resolves demodulator phase ambiguities and establishes node synchronization without manual intervention.

[2] The Linkabit LS4157 sequential decoder is capable of operation at data rates up to $R = 50$ Mbit/s. It uses a constraint length $K = 41$, rate 1/2 code and accepts only hard quantized data. The decoder is fully self-synchronizing. The coding advantage over uncoded data is 4.4 dB at $P_e = 10^{-5}$ at $R = 50$ Mbit/s and greater than 6 dB at $P_e = 10^{-8}$. The coding advantage is larger at lower data rates.

A. Error Probability

It should be recalled that, since the complexity of sequential decoders is relatively independent of constraint length, the constraint length is typically made quite large to provide a very small probability of undetected error. Usually the important contributor of errors is received data buffer overflow due to a computational overload. Such an event causes a long burst of rather noisy output data until the decoder reestablishes code synchronization. During this burst, the probability of bit error is that of the raw channel, perhaps $P_e = 3 \times 10^{-2}$.

Error from a Viterbi decoder occurs in short bursts of length at most 10 to 20. Systems that are sensitive to long bursts of errors should thus use Viterbi decoding. Systems that can tolerate occasional long bursts, with an error indication provided if desired by the decoder, should consider sequential decoding.

The curve of error probability versus E_b/N_0 tends to be much steeper for a sequential decoder than for a Viterbi decoder because of the difference in K. Thus, the sequential decoding advantage tends to increase as lower probabilities of bit error are demanded, although, as before, many errors tend to come in widely separated noisy bursts.

B. Decoder Delay

Sequential decoders tend to require long buffers of at least 200 bits and as much as several thousand bits to smooth out the variations in computational load. Viterbi decoders require a path memory of at most 64 bits. Thus the decoding delay differs by up to two orders of magnitude.

C. Long Tail Required to Terminate Sequences

In time-division multiplexed systems, bursts of separately encoded data may be received at the same decoder from different sources. In these instances, it may be desirable to time share the decoder. As noted in Section III, termination of encoding can be achieved by transmitting a known sequence of length $K - 1$, thus causing the encoder to enter a known state. Since K is typically larger for sequential decoding, the "tailing off" of the encoded sequence can cause a significant degradation in system efficiency. The tailing off of the short constraint length codes for Viterbi decoding causes a much smaller degradation.

If time and implementation permit the storage of the decoder state without code termination, then the cost of tailing off can be ignored. The design of such a time-shared sequential decoder remains for future work.

D. Rates Other than 1/2 and Soft Quantization

Viterbi decoders for rate 1/3 and 8-level quantization are not significantly more complex than those for rate 1/2 and 4- or 2-level quantization. The chief costs occur in the input section of the decoder as discussed in Section VI. In particular, the soft quantized data are processed in the input section and then incorporated in the branch and state metrics. No storage is required. A sequential decoder, on the other hand, must store several thousand branches of received data, each branch containing $\log_2 Q/R_N$ bits for rate R_N and Q level quantization. Although the possibility exists of gaining 0.4 dB by using rate 1/3 rather than rate 1/2 and of gaining 2 dB by using soft decisions rather than hard, these advantages are bought in sequential decoding at a formidable storage and processing cost. In general, then, practical high-rate sequential decoders are limited to rate 1/2 and hard decisions. (It is conceivable that this cost could be minimized by operating the decoder at a very high ratio of computation rate to average bit rate, thereby minimizing the number of branches required in the buffer.)

A second argument against soft quantization with sequential decoding involves the sensitivity of the probability of buffer overflow to channel variations. In Fig. 10, it was demonstrated that changes in receiver AGC of ± 20 percent had negligible effect on the performance of a Viterbi decoder. The degradation is much more pronounced for sequential decoding, since the computational load is very sensitive to changes in channel parameters. Thus, part of the 2-dB gain anticipated for soft decisions might be lost unless great care was exercised in controlling receiver AGC precisely.

In comparing sequential decoding and Viterbi decoding, it thus appears fair to consider soft decisions only for the Viterbi decoder. Under these conditions, the efficiency advantage of a long constraint length sequential decoder is considerably diluted. Consequently, performance of a rate 1/2, $K = 41$ sequential decoder is no better than a rate 1/2 Viterbi decoder of constraint length 5 to 7 (depending on the speed factor, that is, the ratio of computation rate to bit rate) at a P_e of 10^{-5}. The sequential decoder does show a distinct advantage for P_e of 10^{-8} or smaller.

On the other hand, building a system without receiver quantization lowers system costs, since a considerably more crude AGC may be used.

E. Sensitivity to Phase Error and Bursty Conditions on the Channel

The performance of Viterbi decoding under slowly fluctuating phase error was presented in Fig. 15. A similar calculation would indicate much greater degradation in the case of sequential decoding, since the error probability curve is much steeper. Furthermore, this estimate would be optimistic in the case of sequential decoding, since the assumption that the phase varied so slowly that errors occurred independently would probably not hold for sequential decoding. Thus, more careful design of the phase-tracking loop is indicated for a system utilizing sequential decoding rather than Viterbi decoding.

VIII. Conclusions

Viterbi decoding has been shown to be a practical method for improving satellite and space communication efficiency by 4–6 dB, at a bit error rate of 10^{-5} The successful implementation of 2-Mbit/s constraint-length-7 Viterbi decoders effectively demonstrates that the technique is well beyond the stage of being a theoretical curiosity. In fact, a major effort has been under way for the past 2–3 years with the aim of modifying and adapting the algorithm for minimum complexity implementation without sacrificing performance significantly.

In addition, Viterbi decoding has been shown to "degrade gracefully" in the presence of adverse channel or receiver conditions. In particular, the error probability does not change precipitously with E_b/N_0 as is the case with coding techniques that use longer codes and/or require variable decoding effort, such as sequential decoding. This ensures that performance degradation due to an imperfect phase or bit timing reference, or a slight correlation between noise samples, will be minimal. Requirements on AGC accuracy, even for soft decisions, were shown to be quite loose.

Finally the results presented here should provide the communication engineer with the information necessary to evaluate the applicability of Viterbi decoding to space and satellite communication systems with a wide range of requirements and constraints.

References

[1] C. E. Shannon, "Communication is the presence of noise," *Proc. IRE*, vol, 37, Jan. 1949, pp. 10–21.
[2] Codex Corp., Final Rep. "Coding system design for advanced solar missions," Contract NAS 2-3637, NASA Ames Res. Cent., Moffett Field, Calif.
[3] A. J. Viterbi, "Convolutional codes and their performance in communication systems," this issue, pp. 751–772.
[4] J. M. Wozencraft and I. M. Jacobs, *Principles of Communication Engineering.* New York: Wiley, 1965.
[5] I. M. Jacobs, "Sequential decoding for efficient communication from deep space," *IEEE Trans. Commun. Technol.*, vol. COM-15, Aug. 1967, pp. 492–501.
[6] A. J. Viterbi, "Error bounds for convolutional codes and an asymptotically optimum decoding algorithm," *IEEE Trans. Inform. Theory*, vol. IT-13, Apr. 1967, pp. 260–269.
[7] J. A. Heller, "Short constraint length convolutional codes," Jet Propulsion Lab., California Inst. Technol., Space Programs Summary 37-54, vol. III, Oct./Nov., 1968, pp. 171–177.
[8] ——, "Improved performance of short constraint length convolutional codes," Jet Propulsion Lab., California Inst. Technol., Space Programs Summary 37-56, vol. III, Feb./Mar. 1969, pp. 83–84.
[9] Linkabit Corp., Final Rep., "Coding systems study for high data rate telemetry links," Contract NAS2-6024, NASA Ames Res. Ctr. Rep. CR-114278, Moffett Field, Calif.
[10] J. P. Odenwalder, "Optimum decoding of convolutional codes," Ph.D. dissertation, Syst. Sci. Dep., Univ. California, Los Angeles, 1970.
[11] A. J. Viterbi, *Principles of Coherent Communication.* New York: McGraw-Hill, 1966.
[12] J. K. Omura, "On the Viterbi decoding algorithm," *IEEE Trans. Inform. Theory* (Corresp.), vol. IT-15, Jan. 1969, pp. 177–179.

3.4.2

The Viterbi Algorithm

G. DAVID FORNEY, JR.

Invited Paper

Abstract—The Viterbi algorithm (VA) is a recursive optimal solution to the problem of estimating the state sequence of a discrete-time finite-state Markov process observed in memoryless noise. Many problems in areas such as digital communications can be cast in this form. This paper gives a tutorial exposition of the algorithm and of how it is implemented and analyzed. Applications to date are reviewed. Increasing use of the algorithm in a widening variety of areas is foreseen.

I. Introduction

THE VITERBI algorithm (VA) was proposed in 1967 [1] as a method of decoding convolutional codes. Since that time, it has been recognized as an attractive solution to a variety of digital estimation problems, somewhat as the Kalman filter has been adapted to a variety of analog estimation problems. Like the Kalman filter, the VA tracks the state of a stochastic process with a recursive method that is optimum in a certain sense, and that lends itself readily to implementation and analysis. However, the underlying process is assumed to be finite-state Markov rather than Gaussian, which leads to marked differences in structure.

This paper is intended to be principally a tutorial introduction to the VA, its structure, and its analysis. It also purports to review more or less exhaustively all work inspired by or related to the algorithm up to the time of writing (summer 1972). Our belief is that the algorithm will find application in an increasing diversity of areas. Our hope is that we can accelerate this process for the readers of this paper.

This invited paper is one of a series planned on topics of general interest—The Editor.

Manuscript received September 20, 1972; revised November 27, 1972. The author is with Codex Corporation, Newton, Mass. 02195.

II. Statement of the Problem

In its most general form, the VA may be viewed as a solution to the problem of maximum *a posteriori* probability (MAP) estimation of the state sequence of a finite-state discrete-time Markov process observed in memoryless noise. In this section we set up the problem in this generality, and then illustrate by example the different sorts of problems that can be made to fit such a model. The general approach also has the virtue of tutorial simplicity.

The underlying Markov process is characterized as follows. Time is discrete. The state x_k at time k is one of a finite number M of states m, $1 \leq m \leq M$; i.e., the state space X is simply $\{1, 2, \cdots, M\}$. Initially we shall assume that the process runs only from time 0 to time K and that the initial and final states x_0 and x_K are known; the state sequence is then represented by a finite vector $\boldsymbol{x} = (x_0, \cdots, x_K)$. We see later that extension to infinite sequences is trivial.

The process is Markov, in the sense that the probability $P(x_{k+1} | x_0, x_1, \cdots, x_k)$ of being in state x_{k+1} at time $k+1$, given all states up to time k, depends only on the state x_k at time k:

$$P(x_{k+1} | x_0, x_1, \cdots, x_k) = P(x_{k+1} | x_k).$$

The transition probabilities $P(x_{k+1} | x_k)$ may be time varying, but we do not explicitly indicate this in the notation.

It is convenient to define the *transition* ξ_k at time k as the pair of states (x_{k+1}, x_k):

$$\xi_k \triangleq (x_{k+1}, x_k).$$

We let Ξ be the (possibly time-varying) set of transitions

Reprinted from *Proc. IEEE*, vol. 61, pp. 268–278, Mar. 1973.

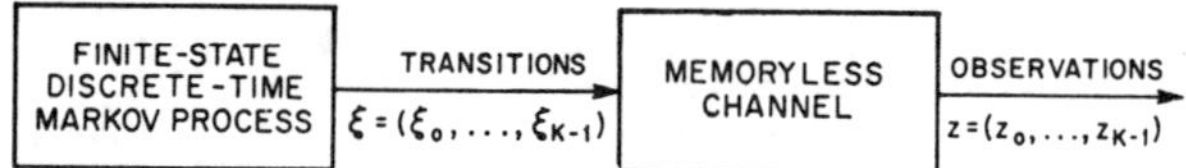

Fig. 1.   Most general model.

$\xi_k = (x_{k+1}, x_k)$ for which $P(x_{k+1}|x_k) \neq 0$, and $|\Xi|$ their number. Clearly $|\Xi| \leq M^2$. There is evidently a one-to-one correspondence between state sequences x and transition sequences $\xi = (\xi_0, \cdots, \xi_{K-1})$. (We write $x \overset{1-1}{\leftrightarrow} \xi$.)

The process is assumed to be observed in memoryless noise; that is, there is a sequence z of observations z_k in which z_k depends probabilistically only on the transition ξ_k at time k:[1]

$$P(z \mid x) = P(z \mid \xi) = \prod_{k=0}^{K-1} P(z_k \mid \xi_k).$$

In the parlance of information theory, z can be described as the output of some memoryless channel whose input sequence is ξ (see Fig. 1). Again, though we shall not indicate it explicitly, the channel may be time varying in the sense that $P(z_k|\xi_k)$ may be a function of k. This formulation subsumes the following as special cases.

1) The case in which z_k depends only on the state x_k:

$$P(z \mid x) = \prod_k P(z_k \mid x_k).$$

2) The case in which z_k depends probabilistically on an output y_k of the process at time k, where y_k is in turn a deterministic function of the transition ξ_k or the state x_k.

Example: The following model frequently arises in digital communications. There is an input sequence $u = (u_0, u_1, \cdots)$, where each u_k is generated independently according to some probability distribution $P(u_k)$ and can take on one of a finite number of values, say m. There is a noise-free signal sequence y, not observable, in which each y_k is some deterministic function of the present and the ν previous inputs:

$$y_k = f(u_k, \cdots, u_{k-\nu}).$$

The observed sequence z is the output of a memoryless channel whose input is y. We call such a process a *shift-register process*, since (as illustrated in Fig. 2) it can be modeled by a shift register of length ν with inputs u_k. (Alternately, it is a νth-order m-ary Markov process.) To complete the correspondence to our general model we define:

1) the state

$$x_k \triangleq (u_{k-1}, \cdots, u_{k-\nu})$$

2) the transition

$$\xi_k \triangleq (u_k, \cdots, u_{k-\nu}).$$

The number of states is thus $|X| = m^\nu$, and of transitions, $|\Xi| = m^{\nu+1}$. If the input sequence "starts" at time 0 and "stops" at time $K - \nu$, i.e.,

$$u = (\cdots, 0, u_0, u_1, \cdots, u_{K-\nu}, 0, 0, \cdots)$$

then the shift-register process effectively starts at time 0 and ends at time K with $x_0 = x_K = (0, 0, \cdots, 0)$.

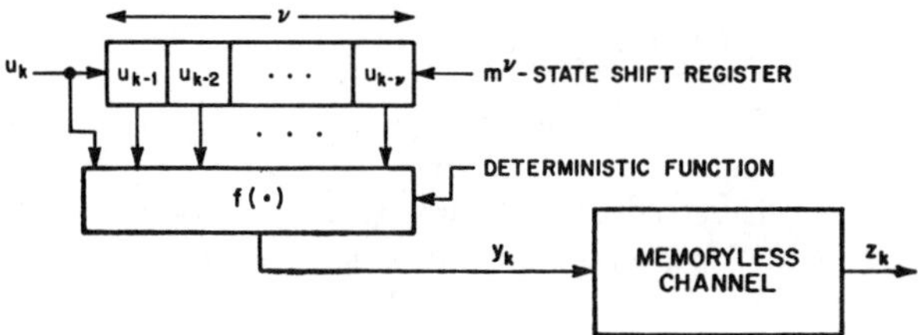

Fig. 2.   Shift-register model.

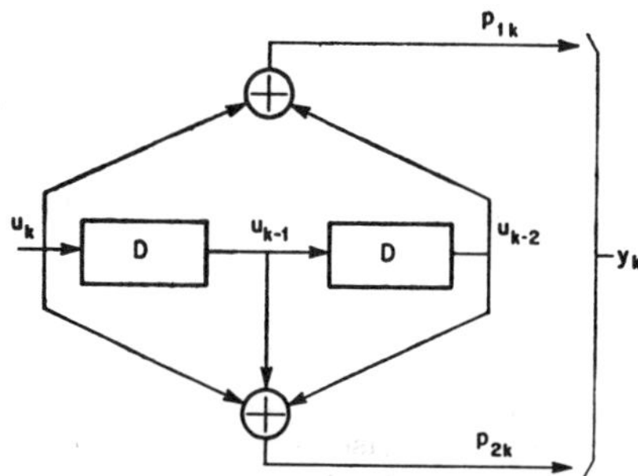

Fig. 3.   A convolutional encoder.

Finally, we state the problem to which the VA is a solution. Given a sequence z of observations of a discrete-time finite-state Markov process in memoryless noise, find the state sequence x for which the *a posteriori* probability $P(x|z)$ is maximum. Alternately, find the transition sequence ξ for which $P(\xi|z)$ is maximum (since $x \overset{1-1}{\leftrightarrow} \xi$). In the shift-register model this is also the same as finding the most probable input sequence u, since $u \overset{1-1}{\leftrightarrow} x$; or also the most probable signal sequence y, if $y \overset{1-1}{\leftrightarrow} x$. It is well known that this MAP rule minimizes the error probability in detecting the whole sequence (the block-, message-, or word-error probability), and thus is optimum in this sense. We shall see that in many applications it is effectively optimum in any sense.

Application Examples: We now give examples showing that the problem statement above applies to a number of diverse fields, including convolutional coding, intersymbol interference, continuous-phase frequency-shift keying (FSK), and text recognition. The adequately motivated reader may skip immediately to the next section.

A. Convolutional Codes

A rate-$1/n$ binary convolutional encoder is a shift-register circuit exactly like that of Fig. 2, where the inputs u_k are information bits and the outputs y_k are blocks of n bits, $y_k = (p_{1k}, \cdots, p_{nk})$, each of which is a parity check on (modulo-2 sum of) some subset of the $\nu + 1$ information bits (u_k, $u_{k-1}, \cdots, u_{k-\nu}$). When the encoded sequence (codeword) y is sent through a memoryless channel, we have precisely the model of Fig. 2. Fig. 3 shows a particular rate-$\frac{1}{2}$ code with $\nu = 2$. (This code is the only one ever used for illustration in the VA coding literature, but the reader must not infer that it is the only one the VA can handle.)

More general convolutional encoders exist: the rate may be k/n, the inputs may be nonbinary, and the encoder may even contain feedback. In every case, however, the code may be taken to be generated by a shift-register process [2].

We might also note that other types of transmission codes (e.g., dc-free codes, run-length-limited codes, and others) can be modeled as outputs of a finite-state machine and hence fall into our general setup [49].

---

[1] The notation is appropriate when observations are discrete-valued; for continuous-valued z_k, simply substitute a density $p(z_k|\xi_k)$ for the distribution $P(z_k|\xi_k)$.

B. Intersymbol Interference

In digital transmission through analog channels, we frequently encounter the following situation. The input sequence u, discrete-time and discrete-valued as in the shift-register model, is used to modulate some continuous waveform which is transmitted through a channel and then sampled. Ideally, samples z_k would equal the corresponding u_k, or some simple function thereof; in fact, however, the samples z_k are perturbed both by noise and by neighboring inputs $u_{k'}$. The latter effect is called intersymbol interference. Sometimes intersymbol interference is introduced deliberately for purposes of spectral shaping, in so-called partial-response systems.

In such cases the output samples can often be modeled as

$$z_k = y_k + n_k$$

where y_k is a deterministic function of a finite number of inputs, say, $y_k = f(u_k, \cdots, u_{k-\nu})$, and n_k is a white Gaussian noise sequence. This is precisely Fig. 2.

To be still more specific, in pulse-amplitude modulation (PAM) the signal sequence y may be taken as the convolution of the input sequence u with some discrete-time channel impulse-response sequence $(h_0, h_1, \cdots)$:

$$y_k = \sum_i h_i u_{k-i}.$$

If $h_i = 0$ for $i > \nu$ (finite impulse response), then we obtain our shift-register model. An illustration of such a model in which intersymbol interference spans three time units $(\nu = 2)$ appears in Fig. 4.

It was shown in [29] that even problems where time is actually continuous—i.e., the received signal $r(t)$ has the form

$$r(t) = \sum_{k=0}^{K} u_k h(t - kT) + n(t)$$

for some impulse response $h(t)$, signaling interval T, and realization $n(t)$ of a white Gaussian noise process—can be reduced without loss of optimality to the aforementioned discrete-time form (via a "whitened matched filter").

C. Continuous-Phase FSK

This example is cited not for its practical importance, but because, first, it leads to a simple model we shall later use in an example, and, second, it shows how the VA may lead to fresh insight even in the most traditional situations.

In FSK, a digital input sequence u selects one of m frequencies (if u_k is m-ary) in each signaling interval of length T; that is, the transmitted signal $\eta(t)$ is

$$\eta(t) = \cos \left[\omega(u_k)t + \theta_k\right], \qquad kT \leq t < (k+1)T$$

where $\omega(u_k)$ is the frequency selected by u_k, and θ_k is some phase angle. It is desirable for reasons both of spectral shaping and of modulator simplicity that the phase be continuous at the transition interval; that is, that

$$\omega(u_{k-1})kT + \theta_{k-1} \equiv \omega(u_k)kT + \theta_k \text{ modulo } 2\pi.$$

This is called continuous-phase FSK.

The continuity of the phase introduces memory into the modulation process; i.e., it makes the signal actually transmitted in the kth interval dependent on previous signals. To take the simplest possible case ("deviation ratio" $= \tfrac{1}{2}$), let the

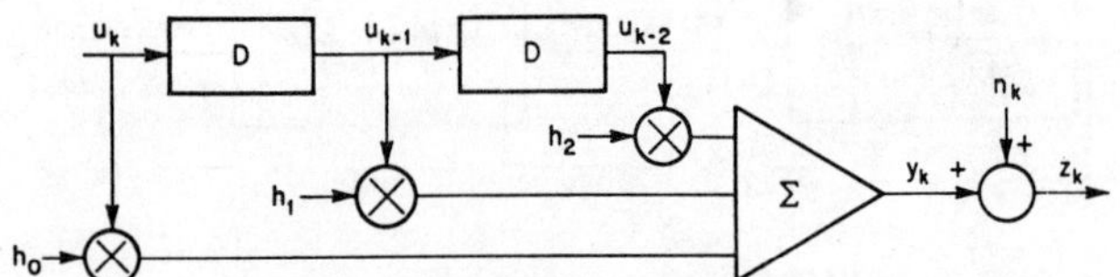

Fig. 4.   Model of PAM system subject to intersymbol interference and white Gaussian noise.

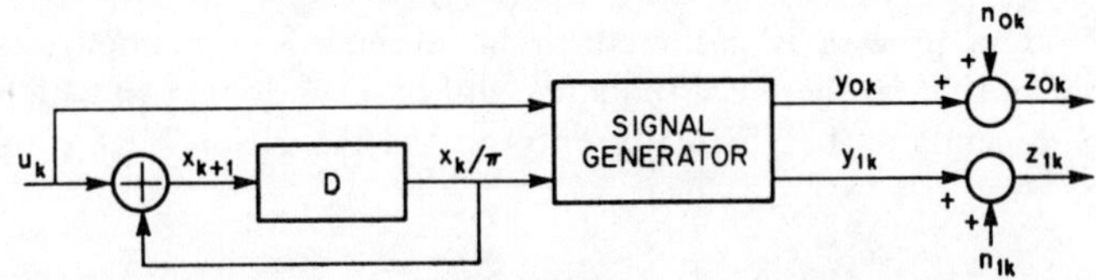

Fig. 5.   Model for binary continuous-phase FSK with deviation ratio $\tfrac{1}{2}$ and coherent detection in white Gaussian noise.

input sequence u be binary and let $\omega(0)$ and $\omega(1)$ be chosen so that $\omega(0)$ goes through an integer number of cycles in T seconds and $\omega(1)$ through an odd half-integer number; i.e., $\omega(0)T \equiv 0$ and $\omega(1)T = \pi$ modulo 2π. Then if $\theta_0 = 0$, $\theta_1 = 0$ or π, according to whether u_0 equals zero or one, and similarly $\theta_k = 0$ or π, according to whether an even or odd number of ones has been transmitted.

Here we have a two-state process, with $X = \{0, \pi\}$. The transmitted signal y_k is a function of both the current input u_k and the state x_k:

$$y_k = \cos \left[\omega(u_k)t + x_k\right] = \cos x_k \cos \omega(u_k)t,$$
$$kT \leq t < (k+1)T.$$

Since transitions $\xi_k = (x_{k+1}, x_k)$ are one-to-one functions of the current state x_k and input u_k, we may alternately regard y_k as being determined by ξ_k.

If we take $\eta_0(t) \triangleq \cos \omega(0)t$ and $\eta_1(t) \triangleq \cos \omega(1)t$ as bases of the signal space, we may write

$$y_k = y_{0k}\eta_0(t) + y_{1k}\eta_1(t)$$

where the coordinates (y_{0k}, y_{1k}) are given by

$$(y_{0k}, y_{1k}) = \begin{cases} (1, 0), & \text{if } u_k = 0, x_k = 0 \\ (-1, 0), & \text{if } u_k = 0, x_k = \pi \\ (0, 1), & \text{if } u_k = 1, x_k = 0 \\ (0, -1), & \text{if } u_k = 1, x_k = \pi. \end{cases}$$

Finally, if the received signal $\xi(t)$ is $\eta(t)$ plus white Gaussian noise $\nu(t)$, then by correlating the received signal against both $\eta_0(t)$ and $\eta_1(t)$ in each signal interval (coherent detection), we may arrive without loss of information at a discrete-time output signal

$$z_k = (z_{0k}, z_{1k}) = (y_{0k}, y_{1k}) + (n_{0k}, n_{1k})$$

where n_0 and n_1 are independent equal-variance white Gaussian noise sequences. This model appears in Fig. 5, where the signal generator generates (y_{0k}, y_{1k}) according to the aforementioned rules.

D. Text Recognition

We include this example to show that the VA is not limited to digital communication. In optical-character-recognition (OCR) readers, individual characters are scanned, salient

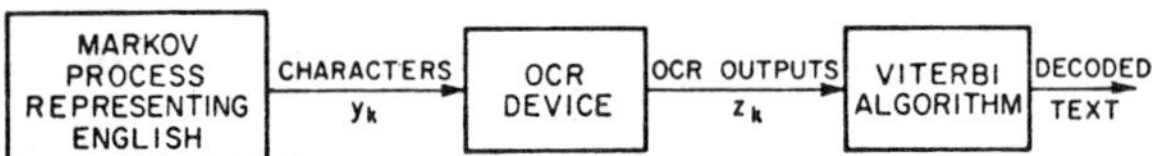

Fig. 6.   Use of VA to improve character
recognition by exploiting context.

features isolated, and some decision made as to what letter or
other character lies below the reader. When the characters
actually occur as part of natural-language text, it has long
been recognized that contextual information can be used to
assist the reader in resolving ambiguities.

One way of modeling contextual constraints is to treat a
natural language like English as though it were a discrete-
time Markov process. For instance, we can suppose that the
probability of occurrence of each letter depends on the ν
previous letters, and estimate these probabilities from the
frequencies of $(\nu+1)$-letter combinations $[(\nu+1)$-grams$]$.
While such models do not fully describe the generation of
natural language (for examples of digram and trigram English,
see Shannon [45], [46]), they account for a large part of the
statistical dependencies in the language and are easily
handled.

With such a model, English letters are viewed as the outputs of an m^ν-state Markov process, where m is the number of
distinguishable characters, such as 27 (the 26 letters and a
space). If it is further assumed that the OCR output z_k is
dependent only on the corresponding input character y_k, then
the OCR reader is a memoryless channel to whose output sequence we may apply the VA to exploit contextual constraints; see Fig. 6. Here the "OCR output" may be anything
from the raw sensor data, possibly a grid of zeros and ones, to
the actual decision which would be made by the reader in the
absence of contextual clues. Generally, the more raw the data,
the more useful the VA will be.

E. Other

Recognizing certain similarities between magnetic recording media and intersymbol interference channels, Kobayashi
[50] has proposed applying the VA to digital magnetic recording systems. Timor [51] has applied the algorithm to a
sequential ranging system. Use of the algorithm in source
coding has been proposed [52]. Finally, Preparata and Ray
[53] have suggested using the algorithm to search "semantic
maps" in syntactic pattern recognition. These exhaust the
applications known to the author.

III. The Algorithm

We now show that the MAP sequence estimation problem
previously stated is formally identical to the problem of finding the shortest route through a certain graph. The VA then
arises as a natural recursive solution.

We are accustomed to associating with a discrete-time
finite state Markov process a state diagram of the type shown
in Fig. 7(a), for a four-state shift-register process like that of
Fig. 3 or Fig. 4 (in this case, a de Bruijn diagram [54]). Here
nodes represent states, branches represent transitions, and
over the course of time the process traces some path from state
to state through the state diagram.

In Fig. 7(b) we introduce a more redundant description of
the same process, which we have called a *trellis* [3]. Here each
node corresponds to a distinct state at a given time, and each
branch represents a transition to some new state at the next

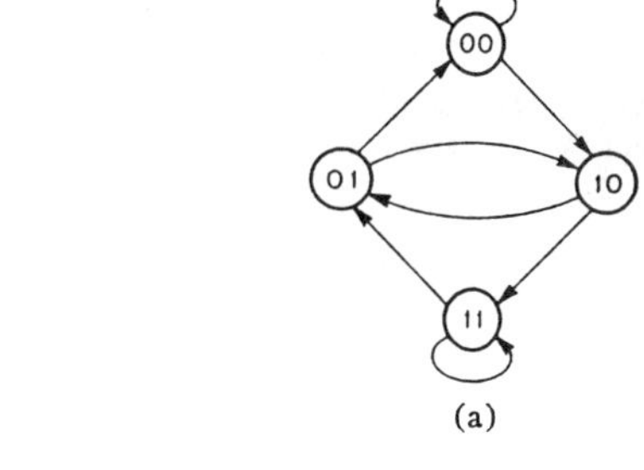

(a)

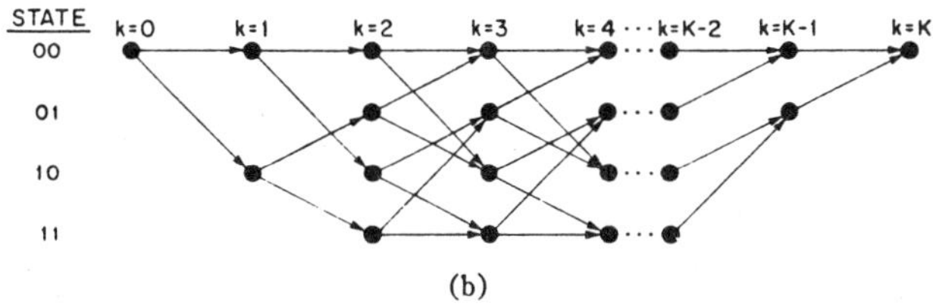

(b)

Fig. 7.   (a) State diagram of a four-state shift-register process. (b) Trellis
for a four-state shift-register process.

instant of time. The trellis begins and ends at the known states
x_0 and x_K. Its most important property is that to every possible state sequence x there corresponds a unique path through
the trellis, and vice versa.

Now we show how, given a sequence of observations z,
every path may be assigned a "length" proportional to $-\ln$
$P(x, z)$, where x is the state sequence associated with that
path. This will allow us to solve the problem of finding the
state sequence for which $P(x|z)$ is maximum, or equivalently
for which $P(x, z) = P(x|z)P(z)$ is maximum, by finding the
path whose length $-\ln P(x, z)$ is minimum, since $\ln P(x, z)$ is
a monotonic function of $P(x, z)$ and there is a one-to-one correspondence between paths and sequences. We simply observe that due to the Markov and memoryless properties,
$P(x, z)$ factors as follows:

$$P(x, z) = P(x)P(z \mid x)$$
$$= \prod_{k=0}^{K-1} P(x_{k+1} \mid x_k) \prod_{k=0}^{K-1} P(z_k \mid x_{k+1}, x_k).$$

Hence if we assign each branch (transition) the "length"

$$\lambda(\xi_k) \triangleq - \ln P(x_{k+1} \mid x_k) - \ln P(z_k \mid \xi_k)$$

then the total length of the path corresponding to some x is

$$- \ln P(x, z) = \sum_{k=0}^{K-1} \lambda(\xi_k)$$

as claimed.

Finding the shortest route through a graph is an old problem in operations research. The most succinct solution was
given by Minty in a quarter-page correspondence in 1957
[55], which we quote almost in its entirety:

The shortest-route problem . . . can be solved very simply
. . . as follows: Build a string model of the travel network,
where knots represent cities and string lengths represent distances (or costs). Seize the knot "Los Angeles" in your left
hand and the knot "Boston" in your right and pull them apart.
If the model becomes entangled, have an assistant untie and
re-tie knots until the entanglement is resolved. Eventually one
or more paths will stretch tight—they then are alternative
shortest routes. . . . It is well to label the knots since after one
or two uses of the model their identities are easily confused.

Unfortunately, the Minty algorithm is not well adapted to modern methods of machine computation, nor are assistants as pliable as formerly. It therefore becomes necessary to move on to the VA, which is also well known in operations research [56]. It requires one additional observation.

We denote by $x_0{}^k$ a segment $(x_0, x_1, \cdots, x_k)$ consisting of the states to time k of the state sequence $x = (x_0, x_1, \cdots, x_k)$. In the trellis $x_0{}^k$ corresponds to a path segment starting at the node x_0 and terminating at x_k. For any particular time-k node x_k, there will in general be several such path segments, each with some length

$$\lambda(x_0{}^k) = \sum_{i=0}^{k-1} \lambda(\xi_i).$$

The shortest such path segment is called the *survivor* corresponding to the node x_k, and is denoted $\hat{x}(x_k)$. For any time $k > 0$, there are M survivors in all, one for each x_k. The observation is this: the shortest complete path $\hat{x}$ must begin with one of these survivors. (If it did not, but went through state x_k at time k, then we could replace its initial segment by $\hat{x}(x_k)$ to get a shorter path—contradiction.)

Thus at any time k we need remember only the M survivors $\hat{x}(x_k)$ and their lengths $\Gamma(x_k) \triangleq \lambda[\hat{x}(x_k)]$. To get to time $k+1$, we need only extend all time-k survivors by one time unit, compute the lengths of the extended path segments, and for each node x_{k+1} select the shortest extended path segment terminating in x_{k+1} as the corresponding time-$(k+1)$ survivor. Recursion proceeds indefinitely without the number of survivors ever exceeding M.

Many readers will recognize this algorithm as a simple version of forward dynamic programming [57], [58]. By this or any other name, the algorithm is elementary once the problem has been cast in the shortest route form.

We illustrate the algorithm for a simple four-state trellis covering 5 time units in Fig. 8. Fig. 8(a) shows the complete trellis, with each branch labeled with a length. (In a real application, the lengths would be functions of the received data.) Fig. 8(b) shows the 5 recursive steps by which the algorithm determines the shortest path from the initial to the final node. At each stage only the 4 (or fewer) survivors are shown, along with their lengths.

A formal statement of the algorithm follows:

Viterbi Algorithm

Storage:

k (time index);

$\hat{x}(x_k), \ 1 \le x_k \le M$ (survivor terminating in x_k);

$\Gamma(x_k), \ 1 \le x_k \le M$ (survivor length).

Initialization:

$$k = 0;$$

$$\hat{x}(x_0) = x_0; \qquad \hat{x}(m) \text{ arbitrary}, \qquad m \ne x_0;$$

$$\Gamma(x_0) = 0; \qquad \Gamma(m) = \infty, \qquad m \ne x_0.$$

Recursion: Compute

$$\Gamma(x_{k+1}, x_k) \triangleq \Gamma(x_k) + \lambda[\xi_k = (x_{k+1}, x_k)]$$

for all $\xi_k = (x_{k+1}, x_k)$.

Find

$$\Gamma(x_{k+1}) = \min_{x_k} \Gamma(x_{k+1}, x_k)$$

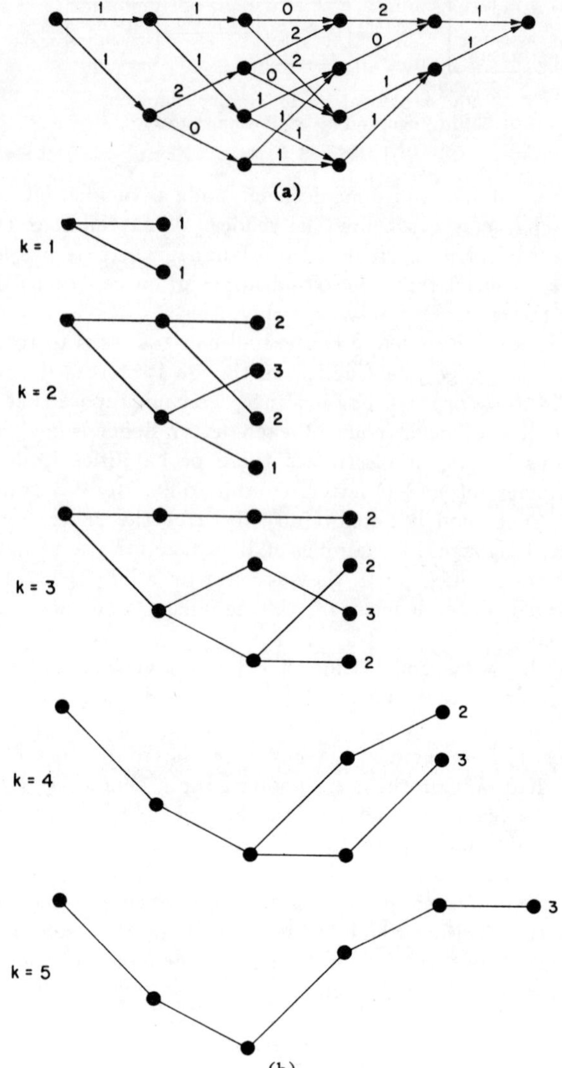

Fig. 8. (a) Trellis labeled with branch lengths; $M = 4$, $K = 5$. (b) Recursive determination of the shortest path via the VA.

for each x_{k+1}; store $\Gamma(x_{k+1})$ and the corresponding survivor $\hat{x}(x_{k+1})$.

Set k to $k+1$ and repeat until $k = K$.

With finite state sequences x the algorithm terminates at time K with the shortest complete path stored as the survivor $\hat{x}(x_K)$.

Certain trivial modifications are necessary in practice. When state sequences are very long or infinite, it is necessary to truncate survivors to some manageable length δ. In other words, the algorithm must come to a definite decision on nodes up to time $k - \delta$ at time k. Note that in Fig. 8(b) all time-4 survivors go through the same nodes up to time 2. In general, if the truncation depth δ is chosen large enough, there is a high probability that all time-k survivors will go through the same nodes up to time $k - \delta$, so that the initial segment of the maximum-likelihood path is known up to time $k - \delta$ and can be put out as the algorithm's firm decision; in this case, truncation costs nothing. In the rare cases when survivors disagree, any reasonable strategy for determining the algorithm's time-$(k - \delta)$ decision will work [20], [21]: choose an arbitrary time-$(k - \delta)$ node, or the node associated with the shortest survivor, or a node chosen by majority vote, etc. If δ is large enough, the effect on performance is negligible.

Also, if k becomes large, it is necessary to renormalize the lengths $\Gamma(m)$ from time to time by subtracting a constant from all of them.

Finally, the algorithm may be required to get started without knowledge of the initial state x_0. In this case it may be initialized with any reasonable assignment of initial node lengths, such as $\Gamma(m) = 0$, all m, or else $\Gamma(m) = -\ln \pi_m$ if the states are known to have *a priori* probabilities π_m. Usually, after an initial transient, there is a high probability that all survivors will merge with the correct path. Thus the algorithm synchronizes itself without any special procedures.

The complexity of the algorithm is easily estimated. First, memory: the algorithm requires M storage locations, one for each state, where each location must be capable of storing a "length" $\Gamma(m)$ and a truncated survivor listing $\hat{x}(m)$ of δ symbols. Second, computation: in each unit of time the algorithm must make $|\Xi|$ additions, one for each transition, and M comparisons among the $|\Xi|$ results. Thus the amount of storage is proportional to the number of states, and the amount of computation to the number of transitions. With a shift-register process, $M = m^\nu$ and $|\Xi| = m^{\nu+1}$, so that the complexity increases exponentially with the length ν of the shift register.

In the previous paragraph, we have ignored the complexity involved in generating the incremental lengths $\lambda(\xi_k)$. In a shift-register process, it is typically true that $P(x_{k+1}|x_k)$ is either $1/m$ or 0, depending on whether x_{k+1} is an allowable successor to x_k or not; then all allowable transitions have the same value of $-\ln P(x_{k+1}|x_k)$ and this component of $\lambda(\xi_k)$ may be ignored. Note that in more general cases $P(x_{k+1}|x_k)$ is known in advance; hence this component can be precomputed and "wired in." The component $-\ln P(z_k|\xi_k)$ is the only component that depends on the data; again, it is typical that many ξ_k lead to the same output y_k, and hence the value $-\ln P(z_k|y_k)$ need be computed or looked up for all these ξ_k only once, given z_k. (When the noise is Gaussian, $-\ln P(z_k|y_k)$ is proportional simply to $(z_k - y_k)^2$.) Finally, all of this can be done outside the central recursion ("pipelined"). Hence the complexity of this computation tends not to be significant.

Furthermore, note that once the $\lambda(\xi_k)$ are computed, the observation z_k need not be stored further, an attractive feature in real-time applications.

A closer look at the trellis of a shift-register process reveals additional detail that can be exploited in implementation. For a binary shift register, the transitions in any unit of time can be segregated into $2^{\nu-1}$ disjoint groups of four, each originating in a common pair of states and terminating in another common pair. A typical such cell is illustrated in Fig. 9, with the time-k states labeled $x'0$ and $x'1$ and the time-$(k+1)$ states labelled $0x'$ and $1x'$, where x' stands for a sequence of $\nu-1$ bits that is constant within any one cell. For example, each time unit in the trellis of Fig. 8 is made up of two such cells. We note that only quantities within the same cell interact in any one recursion. Fig. 10 shows a basic logic unit that implements the computation within any one cell. A high-speed parallel implementation of the algorithm can be built around $2^{\nu-1}$ such identical logic units; a low-speed implementation, around one such unit, time-shared; or a software version, around the corresponding subroutine.

Many readers will have noticed that the trellis of Fig. 7(b) reminds them of the computational flow diagram of the fast

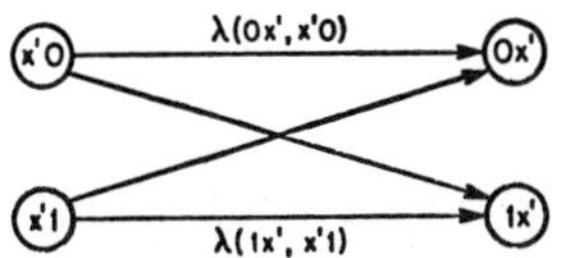
Fig. 9. Typical cell of a shift-register process trellis.

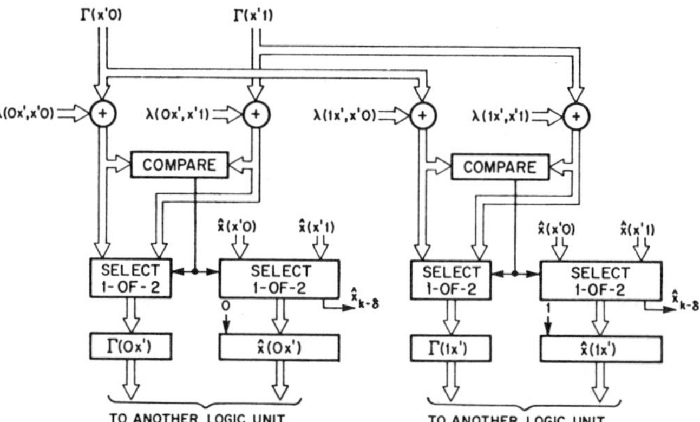
Fig. 10. Basic VA logic unit for binary shift-register process.

Fourier transform (FFT). In fact, it is identical, except for length, and indeed the FFT is also ordinarily organized cellwise. While the add-and-compare computations of the VA are unlike those involved in the FFT, some of the memory-organization tricks developed for the FFT may be expected to be equally useful here.

Because of its highly parallel structure and need for only add, compare, and select operations, the VA is well suited to high-speed applications. A convolutional decoder for a $\nu = 6$ code ($M = 64$, $|\Xi| = 128$) that is built out of 356 transistor–transistor logic circuits and that can operate at up to 2 Mbits/s perhaps represents the current state of the art [22]. Even a software decoder for a similar code can be run at a rate the order of 1000 bits/s on a minicomputer. Such moderate complexity qualifies the VA for inclusion in many signal-processing systems.

For further details of implementation, see [22], [23], and the references therein.

IV. Analysis of Performance

Just as important as the straightforwardness of implementation of the VA is the straightforwardness with which its performance can be analyzed. In many cases, tight upper and lower bounds for error probability can be derived. Even when the VA is not actually implemented, calculation of its performance shows how far the performance of less complex schemes is from ideal, and often suggests simple suboptimum schemes that attain nearly optimal performance.

The key concept in performance analysis is that of an error event. Let x be the actual state sequence, and $\hat{x}$ the state sequence actually chosen by the VA. Over a long time x and $\hat{x}$ will typically diverge and remerge a number of times, as illustrated in Fig. 11. Each distinct separation is called an error event. Error events may in general be of unbounded length if x is infinite, but the probability of an infinite error event will usually be zero.

The importance of error events is that they are probabilistically independent of one another; in the language of probability theory they are *recurrent*. Furthermore, they allow us to calculate error probability per unit time, which is neces-

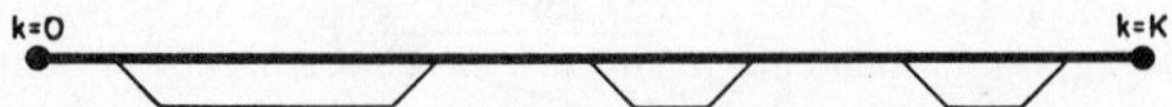

Fig. 11.   Typical correct path *x* (heavy line) and estimated path $\hat{x}$ (lighter line) in the trellis, showing three error events.

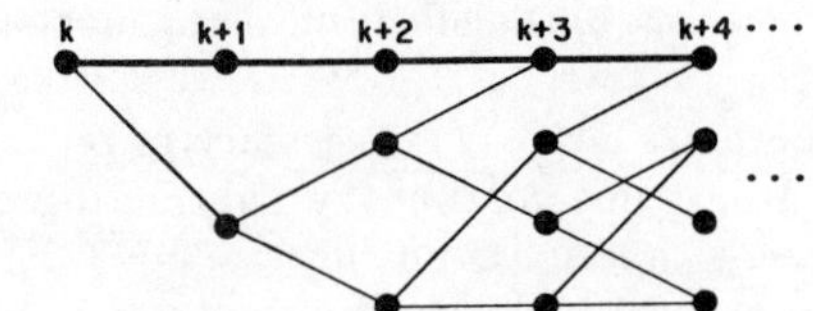

Fig. 12.   Typical correct path *x* (heavy line) and time-*k* incorrect subset for trellis of Fig. 7(b).

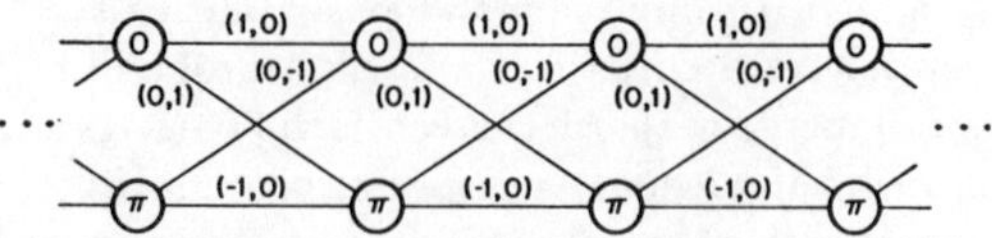

Fig. 13.   Trellis for continuous-phase FSK.

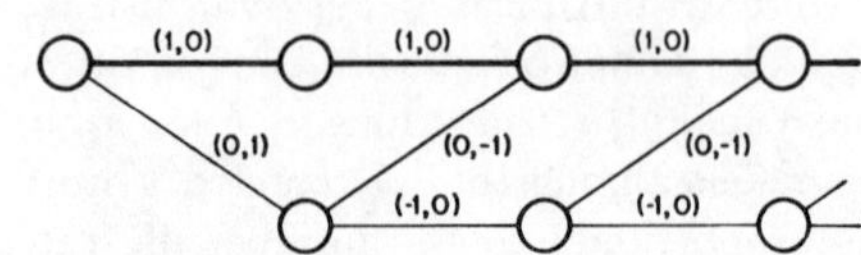

Fig. 14.   Typical incorrect subset. Heavy line: correct path. Lighter lines: incorrect subset.

sary since usually the probability of *any* error in MAP estimation of a block of length K goes to 1 as K goes to infinity. Instead we calculate the probability of an error event starting at some given time, given that the starting state is correct, i.e., that an error event is not already in progress at that time.

Given the correct path *x*, the set $\mathcal{E}_k$ of all possible error events starting at some time k is a treelike trellis which starts at x_k and each of whose branches ends on the correct path, as illustrated in Fig. 12, for the trellis of Fig. 7(b). In coding theory this is called the incorrect subset (at time k).

The probability of any particular error event is easily calculated; it is simply the probability that the observations will be such that over the time span during which $\hat{x}$ is different from *x*, $\hat{x}$ is more likely than *x*. If the error event has length τ, this is simply a two-hypothesis decision problem between two sequences of length τ, and typically has a standard solution.

The probability $P(\mathcal{E}_k)$ that *any* error event in $\mathcal{E}_k$ occurs can then be upper-bounded, usually tightly, by a union bound, i.e., by the sum of the probabilities of all error events in $\mathcal{E}_k$. While this sum may well be infinite, it is typically dominated by one or a few large leading terms representing particularly likely error events, whose sum then forms a good approximation to $P(\mathcal{E}_k)$. True upper bounds can often be obtained by flow-graph techniques [4], [5], [29].

On the other hand, a lower bound to error-event probability, again frequently tight, can be obtained by a genie argument. Take the particular error event that has the greatest probability of all those in $\mathcal{E}_k$. Suppose that a friendly genie tells you that the true state sequence is one of two possibilities: the actual correct path, or the incorrect path corresponding to that error event. Even with this side information, you will still make an error if the incorrect path is more likely given *z*, so your probability of error is still no better than the probability of this particular error event. In the absence of the genie, your error probability must be worse still, since one of the strategies you have, given the genie's information, is to ignore it. In summary, the probability of any particular error event is a lower bound to $P(\mathcal{E}_k)$.

An important side observation is that this lower bound applies to any decision scheme, not just the VA. Simple extensions give lower bounds to related quantities like bit probability of error. If the VA gives performance approaching these bounds, then it may be claimed that it is effectively optimum with respect to these related quantities as well [30].

In conclusion, the probability of any error event starting at time k may be upper- and lower-bounded as follows:

$$\max P(\text{error event}) \leq P(\mathcal{E}_k) \leq \max P(\text{error event})$$
$$+ \text{ other terms.}$$

With luck these bounds will be close.

Example

For concreteness, and because the result is instructive, we carry through the calculation for continuous-phase FSK of the particularly simple type defined earlier. The two-state trellis for this process is shown in Fig. 13, and the first part of

a typical incorrect subset in Fig. 14. The shortest possible error event is of length 2 and consists of a decision that the signal was $\{\cos\omega(1)t, -\cos\omega(1)t\}$ rather than $\{\cos\omega(0)t, \cos\omega(0)t\}$, or in our coordinate notation that $\{(0, 1), (0, -1)\}$ is chosen over $\{(1, 0), (1, 0)\}$. This is a two-hypothesis decision problem in a four-dimensional signal space [59]. In Gaussian noise of variance σ^2 per dimension, only the Euclidean distance d between the two signals matters; in this case $d = \sqrt{2}$, and therefore the probability of this particular error event is $Q(d/2\sigma) = Q(1/\sigma\sqrt{2})$, where $Q(x)$ is the Gaussian error probability function defined by

$$Q(x) \triangleq \int_x^\infty \frac{1}{\sqrt{2\pi}} e^{-y^2/2} dy.$$

By examination of Fig. 14 we see that the error events of lengths 3, 4, $\cdots$ lie at distances $\sqrt{6}$, $\sqrt{10}$, $\cdots$ from the correct path; hence we arrive at upper and lower bounds on $P(\mathcal{E}_k)$ of

$$Q(\sqrt{2}/2\sigma) \leq P(\mathcal{E}_k) \leq Q(\sqrt{2}/2\sigma) + Q(\sqrt{6}/2\sigma)$$
$$+ Q(\sqrt{10}/2\sigma) + \cdots.$$

In view of the rapid decrease of $Q(x)$ with x, this implies that $P(\mathcal{E}_k)$ is accurately estimated as $Q(\sqrt{2}/2\sigma)$ for any reasonable noise variance σ^2. It is easily verified from Fig. 13 that this result is independent of the correct path *x* or the time k.

This result is interesting in that the best one can do with coherent detection in a single symbol interval is $Q(1/2\sigma)$ for orthogonal signals. Thus exploiting the memory doubles the effective signal energy, or improves the signal-to-noise ratio by 3 dB. It may therefore be claimed that continuous-phase FSK is inherently 3 dB better than noncontinuous for deviation ratio $\frac{1}{2}$, or as good as antipodal phase-shift keying. (While we have proved this only for a deviation ratio of $\frac{1}{2}$, it holds for nearly any deviation ratio.) Even though the VA is quite simple for a two-state trellis, the fact that only one type of error event has any significant probability permits still simpler suboptimum schemes to achieve effectively the same performance [43], [44].[2]

V. Applications

We conclude with a review of the results, both theoretical and practical, obtained with the VA to date.

A. Convolutional Codes

It was for convolutional codes that the algorithm was first developed, and naturally it has had its greatest impact here.

The principal theoretical result is contained in Viterbi's original paper [1]; see also [5]–[7]. It shows that for a suitably defined ensemble of random trellis codes and MAP decoding, the error probability can be made to decrease exponentially with the constraint length ν at all code rates R less than channel capacity. Furthermore, the rate of decrease is considerably faster than for block codes with comparable decoding complexity (although the same as that for block codes with the same decoding delay). In our view, the most telling comparison between block and convolutional codes is that an effectively optimum block code of any specified length and rate can be created by suitably terminating a convolutional (trellis) code, but with a lower rate for the block code of course [7]. In fact, if convolutional codes were any better, they could be terminated to yield better block codes than are theoretically possible—an observation which shows that the bounds on convolutional code performance must be tight.

For fixed binary convolutional codes of nonasymptotic length on symmetric memoryless channels, the principal result [6] is that $P(\mathcal{E}_k)$ is approximately given by

$$P(\mathcal{E}_k) \approx N_d 2^{-dD}$$

where d is the free distance, i.e., the minimum Hamming distance of any path in the incorrect subset $\mathcal{E}_k$ from the correct path; N_d is the number of such paths; and D is the Bhattacharyya distance

$$D = \log_2 \sum_{z} P(z \mid 0)^{1/2} P(z \mid 1)^{1/2}$$

where the sum is over all outputs z in the channel output space Z.

On Gaussian channels

$$P(\mathcal{E}_k) \approx N_d \exp\left(- dRE_b/N_0\right)$$

where E_b/N_0 is the signal-to-noise ratio per information bit. The tightness of this bound is confirmed by simulations [8], [9], [22]. For a $\nu = 6$, $d = 10$, $R = \frac{1}{2}$ code, for example, error probabilities of 10^{-3}, 10^{-5}, and 10^{-7} are achieved at $E_b/N_0 = 3.0$, 4.3, and 5.5 dB, respectively, which is within a few tenths of a dceibel of this bound [22].

Channels available for space communications are frequently accurately modeled as white Gaussian channels. The VA is attractive for such channels because it gives performance superior to all other coding schemes save sequential decoding, and does this at high speeds, with modest complexity, and with considerable robustness against varying channel parameters. A number of prototype systems have been implemented and tested [21]–[26], some quite original, and it seems likely that Viterbi decoders will become common in space communication systems.

Finally, Viterbi decoders have been used as elements in very-high-performance concatenated coding schemes [10]–[12], [27] and in decoding of convolutional codes in the presence of intersymbol interference [35], [60].

B. Intersymbol Interference

Application of the VA to intersymbol interference problems is more recent, and the main achievements have been theoretical. The principal result, for PAM in white Gaussian noise, is that $P(\mathcal{E}_k)$ can be tightly bounded as follows:

$$K_L Q(d_{\min}/2\sigma) \leq P(\mathcal{E}_k) \leq K_U Q(d_{\min}/2\sigma)$$

where K_L and K_U are small constants, $Q(x)$ is the Gaussian error probability function defined earlier, σ^2 is the noise variance, and $d_{\min}$ is the minimum Euclidean distance between any two distinct signals [29]. This result implies that on most channels intersymbol interference need not lead to any significant degradation in performance, which comes as rather a surprise.

For example, with the most common partial response systems, the VA recovers the 3-dB loss sustained by conventional detectors relative to full-response systems [29], [32]. Simple suboptimum processors [29], [33] can do nearly as well.

Several workers [34], [35], [42] have proposed adaptive versions of the algorithm for unknown and time-varying channels. Ungerboeck [41], [42] and Mackechnie [35] have shown that only a matched filter rather than a whitened matched filter is needed in PAM.

It seems most likely that the greatest effect of the VA on digital modulation systems will be to reveal those instances in which conventional detection techniques fall significantly short of optimum, and to suggest effective suboptimum methods of closing the gap. PAM channels that cannot be linearly equalized without excessive noise enhancement due to nulls or near nulls in the transmission band are the likeliest candidates for nonlinear techniques of this kind.

C. Text Recognition

An experiment was run in which digram statistics were used to correct garbled text produced by a simulated noisy character recognizer [47]. Results were similar to those of Raviv [48] (using digram statistics), although the algorithm was simpler and only had hard decisions rather than confidence levels to work with. It appears that the algorithm may be a useful adjunct to sophisticated character-recognition systems for resolving ambiguities when confidence levels for different characters are available.

VI. Conclusion

The VA has already had a significant impact on our understanding of certain problems, notably in the theories of convolutional codes and of intersymbol interference. It is beginning to have a substantial practical impact as well in the engineering of space-communication links. The amount of work it has

---

[2] De Buda [44] actually proves that an optimum decision on the phase x_k at time k can be made by examining the received waveform only at times $k-1$ and k; i.e., $(z_{0,k-1}, z_{1,k-1})$, (z_{0k}, z_{1k}). The proof is that the log likelihood ratio

$$-\ln \frac{P(z_{0,k-1}, z_{1,k-1}, z_{0k}, z_{1k} \mid x_{k-1}, x_k = 0, x_{k+1})}{P(z_{0,k-1}, z_{1,k-1}, z_{0k}, z_{1k} \mid x_{k-1}, x_k = \pi, x_{k+1})}$$

is proportional to $-z_{0,k-1} + z_{1,k-1} - z_{0k} - z_{1k}$ for any values of the pair of states (x_{k-1}, x_{k+1}). For this phase decision (which differs slightly from our sequence decision) the error probability is exactly $Q(\sqrt{2}/2\sigma)$.

inspired in the intersymbol interference area suggests that here too practical applications are not far off. The generality of the model to which it applies and the straightforwardness with which it can be analyzed and implemented lead one to believe that in both theory and practice it will find increasing application in the years ahead.

Appendix
Related Algorithms

In this Appendix we mention some processing structures that are closely related to the VA, chiefly sequential decoding and minimum-bit-error-probability algorithms. We also mention extensions of the algorithm to generate reliability information, erasures, and lists.

When the trellis becomes large, it is natural to abandon the exhaustive search of the VA in favor of a sequential trial-and-error search that selectively examines only those paths likely to be the shortest. In the coding literature, such algorithms are collectively known as sequential decoding [13]–[16]. The simplest to explain is the "stack" algorithm [15], [16], in which a list is maintained of the shortest partial paths found to date, the path on the top of the list is extended, and its successors reordered in the list until some path is found that reaches the terminal node, or else decreases without limit. (That some path will eventually do so is ensured in coding applications by the subtraction of a bias term such that the length of the correct path tends to decrease while that of all incorrect paths tends to increase.) Searches that start from either end of a finite trellis [17] are also useful.

In coding applications, sequential decoding has many of the same properties as Viterbi decoding, including the same error probability. It allows the decoding of longer and therefore more powerful codes, at the cost of a variable amount of computation necessitating buffer storage for the incoming data z. It is probably less useful outside of coding, since it depends on the decoder's ability to recognize when the best path has been found without examining other paths, and therefore requires either a finite trellis or a very large distance between the correct path and possible error events.

In the intersymbol interference literature, many of the early attempts to find optimum nonlinear algorithms used bit-error probability as the optimality criterion. The Markov property of the process leads to algorithms that are manageable but less attractive than the Viterbi [36]–[42].

The general principle of several of these algorithms is as follows.[3] First, we calculate the joint probability $P(x_k, z)$ for every state x_k in the trellis, or alternately $P(\xi_k, z)$ for every transition ξ_k. This is done by observing that

$$P(x_k, z) = P(x_k, z_0^{k-1}) P(z_k^K \mid x_k, z_0^{k-1})$$
$$= P(x_k, z_0^{k-1}) P(z_k^K \mid x_k)$$

since, given x_k, the outputs z_k^K from time k to K are independent of the outputs z_0^{k-1} from time 0 to $k-1$. Similarly

$$P(\xi_k, z) = P(x_k, x_{k+1}, z)$$
$$= P(x_k, z_0^{k-1}) P(x_{k+1}, z_k \mid x_k, z_0^{k-1})$$
$$\cdot P(z_{k+1}^K \mid x_{k+1}, x_k, z_0^k)$$
$$= P(x_k, z_0^{k-1}) P(x_{k+1}, z_k \mid x_k) P(z_{k+1}^K \mid x_{k+1}).$$

[3] The author is indebted to Bahl *et al.* [18] for a particularly lucid exposition of this type of algorithm.

Now we note the recursive formula

$$P(x_k, z_0^{k-1}) = \sum_{x_{k-1}} P(x_k, x_{k-1}, z_0^{k-1})$$
$$= \sum_{x_{k-1}} P(x_{k-1}, z_0^{k-2}) P(x_k, z_{k-1} \mid x_{k-1})$$

which allows us to calculate the M quantities $P(x_k, z_0^{k-1})$ from the M quantities $P(x_{k-1}, z_0^{k-2})$ with $|\Xi|$ multiplications and additions using the exponentiated lengths

$$e^{-\lambda(\xi_{k-1})} = P(x_k \mid x_{k-1}) P(z_{k-1} \mid \xi_{k-1}).$$

Similarly we have the backward recursion

$$P(z_k^K \mid x_k) = \sum_{x_{k+1}} P(z_k^K, x_{k+1} \mid x_k)$$
$$= \sum_{x_{k+1}} P(z_k, x_{k+1} \mid x_k) P(z_{k+1}^K \mid x_{k+1})$$

which has a similar complexity. Completion of these forward and backward recursions for all nodes allows $P(x_k, z)$ and/or $P(\xi_k, z)$ to be calculated for all nodes.

Now, to be specific, let us consider a shift-register process and let $S(u_k)$ be the set of all states x_{k+1} whose first component is u_k. Then

$$P(u_k, z) = \sum_{x_{k+1} \in S(u_k)} P(x_{k+1}, z).$$

Since $P(u_k, z) = P(u_k \mid z) P(z)$, MAP estimation of u_k reduces to finding the maximum of this quantity. Similarly, if we wish to find the MAP estimate of an output y_k, say, then let $S(y_k)$ be the set of all ξ_k that lead to y_k and compute

$$P(y_k, z) = \sum_{\xi_k \in S(y_k)} P(\xi_k, z).$$

A similar procedure can be used to estimate any quantity which is a deterministic function of states or transitions.

Besides requiring multiplications, this algorithm is less attractive than the VA in requiring a backward as well as a forward recursion and consequently storage of all data. The following amended algorithm [39], [48] eliminates the latter ugly feature at the cost of suboptimal performance and additional computation. Let us restrict ourselves to a shift-register process with input sequence u and agree to use only observations up to time $k+\delta$ in estimating u_k, say, where $\delta \geq \nu - 1$. We then have

$$P(u_k, z_0^{k+\delta}) = \sum_{u_{k+1}} \cdots \sum_{u_{k+\delta}} P(u_k^{k+\delta}, z_0^{k+\delta}).$$

The $m^{\delta+1}$ quantities in the sum can be determined recursively by

$$P(u_k^{k+\delta}, z_0^{k+\delta}) = \sum_{u_{k-1}} P(u_{k-1}^{k+\delta}, z_0^{k+\delta})$$
$$= \sum_{u_{k-1}} P(u_{k-1}^{k+\delta-1}, z_0^{k+\delta-1})$$
$$\cdot P(u_{k+\delta}, z_{k+\delta} \mid u_{k+\delta-1}, \cdots, u_{k+\delta-\nu}).$$

While the recursion is now forward only, we now must store $m^{\delta+1}$ quantities rather than m^ν. If δ is large, this is most unattractive; if δ is close to $\nu - 1$, the estimate may well be decidedly suboptimum.

These variations thus seem considerably less attractive than the VA. Nonetheless, something like this may need to be hybridized with the VA in certain situations such as tracking a finite-state source over a finite-state channel, where only the state sequence of the source is of interest.

Finally, we can consider augmented outputs from the VA. A good general indication of how well the algorithm is doing is the depth at which all paths are merged; this can be used to establish whether or not a communications channel is on the air, in synchronism, etc. [11], [28]. A more selective indicator of how reliable particular segments are is the difference in lengths between the best and the next-best paths at the point of merging; this reliability indicator can be quantized into an erasure output. Lastly, the algorithm can be altered to store the L best paths, rather than the single best path, as the survivors in each recursion, thus eventually generating a list of the L most likely path sequences.

References

Convolutional Codes

[1] A. J. Viterbi, "Error bounds for convolutional codes and an asymptotically optimum decoding algorithm," *IEEE Trans. Inform. Theory*, vol. IT-13, pp. 260–269, Apr. 1967.

[2] G. D. Forney, Jr., "Convolutional codes I: Algebraic structure," *IEEE Trans. Inform. Theory*, vol. IT-16, pp. 720–738, Nov. 1970.

[3] ——, "Review of random tree codes," NASA Ames Res. Cen., Moffett Field, Calif., Contract NAS2-3637, NASA CR 73176, Final Rep., Dec. 1967, appendix A.

[4] G. C. Clark, R. C. Davis, J. C. Herndon, and D. D. McRae, "Interim report on convolution coding research," Advanced System Operation, Radiation Inc., Melbourne, Fla., Memo Rep. 38, Sept. 1969.

[5] A. J. Viterbi and J. P. Odenwalder, "Further results on optimal decoding of convolutional codes," *IEEE Trans. Inform. Theory* (Corresp.), vol. IT-15, pp. 732–734, Nov. 1969.

[6] A. J. Viterbi, "Convolutional codes and their performance in communication systems," *IEEE Trans. Commun. Technol.*, vol. COM-19, pp. 751–772, Oct. 1971.

[7] G. D. Forney, Jr., "Convolutional codes II: Maximum likelihood decoding," Stanford Electronics Labs., Stanford, Calif., Tech. Rep. 7004-1, June 1972.

[8] J. A. Heller, "Short constraint length convolutional codes," in *Space Program Summary 37-54*, vol. III. Jet Propulsion Lab., Calif. Inst. Technol., pp. 171–177, Oct.–Nov. 1968.

[9] ——, "Improved performance of short constraint length convolutional codes," in *Space Program Summary 37-56*, vol. III. Jet Propulsion Lab., Calif. Inst. Technol., pp. 83–84, Feb.–Mar. 1969.

[10] J. P. Odenwalder, "Optimal decoding of convolutional codes," Ph.D. dissertation, Dep. Syst. Sci., Sch. Eng. Appl. Sci., Univ. of California, Los Angeles, 1970.

[11] J. L. Ramsey, "Cascaded tree codes," MIT Res. Lab. Electron., Cambridge, Mass., Tech. Rep. 478, Sept. 1970.

[12] G. W. Zeoli, "Coupled decoding of block-convolutional concatenated codes," Ph.D. dissertation, Dep. Elec. Eng., Univ. of California, Los Angeles, 1971.

[13] J. M. Wozencraft, "Sequential decoding for reliable communication," in *1957 IRE Nat. Conv. Rec.*, vol. 5, pt. 2, pp. 11–25.

[14] R. M. Fano, "A heuristic discussion of probabilistic decoding," *IEEE Trans. Inform. Theory*, vol. IT-9, pp. 64–74, Apr. 1963.

[15] K. S. Zigangirov, "Some sequential decoding procedures," *Probl. Pered. Inform.*, vol. 2, pp. 13–25, 1966.

[16] F. Jelinek, "Fast sequential decoding algorithm using a stack," *IBM J. Res. Develop.*, vol. 13, pp. 675–685, Nov. 1969.

[17] L. R. Bahl, C. D. Cullum, W. D. Frazer, and F. Jelinek, "An efficient algorithm for computing free distance," *IEEE Trans. Inform. Theory* (Corresp.), vol. IT-18, pp. 437–439, May 1972.

[18] L. R. Bahl, J. Cocke, F. Jelinek, and J. Raviv, "Optimal decoding of linear codes for minimizing symbol error rate" (Abstract), in *1972 Int. Symp. Information Theory* (Pacific Grove, Calif., Jan. 1972), p. 90.

[19] P. L. McAdam, L. R. Welch, and C. L. Weber, "M.A.P. bit decoding of convolutional codes," in *1972 Int. Symp. Information Theory* (Pacific Grove, Calif., Jan. 1972), p. 91.

Space Applications

[20] Linkabit Corp., "Coding systems study for high data rate telemetry links," Final Rep. on NASA Ames Res. Cen., Moffett Field, Calif., Contract NAS2-6024, Rep. CR-114278, 1970.

[21] G. C. Clark, "Implementation of maximum likelihood decoders for convolutional codes," in *Proc. Int. Telemetering Conf.* (Washington, D. C., 1971).

[22] J. A. Heller and I. M. Jacobs, "Viterbi decoding for satellite and space communication," *IEEE Trans. Commun. Technol.*, vol. COM-19, pp. 835–847, Oct. 1971.

[23] G. C. Clark, Jr., and R. C. Davis, "Two recent applications of error-correction coding to communications system design," *IEEE Trans. Commun. Technol.*, vol. COM-19, pp. 856–863, Oct. 1971.

[24] I. M. Jacobs and R. J. Sims, "Configuring a TDMA satellite communication system with coding, in *Proc. 5th Hawaii Int. Conf. Systems Science.* Honolulu, Hawaii: Western Periodicals, 1972, pp. 443–446.

[25] A. R. Cohen, J. A. Heller, and A. J. Viterbi, "A new coding technique for asynchronous multiple access communication," *IEEE Trans. Commun. Technol.*, vol. COM-19, pp. 849–855, Oct. 1971.

[26] D. Quagliato, "Error correcting codes applied to satellite channels," in *1972 IEEE Int. Conf. Communications* (Philadelphia, Pa.), pp. 15/13–18.

[27] Linkabit Corp., "Hybrid coding system study," Final Rep. on Contract NAS2-6722, NASA Ames Res. Cen., Moffett Field, Calif., NASA Rep. CR114486, Sept. 1972.

[28] G. C. Clark, Jr., and R. C. Davis, "Reliability-of-decoding indicators for maximum likelihood decoders," in *Proc. 5th Hawaii Int. Conf. Systems Science.* Honolulu, Hawaii: Western Periodicals, 1927, pp. 447–450.

Intersymbol Interference

[29] G. D. Forney, Jr., "Maximum-likelihood sequence estimation of digital sequences in the presence of intersymbol interference," *IEEE Trans. Inform. Theory*, vol. IT-18, pp. 363–378, May 1972.

[30] ——, "Lower bounds on error probability in the presence of large intersymbol interference," *IEEE Trans. Commun. Technol.* (Corresp.), vol. COM-20, pp. 76–77, Feb. 1972.

[31] J. K. Omura, "On optimum receivers for channels with intersymbol interference," abstract presented at the IEEE Int. Symp. Information Theory, Noordwijk, The Netherlands, June 1970.

[32] H. Kobayashi, "Correlative level coding and maximum-likelihood decoding," *IEEE Trans. Inform. Theory*, vol. IT-17, pp. 586–594, Sept. 1971.

[33] M. J. Ferguson, "Optimal reception for binary partial response channels," *Bell Syst. Tech. J.*, vol. 51, pp. 493–505, Feb. 1972.

[34] F. R. Magee, Jr., and J. G. Proakis, "Adaptive maximum-likelihood sequence estimation for digital signaling in the presence of intersymbol interference," *IEEE Trans. Inform. Theory* (Corresp.), vol. IT-19, pp. 120–124, Jan. 1973.

[35] L. K. Mackechnie, "Maximum likelihood receivers for channels having memory," Ph.D. dissertation, Dep. Elec. Eng., Univ. of Notre Dame, Notre Dame, Ind., Jan. 1973.

[36] R. W. Chang and J. C. Hancock, "On receiver structures for channels having memory," *IEEE Trans. Inform. Theory*, vol. IT-12, pp. 463–468, Oct. 1966.

[37] K. Abend, T. J. Harley, Jr., B. D. Fritchman, and C. Gumacos, "On optimum receivers for channels having memory," *IEEE Trans. Inform. Theory* (Corresp.), vol. IT-14, pp. 819–820, Nov. 1968.

[38] R. R. Bowen, "Bayesian decision procedures for interfering digital signals," *IEEE Trans. Inform. Theory* (Corresp.), vol. IT-15, pp. 506–507, July 1969.

[39] K. Abend and B. D. Fritchman, "Statistical detection for communication channels with intersymbol interference," *Proc. IEEE*, vol. 58, pp. 779–785, May 1970.

[40] C. G. Hilborn, Jr., "Applications of unsupervised learning to problems of digital communication," in *Proc. 9th IEEE Symp. Adaptive Processes, Decision, and Control* (Dec. 7–9, 1970).
C. G. Hilborn, Jr., and D. G. Lainiotis, "Optimal unsupervised learning multicategory dependent hypotheses pattern recognition," *IEEE Trans. Inform. Theory*, vol. IT-14, pp. 468–470, May 1968.

[41] G. Ungerboeck, "Nonlinear equalization of binary signals in Gaussian noise," *IEEE Trans. Commun. Technol.*, vol. COM-19, pp. 1128–1137, Dec. 1971.

[42] ——, "Adaptive maximum-likelihood receiver for carrier-modulated data transmission systems," in preparation.

Continuous-Phase FSK

[43] M. G. Pelchat, R. C. Davis, and M. B. Luntz, "Coherent demodulation of continuous-phase binary FSK signals," in *Proc. Int. Telemetry Conf.* (Washington, D. C., 1971).

[44] R. de Buda, "Coherent demodulation of frequency-shift keying with low deviation ratio," *IEEE Trans. Commun. Technol.*, vol. COM-20, pp. 429–435, June 1972.

Text Recognition

[45] C. E. Shannon and W. Weaver, *The Mathematical Theory of Communication.* Urbana, Ill.: Univ. of Ill. Press, 1949.

[46] C. E. Shannon, "Prediction and entropy of printed English," *Bell Syst. Tech. J.*, vol. 30, pp. 50–64, Jan. 1951.

[47] D. L. Neuhoff, "The Viterbi algorithm as an aid in text recognition," Stanford Electronic Labs., Stanford, Calif., unpublished.

[48] J. Raviv, "Decision making in Markov chains applied to the problem of pattern recognition," *IEEE Trans. Inform. Theory*, vol. IT-13, pp. 536–551, Oct. 1967.

Miscellaneous

[49] H. Kobayashi, "A survey of coding schemes for transmission or recording of digital data," *IEEE Trans. Commun. Technol.*, vol. COM-19, pp. 1087–1100, Dec. 1971.

[50] ——, "Application of probabilistic decoding to digital magnetic recording systems," *IBM J. Res. Develop.*, vol. 15, pp. 64–74, Jan. 1971.

[51] U. Timor, "Sequential ranging with the Viterbi algorithm," Jet Propulsion Lab., Pasadena, Calif., JPL Tech. Rep. 32-1526, vol. II, pp. 75–79, Jan. 1971.

[52] J. K. Omura, "On the Viterbi algorithm for source coding" (Abstract), in *1972 IEEE Int. Symp. Information Theory* (Pacific Grove, Calif., Jan. 1972), p. 21.

[53] F. P. Preparata and S. R. Ray, "An approach to artificial nonsymbolic cognition," *Inform. Sci.*, vol. 4, pp. 65–86, Jan. 1972.

[54] S. W. Golomb, *Shift Register Sequences*. San Francisco, Calif.: Holden-Day, 1967, pp. 13–17.

[55] G. J. Minty, "A comment on the shortest-route problem," *Oper. Res.*, vol. 5, p. 724, Oct. 1957.

[56] M. Pollack and W. Wiebenson, "Solutions of the shortest-route problem—A review," *Oper. Res.*, vol. 8, pp. 224–230, Mar. 1960.

[57] R. Busacker and T. Saaty, *Finite Graphs and Networks: An Introduction with Applications*. New York: McGraw-Hill, 1965.

[58] J. K. Omura, "On the Viterbi decoding algorithm," *IEEE Trans. Inform. Theory*, vol. IT-15, pp. 177–179, Jan. 1969.

[59] J. M. Wozencraft and I. M. Jacobs, *Principles of Communication Engineering*. New York: Wiley, 1965, ch. 4.

[60] J. K. Omura, "Optimal receiver design for convolutional codes and channels with memory via control theoretical concepts," *Inform. Sci.*, vol. 3, pp. 243–266, July 1971.

Section 3.5
Terrestrial Interface

The characteristics of the interface between the user and the earth terminal will vary widely in different systems. In the INTELSAT system, the earth terminal interfaces with a switch which is part of the terrestrial communications system. In other cases, individual or small groups of users, e.g., a PBX, will access the earth terminal through access lines obtained from the terrestrial communication plant. In still other cases, the physical distances will be small enough to justify direct user-owned access, e.g., cable or microwave, to the earth terminal. In all cases, an efficient matching of the satellite system to the user environment is a key part of the communications system design problem. The reprint paper and papers in the Bibliography discuss three aspects of the problem:

1) Echo Control: Reprint Paper 3.5.1 explains the echo problem and discusses a particular type of echo canceller. References [1]-[4] discuss various aspects of echo control.

2) TDM/FDM Transmultiplexers: These devices are designed to convert an FDM signal arriving at the earth station into a TDM signal for transmission over the satellite and vice versa. Reference [5] provides a good introduction to the topic, and subsequent papers in that issue of the transactions complete the development.

3) Multiplexing and Synchronization: Frequently, digital signals will arrive at the earth station at different bit rates, and they must be multiplexed into a single bit stream for transmission. References [6]-[9] discuss various facets of this problem.

REFERENCES

[1] J. W. Emling and D. Mitchell, "The effects of time delay and echoes on telephone conversations," *Bell Syst. Tech. J.,* Nov. 1963.

[2] R. D. Gitlin and J. S. Thompson, "A technique for adaptive phase compensation in echo cancellation," *National Telecommunications Conf.,* 1977.

[3] G. K. Helder, "Digital echo suppressors and the CCITT," *Int. Conf. Communications,* 1976.

[4] O. A. Horna, "Echo canceller utilizing pseudo-logarithmic coding," *National Telecommunications Conf.,* 1977.

[5] R. B. Kieburtz, "Introduction to papers on TDM/FDM transmultiplexers," *IEEE Trans. Commun.,* vol. COM-26, no. 5, pp. 697-698, May 1978.

[6] D. R. Doll, "Multiplexing and concentration," in *Proc. IEEE,* Nov. 1972.

[7] J. R. Pierce, "Synchronizing digital networks," *BSTJ,* pp. 615-636, Mar. 1969.

[8] W. D. White, "Theoretical aspects of asynchronous multiplexing," *Proc. I.R.E.,* pp. 270-275, Mar. 1950.

[9] S. Nash, "Multiplexer survey—Comments on frequency and time division multiplexing," *Telecommunications,* pp. 36-39, Jan. 1970.

ECHO CONTROL IN TELEPHONE COMMUNICATIONS*

H. G. Suyderhoud   M. Onufry   S. Joseph Campanella

COMSAT Laboratories
Clarksburg, Maryland 20734

ABSTRACT

Echo control is increasingly important in modern telephone communications due to the long propagation delay encountered in satellite networks, improvements in circuit performance in terms of lower overall loss and reduced channel noise, and the introduction of speech interpolation methods. The current method of echo control based on the echo suppressor concept relies on the switched insertion of high loss into the send path of the 4-wire system, leading to a natural conflict between requirements for simultaneously blocking echo and passing the wanted speech signal.

To overcome this conflict, echo cancellation methods (which stop echo but are transparent to wanted speech) have been under study for over 10 years and are now being implemented in a form that promises to be economically attractive. This paper defines the echo control problems encountered in modern telephone networks, reports on experience obtained in field tests of experimental echo cancellers, and briefly describes a recently developed economical echo canceller implementation.

INTRODUCTION

Echo in the long-distance telephone plant is a well-known albeit undesirable phenomenon.[1] Its cause, equally well-known, is the lack of impedance matching necessary for reflection-free transmission. This effect is manifested particularly at the hybrid coil, a circuit device which converts local, 2-wire, bidirectional transmission to long-distance, 4-wire, unidirectional transmission for send and return paths, as shown in Fig. 1.

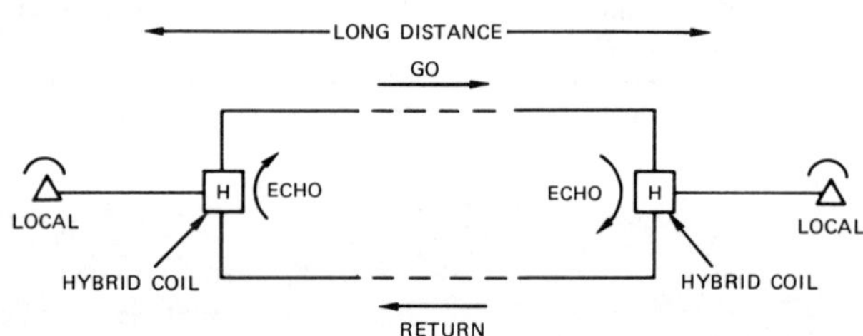

Fig. 1. Echo Occurrences in Long-Distance Telephone Circuits

The hybrid coil, which is an integral part of the 4-wire long-distance circuit, is required to switch a large variety of 2-wire facilities to its 2-wire port. For reflection-free transmission, this port must be perfectly balanced for every 2-wire facility. Since this is not economically feasible, an average echo return loss of only 11—15 dB is obtainable in practice.

Echo, by definition, is a signal, usually reduced in strength, returning to its source after a time delay. Although reflections also occur on local and short-haul connections, the very short delay (<10 ms) renders echo imperceptible. However, on long-haul circuits of over 1500 miles, for example, it can no longer be ignored and therefore requires special attention.

Present echo control measures consist of increasing loss allocation in 4-wire circuits up to approximately 1500 miles and applying echo suppressors to longer circuits. In many instances, echo suppressors introduce speech distortion in the form of chopping of parts of words. In other instances, they do not provide adequate echo control on long propagation delay circuits due to the inherent conflict in design requirements for simultaneously blocking echo and passing speech. These requirements become too severe when echo return loss is insufficient.

Echo cancellation was proposed in the early sixties; preliminary experiments showed that it was, at least potentially, a superior method of echo control.[2] It first appeared that this technique was not economically feasible for large-scale implementation. However, subsequent developments have proved that economy can be realized without sacrificing the basic advantage of the cancellation technique. This paper will review briefly the overall echo problem, duscuss improved results obtained with experimental echo cancellers, and finally present a recent design of an economically feasible echo canceller.

ECHO CONTROL CONSIDERATIONS

In the modern telephone communications system, with trends toward reduced circuit loss and distortion, improved signal-to-noise ratio, introduction of speech interpolation, and increased satellite use, echo control is becoming increasingly important as it becomes more easily perceived. This implies that there is an interaction of effects caused by echo and other possible degradations such as excessive loss, noise, distortion, and delay.[3—5]

To unify these variables, a "grade of service" concept has been developed based on extensive experience and testing both in the laboratory and in actual operation. Its purpose is to establish objectives for performance in terms of circuit quality determined by customer opinion within the constraints of the telephone plant. Grade of service is a combination of two inputs. On the one hand, subjective experiments determine the telephone user's reaction to conditions which may be encountered in the telephone circuit. The results of these experiments are in the form of opinion curves for the condition of interest. The other input consists of a survey of the telephone plant which provides distributions of its objective measurable performance. The distribution of the opinion data is combined with the distribution of the plant performance to obtain a grade of service distribution from which a performance objective may be selected.

An example of such an objective is a grade of service which results in the rating of "good or better" for 95 percent of all telephone calls. Once this has been established, it is possible to allocate objectives to

---

*This paper is based upon work performed in COMSAT Laboratories under the sponsorship of the International Telecommunications Satellite Organization (INTELSAT). Views expressed in this paper are not necessarily those of INTELSAT.

Reprinted from *National Telecommun. Conf.*, vol. I, Nov. 29–Dec. 1, 1976, pp. 8.1-1–8.1-5.

each of the measurable variables, such as noise, cross-talk, echo, and received volume.

The echo performance objective in the Bell System is stated as follows: "The probability of customer dissatisfaction (with the echo performance) of a circuit shall be held to 1 percent or less."[6] Other telephone administrations may select different objectives depending on the desired grade of service and the physical and economic constraints of their telephone plants. For example, Recommendation G.131 of the C.C.I.T.T. contains rules concerning the use of echo suppressors.[7] Rule A states that "the probability that an international connection between any two subscribers will exhibit an objectionable echo should not be greater than 1%. If the probability is greater, an echo suppressor must be provided." This ideal rule is tempered by a practical rule, E, which permits the probability of encountering objectionable echo to increase to 10 percent upon approval by the administrations involved.

The effects of echo and propagation delay have probably been responsible for more subjective experiments than any other single parameter in the telephone plant. To provide some additional insight into the complexities of the echo control problem, the interactive effects of propagation delay, sidetone, circuit loss, and circuit noise on subscriber reaction to echo will be briefly considered.

Sidetone is caused by the mouth-to-ear coupling of a telephone handset. Since it involves no delay, it is a degenerate form of echo. However, it is an important parameter for two reasons. First, it gives the telephone user the impression that the circuit is "alive". Tests at Bell Laboratories[3] have indicated that the preferred sidetone level is quite high, i.e., comparable to the volume which the talker would like to receive from the distant party. Hence, it masks the weaker echo signals which occur with very short delay during active speech. The second function of sidetone is to provide a feedback mechanism which assists the talker in controlling the level of his own speech. This directly influences the echo control problem, since the echo level increases as the transmit speech becomes louder and vice versa.

As the delay increases, the telephone user begins to hear a disturbing hollow effect due to echo. For even larger delays, the echo occurs during silent intervals of speech where it becomes quite noticeable and annoying. Further increases in delay may make it impossible to carry on a conversation.

Noise tends to mask echo in the telephone system. On a very high-quality circuit, the dynamic range of the human ear permits low-level interference such as echo to be heard. Terrestrial circuit noise is additive and proportional to circuit mileage. This is advantageous in terms of the echo control problem because of the masking effect of noise on low-level echo. However, current international practice tends toward lowering the noise objectives and hence increasing the importance of proper echo control.[8]

Circuit loss is also used to control echo. Since echo becomes more apparent with increasing delay, it is reasonable to expect that, as the delay increases, increased loss is required to render echo tolerable. Early work in this area was done by Clark and Mathes[9] in 1925. In 1953 Phillips[4] repeated the tests, studying the relationship between loss and delay up to 85 ms. H. R. Huntly[10] extended the study to a round trip propagation delay of 100 ms; his results are shown in Fig. 2. Figure 3, taken from C.C.I.T.T. Recommendation G.131,[7] shows overall reference equivalent versus round trip propagation time for three probabilities of

encountering objectionable echo. Thus, this information is expressed in terms of the grade of service concept previously discussed. Similar grade of service curves have been developed by D. L. Richards[11] of the United Kingdom Post Office.

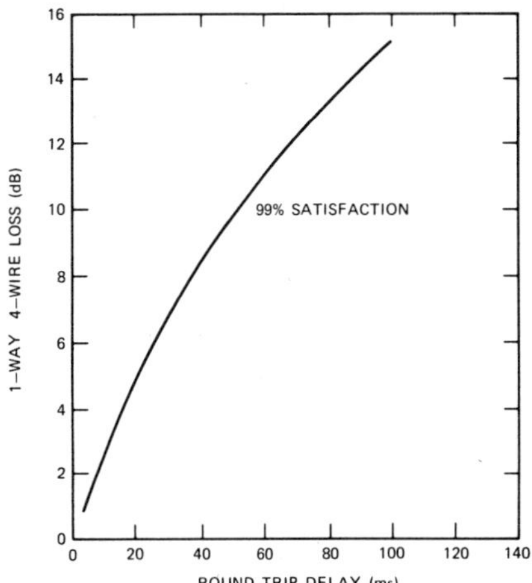

Fig. 2. Echo Tolerance Reported by Huntly[10]

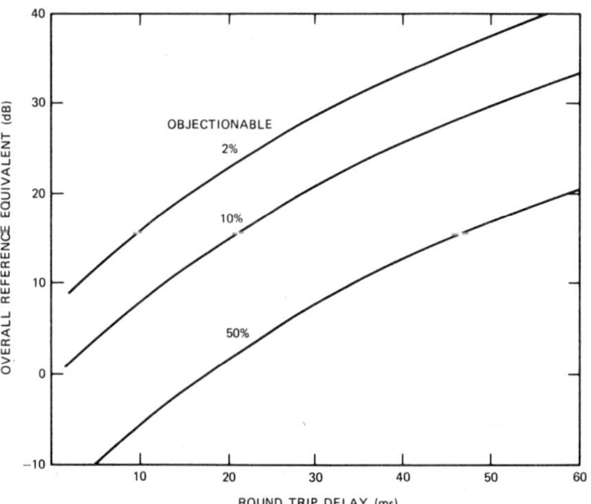

Fig. 3. C.C.I.T.T. Echo Tolerance[8]

<u>CURRENT METHOD OF ECHO CONTROL</u>

Inserting loss in the transmission path is a viable but restricted method of controlling echo in a telephone system in use today. A concept known as "via net loss"[10] is used to systematically distribute the loss through the telephone system as a function of circuit distance for circuits assembled according to the hierarchy of the switching system. This method is obviously limited, since there is a limit to the amount of loss which may be inserted in the transmission path. That limit is reached when the subscriber will no longer tolerate a reduced receive volume to gain increased echo attenuation. At that point echo suppressors are introduced into the circuit to provide echo control.

Echo suppressors, which are currently used to control echo, yield circuits which are acceptable to users in the majority of cases. Nevertheless, due to speech distortions introduced by the basic design limitations of the echo suppressor, the quality of such a circuit is not the same as that of an echo-free circuit. Figure 4 shows a typical echo suppressor application, including a block diagram of a suppressor. As shown in Fig. 4, an echo suppressor in each CT in the the 4-wire circuit employs a speech-operated switched loss circuit to prevent B-to-A transmission when A talks. However, if B talks at the same time and B's speech power becomes equal to or greater than that of A, the switched loss circuit is removed as a result of comparison of the two signals in a differential circuit. Upon removal of the loss, A's echo is permitted to return and become mixed or interspersed with B's speech. This situation may also result in severe chopping of syllables when B's speech power is less than that of A, and loud echo bursts can occur after the loss is removed.

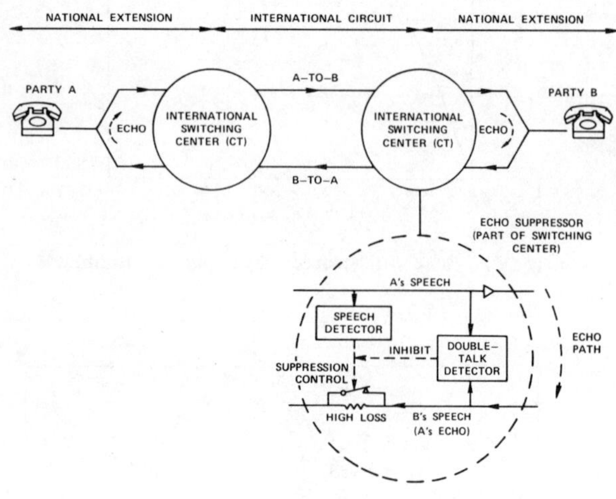

Fig. 4. International Telephone Circuit Including the Echo Suppressor

It should be noted that, if the echo return loss were high, resulting in a very low echo signal, an echo suppressor would be satisfactory. This is not true of the majority of circuits, however, especially international circuits. Continual improvement in terminal balance in the U.S. has resulted in the use of echo suppressors only on circuits of more than 1800 miles.[6] Of course this has economic benefits, since it reduces the number of echo suppressors which are required. It is an example of extension of the grade of service distribution to take into account plant improvements.

As technological advances made long-distance calls via microwave links, undersea cable, and synchronous satellites an increasingly common occurrence, the effects of the increased propagation delay of such circuits aroused increased interest. This led to numerous studies, notably by Bell Laboratories,[6] BPO, INTELSAT, COMSAT, and administrations around the world. Several studies included propagation times extending beyond 1200 ms to assess the effect of double-hop synchronous satellite circuits.[12,13] In general these studies indicated that the effect of propagation delay alone is not the major contributor to transmission impairment. Instead, the dominant contributor is the combination of propagation delay and speech mutilations introduced by the echo suppressors.

Although many techniques have been implemented and succeeded in improving the performance of the echo suppressor,[14] the suppression technique is inherently incapable of good echo control during double talk. In theory, echo cancellation, which emerged during the early sixties, soon became recognized as the ideal

technique of echo control. As shown in Fig. 5, it operates on the principle of adaptive modeling of the echo signal and subsequent subtraction of the model from the real echo, thus leaving the speech paths unaffected for through transmission. Hence, it overcomes the intrinsic shortcoming of the echo suppressor. By means of adaptive feedback processing, a model of the echo path impulse response is obtained and stored in a memory. Subsequently, the incoming signal is processed by convolution with the stored impulse response to obtain a close replica of the real echo signal. This replica is then subtracted from the real echo signal on the send side. Thus, echo is selectively removed without otherwise interfering with the send side signal.

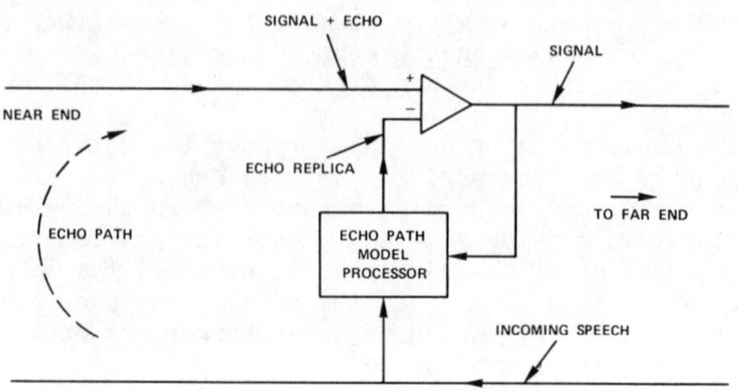

Fig. 5. Principle of Echo Cancellation

DIGITAL ECHO CANCELLER RESULTS

Adaptive echo cancellers require much more elaborate and precise processing than echo suppressors. The component cost alone of early models was about $20,000. However, the recent avalanche of digital circuit components with intensive miniaturization has reduced this figure by a factor of 100 without requiring custom LSI development.

COMSAT Labs built and tested two echo cancellers which were installed on various international commercial satellite circuits.[15] In two tests lasting several months, echo cancellers demonstrated notable subjective quality improvements relative to echo suppressors. Equal performance was obtained in four other tests. However, for these tests it was known that relatively high echo return loss existed. As explained previously, when high echo return loss exists, echo suppressors will provide satisfactory echo control, and the use of cancellers should be expected to produce little or no improvement.* Figure 6, upon which these results are based, shows the composite results of call-back interviews with customers who had completed calls on satellite circuits equipped with either cancellers or suppressors.

ECHO CANCELLER REQUIREMENTS

This section is not intended to present a complete set of echo canceller specifications, but will discuss some general requirements which have emerged from the limited operational experience gained so far. The field trial experience discussed previously is not sufficient to indicate that the majority of circuits in the world have better than 16-dB echo return loss. On the contrary, the circuits on which cancellers were

---

*For completeness, it should be noted that three tests produced negative results with cancellers. In these three cases, however, it was found that the negative results were caused by non-echo-related problems on the circuits on which the cancellers were installed.

tested represented only a fractional cross section of
all terrestrial tail circuits which cause echo. Moreover, it is always possible that a poor echo return
loss circuit will interconnect with an international
circuit requiring echo control. Thus, there is a
potentially widespread need for echo cancellers and
hence proper canceller specifications.

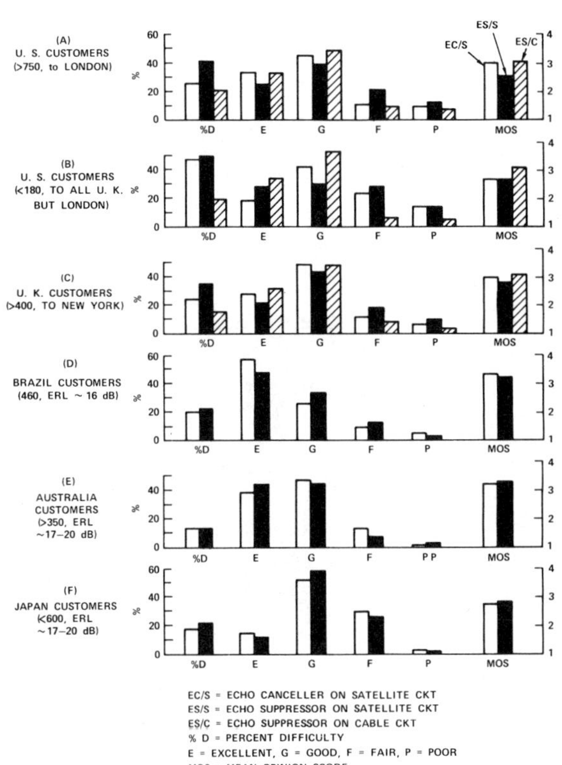

Fig. 6. Distribution of Customer Interview
Responses for Six Populations

Studies of echo tolerance versus delay include
those cited previously, in particular, those by Phillips
of Bell Labs and Hutter in England. Under INTELSAT
sponsorship, COMSAT issued two study contracts to assess
the amount of echo return loss required to obtain satisfactory performance[16] on long delay circuits without
echo suppressors.

It has been concluded from these studies that, if
echo return loss can be maintained at 35 dB or better,
good performance can be expected from such circuits.
This, coupled with the fact that echo return loss on
some circuits may be as poor as 5 dB, indicates that
the echo return loss enhancement required from an echo
canceller is of the order of 30 dB. This is a difficult
requirement to meet economically. Fortunately, however,
the technique known as center clipping can substantially
reduce the 30-dB requirements. A center clipper is a
switch which acts only when the signal power drops below
a given level. It can be invoked to remove all residual
echo after only 20 dB of echo return loss enhancement
is obtained by the canceller. Thus, an important 10-dB
advantage can be achieved by using the center clipper.

During double talk it is necessary to remove the
center clipper to prevent speech distortion. However,
the level of the echo signal mixed with the speech is
sufficiently far down (at least 25 dB) that it is totally
masked by the speech in any case. Thus, a subjectively
transparent echo-free circuit is obtained.

Another requirement is that the canceller must
maintain its cancellation enhancement when time variance is present in the echo path. This latter phenomenon is known to exist in varying degrees on a small
fraction of circuits around the world. The ability of
a canceller to cope with time variance is directly related to its initial speed of convergence, i.e., the
speed required to obtain a given amount of echo return
loss enhancement. With the aid of a center clipper, a
speed of convergence of 0.13 dB/ms will probably suffice, using white noise as a test signal. This means
that, after 150 ms of applying the test signal, 20 dB
of echo return loss enhancement is obtained. In actual
practice, one or two words must be said over the circuit to obtain sufficient cancellation.

There may be circuits with excessive time variance
for which this speech requirement is not quite adequate.
However, present indications are that those conditions
are rather isolated. Whether the extra expense and
complexity required to accommodate those few conditions
is justifiable is uncertain.

A third echo canceller requirement is that it must
be able to accommodate varying amounts of end delay in
the echo path. This is largely an operational question. If all cancellers could always be placed close
to the source of the echo, important economic advantages could be gained from the design simplicity. However, since this is not always possible, the canceller
may be required to accommodate echo path delays up to
40 ms.

Basically, end delay is a signal storage problem.
The difficulty lies not so much in the storage itself
as in the rapid and accurate determination of its values, which may vary significantly from circuit to circuit. Although echo cancellers are becoming economically viable, end delay requires improvement and
innovation.

DESCRIPTION OF A SMALL ECHO CANCELLER

A small and cost-competitive log echo canceller
has been developed at COMSAT Laboratories under
INTELSAT sponsorship. Similar to the first models developed at COMSAT Labs and elsewhere, this echo canceller is based on time-domain convolutional processing
and storing and adaptively updating the echo path impulse response.[17] Its specifications are as follows:

    a. digital convolution processing,
    b. 32-ms total storage capacity,
    c. 110-ms time to convergence, and
    d. variable threshold center clipper.

One of the major simplifications of this echo
canceller is due to its novel adaptive strategy and the
use of pseudologarithmic signal encoding for convolution processing. Analog-to-digital conversion is performed directly into a 7-bit A-law code by inexpensive hardware utilizing three stages of comparators
for sign, exponent, and mantissa. Figure 7 is a block
diagram of the canceller. Note that both the actual
echo subtraction and the center clipping are performed
in analog. Reference 18 describes the entire device.

The echo canceller has been tested in the laboratory in simulated satellite circuits including asym-
metry and phase roll* in the echo path.[19] Excellent
performance has been obtained with phase roll rates up
to 5.7 rad/s, which is the limit of the phase roll

____

*Phase roll is the phenomenon of cyclic variation of
 the echo path impulse response due to carrier fre-
 quenty offsets in FDM systems.

simulation equipment. In a multiple comparison test including six echo suppressors meeting G.161 specifications, the log echo canceller demonstrated significantly higher quality. This echo canceller has shown that echo cancellation can indeed become a viable method of echo control since its size does not exceed that of several present echo suppressors. Moreover, the parts cost is not expected to exceed $200 when large quantities are manufactured.

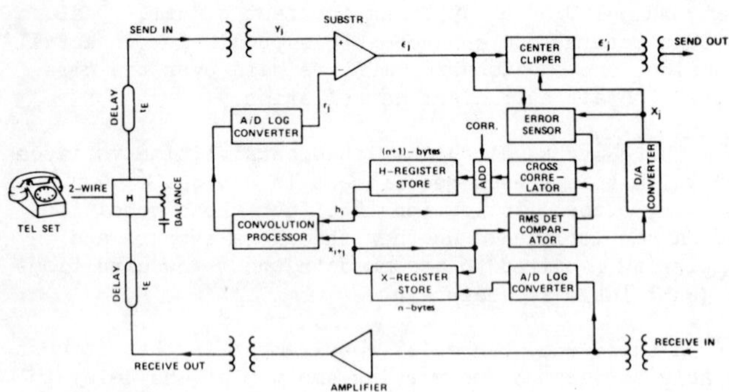

Fig. 7. Block Diagram of Log Echo Canceller

CONCLUSIONS

Echo has been a problem in telecommunications since long-distance telephony came into being about 50 years ago, when the first studies were made to assess the effects of echo and delay. Recently, with the advent of satellite communications, the echo problem has again been brought to the international forefront by study, debate, and standardization, notably in the C.C.I.T.T. It has been widely concluded that in many instances the echo suppressor will be inadequate to overcome the echo problem due to its inherent principle of operation, involving voice switching of the transmission paths.

Although it is much more complex than the suppression technique, the method of echo cancellation which has recently emerged has the potential to finally overcome the echo control problem in telecommunications. One particular design, utilizing log signal formats, may become available soon at a price and size comparable to those of present echo suppressors. The problem of end delay of the echo path still requires attention and innovation, primarily where it will be excessive, as in large countries such as the U.S. or Australia.

REFERENCES

1. J. W. Emling and D. Mitchell, "The Effects of Time Delay and Echos on Telephone Conversations," _Bell System Technical Journal_, Vol. 42, November 1963, pp. 2869–2891.

2. M. M. Sondhi, "An Adaptive Echo Canceller," _Bell System Technical Journal_, Vo. 46, No. 3, March 1967.

3. A. M. Noll, "Subjective Effects of Sidetone During Telephone Conversations," _IEEE Communications and Electronics_, May 1964.

4. G. M. Phillips, "Echo and Its Effects on the Telephone User," _Bell Laboratories Record_, No. 32, August 1954, pp. 281–284.

5. D. L. Richard and G. A. Buck, "Telephone Echo Tests," _Proc. IEE_, Vol. 107, Pt. B, pp. 553–556, 1960.

6. _Transmission Systems For Communications_, Fourth Edition, Bell Telephone Laboratories, Inc., 1970, p. 60.

7. "Line Transmission," _C.C.I.T.T. Green Book_, Vol. III-1, 1973, pp. 48–54.

8. "Telephone Transmission Quality," _C.C.I.T.T. Green Book_, Vol. V, 1973, pp. 172–174.

9. A. B. Clark and R. C. Mathes, "Echo Suppressors for Long Telephone Circuits," Bell Telephone Laboratories, Monograph 165, June 1925.

10. H. R. Huntly, "Transmission Design of Intertoll Telephone Trunks," _Bell System Technical Journal_, Vol. 32, September 1953, pp. 1019–1036.

11. D. L. Richards, _Telecommunications by Speech_, London: Butterworth, 1973.

12. S. J. Campanella, H. G. Suyderhoud, and M. Onufry, "Subjective Evaluation of Dedicated Multiple-Hop Satellite Communications for Government and Military Users," _IEEE Transactions on Communication Technology_, Vol. COM-18, No. 5, October 1970.

13. E. T. Klemmer, "Subjective Evaluation of Transmission Delay in Telephone Conversations," _Bell System Technical Journal_, Vol. 46, July-August 1967.

14. J. E. Unrue, Jr., "Echo Suppressor Design Considerations," _IEEE Transactions on Communication Technology_, Vol. COM-16, No. 4, August 1968.

15. H. G. Suyderhoud, S. Campanella, and M. Onufry, "Results and Analysis of Worldwide Echo Canceller Field Trial," _COMSAT Technical Review_, Vol. 5, No. 2, Fall 1975, pp. 253–275.

16. G. Williams and L. S. Moye, "The Subjective Evaluation of Unsuppressed Echo in Simulated Long-Delay Telephone Communications," _Proc. IEE_, Vol. 118, No. 3/4, March/April 1971, pp. 401–408.

17. H. G. Suyderhoud and M. Onufry, "Performance of a Digital Adaptive Echo Canceller in a Simulated Satellite Circuit Environment," Vol. 33, _AIAA Progress in Astronautics and Aeronautics: Communications Satellite Technology_, edited by P. Bargellini, Cambridge, Mass.: M.I.T. Press, 1974.

18. O. A. Horna, "Transversal Filter Design for Echo Cancellation in Satellite Communications," submitted for publication in _IEEE Transactions on Communication Technology_.

19. H. Brueggemann et al., "Experiments on Adaptive Echo Cancellation," _Australia Telecommunications Research_, Vol. 4, No. 1, May 1970, pp. 14–15.

Section 3.6
Multiple Access

The multiple-access and broadcast capabilities of a communications satellite lead to problems and techniques that are unique to the satellite area. In this section, the multiple access area is developed in detail.

Subsection 3.6.1 contains papers with general discussions of multiple-access techniques. Reprint Paper 3.6.1.1 discusses various multiple access techniques for a hard-limiting satellite repeater. References [1]-[6] also provide general discussions of the multiple-access problem. These references consider various access techniques and analyze their performance for different system. References [7] and [8] explore the theoretical basis of the multiple access and develop capacity expressions.

Possible multiple-access techniques in the time-frequency domain are shown in Fig. 1. In frequency division multiple access (FDMA), the earth stations are assigned frequency bands for transmission of their signals. The assigned band are separated by guard bands to reduce interference. The satellite power is divided among the carriers that are accessing the satellite at any given time. Two common FDMA schemes are FDM/FM/FDMA and SCPC/FDMA. The first technique was introduced in the analog modulation section (3.2). The second technique is single channel per carrier. As the name implies, each carrier is utilized for a single analog or digital channel. Typical modulations are companded FM for voice or QPSK for digital signals. Subsection 3.6.2 discusses FDMA. Reprint Paper 3.6.2.1 discusses the application of FDM/FM/FDMA and SCPC/FDMA in the INTELSAT system. Reprint Paper 3.6.2.2 discusses the design procedures for SCPC and develops designs for several different types of systems. References [9]-[12] discuss various aspects of the FDMA area. Table I summarizes some of the characteristics, advantages, and disadvantages of FDMA which are developed in that section.

In time division multiple access (TDMA), the earth stations are assigned time slots which are separated by guard bands for the transmission of their signals. During its assigned time slot, the earth station has the entire transponder bandwidth and power available. TDMA development started in the mid-1960's, but the first operational usage was in the Telesat system in 1976. Subsection 3.6.3 discusses TDMA. Reprint Paper 3.6.3.1 discusses the principles of TDMA, its advantages and disadvantages, and operating modes. Reprint Paper 3.6.3.2 summarizes the current state of the art in TDMA. Reference [13]-[18] discuss various aspects of TDMA. Table II summarizes some of the characteristics, advantages, and disadvantages of TDMA.

In code division multiple access (CDMA), the transmissions of each earth station are spread over the time-frequency plane by a coded transformation. One method is direct sequence modulation using a pseudonoise (PN) sequence as shown in the simplified diagram in Fig. 2 (from [19]). The RF carrier is modulated by the code sequence of bandwidth B_c using phase-shift keying. In practice, the RF carrier would have already been modulated by a narrowband (B_I) information signal, but it is shown unmodulated for simplicity. Typically B_c is 10 to 100 times greater than B_I so that the signal is spread over a much wider band as shown in the figure. The receiver multiplies the received waveform by a synchronized replica of the PN sequence to obtain the original information bearing RF carrier. The link performance is not affected by the PN transformations. A number of earth stations can operate simultaneously in the same frequency band but utilizing different codes. If these codes were orthogonal, then the stations would not interfere with each other. In fact, there is some interference. A second method is the frequency hopping system shown in Fig. 3 (from [20]). The code sequence shifts the frequency of the information-bearing carrier. The receiver reverses the process.

Both of these techniques are referred to as *spread spectrum* systems. In addition to their multiple access capabilities, they are useful in combatting jamming and are principally used in military systems. CDMA and spread spectrum systems are discussed in Subsection 3.6.4. The IEEE Press Reprint Book in Spread Spectrum Techniques [20] provides an excellent overview of the area. Reprint Paper 3.6.4.1 discusses performance criteria for antijam and anti-intercept systems. Reprint Paper 3.6.4.2 analyzes the performance of various spread spectrum waveforms in multitone CW and noise jamming environments. References [21]-[23] describe various aspects of the CDMA/spread spectrum problem.

The next three subsections of the multiple-access sec-

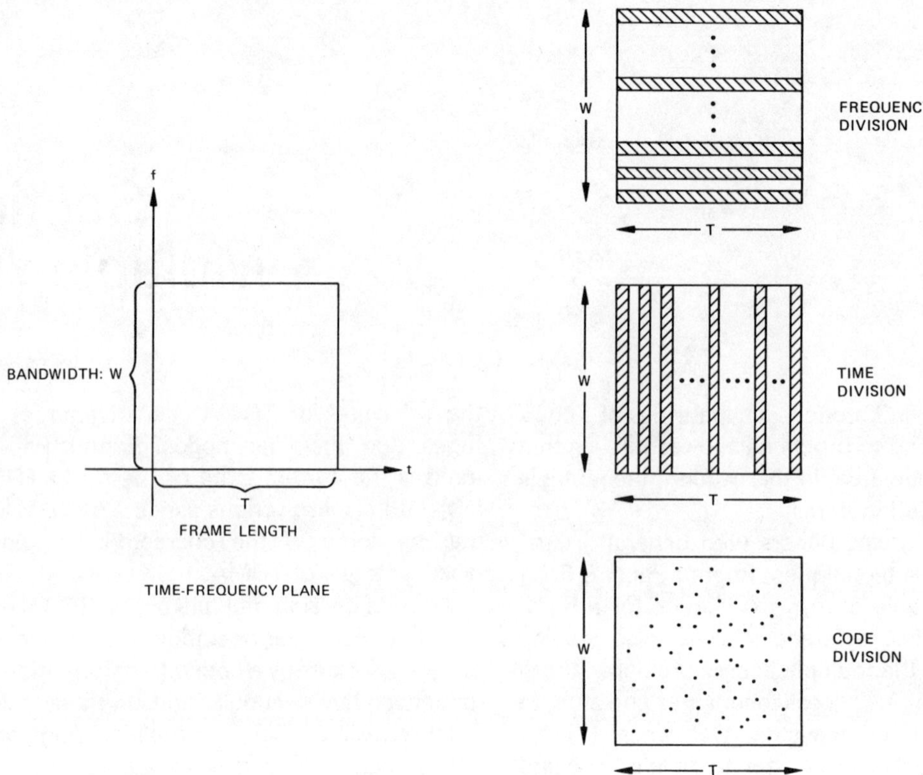

Fig. 1.   Division of time-frequency space.

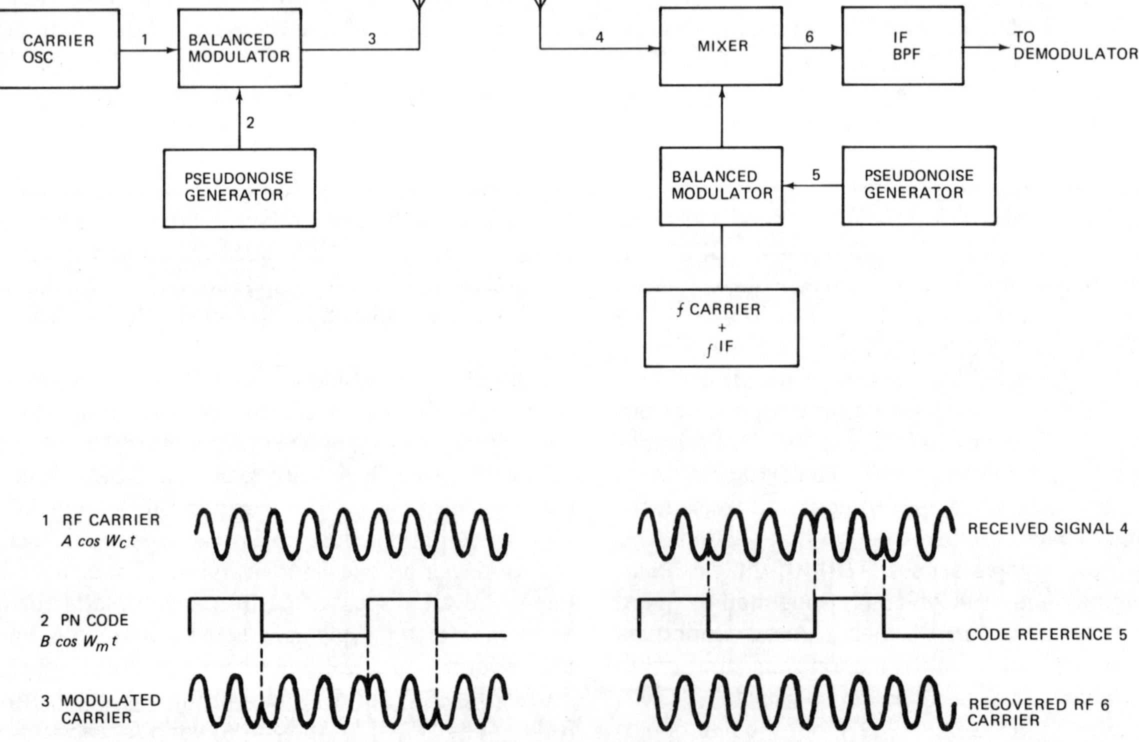

Fig. 2.   Direct sequence (ref. 19).

TABLE I
FDMA

CHARACTERISTICS	ADVANTAGES	DISADVANTAGES
CONSTANT ENVELOPE SIGNALS	USES EXISTING HARDWARE	INTERMODULATION IN REPEATER
SIGNALS CONFINED TO NON-OVERLAPPING FREQUENCY BANDS	NO NETWORK TIMING	REQUIRES UPLINK POWER CONTROL
MULTIPLE ACCESS DEMULTIPLEX-ING BY FILTERING		
MODULATION CAN BE EITHER DIGITAL OR ANALOG		

TABLE II
TIME DIVISION MULTIPLE ACCESS

ADVANTAGES	DISADVANTAGES
1. MAXIMUM USE OF SATELLITE POWER	1. REQUIRES NETWORK TIMING
2. UPLINK POWER CONTROL NOT REQUIRED	2. ANALOG SIGNALS MUST BE CONVERTED TO DIGITAL FORM
3. TRANSMISSION PLAN IS STRAIGHTFORWARD TO CONSTRUCT AND MODIFY	3. INTERFACE WITH FDM TERRESTRIAL PLANT IS EXPENSIVE
4. DIGITAL FORMAT IS COMPATIBLE WITH: (A) SOURCE CODING (B) ERROR CONTROL CODING (C) DIGITAL TERRESTRIAL INTERFACE (D) DEMAND ACCESS ALGORITHMS	
5. CAN WORK WITH A REASONABLE MIX OF EARTH TERMINALS	

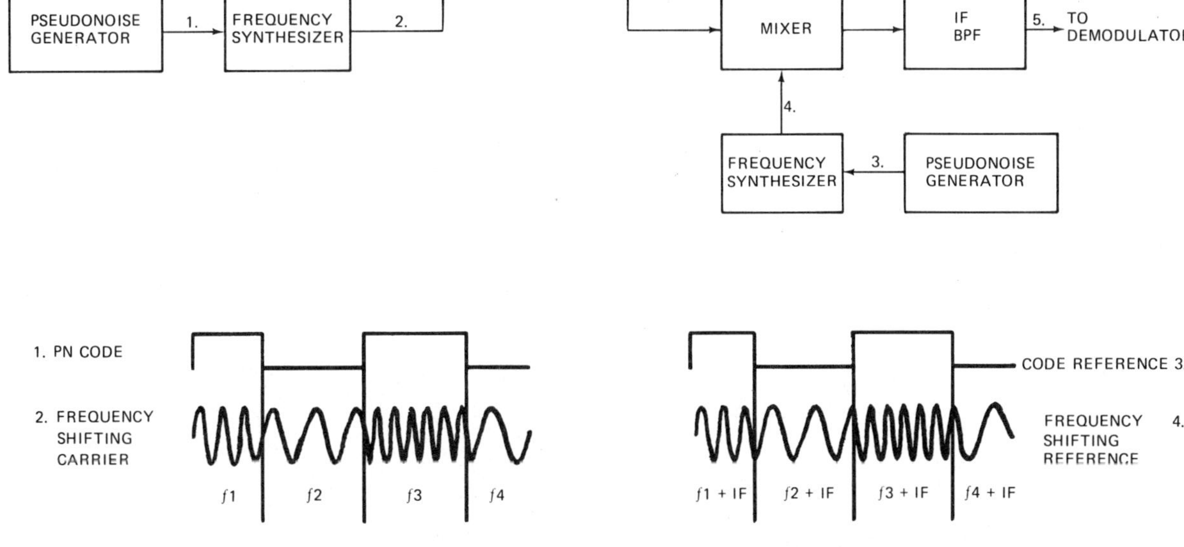

Fig. 3.   Direct hopping system (ref. 19).

tion have separate introductions. Subsection 3.6.5 discusses demand access for voice and data traffic. Subsection 3.6.6 discusses on-board processing. The final Subsection 3.6.7 discusses intersatellite links.

REFERENCES

[1] J. G. Puente, W. G. Schmidt, and A. M. Werth, "Multiple-access techniques for commercial satellites," *Proc. IEEE—Special Issue on Satellite Communications,* vol. 59, no. 2, pp. 218-229, Feb. 1971.

[2] H. Haberle and P. Hein, "Combinations of frequency, time and space division multiple access in multitransponder satellite communications," *Telecommunications Numeriques par satellite,* Paris, France, Nov. 1972.

[3] D. N. McGregor, M. E. Jones, and P. E. O'Neill, Jr., "Comparison of several demand assignment multiple access/modulation techniques for satellite communications," *Int. Conf. Communications Record,* 1971.

[4] D. J. Magill, "Multiple-access modulation techniques," in *Communications Satellite Systems Technology,* New York: Academic, pp. 667-680, 1966.

[5] J. H. Wittman, "Categorization of multiple-access random access modulation techniques," *IEEE Trans. Commun. Technol.,* vol. 15, no. 5, pp. 724-725, Oct. 1967.

[6] J. H. Wittman, "A comparison of satellite multiple access techniques," *IEEE EASCON Conf. Record,* pp. 165-172, 1967.

[7] T. M. Cover, "Broadcast channels," *IEEE Trans. Inform. Theory,* vol. IT 18., no. 1, Jan. 1972.

[8] A. D. Wyner, "Recent results in the Shannon Theory," *IEEE Trans. Inform. Theory,* vol. IT-20, no. 1, Jan. 1974.

[9] P. C. Jain, V. E. Hatfield, J. K. Leung, and D. T. Magill, "Investigation of the use of frequency-division multiple access for application with a mix of user terminals," *Multiple-Access Techniques,* vols. 1-3, Stanford Research Institute, Dec. 1973.

[10] R. M. Gagliardi, "Optimal channelization in FDMA communications—Frequency division multiplexing access in satellite system," *IEEE Trans. Aerosp. Electron. Syst.,* vol. AES-10, pp. 867-870, Nov. 1974.

[11] R. B. McClure, "Analysis of intermodulation distortion in an FDMA satellite communications system with a bandwidth constraint," *IEEE Int. Conf. Communications,* San Francisco, CA, June 8-10, 1970.

[12] H. L. Weinberger and E. M. Kanehira, "Single channel per carrier satellite repeater channel capacity," *IEEE Trans. Aerosp. Electron. Syst.,* vol. AES-11, pp. 805-813, Sept. 1975.

[13] R. M. Hultbert, F. H. Jean, and M. E. Jones, "Time division access for military communications satellites," *IEEE Trans. Aerosp. Electron. Syst.,* vol. AES-1, no. 3, Dec. 1965.

[14] J. G. Puente and T. Sekimoto, "A satellite time-division multiple-access experiment," *IEEE Trans. Commun. Technol.,* vol. COM-16, no. 4, pp. 581-588, Aug. 1968.

[15] R. K. Kwan, "A TDMA application in the Telesat satellite system," *Proc. National Telecommunications Conf.,* pp. 31E-1-31E-6, Nov. 1973.

[16] W. G. Schmidt, "The application of TDMA to the INTELSAT IV satellite series," *COMSAT Tech. Rev.,* vol. 3, no. 2, pp. 257-276, Fall 1973.

[17] I. M. Jacobs, "Integration of demodulation, decoding, buffering and control for TDMA demand assignment systems," *Telecommunications Numeriques par Satellite,* pp. 287-296, Nov. 28-30, 1972, Paris, France.

[18] P. P. Nuspl, K. E. Brown, W. Steenaart, and B. Ghicopoulos, "Synchronization methods for TDMA," *Proc. IEEE,* vol. 65, no. 3, pp. 434-444, Mar. 1977.

[19] R. C. Dixon, *Spread Spectrum Systems.* New York: Wiley, 1976.

[20] R. C. Dixon, Ed., *Spread Spectrum Techniques.* New York: IEEE Press, 1976.

[21] I. L. Lebow, K. L. Jordan, and P. R. Drouilhet, "Satellite communications to mobile platforms," *Proc. IEEE,* vol. 59, no. 2, pp. 139-159, Feb. 1971.

[22] R. Gold, "Study of correlation properties of binary sequences," *Technical Report TR-66-234,* A. F. Avionics Lab., Wright-Patterson Air Force Base, OH.

[23] D. R. Anderson and P. A. Wintz, "Analysis of a spread-spectrum multiple-access system with a hard limiter," *IEEE Trans. Commun. Technol.,* vol. COM-17, no. 2, pp. 285-290, Apr. 1969.

Modulation Techniques for Multiple Access to a Hard-Limiting Satellite Repeater

J. W. SCHWARTZ, MEMBER, IEEE, J. M. AEIN, AND J. KAISER, SENIOR MEMBER, IEEE

Abstract—This paper considers the problem of devising modulation techniques that allow a large number of earth stations to use, simultaneously, a satellite repeater with the transfer characteristics of a wide-band hard-limiting amplifier. Four distinct classes of modulation techniques and their singular properties are described. Three important design considerations are discussed: repeater bandwidth and power sharing, network timing, and operational considerations.

The paper provides a framework for further study of the multiple access problem. The hard-limiting repeater is shown to accommodate a wide variety of signal designs. No single class of multiple-access modulation techniques is found to be uniformly best in satisfying all needs of the diverse networks which can use communication satellites.

INTRODUCTION

COMMUNICATION satellites presently in orbit demonstrate that satellite repeaters can provide a communication capacity equal or superior to that available by other means, and that a single communication satellite can provide many communication channels between widely separated points [3]. However, the development of communication satellite technology has so far been directed toward providing a large communication capacity between only one pair (or at most a few pairs) of earth stations at any one time. For military applications, means are needed for allowing simultaneous access to a single satellite repeater by a number of widely separated stations. With satellite orbits at medium altitudes or above (higher than 2000 nmi), the repeater is visible over large portions of the earth's surface and hence is potentially able to interconnect a large number of far-flung points. Ideally it would be desirable to permit any station in view of the satellite to communicate with one or more other stations at will and with as little interaction as possible with other users of the repeater. The technological and theoretical difficulties involved in providing a system with this characteristic are the portions of the multiple-access problem considered here. The problems of baseband multiplexing[1] and switching (which arise in a central terminal at the interfaces between the satellite

links and the ground links from the subscribers) are not dealt with here.

The size and weight limits on all satellites presently in orbit, and those to be put into orbit in the near future, impose severe limits on the amount of power generated on board the satellite. Therefore, satellite repeaters use traveling-wave tube (TWT) amplifiers, which are the most efficient and reliable devices for wide-band microwave power amplification yet developed. To make maximum use of the dc power consumed by the TWT, it must be operated near the point at which maximum RF output power is obtained—i.e., near saturation. But when a TWT is operated in saturation, it exhibits undesirable properties which include a nonuniform gain-vs.-frequency characteristic and phase distortion. These properties can cause crosstalk among signals sharing the repeater. Therefore, means are needed to ensure that the TWT is operated near, but not in, saturation.

One design is to insert a hard-limiting amplifier (an infinite clipper) with a large bandwidth (10 Mc/s or more) prior to the TWT. The output from the hard-limiter is a constant-envelope signal whose amplitude can be set so that the TWT always operates at a well-chosen point near saturation. The ensuing discussions will demonstrate that the wide-band hard-limiting repeater has great versatility, accommodating many signal designs, and achieves reasonable communication efficiencies. These properties, along with the need to specify only two design parameters (repeater power output and bandwidth), make the hard-limiting repeater a convenient benchmark against which other repeater configurations can be measured.

We shall therefore consider that aspect of the multiple-access problem which involves devising signal modulation techniques that allow simultaneous access to a satellite repeater with the characteristics of a wide-band hard-limiting amplifier.

SIGNAL FLOW

Each user station transmits an RF signal to the satellite with two kinds of modulation. The first kind is *multiple-access* modulation which is designed to accomplish two main purposes: 1) it must make possible separate reception of each transmitted message by any receiving station possessing the requisite demodulator; and 2) it must interfere as little as possible with the transmissions of other user stations in passing through the satellite repeater. The second kind is *message modulation* which conveys the message information. In the fol-

Manuscript received October 18, 1965; revised February 14, 1966. This paper is abstracted from reports of the Institute for Defense Analyses, Research and Engineering Support Division, Arlington, Va., under Contract SD-50, from the Office of the Director of Defense Research and Engineering [1], [2].

The authors are with the Institute for Defense Analyses, Arlington, Va. J. M. Aein is presently on leave to the University of California, Berkeley, Calif.

[1] In this paper, the term *multiple access* applies to the mixing of RF signals. The term *multiplex* applies to the mixing of signals at baseband frequency.

Reprinted from *Proc. IEEE*, vol. 54, pp. 763-777, May 1966.

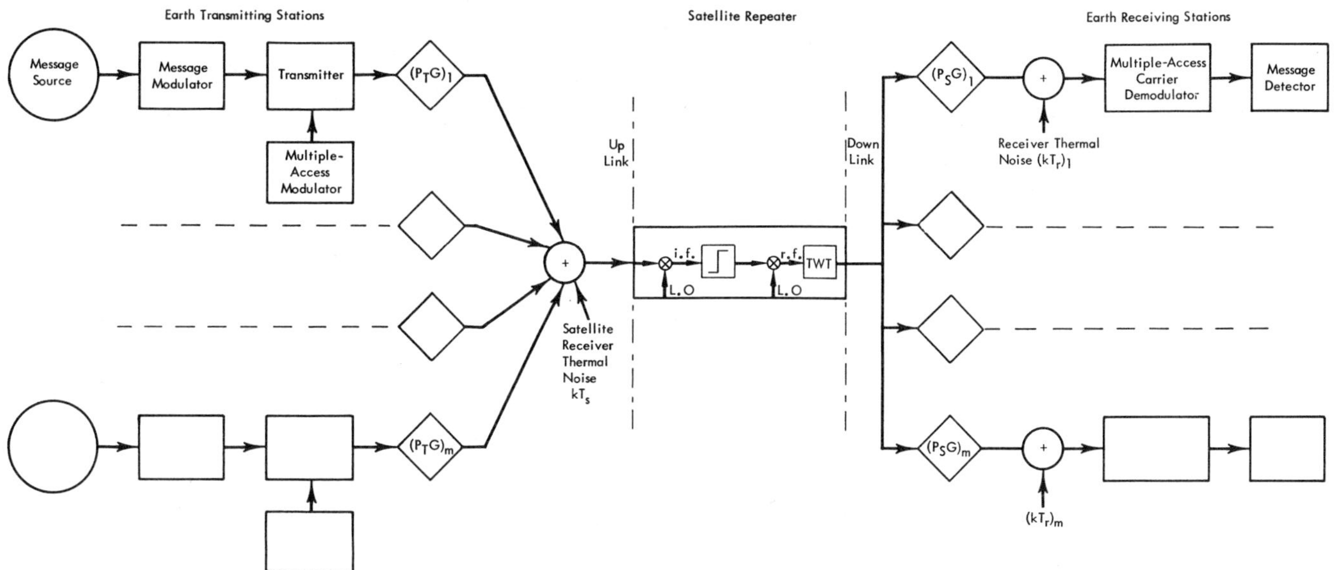

Fig. 1.   Multiple-access signal flow.

lowing discussion, the RF signal with multiple-access modulation but without message modulation is referred to as a *multiple-access carrier* or simply a *carrier*. The allocation of repeater capacity given to a multiple-access carrier with its message modulation is referred to as a *channel*. Two stations communicating in one direction (one transmitting and the other receiving on a given channel) comprise a *link*.

The signal flow in a satellite communications network is shown in Fig. 1. For simplicity, paired communication links are shown; however, in some circumstances one transmitter may wish to communicate with more than one receiver.

The message waveforms must be impressed into each multiple-access carrier waveform. The transmitted signal is generated with a characteristic multiple-access modulation which "addresses" the signal to the receiver, and provides the means by which each receiver can obtain its desired signal from the composite signal emanating from the satellite repeater. For example, in time-division multiple access (TDMA), the multiple-access modulation is a time-gating function which correctly locates each transmission burst relative to those of the rest of the network. In constant-envelope spread-spectrum multiple access (SSMA), a wide-band phase or frequency coding is inserted to provide carrier addressing. In pulse-address multiple access (PAMA), the multiple-access modulation consists of a characteristic pattern of pulses. In frequency-division multiple access (FDMA), the multiple-access modulation is not really a modulation in the strict sense; it is an a priori choice of carrier center frequency. In all cases the RF signal is obtained by two operations: a multiple-access modulation is impressed on the carrier, and a message signal is inserted.

Each transmitted signal is radiated through an an-

tenna, is attenuated by the space loss factor, and arrives at the satellite repeater input. This process is symbolically denoted in Fig. 1 by a diamond enclosing a generic term $(P_T G)$ which represents the effective received power at the satellite repeater input. The transmitted signals from the user stations, combined with thermal noise, are retransmitted by the satellite repeater after wide-band hard-limiting. As shown in Fig. 1, the composite signal at the repeater input is heterodyned down to an intermediate frequency convenient for limiter design, and then heterodyned up to the output radio frequency. The input and output frequencies are offset to prevent the signal from recirculating.

In most cases, the power of a signal from a user station will be much greater than the thermal noise power at the repeater input. Moreover, when sharing the repeater with thermal noise becomes a problem, sharing it with other user stations becomes less important. Therefore, unless otherwise stated, when one or more signals are present at the repeater input, thermal noise at the repeater input is assumed to have no effect on the repeater output signal. However, when no signal is present at the repeater input, the output signal is due to input noise and is assumed Gaussian with a flat spectrum over the repeater band.

The composite repeater output signal is radiated with a power P_s and is received at each receiver with a power $P_r = P_s G$. At the receiver input, receiver thermal noise power per unit bandwidth, $N_0 = kT_r$, combines with the signal received from the satellite. Usually the repeater will have enough bandwidth to make the quotient P_r/N_0 the ultimate limit on communication capacity. This quotient has the dimensions of bandwidth, and for each receiving station it limits the RF signal bandwidth W for a required signal-to-noise power ratio at the receiver input according to the relation P_r/N_0

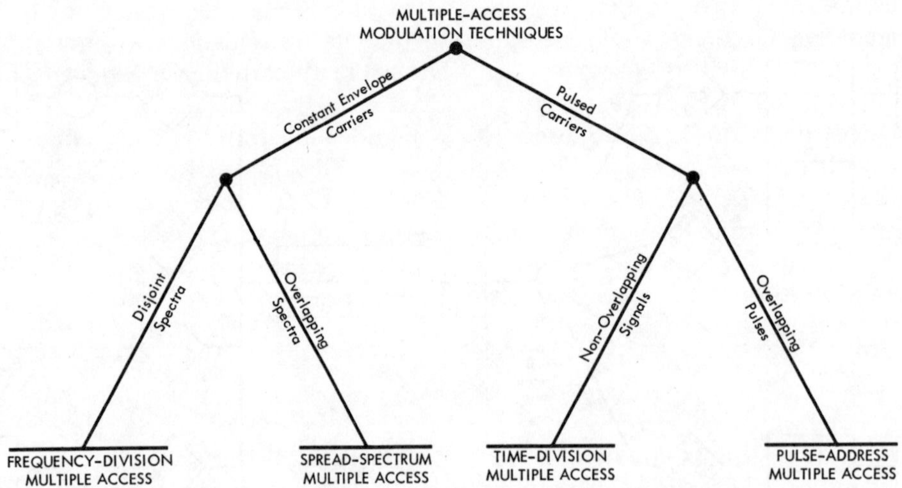

Fig. 2.   Categorization of multiple-access modulation techniques.

$=(S/N)W$. In a multiple-access mode, the desired signal at each receiving station will have some fraction of P_r. An illustrative calculation of P_r/N_0 appears in the Appendix.

The receiving user station demodulates the incoming signals by "tuning" to the desired signal with a multiple-access demodulator. The multiple-access demodulator tends to fall in the same class of devices as the multiple-access modulator. For example, in TDMA the demodulator is a positioned time gate. In SSMA, a correlator with a copy of the phase or frequency coding serves as the demodulator. In PAMA gates matching the pulse pattern can perform the demodulation operation. In FDMA, tuning of the local oscillator frequency provides multiple-access demodulation.

Multiple-Access Modulation Techniques

For practical purposes, a hard-limiting repeater destroys amplitude modulation (except on-off gating) of signals passing through it. Phase and frequency information, on the other hand, tend to be preserved. Therefore, multiple-access modulation techniques for use with a hard-limiting repeater make use of three signal properties: on-off gating, frequency, and phase. In this paper, the various multiple-access modulation techniques are grouped into four categories: frequency-division multiple access (FDMA); constant-envelope, spread-spectrum multiple access (SSMA); time-division multiple access (TDMA); and pulse-address multiple access (PAMA). This categorization is useful for discussion because it groups together systems with similar properties. Figure 2 illustrates the manner by which this categorization is reached.

Frequency-Division Multiple Access

In frequency-division multiple access, the repeater bandwidth is divided into a number of nonoverlapping frequency bands which constitute access channels. Each link is assigned an access channel. For the wide-band hard-limiting repeater, the arrangement of channels is not set by the repeater; a particular arrangement is effected by specifying the spectrum of the signal from each transmitting user station.

Assignment of distinct frequency bands as a means of separating radio signals is a conventional communication practice. Certain complications arise, however, when the signals are mixed in a hard-limiting repeater [4]–[9].

When several signals are simultaneously present at the repeater input, they interact on passing through the repeater and suffer some impairments such as intermodulation noise and intelligible crosstalk. The major source of these impairments is the hard-limiting characteristic of the repeater, which causes intermodulation among signals simultaneously present at the repeater input. This intermodulation noise may be out-of-band noise which only subtracts from the output power available to the desired signals or it may be within the band of a desired signal. To allow the channels to be spaced so that, effectively, all intermodulation noise falls outside the signal bands, much of the bandwidth of the repeater must go unused [10].

Whenever two or more independent signals pass through a hard-limiting repeater, the power ratio of any two component signals at the output will generally differ from the power ratio of the same two components at the input, the change favoring the stronger signal at the expense of the weaker one. The worst case of suppression involves two constant-envelope signals, one greater than the other by a factor of four or more. Then the relative output power of the smaller signal is reduced by another factor of four (6 dB) below its relative power at the repeater input. In contrast to this worst case, if the larger of the two signals can be characterized as a Gaussian process, then the small signal is reduced by, at most, 1 dB below its input relative power. When many large constant-envelope signals arrive simultaneously at the repeater intput with nearly equal power, the effect is like the Gaussian signal case, so that any one constant-envelope signal among them would be suppressed about 1 dB.

If the receiving stations linked to the repeater are all of the same size and sensitivity, the up-link powers of the corresponding transmitting stations should be equal if the data rate requirements are equal. The degree to which up-link power can be coordinated is not known, but the effect of a wide disparity between up-link powers is to reduce the efficiency with which the repeater's output power is used. Assume that the repeater's power and bandwidth have been divided equally among M links. If the up-link powers are not kept equal at the repeater input, the repeater output power does not divide equally, and some stations will receive more power than they require, others less. This situation requires extra margin in the power requirement for each station, thus reducing the total number of stations able to use the repeater simultaneously. Moreover, if one signal at the repeater input is more than four times larger than the sum of the other signals, the strong signal controls a disproportionate amount of the repeater output power, thus robbing the other signals of output power in excess of the input power relationships. For unequal requirements, the up-link power coordination becomes more complicated since the assigned power ratios between signals are unequal.

An advantage of FDMA is the compatibility of the user-station modulation equipment with much existing hardware. For analog FM message modulation, only FM transmitters and receivers are required, and the inputs and outputs of the links appear in analog form, usable for connection into ordinary telephone equipment.

Time-Division Multiple Access (TDMA)

In TDMA, the transmissions in different links do not overlap in the repeater. Each link is assigned exclusive use of the repeater during specified time slots. In this manner, the signals pass through the repeater with virtually no interaction. An access channel in TDMA designates a particular sequence of time slots. The simplest such sequence would provide that the time slots of a channel occur periodically at a definite repetition frequency which may be termed the "frame repetition rate" of the system. All other channels in this arrangement would have the same frame rate but different times of occurrence, and possibly different slot durations [11].

If one signal emitted by the satellite is designated as a time reference, or satellite clock, the signals can be kept separated in time and can be kept in a specified sequence by maintaining specified time intervals relative to the satellite clock. This can be done to within 0.1 μs, and possibly less, by using a straightforward closed-loop timing technique [12]. This procedure does not require a given transmitting station to consider the path delay of any other station. Moreover, because it is

a closed-loop operation, the procedure allows a station to time its transmissions without explicitly calculating either the absolute phase and frequency of the satellite clock or its own path delay.

Figure 3 illustrates this procedure. Suppose that time marks are transmitted by the satellite at intervals of T seconds and that *Station A* wishes to transmit a message to *Station C* via pulses to be received t_1 seconds after each time mark. To set its local clock, *Station A* transmits a series of pulses at intervals nominally T seconds in length. *Station A* then compares the repetition rate of these pulses, when they are received after retransmission by the satellite, with the repetition rate of the time marks. *Station A* then changes the repetition rate of its local clock by exactly the difference in the repetition rates of the received signals. *Station A* also compares the interval between a time mark and the succeeding message pulse with $(T_A/T)t_1$ (in most cases, T_A will be equal to T to measurement accuracy), and adjusts the phase of its local clock to eliminate any difference. When these two comparisons yield no differences, the local clock is properly set.

Note that this procedure is not affected by the identity of the intended receiving station—it could have been *Station B* or *D* instead of *C*. Moreover, the procedure is not affected by satellite motion: the repetition rate of the local clock is always adjusted to match the repetition rate of the returning signal with that of the received time marks; the phase comparison is made between the interval from the time mark to the message pulse and $(T_A/T)t_1$.

The repeater bandwidth must be large enough to keep the signal in one time slot from affecting signals in other time slots. The user stations can transmit message information in any manner they desire, provided only that the message waveforms are suitable for passing through the repeater and do not overlap their assigned time slots. It is important to note that TDMA practically eliminates the need for up-link power coordination. The only way that power coordination can affect system performance is in the event that time sidelobes in the pulse modulation of a strong transmitter interfere with a weak signal in an adjacent channel. For the disparity of transmitter powers anticipated with satellite communication systems, it is expected that the effect of time sidelobes will be negligible.

There are two principal choices to be made in specifying the access channels. First, should all time slots be equal, so that the users in each link will have to adjust their message rate to what can be accommodated in the access channel, or should they be unequal to accommodate specific rates for each link? It is possible to accommodate specific data rates for links with receiving antennas of different sizes by assigning different time slot sizes. Adjusting the size of one time slot will, in general, require adjusting the size of others if waste of transmitted signal power is to be avoided.

The second choice is that of time frame length T. A

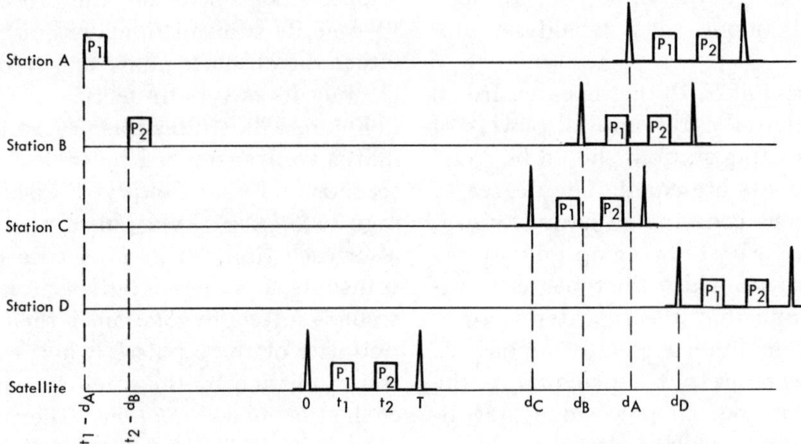

Fig. 3.   Relationships among time scales. The symbol d_x($X = A$, B, C or D) represents
the one-way path delay between *Station x* and the satellite.

great deal of freedom is possible in the choice of this parameter. A simple mode of operation is to choose the frame repetition rate to correspond with the repetition rate of the satellite clock, both rates being as high as the highest sampling frequency required by the message modulation. In this mode each station is allocated one time slot per frame. For real-time voice transmission, the frame rate could be equal to a convenient sampling rate for voice signals, 8 kc/s; ie., $T = 125$ μs. Another mode is a store-and-forward operation in which many message samples are stored and transmitted together in bursts.

Thus TDMA can theoretically provide the highest channel capacity because, in theory, the only inefficiency arises from the guard times between adjacent slots, and these can be made very small relative to the slots themselves. However, TDMA requires that analog message signals such as voice be transmitted in bursts or pulses, and this can necessitate analog-to-digital conversion. In most cases this conversion introduces inefficiencies, so that it can be more efficient, for the overall system, to accept the "inefficiencies" of other multiple-access techniques. Moreover, in some cases stations with very small antennas might be unable to receive their own transmissions or the clock signal. This would make TDMA unworkable.

Constant-Envelope Spread-Spectrum Multiple Access (SSMA)

In SSMA, each multiple-access carrier signal usually occupies the entire repeater bandwidth. An access channel consists of a carrier waveform characterized by a distinct wide-band phase or frequency modulation. Wide-band phase or frequency coding are used to "address" the carrier for the desired receiver. The message-modulation bandwidth is small relative to the RF bandwidth and is inserted into the carrier as an additional phase or frequency variation. The message modulation can take the form of an analog frequency or phase-

modulated signal, or it can take a digital form such as phase or frequency shift keying. The multiple-access modulation can be either analog (for example, a chirp signal) or digital (such as frequency tone hopping or rapid phase keying). The addressed receiving station employs a copy of the multiple-access phase or frequency code to separate the desired signal from the complex signal structure transmitted by the repeater to the user station.

Since greatest consideration to date has been given to digital techniques, an example using rapid phase-shift keying techniques will be described here [13]. Moreover, to simplify the discussion, binary (180°) phase keying will be used to synthesize a constant-envelope spread-spectrum multiple-access carrier. For the most part the descriptive part of the discussion considers a "classical" pseudonoise signaling technique which may be familiar to many readers [14]. For convenience the technique will be described in terms of functional components.

With reference to Fig. 4(A), let us for the moment short-circuit the input to the voltage-controlled oscillator (VCO). Thus a pure sine wave is generated at some convenient IF frequency, say 50 Mc/s. This VCO output is multiplied by a pseudorandom (binary) key stream generator running at a basic frequency one half the desired spread-spectrum bandwidth. The function of the key stream generator is to switch the phase of the VCO output by 180° in accordance with the key stream. The pseudorandom key stream generator can be described as a square-wave generator in which the orderly alternating progression of plus to minus ones is changed to a pseudorandomly alternating progression. The transmitter and the receiver in a given link must generate the same key stream and this must be distinct from the key streams generated in the other links. Thus the key stream generators perform the function of addressing as well as that of spreading the spectrum.

The output from the VCO with the pseudorandom

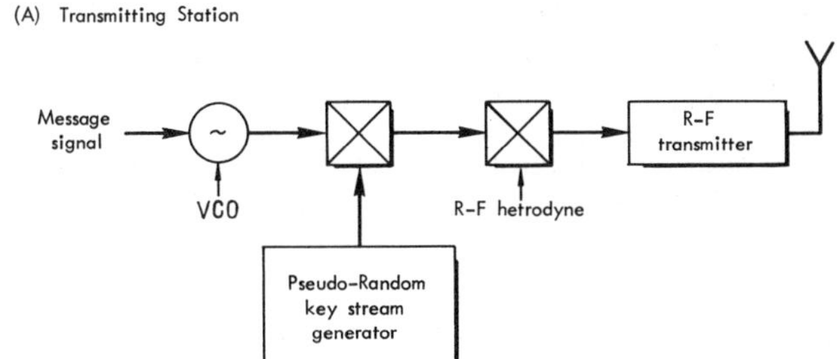

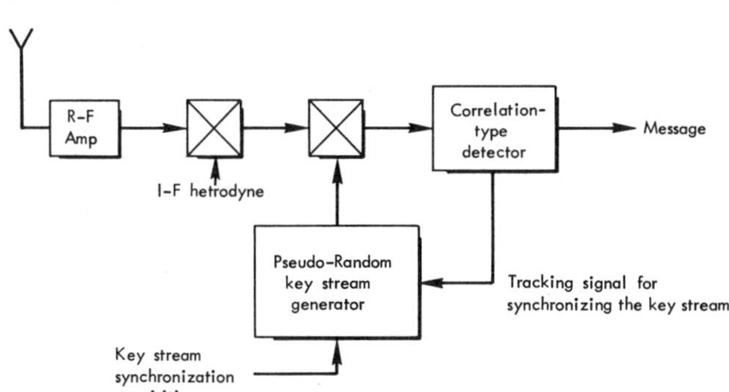

Fig. 4. Functional diagram of a transmitting and receiving station for spread spectrum multiple access. (A) Transmitting station. (B) Receiving station.

square wave is then heterodyned up to the RF frequency, say 8 kMc/s, and radiated to the satellite. Message information is inserted by driving the VCO with a message signal with a predetermined baseband upper frequency of, say, 8 kc/s.

The mathematical description of such a signal as it arrives at the repeater input is given by

$$s(t) = \sqrt{2(P_T G)} \cos\left[\omega_0 t + \Phi(t) + m(t) + \Psi\right]$$

where

$(P_T G) =$ average power in $s(t)$ at the repeater input

$\omega_0 =$ carrier center frequency

$\Phi(t) =$ binary phase coding obtained by multiplying the cosine wave with the key stream generator

$m(t) =$ message-bearing signal; phase or integral of frequency

$\Psi =$ statistically random (but constant in time) RF reference phase associated with path delay and transmitter start-up phase.

The message-bearing angle $m(t)$ may be of the form associated either with more common phase or frequency modulation or that associated with phase-shift keying. In any event, the bandwidth of $m(t)$ is much smaller than that of $\Phi(t)$, typically 8 kc/s, compared to 5 Mc/s for $\Phi(t)$.

Signals from all of the active links arrive simultaneously at the repeater input and occupy the same frequency band W: for example, 5 Mc/s at a center fre-

quency of 8 kMc/s. The repeater sums these inputs, heterodynes them down to IF, and amplifies them with a hard-limiting IF amplifier. The output of the IF amplifier is then heterodyned up to RF, say 7 kMc/s, amplified by a TWT, and transmitted to earth.

An earth receiving station must employ some form of a correlating receiver as shown in Fig. 4(B). After RF amplification and conversion to the IF of the receiver, the composite signal received from the satellite is beat against a pseudorandom square-wave generator whose output is an exact replica of the key stream at the transmitter of the desired signal. (This requires synchronization of the receiver and transmitter.) The resultant is used to drive a correlating message detector. It is the use by the receiver of an exactly matching replica of the desired pseudorandom key stream which allows the receiver to withdraw a useful output from the complicated composite signal radiated by the satellite.

The detector portion of the receiver can take two forms, depending on whether digital or analog messages are to be demodulated. For digital demodulation the detector function is shown in Fig. 5(A). The IF frequency is removed in phase lock with the RF reference phase and the resultant integrated in synchronism with the message digits. The output of the integrating circuit feeds a digital decision circuit which, for simple binary signaling, would be a polarity slicer. Note that the receiver must maintain synchronism by using information from the received message digits. The analog detector shown in Fig. 5(B) is in the form of a phase-locked loop which can be used to detect either phase or frequency modulation. It might also be possible to implement a compressive FM detector.

The fact that the receiver must use an exactly matching replica of the transmitted key stream presents the problem of synchronizing the receiver before message transmission can begin. The receiver's copy of the transmitted pseudorandom square wave must be in very close time synchronization with reception of the desired transmitted wave. A receiver will not know ahead of time when it will be called and, moreover, the propagation delay time over the transmitter-satellite-receiver link may not be known. This circumstance requires an initial synchronization searching procedure by the receiver so that the receiver's local key stream generator can be put in step with the received phase code effected by the transmitter's key stream generator. The link synchronization requirement represents one of the critical aspects of SSMA using pseudonoise carriers. If a 5-Mc/s spread-spectrum bandwidth is used, the receiver local key stream generator must be locked well within 0.2 μs of the epoch of the received signal.

A useful quantity in measuring the performance of communication satellite systems using phased-coded spread-spectrum techniques is the detectability d^2 which can be interpreted as the signal-to-noise ratio at the correlating receiver output. For digital messages,

(A) Detector for Digital Messages

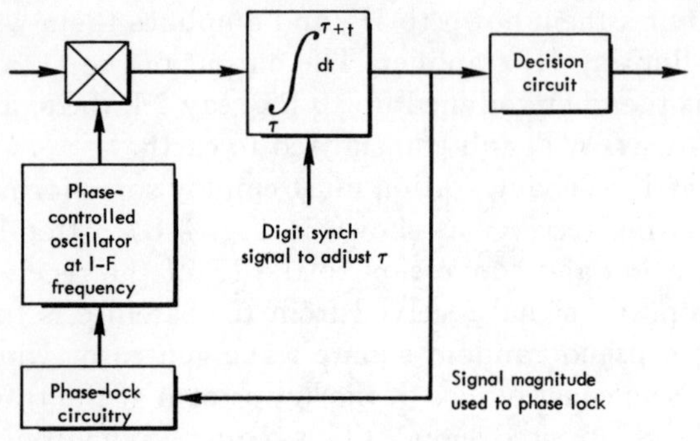

(B) Detector for Analog Messages

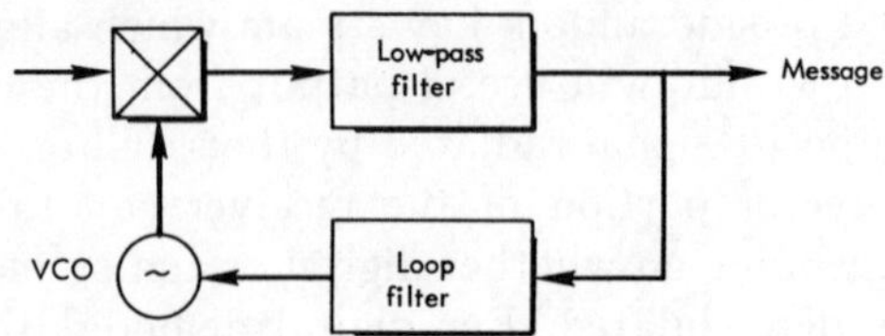

Fig. 5. Detectors for a spread spectrum multiple access receiver. (A) Detector for digital messages. (B) Detector for analog messages.

the detectability can be used to calculate error probabilities; for analog phase or frequency message modulation, the demodulated output quality expressed as a signal-to-noise ratio can be expressed (above threshold) as a factor dependent on the modulation index and the detectability.

A significant feature of spread-spectrum phase-coding techniques is that, due to the beneficial effects of correlation, the receiver outputs behave as if the desired signal were imbedded in Gaussian noise. This effect pertains even though the signals are subjected to the severely nonlinear effects of the hard-limiting repeater. (The presence of the limiting repeater is manifested in the calculation of detectability.) An immediate consequence is that, to a very good approximation, the expression for error probability associated with digital signals in Gaussian noise is applicable, and the output signal-to-noise ratio for analog message modulation can be computed (above threshold) simply as the mod index improvement factor times detectability. For example, if the binary digital message signaling is used, then the bit error probability (assuming equally probable transmitted message bits) is given by

$$P_e = \frac{1}{\sqrt{2\pi}} \int_d^\infty e^{-x^2/2} \, dx$$

where $d^2 =$ detectability. On the other hand, if FM is employed

$$\left(\frac{S}{N}\right)_{\text{out}} = g_{FM} \, d^2$$

where

$g_{FM} = FM$ improvement factor associated with large modulation index FM and phase-lock or frequency-compressive receivers.

For the case of a signal in white Gaussian noise, detectability is given by twice the ratio of received signal power to the product of message bandwidth and noise power density. (The factor of two arises because the receiver is coherent.) This result can be applied if the repeater input is not dominated by a singularly powerful signal and sufficient processing gain is provided. Then the expression for detectability can be obtained in the following manner.

First, assume that the hard-limiting repeater is replaced by an average-power-limited linear repeater, that M signals of equal power impinge on the repeater input, and that each signal has a flat spectral density over the bandwidth W. Then the useful signal power at a user station receiving one of these M signals is P_r/M, where P_r is the power of the composite repeater signal. The effective noise power density is given by the sum of the receiver noise power density N_0 and the power density of the noise generated by the other $M-1$ signals in the composite satellite signal. Since these signals are spread uniformly over the repeater band W, the power density of their sum is

$$\left(\frac{M-1}{M}\right)\left(\frac{P_r}{W}\right),$$

or approximately P_r/W if M is large.

Thus for an ideal average-power-limited linear repeater, the desired signal power received by each user station would be P_r/M and the effective noise power density would be the sum of the receiver noise density N_0 and the noise density in the satellite signal P_r/W. The effect of a hard-limiting repeater is to reduce the useful signal power at each user station by a factor of $\pi/4$ (1 dB). The noise terms are essentially unchanged. Thus the received signal power is $(\pi/4)(P_r/M)$ and the noise power density is N_0+P_r/W. For the case of M equal power signals, detectability is then given by

$$d^2 = \frac{2\left(\frac{1}{b}\right)\frac{\pi}{4}\left(\frac{P_r}{M}\right)}{N_0 + P_r/W}$$

$$= \frac{\pi}{2} \frac{P_r}{MbN_0}\left[\frac{W/W_0}{1 + W/W_0}\right]$$

where

$P_r =$ received power from the repeater
$N_0 =$ user station thermal noise density
$W_0 = P_r/N_0$
$W =$ system bandwidth
$b =$ message bandwidth
$M =$ number of signals, all assumed to be of equal power, present at repeater input.

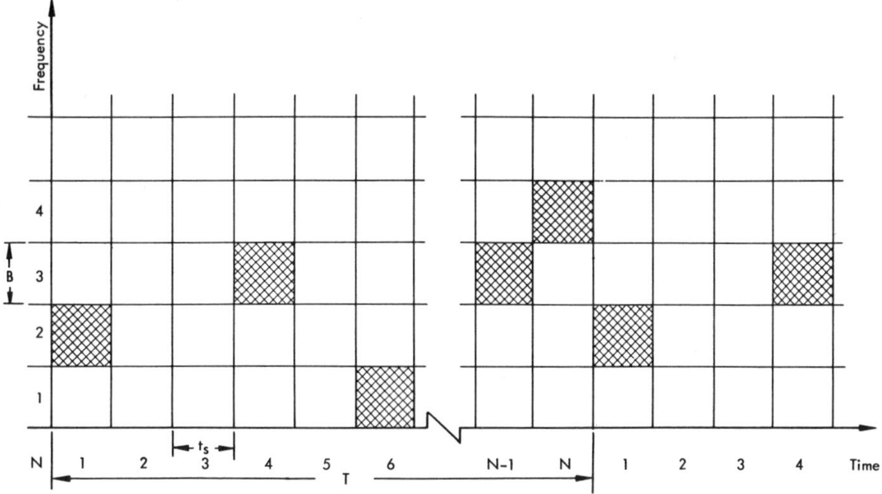

Fig. 6. Time-frequency space in matrix form. The cross-hatched squares illustrate the access channel 2, 0, 0, 3, 0, 1, $\cdots$, 3, 4.

Assuming that all ground stations operating in the spread spectrum mode achieve the same value of $W_0 = P_r/N_0$, the above expression for detectability shows that in order to maximize the multiple-access capability, an intelligent choice for W is a W larger than W_0; however, very much larger values yield marginal improvement, and the importance of the key stream generator synchronization requirements, which provide a practical upper bound on the choice of W, should be kept in mind. Values of W two to four times W_0 seem indicated.

The use of constant-envelope spread-spectrum techniques provides certain desirable features at the expense of requiring link synchronization. The phase or frequency code used to spread the spectrum of the multiple-access carrier also serves to address the transmitted carrier to the desired receiver automatically. In general, owing to the length of the codes used, an extremely large number of distinctly different addresses are available, many more than the number of active links a satellite repeater can support. The phase-coded spread-spectrum multiple-access carrier also has the desirable effect, with one important exception, of making the actual hard-limiting repeater appear, with regard to signal-to-noise ratio calculations, the equivalent of an ideal average-power-limited linear repeater. The one important exception to this feature occurs when one signal at the satellite repeater input dominates the sum of all the other repeater input signals. The strong signal then "captures" the limiter, suppressing the other signals by a factor of four in addition to the signal ratio at the repeater. The capture phenomenon, however, is a characteristic of constant-envelope signals and not just peculiar to spread-spectrum techniques; that is to say, FDMA also suffers capture and, to a lesser degree, PAMA can be degraded by capture.

Pulse Address Multiple Access (PAMA)

PAMA techniques are many and varied, giving rise to systems with diverse properties. A technique can be

devised with almost any single desirable property.

A pulse-address carrier waveform is a series of pulses within specific frequency bands. The pulses and the intervals between the pulses in a given carrier waveform constitute a distinct pattern. The duty factor of each carrier is small; i.e., the amount of time during which pulse is present is much less than the amount of time no pulse is present. In typical systems of this class, the same pattern is used to carry each sample of a message [15]–[18].

Each waveform can be specified by dividing the time-frequency plane into a matrix as shown in Fig. 6. The time axis is divided into frames of length T, each of which is divided into N slots of length t_s. The frequency spectrum is divided into numbered bands of width B. One message sample is transmitted in T seconds, the minimum pulse length is t_s, and B must be at least as great as the bandwidth of a pulse. An access channel specifies a distinct pattern of pulses in time and frequency, and is specified itself by a sequence of numbers. For example, the access channel 2, 0, 0, 3, 0, 1, $\cdots$, 3, 4 is illustrated in Fig. 6.[2]

If there is no network timing, the time scale cannot be marked off in precise slots and frames. Then an access channel specifies the relative positions of pulses in a carrier waveform. For example, the access channel 2, 0, 0, 3, 0, $\cdots$, 3, 4 specifies a carrier waveform in which a pulse in frequency band 2 is followed after $2t_s$ seconds by a pulse in frequency band 3 and that is followed after t_s seconds by a pulse in frequency band 1, etc. Note that if a carrier repeats the same pattern in every frame, the waveform with the pattern 2, 0, 0, 3, 0, 1, $\cdots$, 3, 4 is not distinct from one with the pattern 0, 0, 3, 0, 1, $\cdots$, 3, 4, 2 nor from one with any other cyclic shift of the basic pattern.

---

[2] In some techniques, the pulses themselves contain multiple-access modulation; then an access channel consists of more than a specification of pulse positions in the time-frequency matrix. For instance, spread-spectrum pulses might be used. Such techniques will be discussed presently.

Message information can be sent in various ways. Using pulse-position modulation (PPM) the pattern 2, 0, 0, 3, 0, 1, $\cdots$, 3, 4 could convey "mark," and the pattern 4, 2, 0, 0, 3, 0, 1, $\cdots$, 3 could convey "space." Or using frequency-shift keying (FSK), the pattern 2, 0, 0, 3, 0, 1, $\cdots$, 3, 4 could convey "mark," and the pattern 3, 0, 0, 4, 0, 2, $\cdots$, 4, 5 could convey "space." By extensions of these procedures, either PPM or FSK could be used to convey multilevel digital data or analog data. Digital information can also be conveyed by using each of a number of different patterns to specify a sample value, including on-off signaling for binary data. There are many other possible forms of message modulation.

Although PAMA systems can operate without network timing, they perform better with it. Consider the problem of specifying waveform patterns. This can be treated as a coding problem. Each sequence specifying a waveform is a code word, and the code words together comprise a code. If there is no system clock, the waveform specified by the code word 2, 0, 0, 3, 0, 1, $\cdots$, 3, 4 cannot be distinguished from a waveform specified by 0, 0, 3, 0, 1, $\cdots$, 3, 4, 2 or any other cyclic shift of the basic sequence. Hence, if the network is not synchronized, each waveform can represent any one of N code words—a basic word and its $N-1$ cyclic shifts. If transmissions are timed relative to a system clock, each cyclic shift can be used to specify a different waveshape so that the number of assignable waveshapes is increased. Moreover, when transmissions are timed relative to a system clock, it is possible to reduce mutual interference by using only some of the cyclic shifts of a basic code word [19], [20].

A link between two users is established in the following manner. Each receiver is "tuned" to respond to a waveform with a particular pattern. When the waveform is acquired, a confirmation is transmitted on another waveform selected for answering. The answering waveform is selected according to operational doctrine. For example, it might be a second waveform assigned to the called station for the purpose of answering, or it might be the waveform assigned to the calling user who identifies himself in his calling signal.

PAMA systems are of interest because they allow a link between user pairs to be established in this convenient manner. They provide a large number of waveforms to be permanently assigned to individual users so that the waveshapes themselves serve as addresses. In most pulse-address systems, transmitter powers do not have to be closely controlled and transmissions do not have to be timed relative to a system clock. The procedure for locking a receiver to an incoming signal is rapid and requires a minimum of equipment. Because of the convenient access they provide, some pulse-address systems are known as random-access discrete-address (RADA) systems. Here, the term *fixed-address system* is preferred—denoting a system in which carrier waveshapes are assigned permanently to particular users.

Fixed-address systems provide a means for rapid and simple access to a network operating below full capacity. However, when the network is overloaded, users will not naturally follow procedures that will yield the best network performance. An operational doctrine is needed to ensure proper operation when the network reaches full loading.

PAMA systems take so many different forms that it is difficult to list advantages and disadvantages valid for all of them. One disadvantage they all seem to share is low data rate. However, a PAMA system can be designed with almost any other single desirable property, e.g., operation without network timing, operation without up-link power coordination, operation with simple ground equipment, resistance to intentional interference, ability to serve a small station, etc. An exception to the general property of low data rate for PAMA systems arises when noise or interference at the input to the repeater is appreciable. Then the carriers with short, high-peak-power pulses have an advantage, relative to constant-envelope carriers, in combating up-link noise.

TDMA is a special kind of pulse technique in which each time slot is assigned exclusively to one access channel. This is possible only with network timing, and it limits the number of access channels that can be assigned on a permanent basis. In PAMA systems, no time slot is assigned exclusively to a single access channel. The waveforms are distinct because each is a unique pattern of pulses; the patterns can be identified at the receiver even though some pulses of one waveform overlap pulses of other waveforms. This allows a PAMA system to operate without network timing and a large number of distinct access channels to be assigned to users, although all access channels cannot be used at one time.

The relatively low data rates attained by PAMA techniques arise from two causes: interference among signals transmitted to the satellite and unused portions of the time-frequency plane. Moreoever, there is a tradeoff between these two factors. A PAMA system which serves as a convenient example has been analyzed by Reiffen [2]. In this system, as in Fig. 6, the time frame is divided into N time slots of length t_s, and the frequency spectrum is divided into bands of width B. In particular, $t_s = 1/B$, and B is the pulse bandwidth; B is chosen to be much less than W, the total repeater bandwidth.

The repeater transmits at a constant power level so that a fixed energy E is transmitted in each time slot. If no pulse from any link is present in a time slot, the input to the limiter is thermal noise and the output energy is spread over the full repeater bandwidth W. The output energy in any band of width B is then EB/W. If a single pulse is present in a time slot and saturates the limiter, the full output energy E in that time slot is confined to the pulse band of width B. A receiver can decide whether a pulse is present in a time slot by measuring the energy received in the pulse band during the time slot.

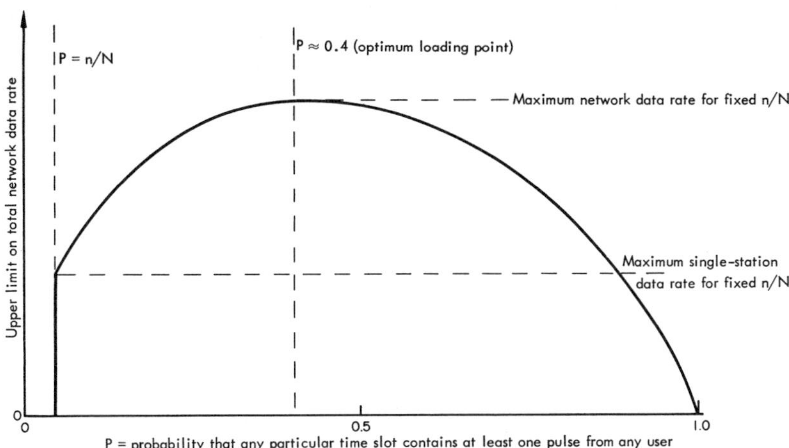

Fig. 7.   Upper limit on total network data rate vs. probability that any one time slot contains at least one pulse from any user (assuming fixed n/N and for the particular PAMA system described).

In this example, all pulses are transmitted at the same center frequency. To transmit binary information, each simplex link is assigned two waveforms: one to transmit *mark* and the other to transmit *space*. Since all signal pulses are transmitted in the same frequency band, the code words specifying the waveforms are binary sequences in which *one* indicates *pulse present*, and *zero* indicates *pulse absent*. Each waveform consists of n pulses per frame so that each generating sequence contains n ones. The duty factor of each wave form is n/N.

The receiver is incoherent. It compares the energies in the first *pulse present* slot of each of its two assigned waveforms, and decides that a pulse is present in the one with the larger energy. It repeats this operation n times, once for each time slot, and by majority rule decides whether *mark* or *space* was sent. Because all transmitters send pulses in the same frequency band, a pulse in a particular time slot looks the same regardless of which transmitter sent it. Therefore, other transmitters are assumed to interfere with a given receiver's operation only by generating false pulses. (This assumption is not strictly valid since overlapping pulses in the same frequency band will generate intermodulation noise which reduces the power in the signal band.)

The highest network data rate is obtained when one user transmits with unity duty factor; i.e., $n/N=1$ (with an appropriate form of message modulation). However, to allow multiple access, the duty factor of each carrier waveform must be small; i.e., $n/N \ll 1$. If only one user is transmitting at a given time, and $n/N \ll 1$, the total network data rate is his data rate, and it is limited to n/N times the data rate he could have if he were transmitting with unity duty factor. At first, the upper limit on network data rate increases if additional transmitters become active, even though the capacity of each link decreases due to mutual interference among pulses; but as more transmitters become active, the interference among users builds up until a

point is reached where additional users would actually reduce the network data rate which can be obtained. Thus a plot of the upper limit on network data rate against the probability in each time slot that at least one pulse is present resembles the plot in Fig. 7. Note that this assumes that each carrier maintains a fixed duty factor of n/N, and the only way to fill more slots is to increase the number of active users.

Similar curves are obtained for other PAMA systems by plotting the upper limit on network data rate against the probability in each element of the time-frequency matrix that a pulse is present. In each case there is a loading point which yields a maximum data rate for fixed n/N. If this loading point is exceeded, the interference is excessive; if it is not reached, the repeater power is being used inefficiently.

With the assumption that $n/N \ll 1$, it turns out that a value of n/N exists to maximize network data rate at the optimum loading point for a particular error rate. However, the network data rate at this point is not very sensitive to changes in n/N, and $(n/N)_{optimum} \approx 1/(2M-1)$, where M is the number of active transmitters (see Cahn [2]). For the system described, the maximum network data rate at this point is 15 dB below the rate possible if a single transmitter sends binary data with a unity-duty-factor carrier. Thus, at best, the example system with $b/N \ll 1$ can transmit at only three percent of the data rate possible with a single transmitter operating at unity-duty factor (or with TDMA assuming negligible guard times).

This analysis assumes that pulses are uniformly distributed in the time frame; i.e., that the probability of "pulse present" is the same in each time slot. This assumption might not be valid, and must be modified if the same patterns are used by each link on each successive frame giving rise to repetitive errors. The analysis also assumes that there is no partial overlap of phases. (Partial overlaps, which will occur if there is no network timing, result in no more than 3 dB additional degradation.) Moreover, at the optimum loading point, the probability of error in each pulse position is high (approximately 25 percent). Coding equipment at the ground stations would be required if the maximum data rate is to be obtained with a suitably low error rate.

The system described requires only simple transmitting and receiving equipment, can operate without network timing, and needs no up-link power control because overlapping signals do not suppress each other. It might have application for a network in which the number of potential users is large, but in which the number of active users at any one time is small and the data sources have low data rates (e.g., teletype). For such a network, convenient access might be most important. Apart from any operational utility, this system serves a function for analysis because a simple and reasonable model for the interference among signals is available.

Other possibilities for PAMA systems can be considered as modifications of this basic system. One modification would assign only one waveform to each link; the message information is conveyed by the center frequency of the signal. All stations transmit *mark* in one frequency band and *space* in another. The interference among overlapping pulses is no longer so easily characterized because pulses overlapping in time can destructively interfere with each other; hence, up-link power disparities can result in the suppression of weak signals. Since this system uses more elements of the time-frequency matrix than does the first system, a greater number of time slots will probably be filled at the optimum loading point. For this reason, the maximum data rate of this system will probably be higher than that of the first system.

This second system can be modified slightly by having different links use different center frequencies. Another variation would have a waveform consisting of pulses in different frequency bands such as in Fig. 6. Unlike the previous examples, these systems reject unwanted signals by frequency filtering as well as time gating. The interference among pulses overlapping in time is more difficult to model in these systems than in the first two examples. They could give somewhat higher maximum data rates than the first examples because they will have more time slots filled at their optimum loading points. They will suffer from effects of intermodulation noise and any lack of up-link power coordination, and might require somewhat more complicated equipment.

Any of the examples given for signal structure in an incoherent PAMA system can be considered for coherent systems. However, maintaining coherence over a pattern of pulses jumping in frequency might not be feasible in practice. Coherent receivers may enhance the data rate by as much as 6 dB relative to an incoherent receiver with the same input-signal-to-noise ratio.

Coherent detection also offers the possibility of another class of systems: pulsed spread-spectrum systems using phase-coded signals with duty factors less than unity [21]. (There are a number of distinct kinds of pulsed spread-spectrum systems. In one, the same pattern of phase reversals is repeated for each pulse and each pulse carries one message sample. In another, a phase-coded spread-spectrum signal with a long, essentially nonrepetitive pattern of phase reversals is time-gated. Similarly, chirp signals can be used.) A time-gated, phase-coded, spread-spectrum system probably has the highest maximum data rate of any PAMA system because it is coherent and employs the spread-spectrum signal to reduce interference among overlapping pulses.

When a pulsed spread-spectrum system is considered, however, the equipment required for the system might be even more complex than for constant-envelope spread-spectrum systems. The motivation to consider on-off gating a spread-spectrum signal is the limiter suppression effect. One concern with constant-envelope systems is that disparity in up-link powers of the various signals might be so extreme that a few strong signals will usurp all the down-link power. This could be countered by using signals with duty factors less than unity.

Pulsed signals are not always an advantage, however, in the presence of disparate up-link powers. Consider the presence of a single strong signal whose up-link power is somewhat less than the combined power of all other users. With constant-envelope signals, that signal will not capture the limiter. If pulsed signals are used, the strong signal, when it is present, will compete with less than the combined power of all other users and will capture the limiter at times. On the other hand, if a single signal has so much up-link power that it captures the limiter even when all other signals are present, pulsed signals yield an advantage. With pulsed signals there will be times when the strong signal is not present, and the other signals present at those times will not be suppressed.

PAMA techniques appear to be most applicable for networks composed of small user stations with relatively low data rate requirements. Preferably, each of these stations will have a short message to transmit at a few widely spaced times. Then PAMA techniques can be especially useful in accommodating networks composed of such stations which have limited transmitter power or limited capability for coordinating their transmitted powers, and which are unable to achieve the network synchronization required for TDMA.

Design Considerations

The wide-band hard-limiting repeater permits a wide choice of modulation systems suitable for multiple access. Characteristics, advantages, and disadvantages of each of the four classes of techniques are listed in Table I. It is clear that a single class will not be appropriate for all applications, and within each class there are many alternatives in choosing a system. Moreover, a possibility which has not been considered here, but one that could be of importance, is sharing the repeater

TABLE I

SUMMARY OF THE CHARACTERISTICS, ADVANTAGES, AND DISADVANTAGES OF THE MULTIPLE-ACCESS TECHNIQUES

	Characteristics	Advantages	Disadvantages
FDMA	Signals have constant envelopes and have spectra confined to nonoverlapping frequency bands (at the repeater input) constituting the access channels. Multiple-access demodulation is accomplished by a filter tuned to the selected frequency band. Compressive techniques might be required to combat intermodulation noise generated by hard-limiting. Message information can be conveyed by any form of angle modulation.	FDMA makes use of existing technology and hardware to a greater extent than the other techniques. Network timing is not required and repeater loading can be determined by relatively simple procedures.	More than one signal present in the repeater at one time produces intermodulation noise which reduces the usable repeater output power. Problems which may arise from different power spectral densities in different portions of the repeater band can make it difficult to operate with disparate receiving sensitivities at the various user stations. Up-link power coordination is required to make full use of the repeater capacity.
TDMA	Carriers are on-off gated in such a way that signals from different links are never present in the repeater at any one instant, but pass through in sequence. Multiple-access demodulation is accomplished by time-gating. Message information can be conveyed by any form of angle modulation within each carrier burst.	This technique avoids the problems of mutual interference among signals passing through a limiter simultaneously. The highest capacity, in theory, can be obtained for a given satellite power due to the absence of intermodulation noise. User stations with different receiving sensitivities or with disparate transmitting powers can be accommodated more readily than when constant-envelope carriers are used. Up-link power need not be coordinated, and each user station can use the highest peak power available to it to counteract interference in the up-link without degrading the other links thereby.	Network timing is required. Analog message information often must be converted to digital form.
SSMA	Carriers have constant envelopes and the spectrum of each extends over most, if not all, of the repeater band. The spectrum of a carrier is spread by angle modulation giving rise to a selected pattern of phase or frequency shifts. Multiple-access demodulation is accomplished by synchronizing the receiver to the desired carrier and duplicating the pattern of phase or frequency shifts. Message information can be conveyed by any form of angle modulation.	The spread spectrum of the carriers provide a processing gain to counteract interference in the up-link. For carriers with wide-band phase codings, the hard-limiting repeater can be made to appear similar to an ideal AGC transponder. The essentially unlimited number of distinct carrier waveforms makes it possible to make fixed address assignments. The network can operate without a central timing source and with a minimum of discipline when the repeater is being used to less than full capacity.	Each carrier occupies a wide bandwidth. Link synchronization is required, and can introduce considerable difficulty. Sophisticated methods are required to determine repeater loading. Up-link power coordination is required to make full use of the repeater capacity.
PAMA	Carriers are low duty cycle pulsed waveforms. A particular carrier is distinguished by the time pattern of the pulses and possibly by their spectra or by phase modulation within the pulses. Multiple-access demodulation is accomplished by synchronizing the receiver to the desired carrier and time-gating, and possibly by frequency filtering or phase demodulation. Message information can be conveyed by a variety of methods.	Pulse waveforms can make use of high peak power to combat up-link noise. The large number of waveforms makes it possible to make fixed address assignments. Network timing is not required and link synchronization is rapid. The network can operate with a minimum of discipline when the repeater is being used to less than full capacity.	Information capacity is relatively low and coding is required to achieve tolerable error rates for maximum data rates.

between one network of users employing one modulation technique, say TDMA, and another network using a different technique, say SSMA.

The remainder of this paper discusses three design considerations: repeater bandwidth and power sharing, network timing, and operational doctrine.

Repeater Bandwidth and Power Sharing

An important problem in the design of a multiple access communication satellite system is that of ensuring compatibility between the repeater and user stations with widely differing antenna aperture, transmitter power, and receiver noise temperatures. The differing user-station capabilities affect the system design in two principal ways. The first is the choice of a repeater RF bandwidth W; the second is the choice of multiple-access modulation and system organization.

Large user stations can radiate a considerable amount of power so that the signal power at the repeater input is large relative to the repeater input thermal noise. This allows the repeater to have a large RF bandwidth and still preserve good up-link signal-to-noise ratio. Large RF bandwidths are desirable for high data rate and good separation of spread-spectrum signals if these are used for multiple access to the repeater. On the other hand, a repeater with a large RF bandwidth may provide poor up-link signal-to-noise ratios for the small user stations. The choice of repeater bandwidth is influenced by the signal-to-noise ratio required at the limiter input.

A network with a mix of user-station capabilities presents problems arising from the unequal apportionment of repeater output power among the active communication links. It is necessary that the relationships between data rate, number of active users, and disparity in user-station capability be established for the various modulation techniques providing multiple access to a hard-limiting communication satellite. The behavior of hard limiters, especially limiter capture by one singularly powerful signal, is of critical importance.

Six factors affecting the design of a communication satellite system with hard-limiting satellite repeaters are as follows:

1) variety of received signal strength at the repeater input
2) thermal noise at the repeater input
3) repeater output power sharing
4) signal interaction in the repeater
5) variety of signal strengths at the user-station receivers
6) thermal noise at the user stations.

Network Timing

Network synchronization can be attained in a network with a communication relay because the time scale at the satellite can be used as a common reference. The procedure described in the discussion of TDMA illustrates a general procedure for setting local clocks so that transmitted signals arrive at the satellite in proper frequency and phase to be transmitted by the satellite in synchronism with a satellite clock signal. The procedure described employed pulses for the message signal and the clock signal; however, either or both signals can have a constant envelope. Accuracy of 0.1 μs, and possibly less, can be obtained.

Network timing can enhance performance with any of the multiple access techniques, with the possible exception of FDMA. TDMA requires it. Advantages for PAMA have been discussed. The critical problem of link synchronization in SSMA can be reduced substantially by timing transmitted signals so that they arrive at the receiving stations in synchronism with the satellite clock signal.

The clock signal can be transmitted from a ground station and be repeated by the satellite in the same manner as any other signal. However, advantages weigh clearly in favor of putting the clock in the satellite, and transmitting the clock signal on a beacon signal which is distinct from the signal coming from the repeater. The principal advantages of locating the clock in the satellite are simplified operating procedures and immunity of the clock signals to interference from signals transmitted to the satellite.

Operational Doctrine

Evaluation of a particular modulation system is not complete until the operational doctrine imposed on the user station has been considered. Conversely, a system designer should determine to what extent desirable communication practices will influence the choice of multiple-access modulation systems.

Access channels may be assigned on a temporary basis for use when needed, *floating address*, or they may be assigned on a permanent basis, *fixed address*. PAMA and SSMA are suited for fixed-address systems because each of the large number of possible carrier waveforms can be considered an address. The number of carrier waveforms, or channels, available in TDMA and FDMA are more restricted, so these modulation techniques are generally more suitable for a floating-address system.

Fixed-address systems, sometimes called *random-access* systems, provide a convenient means of access when the network is operating below full capacity; however, procedures for determining system loading are required. When the system approaches full loading, users will not naturally follow procedures which yield best system performance. In most multiple-access modulation techniques, which allow fixed address systems, the capacity of an access channel is affected by the up-link power used. Thus when the system approaches full loading, power struggles can result; if so, the weaker transmitters will lose. This is not necessarily desirable, so doctrine is required to establish conditions under which links must be dissolved, and how much time must pass before they are re-established.

A doctrine for the operation of military communication satellite systems should specify at least the following: procedures for obtaining an access channel, operations required before transmitting, calling and answering procedures, handling traffic according to priorities, traffic volume concealment, and procedures required in the presence of jamming.

Appendix

Sample Power Budget Calculations

The quotient P_r/N_0 (in cycles per second) relates the signal power and the thermal noise spectral density at the input to a ground station receiver.

The symbol N_0 denotes the spectral density of the thermal noise in the band of the ground station receiver. Its value is computed from the relation

$$N_0 = kT_r, \text{ watts}$$

where

k = Boltzmann's constant, watt-sec/cycle-°K

and

T_r = receiver noise figure, °K.

The symbol P_r denotes the signal power received from the satellite at a given ground station. It is computed from the relation

$$P_r = \frac{P_t G_t G_r \lambda^2 L}{(4\pi D)^2}, \text{ watts}$$

TABLE II

DOWN-LINK POWER BUDGET SUMMARY

Station size	Antenna diameter, ft.	Receiver noise temperature, °K	Communication capacity quotient, P_r/N_0, c/s
Large	60	80	10^8
Small	30	200	10^7
Very small	4	400	10^5

TABLE III

DOWN-LINK POWER BUDGET CALCULATIONS

	Symbol	Units*	Station Size		
			Large	Small	Very Small
Transmitter power (4 watts)	P_t	dBW	+ 6	+ 6	+ 6
Transmitter antenna gain	G_t	dB	0	0	0
Path loss	$(\lambda/4\pi D)^2$	dB	−193	−193	−193
Other losses	L	dB	− 3	− 3	− 3
Receiver antenna gain	G_r	dB	+ 60	+ 54	+ 37
Received power	P_r	dBW	−130	−136	−153
Noise power density	$N_0 = kT_r$	dB(W/c/s)	−210	−206	−203
Capacity quotient	P_r/N_0	dB(c/s)	80	70	50

* dBW denotes decibels with respect to a reference of one watt. dB(c/s) denotes decibels with respect to a reference of one cycle per second. dB(W/c/s) denotes decibels with respect to a reference of one watt per cycle per second.

where

P_t = power transmitted by the satellite repeater, watts

G_t = satellite antenna gain

G_r = receiver antenna gain

λ = wavelength (≈ 3.7 cm, corresponding to a frequency of 8 kMc/s)

L = miscellaneous losses (other than path loss)

D = distance between satellite and receivers.

As an example, let us calculate the value of P_r/N_0 for three receiving-station configurations and a particular satellite configuration. The satellite is in polar orbit at an altitude of 6000 nmi. The repeater transmits 4 watts of RF power into an antenna system whose effective gain in the direction of the user stations is 0 dB (isotropic). Miscellaneous losses are assumed to occur in the downlink between the satellite antenna and the user station receiver. These losses can arise from rainfall, polarization mismatch, and tracking inaccuracies. They are lumped together and are taken to have a value of −3 dB. The path loss is computed for the maximum slant range of 8000 nmi and a frequency of 8 kMc/s. The three receiving station configurations and the corresponding values of P_r/N_0 are shown in Table II. The calculations are shown in Table III. The illustration in Table III is only one example of a satellite system.

The effect of the up-link signal-to-noise ratio (at the satellite) on the final received signal-to-noise ratio is usually negligible, provided the input signal-to-noise ratio at the repeater is greater than about 6 dB. Thus up-link noise can sometimes be ignored; however, for a repeater having a very wide band, up-link noise might have to be considered. Calculation of the signal-to-noise ratio at the satellite repeater input can be made in a manner similar to the calculation for the down-link. To obtain the up-link signal-to-noise ratio (dimensionless), the quotient relating the up-link signal power to the input thermal noise spectral density at the repeater input divided by the repeater bandwidth.

ACKNOWLEDGMENT

This paper is a result of a continuing study of communication satellites by the Institute for Defense Analyses which began in 1958. The multiple-access problem in particular has been the subject of much attention at IDA since 1960. Invaluable contributions were provided by the IDA Multiple-Access Summer Study of July, 1964. Participants in that study, in addition to the authors, were H. L. Blasbalg, IBM Federal Systems Division; W. E. Bradley, IDA; C. R. Cahn, Magnavox Research; W. Doyle, private consultant; B. Elspas, Stanford Research Institute; M. J. E. Golay, Perkin Elmer; I. Jacobs, Bell Telephone Laboratories; J. G. Lawton, Cornell Aeronautical Laboratory; B. Reiffen, MIT Lincoln Laboratory; J. Salz, Bell Telephone Laboratories; G. R. Welti, Deco Electronics; and J. K. Wolf, New York University. The papers written at the summer study are available [2].

The authors are especially pleased to acknowledge the encouragement and contributions of W. E. Bradley in the preparation of this paper.

REFERENCES

[1] J. Kaiser, J. W. Schwartz, and J. M. Aein, "Multiple access to a communication satellite with a hard-limiting repeater: modulation techniques and their applications," Inst. for Defense Analyses, Arlington, Va., IDA rept. R-108, DDC AD457945 category I, January 1965.

[2] J. M. Aein and J. W. Schwartz, Eds., "Multiple access to a communication satellite with a hard-limiting repeater: proceedings of the multiple access summer study," Inst. for Defense Analyses, Arlington, Va., vol. II, IDA rept. R-108, DDC AD465789 category I, April 1965.

[3] E. Bedrosian, N. Feldman, G. Northrop, and W. Sollfrey, "Multiple access techniques for communcation satellite: I. Survey of the problem," RAND RM-4298, September 1964.

[4] W. B. Davenport, Jr., "Signal-to-noise ratios in band-pass limiters," *J. Appl. Phys.*, vol. 24, pp. 720–727, June 1953.

[5] C. R. Cahn, "Crosstalk due to finite limiting of frequency-multiplexed signals," *Proc. IRE*, vol. 48, pp. 53–59, January 1960.

[6] J. J. Jones, "Hard-limiting of two signals in random noise," *IEEE Trans. on Information Theory*, vol. IT-9, pp. 34–42, January 1963.

[7] J. L. Sevy, "The effect of multiple CW and FM signals passed through a hard limiter or TWT," Aerospace Corp., El Segundo, Calif., rept. ATM-63(3111)-1, March 1963.

[8] W. Doyle, "Crosstalk of frequency-multiplexed signals in saturating amplifiers," The RAND Corp., Santa Monica, Calif., RM-3576-NASA, April 1963.

[9] P. D. Shaft, "Limiting of several signals and its effect on communication system performance," *IEEE Trans. on Communication Theory*, vol. COM-13, pp. 504–512, December 1965.

[10] W. C. Babcock, "Intermodulation interference on radio systems," *Bell Sys. Tech. J.*, vol. 32, pp. 63–73, January 1953.

[11] R. M. Hultberg, F. H. Jean, and M. E. Jones, "Time division access for military communications satellites," *IEEE Trans. on Aerospace and Electronic Systems*, vol. AES-1, pp. 272–282, December 1965.

[12] J. W. Schwartz, "Network timing for communication satellites," Inst. for Defense Analyses, Arlington, Va., IDA research paper P-159, DDC AD456159 category I, November 1964.

[13] J. M. Aein, "Multiple access to a hard-limiting communication-satellite repeater," *IEEE Trans. on Space Electronics and Telemetry*, vol. SET-10, pp. 159–167, December 1964.

[14] S. W. Golomb et al., *Digital Communications with Space Applications*, Information Theory Ser. Englewood Cliffs, N. J.: Prentice Hall, 1964.

[15] J. E. Taylor, "Asynchronous multiplexing," *AIEE Trans. (Communications and Electronics)*, vol. 78, pp. 1054–1062, January 1960.

[16] J. R. Pierce and A. L. Hopper, "Nonsynchronous time division with holding and with random sampling," *Proc. IRE*, vol. 40, pp. 1079–1088, September 1952.

[17] W. D. White, "Theoretical aspects of asynchronous multiplexing," *Proc. IRE*, vol. 38, pp. 270–275, March 1950.

[18] M. R. Schroeder, U. S. Patent 3160711, December 8, 1964.

[19] E. N. Gilbert, "Cyclically permutable error-correcting codes," *IEEE Trans. on Information Theory*, vol. IT-9, pp. 175–182, July 1963.

[20] J. W. Schwartz, "A note on asynchronous multiplexing," *IEEE Trans. on Information Theory*, to be published.

[21] H. Blasbalg, "A comparison of pseudo-noise and conventional modulation for multiple-access satellite communications," *IBM J. Res. Dev.*, vol. 9, July 1965.

BIBLIOGRAPHY

[1] J. M. Aein and W. Doyle, "A note on cascaded limiters," *IEEE Trans. on Space Electronics and Telemetry*, (Correspondence), vol. SET-11, pp. 47–49, March 1965.

[2] F. Assadourian, "Intermodulation distortion and efficiency analysis of multicarrier repeaters," *IRE Trans. on Communications Systems*, vol. CS-8, pp. 68–71, March 1960.

[3] W. B. Davenport, Jr., and W. L. Root, *Random Signals and Noise*, Lincoln Lab. Publication. New York: McGraw-Hill, 1958.

[4] W. Doyle, "Band-pass limiters and random walks," *IRE Trans. on Information Theory (Correspondence)*, vol. IT-8, pp. 380–381, October 1962.

[5] ——, "Elementary derivation for band-pass limiter S/N," *IRE Trans. on Information Theory (Correspondence)*, vol. IT-8, p. 259, April 1962.

[6] S. F. George and J. W. Wood, "Ideal limiting," pts. I and II, U. S. Naval Research Lab., Washington, D. C., October 1961.

[7] I. Jacobs, "The effects of video clipping on the performance of an active satellite PSK communication system," *IEEE Trans. on Communication Technology*, vol. COM-13, pp. 195–201, June 1965.

[8] W. J. Judge, "Multiplexing using quasiorthogonal binary functions," *AIEE Trans. (Communications and Electronics)*, vol. 81, pp. 81–83, May 1962.

[9] B. E. Love, "Intermodulations effects in FM and PM systems," *AIEE Trans. (Communications and Electronics)*, vol. 79, pp. 245–255, July 1960.

[10] B. Reiffen and H. Sherman, "Parametric analysis of jammed active satellite links," *IEEE Trans. on Communication Systems*, vol. CS-12, pp. 102–103, March 1964.

[11] J. A. Stewart and E. A. Huber, "Comparison of modulation methods for multiple access satellite communication systems," *Proc. IEE (London)*, vol. III, pp. 534–546, 1964.

Subsection 3.6.2
Frequency Division Multiple Access

FREQUENCY DIVISION MULTIPLE ACCESS (FDMA) FOR SATELLITE COMMUNICATION SYSTEMS

Jack L. Dicks and Martin P. Brown, Jr.*
Communications Satellite Corporation
Washington, D.C.

Abstract

Many methods of multiple access have been proposed for satellite communications; however, the primary mode found in existing systems such as INTELSAT, TELESAT, DCSC, and Molniya/Orbita is Frequency Division Multiple Access (FDMA). This paper discusses the basic engineering fundamentals of FDMA, its attributes and problems, and specific examples of its application. Both multichannel and single-channel-per-carrier (SCPC) systems are described. Comments on future trends are also provided.

Introduction

To date the development of satellite communications has been related to their use in achieving interconnection between a number of earth stations which are either separated by vast distances or displaced by geographical barriers which inhibit the construction of terrestrial radio relay systems. In this type of configuration, the satellite repeater is the nodal point where the problem of "multiple access" between earth stations requires a high degree of flexibility while minimizing the loss of repeater channel capacity. Terrestrial radio relay systems do not have this added complexity since they normally have a single point-to-point access.

The first satellites brought into service in the mid-1960's were not operated in a multiple access mode due to spacecraft power limitations. Later generations, however, have relied heavily on FDMA for its simplicity, proven technology, low cost, and easy adaptation to varying international networks. The advantages of FDMA have been apparent in the years where traffic demand on available satellite power and allocated bandwidth did not restrict system design to the extent that more advanced forms of multiple access were warranted. As bandwidth becomes increasingly precious, more complex and costly techniques will be introduced to make more efficient use of this commodity. Figure 1 illustrates this dilemma with a plot of available power, allocated bandwidth, and, as an example, INTELSAT traffic growth against time.

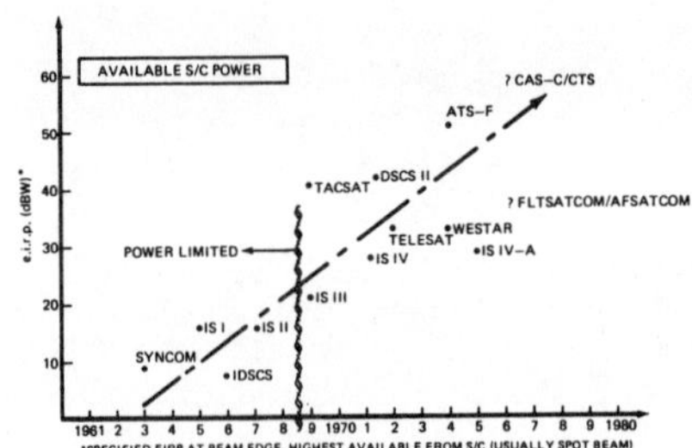

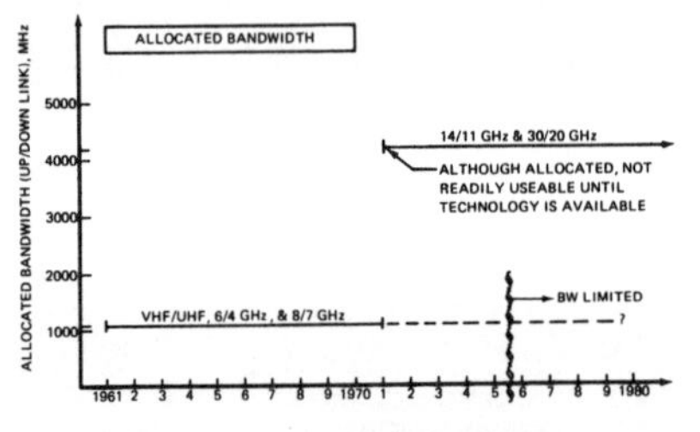

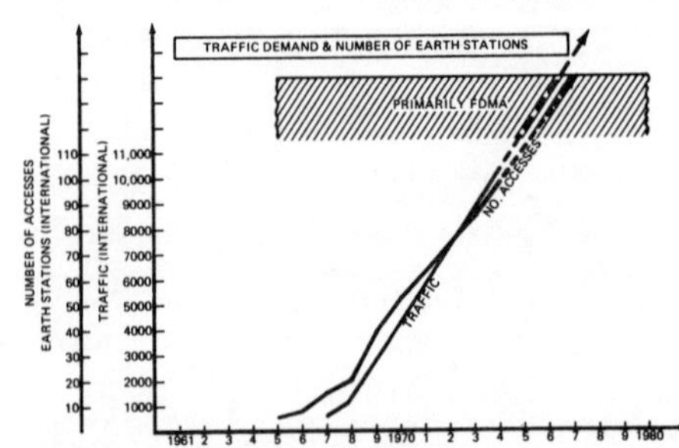

Figure 1. Comparison of Available Spacecraft Power, Allocated Bandwidth, Traffic Demand & No. of Earth Stations

*J. L. Dicks, Director, and M. P. Brown, Sr. Systems Engineer, Communications Systems Requirements Directorate, International Systems Division, COMSAT.

Reprinted from the *Record of the IEEE Electron. & Aerosp. Syst. Conv., (EASCON),* Oct. 7-9, 1974, pp. 167-178.

Frequency Division Multiple Access (FDMA)

In an FDMA system, each up-link radio frequency carrier occupies a discrete frequency allocation and is assigned a specific location within a multicarrier transponder. Receiving earth stations select a desired carrier by means of RF filtering which allows the carrier to be separated, demodulated, and finally de-multiplexed back into individual telephone channels.

Two FDMA techniques are in general operational service today:

- Multichannel-per-Carrier Transmission where the transmitting earth station frequency division multiplexes (FDM) several Single Sideband Suppressed Carrier telephone channels into one carrier baseband assembly which frequency modulates (FM) a carrier and is then applied to an FDMA satellite network. This type of operation is referred to as FDM-FM-FDMA.

- Single-Channel-per-Carrier (SCPC) transmission where each individual telephone channel independently modulates a separate radio frequency carrier. This type of operation is referred to as SCPC-FDMA and can be accomplished by transmitting either a digital or FM carrier.

The majority of satellite circuits in use today employ FDM-FM carriers, either directed to a single destination or as multidestinational carriers. In a system with many points to be interconnected, it is not normally feasible to provide a separate repeater for each carrier. This has resulted in the simultaneous use of a common non-linear amplifier, such as a traveling wave tube (TWT), by a number of independently modulated carriers. Initially, satellite power restrictions limited the number of TWT's to one or two to cover an entire 500 MHz of allocated bandwidth, but today the designer has considerably more power available for use in selecting transponder bandwidth tailored to system requirements.

One type of Single-Channel-per-Carrier system now in international operation was developed specifically to maximize space segment efficiency while at the same time rendering satisfactory service to small users. The digital SPADE* system solves this problem with a pool of 800 channels (400 circuits) which are shared with all earth stations in common view of the satellite.[1] The circuits are assigned on demand and form a temporary connection between any two earth stations accessing the system. When this path is disconnected, the circuits are returned to the demand assignment pool.

A growing number of potential domestic satellite systems have focused on FM-SCPC for its attractiveness in terms of cost, simplicity, and direct relation to small demand applications. FM-SCPC hardware can consist of a small antenna and a minimum amount of easily maintained communication equipment suitable for use with remote weather outposts, oil rigs, pipeline monitoring systems, and other small user networks which would require only one or two channels. A network of these stations can be easily assigned to one satellite transponder on an FDMA basis.

A method of multichannel-per-carrier transmission employing digital techniques is now under development and should be mentioned. In this system, several telephone channels are converted to individual Pulse-Code-Modulation (PCM) bit streams which are multiplexed and, in turn, used to Phase-Shift-Key (PSK) a radio frequency carrier where it is applied to a transponder operating in an FDMA mode. This type of operation is referred to as PCM-PSK-FDMA. Its advantage is that channel capacity can be substantially increased by introducing Digital-Speech-Interpolation (DSI), which requires a digital baseband format.

Overview of FDMA System Constraints

Several transmission impairments must be considered in designing a multiple access system using satellite TWT's. Depending upon the method of modulation employed, amplitude and phase non-linearities must be considered to minimize intermodulation, intelligible crosstalk, and other interfering effects by taking into account the number and size of carriers expected to access the repeater. These impairments are maintained at acceptable limits by operating the TWT at the minimum input back-off necessary to ensure performance objectives are met. Unfortunately, this method of operation results in less available channel capacity compared to a single access mode.

---

*SPADE is an acronym used by INTELSAT for Single-Channel-per-Carrier, Pulse-Code-Modulation, Multiple Access, Demand-Assignment Equipment.

In early systems, intermodulation created by TWT amplitude non-linearity was the dominant factor in limiting system operation. However, as greater power became available and narrow bandwidth transponders capable of operating with only a few carriers near saturation became practical, maximum capacity became dependent on a carefully evaluated trade-off analysis of a number of parameters which include the following:

1) Satellite TWT impairments

   - In-band intermodulation produced by both amplitude and phase non-linearities

   - Intelligible crosstalk caused by FM-AM-PM conversion under multicarrier operation

2) FM transmission impairments not directly associated with the satellite TWT

   - Adjacent carrier interference caused by frequency spectrum overlap between adjacent carriers giving rise to convolution and impulse noise in the baseband

   - Dual path distortion between transponders

   - Interference due to adjacent transponder intermodulation

   - Earth station RF out-of-band emission

   - Frequency reuse co-channel interference

3) General constraints

   - Available power and allocated bandwidth

   - Up-link power control

   - Frequency coordination

   - Vulnerability to interference

In today's complex networks, system designers must consider each of these categories in relation to the requirements of a particular FDMA system, whether FDM-FM or SCPC. The next section on system design will address each item individually.

System Design and Application

As a means of presenting the reader with a general idea of the interdependency of various factors relating to the system design of each FDMA technique, this section has been developed to provide a brief description of each problem, along with applicable references and some historical background material.

FDM-FM-FDMA System Design

Basic link parameters for thermal noise (up- and down-link) and in-band intermodulation are normally considered in FDM-FM-FDMA system design as the major factors establishing overall performance. The total carrier-to-noise ratio (C/N) available at the input to an earth station demodulator for each carrier is given by the equation:[2]

(1)
$$\frac{C}{N}\Big|_{total} = \frac{C}{N}\Big|_{u} + \frac{C}{N}\Big|_{I} + \frac{C}{N}\Big|_{d}$$

where each term is expressed in decibels (dB) and added logarithmically. Each parameter is given by the following link equations:

(2)
$$\frac{C}{N}\Big|_{u} = \psi_s + \frac{G}{T}\Big|_{sat} - BO_i - 10 \log k + 10 \log B - IM^2$$

(3)
$$\frac{C}{N}\Big|_{I} = \text{(See section on intermodulation)}$$

(4)
$$\frac{C}{N}\Big|_{d} = \text{Satellite e.i.r.p.}\Big|_{sat} - PL\Big|_{d} + \frac{G}{T}\Big|_{e.s.} + 10 \log B - 10 \log k + G.A.$$

After determination of the total $\frac{C}{N}$ available from the satellite at the receiving earth station, the transmission parameters for each carrier can be calculated from the FM Equation and Carson's Rule Bandwidth:

(5)
$$\frac{S}{N} = \frac{C}{N} \frac{B}{b} \left(\frac{f_{tt}}{f_{max}}\right)^2 P W, \quad \text{(FM Equation)}$$

or, specifically, in dB

(6)
$$\frac{S}{N}\Big|_{w} = \frac{C}{N} + 10 \log B - 10 \log b + 20 \log \left(\frac{f_{tt}}{f_{max}}\right) + P + W$$

where bandwidth in Hz is:

(7)
$$B = 2(f_{tt} \ell g + f_{max}), \quad \text{(Carson's Rule Bandwidth)}$$

Once the number of channels per carrier and the desired $\frac{S}{N}$ are selected, the $\frac{C}{N}$ can be traded for RF bandwidth to maximize total transponder capacity.

From these relationships, an optimum transponder operating point ("back-off") can be derived by a trade-off between up-path noise, intermodulation noise (treated as thermal), and down-path noise contributions as shown in Figure 2. By controlling the amount of attenuation just before the input to the transponder (satellite gain steps), the up-link $\frac{C}{N}$ ratio can be varied while at the same time maintaining a constant input level to the TWT. The $\frac{C}{N})_{up}$ is thus varied by adjusting the carrier power at the earth station without effecting the $\frac{C}{N}$ due to intermodulation and down-path thermal. Control over most of the remaining impairments is mainly achieved by equipment design factors.

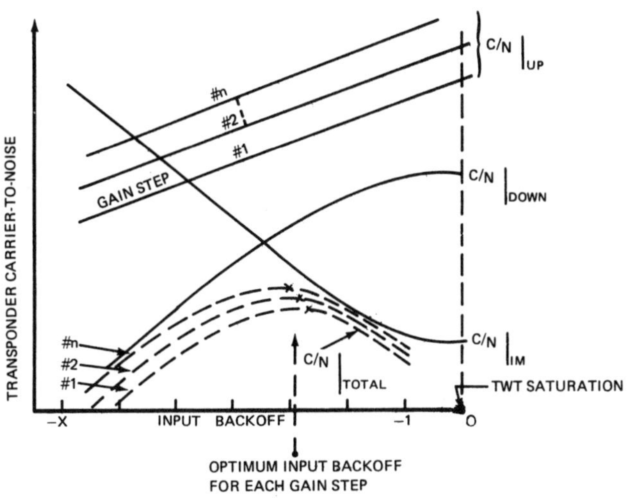

Figure 2. System Optimization Characteristics

Satellite TWT Impairments

<u>Intermodulation</u> -- Considerable study has been undertaken to determine the effect of TWT intermodulation products such as those shown in Figure 3 on modulated signals occupying the same bandwidth. These studies followed the procedure of first computing the magnitude and phase of RF products generated within the non-linear TWT and then predicting the amount of baseband distortion. Initial studies estimated amplitude distortion by expressing the non-linear TWT voltage transfer characteristic as either a polynomial expansion[3] or as a Fourier series.[4] Later, a rigorous mathematical model was developed which considered both phase and amplitude in which the complex output power envelope was expressed as a function of the input power envelope.[5]

In estimating the effect of an intermodulation spectrum on a modulated carrier, it has been customary to assume that the RF intermodulation products are statistically independent of the generating carriers and can therefore be replaced by RF noise of equivalent spectral density when calculating the resultant baseband noise. It is also assumed the equivalent RF spectral density has a Gaussian shape. While these assumptions

have proved adequate for estimating distortion in wideband transponders with many carriers, the use of narrowband transponders with only 3 or 4 carriers operating near saturation has required a re-evaluation of these assumptions. As a result, the mathematical model developed in Reference 5 was expanded to include the features listed below.

- Intermodulation products are treated as interfering carriers.

- Statistical dependence of intermodulation products with respect to the carrier being demodulated is taken into account.

- The relative phase between intermodulation products which appear as symmetrical sidebands on the carrier being demodulated is considered.

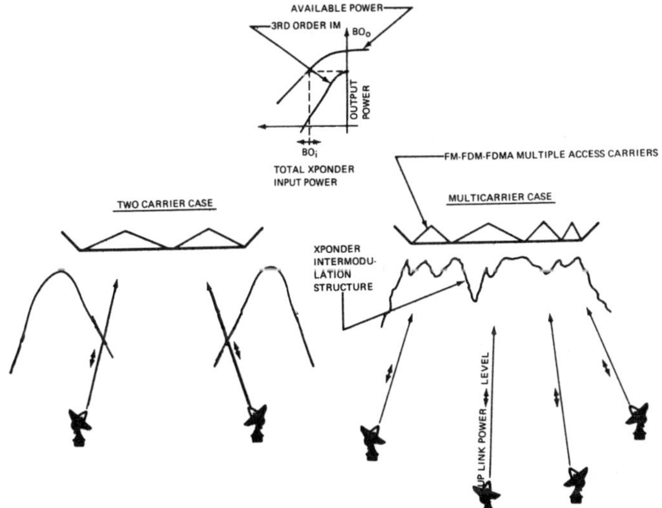

Figure 3. Transponder Intermodulation Patterns

This latest model has shown accurate correlation with measured data and is employed on a day-to-day basis to evaluate frequency plans in the INTELSAT system.[6]

Typical $\frac{C}{N})_{IM}$ values for a narrowband transponder are shown in Figure 4 where the variation of intermodulation is plotted against input back-off for several transponder carrier configurations.

Experience to date with FDM-FM-FDMA indicates that the loss of transponder capacity generally follows the pattern depicted in Figure 5 due to two principle factors:

- As the number of accessing carriers increases, more intermodulation products fall in-band. The buildup of in-band intermodulation requires moving the transponder operating point back from saturation to maintain the same overall intermodulation level. Increasing the input back-off means less power is available to support channel capacity.

As the number of carriers in a given transponder increases, relative size decreases. Smaller size carriers are less efficient in bandwidth utilization and result in less channel capacity.

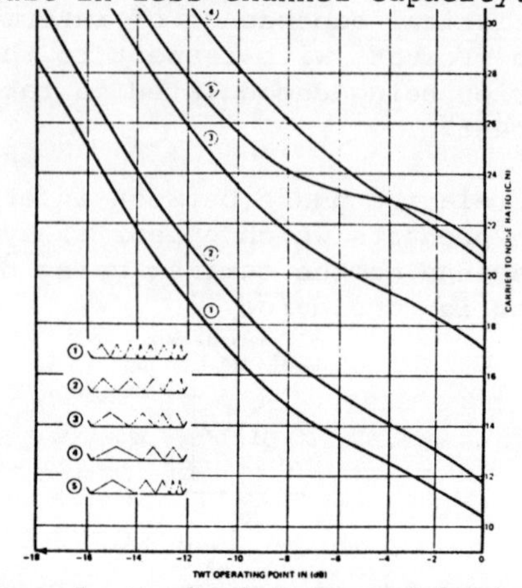

Figure 4. Typical Intermodulation $\frac{C}{N}$ Values for Various Transponder Carrier Configurations

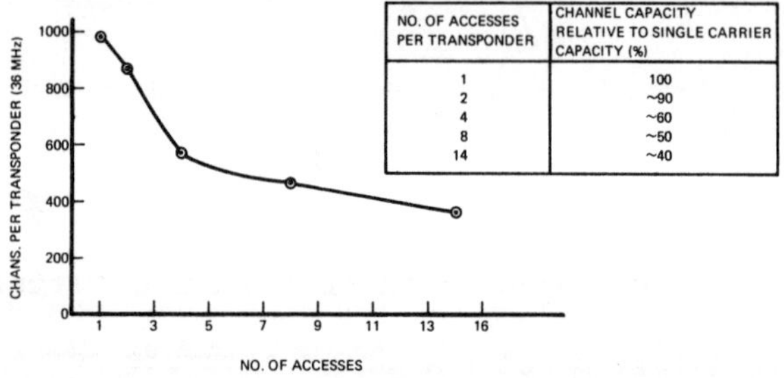

Figure 5. Transponder Capacity As a Function of No. of Accesses

• <u>Bandwidth Utilization</u> -- Further discussion of this last point is important in understanding a limitation of FDMA. Figure 6 shows the variation of RF bandwidth utilization as a function of carrier size. Since channel loading will reflect on carrier bandwidth requirements, it can be seen from the CCIR loading formulas below that carrier sizes above 240 channels are related to a power function (10 $\log$ n) resulting in a constant bandwidth utilization. On the other hand, carrier sizes below 240 channels have a significant penalty in terms of bandwidth utilization for a given $\frac{C}{N}$.

(8) $\ell = -1 + 4 \log_{10} n, \qquad n < 240$ channels

(9) $\ell = -15 + 10 \log_{10} n, \quad n \geq 240$ channels

The carrier sizes shown in Table 1 are employed in the INTELSAT system and illustrate the principle of RF power and bandwidth trade-off.

Carrier Bandwidth (MHz)	Number of Channels		
	$\frac{C}{N} \approx 13$	$\frac{C}{N} \approx 15$	$\frac{C}{N} \approx 20$
2.5	24	36	60
5.0	60	96	132
7.5	96	132	192
10.0	132	192	252

Table 1. Typical INTELSAT Carrier Parameters

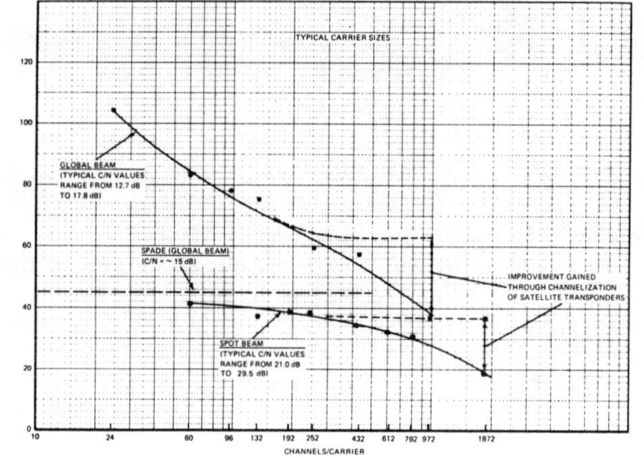

Figure 6. Required RF Bandwidth Per Channel As a Function of Carrier Size

<u>Crosstalk</u> -- In multicarrier FM systems, crosstalk results from a sequence of two phenomena:

1. An amplitude response that varies with frequency-producing amplitude modulation coherent with the original frequency modulation of an RF carrier (FM/AM transfer), and

2. A coherent amplitude modulation which phase-modulates all carriers occupying the same TWT amplifier due to AM/PM conversion.

That is, as a carrier passes through a TWT amplifier, it may produce amplitude modulation from gain slope anomalies in the transmission path. Another carrier passing through the same TWT will vary in phase at the same rate as the AM modulation and thereby pick up intelligible information from any carrier sharing the transponder. Provisions must therefore be made in specifying TWT characteristics to insure AM/PM conversion and gain slope variation are suitable for system requirements. Present day satellite systems have little trouble meeting the CCITT Recommendation (Geneva 1964, Vol. III) for intelligible crosstalk to be 58 dB or better.

An early analysis has given satisfactory results when the gain slope is small.[7] The analysis is basically characterized by a linear coefficient (dB/MHz) and a phase shift vs. drive term, which is represented by an average conversion coefficient (degrees/dB). However, in a complex system, deviations from these assumptions are encountered. For example, if a carrier is located near the transponder bandedge, it experiences a high order unsymmetrical gain slope, particularly at the filter corners. Assuming an average cord in the region of this corner will result in gross error.

Revisions to this model have produced close correlation between predicted and measured results.[8] The new model separates its analysis into two parts, one representing amplitude and group delay frequency response and the other an AM/PM conversion device. The crosstalking carrier is assumed to be modulated with a multichannel baseband while the wanted carrier is represented by an unmodulated carrier. Intelligible crosstalk therefore originates from the baseband channel of the modulated carrier and is transferred to the wanted carrier. Since crosstalk increases proportionally with baseband frequency, the analysis is performed at the top baseband frequency.

A simplified equation for crosstalk ratio (XTR) which can be usefully employed under normal multicarrier operation (-10 to -20 dB input back-off) and small gain slopes is:

$$(10) \quad XTR = 20 \log_{10} \left(\frac{57.3}{K_p S f_{max}} \right) \left(\frac{f_w}{f_i} \right) \left(\frac{P_t}{P_i} \right)$$

FM Transmission Impairment Not Directly Associated with the Satellite TWT

• Impulse and Convolution Noise -- To obtain the maximum channel capacity in a bandwidth-limited transponder, carriers are placed as close together as possible without causing interference from spectrum overlap. Adjacent channel interference of this form has been broken into two subdivisions for analytical purposes:

1. Convolution Noise -- is produced when the amplitude of the desired carrier is greater than the interfering carrier. It is evident in the channel baseband as an increase in noise level.

2. Impulse Noise -- is produced when the amplitude of an adjacent carrier is greater than that of the desired carrier. It is evident in the form of "clicks" or, in data service, as energy bursts.

Several complex mathematical models have been developed to predict impulse noise; however, further refinement is still in progress.[9,10,11,12]

In most cases, impulse noise is the dominant impairment and can be controlled to acceptable levels by providing a suitable guardband on each side of the carrier's occupied bandwidth. CCITT requirements for data service recommend that the number of counts of impulsive noise at a level of -18 dBM0 should not exceed 18 in 15 minutes. Experience to date indicates a guardband of about 10% of the total allocated carrier bandwidth gives sufficient protection to achieve these objectives.

• Satellite Filter Intrinsic Group Delay -- A distortion-producing property found in most satellite filters is group delay. This intrinsic impairment is prominent at the bandedges and can cause significant distortion for very small and very large carrier sizes which deviate in and across this region of the transponder. The amount of group delay distortion caused in the system is a function of filter choice and carrier deviation.

Careful filter design in relation to system carrier parameters, such as occupied bandwidth, can keep this distortion-producing problem in the neighborhood of 500 pW0p if group delay equalizers are used either in the satellite or at the earth station as a compensating device. The INTELSAT system has used this design philosophy with both Chebyshev and Elliptic function satellite filters.

Group delay distortion has been treated by segmenting the filter delay response into linear, parabolic, and ripple components. Generally, only larger carriers will be disturbed by ripple components, which can usually be neglected. Linear and parabolic components have been approximated by the following equations in terms of noise-power-ratio:

$$(11) \quad NPR\big|_{linear} = 20 \log_{10}\left(\frac{10^3}{f_{max}f_{rms}T_{DL}}\right) + 5.2 \text{ dB (Pre-Emphasis)}$$

$$(12) \quad NPR\big|_{parabolic} = 20 \log_{10}\left(\frac{860}{f_{max}f_{rms}^2 T_{DP}}\right) + 3.1 \text{ dB (Pre-Emphasis)}$$

NPR can then be converted to baseband signal-to-noise $\left(\frac{S}{N}\right)$ by:

$$(13) \quad \frac{S}{N}\Big|_w (dB) = NPR + 10 \log_{10}\left(\frac{f_{max}-f_{min}}{b\,\ell^2}\right) + 2.5 \text{ (Psophometric Weighting)}$$

$$(14) \quad \frac{S}{N} (pW0p) = 10^{\frac{(90-S/N)}{10}}$$

• __Dual Path Distortion Between Transponders__ -- The opportunity for a carrier to travel through an indirect path and be recombined at the receiving earth station is also found near the bandedges of each transponder where the skirts of their filters overlap. A dual path is thus available to a carrier where it can be transmitted in part through the adjacent transponder. This impairment source will introduce phase non-linearities in the wanted signal and can be treated as group delay noise, which is primarily generated by the linear componenets of the total group delay response of the satellite transmission path. Since this impairment is dependent upon filter characteristics, a trade-off exists between weight considerations and the kind and degree of filtering required to meet system noise objectives.

With proper transponder guard bands and filter skirts, this impairment can be held to negligible interference and can be controlled as a lump problem with the same equalizers used for intrinsic group delay distortion.

• __Interference Due to Adjacent Transponder Intermodulation__ -- Transponder intermodulation products generated in a multicarrier TWT are transmitted together with the desired signals. Unless the out-of-band products are attenuated by the transponder output filter, they can be of sufficient magnitude to cause noticeable interference to signals transmitted in an adjacent transponder. The magnitude of this interference is dependent on the attenuation provided by the output filter skirts and the level of the in-band intermodulation. The degree of interference is a trade-off between the number of sections of output filtering (weight), the input back-off expected under normal operation, and the amount of interference which can be tolerated in the total system noise budget.

For example, negligible interference has been noticed in the INTELSAT IV satellite system, which has a 6-section Chebyshev output filter for each transponder. On the other hand, system design for the INTELSAT IV-A satellite allows several hundred pico-watts of adjacent transponder interference with 4-section Chebyshev output filters.

• __Earth Station RF Out-of-Band Emission__ -- A trade-off exists between the amount of usable transmitter output power and the amount of earth station emission noise the system can tolerate.[13] When a high power wideband TWT is operated at saturation, its full output power can be realized, but it can also produce severe intermodulation emission to the up-path of other carriers in the FDMA system. This impairment alone will account for the loss of several kilowatts of power due to the need to operate multicarrier wideband HPA's below a given output back-off in order to keep emission noise within a limited budget. Some systems consider up to 7 dB output back-off as a typical upper limit for this purpose. For a 12 kW wideband TWT, this would mean a loss of 9.6 kW just to compensate for out-of-band emission.

RF out-of-band emission (OBE) is normally expressed as a given amount of interference power over a 4 kHz telephone channel (dBW/4 kHz). As can be seen in the equation below, it is dependent upon the transmitting earth station's HPA intermodulation level and the total up-link power of all carriers producing that intermodulation:

$$(15) \quad OBE = IM_{TWT} + P_{e.s.} + \text{Antenna Gain} \quad , \text{in dBW/4 kHz}$$

A limit of 23 to 26 dBW/4 kHz has been established for the particular noise budget found in the INTELSAT system which has been designed to control this impairment to less than 500 pW0p.

• __Frequency Reuse Co-Channel Interference__ -- By making judicious use of frequency band allocations, bandwidth can be expanded by "frequency reuse" techniques such as those achieved by employing either dual polarization or spatial antenna isolation, or both. Depending upon the complexity of the satellite antenna, the degree of isolation actually realized is of a magnitude that can have considerable impact on frequency plan implementation. Figure 7 shows the effect of co-channel interference for a given isolation and carrier level difference. Note also that co-channel interference decreases as the separation between the two carrier center frequencies increase.

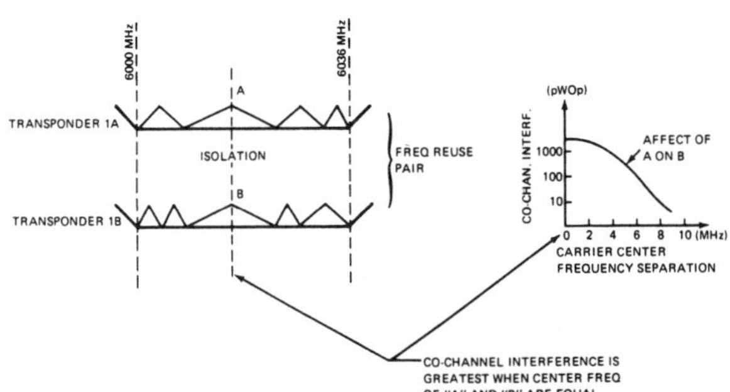

Figure 7. Frequency Reuse Co-Channel Interference

In optimizing a transponder in a frequency reuse system, the additional distortion caused by co-channel interference must be considered in the total noise budget. If a frequency plan is developed in a manner such that carrier center frequencies in frequency reuse transponders are offset as much as possible, this contribution can be kept at a reasonable limit, e.g., 500 pW0p.[14,15]

A rule of thumb for frequency reuse transponders operating in a multicarrier mode is to treat this impairment as thermal noise (assuming that sufficient frequency offset has been employed). In those cases where carriers with a high power density are paired with ones with a low power density, a more rigorous analysis is necessary.[14]

General Constraints

• Available Power and Bandwidth -- The situation where satellite power is the limiting commodity is not as dominant today as it was in the mid-1960's; the main problem is bandwidth. Although more bandwidth has been made available by the 1971 Geneva World Administrative Radio Conference on Space Telecommunications for Fixed-Satellite Service (WARC-ST), the new allocations are in the higher frequency bands which suffer severely from propagation anomilies (see Figure 8). In addition to the requirement to overcome these effects with much higher powered satellites (10-20 dB more e.i.r.p.), new equipment technology is also required, which will prolong the time period before this new bandwidth source can be employed. It can be expected that greater application will be made in the near future of frequency reuse techniques in the 6/4 and 8/7 GHz bands rather than the higher frequency bands.

• Up-Link Power Control -- In an FDMA system, each carrier's up-link power level requires close control to keep transmission impairments for each carrier within the total noise budget. One method of meeting these objectives is to study each transponder carrier configuration on a case-by-case basis before it is actually implemented on the satellite. Once the proper theoretical operating values are determined, each earth station can be requested to provide this exact value. Further refinement can then be obtained through the use of a few strategically located power monitoring stations which measure the down-link of each carrier in the system. Theoretical levels can then be compared to both the reported up-link and measured down-link.

• Frequency Coordination -- Careful frequency coordination is necessary in an FDMA system, especially if most of the carriers are of a multidestinational nature to insure proper clearance with terrestrial radio relay networks sharing the same frequency band.

• Vulnerability to Interference -- Since close power control of each carrier is required, FDMA carriers with generally low modulation indices are susceptible to inteference, especially from an intentional source. A form of "jamming" can be achieved simply by placing a modulated carrier in the FDMA system at the same center frequency as the desired carrier. As the level of this carrier is increased, the level of interference will also increase. If this level is increased significantly, the transponder input back-off can be changed enough to cause disruption to all carriers.

Design Summary

As an example design summary, the INTELSAT IV and IV-A system noise budgets have been distichously displayed in Table 2 to illustrate the weighting placed on each impairment in an international system. The INTELSAT IV system has been in global operation since 1971 in the 6/4 GHz frequency band with approximately 432 MHz of usable bandwidth. The INTELSAT IV-A system will be introduced in the Atlantic Ocean Region in 1975 and will operate in the 6/4 GHz frequency band employing frequency reuse through spatial antenna isolation with approximately 720 MHz of usable bandwidth.

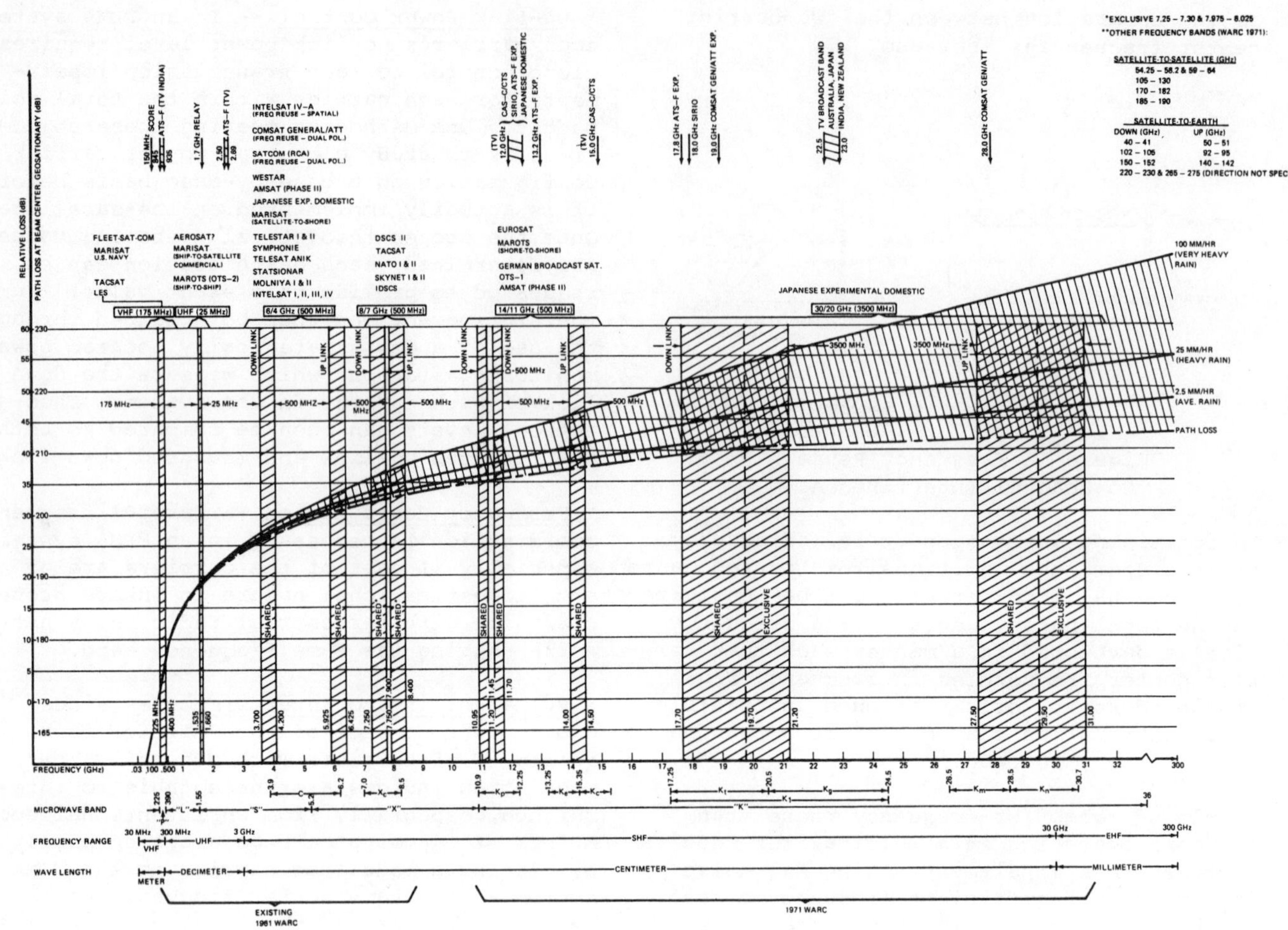

Figure 8.
Communications Satellite Allocated Frequency Bands

IMPAIRMENT	INTELSAT IV (pW0p)	INTELSAT IV-A (pW0p)
SPACE SEGMENT*	Objective	
1. Thermal Noise (up- and down-link)	} 7,500	} ≈ 6,500
2. Transponder Intermodulation		
3. Co-Channel Interference	--	≈ 500 } 7,500
4. Adjacent Transponder Interference due to Intermodulation	--	≈ 500
EARTH SEGMENT	Designated	
1. Earth Station RF Out-of-Band Emission	500	500
2. Earth Station Transmitter Noise, Excluding Multicarrier Intermodulation and Group Delay Noise	250	250
3. Noise due to Total System Group Delay and Dual Path Distortion After Any Necessary Equalization	500 } 2,500	500 } 2,500
4. Other Earth Station Receive Noise	250	250
5. Interference from Terrestrial Systems Sharing the Same Frequency Bands	1,000	1,000
Total System per Channel Noise Budget (CCIR)	10,000	10,000
GENERAL SATELLITE CAPABILITY		
1. Transponder Bandwidth (MHz)	36	36
2. Number of Transponders	12	20
3. Total Bandwidth (MHz)	432	720
4. Nominal Channel Capacity (Atlantic Primary) (plus one transponder each for SPADE and TV)	7,500	12,500

*Objectives may be exceeded as long as total equals 7,500 pW0p.

Table 2. Comparison of INTELSAT IV and
IV-A FDM-FM-FDMA System Noise Budgets

SCPC-Digital-FDMA System Design (SPADE)

The SPADE system is used in a 36 MHz transponder of the INTELSAT system to obtain maximum channel capacity for low demand users as shown in Figure 9. System design is generally the same as the FDM-FM-FDMA technique described above; however, in calculating the $\frac{C}{T}$ required to support 800 channels, three factors are considered: thermal noise, adjacent channel interference, and intermodulation noise. Adjacent channel interference is a function only of the characteristics of the transmit channel low-pass filter and the receive bandpass filter.

In the SPADE system, the channel carrier is voice-activated, that is, the carrier is on only if voice is detected. Since the average talker speaks only 40% of the time, the carrier for any given channel is on only 40% of the time or, in other terms, only 40% of the total carriers will be on. This means that for an 800-channel system, a maximum of 320 channels need to be supported by satellite e.i.r.p. at any given instant.

ments. Figure 10 shows this point graphically.

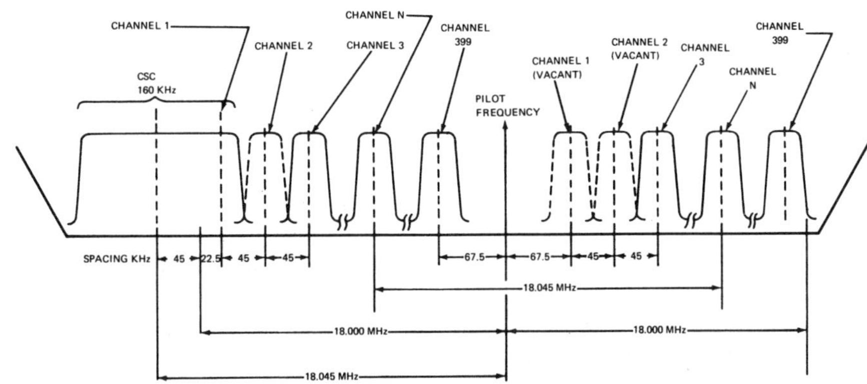

Figure 9.
SPADE Transponder Frequency Plan

When considering intermodulation noise, it is important to note that since all channels are equally spaced, the products created by odd-order amplitude and phase non-linearities fall exactly at allocated channel frequencies. It is then possible to calculate the center frequencies of all third-order products created by a group of 320 carriers randomly allocated among 800 channels and their resulting distortion to each channel determined from this information.[16]

Link equations for digital SCPC operation are essentially the same as FDM-FM. However, because digital SCPC operation is normally power-limited, it is important to determine the threshold $\frac{C}{N}$ required by the demodulator to produce a specific bit-error rate. For the SPADE system, the following equation has been used:

$$(16) \quad \frac{C}{N}_{\text{threshold}} = \frac{E}{N_o} + R - B + M$$

Employing the values provided in the Glossary this will yield a threshold $\frac{C}{N}$ of 13 dB.

By applying up- and down-link equations together with transponder intermodulation, a trade-off analysis can be carried out to determine the optimum transponder operating point. Once the proper channel bandwidth has been selected (and thereby the maximum channel capacity), the appropriate up- and down-link margins should be checked to see if they are adequate to meet system require-

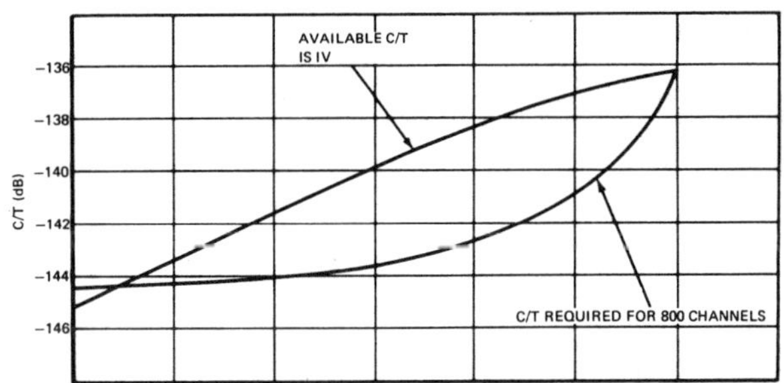

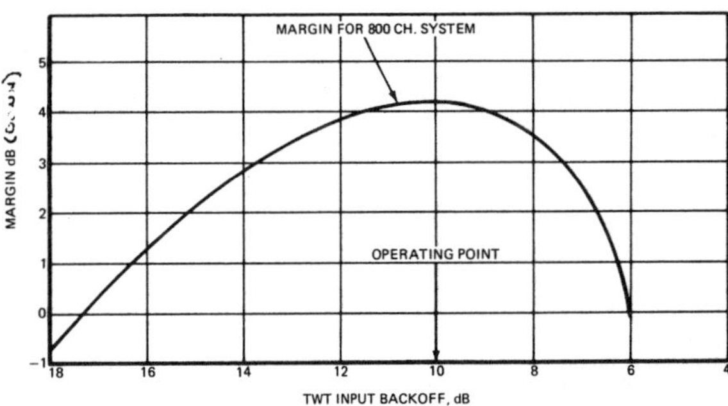

Figure 10. Margin and $\frac{C}{T}$ vs. TWT Input Back-off (800-Channel SPADE System)

SCPC-FM-FDMA System Design

System design of an SCPC-FM-FDMA system is similar to the SPADE system described above in its treatment of intermodulation spectra within the transponder. Once the intermodulation spectrum has been determined for a given carrier arrangement, an optimum tran-

sponder back-off can be derived using the
principles described in the section on FDM-
FM-FDMA.

Because SCPC-FM-FDMA normally finds its
application for small earth terminals, it is
usually operated close to demodulator threshold and total noise budgets exceeding 10,000
pWOp. Improvement in quality has been experienced by using compandors in these systems to improve $\frac{S}{N}$. Although compandors are
not inexpensive, for a weighted $\frac{S}{N}$ greater
than 30 dB approximately 18 dB of improvement
can be achieved.

Future Trends

Figure 11 has been developed as an aid in
tracing the historical and future use of
FDMA in communications satellite systems.
Several conclusions can be drawn from the
information presented on this chart and
the contents of this paper:

* Multicarrier FDMA has been the primary
  mode of multiple access since the beginning of satellite communications and
  will remain so throughout the rest of
  this decade. Both digital and FM methods
  of SCPC-FDMA will become established
  systems before the end of the decade.

* More efficient forms of multiple access
  will also be introduced in the late
  1970's. TDMA will probably come into
  widespread use by the end of the 1980's.
  PCM-PSK-FDMA with DSI may come into
  operational use by the early 1980's.

* In view of propagation difficulties in
  the 14/11 and 30/20 GHz bands, satellite
  traffic demands in the late 1970's and

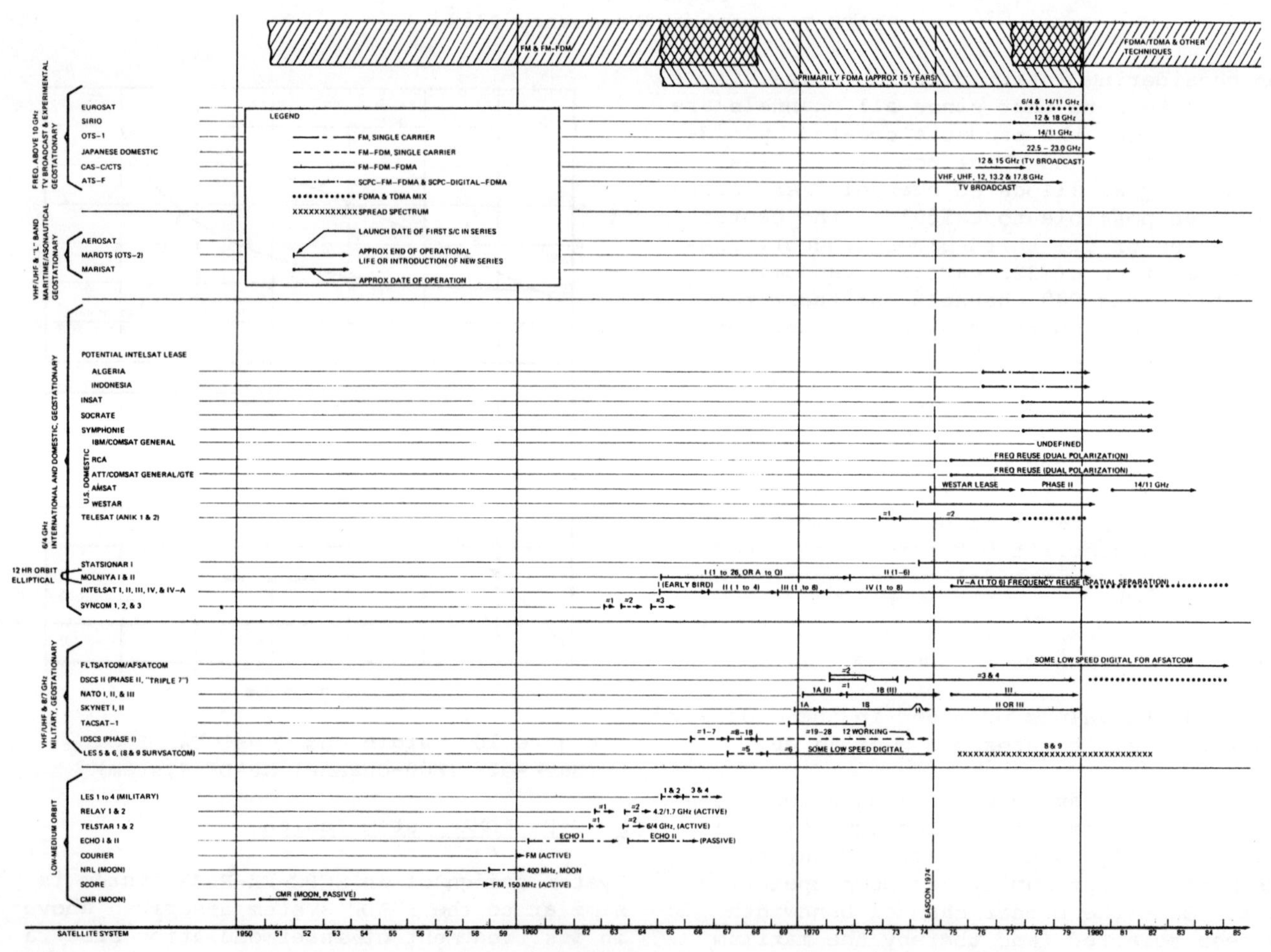

Figure 11.
Future Communication Satellite Multiple Access Trends

early 1980's will force new techniques to be introduced in the 6/4 and 8/7 GHz bands. This will include continued use of FDMA with frequency reuse as a means of bandwidth expansion.

Conclusions

The advantages of FDMA have resulted in its use as the primary method of multiple access for almost a decade. When compared to other forms of multiple access, FDMA is preferred due to its simplicity, proven technology, low cost, and easy adaptation to varying international conditions. These advantages have been true in the years where traffic demands on available satellite power and allocated bandwidth have not restricted system design enough to warrant more advanced forms of multiple access to be introduced. As bandwidth becomes increasingly precious, more complex and costly techniques will be introduced to make more efficient use of this commodity.

This paper has presented a general overview of design considerations for the two major FDMA techniques in operation today: FDM-FM and SCPC (SPADE). Two other techniques, SCPC-FM and PCM-PSK, have also been covered.

Acknowledgements

The authors would like to express their sincere appreciation to Ms. Joyce Strauss who edited and typed this document.

Information in this paper was derived from INTELSAT and COMSAT corporate-sponsored projects; however, the views expressed are not necessarily those of either organization.

References

(1) A. M. Werth, "SPADE: A PCM FDMA Demand Assignment System for Satellite Communications," presented at the London Conference on Digital Satellite Communications, Nov. 1969.
(2) J. L. Dicks, P. H. Schultze, and C. H. Schmitt, "INTELSAT IV System Planning," COMSAT Technical Review, Vol. 2, No. 2, Fall 1972.
(3) R. J. Westcott, "Investigation of Multiple FDM/FM Carriers Through a Satellite TWT Operating Near Saturation," Proceedings of the I.E.E. (UK), Vol. 114, No. 6, June 1967, pp. 726-750.
(4) A. L. Berman and C. E. Mahle, "Non-linear Phase Shift in Traveling Wave Tubes As Applied to Multiple-Access Communications Satellites," I.E.E.E. Transactions on Communications Technology, COM-18, No. 1, Feb. 1970, pp. 37-48.
(5) O. Shimbo, "Effects of Intermodulation, AM-PM Conversion and Additive Noise in Multicarrier TWT Systems," Proceedings of the I.E.E.E., Vol. 59, No. 2, Feb. 1971, pp. 230-238.
(6) N. K. M. Chitre and J. C. Fuenzalida, "Baseband Distortion Caused by Intermodulation in Multicarrier FM Systems," COMSAT Technical Review, Vol. 2, No. 1, Spring 1972.
(7) R. C. Chapman, Jr., and J. B. Millard, "Intelligible Crosstalk Between Frequency Modulated Carriers Through AM-PM Conversion," I.E.E.E. Transactions on Communications Technology, COM-12, No. 2, June 1964, pp. 160-166.
(8) A. L. Berman, C. Mahle, and M. R. Wachs, "INTELSAT IV Transmission Modeling," COMSAT Technical Review, Vol. 2, No. 2, Fall 1972.
(9) R. G. Medhurst, "FM Interfering Carrier Distortion: General Formula," Proceedings of the I.E.E. (UK), Vol. 109B, No. 44, March 1962, pp. 149-150.
(10) R. Hamer, "Radio-Frequency Interference in Multichannel Telephony FM Radio Systems," Proceedings of the I.E.E. (UK), Vol. 108B, No. 37, Jan. 1961.
(11) M. Wachs, "Analysis of Adjacent Channel Interference in a Multicarrier FM Communications System," COMSAT Technical Review, Vol. 1, No. 1, Fall 1971.
(12) T. Furuya, A. Fujii, and K. Tezuka, "Impulse Noise in FM Receivers in the Presence of Adjacent Channel Interference and Thermal Noise," I.C.C., June 1972, Cat. No. 72-CHO-622-1-COM.
(13) M. P. Brown, Jr., "Review of INTELSAT Earth Station RF Out-of-Band Emission Criteria," COMSAT Technical Review, Vol. 4, No. 1, Spring 1974.
(14) B. A. Pontano, J. C. Fuenzalida, and N. K. M. Chitre, "Interference into Angle-Modulated Systems Carrying Multichannel Telephony," I.E.E.E. Transactions on Communications, June 1973.
(15) J. L. Dicks and M. P. Brown, Jr., "INTELSAT IV-A Satellite Transmission Design," A.I.A.A. Paper No. 74-474, April 1974.
(16) R. B. McClure, "Analysis of Intermodulation Distortion in an FDMA Satellite Communications System with a Bandwidth Constraint," I.E.E.E. 1970 International Conference on Communications Record.

Glossary

b = Channel bandwidth (normally 3.1 kHz), in Hz (6), in MHz (13)

B' = Channel noise bandwidth = 45.8 dBHz

B = Carson's Rule Bandwidth, in Hz

BO_i = Satellite transponder TWT input back-off, in dB.

BO_o = Satellite transponder TWT output back-off, in dB

$\left.\frac{C}{N}\right|_u$ = Up-link carrier-to-noise ratio, in dB

$\left.\frac{C}{N}\right|_I$ = Carrier-to-intermodulation noise ratio, in dB

$\left.\frac{C}{N}\right|_d$ = Down-link carrier-to-noise ratio, in dB

$\frac{E}{N_0}$ = Ratio of energy-per-bit-to-noise density = 8.4 dB

f_{tt} = r.m.s. deviation caused by a 0 dBM0 test tone signal, in Hz (6) and (7)

f_{max} = Top baseband frequency, in Hz (6), in MHz (10), (11), (12), and (13)

f_{rms} = r.m.s. multichannel deviation, in MHz (11), (12), and (13)

f_{min} = Minimum baseband frequency, in MHz (normally .012)

$\frac{f_w}{f_i}$ = Ratio of wanted carrier peak deviation per channel to crosstalking carrier peak deviation per channel

g = Peak-to-r.m.s. voltage ratio (also known as the peak factor) of a multichannel load, INTELSAT = 10 dB for an FMFB demodulator

$\left.\frac{G}{T}\right|_{sat}$ = Satellite gain-to-noise temperature ratio at the up-link frequency, in dB/K

$\left.\frac{G}{T}\right|_{e.s.}$ = Earth station gain-to-noise temperature ratio at the down-link frequency, in dB/K

$G.A.\left.\right|_d$ = Average down-link geographic adjustment factor for path loss and antenna gain when compared to beam edge, in dB

K_p = The average value of AM/PM transfer coefficient in deg./dB for a given back-off

$\frac{1}{k}$ = 1/Boltzman's constant = $\dfrac{1}{1.38 \cdot 10^{-23} \text{joule/K}}$ = $-10 \log k$ = 228.6 dBW/K-Hz

ℓ = Multichannel CCIR load factor (see section on "bandwidth utilization" and equations (8) and (9)

M = Implementation Margin = 2.5 dB

$1M^2$ = Effective gain over isotropic of a 1-M^2 antenna, = 37 dB at 6 GHz

P = Psometric weighting factor = 2.5 dB

$P_{e.s.}$ = Power of earth station transmitter at saturation, in dBW

$\frac{P_t}{P_i}$ = Ratio of total transponder power to power of the crosstalking carrier

$PL\left.\right|_d$ = Path loss on down-link at beam edge, = 196.7 dB at 4 GHz

R = Channel bit rate = 48.1 dB for a 10^{-4} bit error rate

$\mathbf{s_s}$ = Flux density for single-carrier transponder saturation at the up-link frequency, in dBW/m^2

S = The linear gain slope over the region occupied by the first two sidebands

$\frac{S}{N}$ = Signal-to-noise ratio at the demodulator output, in dB

T_{DL} = Linear delay component (nanosec/MHz) determined by a cord connecting the end points of the intersection of the carrier's occupied bandwidth and the group delay curve for the transmission path.

T_{DP} = Parabolic delay component (nanosec/MHz^2) determined by the vertical segment bisecting the linear component and the group delay curve.

W = Pre-emphasis weighting factor = 4 dB

$e.i.r.p.\left.\right|_{sat}$ = Satellite transmit e.i.r.p. for single-carrier saturation at beam edge, in dBW

Antenna Gain = Earth station antenna gain, in dB

IM_{TWT} = Multicarrier intermodulation level from the earth station TWT referenced to single carrier saturation, in dB/4 kHz

DESIGN OF FM SINGLE-CHANNEL-PER-CARRIER SYSTEMS

Myron E. Ferguson

California Microwave, Inc.
Sunnyvale, California

ABSTRACT

The paper discusses the factors which impact on the design of FM SCPC communications satellite systems. Stress is placed on the SCPC equipment parameters which are important to the system design. A number of design formulas are given and seven different systems designs are shown ranging in size from a bush telephone system to a high-density domestic satellite system.

INTRODUCTION

A large FM single-channel-per-carrier (SCPC) satellite communication system is scheduled for service this year and another for late 1976.[1] It is expected that such systems will be in wide usage in the future. This paper discusses systems design considerations and presents several possible systems which can be designed using FM SCPC.

Depending upon the effective isotropic radiated power (EIRP) of the satellite and upon the antenna gain-to-temperature ratio (G/T) of the earth station, systems range from high density (1333 voice channels) in which the satellite capacity is bandwidth limited to low density (less than 100 channels) in which the satellite capacity is power limited. The flexibility of FM systems in being readily adaptable to a wide range of applications, the implementation simplicity, and the conservation of satellite power and/or bandwidth in most cases are attractive parameters in considering FM for satellite SCPC applications.

While this paper addresses itself specifically to FM SCPC systems in which emphasis and compandors are used, the considerations and approach to system design is applicable to other FM SCPC configurations as well.

SYSTEMS CONSIDERATIONS

The FM SCPC equipment has applications extending from a single voice channel remote (bush) type terminal to a high-density multiple access system. The flexibility of FM SCPC equipment is evident in the wide potential ranges of applications. In some applications, the FM SCPC equipment will be used in the same transponder with a video carrier for domestic distribution in a country. In these applications, the low power requirement of the FM equipment is of paramount importance. In other applications, the narrow bandwidth requirement of the FM SCPC system will permit 1333 or more voice channels in a single 30-MHz satellite transponder.

The satellites with which the FM SCPC system can be used include Intelsat IV and IV–A satellites, Canadian ANIK, Western Union's WESTAR [4], RCA's forthcoming domestic satellite [5], the Indonesia domestic satellite (similar to WESTAR), and other future potential satellites whether of the global-beam Intelsat type or the narrow-beam domestic satellite type.

With the wide variety in satellites which may be used with the FM SCPC system, there is a comparable diversity of earth stations which are suitable. These range from the standard Intelsat earth station with its 30-meter diameter antenna to the single-channel bush telephone terminal with a 5-meter antenna. The comparable range of gain-to-temperature ratios is from over 40.7 dB for the Intelsat stations to less than 20 dB for some bush telephone stations. The FM system permits a wide range of tradeoffs in terms of satellite capacity, earth station size, and grade of performance.

Satellite Characteristics

Those characteristics which are of importance in communications systems are satellite gain, satellite EIRP, G/T, bandwidth, transponder translation oscillator drift, intermodulation, and FM noise.

Satellite gain may be expressed as output EIRP divided by the effective isotropic receive power. EIRP includes satellite antenna gain and so varies as a function of the location of the individual earth terminal. When a multiplicity of carriers is present in a given transponder, as is the case with SCPC siganls, it is necessary to operate the satellite transponder at less than full output to reduce the level of intermodulation products among the carriers present in the output amplifier of the transponder. This reduction in output (output backoff), usually expressed in decibels, represents the reduction in total output power at the operating point relative to saturation.

G/T is a measure of the sensitivity of the receiving side of the satellite transponder. It is of importance in system design since it is a key characteristic in determining the noise contribution of the uplink. The 36-MHz bandwidth of the satellite transponder is fairly uniform among all satellites used today for satellite communications. Note, however, that due to the design of the RF filters the WESTAR satellite has only about 30 MHz which can be used for SCPC and the RCA satellite will have only about 34 MHz which can be used.

Table 1 shows the various parameters of a number of satellites which can be used for SCPC systems operation. Figure 1 gives two relevant satellite characteristics for typical satellites; the gain and total power to intermodulation-noise density ratio as a function of the backoff of total output power below saturation. The curves are a composite of available data on several satellites, and are typical of satellites using similar TWTA's. [2, 3]

Earth Station "Sizes"

As noted above, the primary receiving characteristic of an earth station is its gain-to-temperature ratio (G/T) which is a figure of merit for the receiving station. A number of factors enter into the gain-to-temperature ratio, with the primary ones being the antenna gain and the noise temperature of the first amplifier in the receiving terminal.

TABLE 1. RELEVANT SATELLITE CHARACTERISTICS

		INTELSAT IV* (Global Beam)	WESTAR[4] (Published)	RCA[5] (Proposed)
G/T	dB/°K	-17.6	-6.0	-4.8
EIRP, Saturated**	dBW	22.5	34.0	33.0
GAIN, Saturated**	dB	132.2	151.1	150.0
Useful Bandwidth	MHz	36.0	30.0	34.0

*   Gain Step 1.

**   Isotropic, over most of useful coverage area

G/T ratios are usually based on system requirements from a communications point of view while the actual antenna gain and low-noise-amplifier temperature are functions of the economic tradeoffs necessary to achieve a given G/T ratio.

Reprinted from *IEEE Int. Conf. on Commun.*, June 16–18, 1975, vol. I, pp. 12-11–12-16.

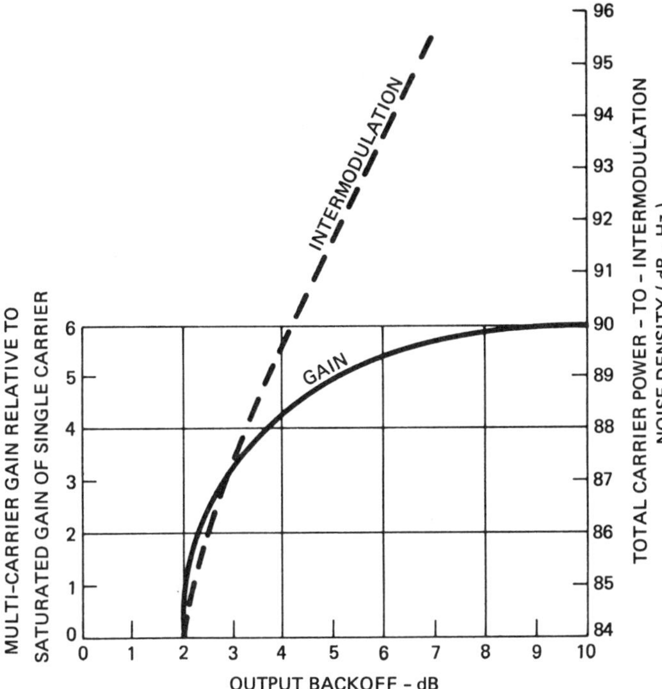

Figure 1. Typical Satellite Characteristics

Besides the satellite G/T, the other interface parameter with the satellite is the earth station transmit EIRP. The EIRP is a function of antenna gain and high-power amplifier capability. Higher gain satellites will decrease earth station EIRP's but will increase up-link thermal noise.

Table 2 gives the relevant parameters for some typical communications satellite earth terminals which could be used in SCPC service.

	LARGE INTELSAT	SMALL INTELSAT	WESTERN UNION	LARGE BUSH	SMALL BUSH***
Antenna Diameter	30M	11M	51'	15'	10'
LNA	Cooled Paramp	Uncooled Paramp	Cooled Paramp	Transistor	Transistor Amp
G/T dB/°K	40.7*	31.7*	37.0**	19.3*	12.9*
Transmit Antenna Gain dBi	63.4	54.6	57.7	46.9	43.4

*   At 5° elevation angle

**   At 30° elevation angle

***   10-foot antennas are of questionable usability because their wide beamwidths make them more susceptible to interference to and from other spacecraft.

Table 2. Earth Station Characteristics

Satellite Loading

Of prime importance in a satellite system design is the number of channels which a given transponder is capable of handling at one time. An advantage of SCPC systems over FDM-FM systems is that it is possible to transmit the carrier only during those intervals during which the voice is present. An activity factor of 40% is used in SCPC service so that the number of channels assigned to a satellite transponder can be increased by 2½ times over that which would be possible if voice operated carrier (VOX) were not used. Note, however, that this loading affects only the satellite power limitations, not the bandwidth limitations.

Noise Objectives

Several considerations must be accounted for when selecting the system operating noise objective. First, if the system uses demand assignment multiple access (DAMA) the design must take into account the stations with the lowest gain-to-temperature ratios and those in the lowest EIRP area of the satellite. Satellite signals do not have the multipath fading that is encountered in line-of-sight microwave systems and the amount of operating margin need be only a few decibels to account for such factors as rain attenuation and antenna pointing error. System operating margin is generally achieved by using an earth station G/T which is greater than the minimum calculated value.

International satellite systems use a noise objective of 10,000 pWp0 for the satellite link. To use a noise objective less than 10,000 pWp0 for the least sensitive station in a network would cause an economic burden which is not necessary; to design the nominal system performance for more than 10,000 pWp0 will not allow adequate margin for operation under unfavorable conditions. In the description of system performances in later sections of this paper, 10,000 pWp0 is used as a system noise objective. When this noise objective is used as a nominal design criterion for the system, fine performance of the FM system below threshold will make it possible to operate the system under degraded conditions without severely jeopardizing the overall quality of the communications system.

Weighting

Psophometric and other noise-weighting networks are commonly used to account for the frequency response of the typical receiving telephone circuit and for the annoying effects of noise at various frequencies to a typical listener.

When compandors are used, the appropriateness of using weighting networks and weighting improvements needs to be re-examined since the noise is present only when speech is present and is absent during the intersyllabic intervals. The use of weighting networks in making uncompanded testtone-to-noise measurements is appropriate, albeit the exact shape of the weighting curve may be subject to question. In making systems calculations and measurements, it is suggested that C-message or psophometric weighting be used, with improvements of 2.2 and 2.5 dB respectively, for flat input noise.

COMPUTATIONAL CONSIDERATIONS

Signal-to-Noise vs. Carrier-to-Noise

The signal-to-noise ratio of an FM system is given in the following equation:

$$(S/N)_1 = \left(\frac{C}{N_0}\right) \cdot \frac{3}{2} \cdot \frac{(\Delta f^2)}{(f_U^3 - f_L^3)} \qquad (1)$$

where $(S/N)_1$ is the signal-to-noise ratio, without VF signal processing,

     C/N_0 is the carrier-to-noise density ratio,

     Δf is the peak deviation of the signal, S,

     f_U is the upper frequency of the receive bandwidth, and

     f_L is the lower frequency of the receive bandwidth.

With the parameters used in voice FM SCPC systems, that is, with an upper frequency of 3400 Hz and a lower frequency of 300 Hz, the signal-to-noise ratio (which is also the testtone-to-noise) (S/N_A) expressed in decibels, is given in equation (2).

$$(S/N)_A \Big|_{dB} = \left[\frac{C}{N_0}\right]_{dB} - 104.2 + 20 \, Log \, (\Delta f) \qquad (2)$$

When 6.3 dB pre-emphasis improvement and 2.5 dB weighting is included, the uncomponded $(S/N)_B$ is given by equation (3).

$$(S/N)_B \Big|_{dB} = \left[\frac{C}{N_0}\right]_{dB} - 95.4 + 20 \, Log \, (\Delta f) \qquad (3)$$

and with a 17 dB compandor improvement,

$$(S/N)_C \Big|_{dB} = \left[\frac{C}{N_0}\right]_{dB} - 78.4 + 20 \, Log \, (\Delta f) \qquad (4)$$

where $(S/N)_C$ is the signal-to-noise ratio with compandor.

Path Losses

For a synchronous satellite and earth stations at the edge of the coverage area with an elevation look angle of $5°$, the path losses between isotropic antennas are:

6 GHz	200.3 dB
4 GHz	196.8 dB

At a look angle of $30°$ these losses are reduced by 0.6 dB.

Satellite Intermodulation

When multiple signals are present in the satellite output TWTA, intermodulation products will arise among the signals. With FM SCPC the definition of the magnitude and nature of these products is different than for either FDM-FM or digital SCPC systems since the FM SCPC carriers are, in general, not modulated as heavily. The result is that there are generally carrier "spikes" present with each carrier as well as modulation sidebands. When several of these carriers plus sidebands intermodulate there will be a resultant carrier spike at the product frequency as well as intermodulation products involving the FM sidebands.

Spectral Distribution of Carriers. With other than the smallest systems the only intermodulation product which is of concern is of the $(f_1 + f_2 - f_3)$ type [8]. The largest number of these products will, in general, occur at the center of the band occupied by the SCPC carriers. Observation of an FM SCPC carrier operating with emphasis and deviations normally to be expected shows that with a median talker level about one-half of the power remains in the carrier spike and one-half in the sidebands. Assuming that the sideband power has a Gaussian distribution whose standard deviation is equal to the rms deviation of a median talker, (-13.9 dBm [7]) the distribution of sideband energy in the $(f_1 + f_2 - f_3)$ product was computed as follows:

12.5% of the total product power will be in the carrier spike,

67.8% of the total product power will be in the sidebands between 300 and 3400 Hz from the carrier,

15.7% of the total product power will be in the sidebands less than 300 Hz from the carrier,

4% of the total product power will be more than 3400 Hz from the carrier.

IM Product Power. Wescott shows that for other than trivially small cases, the number of $(f_1 + f_2 - f_3)$ products falling at the center of a uniformly loaded band of n carriers will be $3n^2/8$. [8] With n carriers distributed randomly in N frequency slots, the number of products falling into the central channel will be reduced by the ratio of (N/n). Using Wescott's relationship that the $(f_1 + f_2 - f_3)$ product has a power which is 5.5 dB greater than the $(2f_1 - f_2)$ product, the relationship that the C/I ratio improves 2 dB for each dB reduction in carrier level, and the empirical relationship that with two carriers at maximum output power the carrier-to-$(2f_1 - f_2)$ product ratio is 12 dB; the SCPC carrier-power to IM-product power in the worst (central) channel with no output backoff will, on the average, be given by:

$$\left. C/I \right]_{dB} = 4.7 + 10 \, \text{Log} \left[N/n \right] \tag{5}$$

For high-density systems which use the entire transponder and a 40% activity factor, the staggering factor, 10 Log (N/n), will be 4.0 dB. For lower density systems the frequency staggering factor will be greater.

With satellite output backoff (OBO), equation (5) becomes

$$\left. C/I \right]_{dB} = 4.7 + 10 \, \text{Log} \left[N/n \right] + 2 \, (OBO) \tag{6}$$

Testtone-to-Intermodulation Noise Ratio. With an FM SCPC system using 6 dB/octave emphasis, the system is effectively phase modulated and the testtone-to-IM ratio is given by:

$$\left. TT/I \right]_{dB} = \left. C/I \right]_{dB} + 20 \, \text{Log (peak testtone dev./testtone freq.)} \tag{7}$$

There are two factors which can help reduce the TT/I noise in a channel, the frequency stability of the terminals should be such that the carrier spike energy in the product will be sub-audio and the terminal should have a reasonably good high-pass filter to reduce the amount of energy below 300 Hz which reaches the receive channel voice output terminals.

Assuming that the carrier spike of the product will remain sub-audio and noting that 67.8% of the total product energy will be demodulated into the band between 300 and 3400 Hz, we can combine equations [6] and [7] to give:

$$\left. TT/I \right]_{dB} = 6.4 + 10 \, \text{Log} \frac{N}{n} + 2(OBO) \tag{8}$$

$$+20 \, \text{Log (peak testtone dev./testtone freq.)}$$

Margins

Two types of margins are of interest in applications of satellite systems; the operating margin with regard to the carrier-to-noise-density ratio and the equipment implementation margin. The operating margin is included to allow for rain attenuation, antenna pointing errors, and other effects which can reduce the C/N_0 at the input to the equipment. The equipment implementation is the practical margin necessary to allow for equipment imperfections when compared to the theoretically computed performance values.

Because of the variable quantity of the compandor improvement and because of its subjective nature, measurements made in assessing the performance of the system must, perforce, be made with the compressor and expandor switched out of the path. Under these conditions the implementation margin can be readily quantified. In the applications discussions of following sections, the uncompanded S/N is computed using testtone as a reference signal and then compandor improvements added, recognizing that, in general, systems measurements are made without the compandors.

IF Bandwidth

The appropriate pre-detection bandwidth for use in the demodulator is given by Vandemark [9] and its derivation is explained there. The IF noise bandwidth is:

$$\text{Bandwidth} = 2.8 \, [(\text{testtone peak deviation}) + 1000] \tag{9}$$

C/N_0 Calculations

Downlink

$$\text{In dB:} \quad \left. \frac{C}{N_0} \right] = \left. EIRP \right]_{SAT} - L_p + \left(\frac{G}{T} \right)_{ES} - K \tag{10}$$

where: $\left. EIRP \right)_{SAT}$ = satellite eirp in dBW

L_p = 196.8 dB at $5°$ elevation angle

= 196.2 dB at $30°$ elevation angle

$(G/T)_{ES}$ = earth station gain-to-temperature ratio in dB/°K

K = Boltzman's Constant = -228.6 dB W/Hz-°K

or: (11)

$$C/N_O \Big]_{dB} = EIRP \Big]_{SAT} + (G/T)_{ES} + 31.8 \text{ at } 5° \text{ elev. angle}$$

Uplink (12)

$$C/N_O \Big]_{dB} = EIRP \Big]_{SAT} - G_{SAT} + (G/T)_{SAT} - K$$

where: $EIRP)_{SAT}$ = satellite eirp in dBW   (13)

G_{SAT} = satellite isotropic gain in dB

$(G/T)_{SAT}$ = satellite receive gain-to-temp. ratio in dB/°K

FM Noise

All beating oscillators between the modulator and the demodulator will contribute FM or phase noise to the carriers. Vandemark [9] gives typical contributions for transmit channel equipment, upconverter, satellite translation oscillator, downconverter, and receive channel equipment. The total FM noise deviation in the 300–3400 Hz band, relative to a 1-Hz deviation, after de-emphasis, is typically 28.2 dB.

TYPICAL SYSTEMS

It is not the purpose of this paper to delve into the economics of the optimum design of the earth station in terms of antenna size, LNA temperature, and power amplifier size. In the systems examples given here, the requisite G/T is given for each application and an example is given as to how the G/T may be obtained. In each case the noise contributions outside of the LNA itself are assumed to be 32°K, which is a reasonably achievable value with good antennas operating at an elevation angle of 5°. An antenna aperature efficiency of 60% is similarly assumed. An exception is the standard Intelsat earth station whose "standard" G/T is 40.7 dB, even though most stations use antenna sizes and LNA's which typically yield G/T's of over 42.0 dB.

The systems described use equipment whose channels are spaced the same as the Intelsat SPADE SCPC Systems, i.e., 45 kHz, or one-half of the SPADE spacing, i.e., 22.5 kHz. FM systems are readily capable of other spacings, however.

The parameters and characteristics of the different systems are given in Table 3.

TABLE 3. TYPICAL FM SCPC SYSTEMS DESIGN

		INTELSAT IV GLOBAL BEAM				DOMESTIC SATELLITE		BUSH SYSTEM	
		VIDEO +SCPC	1200 CHAN	800 CHAN	200 CHAN	1333 CHAN	667 CHAN	BUSH TERM.	CENTRAL TERM.
TEXT REF		A	B	C	D	E	F	G	H
SATELLITE		ITS–IV	ITS–IV	ITS–IV	ITS–IV	WESTAR	WESTAR	RCA	
OUTPUT BACKOFF	DB	12.9*	6	4.4	2	8	4.4	2.5	
GAIN	DB	133.7	137.7	136.8	132.5	157.0	155.8	149 **	
TOTAL EIRP (SCPC)	DBW	9.6	16.5	18.1	20.5	26.0	29.6	25.5 ***	
NO. OF SCPC CHANNELS		200	1200	800	200	1333	667	240	120
ACTIVITY FACTOR	%	40	40	40	40	40	40	40	40
NO. OF SCPC ACTIVE CHANNELS	(n)	80	480	200	50	533.2	267	96	48
EIRP/SCPC CHANNEL	DBW	-9.4	-10.3	-4.9	3.5	-1.3	+5.3	5.1	0.0
G/T	DB/°K	-17.6	-17.6	-17.6	-17.6	-6.0	-6.0	-4.8	
EARTH STATION									
G/T		31.7	40.7	29.3	21.3	37.0	19.1	19.3	30.9
TYPICAL ANT DIA	M/FT	11/36.1	30/98.4	10/32.8	5/16.4	15.5/51	5/16.4	4.6/15	10/32.8
LNA TEMP/NF	°K/DB	55/0.75	20/0.29	90/1.2	162/1.9	20/0.29	263/2.8	225/2.5	55/0.75
SCPC SYSTEM									
RMS TESTTONE DEV	Hz	4600	3300	4600	4600	3300	4600	4600	4600
IF NOISE BANDWIDTH	kHz	25.7	18.75	25.7	25.7	18.75	25.7	25.7	25.7
CHANNEL SPACING	kHz	45	22.5	45	45	22.5	45	45	45
NO. OF CHANNEL SLOTS	(N)	200	1600	800	800	1333	667	1511	1511
DOWNLINK C/N_O	DB-Hz	54.1	62.2	56.2	56.6	67.5	56.2	56.2	62.7
UPLINK C/N_O	DB-Hz	67.8	63.0	73.9	82.0	64.3	71.3	79.9	74.8
TOTAL C/N_O	DB-Hz	53.9	59.6	56.1	56.6	62.6	56.1	56.2	62.4
C/N	DB	10.0	16.8	12.0	12.5	19.9	12.0	12.1	18.2
WEIGHTED, UNCOMPANDED									
TT/THERMAL NOISE	DB	34.7	37.6	37.1	37.6	40.3	37.1	37.2	43.3
WEIGHTED, UNCOMPANDED									
TT/INTERMOD	DB	54.9	39.5	39.9	41.2	42.3	37.9	39.6	34.5
WEIGHTED, UNCOMPANDED									
TT/FM NOISE	DB	45.0	42.2	45.0	45.0	42.2	45.0	45.0	45.0
WEIGHTED, UNCOMPANDED									
TT/TOTAL NOISE	DB	34.3	34.6	34.8	35.5	36.7	34.1	34.8	34.1
COMPANDED EQUIVALENT TT/NOISE									
FOR MEDIAN TALKER	DB	51.3	51.6	51.8	52.5	53.7	51.1	51.8	51.1
COMPANDED EQUIVALENT TT/NOISE									
FOR LOW-LEVEL TALKER	DB	57.1	57.4	57.6	58.3	59.5	56.9	57.6	56.9
VIDEO SYSTEM									
EIRP	DBW	20.1							
C/N_O	DB-Hz	83.4							

* FOR SCPC CHANNELS

** TO REMOTE STATIONS ON EDGE OF ANTENNA PATTERN

*** TO REMOTE STATIONS ON EDGE OF ANTENNA PATTERN WHERE SATURATED EIRP IS 28 DB

An Intelsat Transponder Transmitting Both Video and SCPC

Domestic satellite applications in small countries may use an Intelsat IV global-beam transponder both for SCPC telephony and for video distribution. One such application is shown in Figure 3.

In this application the video carrier takes most of the satellite EIRP and occupies one-half of the transponder bandwidth as shown in Figure 4. The size of the receiving earth terminals are dictated by the need to receive the video carrier with some margin above threshold. Such stations have a G/T ratio of at least 31.7 dB. When the video carrier is present, the satellite transponder will be driven well towards saturation and the satellite gain will be lowered for the SCPC signals. The pertinent characteristics of this system are given in Table 3 in column "A". With 45-kHz centers, 200 channels are possible on a bandwidth basis in the 9 MHz allocated for this purpose in the frequency plan of Figure 4. If more power were available for the SCPC system, it would be possible to increase the number of available channel bandwidths beyond 200.

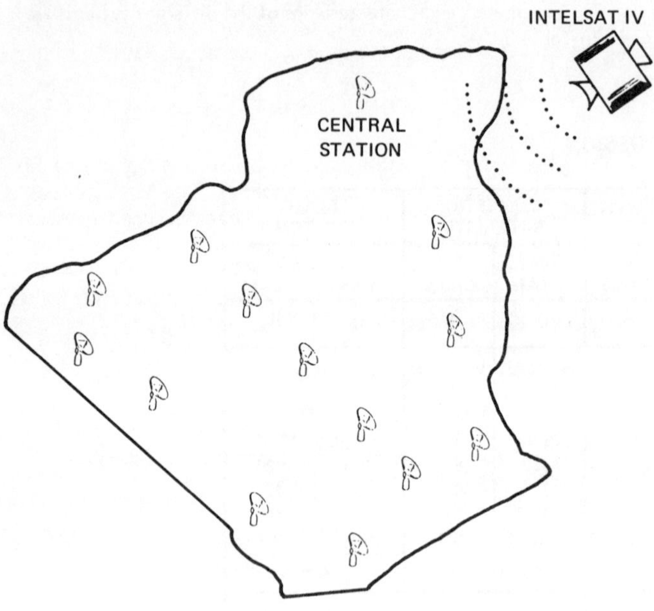

VIDEO: FROM CENTRAL STATION TO ALL REMOTE STATIONS   SCPC: TRANSMIT/RECEIVE ALL STATIONS

FIGURE 3. SMALL COUNTRY VIDEO/SCPC SYSTEM

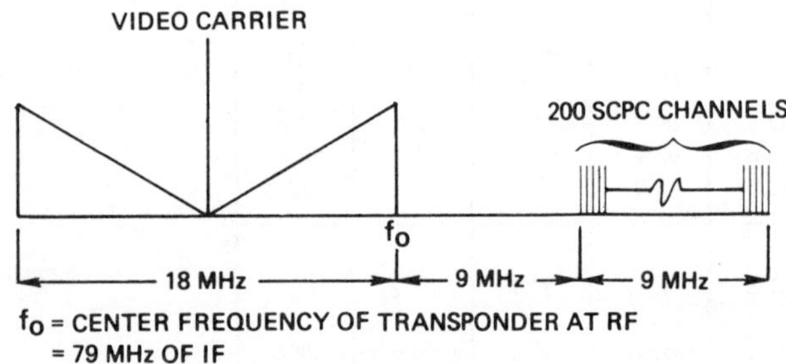

f_0 = CENTER FREQUENCY OF TRANSPONDER AT RF
= 79 MHz OF IF

FIGURE 4. FREQUENCY PLAN, VIDEO + FM SCPC

An Intelsat IV Global-Beam Transponder Transmitting FM SCPC

If the Intelsat IV global-beam transponder is used only for transmitting SCPC signals, three systems are shown; one using standard Intelsat earth stations yielding a total of 1200 channel allocations, one using smaller earth stations yielding a total of 800 channels of SCPC, and one using 5-meter earth stations yielding a total of 200 channels of SCPC.

Intelsat IV Global-Beam Transponder with 1200 Channels. For this application, it is assumed that standard Intelsat earth stations with G/T ratios of 40.7 dB or greater will be used. In this case, the Intelsat satellite is operated with 6 dB of output backoff. The computed resulting performance is shown in Table 3 column "B". Of note in this application arrangement is the relatively high carrier-to-noise ratio (16.8 dB) which is achieved.

Intelsat IV Global-Beam Transponder with 800 Channels. When the use of the FM SCPC equipment is optimized for a channel capacity of 800 channels, i.e., when using 45-kHz allocated-channel bandwidth, it is possible to optimize the system to minimize the G/T ratio required of the earth station. This is done in the arrangement shown in column "C" of Table 3.

In this case, note that the G/T of the earth station is only 29.3 dB/°K. This G/T can be achieved in several ways, one of which is shown on the table, a 10 meter antenna with a 90°K LNA.

Intelsat IV Global-Beam Transponder with 200 Channels. Column "D" of Table 3 shows a system design with even smaller earth terminals using 5-meter antennas and capable of 200 channels. This is a case where the satellite is definitely power limited. In this case the staggering factor is used to advantage so that only 2 dB of output backoff is required.

Domestic Satellite Systems

When domestic systems such as the Western Union WESTAR is used with its higher gain and greater EIRP, a different set of performance parameters become possible. For example, when used with the standard Western Union earth station, a total of 1333 channels can be used with some margin in the companded, equivalent testtone-to-noise ratio or a lower G/T earth station may be used while still achieving 1333 channels of 10,000-picowatt voice performance. A lower density (800 channels) system is feasible using even less sensitive earth stations.

High-Density Domestic Systems. When a satellite such as the Western Union WESTAR is used for SCPC applications with the earth terminals which are presently in operation by Western Union, the satellite is bandwidth limited, i.e., the capability of the satellite is limited by the 1333 channels that are available at 22.5-kHz spacings. Even under those high-density conditions, the performance of the system will be better than the nominal 10,000-picowatt objective. This is shown in column "E" of Table 3 where the companded, equivalent testtone-to-noise ratio for the average talker is shown to be 53.7 dB.

Low-Density Domestic Systems. If the same WESTAR domestic satellite is used in a system arrangement which trades the capability of the satellite for less-expensive earth stations and the number of allocated channels is reduced to 667, the resulting system can operate with earth station G/T ratios of as low as 19.1 dB/°K as shown in column "F" of Table 3. Here toll quality performance is achieved with an antenna 15 feet in diameter and a low-noise amplifier whose noise figure is 2.8 dB such as can be achieved readily with transistor amplifiers.

The low cost of such stations clearly opens the way for satellite communications in areas which up to this point have been considered suitable only for large-centralized earth terminals.

Bush-Type Telephone Systems

In some parts of the world applications exist for communications into remote villages which are not feasible to be served in any way other than by satellite communications. In some of these cases, the available domestic satellite may not have as high an antenna gain in the direction of the bush villages as in the main coverage area of the domestic satellite. Such a problem, for example, exists in Alaska and in northern Canada. (See Figure 5.) In this case, the satellite definitely becomes power limited rather than bandwidth limited because the lower EIRP and lower gain of the satellite looking toward the remote bush station cannot, economically, be overcome by the use of larger bush terminals.

In a bush-type telephone system there will generally be one, or a few, larger terminals which are used to carry higher-density traffic. These central terminals will contain the demand-access control equipment and will generally be the terminus of calls from the bush stations. In the examples shown in columns "G" and "H" of Table 3, it is assumed that two-thirds of the traffic will be between a bush terminal and a central terminal and one-third of the traffic will be between the bush terminals themselves. In this case, two different satellite transmit powers per SCPC carrier will be used, one in transmitting to the bush terminal and a lower value in transmitting to the central terminal. This, of course, requires that the SCPC equipment, if fixed-allocations of frequencies are used, be adjusted so that those carriers going to the bush terminals have different output powers than the carriers going to the central terminal. When demand assignment is used, this change in bush terminal and central terminal output power is achieved by control by the DAMA equipment. (The SCPC transmit-channel unit should be capable of changing output power by external command.)

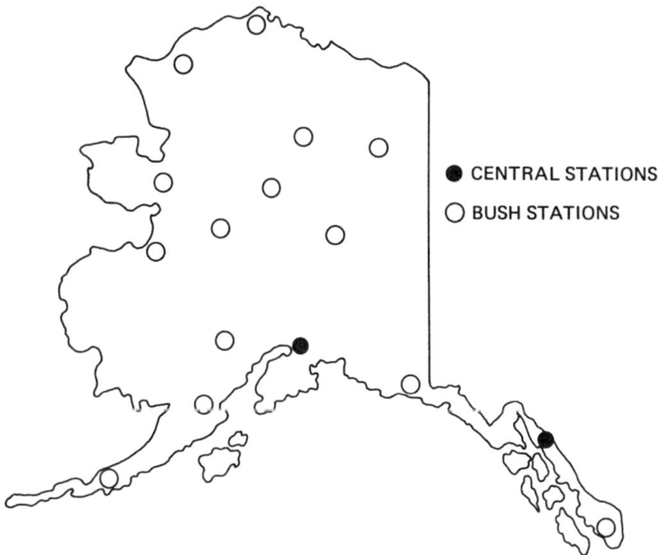

FIGURE 5. BUSH TELEPHONE SYSTEM

The G/T of the central terminal is assumed to be 30.9 dB/°K although this could be reduced with little impact on system performance since the downlink to the central station is seen to take very little satellite power. The bush stations use an antenna with a diameter of 15 feet and an LNA noise figure of 2.5 dB. (It is not likely that antenna diameters smaller than about 5 meters of 15 feet should be used in satellite service because of the potential interference problem existing among satellites in the synchronous orbit locations.) With a total EIRP from the satellite of 23 dBW (some 5 dB below the EIRP of the satellite in its nominal coverage range) and with the 15-foot antenna, and 2.5 dB noise figure LNA, the system will support 240 channels to bush terminals and 120 channels to central terminals. Thus the system will support 180 simultaneous calls approximately one-third of which are bush-to-bush and two-thirds are bush-to-central.

ECONOMICS OF THE SCPC APPROACH

The economics of the FM SCPC approach are viable only under certain applications conditions. The high-channel capacity capability of the SCPC used with either high-gain satellites or with high-sensitivity earth stations is attractive in terms of the utilization of the satellite capability. However, with satellites such as the proposed 24-transponder RCA satellite, transponder costs may not predominate in the total operating costs of a system. For such systems, the multiple-access capability of the SCPC approach will greatly reduce both the satellite transponder and the earth station terminal costs.

It is assumed, of course, that a low-cost demand-assignment control system can be implemented. For thin-route and low-density systems, Whitworth and Potano [9] describe one approach to the demand-access problem. Their approach for demand access if fully compatible with the SCPC system described by Vandemark [6]. Alternate demand-assignment control systems offer the possibility of economic central control for demand assignment and multiple access for high-density systems.

Even without the use of DAMA, the FM SCPC system should prove economic in those arrangements involving a large number of relatively small-capacity terminals. Here the more efficient use of satellite power through the use of voice-operated carrier may make the SCPC system quite competitive to FDM-FM systems approaches. Another economic application of the FM SCPC is in conjunction with portable terminals which must be available for installation to provide a few channels of voice communication in a short period of time. Such systems can be installed and integrated with other SCPC carriers in the same transponder on an economic and efficient basis.

CONCLUSIONS

The paper has described a wide variety of applications of FM SCPC using both Intelsat and domestic satellites. The relatively low cost of the FM SCPC and its usefulness with relatively small, inexpensive earth terminals make available to the communications fraternity a new tool for providing services in ways that are more economic than others and, frequently, are able to provide services in areas which could not be reached in other ways.

REFERENCES

[1] A. Bairi and J. Leonhard, "A Domestic Satellite System for Algeria," 1975 IEEE International Conference on Communications Record.

[2] "INTELSAT IV System: Some Transmission Considerations," ICSC/T-30-11E W/3/69, Intelsat, Washington D.C.

[3] "INTELSAT IV Channel Capacity Versus Earth Station G/T," ICSC/T-36-13E W/5/70, Intelsat, Washington D.C.

[4] S.N. Verma, "WESTAR Communications Characteristics," IEEE 1974 National Telecommunications Conference Record, pp 108–113.

[5] "Domestic Communications Satellite," p/o FCC filing for Domestic Satellite System, Revision A; Nov. 8, 1973, RCA Gov't and Com'l Systems Astro-Electronics Div., Princeton, N.J.

[6] G. Vandemark, "A SCPC Terminal for Satellite Communications," 1975 IEEE International Conference on Communications Record.

[7] P.M. Boudreau, and N.G. Davies, "Modulation and Speech Processing for Single-Channel-Per-Carrier Satellites Communications," 1971 IEEE International Conference and Communications Record, pp 19–9/19–15.

[8] R.J. Wescott, "Investigation of Multiple FM/FDM Carriers through a Satellite TWT Operating Near to Saturation," Proc. IEE, vol 114, No. 6, June 1967.

[9] J.E. Whitworth and B.A. Potano, "Low Cost Satellite Earth Terminals for Thin Route and Remote Subscriber Applications," IEEE 1974 National Telecommunications Conference Record, pp 128–133.

Subsection 3.6.3
Time Division Multiple Access

3.6.3.1

TIME-DIVISION MULTIPLE ACCESS

O. GENE GABBARD and PRADMAN KAUL

Digital Communications Corporation, Gaithersburg, Maryland

Abstract

Time-Division Multiple Access (TDMA) is an effective method of significantly increasing satellite channel capacity and improving satellite system flexibility. As a result, there is increasing worldwide interest and a stepped-up hardware development pace in TDMA. This paper discusses the principles of TDMA, its advantages, disadvantages, and operating modes. Also presented are typical link equations for satellite TDMA transmission.

INTRODUCTION

Satellite time-division multiple access (TDMA) is a technique whereby a multiplicity of satellite earth stations share a common transponder by transmitting bursts of carrier signals which arrive at and pass through the transponder in a sequential, non-overlapping mode. Multiple transponder operation is also possible from single TDMA terminals using time-switched earth station up- and down-converters. Recently, TDMA has received worldwide attention because of its high efficiency. Various experimental TDMA systems operating at bit rates from 6 Mbps to 60 Mbps have been developed and successfully tested [1, 2, 3, 4].

If more than one carrier is amplified simultaneously by a single transponder, the total channel capacity of that repeater is reduced because of the non-linear characteristics of typical satellite transponders when operated at high output power levels. Moreover, operating in a frequency-division multiple access (FDMA) mode complicates the establishment of a frequency plan and the bandwidth-power relationship of individual carriers. Changes to existing links are generally difficult to achieve. These may affect all users, decrease overall channel capacity, and restrict system flexibility. By contrast, a multiple access system time-sharing the satellite does not suffer from these difficulties since only one carrier is present at any one instant of time in a given satellite transponder.

The difference in efficiency can be illustrated comparing approximate channel capacities of an INTELSAT IV global beam transponder operating with normal INTELSAT-standard 30 meter earth stations, using FDMA and TDMA, respectively. Assuming ten accesses, typical capacity using FM/FDMA is approximately 450 one-way voice channels. With TDMA, using standard 64 kbps voice frequency PCM encoding, the capacity of the same transponder is about 900 channels. If Digital Speech Interpolation (DSI) is used to process the PCM bit streams, the capacity is further increased to approximately 1800 channels.

TDMA GENERAL DESCRIPTION

A pictorial representation of TDMA operation is shown in Figure 1. One earth station acts as reference and transmits periodic bursts without closed loop control. The other stations in the network use closed loop synchronization through the satellite to place burst transmissions within their assigned time slots (assumes global beam operation). In the INTELSAT TDMA system, a separate non-traffic bearing reference synchronization burst is radiated by an earth station which subsequently also radiates a normal data burst. If for some reason, this station fails to transmit the synchronization bursts, a back-up station automatically assumes the reference station responsibility for the network, and places a synchronization burst in the same time location in the TDMA frame. Any burst movement or re-location is, therefore, avoided in the frame.

Figure 2 provides typical TDMA frame and burst formats. For simplicity a four-access network (Stations A through D in one frame) is shown. However, up to about 60 could be accommodated in an actual system. Typically, 5 to 15 accesses per frame are expected in operating networks. The burst lengths are normally different, since different gross traffic loads are transmitted by each station. Reconfiguring of the traffic load is easily accomplished by re-allocating the burst lengths assigned to each station. Such re-allocation can be accomplished manually or with the aid of a miniprocessor since digital control is used.

A typical burst is expanded in the lower portion of Figure 2. The guard time, normally 100 to 200 nanoseconds, is provided to assure that overlap between successive bursts never occurs. This is followed by a carrier-recovery/bit-timing-recovery (CR/BTR) sequence, which is used by the PSK demodulator in each receiver to recover local carrier and digital clock for coherent demodulation. For example, the CR/BTR sequence is typically 60 bits in a 60 Mbps system or about one microsecond per access.

Reprinted from the *Record of the IEEE Electron. & Aerosp. Syst. Conv., (EASCON),* Oct. 7-9, 1974, pp. 179-184.

The minimum realistic number of CR/BTR bits provided in the preamble for burst-operated systems is essentially the same for a wide range of bit rates. This arises from the fact that carrier recovery and bit timing recovery at the receiver must be accomplished by realizable phase-lock loops and/or filters which are sufficiently narrow-band to provide good output signal-to-noise performance. Generally, CR/BTR loops which have bandwidths of from 0.5 to 2 percent of the bit rate provide a good compromise between acquisition time and demodulation bit error rate performance degradation resulting from finite loop output signal-to-noise ratios. Adaptive phase-lock loops which acquire in wideband mode and track with a narrower bandwidth can be used to reduce the amount of CR/BTR overhead. However, such adaptive techniques have not yet been developed for high bit rate demodulators.

After CR/BTR, a unique word (UW) of about 20 bits is transmitted, which, when detected at the receiver, is used to establish an accurate time reference in the received burst. The location of each data bit in the burst can then be determined by counting clock relative to the detection time of the UW. The UW can also be used to identify the transmitting station. Alternatively, a station identification code (SIC) of about six bits can follow the UW to provide the identification function. The planned INTELSAT TDMA system (frame and burst formats are shown in Figure 3) uses this latter technique.

The probability of missing or falsely detecting the UW is of primary importance in practical TDMA systems. If the UW is missed, usually an entire data burst is lost. This causes impulses or "clicks" in voice traffic being carried. For data, large blocks are lost. A falsely detected UW causes "skew" in the demultiplexed data producing essentially the same effects. Generally, a probability of miss or false UW detection of 1×10^{-8} or better is required at the threshold bit error rate (i.e., 1×10^{-4}). This will provide a mean time to UW miss or false detection of several hours for frame lengths in the one millisecond region.

In order to achieve realistic miss detection performance, UW's which contain a reasonable number of bit errors must be accepted by the detector as being valid. For example, a 20-bit UW which contains four or less errors may be accepted, while all words containing more than four errors are rejected (or missed). For this example, the "detection threshold" is four bits. Allowing such a detection threshold increases the probability of false detection (i.e., that random data patterns will meet the criteria for detection). This can be combated by using "aperture gating". A narrow window or gate is established in the immediate vicinity where the UW is expected. Typically, apertures of 10 to 20 bits in width are used. Assuming a TDMA frame containing 45,000 bits, an improvement ratio of about 45,000/10 or 4,500 is obtained in the false detection probability.

A great deal of analysis has been done in the area of UW detection [5]. Unfortunately, it is often very difficult to achieve hardware UW miss detection performance which is near theoretical predictions. Typically, measured performance has been two to five orders of magnitude from calculated values. Such large departure from theoretical is due primarily to intersymbol interference, and carrier/bit timing acquisition problems in practical demodulators. Near theoretical performance can be obtained if filter type CR/BTR loops are used in the demodulator [6, 7].

The detected UW's are also used to establish and maintain accurate burst synchronization in TDMA. For systems operating over satellites in synchronous orbit, the burst synchronizer is usually implemented as a digital phase-lock loop with a tracking bandwidth of approximately 3 Hz and a phase resolution of one symbol (two bits using four-phase PSK modulation). The detected reference (or synchronization) UW serves as the reference input to the loop, while the local station detected UW becomes the feedback input to the phase-lock loop through the satellite. Early TDMA experiments show that 100 nanosecond guard times were practical using the described technique [8].

In some cases, particularly for higher order PSK modulation (i.e., eight-phase), it is desirable to restrict the effective modulation to two-phase in at least the CR/BTR and UW portions of the preamble. Two-phase $(0, \Pi, 0, \Pi, \ldots)$ modulation for CR/BTR provides high energy at the symbol rate, and usually has less intersymbol interference (influence of the numerical value of one symbol on adjacent symbols). More optimum CR/BTR circuits can, therefore, be realized. Higher effective energy per bit to noise spectral density (E_b/N_0) is also achieved for a two-phase modulated UW than in four- (3 dB gain) or eight-phase (6 dB gain) PSK systems. This decreases the bit error rate in the received UW and substantially improves the UW miss detection probability [9].

Other housekeeping functions such as control signalling channels, transmission bit error rate (BER) sequences, voice order wires, teletype order wires, test channels, etc., may be inserted between the UW and data portion of the burst. Taking all of these into account, the total preamble (all overhead bits) will usually be 100 to 200 bits per burst frame.

For high capacity TDMA, frame lengths from 125 microseconds to 1 or 2 milliseconds are most practical. If a frame length of 750 microseconds were used with 15 accesses and 150 bits of overhead each, at 60 Mbps transmission rate, the total overhead to information bits per frame would be 2,250/45,000. The resulting percentage of overhead to total bits per frame is five percent, and the transmission efficiency is 95 percent. The total time allocated to guard time would be 15 times 150 nanoseconds, or 2.25 microseconds per frame, amounting to only 0.3 percent of the total frame time.

As can be seen, the frame length, preamble length, and number of accesses or bursts per frame determine the transmission efficiency of a TDMA system. Frame length is directly proportional to the efficiency. As the length increases, the number of overhead bits per unit time decrease. Thus, the efficiency is increased. The limiting constraint is that the size of the buffer memories used for compression and expansion of the data to form the TDMA bursts also increases with the frame length, thus increasing the complexity and cost of the hardware.

To aid in understanding of TDMA, a block diagram of the currently proposed INTELSAT TDMA terminal equipment is shown in Figure 4. Information to be transmitted is processed by the appropriate Terrestrial Interface Module (TIM). Such modules will include PCM encoding of voice (VF PCM Codec), FDM supergroup PCM encoding, direct digital interface (DDI), DSI, etc., as shown. These TIM's work into flexible multiplexers/demultiplexers under control of the Common TDMA Terminal Equipment (CTTE). Within the common equipment are located redundant minicomputers which control the multiplexers/demultiplexers, initial time slot acquisition, order wire signalling and switching, etc. These processors also monitor the overall functioning of the CTTE, and cause switchover when a failure occurs.

The multiplexed data on the transmit side is combined with the preamble and then scrambled by half-adding with a pseudo-random sequence of 32,767 bits length. This operation guarantees a reasonable number of bit transitions during portions of the transmitted burst that are lightly loaded (all zeros data). Reasonably continuous bit transitions are required at the demodulator in order to retain coherent bit timing recovery. Scrambling also effectively produces a nearly white noise frequency spectrum at the output of the PSK Modulator, thus keeping the radiated satellite energy per 4-kHz of bandwidth low to avoid interference with terrestrial microwave facilities operating in the same band [10]. Before modulation, the data is also differentially encoded. This process is used to remove carrier phase ambiguity, which can invert or bit position translate the data, at the PSK demodulator.

Simpler forms of TDMA hardware are more optimum for systems other than INTELSAT. For example, TDMA presently under procurement by TELESAT Canada for domestic operation does not incorporate minicomputers or highly flexible multiplexers/demultiplexers [11]. Instead, the PCM encoding euqipment is hardwired to the common control equipment. Changes in format, interface equipment, etc., are accomplished by a build-on capability and reconfiguring of wire-wrappable back planes.

CHANNEL CAPACITY

For bandwidth limited cases, the nominal capacity of a satellite transponder using TDMA can be approximated by the following expression [12].

$$R_b = W + B - C_W$$

where R_b = the satellite transmission link bit rate, dB (band-limited case)

W = bandwidth of the satellite transponder

B = bit-rate-to-symbol-rate ratio (1 for two-phase PSK, 2 for four-phase PSK, and 3 for eight-phase PSK)

C_W = the ratio of the transponder bandwidth to the possible band-limited symbol rate through the transponder (typically 1.2).

For an INTELSAT IV transponder,

W = 36 MHz = 76 dB

B = 3 dB

$C_W \cong 0.8$ dB

and $R \cong 78$ dB $\cong 60$ Mbps. The maximum bit rate through the transponder using four-phase PSK modulation strictly from a bandwidth point of view, is 60 Mbps.

If the satellite channel is power-limited on the down link, the following equation holds:

$$R_p = e.i.r.p. - P_L + G/T - K - E_b/N_o - M$$

where R_p = the satellite transmission link bit rate, dB (power-limited case)

e.i.r.p. = equivalent isotropically radiated power of the satellite per transponder

P_L = transmission path loss at 4 GHz (197 dB)

G/T = gain/temperature ratio for the receiving earth station

K = Boltzmann's constant (-228.6 dBW/Hz°K)

E_b/N_o = energy per transmitted bit to the noise power density ratio at a probability of bit error of 1 x 10⁻⁴ (8.4 dB for four-phase coherent PSK modulation)

M = total system margin.

For an INTELSAT IV global beam, e.i.r.p. = 22.5 dBW. The G/T for a standard INTELSAT earth station is 40.7 dB/°K. If a margin of 8 dB is chosen (2 dB for hardware implementation, 2 dB for earth station and satellite effects, and 4 dB for fade, tracking, etc.), then the power-limited bit rate, using four-phase coherent PSK, is,

$$R_p = 22.5 - 197 + 40.7 + 228.6 - 8.4 - 8 \text{ (dB)}$$

$$= 78 \text{ dB or} \approx 60 \text{ Mbps.}$$

In this case, the power-limited and band-limited
link bit rates are approximately equal, achieving
an optimum operating point. Actually, 60 Mbps is
conservative. By using careful channel equalization, up to 67.2 Mbps has been transmitted through
a 36 MHz bandwidth transponder using four-phase
PSK [13].

If the power-limited link bit rate were less than
the band-limited rate, then forward acting error
correction techniques, such as Rate-3/4 or Rate-7/8
convolutional encoding with threshold or soft decision decoding, could be used [14, 15]. These techniques can provide from 2 to 5 dB improvement. On
the other hand, if the channel is significantly
band-limited, higher order PSK modulation can be
used. Eight-phase PSK allows an increase of up to
50 percent (1.7 dB improvement) in bit rate through
a band-limited channel [9].

The voice channel capacity, C, of a TDMA system
using digital voice encoding can be expressed as:

$$C = \frac{1}{V}\left[R - \frac{NP}{T} \right]$$

where R = satellite transmission link bit rate

       V = bit rate for one voice channel

       N = number of bursts in a frame

       P = number of bits in the preamble

       T = the frame period

As an example, for the INTELSAT IV satellite transponder, some typical values would be:

     R = 60 Mbps
     V = 64 kbps
     N = 10
     P = 150
     T = 750 microseconds
and $C \cong$ 900 channels.

The channel capacity for direct digitally encoded
voice can be doubled by using Digital Speech Interpolation (DSI) [16, 17]. DSI systems are similar
to Time-Assignment Speech Interpolation (TASI) systems used over cables. However, the use of digital
processing significantly improves speech clipping
and freeze-out performance. The DSI principle is
to use the inactivity periods of one speech channel to carry the channel data of other active channels. For telephone conversations, one party is
normally listening while the other is speaking with
pauses; therefore, the mean activity of speech is
typically 30 to 40 percent. The ratio between the
number of input speech channels and the time slots
used for actual transmission is called DSI gain.
This gain is inversely proportional to the mean
activity of the input channels. Conservatively, a
DSI gain of two is obtainable. DSI can also be
implemented to accommodate multi-destination
operation.

ADVANTAGES AND DISADVANTAGES OF TDMA

The main advantages of TDMA are:

(1) The transponder can be operated at saturation, making maximum use of available satellite
power.

(2) TDMA can be used for voice, data, and
video transmission and can carry directly the new
digital carrier formats, such as T-carrier. In
more versatile TDMA common equipment designs, various interface modules can be used, such as supergroup PCM codecs, direct voice encoding PCM channel banks, DSI equipment, and direct digital
interfaces (DDI).

(3) The capacity of each earth station can be
changed more easily by changing burst formats.
This provides improved flexibility.

(4) All earth stations have the same transmit
and receive frequency. Hence, the frequency plan
for the earth stations is greatly simplified.

(5) In bandwidth limited applications, TDMA
can increase the information rate by using higher
order modulation. For example, by using an eight-
phase PSK modem, the information rate can be increased by approximately 50 percent over a four-
phase PSK modem. Generally, very efficient bandwidth to power trade-offs can be obtained with
digital transmission.

(6) In power-limited applications, forward
acting error correcting techniques can be used to
gain 2 to 6 dB in capacity and/or bit error rate
performance.

(7) TDMA is directly compatible with developing terrestrial digital voice and data networks.

Some disadvantages of TDMA are:

(1) At the present time, FM is generally less
expensive and more efficient for cases of a single
large carrier (i.e., 960 channel) accessing a
transponder.

(2) In general, TDMA is not effective for
earth stations with very light total transmit traffic. TDMA is usually an optimum technique for
earth stations radiating a total of 20 channels or
more.

(3) Today, there are few sources of commercial
TDMA hardware. There are many suppliers of FM/FDMA
equipment.

(4) FM/FDMA is directly compatible with existing FDM terrestrial telephony transmission
equipment.

FUTURE OF TDMA

TDMA is receiving worldwide attention for application to domestic, international and military
satellite systems. The two access TDMA system

being implemented by TELESAT Canada will go into commercial service in 1975 between Allan Park, Ontario and Harrietsfield, Nova Scotia. This message link will have a capacity of 800 voice channels and will extend traffic from the CANTAT II Trans-Atlantic cable. Members of INTELSAT are in the process of procurring TDMA hardware for field trials in the Atlantic and Pacific Ocean areas starting in 1976. Testing is expected to continue for one or two years with commercial operating beginning in the 1978 to 1980 time frame. This international TDMA system will include multi-transponder operation through at least the INTELSAT IV/IV-A satellite era due to the heavy traffic volume in relationship to the capacity of each satellite transponder.

The power-bandwidth efficiency of TDMA, augmented by DSI, offers increases in current satellite transmission capacities of as much as four-fold. The economic importance of this increase in capacity, coupled with the tremendous advances in high-speed semiconductor technology during the past few years, has brought TDMA techniques to the "ready for production" stage. Four-phase PSK modems for TDMA operation in the 60 Mbps region are available virtually off the shelf with performance within 2 dB of theoretical, and experimental 1 Gbps units operating within 4 dB of theoretical have already been constructed. Implementation of complex digital processing functions needed for TDMA hardware is now routine to 100 Mbps and reasonably practical to 500 Mbps [18]. PCM encoders of many types, including direct encoding of FDM supergroup and mastergroup signals, are readily available and DSI is in the final stages of development. Production DSI equipment for TDMA is expected to be available by mid-1975 from at least two or three sources. Many manufacturing companies and future users of TDMA are placing development funds into TDMA technology on a worldwide basis.

Looking to the 1980's, further advances in technology now being pursued in the laboratory environment are expected to further expand the use of time-division digital transmission by satellite. Single satellite capacities of 200,000 to 500,000 voice circuits are foreseen using Satellite-Switched Time-Division Multiple Access (SDMA/TDMA) techniques [19, 20]. Operating in this mode, satellites are expected to have fewer transponders (4 to 10) of greater bandwidth (200 to 600 MHz each), and may decrease in gross launch weight. If SDMA/TDMA technology continues to be actively pursued, systems of this type could be operational by 1985.

REFERENCES

[1] T. Sekimoto and J. G. Puente, "A Satellite Time-Division Multiple Access Experiment," IEEE Transactions on Communications Technology, Volume COM-16, August 1968, pp. 581-588.

[2] W. G. Schmidt, O. G. Gabbard, E. R. Cacciamani, W. G. Maillet, and W. W. Wu, "MAT-1: INTELSAT's Experimental 700-Channel TDMA/DA System," Proceedings of the 1969 INTELSAT/IEE Conference on Digital Satellite Communications, London, England, pp. 428-440.

[3] W. Maillet, "INTELSAT's 50 Mbit/s TDMA-2 System," Proceedings of 1972 International Conference on Digital Satellite Communications, Paris, France, pp. 26-32.

[4] K. Nosaka, "TTT System - 50 Mbps PCM/TDMA System with Time Preassignment and TASI Features," Proceedings of 1969 INTELSAT/IEE International Conference on Digital Satellite Communications, London, England, pp. 83-94.

[5] W. Schrempp and T. Sekimoto, "Unique Word Detection in Digital Burst Communications," IEEE Transactions on Communications Technology, Vol. COM-16, August 1968, pp. 597-605.

[6] A. Ogawa, I. Muratani, M. Okawa and K. Nosaka, "Analysis and Satellite Field Test Results of PSK Modem for PCM/TDMA System," Proceedings of International Colloquium on Space and Communications, Paris, France, 1971.

[7] Y. Matsuo, S. Sugimoto and S. Yokoyama, "A Study on Carrier and Bit-Timing Recovery for Ultrahigh Speed PSK/TDMA Systems," Proceedings of 1972 International Conference on Digital Satellite Communications, Paris, France, pp. 277-286.

[8] O. G. Gabbard, "Design of a Satellite Time-Division Multiple Access Burst Synchronizer," IEEE Transactions on Communications Technology, Vol. COM-16, August 1968, pp. 589-596.

[9] A. Ogawa and M. Ohkawa, "A New Eight-Phase PSK Modem System for TDMA," Proceedings of 1972 International Conference on Digital Satellite Communications, Paris, France, pp. 256-265.

[10] A Tomozawa, "Study of Spectral Energy Dispersal for Use with TDMA," Proceedings of 1972 International Conference on Digital Satellite Communications, Paris, France, pp. 297-305.

[11] R. K. Kwan, "A TDMA Application in the Telesat Satellite System," Proceedings of NTC, 1973.

[12] J. G. Puente, W. G. Schmidt, A. M. Werth, "Multiple Access for Commercial Satellites," Proceedings of IEEE, February 1971.

[13] COMSAT Press Release, "Telesat and COMSAT Test New TV System" June 10, 1974 (appeared in Defense Space Business Daily).

[14] W. W. Wu, "Applications of Error-Coding Techniques to Satellite Communications," COMSAT Technical Review, Vol. 1, No. 1, Fall 1971, pp. 183-219.

[15] I. M. Jacobs, "Integration of Demodulation, Decoding, Buffering, and Control for TDMA Demand-Assignment Systems," Proceedings of 1972 International Conference on Digital Satellite Communications, Paris, France, pp. 288-296.

[16] I. Poretti, G. Monty, A. Bagnoli, "Speech Interpolation Systems and Their Applications In TDM/TDMA Systems," _Proceedings of 1972 International Conference on Digital Satellite Communications_, Paris, France, pp. 348-357.

[17] J. A. Sciulli and S. J. Campanella, "A Speech Predictive Encoding Communications System for Multichannel Telephony," _IEEE Transactions on Communications_, Vol. COM-21, July 1973.

[18] P. Kaul, O. G. Gabbard, "Adapting High Speed Logic to Data Communications," _Electronics_, July 17, 1972, pp. 72-76.

[19] W. G. Schmidt and R. Cooperman, "A Satellite-Switched SDMA/TDMA System for a Wideband Multibeam Satellite," _Proceedings of International Communications Conference_, 1973.

[20] T. Muratani, "Satellite-Switched Time-Division Multiple Access," _Proceedings of EASCON_, 1974, Washington, D. C.

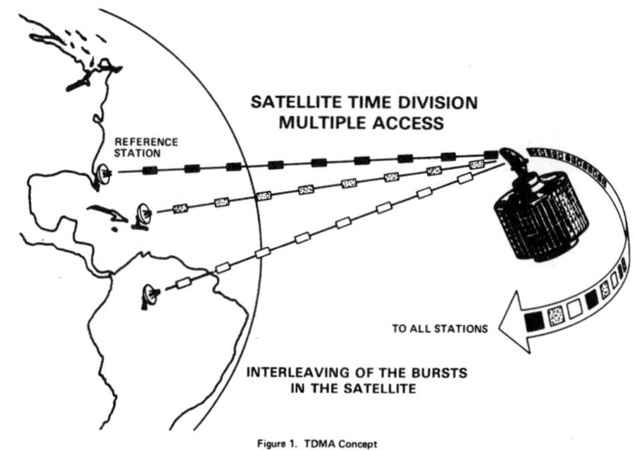

Figure 1. TDMA Concept

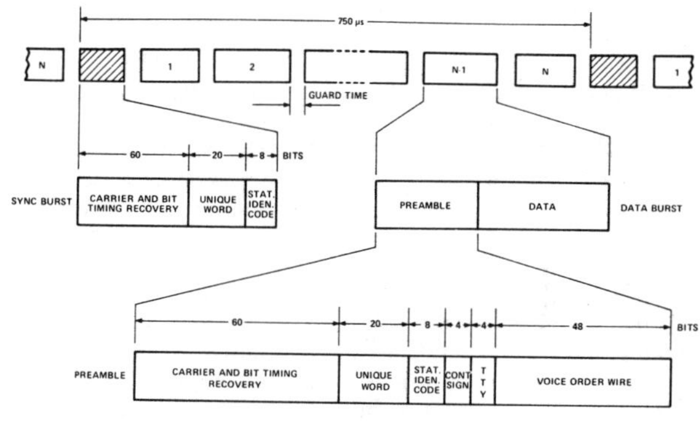

Figure 3. Frame and Burst Formats for Intelsat TDMA

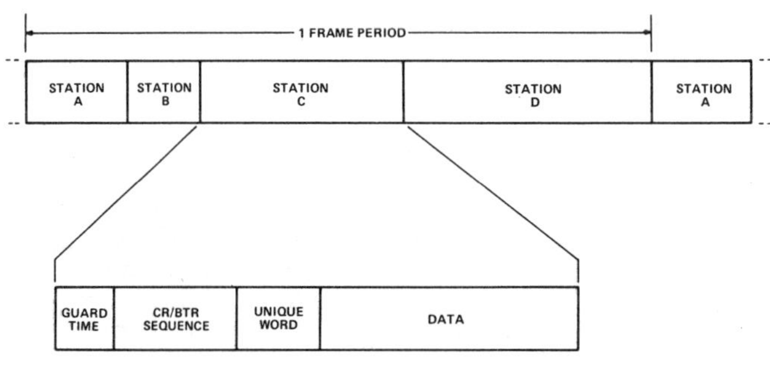

Figure 2. Typical TDMA Frame and Burst Formats

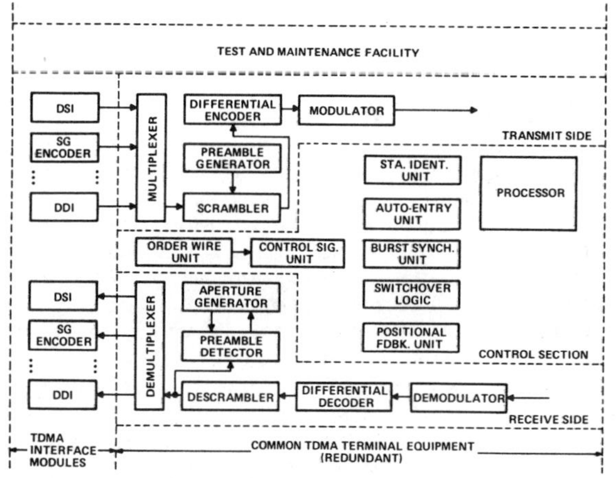

Figure 4. Block Diagram of Intelsat TDMA Terminal

3.6.3.2

TMDA, THE STATE-OF-THE-ART*

GEORGE D. DILL

COMSAT Laboratories, Clarksburg, Maryland

Abstract

Since its initial development, significant
advances have been made in the state-of-
the-art of time-division multiple access
(TDMA), primarily in the areas of acquisition and synchronization, modulation,
transmission link encoding, baseband processing, and implementation techniques.
These advances have resulted in an improved capability as well as lower implementation costs. Consequently, TDMA is
now being implemented in operational communications satellite systems. Most notable among these are the Canadian Telesat
system and the forthcoming Satellite Business Systems (SBS) system. This paper
summarizes the state-of-the-art of selected TDMA subsystem developments.

Introduction

Time-division multiple access (TDMA),
which is a relatively recent development
in telecommunications systems,[1-3] is most
applicable to satellite communications
networks. Since the mid-1960s when its
development was initiated, many laboratory
models have been developed and tested.
Several TDMA systems have already been put
into operational service; still more are
being planned. TDMA's chief attribute is
a greater satellite channel capacity. In
addition, it is compatible with modern
digital communications systems, and it offers a greater flexibility in accommodating transmission system growth. Recent
implementation techniques promise lower
investment cost and greater reliability
than other types of multiple-access systems. This paper summarizes the state-of-
the-art of selected TDMA subsystem developments which significantly enhance its
channel capacity, its performance, and its
economy.

TDMA Concept

A communications satellite in a geostationary orbit provides two unique features: it is within the view of a large
portion of the earth's surface (nearly one
third) and it is continuously in view of
essentially the same portion of the
earth.[4] These two features may be used to
simplify the provision of long-distance
telecommunications circuits. For example,
calls may be routed directly from any one
switching center to any other within the
coverage area of the satellite (without
transiting), and all transmissions may be
conducted via a single transponder (as
opposed to many repeaters in tandem).

In practice, a global beam (a beam sub-
tending nearly one third of the earth's
surface), a zone beam (a beam covering a
large geographical area but less than the
entire globe), or a combination of several
zone or spot beams covering selected locations on the earth's surface may be used
to relay the signals to and from a large
number of earth stations having TDMA terminals. The satellite also contains one
or more signal chains (receivers, filters,
frequency translators, and power amplifiers) to amplify and translate the received signal before retransmission.
These features permit many earth stations
to relay many transmission carriers simultaneously via the same satellite and even
the same transponder. The concept in
which more than one earth station operates
via the same transponder is called "multiple access." That in which a transponder
is sequentially shared, one carrier at a
time, is called time-division multiple
access (TDMA).

In a TDMA network, each earth station
transmits information (digitized samples
of voice, digital data, or digitized

---

*This paper is based upon work performed in COMSAT Laboratories under the sponsorship
 of the International Telecommunications Satellite Organization (INTELSAT). Views
 expressed in this paper are not necessarily those of INTELSAT.

Reprinted from the *Record of the IEEE Electron. & Aerosp. Syst. Conv., (EASCON)*, Sept. 26-28, 1977, pp. 31-5A-31-5I.

video) in short bursts. Each burst is synchronously transmitted at a specific time interval so that it will not overlap any other TDMA burst in the satellite transponder. The bursts are individually amplified by the satellite transponder and retransmitted in the appropriate down-beam. The respective TDMA terminals in each down-beam systematically detect and demultiplex the appropriate bursts and feed the information to the respective terrestrial interface trunks. Figure 1 is a diagram of the TDMA process.

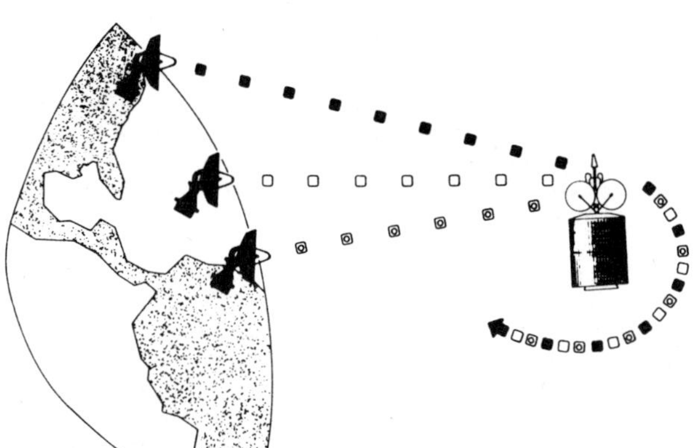

Figure 1. TDMA System Concept

As shown in Figure 2, each TDMA terminal consists of one or more terrestrial interface modules (TIMs), as well as TDMA control units, PSK modems, and a TDMA maintenance unit. The information to be transmitted is normally received from the respective switching centers (CTs) via a terrestrial link. The TIMs provide baseband processing to convert the received signals from analog to digital format, compress and expand the data, and provide the sampled data (either in composite or per-channel format) either to the DSI unit or directly to the TDMA control unit for transmission. The TDMA control unit provides for both the TDMA system control and TDMA acquisition and synchronization. Depending on the design, it may also provide transmit and receive TIM multiplexing and data compression and expansion buffering.

In most TDMA systems, a reference burst is transmitted within an arbitrary time interval defining a TDMA frame. A typical frame interval may range from $\tau = 125$ μs to $\tau = 15$ ms for most heavy traffic networks and up to several seconds for some low-speed data networks. Each TDMA terminal transmits its information burst frame in synchronism with the reference bursts. While the reference burst contains only a preamble and system control data, each information burst contains three types of information: preamble; system control data; and digitized voice, data, or video information. The preamble,

which is used by each demodulator to recover the carrier phase and symbol clock, also contains a unique word which identifies the beginning of the system control data. The system control data may include such functions as station identification code, terminal operating mode data, bit-error rate test data, and voice or tele-typewriter order wire. Typical TDMA superframe, frame, and burst structures are shown in Figure 3.

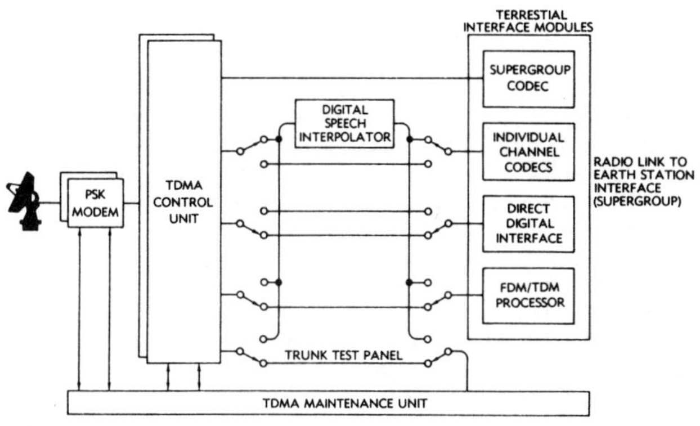

Figure 2. Simplified TDMA Terminal Block Diagram

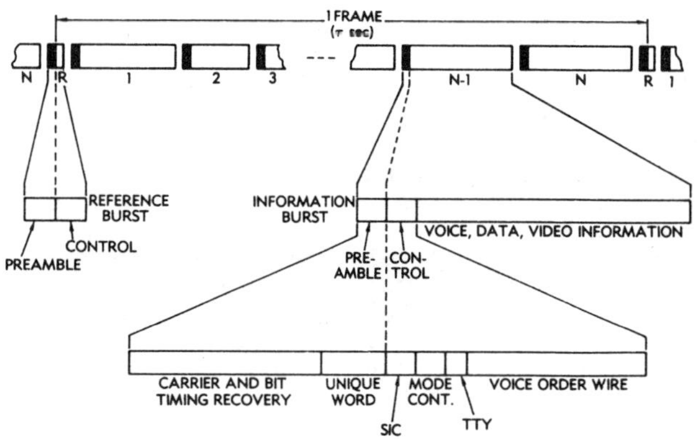

Figure 3. Typical TDMA Frame and Burst Structures

Since its conception, significant developments have improved the performance of TDMA. These developments fall into five general categories:

    a. acquisition and synchronization techniques,
    b. modulation techniques,
    c. transmission link encoding techniques,
    d. baseband processing techniques, and
    e. implementation techniques.

In addition, studies, developments, and tests have shown that, with a switch on-board the satellite (SS-TDMA), TDMA can provide an additional n-fold increase in satellite channel capacity.[5] The following sections summarize the key developments in each of these five areas of technology.

Acquisition and Synchronization

Two general types of acquisition and synchronization have been developed for TDMA systems. One is a burst-synchronous type which maintains time synchronization of all bursts through the satellite via a closed- or open-loop acquisition and synchronization control algorithm. The second is a bit-synchronous type which maintains time synchronization of all bits through the satellite via a closed-loop acquisition and synchronization control algorithm. Most TDMA systems developed to date are of the former type. Only one type of these is known to be fully developed to date, the U.S. Army's Converter Group OU-98 system.[6] Because of their predominance, this paper summarizes only the frame-synchronous-type TDMA systems.

Perhaps the most significant development in TDMA systems was the realization of efficient and reliable acquisition and synchronization techniques. Each TDMA terminal is at a different slant range with reference to a geostationary satellite, and a geostationary satellite has some relative motion with respect to each location on the earth's surface. Therefore, each TDMA terminal must be capable of first locating and then controlling its transmit burst time phase so that each burst always arrives at the transponder at a prescribed time relative to all other bursts, hence ensuring that no two bursts overlap and that the guard time between any two bursts is small to obtain high transmission efficiency. The acquisition process comprises the initial location or relocation of a burst within a prescribed time slot of a TDMA frame prior to establishing synchronization. Synchronization is the continuing process of controlling the TDMA burst so that it remains within the prescribed time slot once the appropriate time slot has been located.

The MATE experiment,[2] performed by COMSAT Laboratories in 1966, was the first major successful development and demonstration of independent TDMA burst acquisition and synchronization. During tests via the INTELSAT I F-1 satellite (Early Bird), two remote TDMA terminals, one located at Andover, Maine (U.S.), and the other at Mill Village (Canada), acquired the appropriate burst slot and maintained synchronization to within one symbol using a

closed-loop ranging technique. With this technique, the time difference between a TDMA terminal's unique word correlation and that of a reference burst is periodically compared (at least once every 300 ms) to the prescribed value and the transmit time phase adjusted by +1, -1, or 0 symbols.

The success of the early TDMA system acquisition and synchronization tests spurred the development of other TDMA systems, such as the Japanese TTT system developed for KDD Laboratories,[7] the TDMA-2 system developed by COMSAT Laboratories for INTELSAT,[8,9] the TDMA common signaling channel system developed for the INTELSAT SPADE system, the Telesat TDMA system developed by Digital Communications Corporation,[10] and the INTELSAT TDMA field trial terminals.[11]

A key element in the development of a closed-loop synchronization system is a unique word correlator. The analog-summing network initially used for this purpose suffered from analog reference voltage instabilities. Subsequently, a digital summing network was developed to provide a stable and repeatable operational performance. More recently, the correlation has been performed using read-only memories (ROMs), which are not only stable, but also consist of many fewer chips.[12]

With the advancement of communications satellite designs, hemispheric, zone, and spot beams have been introduced to increase the useful spectrum available to commercial communications satellites. Closed-loop acquisition and synchronization techniques may not be applicable to these new satellites since no TDMA terminal can see its own burst. Several new developments have been proposed for spot beam operation and some have been tested. These new acquisition and synchronization techniques fall into three general categories: open-loop, feedback, and subnetwork.[13,14]

In feedback acquisition and synchronization a transmitting TDMA terminal's relative burst position is processed by three other TDMA terminals, each returning an appropriate control signal to the transmitting terminal. These signals are used singly or severally by the transmitting TDMA terminal to adjust the transmit burst time phase. Three or more cooperating TDMA terminals are required to ensure that the failure of one cooperating terminal does not cause the transmitting TDMA terminal to fail. Feedback-type acquisition and synchronization has several drawbacks:

a. each TDMA terminal depends on one or more other terminals,

b. each TDMA terminal must have extra hardware to support at least three other TDMA terminals;

c. depending on the number of TDMA terminals per beam, some TDMA terminals may be required to support more than three other TDMA terminals; and

d. the start-up procedure is more complicated than that of closed-loop or open-loop acquisition and synchronization.

Subnetwork acquisition and synchronization is the same as closed-loop operation except that all of the terminals in each beam are synchronized together, one beam is designated as the master, and the reference terminals in each subnetwork phase lock their frame periods to the master reference terminal. For a system in which there are only a few beams and each terminal has an interbeam operation capability, this technique is nearly the same as closed-loop operation. However, in some instances, not all TDMA systems have an interbeam operation capability so that this technique may not be applicable.

The open-loop techniques depend on a central facility to determine the satellite location, compute the predicted slant range from each TDMA terminal in the system to the satellite, and distribute the predicted values to each of the TDMA terminals. Perhaps the most advanced development of the open-loop technique is the CENSAR system developed in Canada for the CTS earth stations.[15]

INTELSAT is planning to conduct open-loop acquisition and synchronization tests during its TDMA field trials in 1978. Open-loop acquisition and synchronization were also used in the TDMA low-speed data ship terminal tests conducted by COMSAT in 1975 for the Maritime Administration.[16] However, because the bit times of this TDMA system are so long (nearly equivalent to the variation in slant range due to satellite motion), the appropriate transmission time phase needed to be computed only once. Table 1 is a summary of the synchronization accuracies achieved by the various techniques discussed herein.

Table 1. Synchronization Technique Accuracies

Technique	Accuracy (symbols)
Closed Loop	±1
Open Loop	±1 to ±10
Feedback	±1 to ±3

Modulation Techniques

Because of the requirement to compress the transmitted information into very short bursts so that several TDMA terminals can sequentially share one transponder, it is most convenient to sample and digitize the information before applying it to a compression buffer and modulator. Although modulation techniques had been studied extensively by the early 1960s, only a few applications required burst mode operation and none of these were required to operate at a rate of tens of Msymbols/s. Several study and development projects were initiated in the 1960s to develop high-speed burst modems. The key problem was the achievement of reliable carrier and bit timing recovery on independent bursts within a reasonable margin from the theoretical curves. These efforts showed that, with the available satellite weight and power budgets, the most appropriate modulation technique for high-channel-capacity TDMA systems in terms of performance, feasibility, and cost would be differentially encoded, 4-phase phase-shift keying (DPSK).

The initial DPSK modems used phase-lock loops for both carrier and bit-timing recovery.[17] However, in most TDMA systems the carrier phase is independently acquired for each burst; therefore, it has a relatively large initial phase error. In this case, the phase-lock loop design sometimes fails to attain phase lock. In addition, it sometimes suffers from occasional phase slips. Gardner, Yokoyama, and others found that a high-Q tank circuit in the carrier phase receiving circuits would eliminate these phase-lock "hangups."[18,19] Figure 4 is a plot of the performance (BER vs E/N_o) of a QPSK burst-type demodulator operating via a nonlinear link. Most TDMA modems now operate within 2 to 3 dB of the theoretical curve.

The Telesat TDMA system, the INTELSAT TDMA field trial system, and the European regional TDMA field trial system all use 4-phase QPSK modulation. Experimental, differentially encoded, 8-phase PSK modems[20] have also been developed and tested, although none are in use today since most TDMA systems are power-limited. In addition, other types of burst modems such as fast frequency shift keying and offset QPSK are now being studied and tested for potential application to future TDMA systems.[21,22]

Transmission Link Encoding Techniques

Assume that there is limited signal power in the transmission link from the satellite to the earth station (due either to limited satellite power or to small-aperture earth stations). For operation in the millimeter wave region, in which signal fades may occur occasionally due to changes in the atmospheric conditions, or

signal interference may occur due to nonlinearities in the channel or other transmitters at the same frequency, it may be
useful to employ transmission link encoding to improve the effective received
carrier-to-noise ratio. Transmission link
encoding is the process of systematically
evaluating the sequence of information
bits and adding error-correcting bits so
that an error may be detected if one or
more bits are improperly received, and
there is sufficient redundancy in the encoded information to restore the original
information bits. Because some additional
"parity bits" are added to the information
bits, this technique represents a tradeoff
of bandwidth for power. Depending on the
encoding technique used and the pre-
encoded condition of the transmission
bits, an effective carrier-to-noise improvement of up to 6 dB may be achieved.

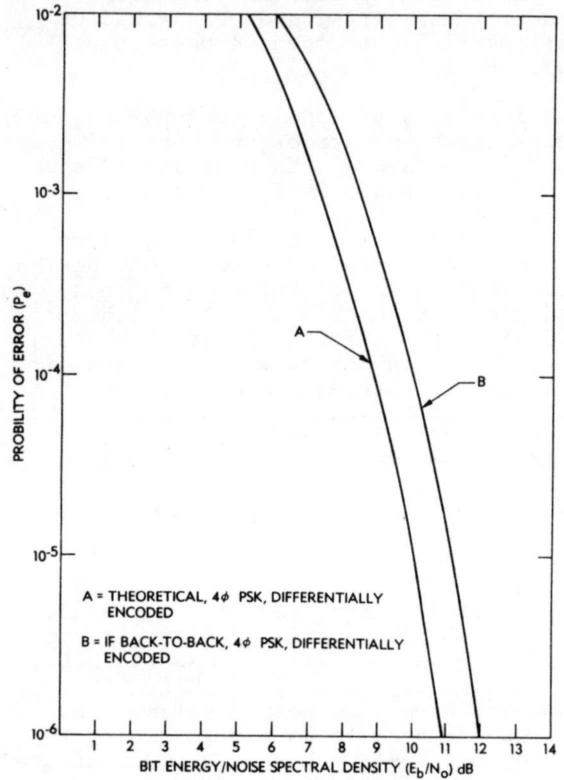

Figure 4. BER vs E_b/N_o for DPSK
Demodulation

Two forward-acting error encoding techniques are in use today: block encoding
and convolutional encoding. Perhaps the
most common type of block codes used are
Bose-Chaudhuri-Hocquenghen (BCH) codes.[23]
In recent years, convolutional codes have
gained prominence; the most notable are
codes suited to Viterbi decoding.[24] However, Viterbi-type decoders are limited
in both the through bit rate and the
coding rate (ratio of information to composite information/parity bit rate) by

the requirement for parallel processing.
At the present time, the maximum through
bit rate is 10 Mbps using coding rates of
up to rate 3/4.

Transmission link encoding gains also
depend on the type of decoder used, i.e.,
hard decision or soft decision. Typically, a 2-dB difference in encoding gain
may be expected, with soft decision providing the greater gain. However, soft
decision decoding does require additional
hardware (a high-speed analog-to-digital
converter) in the demodulation.

Threshold-type decoding is perhaps the
most prominent error decoding technique
in use because it is generally simpler to
implement and can perform at higher data
and encoding ratios. However, it generally provides a slightly smaller decoding
gain than soft decision decoding. Majority decoding (a class of threshold decoding) is a case in point.[25] A high-speed
forward error encoder/decoder of this
type capable of operating at data rates
of 120 Mbps with a coding rate of 3/4 is
presently being developed by KDD Laboratories of Japan under INTELSAT
sponsorship.

Modern digital signal processing techniques are increasing the through bit
rate of both block and convolutional
decoding. One application is the use of
ROMs to provide high-speed recovery of
the error pattern based on the real-time
syndrome. In addition, with the use of
standard MSI and LSI integrated circuit
chips such as the 100K series ECL and
Schottky TTL chips, real-time processing
is more practical at very high speeds.
It is anticipated that the speeds of new
transmission link encoding schemes will
be limited only at processing speeds
above 120 Mbps and at coding rates above
9/10.

Baseband Processing Techniques

Several baseband processing techniques
which improve the performance or reduce
the cost of TDMA systems have been
developed.

Sampling and Compression

When several TDMA terminals use a common
transponder one carrier at a time, it is
most convenient for the information to be
sampled (8-bit PCM or delta encoding/
decoding) and compressed so that several
earth stations can sequentially use the
same transponder once each frame period.
When TDMA system developments were initiated, PCM voice and computer data
transmission and signal compression techniques had already been developed and
were in operation over a limited number

of digital terrestrial transmission
facilities (microwave and cable systems).
As a result of the available technologies,[7,9,10] the initial baseband interfaces to TDMA systems were direct digital
interfaces, directly encoded supergroups,
and individually encoded voice channels.
Figures 5a and 5b are diagrams of supergroup encoding and individual channel
encoding baseband processing techniques,
respectively.

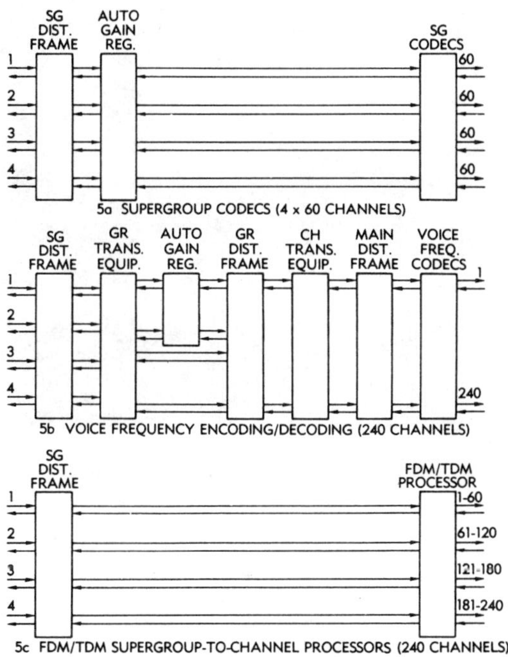

Figure 5.   TDMA Baseband Processing
Techniques

As a result of significant advancements
in digital logic function developments
and digital signal processing studies, it
became evident that direct conversion of
FDM/FM groups and supergroups to and from
12 or 60 individual time-division-
multiplexed 8-bit PCM channels could be
made practical. Figure 5c is a simplified block diagram of a 60-channel FDM/
TDM processor (transmultiplexer). COMSAT
Laboratories and others have successfully
developed such a processor.[26,27] These
types of processors may be used to simplify the interface of TDMA terminals to
FDM/FM terrestrial plants. Because they
provide the signals in an individual
digital channel format, they are also
applicable to channel reordering, digital
speech interpolation, and channel monitoring in TDMA systems. They may be used
in the reverse direction when some FDM/FM
carriers are received via the satellites
and the terrestrial plant is digital.

Speech Interpolation

Another available baseband processing
technology was speech interpolation.[28,29]
Because satellites are in view of many
earth stations, common path routing may
be used to bring together larger numbers
of channels per trunk group, thereby
significantly enhancing the application
of speech interpolation to satellite
trunk groups (reducing the probability
of speech clipping). As a result several
new digital speech interpolation systems
were developed and tested:

    a. INTELSAT TDMA/DSI: This equipment
is specified in the INTELSAT Prototype
TDMA System Specification, BG-1-18
(Rev. 2). It includes a single bit reduction capability during overload conditions. Equipment is currently being
fabricated by SIT Siemens (Italy) and SAT
(France) for evaluation in the INTELSAT
TDMA field trial.[30]
    b. CELTIC: This equipment was devel-
oped in France for use in submarine cable
circuits; field trials are being conducted in 1977. Although processing is
performed digitally, interpolated speech
is converted to analog format for transmission over the cable.[31]
    c. SPEC: This equipment was developed
by COMSAT Laboratories. It employs a
variable threshold, first-order predictor
to avoid competitive front-end signal
clipping that occurs on the preceding
types of speech interpolation systems.[32]

These systems all achieved an effective
channel improvement gain of 2:1 or more.
Other digital speech interpolation systems
are being studied and developed for TDMA
applications, e.g., speech interpolation
on delta encoded channels.

Demand Assignment

Various types of demand-assignment processing have been considered for TDMA system applications. For example, both call
and variable burst length/burst position
demand assignment were developed and
tested on the INTELSAT TDMA-2 develop-
mental system.[9] Because call-type demand
assignment is most applicable to low-
intensity trunks, it was not included in
the INTELSAT TDMA field trial system,
which is designed primarily for large
trunk groups.

Some additional satellite channel capacity can be gained by using variable burst
length/burst position demand assignment.
In this case, as the number of channels
active in a trunk group diminishes, the

burst length is shortened; when the number
of active channels per trunk group increases, the burst length is increased.
If each TDMA terminal's burst length is
allowed to change accordingly, and if the
burst position of each TDMA terminal can
be changed dynamically, those terminals
with heavy activity may utilize capacities
unused by those with low activity.

This capability would be most effective
in a system that encompasses a relatively
large geographic area such as the Atlantic
Ocean region, where a total of five time
zones are spanned from east to west and
there is significant north-south traffic
in both the Euro-African hemisphere and
the North-South American hemisphere.
However, zone and spot beam satellites are
to be used in the INTELSAT system. Because the east-west and west-east transponders are independent, they cannot be
dynamically switched. Therefore, burst
length and burst position demand assignment would provide little, if any, improvement. Demand assignment can also be
expected to yield little improvement for
national or regional networks of limited
geographical size, since the busy hour of
most TDMA terminals will be simultaneous.
However, Satellite Business System's TDMA
system is presently capable of providing
this service.[33] With the availability of
an onboard dynamic switch such as that
which would be used for SS-TDMA, burst
length/burst position demand assignment
may be useful.

Call-type demand assignment has been considered practical for the single-channel-
per-burst TDMA system. In this case, the
system is designed to serve many low-
intensity trunks by transmitting information for only a single channel in each
burst. If more than one channel is
active, multiple bursts are used. A reasonable efficiency is achieved by using
relatively long frames (20 ms or more).
COMSAT Laboratories has recently completed
a system design study and specification of
call-type demand assignment for the U.S.
Army Satellite Communications Agency.[34]

Implementation Techniques

The initial TDMA developments by COMSAT
for the MATE experiment included the construction of flip-flops from individual
transistors to achieve the desired operating speeds. A single flip-flop required
a physical space as large as 3 by 5
inches. However, with the rapid development of the integrated circuit transistor
and transistor logic (TTL) and emitter
coupled logic (ECL) systems in the late
1960s and early 1970s, the development of
less costly TDMA terminals became possible. Figure 6 is a photograph of an
INTELSAT type TDMA terminal. This system,

developed in 1976, operates at 60 Mbps.
The INTELSAT and European regional TDMA
field trial terminals and the Telesat TDMA
system terminals are very similar in
construction.

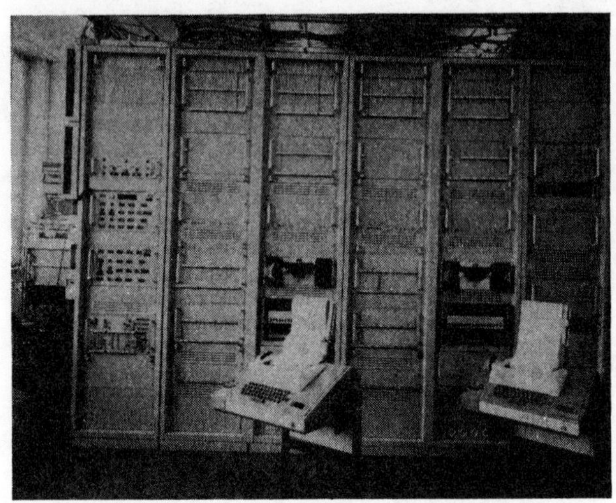

Figure 6. INTELSAT-Type TDMA
Terminal

Later developments by COMSAT Laboratories
have resulted in 120-Mbps TDMA terminals
which can meet the INTELSAT Prototype TDMA
Specification, BG-1-18 (Rev. 2), are much
smaller, use fewer chips, and should be
much more reliable (with a much lower FIT
count). These terminals were achieved by
developing more capable algorithms using
standard off-the-shelf integrated circuit
chips to perform the prescribed functions.
A breadboard of the newly designed
transmit/receive synchronizers and the
TDMA control unit processor is now in
operation. A low-cost breadboard of a
QPSK modem operating at 44 Mbps has
already been built and will be modified to
operate at 120 Mbps. A fully redundant
TDMA control unit and 4-phase QPSK modem
built with these new concepts contains
less than 1000 integrated circuit chips
and may be housed in a single shelf which
is 19 x 10.5 x 20 inches.

Summary

Time-division multiple access (TDMA) is a
useful communications system. All of the
required technologies and operating subsystems have been developed and tested.
It can provide a greater channel capacity
than the conventional FDM/FM/FDMA systems.
With the recent developments in implementation technology, the cost per channel is
competitive, and less physical space is

required to house each channel. In addition, because it consists of fewer active components, it is expected to be more reliable than existing systems.

Jn view of traffic projections for most networks (national, regional, and international), the satellite channel capacities for most systems will have to be significantly increased (by more than one order of magnitude) relative to those of the 1980s. As a result of the limited available spectrum and the limited physical plant size for most earth stations, a more efficient modulation and multiple-access technique such as TDMA, combined with more beams per satellite to increase the effective usable spectrum, appears highly desirable for the future. It has been shown that the initial application of TDMA and subsequent incorporation of a dynamic satellite switching matrix (SS-TDMA) will result in a much more capable system.[35] Thus, with an appropriate design, TDMA can provide a more economical multiple-access capability than FDMA; it can provide a greater channel capacity than FDMA; and it can evolve into an SS-TDMA system with even greater capacity, thereby providing a longer useful life for the multiple-access system hardware.

References

1. S. Metzger and N. J. Bergenfield, "Multiplex Branch Repeater Station," Patent No. 2,861,128, U.S. Patent Office, November 1958.

2. T. Sekimoto and J. G. Puente, "A Satellite Time Division Multiple Access Experiment," IEEE Transactions on Communications Technology, Vol. COM-16, August 1968.

3. Y. Inoue, "Acquisition in TDMA Satellite Communications System—SMAX," Electronic Communication Laboratory Review, Vol. 17, August 1960.

4. L. J. Carter, Communications Satellites, Cambridge: Academic Press, 1962.

5. G. D. Dill, T. Muratani and R. Cooperman, "Application of SS-TDMA to INTELSAT Networks," INTELSAT/IECE/IEE Third International Conference on Digital Satellite Communications, Kyoto, Japan, November 1975.

6. K. Coffrin and G. Gabaui, "TDMA for Defense Satellite Communications Systems," IEEE INTERCON Conference Record, 1973.

7. N. Nosaka, "TTT System—50 Mbps PCM-TDMA System with Time Preassignment and TASI Features," INTELSAT/IEE First International Conference on Digital Satellite Communication, London, England, November 1969.

8. W. G. Schmidt et al., "MAT-1: INTELSAT's Experimental 700-Channel TDMA/DA System," INTELSAT/IEE First International Conference on Digital Satellite Communication, London, England, November 1969.

9. W. G. Maillet, "INTELSAT's 50-Mbps TDMA-2 System," INTELSAT/IEE Second International Conference on Digital Satellite Communications, Paris, France, 1972.

10. R. K. Kwan, "The Telesat TDMA System," International Conference on Communications, San Francisco, California, 1975.

11. G. D. Dill, J. Deal, and W. Maillet, "INTELSAT Prototype TDMA System," IEEE INTERCON Conference Record, 1975.

12. B. Warner, "Build High Speed Sync Pattern Detectors," Electronic Design, Vol. 21, October 1975.

13. G. D. Dill and A. Tomazawa, "Time Division Multiple Access Frame and Burst Synchronization for Spot Beam Satellites," IEEE Electronic and Aerospace Systems Convention, October 1974.

14. P. Nuepl et al., "Synchronization Methods for TDMA," IEEE Proceedings, Special Issue on Satellite Communications, Vol. 65, No. 3, March 1977.

15. P. Nuepl, N. Davies, and R. Olsen, "Ramping and Synchronization Accuracies in a Regional TDMA Experiment," INTELSAT/IECE/ITE Third International Conference on Digital Satellite Communications, Kyoto, Japan, November 1975.

16. R. B. McClure, J. M. Husted, and F. Giorgio, "Maritime Satellite Communications Terminal Implementation," INTELSAT/IEEE/ITE Third International Conference on Digital Satellite Communications, Kyoto, Japan, November 1975.

17. C. J. Wolejsza, A. M. Walker, and A. M. Werth, "PSK Modems in Satellite Communications," INTELSAT/IEEE First International Conference on Digital Satellite Communication, London, England, November 1969.

18. Y. Matsuo, S. Sugimoto, and S. Yokoyama, "A Study on Carrier and Bit Timing Recovery for Ultra-High-Speed PSK-TDMA Systems," INTELSAT/IEE Second International Conference on Digital Satellite Communications, Paris, France, 1972.

19. F. M. Gardner, "Comparison of QPSK Carrier Regenerator Circuits for TDMA Application," International Conference on Communications, Minneapolis, Minnesota, June 1974.

20. A. Ogawa and M. Ohkawa, "A New Eight-Phase PSK Modem System for TDMA," INTELSAT/IEE Second International Conference on Digital Satellite Communications, Paris, France, 1972.

21. D. P. Taylor, S. T. Ogletree and S. S. Kaykim, "A Prototype 60 Mbps Fast Frequency Shift Keying Modem," IEEE International Conference on Communications, Philadelphia, Pennsylvania, June 1976.

22. S. A. Rhodes, "Carrier Synchronization Techniques for Off-Set QPSK Signals," National Telecommunications Conference, San Diego, California, December 1974.

23. W. R. Bennett and J. R. Davey, Data Transmission, New York: McGraw-Hill, 1965.

24. A. J. Viterbi, "Convolutional Codes and Their Performance in Communication Systems," IEEE Transactions on Communications Technology, Vol. COM-19, October 1971.

25. J. L. Massey, Threshold Decoding, Cambridge, M.I.T. Press, 1963.

26. F. Takahata et al., "Development of TDM/FDM Transmultiplexer," IEEE National Telecommunications Conference, Los Angeles, California, December 1971.

27. S. L. Freeny et al., "Design of Digital Filters for an all Digital Frequency Division Multiplex-Time Division Multiplex Translator," IEEE Transactions on Circuit Theory, Vol. CT-18, November 1971, pp. 702–711.

28. J. M. Fraser, D. B. Bullock and H. G. Lin, "Overall Characteristics of a TASI System," Bell System Technical Journal, Vol. 41, No. 4, July 1962.

29. S. J. Campanella, "Digital Speech Interpolation," COMSAT Technical Review, Vol. 6, No. 1, Spring 1976.

30. I. Poretti, G. Minty, and A. Bagnoli, "Speech Interpolation Systems and Their Application in TDM/TDMA Systems," INTELSAT/IEE Second International Conference on Digital Satellite Communications, Paris, France, 1972.

31. F. D. Daynonnet, A. Jourset, and A. Profit, "Le Celtic: Concentrateur Exploitant les Temps d'Inactivite des Circuits," L'Onde Electrique, Vol. 42, No. 426, September 1962.

32. S. J. Campanella and J. A. Sciulli, "Speech Predictive Encoded Communication System for Multichannel Telephony," IEEE Transactions on Communications, Vol. COM-21, No. 7, July 1973.

33. C. Kittiver and F. Zitzmann, "The SBS System—An Innovative Domestic Satellite System for Private Line Networks," AIAA/CASI Sixth Communications Satellite Systems Conference, Montreal, Canada, April 1976.

34. J. Deal and J. Buegler, "A Demand Assigned, Time Division Multiple Access System for Military Tactical Application," AIAA/CASI Sixth Communications Satellite Systems Conference, Montreal, Canada, April 1976.

35. G. D. Dill, Y. Tsuji, and T. Muratani, "Applications of SS-TDMA in a Channelized Satellite," IEEE International Conference on Communications, June 1973, ICC Conference Record, Vol. 3.

Subsection 3.6.4
Code Division Multiple Access Spread Spectrum

Performance Criteria for Spread Spectrum Communications

MARLIN P. RISTENBATT, SENIOR MEMBER, IEEE, AND JAMES L. DAWS, JR., MEMBER, IEEE

Abstract—The criteria for antijam (AJ) and anti-intercept (AI) systems are described, in terms of the appropriate action taken by the jammer or interceptor. Avoiding pseudonoise (PN) sequences which can be partially or totally predicted is a foremost criteria, especially for AJ, and use of nonlinear feedforward logic (NFFL) with long-period linear maximal sequences appears attractive.

A frequency-hopping (FH) system must anticipate a multitone jamming signal, and an error-control code is necessary. A novel method for generating a multitone signal using repeating maximal sequences is described. PN systems must anticipate a tone jammer, and now an error-control code may be needed to assure that sporadic PN sequence correlation with the tone does not reduce the processing gain.

Any AI system must anticipate that the interceptor may, if advantageous, integrate his decision energy over many communicator symbols (up to the message length). Now spreading both in frequency and in time may be valuable.

Finally, the use of an automatic adaptive data rate is suggested to realize flexibly either AJ or AI objectives.

I. INTRODUCTION

IN spread spectrum systems either the RF bandwidth (W) and/or the total time epoch of the signal is "spread" beyond that required for basic signaling to provide special functions or to cope with unintended parties. Special functions, which may be of interest to civilian as well as military users, are ranging, multiple-access, selective addressing, antimultipath, and minimum power-density signaling.[1]

Military communicators, who are the chief developer of spread spectrum techniques, build antijam (AJ) systems if the unintended party is a jammer and anti-intercept[2] (AI) systems if avoidance of signal intercept is important. There is also interest in antispoof (IFF)[3] signaling, but this will not be treated here.

This paper describes quantitatively the system performance criteria for AJ and AI systems in terms of the appropriate action taken by the unintended party. The potential availability of new spectrum regions, made accessible by device developments, now makes impressive processing gains achievable by spread spectrum system designers. The same new technology is, of course, available to the unintended party.

The major purpose of this paper is to note the quantitative relation between the AJ and/or AI system goal and the goal of the corresponding unintended party. A secondary purpose is to relate the pseudonoise (PN) sequence properties, which are used to achieve the spreading, to the AJ system goals. A final purpose is to advance certain new concepts and ideas for consideration.

Section II describes the common spread spectrum approaches in general terms. Section III describes the PN code sequence properties that are influential in practical systems. Sections IV and V treat AJ and AI systems, respectively. The final section proposes a concept for flexible AJ and AI operation.

II. GENERAL SPREAD SPECTRUM ASPECTS

We will assume that the information to be communicated is digital, stemming either from a data terminal or digitized voice or video. Figure 1 shows a general block diagram of a digital spread spectrum system. A digital source of rate R b/s is first encoded via a block or convolutional code; a block error-*detecting* code will be used if the link is involved in retransmission (ARQ), and a convolutional code will be used if forward-only transmission is involved (see Ref. 1). Specific reasons for coding with spread spectrum are treated later. The rate entering the spread spectrum modem is the product of bit stream R and the code-rate. The modem output is spread over a total RF bandwidth W, and the time epoch may be expanded if the message is finite-length. The common spread spectrum systems are: frequency hopping (FH), direct-sequence pseudonoise (PN), time-hopping (TH), and chirp, plus combinations of these. Specific block diagrams of these systems are available elsewhere (Refs. 2, 3, and this issue) and will not be repeated here.

In FH systems, a single hop of bandwidth w_h and length $t_h (w_h t_h \approx 1)$ is pseudorandomly hopped over the total bandwidth W, under the control of k-tuples from a PN-sequence used as a random-number generator, as suggested in Fig. 2. Usually frequency-shift-keying (FSK), either binary or M-ary, is used to frequency-offset the hop-stream produced by the random-number generator.

In direct-sequence PN systems, the PN sequence forms a series of "chips" that phase-shift-key (PSK) a carrier; this code modulation can be either binary or quaternary. Then binary data sequentially phase-shifts successive k-tuples of such sequence chips (Fig. 3). The resulting PN signal covers the total bandwidth W, at all times, as indicated in Fig. 3(d).

In a TH system, the RF signal is dispersed in time, with PN code-determined time gaps between signal elements. This forms a type of intermittent signal which is useful for AI signaling, and may be useful for fall-back AJ signaling. In a chirp system, one possibility is to use an up-chirp for a (data) binary "one" and a down-chirp for a binary "zero" forming a series of up- or down-sweeps in FT space (Ref. 4). For AJ, one adds FH to the chirp start-frequency, to afford protection against a chirp jammer.

Manuscript received January 19, 1976; revised April 6, 1977.

The authors are with the Cooley Electronics Laboratory, Department of Electrical and Computer Engineering, University of Michigan, Ann Arbor, MI 48104.

[1] This refers to (a) the use of data rate reduction to allow communication with devices having very limited power capability, or (b) frequency spreading to afford negligible interference to nearby (relatively) narrowband receivers.

[2] AI systems are frequently called "low probability of intercept (LPI)" systems or "covert communications."

[3] Identify Friend or Foe (IFF) refers to the process of using a private "password" signal to identify oneself as being a member of the defined group.

Reprinted from *IEEE Trans. Commun.*, vol. COM-25, pp. 756–761, Aug. 1977.

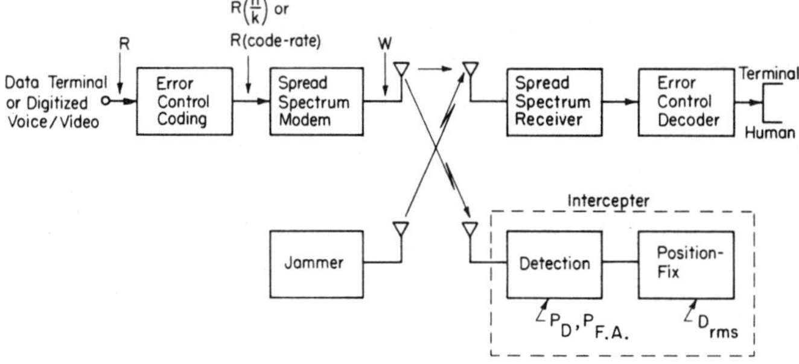

Fig. 1.   Spread spectrum system and unintended parties.

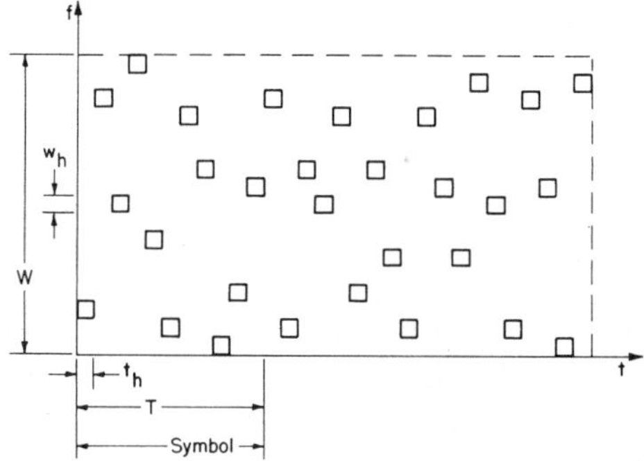

Fig. 2.   Frequency versus time sketch for FH.

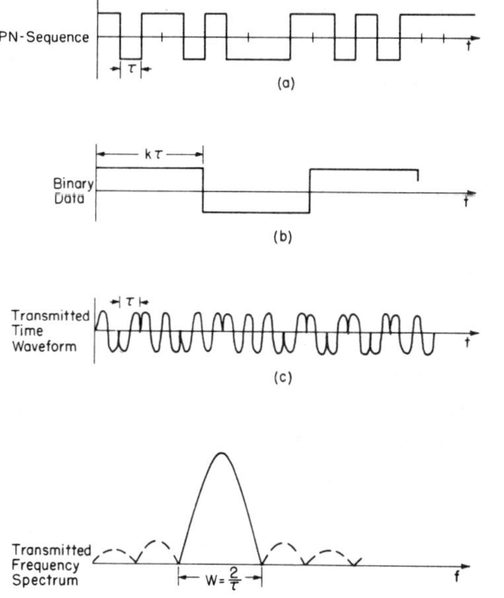

Fig. 3.   PN Signal formation. (a) PN sequence. (b) Binary data. (c) Chip and data modulated time waveform. (d) Corresponding (filtered) spectrum.

Each type of spread spectrum system requires PN code sync; hence one must acquire sync either from a "cold start" or within an *a priori* uncertainty interval if continuous-clocking is used. The *a priori* uncertainty interval is determined by the maximum possible path-delay.

Spread spectrum applications are sensibly divided into the cases where (a) the links are relatively "isolated"; or (b) the links are relatively "dense" and interacting. The isolation may stem either from physical separation of platforms (remote ship or airplane) or from antenna-beaming (satellite uplinks).

Traditional dense applications occur in omnidirectional communication links among tactical mobile vehicles or aircraft. In dense applications, where many radios are within range of each other, frequency-management and self-interference (near-far) must be handled, as well as unintended parties. In isolated applications, usually a single radio link must cope only with the unfriendly jamming or with the possible signal exploitation after interception.

When adopting spread spectrum techniques in isolated cases, the major contest is between platforms, and the "processing gain" of the system versus that of the unintended party. In dense cases, however, spread spectrum techniques offer yet another dimension—the denial of obvious frequency channelization to the unintended party. Traditionally this channelization provides the underlying basis for intercept-then-jam strategies. Potential jammers are able to monitor the spectrum activity in dense environments and relate specific spectrum activity to particular types of vehicles or organizations, for selective jamming. Adoption of spread spectrum in a frequency region should significantly alter the ability of unintended parties to make sense out of the spectrum activity.

Finally, we note that spread spectrum system design is heavily influenced by the particular frequency region, both because the physical channel has characteristic propagation and noise properties, and because electronic transceivers vary somewhat for the different frequency regions. It is sensible to divide the major candidate channels into the categories of: (1) extremely low frequency (ELF)/very low frequency (VLF); (2) high-frequency (HF); (3) line-of-sight (LOS)—VHF/UHF/SHF; and (4) satellite-relay. ELF/VLF appears useful for long-distance, very low data-rate systems; HF features long distance, variable propagation, and moderate data rates. The VHF/UHF frequencies are typical for military LOS applications, and these channels would form a dense application if spread spectrum is adopted. SHF is not yet crowded with many radios, and may be the spectrum region where spread spectrum is adopted for dense and isolated environments. The satellite link has a unique, long distance capability, either broadcast or point-to-point, and can support a high processing gain. Optical links (not listed above) achieve substantial gain against an unintended party by virtue of the extreme radiation directivity, and may not need spread spectrum except in rare cases.

III. PN SEQUENCE ASPECTS

The PN sequences that are used for the spreading in any system must meet the two critical criteria of: (1) denying any information about future sequence k-tuples to the unintended party, and (2) permitting practical implementation, including convenient code changes. Sometimes it is desirable for the sequence autocorrelation behavior to have a high peak-to-sidelobe ratio, for acquisition sync purposes.[4] Finally, proper k-tuple statistics are desirable, depending on the type of system (see below).

Denying future information is probably the foremost

---

[4] A sequence specially suited to acquisition sync may be used first, followed by a long code.

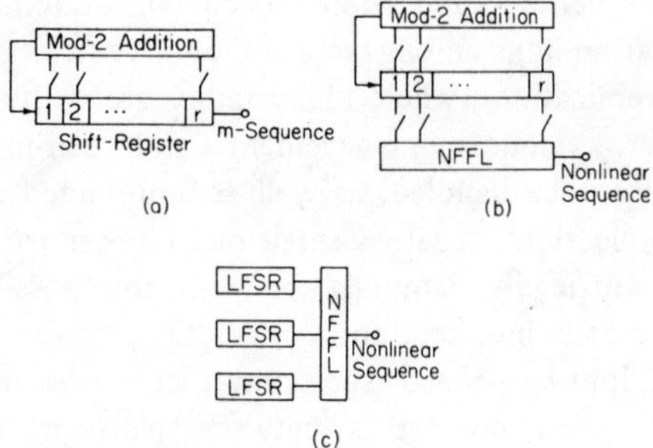

Fig. 4. Sequence generators. (a) LFSR. (b) NFFL using single LFSR. (c) NFFL with multiple LFSR's.

criteria; if future sequence values are not totally uncertain, the unintended party should be able to reduce the intended system processing gain, at least in AJ systems. This means that the PN code period must exceed the time between code changes, and the code must not be "crackable" in the encryption sense.

Practical and efficient implementation techniques for PN sequences center around use of shift-registers (Refs. 5, 6). Since crackability is such an important issue, it appears sensible to categorize PN code generators, according to increasing complexity, into the classes of: (a) linear feedback shift-registers (LFSR); (b) one or more LFSR's with nonlinear feedforward logic (NFFL); and (c) general nonlinear feedback shift-registers (NLFSR). Other ad hoc cases are possible, but are ignored here.

Linear sequences are the least complex and limit the logic used in the generator to modulo-two addition. LFSR's can generate linear maximal sequences, m-sequences, using selected stages of an r-stage shift register as shown in Fig. 4(a). The correct selection of the feedback stages will result in a maximal sequence of length $L = 2^r - 1$. Although m-sequences have been thoroughly studied (Ref. 5 through Ref. 12), they are practically useless when unintended parties are involved (at least for AJ). The classical properties of m-sequences are: (1) maximal length for an r-stage generator; (2) maximum possible peak/side-lobe autocorrelation over the period L; (3) large number of possible maximal connections (codes) per r-register; and (4) balanced k-tuple statistics. The major problem is that an unintended party need only $2r$ successive sequence bits to crack the sequence: determine the feedback connections and the initial state of the r-register.

In 1973 we found (Ref. 10) that the relatively simple procedure of adding nonlinear logic (AND, OR, INVERT, etc.) to an LFSR in a feedforward fashion, as in Fig. 4(b), significantly increases the complexity of cracking. Our measure of complexity here is the number of consecutive sequence bits that must be known for cracking.[5] This was found by evaluating the register length of the equivalent linear generator. This length, in turn, was found using Massey's algorithm (Ref. 11), which is a general solution to synthesize the shortest linear generator capable of generating a prescribed finite sequence of bits. We found that the number of stages

<hr>

[5] Another measure of complexity has recently been proposed, based on the number of independent k-tuples—see Ref. 13.

$(_rN_m)$ in the linear equivalent generator of an r-stage maximal generator to which an m-order nonlinearlity is added, is practically always given by:

$$_rN_m = \sum_{i=0}^{m} \binom{r}{i} \qquad (1)$$

where

m = maximum number of inputs to any single nonlinear gate

r = number of stages in the LSRG

$\binom{r}{i}$ = binomial coefficient.

Since $2(_rN_m)$ consecutive error-free bits are required to crack the sequence, the cracking complexity increases very rapidly with r and m. For example, a 17-long LFSR has an $_rN_m$ of 3,214 for $m = 4$ and 131,072 for $m = 17$. In certain rare cases Eq. (1) is violated (see Refs. 10, 12).

Another type of NFFL uses multiple LFSR's with a nonlinear feedforward logic, such as in Fig. 4(c). A generator proposed by Geffe (Ref. 14) uses three LFSR's with two 2nd order AND gates. Groth (Ref. 15) has proposed a yet-more-complex NFFL arrangement.

The basic properties of the general NLFSR, which also have an equivalent linear generator, are treated in Ref. 6; crackability studies of such generators are presumably done only by official encryption parties.

Special sequence generators aimed at improved code acquisition sync properties have been (Refs. 16, 17) and continue to be pursued (Refs. 18, 19). Any special techniques must withstand exploitation by an unintended party.

The possible use of code division multiple-access (CDMA, Ref. 1) motivated examination of cross-correlation behavior between different sequences. The so-called Gold sequences (Ref. 8) which are generated by modulo-2 addition of selected m-sequences, provide a set of sequences (one for each value of relative phase shift between the m-sequences) that have a stated bound on the pair-wise cross-correlation, taken over the full period. Since actual systems are integrating over a k-tuple which is small compared with full-period ($k \ll L$), it is the partial-period cross-correlation results that are of interest. Both Gold (Ref. 20) and Daws (Ref. 7) have shown that the average partial period cross-correlation between two maximal sequences of the same length for a fixed value of relative offset is equal to the full period cross-correlation for that value of offset. The Gold codes, with bounded full-period cross-correlation will, for the partial-period case, have a bounded mean value of cross-correlation and the same spread (variance) as any other sequences of the same length. Thus the Gold codes result in a partial-period correlation that is statistically as low as possible for codes of a given length. The practical utility of these results are currently uncertain, both because they apply to linear sequences and because the comparative value of CDMA is not yet assured.

A remaining sequence issue is the k-tuple behavior. This is best postponed to the next AJ section.

IV. ANTIJAM PERFORMANCE CRITERIA

FH

If FH with FSK modulation is the AJ technique, then the first system concern is to assure sufficiently rapid hopping so as to deny any possibility of follower-jamming for the geometry-determined path delays. We may note that surface-acoustic-wave (SAW) devices form the basis of compressive receivers capable of very rapid frequency-determination (Ref. 21). We recently designed and built (Ref. 22) a breadboard frequency-follower, capable of locating a lone frequency-hopping signal within ± 100 Hz over a 100 MHz band (650 to 750 MHz) in a period of 1.5 ms. The follower design centered around a recent SAW dispersive delay line[6] having a compression ratio of 10^4 (250 MHz bandwidth, 40 μs differential delay).

Assuming sufficiently rapid hopping, the system design should then anticipate a multitone jamming signal. To justify this we need to distinguish the hop-decision error-rate (P_E') from the final error-rate (P_E) after the error-control encoding/decoding is included. The minimum jammer power is that power required to cause an error-rate higher than the specified P_E.

The jammer's best strategy is to space a series of tones across the bandwidth (W), where the (received) power per jammer tone slightly exceeds the system's received power per hop. Whenever a hopping receiver's instantaneous bandwidth (Mw_h) lands on a jamming tone, the conditional error-probability is:

$$P'(E|\text{hit}) = \frac{M-1}{M}$$

where

$P'(E \mid \text{hit})$ = error-probability conditional on a jammer tone being in a receiver bandwidth

M = the level of FSK modulation; $M = 2$ for binary.

The jammer must use as many such tones as required to exceed the system P_E requirement. Further, the tones will be spaced no closer than Mw_h so that no more than one tone can appear in the modulation bandwidth. Assume that both the FH and the FSK positions are spaced at the orthogonal $(1/t_h)$ nonoverlap positions. Then:

$$W = N_f w_h$$

where

W = total RF bandwidth
N_f = number of orthogonal bins in total W
w_h = bandwidth of single hop

[6] A 10,000: 1 Linear FM Reflective Array Dispersive Filter developed and built by Hughes Aircraft Company.

Then the hop-decision P_E' would be:

$$P_E' = \frac{M-1}{M} \frac{N_j}{N_f/M} = (M-1)\frac{N_j}{N_f} \qquad (2)$$

where

N_j = number of jamming tones
N_f = number of orthogonal frequencies.

Equation (2) allows us to observe that, if no error-control coding were used, the minimum required jammer power is heavily dependent on the specified error-rate. A typical error-rate of 10^{-3} or less would permit a jammer to use a power much smaller than predicted by the processing gain (W/w_h). Error-control coding (see Fig. 1) serves to make the symbol decision span a series of hops. P_E now depends on the code and on the P_E'. In general, a good code causes the required jammer power to rise close to that value predicted by the white-Gaussian noise (WGN) model:

$$J = j_0 W$$

where

J = total jammer power
j_0 = effective "noise power density"
    = J/W.

For example, if a binary FSK with a given error-control code can be approximated as constituting incoherent orthogonal signals, with symbol energy E_s, then the P_E is approximated by (Ref. 24)

$$P_E \cong 0.5 e^{-E_s/2j_0}$$

where

E_s = ST
S = received signal power
T = symbol length

In FH systems the PN sequence is used as a random number generator. In this case the possible frequencies, and hence the k-tuples, should be approximately equally likely. The k will be less than r, since the required large $L = 2^r - 1$ exceeds the 2^k hop possibilities even for impressive processing gains. This $k < r$ factor contributes to the difficulty of cracking the PN sequence. Since all linear m-sequences will have equally likely k-tuples $(k \leqslant r)$ over a period, addition of an NFFL can be arranged to exhibit a similar property in the nonlinear output sequence.

Since multitone signals are the optimum nonadaptive jammer against FH, it is interesting to note a particularly simple way to generate flexible, multitone signals. A readily available, near-optimum digital technique consists of using periodic repetitions of a linear m-sequence. The amplitude spectrum of a periodic series of impulses having a plus or minus sign corresponding to an m-sequence is a series of equal-height lines, extending indefinitely in frequency, with spacing between the lines given by:

$$\Delta w_j = \frac{1}{\text{period}} = \frac{f_c}{L}$$

where

Δw_j = tone spacing
f_c = clock frequency of shift-register
L = sequence length in bits.

Since a jammer prefers a near-constant envelope signal, one starts with the rectangular-wave form of an m-sequence (the usual form). The equally-spaced spectrum lines now have a $\sin x/x$ amplitude behavior, with the first zero occurring at f_c. One can obtain a near-constant tone amplitude by using a digital filter which both: (1) passes only the portion out to about $f_c/2$, and (2) compensates in amplitude, so that the output has essentially flat tones.

We constructed a breadboard of a multitone jammer which has a total bandwidth of 100 MHz, with tones spacings as low as 1 kHz. The m-sequences were clocked at 150 MHz, and used to PSK an RF carrier. The digital filter has a passband of 50 MHz. When the result was balanced-modulated up to a higher frequency, the bandwidth was automatically doubled.

PN

If a system uses direct-sequence PN for AJ, then it should anticipate a (single) tone jammer, placed near or at the PN carrier frequency. A tone jamming signal is the best practical nonadaptive interference because: (a) it stresses the carrier suppression issue of balanced-modulators (if used)—see Ref. 3; and (b) it benefits from any unbalance in the k-tuples.

For PN systems the number of sequence bits per information symbol (k) is determined by the desired bandspreading factor and is typically much larger than the length of the shift-register generating the sequence. It is interesting to note that one would like every sequence k-tuple to have an equal number of ones and zeros (be balanced) since serious unbalance increases the correlation with the jamming tone. Serious and frequent unbalance would cause significant correlation with the jamming tone, and reduced processing gain.

If the PN code were a linear m-sequence, we know (Ref. 9) that the k-tuple balance when $k > r$ (but $k \ll L$) does not differ significantly from that of truly random sequences. Although slightly nonoptimum against a tone, they would be optimum against a broadband jammer.

Nonlinear sequences using NFFL are potentially very unbalanced since the individual nonlinear AND or OR circuits have an unbalanced output for balanced inputs. Therefore some care must be taken in the design of the NFFL to get reasonable balance properties. One common technique for achieving the desired balance is to add (modulo-2) the original maximal sequence to the output of the nonlinear elements. In this case it may be noted that if the output of the nonlinearity is very unbalanced (for example, very few ones) the resulting sequence is the original maximal sequence with errors. Although the linear equivalent generator would be quite long (as noted above), the original linear maximal may be a good approximation for the output sequence.

We note, however, that as the processing gain increases, k becomes increasingly greater than r, and the k-tuples from the LFSR approach better balance. If reasonable care is taken in designing the NFFL, the same tendency should hold. Further, the error-control code (Fig. 1) will smooth over occasional tone correlations produced by unbalanced k-tuples.

V. ANTI-INTERCEPT PERFORMANCE CRITERIA

Spread spectrum techniques can implement an AI (or covert) communication system, where the requirement is to minimize the exploitation of the communication signal for purposes of detecting and position-fixing the signal emitting platform. Keeping any required communication covert is especially important for those platforms that are difficult to detect and position-fix by other means. Since minimum transmitted energy contributes to AI, covert signals are usually minimum data rate (R) continuous signals, or relatively short, finite digital messages. It is desirable for the communicator to estimate the propagation loss via feedback techniques and use only the correct power.

AI seems to make sense for isolated (as opposed to dense) cases, dealing with a single or group of remote platforms. The interceptor usually has two goals: (1) detect presence of signal of interest in midst of natural noise and RFI; and (2) locate position of the emitter (see Fig. 1). The detection portion is evaluated in terms of detection probability (P_D) and false-alarm probability ($P_{F.A.}$) In general, the detection step involves integration of received power over either the entire message (if finite), or over the maximum practical value for continuous signals. Often detection includes a recognition step, using some recognizable feature, to confirm that the signal is of interest. The position-fixing is evaluated in terms of the "rms-error-distance" (D_{rms}).

In LOS and satellite channels, the detector background should be natural noise, with little propagation from distant RFI sources. Then a successful detection may amount also to recognition. In the HF case, however, the interceptor receives propagation from quite distant sources, and may have a "needle in a haystack" problem. In such a case, the interceptor will usually divide the detection step into two phases: (1) initial sort; and (2) recognition. The initial sort or editing processing attempts to sieve, both in frequency and in time, so as to pass that part of the T-W space that appears most likely to contain a signal. A following recognition step continuously examines the initial sort output and makes any "alarm" decisions.

It appears desirable for the communicator to choose a signal as close in appearance to the background (noise or RFI) as possible since this maximizes the difficulty of recognizing the signal. Thus, in satellite and LOS-links, where the background is white Gaussian noise, the covert signals might use filtered direct-sequence PN signals, which have time and frequency properties approximating those of white Gaussian noise.

It is instructive to compare the communicator and intercept performance equations for a finite message. Two important communicator requirements are the total information bits (N) in the message, and the error-probability (P_E). The total number of transmitted symbols (N_t) in the message is determined by the level of M chosen for the signaling:

$$N_t = \frac{N}{\log_2 M} \; .$$

The symbol error-probability is determined by the received energy-to-noise-density ($E_s/n_0 = ST/n_0$), and is a function of the M-level and the coding method used.

While the communicator must make N_t (M-ary) decisions during the message epoch, the interceptor can base his detection and position-fix accuracy on the entire message (a single decision per message). In principle, the detection capability is limited by the total energy (to noise density) received by the interceptor. For weak ($S_I/n_0 W_I \ll 1$) unknown signals in white Gaussian noise the energy detector, usually called a radiometer at microwave frequencies and above, is the optimum.

If the signal energy is spread or dispersed evenly over a total time (T_I) and a total bandwidth W, then the interceptor can integrate only up to T_I, and use an intercept bandwidth $W_I \le W$. Then the "detectability index (d)" (Ref. 23, 24), which determines the detection-probability (P_D) and accompanying false-alarm probability ($P_{F.A.}$), is given by

$$d^2 = \frac{1}{T_I W_I} \left(\frac{E_I}{n_0} \right)^2 = \frac{(N_t T)^2}{T_I W_I} \left(\frac{S_I}{n_0} \right)^2 \qquad (3)$$

where

$$\begin{aligned}
T &= \text{length of } M\text{-ary symbol or signal} \\
S_I &= \text{average signal power received by interceptor} \\
N_t &= \text{number of transmitted symbols in message} \\
n_0 &= \text{noise power density.}
\end{aligned}$$

It is seen that detectability (d) is minimized if N_t is minimized and T_I and W_I are maximized. The S_I will be controlled by the communicator's $E_s = ST$ requirement, and by the interceptor distance from the emitter, relative to that of the communication receiver.

Hence, we see that spectrum spreading, including time-hopping, using pseudorandomized gaps, if the message is finite, is desirable since the gap pattern is known to the communicator but unknown to the interceptor. Although Eq. (3) specifies the theoretical detection constraint, in practice one is often limited by both: (1) ability to find a relevant noise-alone T-W space to measure background noise for setting the decision threshold, and (2) the (message-long) signal processing limitation of a given complexity level.

While the interceptor detection objective is to maximize the P_D for a fixed (small) $P_{F.A.}$, the position-fix objective is to minimize the D_{rms}. Position-fixing requires finding at least two "directions" or line-of-bearings. If the S_I/n_0 is adequate and one has a carrier component available, direction-finding (DF) can be measured in terms of the period of the carrier-cycle, assuming the cycle ambiguity can be resolved in some fashion. Another DF approach, also requiring moderate SNR, is to combine antenna elements so as to form multiple-lobes. For weak, wide-band signals the optimum approach is to estimate the relative time delay of the signal as received at two or more positions. This time delay is estimated by correlating the two receptions.

The accuracy of position-fixing is discussed in Ref. 23. Although the position-fixing accuracy theoretically goes up as $T_I W_I$ increases, it is important to recognize that this theoretical accuracy can only be achieved if the S_I/n_0 is above a certain minimum threshold.

VI. AUTOMATIC DATA RATE FOR AJ/AI

Finally, we propose consideration of the concept of automatic data rate variation. One increases both AJ and AI as data rate decreases. AI increases both because intercepted S_I/n_0 decreases and time hopping can be used. For AJ, a decreasing rate permits a larger receiver integration, T, and hence a higher symbol energy E_s.

Since either AJ or AI objectives may be as important as data rate, at various times, it appears desirable to provide for automatic data rate variation so that the communicator can cope with what he perceives to be the greatest current threat (data rate, AJ, or AI). We visualize a link where two terminals are operating either half-duplex or full-duplex. The forward-link data rate is to be automatically changed as necessary with no visible signal change, to operate normally without repeats (or retransmission) as jamming or intercept conditions change. Provision of automatic rate changes means that the communicator need only set in the criteria to either: (1) meet whatever jamming threat he encounters so long as he is able to tolerate data rate reduction, or (2) minimize his exposure to an envisaged detection threat by appropriately restraining power output.

Such a flexible mode of operation is attractive from a number of different viewpoints. First of all it tends to change the basic "game" between the communicator and the unintended party. An unintended party considering construction of a jammer will face the proposition that: (1) he may not be able to deny communication, but only minimize the rate, and (2) he will not know how well he is doing. An equivalent position occurs for the potential interceptor. He may face an intercepted S_I/n_0 below his threshold.

A flexible data rate is sensible from another point of view. Field-dispersed equipment constitutes an investment to accomplish certain communication goals. Automatic variable data rate implements a more flexible utilization of that investment than any other method.

A final, significant advantage of this automatic approach should be an increased reliability since any unexpected system losses will result in a reduced rate rather than in total disabling.

ACKNOWLEDGMENT

The authors would like to acknowledge the many helpful discussions with Ervin Holland-Moritz.

GLOSSARY OF SYMBOLS

d	detectability index
E_I	energy integrated by interceptor
E_s	energy per communicator information symbol
f_c	clock frequency of a shift-register generator
j_0	single-sided jammer power density
J	total jammer power
k	number of PN code bits in a symbol (either direct-sequency or frequency hop), or number of information bits in a block codeword
L	length of PN code or sequence in bits
LFSR	linear feedback shift register
m	order of nonlinear logic
M	number of alternative information modulation symbols
N	number of information bits in finite message
N_f	number of orthogonal frequencies in total W
N_j	number of jamming tones in total W
NFFL	nonlinear feed forward logic
NLFSR	nonlinear feedback shift register
N_t	number of transmitted symbols in finite message
$_rN_m$	number of stages in the equivalent linear generator of an NFFL generator
n	total number of bits, information plus check, in a block codeword
P_E	final error rate, including error-control coding/decoding
$P_E{}'$	hop-decision error rate
r	number of shift-register stages
R	source data rate in b/s
S	received power from system transmitter
S_I	power received by interceptor, from system transmitter
T	symbol time for system
t_h	time per hop in FH
T_I	largest intercept integration time
w_h	bandwidth per hop in FH
W	total bandspread bandwidth
W_I	intercept bandwidth ($W_I \leqslant W$)
Δw_j	tone spacing of multitone jammer

REFERENCES

[1] M. P. Ristenbatt, "Alternatives in Digital Communications," *Proc. IEEE,* Invited Paper, June 1973, pp. 703–721.

[2] C. R. Cahn, *Spread Spectrum Applications and State-of-the Art Equipments,* Magnavox Communications and Navigation, Torrance, California, January 1973.

[3] R. C. Dixon, *Spread Spectrum Systems,* Wiley & Sons, New York, 1976.

[4] J. Otto, "Chirping RPV Data Links for ECM Protection," *Microwaves,* December 1974, pp. 54-60.

[5] T. G. Birdsall and M. P. Ristenbatt, *Introduction to Linear Shift-Register Generated Sequences,* Cooley Electronics Laboratory TR 90, University of Michigan, Ann Arbor, Michigan, October 1958.

[6] S. W. Golomb, *Shift Register Sequences,* Holden-Day, San Francisco, California, 1967.

[7] J. L. Daws, Jr., *Partial-Period Correlation of Pseudo-Random Binary Sequences,* Cooley Electronics Laboratory TR202, University of Michigan, Ann Arbor, Michigan, May 1970.

[8] R. Gold, "Optimal Binary Sequences for Spread Spectrum Multiplexing," *IEEE Trans. Information Theory,* Vol. IT-13, October 1967.

[9] J. H. Lindholm, "An Analysis of the Pseudo-Randomness Properties of Subsequences of Long *m*-Sequences," *IEEE Trans. Information Theory,* Vol. IT-14, July 1968.

[10] M. P. Ristenbatt, J. L. Daws, Jr., and H. M. Pearce, "Crack Resistant Sequences for Data Security," *IEEE National Telecommunications Conference Record,* Vol. 1, November 1973, pp. 15F1-15F5.

[11] J. L. Massey, "Shift-Register Synthesis and BCH Decoding," *IEEE Trans. Information Theory,* Vol. IT-15, No. 1, January 1969.

[12] E. L. Key, "An Analysis of the Structure and Complexity of Nonlinear Binary Sequence Generators," *IEEE Trans. Information Theory,* Vol. IT-22, No. 6, November 1976, pp. 732-736.

[13] A. Lempel and J. Ziv, "On the Complexity of Finite Sequences," *Trans. Information Theory,* Vol. IT-22, No. 1, January 1976, pp. 75-81.

[14] P. R. Geffe, "How to Protect Data with Ciphers That Are Really Hard to Break," *Electronics,* January 4, 1973.

[15] E. J. Groth, "Generation of Binary Sequences with Controllable Complexity," *IEEE Trans. on Information Theory,* Vol. IT-17, No. 3, May 1971.

[16] "Communication Techniques–Deep Space Range Measurement," *Jet Propulsion Laboratory Research Summary* No. 36-1, Vol. 1, California Institute of Technology, Pasadena, California, February 15, 1960, pp. 39-46.

[17] M. P. Ristenbatt, "Pseudo-Random Binary Coded Waveforms," Chapter 4, *Modern Radar Analysis, Valuation and System Design,* R. S. Berkowitz, Ed., Wiley & Sons, 1965.

[18] N. B. Chakrabarti and M. Tomlinson, "Design of Sequences with Specified Autocorrelation and Cross Correlation," *IEEE Trans. Communications,* Vol. COM-24, No. 11, November 1976, pp. 1246-1251.

[19] W. P. Baier, "On Parasitic Correlation Peaks in Cross-Correlation Circuits for Binary Pseudorandom Sequences," *IEEE Trans. Communications,* Vol. COM-24, No. 10, October 1976, pp. 1143-1147.

[20] R. Gold, *Code Synthesis Study,* Final Report, Vol. II Robert Gold Assoc., Los Angeles, California, 15 June 1973.

[21] J. B. Harrington and R. B. Nelson, "Compressive Intercept Receiver Uses SAW Devices for Signal Sorting with Stability and Flexibility," SAW Series No. 5, *Microwave J.,* September 1974, pp. 57-62.

[22] E. K. Holland-Moritz, J. L. Daws, Jr., G. W. McClure, and M. P. Ristenbatt, "A Transponder Against FH Signals Using a SAW Device," Abstract, paper presented Adhoc Symposium on Spread Spectrum Communications, 7-9 September 1976, Monterey, California, p. 10.

[23] M. P. Ristenbatt, "Estimating Effectiveness of Covert Communications," *Proc. 1973 Symposium on Spread Spectrum Communications,* Vol. 1, TD 271, 13-16 March 1973, Naval Electronics Laboratory Center, San Diego, California, pp. 113-118.

[24] A. D. Whalen, *Detection of Signals in Noise,* Academic Press Electrical Science Series, New York, 1971.

MODULATION TECHNIQUES FOR COMMUNICATION, PART 1: TONE AND NOISE JAMMING PERFORMANCE OF SPREAD SPECTRUM M-ARY FSK AND 2, 4-ARY DPSK WAVEFORMS

SAM W. HOUSTON

TRW Systems
One Space Park
Redondo Beach, CA. 90278

Abstract

The equivalent bit error performance of a variety of spread spectrum waveforms in both multitone CW and noise jamming environments is presented. The organization of the paper follows a complete and rather detailed derivation of the results which are summarized and given in tabular form at the conclusion. Spread Spectrum waveforms considered are frequency hopping (FH), pseudo random noise modulation (PN), and hybrid (PN/FH) combined with M-ary frequency shift keying and 2, 4-ary differential phase shift keying. The consideration of multiple chips (frequency hops) per bit and repeat jamming threats is excluded.

The performance summary for the class of waveform/jammers considered may be used to determine AJ margins directly for the uncoded waveforms or for block or convolution coded waveforms assuming hard decision demodulation.

1. INTRODUCTION

This correspondence presents the derivation of the equivalent bit error performance of a class of spread spectrum waveforms for both a noise and tone-jamming environment. The modulation techniques considered are M-ary Frequency Shift Keying (MFSK), and 2 and 4-ary Differential Phase Shift Keying (DPSK). The spread spectrum techniques considered are Frequency Hopping (FH) and Pseudo-noise Modulation plus Frequency Hopping (Hybrid PN/FH). The particular waveform under consideration is defined by stating both the modulation and spread spectrum technique employed (e.g., MFSK/FH). Incidentally, the performance of M-ary Code Shift Keying (MCSK) is also determined since it is identical to that of MFSK/PN-FH.

The noise jamming threat is considered for all waveforms in Section 2, and the Multitone CW Jamming threat is treated in Section 3. In each section the jamming strategy is described and the performance analysis carried as far as possible for a general waveform. The remainder of Sections 2 and 3 treats each considered waveform individually, as required. Section 4 summarizes the results and identifies the worst-case of the jam threats considered, except in the instances where performance is identical.

2. PARTIAL BAND NOISE JAMMING

We first consider a jamming strategy in which the jammer uses his available power, J, to emit noise of one-sided spectral density, N_J, over a fraction α of the spread spectrum bandwidth, W. Suppose that quiescent noise density is N_o, then the effective noise density is

$$N_e = \begin{cases} N_J + N_o \ , & \text{with probability } \alpha \\ \\ N_o \ \ \ \ \ \ \ , & \text{with probability } (1-\alpha) \end{cases} \tag{1}$$

where $N_J = J/\alpha W$. For this jam strategy a result is obtained which is valid for all of the waveforms considered. The symbol error probability for the general M-ary modulation can be written in general as a function of M and $X = (S/J)(W/R_c)$, the channel bit energy to jam noise density ratio, as

$$P_E(M, X) = \alpha P_E\left(M, \frac{E_c}{N_J+N_o}\right) + (1-\alpha) \ P_E\left(M, \frac{E_c}{N_o}\right) \tag{2}$$

Assuming that the quiescent noise contribution is negligible, we have

$$P_E(M, X) = \alpha P_E\left(M, \frac{E_c}{N_J}\right) \tag{3}$$

Note that

$$\frac{E_c}{N_J} = (S/R_c)/(J/\alpha w) = \alpha X \tag{4}$$

The relation between the bit and symbol error rate is

$$P_B(X) = \left[\frac{M}{2(M-1)}\right] P_E(M, X) \tag{5}$$

In general, (5) holds as an equality only for orthogonal signal sets. Hence, for the DPSK waveforms the right-hand side of (5) is an upper bound to $P_B(X)$, which for low SNR becomes a tight upper bound. For binary DPSK, however, the equality of (5) does hold since $P_B = P_E$.

We now combine (3) - (5) to obtain

Reprinted from *IEEE 1975 National Aerosp. & Electron. Conf.*, June 10-12, 1975, pp. 51-58.

$$P_B(X) = \frac{M}{2(M-1)X} \left(\frac{E_c}{N_J}\right) P_E\left(M, \frac{E_c}{N_J}\right) \qquad (6)$$

The jammer will distribute his available power, J, optimally to maximize P_B. His only control variable to accomplish this purpose is α, hence, he seeks the optimal value of this parameter, α_o, subject to the constraint, $0 < \alpha_o \leq 1$. Since the parameter α determines E_c/N_J an equivalent means of maximizing $P_B(X)$ may be found by differentiating (6) with respect to E_c/N_J and setting the result = 0. We obtain

$$\frac{\partial P_B}{\partial\left(\frac{E_c}{N_J}\right)} = \frac{M}{2(M-1)X} P_E\left(M, \frac{E_c}{N_J}\right) + \frac{M}{2(M-1)X}$$

$$\left(\frac{E_c}{N_J}\right) \frac{\partial P_E\left(M, \frac{E_c}{N_J}\right)}{\partial\left(\frac{E_c}{N_J}\right)}$$

$$(7)$$

$$\Rightarrow -\frac{\partial P_E}{P_E} = \frac{\partial\left(\frac{E_c}{N_J}\right)}{\left(\frac{E_c}{N_J}\right)}$$

The general solution to the differential equation (7) is

$$\left(\frac{E_c}{N_J}\right)_o = \frac{k}{P_E\left[M, \left(\frac{E_c}{N_J}\right)_o\right]} \qquad (8)$$

where k = constant determined for each type modulation. Substituting $(E_c/N_J)_o = X_o$ from (8) in (6)

$$P_B(X) = \frac{Mk}{2(M-1)X} \qquad (9)$$

Recall that (9) is the expression for bit error probability only for $\alpha < 1$ or $X > X_o$. For $X \leq X_o$, we set $\alpha = 1$ and obtain from (6) the applicable $P_B(X)$ as

$$P_B(X) = \frac{M}{2(M-1)} P_E(M, X) \qquad (10)$$

In summary then, $P_B(X)$ for a partial or full-band noise jammer is

$$P_B(X) = \begin{cases} \dfrac{Mk}{2(M-1)X} & , \ X > X_o \\[3mm] \dfrac{M}{2(M-1)} P_E(M, X), & X \leq X_o \end{cases} \qquad (11)$$

where $k = X_o \, P_E\left[M, X_o\right]$.

Note that (11) was obtained independent of whether the particular waveform was pure frequency-hopped or hybrid (PN/FH), and is, therefore, valid for all the waveforms considered in the noise-jamming environment. Relation (11) is particularly useful since it will be shown in the sequel that it also holds for the hybrid waveforms in the case of multitone CW jamming as well. We now evaluate (11) for the various waveforms under consideration.

2.1 M-ary FSK (or CSK)/FH or Hybrid (PN/FH)

For orthogonal noncoherent MFSK, the bit error probability is

$$P_B(X) = \frac{1}{2(M-1)X} \left(\frac{E_c}{N_J}\right) \sum_{n=2}^{M} (-1)^n \binom{M}{n} e^{-\frac{\ell E_c}{N_J}\left(\frac{n-1}{n}\right)} \qquad (12)$$

where $\ell = \log_2 M$, from (4), we may write (12) as

$$P_B(X) = \frac{\alpha}{2(M-1)} \sum_{n=2}^{M} (-1)^n \binom{M}{n} e^{-\ell\alpha X\left(\frac{n-1}{n}\right)} \qquad (13)$$

Differentiating (13) with respect to α and equating to zero we obtain

$$\frac{\displaystyle\sum_{n=2}^{M} (-1)^n \binom{M}{n} \frac{1}{n} e^{y/n}}{\displaystyle\sum_{n=2}^{M} (-1)^n \binom{M}{n} e^{y/n}} = \frac{y-1}{y} \qquad (14)$$

where we have introduced the parameter $y = \ell \alpha X$, Figure 1 shows the solution of (14), obtained by computer, for $2 \leq M \leq 32$. The value of y_o corresponding to a given M maximizes $P_B(X)$ in (12). The general result for M-ary FSK for the noise jammer becomes

$$P_B(X) = \begin{cases} \dfrac{k'}{X} & , \ X > X_o \\[3mm] \dfrac{1}{2(M-1)} \displaystyle\sum_{n=2}^{M} (-1)^n \binom{M}{n} e^{-\ell X\left(\frac{n-1}{n}\right)} , & X \leq X_o \end{cases} \qquad (15)$$

Several of the M-ary cases were evaluated by computer. The results are

$$\begin{aligned} M &= 2, & k' &= 0.3679 \ ; & X_o &= 2.000 \\ M &= 4, & k' &= 0.2329 \ ; & X_o &= 1.170 \\ M &= 8, & k' &= 0.1954 \ ; & X_o &= 0.930 \\ M &= 16, & k' &= 0.1803 \ ; & X_o &= 0.872 \\ M &= 32, & k' &= 0.1746 \ ; & X_o &= 0.798 \end{aligned} \qquad (16)$$

2.2 Binary DPSK/FH or Hybrid (PN/FH).

For binary DPSK the symbol error probability is

$$P_E\left(2, \frac{E_c}{N_e}\right) = \frac{1}{2} e^{-\frac{E_c}{N_e}} \qquad (17)$$

Its noise-jamming performance, from (11), is

$$P_B(X) = \begin{cases} \dfrac{k}{X} & , \ X > X_o \\[2em] \dfrac{1}{2} e^{-X} & , \ X \leq X_o \end{cases} \qquad (18)$$

where $k = X_o \, P_E \, [M, X_o]$. To find k and X_o write, from (6)

$$P_B(X) = \frac{1}{X} \left(\frac{E_c}{N_J} \right) P_E \left(2, \frac{E_c}{N_J} \right) \qquad (19)$$

Setting $\partial P_B / \partial (E_c/N_J) = 0$, we obtain, using (19), $X_o = 1$. Therefore,

$$P_B(X) = \begin{cases} \dfrac{1}{2eX} & , \ X > 1 \\[2em] \dfrac{1}{2} e^{-X} & , \ X \leq 1 \end{cases} \qquad (20)$$

2.3 <u>4-ary DPSK/FH or Hybrid (PN/FH)</u>. From (11) we write the bit error performance of 4-ary DPSK as

$$P_B(X) = \begin{cases} \dfrac{2}{3} \, P_E(4, X) & , \ X \leq X_o \\[2em] \dfrac{2k}{3X} & , \ X > X_o \end{cases} \qquad (21)$$

Useful approximations exist for the DPSK 4-ary symbol error probability, $P_E(4, X)$, but these are valid only for relatively high SNR (>5 dB), which is above our present range of interest. Hence, we shall use numerical methods to find $(E_c/N_J)_0$. Using a table provided by Lindsey* of exact 4-ary DPSK symbol error probabilities versus the SNR, E_c/N_J, at 1 dB increments, we obtain the curve of Figure 2, which shows the product $(E_c/N_J) \, P_E(4, \ E_c/N_J)$ as a function of E_c/N_J. The maximum value of this curve is $k = (E_c/N_J)_0 \, P_E \, [4, (E_c/NJ)_0] \cong 0.318$ at $(E_c/N_J)_0 \approx 1.26$, hence, from (21)

$$P_B(X) = \begin{cases} \dfrac{0.212}{X} & , \ X > 1.26 \\[2em] \dfrac{2}{3} \, P_E(4, X) & , \ X \leq 1.26 \end{cases} \qquad (22)$$

3. MULTITONE CW JAMMING

The second type of jam threat considered is generally called multitone CW jamming. This jammer strategy distributes the total jam power, J, equally in q separate random phase jam tones. In this case the performance of the pure frequency-

hopped and hybrid systems must be analyzed separately. Let us first consider the hybrid waveforms. For the hybrid waveforms the expanded hop bandwidth is 1/P, where P = PN chip duration. The number of frequency bins, or hop bandwidths within the total spread spectrum bandwidth, W, is n = WP. The probability that the jammer hits a frequency bin in any one hop is simply $P_H = q/n = q/WP \leq 1$. Assume that the jammer places at most one tone centered in the transmitted hop band. The resulting jamming noise spectral density is $N_J = JP/q$. The expression for the symbol error rate may be written in general as before, treating the quiescent noise density, N_o, as negligible as

$$P_E(M, X) = P_H \cdot P_E \left(M, \frac{E_c}{N_J} \right) \qquad (23)$$

This expression is identical to (3), with $P_H = \alpha$. The jammer seeks to maximize (23) by an optimal distribution of his total power, J. The parameter which he can vary is q, the number of jam tones $= \alpha J/S$. Since $P_H = 1/X \cdot E_c/N_J$ the expression for the symbol error probability (23), becomes

$$P_E(M, X) = \left(\frac{1}{X} \right) \left(\frac{E_c}{N_J} \right) P_E \left(M, \frac{E_c}{N_J} \right) \qquad (24)$$

and from (5) we may write the bit error probability as

$$P_B(X) = \frac{M}{2(M-1)X} \left(\frac{E_c}{N_J} \right) \cdot P_E \left(M, \frac{E_c}{N_J} \right) \qquad (25)$$

Note that (25) is identical to (6). The optimal value, $(E_c/N_J)_0$, is found just as before by varying the value of P_H or α, in the range (0, 1). Hence, the performance of the hybrid waveforms is identical for both the noise and tone jammers. The tone-jamming performance of the pure frequency-hopped waveforms under consideration must be evaluated individually. This evaluation follows in the sequel.

3.1 <u>M-ary FSK/FH: Jam Tone Spacing = $R_c/\log_2 M$</u>. The MFSK/FH waveform hops over the total spread spectrum bandwidth, W. The frequency spacing between the M-ary symbol keying tones is $R_c/\log_2 M$ Hz, and hence the hopping bandwidth is $M \cdot R_c/\log_2 M$ Hz. We assume that the jammer has perfect knowledge of system operation, except for the hopping sequence. He does not know, a priori, where within W, the spread spectrum bandwidth, to concentrate his jam tones. He can, however, seek an optimum distribution of his total available power, J, by varying the number of jam tones. In this section we assume that whatever number of jam tones he employs are contiguous and spaced at frequency intervals equal to $R_c/\log_2 M$, the spacing between the M-ary symbol keying tones. The probability that the jammer will hit a symbol keying tone is

---

*Lindsey, W.C. and Simon, M.K., <u>Telecommunication System Engineering</u>, Prentice-Hall, 1973, p. 250.

$$p = \frac{qR_c}{W \log_2 M} = \frac{\alpha}{\ell X} \ , \ 0 \leq \alpha \leq 1 \qquad (26)$$

Now, assume that the jammer causes a symbol error if he hits one of the $(M-1)$ unkeyed symbol tones. The correct symbol error probability, $P_C(M, X) = 1 - P_E(M, X)$, becomes

$$P_C(M, X) = (1-p)^M \left\{ \begin{array}{l} \text{probability he hits none of} \\ \text{the M tones searched} \end{array} \right\} \atop + p \left\{ \begin{array}{l} \text{probability he hits keyed} \\ \text{tone} \end{array} \right\} \quad (27)$$

and the bit error probability from (5) and (26) is

$$P_B(X) = \frac{M}{2(M-1)} \left[1 - \left(\frac{\alpha}{\ell X} \right) - \left(1 - \frac{\alpha}{\ell X} \right)^M \right] \qquad (28)$$

We now maximize $P_B(X)$ with respect to α, setting $\partial P_B / \partial \alpha = 0$, to obtain

$$\alpha_o = \ell X \left[1 - M^{-\left(\frac{1}{M-1} \right)} \right] \qquad (29)$$

This value of α_o is optimum only if it is less than unity which requires a minimum ratio, J/S. Namely, for $\alpha_o < 1$

$$\ell X \left[1 - M^{-\left(\frac{1}{M-1} \right)} \right] < 1 \Rightarrow X < \left\{ \ell \left[1 - M^{-\left(\frac{1}{M-1} \right)} \right] \right\}^{-1} \quad (30)$$

For the range of X satisfying (30), we have, substituting (29) in (28)

$$P_B(X) = \frac{1}{2} M^{-\left(\frac{1}{M-1} \right)} \ , \ X < \left\{ \ell \left[1 - M^{-\left(\frac{1}{M-1} \right)} \right] \right\}^{-1} \quad (31)$$

Otherwise $\alpha_o = 1$, for which (28) is

$$P_B(X) = \frac{M}{2(M-1)} \left[1 - \left(\frac{1}{\ell X} \right) - \left(1 - \frac{1}{\ell X} \right)^M \right] \qquad (32)$$

for $X > \left\{ \ell \left[1 - M^{-\left(\frac{1}{M-1} \right)} \right] \right\}^{-1}$

3.2 M-ary FSK: Jam-Tone Spacing = $MR_c / \log_2 M$.

In this strategy, the jammer spreads his tone spacing seeking to hit only one of the contiguous M symbol keying tones. Note that there are $b = W\ell / MR_c$ banks of M contiguous symbol keying tones within the spread spectrum bandwidth, W. The probability of symbol error becomes

$$P_E(M, X) = \frac{q}{b} \left\{ \begin{array}{l} \text{probability he hits} \\ \text{the M-ary tone bank} \end{array} \right\} \atop \text{times} \ \frac{M-1}{M} \left\{ \begin{array}{l} \text{probability he does not} \\ \text{hit the keyed tone} \end{array} \right\} \quad (33)$$

where $q = \alpha J/S$, as before. From (5), the bit error probability becomes $P_B(X) = M\alpha / 2\ell X$. Since $P_B(X)$ is directly proportional to α, the maximum value of $P_B(X)$ occurs for the maximum possible value of α, namely $\alpha_o = 1$, for which

$$P_B(X) = \begin{cases} \dfrac{M}{2\ell X} \ , & X \geq \dfrac{M}{\ell} \\[3mm] \dfrac{1}{2} \ , & X \leq \dfrac{M}{\ell} \end{cases} \qquad (34)$$

The restriction $X \geq M/\ell$ is necessary since $P_B \leq 1/2$.

3.3 Binary DPSK/FH.

The derivation of binary DPSK/FH performance with a multitone CW jammer relies largely on certain geometric relations for which Figure 3 is useful. In binary DPSK, the transmission of one of the two biphase vectors conveys differential change-of-phase, $\Delta\theta$, information from bit-to-bit. In this analysis we assume the differentially encoded PSK signal vectors to be $\mathbf{P}_1 = A/\underline{0}$ and $\mathbf{P}_2 = A/\underline{\pi}$. The DPSK receiver estimates the phase difference between each consecutive signal pair, $\widehat{\Delta}\theta$.

$$\widehat{\Delta}\theta = \underline{/}(\mathbf{P}_k + \mathbf{J}) - \underline{/}(\mathbf{P}_{k-1} + \mathbf{J}) \qquad (35)$$

Further, we assume that the tone-jamming signal vector, $\mathbf{J} = I/\underline{\phi}$, is constant, in phase and magnitude, during the hopping interval, T_H. Two types of errors (caused by the CW interference) can occur: 1) $\widehat{\Delta}\theta = 0$ if $\Delta\theta = \pi$ and 2) $\widehat{\Delta}\theta = \pi$ if $\Delta\theta = 0$. Note, however, that by our constant jamming vector assumption, and in the absence of noise, a Type 2) error cannot occur since for $\mathbf{P}_k = \mathbf{P}_{k-1}$ in (35), $\widehat{\Delta}\theta$ is always identically equal to zero. Hence, Pr [Type 2) error] = 0. A Type 1) error occurs if the angle α between R_1 and R_2 in Figure 3 is between $0°$ and $90°$ or between $270°$ and $360°$. This is equivalent to $|\alpha - \pi| > \pi/2$ or $\cos\alpha > 0$, hence, the bit error probability may be written as

$$P_B = Pr[\Delta\theta \doteq \Pi] \cdot Pr[\widehat{\Delta}\theta = 0] = \frac{1}{2} Pr[\cos\alpha > 0] \qquad (36)$$

From Figure 3, we obtain the relation

$$\cos\alpha = \frac{1}{R_1 R_2} \left[I^2 - A^2 \right] \qquad (37)$$

Therefore, if $I^2 > A^2$, $P_B = 1/2$. Assume now that power in each of the q jam tones is greater than A^2, then

$$P_B = \frac{1}{2} Pr[I^2 > A^2] = \frac{1}{2} P_H \qquad (38)$$

where $P_H = \alpha/X$ since α_o is obviously = unity and (38) becomes

$$P_B(X) = \begin{cases} \dfrac{1}{2X} \ , & X > 1 \\[3mm] \dfrac{1}{2} \ , & X \leq 1 \end{cases} \qquad (39)$$

3.4 4-ary DPSK/FH.

The signal and interference

vectors for 4-ary DPSK are illustrated in Figure 4. In this analysis we adopt the convention that the transmission of $\mathbf{P}_k \Rightarrow (\theta = k$ radians), where $k = 0, \pi, \pm \pi/2$. In practice these signal vectors are usually rotated $\pi/4$ radians to avoid the axes, but for convenience we adopt this convention. Note that four types of errors may occur. Denote these as Q_k = (Probability of incorrectly detecting a phase transition of k radians). The 4-ary DPSK symbol error probability may be written as

$$P_E(4) = \frac{P_H}{4}\left[Q_o + Q_\pi + Q_{\frac{\pi}{2}} + Q_{-\frac{\pi}{2}}\right] \quad (40)$$

Again we assume the CW jam signal vector, $\mathbf{J} = I\underline{/\phi}$, to be constant both in phase and magnitude, hence $Q_o = 0$. Since ϕ is uniformly distributed in the interval $(0, 2\pi)$, we argue that $Q_{\pi/2} = Q_{-\pi/2}$ and using (5), we write the bit error probability as

$$P_B = \frac{P_H}{6}\left[Q_\pi + 2Q_{\frac{\pi}{2}}\right] \quad (41)$$

Determination of $Q_{\pi/2}$. Assume that $\Delta\theta = \pi/2$. Note that if $\hat{\Delta\theta} = \Delta\theta + \phi = \psi$ differs in phase from $\pi/2$ by more than $\pi/4$, the phase transition of $\pi/2$ radians is incorrectly detected, hence

$$Q_{\frac{\pi}{2}} = Pr\left\{\left|\psi - \frac{\pi}{2}\right| > \frac{\pi}{4}\right\} = Pr\left\{\sin\psi < |\cos\psi|\right\} \quad (42)$$

Using the trigonometric identities for $\sin(\theta_3 - \theta_1)$ and $\cos(\theta_3 - \theta_1)$ and letting $\phi' = \phi - \pi/4$ and $\rho = I/A$, we find that the condition $\sin\psi < |\cos\psi|$ is equivalent to

$$\frac{1}{\rho} + \sqrt{2}\cos\phi' < |\rho + \sqrt{2}\cos\phi'| \quad (43)$$

This inequality holds for all ϕ' if $\rho > 1$. Hence, if $I > A$, $Q_{\pi/2} = 1$. In the region $0 < \rho < 1$, the inequality (43) holds only for a certain range of ϕ'. Note that for $\rho < 1$, the inequality holds only for $|\rho + \sqrt{2}\cos\phi'| = -(\rho + \sqrt{2}\cos\phi')$, hence, for $\rho < 1$, (43) may be written

$$\cos\phi' < \frac{-(\rho^2 + 1)}{2\sqrt{2}\,\rho}, \quad \rho < 1 \quad (44)$$

Note that the inequality (44) holds for no value of ϕ' if $\rho < \sqrt{2} - 1$. Let $\phi' = \beta + \pi$, then (44) becomes

$$\cos\beta > \frac{\rho^2 + 1}{2\sqrt{2}\,\rho}, \quad \sqrt{2} - 1 < \rho < 1 \quad (45)$$

We now observe that the range of ϕ' satisfying the inequality (45) is 2β where, $\beta = \cos^{-1}(\rho^2 + 1)/(2\sqrt{2}\,\rho)$, $\sqrt{2} - 1 < \rho < 1$. Hence, $Q_{\pi/2} = 2\beta/2\pi = \beta/\pi$, $\sqrt{2} - 1 < \rho < 1$, and we have

$$Q_{\frac{\pi}{2}} = \begin{cases} 0 & , \rho < \sqrt{2} - 1 \\ \frac{1}{\pi}\cos^{-1}\left(\frac{\rho^2 + 1}{2\sqrt{2}\,\rho}\right), & \sqrt{2} - 1 < \rho < 1 \\ 1 & , \rho > 1 \end{cases} \quad (46)$$

Determination of Q_π. Assume that $\Delta\theta = \pi$. Note that if $\hat{\Delta\theta} = \Delta\theta + \phi = \psi$ differs in phase from π by more than $\pi/4$, the phase transition of π radians is incorrectly detected, hence

$$Q_\pi = Pr\left\{|\psi - \pi| > \frac{\pi}{4}\right\} = Pr\left\{\cos\psi > -|\sin\psi|\right\} \quad (47)$$

By the trigonometric properties of Figure 4

$$\sin\psi = \frac{2AI\sin\phi}{R_1 R_2} \quad (48)$$

and using (37) from the 2-ary development, we may now write (47) as

$$Q_\pi = Pr\left\{I^2 - A^2 > -|2AI\sin\phi|\right\} \quad (49)$$

Again letting $\rho = I/A$, we write the inequality as

$$|\sin\phi| > \frac{1 - \rho^2}{2\rho} \quad (50)$$

Note that for $\rho > 1$, the inequality (50) holds for all values of ϕ, and for $\rho > \sqrt{2} - 1$ (50) holds for no value of ϕ. For $\sqrt{2} - 1 < \rho < 1$, $0 \le \phi \le \pi$, (50) may be written

$$\sin\psi \left(\frac{1 - \rho^2}{2\rho}\right) \quad (51)$$

Let $\phi = \beta + \pi/2$, then (51) gives

$$\beta = \cos^{-1}\left(\frac{1 - \rho^2}{2\rho}\right) \quad (52)$$

Note that the range of ϕ which satisfies the inequality (51) is 2β and we have, from (49)

$$Q_\pi = \begin{cases} 0 & , \rho < \sqrt{2} - 1 \\ \frac{2}{\pi}\cos^{-1}\left(\frac{1 - \rho^2}{2\rho}\right), & \sqrt{2} - 1 < \rho < 1 \\ 1 & , \rho > 1 \end{cases} \quad (53)$$

We not gather our results and evaluate the bit error probability, (41). Note that

$$\frac{\frac{J}{q}\,(\text{jam tone power})}{S} = \frac{I^2}{A^2} = \rho^2 \quad (54)$$

which implies that the number of jam tones is $q = J/\rho^2 S$. Since the number of frequency bins, or 4-ary DPSK symbol tones in the total spread spectrum bandwidth, W is n = 2W/R$_c$, we find that $P_H = q/n = 1/2X\rho^2$. Using this result and noting that $P_H = q/n = R_c J/2\rho^2 SW = 1/2X\rho^2$, (41) becomes

$$P_B(X) = \frac{1}{12X\rho^2} \left[Q_\pi + 2Q_{\frac{\pi}{2}} \right] \qquad (55)$$

Summarizing our results from (46) and (53), the bit error probability for 4-ary DPSK/FH with tone-jamming is

$$P_B(X) = \begin{cases} 0 & , \rho < \sqrt{2} - 1 \\[2mm] \dfrac{1}{6\pi\rho^2 X} \left[\cos^{-1}\left(\dfrac{1-\rho^2}{2\rho}\right) \right. & \\[2mm] \left. + \cos^{-1}\left(\dfrac{1+\rho^2}{2\sqrt{2}\,\rho}\right) \right] & , \sqrt{2} - 1 < \rho < 1 \\[2mm] \dfrac{1}{4\rho^2 X} & , \rho > 1 \end{cases}$$

It may be verified that the maximum value of $P_B(X) = 0.2592$ occurs for the optimal jam stragegy of setting $\rho = I/A = 0.52$.

4. SUMMARY OF RESULTS

The performance summary for the class of waveform/jammers is summarized in Table 1. The results of the table may be used to determine AJ margins directly, assuming hard decision demodulation, for a given channel bit error performance, spread spectrum bandwidth, W, and channel data rate, R$_c$. For the hybrid (PN/FH) waveforms, performance is identical for both noise and tone-jamming. For the M-ary FSK (or CSK)/FH systems, the multitone CW jammer with tone spacing equal to M R$_c$/log$_2$ M Hz represents the worst-case jamming threat. For 4-ary DPSK/FH, the worst-case jammer is Multitone CW with R$_c$/2 tone spacing with ρ^2 = ratio of jam tone power to symbol tone power = 0.52^2. Figures 5 and 6 plot the performance, for worst-case jamming, of the FH and Hybrid PN/FH Waveforms, respectively. Note that increasing the order of M degrades the performance of the pure FH waveforms, while the reverse is true for the hybrid PN-FH waveforms.

Acknowledgement

The author appreciates and acknowledges the contributions and helpful comments made by D. Brown and F. R. Skalbania of TRW Systems.

Table 1. Channel Bit Error Probabilities for Jammed Waveforms

WAVEFORM	JAMMING SIGNAL	CHANNEL BIT ERROR PERFORMANCE, $P_B(X)$
MFSK/FH	Multitone CW with $\frac{M R_c}{\ell}$ Hz Tone Spacing (Worst Case)	$P_B(X) = \begin{cases} \frac{M}{2\ell X} & , X > \frac{M}{\ell} \\ \frac{1}{2} & , X \le \frac{M}{\ell} \end{cases}$
	Multitone CW with $\frac{R_c}{\ell}$ Hz Tone Spacing	$P_B(X) = \begin{cases} \frac{M}{2(M-1)}\left[1 - \left(\frac{1}{\ell X}\right) - \left(1 - \frac{1}{\ell X}\right)^M\right], & X > \left\{\ell\left[1 - M^{-\left(\frac{1}{M-1}\right)}\right]\right\}^{-1} \\ \frac{1}{2} M^{-\left(\frac{1}{M-1}\right)}, & X \le \left\{\ell\left[1 - M^{-\left(\frac{1}{M-1}\right)}\right]\right\}^{-1} \end{cases}$
MFSK/Hybrid (PN/FH) MCSK/Hybrid (PN/FH)	Partial Band Noise Multitone CW with $\frac{1}{P}$ Hz Tone Spacing or Partial Band Noise	$P_B(X) = \begin{cases} \frac{k'}{X} , & X > X_o \\ \frac{1}{2(M-1)} \sum_{n=2}^{M} (-1)^n \binom{M}{n} e^{-\ell X\left(\frac{n-1}{n}\right)} , & X \le X_o \end{cases}$ for $M = 2$, $k' = 0.3679$; $X_o = 2.000$ $M = 4$, $k' = 0.2329$; $X_o = 1.170$ $M = 8$, $k' = 0.1954$; $X_o = 0.930$ $M = 16$, $k' = 0.1803$; $X_o = 0.872$ $M = 32$, $k' = 0.1746$; $X_o = 0.798$
2-ary DPSK/FH	Multitone CW with R_c Hz Tone Spacing (Worst Case)	$P_B(X) = \begin{cases} \frac{1}{2X} & , X > 1 \\ \frac{1}{2} & , X \le 1 \end{cases}$
2-ary DPSK/Hybrid (PN/FH)	Partial Band Noise Multitone CW with $\frac{1}{P}$ Hz Tone Spacing or Partial Band Noise	$P_B(X) = \begin{cases} \frac{1}{2eX} & , X > 1 \\ \frac{1}{2} e^{-X} & , X \le 1 \end{cases}$
4-ary DPSK/FH	Multitone CW with $\frac{R_c}{2}$ Hz Tone Spacing (Worst Case with $\rho = 0.52$)	$P_B(X) = \begin{cases} 0 & \rho < \sqrt{2} - 1 \\ \frac{1}{6\pi\rho^2 X}\left[\cos^{-1}\left(\frac{1-\rho^2}{2\rho}\right) + \cos^{-1}\left(\frac{1+\rho^2}{2\sqrt{2}\,\rho}\right)\right], & \sqrt{2} - 1 < \rho < 1 \\ \frac{1}{4\rho^2 X} & , \rho > 1 \end{cases}$ Worst Case: with $\rho = 0.52$, $P_B(X) = \frac{0.2592}{X}$
4-ary DPSK/Hybrid (PN/FH)	Partial Band Noise Multitone CW with $\frac{1}{P}$ Hz Tone Spacing or Partial Band Noise	$P_B(X) = \begin{cases} \frac{0.212}{X} & , X > 1.26 \\ \frac{2}{3} P_E(4, X) & , X \le 1.26 \end{cases}$

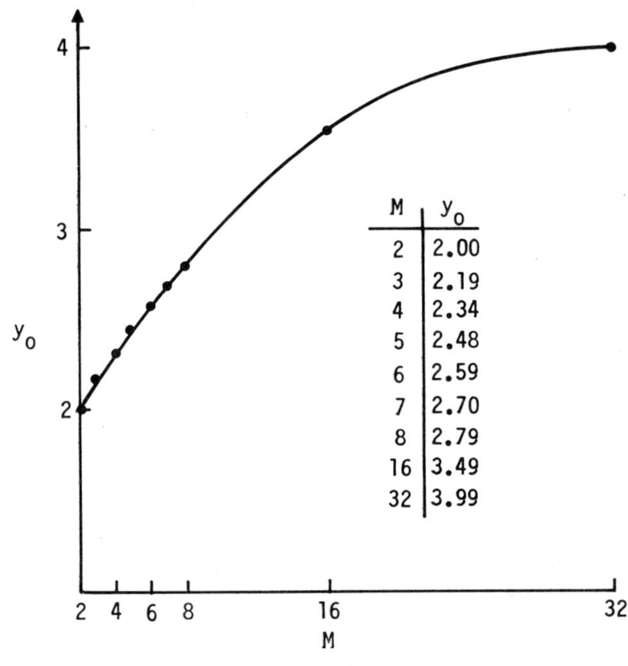

SOLUTION OF (16) FOR y_0 AS M VARIES

Figure 1.

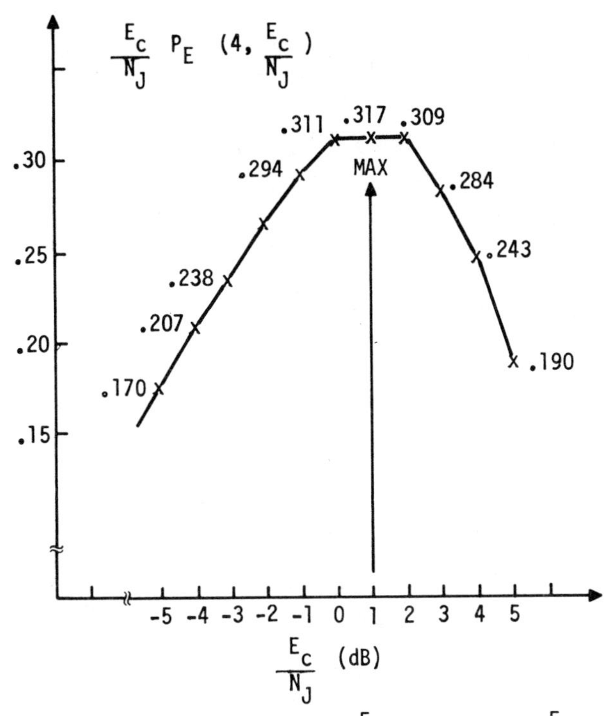

NUMERICAL SOLUTION FOR $k = \left(\dfrac{E_c}{N_J}\right)_0 P_E \left[4, \left(\dfrac{E_c}{N_J}\right)_0\right]$

Figure 2.

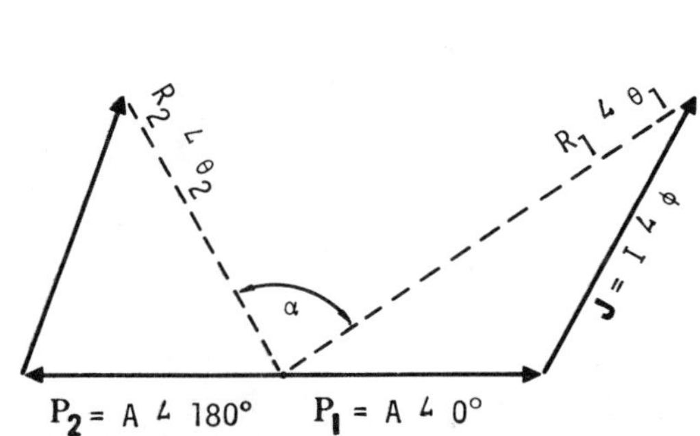

VECTOR DIAGRAM SHOWING RESULTANT OF DIFFERENTIALLY ENCODED 2-ARY PSK SIGNALS, P1 AND P2, AND CONSTANT TONE JAM INTERFERENCE SIGNAL, J.

Figure 3.

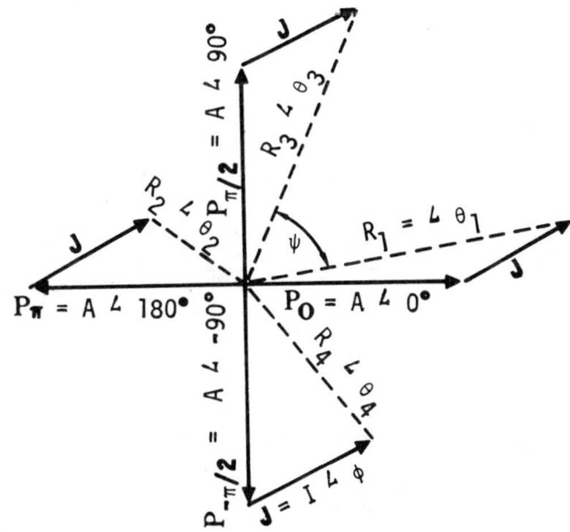

VECTOR DIAGRAM SHOWING RESULTANT OF DIFFERENTIALLY ENCODED 4-ARY PSK SIGNALS $\{P_k\}_{k=1}^4$ AND CONSTANT TONE INTERFERENCE SIGNAL J.

Figure 4.

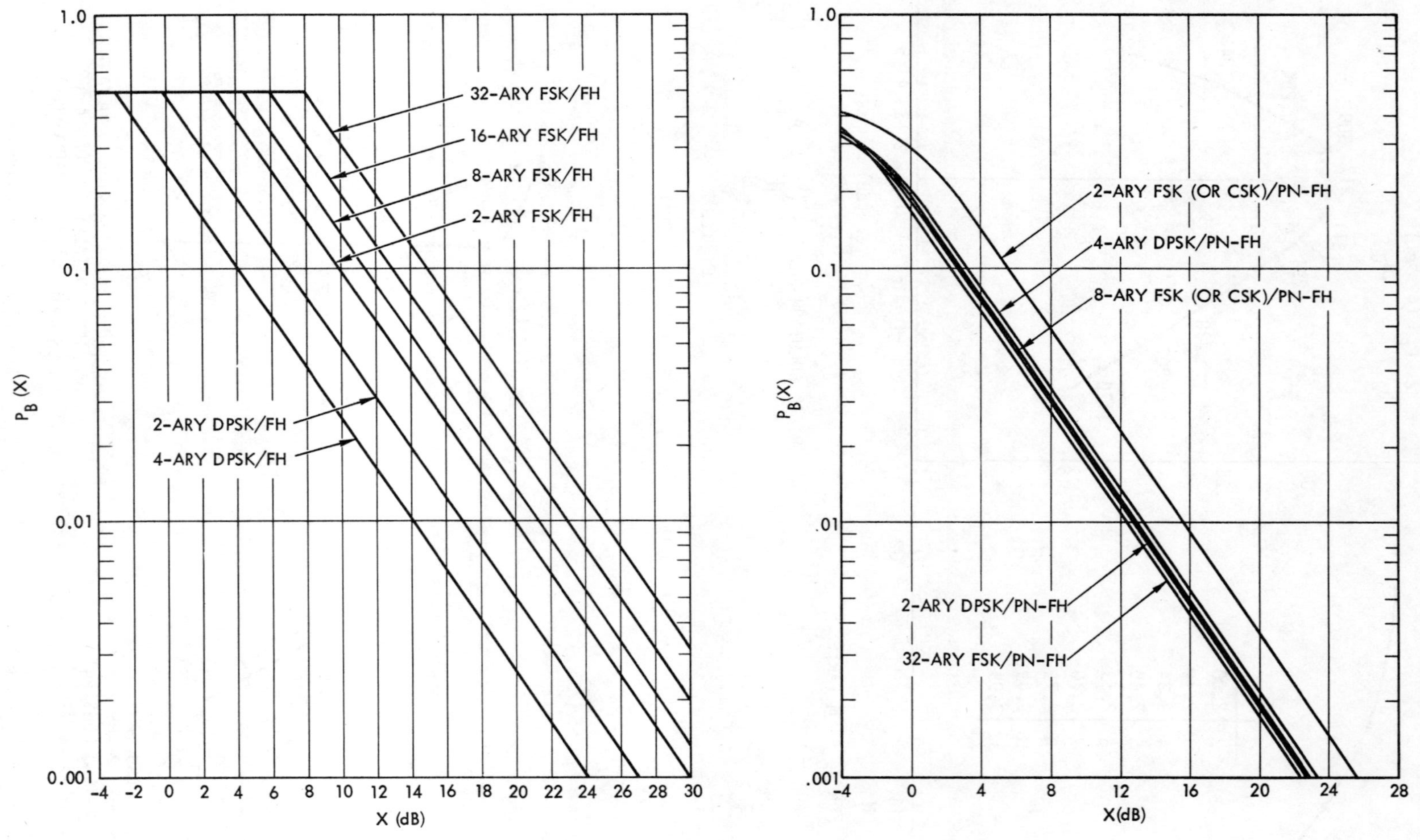

AJ PERFORMANCE FOR FH SYSTEMS. CURVES SHOWN ARE FOR THE WORST OF THE JAMMING THREATS CONSIDERED.

Figure 5

AJ PERFORMANCE FOR THE PN-FH SYSTEMS.

Figure 6

In the previous subsections, multiple-access schemes have been studied as a function of the division of the time-frequency plane. Specifically, FDMA, TDMA, and CDMA were analyzed. In this section, the question of how the channels are assigned to the users is discussed. The two basic alternatives are fixed assignment of the channels or dynamic (or demand) assignment of the channels. Three types of traffic inputs are of interest: voice, digital data, and combinations of voice and data.

One of the key characteristics of voice traffic is that conversations have an average duration of several minutes. The overhead time to connect and disconnect a circuit is typically one second, so that channel assignment schemes are efficient. Some of the issues of importance in demand assignment for voice traffic are

1) user requirements (Erlangs, number of destinations, grade of service),
2) capacity improvements due to demand assignment,
3) assignment algorithms (centralized versus distributed control),
4) equipment cost and complexity.

A simple example to demonstrate the potential improvement due to demand assignment is useful. Consider a satellite system in which an earth station is required to communicate to 40 destinations, the traffic to each destination is 0.5 Erlangs, and the required blocking probability is $P_B = 0.01$. If the traffic is modeled such that the call arrivals are a Poisson process and the call holding times have an exponential distribution, then, for a trunk group with N channels in which blocked calls are cleared, the blocking probability is given by the Erlang B distribution,

$$P_B = \frac{\alpha^N/N!}{\displaystyle\sum_{n=o}^{N} \alpha^n/n!} \tag{1}$$

where α is the intensity in Erlangs. If the system is designed so that the channels are pre-assigned to each destination, then it will require four channels per destination to achieve $P_B = 0.01$. (This is the smallest value of N that makes the right side of (1) less than 0.01). Since there are 40 destinations, 160 channels are required for pre-assigned operation. In the demand-assigned mode of operation, the total traffic is considered since any channel can be assigned-to any user. Thus, the total traffic load is 20 Erlangs and

the P_B formula gives a requirement of 30 channels. Thus, demand assignment has improved the efficiency of the system by a factor of $160/30 = 5.3$. Table I (from [1]) shows the comparison for various Erlang loads and numbers of destinations. A more general discussion of the grade of service in demand-assigned systems is given in Reprint Paper 3.6.4.1.

TABLE I

PRE-ASSIGNMENT VERSUS DEMAND-ASSIGNMENT SATELLITE CHANNEL REQUIREMENTS BASED ON GRADE OF SERVICE $P_B = 0.01$

NUMBER OF DESTINA-TIONS, d	CHANNEL REQUIREMENT ($\alpha = 0.1$)		CHANNEL REQUIREMENT ($\alpha = 0.5$)		CHANNEL REQUIREMENT ($\alpha = 1.0$)	
	PRE-ASSIGN-MENT	DEMAND-ASSIGN-MENT	PRE-ASSIGN-MENT	DEMAND-ASSIGN-MENT	PRE-ASSIGN-MENT	DEMAND-ASSIGN-MENT
1	2	2	4	4	5	5
2	4	3	8	5	10	7
4	8	3	16	7	20	10
8	16	4	32	10	40	15
10	20	5	40	11	50	18
20	40	7	80	18	100	30
40	80	10	160	30	200	53

The potential improvement in the communication system capacity is the primary motivation for demand-assigned systems. Reprint Paper 3.6.4.2 discusses a demand-assigned system for INTELSAT, called SPADE (*s*ingle-channel-per carrier, *P*CM multiple-*a*ccess *d*emand assignment *e*quipment). Reference [2] also describes the SPADE system. A demand-assigned TDMA system is described in [3]. Analyses of demand-assignment schemes are given in [4]-[6]. Other demand-assignment schemes for voice traffic are discussed in papers listed in the Bibliography.

The second type of input of interest is the data traffic generated in computer communications systems. The traffic that is generated in computer–computer and computer–terminal applications (e.g. time-sharing and inquiry-response applications) is characterized by a large variance in the interarrival times of messages and message lengths. In addition, the user frequently imposes a message delay that is small compared to the message interarrival time. If the channel assignment schemes used for voice traffic were used for this type of data traffic, two problems would result. The time to connect and disconnect a channel would be comparable to the time the channel was in use for transmitting the message and the resulting overhead would be unacceptable. The time to establish the connection might lead to excessive delay. As a result of the limitations of the classical channel demand-assignment schemes, a large variety of random access and packet reservation schemes have been developed in the last several years. Table II is a representative list of multiple access schemes.

TABLE II
REPRESENTATIVE MULTIPLE-ACCESS SCHEMES FOR DATA TRAFFIC

```
1)  CHANNEL RESERVATION SCHEMES

    (A)  FIXED ASSIGNMENT
    (B)  DEMAND-ASSIGNED FOR SESSION
    (C)  DEMAND-ASSIGNED FOR MESSAGE
    (D)  POLLING SYSTEMS

2)  RANDOM ACCESS SCHEMES

    (A)  ALOHA
    (B)  SLOTTED ALOHA
    (C)  RESERVATION ALOHA

3)  PACKET RESERVATION SCHEMES

    (A)  FIFO RESERVATION (ROBERTS)
    (B)  ROUND ROBIN (BINDER)
    (C)  CONTENTION-BASED DEMAND-ASSIGNMENT PROTOCOL (CPODA)
```

Reprint Paper 3.6.5.3 provides a unified presentation of packet broadcasting theory. Reprint Paper 3.6.5.4 compares various satellite multiple-access schemes for data traffic and analyzes their delay-throughput performance. References [7]–[23] discuss various aspects of demand assignment for data.

REFERENCES

[1]   G. D. Dill, "Comparison of circuit call capacity of demand-assignment and preassignment operation," *COMSAT Tech. Rev.,* vol. 2, no. 1, pp. 243–256, Spring 1972.
[2]   B. I. Edelson and A. Werth, "SPADE system progress and application," *COMSAT Tech. Rev.,* vol. 2, no. 1, Spring 1972.
[3]   J. H. Deal and J. Buegler, "A demand-assignment time-division multiple-access system for military tactical application," *AIAA Paper 76-270, AIAA/CASI 6th Communications Satellite Systems Conf.,* Apr. 1976.
[4]   J. M. Aein and O. S. Kosovych, "Satellite capacity allocation," *Proc. IEEE,* vol. 65, no. 3, Mar. 1977.
[5]   J. M. Aein, "A multi-user-class, blocked-calls-cleared, demand access model," *IEEE Trans. Commun.,* vol. 65, no. 3, Mar. 1977.
[6]   Collins Report, "Demand Assignment Techniques Study for Military Satellite Communications Applications," June 25, 1977, prepared for Milsatcom Office, Defense Communications Agency.
[7]   P. E. Jackson and C. D. Stubbs, "A study of multi-access computer communications," *Spring Joint Computer Conf., AFIPS Conf. Proc.,* vol. 34, pp. 491–504, 1969.
[8]   L. G. Roberts, "Dynamic allocation of satellite capacity through packet reservations," *Proc. 1973 National Computer Conf.,* pp. 711–716.
[9]   L. Kleinrock and S. S. Lam, "Packet-switching in a slotted satellite channel," *Proc. 1973 National Computer Conf.,* pp.703–710.
[10] R. Binder, "A dynamic packet switching system for satellite broadcast channels," *Proc. Int. Conf. Communications,* San Francisco, CA, pp. 41-1-41-5, June 1975.
[11] W. Crowther, R. Rettberg, D. Walden, S. Ornstein, and F. Heart, "A system for broadcast communication: Reservation ALOHA," *Proc. 6th Hawaii Int. Conf. System Sciences,* pp. 371–374, Jan. 1973.
[12] R. Binder, J. McQuillan, and R. Rettberg, "The impact of multi-access satellites on packet switching networks," *EASCON '75, Convention Record.* New York: IEEE, 1975, pp. 63-A-63-F.
[13] S. S. Lam, "A new measure for characterizing data traffic," *IBM T. J. Watson Research Center, Yorktown Heights, NY, Research Report RC6242,* Oct. 1976.
[14] ——"Delay analysis of a packet-switched TDMA system," *NTC '76 Conf. Rec.* New York: IEEE, pp. 16.3-1–16.3-6.
[15] L. Kleinrock and S. S. Lam, "Packet switching in a multi-access broadcast channel: Performance evaluation," *IEEE Trans. Commun.,* vol. COM-23, pp. 410–423, Apr. 1975.
[16] S. S. Lam and L. Kleinrock, "Packet switching in a multi-access broadcast channel: Dynamic control procedures," *IEEE Trans. Commun.,* vol. COM-23, pp. 891–904, Sept. 1975.
[17] A. B. Carleial and M. E. Hellman, "Bistable behavior of ALOHA-type systems," *IEEE Trans. Commun.,* vol. COM-23, pp. 401–410, Apr. 1975.
[18] R. L. Pickholtz and W. F. Vogelzang, "Dynamic satellite slot allocations for data packets via Adaptive TDMA," *Proc. National Telecommunications Conf.,* Dec. 1977.
[19] L. Lee and I.M. Jacobs, "A priority-oriented demand assignment (PODA) protocol and an error recovery algorithm for distributively controlled packet satellite communication networks," *EASCON '77.*
[20] I. Rubin, "A group random access procedure for multi-access communication channels," *Proc. National Telecommunications Conf.,* Dec. 1977.
[21] ————"On message delays in FDMA and TDMA communication channels," presented at Conference on Information Sciences and Systems, John Hopkins Univ., Mar. 1978.
[22] I. Jacobs *et al,* "CPODA—A demand assignment protocol for SATNET," *5th Data Communications Symp.,* Snowbird, UT, pp. 2-5-2-9, Sept. 1977.
[23] I. M. Jacobs, R. Binder, and E. Hoversten, "General purpose packet satellite networks," *Proc. IEEE,* vol. 66, no. 11, pp. 1448-1467, Nov. 1978.

3.6.5.1

The Grade of Service in Multiple-Access Satellite Communications Systems with Demand Assignments

GABRIEL FRENKEL, MEMBER, IEEE

Abstract—Demand assignment of satellite power on a call-by-call basis results in increased system capacity. The probability that a call is blocked is not readily available, and standard formulas commonly used for the Poisson and Erlang B traffic model cannot be directly applied. Expressions for blocking probabilities are derived here for fully variable, variable-source, and variable-destination demand assignments. A numerical example is given for their application in a network of terminals of varying sizes and known point-to-point traffic.

I. INTRODUCTION

Satellite communication systems which provided telephone service in the past made use of point-to-point dedicated circuits. Recently, however, the concept of sharing the satellite resources among a number of links is gaining progressively wider application [1]–[3]. Demand assignment, achieved by pooling the traffic on a number of links, may be carried out in one of three ways leading to a fully variable, variable-source, or variable-destination demand-assignment system. First, in a fully variable system all the traffic in the network is pooled, and all the satellite resources are pooled and assigned to specific calls as the demand arises. Secondly, in a variable-destination system a portion of the satellite resources is permanently dedicated to one-way calls leaving a particular earth terminal, and is assigned to specific calls as the demand arises. Thirdly, in a variable-source system a portion of the satellite resources is permanelty dedicated to one-way calls entering a particular earth terminal, and is assigned to specific calls as the demand arises.

The method whereby a portion of the satellite resources is assigned to a call depends on the multiple-access technique. The two such techniques particularly suited for demand assignment are Time-Division Multiple Access (TDMA) and Frequency-Division Multiple Access (FDMA), of which the SPADE system is the best publicized example [3]–[6]. In a TDMA system a call is serviced by allocating a time slot or a set of time slots within the TDMA frame format. If the terminals do not have equal figures of merit (ratio of antenna gain to receiver noise temperature), in order to insure equal voice quality, the fraction of satellite power dedicated to the one-way voice channel must be inversely proportional to the figure of merit of the receiving terminal. This is accomplished by assigning a different number of time slots to a one-way voice channel according to the figure of merit.

In a FDMA system such as SPADE, a separate carrier is assigned to each one-way voice channel, and satellite power allocation to earth terminals of unequal figures of merit may be accomplished by transmitter power control of earth terminals. This results in unequal signal levels at the satellite receiver input and hence unequal signal output powers from the satellite for the various signals. This mode of operation implies a power-limited system in which the available satellite power must be husbanded and allocated to insure acceptable or equal

voice quality. A call is blocked if the available satellite power has been completely allocated and none is left to establish a two-way voice channel for it.

A similar situation exists when blocking occurs primarily because all the available channels have been allocated. In the case of an FDMA system, this would mean that all the available carriers are exhausted. The total number of available carriers depends on the available bandwidth and on the carrier separation, hence in this case a bandwidth-limited system is implied. A bandwidth-limited TDMA configuration implies time slots of equal duration for the channel and signaling rates (burst rates), dictated by the available bandwidth. In this case, blockage occurs when no more time slots are available for an incoming call.

Special techniques, namely voice-controlled carrier (VOX) [4] in the case of SPADE and digital speech interpolation (DSI) are generally used to increase satellite capacity. The effects of these techniques are not included in the analyses of this concise paper.

In addition to multichannel telephony, demand assignment is applicable to systems consisting of a large number of small terminals operating at low duty cycle and sharing a much smaller number of channels. A typical example is an air traffic-control system in which a large number of aircraft are in contact with a control terminal. If the small terminals are grouped according to figure of merit, each group may be considered the equivalent of a multichannel terminal, and the number of active users in any one group is the equivalent to the number of active calls in the multichannel case. The derivations of this concise paper are directly applicable to such a configuration.

In a telephony network utilizing satellite links with dedicated circuits, the grade of service may be determined through the customary techniques used for terrestrial links, and the well-known blocking formulas, such as the Poisson and Erlang B formulas, apply. In a demand-assignment system, those techniques are in general not applicable. This concise paper presents extensions of the analytic methods used for obtaining the blocking probability on dedicated circuits to power-limited and bandwidth-linked demand-assignment systems.

II. SYSTEM CONFIGURATION

The system parameters of primary interest for demand assignment are the matrix elements of point-to-point traffic (given in call seconds/second for example) between a set of ground stations. The figures of merit of the stations may be different. The traffic matrix is time varying, and because of the large geographic coverage through satellites, the peak traffic may occur at different times of the day on different links. The objective considered here for demand assignment is to insure the minimum acceptable voice quality for all calls which are served while minimizing the probability of blockage. If a call may be served only at the cost of degrading the calls already in progress below the acceptable quality, the call is blocked.

All three types of demand assignments will be considered here for both TDMA and FDMA. In spite of the ease of implementing a fully variable demand-assignment system for SPADE-type multiple access, variable source and variable destination implementations may still be considered because they do not require network control. This is so because the power to be allocated is dedicated to one of the terminals of the call to be established, hence only signaling between the two terminals involved in the call is necessary. Power-limited operation

Paper approved by the Associate Editor for Communication System Disciplines of the IEEE Communications Society for publication without oral presentation. Manuscript received November 20, 1973; revised March 11, 1974. This work was performed for the Defense Communications Agency.

The author is with the Computer Sciences Corporation, Falls Church, Va. 22046.

for SPADE-type systems would be accomplished by transmitting the various carriers at different powers according to the figure of merit of the destination terminal. Blocking occurs when the total transmitter power exceeds a preset value.

In the case of TDMA, power control and power allocation imply the allocation of a varying number of time slots, which may be accomplished in one of two ways. In one case, as soon as a call is terminated the vacant slot is filled by displacing the remaining call. In this way vacant slots will always be adjacent [7]. In an alternate method, the occupied slots are not reorganized when a vacancy occurs. This leads to the phenomenon referred to as fragmentation. A new call may be assigned a number of time slots which are not adjacent. If each such time slot would have to be synchronized anew, the preamble time devoted to such synchronization could become excessive. A TDMA frame-to-frame coherent system is an alternative.

III. THE GRADE OF SERVICE IN TELEPHONY NETWORKS WITH PERMANENTLY ASSIGNED TRUNKS

The grade of service, defined as the probability that a call will not find an idle server immediately, may be obtained for demand-assignment networks through the extension of the customary analysis carried out for networks with permanently assigned trunks. The following is a summary of some basic concepts and derivations (see [8] for details). The discussion will be limited to the two major blocking formulas: Poisson and Erlang B. In both cases, the underlying assumption is that the arrival time of the calls has the Poisson distribution and holding times are distributed exponentially. In the Poisson traffic model, used for final routes (in which a blocked call does not have an alternate route available), blocked calls are held for a time equal to their intended holding time. If a channel becomes idle prior to the expiration of this holding time, the call will seize it and use it for the remaining part of its holding time. The Erlang B model, used for routes for which alternate routes are available, is based on the assumption that the blocked calls are cleared and do not reappear at some future time.

The derivation for blocking probability in the Poisson model is based on the stationarity condition which dictates that the rate of change from x to $x + 1$ calls in progress, caused by incoming calls, equals the rate of change from $x + 1$ to x calls, caused by calls being terminated.

$$np(x) = \frac{(x + 1)p(x + 1)}{h} \tag{1}$$

where

n expected number of calls/second;
h expected holding time of a call;
$p(x)$ probability function of x.

Since waiting blocked calls behave like calls in progress, the random variable x is the sum of busy trunks and waiting calls. The probability of blockage for a call arriving at random is given by

$$P(c,a) = P[x \geq c] = \sum_{x=c}^{\infty} p(x) \tag{2}$$

where c is the number of available trunks, and a, the number of call seconds/second, is given by

$$a = nh. \tag{3}$$

From (1) repeated substitution yields

$$p(x) = \frac{a^x}{x!} p(0) \tag{4}$$

$$\sum_{0}^{\infty} p(x) = p(0)e^a = 1 \tag{5}$$

hence

$$p(0) = e^{-a}$$

and

$$p(x) = e^{-a} \frac{a^x}{x!}.$$

From (2)

$$P(c,a) = \sum_{x=c}^{\infty} e^{-a} \frac{a^x}{x!}. \tag{6}$$

In the case of the Erlang B model, the recursion formula (1) as well as (4) still applies. The random variable x in this case is the number of calls in progress, and in contrast to the Poisson model it has the range of zero through c, hence (5) becomes

$$\sum_{x=0}^{c} p(x) = \sum_{x=0}^{c} \frac{a^x}{x!} p(0) = 1$$

yielding

$$p(0) = \left[\sum_{x=0}^{c} \frac{a^x}{x!} \right]^{-1} \tag{7}$$

and

$$p(x) = a^x \bigg/ x! \sum_{i=0}^{c} \frac{a^i}{i!}. \tag{8}$$

A call will be blocked if it finds c trunks busy, hence for the Erlang B model, the probability of blocking is given by

$$B(c,a) = a^c \bigg/ c! \sum_{i=0}^{c} \frac{a^i}{i!}. \tag{9}$$

IV. THE GRADE OF SERVICE FOR DEMAND ASSIGNMENT ON POWER-LIMITED SATELLITE LINKS

A. Blocked Calls Held in a Fully Variable Demand-Assignment System

In an extension of the derivation of Section III, the following symbols are defined:

x_i the sum of busy trunks and waiting calls on the ith link, $i = 1,2,\cdots,M$;
n_i the expected number of calls/second on the ith link, $i = 1,2,\cdots,M$;
M the total number of links in the network;
W_a available satellite power;
w_i satellite power to be allocated for a two-way trunk of the specified voice quality on the ith link, $i = 1,2,\cdots,M$.

The stationarity condition yields the expression analogous to (1) for the joint probability function of the discrete random variables $x_1,x_2,\cdots,x_M$

$$n_i p(x_1,x_2,\cdots,x_i,x_{i+1},\cdots,x_M)$$
$$= \frac{(x_i + 1)p(x_1,x_2,\cdots,x_i + 1,x_{i+1},\cdots,x_M)}{h} \tag{10}$$

yielding

$$p(x_1,x_2,\cdots,x_i,\cdots,x_M) = \frac{a_i^{x_i}}{x_i!} p(x_1,x_2,\cdots 0,x_{i+1},\cdots,x_M) \tag{11}$$

where $a_i = n_i h$ is the load offered on the ith link. Repeated application of (11) yields

$$p(x_1,x_2,\cdots,x_M) = \prod_{i=1}^{M} \frac{a_i^{x_i}}{x_i!} p(0,0,\cdots,0). \tag{12}$$

$p(0,0,\cdots,0)$, the probability that no trunk is busy on any of the links, may be obtained by normalizing the probability function

$$\sum_{x_1=0}^{\infty} \sum_{x_2=0}^{\infty} \cdots \sum_{x_M=0}^{\infty} \prod_{i=1}^{M} \frac{a_i^{x_i}}{x_i!} p(0,0,\cdots,0) = \prod_{i=1}^{M} e^{a_i} p(0,0,\cdots,0) = 1 \tag{13}$$

or

$$p(0,0,\cdots,0) = \exp\left(-\sum_{i=0}^{M} a_i\right) \tag{14}$$

and

$$p(x_1,x_2,\cdots,x_M) = \prod_{i=1}^{M} \frac{e^{-a_i}a_i^{x_i}}{x_i!}. \tag{15}$$

The satellite power needed to support a particular set of calls $x_1,x_2,\cdots,x_M$ on the M links is given by

$$W(x_1,x_2,\cdots,x_M) = \sum_{i=1}^{M} w_i x_i. \tag{16}$$

A call on the ith link will be blocked if the satellite power already allocated to calls in progress exceeds $W_a - w_i$; hence the blocking probability is given by

$$P_i(W_a;a_1,a_2,\cdots,a_M) = 1 - \sum_{A(W_a-w_i)} \prod_{j=1}^{M} \frac{e^{-a_j}a_j^{x_j}}{x_j!} \tag{17}$$

where $A(W_a - w_i)$ is the totality of all points in the $x_1,\cdots,x_M$ space given by

$$A(W_a - w_i) = [x_1,x_2,\cdots,x_M; \sum_{j=1}^{M} w_j x_j \leq W_a - w_i]. \tag{18}$$

Since $P_i(W_a;a_1,a_2,\cdots,a_M)$ is a monotonic, nondecreasing function of w_i, calls on links requiring more satellite power have an equal or higher probability of being blocked; as w_i is increased, some points in the domain defined by (18) are eliminated, eliminating some probability elements in the sum in (17). If the system design goal is a maximum acceptable blocking probability for all traffic, inserting the largest value of w_i in the network into (17) yields the probability of blockage on the link with the highest such probability; other links will perform better. With $F(x)$ as the distribution function of satellite power, (17) equals $1 - F(W_a - w_i)$. The right-hand side is the probability of blocking at the power level $W_a - w_i$.

B. Blocked Calls Rerouted in a Fully Variable Demand-Assignment System

When blocked calls are rerouted, the recursive formula (10) yields (12), with $p(0,0,\cdots,0)$ obtained through normalization. The random variable x_i is the number of calls in progress on the ith link. The sample space of $x_1,x_2,\cdots,x_M$ is $A(W_a)$ defined by (18) yielding the normalizing equation

$$p[x_1,x_2,\cdots,x_M \subset A(W_a)] = \sum_{A(W_a)} \sum_{i=1}^{M} \frac{a_i^{x_i}}{x_i!} p(0,0,\cdots,0) = 1$$

hence

$$p(0,0,\cdots,0) = \left[\sum_{A(W_a)} \prod_{i=1}^{M} \frac{a_i^{x_i}}{x_i!} \right]^{-1} \tag{19}$$

and

$$p(x_1,\cdots,x_M) = \prod_{i=1}^{M} \frac{a_i^{x_i}}{x_i!} \bigg/ \sum_{a(W_a)} \prod_{i=1}^{M} \frac{a_i^{x_i}}{x_i!}. \tag{20}$$

A call on the ith link will be blocked if the satellite power already allocated to calls in progress exceeds $W_a - w_i$, hence the blocking probability is given by

$$B_i(W_a;a_1,a_2,\cdots,a_k) = 1 - \sum_{A(W_a-w_i)} p(x_1,\cdots,x_M)$$

$$= 1 - \sum_{A(W_a-w_i)} \prod_{i=1}^{M} \frac{a_i^{x_i}}{x_i!} \bigg/ \sum_{A(W_a)} \prod_{i=1}^{M} \frac{a_i^{x_i}}{x_i!}. \tag{21}$$

From (17)

$$B_i(W_a;a_1,\cdots,a_M) = 1 - \frac{1 - P_i(W_a;a_1,\cdots,a_M)}{1 - P_i(W_a + w_i;a_1,\cdots,a_M)}$$

$$= \frac{P_i(W_a;a_1,\cdots,a_M) - P_i(W_a + w_i;a_1,\cdots,a_M)}{1 - P_i(W_a + w_i;a_1,\cdots,a_M)}. \tag{22}$$

This is a generalization of the well-known formula for conventional trunks

$$B(c,a) = \frac{P(c,a) - P(c+1,a)}{1 - P(c+1,a)} \tag{23}$$

where $P(c,a)$ and $B(c,a)$ are given by (6) and (9).

As in the case of the Poisson traffic model, the link with the largest value of w_i has the poorest service, and the corresponding value of the blocking probability in (22) may serve as an upper bound for the other links.

C. The Grade of Service in Variable-Source and Variable-Destination Demand-Assignment Systems

In a variable-source or variable-destination demand-assignment system operating under power-limited conditions, blocking occurs when the power assigned to a terminal is exhausted. For example, in a variable-destination TDMA system an earth terminal establishes a one-way link by transmitting in a number of slots assigned to the call, from within the burst permanently or semipermanently allocated for transmission to the terminal within the TDMA format. The number of time slots assigned by the terminal to a call, which is proportional to the average power allocated to the call, will depend on the figure of merit of the receiving station and on the voice quality desired. Blocking occurs when a sufficient number of slots or equivalently sufficient power is not available within the burst allocated to the terminal for a new call. Similarly, in a variable-source TDMA system an earth terminal establishes a one-way link by transmitting in a number of slots located within the burst permanently or semipermanently allocated for reception to the destination terminal.

Each call may be blocked twice, and since in general sufficient information about the dependence of these probabilities is not available, one may conservatively take the sum of blocking probabilities of the one-way trunks as the probability that the call is blocked.

The derivation of these probabilities is identical in all respects to the one given in Section III for a fully variable system. In a variable-destination demand assignment system for blocked calls held, the probability that the one-way trunk cannot be established from the rth to the r_ith terminal is given by

$$P_{ir}(W_{ar};a_{r1},a_{r2},\cdots,a_{ik_r}) = 1 - \sum_{A'(W_{a_r}-v_{ri}')} \prod_{j=1}^{k_r} \frac{e^{-a_{rj}}a_{rj}^{x_{rj}}}{x_{rj}!} \tag{24}$$

where

$$A'(W_{ar} - w_{ri}') = [x_{r1},x_{r2},\cdots,x_{rk_r}; \sum_{j=1}^{k_r} w_{rj}' x_{rj} \leq W_{ar} - w_{ri}'] \tag{25}$$

W_{ar} satellite power allocated to the rth terminal for one-way transmission;

w_{ri}' satellite power needed to establish a one-way trunk of desired quality from the rth to the r_ith terminal ($i = 1,2,\cdots,k_r$);

k_r number of links from the rth terminal;

a_{rj} the load offered between the rth and r_jth terminal;

x_{rj} the number of one-way trunks from the rth to the r_jth terminal.

For rerouted blocked calls, the probability that the one-way trunk cannot be established from the rth to the r_ith terminal is given by

$$B_{ir}(W_{ar};a_{r1},a_{r2},\cdots,a_{rk_r})$$

$$= \frac{P_{ir}(W_{ar};a_{r1},\cdots,a_{rk_r}) - P_{ir}(W_a + w_{ri}';a_{r1},\cdots,a_{rk_r})}{1 - P_{ir}(W_a + w_{ri}',a_{r1},\cdots,a_{rk_r})}. \tag{26}$$

For a variable-source demand-assignment system, if the satellite power allocated to the rth terminal for reception equals W_{ar} and the satellite power needed to establish a one-way trunk of desired quality from the ith terminal equals w_{ri}' expressions for one-way blocking probabilities identical to (24) and (26) are obtained. For both variable-source and variable-destination systems, the total satellite power needed for a given one-way probability of blocking (assuming the same probability of blocking on all links) is given by

$$W_a = \sum_{r=1}^{N} W_{ar} \tag{27}$$

where N is the total number of terminals, and W_{ar} is obtained from (24) or (26) for the desired one-way blocking probability.

TABLE I
NORMALIZED POWER FOR ONE-WAY AND TWO-WAY LINKS

Link No.	Link Traffic CCS	Normalized Power for Two-Way Trunk, w_i	Normalized Power for One-Way Trunk into Terminal		Normalized Power for One-Way Trunk into Terminal	
			Terminal No.	w'_{ri}	Terminal No.	w'_{ri}
1	19.0	2.26	1	1.26	3	1.00
2	39.6	4.42	1	1.26	6	3.16
3	168.0	2.00	2	1.00	3	1.00
4	88.0	4.16	2	1.00	4	3.16
5	311.0	2.00	2	1.00	5	1.00
6	168.0	4.16	2	1.00	7	3.16
7	19.1	4.16	2	1.00	9	3.16
8	19.1	4.16	3	1.00	4	3.16
9	63.0	2.00	3	1.00	5	1.00
10	19.1	4.16	3	1.00	8	3.16
11	585.0	4.16	4	3.16	5	1.00

D. The Grade of Service With Terminals of Equal Figures of Merit

When all the terminals have the same figure of merit, the allocation for each one-way circuit expressed in terms of satellite power is the same for all links. The number of available trunks in the system is fixed and the blocking probability is obtained from the conventional blocking formulas (6) and (9) with values for c and a dependent on the type of demand assignment and semipermanent allocation of circuits. Thus in a fully variable system a is the total traffic load and c is the number of circuits; in a variable-distinction or variable-source system a is the traffic load into a terminal and c is the number of one-way circuits dedicated to it. Since the case of identical terminals is a special case of a more general configuration, one would expect that substituting equal weights w into (17) and (22) would yield (6) and (9). The domain in (18) becomes

$$\sum_{i=1}^{M} x_i \leq c - 1$$

where c is the number of available circuits, and utilizing repeatedly the equality

$$\sum_{x_i=0}^{x} \frac{a_i^{x_i}}{x_i!} \cdot \frac{a_j^{x-x_i}}{(x-x_i)!} = \frac{(a_i + a_j)^x}{x!} \tag{28}$$

yields (6) and (9) with

$$a = \sum_{i=1}^{M} a_i \tag{29}$$

$$c = \text{int} \left[\frac{W_a}{w_i} \right]. \tag{30}$$

Similarly for terminals with equal figures of merit in a variable-source and variable-destination system, (24) and (26) reduce to (6) and (9).

V. PROBABILITY OF BLOCKING IN A BANDWIDTH-LIMITED SYSTEM

In a demand-assignment system which is bandwidth limited, the probability of blocking will depend on the number of available circuits, and is obtained from (6) and (9). The value of traffic load to be used is the one dictated by the type of demand assignment. It is the total traffic for a fully variable system, and the traffic associated with a particular terminal for the variable-source and variable-distinction systems. Thus in this case the results given previously for a network of terminals with equal figures of merit apply.

VI. ILLUSTRATIVE EXAMPLE

Table I shows the network configuration selected for the example. The power needed to yield a one-way channel of the desired quality into the largest terminal is taken equal to one unit; power requirements for the other two terminal sizes in the network are normalized and equal 1.26 and 3.16. The power needed to establish a duplex link equals the sum of the two one-way power requirements associated with the two terminals of the link. The traffic on each of the eleven links is shown in the second column.

Table II shows the satellite power needed for the probability of the blocking of 0.1 and 0.05 for the various methods of demand assignment. Since the expressions previously developed yield blocking probability as a function of satellite power, the satellite power associated with the desired blocking probabilities was obtained by computing the blocking for many small incremental steps of satellite power in the region of interest. The last column of Table II tabulates the power (in the units defined previously) under permanent assignment of power to each link. The case of equal-sized terminals refers to the configuration in which all terminals require one unit of power for establishing a one-way link. This case was included because the effectiveness of demand assignment is reduced if one link requires a substantial portion of the total satellite power. Such would be the case if the particular link is comprised of small terminals. Thus in the case of identical terminals the effects of this imbalance are eliminated. As pointed out in Section IV, the larger the power needed to complete the call, the higher the probability of blocking. The entries in the "small-terminals" row show the power required when the call on the link requiring most power has the blocking probability of 0.1 or 0.05; the blocking probability for calls on other links is lower. Conversely, the entries in the "large-terminals" row imply that links requiring the least power will have the shown probability of blocking; on others the blocking probability will be higher.

All satellite-power requirements are given in percent of the power needed to achieve a blocking probability of 0.1 with permanent assignment. The actual power (in units as defined previously) is obtained by multiplying by the entry in the last column.

The entries in Table II for a fully variable demand assignment were computed from (17) and (22). These equations yield blocking probability versus satellite power W_a. The entries for variable-destination demand assignment were computed from (24) and (26). These equations give the blocking probability versus satellite power W_{ar} allocated to the rth station. By summing over all stations the power needed for a particular blocking probability, the total power required is obtained. The entries for variable-source demand assignment were obtained from (6) and (9) to which (24) and (26) are reduced, as shown in Section IV-C. As discussed previously, these equations may be used to compute the one-way probability of blocking as a function of the power allocated to the terminal, and again summing over all terminals, the total power required is obtained.

As expected, fully variable demand assignment results in maximum power saving. Caution should be used in drawing conclusions from the entries in Table II. These show reductions in power requirements for the traffic matrix of Table I. There are only eleven links in the network, four of which carry relatively little traffic. Pooling the traffic through demand assignment is particularly effective when the number of links is large. The actual improvement that results can be evaluated through calculations similar to those shown here.

TABLE II
Normalized Power in Percent for Blocking Probabilities of 0.1 and 0.05

Type of Demand Assignment		Fully Variable		Variable Source		Variable Destination		Permanently Assigned		Normalizing Factor W_a
Blocking Probability		0.1	0.05	0.1	0.05	0.1	0.05	0.1	0.05	
Large Terminals	Poisson	67	71	92	102	86	93	100	115	254.9
	Erlang B	57	65	98	105	76	89	100	120	223.9
Small Terminals	Poisson	68	71	92	102	89	96	100	115	254.9
	Erlang B	67	73	98	105	93	106	100	120	223.9
Equal Sized Terminals	Poisson	67	71	87	95	87	95	100	115	152.0
	Erlang B	64	71	90	99	90	99	100	120	134.0

ACKNOWLEDGMENT

The author wishes to thank the reviewers for many helpful suggestions which resulted in a more informative presentation.

REFERENCES

[1] F. Auguerra, "Traffic considerations for demand assigned telephone through a satellite system," in *Proc. 5th Int. Teletraffic Congr.* (Rockefeller Univ., New York, June 14–20, 1967.

[2] J. G. Puente, "A PCM-FDMA demand assignment satellite multiple access experiment," presented at the AIAA 2nd Communications Satellite Systems Conf., Apr. 1968.

[3] A. M. Werth, "SPADE: A PCM/FDMA demand assignment system for satellite communications," presented at the Intelsat/IEE Int. Conf. Digital Satellite Communications, London, England, Nov. 25–27, 1969.

[4] E. R. Cucciamani, Jr., "The SPADE system as applied to data communications and small earth station operation," COMSAT Tech. Rev., vol. 1, pp. 171–182, Fall 1971.

[5] J. G. Puente and A. M. Werth, "Demand-assigned service for the Intelsat Global Network," *IEEE Spectrum*, vol. 8, pp. 59–69, Jan. 1971.

[6] J. G. Puente, W. G. Schmidt, and A. M. Werth, "Multiple-access techniques for commercial satellites," *Proc. IEEE (Special Issue on Satellite Communications)*, vol. 59, pp. 218–229, Feb. 1971.

[7] G. Eckhardt, S. Reidel, and H. Rupp, "A TDMA system proposed for the experimental German-French communication satellite Symphonie," presented at the Intelsat/IEE Int. Conf. Digital Satellite Communication, London, England, Nov. 25–27, 1969.

[8] *Switching Systems*, American Telephone and Telegraph Co., 195 Broadway, New York, 1961.

Demand-assigned service for the Intelsat global network

In April 1970 Intelsat decided to allow earth-station owners to access the Atlantic Intelsat IV satellite via the SPADE system. It is expected that by mid-1971 another milestone will have been achieved with the introduction of the first operational demand-assignment system

J. G. Puente, A. M. Werth   Comsat Laboratories

Communications satellites have been in commercial operation for more than five years. With the growth of traffic requirements, particularly those of the developing nations, the early methods used in the Intelsat network for multiple access have had to be updated. The full potential of satellite communications can be realized by using the satellite to provide connections between any two users on demand. This article traces the development of the multiple-access techniques employed with Intelsat from the two access methods associated with Early Bird to the demand-assignment technique planned for Intelsat IV. This latter method, SPADE, is to be the first operational demand-assignment multiple-access system and is scheduled for use in 1971. Details of the SPADE system and its use for other applications are reviewed.

The launch of the Early Bird satellite in April 1965[1] initiated a new era in telecommunications. Its 240 voice circuits provided a greater capacity than all of the cables laid between the United States and Europe over the previous ten years. In addition, most of Western Europe was in sight of Early Bird and, for the first time, it was technically possible to provide direct, reliable, and instantaneous high-quality voice, television, and data communication between distant points.

Early Bird was a modified version of Syncom,[2] the first of the synchronous satellites successfully constructed by Hughes Aircraft Corp. and launched by NASA. Early Bird featured a hard-limiting transponder that used frequency-division multiple access, and thus restricted the number of earth stations accessing the satellite to two if channel capacity were to be maintained. Because of this

FIGURE 1. Early satellite operation. (Dotted lines show conversations in progress.)

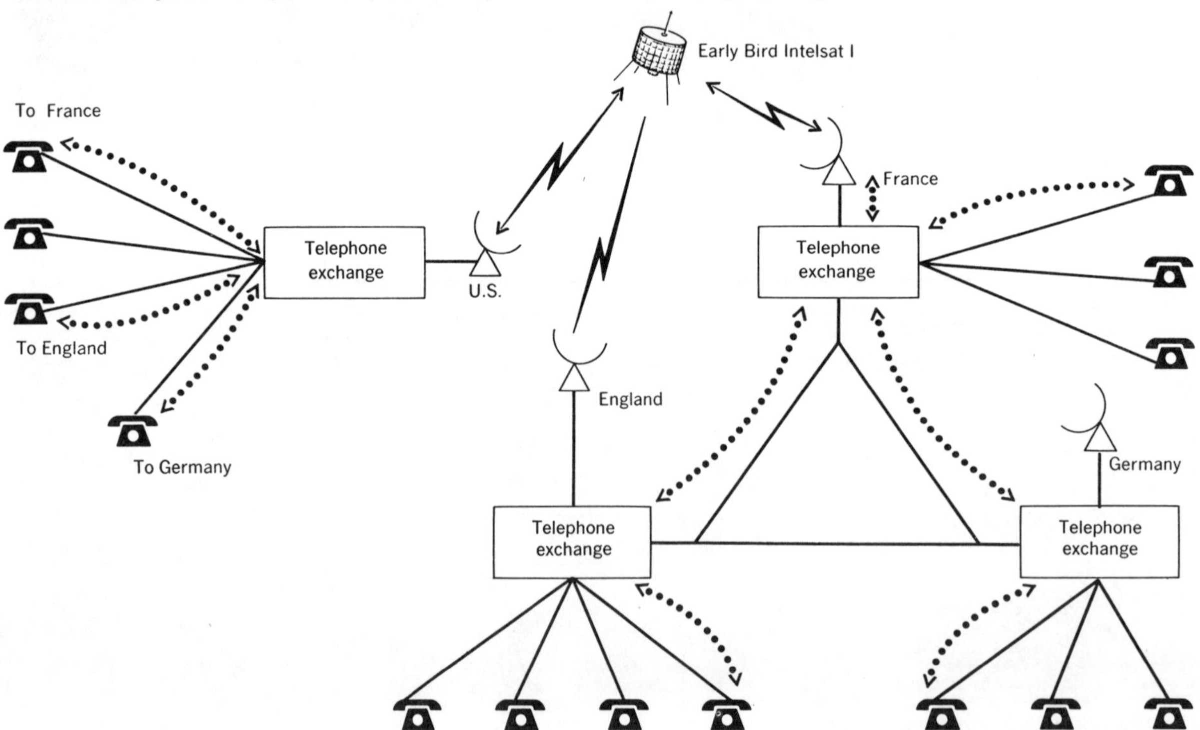

Reprinted from *IEEE Spectrum*, vol. 8, pp. 59–69, Jan. 1971.

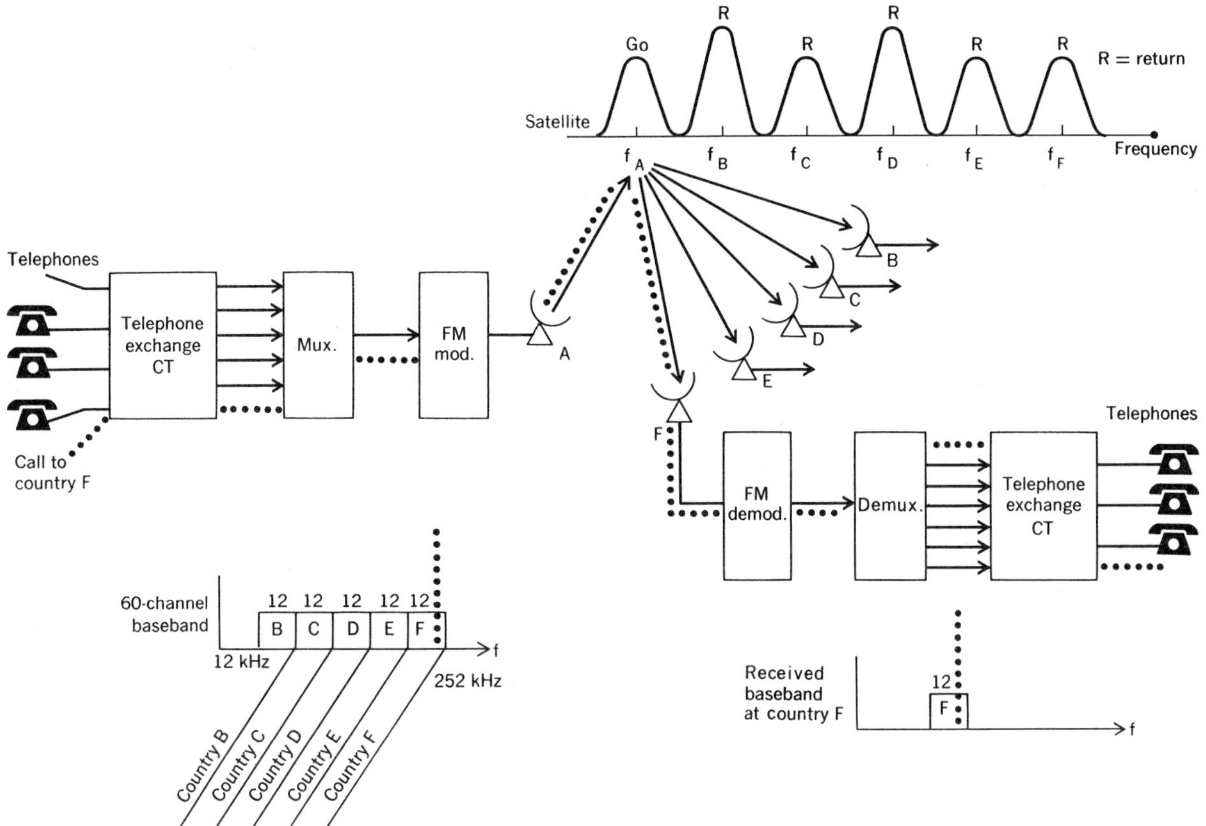

FIGURE 2. Preassigned multidestination FDM/FM carriers.

FIGURE 3. Atlantic Basin interconnection of earth stations in early 1970s.

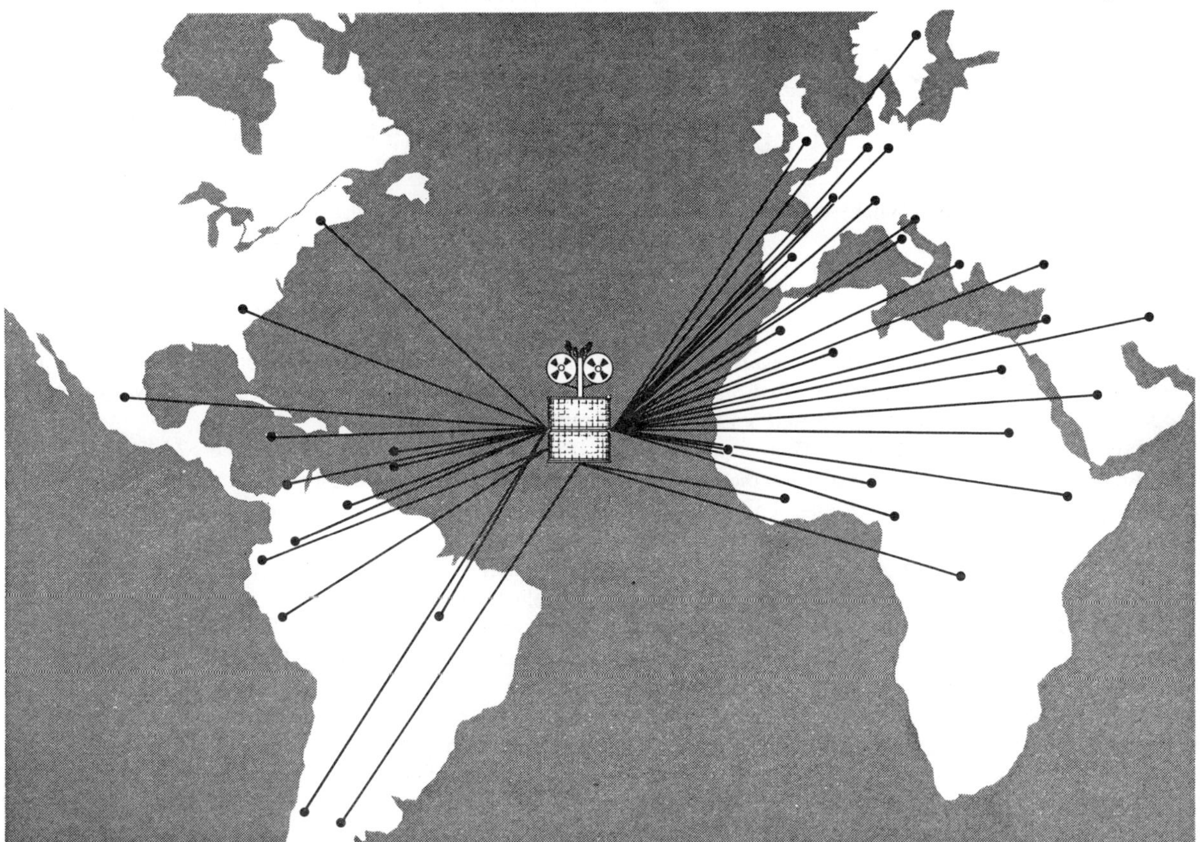

IEEE spectrum JANUARY 1971

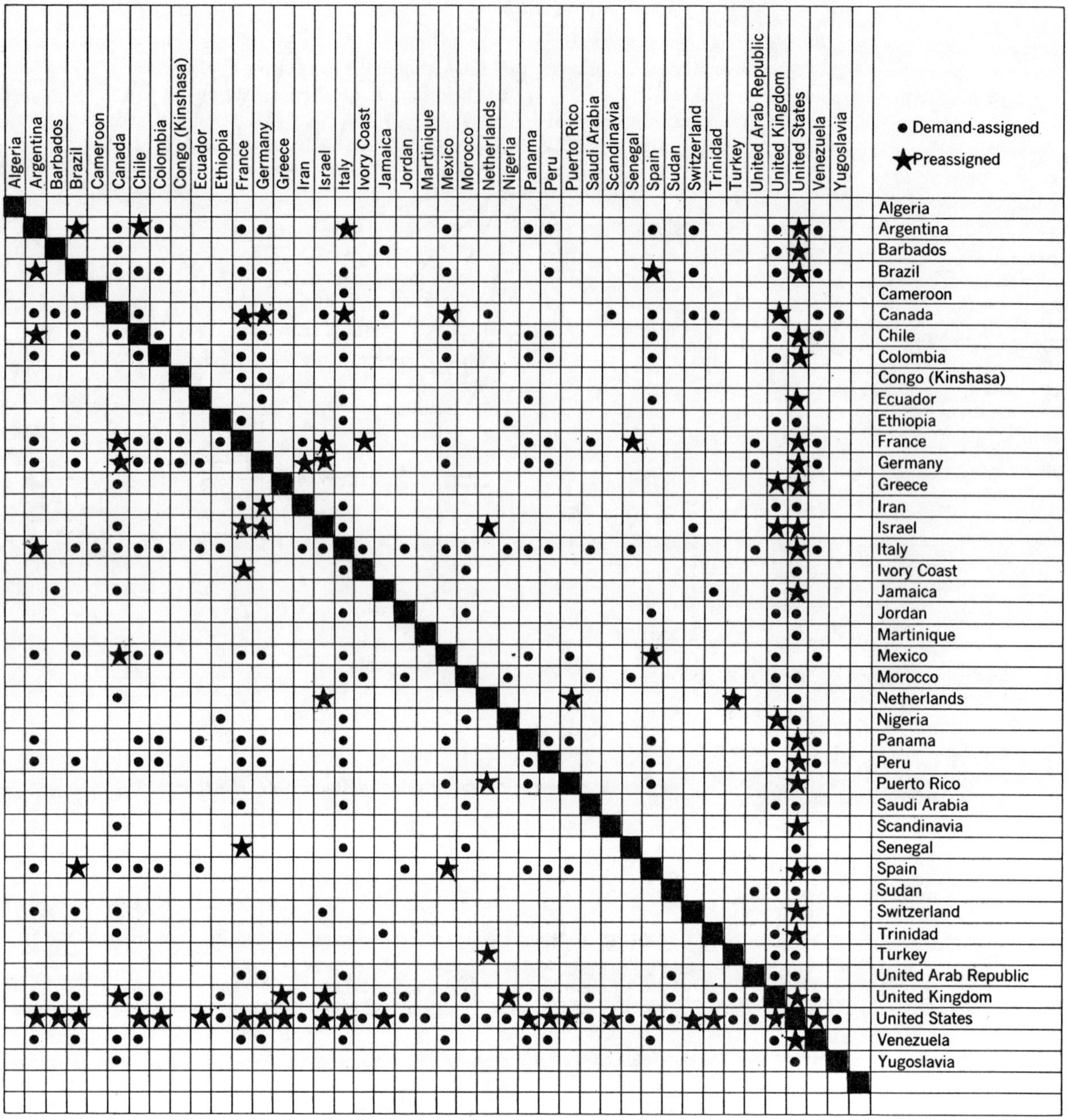

FIGURE 4. Summary of traffic density among Intelsat stations in early 1970s.

limitation, commercial satellite operation began as shown in Fig. 1. The traffic was between the United States and Europe; therefore a U.S. earth station and a European earth station could access Early Bird. There were three European earth stations capable of accessing the satellite, each having traffic with the United States. To allow each European station to operate with the satellite, a triangular terrestrial communications network was interconnected among the European users. Only one European earth station would communicate with the U.S. during any one period of time (e.g., one month), and would distribute traffic to the other countries connected at the corners of the triangle. Figure 1 shows France receiving the satellite signal from the U.S. for three calls: one to France, one to England, and one to Germany. This mode of operation provided a great deal of experience in general satellite operations but, because of satellite access limitations, failed to take advantage of that mode's unique capability—direct communication between two or more countries.

Frequency-division multiple access (FDMA) was accommodated more readily on Intelsat II and Intelsat III by operating their traveling-wave-tube (TWT) power amplifiers in the linear region as contrasted to the use of a hard limiter in Early Bird. Thus many FM carriers from a multiplicity of earth stations could access the satellite.[3] (The operating point of the satellite TWT power amplifier is determined by the illumination of transmitted signals from the earth. All earth stations transmitting to the satellite control and monitor their output power so that the satellite power amplifier continues to operate in its linear region. There are no power-level controls on the satellite. As the received power at the satellite is decreased, the TWT is "backed off" from saturation and its transfer characteristic becomes more linear. Other factors then become important, such as the conversion of amplitude to phase modulation in traveling-wave tubes.)

As the number of accesses increased, the satellite channel capacity still decreased, but not nearly as quickly as for a hard limiting satellite such as Early Bird. For this reason Intelsat II and III have normally been regarded as multiple-access satellites when operating with FM-FDMA. The objective in optimizing the satellite channel capacity was to have a minimum number of carriers entering the satellite, that is, one carrier from each earth station, carrying traffic to a number of distant destinations.

Thus evolved the multidestination FM carrier diagrammed in Fig. 2. The figure has been simplified to bring out the salient points. Long-distance telephone calls originating in country A enter a telephone exchange or transit center (CT) and are multiplexed as shown in the 60-channel baseband. Country A transmits on a single FM carrier f_A with a 60-channel capacity. It has preassigned the 60 channels in groups of 12 channels to be received by five other countries. Therefore, countries B, C, D, E, and F require an FM receive chain operating on frequency f_A in order to receive telephone traffic from A. Conversely, country A must have a receive chain for each of the countries in its baseband. Thus a receive chain is required for every country with whom communications are desired. This will become more important in our later discussion.

In tracing a telephone call from country A to country F, the CT switches the call to the proper slot in the baseband—i.e., one of the 12 channels allocated to station F in the baseband. The baseband continuously modulates an FM carrier on f_A and the carrier is transmitted to the satellite to be amplified and retransmitted to the earth. Stations B, C, D, E, and F receive the carrier f_A. At earth station F the received signal is demultiplexed, selecting only those 12 channels preassigned for it on the f_A carrier. The CT in country F then makes the connection to the party called via terrestrial facilities, and the connection is complete. Of course the return connection to country A from country F is accomplished in a similar manner via carrier frequency f_F.

Note that in this mode of satellite communications, if the 12 channels to country F in country A's baseband are being used, a new call entering the CT would receive a busy signal from the satellite system even though the remaining 48 channels in the baseband might be unoccupied. We can see that multidestination carriers are a step toward improving the utilization of satellite capacity, but, as shown above, they are not fully utilizing the satellite's capability from the point of view of occupancy. Nevertheless, for large-capacity links between the two points, the statistics of call placement are such that very few calls are lost due to overload.

Consider next the case where connections are required between an earth station and many other users, but the total traffic between any paired earth station link is very low. To improve the efficiency of this model, the satellite capacity should somehow be shared in such a way that when a new call comes in and there is any unused capacity in the satellite, the connection can be established. The assignment of satellite capacity in preassigned, fixed blocks between two users prevents reallocation of unused channels. Therefore, for a network of light traffic links, satellite channels should be assigned on demand—that is, when a call comes into the CT. When the call is completed, the satellite channel is made available to any other pair of users. The introduction in the Intelsat network of demand assignment (DA)[4] represents the third major innovation in commercial satellite communications since 1965.

The need for demand assignment

Figure 3 illustrates the configuration of earth stations predicted for the Atlantic Basin in the early 1970s. The present projection of operating earth stations having access to an Intelsat IV satellite poses serious problems because of the increasing number of connections required among various earth stations. Potentially, a total of 820 links are possible for the 41-station Atlantic network. Considering that a large proportion of any such connections would carry light traffic, the satellite, operating as it does today, in an FDM/FM/FDMA (frequency-division multiplex, frequency-modulated, frequency-division multiple access) multidestination mode as described earlier, could not efficiently accommodate the increasing number of connections.

Figure 4 presents a summary of traffic data[5] and shows

I. Traffic for Argentina

Traffic Between Argentina and	Number of Circuits	Paid Minutes per Day
Brazil	**29***	3022
Canada	2	70
Chile	**35**	12 750
Colombia	4	15
France	7	350
Germany	6	350
Italy	**15**	824
Mexico	4	84
Panama	1	42
Peru	6	378
Spain	10	196
Switzerland	3	280
United Kingdom	10	220
United States	**52**	8882
Venezuela	4	70

* Boldface numbers indicate preassigned circuits.

II. Total traffic, Atlantic Basin, 1973

Preassigned voice channels	3618
Preassigned record data (or alternate voice data) channels	1544
Total channels preassigned	5162
Voice-channel candidates for demand assignment	1158
Total equivalent channels required	6320
Percentage DA of total	18.3%
Number of paths (preassigned)	84
Number of paths (demand-assigned)	266
Total paths	350
Percentage DA of total	76.0%

IEEE spectrum JANUARY 1971

the density expected for each of the known predicted links. The black dots at the intersections of two countries represent traffic in excess of 12 circuits, the colored dots indicate traffic less than 12 circuits. (A figure of 12 circuits has been taken arbitrarily for illustrative purposes since it represents a standard group.)

It is evident from Fig. 4 that the majority of the presently predicted traffic between these links is relatively light. To illustrate this fact, Table I examines in more detail the traffic for Argentina.[6] Note that of 15 links for Argentina, only 4 have a requirement for more than 12 circuits. Similar analyses of traffic for most other nations result in nearly the same distribution.

Table II summarizes the total predicted traffic for the Atlantic Basin in 1973. Of the 6320 total channels required, 1158 (or 18.3 percent) are candidates for demand assignment. Of particular interest in these statistics is

that there is a predicted total of 350 paths. Using the guideline that 12 circuits or less should be served by demand assignment, 266 (or 76 percent) of these paths fall into this category. Furthermore, it should be noted that in the matrix of Fig. 4 there is a sizable number of boxes with no dots. Although these blanks may represent paths for which no predictions are available, they generally represent paths over which traffic is so light that it is not economically attractive to service or even plan these links. It follows that the growth of the Intelsat network will occur not only between those areas indicated by dots, but also between those locations where economical communications links have previously been impossible.

Experience in satellite communications shows that there will be a need for accommodating the heavy trunks on a preassigned basis between certain locations. These links are economically justifiable because they provide efficient service. It is evident, however, that if all desired links were to be handled on this preassigned basis, the capacity of the satellite would quickly be exhausted. Therefore, it appears that the answer to the problem of providing efficient global service consists of implementing a mixture of preassigned and demand-assigned services.

Figure 5 illustrates methods of assigning multidestination satellite channels. In a preassigned (PA) system, if earth station A has traffic to stations B through F, it can multiplex this traffic on a single carrier and transmit it through the satellite. Then each destination earth station would receive station A's carrier and extract the communications addressed to that particular destination. The lower half of Fig. 5 shows that, using demand assignment, earth station A would connect to earth station B on a demand basis over channel 3, etc. After each of these communications, the channels would be available to other pairs of users. The fundamental difference between PA and DA is that for DA the channels are never permanently assigned between any pair of stations.

It is appropriate at this point to summarize briefly the key goals Intelsat expects to achieve with the implementa-

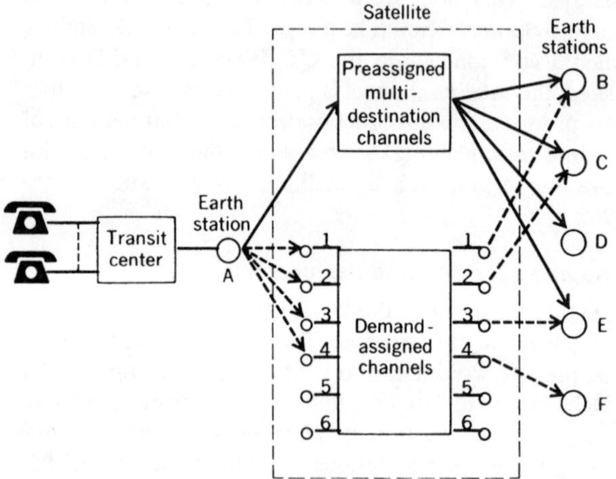

FIGURE 5. Preassigned and demand-assigned operation of satellite channels.

FIGURE 6. Functional earth-station requirements of demand-assigned satellite systems.

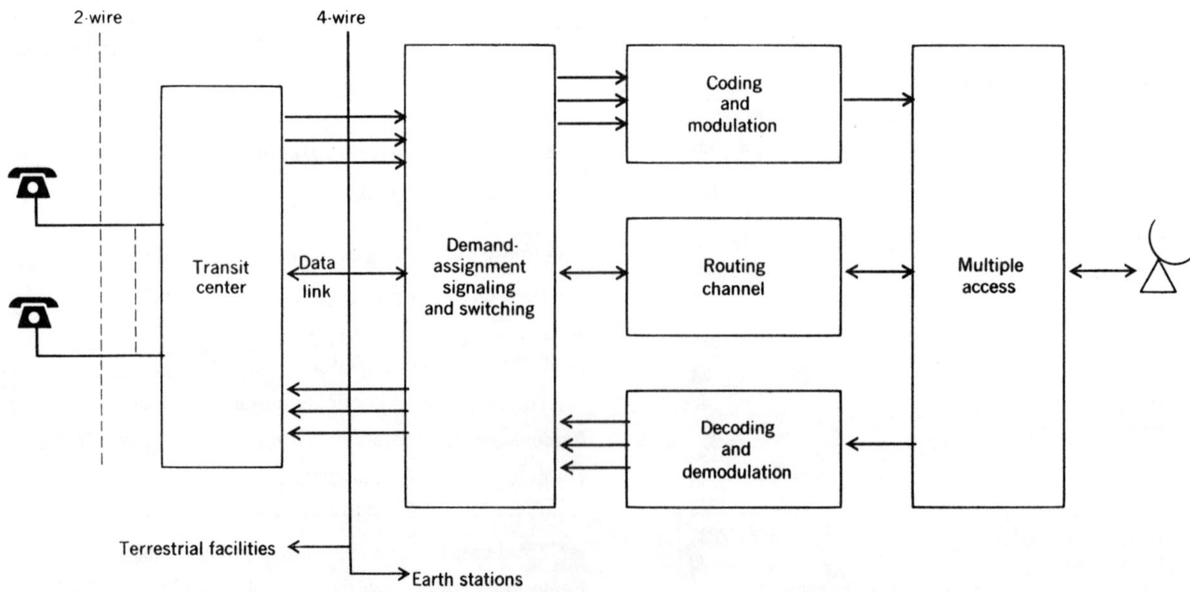

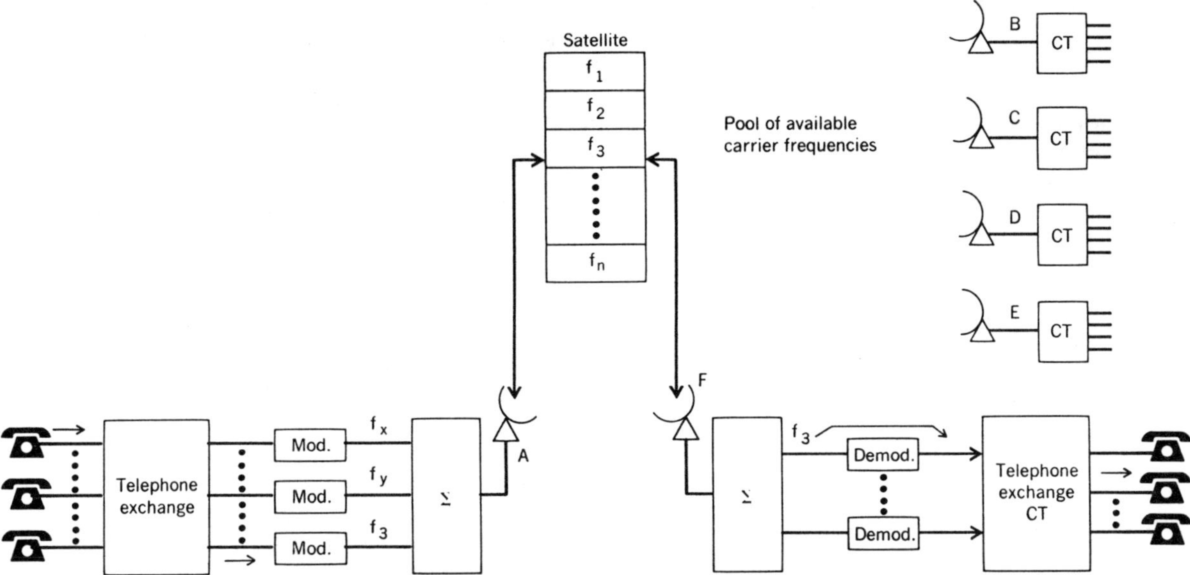

FIGURE 7. Voice-signal flow in a demand-assignment FDMA system. (f_y = carrier frequency; x = 1, 2, 3, . . ., n.)

FIGURE 8. Signaling flow in demand-assignment system.

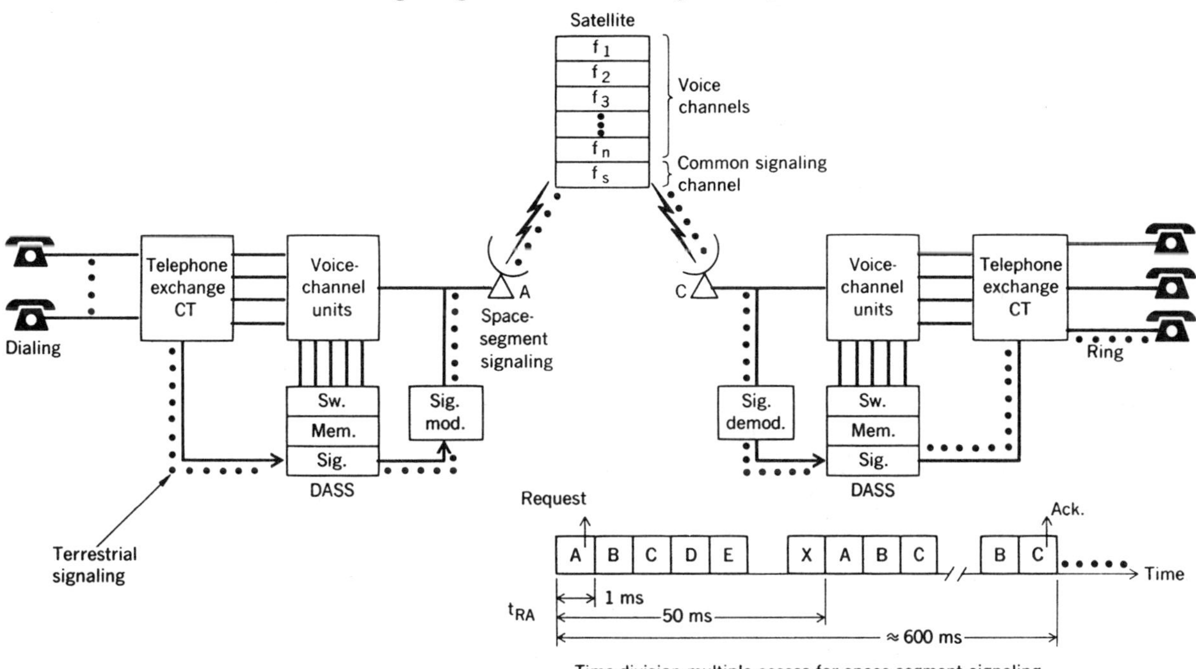

tion of a demand-assignment system. They are:

1. To provide efficient service to light traffic links.
2. To handle overflow traffic from medium-capacity preassigned links.
3. To allow establishment of a communication link from any earth station to any other earth station within the same zone on demand.
4. To utilize satellite capacity efficiently by assigning circuits individually.
5. To make optimum use of existing earth-station equipment.

The increased total number of terrestrial circuits that might be served by a DA pool through a more efficient use of the space segment could delay the necessity of an additional satellite in a given region. This not only would provide a space-segment cost saving, but also could provide an additional saving to some earth-station owners by deferring for a substantial time period the need for an additional antenna.

It is also important to note that DA channels have a greater equivalent traffic-carrying capability than PA channels. From the point of view of traffic, DA channels,

III. Some technical characteristics of the SPADE system

Channel encoding	PCM
Modulation	4-phase PSK (coherent)
Bit rate	64 kb/s
Bandwidth per channel	38 kHz
Channel spacing	45 kHz
Stability requirement	±2 kHz (with AFC)
Bit error rate at threshold	10^{-4}
TDMA common signaling channel:	
Bit rate	128 kb/s
Modulation	2-phase PSK
Frame length	50 ms
Burst length	1 ms
Number of accesses	50 (49 stations + 1 reference)
Bit error rate	10^{-7}

since they are not permanently connected between two points, can serve a larger number of terrestrial circuits than can PA channels. It is estimated that a two- to three-fold increase in circuit utilization can thus be effected.[7]

In an effective DA system it is possible to establish communications between any two stations in common view of a satellite; therefore, even the least dense traffic links can be accommodated with no penalty to the system. With this capability the loading on these links would be allowed to grow in gradual steps as the availability of the service attracts new traffic.

It should be recognized that the advantages of DA apply most effectively to light-to-medium-capacity multidestination traffic. As noted earlier, the efficiency, both in terms of equipment and spectrum utilization, of high-density limited-destination trunks requires that a mix of preassignment and demand assignment be maintained.

Characteristics of demand-assignment systems

Figure 6 illustrates the functional requirements for any type of demand-assignment system. There are three key functions:

1. *Multiple-access equipment.* The equipment necessary to provide access to the satellite—for example, FDMA and time-division multiple access (TDMA).

2. *Coding and modulation equipment.* The equipment necessary to modulate/encode and demodulate/decode the incoming and outgoing analog signals, respectively.

3. *Demand-assignment signaling and switching (DASS).* The equipment necessary to connect with existing terrestrial equipment and the other users in the demand-assignment pool through the routing channel. All channel requests, busy signals, terminating functions, system status reports, etc., are received and transmitted via the routing channel under the control of DASS.

The transit center is not part of the demand-assignment multiple-access system. It is normally located at major metropolitan areas and is connected to the earth station via cable or microwave links. Demand-assign-

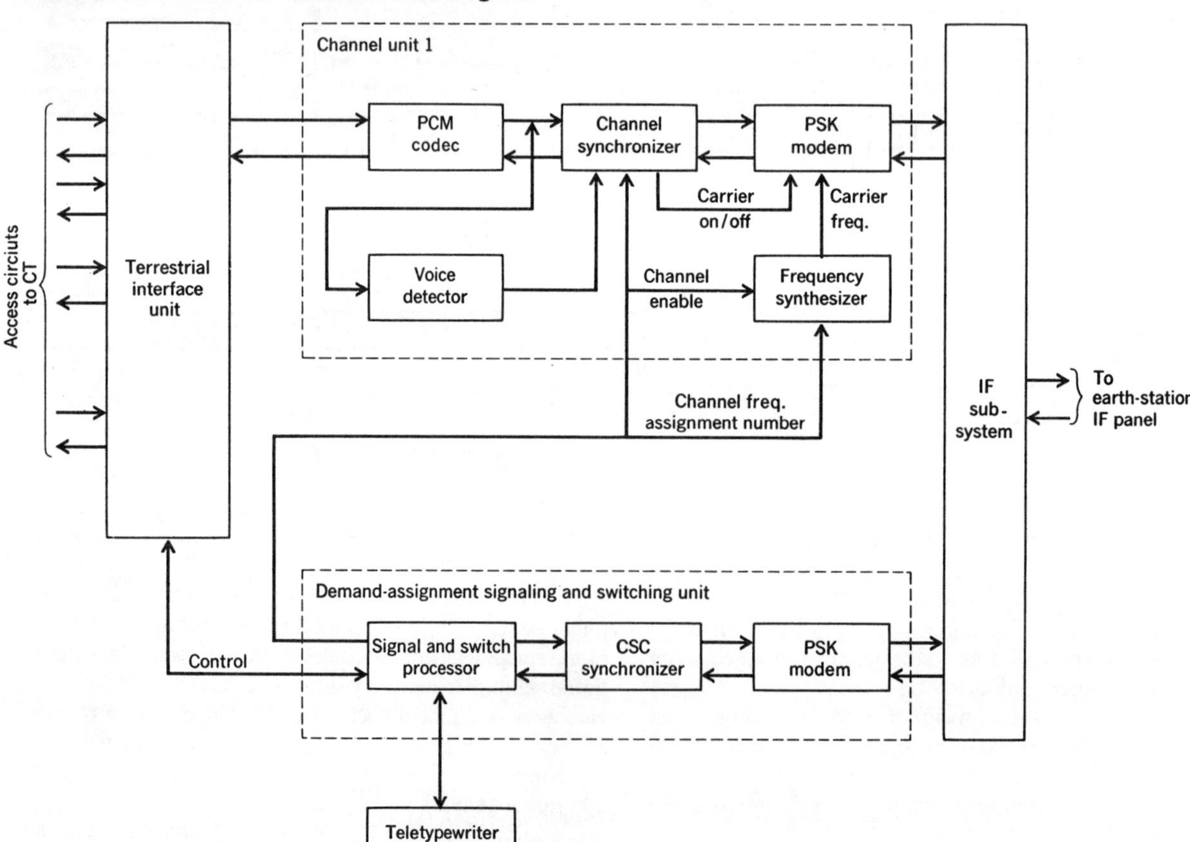

FIGURE 9. SPADE terminal functional block diagram.

ment systems, through the flexibility of a stored programmed DASS,[8,9] can be designed to interface with all of the standard CCITT signaling systems, such as CT1, R-2, CT5, CT5 bis, and the forthcoming CT6.

Two demand-assignment multiple-access systems developed by Intelsat are described in Refs. 10 and 11. For the Intelsat system, where there are a large number of low-traffic sources desiring interconnectivity, a single-channel-per-carrier FDMA system is efficient. On the other hand, where there is a relatively lower number of accesses carrying medium-to-heavy traffic (12–60 circuits) per link, TDMA appears to offer a more efficient solution. As the number of channels per link increase, the effectiveness of the application of demand-assigned channels decreases.[12]

The particular demand-assignment system developed by Intelsat, which is to be implemented in mid-1971, uses FDMA and is described in the following section.

Principles of FDMA demand assignment[3]

The colored lines in Fig. 7 show a typical path allocation for a single voice link in a demand-assignment FDMA system. A subscriber requesting a call to a DA destination is connected at the transit center to one of a number of terrestrial access circuits terminating in DA equipment at earth station A. Note that, regardless of the ultimate call destination, any access circuit may be used, thus allowing more efficient use of the CT-to-earth-station link. In the earth-station DA equipment, the call is routed to modulating equipment and a particular carrier frequency pair (f_3 in this case) out of the pool of available network frequencies is assigned to the modulator and to the distant-end channel. For the duration of that connection frequency, f_3 remains assigned to that circuit and "is busied out" to other requesters. At the end of the communication, f_3 is returned to the network pool for subsequent reallocation. At earth station F, the call arriving on frequency f_3 is demodulated in the particular unit to which circuit f_3 has been assigned and is forwarded, via the CT, to the subscriber over the terrestrial link. Again, the receiving unit and its associated terrestrial connection are assigned arbitrarily, and only the allocated frequency determines which particular circuit the receiver will process. The important factor is that the equipment assignment is independent of either call source or destination.

The method used to establish calls among users of the DA network will now be examined. Figure 8 illustrates a typical signaling and allocation method. Automatic dialing, which may originate either from the subscriber or from an operator at the CT, is assumed. The colored dots show the signaling proceeding from the subscriber through the CT to earth station A. There the DASS unit routes the signaling via a common signaling channel (CSC), operating in a time-division multiple-access broadcast mode, to all participating earth stations. At the initiation of this routing, station A requests allocation of a particular frequency for this call from a busy–idle table of frequencies continually updated via the CSC. Station C, for which the call is destined, monitors the common signaling channel and notes the arrival of station A's request and the allocated frequency. If no other station has requested that frequency before station A, station C transmits an acknowledgment of the call request, assigns the frequency to the channel equipment, and proceeds to verify continuity and establish ultimate connection to the subscriber via the CT.

Station A monitors the common signaling channel between the time it requests a frequency and the time it receives its own request (about 240 ms). If, during that time, another station had requested that frequency, both stations A and C would register a busy to that allocation on their busy–idle tables and station A would initiate a new request. To minimize conflicting requests for simultaneous frequency assignment, each station chooses randomly from a list of available frequencies stored in DASS.

Note on the time scale of Fig. 8 that access to the common signaling channel occurs once every 50 ms for each station and the access duration is 1 ms. Station A, therefore, may initiate other calls while awaiting its own and station C's response.

Having established the connection between the subscribers using stations A and C, the DASS and CSC proceed with processing of other incoming and outgoing requests. At the end of the call, the disconnect signal is broadcast over the CSC and all stations note that this frequency is again available for assignment. The DASS unit may also note the duration of the allocation and other

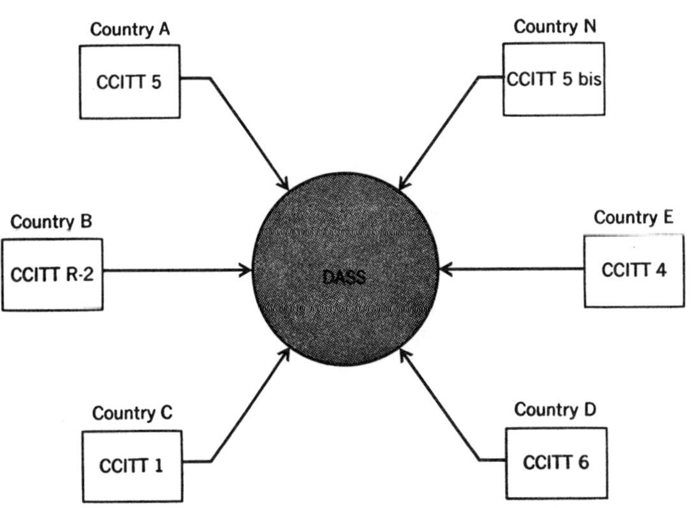

FIGURE 10. Signaling systems SPADE can interface.

IV. Intelsat IV
transmission characteristics (global beam)

RF Bandwidth per Accessing Carrier, MHz	Channels per Carrier	Total Accesses per Transponder	Total Channels per Transponder
FDM/FM/FDMA (multichannel carriers):			
2.5	24	14	336
5	60	7	420
10/5*	132	4	456
36	900	1	900
PCM/PSK/FDMA (SPADE):			
0.045	1	800	800

* Three carriers at 10 MHz and one carrier at 5 MHz.

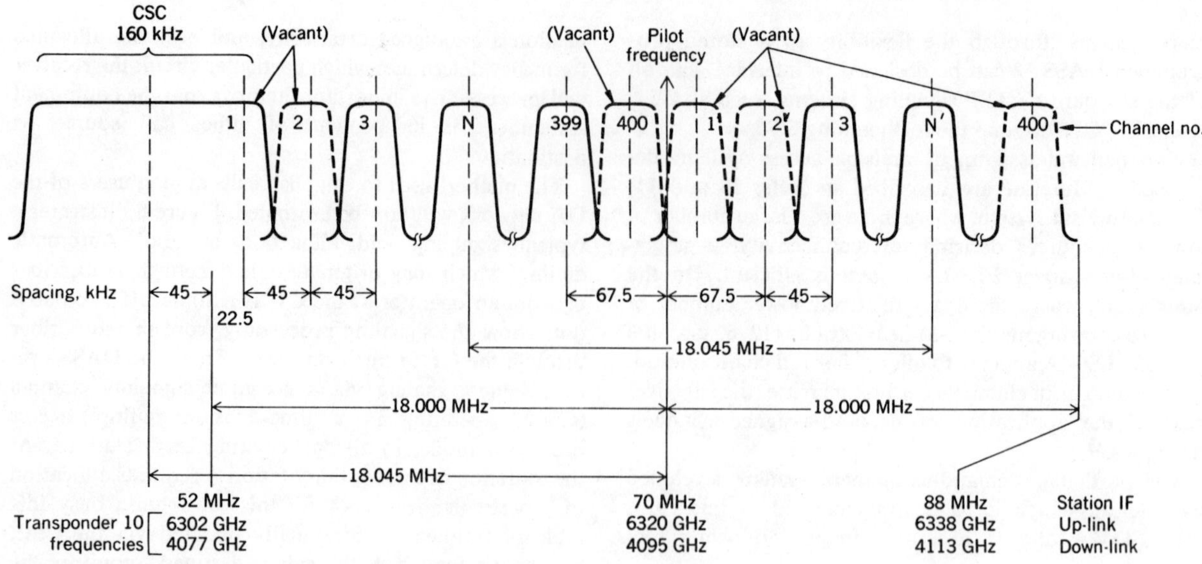

FIGURE 11. SPADE multichannel frequency-allocation spectrum.

vital statistics pertinent to each call.

This DA/FDMA method, which at present is being implemented by Intelsat, is known as the SPADE system. (The term SPADE is derived from *s*ingle-channel-per-carrier, *p*ulse code modulation; multiple-*a*ccess, *d*emand-assignment *e*quipment.) It is particularly timely in view of the rapid increase in the number of earth stations accessing Intelsat satellites, and the associated evolution of new traffic patterns. Further, it interfaces easily with existing earth-station facilities and is compatible with present methods of operations.

The SPADE demand-assignment system

Some of the technical characteristics of the SPADE system are summarized in Table III. This is a single-channel-per-carrier FDMA system providing demand assignment without central control. Pulse code modulation (PCM) is used for channel encoding and four-phase coherent phase-shift keying (PSK) is used for modulating each carrier. Other significant parameters of the system, including the characteristics of the common signaling channel, are also shown in Table III.

Figure 9 is a block diagram of a typical installation. Telephone circuits from the local transit center are linked to the SPADE terminal via the terrestrial interface unit. This equipment provides the interface for call signaling as required to initiate, supervise, and terminate all calls systematically. When a call request is received, the DASS unit automatically selects a frequency pair from the pool of available frequencies and alerts the destination station of an incoming call and the frequency assignment for response. All DASS units utilize the signaling information disseminated by the CSC to update a channel table such that the frequencies just assigned are unavailable for new calls.

The frequency selected is provided to the channel unit by means of a frequency synthesizer that is capable of generating any of the 800 discrete pool frequencies using digital codes provided by DASS. This frequency is used both for the outgoing carrier and the received-signal local oscillator. Channel pairings are based on the common use of the synthesizer for received and transmit signals.

Upon turn-on of the modem, the DASS unit conducts a two-way circuit continuity check. Once the call has been established, the analog signal received by the channel unit is sent to a PCM codec, which transforms it to a digital signal for outgoing transmission and transforms returning signals from digital to analog form.

The content of the voice channel coming from the CT is detected by a voice detector, which is used to gate the channel carrier on or off. This conserves satellite power as a function of talker activity. The digital bit streams in and out of the voice encoder/decoder are synchronized by the transmit–receive synchronizer, where timing, buffering, and framing functions are performed. The PSK modem modulates the assigned carrier frequency with the outgoing bit stream and coherently demodulates the incoming bursts by recovering carrier and bit timing associated with the received signals. The modulated carriers, both outgoing and incoming, are passed through a common intermediate-frequency (IF) subsystem, which interfaces with the earth station up- and down-converters at IF. The carrier used for the CSC modem is also passed through the IF subsystem.

When the call is completed, a control signal from the CT allows DASS to return that circuit to the frequency pool for reassignment. This information is passed to all stations via the CSC.

The design of the SPADE system has taken into account the fact that a variety of signaling and switching systems will interface with each SPADE terminal, as illustrated in Fig. 10. The terrestrial interface unit and the DASS in the SPADE terminal provide the required "universal interface" to each of these CCITT signaling and switching systems. Communications between countries having different signaling and switching systems can therefore be established using the SPADE DASS to provide compatibility.

System capacity with Intelsat IV satellites

The Intelsat IV satellite consists of 12 independent transponders. The allocated satellite frequency bandwidth

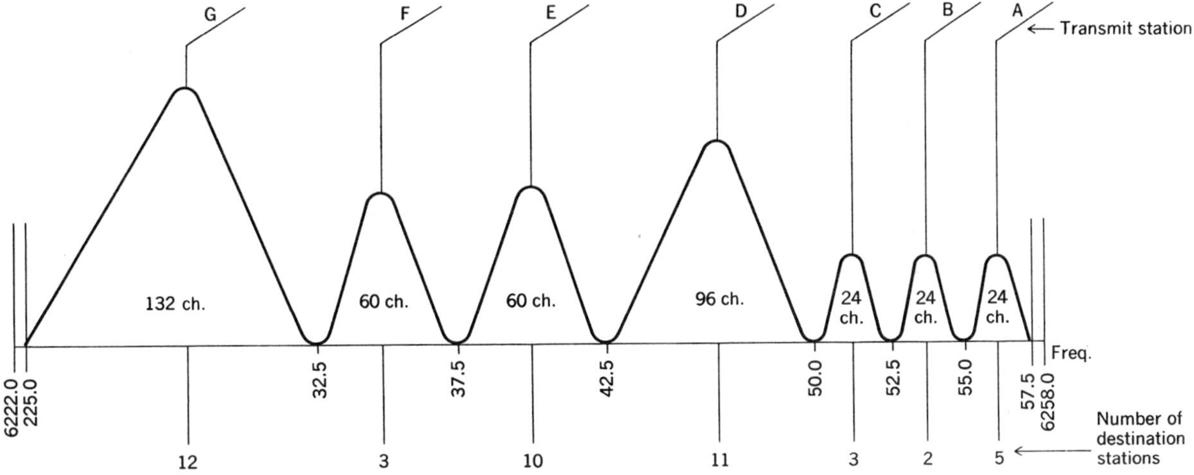

FIGURE 12. Typical FM/FDMA frequency-allocation spectrum.

FIGURE 13. Transponder capacity in mixed-size earth-station network.

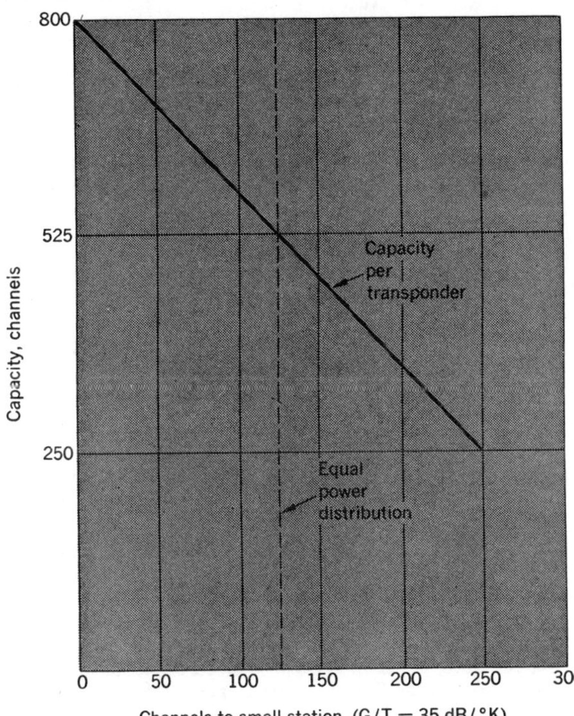

ent sizes by specifying both the occupied RF bandwidth and the number of 4-kHz channels per accessing carrier. Some of these standard sizes are shown in Table IV. Since the smaller-capacity carriers are less efficient of bandwidth, the capacity of the transponder drops as the number of accesses increase. Thus, in Table IV, 14 accessing carriers, each occupying 2.5 MHz and each carrying 24 channels, fill the 36-MHz transponder bandwidth with a total of only 336 channels. Although various-size carriers can be transmitted from the same earth station, each carrier constitutes an access.

For SPADE, using PCM/PSK/FDMA, each RF carrier accommodates a single channel and occupies a bandwidth of 45 kHz. Thus, the 36-MHz bandwidth can be filled with 800 individual carriers, each representing an access to the satellite even though, again, several carriers emanate from the same earth station.

Since each carrier is associated with a voice channel, it need not be transmitted unless voice is present on that channel. Thus, there is sufficient satellite power to support 800 voice channels, which are on or off as a function of talker activity. This feature is not to be confused with time-assignment speech interpolation (TASI) where an idle channel is actually reallocated to another user during pauses in speech. In SPADE, the channel allocation remains fixed for the duration of the call, but during that call the carrier is turned on and off by talker activity to conserve satellite power.

Figure 11 shows the distribution of SPADE carriers over the 36-MHz band of transponder 10 and the corresponding band centered at the 70-MHz IF interface between the SPADE terminal and the earth station up and down frequency converters. At the lower end of the band, space is allocated for the common signaling channel and a system pilot is located at band center.

By contrast, Fig. 12 shows global beam transponder 8 being accessed by three 24-channel carriers, two 60-channel carriers, one 96-channel carrier, and one 132-channel carrier, resulting in a capacity of 418 channels. The number of destinations for each carrier is also given in this illustration.

of 500 MHz at 6 GHz and 4 GHz is distributed equally to each transponder so that each has an RF bandwidth of 36 MHz.

The SPADE system has been assigned to transponder 10, which lies between 6.302 and 6.338 GHz for reception and 4.077 and 4.113 GHz for transmission. Of the 12 transponders, eight can be selected to transmit either via narrow-spot beam antennas or a global coverage antenna. The four others are permanently assigned to the global earth coverage satellite antenna. Transponder 10 falls into the latter category.

The remaining transponders are accessed using conventional FDM/FM/FDMA multidestination, multichannel carriers. Intelsat has standardized these carriers to differ-

The capacity of 800 channels[14] shown in Table IV assumes that the operating network is accessed by standard earth stations—i.e., earth stations with gain-to-temperature ratios (G/T)* of at least 40.7 dB/°K. At the expense of capacity per transponder, it is possible to operate the SPADE system in a network of mixed earth-station sizes. Figure 13 is a plot of the channels destined for small stations with G/T of 35 dB/°K. It can be seen from this illustration that if 125 channels are destined for the small earth stations, then the total network capacity is reduced to 525 channels. This represents the point at which half the available satellite power is being used to service the channels destined to the smaller stations and the remaining half is used to service channels terminating at the large standard stations. This curve is presented to illustrate the effect on network capacity occasioned by the introduction of small earth stations.[15] The main parameter being varied is the satellite-to-earth-station carrier power. All carriers destined to small stations must exceed the level of carriers destined to large stations by the difference in G/T between the large and small terminals. At the time the frequency is assigned for a call, each transmitting station is apprised of the size of the destination terminal and the level of the outgoing carrier is set accordingly. The implementation of the SPADE system, which allows the use of a single channel per carrier together with the DASS assignment feature, makes the penalties of mixing earth-station sizes less severe than for presently known frequency- or time-division multiple-access methods.

Conclusion

The technique of demand assignment will enhance the usefulness of satellite communications networks. In particular, it is advantageous for interconnection of countries requiring relatively few channels and for overflow traffic from medium preassigned trunks. The SPADE system, which is the specific system that Intelsat plans to use, is more efficient of power and bandwidth per channel than present methods.

* Gain-to-temperature ratio is the figure of merit applied to the receiving sensitivity of a given earth station without reference to its specific size or noise figure. Intelsat has ruled that the minimum G/T for a standard earth station in its network shall be at least 40.7 dB/°K.

This article is based upon work performed at Comsat Laboratories under Corporate sponsorship and under the sponsorship of the International Telecommunications Satellite Consortium (Intelsat). The views expressed herein are not necessarily those of Intelsat.

The authors gratefully acknowledge the assistance of the many members of the technical staff at Comsat Corporation during the course of this program. Further acknowledgment is due to Intelsat, under whose auspices the system was developed, and to the delegates of the various SPADE working groups, whose diligence contributed notably to the successful embodiment of an R&D concept into an operational system.

REFERENCES

1. Votaw, M. J., "The Early Bird project," *IEEE Trans. Communications Technology*, vol. COM-14, pp. 507–511, Aug. 1966.

2. Murphy, C. G., "A Syncom satellite program," Paper 63-264, presented at the AIAA Summer Meeting, Los Angeles, Calif., June 1963.

3. Berman, A. L., and Mahle, C. E., "Nonlinear phase shift in traveling-wave tubes as applied to multiple access communication satellites," *IEEE Trans. Communication Technology*, vol. COM-18, pp. 37–48, Feb. 1970.

4. Lutz, S. G., "A traffic study of multiple access satellite communication based on an Atlantic system model," Hughes Research Rept. 376, Aug. 1967.

5. CCITT/CCIR, "General plan for the development of the interregional telecommunication network (Mexico 1967-1970-1975)," ITU, Geneva, 1968.

6. Report of the Atlantic Region Traffic Subgroup Meeting, Washington, D.C., May 1969, ICSC Doc. 39-32.

7. Anguera, F., "Traffic considerations for demand-assigned telephone service through a satellite system," presented at the International Teletraffic Congress, Mexico City, Apr. 1967.

8. Dill, G., and Shimasaki, N., "Signaling and switching for demand-assigned satellite communications," *Proc. Intelsat/IEE Conf. Digital Satellite Commun.*, pp. 297–307, 1969.

9. Dill, G., and Shimasaki, N., "The terrestrial interface at SPADE terminals," Paper 70-413, presented at the AIAA 3rd Communications Satellite Systems Conference, Los Angeles, Calif., Apr. 6–8, 1970.

10. Werth, A. M., "SPADE: A PCM FDMA demand assignment system for satellite communications," IEEE Paper 70-CP-423-COM, presented at the 1970 International Conference on Communications, San Francisco, Calif., June 1970.

11. Schmidt, W. G., et al., "MAT-1: Intelsat's experimental 700-channel TDMA/DA system," *Proc. Intelsat/IEE Conf. Digital Satellite Commun.*, pp. 428–440, 1969.

12. "BTR special study for Intelsat on demand assignment techniques for satellite communications," British Telecommunications Research Ltd. Rept. 97/69/04, Jan. 1969.

13. Morita, M., Tukami, T., and Yamato, S., "STAR system. I—General description," *Nippon Elec. Co. Res. Develop. J.*, Oct. 1966.

14. McClure, R. B., "The effect of earth station and satellite parameters on the SPADE system," presented at the IEEE Conference on Earth Station Technology, London, Oct. 1970.

15. Cacciamani, E. R., "The SPADE concept applied to a network of large and small earth stations," Paper 70-420, presented at the AIAA 3rd Communications Satellite Systems Conference, Los Angeles, Calif., Apr. 6–8, 1970.

The Throughput of Packet Broadcasting Channels

NORMAN ABRAMSON, FELLOW, IEEE

Abstract—Packet broadcasting is a form of data communications architecture which can combine the features of packet switching with those of broadcast channels for data communication networks. Much of the basic theory of packet broadcasting has been presented as a byproduct in a sequence of papers with a distinctly practical emphasis. In this paper we provide a unified presentation of packet broadcasting theory.

In Section II we introduce the theory of packet broadcasting data networks. In Section III we provide some theoretical results dealing with the performance of a packet broadcasting network when the users of the network have a variety of data rates. In Section IV we deal with packet broadcasting networks distributed in space, and in Section V we derive some properties of power-limited packet broadcasting channels, showing that the throughput of such channels can approach that of equivalent point-to-point channels.

I. INTRODUCTION

A. Packet Switching and Packet Broadcasting

THE transition of packet-switched computer networks from experimental [1] to operational [2] status during 1975 provides convincing evidence of the value of this form of communications architecture. Packet switching, or statistical multiplexing [3], can provide a powerful means of sharing communications resources among large number of data communications users when those users can be characterized by a high ratio of peak to average data rates. Under such circumstances, data from each user are buffered, address and control information is added in a "header," and the resulting bit sequence, or "packet," is routed through a shared communications resource by a sequence of node switches [4], [5].

Packet-switched networks, however, still employ point-to-point communication channels and large multiplexing switches for routing and flow control in a fashion similar to conventional circuit switched networks. In some situations [6]–[10] it is desirable to combine the efficiencies achievable by a packet communications architecture with other advantages obtained by use of broadcast communication channels. Among these advantages are elimination of routing and network switches, system modularity, and overall system simplicity. In addition, certain kinds of channels available to the communications systems designer, notably satellite channels, are basically broadcast in their structure. In such cases use of these

Manuscript received January 19, 1976; revised June 11, 1976. This work was supported by The ALOHA System, a research project at the University of Hawaii which is supported by the Advanced Research Projects Agency of the Department of Defense and monitored by NASA Ames Research Center under Contract NAS2-8590. The views and conclusions contained in this paper are those of the author and should not be interpreted as necessarily representing the official policies, either expressed or implied, of the Advanced Research Projects Agency of the United States Government.

The author is with The ALOHA System, University of Hawaii, Honolulu, HI 96822.

channels in their natural broadcast mode can lead to significant system performance advantages [11], [12].

B. Outline of Results

Packet broadcasting is a form of data communications architecture which can combine the features of packet switching with those of broadcast channels for data communication networks. Much of the basic theory of packet broadcasting has been presented as a byproduct in a sequence of papers with a distinctly practical emphasis. In this paper we provide a unified presentation of packet broadcasting theory.

In Section II we introduce the theory of packet broadcasting as implemented in the ALOHA System at the University of Hawaii; also in Section II we explain a modification of the basic ALOHA method, called slotting. In Section III we provide some theoretical results dealing with the performance of a packet broadcasting channel when the users of the channel have a variety of data rates. In Section IV we deal with packet broadcasting networks distributed in space, and present some incomplete results on the theoretical properties of such networks. Finally, in Section V we derive some properties of power limited packet broadcasting channels showing that the throughput of such channels can approach that of equivalent point-to-point channels. This result is of importance in satellite systems using small earth stations since it implies that the multiple access capability and the complete connectivity (in the topological sense) of packet broadcasting channels can be obtained at no price in average throughput.

II. PACKET BROADCASTING CHANNELS

A. Operation of a Packet Broadcasting Channel

Consider a number of widely separated users, each wanting to transmit short packets over a common high-speed channel. Assume that the rate at which users generate packets is such that the average time between packets from a single user is much greater than the time needed to transmit a single packet. In Fig. 1 we indicate a sequence of packets transmitted by a typical user.

Conventional time or frequency multiplexing methods (TDMA or FDMA) or some kind of polling scheme could be employed to share the channel among the users. Some of the disadvantages of these methods for users with high peak-to-average data rates are discussed by Carleial and Hellman [13]. In addition, under certain conditions polling may require unacceptable system complexity and extra delay.

In a packet broadcasting system the simplest possible solution to this multiplexing problem is employed. Each user transmits its packets over the common broadcast channel in a completely unsynchronized (from one user to another) manner. If each individual user of a packet broadcasting chan-

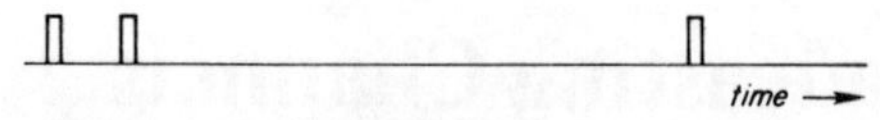

Fig. 1.   Packets from a typical user.

Fig. 2.   Packets from several users on an ALOHA channel.

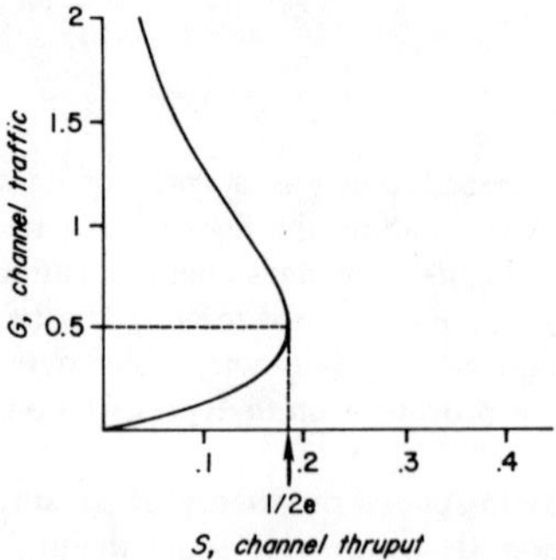

Fig. 3.   Channel throughput versus channel traffic for an ALOHA channel.

nel is required to have a low duty cycle, the probability of a packet from one user interfering with a packet from another user is small as long as the total number of users on the common channel is not too large. As the number of users increases, however, the number of packet overlaps increases and the probability that a packet will be lost due to an overlap also increases. The question of how many users can share such a channel and the analysis of various methods of dealing with packets lost due to overlap are the primary concerns of this paper. In Fig. 2 we show a packet broadcasting channel with two overlapping packets. Since the first packet broadcasting channel was put into operation in the ALOHA System radio-linked computer network at the University of Hawaii [6], they have been referred to as ALOHA channels.

B. ALOHA Capacity

A transmitted packet can be received incorrectly or lost completely because of two different types of errors: 1) random noise errors and 2) errors caused by packet overlap. In this paper we assume that the first type of error can be ignored, and we shall be concerned only with errors caused by packet overlap. In Section II-D we describe several methods of dealing with the problem of packets lost due to overlap, but first we derive the basic results which tell us how many packets can be transmitted with no overlap.

Assume that the start times of packets in the channel comprise a Poisson point process with parameter λ packets/second. If each packet lasts τ seconds, we can define the normalized channel traffic G where

$$G = \lambda\tau. \tag{1}$$

If we assume that only those packets which do not overlap with any other packet are received correctly, we may define $\lambda' < \lambda$ as the rate of occurrence of those packets which are received correctly. Then we define the normalized channel thruput S by

$$S = \lambda'\tau. \tag{2}$$

The probability that a packet will not overlap a given packet is just the probability that no packet starts τ seconds before or τ seconds after the start time of the given packet. Then, since the point process formed from the start times of all packets in the channel was assumed Poisson, the probability that a packet will not overlap any other packet is $e^{-2\lambda\tau}$, or e^{-2G}. Therefore

$$S = Ge^{-2G} \tag{3}$$

and we may plot the channel throughput versus channel traffic for an ALOHA channel (Fig. 3).

From Fig. 3 we see that as the channel traffic increases, the throughput also increases until it reaches its maximum at $S = 1/2e = 0.184$. This value of throughput is known as the capac-

ity of an ALOHA channel, and it occurs for a value of channel traffic equal to 0.5. If we increase the channel traffic above 0.5, the throughput of the channel will decrease.

C. Application of an ALOHA Channel

In order to indicate the capabilities of such a channel for use in an interactive network of alphanumeric computer terminals, consider the 9600 bits/s packet broadcasting channel used in the ALOHA System. From the results of Section II-B we see that the maximum average throughput of this channel is 9600 bits/s times 1/2e, or about 1600 bits/s. If we assume the conservative [14] figure of 5 bits/s as the average data rate (including overhead) from each active[1] terminal in the network, this channel can handle the traffic of over 300 active terminals and each terminal will operate at a peak data rate of 9600 bits/s. Of course, the total number of terminals in such a network can be much larger than 300 since only a fraction of all terminals will be active and a terminal consumes no channel resources when it is not active.

D. Recovery of Lost Packets

Since the packet broadcasting technique we have described will result in some packets being lost due to packet overlaps, it is necessary to introduce some technique to compensate for this loss. We may list four different packet recovery techniques for dealing with the problem of lost packets. The first three make use of a feedback channel to the packet transmitter and the repetition of lost packets, while the fourth is based on coding.

1) Positive Acknowledgments (POSACKS): Perhaps the most direct way to handle lost packets is to require the receiver of the packet to acknowledge correct receipt of the packet. Each packet is transmitted and then stored in the transmitter's buffer until a POSACK is received from the receiver. If a POSACK is not received in a given amount of time, the transmitter can repeat the transmission and continue to repeat until a POSACK is received or until some other criterion is met. The POSACK can be transmitted on a sepa-

---

[1] A terminal is defined as active from the time a user transmits an attempt to log on until he transmits a log off message.

rate channel (as in the ALOHANET [6]) or transmitted on the same channel as the original packets (as in the ARPA packet radio system [15]). An error detection code and a packet numbering system can be used to increase the reliability of this technique.

2) Transponder Packet Broadcasting: Certain communication channels—notably communication satellite channels—transmit packets on one frequency to a transponder which retransmits the packets on a second frequency. In such cases all units in a packet broadcasting network can receive their own packet retransmissions, determine whether a packet overlap has occurred, and repeat the packet if necessary. This technique has been employed in ATS-1 satellite experiments in the Pacific Educational Computer Network (PACNET) [16] and in the ARPA Atlantic INTELSAT IV packet broadcasting experiments [17].

3) Carrier Sense Packet Broadcasting: For ground-based packet broadcasting networks where the signal propagation time over the furthest transmission path is much less than the packet duration, it is feasible to provide each transmission unit with a device to inhibit packet transmission while another unit is detected transmitting. A carrier sense capability can increase the channel throughput, even if these conditions are not met, when used in conjunction with other packet recovery methods. Carrier sense systems have been analyzed by Tobagi [18] and by Kleinrock and Tobagi [19]. A comprehensive yet compact analysis of such systems is provided in [42].

4) Packet Recovery Codes: When a user employs a packet broadcasting channel to transmit long files by breaking them into large numbers of packets, it is possible to encode the files so that packets lost due to broadcasting overlap can be recovered. It is clear that some of the existing classes of multiple burst error-correcting codes [20] and cyclic product codes [21] can be used for packet recovery in transmissions of long files. It is also clear that these codes are not as efficient as possible for packet recovery and that considerable work remains to be done in this area.

E. Slotted Channels

It is possible to modify the completely unsynchronized use of the ALOHA channel described above in order to increase the maximum throughput of the channel. In the pure ALOHA channel each user simply transmits a packet when ready without any attempt to coordinate his transmission with those of other users. While this strategy has a certain elegance, it does lead to somewhat inefficient channel utilization. If we establish a time base and require each user to start his packet only at certain fixed instants, it is possible to increase the maximum value of the channel thruput. In this kind of channel, called a slotted ALOHA channel, a central clock establishes a time base for a sequence of "slots" of the same duration as a packet transmission [41]. Then when a user has a packet to transmit, he synchronizes the start of his transmission to the start of a slot. In this fashion, if two messages conflict they will overlap completely, rather than partially.

To analyze the slotted ALOHA channel, define G_i as the probability that the ith user will transmit a packet in some slot. Assume that each user operates independently of all other

users, and that whether or not a user transmits a packet in a given slot does not depend upon the state of any previous slot. If we have n users, we can define the normalized channel traffic for the slotted channel G where

$$G = \sum_{i=1}^{n} G_i. \tag{4}$$

Note that G may be greater than 1.

As before, we can also consider the rate at which a user sends packets which do not experience an overlap with other user packets. Define $S_i \leqslant G_i$ as the probability that a user sends a packet and that this packet is the only packet in its slot. If we have n users, then we define the normalized channel throughput for the slotted channel S where

$$S = \sum_{i=1}^{n} S_i. \tag{5}$$

Note that S is less than or equal to 1 and $S \leqslant G$.

For the slotted ALOHA channel with n independent users, the probability that a packet from the ith user will not experience an interference from one of the other users is

$$\prod_{\substack{j=1 \\ j \neq i}}^{n} (1 - G_j).$$

Therefore we may write the following relationship between the message rate and the traffic rate of the ith user:

$$S_i = G_i \prod_{\substack{j=1 \\ j \neq i}}^{n} (1 - G_j). \tag{6}$$

If all users are identical, we have

$$S_i = \frac{S}{n} \tag{7}$$

and

$$G_i = \frac{G}{n} \tag{8}$$

so that (6) can be written

$$S = G \left(1 - \frac{G}{n}\right)^{n-1} \tag{9}$$

and in the limit as $n \to \infty$, we have

$$S = Ge^{-G}. \tag{10}$$

Equation (10) is plotted in Fig. 4 (curve labeled "slotted ALOHA"). Note that the message rate of the slotted ALOHA channel reaches a maximum value of $1/e = 0.368$, twice the capacity of the pure ALOHA channel.

This result for slotted ALOHA channels was first derived by Roberts [41] using a different method.

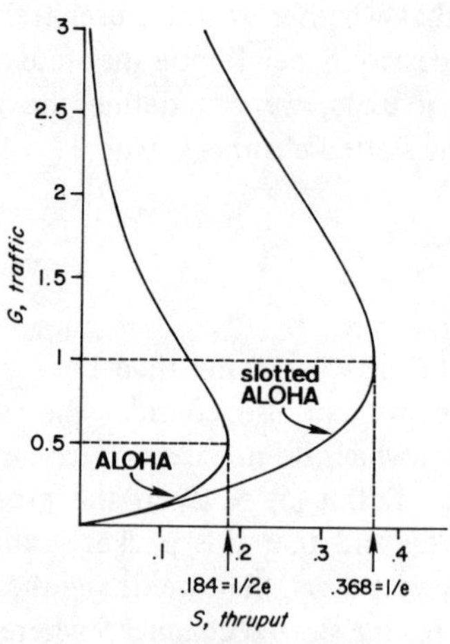

Fig. 4. Traffic versus throughput for an ALOHA channel and a slotted ALOHA channel.

III. PACKET BROADCASTING WITH MIXED DATA RATES

A. Unslotted Case: Variable Packet Lengths

In Section II we were concerned with the analysis of ALOHA channels carrying a homogeneous mix of packets. If some channel users have a higher average data rate than others, however, the high rate users must either transmit packets more frequently or transmit longer packets. In this section we shall analyze the unslotted ALOHA channel when carrying packets of different lengths, and we shall analyze the slotted ALOHA channel when the probability of transmitting in a given slot varies from user to user.

Let us assume an unslotted ALOHA channel with two different possible packet durations, τ_2 and τ_1. Assume $\tau_2 \geqslant \tau_1$, and therefore we refer to the two different length packets as long packets and short packets, respectively. Assume also the start times of the long packets and short packets form two Poisson point processes with parameters λ_2 and λ_1 packets/second, and that the two Poisson point processes are mutually independent. Then we can define the normalized channel traffic for those packets of duration τ_i:

$$G_i = \lambda_i \tau_i, \qquad i = 1, 2. \tag{11}$$

Again assume that only those packets which do not overlap with any other packet are received correctly and define $\lambda_i' < \lambda_i$ as the rate of occurrence of those packets of duration τ_i which are received correctly. Define the normalized throughput of packets of duration τ_i as

$$S_i = \lambda_i' \tau_i, \qquad i = 1, 2. \tag{12}$$

Since we assumed two independent Poisson point processes, the probability that a short packet will be received correctly is

$$\exp\left[-\lambda_1(2\tau_1) - \lambda_2(\tau_1 + \tau_2)\right] \tag{13}$$

and if we define

$$G_{12} \triangleq \lambda_1 \tau_2 \tag{14a}$$

$$G_{21} \triangleq \lambda_2 \tau_1 \tag{14b}$$

(13) becomes

$$\exp\left[-2G_1 - G_{21} - G_2\right]. \tag{15}$$

Therefore

$$S_1 = G_1 \exp\left[-2G_1 - G_{21} - G_2\right] \tag{16a}$$

and, by a similar argument, the throughput of long packets is

$$S_2 = G_2 \exp\left[-G_{12} - G_1 - 2G_2\right]. \tag{16b}$$

For any given values of λ_1 and λ_2 we may calculate G_1, G_2, G_{12}, and G_{21}; substitution of these values into (16a) and (16b) will allow calculation of the throughputs S_1 and S_2. Therefore (16a) and (16b) may be used to define an allowable set of throughput pairs (S_1, S_2) in the (S_1, S_2) plane.

To determine the boundary of this region we define

$$\alpha \triangleq \frac{\tau_2}{\tau_1}. \tag{17}$$

Note that $\alpha \geqslant 1$. We may rewrite (16a) and (16b) in terms of α, the ratio of long packet duration to short packet duration:

$$S_1 = G_1 \exp\left[-2G_1 - \left(1 + \frac{1}{\alpha}\right)G_2\right] \tag{18a}$$

$$S_2 = G_2 \exp\left[-(1 + \alpha)G_1 - 2G_2\right]. \tag{18b}$$

The boundary of the set of allowable (S_1, S_2) pairs in the (S_1, S_2) plane is defined by setting the Jacobian

$$J = \left| \frac{\partial S_i}{\partial G_j} \right|, \qquad i, j = 1, 2 \tag{19}$$

equal to zero. A simple calculation shows that the Jacobian is zero when

$$G_2 = \frac{1 - 2G_1}{\left(\dfrac{(\alpha - 1)^2}{\alpha} G_1 + 2\right)}. \tag{20}$$

Note that this checks for $G_1 = 0$ and for $\alpha = 1$.

We need only substitute this expression for G_2 into (18a) and (18b) to obtain two equations for S_1, the short packet throughput, and S_2, the long packet throughput, in terms of the single parameter G_1; and as G_1 varies from 0 (all long packets) to 1/2 (all short packets), we will trace out the boundary of the achievable values of throughput in the (S_1, S_2) plane. These achievable throughput regions are indicated for several values of α in Fig. 5.

The basic conclusion of this analysis is that the total channel throughput can undergo a significant decrease if all packets are not of the same length. Thus if the two different

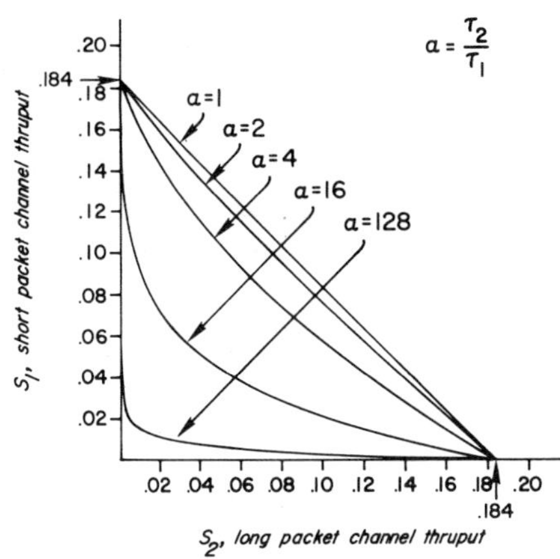

Fig. 5. Achievable throughput regions in an unslotted ALOHA channel.

packet lengths differ by a large factor, it is often preferable to break up long packets into many shorter packets as long as the overhead necessary to transmit the text in each packet is small. Ferguson [23] has generalized these results to show that channel throughput is maximized over all possible packet length distributions with fixed length packets.

In view of this discouraging result, we might conclude that an inhomogeneous mix of users inevitably leads to a decrease in the maximum value of channel throughput. Surprisingly, this conclusion is not warranted, and we shall show in Section III-B that a mix of users of varied data rates can lead to an increase in the maximum values of channel throughput.

B. Slotted Case: Variable Packet Rates

In the section we shall consider a slotted ALOHA channel used by n users, possibly with different values of channel traffic G_i. From (6) we have a set of n nonlinear equations relating the channel traffics and the channel throughputs for these n users:

$$S_i = G_i \prod_{\substack{j=1 \\ j \neq i}}^{n} (1 - G_j), \qquad i = 1, 2, \cdots, n. \tag{21}$$

Define

$$\alpha = \prod_{j=1}^{n} (1 - G_j); \tag{22}$$

then (21) can be written

$$S_i = \frac{G_i}{1 - G_i} \alpha, \qquad i = 1, 2, \cdots, n. \tag{23}$$

For any set of n acceptable traffic rates $G_1, G_2, \cdots, G_n$, these n equations define a set of channel throughputs $S_1, S_2, \cdots, S_n$ or a region in an n-dimensional space whose coordinates are the S_i. In order to find the boundary of this region, we calculate the Jacobian:

$$J = \left| \frac{\partial S_j}{\partial G_k} \right|, \qquad j, k = 1, 2, \cdots, n. \tag{24}$$

Since

$$\frac{\partial S_j}{\partial G_k} = \begin{cases} \displaystyle\prod_{\substack{i=1 \\ i \neq j}}^{n} (1 - G_i), & j = k \\[2em] \displaystyle -G_j \prod_{\substack{i=1 \\ i \neq j,k}}^{n} (1 - G_i), & j \neq k \end{cases} \tag{25}$$

after some algebra we may write the Jacobian as

$$J = \alpha^{n-2} \begin{vmatrix} (1 - G_1) & -G_1 & -G_1 \\ -G_2 & (1 - G_2) & -G_2 \\ -G_3 & -G_3 & (1 - G_3) \\ & \vdots & \end{vmatrix} \cdots$$

$$= \alpha^{n-2} [1 - G_1 - G_2 - \cdots - G_n]. \tag{26}$$

Thus the condition for maximum channel throughputs is

$$\sum_i G_i = 1. \tag{27}$$

This condition can then be used to define a boundary to the n-dimensional region of allowable throughputs $S_1, S_2, \cdots, S_n$.

Consider the special case of two classes of users with n_1 users in class 1 and n_2 users in class 2:

$$n_1 + n_2 = n. \tag{28}$$

Let S_1 and G_1 be the throughputs and traffic rates for users in class 1, and let S_2 and G_2 be the throughputs and traffic rates for users in class 2. Then the n equations (21) can be written as the two equations

$$S_1 = G_1 (1 - G_1)^{n_1 - 1} (1 - G_2)^{n_2} \tag{29a}$$

$$S_2 = G_2 (1 - G_2)^{n_2 - 1} (1 - G_1)^{n_1}. \tag{29b}$$

For any pair of acceptable traffic rates G_1 and G_2, these two equations define a pair of channel throughputs S_1 and S_2 or a region in the (S_1, S_2) plane.

From (27) we know that the boundary of this region is defined by the condition

$$n_1 G_1 + n_2 G_2 = 1. \tag{30}$$

We can use (30) to substitute for G_1 in (29a) and (29b) and obtain two equations for S_1 and S_2 in terms of a single parameter G_2. Then as G_2 varies from 0 to 1, the resulting (S_1, S_2) pairs define the boundary of the region we seek. These achievable regions are indicated for various values of n_1 and n_2 in Figs. 6 and 7.

The important point to notice from Figs. 6 and 7 is that in a lightly loaded slotted ALOHA channel, a single large user can transmit data at a significant percentage of the total channel data rate, thus allowing use of the channel at rates well above the limit of $1/e$ or 37 percent obtained when all users have the same message rate. A throughput data rate above the $1/e$ limit

$$S_1 = G_1{}^2 \tag{31a}$$

$$S_2 = (1 - G_1)^2. \tag{31b}$$

2) For $n_2 \to \infty$:

$$S_1 = G_1(1 - G_1)^{n_1 - 1} \cdot \exp\left[-(1 - n_1 G_1)\right] \tag{32a}$$

$$S_2 = (1 - n_1 G_1)(1 - G_1)^{n_1 - 1} \cdot \exp\left[-(1 - n_1 G_1)\right]. \tag{32b}$$

3) For $n_1 = n_2 \to \infty$:

$$S_1 = \frac{G_1}{e} \tag{33a}$$

$$S_2 = \frac{1 - G_1}{e}. \tag{33b}$$

Additional details dealing with excess capacity and the delay experienced with this kind of use of a slotted ALOHA channel may be found in [11] and [25]. A different view of the use of a slotted packet broadcasting for different sources may be found in [43].

IV. SPATIAL PROPERTIES OF PACKET BROADCASTING NETWORKS

A. Packet Repeaters

In this section we deal with certain spatial properties of packet broadcasting networks. Not long after the initial units of the ALOHA System went into operation, it was realized that the range of the network could be extended beyond the range of a single radio link in the network (about 200 km) by the use of packet repeaters. A packet repeater operates in much the same manner as a conventional radio repeater with one major exception. Since radio transmission in a packet broadcasting network is intermittent, a packet repeater can receive a packet and retransmit that packet in the same frequency band by turning off its receiver during a retransmission burst. Thus a packet repeater can sidestep many of the frequency allocation and spatial cell problems [26] of conventional land-based repeater networks.

The use of packet repeaters leads to the consideration of packet broadcasting networks with more than one central station distributed over very large areas. Users transmit a packet, and if the packet cannot be received directly by its destination, it is forwarded to its destination by one or more packet repeaters according to some routing algorithm [27]. The study of such networks has led to the analysis of two communication theory issues related to the performance of the networks: 1) capture effect and 2) the distribution of packet traffic and packet throughput in space.

B. Capture Effect

Up to this point we have analyzed packet broadcasting channels under the pessimistic assumption that if two packets overlap at the receiver, both packets are lost. In fact, this assumption provides a lower bound to the performance of

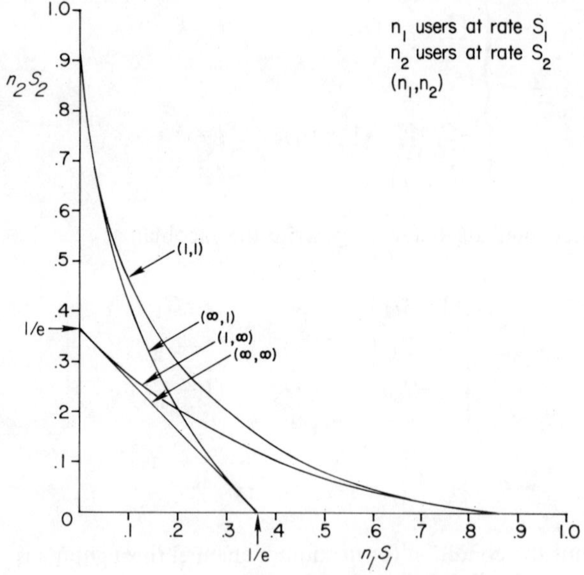

Fig. 6. Allowable channel throughputs.

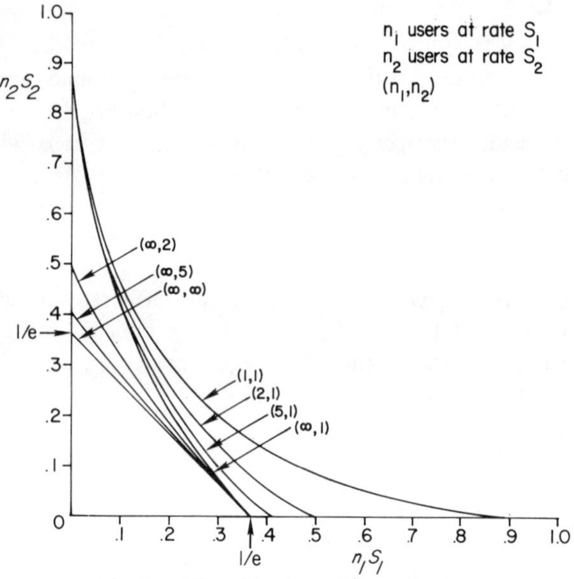

Fig. 7. Allowable channel throughputs.

has been referred to as "excess capacity" [24]. Excess capacity is important for a lightly loaded packet broadcasting network consisting of many interactive terminal users and a small number of users who send large but infrequent files over the channel. Operation of the channel in a lightly loaded condition, of course, may not be desirable in a bandwidth-limited channel. For a communications satellite where the average power in the satellite transponder limits the channel, however, operation in a lightly loaded packet-switched mode is an attractive alternative. Since the satellite will transmit power only when it is relaying a packet, the duty cycle in the transponder will be small and the average power used will be low (see Section V).

Finally, we note that it is possible to deal with certain limiting cases in more detail, to obtain equations for the boundary of the allowable (S_1, S_2) region.

1) For $n_1 = n_2 = 1$:

Upon using (30) in (29), we obtain

real packet broadcasting channels, since in many receivers the stronger of two overlapping packets may capture the receiver and may be received without error. Metzner [40] has used this fact to derive an interesting result, showing that by dividing users into two groups—one transmitting at high power and the other at low power—the maximum throughput can be increased by about 50 percent. This result is of importance for packet broadcasting networks with a mixture of data and packetized speech traffic.

In order to include the effect of capture in a packet broadcasting network, we consider a distribution of packet generators over a two-dimensional plane and a single packet broadcasting receiver which receives packets from these generators [41]. The receiver then may be viewed as a "packet sink" and the packet generators as a distribution of "packet sources" in the plane. We assume that the rate of generation of packets in a given area depends only on r, the distance from the packet sink, and is independent of direction θ.

Then we may define a traffic density and a throughput density analogous to the normalized traffic G and normalized thoughput S defined in Section II-B.

$G(r)$ = normalized packet traffic per unit area at a distance r.

$S(r)$ = normalized packet thruput per unit area at a distance r.

The traffic due to all packet generators in a differential ring of width dr at a radius r is

$$G(r)\,2\pi r\,dr. \tag{34}$$

We assume that packets from different users are generated so that the packet starting times of all packets generated in the differential ring constitute a Poisson point process. Then since the sum of two independent Poisson processes is a Poisson point process, if users in different rings are independent, the start times of all packets generated in a circle of radius r also constitute a Poisson point process, and the total traffic generated by all users within a distance r of the center is

$$\int_0^r G(x)\,2\pi x\,dx. \tag{35}$$

If we assume that a packet from a user at a distance r from the center will be received correctly unless it is overlapped by a packet sent from a user at a distance ar or less ($a \geqslant 1$), then using the results of Section II-B the probability that such a packet will be received correctly is

$$\exp\left[-4\pi \int_0^{ar} G(x)x\,dx\right]. \tag{36}$$

Any packet generated from a packet source in the circle of radius ar shown in Fig. 8 will interfere with packets generated from a source in the circle of radius r. A packet generated outside the circle of radius ar will not interfere with packets generated from a source in the circle of radius r.

We can relate the normalized packet throughput to the normalized packet traffic in the usual way:

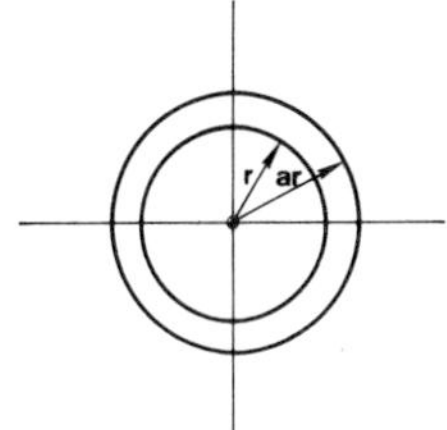

Fig. 8.   Regions of interfering packets.

$$2\pi r S(r)\,dr = 2\pi r G(r) \exp\left[-4\pi \int_0^{ar} G(x)x\,dx\right] dr$$

or

$$S(r) = G(r) \exp\left[-4\pi \int_0^{ar} G(x)x\,dx\right]. \tag{37}$$

If we take a derivative of (37) with respect to r and use (37) to substitute for the exponential, we get

$$S'(r)G(r) = G'(r)S(r) - 4\pi r a^2 S(r)G(r)G(ar). \tag{38}$$

We have not found a general solution of (38) for relating $S(r)$ to $G(r)$ in the presence of capture. We have been able to analyze two special cases, however.

C. Two Solutions

In the first of these special cases we assume a constant traffic density $G(r)$. We can then show that the throughput density $S(r)$ has a Gaussian form, due to the fact that those packets generated further from the receiver will be received correctly less frequently than those packets generated close to the receiver.

In the second special case analyzed we assume a constant packet throughput density $S(r)$ and perfect capture ($a = 1$). Under these assumptions, the packet traffic density will increase as the distance from the receiver increases. We show that there exists a radius r_0 such that the packet traffic density is finite within a circle of radius r_0 around the receiver, while the packet traffic density becomes unbounded on the circle of radius r_0.

For the important case of a packet broadcasting channel distributed over some geographical area and using a packet retransmission policy (Section II-D), this result has an interesting interpretation. In such a situation any packet transmitted from a terminal located within the circle of radius r_0 will be received correctly with probability one (after a finite number of retransmissions), while the expected number of retransmissions required for a packet transmitted from a terminal further from the center than r_0 will be unbounded. Thus there exists a circle of radius r_0 such that terminals transmitting from within this circle can get their packets into the central receiver, while terminals transmitting from outside this circle spend all their time retransmitting their packets in vain. We call r_0 the Sisyphus distance of the ALOHA channel.

1) Constant Packet Traffic Density: Assume the density of normalized packet traffic is constant over the plane

$$G(r) = G_0 \tag{39}$$

and define the distance r_1 as the radius of a circle within which the total packet traffic is unity:

$$\pi r_1{}^2 G_0 \triangleq 1. \tag{40}$$

Then (38) reduces to

$$S'(r) = -\frac{4ra^2}{r_1{}^2} S(r) \tag{41a}$$

with the boundary condition

$$S_0 = G_0 \tag{41b}$$

so that the packet throughput density is

$$S(r) = G_0 \exp\left[-2a^2 \left(\frac{r}{r_1}\right)^2 \right] \tag{42}$$

and the total normalized packet thruput from a circle of radius r is

$$S = \int_0^r S(r')2\pi r' \, dr'$$

$$= \frac{1}{2a^2} \left\{ 1 - \exp\left[-2\left(\frac{ar}{r_1}\right)^2 \right] \right\} \tag{43}$$

and

$$\lim_{r \to \infty} S = \frac{1}{2a^2}. \tag{44}$$

Note that a total throughput which can be supported by a single packet sink with "perfect capture" ($a = 1$) is equal to one half.

2) Constant Packet Throughput Density: Another case of interest where we have found a solution for (38) is that of constant packet throughput density in the plane. Assume

$$S(r) = S_0 \tag{45}$$

over the region in the plane where $S(r)$ and $G(r)$ are bounded.

Then (38) becomes

$$G'(r) = 4\pi r a^2 G(r)G(ar). \tag{46}$$

For the case of $a = 1$ (perfect capture), (46) becomes

$$G'(r) = 4\pi r G^2(r) \tag{47}$$

with the boundary condition

$$G(0) = S_0 \tag{48}$$

so that

$$G(r) = \frac{S_0}{1 - 2\pi r^2 S_0}. \tag{49}$$

Note that the normalized packet traffic per unit area is finite

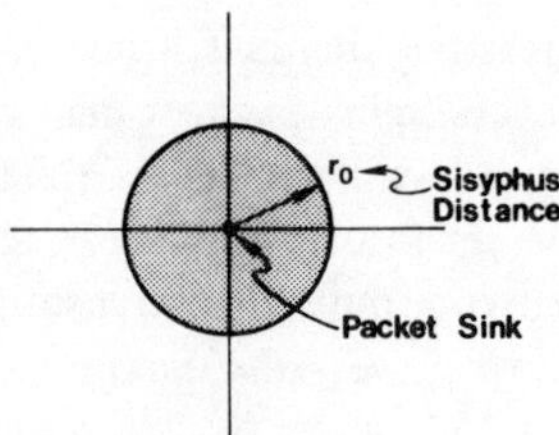

Fig. 9. Region of constant packet throughput S_0 for a single packet sink.

for

$$0 \leqslant r < r_0 \tag{50}$$

where

$$r_0 \triangleq [2\pi S_0]^{-1/2} \tag{51}$$

and r_0 is the Sisyphus distance mentioned in Section IV-C. Note that the Sisyphus distance also has the property that

$$\pi r_0{}^2 S_0 = \frac{1}{2}. \tag{52}$$

As in the previous case, the total packet throughput which can be supported by a single packet sink operating with perfect capture is one half.

V. PACKET BROADCASTING WITH AVERAGE POWER LIMITATIONS

A. Satellite Packet Broadcasting

In previous sections we have analyzed the performance of packet broadcasting channels and compared the performance of these channels to that of conventional point-to-point channels operating at the same peak data rate. Such a comparison is of interest in the case of channels limited by multiple access interference rather than noise, since an increase in the transmitted power of such channels will not lead to improved performance. But just as the average data rate of a packet broadcasting channel can be well below its peak data rate when it is operated at a low duty cycle, the average transmitted power of a packet broadcasting channel can be well below its peak transmitted power.

In this section we analyze the throughput of a packet broadcasting channel when compared to that of a conventional point-to-point channel of the same average power. This analysis is of interest in the case of satellite information systems employing thousands of small earth stations. For a satellite system the fundamental limitation in the downlink is the average power available in the satellite transponder rather than the peak power. Our results show that in the limit of large numbers of small earth stations, the packet throughput approaches 100 percent of the point-to-point capacity. Thus the multiple access capability and the complete connectivity (in the topological sense) of an ALOHA channel can be obtained at no price in average throughput. Furthermore, since our results suggest the use of higher peak power in the satellite transponder (while the average power is kept constant), the small earth stations may use smaller antennas and simpler receivers and modems than would be necessary in a conventional system.

In existing satellite systems the TWT output power in each transponder cannot be varied dynamically. In such systems the advantages implied by our analysis may be realized by frequency-division sharing a single transponder among several voice users and a single channel, operating in an ALOHA mode or some other burst mode, and occupying a frequency band equivalent to one or more voice users. The type of operation implied by our analysis also suggests investigation of high peak power satellite burst transponders (perhaps employing power devices similar to those used in radar systems) for use in information systems composed of large numbers of ultra-small earth stations.

B. Burst Power and Average Power

The capacity of a satellite channel can be calculated by the classical Shannon equation

$$C = W \log\left(1 + \frac{P}{N}\right) \tag{53}$$

where C is the capacity in bits (if the log is a base two logarithm), W is the channel bandwidth, P is the average received signal power at the earth station, and N is the average noise power at the earth station. Equation (53) expresses the capacity of the satellite channel under the assumption that the transponder transmits continuously.

If the channel is used in burst mode the transponder will emit power only when a data burst occurs, and the average power out of the transponder will be less than the burst power. Let D be the ratio of the average power transmitted to the power transmitted during a data burst. For a linear transponder D will equal the channel traffic G, and for a hard-limiting transponder D will equal the duty cycle of the channel. For both the unslotted and slotted ALOHA channel the duty cycle is $1 - e^{-G}$. Thus for a linear transponder[2]

$$D = G, \tag{54a}$$

while for a hard-limiting transponder

$$D = 1 - e^{-G}. \tag{54b}$$

Note that in the case of a hard-limiting transponder with small values of channel traffic, the duty cycle approaches that of a linear transponder.

If we retain P as the notation for the average signal power received at the earth station, the power received during a data burst will be P/D. Thus (53) should be modified in two ways.

<hr>

[2] Our analysis is of significance only for $G < 1$. The analysis is formally correct, however, for all G, even though the designation of the power transmitted during bursts as "peak power" becomes inappropriate for the linear transponder case when $G > 1$. (In such a situation the "peak power" is less than the average power.)

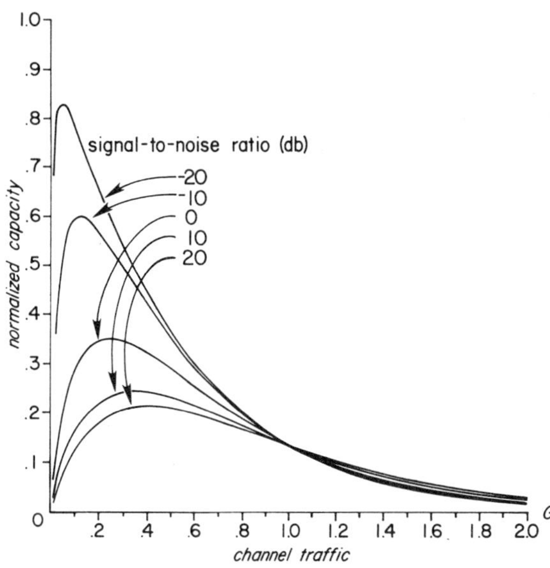

Fig. 10.   Linear transponder; unslotted channel.

1) We replace W by SW to account for the fact that the channel is only used intermittently.

2) We replace P in (53) by P/D to keep the average power of the channel fixed at P.

We should note that when we make these changes, we are assuming that the packet length of the system is long enough so that the asymptotic assumptions which are used to derive (53) still apply. In practice, this is not a problem.

With these two changes then, we have four different cases.

1) Unslotted channel, linear transponder:

$$C_1 = Ge^{-2G} W \log\left(1 + \frac{P}{GN}\right). \tag{55a}$$

2) Unslotted channel, limiting transponder:

$$C_2 = Ge^{-2G} W \log\left(1 + \frac{P}{(1 - e^{-g})N}\right). \tag{55b}$$

3) Slotted channel, linear transponder:

$$C_3 = Ge^{-G} W \log\left(1 + \frac{P}{GN}\right). \tag{55c}$$

4) Slotted channel, limiting transponder:

$$C_4 = Ge^{-G} W \log\left(1 + \frac{P}{(1 - e^{-G})N}\right). \tag{55d}$$

We have calculated the normalized capacities C_i/C for $i = 1, 2, 3, 4$ for different values of P/N, the signal-to-noise ratio of the earth station when the transponder operates continuously. The normalized capacities are plotted in Figs. 10, 11, 12, and 13 for P/N equal to -20, -10, 0, 10, and 20 dB. Of particular interest in these curves is the fact that the highest values of C_i/C occur just where we would want them to occur—for small values of channel traffic (G) and for small earth stations (low P/N). In the limit we have (for a fixed value of G)

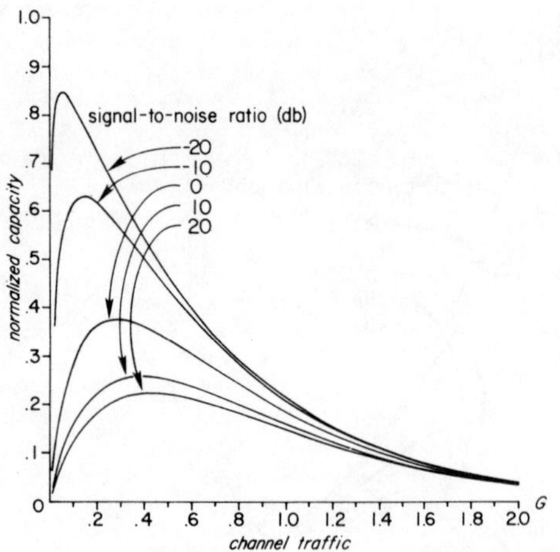

Fig. 11.   Limiting transponder; unslotted channel.

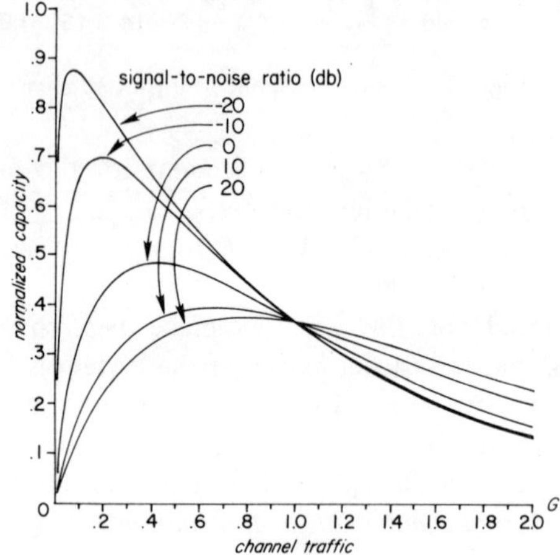

Fig. 12.   Linear transponder; slotted channel.

$$\lim_{\frac{P}{N}\to 0} \frac{C_i}{C} = \frac{S}{D}, \qquad i = 1, 2, 3, 4 \tag{56}$$

so that

1) unslotted channels, linear transponder

$$\lim_{\frac{P}{N}\to 0} \frac{C_1}{C} = e^{-2G} \tag{57a}$$

2) unslotted channels, limiting transponder

$$\lim_{\frac{P}{N}\to 0} \frac{C_2}{C} = \frac{Ge^{-2G}}{(1 - e^{-G})} \tag{57b}$$

3) slotted channel, linear transponder

$$\lim_{\frac{P}{N}\to 0} \frac{C_3}{C} = e^{-G} \tag{57c}$$

4) slotted channel, limiting transponder

$$\lim_{\frac{P}{N}\to 0} \frac{C_4}{C} = \frac{Ge^{-G}}{(1 - e^{-G})} \tag{57d}$$

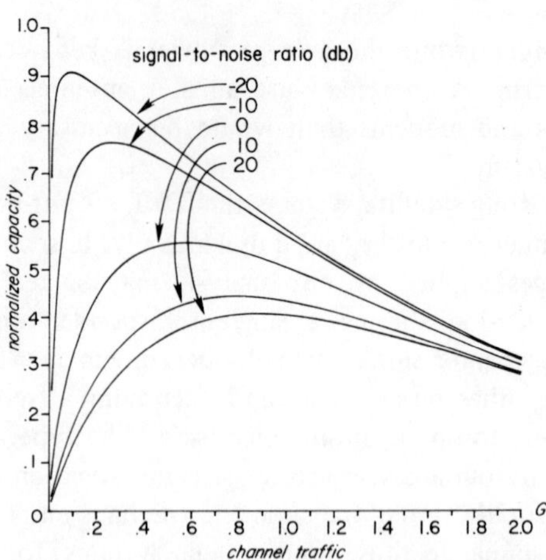

Fig. 13.   Limiting transponder; slotted channel.

and in all cases

$$\lim_{G\to 0} \ \lim_{\frac{P}{N}\to 0} \frac{C_i}{C} = 1. \tag{58}$$

Thus this multiplexing technique allows a network of small inexpensive earth stations to achieve the maximum value of channel capacity, at the same time providing complete connectivity and multiple access capability.

VI. BACKGROUND AND ACKNOWLEDGMENT

The term "packet broadcasting" was first coined by Robert Metcalfe in his Ph.D. dissertation [28]. As is often the case with simple ideas, the concept of combining burst transmission and Poisson user statistics to provide random access to a channel has occurred independently to a number of investigators. The first attempt at an analysis of such a system of which I am aware is contained in an internal Bell Laboratories memorandum by Schroeder [29], suggested by an earlier paper by Pierce and Hopper [30]. Two other early related papers were written by Costas [31] and Fulton [32]. Of course, a theoretical analysis is not necessary in order to build such a system, and anyone who has sat in a taxi listening to the staccato voice bursts of a radio dispatcher and a set of taxi drivers sharing a single voice channel will recognize the operation of a voice packet broadcasting channel using a carrier sense protocol. And even after an analysis is available, the concept of packet broadcasting may be suggested without reference to the theory [33].

The first papers analyzing packet broadcasting in the form implemented in the ALOHA System [6] assumed fixed packet throughput and a retransmission protocol as described in Section II-D-1). This approach leads to a number of questions involving optimum retransmission policy [28], the behavior of the channel with a finite number of users [39], stability of the channel [13], and transmission of long files by means of various reservation schemes [34], [44]. A comprehensive treatment of these as well as other interesting packet broadcasting questions may be found in Kleinrock [42]. In this paper we have taken a different approach by assuming a given packet traffic rather than throughput. With such a starting point, the questions mentioned above do not assume key importance in the theory, although their practical importance is not diminished.

Much of the theory of packet broadcasting was developed in two working groups sponsored by the Advanced Research Projects Agency of the Department of Defense. These groups circulated a private series of working papers—the ARPANET Satellite System notes (ASS notes) and the Packet Radio Temporary notes (PRT notes)—where many of the theoretical results described or referenced in this paper appeared for the first time. Unfortunately, the several references to ASS notes in papers subsequently published in the open literature may have produced some confusion in the minds of those trying to trace the references. Among the most significant of the ASS note and PRT note results was the first derivation of the capacity of a slotted ALOHA channel and the first analysis of the use of the capture effect in packet broadcasting, both by Larry Roberts. That note has since been republished in the open literature [41].

The results of Section III-A dealing with two different packet lengths were suggested by an ASS note written by Tom Gaarder, and the results of Section III-B dealing with the excess capacity of a slotted channel were suggested by an ASS note written by Randy Rettberg. Other problems which were first analyzed in ASS notes or PRT notes but not emphasized in this paper include various packet broadcasting reservation systems [22], [35], [36], carrier sense packet broadcasting [18], [19], and questions dealing with packet routing and protocol issues in a network of repeaters [37]. The reader interested in theoretical network protocol questions should also see Gallagher [38], although this work did not originate in an ASS note or PRT note.

The first system to employ packet broadcasting techniques was the ALOHA System computer network at the University of Hawaii in 1970. Subsequently, packet repeaters were added to the network and packet broadcasting by satellite was demonstrated in the system. Some of the people involved in the implementation and development of the system were Richard Binder, Chris Harrison, Alan Okinaka, and David Wax.

The historical relevance of [29] and [32] was pointed out to me by Joe Aein, to whom I am indebted, in spite of my embarassment at having forgotten I was thesis supervisor on the second of these papers.

REFERENCES

[1] L. G. Roberts and B. D. Wessler, "Computer network development to achieve resource sharing," in *1970 Spring Joint Comput. Conf., AFIPS Conf. Proc.,* vol. 36. Montvale, NJ: AFIPS Press, 1970, pp. 543–549.

[2] L. G. Roberts, "Data by the packet," *IEEE Spectrum,* vol. 11, pp. 46–51, Feb. 1974.

[3] W. W. Chu, "A study of asynchronous time division multiplexing for time-sharing computer systems," in *1969 Spring Joint Comput. Conf., AFIPS Conf. Proc.,* vol. 35. Montvale, NJ: AFIPS Press, 1969, pp. 669–678.

[4] F. Heart, R. Kahn, S. Ornstein, W. Crowther, and D. Walden, "The interface message processor for the ARPA computer network," in *1970 Spring Joint Comput. Conf., AFIPS Conf. Proc.,* vol. 36. Montvale, NJ: AFIPS Press, 1970, pp. 551–567.

[5] H. Frank, I. Frisch, and W. Chou, "Topological considerations in the design of the ARPA computer network," in *1970 Spring Joint Comput. Conf., AFIPS Conf. Proc.,* vol. 36. Montvale, NJ: AFIPS Press, 1970, pp. 581–587.

[6] N. Abramson, "The ALOHA system—Another alternative for computer communications," in *1970 Fall Joint Comput. Conf., AFIPS Conf. Proc.,* vol. 37. Montvale, NJ: AFIPS Press, 1970, pp. 281–285.

[7] R. E. Kahn, "The organization of computer resources into a packet radio network," in *Nat. Comput. Conf., AFIPS Conf. Proc.,* vol. 44, May 1975, pp. 177–186.

[8] N. Abramson and E. R. Cacciamani, Jr., "Satellites: Not just a big cable in the sky," *IEEE Spectrum,* vol. 12, pp. 36–40, Sept. 1975.

[9] I. T. Frisch, "Technical problems in nationwide networking and interconnection," *IEEE Trans. Commun.,* vol. COM-23, pp. 78–88, Jan. 1975.

[10] R. M. Metcalfe and D. R. Boggs, "ETHERNET: Distributed packet switching for local computer networks," *Commun. Ass. Comput. Mach.,* to be published.

[11] N. Abramson, "Packet switching with satellites," in *Nat. Comput. Conf., AFIPS Conf. Proc.,* vol. 42, 1973, pp. 695–702.

[12] R. D. Rosner, "Optimization of the number of ground stations in a domestic satellite system," in *EASCON'75 Rec.,* Sept. 29–Oct. 1, 1975, pp. 64A–64J.

[13] A. B. Carleial and M. E. Hellman, "Bistable behavior of ALOHA-type systems," *IEEE Trans. Commun.,* vol. COM-23, pp. 401–410, Apr. 1975.

[14] P. E. Jackson and C. D. Stubbs, "A study of multiaccess computer communications," in *1969 Spring Joint Comput. Conf., AFIPS Conf. Proc.,* vol. 34. Montvale, NJ: AFIPS Press, 1969, pp. 491–504.

[15] T. J. Klein, "A tactical packet radio system," in *Proc. Nat. Telecommun. Conf.,* New Orleans, LA, Dec. 1975.

[16] K. Ah Mai, "Organizational alternatives for a Pacific educational computer-communication network," ALOHA Syst. Tech. Rep. CN74-27, Univ. Hawaii, Honolulu, May 1974.

[17] R. Binder, R. Rettberg, and D. Walden, *The Atlantic Satellite Packet Broadcast and Gateway Experiments.* Cambridge, MA: Bolt Beranek and Newman, 1975.

[18] L. Kleinrock and F. A. Tobagi, "Packet switching in radio channels: Part I—Carrier sense multiple-access modes and their throughput-delay characteristics," *IEEE Trans. Commun.,* vol. COM-23, pp. 1400–1416, Dec. 1975.

[19] F. A. Tobagi and L. Kleinrock, "Packet switching in radio channels: Part II—The hidden terminal problem in carrier sense multiple-access and the busy-tone solution," *IEEE Trans. Commun.,* vol. COM-23, pp. 1417–1433, Dec. 1975.

[20] R. T. Chien, L. R. Bahl, and D. T. Tang, "Correction of two erasure bursts," *IEEE Trans. Inform. Theory,* vol. IT-15, pp. 186–187, Jan. 1969.

[21] N. Abramson, "Cyclic code groups," *Problems of Inform. Transmission,* Acad. Sci. USSR, Moscow, vol. 6, no. 2, 1970.

[22] L. G. Roberts, "Dynamic allocation of satellite capacity through packet reservation," in *Nat. Comput. Conf. AFIPS Conf. Proc.,* vol. 42, June 1973, pp. 711–716.

[23] M. J. Ferguson, "A study of unslotted ALOHA with arbitrary message lengths," in *Proc. 4th Data Commun. Symp.,* Quebec, Canada, Oct. 7–9, 1975, pp. 5-20–5.25.

[24] R. Rettberg, "Random ALOHA with slots-excess capacity," ARPANET Satellite Syst. Note 18, NIC Document 11865, Stanford Res. Inst., Menlo Park, CA, Oct. 11, 1972.

[25] N. Abramson, "Excess capacity of a slotted ALOHA channel (continued)," ARPANET Satellite Syst. Note 30, NIC Document 13044, Stanford Res. Inst., Menlo Park, CA, Dec. 6, 1972.

[26] L. Schiff, "Random-access digital communication for mobile radio in a cellular environment," *IEEE Trans. Commun.,* vol. COM-22, pp. 688–692, May 1974.

[27] H. Frank, R. M. Van Slyke, and I. Gitman, "Packet radio network design—System considerations," in *Nat. Comput. Conf., AFIPS Conf. Proc.,* vol. 44, May 1975, pp. 217–232.

[28] R. M. Metcalfe, "Packet communication," Rep. MAC TR-114, Project MAC, Massachussetts Inst. Technol., Cambridge, July 1973.

[29] M. R. Schroeder, "Nonsynchronous time multiplex system for speech transmission," Bell Lab. Memo., Jan. 19, 1959.

[30] J. R. Pierce and A. L. Hopper, "Nonsynchronous time division with holding and random sampling," *Proc. IRE,* vol. 40, pp. 1079–1088, Sept. 1952.

[31] J. P. Costas, "Poisson, Shannon, and the radio amateur," *Proc. IRE*, vol. 47, p. 2058, Dec. 1959.

[32] F. F. Fulton, Jr., "Channel utilization by intermittent transmitters," Tech. Rep. 2004-2, Stanford Electron. Lab., Stanford Univ., Stanford, CA, May 12, 1961.

[33] K. D. Levin, "The overlapping problem and performance degradation of mobile digital communication systems," *IEEE Trans. Commun.*, vol. COM-23, pp. 1342–1347, Nov. 1975.

[34] R. Binder, "A dynamic packet-switching system for satellite broadcast channels," in *Conf. Rec., Int. Conf. Commun.*, vol. III, June 1975, pp. 41-1-41-5.

[35] S. S. Lam and L. Kleinrock, "Packet switching in a multiaccess broadcast channel: Dynamic control procedures," *IEEE Trans. Commun.*, vol. COM-23, pp. 891–904, Sept. 1975.

[36] L. Kleinrock and S. S. Lam, "Packet switching in a multiaccess broadcast channel: Performance evaluation," *IEEE Trans. Commun.*, vol. COM-23, pp. 410–423, Apr. 1975.

[37] I. Gitman, "On the capacity of slotted ALOHA networks and some design problems," *IEEE Trans. Commun.*, vol. COM-23, pp. 305–317, Mar. 1975.

[38] R. G. Gallager, "Basic limits on protocol information in data communication networks," *IEEE Trans. Inform. Theory*, vol. IT-22, pp. 385–398, July 1976.

[39] M. J. Ferguson, "On the control, stability, and waiting time in a slotted ALOHA random-access system," *IEEE Trans. Commun.*, vol. COM-23, pp. 1306–1311, Nov. 1975.

[40] J. J. Metzner, "On improving utilization in ALOHA networks," *IEEE Trans. Commun.*, vol. COM-24, pp. 447–448, Apr. 1976.

[41] L. G. Roberts, "ALOHA packet system with and without slots and capture," *Comput. Commun. Rev.*, vol. 5, pp. 28–42, Apr. 1975.

[42] L. Kleinrock, *Queueing Systems, Volume 2: Computer Applications.*   New York: Wiley, 1976, pp. 360–407.

[43] I. Gitman, R. M. Van Slyke, and H. Frank, "On splitting random accessed broadcast communication channels," in *Proc. 7th Hawaii Int. Conf. Syst. Sci.–Suppl. on Comput. Nets,* Western Periodicals, Jan. 1974, pp. 81–85.

[44] W. Crowther, R. Rettberg, D. Walden, S. Ornstein, and F. Heart, "A system for broadcast communication: Reservation ALOHA," in *Proc. 6th Hawaii Int. Conf. Syst. Sci.,* Western Periodicals, Jan. 1973, pp. 371–374.

3.6.5.4
Satellite Multiaccess Schemes for Data Traffic

Simon S. Lam

IBM Thomas J. Watson Research Center
Yorktown Heights, New York 10598

ABSTRACT

Satellite systems have traditionally been designed for voice traffic. In this paper, satellite multiaccess schemes for data traffic are considered. Three general categories of schemes are identified and described: channel reservation schemes, random access schemes and packet reservation schemes. Bounds on achievable channel throughput are presented for the three classes of schemes. Finally, numerical examples are used to compare the delay-throughput performance of several specific schemes for a variety of data traffic environments.

1. INTRODUCTION

We are witnessing two rapidly growing fields in the world of communications: distributed computer networks and satellite communications. The growth in computer data traffic is evidenced by the development of packet-oriented public data networks in several countries [1]. The pace of development of communication satellite systems has accelerated markedly and satellite system costs have come down significantly [2]. In addition to potential cost reductions, satellites offer special capabilities which may be used to great advantage by communication users, in particular, data communication users.

Computer data traffic has more diverse characteristics and transmission requirements than voice traffic. The purpose of this paper is to present a taxonomy of satellite multiaccess schemes and examine their delay-throughput performance. We begin by reviewing some unique satellite characteristics. A model and a new measure for characterizing computer data traffic are then given. Next, three categories of satellite multiaccess schemes (channel reservation schemes, random access schemes and packet reservation schemes) are identified and described. Bounds on achievable channel throughput are then given. The delay-throughput performance of these schemes are compared for a variety of data traffic environments.

2. SATELLITE CHARACTERISTICS

Satellites have special capabilities which can be used to great advantage in a computer communication network. A satellite can receive signals from any earth station (multiaccess) and transmit signals to all earth stations (broadcast) in its antenna pattern. This allows implementation of schemes for dynamic allocation of the satellite transmission capacity, thus achieving statistical averaging of traffic loads over a large number of users who are geographically distributed. The multiaccess capability is especially beneficial to computer data traffic sources which are typically much more "bursty" than voice traffic. The broadcast capability also permits routing via a destination address in each transmission burst. This would result in a fully interconnected network with direct "logical" connections between all earth station pairs.

On the other hand, the satellite channel propagation delay of approximately 0.27 second will impact the computer-communication network environment in many ways. It will require modifications in error and flow control protocols at the link level, the source-to-destination level as well as any intermediate level requiring these protocols. It will increase buffer requirements not only at the two nodes connected by a satellite link but also throughout the computer-communication network. Some of these issues are discussed in [3].

The throughput of a satellite channel is defined as follows. Let C be the channel transmission rate in bits per second (bps) and let there be on the average P bits in a transmitted data block. The *channel throughput* is then defined to be the ratio of the rate of successfully transmitted data blocks multiplied by P to the rate C. Note that this definition includes as useful throughput all overhead bits in a data block as well as any unused bits in fixed length blocks (packets) obtained from segmenting variable length messages. The *channel capacity* is defined to be the maximum possible channel throughput.

We note that although channel throughput is an important factor in determining the satellite cost per unit of transmitted data, it is by no means the only criterion for evaluating a satellite multiaccess scheme.

(In fact, for a power-limited FDMA satellite system, such as SPADE [4], it may be possible to take advantage of the knowledge of a low channel activity factor to increase the effective transponder capacity, in number of channels.) Since an overriding objective is to minimize the total system cost, other factors such as earth station cost and the satellite system's impact on the overall computer-communication network environment, are important and should also be considered.

3. DATA TRAFFIC CHARACTERISTICS

Data traffic streams generated in data processing applications (time-sharing, data base inquiry-response, etc.) typically have large variability in their transmission requirements. The length of messages ranges from a single byte to thousands of bytes. One such message is often made available instantly by some control signal (e.g., carriage return of a terminal) and must be transported from source to destination within a specified delay constraint.

A Model

The following model will be used to represent the traffic characteristics and transmission requirements of a computer-communication network. Consider a finite number of data traffic sources in the network. Each source is modeled as a point process with instants of message "arrivals" being the points of interest. Each message may be segmented into one or more fixed size blocks called packets. Messages generated by a traffic source may belong to one or more classes with different message interarrival time and message length (number of packets) statistics. Source-destination delay constraints are specified for individual classes of messages. Only average delay constraints are considered in the following.

How Bursty is Data Traffic?

Computer data traffic sources are often described to be "bursty." A quantitative measure of how bursty a given data traffic source is, called *bursty factor*, is presented below [5].

The bursty nature of a data traffic source stems from more than just the randomness in message generation time and size. User-specified message delay constraints to be met for these traffic sources are actually the single most important factor in determining if data traffic sources behave in a bursty manner. Suppose we are given a data traffic source with

T = average interarrival time between messages

and

δ = average message delay constraint (excluding channel propagation delays).

The bursty factor β of the traffic source is defined to be

$$\beta = \delta/T \tag{1}$$

Note that, unlike usual measures, such as the ratio of peak-to-average data rates (PAR), this new measure is not dependent upon a specific design. It also exposes the role played by the delay constraint explicitly. Now consider a traffic source which is formed by merging together N sources with different statistics and delay constraints. The bursty factor of the aggregate source is defined to be the sum of the bursty factors of the individual sources, namely

$$\beta = \beta_1 + \beta_2 + \cdots + \beta_N \tag{2}$$

The usefulness of β is due to the following results. Suppose a channel is fixed assigned to a data traffic source with bursty factor β and all delay constraints are met. Then the resulting peak-to-average ratio satisfies

$$\text{PAR} \geq 1/\beta \tag{3}$$

Reprinted from *1977 Int. Conf. Commun.*, vol. III, June 12–15, 1977, pp. 37.1-19–37.1-24.

and the channel throughput S satisfies

$$S \leq \beta \qquad (4)$$

In other words, the bursty factor gives an upper bound on the "duty cycle" of a traffic source.

4. SATELLITE MULTIACCESS SCHEMES

For purposes of this paper, satellite multiaccess schemes for data traffic can be classified into three general categories:
(1) channel reservation schemes,
(2) random access schemes,
(3) packet reservation schemes.
All of the above are based upon a satellite channelization using FDMA or TDMA techniques. Other techniques such as code division and space division multiple access are not considered. The three categories of schemes are described below.

4.1 Channel Reservation Schemes

Included in this category are conventional satellite multiaccess schemes (for voice traffic). The satellite transponder is channelized using FDMA, TDMA or a combination FDMA-TDMA technique. The channels (forming a common pool) can then be either fixed assigned for periods of hours/days or demand assigned for periods of seconds/minutes. The traffic sources may represent individual data ports within earth stations or entire earth stations acting as data concentrators. Channels may be assigned between source-destination pairs of users as well as to traffic sources which can send packets to different destinations by attaching a destination address to each packet.

Typically, a channel is set aside for signalling among the earth stations. Access to the signalling channel can be achieved by having time slots (within a frame) fixed assigned to earth stations. Demand assignment can then be accomplished with a central controller at a master earth station or with decentralized controllers at all earth stations (e.g., SPADE [4]).

Channel reservation schemes are very efficient for voice traffic since voice conversations typically are several minutes long. The total allocation-deallocation time overhead for circuit-switching is on the order of one second---an insignificant fraction of the connect time.

The efficiency of channel reservation schemes for data depends upon the specific traffic environment. Three channel reservation schemes are distinguished for data traffic:
(1) FA -- channels are fixed assigned to traffic sources
(2) DA/Session -- channels are demand assigned for each "computer session"
(3) DA/Message -- channels are demand assigned for each message.

We next consider upper bounds on channel throughput for the three channel reservation schemes. For the scheme DA/Session, the allocation-deallocation time overhead is typically insignificant compared to the channel connect time for a session and can be ignored. For purposes of this paper, no distinction will be made between FA and DA/Session. For these schemes, the channel throughput is bounded above by the bursty factor of the given traffic source. Thus

$$S \leq \beta = \delta/T \qquad (5)$$

For the scheme DA/Message, suppose the average channel allocation and deallocation times are t_a and t_d seconds, respectively. Assume that during allocation and deallocation times the channel is idle. Let t be the average channel transmission time for a message. The channel throughput S must then satisfy

$$S \leq t/(t + t_a + t_d)$$

The message delay (assuming zero blocking probability for allocation requests) is $t + t_a$ and must satisfy

$$t + t_a \leq \delta$$

From the above two inequalities, we get the following bound

$$S \leq (\delta - t_a)/(\delta + t_d) \qquad (6)$$

The above bounds on channel throughput are plotted in Fig. 1 as a function of δ and for different values of T, t_a and t_d.

Note that Eq. (6) does not depend on the average message interarrival time T . Therefore, summing traffic sources into a single stream has no effect on the throughput bound of DA/Message. On the other hand, FA (and also DA/Session) will have a very low channel throughput if T is large compared to the delay constraint δ . In such case, the channel throughput can be improved by merging many traffic sources into the same channel.

The curves in Fig. 1 represent upper bounds on channel throughput. The actual throughput at the system design point must be smaller due to (1) queueing delays, and (2) the fact that channel speeds are available only at certain discrete values. For a demand assigned system, the difference between the actual throughput and the upper bound also depends upon the blocking probability that can be tolerated for allocation requests.

Formulas for the average message delay of FIFO and priority disciplines using a fixed assigned channel are given in [6].

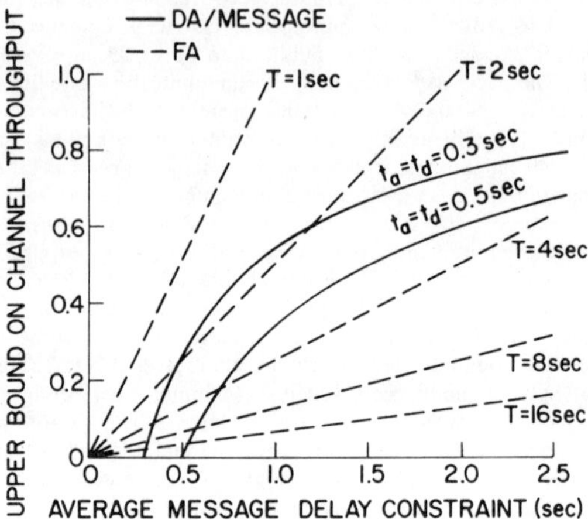

Fig. 1 Upper bounds on channel throughput for channel reservation schemes.

4.2 Random Access Schemes

Multiaccess schemes included in this category are: (1) ALOHA, which had its origin in the ALOHA System at the University of Hawaii [7], (2) slotted ALOHA and (3) Reservation-ALOHA [8]. It is assumed that a common FDMA satellite channel is shared by a population of users (earth stations). Whenever packet transmissions from different users overlap in time at the satellite, it is assumed that neither one can be correctly received (destructive interference).

ALOHA

In the ALOHA scheme, channel users are not synchronized in any way. Each user transmits a packet whenever one is ready. Each packet contains parity bits for error detection. Since a satellite channel has the broadcast capability, a user can actually monitor his own packet transmissions on the down link. If he receives a correct copy of a previously transmitted packet, he can assume that the intended receiver has also received it correctly (given that the channel has a very low error rate). At the same time, a positive acknowledgment scheme can also be used for the detection of transmission errors. In the event that two or more packets interfere with each other at the satellite (a collision), each of the users involved will detect the collision. Each will then retransmit the packet involved in the collision after a randomized delay. This randomization procedure turns out to be crucial to the throughput, delay and stability performance of such random access channels [9-11].

In [7], Abramson first derived the capacity of an ALOHA channel shared by a large number N of low-rate users (very bursty). All messages are assumed to consist of single packets only. In the limit of an infinite user population ($N \to \infty$ and for each user $\beta \to 0$), the packet birth process is a Poisson process. Abramson then made the assumption that the sum of new transmissions and retransmissions in the channel (called *channel traffic*) can be approximated by a Poisson process, which gives rise to the following relationship

$$S = Ge^{-2G} \qquad (7)$$

where

S = aggregate channel throughput in packets/packet time,
G = aggregate channel traffic in packets/packet time.

From the above equation, the maximum possible ALOHA channel throughput (under the above assumptions) is obtained at $G = 0.5$. Thus the ALOHA channel capacity for an infinite population model is

$$C_A = 1/e = 0.184$$

Slotted ALOHA

The slotted ALOHA scheme was first proposed and studied by Roberts [12]. In this case, channel users are required to synchronize their packet transmissions into fixed length channel time slots. The protocols of slotted ALOHA are just like ALOHA. However, by requiring synchronization of packet start times, packet collisions due to partial overlaps are avoided. Under the same assumptions given above for ALOHA, the relationship between channel throughput and channel traffic is given by

$$S = Ge^{-G} \qquad (8)$$

where S is maximized at $G = 1$. The resulting slotted ALOHA channel capacity for an infinite population model is twice that of the unslotted case, namely

$$C_{SA} = 1/e = 0.368$$

In [11], it was shown via simulations that the above equation is quite accurate for as few as 10 users with balanced traffic. When the traffic distribution is unbalanced with a combination of high-rate and low-rate users, the channel throughput can be considerably improved [13].

Reservation-ALOHA

The traffic environment suitable for ALOHA and slotted AL-OHA is that of a large population of low-rate bursty users and short messages (single packets). The Reservation ALOHA scheme was proposed by Crowther et al. [8] for less bursty users. In addition to time slotting, the slots are organized into frames. The duration of a frame is assumed to be greater than the satellite channel propagation delay so that each user is aware of the usage status (mine, others', unused) of time slots one frame ago. A slot is unused if it is empty or contains a collision. Those slots in the last frame which were unused are available for random access by all users just as in slotted ALOHA. A slot which had a successful transmission by user X in the last frame is off limits to everyone except user X. However, if user X fails to use it in the current frame, then that slot becomes available again for general contention in the next frame. Thus we see that this scheme will achieve very good channel throughput for users with long messages or steady arrival streams.

Under the assumption of equilibrium conditions, the throughput S_{RA} of a Reservation-ALOHA channel can be expressed in terms of the slotted ALOHA throughput S_{SA} for the contention portion of the channel [14]

$$S_{RA} = S_{SA}/(S_{SA} + 1/L) \qquad (9)$$

where L = average number of packets transmitted before a user releases his "captured" slot. The above equation is derived in [14] under certain assumptions. An example that satisfies these assumptions is an infinite population model with geometrically distributed message length (in number of packets). From Eq. (9), the capacity of a Reservation-ALOHA channel is

$$C_{RA} = C_{SA}/(C_{SA} + 1/L) \qquad (10)$$

For an infinite population model such that $C_{SA}=1/e$, we then have

$$C_{RA} = 1/(1 + e/L) \qquad (11)$$

which gives

$$1/(1+e) \le C_{RA} \le 1$$

for L ranging from 1 to ∞.

Note that in the worst case, the Reservation-ALOHA channel capacity is actually less than than of slotted ALOHA. This deficiency can be remedied if an end-of-use flag is included in the last packet before a user stops using his captured slot. Slots containing packets with end-of-use flags can then be treated as if they are unused and are available for general contention in the next frame. In this case, the Reservation-ALOHA channel throughput is [14]

$$S_{RA} = S_{SA}/(S_{SA} + (1-S_{SA})/L) \qquad (12)$$

We then have

$$C_{RA} = C_{SA}/(C_{SA} + (1-C_{SA})/L) \qquad (13)$$

and

$$C_{SA} \le C_{RA} \le 1$$

The channel capacity of Reservation-ALOHA is shown in Fig. 2 for the infinite population model with and without employing end-of-use flags.

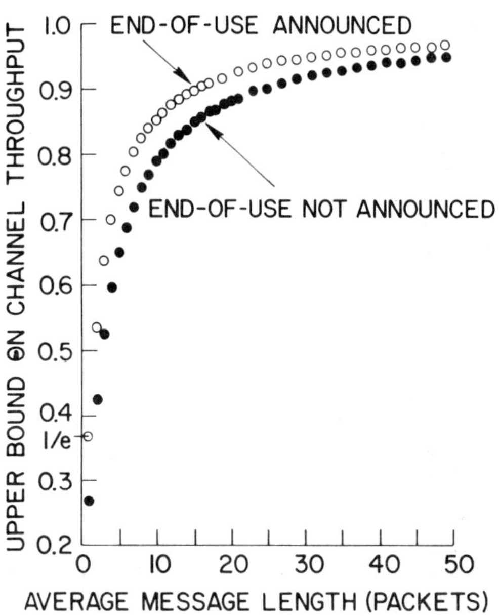

Fig. 2 Upper bounds on channel throughput for Reservation-ALOHA.

Other Performance Criteria

Apart from channel capacity, it is necessary to characterize the delay performance of any given multiaccess scheme. In addition, all random access schemes suffer from potential instability behavior and may require some form of adaptive control.

For slotted ALOHA, the delay-throughput tradeoff was first shown by Kleinrock and Lam [9]. A theory and a stability definition for characterizing stable and unstable channels for a finite population of users is given in [10] where a measure of instability for unstable channels is also introduced. A comprehensive theoretical treatment of adaptive control using a Markov decision model is given in [15]. Various heuristic schemes and their performance are presented in [16].

For Reservation-ALOHA, the tradeoff between average message delay and channel throughput is treated in [14]

4.3 Packet Reservation Schemes

Packet reservation schemes, like random access schemes, are intended for sharing a single multiaccess channel among a population of geographically distributed users (earth stations). In packet reservation schemes, the channel transmission capacity is demand assigned to individual packets or groups of packets. Because of demand assignment, a disadvantage of such schemes is that the average message delay is at least twice the satellite channel propagation delay (≥ 0.54 second).

Packet reservation schemes typically require mechanisms for making reservations and queue management, queue disciplines, as well as protocols for error control and recovery. Below we describe two packet reservation schemes [17,18] both based upon a distributed queue management method. They serve to illustrate the concepts and protocols involved. The possibility of centralized queue management using an intelligent satellite has been proposed [19] but not considered here.

A FIFO Scheme

We describe below a FIFO scheme first proposed and studied by Roberts [17]. The satellite channel is divided into time slots. For a certain number of slots used for data packets, one slot is subdivided into V small slots. The small slots are for reservation packets (as well as possibly positive acknowledgment and small data packets) to be used on a contention basis with the slotted ALOHA technique.

A single global queue consisting of messages holding reservations is maintained via a distributed queue management method. It makes use of the satellite broadcast capability such that a reservation packet successfully transmitted with no interference can be received by all users. Each user maintains his copy of the global queue status. It is sufficient for each user to record only the queue size and the queue positions of his own reserved messages. The queue discipline proposed by Roberts is FIFO according to the order reservation packets are received.

For a currently inactive user who wants to join the queue, it is necessary for him to first acquire sufficient queue status information (queue synchronization). In this scheme, the current queue length information may be supplied in the header of each data packet transmitted. Alternatively, it may be announced periodically by a master earth station. Note that such queue length information is one propagation delay old when received. In order for a currently inactive user to acquire queue synchronization, he must update the queue length information with reservation packets received within a satellite propagation time just prior to receiving the queue length information.

To maintain coordination between all users, it is necessary and sufficient that each reservation packet which is received correctly by any user is received correctly by all users. This can be assured by properly encoding the reservation requests. A simple strategy proposed by Roberts is to send parity-checked copies of each reservation request in triplicate within a reservation packet.

Due to the decentralized nature of queue management, the impact of any error in a user's queue status information is merely to delay some data packets momentarily. Error recovery can be accomplished by requiring any user who receives a reservation packet with error or who has been involved in a collision in a reserved data slot to discard his acquired reservations and reacquire queue synchronization.

A Round-Robin Scheme

We describe here a round-robin scheme first proposed and studied by Binder [18]. The satellite channel is divided into time slots and organized into frames of M slots each. The frame time is required to be longer than the satellite propagation time. Also, the number N of users is required to be less than or equal to M. Each user is fixed assigned a time slot within the frame. Each user sends information concerning his current queue length in the header of the data packet transmitted into his fixed assigned slot.

Distributed queue management such as described above is assumed. The global queue status consists of the queue lengths of the active users. Any unassigned slots as well as unused slots (assigned to currently inactive users) within a frame are used by the active users in a round-robin fashion.

A user who has been inactive can transmit into his fixed assigned slot to deliberately generate a conflict. Such conflict is detected by all users and the protocol dictates that following a collision only the owner of the slot can use it in the next frame. The queue status is announced by a master earth station at the beginning of each frame. Thus a previously inactive user can acquire queue synchronization by just listening in for one complete frame.

Error control and recovery can be done in a similar fashion as described above. A special feature of this scheme is that even if a user is in the process of acquiring queue synchronization, he can use his

fixed assigned slot in the mean time. Thus in the worst case, his throughput is equal to that of a single TDMA channel.

Other Scheduling Disciplines

Apart from the FIFO and round-robin queue disciplines given above, other scheduling disciplines can be used to achieve response time characteristics suitable for a given traffic environment and specific delay requirements. For example, head-of-the-line priority can be imposed to give shorter delays to control and interactive traffic at the expense of longer delays for other messages.

Reservation Overhead

The maximum channel throughput of a packet reservation scheme is $1-\alpha$, where α is the minimum fraction of channel capacity needed to accommodate the reservation request traffic. As illustrated in the above schemes, two methods (slotted ALOHA or fixed assignment) can be employed for accessing the reservation channel. With slotted ALOHA, each message requires one reservation packet. The minimum overhead is thus

$$\alpha = 1/(1+C_{SA}VL) \tag{14}$$

where C_{SA} is the slotted ALOHA channel capacity, V is the ratio of data packet size to reservation packet size and L is the average message length (in number of packets). We make two observations: (1) α decreases as L decreases, and (2) α is independent of the number of users. The first observation says that a slotted ALOHA reservation channel incurs a large overhead when the data traffic consists of mostly short messages (e.g., single packets). The second observation says that it can be used for a very large user population without any performance degradation.

With a fixed assigned reservation channel, the minimum overhead is

$$\alpha = N/(MV) \tag{15}$$

where the number N of users is required to be smaller than the frame size M. Observe that α increases with N but is independent of the average message length. A fixed assigned reservation channel has other advantages over slotted ALOHA. First, the reservation packets suffer no contention delay. Second, unlike slotted ALOHA, adaptive channel control is not required. Equations (14) and (15) are shown in Fig. 3 in which $1-\alpha$ is plotted as a function of the number of users with L and M as parameters.

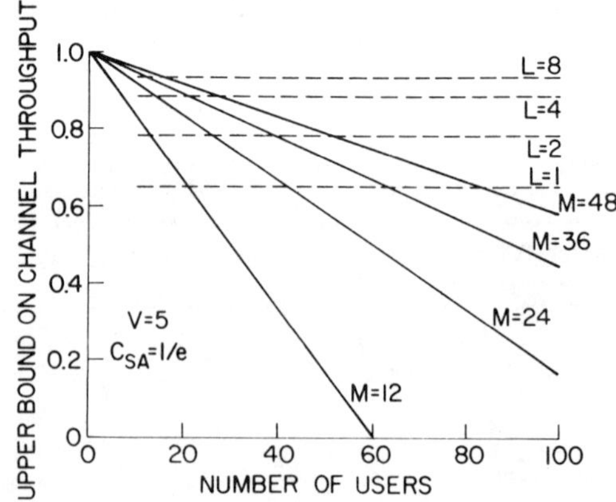

Fig. 3 Upper bounds on channel throughput for packet reservation schemes.

5. PERFORMANCE COMPARISONS

The applicability of a satellite multiaccess scheme depends upon the particular characteristics and delay requirements of the traffic environment under consideration. In this section, we shall compare, via some numerical examples, the 3 categories of satellite multiaccess schemes for a variety of data traffic environments.

Delay formulas for fixed assigned TDMA and slotted ALOHA are taken from [6] and [9] respectively. The expected delay of a message in a packet reservation scheme is the sum of the expected delay D_1 incurred by its reservation request and the expected delay

D_2 incurred by the data message itself after the reservation has been made. To calculate D_1 and D_2 we shall regard the channel (with capacity C bps) to be split up into two channels: a data channel with capacity $C\gamma$ bps and a reservation channel with capacity $C(1-\gamma)$ bps. D_1 and D_2 are then calculated separately.

In Fig. 4, we show the delay-throughput performance of representatives from the three categories of multiaccess schemes: (1) slotted ALOHA, (2) FIFO with a fixed assigned reservation channel, and (3) a fixed assigned TDMA channel; they are labeled as ALOHA, FIFO and TDMA respectively. We assume that there are 10 users sharing a 50 KBPS satellite channel with a channel propagation delay of 0.27 second. All messages consist of single packets of 1125 bits each and in the case of FIFO, a frame size of 12 slots is assumed with $\gamma=1/6$ and $V=5$. In this case, we observe from Fig. 4 that the performance of fixed assigned TDMA is the best, except when the channel throughput is less than 0.2 where slotted ALOHA gives a smaller delay.

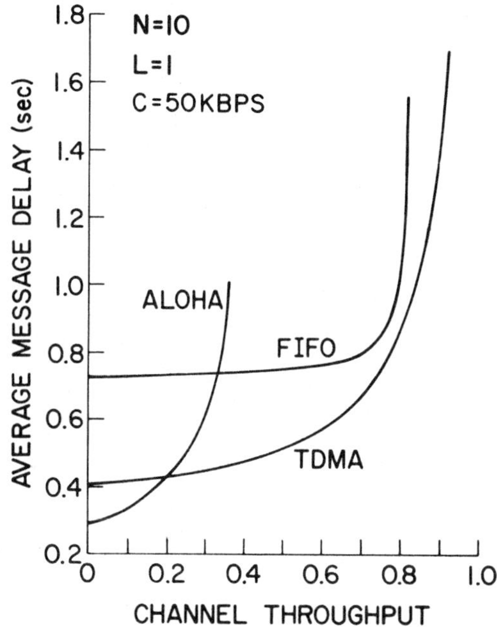

Fig. 4 Delay-throughput tradeoff for 10 users and
short messages.

In Fig. 5, we show the effect of decreasing the data rate of each user while the number of users is increased to 50 so that the total data rate remains the same as before. Note that the bursty factor of each user is now $1/5$ of the bursty factor in the previous case. This predicts the significantly degraded performance of fixed assigned TDMA shown in Fig. 5. FIFO with a fixed assigned reservation channel (frame size $= 24$, $\gamma = 5/12$) also has poor performance due to the large reservation overhead needed for a large number of users. FIFO with a slotted ALOHA reservation channel (labeled as FIFO* in Fig. 5) gives better performance but the reservation overhead is still quite large; $\gamma=0.4$ assuming a value of 0.3 for C_{SA} in Eq. (14). The delay performance of slotted ALOHA is independent of the increase in the number of users and is most suitable for this environment if a throughput of about 0.3 or less is acceptable.

Now suppose we are back to having 10 users but the data traffic consists of single-packet and eight-packet messages in equal number. The average message length L is increased to 4.5 packets. This implies that the bursty factor of each user is $1/4.5$ of that in Fig. 4 and predicts the poor performance of TDMA compared to FIFO in Fig. 6. The delay curve for slotted ALOHA is plotted using the delay model for multi-packet messages in [11]. We see that it performs much worse than FIFO in this traffic environment except for a throughput of less than 0.2 . Simulations also seem to indicate that when there are more multi-packet messages in the input traffic, the slotted ALOHA channel becomes more "unstable" [11].

A large disparity in delay requirements which often exists between different classes of messages can be capitalized upon to improve

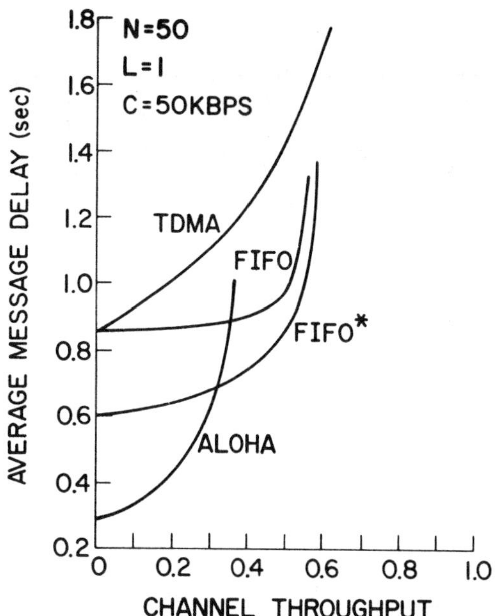

Fig. 5 Delay-throughput tradeoff for 50 users and
short messages.

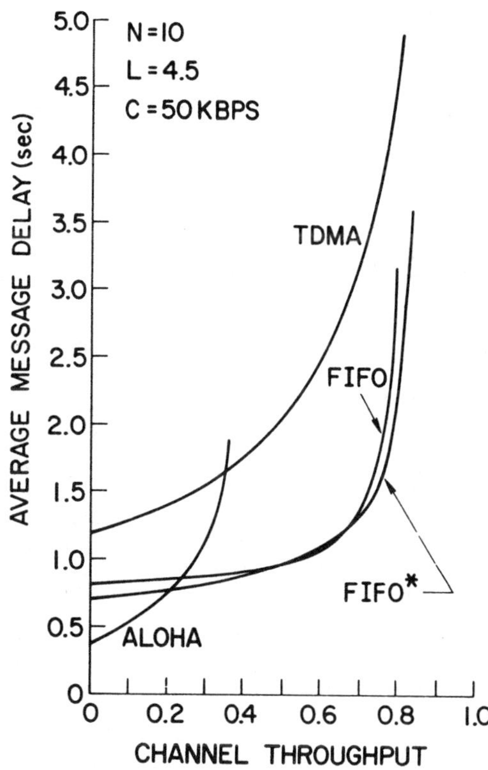

Fig. 6 Delay-throughput tradeoff for 10 users and
long messages.

channel throughput through use of an appropriate scheduling discipline in a packet reservation scheme. We shall illustrate this point by considering a traffic environment in which 20 percent of the messages are single packets while messages having 2, 3, 4, 5, 6, 7, 8 or 16 packets make up 10 percent each of the total input. In Fig. 7, we have shown the expected message delay versus message length at a throughput S=0.2, 0.6 and 0.8 for two scheduling disciplines: FIFO and shortest-message-first (SMF). Notice that the delay of short messages in SMF is significantly lower than that for FIFO. This is at the expense of a long delay (10 seconds) for messages which are 16 packets long. If a 10 second delay for these long messages is acceptable to the users, then SMF is far superior to FIFO for this traffic environment.

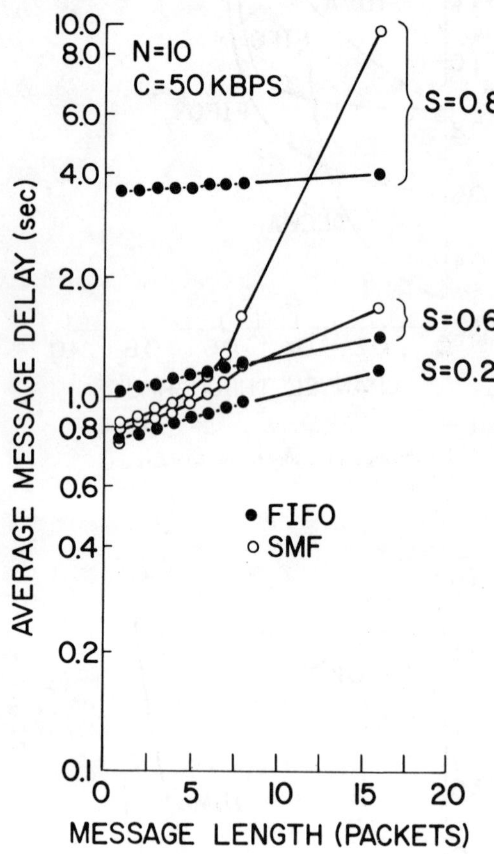

Fig. 7 Delay vs message length for two scheduling schemes.

6. CONCLUSIONS

We have identified and described three general categories of satellite multiaccess schemes that can be used for data traffic: channel reservation schemes, random access schemes and packet reservation schemes. A new measure called bursty factor was introduced for characterizing data traffic sources. Bounds on achievable channel throughput were presented for the three classes of schemes. Numerical examples were used to explore the suitability of specific multiaccess schemes in a variety of traffic environments.

It was found that for nonbursty traffic sources (large bursty factor), fixed assigned channels are adequate (see Fig. 4). For bursty traffic sources (small bursty factor), packet reservation schemes are suitable for multi-packet messages (see Fig. 6); slotted ALOHA is suitable for short single-packet messages and is required when the average delay constraint is less than 0.6 second (see Fig. 5). Finally, if the traffic sources are bursty but the average delay constraint is large (several seconds), then the scheme DA/Message is also applicable (see Fig. 1).

Our main emphasis has been the achievable satellite channel throughput. We should, however, keep in mind that although achievable channel throughput is an important factor in determining the satellite cost per unit of data transmitted, it is not the only criterion for evaluating multiaccess schemes. If the total system cost is to be optimized, then factors such as earth station cost (as a function of antenna size, intelligence, and power requirement etc.) and the impact of the multiaccess scheme on the computer-communication network environment should also be evaluated.

REFERENCES

[1] *Proceedings of the Third International Conference on Computer Communication*, Toronto, August 1976.

[2] Abramson, N. and E. Cacciamani, "Satellites: not just a big cable in the sky," *IEEE Spectrum*, Sept. 1975, pp. 36-40.

[3] Binder, R., J. McQuillan and R. Rettberg, "The Impact of Multi-Access Satellites on Packet Switching Networks," *EASCON'75 Convention Record*, IEEE, New York, 1975, pp. 63-A to 63-F.

[4] Edelson, B. and A. Werth, "SPADE system progress and application," *COMSAT Technical Review*, Vol. 2, No.1, 1972, pp. 221-242.

[5] Lam, S.S., "A New Measure for Characterizing Data Traffic," IBM Thomas J. Watson Research Center, Yorktown Heights, N.Y., Research Report RC6242, Oct. 1976.

[6] Lam, S.S., "Delay Analysis of a Packet-Switched TDMA System," *NTC 76 Conf. Rec.*, IEEE, New York, pp. 16.3-1 to 16.3-6.

[7] Abramson, N., "THE ALOHA SYSTEM--Another alternative for computer communications", *AFIPS Conf. Proc.*, Vol. 37, AFIPS Press, Montvale, N.J., 1970, pp. 281-285.

[8] Crowther, W. et al., "A System for Broadcast Communication: Reservation-ALOHA," *Proc. 6th HICSS*, University of Hawaii, Honolulu, January 1973.

[9] Kleinrock, L and S.S. Lam, "Packet-switching in a slotted satellite channel," *AFIPS Conf. Proc.*, Vol. 42, AFIPS Press, Montvale, N.J., 1973, pp. 703-710.

[10] Kleinrock, L. and S.S. Lam, "Packet Switching in a Multiaccess Broadcast Channel: Performance Evaluation, " *IEEE Trans. on Commun.*, Vol. COM-23, April 1975, pp. 410-423.

[11] Lam, S.S., "Packet Switching in a Multi-Access Broadcast Channel with Application to Satellite Communication in a Computer Network," Ph.D. Dissertation, Computer Science Dept., Univ. of California, Los Angeles, March 1974.

[12] Roberts, L., "ALOHA Packet System with and without Slots and Capture", ARPA Network Information Center, Stanford Research Institute, Menlo Park, California, ASS Note 8 (NIC 11290), June 1972.

[13] Abramson, N. "Packet switching with satellites," *AFIPS Conf. Proc.*, Vol. 42, AFIPS Press, Montvale, N.J., 1973, pp. 695-702.

[14] Lam, S.S., "Performance of a Multiaccess Scheme for Satellite Packet Switching: Reservation-ALOHA," to appear.

[15] Lam, S.S. and L. Kleinrock, "Packet Switching in a Multiaccess Broadcast Channel: Dynamic Control Procedures," *IEEE Trans. on Commun.*, Vol. COM-23, Sept. 1975, pp. 891-904.

[16] Lam, S.S. and L. Kleinrock, "Dynamic control schemes for a packet switched multi-access broadcast channel," *AFIPS Conf. Proc.*, Vol. 44, AFIPS Press, Montvale, N.J., 1975, pp.143-153.

[17] Roberts, L., "Dynamic allocation of satellite capacity through packet reservation," *AFIPS Conf. Proc.*, Vol. 42, AFIPS Press, Montvale, N.J., 1973, pp. 711-716.

[18] Binder, R., "A Dynamic Packet-Switching System for Satellite Broadcast Channels," *ICC 75 Conf. Rec.*, IEEE, New York, 1975, pp. 41-1 to 41-5 .

[19] Ng, S.F.W. and J.W. Mark, "A Multiaccess Model for Packet Switching with a Satellite Having Some Processing Capability," *IEEE Trans. on Commun.*, Vol. COM-25, Jan. 1977, pp. 128-135.

Most present satellite communication systems provide a transparent repeater channel to the users of the system. The channel characteristics are static and play a passive role in the communications link. Such links are only reconfigurable by the activation of a nonreal time-switching operation. On-board processing represents a set of generic techniques, involving real or near-real time processing or switching, that can be used on the satellite to provide an improved performance level in the total communications system. There are several potential benefits from on-board processing:

1) Increased Connectivity: This can be achieved by changing the transmission paths or dynamically rerouting messages.

2) Increased Capacity: This is accomplished by improving the satellite fill factor.

3) Increased Communications Link Efficiency: This can be achieved by separating the uplink and downlink designs.

4) Increased Flexibility: This is achieved by the dynamic reconfiguration capability.

There are three generic types of processors: RF, bit stream, and full baseband. Each serves a different role, has different design factors, and provides increasing levels of flexibility. The applications to be served will play a significant role in determining the type of processing that is most suitable for a particular system design. Reference [1] provides an introduction to on-board processing.

A. RF Processor

An RF processor deals directly with the uplink RF signal and does all processing at RF. No translations of the signal to a low frequency IF or baseband need be performed, although a simple frequency translation may be used to achieve operation at a convenient frequency. In its simplest form, an RF processor would be a satellite switch matrix having n inputs and m outputs with the capability of connecting any input to any output. Such a switch could be used to improve the connectivity in a TDMA system.

The basic design factors for such a processor are as follows.

1) Frequency: The frequency at which the RF signal enters and leaves.

2) Bandwidth: This is related to the data rate or bandwidth of the signal being switched.

3) Mass: A lightweight design is desirable.

4) Reliability: Such a processor should have a long life and should have a fail-safe mode; that is, if failure occurs, it should not cause major degradation in system performance.

5) Power: The process should use a minimum of power and be active only when processing the signal.

6) Speed: This relates to the rate at which the switch can change state.

Such processors still provide a transparent link and the uplink and downlink are coupled as in a satellite with a static switch.

The most common form of an RF processor is satellite switched TDMA (SS-TDMA). Reprint Paper 3.6.6.1 and [2]–[9] discuss various aspects of SS-TDMA. There are currently no operational SS-TDMA systems, but a 4×4 switch is planned for Advanced Westar [6].

B. Bit Stream Processor:

The bit stream processor has the capability of taking the RF signal, translating it to IF, demodulating the digital bit stream, and then using the demodulated bit stream to remodulate a carrier at IF or RF. The remodulation operation is done on a symbol by symbol basis using hard decision logic. No use is made of the signal structure to control the processing, and a bit stream processor has essentially no storage capability. These processors are often referred to as regenerative repeaters. Such a scheme could, for example, employ a DPSK uplink and a coherent PSK downlink. The major difference between this form of processor and the RF processor is that the uplink and downlink are uncoupled. The TWTA transmits just the remodulated signal and not the uplink noise. The uplink noise effects result in decision errors. Separating the uplink and downlink allows an independent optimization of each link and a significant improvement in the overall system performance. An example of the possible improvement is shown in Fig. 1 of Reprint Paper 3.6.6.2. Reprint

Paper 3.6.6.2 and [10]-[15] discuss various aspects of on-board repeaters. Two military satellites have on-board processors in orbit. The LES 8/9 satellite processors perform a demodulation, formatting, and modulation function.

C. Full Baseband Processor

The most complicated form of processor is a full baseband processor. In the general case of this realization, the RF signal would be demodulated, decoded using soft decisions, stored in buffers, reformatted, re-encoded and remodulated. The processor could use information in the decoded signal to route the messages in a dynamic fashion or otherwise specify the processing.

One potential application of this form of processor would be in a mobile satellite system, where there are many users with low-gain antennas and low-power transmitters. In this case, the uplink could be a random-access mode with low carrier power to noise spectral density (C/N_0). By using the processor, the efficiency of the access scheme could be increased, and an optimal use of uplink C/N_0 could be made. In addition, if the downlink is to a few large earth stations, TDM may be employed via multiple-beam antennas. The processor would provide the routing and provide the required change in modulation.

Another scheme involves the use of FDMA on the uplink and TDMA on the downlink. This would simplify the earth station complexity, make optimum use of the satellite transmission capability, and timing could be centrally provided by the satellite.

A third area of application would be in computer communications. In this case, the processor would act as a message switch or processor for a packet-type network.

Currently there are no full baseband processors in operation. Lincoln Laboratory is developing a processor for UHF anti-jam communications [17]. Chapter V of [18] discusses current activities in the on-board processing area. With the rapid developments in LSI technology, microprocessors, and surface acoustic wave (SAW) and charge coupled devices (CCD), an increased technology base will be available to develop both bit-stream and full baseband processors, and they will become more widely used in specialized systems.

REFERENCES

[1] R. Y. Huang and P. Hooten, "Communications satellite processing repeaters," *Proc. IEEE,* vol. 59, no. 2, pp. 238-252, Feb. 1971.

[2] X. Rozec and F. Assal, "Microwave switch matrix for communications satellites," *Proc. Int. Communications Conf.,* pp. 35-13-17, June 1976.

[3] Y. Ito, Y. Urano, T. Muratani, and M. Yamaguchi, "Analysis of a switch matrix for SS/TDMA system," *Proc. IEEE,* pp. 411-419, Mar. 1977.

[4] G. Dill, Y. Tsuji, and T. Muratani, "Application of SS/TDMA in a channelized satellite," *Proc. Int. Communications Conf.,* pp. 51-1-51-5, June 14-16, 1976.

[5] W. G. Schmidt, "Satellite-switched TDMA: Transponder-switched or beam-switched?" AIAA Paper No. 74-460, Los Angeles, CA, Apr. 22-24, 1974.

[6] J. Ramasastry, W. Heiser, E. Levine, and R. Markham, "Western Union's satellite-switched TDMA advanced Westar system," *Proc. 7th AIAA Communications Satellite Systems Conf.,* San Diego, CA, pp. 497-506, Apr. 24-27, 1978.

[7] R. P. Liccini, "Hardware tradeoffs for a baseline spacecraft switched time division multiple access (SSTDMA) communication system," *EASCON '77.*

[8] N. Shimasaki and R. Rapuano, "Synchronization for a communications distribution center on-board a satellite," *Conf. Record Int. Conf. Communications,* pp. 42-20-42-25, 1971.

[9] D. P. Dobson and A. E. Ring, "An LSI controller for satellite switched TDMA," *Int. Telemetering Conf.,* Los Angeles, CA, pp. 56-70, Sept. 28-30, 1976.

[10] S. Campanella, F. Assal, and A. Berman, "On board regenerative repeater," *Proc. Int. Communications Conf.,* pp. 121-125, June 1977.

[11] J. T. Chiao and F. Chetnik, "Satellite regenerative repeater study," *Canadian Proc. Communication and Power Conf.,* pp. 222-225, Oct. 1976.

[12] L. Cuccia, "Modern transponder technology for baseline designs of data processing and switching communications satellites," *IAF,* 1976.

[13] L. Cuccia, R. Davies, and W. Matthews, "Baseline considerations of beam switched SS-TDMA satellites using baseband matrix switching," *Proc. Int. Communications Conf.,* pp. 126-131, June 1977.

[14] J. Katz and K. Schneider, "Processing methods to suppress effects of uplink multipath in satellite systems," *Proc. Int. Communications Conf.,* pp. 138-143, June 1977.

[15] K. Nosaka and T. Muratani, "New satellite transponder with regenerative repeater and higher order DPSK for TDMA/SDMA system," *2nd Int. Conf. Digital Satellite Communications,* Paris, France, 1972.

[16] L. J. Collins, L. R. Jones, D. R. McElroy, D. A. Siegel, W. W. Ward, and D. K. Willim, "LES 8/9 communication system test results," *Proc. AIAA 7th Communication Satellite Systems Conf.,* San Diego, CA, pp. 471-478, Apr. 24-27, 1978.

[17] L. Metzger, "On-board satellite signal processing," vol. 1, Technical Note 1978-2, vol. 1, Lincoln Laboratory, Jan. 1978.

[18] J. B. Woodford, Ed., "Communications satellite technology: State of the art and development opportunities," Aerospace Report ATR-78 (7735)-1, July 1978.

3.6.6.1

SATELLITE-SWITCHED TIME-DOMAIN MULTIPLE ACCESS

Takuro Muratani

COMSAT Laboratories, Clarksburg, Md.

ABSTRACT

Three key elements of the SS-TDMA system have been discussed. For the microwave switch matrix various types of redundancy have been presented and evaluated. The distribution control unit, consisting of a small number of MOS/LSI chips, will be able to achieve the reliability objective with small weight and power consumption. The frame synchronization techniques have also been categorized and evaluated.

A capacity comparison of SS-TDMA and FDMA/TDMA satellites shows an increase of over 30 percent in available capacity through the adoption of the SS-TDMA system.

INTRODUCTION

A primary problem in all telecommunications systems is to obtain sufficient frequency spectrum in which to carry the traffic. Multiple spot-beam satellite antennas and orthogonal polarization are two practical means of increasing the usable spectrum. However, when multiple spot-beam antennas are used, care must be exercised in the system design to maintain the interconnectivity among the earth stations. Two methods of providing this interconnectivity have been considered.

One solution is to divide the whole frequency spectrum into sub-bands by means of filters. The required interconnectivity is obtained by arranging the connection from these filters to the output beams. This approach requires a number of channelizing filters at least equal to the square of the number of beams.

A second solution is to use satellite-switched time-domain multiple access (SS-TDMA). Excellent interconnectivity may be provided by cyclically interconnecting the TDMA signals among beams in rapid sequence via an onboard time-domain switch matrix. The space segment of this system comprises a microwave switch matrix and a programmable distribution control unit. The microwave switch matrix is controlled by a stable time base according to the programmable memory contents in the distribution control unit. The earth stations in the network synchronize their transmissions to the onboard switch sequence and communicate with those in other beams by transmitting TDMA bursts in proper timing to the sequence.

Since the SS-TDMA system[1] was first proposed, several detailed configurations of this system have been analyzed and proposed.[2-8] In addition to these efforts, laboratory analysis has also been performed on several variations of the SS-TDMA system and its future trends.

SS-TDMA CONCEPT

There are three key operational elements in this system. The first is the microwave switch matrix, which will perform the actual switching operations used to interconnect the various TDMA signals among the multiple antenna beams. The second element is the distribution control unit, which systematically controls the distribution path of the TDMA signals passing through the switching matrix. The third element is the network acquisition and synchronization control.

An important requirement for proper functioning of the system is the derivation of synchronization information directly from the satellite. The size of the network and high data rates impose stringent performance characteristics on the earth station synchronizer. The basic idea of SS-TDMA is simply shown in Figure 1. Figure 2 shows the transmit frame format. TDMA signals from a zone are cyclically interconnected to other beams or zones so that the set of transponders appears to

This paper is based upon work performed in COMSAT Laboratories under the sponsorship of the International Telecommunications Satellite Organization (INTELSAT). Views expressed in this paper are not necessarily those of INTELSAT.

have beam-hopping capability. A sync
window or a reference window is usually
required to synchronize the TDMA signals
to the onboard switch sequence.

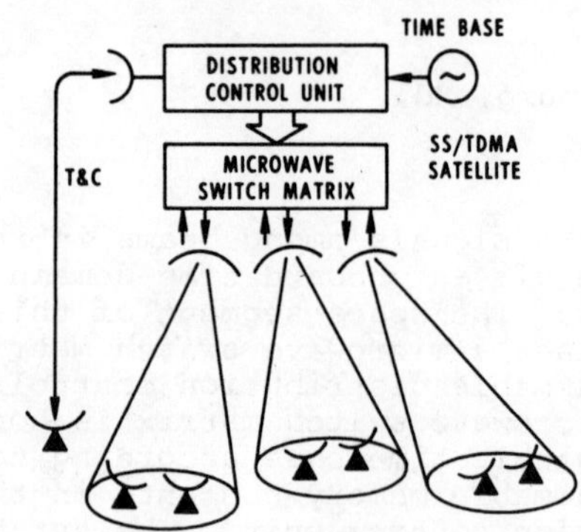

Figure 1. SS-TDMA System

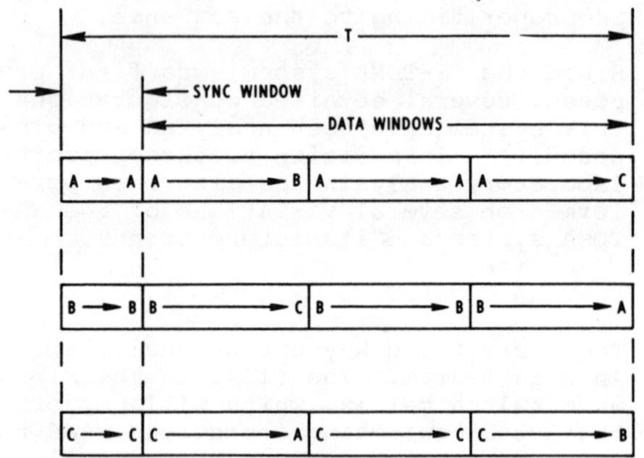

Figure 2. Transmit Frame Format

TIME SLOT ASSIGNMENT

The frame format of the onboard switch
matrix shall be controlled in the manner
which provides a maximum traffic handling
capability.

Figure 3 is a traffic matrix among
N spot-beam coverages. In this figure
t_{ij} is the traffic from the ith zone to
the jth zone, expressed in terms of the
number of telephone channels; S_i is the
total traffic originated from the ith
zone; and R_j is the received traffic
destined to the jth zone. The respective
outgoing and incoming traffic per zone
are

$$S_i = \sum_{j=1}^{N} t_{ij} \qquad (1)$$

$$R_j = \sum_{i=1}^{N} t_{ij} \qquad (2)$$

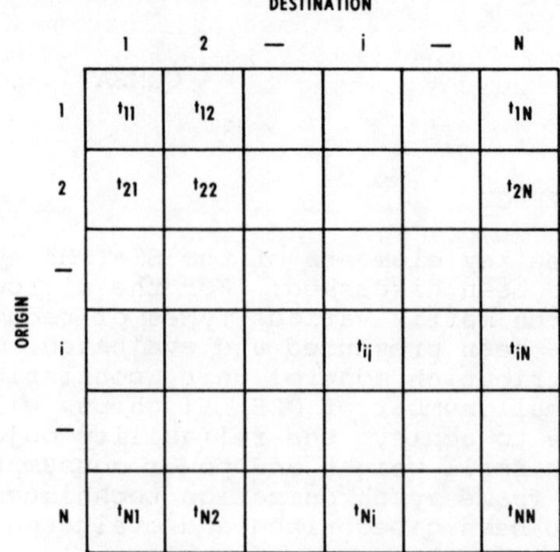

Figure 3. Traffic Matrix

When the traffic is transmitted via the
non-blocking switch matrix, a k-second
time slot is assigned to each channel in
the T-second TDMA frame. To efficiently
use the time frame, all the traffic which
is expressed in Figure 3 shall be transmitted within a time T_R which should be
made as short as possible. Clearly T_R is
bounded as follows:

$$T_R \geq k[\max(S_1, \ldots, S_i, \ldots, S_N; R_1,$$

$$\ldots, R_j, \ldots, R_N)] \qquad (3)$$

Now the question is whether or not an
equal sign can be obtained in equation (3) for a given traffic matrix among
beams. Fortunately this is always possible, and the following statement[9] can be
verified:

> The necessary and sufficient time
> to communicate all of the traffic
> in the traffic matrix is k × max
> $[(S_1, \ldots, S_i, \ldots, S_N;$
> $R_1, \ldots, R_j, \ldots, R_N)]$.

This result guarantees that the most efficient format can always be achieved,
since it is impossible to communicate all
of the traffic in a period shorter than
the required transmitting or receiving
time duration for the spot-beam zones
with the maximum traffic. Also, this
result implies that onboard storage and
forward memory are not required to
achieve the maximum traffic handling
capacity.

An example of the time slot assignment
required to meet long-range predictions
of future traffic is shown in Figure 4.
This frame format has been derived from
the particular traffic matrix in Figure 5. There is no vacant time slot in
the transmit format originating from
zone 4, which carries the greatest amount
of traffic among all zones.

1. EAST HEMI.	TO 4	TO 5	TO 1	TO 2	TO 2 / TO 5	TO 3
2. WEST HEMI.	TO 2	TO 3	TO 5	TO 1	TO 1 / TO 4	TO 4
3. NORTH EUROPE	TO 5	TO 2	TO 4	TO 4		TO 1
4. NORTH AMERICA-1	TO 1	TO 4	TO 3	TO 5	TO 3 / TO 2	TO 2
5. NORTH AMERICA-2	TO 3	TO 1	TO 2	TO 4	TO 1	

(overall frame length = 263)

Figure 4.   Transmit Frame Format

	DESTINATION					
ORIGIN	1	2	3	4	5	SUBTOTAL
ZONE 1	44	40	21	71	71	247
ZONE 2	40	48	56	45	45	234
ZONE 3	21	56	0	63	63	203
ZONE 4	71	45	63	56	28	263
ZONE 5	71	45	63	28	0	207
SUB-TOTAL	247	234	203	263	207	1154

Figure 5.   Atlantic Traffic Matrix
(one unit = 100 channels at
end 1980)

SWITCH MATRIX

The microwave switch matrix is one of
the key elements in the SS-TDMA system.
The choice of matrix configuration shall
consider its reliability, insertion
loss, isolation, and transmission
characteristics.

Matrix Reliability

To achieve the required matrix reliability, it is essential to use a redundant
design which is categorized as follows:

   a. <u>diode redundancy</u>. A switch matrix
is composed of a set of basic switching
elements such as SPST (single-pole,
single-throw), SPDT (single-pole, double-
throw), or DPDT (double-pole, double-
throw) switches. These switching
elements, made with PIN diodes, seem
promising because of their excellent ON/
OFF insertion loss ratios. Figure 6a
shows a SPST switch formed via a quadrant of diodes. Figure 6b shows an example of a SPDT switch, and Figure 6c
shows an example of a DPDT switch. These

switches are operable against any one
fixed state diode failure. However, to
achieve the transmission characteristics
required to accommodate this type of
redundancy, extensive work must be
undertaken.

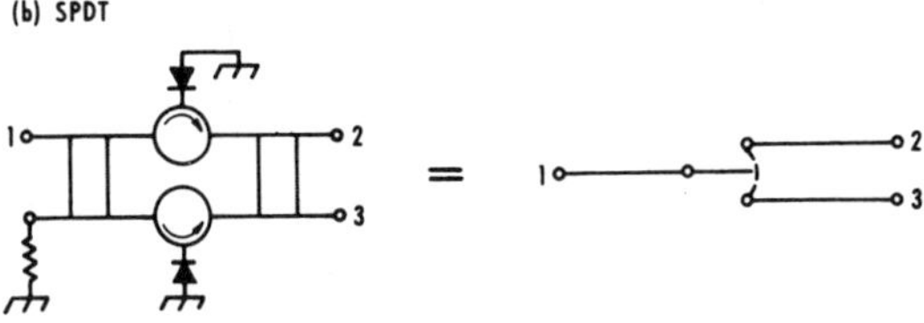

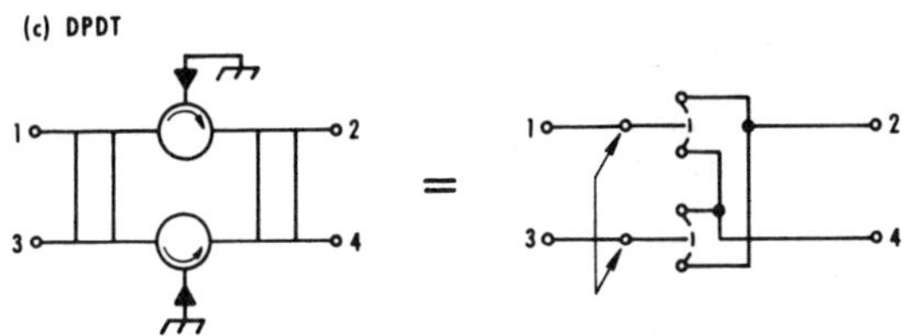

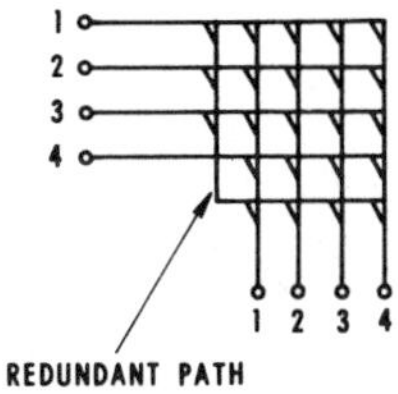

Figure 6.   Examples of Diode
Level Redundancy

   b. <u>element level redundancy</u>. A switch
matrix can be composed of redundant
switch elements. For instance, an
(N + 1) x (N + 1) cross-bar-type matrix
is shown in Figure 7a. It can be operated
as an N x N switch matrix which can overcome fixed state failure of any one
element. Another example, shown in
Figure 7b, is a rearrangeable switch
matrix with β-elements (DPDT switch)
which provides an efficient redundancy
technique by adding only N/2 extra β-
elements to the basic nonredundant N x N
matrix.

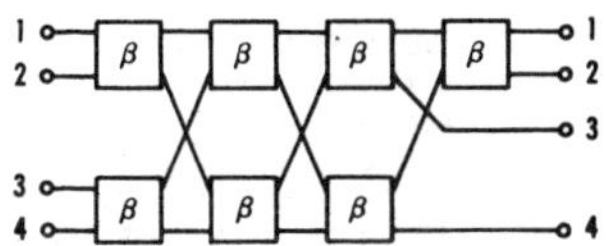

Figure 7.   4 x 4 Matrix with
Redundant Element

c. <u>matrix level redundancy</u>. Multiple N x N matrices used in parallel or in cascade are another method of achieving redundancy. However, the parallel scheme requires the use of another set of switches at the input and/or output of the matrix. These added switches must be extremely reliable and they contribute additional weight to the satellite. An additional disadvantage of the cascade scheme is that it is incapable of surviving intermediate state failures of the switching element; it also contributes additional weight to the satellite.

For fixed state failures of switching elements, the total failure rate is

$$\text{failure rate} \cong \left[\binom{N_o}{p+1} \left(1 - \frac{1}{\exp \lambda \pi} \right)^{p+1} \right]^q$$

where

λ = fit number of a switching element

N_o = number of switching elements in a matrix

p = number of element failures allowable in a matrix

q = number of multiple matrices in parallel or in cascade.

Examples of reliability as a function of switch age are shown in Figure 8. These cases assume only element level redundancy, which permits any 1-element failure.

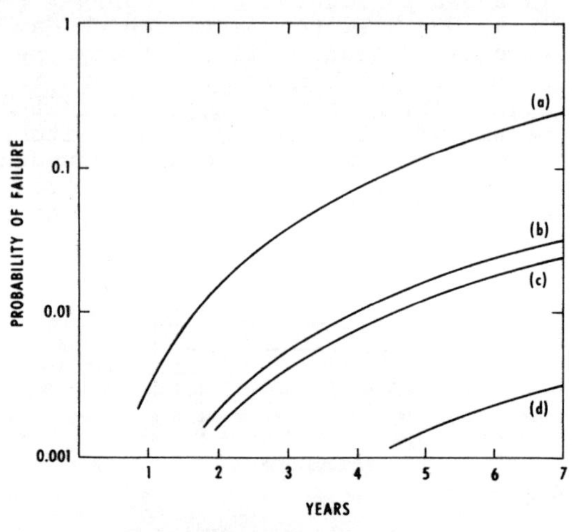

Figure 8. Failure Rate of Microwave Switch Matrix

Insertion Loss

The insertion loss of an N x N switch matrix depends largely on its configuration. For an N x N cross-bar-type matrix without redundant elements, the insertion loss (defined here as the largest power loss in any possible route through the matrix) is

$$\text{insertion loss} = (2N - 1) \times \alpha \quad [\text{dB}]$$

where α is the insertion loss of a cross-point. For a rearrangeable N x N switch matrix the insertion loss is

$$\text{insertion loss} = (2 \log_2 N - 1) \times \gamma \quad [\text{dB}]$$

where γ is the insertion loss of single β-element.

Isolation

Isolation is defined as the lowest ratio of the desired signal to the total interference power. If it is assumed that the input power levels are adjusted to give maximum isolation, the isolation is

$$\text{isolation} = I_1 - 10 \log_{10} N \quad [\text{dB}]$$

for a cross-bar-type matrix, where I_1 = single-element isolation, and

$$\text{isolation} = I_2 - 10 \log_{10}$$
$$\times (2 \log_2 N - 1) \quad [\text{dB}]$$

for a rearrangeable switch matrix, where I_2 = single β-element isolation.

DISTRIBUTION CONTROL UNIT

It is envisioned that MOS/LSI technology may be fully adopted in the implementation of the distribution control unit (DCU) to reduce the required number of chips and the power consumption. The use of MOS/LSI technology will also result in a high subsystem reliability.

The DCU is composed of the following functional elements (see Figure 9):

a. <u>T&C interface</u>. The T&C interface loads and reads out the memory contents and status of the DCU.

b. <u>programmer</u>. The programmer generates the cyclic control signals for opering the memories, output circuits, and timing circuits.

c. <u>control memory and counter</u>. Two memories will be required in a DCU. While one is operating to control the microwave switch matrix, the other is either standing by to update and monitor the memory contents or running at the same speed as the on-line unit to ensure that its contents are coincident with those of the operating memory.

d. <u>output circuit</u>. The output circuit reads the operating memory contents and transforms them to a signal format for control of the microwave switch matrix.

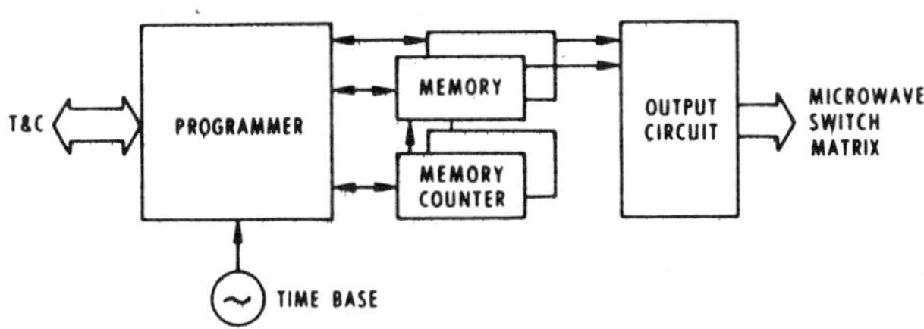

Figure 9. DCU Block Diagram

Memory Size

The frame period is subdivided into n_o frame units. The frame unit is the minimum time period during which the switch elements are maintained in the same pattern.

For an N x N cross-bar matrix, where N is a power of two, N switching elements in a row can be controlled by ($\log_2 N$) bits of signals. Therefore, the total memory size is

$$(\text{memory size})_{CB} = n_o(N \log_2 N) \quad [\text{bits}]$$

For an N x N rearrangeable switch matrix the required number of elements is

$$\text{number of elements} = N \log_2 N - N + 1$$

These elements shall be controlled independently; therefore the total memory size is

$$(\text{memory size})_{RS} = n_o(N \log_2 N - N + 1) \quad [\text{bits}]$$

FRAME SYNCHRONIZATION

In the SS-TDMA system, the transmitting TDMA bursts must be synchronized to the onboard switching sequence. Several options are available as indicated:[4-8]

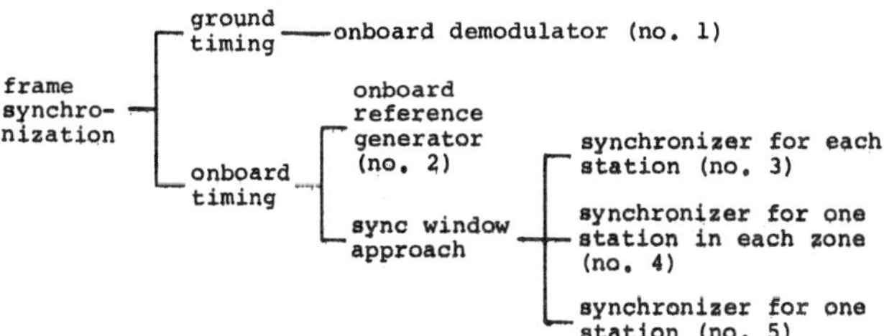

First the scheme is divided into two categories, depending on whether the satellite or the earth station is to provide the basic frame timing source for the satellite switch sequence. The first method requires a demodulator to detect code words on the satellite. Therefore, it requires more onboard equipment although it may provide a highly accurate frame frequency.

The second method requires an onboard modulator for broadcasting the reference signal from the satellite. The earth stations can synchronize their transmission to the satellite by comparing the received timing of the reference signal and the looped back signal in a loop back time slot assigned for this purpose. In this scheme, no special equipment is required in each earth station, and all earth stations are identical. A disadvantage is the requirement for an onboard satellite reference signal generator, which involves additional satellite power and weight as well as decreased system reliability.

The third and fifth methods do not require extra onboard hardware. In these methods, a short loop back time slot (sync window) is used to provide the reference timing of the satellite switch sequence. The earth station puts a burst in this window and measures the portion of the burst cut away by the edge of the window. Incrementally adjusting the transmitting time of the sync window burst makes it possible to adjust the earth station timing so that it is synchronous with the network.

The difference between methods 3 and 5 depends on the number of stations requiring special equipment to perform the network synchronization function. In the third method, each station has the synchronizer; therefore all earth stations are identical and independent. This method does not require an additional loop back time slot; hence, it results in a slightly more efficient use of the frame for data transmission.

In the fifth method, only one station at a time is responsible for synchronizing its transmission to the sync window, and upon achieving synchronization, that station transmits its reference bursts to all zones on a time-division basis. The other stations can synchronize their transmission to the satellite switch sequence indirectly by synchronizing it to the reference burst broadcasted from the reference station. The advantage of this method is the small number of special synchronizers required. All of the stations can use the conventional TDMA terminals even in the SS-TDMA operation.

The fourth method requires one reference
station in operation at a time in each
zone. This method seems to have no significant operational advantage over the
third or fifth method.

OVERHEAD TIME PER FRAME

Some overhead time is required in SS-TDMA
operation on account of several factors.
To estimate the available communications
capacity of a SS-TDMA system, the receiving or transmitting frame which carries
the maximum traffic shall be analyzed.
The following notation will be used:

n   Number of data windows in the
    frame in which maximum traffic is
    carried.

n_i Number of bursts carrying traffic
    in the ith data window (i = 1,
    . . ., n)

T_s Maximum time required for switch-
    ing the microwave switch matrix
    including its variations.

T_w Peak-to-peak sync accuracy of the
    ground synchronizer to the sync
    window.

T_R Peak-to-peak sync accuracy of the
    TDMA terminals to the reference
    burst.

T_d Peak-to-peak jitter of the switch-
    ing timing generator of the
    microwave switch matrix.

T_u Duration of the minimum switching
    segment.

n_ℓ Number of earth stations which
    have no traffic between any sta-
    tion in the zone which carries
    maximum traffic.

T_p Duration of shortest burst or
    duration of preamble words.

The overhead time in a frame differs depending on these parameters and the frame
synchronization scheme. It is calculated
for each of the frame synchronization
schemes described in the previous section, and the results are summarized
as follows for methods 1 through 5,
respectively:

$$\Delta T_1 = n(T_s + T_d) + \left(n_\ell + \sum_{i=1}^{n} n_i\right)(T_R + T_p)$$

$$\Delta T_2 = nT_s + \left(n_\ell + \sum_{i=1}^{n} n_i\right)(T_R + T_p)$$

$$\Delta T_3 = nT_s + (T_w + T_s + T_p) \sum_{i=1}^{n} n_i + T_u$$

$$\Delta T_4 = T_u + n(T_s + T_w + T_R) + (T_R + T_p) \times \left(\sum_{i=1}^{n} n_i - n + n_\ell\right)$$

$$\Delta T_5 = n(T_s + T_w + T_R) + (T_R + T_p) \times \left(\sum_{i=1}^{n} n_i - n + n_\ell\right)$$

The typical values of the parameters are

T_s = 200 ns

T_w = 100 ns

T_R = 100 ns

T_d = 100 ns

T_u =   6 µs

T_p = 500 ns (demodulator sync word and
        unique word plus station
        identification bits)

and n and n_ℓ are assumed to be 8 and 5,
respectively.

Figure 10 shows the total overhead time
per frame for each of the five methods of
frame synchronization as a function of
the number of data bursts in a frame (sum
of n_i).

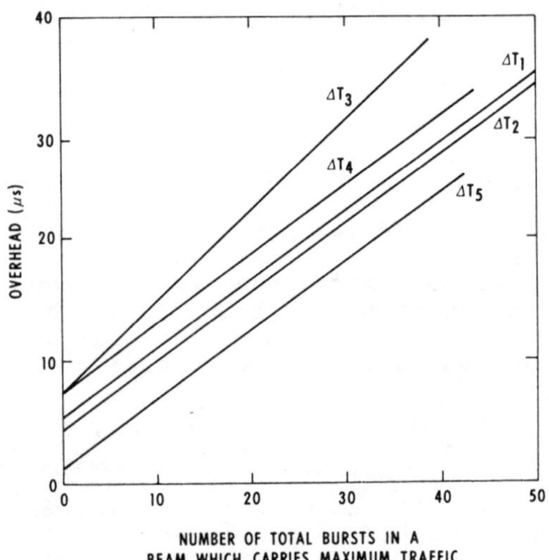

Figure 10. Overhead Time per Frame

EXAMPLE

Analysis of a SS-TDMA system should consider not only the weight and power requirements of the onboard subsystems, but
also the ultimately available communications capacity. The following is a sample communications capacity comparison of
FDMA/TDMA and SS-TDMA.

FDMA/TDMA

A multibeam channelized satellite provides the interconnectivity among beams
by using a number of filter channels
equal to the number of beams. This is
the minimum requirement to communicate
among stations in different zones. The
addition of filter channels to this minimum requirement results in an increased
total capacity. If the number of filter
channels equals the number of frame units
in a SS-TDMA system, the same capacity
can be obtained in a FDMA/TDMA satellite
as in a SS-TDMA satellite. However, in
this case, the required number of channelizing filters becomes too large for
practical implementation. For example,
if the number of zones is five, to obtain
the same capacity provided by a SS-TDMA
system having 124 frame units, the required number of channelizing filters
would be 620 in a FDMA/TDMA system.

When the number of equal-bandwidth filter
channels of a FDMA satellite is the same
as the number of beams, the maximum communications capacity for a traffic pattern such as that shown in Figure 3 is

$$\text{(maximum capacity)}_{\text{FDMA}} = \left(\sum_{i=1}^{N} S_i \right)$$

$$\times \frac{\text{capacity/beam}}{N[\max (t_{ij})]}$$

SS-TDMA

The maximum communications capacity of
the SS-TDMA satellite can be directly derived from the results of the third section of this paper:

$$\text{(maximum capacity)}_{\text{TDMA}} = \left(\sum_{i=1}^{N} S_i \right)$$

$$\times \frac{\text{capacity/beam}}{\max (S_i, R_j)}$$

The results of a comparison of FDMA/TDMA
and SS-TDMA satellites with the traffic
pattern of Figure 5 appear in Table 1.
The difference in maximum total capacity

(over 30 percent) results from the excellent interconnectivity of the SS-TDMA
approach. The satellite weight will be
clearly reduced by using the SS-TDMA
approach because of the much smaller number of filters required.

Table 1. Comparison of FDMA/TDMA
and SS-TDMA

Item	FDMA/TDMA	SS-TDMA
Available Total Band (MHz)	400	400
Number of Frequency Channels	5	1
Modulation	60 Mbd, 4-phase PSK	300 Mbd, 4-phase PSK
Number of Beams	5	5
Number of Channelizing Filters	25	5
Number of Voice Channels per Beam (64 kbps/ch)	9,300	9,300
TDMA Frame Length (μs)	750	750
Overhead Time (μs)	9[a]	20[b]
Total Capacity (ch)	29,870	39,700

[a] 7 bursts, each with a 72-symbol preamble.
[b] 32 bursts, fifth method.

CONCLUSIONS

Three key elements of the SS-TDMA system
have been discussed. It is vital for the
microwave switch matrix to use a redundant design. Various types of redundancy
have been discussed and it has been shown
that large matrices with high reliability
can be implemented. The distribution
control unit can exploit the recent developments of MOS/LSI technology. Since
only a few chips will be required for a
unit, the reliability objective can be
achieved with small weight and power
consumption.

The frame synchronization techniques between the onboard switch sequence and the

earth station TDMA transmission have been categorized and evaluated in terms of system reliability and required overhead time. A sync window approach which does not require additional onboard hardware is recommended.

A capacity comparison of SS-TDMA and FDMA/TDMA satellites shows that an increase of over 30 percent in available capacity can be obtained through the adoption of the SS-TDMA system.

REFERENCES

1. W. Schmidt, "An On-Board Switched Multiple Access System for Millimeter-Wave Satellites," INTELSAT/IEE Conf. on Digital Satellite Communications, 1969.

2. W. Schmidt and R. Cooperman, "A Satellite-Switched SDMA/TDMA for a Wideband Multibeam Satellite," ICC 1973.

3. W. Schmidt, "Satellite-Switched TDMA: Transponder-Switched or Beam-Switched," AIAA Paper No. 74-460, 1974.

4. K. Nosaka and T. Muratani "New Satellite Transponder with Regenerative Repeater and Higher Order DPSK for TDMA/SDMA System," 2nd International Conf. on Digital Satellite Communications, Paris, 1972.

5. N. Shimasaki and R. Rapuano, "Synchronization for a Communications Distribution Center On-Board a Satellite," ICC 1971.

6. R. Rapuano and N. Shimasaki, "Synchronization of Earth Stations to Satellite-Switched Sequences," AIAA Paper No. 72-545, 1972.

7. M. Asahara et al., "Synchronization and Acquisition in SDMA Satellite Communication System," ICC 1974 (43E-1).

8. C. Carter and S. Haykin, "Precision Synchronization to a Switched Satellite Using PSK Signals," ICC 1974 (43D-1).

9. Y. Ito, Y. Urano, and T. Muratani, "Optimal Traffic Assignment for an On-Board TDM Switching System," to be presented in IECE of Japan.

On-Board Regenerative Repeaters Applied to Digital Satellite Communications

KEIICHIRO KOGA, TAKURO MURATANI, AND AKIRA OGAWA

Abstract—In pulse-code modulation/time-division multiple access (PCM/TDMA) satellite communications, though the number of repeating stage is usually only one, the regenerative repeater will be worthwhile enough to be introduced on board the satellite. This is because the improvement of the communication quality by the increase of the transmission power is difficult in the satellite link and the regenerative repeater is very effective for removing cochannel interferences which will become predominant factor of signal degradation in the future and it will also provide great facility for various kinds of on-board processings. This paper first presents basic configurations of the on-board regenerative repeater, and discusses about simplification of the repeater. Next various kinds of on board processings are discussed in connection with the regenerative repeating. Then a concrete regenerative repeater applied to satellite switched/TDMA (SS/TDMA) satellite system which augments the communication capacity by frequency reuse is presented and compared with the conventional frequency translating type in terms of link budget, power dissipation, and weight. Lastly the result of the development of an experimental regenerative repeater is described.

I. INTRODUCTION

THE rapid increase of the communication demand has not only prompted the introduction of large capacity communication satellite and the augmentation of the number of the satellites, but also raised a requirement of using the communications bandwidth more efficiently. Digital satellite communications, especially pulse-code modulation/time-division multiple access (PCM/TDMA), will provide more than two times of communication capacity compared to the frequency modulation/frequency division multiple access (FM/FDMA) system using the same bandwidth. Furthermore, it will provide much better flexibility to accommodate changes of traffic demands. The introduction of the PCM/TDMA has already begun in some countries [1]. Under these circumstances, it will be worthwhile to raise a question of whether regenerative repeaters should be incorporated in the satellite transponders for TDMA operation. Usually the regenerative repeating is considered to be advantageous when the number of repeating stages becomes large. Because, when the repeating number is small, the nonregenerative repeating system will be able to attain the same transmission quality as the regenerative one's only by increasing the transmission power. In the satellite communication, however, even the one-stage regenerative repeating will be worthwhile enough to be employed. The reasons for this are the following.

1) In the satellite link, especially in the down-link, the increase of transmission power is difficult.

2) Isolation of up-link and down-link by regeneration on board will prevent the accumulation of thermal noise and interferences. In the future satellite communications system,

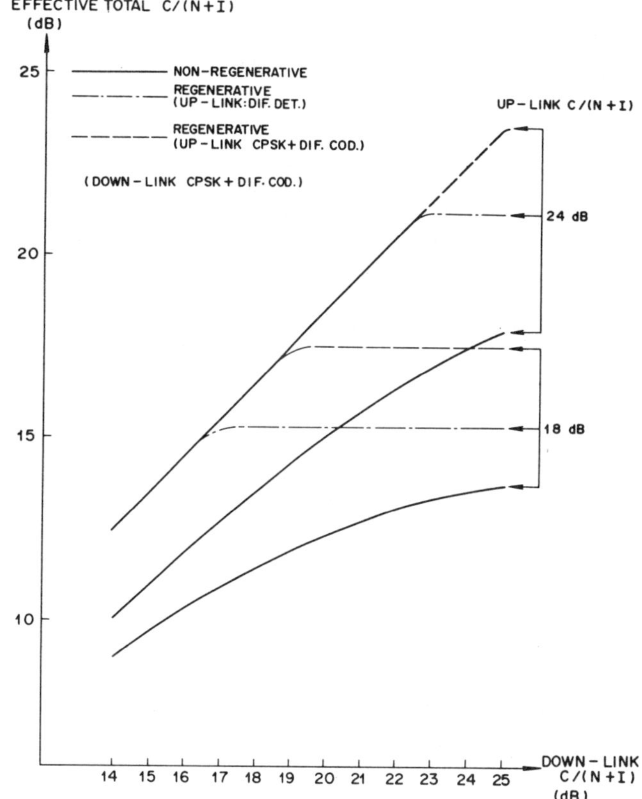

Fig. 1. Comparison of regenerative and nonregenerative system.

co-channel interference will become the predominant factor of signal degradation because of the measures taken to augment the satellite communication capacity, such as frequency reuse by multiple spot beams or increase of satellite number in a given orbital arc.

3) Another merit of isolating the up-link and the down-link is that it makes the independent optimization of each link possible. For instance the modulation format of the down-link need not be the same as the up-link's one.

4) With the regenerative repeating, it is possible to avoid the signal degradation due to the nonlinearities of the travelling wave tube amplifiers by substituting them with high-power direct modulators.

5) The regeneration of the baseband signal in the repeater provides great facility for various kinds of baseband-signal processings on board the satellite. Processings on board the satellite have two major purposes. One is to provide new conveniences to the users. The other is to make more efficient use of the potential communication capacity of the satellite.

Fig. 1 shows the effective total carrier to noise plus interference power ratio $[C/(N+I)]$ versus the down-link $C/(N+I)$

Manuscript received May 21, 1976; revised September 7, 1976.
The authors are with Kokusai Denshin Denwa (KDD) Laboratories, 2-1-23 Nakameguro, Meguro-ku, Tokyo 153, Japan.

taking the up-link $C/(N+I)$ as a parameter. Here, phase-shift keying (PSK) is assumed. In the regenerative system, miscellaneous losses of 0.5 dB on up-link and 1.5 dB on down-link are assumed. The large down-link loss is due to the possible degradation by the high-power direct PSK modulator in the repeater. In the nonregenerative system, total miscellaneous losses of 1.0 dB and satellite TWT nonlinear loss of 2.5 dB [2] are assumed. As can be seen in the figure, very efficient system design becomes possible by adopting the regeneration on board the satellite. For example, in the generative system, the effective total $C/(N+I)$ of 15 dB can be attained with the up-link $C/(N+I)$ of 18 dB, and the down-link $C/(N+I)$ of 17.5 dB. However, in the nonregenerative system the up-link $C/(N+I)$ of 24 dB and the down-link $C/(N+I)$ of 20.5 dB are required to attain the same effective total $C/(N+I)$. In this case the regenerative repeater can save 6 dB on the up-link and 3 dB on the down-link.

From the reasons mentioned above, the trend will be to adopt a regenerative repeater in place of the conventional frequency-translating repeater. It will be very desirable for the digital satellite communication.

This paper first discusses basic configurations of the on-board regenerative repeater. Descriptions will be made on the fundamental functions of the regenerative repeater and the implementation of them by rather simple configurations.

Next various types of on-board processing are discussed in connection with the regenerative repeating. Though the concept of the "on-board processing" is not necessarily connected with the regeneration, many of the processings are possible only in the regenerative repeater and some processings are far more easily performed in the regenerative repeater than in the nonregenerative one.

A rather concrete satellite system of satellite switched/TDMA (SS/TDMA) which augments the system communication capacity by frequency reuse will then be presented. The link and payload budgets of the regenerative type and the nonregenerative type satellite are compared and the merits of the regenerative type are displayed.

Lastly the result of development of an experimental model regenerative repeater will be presented. The model, which comprises a four phase PSK demodulator and a high-power microwave PSK modulator will be useful to evaluate the hardware feasibility of on-board regenerative repeater by state of the art.

II. CONFIGURATIONS OF REGENERATIVE REPEATERS FOR PSK TRANSMISSION

Among various modulations, PSK will be one of the most promising modulations, which has been widely adopted in the recent digital communications applications.

This section will describe regenerative-repeater configurations to be applied to wideband communications satellite links. A PCM/TDMA link is an example of wide-band communication link which utilizes burst mode PSK signals. When regenerative repeaters are incorporated in the satellite for PCM/TDMA signals, the on-board repeater receives the signal bursts successively which are transmitted from earth stations in nonoverlapping basis (see Fig. 2).

Though the carrier frequencies or clock frequencies are quite close among all bursts, coherency of the carrier or the clock phases between bursts may not be anticipated.

Consequently in order to regenerate the baseband signals, carrier and clock recoveries should be done at the preamble of

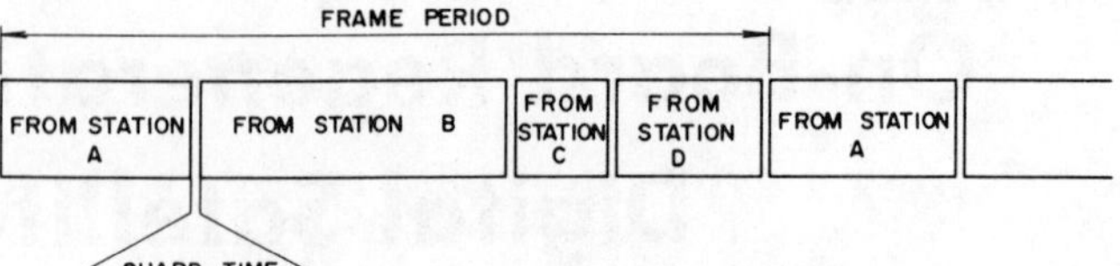

Fig. 2. TDMA frame structure.

each burst, and they must be performed quickly to maintain the high communication efficiency.

In recovering the carrier, the attendant problem is that of properly resolving the correct phase of the recovered carrier [3]. There are two types of solutions to this phase-ambiguity problem. One solution is to really determine the correct phase by some methods, for instance by inserting a code word in the transmitted bit stream at regular intervals or by error-correcting coding the transmitted bit stream. At the receive side, a phase of the recovered carrier, by which the specific code is demodulated correctly or by which a error-correcting decoder receives the demodulator output whose error rate is not abnormally high, is selected. Another solution is to remove the effect of selecting the false phase by differentially encoding the transmitted bit stream. Though the latter way is simpler than the former, the cost to be paid for the simplicity is the doubling of the error rate for the same C/N.

Following are the discussions on various types of regenerative repeaters to be applied to PCM/PSK/TDMA signals.

A. Ideal Type

An ideal type of the PSK regenerative repeater will have the following functions (see Fig. 3(a)):

1. carrier recovery
2. coherent detection
3. clock recovery
4. decision
5. differential decoding
6. data processing
7. differential encoding
8. carrier generation
9. modulation.

Since the repeater on board the satellite is strongly required to be high in reliability, light in weight, and low in power consumption, the above functions must be realized fully considering these requirements. Though these functions will be realized with sufficient reliability to be on board in the future, fast carrier-recovery circuits and fast clock-recovery circuits are critical elements and may not have desirable application. The omission of these elements simplifies the repeater and increases the reliability. The regenerative repeaters in which either one of them or both of them are omitted are shown and scrutinized subsequently.

B. Omission of Carrier Recovery

Carrier recovery may be the most difficult to be realized with sufficient stability. Consequently the omission of it makes the repeater reliable. The adoption of the differentially coherent detector fits the purpose. Though differentially coherent detector needs approximately 0.5 dB, 2.5 dB, and 3 dB higher C/N's for 2-phase, 4-phase, and 8-phase PSK signals, respectively, than those for the coherent detector, it is rather easy to compensate for it by increasing the earth-station transmission power. As is seen in Fig. 1, in the region where

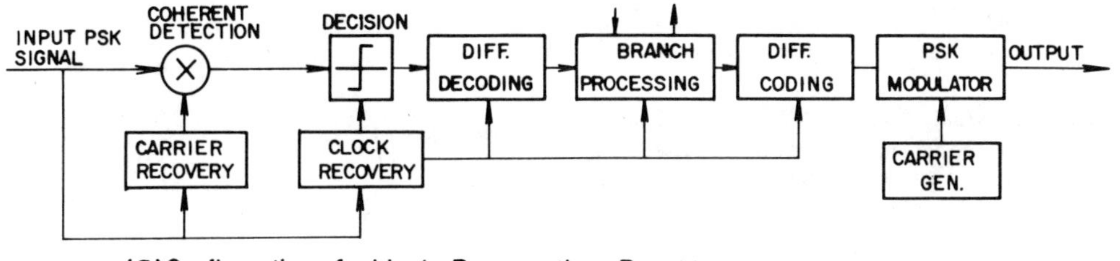

(a) Configuration of ideal Regenerative Repeater

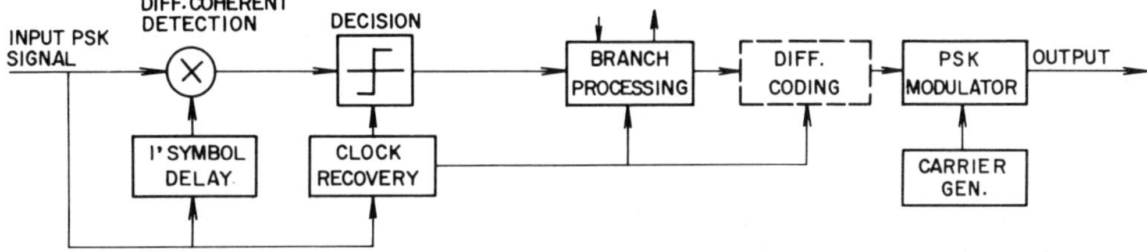

(b) Configuration of Regenerative Repeater without Carrier Recovery

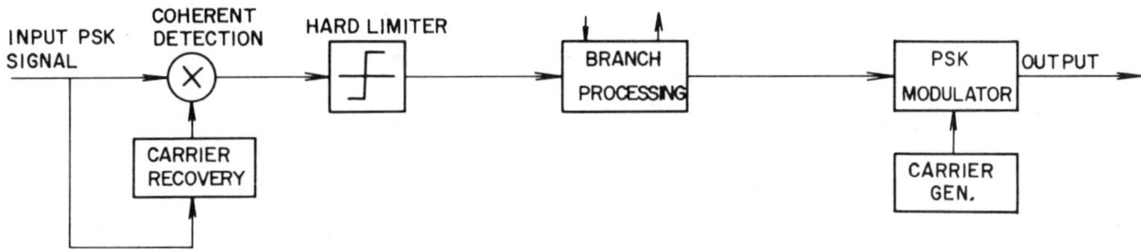

(c) Configuration of Regenerative Repeater without Clock Recovery

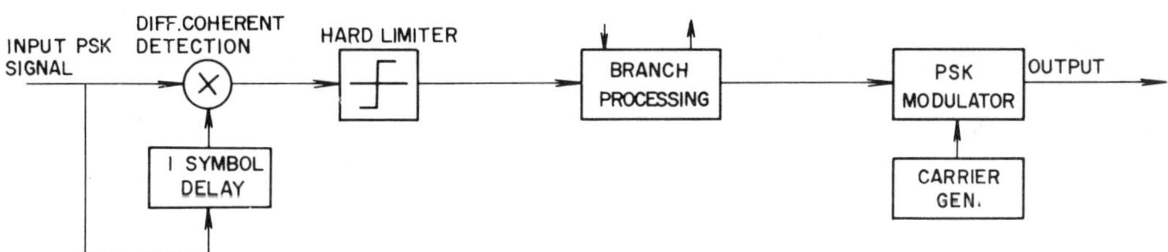

(d) Configuration of Regenerative Repeater without Carrier Recovery or Clock Recovery

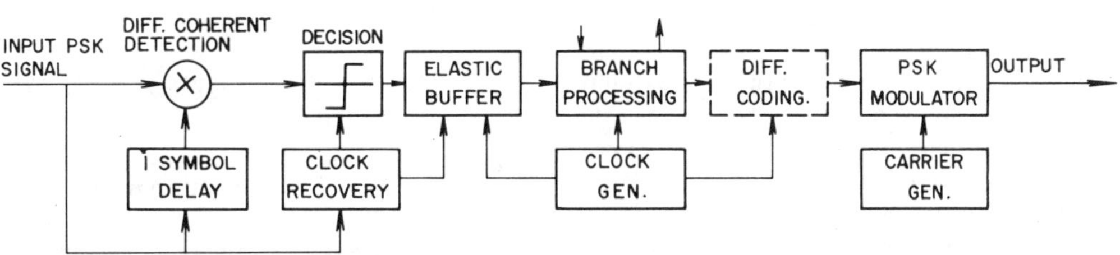

(e) Configuration of Regenerative Repeater with Retiming by the Clock Oscillated On – board the Satellite

Fig. 3. Basic configurations of a regenerative repeater.

down-link error is predominant, the adoption of the differentially coherent detector on board the satellite makes little difference to the total error rate. Furthermore, recently a method by which the characteristic of the differentially coherent detector is made almost same as that of the coherent detector has been proposed [4], therefore the difference of the required C/N's is not a big problem.

One of the most important problems in implementing the differentially coherent detector may be the stability of the one-symbol delay element. The delay-time variation results in the phase error of the reference signal and deteriorates the error rate. The phase error ΔP is expressed as

$$\Delta P = \frac{f_c}{f_s} E_t \times 360 \text{ (degrees)} \tag{1}$$

where f_c is the carrier frequency, f_s is the symbol frequency, and E_t is the delay-time variation. E_t mainly depends on the temperature coefficient of variation of electrical length of the delay element and on the amount of temperature variation in the repeater. The temperature coefficient of $5 \times 10^{-5}/°C$ has been attained for commercially available SJ coaxial cable. When the stability of the delay element is not sufficient, it will be needed to lower the signal frequency, that is, to IF frequency at the cost of the increase of weight and power consumption of the repeater.

The error pattern somewhat varies with the detection methods. In the case of coherent detection followed by differential decoding, most of single-symbol error results in double-bit errors because of the differential decoding. (Here, of course, Gray coding is assumed.) In the case of differentially coherent detection, though many of the errors are single-bit errors, the probability of double-bit errors is not negligible [5]. The attention must be paid to the error pattern especially when error correction is introduced.

Since the differentially coherent detector has a function of differential decoding, the differential decoding circuit is not necessary. If the signal is doubly differential encoded at the transmit stations, even the differential encoder can be omitted in the satellite [6], [7] [see Fig. 3(b)]. However in this case not only the down-link errors but also the up-link errors are doubled by the differential decoder at the receive station. Consequently when the effect of doubling the up-link errors is not negligible, or when some on-board processings are performed for the signal which is not differential encoded, the double-differential encoding scheme at the transmit station cannot be adopted.

Since when the regenerative repeater is adopted the carrier becomes coherent over all TDMA bursts on the down-link, the coherent PSK can be rather easily applied to it. Exact phase of the recovered carrier is determined so that a specific pattern in a burst may be reproduced. Though some amount of hardware must be added to the earth station receiver in this case, the differential encoder in the satellite can be omitted and the down-link error rate becomes the smallest.

C. Omission of Clock Recovery

When the clock recovery is omitted, retiming, differential encoding, decoding, or other processings which need the timing clock cannot be performed (see Fig. 3(c)).

However, timing jitter caused by the omission of the retiming will be tolerable if the ratio of the transmission bandwidth to the symbol rate is not very small. The phase ambiguity problems of the recovered carriers, both at the satellite repeater and at the earth station receiver, can be solved at the same time by a differential encoding at the transmit station and a differential decoding at the receive station. Consequently the retiming, differential encoding, or decoding is not always needed in this configuration. The output of the demodulator in the repeater, however, is not always the exact replica of the transmitted data from the earth station because the phase ambiguity of the recovered carrier is not removed in the repeater.

This configuration does not seem attractive because while the carrier recovery circuit which is the most critical remains, only the clock-recovery circuit is omitted.

D. Omission of Carrier Recovery and Clock Recovery

If both the carrier recovery and clock recovery are omitted, the configuration of the repeater becomes very simple and provides reliable performance (see Fig. 3(d)). The problem is that the differentially coherent detector is essentially accompanied with large timing jitter, that is, the fluctuation of the zero-cross time of the signal, at its demodulated output except in the case of two-phase PSK. Fig. 4 illustrates how the jitter is caused taking the differentially coherent detection of four-phase PSK for example. In the figure, the phase of the input

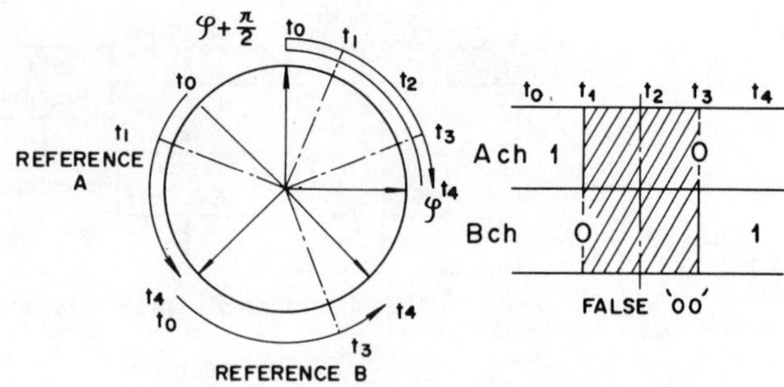

Fig. 4. Jitter generation by differentially coherent detection.

TABLE I

PHASE TRANSITION OF INPUT SIGNAL AND REFERENCE SIGNALS

Phase Transition		
Input Signal	$\varphi \longrightarrow \varphi + \pi/2 \longrightarrow \varphi$	
A Ch. Reference	$\varphi + 3\pi/4 \longrightarrow \varphi + \pi/2 + 3\pi/4$	
B Ch. Reference	$\varphi + 5\pi/4 \longrightarrow \varphi + \pi/2 + 5\pi/4$	

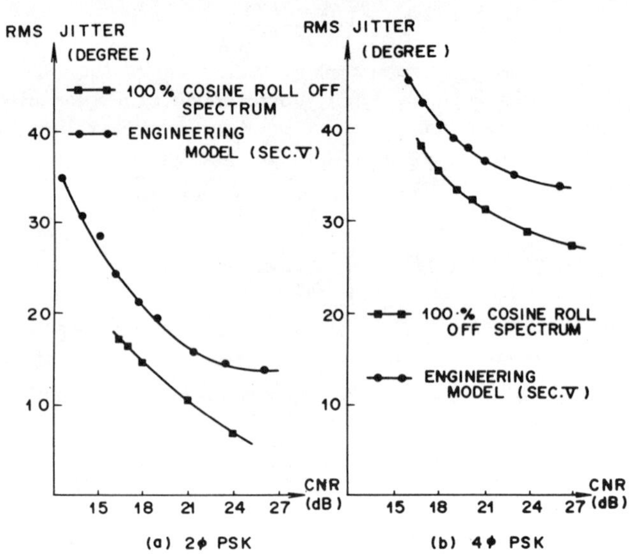

Fig. 5. Jitter at the output of the differentially coherent detector (simulation result).

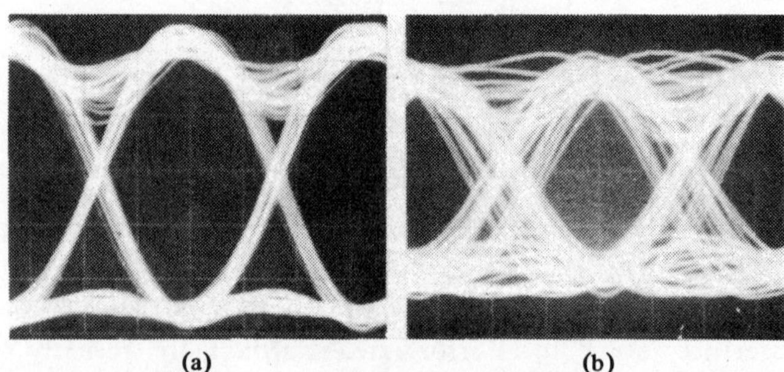

Fig. 6. Eye patterns at the output of the differentially coherent detector. (a) 2ϕPSK. (b) 4ϕPSK.

signal changes as $\varphi \to \varphi + \pi/2 \to \varphi$. During that time the phases of the reference signals, which are the phase-shifted input signals of one-symbol time before, change as in Table I. When the signal is not band-limited, it does not take any time for the phase of the signal to change from $\varphi + \pi/2$ to φ, and no

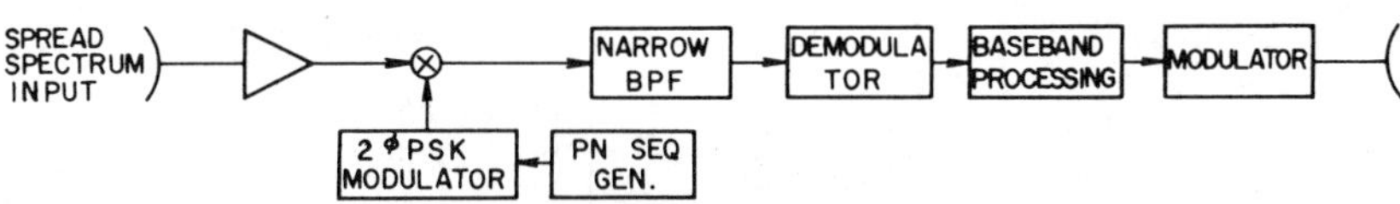

Fig. 7. An example of an anti-jam repeater.

jitter is produced. However when the signal is band-limited, it takes some time, that is from t_0 to t_4 in the figure, to change the phase of the signal. At time point t_1 which is not the center between adjacent symbols, the input signal and the A-channel reference signal become at quadrature. At time t_3 the input signal and the B-channel reference become at quadrature. Consequently A-channel modulator output changes from "1" to "0" at time point t_1 and B-channel output changes from "0" to "1" at time point t_3. Usually the retiming removes the timing jitter. However in this case, since the retiming is not performed, the timing jitter is carried over to the down-link without being suppressed.

In Fig. 5 the amount of jitter at the output of the differentially coherent detector versus the input C/N obtained by the computer simulation is shown. From the above-mentioned cause, the jitter for four-phase PSK signal exceeds the limit for practical use even when the full-cosine roll-off spectrum in which intersymbol interference becomes zero not only at the center of each pulse but also at the middle of the adjacent pulses is adopted and the C/N is infinite. However in two-phase PSK, there occurs no essential jitter mentioned above, so this simple configuration will be applicable when the wide transmission bandwidth is available. Fig. 6 shows the eye-patterns at the demodulator output in the experimental regenerative repeater of this configuration which is mentioned in Section V. The jitters in these eye-patterns agree with the above-mentioned simulation results.

Since differential encoding cannot be performed in the repeater because of the lack of the timing clock, the doubly differential encoding at the transmit station or the coherent PSK on the down-link must be employed.

E. Retiming by the Clock Generated in the Satellite

When the clock recovery and retiming is performed, the second retiming of all TDMA bursts by the clock oscillated in the satellite becomes attractive. This configuration of the regenerative repeater is illustrated in Fig. 3(e).

Since the down-link clock becomes coherent over all TDMA bursts, not only the earth-station receiver is simplified, but also the communication efficiency is improved because of the unnecessity of the dummy symbols which are assigned at the top of the each burst for the clock recovery at the earth station. Also in the satellite, signal processings become easy because of the coherency of the clock over all TDMA bursts.

In order to absorb the slight difference of the recovered clock frequency and the on-board oscillated clock frequency, an ellastic buffer is used. Before each TDMA burst begins, the ellastic buffer must be reset to prevent its overflowing or becoming empty during the burst because of the clock frequency difference.

There are several ways to know the reset timing. One way is to detect the frame-synchronization word at the top of each burst. However the complexity of the equipment needed on the satellite may make it unpractical. Another way is to know the interval between TDMA bursts by envelope detection of the received signal. This method will require a little longer guard time than usual. In the SS/TDMA repeater which will

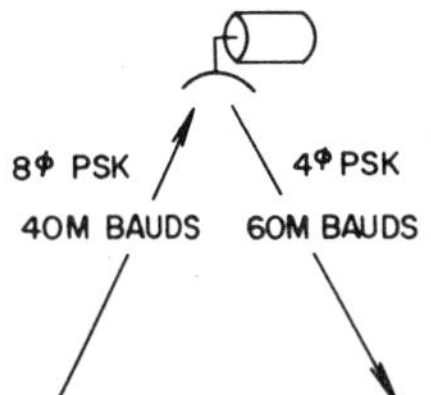

Fig. 8. Modulation conversion from 8φ PSK to 4φ PSK (120 Mbit/s).

be mentioned in Section IV, the timing source is contained originally to know the switching timing. According to the on-board timing source, the repeater sends reference burst to the earth stations to inform the TDMA frame timing and performs switching. With a little modification, the timing source can also provide the reset timing of the ellastic buffer.

III. PROCESSING ON BOARD THE SATELLITE

Although an on-board repeater is required to be simple, light in weight, and reliable, the improvements of electronic devices have gradually made it possible to answer the request for more complicated repeaters which provides larger communication capacity, or the various kinds of services. And various proposals for "processing repeaters" in place of the conventional frequency translating repeater has been made [8]–[10]. The regenerative repeater is a kind of processing repeater in itself. Though there are many kinds of nonregenerative processing repeaters, the regeneration provides facilities for various kinds of other processings. One reason is that the receive side and the transmit side of the repeater can be individually designed, another is that the baseband signal is easy to be processed.

The independency of the receive side and the transmit side allows the modulation conversion. An example of the modulation conversion is the spread-spectrum modulation on the up-link and the usual modulation, for instance PSK, on the down-link [7], [8] (see Fig. 7). The spread-spectrum modulation not only removes jam or other interferences, but also enables only the specific earth stations [which use the key pseudonoise (PN) sequences] to access the repeater. And the additional baseband processings help the function. However this kind of repeater is not suitable for commercial communications because the transmission efficiency is very low.

The modulation conversion may be applied to the commercial communications, if the saving of the cost by the improvement of the system efficiency more than makes up for the extra cost required for the modulation disparity. In many cases the up-link C/N and the down-link C/N are different. In the communication system it is desirable to balance the two capacity-limiting factors, that is, the transmission power and the frequency bandwidth, and it leads to the high system efficiency. Consequently, the independent selection of the up-link modulation and down-link modulation makes more efficient use of the satellite communication capacity possible. Take the eight-phase PSK on the up-link and the four-phase PSK on the down-link, for example, because the up-link C/N is usually higher than the down-link one. Though in order to attain the same error rate on both links, the required C/N for

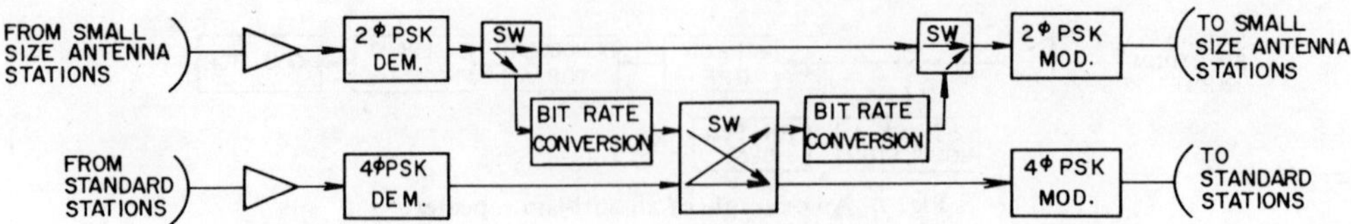

Fig. 9. Connection of the 2φ PSK transponder and the 4φ PSK transponder.

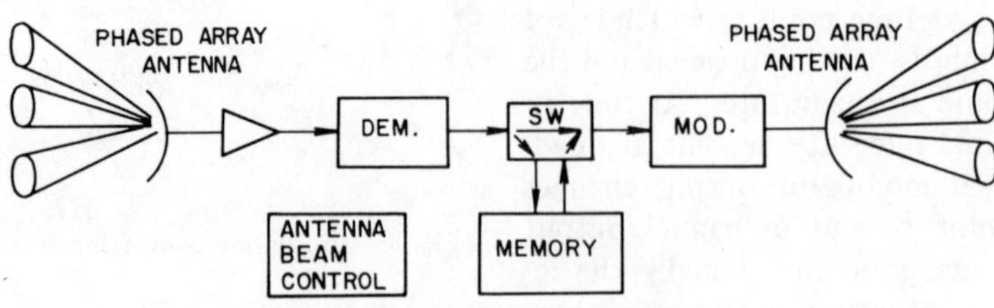

Fig. 10. Beam-hopping repeater.

the up-link is approximately 5 dB higher than that for the down-link, the required bandwidth for the up-link is two thirds of that for the down-link (see Fig. 8). Since the eight-phase PSK is relatively subject to the effect of intersymbol interferences, the nonlinearity of the traveling wave tube amplifier (TWTA) causes much deterioration in error rate. However in regenerative repeating, there is no need to consider the effect of the TWTA nonlinearity in the satellite. Consequently if the TWTA amplifier in the earth station is used with sufficient backoff or some nonlinearity compensation is performed, the eight-phase PSK is applicable on the up-link.

It is desirable to prepare a transponder for earth stations with small size antennas. Because of the insufficiency of their antenna gain, two-phase PSK, for example, may be used. Fig. 9 shows a satellite-system configuration which comprises one standard transponder and one transponder for small-size antenna stations. In order to connect the standard antenna stations and the small antenna stations, some of the TDMA bursts from standard stations are retransmitted from the transponder for the small-size antennas after the modulation conversion and vice versa. This is a kind of satellite switching. Since the bit rates of the signals for standard stations and the ones for small size antenna stations are different, the bit-rate conversion is also required, and a considerable amount of memories are needed for this purpose. The satellite switching and the bit-rate conversion are, of course, done at baseband stage.

Because of the baseband stage, the regenerative repeater has a great freedom in on-board processing. For instance, storing the signal in the memory or error correction is possible only when the baseband signal is obtained. The switching of the baseband signal is far easier than that of the microwave signal.

On-board memory can be used for TDMA bursts rearrangement as well as for bit-rate conversion, as mentioned above. Beam hopping [11] is a technique in which a transponder is time shared by multiple beams. By this, a transponder and a phased-array beam antenna can accommodate many spot-beam areas with small traffics. The connection among the beam spots is performed by storing the signal in the memory as shown in Fig. 10.

Error correction is a very effective means to reduce transmission errors [12], [13]. Though it is not appropriate to introduce it into bandwidth-limited system, it shows its effect when applied to power-limited system. Consequently, the

TABLE II
EXAMPLES OF CODING GAIN (k: CONSTRAINT LENGTH)

Code Decoding Method	Coding Rate		C/N Gain (dB)	E/N$_0$ Gain (dB)
Wyner - Ash Single		1/2	2.4	-0.6
Error Correcting		3/4	1.7	0.5
Code		7/8	1.5	0.9
Self-Orthogonal		1/2	3.5	0.5
Code		3/4	2.5	1.2
(Double Error Correction)		7/8	1.8	1.2
Nonsystematic	k=8	1/3	10.2	5.4
Convolutional Code	k=7	1/2	8.1	5.1
Viterbi Decoding	k=6	2/3	6.9	5.1
(Soft Decision)	k=9	3/4	4.7	3.5
Sequential Decoding		1/2	6.6	3.6
k : Constraint Length.				

error correction should be taken into consideration, when modulation method is selected, for the combination of them always makes more efficient transmission possible than the modulation only. For instance, when a transmission system is bandwidth-limited for two-phase PSK and power-limited for four-phase PSK, the combination of the four-phase PSK and the error correction will be the best solution. The error correction technique is particularly suitable for satellite link, for basically errors occur randomly on the satellite link and random error correction is rather easier than burst error correction, which usually must be done on terrestrial links.

Of course even on the satellite link, the coherent detection plus differential encoding or the differentially coherent detection, as is remarked in Section II, makes double-bit errors. However they can be easily coped with by simple bit interleaving.

Here only the forward error correction is considered because of the large transmission delay of the satellite link and the high-speed operation.

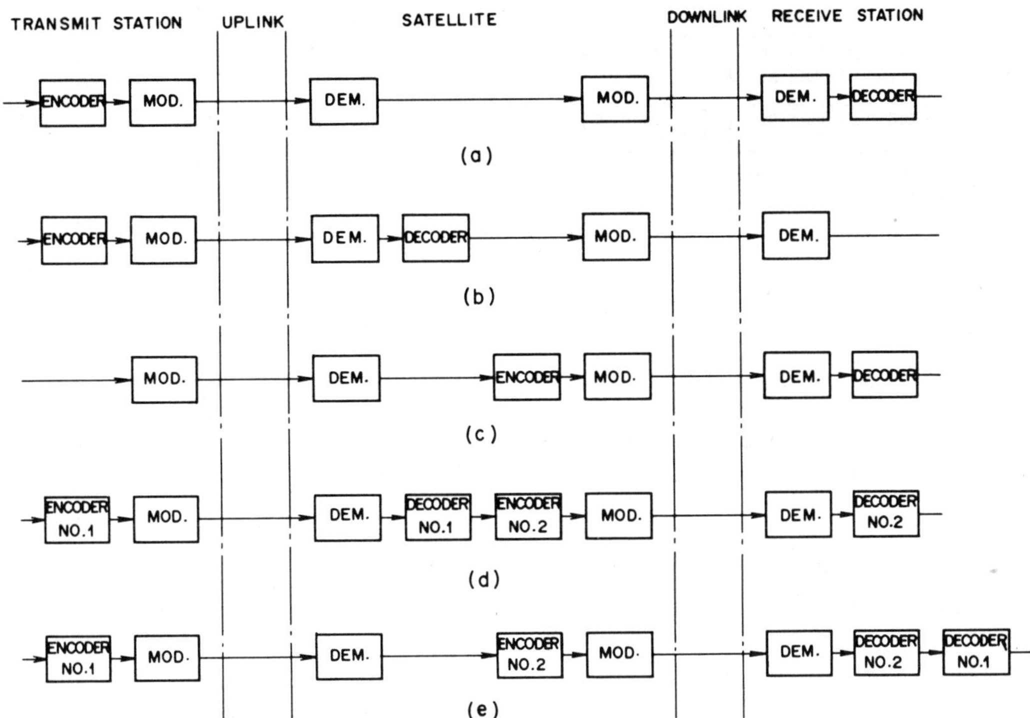

Fig. 11. Satellite communication system configurations with error correction.

The coding gain depends on the code, the coding rate, and the decoding method. Typical figures are shown in Table II. As is seen in the table, the introduction of error correction makes it possible to trade off the transmission bandwidth and the required E/N_0 (Energy per bit to noise-power density ratio) fairly freely, and the freedom in the system design is increased.

In Fig. 11 various types of satellite communication-system configurations with error correction are shown. In configuration (a), the encoding and the decoding are performed at the earth stations. This configuration can be applied to either nonregenerative or regenerative repeaters. In configuration (b), error correction is introduced only into the up-link. In this case the decoder, which is far more complicated than the encoder, must be located in the satellite. Consequently, it will not be practical to adopt the decoder on board the satellite except when the coding rate is low. In configuration (c), error correction is introduced only into the down-link. Since the encoder is usually simpler than decoder, this configuration may be easily realized. In configuration (d), individual error correction is performed on each link. Though optimal error correction can be selected on each link, the error-correcting decoder is to be in the satellite. In configuration (e), only an encoder is located in the satellite, and the signal is encoded singly on the up-link and doubly on the down-link. When the down-link needs more protection than up-link, this configuration has almost the same effect as the configuration of (d) and is easier to implement. However, in this case, the special consideration must be given to the selection of codes, for the decoding error of the inner code must be corrected by the outer-code decoder.

IV. SDMA/SS/TDMA REGENERATIVE REPEATER

In this section a rather concrete satellite system is discussed, and the merit of the regenerative repeater is displayed, taking SS/TDMA satellite system as an example.

Space-division multiple access/satellite switched/TDMA (SDMA/SS/TDMA) is a very efficient satellite communication system, in which multiple spot-beam antennas cover each area using the same frequency band and the interconnection among the spots are made by time-division switching.

The SS/TDMA satellite system can, of course, be realized by conventional nonregenerative type repeater. Many papers concerning this type of SS/TDMA repeater have been published [9], [14]–[17]. However, when all communications are digitalized, the regenerative type SS/TDMA repeater becomes very attractive as described below.

An example of the regenerative type SS/TDMA satellite is shown in Fig. 11. The number of the spot beams is eight, and the 8×8 switch matrix does the routing among the spot beams. The input signal from each spot area is detected by the differentially coherent detector. After the decision by the recovered clock, the received signal is retimed by the common clock oscillated in the satellite. For this purpose an ellastic buffer is needed. The ellastic buffer must be reset at the top of each TDMA burst. This can be done by the switch-matrix control unit, which is mentioned later. The switch matrix can be easily realized by conventional integrated circuits, for it handles baseband signals. The switch-matrix control unit controls the matrix in accordance with the command from the control station so as to meet the traffic distributions.

One of the problems in SS/TDMA is the synchronization of the on-board switching with the transmission of the TDMA burst from each earth station. If the basic timing is provided by an earth station, it will require a lot of extra on-board equipment to receive the timing signal and set the control by it. Two types of on board timing generation will be possible. One is the generation of reference bursts on board, and the other is the sync window approach [16]. While the latter requires earth stations to have special synchronizers, the former requires a reference-code generator, a modulator, and a carrier source on board the satellite if the nonregenerative

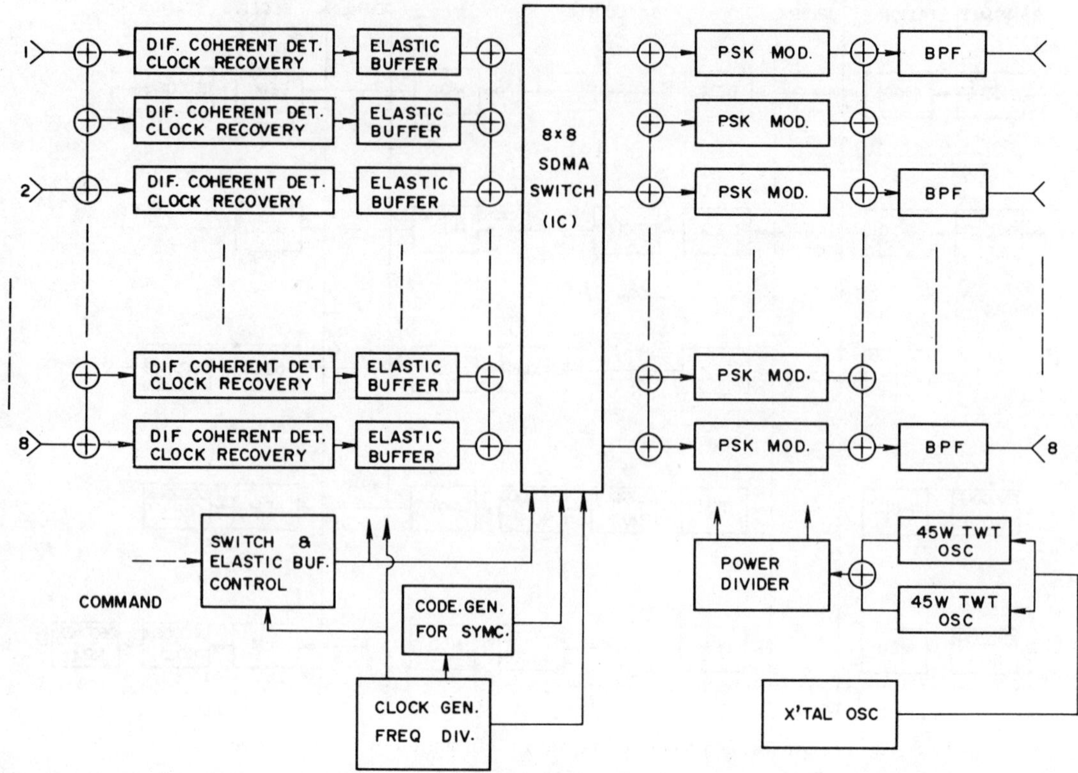

Fig. 12. Regenerative SS/SDMA repeater.

repeater is adopted. However in the regenerative repeater, which originally contains modulators and a carrier source, only a reference-code generator is required, and its output is mixed with received signals by the switch matrix.

The output signals from the switch matrix are fed to the PSK modulators prepared for respective spot beams. A single high-power oscillator controlled by a crystal oscillator supplies carriers to all PSK modulators. Since the modulator outputs are directly transmitted to the earth stations, a large number of traveling wave tube amplifiers which are needed in the conventional systen can be omitted, and the weight of the repeater is decreased remarkably.

As is seen in Fig. 12, in the major part of the repeater a spare circuit is provided for two active circuits to keep the reliability. The comparisons of the regenerative-type repeater and the nonregenerative-type repeater (Fig. 13) in the link budget, the weight, and the power dissipation are shown in Table III and Table IV under the condition that two of communication subsystems shown in Fig. 11 and in Fig. 12, except for the commonly usable parts such as antennas, are equipped in the satellite to compose a system as follows:

1) voice-channel capacity: 64 000 ch (32 000/subsystem);
2) antenna: eight adjustable beams;
3) frequency allocation: Up-path = 14.05 ~ 14.25 GHz and 14.30 ~ 14.50 GHz Down-path = 11.00 ~ 11.20 GHz and 11.50 ~ 11.70 GHz;
4) modulation: 4-phase PSK;
5) number of carriers: 8 × 2 = 16;
6) symbol rate/carrier: 160 MBd (320 Mbit/s).

If an earth station accesses multiple TDMA transponders, it is desirable to reduce the required number of TDMA terminals

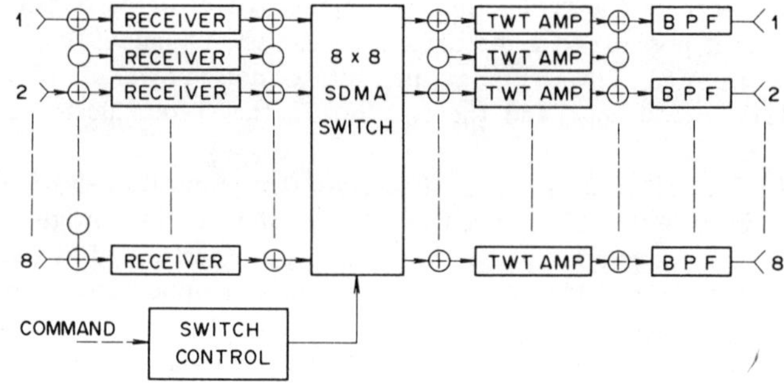

Fig. 13. Nonregenerative SS/SDMA repeater.

by time-sharing use of the terminals, namely by transponder hopping [18]. In order to time share the TDMA terminal, burst positions are assigned for each earth station to avoid the simultaneous transmission to multiple transponders or simultaneous reception from multiple transponders. In SS/TDMA system, in return for its high efficiency, not only the burst scheduling but time-slot scheduling must be considered [19]. It may make it difficult to make the convenient burst scheduling for the transponder hopping. With memories on board, however, store-and-forward instead of real-time circuit switching becomes possible. The rearrangement of the bursts in the satellite makes the burst scheduling independent of the time-slot scheduling, and it not only makes the both schedulings easy but also makes the efficient use of TDMA terminals possible.

V. EXPERIMENTAL REGENERATIVE REPEATER

A regenerative repeater without carrier recovery or clock recovery has been developed in order to examine the possibility of the simple configuration and the feasibility of high-power microwave direct PSK modulator. Outlines of the

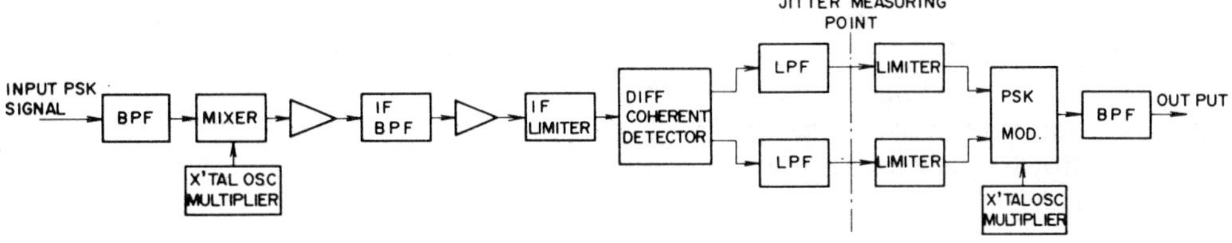

Fig. 14. Engineering model configuration.

TABLE III
PER CARRIER LINK BUDGET (ELEVATION: 10°)

Link Parameters		Ordinary System (Fig. 13)	Regenerative System (Fig. 12)
Threshold Bit Error Rate		10^{-4}	10^{-4}
Threshold Carrier/Noise	(dB)	14.0	14.0
Up-link Budget	(14 GHz)		
Earth Station EIRP	(dBw)	85.0	85.0
Path Loss	(dB)	207.5	207.5
(G/T) sat.	(dB/°K)	5.5	5.5
Carrier/Thermal Noise			
in 160 MHz Bandwidth	(dB)	29.6	29.6
Carrier/Interference	(dB)	29.0	29.0
Carrier/Total Noise	(dB)	26.5	26.5
Detection Loss*	(dB)		3.0
Up-link Carrier/Noise	(dB)		23.5
Down-link Budget	(11 GHz)		
Satellite EIRP	(dBw)	43.0	43.0
Path Loss	(dB)	205.5	205.5
(G/T) E.S.	(dB/°K)	40.0	40.0
Carrier/Thermal Noise	(dB)	24.1	24.1
Carrier/Interference	(dB)	27.0	27.0
Carrier/Total Noise	(dB)	22.3	22.3
Overall Carrier/Noise	(dB)	20.9	
Miscellaneous Losses	(dB)	1.0	1.5
Down-link Carrier/Noise	(dB)		20.8
Sat. TWT Nonlinear Loss	(dB)	2.5	
Effective Total Carrier/Noise	(dB)	17.4	20.8

* Miscellaneous losses 0.5 dB plus 2.5 dB loss by the adoption of differential coherent detection.

TABLE IV
SS/SDMA SATELLITE SYSTEM WEIGHT AND POWER (NONREGENERATIVE SYSTEM/REGENERATIVE SYSTEM)

	Number	Weight Each (kg)	Weight Total (kg)	Number of ON	Watts Each	Watts Total
14/11 GHz Spot Antenna	1/1	15/15	15/15			
14 GHz Receiver	24/24	1.0/1.5	24/36	16/16	4/5	64/80
SDMA Switch with Controller	2/2	4.0/0.5	8/1	2/2	25/5	50/10
11 GHz TWT Amp. (5 W)	24/0	1.5/0	36/0	16/0	25/0	400/0
11 GHz TWT Amp. (45 W)	0/4	0/3.0	0/12	0/2	0/90	0/180
11 GHz Up Converter and Filter	24/0	1.5/0	36/0	16/0	5/0	80/0
11 GHz PSK Mod. Driver and Filter	0/24	0/1.5	0/36	0/16	0/5	0/80
Total			119/100			594/360

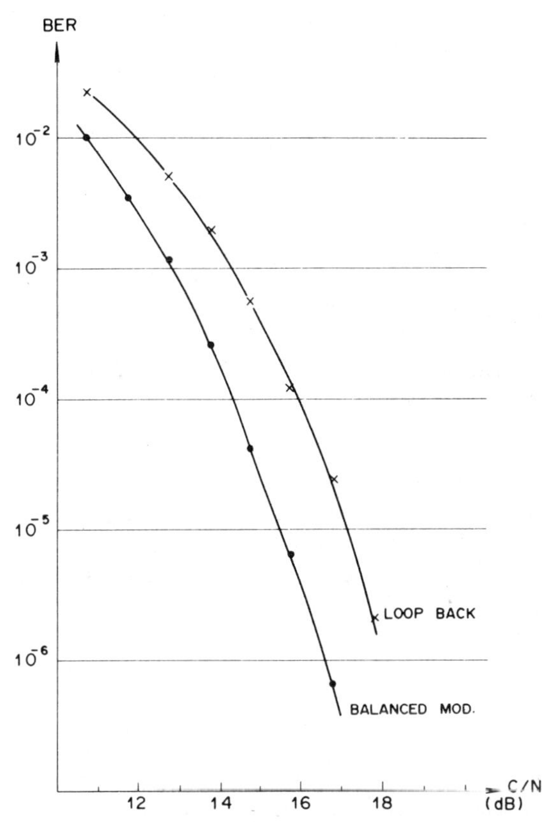

Fig. 15. Error-rate performance of an experimental model.

model are as follows:

1) input-signal frequency: 6030 MHz;
2) modulation: two-/or four-phase PSK;
3) modulation rate: 30 MBd;
4) Input-BPF bandwidth: 120 MHz (3 dB down);
5) IF BPF: Gauss filter, 140 ± 20 MHz (3 dB down);
6) detection: Differential coherent detection;
7) modulator: Reflection type (using p-i-n diodes);
8) output-signal frequency: 3805 MHz;
9) output-BPF bandwidth: 36 MHz (3 dB down).

The configuration of the model is shown in Fig. 14.

Since the ratio of the carrier frequency to the symbol frequency is too large to attain the stable reference, the signal was converted to IF signal of 140 MHz before detection.

Fig. 15 shows the error-rate performance of the experimental model. The circuit was cut off at the detector output. Data were fed to the modulator through LPF, and the repeater output was fed to the repeater input through 4/6 GHz converter. Noise was added at IF stage. The characteristic of the high-power reflection-type modulator is a little inferior to the usual low-power modulators such as ring modulator. In order to evaluate the degradation by the modulator, one more curve is shown in the Fig. 15. This is the performance when 140-MHz balanced-type modulator output was directly fed to IF BPF of the experimental model. As can be seen in the figure, about 1.5 dB degradation exists at the error rate of 10^{-4}, though this figure includes the bandlimit effect by the output BPF.

One of the important factors which determines the realizability of the regenerative repeater mentioned in Section IV is the high-power PSK modulator. In order to obtain the effective isotropic-radiation power (EIRP) of 43 dBw, modulator output of 2 W is needed with the antenna gain of 40 dB. In the experimental model, a reflection-type modulator using p-i-n diodes is adopted and the output of 2.7 W is obtained for 4 W input [20].

The power dissipation of the modulator driver is approximately 5 W. And the transition time of the modulator output is approximately 5 ns. The development of devices and the improvement of the circuit will make the modulator of 160 M bands possible to operate with the power dissipation within 5 W in the near future.

VI. Conclusions

In PCM/TDMA satellite communications, the regenerative repeater on board the satellite is very effective to improve the communication quality and provides facilities for on-board processings. This paper first developed discussions on basic configurations of the on-board regenerative repeater. Simplification of the repeater by omission of carrier-recovery circuit or clock-recovery circuit, or both of them, were scrutinized. Then various kinds of on-board processings were presented in connection with regenerative repeating. Isolation of the up-link and the down-link by the regeneration, and the baseband stage in the repeater facilitate on-board processings. A regenerative repeating SS/TDMA satellite system was proposed to show the merits of the regenerative repeating concretely. The regenerative repeating is very appropriate to be applied to SS/TDMA system from the following reasons. The satellite switching can be easily performed at baseband stage using conventional integrated circuits. The retiming of all the received signals by the clock oscillated in the satellite can be done easily in the regenerative repeater and it will improve the communication efficiency and simplify the earth station receiver. Many TWTA in the repeater can be omitted by adopting high-power PSK modulator, and it will reduce the weight of the repeater remarkably. The synchronization of the satellite switching and the transmission of the TDMA bursts from earth stations can be easily attained by generating reference bursts in the satellite. Lastly the result of the development of an experimental regenerative repeater, which was manufactured especially in order to examine the feasibility of the high-power PSK direct modulator, was presented.

It is expected that the regenerative repeater will be introduced into the future satellite communications to provide high quality, high efficiency, and many operational flexibilities.

Acknowledgment

The authors wish to thank Dr. S. Oshima, Mr. N. Sasamoto, Dr. Y. Nakagome, Dr. H. Kaji, Dr. H. Yokoi, Dr. K. Nosaka, and Dr. M. Yamaguchi for their guidance and encouragement. Thanks are also due to H. Sasaki and colleagues in radio communications research group for their helpful discussions.

References

[1] R. K. Kwan, *et al.*, "The TELESAT TDMA system," presented at Proc. Inter. Conf. on Communications, Session 44, June 1975.

[2] M. R. Wachs and D. E. Weinreich, "A laboratory study of the effects of CW interference on digital transmission over nonlinear satellite channels," in *Proc. INTELSAT/IECE/ITE 3rd Int. Conf. Digital Satellite Commun.*, 1975, pp. 65–72.

[3] E. R. Cacciamani and C. J. Wolejsza, Jr., "Phase ambiguity resolution in a four-phase PSK communications system," *IEEE Trans. Commun. Technol.*, vol. COM-19, pp. 1200–1210, Dec. 1971.

[4] M. Kato and H. Inose, "DDM differentially coherent detection scheme of DPSK signals," *Trans. IECE (Japan)*, vol. J59-A, no. 2, pp. 133–140, Feb. 1976.

[5] J. Salz and B. R. Saltzberg, "Double-error rates in differentially coherent phase systems," *IEEE Trans. Commun. Syst.*, vol. CS-12, pp. 202–205, June 1964.

[6] K. Nosaka and T. Muratani, "New satellite transponder with regenerative repeater and higher order DPSK for TDMA/SDMA system," in *Proc. INTELSAT/SEE 2nd Int. Conf. Digital Satellite Commun.*, 1972., pp. 460–468.

[7] K. Koga, T. Muratani, A. Ogawa, M. Yamaguchi, and K. Nosaka, "A satellite regenerative repeating system using higher order DPSK," to be published in *Trans. IECE (Japan)*, 1976.

[8] D. P. Sullivan, "Future trends in military communication satellite repeaters," *IEEE Trans. Aerosp. Electron. Syst.*, vol. AES-6, pp. 129–136, Mar. 1970.

[9] R. Y. Huang and P. Hooten, "Communication satellite processing repeaters," *Proc. IEEE*, vol. 59, pp. 238–252, Feb. 1971.

[10] A. D. Perry, Jr., "Electronics-communications," *Space/Aeronaut.*, pp. 81–84, July 31, 1968.

[11] W. G. Schmidt, "Satellite-switched TDMA: transponder-switched or beam-switched," *AIAA*, paper no. 74–460, 1974.

[12] G. D. Forney, Jr., "Coding and its application in space communications," *IEEE Spectrum*, vol. 7, pp. 47–58, June 1970.

[13] W. W. Wu, "Applications of error-coding techniques to satellite communications," *Comsat Tech. Rev.*, vol. 1, no. 1, pp. 183–219, Fall 1971.

[14] W. G. Schmidt, "An on-board switched multiple-access system for millimeter-wave satellites," in *Proc. INTELSAT/IEE 1969 Int. Conf. Digital Satellite Commun.*, pp. 399–407.

[15] W. G. Schmidt and R. Cooperman, "A satellite-switched SDMA/TDMA system for a wideband multibeam satellite," presented at Proc. Inter. Conf. Commun., June 1973.

[16] T. Muratani, "Satellite-switched time-domain multiple access," in Conf. Rec., 1974 *IEEE Eascon* pp. 189–196.

[17] Y. Ito, M. Kyogoku, and M. Yamaguchi, "Test result of experimental microwave on-board switching system," *Trans. IECE (Japan)*, vol. J59-B, no. 2, pp. 122–129, Feb. 1976.

[18] W. G. Schmidt, "The application of TDMA to the Intelsat IV satellite series," *Comsat Tech. Rev.*, vol. 3, no. 2, pp. 257–273, Fall 1973.

[19] Y. Ito, Y. Urano, T. Muratani, and M. Yamaguchi, "Analysis of a switch matrix for an SS/TDMA System," this issue, pp. 411–419.

[20] Y. Takayama and Y. Higo, "3.8 GHz 4 W PSK modulator using p-i-n diodes," *NEC Res. Dev.*, nos. 37/38, pp. 30–38, Apr./July 1975.

An intersatellite link (ISL) is a two-way communication channel between two or more satellites. It provides a connection between users who access two different satellites in the same or different orbits. In so doing, it can significantly reduce the number of additional accesses needed, or provide connectivity between users who cannot see the same satellite.

The first reported intersatellite link was made in January 1975 between Oscar 6 and Oscar 7, the two VHF amateur radio satellites [1]. Later that year, a link was established between SPACELAB and ATS-6. In February 1976, the LES 8 and LES 9 (Lincoln Experimental Satellites, Massachusetts Institute of Technology, Lincoln Laboratory) established a link between two three-axis stabilized synchronous satellites at 36 GHz. An intersatellite link is planned on the TDRSS satellite [2].

An ISL capability can impact significantly upon system performance. The following system characteristics are most directly affected.

1) Connectivity: The ISL can be used to provide flexibility in satellite system design while still achieving a required level of interconnectivity within a community of users. Typically, the total community of users will be composed of subgroups which form strong communities of interest, i.e., have a lot of internal traffic with a more modest, but nonzero, traffic requirement between these subgroups. Normally, if full interconnectivity is to be provided, then the total community must access a single large-capacity satellite. An ISL capability allows subgroups of the total community to access separate lower capacity satellites, with the ISL accommodating the intercommunity traffic so that the capability for full connectivity is retained. The system benefits of intersatellite links in the future INTELSAT system are described in Reprint Paper 7.1.

2) Coverage: If a satellite system is to provide global coverage, then users in one region must be capable of talking to those in regions not covered by their set of satellites. This may be accomplished by terrestrial means or by ''double hopping'' the signal; that is, sending it up to one satellite and down and then repeating the process. This double hopping requirement can be eliminated by use of an ISL, with a resultant decrease in the transmission delay.

3) Orbit Utilization: By allowing more flexibility in orbit position without large connectivity and capacity penalties, the ISL can significantly increase orbit utilization efficiency.

4) Cost: The cost of the earth segment can potentially be significantly reduced by the increased connectivity provided by the ISL. This results from a reduction in the number of earth stations required.

An ISL system is composed of four generic subsystems: receiver, transmitter, acquisition and tracking, and an antenna/lens system. An optimized design is developed by trade-offs among these subsystems. The subsystems perform the following functions.

1) Receiver: The receiver, in the most general case, detects the signal and demodulates and decodes the received information. It also provides the interface between the ISL and the normal mode of satellite transmissions, including any required format conversation. In some realizations, there will be no decoding and minimal format conversion.

2) Transmitter: In the general case, the transmitter selects those information signals that are to be sent across the link, encodes them, and then provides modulation, frequency translation, and amplification. In some systems, the transmitter may simply remodulate an appropriate high-power source.

3) Tracking and Acquisitions: When the two satellites are in their respective orbits, the two systems must acquire each other and then track to an appropriate accuracy ($\approx$ 0.1 of the beamwidth) as the information is transmitted across the link. Such techniques as raster scan and monopulse can be used in this subsystem. It should be noted that it is not necessary to orient the entire spacecraft to the required accuracies, but only a portion of the ISL subsystem, e.g., a lens, reflector, or antenna feed.

4) Antenna/Lens Subsystem: The aperture used to interface with free space depends upon the wavelengths used for transmission. For standard microwave or millimeter-wave transmissions, an antenna system such as a reflector or lens would be adequate. For optical wavelengths, an optical reflector or lens system would be required. The key differences are, of course, aperture sizes, surface tolerances, and materials.

There are two frequency ranges that appear attractive for ISL operation—millimeter wave and optical. In both of these ranges there are sources available, although there is a significant need for source technology improvement. In the millimeter band of 55–60 GHz, allocated to ISL's, there are TWTA's that can be used in a standard fashion. The tubes have in excess of 10-W peak output with about 10–15% efficiency. In order to provide low noise reception, a cooled paramp, which can yield a receive noise temperature of 750 K, is available. Present uncooled receiver noise temperatures for the 55–60 GHz range are less than 2500 K. However, using a passive cooler, this can be reduced. In the higher millimeter ISL bands, above 100 GHz, no strong technology base exists.

The optical frequency range is currently under development. For wavelengths in the range of 0.5 to 10.6 μm, increased antenna gains, increased path losses, and changes in sources and detectors occur. The most promising sources in these frequency ranges are all lasers with coherent emissions. At the present time, there are four main candidates.

1) CO$_2$ (10.6 μm): The source here is a gas laser in the infrared region at 10.6 μm. This laser, which has undergone considerable development, is capable of delivering several watts of power and has high efficiency ($\approx$ 10%). Typical modulators use cadmium telluride crystals, often in an intracavity configuration to reduce the required driver power. Large bandwidths are available, on the order of 300 MHz, and detection is performed by heterodyning with another laser and using a diode type detector, which must be cooled to approximately 110 K.

2) Nd:YAG (1.06 μm): The neodinium yttrium aluminum garnet (YAG = $Y_3Al_5O_{12}$) laser is similar to the CO$_2$ in that it has a high power output, but it is less efficient and is a solid rather than a gas. The Nd:YAG laser is externally pumped by a light source in an appropriate frequency band. Pump sources include the sun, low-pressure discharge lamps, and solid-state light-emitting diodes and lasers.

3) HeNe (0.63 μm): This is the oldest source. It is a gas laser with a very long lifetime, but low output power (on the order of milliwatts) and low efficiency ($\approx$ 0.01%). It is useful for low data rate (<100 Mbit/s) links that span reasonably short paths (<10° of synchronous arc).

3) GaAs (0.9 μm): Gallium arsenide or GaAlAs is a solid-state laser of high efficiency, potentially long lifetime, and modest output power, at least at present. Because it can be directly modulated, tuned to various wavelengths by doping, and is amenable to direct detection techniques, GaAs is an attractive source for those applications where it has sufficient power.

A direct comparison of optical and microwave receiver performance is not straightforward because millimeter receivers have standard kT noise; whereas, the fundamental noise for optical receivers is quantum in nature and depends on the detection technique [3]. Thus, care must be taken in comparing optical and microwave systems.

References [4]–[6] discuss various optical systems. Reprint Paper 3.6.7.1 describes the intersatellite link on the LES 8/9 satellites.

References

[1] P. I. Klein and R. Soifer, "Intersatellite communication using the AMSAT-OSCAR 6 and AMSAT-OSCAR 7 radio amateur satellites," *Proc. IEEE,* pp. 1526–1527, Oct. 1975.

[2] W. M. Holmes, "The tracking and data relay satellite system," *Proc. AIAA 7th Communications Satellite Systems Conf.,* San Diego, CA, pp. 158–164, Apr. 1978.

[3] E. V. Hoversten, "Optical communication theory," *Laser Handbook,* Eds. F. T. Arecchi and E. O. Schulz-Dubois, New York: North-Holland, 1972, pp. 1805–1862.

[4] J. D. Barry *et al.,* "1000-Mbits/s intersatellite laser communications system technology," *IEEE Trans. Commun.* pp. 470–478, Apr. 1976.

[5] W. L. Ohern and C. A. Rodenberger, "Laser data links for communications satellites," *Proc. 5th AIAA Communications Satellite Systems Conf.,* Los Angeles, CA, Apr. 22–24, 1974.

[6] D. C. Forster, F. E. Goodwin, and W. B. Bridges, "Wide-band laser communications in space," *IEEE J. Quant. Elec.,* vol. QE-8, pp. 263–272, Feb. 1972.

SATELLITE-TO-SATELLITE DATA TRANSFER AND CONTROL*

DAVID M. SNIDER and DAVID B. COOMBER

Lincoln Laboratory, Massachusetts Institute of Technology
Lexington, Massachusetts

ABSTRACT

The LES-8 and -9 satellites are three-axis
stabilized earth-orbit spacecraft which provide
communications at UHF and K-band frequencies with
the capability of simultaneously establishing and
automatically maintaining a K-band (36-38 GHz)
satellite-to-satellite data link. Narrow beam (1
deg) antennas are used which incorporate flat
reflecting plates for beam pointing in such a way
that slip rings or waveguide rotary joints are
not required. The antenna beams are electrically
lobed by sequentially switching offset feedhorns
to produce error signals for autotracking.
Spatial acquisition is accomplished by a digital
pointing control which scans the spatial uncertainty while an autotrack receiver eliminates
frequency offsets and doppler. Link acquisition
is accomplished when the antenna boresights are
aligned and the received signals are in the
receiver passbands. The link can support a data
rate of 10 or 100 kb/s, depending on satellite
separation, while maintaining antenna beam
alignment of 0.04 degrees. Differences in
spacecraft timing and doppler effects are accommodated by an "elastic" shift register which
inserts a variable delay to allow synchronization
of the crosslink frame to the nominal downlink
frame so that simple time-multiplexing will yield
a downlink containing both crosslink and uplink
data streams.

The inter-satellite data "crosslink" extends
the earth coverage over that which a single
satellite can provide and demonstrates how a
global communications system might be realized
without the restrictions imposed by ground
relay. Crosslink antenna pointing angles have
been used as a reference for the Attitude Control
system to maintain satellite spatial orientation
without any earth references. The link also
includes capability to relay Command and Telemetry
so that control of a spacecraft not nominally in
range of any ground station is achieved.

The paper describes the operation and
characteristics of the lobing antenna and pointing
control, the solid state transmitters and receivers, the autotrack receiver and elastic shift
register. Ground based and post-launch test
results of pointing and communication systems are
presented. Also, a user-oriented system is
described which, among other features, allows an

operator to easily establish crosslink from the
ground terminal using pointing angles calculated
by a small computer using spacecraft derived
earth pointing parameters.

INTRODUCTION

The LES-8 and -9 satellites are three axis
stabilized spacecraft which were placed into geosynchronous orbit on March 14, 1976. Each satellite provides earth communications in the military
UHF band through a 10-dB gain dipole array and at
K-band (37 GHz) through a 25-dB gain horn antenna.
Both spacecraft are also capable of simultaneously
establishing and automatically maintaining a K-
band satellite-to-satellite data link. The
inter-satellite data "crosslink" extends the
earth coverage over that which a single spacecraft
can provide by allowing the retransmission of
information initially received by one satellite
on the downlink of the other satellite. The
crosslink can support a data rate of 2.4 to 100
Kb/s while providing earth coverage for the up
and downlinks determined by the UHF and K-horn
antenna beamwidths and the satellite-to-satellite
central separation angle of up to 100 degrees
(see Fig. 1).

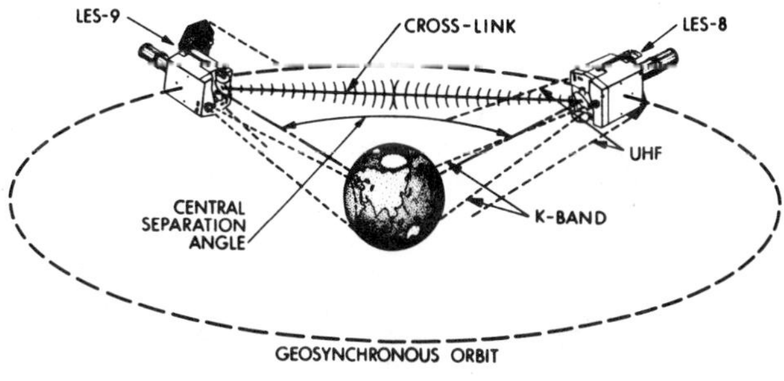

Figure 1. ANTENNA COVERAGE

Each spacecraft employs solid state components
in redundant configurations wherever possible to
enhance satellite reliability and lifetime. In
keeping with this philosophy, four silicon Impatt
diodes are combined in the output of the K-band
crosslink transmitters to reliably generate
0.5watt BPSK modulated signal. The crosslink
receivers each incorporate two gallium arsenide
diodes in a balanced mixer design which provides

*This work was sponsored by the United States Air
Force and the United States Navy.

a 7-dB front-end noise figure. The desired
satellite-to-satellite data rate of 100 Kbps
together with the available transmitter output
power and receiver noise figure dictate crosslink
antenna gains of about 42-dB for a satellite-
to-satellite separation of 40,000 km. (See Link
Calculation, Fig. 2)

Required E_b/N_o	=	10 db (BER = 10^{-6})
Desired data rate	=	100 Kbps (50 dB-Hz)
Required P_r/N_o	=	60 dB-Hz
Available transmitter Po(rf)	=	+28 dbm
Tx filter and polarizer losses	=	−2 db
Path Loss (40,000 Km, 37 GHz)	=	−216 db
Rx filter and polarizer losses	=	−1 db
K.T.N.F. (noise figure = 7 db)	=	−167 dbm-Hz
Resulting P_r/N_o	=	−24 dB-Hz
Required antenna gain (60 + 24)	=	84 dbI

Figure 2. LINK CALCULATION

The antenna gain is provided by a pair of
18-inch parabolic dishes (one per spacecraft)
but the resulting narrow beamwidths (1 degree)
require the alignment of the antenna boresights
along the line of sight between the spacecraft.
Antenna pointing is provided by controlling the
orientation of flat reflecting plates to redirect
the beams. This technique has the advantage
that no rf rotary joints are needed. The plates
are supported by elevation over azimuth positioners which incorporate integral resolvers that
have been aligned to the satellite physical axes
so that the antenna beams can be pre-pointed in
any desired direction over a range of ± 10
degrees in elevation and ± 52 degrees azimuth.
Brushless dc tachometers and torque motors are
used in conjunction with wire crossovers to
provide damping and pointing torque without
resorting to slip rings. The pointing control
also accepts pointing error signals produced by
electrically lobed feedhorns and autotrack
receivers such that link acquisition is assured
if the antenna boresights are within 0.7 degrees
of each other and the boresight P_r/N_o > 46 dB.

If the antenna boresights are not within
0.7 degrees of each other the pointing controls
have available (by Command) a SCAN mode whereby
the antenna beams are automatically stepped over
a spatial uncertainty of 2.1 degrees x 3.5
degrees in 0.7 degree steps while the autotrack
receivers sweep a frequency uncertainty of ± 5
KHz. The spatial scan range was sized to overcome
the 3σ uncertainty in pointing direction due to
tolerances in attitude control, orbit fit and
spacecraft structural deflections while the
frequency sweep accommodates doppler and crystal
oscillator drift. Our post-launch experience
has been that pre-acquisition pointing angles
are usually within 0.5 degrees of antenna boresight.

Due to the many modes of LES-8/9 and the
large number of potential users, the basic timing
was selected to run off real time, or UTC, so
that uplink and downlink synchronization for
several modes can be achieved with only a knowledge of UTC. The reception of crosslink data
from another satellite is therefore complicated
because the instantaneous data rate over the
crosslink will, in general, be slightly different
from the downlink data rate of that same satellite due to the variable satellite-to-satellite
separation. To overcome this basic difficulty we
have employed an elastic shift register (ESR)
which is a variable length buffer capable of
slightly different input and output rates. Provision is made for the long-term effect of higher
or lower relative clock rates at input and output
by allowing the ESR to delete or repeat frames of
50 bits. Using entire 50-bit frames avoids
disturbing the frame sync in the ground terminals,
while in addition, 1 second warning is given of
an impending frame deletion or repetition by
modification of a communications sync code and by
Telemetry.

The synchronizer following the ESR establishes
where to tap the buffer so that a crosslink frame
will coincide with a downlink frame to allow
easily demodulated TDM. In its acquisition mode,
the synchronizer tests subsequent bit positions
for fewer than 7 frame sync failures in 32 frames.
Achieving this at some point, the monitor mode is
entered which must find fewer than 103 out of
256 possible frame sync failures to maintain this
tap position. Average time to acquire frame sync
at 100 kb/s is 0.18 msec. After frame sync is
established, the system must acquire parity sync
on the low rate (200 bps) data multiplexed into
the crosslink so that telemetry and low rate
uplink data received by the other satellite is
successfully routed to the 8-ary UHF downlink.
Worst case acquisition time for this mode is about
10 seconds.

In view of the complex nature of the crosslink communications system on LES-8/9, a comprehensive, yet manageable, control center for these
functions was a logical result of a desire to
fully understand and exercise the various satellite links. This has been realized by integrating the Command and Telemetry systems into a
set of display panels which show the configuration of the communications modes of each satellite, indicating binary switch positions, signal
levels at various points in dB, real time clock
monitoring, dish pointing angles, attitude control corrections, and similar functions. An
integral part of the display panels is the command generator which has push buttons and data
switches located on the panel adjacent to related

telemetry. Key guarding is provided to prevent
unauthorized commanding and the command generator
is designed so that only communications-related
commands can be output.

SPATIAL ACQUISITION

In order to establish the satellite-to-
satellite crosslink, both satellite antenna boresights must first be simultaneously brought to
within 0.35 degrees of the line of sight (LOS)
between the spacecraft so that spatial pull-in
can occur. Simultaneity is a requirement since
each spacecraft is the beacon for the other to
pull-in and track on. Once spatial acquisition
has been accomplished, the full antenna gain is
available to enhance frequency acquisition.

The cartoon shown in Figure (3) depicts how
the spatial and frequency uncertainties are
resolved automatically. The satellites are shown
with antenna beams pre-pointed to lie along the
calculated LOS. Tolerances on orbit fit, Attitude Control and deformation of the satellite
structure due to thermal and "zero-G" stress
total to about ±1 degree so that these initial
pre-point angles can be in error by that amount.
To overcome these errors, the satellites are
commanded to begin incrementally searching a 2.1
x 3.5 degree area centered about the direction of
best available information in 0.7 degree steps.
The spatial scans of both spacecraft are not
synchronized so that one spacecraft scans at a
rate sixteen times slower than the other to allow
the fast scan satellite to complete its 15 position scan before the slow scan satellite increments to a new position. The fast scan spacecraft takes 0.5 seconds to reach a new pointing
position and maintains that direction for 2.3
seconds. In this way adequate time is provided
for the receivers to sweep the full frequency
uncertainty without synchronizing the frequency
sweeps to either spatial scan. Should the boresight P_r/N_o degrade, both spacecraft can be
scanned at half the normal rate (by Command)
allowing time for two frequency sweeps with a
resulting improvement in the probability of
spatial and frequency acquisition.

CROSSLINK BLOCK DIAGRAM

A simplified functional block diagram of the
crosslink system is presented in Figure (4). The
required transmit/receive antenna gain is provided by an 18-inch parabolic reflector illuminated by a circular-polarized feedhorn. The beam
is electronically "squinted" 0.07 degrees by
sequentially energizing pairs of bi-phase modulators which are connected to offset feeds that
have 20 dB coupling to the main feed. The beam
squinting impresses amplitude modulation on the
K-band received signal for sources off boresight.
After mixing down to IF (3 MHz) an AGC amplifier
normalizes the AM waveform so that variations in
signal level due to the antenna pattern and
crosslink range are eliminated. The AM signal is
then bandpass filtered and detected in synchronism with the modulator drivers to produce
azimuth and elevation error signals. These

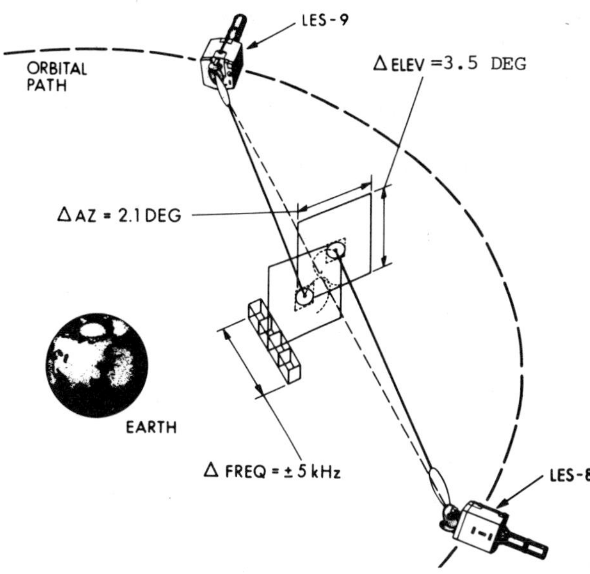

Figure 3. SPATIAL ACQUISITION

signals are used in a closed loop fashion with the
signal from the other spacecraft as a reference to
provide satellite-to-satellite autotracking. Amplitude modulation of the transmit signal caused by
bi-phase modulator effects at the transmit frequency is indistinguishable from real mispointing
and is eliminated by including notch (band-stop)
filters between the bi-phase modulators and the
offset feeds.

The antenna pointing direction is controlled
by the position of a flat reflecting plate placed
in front of the parabolic dish to re-direct the
focused beam (Fig. 5). The flat reflector is
mounted to a gearless bi-axial drive which
utilizes resolvers as a reference in the pointing
control feedback loop to pre-point the antenna.
Shaft torque and damping are provided by brushless dc torquemotors and tachometers while an
azimuth axis wire cross-over eliminates the need
for slip rings. Since the antenna feed network
does not rotate with the reflecting plate, the
lobing axes change with plate orientation so that
the pointing error signals must be put through a
coordinate rotation to realign the electrical
lobing axes with the non-rotating feed axes. The
functions of error signal electronic rotation,
beam pre-pointing, spatial scan control and
autotracking are all provided by the Pointing/
Scan control.

Lobing Feed

The five horn feed assembly is shown in Fig.
(6). The mechanical support and alignment mechanism utilizes slant and slotted ways to adjust
the feed horn assembly to within ±0.002 inches
laterally and ±0.10 degrees angularly. The
center feed design incorporates a crossed septum
which increases the lobing error slope by increas-

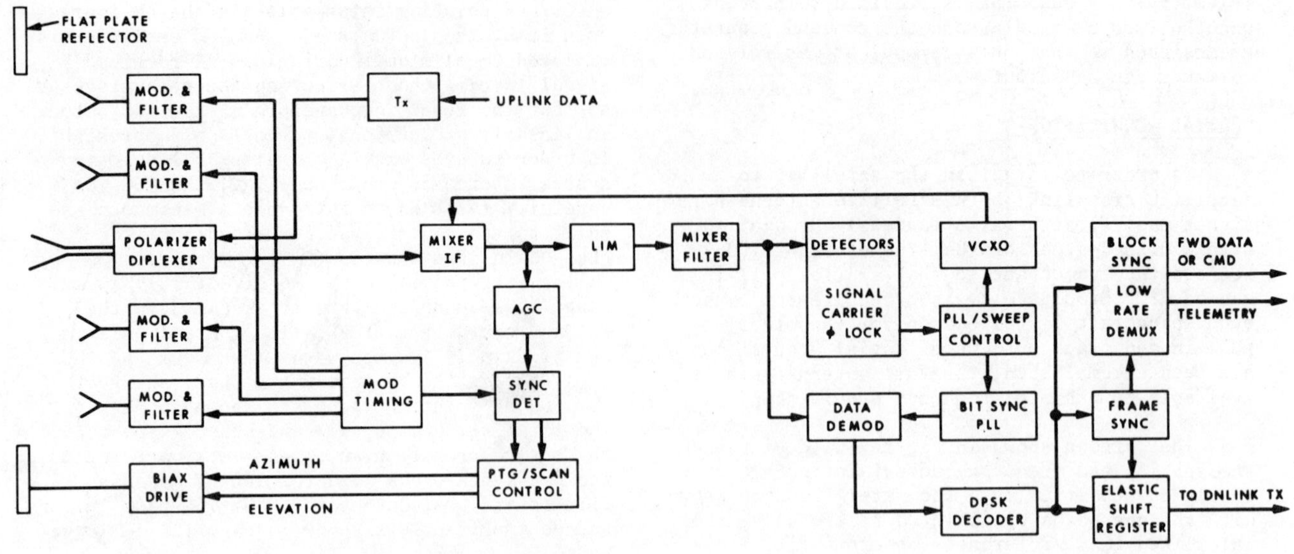

Figure 4. CROSSLINK SYSTEM BLOCK DIAGRAM

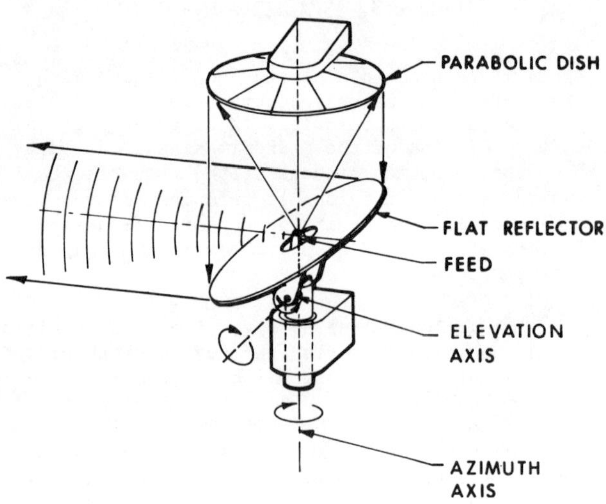

Figure 5. CROSSLINK ANTENNA SYSTEM

ing the coupling between the main and offset feeds without reducing the antenna efficiency. The bi-phase modulators require 65 ma at 3 volts to provide 180 degrees RF phase shift and are driven by 100 Hz square waves shifted by 90 degrees to each other such that 360 degrees of antenna lobing occurs in four steps at a 100 Hz rate. The effect of energizing the phase modulators is shown clearly in the antenna patterns in Fig. (7). Here opposite pairs of modulators have been energized prior to recording a pure azimuth pattern with a resulting "squint" in the pattern of about ±0.07 degrees. The antenna lobing characteristics are presented in Fig. (8). The concentric lines are contours of constant

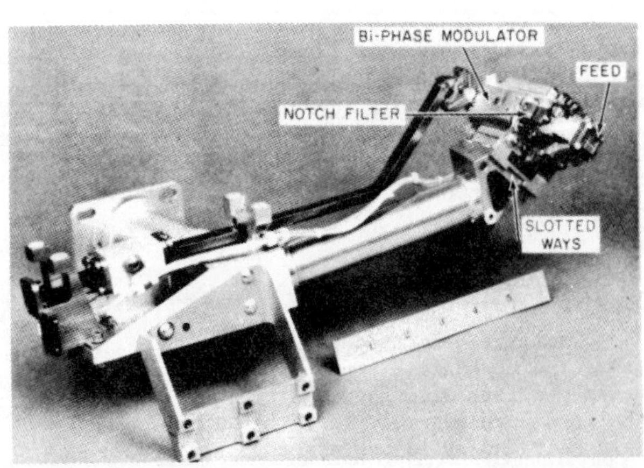

Figure 6. LOBING FEED ASSEMBLY

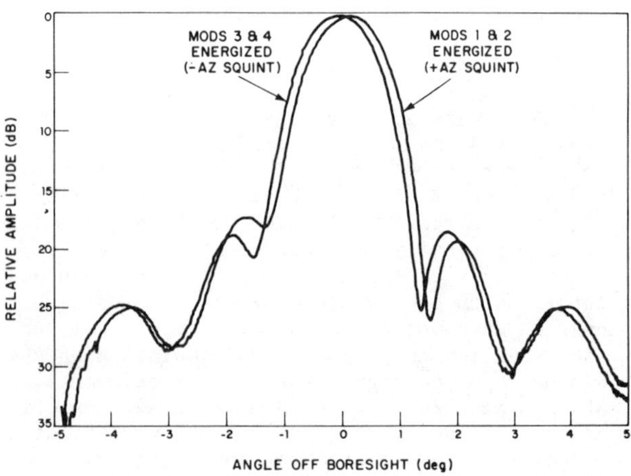

Figure 7. LOBED ANTENNA PATTERNS

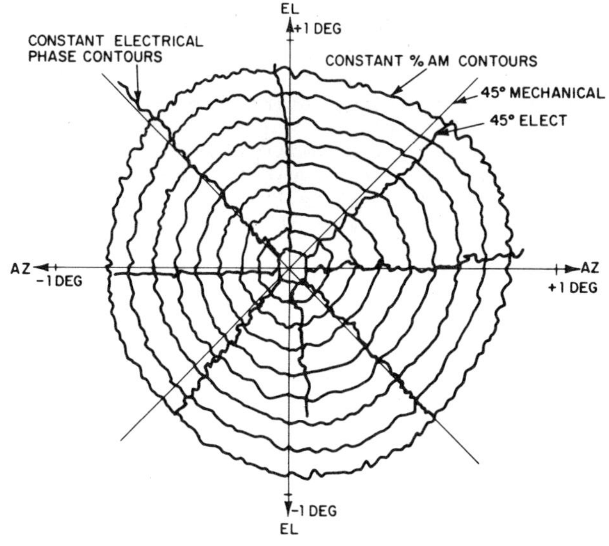

Figure 8. LOBING SENSITIVITY AND ORTHOGONALITY

percentage AM and show the variation in lobing sensitivity to be less than ±10 percent within one degree of boresight. The radial lines are contours of constant AM phase and show the electrical lobing axes to be within ten degrees of orthogonality. Reducing these imperfections to be within the total autotrack amplitude and phase variation stability budget of ±50 percent and ±40 degrees requires the axial ratio of each feed to be fine-tuned to better than 0.8 dB. In addition, ferrite isolators are required in the waveguide runs leading to the transmitter and receiver so that any reflection from the diplexer bandpass filters will not upset the axial ratio.

The K-Band receiver consists of a two diode balanced mixer followed by a 2 GHz transistor amplifier providing a 7 dB front end noise figure. In order to minimize waveguide and circuit losses and reduce the probability of radiative interference, the receiver is built as an integrated mechanical unit physically attached to the antenna feed assembly. After frequency conversion to 3 MHz, the signal is passed through an attentuator which eliminates the possibility of sidelobe acquisition by reducing the signal-to-noise ratio at IF due to sidelobe radiation below the S/N required to spatially pull-in. The attentuator is adjustable by Command, prior to spatial acquisition, in 5 dB steps over a 75 dB range such that boresight autotracking is assured for any central separation angle greater than 1 degree.

CROSSLINK AUTOTRACK RECEIVER

The input to the autotrack receiver is a DPSK modulated 3 MHz signal from the step attenuator which also contains satellite to satellite

autotrack pointing information in the AM impressed upon it by the lobing feed. An AGC amplifier is employed to eliminate variations in received signal level without affecting the percentage AM so that the resulting waveform at the AGC output is linearly proportional to angle off boresight. In order to stay within the allowable pointing system inter-axis cross-coupling, the AGC was carefully tailored to introduce a minimum phase shift to the 100 Hz tracking signal due to changes in input levels (AM to PM conversion) while still providing the required 55 dB dynamic range and 350 dB/sec rate to accommodate the antenna characteristics during spatial pull-in and tracking. The measured effects of the AGC amplifier on the amplitude and phase shift of the 100 Hz pointing signal is shown in Fig. (9). The total servo amplitude and phase response of the entire crosslink autotrack receiver, including the pre-synchronous detector 100 Hz bandpass filters and post-detector 3.5 Hz low-pass filters and dc amplifiers is shown in Fig. (10). Interpretation of Figs. (9) and (10) is significant

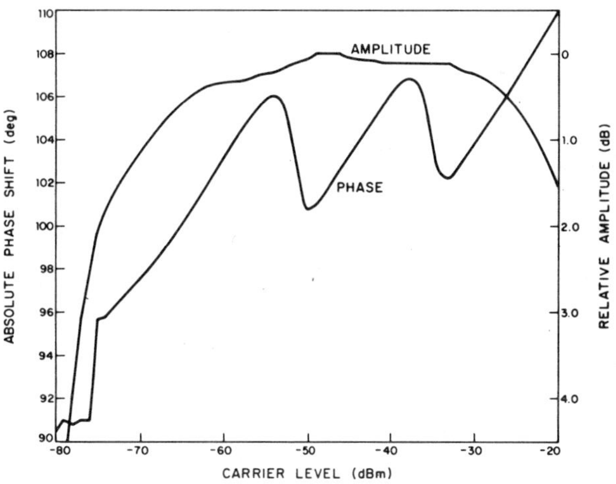

Figure 9.
AMPLITUDE AND PHASE SHIFT OF 100 Hz MODULATION

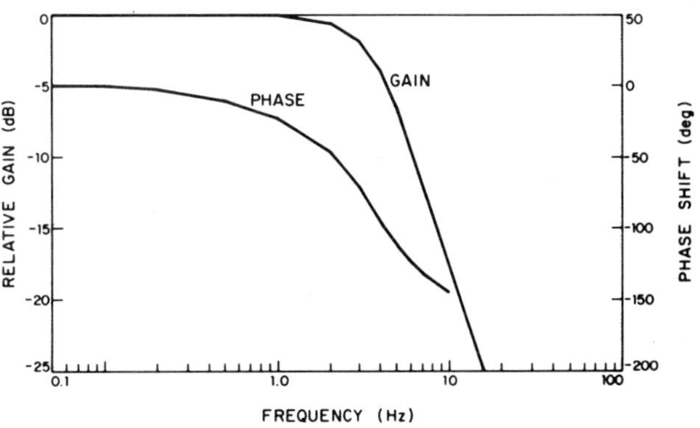

Figure 10. SERVO AMPLITUDE AND PHASE RESPONSE

because the difference in stability reducing mechanisms between the AGC loop and the rest of the crosslink receiver is evident. The AGC loop reduces system stability by shifting the phase of the 100 Hz tracking signal and thereby introducing channel cross-coupling prior to synchronous detection. The receiver filters insert an amplitude and phase response in series with the Azimuth and Elevation channel servo loops leading to a direct reduction in stability margin.

The outputs from the crosslink autotrack receiver are dc voltages proportional to the antenna beam misalignment from the L.O.S. which have the appropriate polarity to cause the reflecting plate to move and reduce the misalignment to zero.

POINTING/SCAN CONTROL

The operating mode of the pointing system is selected by ground Command and implemented by the Pointing/Scan Control. A simplified block diagram of the Azimuth axis portion of the Control is shown in Fig. (11). In the Acquisition Mode (ACQ) closed loop error signals to pre-point the antenna are derived from the differences between the desired pointing direction and the actual pointing angles provided by resolvers located in the biax drive. The ACQ mode is initiated by placing the desired pointing angles into the command registers via the Command system and verifying correct reception via Telemetry and executed by transmission of the ACQ command. In the SCAN mode the antenna is spatially scanned by digitally adding ROM (read only memory) generated pointing angles to the ACQ mode signals. The SCAN is an incremental spiral about the angles

stored in the command registers and is initiated by transmission of the SCAN START command. Four scan rates are available from 2.8 to 89.6 seconds per step. The crosslink autotrack mode (TRACK) uses the error signals derived by the autotrack receiver to control the position of the antenna. The coordinate rotation required to unravel the cross-coupling caused by having a feed system which does not rotate with the antenna is given by the following equations:

$$AZ \text{ error} = Z_1 \cos AZ + Z_2 \sin AZ$$

$$EL \text{ error} = Z_2 \cos AZ - Z_1 \sin AZ$$

where Z_1 and Z_2 are the azimuth and elevation channel error signals from the autotrack receiver.

Digital integration is included in the pointing control to overcome friction drag in the biax and to compensate for mechanical, magnetic, and electrical biases in the system. 350 IC's mounted on 16 multilayer P.C. boards are required to build this equipment, which also includes back-up modes, telemetry points, and biax overtemperature safety circuits not detailed in this paper.

BIAX DRIVE

A cross section of the azimuth portion of the biax drive is shown in Fig. (12). The mechanical design accommodates orienting the tachometer and torquemotor permanent magnets to minimize their net magnetic moment so that the torque exerted on the spacecraft due to the interaction with the earth's magnetic field is small compared to the other disturbances affecting the Attitude Control system.

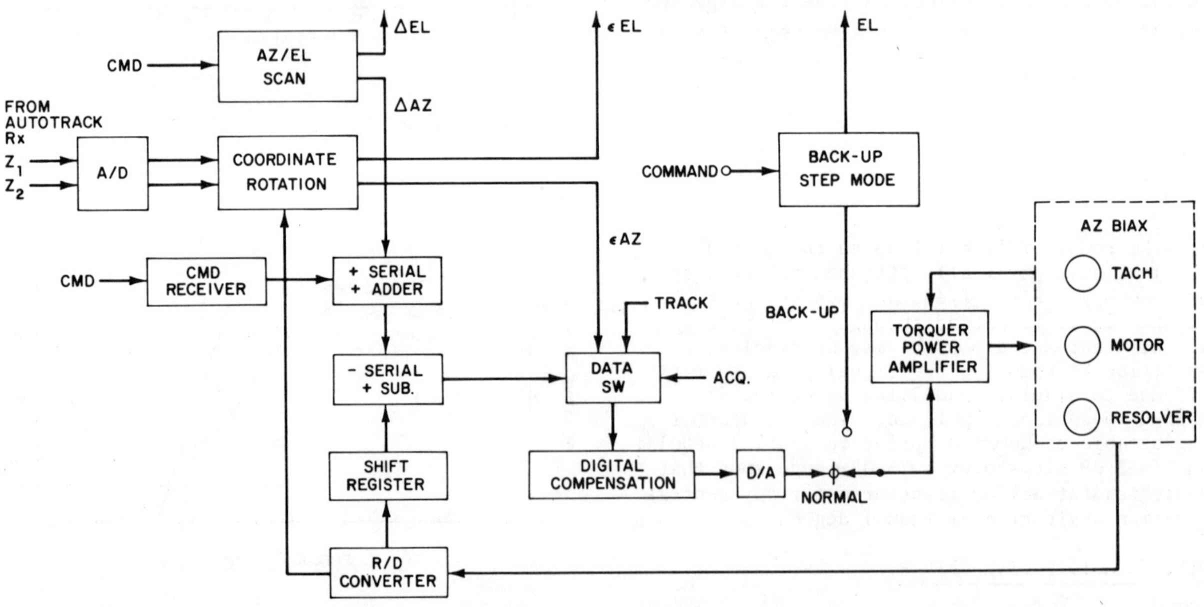

Figure 11. POINTING/SCAN CONTROL BLOCK DIAGRAM

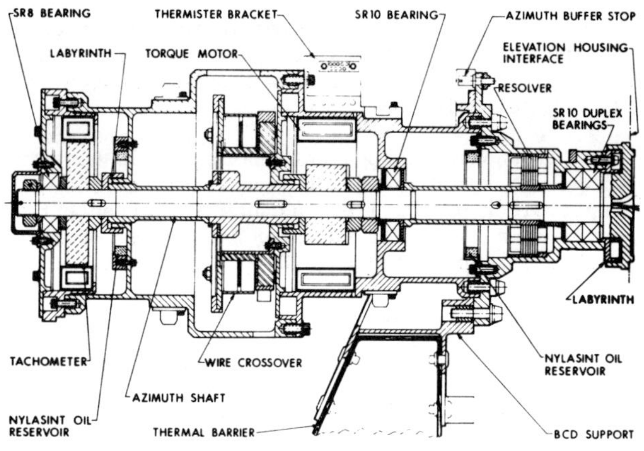

Figure 12. *CROSS-SECTION AZIMUTH DRIVE*

The small clearances associated with the tachometer, torquemotor, resolve and labrynth require a ball bearing supported Rene' steel shaft to provide adequate mechanical stiffness to prevent physical contact of internal parts during launch. The labrynth is a 0.005 inch gap about one-half inch long which controls the outgassing rate of the ball bearing lubrication. Since this is a direct-drive, low torque system, the drag (torque load) of a wire harness cannot be tolerated. The wire crossover is a thirty-six conductor device which provides electrical connection through the azimuth shaft and to the axis in a low torque fashion. The design and construction of the elevation axis is similar to the azimuth except that a wire crossover is not needed since the angular range is less (10 degrees vs. 104 degrees).

Since the biax assembly is by nature of its function external to the main spacecraft body, a separate thermal control system is required to prevent excessive lubrication depletion by outgassing and to minimize thermally induced component stress. Proportional control heaters maintain body temperature at +10°C during shadow, while thermal radiators are sized to limit the maximum temperature during solar heating to +35°C. Thermal cans and multilayer thermal blankets are utilized extensively to minimize the heater power requirements and radiator size.

SIGNAL ACQUISITION

After spatial acquisition of the crosslink antennas, the crosslink signal must be acquired. The signal between LES-8 and LES-9 is a carrier DPSK modulated at command-selectable rates of either 10 or 100 kb/s. The data format is the same for both rates and timing is synchronized to the on-board real-time-clock. The frame format is described in Fig. (13). The first bit in every frame, always a one, is used for establishing frame synch, to be described later; the second bit is low-rate data which is command-selectable from one of four sources; the third bit is multiplexed between telemetry data and slip/sequence data used in ground monitoring of crosslink sync;

the fourth bit is telemetry data from the crosslink satellite; the remaining 46 bits are user bits.

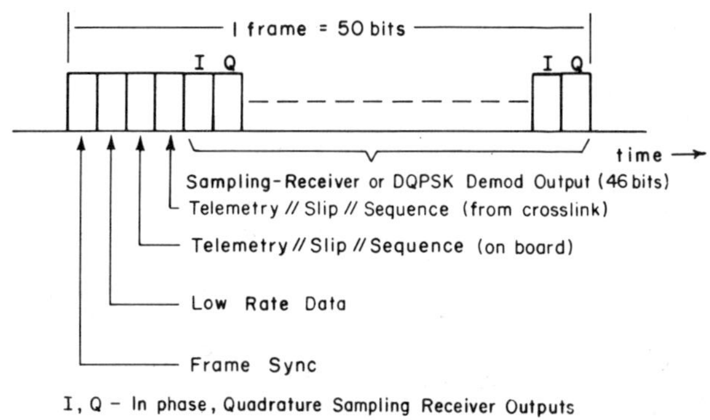

Bit Rate	Bit Duration	Frame Duration	Frame Rate
10 kbit/sec	100 μsec	5 msec	200/sec
100 kbit/sec	10 μsec	0.5 msec	2000/sec

Figure 13. *CROSSLINK DATA FORMAT*

Frequency uncertainties due to angular separation (one degree per day causes over one kilohertz Doppler shift) and eccentricity (0.01 eccentricity causes almost ±4 kilohertz Doppler shift) are accommodated by a ±5 KHz search range of a 2 GHz voltage-controlled crystal oscillator (VCXO). The resulting 3 MHz IF is limited and mixed down to baseband inphase and quadrature components. The box labelled "DETECTORS" in the block diagram determines the presence of system phase lock, unmodulated signal or modulated carrier. The receiver is closed as a second-order phase-lock loop which holds lock for $P_r/N_o > 38$ dB-Hz. After acquiring frequency and phase, the "BIT SYNC PLL" adjusts bit timing in 1/3 microsecond steps to optimally sample the data matched filters.

Since the first bit in each crosslink frame is a one, frame sync is found by testing successive bit positions for the presence of a predominance of ones in sequential frames. Two modes are used, acquisition and monitor. In the initial acquisition mode, a position is tested in 32 successive frames. If seven or more positions are not ones, the timing slips one bit and testing resumes. Eventually frame acquisition will switch the system into the monitor mode. The system still counts zeroes but will reject the sync point only if more than 102 zeroes are found in 256 consecutive frames. Loss of sync and switching to the acquisition mode uses a criterion almost twice as strict as that for entering the monitor mode.

A block consists of four Low Rate Data bits, noted in Fig. (13), accumulated from four consecutive frames. Block sync is established by identifying the parity bit corresponding to the three previous data bits. The crosslink frame sync process averages 2.3 seconds in the 10 Kb/s mode (0.68 seconds in 100 Kb/s), while block sync takes 5 to 10 seconds.

In order that the data bit stream from the crosslink can be interleaved with the data bit stream from the satellite's own uplink, synchronization requires buffering the crosslink data stream. In addition, the buffering will not be a constant delay due to relative motion of the satellites. Adjustable buffering is provided by an "elastic shift register" into which crosslink data bits are inserted at a time varying rate determined by the crosslink data rate and the crosslink delay, and from which data bits are extracted at the downlink rate.

ELASTIC SHIFT REGISTER

A block diagram of the elastic shift register is shown in Fig. (14). The Up/Down Counter controls which bit of the 64 bit shift register is selected to appear at the Data Output. For example, if the number in the Up/Down Counter is 35, then A35 is selected for the Data Output. Whenever an Input Clock pulse is received, two things happen:

1. The Input Data bit is shifted into Q0 and all other bits in the Shift Register move over one bit, e.g., the information bit stored in Q35 moves to Q36.

2. The Up/Down Counter increases its count by one and causes the selector to move over one bit, e.g., if the selector was selecting Q35, it is now selecting Q36.

It should be apparent that there is no net change in the Data Output as a result of the Input Clock pulse, i.e., as an information bit moves down the Shift Register, the selector moves with it, keeping the Data Output constant.

When an Output Clock pulse is received, it decrements the Up/Down Counter by one and thus causes the selector to move one bit to the left, e.g., if the selector was selecting Q36, it is now selecting Q35, which contains the next information bit. It does not affect any information stores in the Shift Register.

Notice that there is no requirement that the Input and Output Clock Pulses alternate and that, ignoring for the moment the problems of end-runover and time-coincidence of Input and Output Clock pulses, the input and output terminals are completely independent and may be operated asynchronously at different frequencies.

The time-coincidence problem is handled by sending the Input Clock pulse through a special synchronizer which delays it if necessary to avoid coincidence with the Output Clock pulse.

Obviously, if the Input Clock rate exceeds the Output Clock rate, more information will be inserted into the ESR than is extracted and a means of "dumping" the excess information is required. Similarly if the output rate exceeds the input rate, the ESR will be drained of infor-

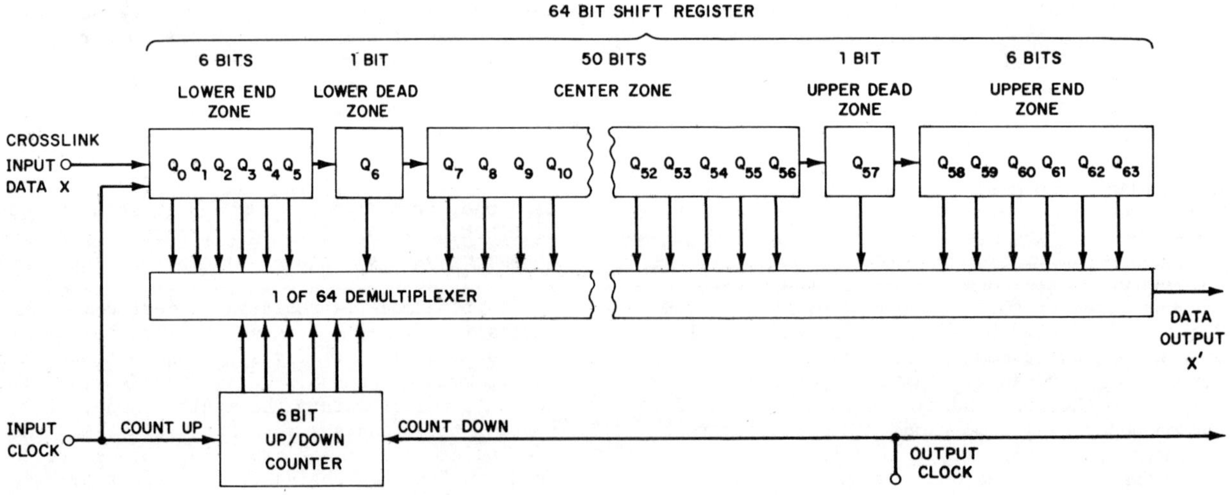

Figure 14. ELASTIC SHIFT REGISTER

440

mation and a means of "creating" more information
is needed. In order to avoid disturbing the
frame sync system of ground terminals, entire 50-
bit frames are deleted, to effect a "dumping" of
information, or repeated, to effect the "creation"
of information. A deletion is accomplished by
decrementing the Up/Down Counter by 50 which
makes the selector skip over 50 bits which are
subsequently lost forever. A repetition is
accomplished by incrementing the Up/Down Counter
by 50 which makes the selector move 50 bits to
the right. In this case the selector jumps back
over the 50 bits which had been previously output
and which will now be output again.

Other system considerations dictate that the
ESR give the ground terminals some warning of an
impending deletion or repetition. The warning
convention adopted is that the ESR will send a
delete or repeat warning message through Telemetry
and through the third frame for the one-second
interval immediately preceding the actual deletion
or repetition which occurs on the one-second
"tick" of the real time clock.

The 64 bits of the ESR may be conceptually
divided up as shown in Fig. (14). If the selector
is found pointing to a bit in the upper-end zone
at the end of a one-second interval, a delete
warning message is initiated and at the end of
the one-second warning, the Up/Down Counter is
decremented by 50, which moves the selector back
to a bit which is, by necessity, in the center
zone. Similarly when the selector is found in
the lower-end zone, a repeat warning message is
initiated after which the Up/Down Counter is
incremented by 50, which moves the selector back
into the center zone. One bit dead zone is added
at either end of the center zone to prevent the
ESR from unnecessarily repeating and deleting
frames under low-Doppler conditions. Because of
the discrete nature of the "BIT SYNC PLL" the
crosslink clock is adjusted in 1/3 microsecond
increments.

The heart of the ESR control system is a
divide-by-50 counter whose count runs from 7 to
56 instead of 0 to 49. This range of number
corresponds to the center zone of the ESR. When
the system is synchronized, the contents of this
counter will indicate the position of the sync
bit in the center zone of the ESR (since the
center zone has 50 bits, there is always one sync
bit in it). Accordingly, this counter operates
off the Input Clock, i.e., when an input data is
inserted into the ESR, all the bits in the shift
register move over one bit, the Up/Down Counter
increments by one, and the sync counter increments
by one. At one-second intervals the contents of
the sync counter are jam loaded into the Up/Down
counter. This means that the up/down counter
and, consequently, the data selector are forced
to select a sync bit at the beginning of a one-
second interval, and this also means that the
beginning of a frame coincides with the beginning
of a one-second interval.

If the ESR was previously operating in the
center zone, there is no change in the Up/Down

Counter. If the ESR is in either dead zone the
jam loading is inhibited. If a slip warning
message has been sent (which implies that the ESR
is in an end zone) the jam loading is enabled and
the ESR is forced to the center zone. If the
system is already synchronized, the ESR is moved
from selecting a sync bit in the end zone to a
sync bit in the center zone, which is a jump of
exactly 50 bits. Thus a slip of 50 bits is
achieved via the modulo-50 property of the sync
counter--no explicit addition or subtraction of
50 to the contents of the Up/Down Counter is
required. If the system is not previously synchronized, the ESR is still forced to a position
which represents the sync counter's tentative
estimate of where the sync bit is. When the sync
counter ultimately achieves synchronization the
ESR will also be synchronized at the next one-
second tick. Because the ESR is updated only at
one-second intervals, the ESR phase telemetry is
not an adequate measure of the progress of the
system in acquiring sync.

Initially, the sync-counter is adjusted to
ultimately contain the position of the sync bit
by making it ignore a single clock pulse whenever
the sync system concludes that the sync counter
does not yet have the correct position of the
sync bit. This effectively decrements the sync
counter by one, which means that the sync counter's
estimate of the position of the sync bit is moved
one bit to the left in the ESR, or in other
words, it is now trying a bit position which is
later in time. When searching for the sync bit,
it is preferable to search from early to late so
that generally the sync bit is encountered and
tested before any of the low-rate bits, which
might possibly contain a preponderance of 1's and
cause a false sync condition.

MONITORING AND CONTROLLING THE CROSSLINK

Early in the development of LES-8/9 it
became apparent that some practical way of keeping
track of the various states of the satellites was
needed. This was not only apparent with single-
satellite tests, but seemed imperative with dual-
satellite tests involving the crosslink. A
system developed to use the existing command and
telemetry links to the satellite soon proved to
be indispensable in extensive exercise of the
communications subsystems of LES-8/9.

A photograph of the major panel displaying
communications command and telemetry is shown in
Fig. (15). This panel links a block diagram of
the satellite communications system with appropriate commands and telemetry so that a user can
note some change that is required, push a button
to flip a relay or set a data source for instance,
and then note the confirming telemetry after the
command has been received in the satellite and
the telemetry has returned to the ground station.
In this manner the user can note possible effects
of commands before actually sending them, since a
block diagram is there to functionally indicate
interactions which may not always be obvious. It
is also an operating convenience to have a light
change state shortly after sending a command to

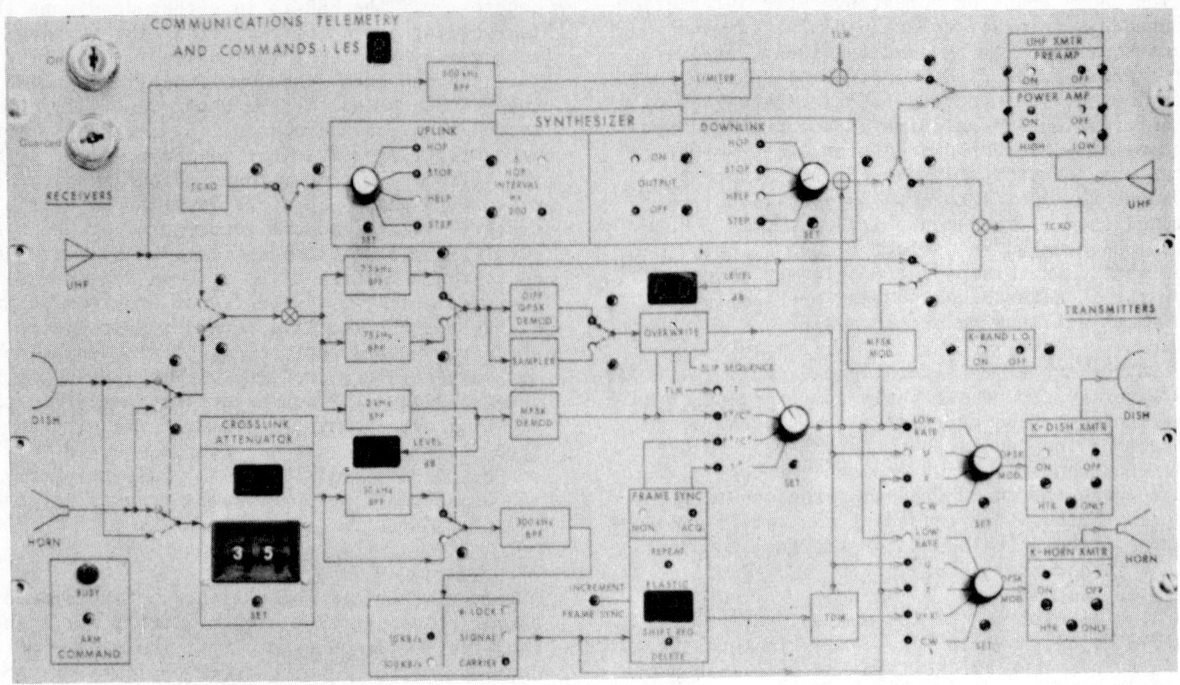

Figure 15. COMMUNICATIONS COMMAND/TELEMETRY PANEL

confirm its proper reception. The entire set of LES-8/9 Communications Command and Telemetry display panels is shown in Fig. (16). This arrangement is particularly well suited to the establishment and maintenance of a crosslink since crosslink pointing angles are converted from raw telemetry to decimal degrees, along with attitude control system roll and pitch information to allow quick interpretation. Associated commands are also interleaved with telemetry on the panels so that logical operation can occur. A close-up of the crosslink pointing panel is shown in Fig. (17). In using this panel to establish a crosslink, the user first initializes a mini-computer, primarily used by the communications modem, with attitude control references as displayed on a panel below the crosslink panel. The mini-computer now can send crosslink pointing command data bytes to the command system input storage in each rack. This transfer is initiated by the user and each command input accepts only pointing information for the satellite it is to command. This stored data is then converted as it would be if sent to the satellite and displayed in two windows labelled AZIMUTH and ELEVATION, Computed Pointing. A toggle switch can select either this data or manual data which in the case of computer failure, for instance, can be sent by pushing either the azimuth or elevation push button command load in the block diagram. The next set of windows to the left, COMMAND DESIGNATE MEMORY, is a telemetry display, suitably converted to decimal degrees, of the received angle command in the satellite just before it will be used to point the antenna in the acquisition mode. This display should match the computed pointing display to verify correct data transfer. Pushing the EXECUTE button transfer this angle data to the dish which moves to the new angle. If the other dish is also pointed correctly the energy light will come on and autotrack can be engaged. The Dish Pointing and COMMAND DESIGNATE MEMORY angles are independent in autotrack and a user can note the reasons for this on the same panel.

While these panels do not set up a link or define an experimental mode by themselves, they do present the operator with a clear picture of the many modes of these satellites and offer a straightforward way to interact with the satellites. It should be noted that some additional flexibility has been incorporated in telemetry display by providing randomly selectable "probes" into the telemetry stream which can access any of the 400 telemetry words and display these 12-bit binary words in a number of formats, a "rule" taking the form of MX + B where X is the raw telemetry value. Using this tool, we can monitor oscillator temperatures in degrees Centigrade, K-band power in milliwatts, or converter outputs in volts or amps, just to give examples of this useful feature.

As an aid in setting up a crosslink, for example, this system allows a crosslink to be establish in less than 15 minutes, even for non-zero roll and pitch. Simple mode changes take a few seconds. Telemetry is always at hand to allow noting system configurations as quickly as possible. In the case of a hardware failure, a single rack can be used with either LES-8 or LES-9 by throwing a single switch.

Figure 16.
LES-8/9 COMMUNICATIONS COMMAND/TELEMETRY SET

COMMAND AND TELEMETRY DISPLAY IMPLEMENTATION

Each display system contains a synchronizer that can accept a raw telemetry stream of 100 b/s or 10 kb/s and convert it to usable telemetry. This allows maximum flexibility in selection of source which may be a straight telemetry downlink or stripped data from a downlink otherwise used for communications, either at UHF or K-band. The data synchronizer used requires only input telemetry data and a 1-MHz oscillator input. Displays note the occurrence of bit sync, frame sync and parity errors. After synchronization, 12-bit data bytes and 10-bit addresses are sent to the processing hardware for formatting and conversion to a usable display. Some conversion is simple logic while some is fairly complicated, such as angle conversion (0.44° per binary bit, two's complement), and uses a microprocessor chip.

The entire telmetry display subsystem uses about 500 integrated circuits. The command subsystem adds about 150 integrated circuits. Typical packages used in shown in Fig. (18).

The command subsystem was treated very carefully to avoid sending a command which could prove harmful to the satellite. A subset of the

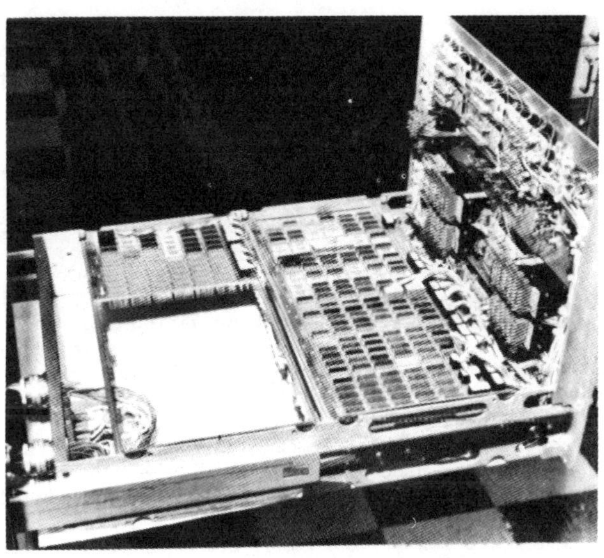

Figure 18.   COMMAND/TELEMETRY PANEL CONSTRUCTION

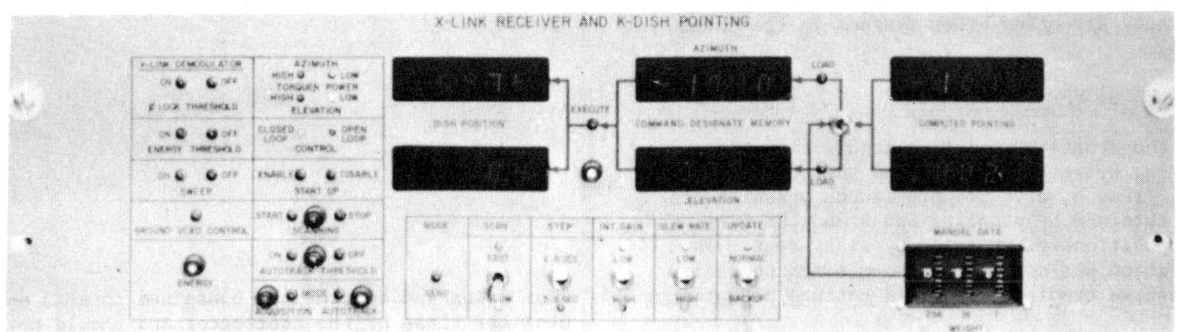

Figure 17.   CROSSLINK POINTING PANEL

total command list was chosen according to the
usefulness of a command in establishing a desired
communications mode. In the satellite itself, a
command can either enter data into a holding
register, execute that associated data, change a
relay, or produce a digital pulse at some input.
Because of the 2^{12} possibilities for a data
command, this command can only be practically
formed with a logical circuit which inputs the
desired data and outputs the 63 bits which make
up a command. Other commands are stored in their
entirety in ROMs and output as requested by the
operator.

To reduce confusion which could cause a mis-
taken command to be sent, the selection of commands is decided by pushing a button located in a
block diagram near the function to be commanded.
No translation from function to name to number
and the associated notebook listing is required.
As mentioned before, a confirming telemetry
display is located nearby to make doubly-sure
that the satellite was properly commanded.

This implementation has also made possible
a simplification that is more subtle. Some
functions require more than the basic 12-bit data
command to properly change. The command implementation used knows which push buttons require
1, 2 or more data transfers, formats the data
from the panel appropriately, and sends as many
commands as necessary. This ability to input
data from the panel and translate it to satellite
usable data is another time-saving feature
available in this system. The system described
is easily learned by new users of LES-8/9. It is
a convenient way of describing various communications modes, by noting switch positions and data
in red on copies of the panels. These copies
then serve to "blueprint" the operation and allow
quick reference to past operations. A keylock
switch can disable the command functions of the
system when used as a telemetry-only panel.
Another keyswitch also serves to disable commands,
such as setting the onboard real-time clock,
which are sensitive to most communication experiments and are not sent without strict knowledge
of how they are to be used.

Experience with these panels has shown that
they have had the desired result of allowing the
conductors of communications experiments to
interact closely with the satellites without
adversely affecting other systems in the satellites.

MEASURED SYSTEM PERFORMANCE

The crosslink pointing acquisition and
tracking system performance is summarized in
Figs. (19a, b, c). The pre-launch measurements
were obtained by mounting LES-8 and LES-9 on two
axis positioners separated by sixty feet. This
separation was considered the minimum necessary
to observe credible far field antenna performance.

The data shown in Fig. (19a) was obtained
by pre-pointing the antenna boresight to generate
worst case antenna alignments during SCAN. The

energy threshold was enabled to allow automatic
switching from SCAN to TRACK if the energy detector
in the autotrack receiver exceeded its threshold
value. The resulting relationship shown in
Fig. (19a) is therefore determined by the threshold setting of the energy detector in the
receiver phase lock loop and not by the pointing
equipment. The curve labelled CW indicates the
improved pull-in performance that is available
from a separate carrier detector in the autotrack
receiver if the link degrades for any reason. It
is evident from Fig. (19b) that the improvement
in P_r/N_o due to pull-in is adequate to ensure
proper tracking.

Extensive use of the satellite-to-satellite
data link since launch has demonstrated complete
reliability in all cases. Both 10 Kb/s and
100 Kb/s links are normally operated in a region
where bit errors due to random noise is less than
a part in 10^4. To compare present performance with
that prior to launch, each dish was pointed to
earth and a crosslink was established between a
satellite and a ground station simulating another
satellite. Data gathered in this way is compared
to pre-launch data and shown in Figs. (20a, b)
for LES-9. The crosslink has now been used to
relay commands, telemetry, low-rate and high-rate
communications simultaneously in both directions.

ACKNOWLEDGEMENTS

The design, integration, and testing of the
systems described in this paper crossed many
technical disciplines during the past five years.
The authors wish to acknowledge the effort and
cooperation of Dr. R. Bauer, Dr. D. Peterson,
D. Hodsdon, and C. Much for integration and
acceptance testing of the crosslink system;
C. Sullivan for the inherently simple biax drive;
D. Nathanson for the biax thermal control system;
Dr. N. Smith for electro-mechanical design and
analysis, and Dr. A. Dion for conceiving and
delivering the lobing feed.

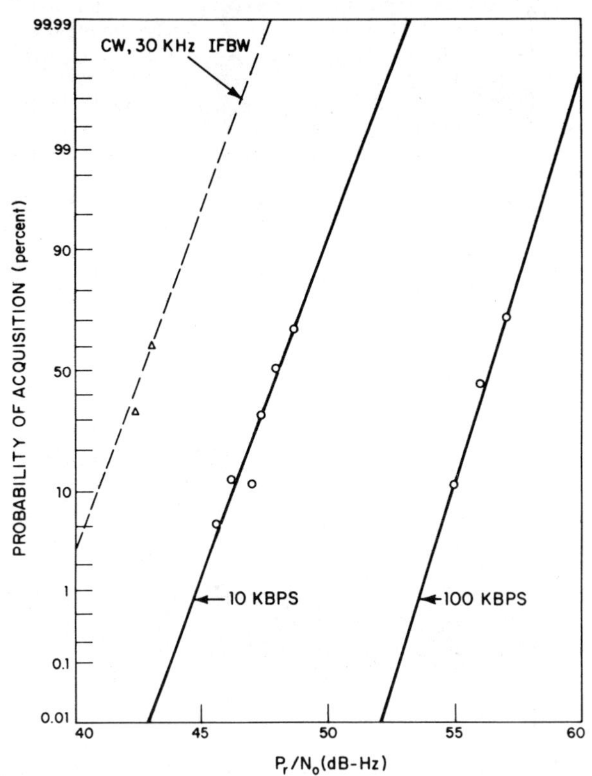

Figure 19a. PROBABILITY OF SPATIAL ACQUISITION

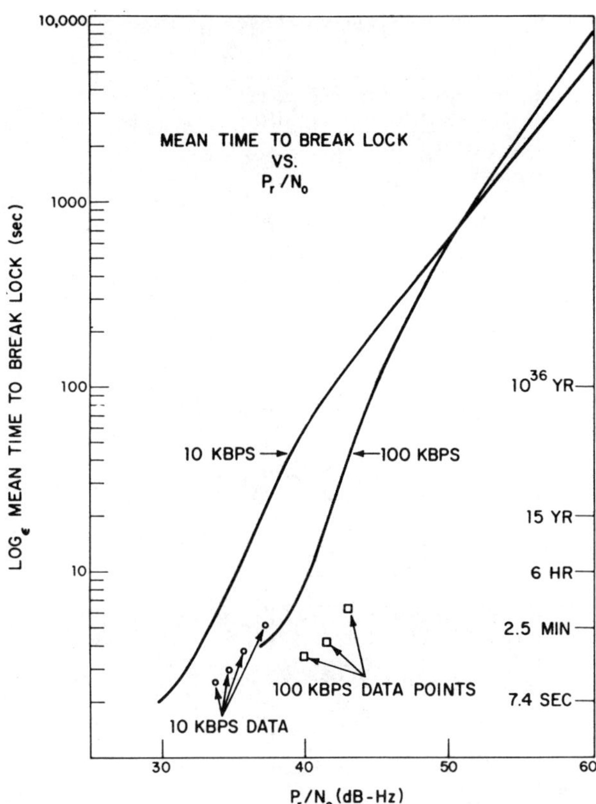

Figure 19b. MEAN TIME TO BREAK LOCK

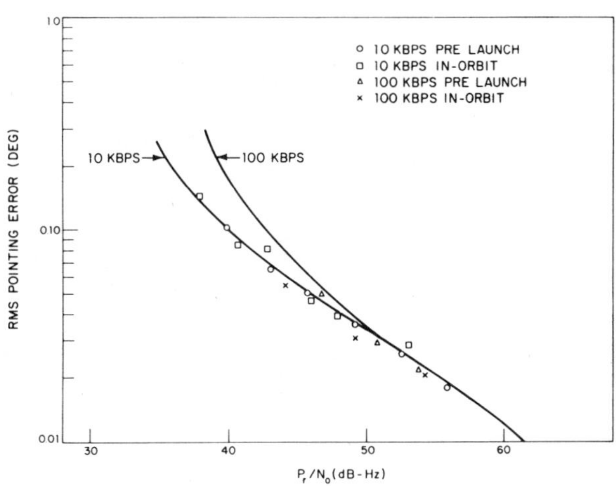

Figure 19c. TRACKING ACCURACY

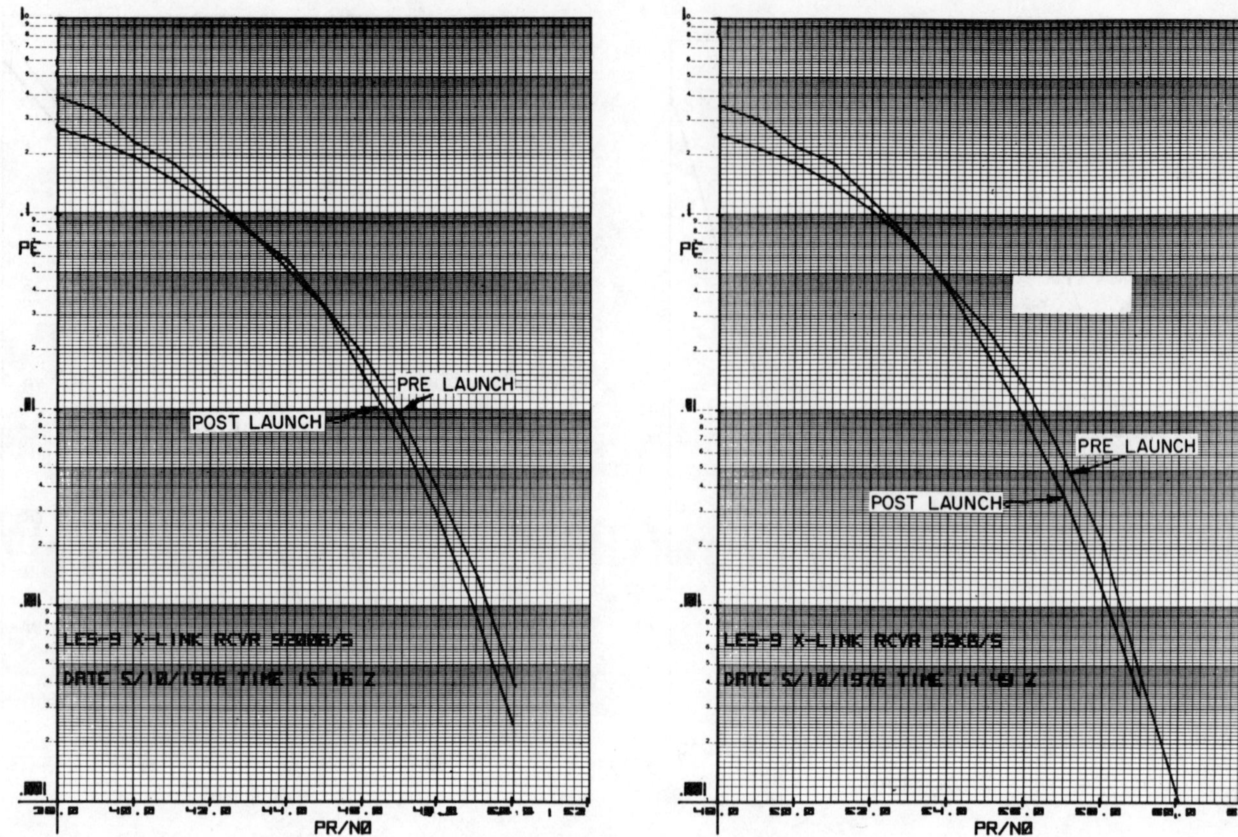

Figure 20a.   CROSSLINK LOW RATE BER

Figure 20b.   CROSSLINK HIGH RATE BER

Part IV

Spacecraft and Communication System Design and Technology

In this part, some of the aspects of the design of communications satellites are discussed. Section 4.1 discusses the design of the various subsystems of the spacecraft which are needed to support the communications missions. Section 4.2 discusses satellite antennas. Section 4.3 discusses satellite receiver and transmitter technology with emphasis on the active components. Section 4.4 discusses the various passive elements in the satellite transponder. These topics are not part of the primary focus of the book, but they provide the necessary background to understand some of the limitations on the satellite communications subsystems.

Section 4.1
Spacecraft Designs

In this section, the design of the spacecraft and the related launcher is discussed. Reprint Paper 4.1 discusses the various design problems in a communication satellite mission. It is convenient to divide the discussion into eight topics:

1) launch procedures and orbital dynamics;
2) stabilization;
3) attitude and orbit control;
4) power generation, conditioning, and storage;
5) thermal control;
6) tracking, telemetry, and control;
7) structures;
8) reliability and testing.

LAUNCH VEHICLES

The presently available set of launch vehicles are the Delta (e.g., 2914, 3914), Atlas Centaur, and Titan III-C. All of the current operational commercial communication satellites have been launched on the Delta or Atlas-Centaur vehicles which are expendable. Reference [1] summarizes the payload capabilities and costs for these launch vehicles. The development of satellite systems in the 1980's, however, will rely upon the space shuttle to place the satellites into geosynchronous orbit. This will increase the maximum payload mass and dimension and reduce the per pound cost in orbit.

The present launch strategy with the expendable vehicles is, first, a launch into a low altitude parking orbit (185 km); than an appropriately times thrusting into an elliptical transfer orbit with apogee and perigee over the equator; and, finally, a positioning in the synchronous orbit with the firing of the apogee kick motor (AKM). Normally, the Titan III-C uses the trans-stage for both the transfer orbit and synchronous orbit maneuvers. However, it is possible to use an AKM with the Titan III-C and achieve a synchronous orbit payload of 4830 lb.

The space shuttle can be used to bring up to 30 000 kg of payload in a 18.3 m by 4.6 m diameter shuttle bay into a parking orbit (300 km). From this parking orbit, an Interim Upper Stage (IUS) will be used to transfer the payload to synchronous orbit—first by placing it in transfer orbit, and then at apogee, kicking it into synchronous orbit.

Reference [2] provides an overview of the present status of the space shuttle program, which has an operational flight scheduled for 1980.

There are several alternative IUS concepts that are being considered for the shuttle era (e.g., [3] or [4]). The first uses a three-axis-stabilized upper stage. This technique is consistent with the Titan III-C launch strategies and will require the development of a set of two new motors. The second method for transfer from parking to synchronous orbit is by use of a spin-stabilized upper stage (SSUS). For this technique, stabilization is obtained by spinning the spacecraft as is presently done in the Delta and Centaur launches. A perigee kick motor (PKM) is used to get from parking orbit to transfer orbit, and an AKM is then used to achieve synchronous orbit.

NASA has proposed a preliminary pricing policy for shuttle launches. For a dedicated launch the cost in 1975 dollars will be $20 M. However, for shared launches the prorated price will be based on a combined length-weight basis. Specifically the ratios of payload length to shuttle length (18.3 m or 60′) and payload weight to maximum shuttle launch weight (30 000 kg or 65 000 lb) will be determined. The largest of these two factors will be used to determine what fraction of the $20 M will be charged. In addition, in most cases the shuttle will not be fully filled by the several users, and since each launch must be self supporting the prices will be increased to compensate for the underfill. For example, with a typical 75% fill the prices will be increased by a 4/3 adjustment factor.

In the 1980's, additional non-U.S. launch vehicles will become operational. The Japanese N-Vehicle has a synchronous orbit payload of 260 lb. The ESA Ariane will have a capability of 1830 lb, close to that of the Centaur, by taking advantage of a near-equatorial launch base (e.g., [5]).

STABILIZATION

The two main types of satellite stabilization are spin stabilization and three-axis (or body) stabilization. Most currently operational communications satellites (e.g., IN-TELSAT IV-A, Westar, Anik, and DSSC-II) are spin stabilized, but a number of three-axis satellites are in operation or planned (e.g., RCA SATCOM and DSSC III). References [6]-[8] discuss various aspects of stabilization.

ATTITUDE AND ORBIT CONTROL

The major issues in attitude and orbit control are outlined in Section IV of Reprint Paper 4.1. References [9]-[10] amplify this discussion. One of the factors determining satellite lifetime is the amount of fuel required for North-South station keeping. Ion engines and thermally augmented hydrazine thrusters will prove useful in future systems (e.g., [11]-[12]).

POWER GENERATION, CONDITIONING, AND STORAGE

Currently, prime power is normally provided by photovoltaic solar arrays, with batteries used during eclipse. Although radioactive thermal generators (RTG) have recently been flown on communication satellites, they are not expected to replace photovoltaic cells. In the future, higher efficiency solar arrays utilizing improved cells, such as the "violet" and nonreflective cells, will provide more prime power per unit area (e.g., [13]-[14]).

In addition, the use of three-axis stabilized satellites with deployable solar-oriented arrays will result in much larger effective solar array areas than have been typical of the predominantly spin-stabilized satellites currently in use. The battery lifetime is a significant factor influencing the satellite lifetime when there is a requirement to function during eclipse. Batteries with longer life potential, such as nickel hydrogen, are expected to increase this lifetime [15].

THERMAL CONTROL

A thermal control system is required to keep the equipment within appropriate temperature limits. The environment experienced by a satellite is described in Section VI of Reprint Paper 4.1. Both passive and active techniques are used to provide thermal control. References [18]-[20] discuss thermal control techniques.

TRACKING, TELEMETRY, AND CONTROL

Telemetry and command and control requirements will increase as satellites become more complex. Monitoring and control of on-board processors, multiple-beam antenna performance, solar-array and battery capabilities, and TWTA performance will be essential to an effective system. It is likely that microprocessors will be employed in performing the monitoring and control and some of the other basic bus functions. Reference [21] and Section VII of Reprint Paper 7.1 discuss the same aspects of the TT&C area.

STRUCTURES

Significant progress has been made in the structural design area by efficient configuration design and the use of lightweight materials. Reference [22] discusses aspects of the structural problem.

RELIABILITY AND TESTING

The importance of testing on spacecraft reliability has been well documented. Reference [23] discusses some of the important issues in spacecraft reliability.

The discussion up to this point has considered various elements of spacecraft design. References [24]-[26] discuss examples of the design of the overall spacecraft and show how the above functions are integrated into a complete design.

REFERENCES

[1] Z. O. Bleviss, "Expendable launch vehicles for synchronous communications satellites," in *Satellite Communications; Future Systems,"* D. Jarett, Ed. New York: AIAA, 1977, pp. 319-346.

[2] T. R. Kloves, "Space shuttle program: An overview," *Ibid.,* pp. 347-356.

[3] W. G. Huber and C. F. Huffaker, "IUS capability for earth orbital spacecraft," *Ibid.,* pp. 357-376.

[4] C. R. Jones, "Cost-competitive STS for geostationary payloads," *Ibid.,* pp. 389-404.

[5] A. F. Lebeau and R. M. Orye, "Ariane launch vehicle: A European program," *Ibid.,* pp. 377-388.

[6] T. B. Garber, "Three-axis or dual-spin stabilization for future synchronous communications satellites—A summary," *Communications Satellites for the 70's: Systems,* vol. 26, AIAA Series, N. E. Feldman and C. M. Kelly, Eds., Cambridge, MA: M.I.T. Press, 1971, pp. 639-653.

[7] J. T. Neer, "INTELSAT IV nutation dynamics," *Communications Satellite Technology,* vol. 33, AIAA, P.L. Bargellini, Ed. Cambridge, MA: M.I.T. Press, 1974, pp. 57-84.

[8] J. E. Keigler, W. J. Lindorfer, and L. Muhlfelder, "Momentum wheel three-axis attitude control for synchronous communications satellites," *Ibid.,* pp. 141-161.

[9] K. L. Lebsock, "High pointing accuracy with a momentum bias attitude control system," *Proc. AIAA 7th Communications Satellite Systems Conf.,* San Diego, CA, pp. 269-277, Apr. 1978.

[10] G. E. Schmidt, "Magnetic attitude control for geosynchronous spacecraft," *Ibid.,* pp. 278-284.

[11] W. C. Isley and K. I. Duck, "Propulsion requirements for communications satellites," *Op.cit.* [7], pp. 165-190.

[12] L. B. Holcomb and D. H. Lee, "Survey of auxiliary propulsion systems for communications satellites," *Op.cit* [7], pp. 191-243.

[13] J. Lindmayer and J. F. Allison, "The violet cell: An improved silicon solar cell," *COMSAT Techn. Rev.,* vol. 3., no. 1, pp. 1-22, Spring 1973.

[14] J. F. Allison, R. Arndt, and A. Meulenberg, "A comparison of the COMSAT violet and non-reflective solar cells," *COMSAT Techn. Rev.,* vol. 5, no. 2, pp. 211-224, Fall 1975.

[15] J. F. Stockel, J. D. Dunlop, and F. Betz, "NTS-2 nickel hydrogen battery performance," *Op.cit.* [9], pp. 66-71.

[16] L. G. Chidester, "Advanced lightweight solar array technology," *Ibid.,* pp. 55-60.

[17] R. H. Sparks, "Nickel-cadmium battery technology advancements for geosynchronous orbit spacecraft," *Ibid.,* pp. 61-65.

[18] A. J. Gopin and A. S. Friedman, "Thermal analysis and control of the Air Force tactical communications satellite," *Communications Satellites for the 70's: Technology (AIAA Progress in Astronautics and Aeronautics,* vol. 25), N. E. Feldman and C. M. Kelly, Eds. Cambridge, MA: M.I.T. Press, 1971.

[19] J. A. Robinson, "Thermal control, the INTELSAT IV spacecraft, *COMSAT Tech. Rev.,* vol. 2, no. 2, pp. 376-388, Fall 1972.

[20] J. W. Lucas, Ed., *Fundamentals of Spacecraft Thermal Design,* vol. 29, AIAA Progress in Astronautics and Aeronautics, M.I.T. Press, 1972.

[21] F. W. Weber, "Telemetry and command subsystems, the INTELSAT IV spacecraft," *COMSAT Tech. Rev.,* vol. 2, no. 2, pp. 341-358, Fall 1972.

[22] L. A. Harris, "Advanced composites, an assessment of the future," *Astronautics and Aeronautics,* pp. 22-33, Mar. 1976.

[23] R. Strauss and J. Owens, "Satellite reliability," IAF Congress, Prague, Czech., 1977.

[24] R. J. Rusch, J. T. Johnson, and W. Baer, "INTELSAT V spacecraft design summary," *Op.cit.* [9], pp. 8-20.

[25] G. E. A. Abutaleb, M. C. Kim, K. F. Manning, J. F. Phiel, Jr., and L. H. Westerlund, "The COMSTAR satellite system," *COMSAT Tech. Rev.,* vol. 7, no. 1, pp. 35-83, Spring 1977.

[26] "INTELSAT IV spacecraft," *COMSAT Tech. Rev.,* vol. 2, no. 2, pp. 271-572, Fall 1972.

Design Problems of Spacecraft for Communication Missions

RENÉ C. COLLETTE AND BERNARD L. HERDAN

Abstract—The process of designing a spacecraft for geosynchronous communications is described from the point where the mission and key payload performance requirements have been defined. Following a description of the launch and injection process, the constraints which this and the communications requirements impose on the satellite platform are described together with their potential design implications. Types of stabilization are compared with emphasis on selection between the dual spin and body stabilized classes. Typical guidelines are given for the design of each major spacecraft subsystem. The paper concludes with more general considerations such as mass and reliability optimization as well as configuration synthesis.

I. LAUNCH AND ORBITAL INJECTION

THE MAJORITY of communication satellites are geosynchronous, i.e. the orbit is nearly circular with a period of 23 h 56 min in mean solar time, so that the satellite's diurnal movements with respect to a given earth

Manuscript received August 16, 1976; revised October 14, 1976.
The authors are with the Telecommunication Satellites Program Office, European Space Agency, Noordwijk, The Netherlands.

meridian are extremely small. Before entering into satellite design considerations, it is instructive to recall the practical aspects related with the achievement of such an orbit.

A. Basic Considerations

Considering first of all the trajectory design, the motions of launch vehicles or satellites about the earth follow, to a first approximation, Newton's classical laws for the motion of two gravitating bodies in isolated space. Proceeding from this, it may be shown that U, the total energy of such a two-body system, is given by

$$U = \frac{1}{2} mv^2 - \frac{GmM}{r}$$

where v is the velocity of the vehicle, m is the mass, r is the distance from the earth center, M is the mass of the earth, and G is the universal gravitational constant.

In the particular case of giving a body sufficient energy to reach geosynchronous orbit, a launching system must be devised to put the satellite at a distance of 42164 km from the earth's center and to accelerate it to a velocity of 3070 m/s. Because of physical limitations, the launcher does not impart this energy to the satellite in a continuous fashion. Indeed, in an equipotential field, the maximum velocity increment Δv of a propulsion system is given by the following relationship:

$$\Delta v = c \ln \frac{1}{1 - m_f/m_0}$$

where c is the effective exhaust velocity of the gas, which is a function of the fuel type and nozzle design, m_0 is the total mass of system at ignition, and m_f is the mass of expanded fuel.

This maximum increase in velocity is, in the limit, constrained by the practicalities of what can be achieved in terms of propulsion, i.e. exhaust velocity and launch vehicle structure technology (m_f/m_0 ratio). It turns out therefore, from the equation indicated above, that a single rocket cannot provide a sufficient velocity increment to inject the satellite directly into orbit. To overcome this limitation, multistage launch procedures are used which permit the required final velocity to be achieved by the sequential operation of each stage after the previous one has burnt to depletion and has been jettisoned.

B. Practical Flight Plan for a Geosynchronous Mission

Without going further into the details of the mathematical and hardware considerations, it may be seen that there is a need for the optimization of the trajectory required for synchronous orbit insertion, which takes into account the practical constraints of launch vehicle design and the use of optimum efficiency techniques, for example the so-called "Hohman" transfer orbit. A typical flight plan for a geostationary mission, with zero inclination with respect to the equatorial plane, is illustrated in Figs. 1 and 2. The main sequence of events would be as follows.

1) Launch from a site near the equator in an easterly direction to take maximum advantage of the earth's rotational velocity. Burn to depletion and dropoff of the vehicle's first stage. Ignition of the second stage and cutoff when an initial circular parking orbit has been achieved, with a typical altitude of 185 to 250 km.

2) Shortly before the crossing of the equator, second burn of the second stage and (if any) burn to depletion of a third stage. The parking orbit is thus modified to an intermediate elliptical orbit known as the transfer orbit having its apogee at geosynchronous altitude, its perigee on the equatorial crossing point, and an inclination such that the mass reaching final orbit is maximized.

3) Separation of the satellite from the launch vehicle followed by attitude manoeuvres during transfer orbit (spacecraft in spinning mode) to provide a suitable alignment prior to the ignition of the apogee kick stage, the so-called apogee motor, which is in general an integral part of the spacecraft.

4) After usually several revolutions in transfer orbit, ignition and burn to depletion of the apogee motor. This operation again modifies the orbit, which becomes near circular at synchronous altitude.

5) Fine corrections, using an auxiliary propulsion system to position the satellite at its required longitude station and correct the orbital period and eccentricity to the required values. At the same time, establishment of the proper satellite operating mode including, if applicable, transition from spin to body stabilization and solar array deployment.

6) Periodic orbit corrections ('stationkeeping') to maintain the satellite at the required location for the mission duration.

II. Overall Satellite Design Constraints

The communications satellite design process starts with the definition of key service characteristics, taking account of customer mission requirements and frequency constraints. A progressive process of iteration is customary, involving in particular the matching of space segment communications capacity to the expected traffic model and a number of major tradeoffs between satellite capabilities and those of the corresponding earth stations. The end result of this initial phase is the characterization of the requirements of the space segment in terms of number of satellites and their communications performance.

The overall set of technical constraints on the satellite design arise primarily from these communications performance requirements, plus constraints from the launch vehicle, the injection sequence, and the space environment. Each of these factors is examined below. Naturally there is invariably the additional condition of having to achieve a cost effective result utilizing mass, volume, power, and development funds in a balanced and optimized manner.

A. Communications Performance Requirements

Once the operating frequencies, number of channels, EIRP requirements and other key parameters have been finalized via a series of trade offs between onboard and earth station capabilities, the configuration of the communications payload can then be established, leading to the following key demands on the satellite platform.

1) Accommodation of equipment in a suitable layout which minimizes signal losses and interconnections.

2) Support of the corresponding total mass.

3) Provision of the required electrical power within specified voltage tolerances.

4) Control of temperatures within limits imposed by the equipment for operating and nonoperating modes.

5) Provision of the necessary telemetry and command services to permit ground monitoring and control of repeater functions.

The definition, in terms of received flux contours, of the required area of ground coverage is, once again, the subject of tradeoffs between the satellite and the ground stations. Depending on the mission, one or more antennas may be needed, and these may be fed in a variety of ways. For large antennas, the launch vehicle envelope may necessitate that they are folded at launch and deployed after acquiring synchronous orbit. All these factors lead to geometric constraints on the satellite configuration. The required coverage area and the communications link budgets lead to the definition of antenna boresight pointing requirements and hence the specification on the platform stabilization accuracy, which may be a key factor in the satellite design definition. For missions employing orthogonal polarization frequency reuse, there are also tight limits on satellite rotation about the satellite–earth line (yaw axis).

Coupled with the definition of antenna coverage tolerances and also depending on the relevant earth station facilities, the question of the need for North-South stationkeeping is a factor which may influence the satellite design in an important manner. While orbit corrections in the East-West direction are relatively inexpensive in terms of fuel usage and are anyway necessary to maintain the satellite longitude and so pre-

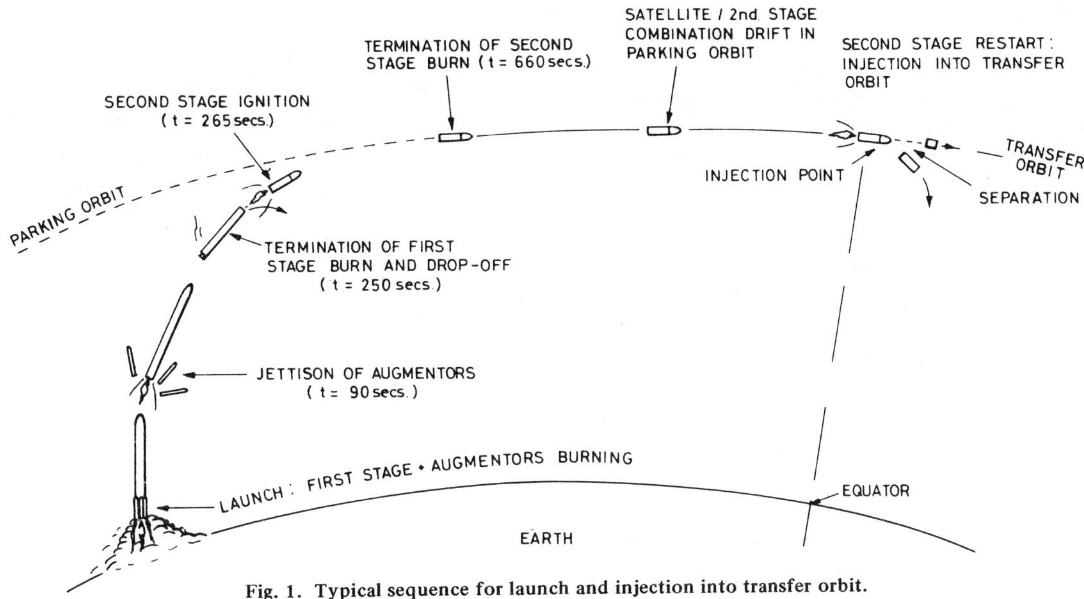

Fig. 1. Typical sequence for launch and injection into transfer orbit.

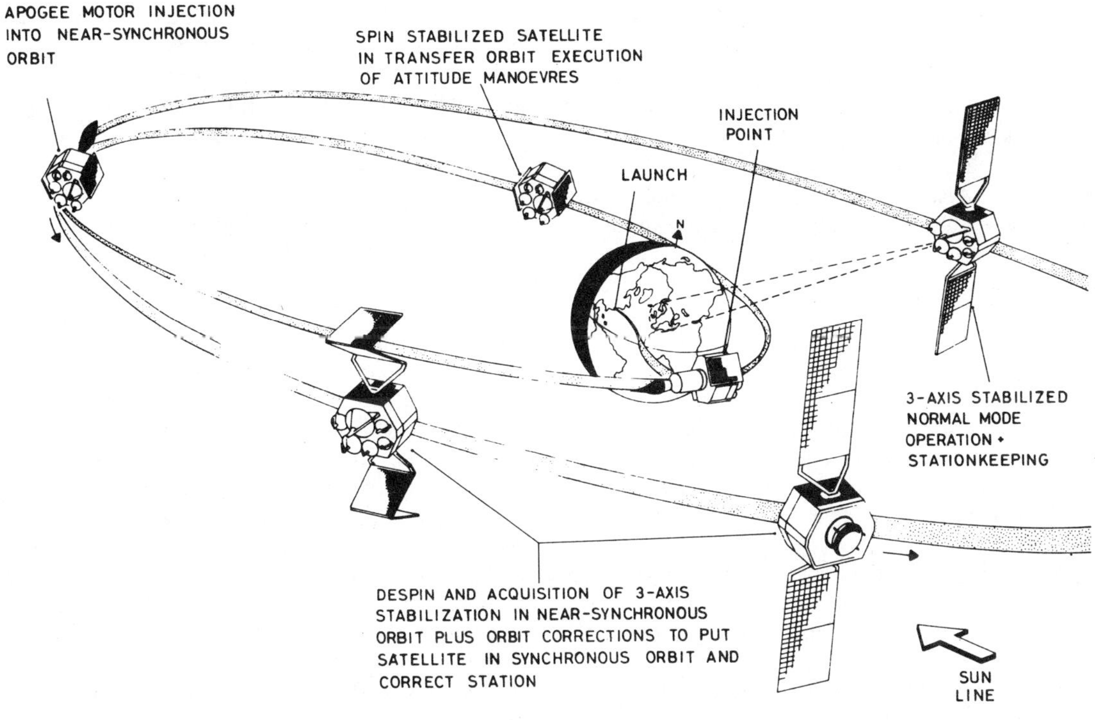

Fig. 2. Typical sequence for injection into synchronous orbit and acquisition of normal mode operation.

serve adequate separation from other geosynchronous satellites, controlling in the North-South direction might not be essential. Indeed latitudinal station keeping costs a great deal of propellant (of the order of 20 kg per year for a 900 kg satellite) due to the magnitude of the sun/moon gravitational perturbing forces which cause the plane of the orbit to rotate at the rate of nearly one degree per year. In general, such N-S station-keeping is necessary in the following cases.

1) Variations in satellite antenna coverage, even small, are unacceptable.

2) Some earth stations are operating at extremely low elevation angles such that the satellite may vanish below the effective station horizon at an extreme of its North-South motion.

3) The earth station antennas have relatively narrow beam-width and cannot be equipped for tracking of the satellite. (Typical example: system employing a large number of low-price receive-only stations, e.g. for T.V. reception.)

In other cases, it might be advantageous to dispense with *N-S* stationkeeping and operate a system in which the orbit inclination slowly drifts, resulting in an apparent daily motion of the satellite.

The communications performance of a satellite must always be keyed to a time duration. This, in turn, imposes clear reliability objectives for the payload and spacecraft platform. As an example, a system requirement that 7 repeater chains be operational after eight years with a probability of 72 percent could, for instance, be met by designing the payload with 10 chains (of which 3 are allowed to fail during the 8 years) achieving therefore a payload reliability of 80 percent, and then allocating a constraint that the platform achieves a reliability of 90 percent. In general, economic factors are encouraging a trend towards longer and longer life satellites, limited only by the aging or wearout of key elements such as traveling wave tubes, bearings, and batteries. Normally 7 to 10 yr are specified nowadays.

As a final point, the requirement of the communication mission for satellite payload operation through eclipse periods has a significant effect on the power system design. A satellite in geostationary orbit experiences, during two 45-day seasons each year, daily periods of up to 72 min during which the sun is eclipsed by the earth. Since all current communications satellites rely on solar power for normal operation, batteries must be carried to support the power demands during eclipse. In some cases, the communications mission can accept a reduction in payload operation during eclipse which always occurs around midnight satellite time; a smaller battery can then be carried and considerable mass saved.

B. Launch Vehicle

The launch vehicle imposes on the satellite the primary constraints of mass and permissible envelope, a variety of detailed electrical and mechanical interfaces, and a very serious environment during the launch sequence itself. The launch vehicle selection is generally made early in a project, based on an estimate of the probable minimum satellite mass and on available funds. Apart from the basic question of economics, some indirect benefits result from the use of certain vehicles. Titan, for example, provides a more accurate injection, three-axis control of the payload right up to injection into synchronous orbit, and (in some cases) a softer ride. The choice of launchers is, of course, very limited and, in fact, an ideal match to the mission requirements is often impossible. The following table (Table I) indicates some of the expendable vehicles which are available for launch of synchronous satellites and illustrates the large steps between them in terms of payload capability. The larger vehicles, in particular the Titan and Titan/Centaur, may be used for multiple satellite launches on a single vehicle.

Although not within the general scope of this paper, mention must be made of the NASA Space Shuttle which will be operational in the 1980's and which should provide the following set of benefits.

1) Relaxation of mass constraints resulting in technical and economical benefits.

2) Larger envelope, hence greater freedom in configuration definition and necessity for deployment mechanisms only for the largest antennas.

3) Because of the larger cylindrical surface available for the installation of solar cells, increased power capabilities for spin-stabilized satellites, hence extended potential applications for this class of spacecraft.

4) Less severe launch environment, both from the viewpoint of mechanical vibration and thermal conditions.

5) Possibility of in-orbit checkout (and repair) before ejection from the shuttle.

It should be noted that the shuttle itself can only bring its payload into a low earth orbit, with an additional stage being needed to put the satellite into the elliptical transfer orbit mentioned in Section I. This shuttle-born stage will be either expendable (developments from existing launcher hardware)

TABLE I

VEHICLES AVAILABLE FOR LAUNCH OF SYNCHRONOUS SATELLITES AND THEIR PAYLOAD CAPABILITIES

Vehicle	Origin	Approximate Payload Into Synchronous Orbit, Excluding Apogee Motor (kg)	Envelope Diameter (m)
Delta (2914 or 3914)	US	340 to 460	2.2
Atlas SLV-3C Centaur	US	870	2.9
Titan III C/Agena Types	US	1470 to 1940	3.1
Titan III E/Centaur DIT	US	3360	3.7
Ariane (from 1980)	Europe	900	3.0
N (from 1977)	Japan	130	1.4
Soyuz SL-4	USSR	7500*	–
Proton SL-9	USSR	18000*	–

*Low earth parking orbit

as foreseen for the period 1980–85 or reusable as foreseen for the period 1985 onwards. Against the technical benefits listed above, the economic conditions of a shuttle launch, although promising, are still unclear. The long term objective is to achieve costs noticeably below those of today's conventional launching systems.

C. Environment and Injection Sequence

The design of a communications satellite is constrained not only by the eventual operating requirements and conditions in synchronous orbit but by the requirements to survive ground handling and the launch environment, and to provide all necessary functions for the injection sequence. For the purposes of ground handling, the satellite must be mechanically compatible with requirements for lifting and manoeuvering, and able to survive transportation shocks, etc. One must also be able to monitor, control, and supply power through hard lines during certain operations. As far as the launch vehicle is concerned and apart from mass and envelope constraints, there are detailed mechanical and electrical interface requirements such as limitations on the center of mass location and balance accuracy, as well as electrical constraints for umbilical connections and RF field levels. During the launch vehicle ascent, the spacecraft has of course to survive a rather severe mechanical and acoustic vibration environment, the most important loads being experienced during lift-off, engine ignitions and cut offs, separation and jettison of main booster elements, and satellite separation from the vehicle.

As described in Section I, the launch vehicle will inject the satellite into a transfer orbit having a perigee between 180 and 500 km and an apogee at a synchronous altitude of about 35.700 km. The satellite has to operate in this elliptical orbit and provide all the necessary functions for injection into synchronous orbit via the use of a supplementary apogee motor and suitable attitude orientation and control system. In the transfer orbit, the operating mode of the satellite is generally very different from that on station. For example, the repeater payload is switched off and must be protected from growing too cold. Except for specific cases where the last stage of the launch vehicle provides three-axis stabilization up to and during apogee firing, the satellite is usually spin stabilized in transfer orbit. For satellite which will also be spin-stabilized on-station, this does not present any new basic design requirements. For satellites which will be body-stabilized on-station, the spin-stabilized transfer orbit mode leads to two essentially separate attitude-control systems being needed as well as radically different design conditions in the thermal, power, telemetry, and command subsystems in the two mission phases.

Once on-station, the satellite has to be designed to operate reliably for a number of years in a different but nevertheless harsh environment. Not only is there a hard vacuum, but different parts of the spacecraft see widely varying temperature extremes depending on the sun-satellite-earth geometry at any particular time of the day and year. In addition, the satellite is exposed to damaging proton and electron fluxes and a variety of environmental attitude and orbit perturbations, yet must still continue to perform perfectly.

III. Selection of Stabilization Type

Probably no aspect of communications satellite system design causes more controversy between the experts than the choice between the different classes of stabilization systems. This controversy is by no means misplaced since the question is a complex one, and the decision on the stabilization technique to be used for a satellite has a strong influence on both the overall configuration and on each subsystem design.

A. Main Classes of Stabilization

Two main classes of satellite stabilization exist: spin stabilized and body stabilized. Spin-stabilized satellites have a gyroscopic stiffness due to rotational motion of all or part of the body, and require only periodic corrections via thrusters to maintain the attitude characterized by the spin axis. There are three subsets of this class relevant to communication missions.

1) Simple spinners which employ a spinning drum plus an electronically phased array to ensure a constant earth-pointing beam.

2) Dual-spin satellites which also employ a spinning drum, plus a despun platform which is maintained by counter rotation to be almost stationary, so that the antennas mounted on it are constantly earth-pointing.

3) Triple-spin satellites which are a derivative of the dual-spin class, but employ a third section from which solar array panels are deployed and rotated once per day, to maintain them sun-pointing.

A stylistic representation of typical versions of each of these types is shown in Fig. 3(a) and 3(b). Only the dual-spin type has seen general application and is therefore singled out for further consideration here as representative of the spin-stabilized class.

Three-axis or body-stabilized satellites are attitude controlled about each axis, either by the explicit use of sensing plus torque about each axis or by the use of momentum devices to provide gyroscopic stiffness about one, two, or three axes, supplemented by sensing and torque as required to maintain attitude in the presence of environmental perturbations. There are again three principal subsets of this class relevant to communication satellites.

1) Satellites without any gyroscopic stiffness which employ sensors and gyros for attitude determination, plus thrusters for control about each axis.

2) Satellites employing a fixed or gimballed momentum

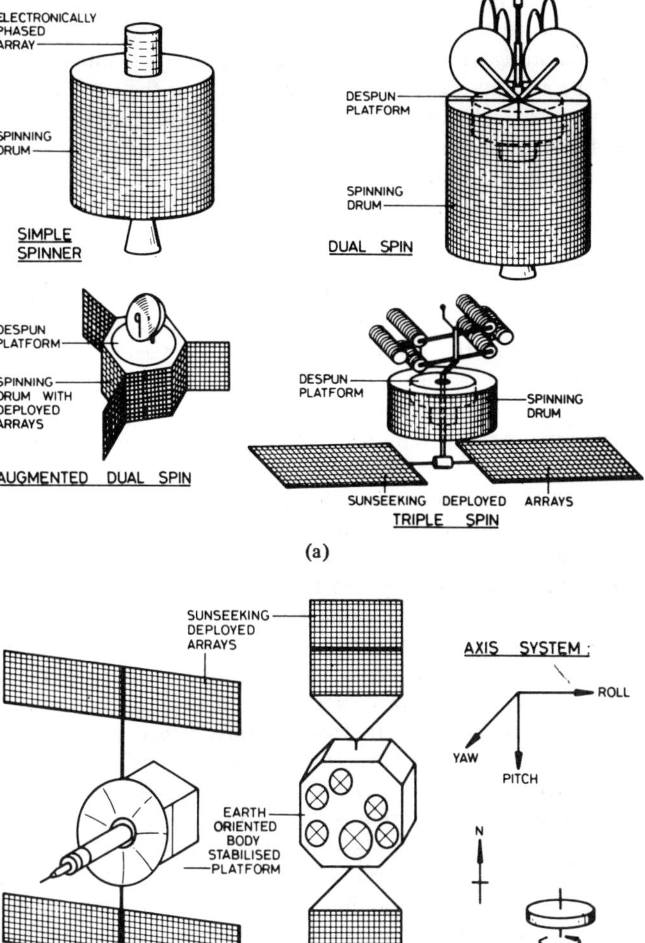

Fig. 3. (a) Various spin-stabilized configurations. (b) Body-stabilized configurations.

wheel with a finite nominal speed which imparts gyroscopic stiffness to one axis, and whose control of attitude is ensured via variations in the nominal speed of the wheel and the occasional use of torque devices (e.g. thrusters and/or wheel gimbals) to maintain the required wheel spin-axis alignment.

3) Satellites employing three reaction wheels, one directed along each axis, to control the attitude via variations in individual wheel speeds which are centered around zero momentum.

A representation of some typical three-axis stabilized satellite configurations is shown in Fig. 3(b). The momentum wheel type of system is currently the most common, and is thus selected here for comparison below with the dual-spin type of stabilization.

B. General Comparative Analysis

1) Overall Configuration: Fig. 3 illustrates the main configurational differences. For the dual-spin type of satellite, the drum is covered with solar cells, and the antennas are generally mounted on the despun platform of a central stalk with their boresights approximately normal to the spin axis. On the other hand, body-stabilized satellites derive their power from sun-tracking deployed solar arrays, which are folded in transfer orbit. In this case it is more usual for the antennas to be mounted flat onto the main satellite body with their boresights approximately parallel to the main axis (yaw) of the satellite. The envelope constraints of the launch vehicle are of particular importance for the dual-spin type of satellite, both because they limit the maximum size of the drum and therefore the available power as well as the height of the antenna farm above the main body.

2) Power Generation: As indicated above, the dual-spin satellite suffers from power limitations as a function of launch vehicle envelope cross-section. Even for the largest envelopes, the maximum power which can be generated by a dual-spin satellite is in the region of 1 kW after 7 years in orbit. A further point is that the mass (and cost) of a body-mounted drum array is much greater than that of a sun-oriented deployable array since it requires about three times the area of solar cells to provide for the same power requirement. To count against these points, the spinner array is inherently simpler and avoids the need for deployment mechanisms. For completeness, it should be mentioned that the triple-spin type of configuration overcomes the dual-spin problem of power limitations, but at the same time introduces new problems. These include mechanical and configurational complexity, excessive unbalanced solar torques and power-transfer difficulties. For these various reasons, it is clear that missions requiring higher powers from a mass-effective system are best supported by a body-stabilized satellite design.

3) Attitude Control: The attitude-control system of a dual spin is much simpler and lighter than that of a body-stabilized satellite. For satellites which are spin stabilized for the injection phase it has already been noted that the body-stabilized class then requires two separate on board attitude systems with the attendant mass, cost, and reliability drawbacks. On the other hand, the dual-spin system contains some inherent features which make it unsuitable for applications where very high pointing accuracy is needed. These drawbacks include nutation due to residual imbalance in the spinning section, torque ripple in the mechanism used to despin the platform carrying the payload, slop due to the results of bearing wear, and thermal and/or mechanical distortion of the "stalk" carrying the antennas.

Spinning satellites have a certain drawback, however, when the configuration is such that the spin axis cannot be the axis of maximum moment of inertia. In such cases where the spin/lateral inertia ratio is unfavorable, the dynamic system is basically unstable, and there is a need for mechanical damping on the despun platform and/or active nutation control, as well as special steps to avoid potential instability caused by energy dissipation in bearings or due to fuel slosh.

4) Thermal Control: The thermal control of a dual-spin satellite suffers, as for power generation, by the geometric envelope limitations. Where high payload dissipations are involved, only the North-facing side of the satellite is able to be used for significant radiation to space, and even this is to some extent blocked by the antenna farm. The body stabilized satellite, which can make available two relatively large and unobstructed North and South faces, is thus preferable from this viewpoint once on-station. The despin mechanism of the dual-spin satellite poses a unique thermal control problem and generally needs heating to prevent it growing too cold and being damaged.

In such a complex matter it is impossible to derive any definitive conclusions in terms of a general preference for one system or the other. Suffice to state that, for applications involving a total power requirement at end of life of a certain maximum value (which depends on the launch vehicle employed and is around 1 kW for Atlas/Centaur) and for antenna-pointing accuracy and thermal control requirements which are not too severe, the dual-spin class of satellite has proved to be advantageous. On the other hand, the body-stabilized design has also many attractive features for this range of applications, and furthermore is the obvious solution where higher power demands or very accurate antenna pointing are involved. Both types of configuration will be reviewed in the following sections which deal with subsystem design aspects.

IV. ATTITUDE AND ORBIT CONTROL

In this section, the implications of the various attitude and orbit control requirements are addressed for both the injection and on-station phases of the mission. Brief mention is made of the design impact of N-S stationkeeping requirements.

A. Injection Phase Functions

As stated in Section II-C, most communication satellites are spin-stabilized in transfer orbit. The following main tasks are to be performed during this phase.

1) Damp any residual nutation stemming from the launcher separation manoeuvre down to a level that permits an attitude reconstitution.

2) Perform the attitude determination of the spin axis at a convenient moment, i.e. when earth and sun are in the field of view of the sensors and the satellite is in contact with a ground station.

3) Process the spin axis to the orientation that is required for the apogee motor firing.

4) Perform a final attitude determination before motor firing to ensure that the thrust axis is correctly aligned and any excessive nutation is damped out.

5) Fire the motor in order to circularize the orbit and to remove most or all of the equatorial inclination.
Each of these functions will now be discussed in more detail.

1) Nutation Damping: The residual nutation of a basically stable satellite, i.e. whose spin axis is the axis of maximum inertia, can be damped out relatively easily. For an efficient and lightweight damping system, an oscillating system such as a pendulum or a ball in a curved tube can be used, the resonance frequency and damping coefficient being tuned to the satellite nutational frequency as determined by the satellite moments of inertias. In the case of a basically unstable configuration, the damping has to be performed by an active control loop consisting of a nutation sensor—a rate gyro or accelerometer—and an actuator (for example, a thruster that has to be activated according to the output signal of the sensor). Dual-spin satellites can be stable with respect to nutation damping even if the spinning drum alone would be unstable. The condition is that the counter-rotating platform must have a low or zero spin-rate with respect to the spinning drum and that the main energy dissipation must be on the platform. In this case therefore the damper has to be mounted on the despun platform.

2) Attitude Determination (Initial and Final): The computation of satellite spin-axis orientation is normally performed on the ground, based on transmitted angle data. The on-board sensors measure spin axis-earth elevation, spin axis-sun elevation, and sun-satellite-earth angles, which data may be used to solve the appropriate spherical triangle and

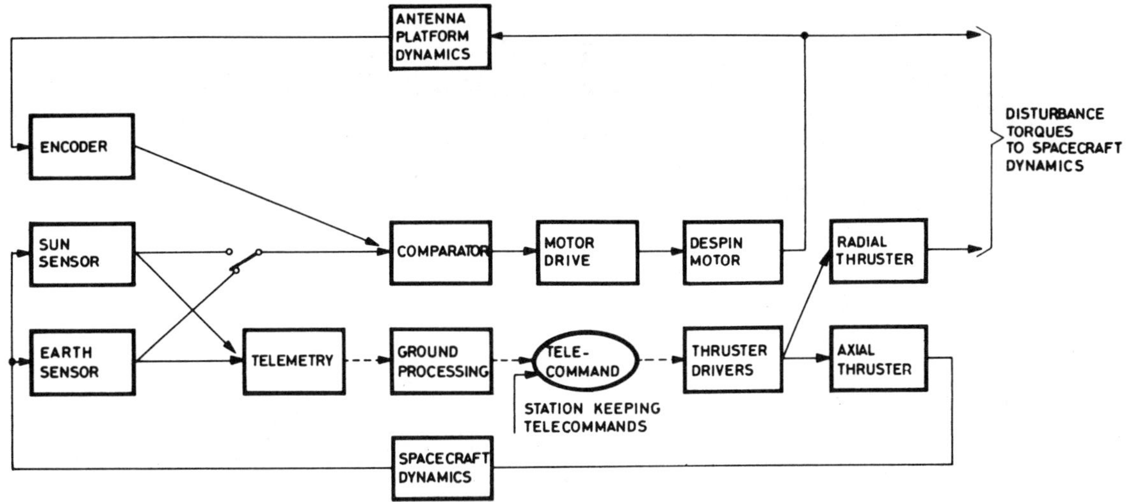

Fig. 4. AOCS functional block diagram for a dual-spin design.

hence derive spin-axis orientation with respect to inertial references.

3) Spin-Axis Orientation: Thruster-activated precession manoeuvres of typically 130 degrees have to be performed early in the transfer orbit and while there is coverage from a ground station. The minimum thrust level is relatively high, of the order of several Newtons, and the manoeuvres may take 1.5 h or more.

4) Apogee Motor Firing: The motor has to be activated at a moment where the satellite is at the correct point of the transfer orbit near apogee, and the spin axis has the correct attitude (with an accuracy of typically better than 1 degree).

Since the mass of the apogee motor fuel represents about 50 percent of the total satellite mass, errors in attitude, timing, or total thrust have a detrimental effect.

5) Acquisition Phase: After apogee injection and prior to positioning at the nominal longitude, the satellite must acquire the correct attitude. This phase is referred to as earth acquisition. In the case of a spin-stabilized configuration, this manoeuvre is quite simple, i.e. the rotation of the spin axis from parallel to perpendicular to the equatorial plane. In the case of a body-stabilized system, the manoeuvres are more complex (see Fig. 2) since the spacecraft must first be despun followed by flywheel run-up, solar array deployment prior to being correctly pointed towards the earth. These manoeuvres as well as their exact sequence depend on the detailed design of the satellite and will not be described here.

B. On-station Functions–Dual-Spin Design

The attitude control requirements for a satellite that is spin stabilized during its normal operation differ little from those during transfer orbit. In fact, the attitude determination of the spin axis is easier because the earth will be continuously available as a target for the pencil-beam infrared sensor. The counter-rotating antenna is controlled so as to be locked onto the coverage zone at all times. This control loop must be kept closed continuously and, in order to avoid communications outages, automatic on-board or ground-based means are necessary to rapidly reacquire despin lock in case of a failure. Attitude-correction manoeuvres around the other two axes are performed by a single axial thruster that is mounted parallel to, but at some distance from, the spin axis. If this thruster is actuated in a pulsed mode, always in the same sector of the spin cycle, then—depending on the sector—control manoeuvres can be executed around any desired axis perpendicular to the spin axis. If the same thruster is actuated continuously, it provides a thrust through the center of mass in the N-S direction, and can be used for orbit-control manoeuvres. Position-control manoeuvres in the east or west direction are performed by a second thruster that is mounted with its axis perpendicular to the spin axis of the spacecraft. The latter thruster has to be pulsed in the appropriate sector during each rotation to execute a maneuver. Finally, a third thruster mounted tangentially at the circumference of the satellite may be used, if necessary, to control the spin rate of the satellite. A typical block diagram of the dual-spin type control system is shown in Fig. 4.

The dual-spin satellite has a high inherent gyroscopic stiffness but may still need frequent attitude corrections. Its inertia about the spin axis is several hundred $kg \cdot m^2$, and the spin rate is in the range of 5 to 10 rad/s. Thus the angular momentum is very high, typically 2000 Nms. However, there is usually a significant offset between the center of solar pressure (CP) and the center of mass (CM) due to basic configurational assymetries and this in turn gives rise to an attitude disturbance torque. The CP/CM offset may be kept to below 5 cm for smaller antenna farms by careful arrangement of equipments within the main drum, but for larger antennas the offset can exceed 60 cm. The sun radiation pressure combined with this CP/CM offset gives rise to an attitude-disturbance torque which can be sizeable. The spin-axis direction must, therefore, be corrected at regular intervals. Depending on the particular spacecraft configuration and the allowable attitude deviation, i.e., the control deadband, such corrections are normally made with a frequency which varies from every 3 days to every 60 days or so. Since the stiffness of the system is high, the thrust level of the control thrusters is not critical. A thrust level in the order of 10 to 20 N can be easily achieved with monopropellant hydrazine or bi-propellant thrusters.

C. On-Station Functions–Body-Stabilized Design

Three-axis stabilization can in principle be achieved without angular momentum storage. However, nongyroscopic systems

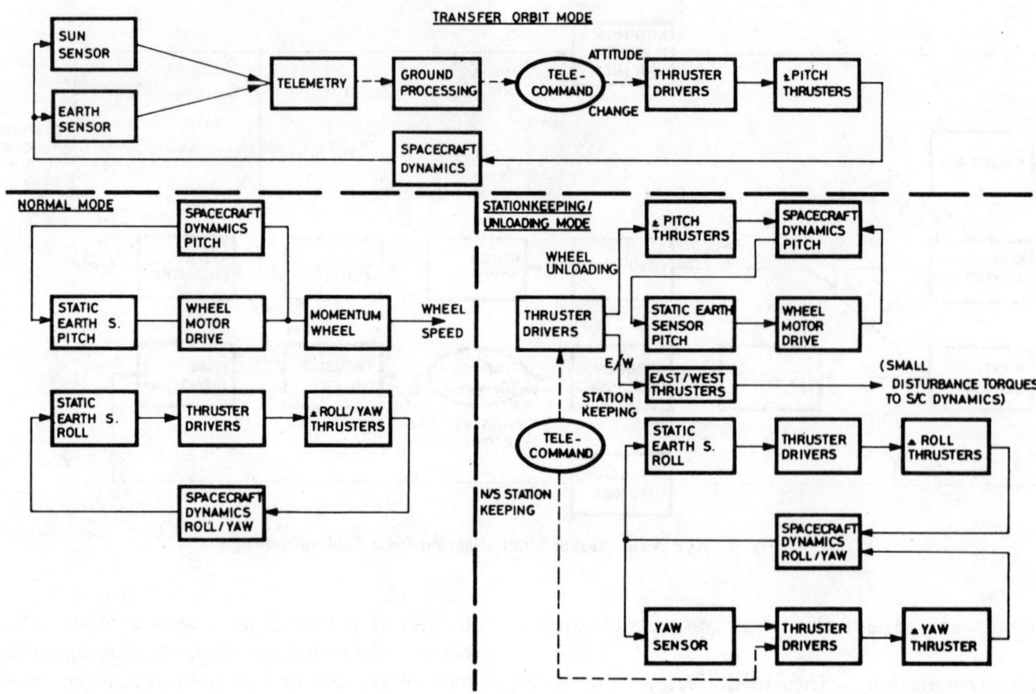

Fig. 5. AOCS functional block diagram for a body-stabilized design (various modes of operation).

require a yaw sensor somewhat difficult to implement on a communications satellite. Indeed, star sensors are obstructed by the solar array configuration, sun sensors cannot be used in orbital positions where the sun is colinear with the earth and gyros can only keep a reference for a restricted period of time and need period updating. A combination of sun sensor and gyro is a possible solution but it proves to be more complicated than a momentum bias system. The latter approach is described hereafter.

The basic element of a momentum bias system is the momentum wheel or flywheel, which has to be sized (in terms of mass and speed) such that the overall system has sufficient gyroscopic stiffness in the presence of environmental torques to avoid unduly high frequency of thruster operations, while at the same time keeping the mass low. The disturbance torques themselves are also minimized by constraining the CP/CM offset to a very low value.

The satellite attitude cannot be determined by the transfer orbit sensor as in case of the dual spinner. For the three-axis stabilization mode, a different sensor is used, typically a static-earth infrared sensor that provides satellite pitch and roll-angle information (axes are defined in Fig. 3). To determine and correct the wheel spin-axis attitude, roll and yaw angles are needed, however the latter is not measured but derived from roll information since a roll angle becomes a yaw angle one quarter of an orbit later during which period the wheel spin axis orientation is sensibly constant. Roll (or roll/yaw) thrusters are usually employed to perform roll-attitude corrections. Control around the spin axis of the momentum wheel is based on the pitch signal of the static infrared sensor, referred to above, whose output acts on the drive motor of the momentum wheel to accelerate or decelerate the wheel accordingly, thus providing reaction torques on the satellite. This loop is always closed on-board the satellite. Since the angular momentum of the wheel is considerably smaller than that of the dual-spinner drum, the

accumulated speed of the wheel is much more sensitive to external torques. For example, a torque of 2.10^{-2} Nm around the spin axis would completely despin the wheel or inversely double its speed, within 3 years or so. In practice, a speed variation of ± 10 percent is acceptable. Hence a speed adjustment manoeuvre, called desaturation, has to be performed at intervals of several weeks using a set of control thrusters around the pitch axis.

Stationkeeping manoeuvres, i.e. orbit control, are performed using thrusters in a continuous (rather than pulsed) mode. East/West manoeuvres use thrusters acting through the CM in east and west direction. N-S manoeuvres could, in principle, be similarly performed by a single thruster either on the north or south face of the satellite. However, these positions are normally occupied by the solar-array drive mechanisms. Consequently, two thrusters have to be provided with symmetric arrangement about the CM and actuated synchronously in north or south direction. Since their thrust levels cannot be kept completely equal, the thrust difference has to be controlled to avoid excessive disturbance torques.

A typical block diagram for the three basic control modes (transfer orbit, normal operations, and stationkeeping/unloading) is presented in Fig. 5 for a system where all main loops are closed on-board. A further system is required to effect the transition from the transfer orbit. Also an additional control loop is required to drive the solar arrays in such a way that they always point to the sun. In fact, this loop can be very simple if the solar arrays are driven by a motor that is actuated by a clock providing the right slewing rate. The proper phase angle can be updated and biased to compensate disturbance torques once per day if a zero reference mark on the drive mechanism is available. An alternative method employs a sun sensor to provide closed loop control of the array motor.

In case of the failure of an element of the control system, an outage may be caused by rapid loss of stabilization. To

Fig. 6. Rigid-panel deployable solar array.

protect against this eventuality, on-board failure sensing and automatic switching to redundant sets of equipment is essential.

D. N-S Stationkeeping

For a mission duration of 7 years, the fuel required to execute full N-S stationkeeping represents 15–20 percent of the useful satellite mass if a conventional hydrazine system is employed. It is therefore very attractive to reduce this fuel mass by one of the two following ways.

1) Suppression of N-S stationkeeping requirements, where this is acceptable for the mission.

2) Increase of the specific impulse by use of a different thruster technology.

If stationkeeping can be suppressed, the maximum savings in mass results, but note that one more degree of freedom is needed on-board so that while the spin axis of the momentum wheel stays parallel to the earth axis, the main body of the satellite is sinusoidally nodded once per day around its roll axis to maintain correct antenna coverage. The simplest way to implement this facility is to gimbal the momentum wheel around one axis (parallel to the roll axis of the satellite) and provide stepper motor actuation. Another possibility is to keep the spacecraft attitude constant and nod the antenna or antennas on a daily basis, which may be acceptable for a satellite with a single antenna, but becomes impractical for more complex arrangements.

If, on the other hand, N-S stationkeeping is essential, methods exist to improve on the specific impulse of conventional thrusters including the use of bi-propellant, power augmented electrothermal hydrazine decomposition, or electrical propulsion thrusters. The specific impulse of the latter is an order-of-magnitude higher than the other and in fact the dry mass of its components, including high-voltage power supply, represents most of the mass of such a system.

V. Power Generation, Conditioning, and Storage

The power subsystem fulfills the function of power generation, regulation or conditioning, distribution to user systems, and storage of energy for periods when the primary power source is not functioning such as during eclipses.

A. Power Generation Function

Power generation for geostationary satellites employs photovoltaics. Not only does this offer most advantages in terms of technological standard, power mass ratio, reliability, and flexibility, but also its selection is determined by the fact that 99 percent of the mission lifetime happens in sunlight. As explained in Section III-B, the two main classes of satellites,

dual-spin and body-stabilized, respectively employ a body-mounted solar array and deployable solar panels or wings. The drum array is essentially power limited due to the geometric configuration and is also heavier since its area must be greater than that of a sun-oriented solar wing by the factor

$$F = \frac{\pi}{1 + \Delta T \, K_{p/t}} \simeq 2.7$$

where ΔT = difference in normal operating temperature between the sun-oriented panel and drum array (the latter running typically 30° cooler), and $K_{p/t}$ = solar-cell efficiency factor, which is about 0.5 percent improvement per degree centigrade of temperature reduction.

A good figure of merit for a solar array is its performance at end of operating life (e.o.l.) expressed in watts per kilogram, including mass of the cells, substrates, supporting structures, deployment and power transfer mechanisms. For a dual-spin satellite drum array, at power levels of around 600–800 W at e.o.l., it is possible to achieve a figure of 9 to 11 W/kg, while the equivalent figure for the rigid-panel deployable array of a body-stabilized satellite is in the range of 18 to 22 W/kg. Thus there is a potential mass advantage for an 800 W body-stabilized satellite of about 40 kg. As an example, Fig. 6 shows a rigid-panel array whose specific mass is 20 W/kg. With larger arrays and employing a flexible substrate for mounting cells, a power specific mass of up to 50 W/kg seems to be achievable. With very large arrays the contribution of solar-cell mass will be about 50 percent of the total array mass. Further improvements are expected from the panel structure technology, and also from the solar cell performance.

Degradation of silicon solar cells is caused by generation of defects in the semi-conductor structure due to electron and proton bombardment in space. The major contributor in geostationary orbit is from protons during solar flares, whose activity is cyclic, thus dependent on the launch date. In order to reduce this damage, solar cells are covered with shields from fused silica or cerium-doped microsheets. As a typical example Fig. 7 shows the degradation of an array output versus satellite lifetime. The seasonal power oscillations are due to changes of solar aspect and variation of solar intensity during the year. The normal increase of solar array power of about 9 percent during equinox is fortuitous, since eclipses occur at this time and more power is necessary for battery recharge. But battery recharge power may be up to 25 percent of the equinox sunlight power requirements, and therefore this 9 percent natural power increase may not be sufficient. In this case additional solar-cell area needs to be provided and the equinox power condition becomes the critical design case. The communications payload requirements must, of course, be satisfied through the whole year.

However, the degradation of the solar array with life can be matched to the degradation of the payload capacity—in other words, the number of surviving channels. This means that at the beginning of life, more channels of a repeater can be operational than a few years later, giving so-called graceful degradation of both the array and payload performance.

B. Power-Conditioning Function

The power delivered from the solar array depends on the selected operating voltage and the extent of irradiation degradation. The cell voltage is also strongly dependent on

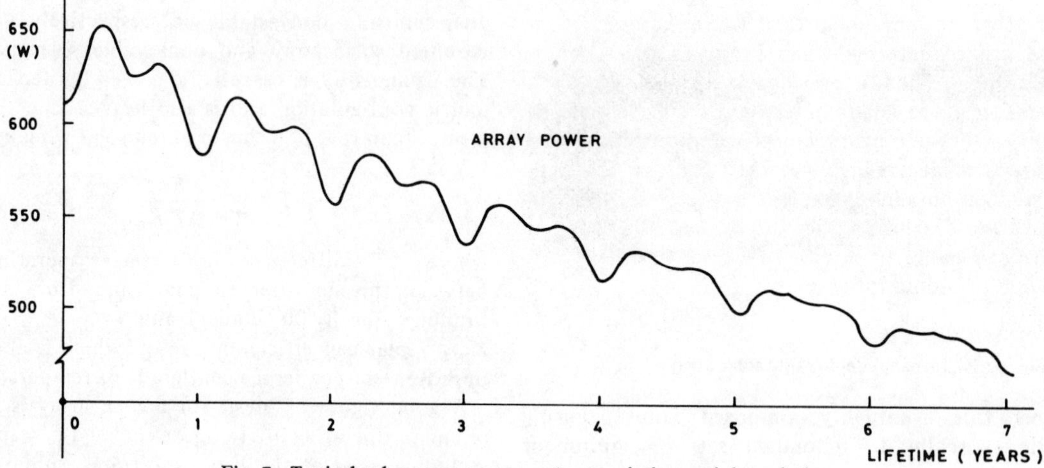

Fig. 7. Typical solar-array power-output variation and degradation.

temperature, and for example, while the array temperature at exit from eclipse rises over a few minutes from $-180°$C to $+60°$C, the array voltage can briefly exceed the nominal operating value by up to 2.5 times. Regulation or damping of the array voltage is therefore necessary to avoid such an overvoltage on the users. With regard to the battery, its voltage must also be matched in some way to the array voltage to allow operation of the users under sunlight and eclipse conditions. Two main types of power systems exist, based on the regulated and the unregulated-bus concepts, as illustrated in Fig. 8.

In the regulated-bus concept, a shunt parallel to the solar-array generator defines the operating voltage which is regulated to 1 or 2 percent. Under e.o.l. conditions, the operating voltage matches the maximum-power point of the solar array.

The following are the main advantages of the regulated-bus concept.

1) Clear definition of supply characteristics under all operating conditions.

2) Simplification of power conditioning for users who need higher or lower voltages than offered from the bus, and direct supply for users with suitable supply voltage requirements.

3) Low impedance and ripple on the supply line.

The next three points describe the disadvantages of the regulated-bus concept.

1) A high reliability requirement for the central power conditioning equipment.

2) The shunt must be able to withstand high variations of supply demand, and for maximum efficiency must be realized in an analogue/digital form to provide sequential switching of solar-array sections.

3) The storage energy loop has significant losses (about 15 percent of the storage power), which means that a regulated bus concept is less advantageous for mission with high eclipse-power requirements.

In the unregulated-bus concept the user-load characteristics and the battery determine the bus voltage. A varying load demand thus changes the input voltage to other users. Also when switching from the sunlight to the eclipse mode of operation, the voltage conditions change significantly. Consequently, the users must be equipped with their own power-regulation equipment to cope for the varying input-voltage conditions. In order to avoid dangerously high voltages at

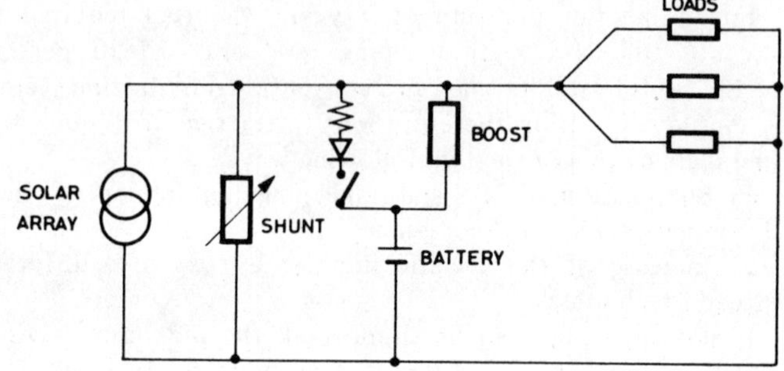

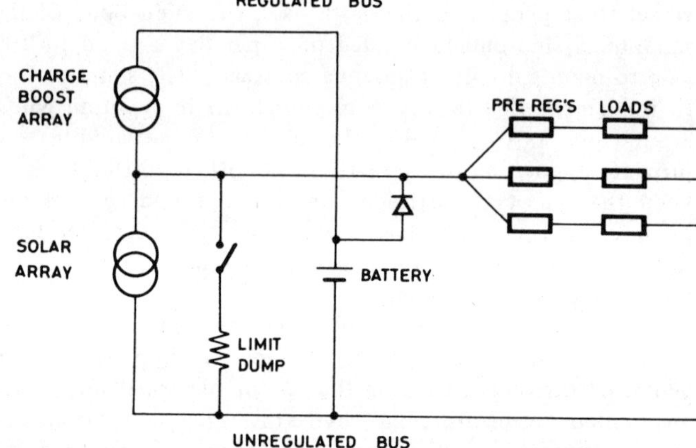

Fig. 8. Regulated and unregulated power-bus concepts.

end of eclipse, some kind of dumping facility (limiter) must be provided.

The following are the main advantages of the unregulated limited-bus concept.

1) Simplicity and lower mass of power subsystem equipment, which has mainly limiting and distribution functions.

2) Overall system mass lower, especially when degradation or loss of individual users is tolerable and when mission demands high eclipse powers.

The disadvantages are:

1) power conditioning in user equipment is more complex;

2) higher bus impedance and ripple;

3) protection necessary to avoid high supply voltages, which could overstress connected equipments.

It should further be mentioned that multiple-bus concepts exist. These are of interest when supply sources can be dedicated to groups of loads, or when the characteristics of user equipment, e.g. payload and support subsystems, are quite different. One such example is a communication satellite which carries a single large payload, e.g. for TV transmission in sunlight mode only, where the very different requirements of the TV transponder and the support functions of the satellite can better be satisfied by a separate bus using conventional conditioning equipments.

C. Energy Storage Function

Geostationary satellites experience 90 eclipses during one year on station with a maximum duration of 72 min per day. This leads to a relatively low number of charge/discharge cycles of the energy storage system and hence allows for the application of a high depth of discharge. This means that 50 to 70 percent of the capacity of a battery can be utilized during each discharge cycle, compared to 10–20 percent for low-orbit satellites which have thousands of cycles a year. The mass of the energy-storage equipment, normally Ni-Cd batteries with their proven high reliability and long lifetime, is high in comparison with the photovoltaic power generation hardware. Considering the battery alone, the power specific mass at 70 percent discharge is about 12 W/kg when calculated for the maximum duration eclipse. In a regulated bus concept, the mass and inefficiency of charge and discharge regulators must then be added, which reduces the power specific mass to below 10 W/kg. Some mass improvements on energy storage devices are likely in the future and would lead to major mass savings for communications satellites. Ni-H$_2$ and Ag-H$_2$ batteries are in development which could offer a mass saving compared to Ni-Cd batteries of about 30 or 60 percent respectively. Application of these batteries in communication satellites, other than for experimental purpose, is unlikely before 1980.

Ni-Cd batteries can safely be operated in a moderate temperature range only, preferably $-5°$ to $+10°$C. Particular care must be given that the storage temperature is favourable ($<15°$C) because the majority of battery life in orbit is in the storage mode. Furthermore, to improve battery performance, a low trickle charge is applied during storage and/or reconditioning (full discharge to zero and recharge) prior to the eclipse season. The latter has been shown to be particularly valuable for long life missions. During the eclipse season about 23 h are available for recharge. Protection based on a safe upper limit of cell voltage of about 1.55 V is necessary to avoid overcharging of the battery. Battery stress is further reduced by limiting the ampere-hours input to the minimum necessary to recharge the battery. During discharge care must be taken to limit the discharge current to avoid capacity loss and to limit minimum battery voltage. Below 1.1 V or so, the cell voltage decreases rapidly and since not all the cells of one battery have exactly the same capacity, the weakest cells could see their polarity reversed while the total battery package still delivers energy, leading to high pressure of hydrogen in the reversed cells and possible destruction.

VI. THERMAL CONTROL

The environment experienced by a satellite, as a function of mission phase, leads to requirements for a thermal-control system to ensure that equipments remain within their operating and nonoperating temperature limits and hence continue to meet their performance objectives throughout the satellite lifetime. The following is a summary of the main environmental factors.

Environmental Parameter	*Value and Implications for Geosynchronous Satellite*
1) Vacuum.	$\simeq 10^{-10}$ torr (mm Hg)
2) Solar radiation (thermal flux in infrared, visible and ultraviolet regions).	$\simeq 1400$ W/m^2, with slight variations according to time of year.
3) Thermal radiation to free space from satellite	Free space has an equivalent temperature of $\simeq 4°$K.
4) Albedo (solar radiation reflected from earth).	Negligible (but must be considered in transfer orbit).
5) Ultraviolet, electron, and proton irradiation.	Negligible as regards thermal flux, but harmful effects on surface finishes.
6) Eclipse.	Leads to rapid external temperature variation.

During pre-launch activities, the satellite will usually be maintained in a specified favourable environment (typically $22°$C $\pm 5°$C), although during transportation and some test phases the limits may be somewhat wider. During the first stage of launch the shroud is heated by aerodynamic friction; the interior can radiate heat to the satellite but the period is relatively short and the only possible impact is on large-area, small-mass, external equipments such as antenna dishes and solar-array panels. After the shroud is jettisoned, the satellite is directly exposed to aerodynamic heating, albedo, earthshine, and direct solar radiation and this may result in requirements for special attitude manoeuvers during ascent and/or launch window limitations. Conditions during transfer orbit are broadly similar to those experienced in the final geosynchronous orbit, except for small extra albedo, earthshine, and aerodynamic thermal loading near the perigee of each orbit. Note, however, that the high-power payload equipments are switched off at this time. The period of apogee motor firing is thermally significant in two ways: during and for some time after the firing there is heat soakback from the motor case, which has to be restricted to acceptable levels, and, during the actual firing, the energy in the rocket plume plus radiation from the hot nozzle itself subject adjacent equipments and surfaces to a high thermal flux. During the drift orbit phase, i.e. after ABM firing until the satellite acquires normal operational status, there may be significant orientation and on-board configuration changes. However, thermal requirements in this phase can usually be derived by analogy with the transfer orbit or on-station phases. Finally, for the on-station phase, the complete satellite including payload has to operate during exposure to winter and summer solstice conditions (which might affect the satellite very differently according to whether or not it is broadly symmetrical about the Earth's equatorial plane), to equinox conditions (with two 45-day eclipse periods per year) and up to end-of-life conditions (affected both by degradation of the thermal control surfaces and by reduction in dissipated power due to degraded solar-array output).

A. Thermal Control Techniques

Achievement of satisfactory thermal control in the environmental conditions outlined above may rely on the use of both

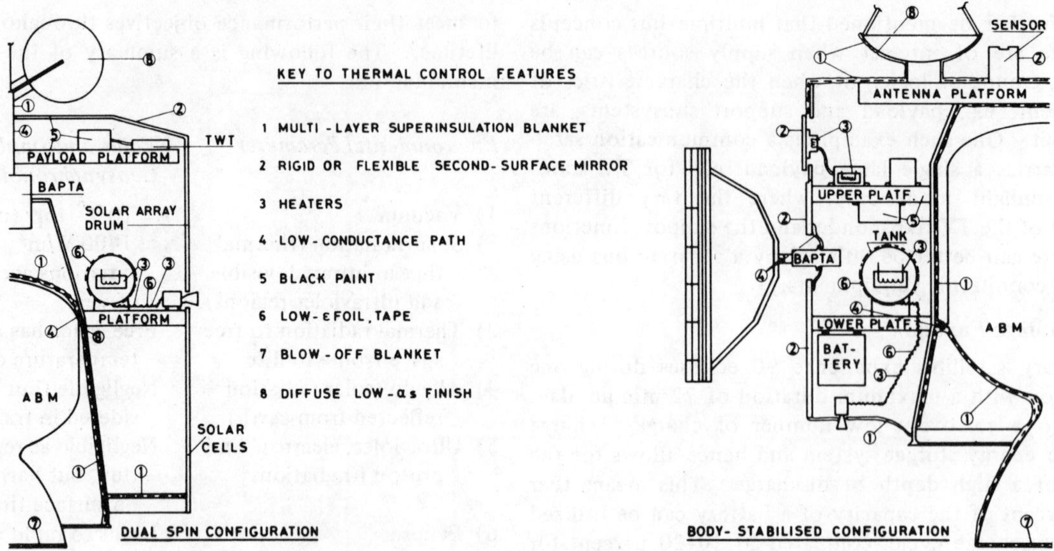

Fig. 9. Thermal control for dual-spin and body-stabilized configurations.

passive and active techniques, with a preference for use of passive techniques wherever feasible, on grounds of design simplicity, cost, and reliability.

1) Passive Techniques: As a first step, satellite and equipment surfaces may be treated to give them a high or low emissivity ϵ (the relative capability of radiating heat away at a given temperature) and a high or low solar absorptivity α_s (the relative capability of absorbing incident solar energy). Some finishes, for example black paint, have high ϵ and α_s, while others, such as polished metal surfaces, have low ϵ and α_s, and so on. The ϵ and α_s parameters take values lying in the range zero to unity, and the ratio α_s/ϵ is of prime importance in determining whether, when exposed to sun, a given surface finish at a particular temperature leads to a net outward radiation. Since communication satellite payloads need to dissipate heat, radiators with a very low α_s/ϵ finishes are employed: typical finishes are special second surface mirrors with α_s/ϵ values down to 0.06, made either from quartz chips or Teflon sheet with silver or aluminium backing. Combined with the use of such surface finishes, most passive thermal control systems also rely on the use of super-insulation blankets composed of many layers of thin Mylar or Kapton film coated with vacuum-deposited aluminium and having extremely low conductance in the space environment. Interface fillers are also utilized to improve conductive heat transfer between equipment bases and their mounting surfaces, particularly for small, high-dissipation units such as traveling wave tubes: typically, a conductance increase of nearly one order of magnitude is obtainable.

2) Active Techniques: As a supplement to passive means, the following active methods of thermal control may be employed as necessary.

a) Electrical heaters, activated either by thermostats or telecommand.

b) Hinged flaps and multiple-blade louvre mechanisms, arranged to expose or to cover high-ϵ radiator areas and activated by temperature-responsive bi-metallic springs or fluid expansion circuits.

c) Heat pipes, which transfer heat from a source to a sink (commonly an external radiator) at a very small temperature difference by vaporizing and recondensing a sealed working fluid in a tube. Gas-controlled heat pipes have in addition the capability of transporting variable quantities of heat from a source, maintaining it at constant temperature by rejecting the heat over varying areas of radiator at the condenser end.

B. Functional Design Approach and Thermal Analysis

The design of the thermal-control system is controlled by and interacts with the overall satellite configuration and the specific equipments within it. Considering first the overall configurational aspects, distinction must be drawn again between dual spin and body stabilized designs. Fig. 9 shows typical thermal-control solutions for the two configurated types.

For a dual-spin satellite, the drum bearing the solar array forms the main environment for most of the internal equipment and stays at a very even temperature throughout the year, typically $20°$ to $25°$ C, except for the two equinox eclipse seasons. The temperature then drops to $-80°C$ or lower, at maximum eclipse, which makes it necessary to thermally isolate many of the equipments from the drum interior. The north face can radiate freely to space, except for the obstruction caused by the antennas, and received solar radiation during the summer equinox season. The south face contains the apogee-motor nozzle and launch-vehicle adaptor ring, and is usually further from the high-dissipation communication equipments which have to be located near the antennas. Hence, it is usually covered by a shield or a multilayer insulating blanket. Thus for a dual-spin satellite, almost all the heat dissipation from the payload is accomplished via the use of radiators on the north face, where these are either mounted on the top of the spinning drum or on the tip of the despun section.

For body-stabilized satellites, the main thermal control is almost independent of the solar-array temperatures, which vary between typically $60°C$ in sunlight and $-160°C$ in eclipse. The spacecraft body, since it rotates about the N-S axis at only one revolution per day, is best covered by superinsulation blankets on all but the north and south faces in order to avoid the extreme daily temperature differences which would affect internal equipments. The north and south faces are used for dissipating heat and the satellite design can be symmetrical

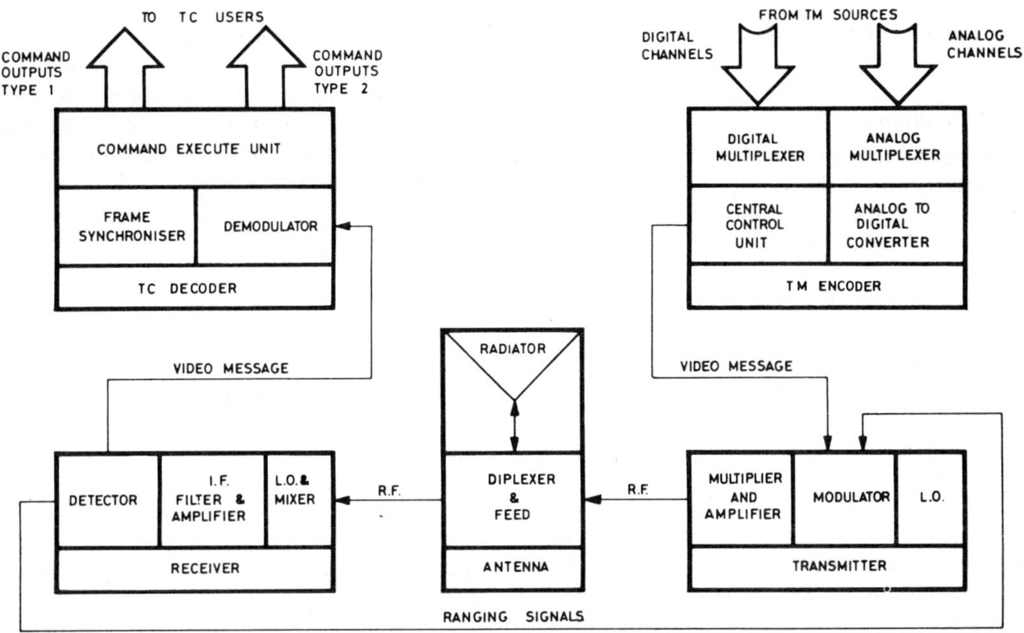

Fig. 10. Example of TTC subsystem block diagram.

about the plane perpendicular to the N-S axis, as each face receives a solar input alternately at winter or summer solstice.

As a necessary tool for arriving at satisfactory thermal design, and for subsequent interpretation of test results, one has to construct an analytical model of the satellite which includes a large number of temperature nodes. Inputs to the model are radiative and conductive couplings between specified nodes, dissipations of equipments, radiative couplings of external nodes to space, solar inputs to external nodes, and masses and mean specific heats of nodes. Computer runs give predicted nodal temperatures and the design is adjusted until a satisfactory result is achieved. When, later in the program, acutal thermal-vacuum simulation tests are performed, correlation is made between thermal-model predictions and test results as a final verification of thermal-design adequacy, also taking into account expected degradation in thermal surfaces over the satellite lifetime and cases where some equipments may have failed.

VII. Telemetry, Tracking, and Command

Telemetry, tracking, and command—commonly abbreviated to TT&C—are key service functions on which all satellites must rely. Telemetry is necessary to monitor and evaluate the satellite performance and to provide ground controllers with sufficient data for routine operation of the satellite and failure diagnosis. The matching telecommand function is necessary for initiation of manoeuvres which are not under closed-loop on-board control and for status or mode changing, either as part of normal operations or to respond to an emergency by corrective actions. Finally, tracking is required to perform corrections or manoeuvres.

In addition to these mission-dependent requirements, the TT&C subsystem design is also strongly constrained in the selection of transmission frequencies. It has to operate within specified bands allocated and controlled by the International Telecommunications Union and interference with other users operating in the chosen band must be avoided. The frequency selection and transmission standards employed will also be constrained through economic considerations to ensure compatibility with an existing and suitable network of ground stations for the launch and injection phase and, of course, the master ground station used for routine control during the operational phase. A typical tradeoff could involve selection between the following.

1) TT&C operation in all phases within the communication payload frequency band, such as the Intelsat satellites which operate continuously at C-band, and use a special network of ground stations during injection.

2) TT&C operation in all phases within one of the standard service bands. (VHF or S-band).

3) A combination of both employing VHF or S-band during injection and the communications band proper once in geosynchronous position.

Fig. 10 shows a typical system block diagram with the various functions described hereafter.

A. Telemetry Function

For communication satellites, the usual telemetry requirements call for a moderate number of small bandwidth channels which are most suitably handled by a time-multiplex system in which each channel is sampled at a fixed time interval, the samples then being multiplexed to form a serial message. The data from the on-board sensors and detectors may be in analog or digital form and conversion of the analog data into digital data is generally performed so that the output message can be more easily handled. The telemetry bit rate is a function of the number of channels which must be sampled and the sampling rate of each channel. Typically, it is in the range 150–1000 bits/s.

Since satellite telemetry requirements may vary between mission phases, it is sometimes beneficial to foresee more than one operating mode for the data handling system—for example by providing a command to change the telemetry format, or the bit rate during the transition from injection to normal operation. For low telemetry bit rates, it is normal to employ a subcarrier prior to modulation of the RF transmitter. While no firm rule can be given, commonly used modulations are phase-shift keyed (PSK) or frequency-shift keyed (FSK) for the subcarrier, followed by phase modulation of the RF carrier.

B. Telecommand Function

Except in those cases where the telemetry and telecommand links form an integral part of the satellite-control operation, the rate at which commands are sent to communication satellites is generally very low, especially when routine operations are established and only a handful of commands are required each month. However, the total number of different commands can be quite high, typically 300 or more, and it is vital that a critical command can always get through and that no false or erroneous command can be executed by the telecommand decoder.

The balancing between the two conflicting requirements of guaranteeing reliable operation at low S/N ratios while offering protection against false commands is usually achieved by designing into the message a number of safety features which the on-board system can check prior to executing the order or, alternatively, by some type of ground-based system checking the on-board-received command and then authorizing it by a separate command. Narrow-band RF frequency response, subcarrier and data frequency discrimination, and well-protected address codes are typical methods used to ensure discrimination against telecommands intended for other satellites. To protect against bit errors due to noise, coding techniques and repetition of data can be employed. On-board, due to the number of command outputs involved, it is common design practice to employ a limited range of standard output interfaces, compatible with loading of registers, relay driving, or logic switching.

C. Tracking Function

To meet the requirements on satellite tracking, the most commonly employed systems involve angular, range, or range-rate measurements. For angular measurements, data can be obtained from the azimuth and elevation angles of the ground antenna tracking the satellite, or interferometry can be used. Both methods necessitate that the satellite provide a beacon signal and the latter method also leads to requirements on the spectrum cleanliness close to the RF carrier. In the interests of efficiency, the transmitter used for telemetry transmission normally acts as the beacon source. Range measurements may be performed by sending a tone to the satellite via the telecommand receiver, and retransmitting the tone back via the telemetry transmitter, computing the time interval for the round trip. Of course, variations in the propagation delay of this tone lead to measurement errors and hence the on-board phase variations must be minimized. An alternative method, based on the same principle but involving range measurements through the SHF payload repeater itself can, where feasible, offer distinct advantages, especially since at higher frequencies the range measurement is less disturbed by ionospheric effects.

D. TTC Antenna

In general, the receive and transmit antenna system is common to all three functions discussed above. Given the potential criticality of satellite-earth links at varying satellite attitudes and positions it is desirable that the maximum angular coverage is provided by these antenna so that normal operations and recovery action are at all times feasible. On the other hand, antenna design complexity and transmitter power requirements increase noticeably with the widening of the coverage angle. No specific design guidelines can be given here, and it can only be stated that the tradeoff, an especially difficult one at higher frequencies (e.g. S-band and above), must be resolved early in the project due to possible interactions with the satellite configuration.

VIII. STRUCTURE

A. Design Techniques and Analysis

The satellite configuration, and hence the structural layout, have to be defined early in a project since the structure is the first element of hardware required in the development program. This, plus economic factors, normally leads to the use of well-known design and fabrication techniques and materials. These consist of stringer-stiffened cylinders and cones, open and closed corrugated shells, tubular struts for platforms and equipment support, and flat honeycomb core panels for equipment-mounting surfaces, fabricated from materials such as aluminium alloys, magnesium alloys, and reinforced plastics.

The integrated structure must be designed to survive a number of specific load conditions. Complex analysis techniques are employed to predict the response of the structure and equipment to expected load inputs, determine the stress levels in the structural parts, and calculate the displacements for all parts of the structure. The most widely used analysis technique is the finite element method which allows the structure to be divided up into small segments called elements the information on elemental geometry, elastic properties, forces, etc. being stored in matrix form. The use of a large digital computer is necessary in order to manipulate the information matrices.

B. Configurational Aspects

A dual-spin satellite requires a structural path which transfers the loads from the despun platform (whose mass may be 40 percent of the total) through the bearing assembly to the launch vehicle interface. A conical spinning structure results which allows for the apogee motor to be carried within the cone. A series of structural members radially positioned around the cone is required to support the solar-array drum and other equipment such as fuel tanks, batteries, etc. The spinning structure is normally relatively heavy and empty, its particular requirement being to support the maximum diameter array drum. The despun section consists basically of a flat payload equipment platform supported at the hub of the platform by the bearing mechanism with support structure provided for the antenna.

A three-axis stabilized satellite has the same basic geometrical constraints as the dual-spin satellite but, in the absence of a central bearing, the payload equipment platform can be integral with the lower half of the structure. This is a configuration which is potentially more mass efficient and gives more scope for controlling the adverse vibration environment imposed on the equipment.

C. Technology Trends

Recent structure designs have achieved remarkably low structure mass to total mass ratios of around 6 percent by adoption of more efficient configurations by better detail design and by more accurate definition of the load environment, as well as by more accurate prediction of the stresses, displacements, and failure criteria of every structural member.

Apart from the use of well-known aluminium and magnesium alloys, carbon-fiber reinforced plastics and beryllium products are now being introduced. Carbon-fiber reinforced plastic is used already for solar-panel substrates and studies into the utilization of this material for facing sheets of honeycomb core of equipment platforms have been underway for some time. Mass savings have also been made using beryllium structural components in military satellites, but the high cost and new manufacturing techniques required exclude the wide use of this material in the commercial satellite world. In fact, from now on it is expected that greater mass savings will be obtained in the future from electronic equipment-packaging optimization than from structural refinements.

IX. Reliability Optimization/System Engineering and Tests

Although these types of activities are not within the scope of the paper, they are obviously of paramount importance. Indeed, the task of producing an overall design for a satellite to meet the various performance requirements and reliability objectives in an optimum manner starts with major decisions on the basic configuration and the allocation of the various budgets to provide guidelines for the design and development of the satellite subsystems. During the subsystem design and development, complex system-engineering tasks revolve around the need to optimize the satellite system reliability, verify the interfaces, monitor and control optimum use of available mass, power, TM and TC channels, and ensure the achievement of required system performances. Later in the program, the various satellite models are to be assembled and a comprehensive series of satellite system tests have to be executed under severe environmental conditions to verify that these required performances will indeed be met in orbit.

Acknowledgment

The design guidelines of this paper reflect not only the views and experience of the authors, but also those of a number of members of the ESA Telecommunication Satellites Program Office. We should like to acknowledge, in particular, the contributions of R. Steels, U. Renner, H. Lechte, D. Howle, and D. Baston, and to thank other members of the team who provided helpful information or criticism.

Section 4.2
Antennas

Spacecraft antennas play a particularly important role in system design. They are used to provide gain for both uplink and downlink signals and to direct energy to desired locations while, at the same time, rejecting energy coming from other locations. Many of the earlier satellites provided global coverage (17.3° in synchronous orbit). To do this using a single polarization was a relatively direct task. However, as discussed in Section 3, as the required satellite capacities increase, some form of frequency reuse is required unless the available bandwidth is increased or new frequency bands are used. This frequency reuse can be obtained by using multiple-beam antennas and/or dual polarization. These new features require more complicated and larger antennas. The antenna must provide beam-to-beam isolation and must also have polarization isolation if two polarizations are used. In this section, the three generic types of antennas—reflectors, arrays, and lenses—are discussed. Reprint Paper 4.2.1 surveys the state of the art in multiple-beam antennas development. Reference [1] provides a good general discussion on communications satellite antennas.

Reflectors are the most commonly used spacecraft antenna. They have been used on the INTELSAT IV and IV-A satellites, and ANIK satellites, WESTAR, and SAT-COM. On ATS-6, a 10 m deployable center-fed paraboloid was used. Reflector antennas are composed of 1) reflecting surfaces, 2) feed systems, and 3) distribution networks.

1) The reflecting surface is generally parabolic, although spherical surfaces have been proposed. In addition, there can be subreflectors as in the Cassegrain antenna. The size of the reflecting surface determines the antenna gain at a given frequency. The surface tolerances required are also related to the operating frequencies, with the higher frequencies requiring more control of the surface tolerances.

2) The feeds for the reflector can be any set of radiating elements (e.g., horns, arrays, or lenses). The feeds will normally be essentially located in the focal plane of the effective optical system, although they may be offset from the optical axis; or the optical axis may be folded through the use of a secondary reflector. Moveable feeds can be employed to steer the separate beams, or the beams may be electronically steered.

3) The distribution network contains phase shifters and provides power distribution to the radiating structures at the feed. Beam shaping and beam steering are performed in this network. For multiple-beam dual-polarization antennas, this distribution system is a critical element in system performance.

Reference [2] provides a good general discussion of reflector antennas. References [3]–[6] address the satellite antennas specifically. Large deployable reflectors are also of interest in some applications (e.g., [18, ch.1]).

Array antennas are composed of three basic components: array elements, phase shifters, and the feed system [7].

1) The array of elements is the interface with free space and is composed of the individual radiating elements. These elements may be individual dipoles, helical elements, open-ended waveguides, or any other set of radiating elements. The element spacing may or may not be uniform.

2) Phase shifters feed each element and are used to point and form the desired beam(s). The phase shifters should not unduly attenuate the signal, and values of 1.0 dB are commonly attained.

3) The feed system distributes the energy to the phase shifters and provides amplitude weighting across the array for beam shaping.

The phase shifters and feed network can be adaptively adjusted to form beams of arbitrary shapes. Techniques of this type are called adaptive beam forming [8]. The phase shifters can be digitally or analog-controlled, with respectively discrete or continuous phase shifts. The power distribution network can use variable power dividers to change the amplitude distribution [9]. This gives the array antenna a great deal of flexibility, both as a primary radiator and as a secondary radiator.

The array has been considered for possible spacecraft operation; however, it is not presently used in its most complete form in any commercial satellite. It has in general been too big and too inefficient (e.g., the feed and phase shifters are too lossy). The MARISAT satellite does use a simple four-element helix array at L-band and a three-element helix array at UHF. In both cases, the phase shifters and feed elements are fixed.

LENS

The lens antenna is a transmission device in contrast to the reflector antenna. The lens acts as a focusing medium for the electromagnetic energy generated by the feed structure behind the lens. In addition to the feed structure, a distribution network of phase shifters and power dividers is also required. There are three generic types of lens structures: 1) metallic, 2) dielectric, and 3) TEM designs.

1) The metallic lens is composed of an array of waveguides of varying length, with the resulting difference in propagation times accounting for its beam shaping property. The waveguides are hollow, and such a lens can be made lightweight. However, these metallic lens antennas have small fractional bandwidths on the order of 5 percent.

2) The dielectric lens is a broadband lens composed of dielectric material. In general, it is quite heavy, but recent work on lightweight dielectrics indicates that this weight factor can be reduced.

3) The TEM lens design consists of TEM lines that do the required phasing of the electromagnetic field. A TEM lens antenna is lighter than a dielectric lens and broader band than a metallic lens. The lens can also be used as a feed for a reflector antenna.

As a spacecraft antenna, the lens has received attention recently for its proposed use on the DSCS-III satellites (see Reprint Paper 2.1.3). For this application, the antenna must provide flexible antenna profiles and be capable of nulling out interference sources on the earth. It does this by controlling the various pencil beams that are formed by the feed network via the amplitude variable power dividers and output phase shifters. The lens has the capability of scanning off boresight quite easily; and, with sufficiently complex feed structures, large nulls can be attained at arbitrary locations.

There are other antennas that may be of use for specific applications. The helix antenna is being employed, in a simple array structure, for the MARISAT satellite because it provides reasonable gain and good circular polarization, which in turn, gives multipath protection. Horn antennas have been extensively used for global coverage, but they do not have the capability to provide the narrow beams and high gain required in advanced systems.

Other references of interest include [10]-[17] as well as a number of papers in the Bibliography at the end of this volume.

REFERENCES

[1] L. J. Ricardi "Communication satellite antennas," *Proc. IEEE,* vol. 65, no. 3, pp. 353-367, Mar. 1977.

[2] P. J. B. Clarricoats and G. T. Poulton, "High-efficiency microwave reflector antennas—A review," *Proc. IEEE,* vol. 65, no. 6, pp. 1470-1504, Oct. 1977.

A. W. Rudge, "Multiple-beam antennas: Offset reflectors with offsetfeeds," *IEEE Trans. Antennas Propagation,* vol. AP-23, no. 3, pp. 317-322, May 1975.

[4] J. Duncan *et al.,* "Dual polarization multiple beam antenna for frequency reuse satellites," *Op.cit.* [5].

[5] C. C. Han, A. E. Smoll, H. W. Bilenko, C. A. Chuang, and C. A. Klein, "A general beam shaping technique—Multiple-feed offset reflector antenna system," *AIAA/CASI 6th Communications Satellite Systems Conf.,* pp. 1-6, Apr. 5-8, 1976, Montreal, Canada.

[6] R. H. Turrin, "A multibeam, spherical reflector satellite antenna for the 20- and 30-GHz bands," *Bell Syst. Tech. J.,* vol. 54, no. 6, pp. 1011-1026, July-Aug. 1975.

[7] L. Stark, "Microwave theory of phased array antennas," *Proc. IEEE,* vol. 62, no. 12, pp. 1661-1701, Dec. 1974.

[8] W. F. Gabriel, "Adaptive arrays—An introduction," *Proc. IEEE,* vol. 64, no. 2, pp. 239-272, Feb. 1976.

[9] A. R. Dion and L. J. Ricardi, "A variable coverage satellite antenna system," *Proc. IEEE,* vol. 59, no. 2, pp. 252-262, Feb. 1971.

[10] S. I. Ghobrial, "Cross-polarization in satellite and earth-station antennas," *Proc. IEEE,* vol. 65, no. 3, Mar. 1977.

[11] R. Kyle, "Phased array antenna fundamentals," *Military Electronics/Countermeasures,* Mar. 1978.

[12] D. O. Reudink and Y. S. Yeh, "A scanning spot-beam satellite system," *Bell Syst. Tech. J.,* vol. 56, no. 8, Oct. 1977.

[13] R. Mittra, V. Galindo-Israel, D. A. Bathker, and W. N. Moule, "A review of performance characteristics of satellite antennas and some observations on future directions in satellite antenna design," *EASCON '77.*

[14] D. T. Nakatani and S. L. Flateau, "Multiple beam antenna design for satellite communication systems," *ICC'77,* pp. 33.3-317-33.3-321.

[15] D. H. Townsend, "A satelliteborne adaptive array employing ground implemented beam formation," *ICC'77,* pp. 33.5-323-33.5-327.

[16] W. A. Imbriale and G. G. Wong, "An S-band phased array for multiple access communications," *NTC '77,* pp. 19:3-1-19:3-7.

[17] P. F. Sielman, "The role of adaptive multibeam systems," *7th AIAA Communications Satellite Systems Conf.,* 77-60, pp. 1-29.

[18] C. B. Coulborn and F. L. Hennessey, "Antennas," Chapter 1 in Aerospace Report No. ATR-78 (7735)-1, July 1978.

ADVANCES IN MULTIBEAM SATELLITE ANTENNA TECHNOLOGY

E. W. MATTHEWS, W. G. SCOTT, and C. C. HAN

Aeronutronic Ford Corporation
Western Development Laboratories Division
Palo Alto, California

Abstract

This paper surveys the state of the art in multibeam antenna development as applicable to communication satellites. It defines three basic forms of multibeam antennas -- those providing space diversity, polarization diversity, and variable-shaped beams. The use of lenses, reflectors, and phased arrays in multibeam configurations is described. Finally, hardware developments in the field are surveyed, including the LES-7 waveguide lens, more recent experimental TEM and waveguide lenses, a dual K-band reflector, and offset-fed reflectors for the INTELSAT IV-A and V satellites.

I. Multibeam Antennas - Forms and Functions

Multiple beam antennas have been considered for nearly a decade now as one answer to the bandwidth constraints affecting multiple users of a common radiating communications system, particularly involving a satellite. If the users (or groups of users) are separated geographically, each may reuse the same frequency band by accessing the satellite through a different antenna beam. Space limitations on a satellite, however, call for a single antenna structure capable of radiating many simultaneous beams which are sufficiently isolated that individual beams do not interfere with each other. This is known as frequency reuse through space diversity.

Additionally, two users (or groups of users) in the same geographical area may both use the same frequency band without interference by employing orthogonal polarizations (right and left circular, or horizontal and vertical). This is known as frequency reuse through polarization diversity, and is limited to a single reuse since only two orthogonal polarizations exist. Space diversity allows considerably more flexibility, being limited only by the number of individual beams implemented in the antenna and by the antenna's resulting size.

A third form of multibeam antenna needs to be recognized and distinguished from the above two; this is actually a variable-shaped beam antenna, such as proposed for the Defense Satellite Communication System (DSCS-III) program, as discussed in another session. This antenna type may be implemented as a multibeam antenna with space diversity, but differing from the above types in that the beams are interconnected to a single common port. The shape of the resulting single beam may be controlled by means of the interconnection network, thus providing two advantageous features: first, available power can be concentrated only in the areas actually in use at a given time, resulting in increased

EIRP and G/T; and second, areas where interference is being experienced (jamming) can be discriminated against by forming nulls in the net antenna pattern in those directions, either by command or adaptively.

II. Types of Multibeam Antennas

Three distinct types of multibeam antennas are examined - lenses, reflectors, and arrays. Each type is discussed, their operating principles and subtypes described, and their relative merits compared.

A. Multibeam Lens Antennas

Multiple beams in space may be created by focusing the energy from a primary array of feed horns through a microwave lens. The axial symmetry of such a configuration allows low aberration for scanned beams, thus resulting in low sidelobes and cross-polarization components, which are important considerations for frequency reuse applications. Two types of lenses have been considered for microwave use.

Refractive Lenses. The classic solid dielectric lens shown in Fig. 1 was studied for multibeam use by Lockheed[1] and Hughes.[2] For a 30 dB sidelobe specification, it was shown[2] that random variations of refractive index must be controlled to within 0.8% around a nominal value of 2.0. Since lightweight low-loss materials of this uniformity were not available, no further effort was applied.

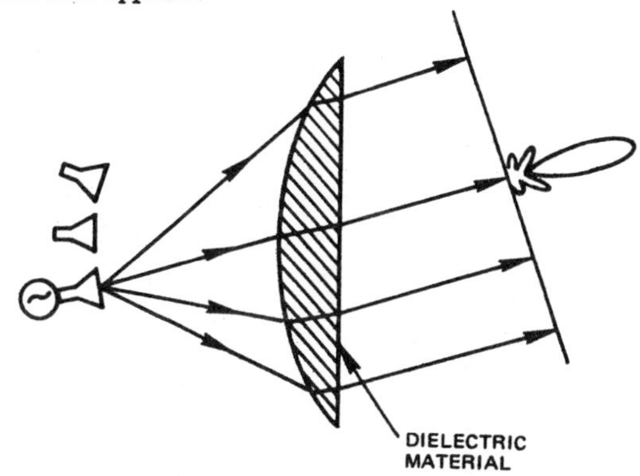

Fig. 1 Refractive Lens

Constrained Lenses. A constrained lens is a form of space-fed array, as shown in Fig. 2. A ray path through the lens is constrained to follow an RF transmission line, interconnecting small pickup and reradiating elements. Optimum low-aberration beam control

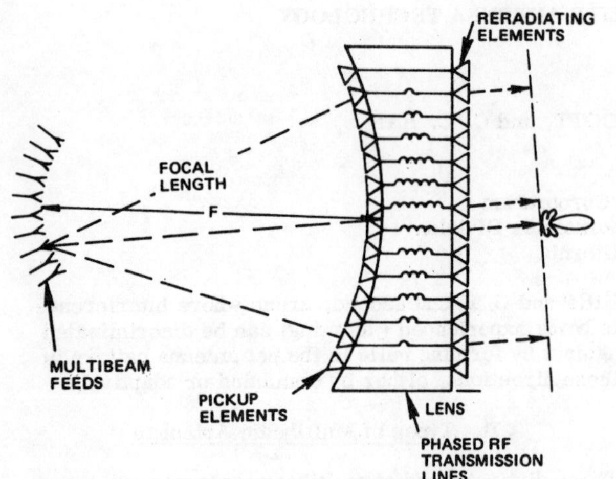

Fig. 2 Constrained Lens

is established by proper choice of line length and element positioning. Characterization of this lens type is generally in terms of the type of wave carried by the RF lines, i.e., waveguide or TEM (bootlace). The latter are non-dispersive and as such usually exhibit greater band-widths; they may utilize coaxial, microstrip, or similar types of delay lines, generally much smaller in cross-sectional dimensions than waveguides. As a consequence, this type of interconnection allows more freedom in choosing the lens optical design to achieve minimum aberrations. The TEM lens should thus be capable of lower sidelobe performance than the waveguide, although it may be somewhat more lossy due to the dissipation in its lines.

B. Multibeam Reflector Antennas

Reflector-type multibeam antennas are potentially attractive because of their design simplicity, inherent bandwidth, ease of construction, light weight, and low cost. However, studies have shown that the usual large multibeam feed structures cause excessive blockage and

consequent high sidelobe levels for all configurations except offset-fed types, such as shown in Fig. 3. These designs usually consist of a section of a larger parabola, whose focal point is located such that the feed does not block directly reflected rays from the section. Feed energy is constrained to illuminate the parabolic section, with proper amplitude taper for desired low side-lobes. Offset-fed types are less well known than sym-metrically fed types (including cassegrainian), and are subject to rapid deterioration of sidelobes and cross-polarization properties for off-axis scanned beams unless special techniques are used to compensate for the usual phase errors, such as reflector shaping and dual-focus designs.

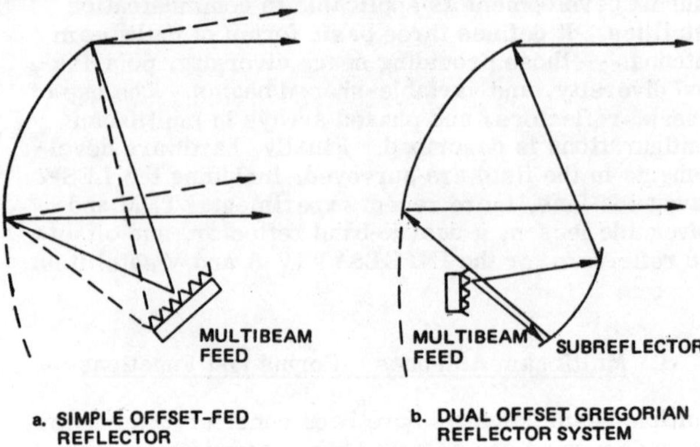

Fig. 3 Reflector Antennas

To illustrate what performance can be expected from an offset-fed reflector system, calculations were run for an 8 GHz design with a reflector diameter of 8 feet (64 wavelengths) and a focal length of 10 feet, producing a 1.3° half-power beamwidth. A parabolic amplitude taper with -28 dB edge illumination was assumed, with constant phase. Scanning was accomplished by dis-placing the feed from the focal point and readjusting feed orientation to maintain the same illumination pat-tern over the reflector. Patterns for scans up to ±10° in the plane of scan are shown in Fig. 4; patterns for

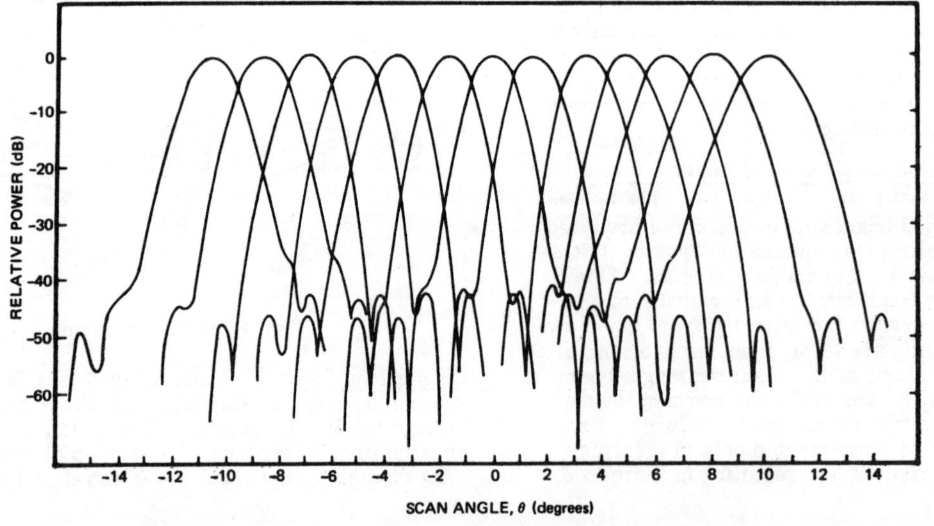

Fig. 4 Computed Vertical Plane Scan Patterns of Offset Paraboloid

orthogonal and 45° cuts through the on-axis beam, as well as those for the extreme scan positions, are shown in Fig. 5. These figures show sidelobes generally maintained below −40 dB over the entire scan range, with little change in gain, and a gradual broadening of the mainlobe for the scanned beams, as shown in Fig. 6. This broadening may actually be more harmful for frequency reuse applications than degraded sidelobes, since the broadened beam will increase the minimum spacing between beams required to maintain a given degree of isolation, which is roughly equal to the distance between nulls of the main beams.

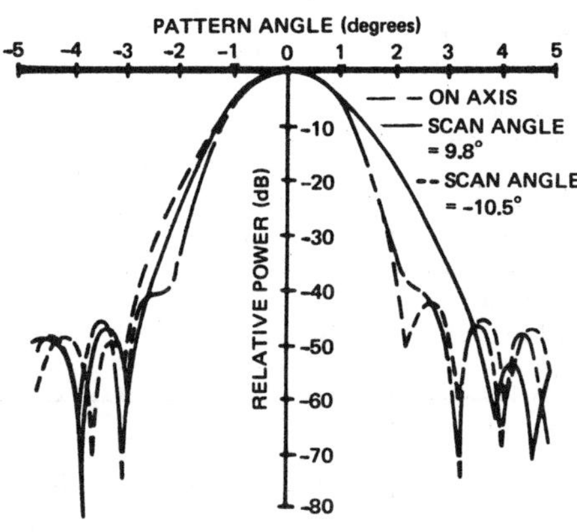

a. 45° Plane

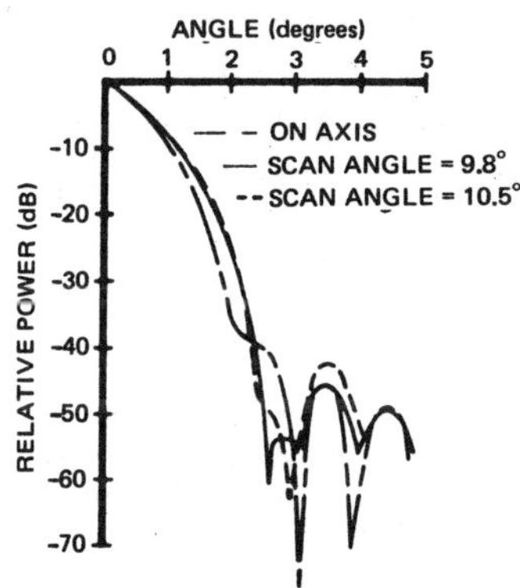

b. Horizontal Plane

Fig. 5 Vertical Plane Patterns for Central and Extreme Angle Scan Positions

The Gregorian dual-reflector system, shown in Fig. 3b and described by Fitzgerald,[3] represents a potential improvement in offset-fed reflector performance, since the second reflector allows an additional degree of freedom available for optimizing off-axis scanning characteristics. Two versions have been studied -- the true Gregorian with multiple individual displaced feeds, and the near-field Gregorian with a small planar array feed, which scans the secondary beam by tilting the phase front of the array feed. The advantage of the array feed is that continuous control of the scanned beam is available with a modest-sized array, whose size is reduced from the radiating aperture diameter by the ratio of the focal lengths (magnification) of the two reflectors.

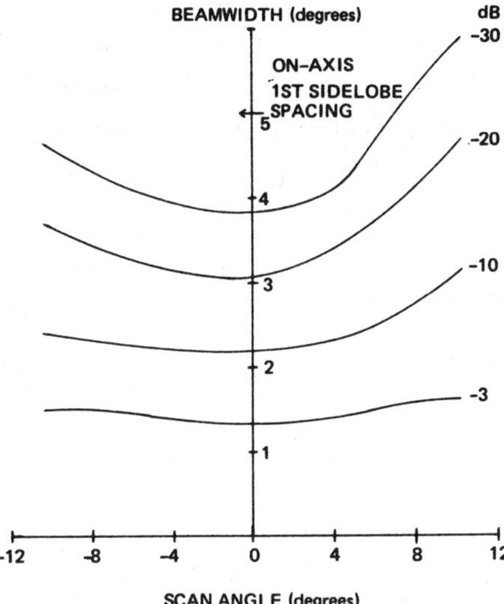

Fig. 6 Beamwidth Scan Broadening

However, this size reduction also requires the array to scan over proportionately wider limits, which becomes difficult for earth-coverage applications with any substantial degree of magnification. Furthermore, true multibeam applications (as contrasted with shaped beams) require multiple phasing networks for the array feeds, as will be mentioned in the next section. Apparently the true Gregorian system has received little study, although its potential for multiple beam use is attractive.

In the array-fed reflector each feed element separately illuminates the reflector to generate a component beam in the far field. By properly exciting feed elements, and thus summing individual component beams, a desired shaped beam may be achieved to serve a specific ground coverage area. Depending on the feed-array element, this system can be operated for any linear or circular polarization. Circular polarization (CP) is attractive for an offset reflector when polarization diversity is used since an offset parabolic reflector does not generate a cross-polarized signal when the feed has perfect CP pattern. Thus for CP beams, good polarization isolation and axial ratio can be obtained by properly designing the element and array configuration.

C. Multibeam Phased Array Antennas

A conventional planar phased array consisting of identical low-gain, uniformly spaced radiators can be adapted for multiple beam use by providing separate beam forming and steering networks for each individual beam, and combining their outputs at each array element, as depicted in Fig. 7. Separate sets of networks separated by diplexers are also required for receive and transmit; or different arrays (or array elements) can be used for the two functions. The weight and complexity of such a system increase in proportion to the number of simultaneous beams, and become a major factor against the choice of such a system for any significant number of beams. The possible use of a Butler matrix to passively form multiple beams in space from

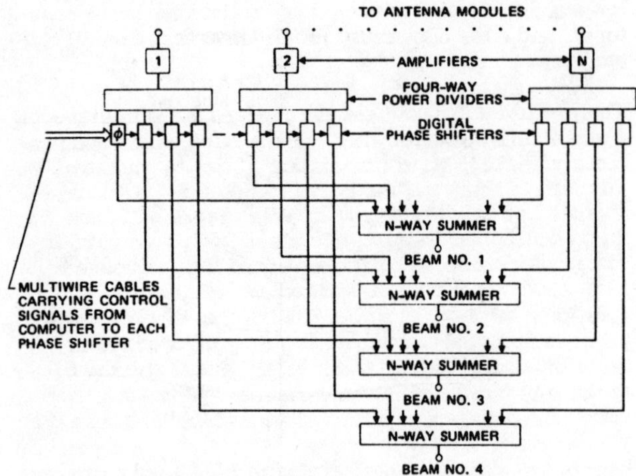

Fig. 7 Typical Multibeam Array Antenna System Schematic Using Four Active Beams

a common array seems appealing, but introduces too great a degree of inflexibility in beam positioning to be of general use.

Another factor weighing heavily against phased array multibeam transmit antennas is the need to provide individual power amplifiers at each radiating element, each of which must handle simultaneous signals from all of the multiple beams. To avoid excessive intermodulation generation, AM/PM conversion, and signal suppression effects due to the presence of the multiple signals, great care must be taken to avoid saturation within these devices, forcing them to operate in their low-efficiency linear regions. This penalizes the overall transmitter efficiency available by a factor of 5 to 10 dB, and thus further increases weight and prime power requirements severely.

As an example of the extent of such compromises, we quote a summary of the characteristics of several alternate multiple-beam phased array antennas configured by Hughes[4] for a general purpose Armed Forces satellite. The antennas were to operate at X-band (7–8 GHz), and have the ability to position 10 to 30 simultaneous beams anywhere on earth from synchronous altitude (±8.7° field of view). The number of discretely driven elements is determined by the minimum beamwidth required to avoid excessive gain loss when scanning to the edge of the earth, and meanwhile maintaining grating lobes outside the earth. This results in a maximum element spacing of about 5 inches, with the total number of elements required for a 5-foot-diameter array being around 91, and a corresponding increase to 472 for a 10-foot array. Other characteristics of these two sizes are summarized in Table 1, based on an assumed radiated power requirement of 10 W per beam, and using TWT power amplifiers with saturated power levels 10 dB greater than actually required, thus assuring linear operation. FET preamplifiers at each receive antenna element were assumed, with 20 to 30 dB gain, and 4-bit diode phase shifters consuming 0.5 W each. Totals for even the smallest 10-beam 5-foot array show an estimated weight of over 700 lb and a prime power requirement of 4.5 kW, for net radiated power of only 100 W. The design would require 910 phase shifters to form the 10 separate beams, and 91 10-watt TWT's each operating at about 10% of

Table 1. Projected Characteristics of an X-Band Multibeam Array Antenna

Parameter	Characteristics			
Array diameter, ft	5		10	
Number of elements	91		472	
Beamwidth, degrees	2		1	
Gain on axis, dB	40		46	
Maximum sidelobe, dB				
On axis	25		25	
9° scan	22		23	
Number of simultaneous beams	10	30	10	30
Weight, lb				
Transmitters	560	900	1250	2300
Receivers	28	33	120	130
BFN's	120	575	330	1640
Totals	708	1508	1700	4170
Power, kW				
Transmitters	4.0	12	4.0	12
Receivers	0.04	0.075	0.14	0.16
BFN's	0.45	1.4	2.3	7.1
Totals	4.5	13.5	6.4	19.3

rated power, with a net efficiency of only 2.5%. In contrast, a TEM waveguide lens of the same size, with a 10-beam feed network and ten 100-watt TWT's, would weigh less than 125 lb.

A more attractive use for a phased-array multibeam antenna is in producing shaped beams, rather than multiple simultaneous beams, as specified for the DSCS-III antenna. A single beam as narrow as 2° is desired anywhere on earth from synchronous altitude, or a variable pattern up to full earth coverage, with the additional ability to produce nulls within this pattern to discriminate against jammers. The phased array is well suited to this application, since it is able to produce high-gain low-sidelobe beams individually or in various combinations merely by adjusting phases over the aperture. With tapered amplitude illumination and careful phase control (typical 5° tolerances), sidelobes as low as −35 dB can be achieved.[5] Multiple phasing networks for multiple beams are not required, since only a single beam of variable shape is to be produced. The overall array size is determined by the narrowest beam desired, resulting in a 5-foot diameter for tapered low-sidelobe illumination at X-band. The number of array elements required is a function of the grating lobe condition as well as the sidelobe level and beam pointing accuracy desired. The 5-foot aperture can be filled with 91 elements, each 3 wavelengths in diameter, arranged in triangular fashion, which will maintain 30 dB sidelobes for scanning to the edge of earth, while eliminating grating lobes from the earth's field of view. Careful control of the transmit array power amplifiers on each element would still be necessary to avoid phase changes with signal levels, and consequent beam squint, but the problem would be considerably eased since multiple users in the same bands would not be present. Furthermore, at least 6-bit digital phase shifters would be necessary to achieve phase accuracies consistent with the low sidelobe requirements. The resulting design would have as much beam and null control capabilities as a 91-beam lens antenna, as shown by corresponding calculated patterns for the two cases.

III. Multibeam Antenna Applications

A. Multibeam Lenses

A true multibeam antenna for spatial frequency reuse produces many separate beams from one common aperture. Typically, 30 dB sidelobes are needed and from 5 to 10 or more beams may be desired. Use of a 7-element feed cluster is necessary to form each low-sidelobe pencil beam while retaining approximately one-beamwidth switched beam positioning resolution. The complete feed array may contain 50 to 100 feed radiators, followed by a set of 7-to-1 BFN's (beam forming networks), one for each beam port (Fig. 8). Each BFN may be switched to different sets of feed elements, if desired, thereby providing each beam with a range of pointing directions to be chosen by ground command. The excellent low-aberration optics of the TEM lens permits the excitation sets for all BFN's to be the same (invariant with beam position), thus permitting a vast simplification in ground command requirements for low-sidelobe beam pointing control.

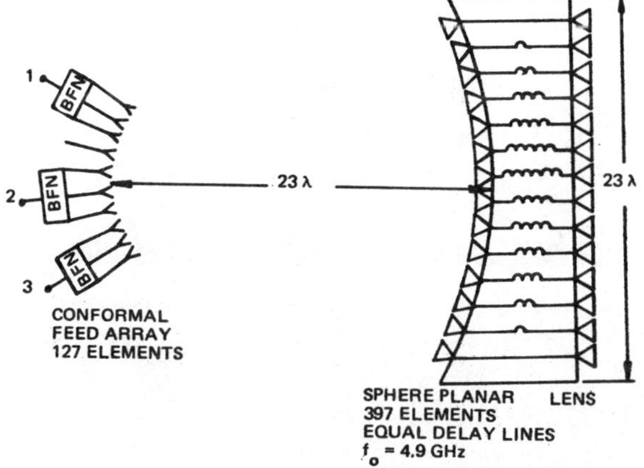

Fig. 8 Multibeam TEM Lens Diagram

Experimental TEM Lens. This application has been the focus of extensive development of low-sidelobe TEM lenses at Aeronutronic Ford. The particular TEM line lens has a spherical inner (S_1) surface and a planar outer (S_2) surface, with all interconnection lines of equal length. Such geometry provides excellent small-angle scan focusing necessary to maintain low sidelobes for all beam positions over the lens FOV (field of view). The use of RF printed circuit techniques (such as open microstrip) for interconnecting lines offers the advantages of

low weight (thin PC cards), high reliability (no connectors), and ease of quantity precision production (photoetching).

The preliminary design of such a lens for simultaneous 4 and 6 GHz operation (Fig. 8) has been described previously. [6] Each triad (consisting of one S_1 radiator, one microstrip line, and the corresponding S_2 radiator) is photoetched onto the two sides of a thin (0.020 inch thick) flat copper-clad dielectric card. Each card is mated with a second orthogonal card into a crossed-card configuration so as to produce lens operation independent of polarization. A collection of such card pairs is mounted through slots in a pair of circular metallic sheets (the S_2 sheet being planar and the S_1 sheet spherical) so that each radiator protrudes "outward" beyond its respective sheet, which acts as a ground plane for the radiators. In effect, the S_1 and S_2 sheets are slotted racks serving to hold card pairs in a circular array matrix. This collection of cards and conducting sheets is the TEM lens (Fig. 9).

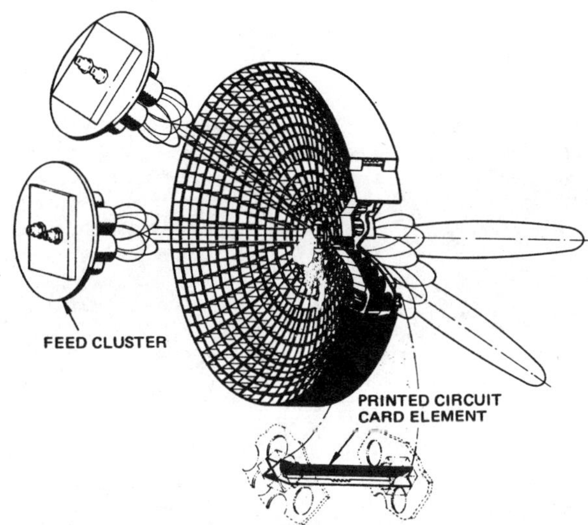

Fig. 9 Printed Circuit Card, TEM
Lens, and Feed Clusters

The feed array was designed to consist of small radiators in a nearly triangular lattice on a spherical focal scan surface. To produce each low-sidelobe pencil beam, a symmetrical cluster of feed elements was driven as an array, using 7 elements at 6 GHz and 13 elements at 4 GHz (Fig. 10). A computerized synthesis technique was developed to calculate optimized phase and amplitude excitation coefficients of the individual clusters. [7]

An electrical test model of this antenna was constructed for experimental pattern testing at C-band (Fig. 11). This model has a diameter of 5 feet, as well as a 5-foot focal length, and uses 396 card pairs in the TEM lens. It includes a 19-element linearly polarized feed cluster on a mechanical mount that permits positioning the cluster to any position within the earth coverage FOV. Either a 7-way (6 GHz band) or 13-way (4 GHz band) power divider could be connected to the 19-element feed cluster for tests, with unconnected elements terminated in matched loads.

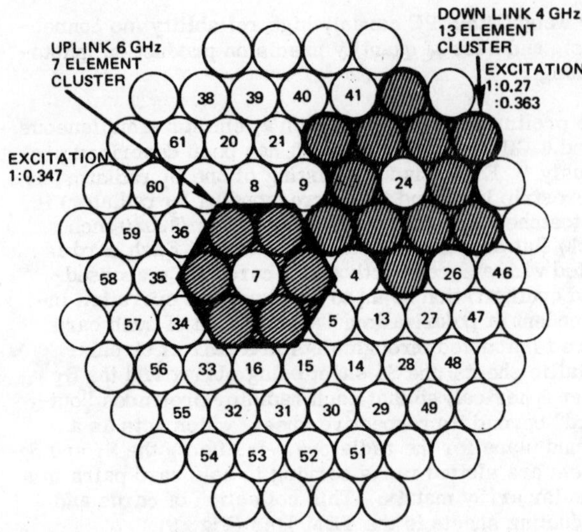

Fig. 10 Feed Cluster Geometry for TEM Lens

Figure 12 shows an H-plane (relative to feed cluster polarization) secondary pattern measured at 4.9 GHz with a single on-axis feed element energized to produce a basic singlet pattern of the lens. Also shown is a calculated pattern for this case, with good agreement. Figure 13 shows similar patterns using a 7-element cluster feed with the synthesized excitations for low sidelobes. For beam positions over the earth's FOV (0 to 8.6° scan), the on-earth sidelobes remained below –30 dB.

Subsequent tests verified wideband performance and indicated that smaller lens elements would be necessary to achieve best efficiency for both transmit (3.7 to 4.2 GHz) and receive (5.9 to 6.4 GHz) operation in the one multibeam antenna. Currently, such smaller elements are being developed via active impedance measurement subarray techniques.

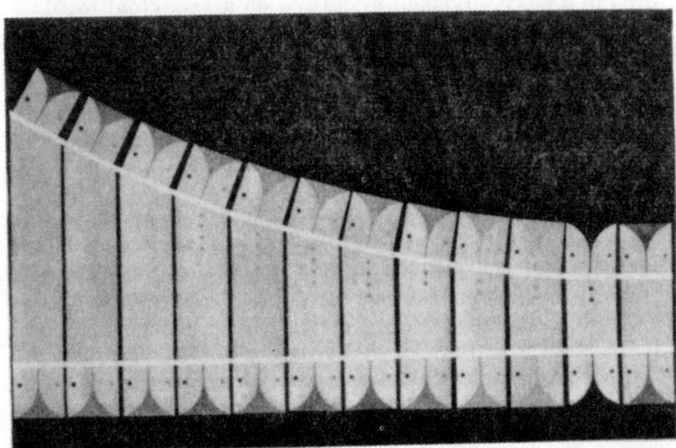

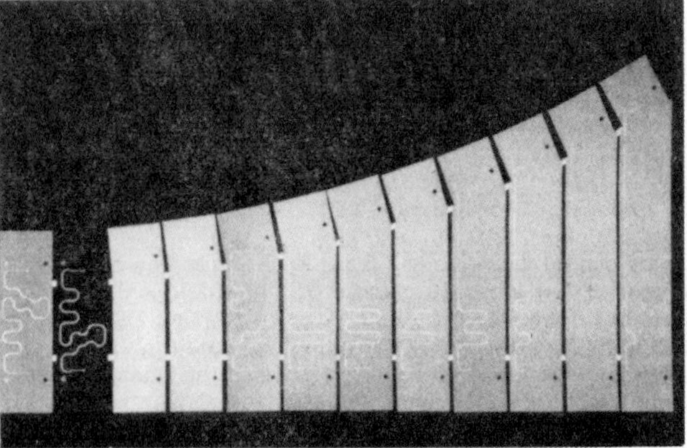

Fig. 11 TEM Lens Model

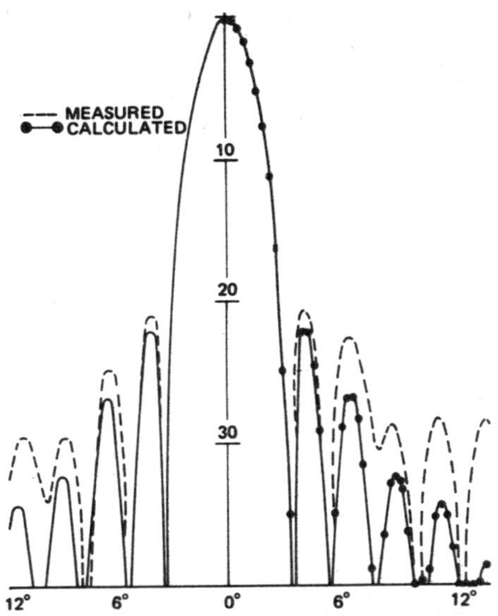

Fig. 12 Measured vs Calculated 4.9 GHz Pattern
of TEM Lens with Single Element Feed

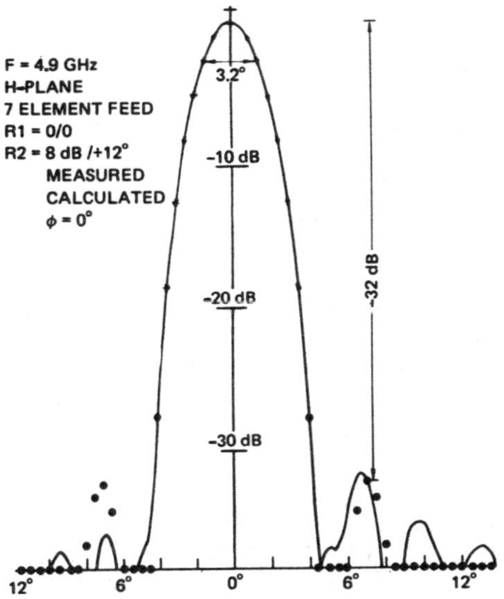

Fig. 13 Measured vs Calculated 4.9 GHz
Pattern of TEM Lens with 7-Element
Feed Cluster (On-Axis)

TEM Lens Beam Isolation Characteristics. In order to explore the system capabilities of the TEM lens antenna, a computer-calculated design was devised, providing the capability of generating 61 low-sidelobe dual-polarized beams over a 17° FOV at both 4 and 6 GHz. The general requirements of isolation between beams, as well as the total coverage provided by the beams, invoke certain restrictions in the design of a multibeam antenna. These generally fall into the following areas:

a. Antenna beamwidths and beam separation angles must be chosen to provide the required isolation as well as coverage.

b. Characteristics of the individual beams, including both beam shape and sidelobe levels, must be properly selected. Sidelobe levels of at least -35 dB are generally required, while a steep falloff of the main beam allows closer spacing of individual beams without interference.

c. Use of polarization diversity requires (for CP) that axial ratios of the individual beams be maintained under 0.5 dB to prevent excessive leakage into the cross-polarized channel.

d. Feed networks that provide excitations for individual beams to achieve the desired shapes and low sidelobes, as well as necessary switching, should be as simple as possible to avoid excessive losses.

Limits on interbeam isolation may be established by any of the following coupling criteria: (1) main beam coupling, (2) sidelobe levels, or (3) cross-polarized levels. These levels must ordinarily be below the desired isolation level between beams, over the entire coverage area of each beam, with sufficient margin to allow for all contributions to the total isolation from other beams. The minimum beam separation to meet typical isolation requirements is usually established by the null-to-null beamwidth of the downlink (broader) beam.

After extensive study of calculated patterns, it was established that a maximum of six beams of the available 61 dual-polarized beams could be excited simultaneously while maintaining an isolation of 27 dB between beams, for a total of 12 times frequency reuse. [8]

B. Variable-Shaped Beam Lenses

An alternate use of a multibeam antenna lies in combining all beams through a ground-controlled RF variable power combiner (or variable power divider), [9] thus producing a single composite beam whose shape may be varied from a single narrow pencil beam to a much wider, flattop composite of up to N beams. N is usually chosen so that the widest flattop conical beam provides full coverage of the earth. Another feature of the variable beam concept is the ability to control the BFN so that a narrow null is placed in a previously established broad coverage beam in the direction of a discrete source of strong interference, thereby improving an otherwise unacceptable signal-to-noise ratio. Both waveguide and TEM lenses have been considered for the variable beam application.

LES-7 Lens. Dion and Ricardi[10] of MIT's Lincoln Laboratory have developed a laboratory model variable beam antenna using a 30-inch-diameter stepped waveguide lens with a 19-horn feed array and a BFN tree of two-to-one ferrite waveguide variable power combiners. They have developed a scalar computer program[11] for calculating radiation patterns of this lens, and have studied application of up to 61-element feed arrays with lens diameters up to 50 inches at X-band.[12, 13]

The 30-inch-diameter lens model is illustrated in Fig. 14 consisting of over 700 square waveguide cells with one ring step in its outer surface. The 19-element linearly polarized feed array is mounted on

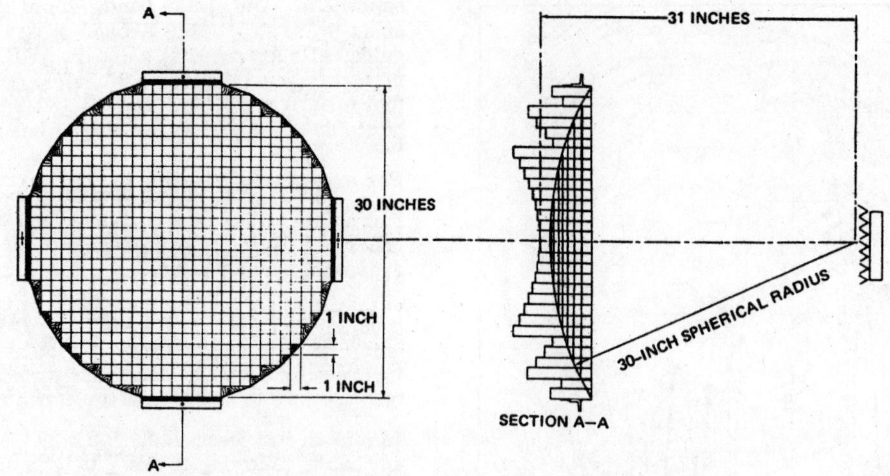

Fig. 14 LES-7 Experimental Waveguide Lens Antenna Design

a flat surface centered at the lens focal point ($f/D = 1$). Figure 15 shows a set of five singlet beams, each produced by separate excitation of one feed horn at a time. Figure 16 shows the composite earth coverage pattern obtained by uniform coherent excitation of all 19 feeds. Figure 17 shows a typical compound coverage contour pattern obtained by excitation of four horns, illustrating the wide flexibility of this concept.

Figure 18 shows a measured earth coverage contour pattern plot with one feed turned off so as to create a narrow deep interference rejection null. The dashed contours are the calculated –6 dB and –12 dB contours for comparison. The dotted area has a null depth of 15 dB or greater. A similar result is obtained when two adjacent feeds are turned off (Fig. 19).

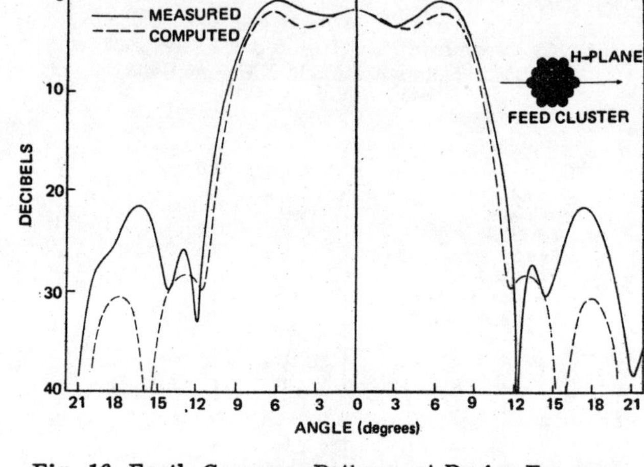

Fig. 16 Earth-Coverage Patterns at Design Frequency

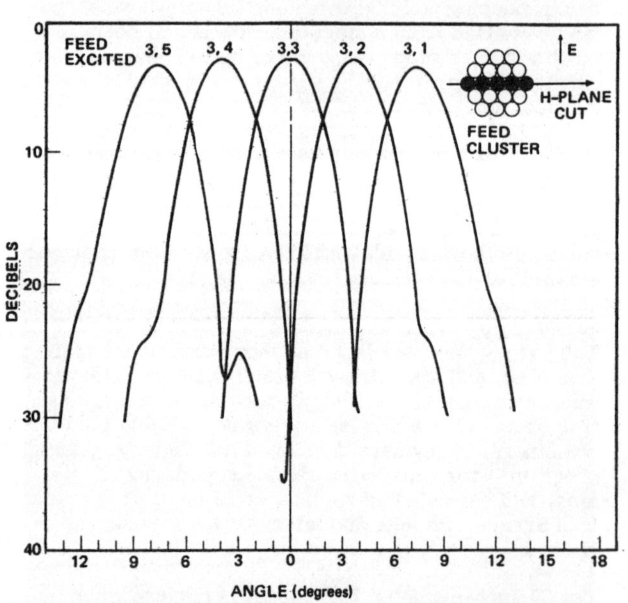

Fig. 15 Superimposed H-Plane Patterns of Center Row Beams Measured at Design Frequency

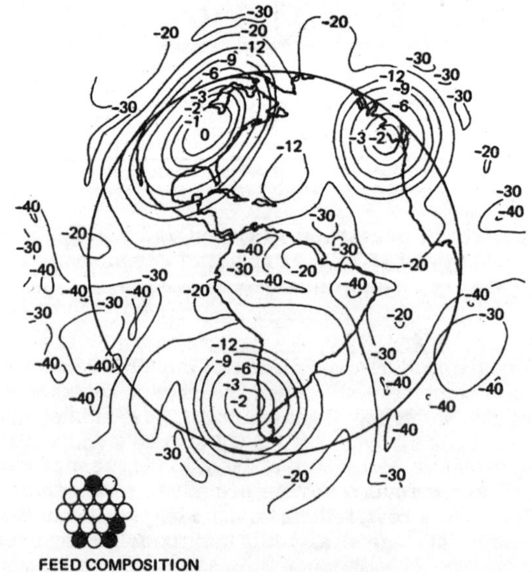

Fig. 17 Compound Coverage Contour Pattern (Measurements in Decibels)

476

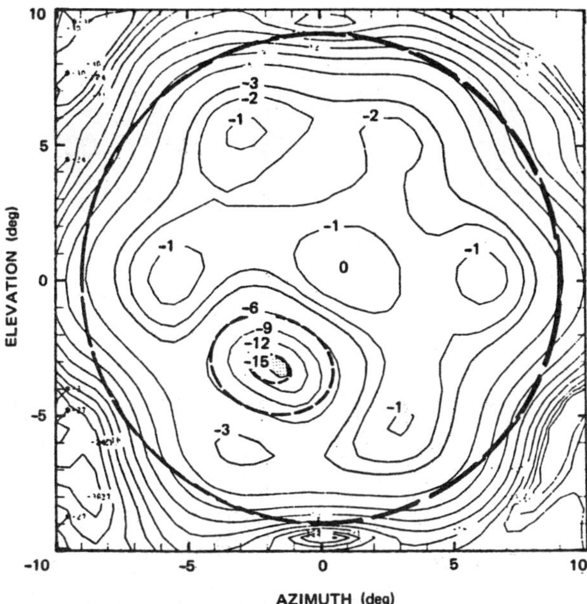

Fig. 18 Measured Earth-Coverage Contour
With Feed 22 Off

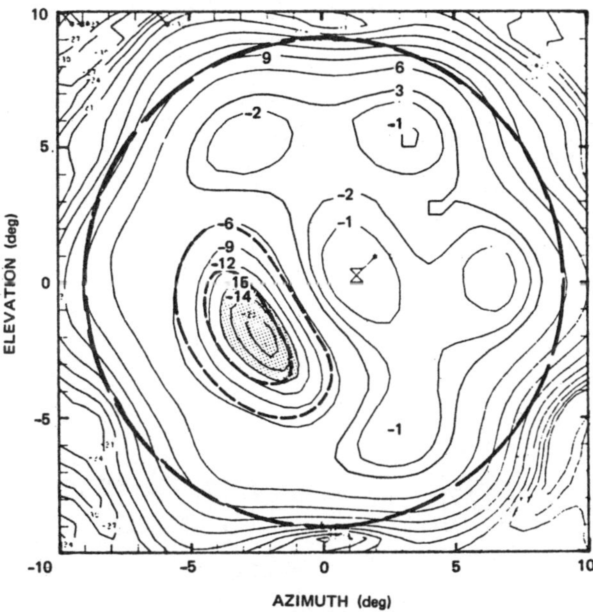

Fig. 19 Measured Earth-Coverage Contour
With Feeds 22 and 32 Off

Recent Experimental Waveguide Lenses. Modifications
of the Dion and Ricardi waveguide lens have been designed and constructed by Lockheed (48 inch diameter)
and by Aeronutronic Ford (46 inch diameter), wherein
model antennas with 61 circularly polarized feed elements have been tested. In the Aeronutronic Ford
antenna, the lens is a matrix of square hollow metallic
waveguides forming a metallic "eggcrate" structure.
S_1 is spherical, with the sphere centered at the lens
focal point ($f/D = 1$) as in the TEM lens. S_2 is elliptical,
with three ring steps (setbacks) to prevent excessive
lens edge thickness and unacceptably narrow frequency
bandwidth. The usable bandwidth of the stepped lens is
about 5% to 10%. This X-band design has 1528 waveguide cells arranged in a square grid. The 46-inch-
diameter lens produces a 2° half-power pencil beam
that may be positioned over an 18° conical FOV. It exhibits 20 dB sidelobes for singlet beam excitation.

For experimental testing the feed array for this antenna
was excited from a 61-to-1 variable BFN formed of
variable power dividers, attenuators, and phase shifters. The test model is illustrated in Fig. 20. The feed
array is positioned on a spherical surface (Fig. 21) for
simplifying shaped beam excitations (no phase corrections are needed for off-axis beams, in contrast to requirements for a flat array). Figure 22 shows the
measured circularly polarized patterns of five adjacent
beams, energized and measured one at a time.

Figure 23 shows a right hand circularly polarized
(RHCP) measurement of the on-axis singlet beam at
midband. Main beam axial ratio remained below 1.5
dB for all positions over a bandwidth of 5.5%.

Figure 24 illustrates measured beam scanning between
adjacent singlet positions by gradual change of the BFN
feeding the two beam feeds. Figure 25 shows the measured RHCP earth coverage pattern with all 61 feeds
equally excited. Axial ratio remained below 1.5 dB

Fig. 20 Multiple Beam Waveguide Lens Antenna

Fig. 21 61-Element Spherical Feed Array

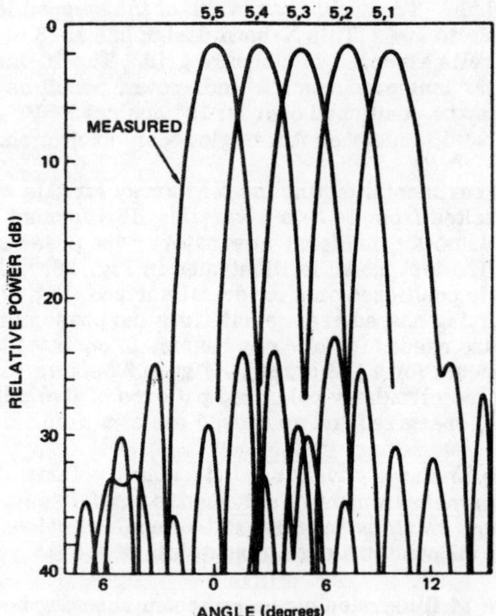

Fig. 22 Adjacent Scanned Beam Patterns, f_0, RHCP

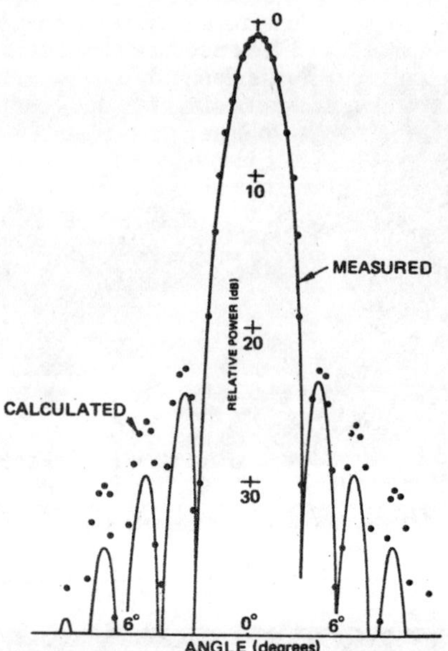

Fig. 23 Singlet Beam Pattern, Central Beam, f_0, RHCP, $\phi = 0°$

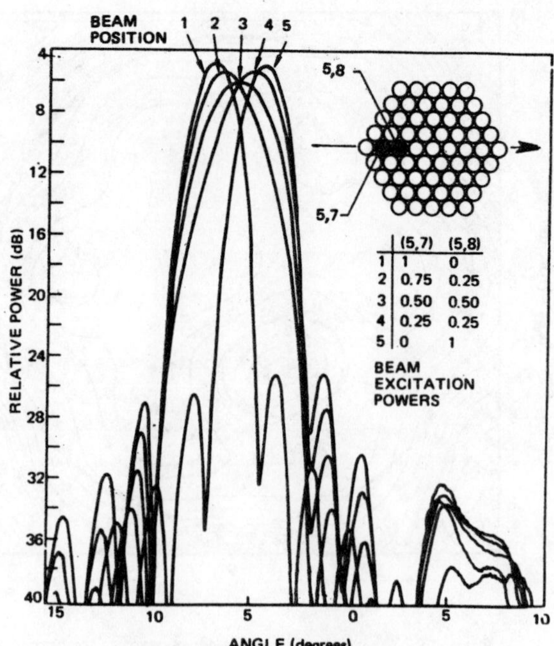

Fig. 24 Scanned Doublet Beam Pattern, (5,7), (5,8), f_0, RHCP, $\phi = 0°$

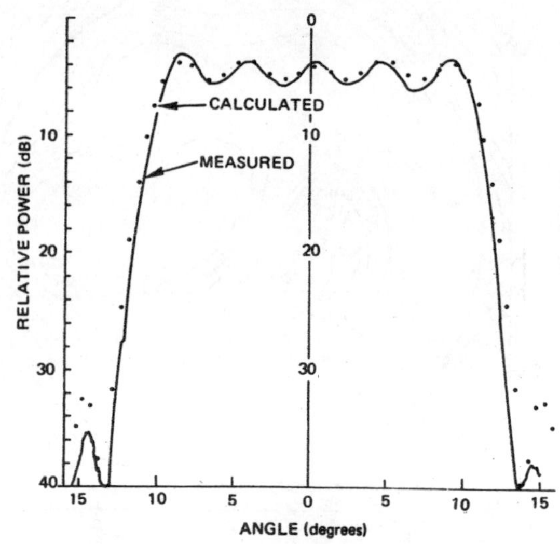

Fig. 25 Earth Coverage Beam Pattern, 61 Feeds, Equal Excitation, f_0, RHCP, $\phi = 45°$

across the 5.5% band for this case. The insertion of a deep null into the earth coverage beam was accomplished by phase reversal and appropriate amplitude control of one feed (Figs. 26 and 27). For the simple turnoff and subtraction null techniques, the null bandwidth was found to be small (due primarily to the dispersive lens defocusing phenomena), about 1% for a -20 dB null.

A 30-inch-diameter, 19-beam transmit antenna similar to the Lincoln Labs model was constructed by Lockheed for initial evaluations of the antenna and its control network. More recently, Lockheed built a 48-inch-diameter receive lens, with 61 feeds and a number of design

optimizations. This lens employs more than 2200 waveguide elements in a two-zone configuration. Although no measured data have been published, results are reported to have been excellent.[14]

One major advantage of using a waveguide lens is the ability to employ unique manufacturing techniques that greatly reduce the weight tolerances required to build a structure that can withstand space applications with no degradation of electrical properties. A considerable amount of research has been done on the use of a variety of carbon-composite structures in forming very strong but lightweight square waveguides.

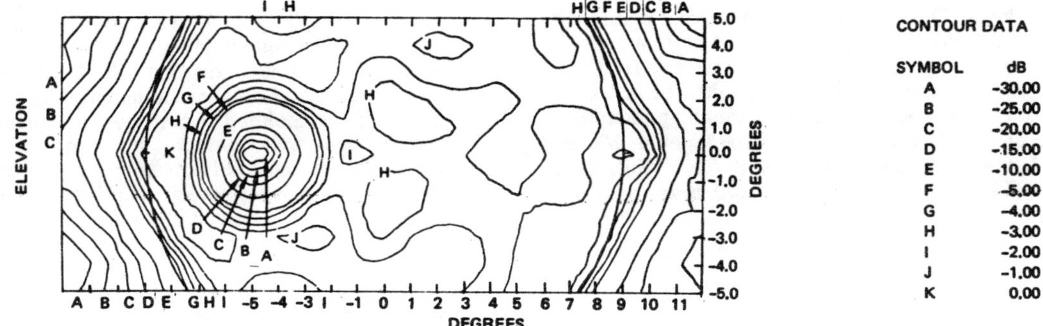

Fig. 26 Earth Coverage Null, Singlet Beam (5,7) Subtraction Digitized Contour
Plot, Measured at f_O, RHCP

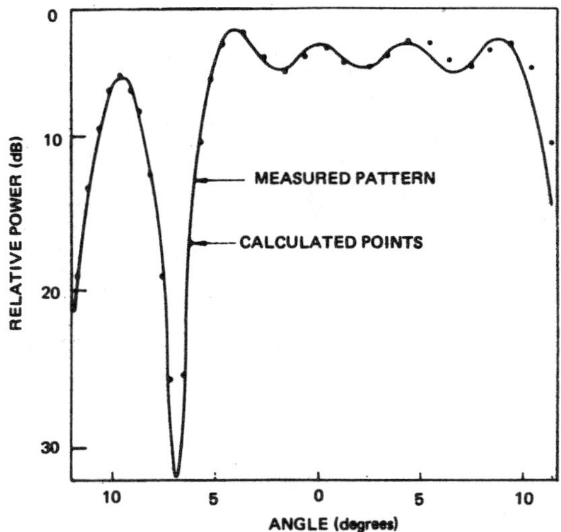

Fig. 27 Earth Coverage Null, Singlet Null
Beam Pattern, (5,8), f_O, RHCP

Lockheed also reports development of a unique power divider assembly and beam-forming network which greatly
reduces the number of levels required in the network,
thus reducing beam-forming network loss while providing a much greater degree of precision in the setting
of power at individual beam ports.

Variable-shaped beam lens antennas are currently under
investigation by two contractors (Hughes and General
Electric) for the Air Force (SAMSO) as part of contract
definition studies for DSCS-III.

TEM Variable Beam Lens. An application of a TEM
lens to the variable beam concept was studied by Binz
and Waineo.[15] The lens was formed of multiple flat
sheets of layered form and printed circuit boards, each
board containing a row of many balanced stripline delay
lines, pickup, and reradiator dipoles. Single-sense circular polarization is achieved from this linearly polarized lens by placing a form of sheet polarizer over the
lens aperture, which converts linear to circular polarization. Four-percent-bandwidth waveguide polarizers
based on ferrite faraday rotators were proposed for use
in the 64-to-1 BFN. A feature of this lens was the calculated 30 dB sidelobe level performance of the pencil
beam formed from a properly excited cluster of feed
elements. No test model has been reported.

C. Multibeam Reflector Antennas

Symmetrical Six-beam K-Band Model. Since aberrations due to offset-fed reflectors generally increase as
the number of beamwidths of scan increase, it is especially attractive to utilize a symmetrical feed structure
for higher-frequency narrow beam applications, such as
a domestic satellite system. If the number of simultaneous beams is limited, feed structure blockage may be
tolerable without producing excessive sidelobes.
Turrin[16] has proposed a novel structure to minimize
such blockage by extending the feed horns through holes
in an auxiliary plane reflector, thus eliminating most
of the support and feed network structures from the
blockage region. His design was for a spherical reflector accommodating six simultaneous beams, pointing
within a 13° area (to cover ground stations from Hawaii
to Puerto Rico), for use at both 20 and 30 GHz. The
basic antenna layout is shown in Fig. 28, with feed

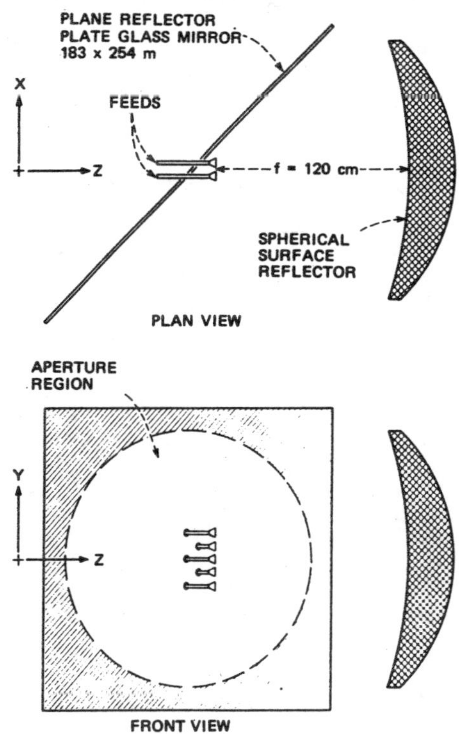

Fig. 28 Multibeam Spherical/Planar Reflector

locations as in Fig. 29. Orthogonal linear polarizations were used for the two bands, obtained from a novel finned feed horn design. A 5-foot-diameter (150 cm) model was built and tested. Beamwidths of approximately 0.7° were obtained in both bands, principally because of different feed illumination tapers. Measured gain and first-sidelobe levels are shown in Fig. 30 as a function of beam pointing direction off-axis; on-axis gains were 47.0 dB at 19 GHz and 49.2 dB at 30.2 GHz, some 3 to 4 dB below theoretical maxima for a 5-foot aperture. Isolation between beams was more than 28 dB, except for two adjacent co-polarized beams designed to cover New York and Atlanta, only 1.3° apart, where isolation was 24 to 26 dB.

<u>INTELSAT IV-A Antenna.</u> The most representative multibeam antennas in use today are of the offset-fed reflector type as employed on the INTELSAT IV-A, the second of the INTELSAT IV series of communication satellites. Increased channel capacity in the IV-A is achieved by reuse of the 500 MHz bands at both 4 and 6 GHz by means of antenna beamshaping, which produces two spatially isolated beams covering eastern and western hemispherical areas, as depicted in Fig. 31.[17] Both Atlantic and Pacific regions are shown, with coverage areas determined by locations of INTELSAT ground stations. Reduced coverage is also required at secondary regions shown as rectangles. The 6 GHz receive beams (LHCP) each cover an entire hemisphere. The 4 GHz transmit beams (RHCP) are further divided into northern and southern quadrants, which are individually available for even- or odd-numbered channels on command.

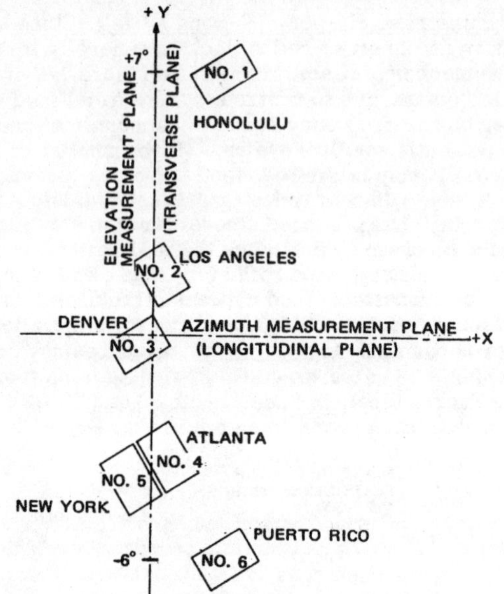

Fig. 29 Feed Horn Positions

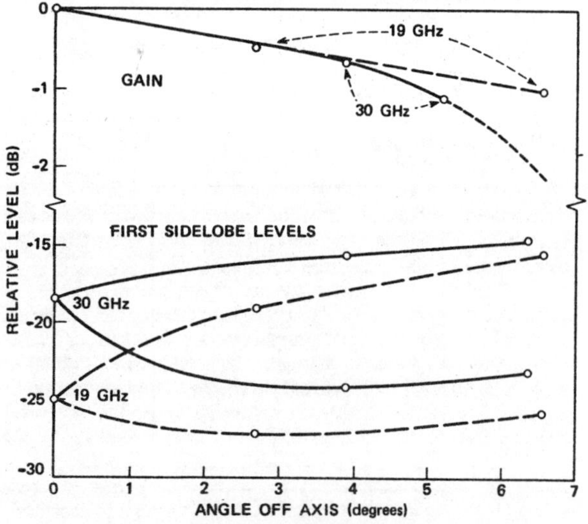

Fig. 30 Single-feed Measurements of Off-axis Performance

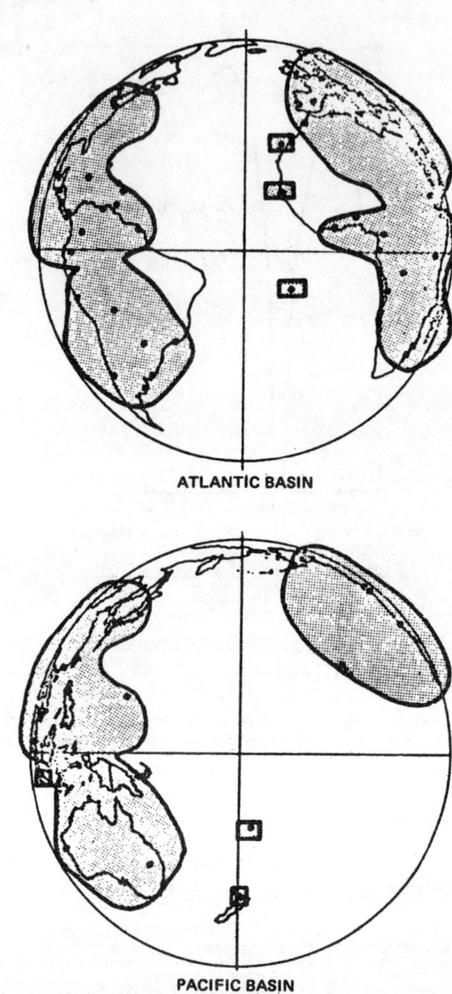

Fig. 31 INTELSAT IV-A Shaped Beam Coverage

The general arrangement of antenna hardware on the IV-A spacecraft mast is shown in Fig. 32. The two 53-inch-square parabolic reflectors at the base of the mast form the even-odd north-south transmit beams, while the east-west beams are provided by separate sets of feeds for each reflector. The shaped-beam receive reflector is located above the two transmit structures. A biconical global-coverage antenna appears at the tip of the mast. Performance characteristics of the antenna system are summarized in Table 2, as extracted from the Hughes report.[17]

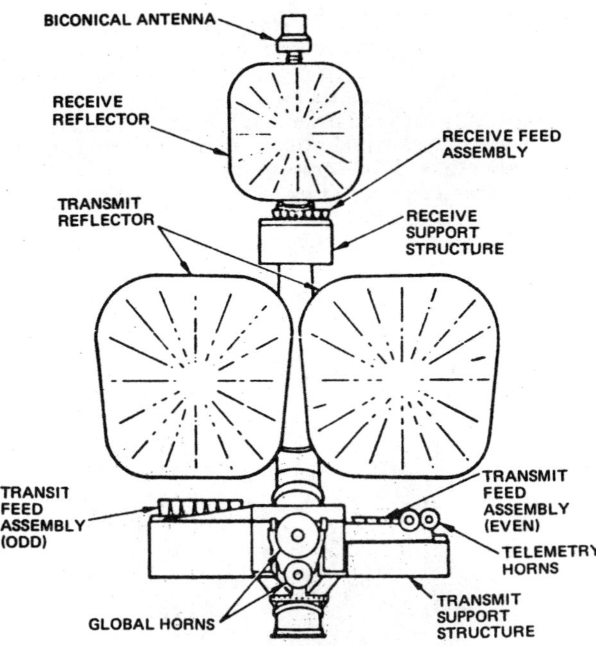

Fig. 32 General Arrangement of Antenna
Hardware, INTELSAT IV-A

Table 2. INTELSAT IV-A Antenna
Performance Characteristics

Coverage:	All stations plus major land mass in Atlantic and Pacific basins (Fig. 31)
Frequencies:	Receive 5932 to 6378 MHz Transmit 3707 to 4153 MHz
Polarization:	LHCP on receive RHCP on transmit Ellipticity $\leq$ 3 dB
Gain:	Receive 22 dB Transmit 24 dB (sector) 21 dB (T mode)
Sidelobe level:	C/I $\leq$ 27 dB
Slope:	$\leq$ 3 dB/degree

The two dominant features of the shaped-beam antenna
are the multihorn feed array and the highly offset square
reflectors. Both features contribute to achieving the
required sidelobe isolation. A feed horn located at the
focal point of a parabolic reflector produces a secondary
pattern coincident with the antena boresight. An identical feed horizontally displaced from the focal point
produces a secondary pattern displaced in azimuth from
the boresight. If the reflector f/D ratio is sufficiently
large, these two secondary patterns will have very
similar shapes.

Three such feed horns, horizontally arrayed about the
focal point, will thus produce three secondary patterns
staggered in azimuth about the antenna boresight. If the
relative beam displacement is equal to the nominal side-

lobe width, the beam arrangement shown in Fig. 33
will be realized. In this configuration, the sidelobes of
beams 1 and 3 are in phase with each other but are 180°
out of phase with sidelobes of the middle beam. This
sidelobe cancellation can be maximized, yielding the
pattern shown in Fig. 33b, if the outer beams are reduced in amplitude relative to the center beam. This
reduction can be accomplished by a feed network power
splitter which provides the required unequal power
levels to the feed horns.

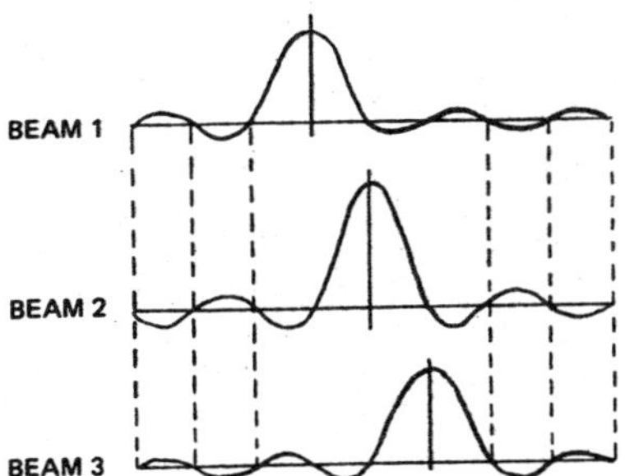

a. Component Beams for Three
Adjacent Feed Horns

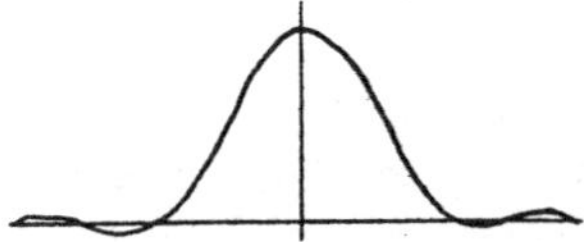

b. Composite Shaped Beam

Fig. 33 Shaped Beam Superposition Concept

The particular arrangement shown in Fig. 33 is a case
of three horns with a symmetric power distribution.
For applications such as INTELSAT IV-A, where sidelobes on only one side of the main beam need be low,
asymmetric power distributions may be used. The
number of horns may also be varied so long as the
power distribution is reoptimized for each case. The
result is a series of two-horn, three-horn, and four-
horn arrays, each with low sidelobe properties but different beamwidths. Several such arrays can be stacked
vertically to achieve the coverage requirements of
INTELSAT IV-A, yet still retain sidelobe performance
in the azimuth direction.

The reflector associated with this feed network is also
designed to enhance sidelobe performance. A focal
length of 50 inches was selected to minimize phase
errors and maximize sidelobe performance. In addition, sidelobes produced by aperture blockage are
eliminated by using a highly offset parabolic reflector
with the bottom edge of the reflector located 12 inches
above the focal axis. Finally, the reflector cross section is designed to be nearly square. This arrangement
minimizes sidelobe interference between the NW and
SE and the NE and SW portions of the coverage regions.

A typical feed-horn array for the odd-channel transmit
antenna consists of 37 horns with integral polarizers
(Fig. 34). This array is energized by quadrants from a
TEM transmission line power division network.

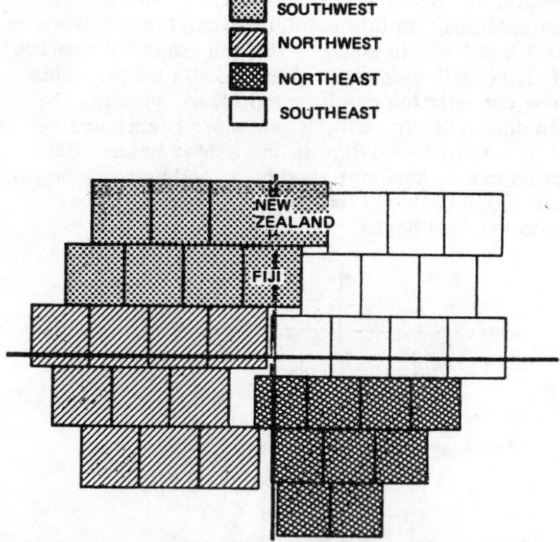

Fig. 34 INTELSAT IV-A Odd-Channel Transmit
Antenna Feed System

<u>INTELSAT V Antenna</u>. The INTELSAT V requirements
are similar to those of INTELSAT IV-A except that
polarization diversity is also required to provide simultaneous coverage of two overlapping regions in each
hemisphere. A configuration common to the Atlantic,
Pacific, and Indian Ocean theaters is desired, with a
minimum amount of switching in the feed networks to
accommodate differences between areas. Steerable spot
beams for the 11/14 GHz bands are also required. This
satellite is scheduled for development for delivery in
1979.

Offset-fed reflectors are also being considered for
INTELSAT V, with considerably tighter control of individual shaped beams necessary to maintain low axial
ratios (under 0.75 dB) to achieve at least 27 dB isolation
between beams. Aeronutronic Ford has built a 9 GHz
(3 ft diameter) scale model of this C-band antenna (Fig.
35). Phase and amplitudes for the 48-horn feed array
shown were determined by computer optimization to yield
the desired coverage pattern.[18]

A measured contour plot of the Western Hemisphere
beam from this model is shown in Fig. 36a for the principal RHCP polarization; ground stations to be covered
are indicated by "+" signs (representing the Indian Ocean
area). Pattern shaping to conform to this desired coverage area is evident, as are the low sidelobe levels in the
Eastern Hemisphere (affording at least 27 dB isolation
from the Eastern Hemisphere beam). A measured cross-
polarized (LHCP) radiation contour plot of the same beam
is given in Fig. 36b; the reference level for this plot is
30 dB below that of the principal-polarized plot. Results
show that axial ratios in the Western Hemisphere coverage area are below the desired 0.75 dB.

Similar results have been reported by TRW[19] in which a
60-inch offset-fed reflector with a 45-element feed was
tested at 4 GHz. Dual circularly-polarized cup-dipole
feed elements were used because of their inherent low
axial ratios and because of their adjustability to compensate for mutual coupling in an array environment, as

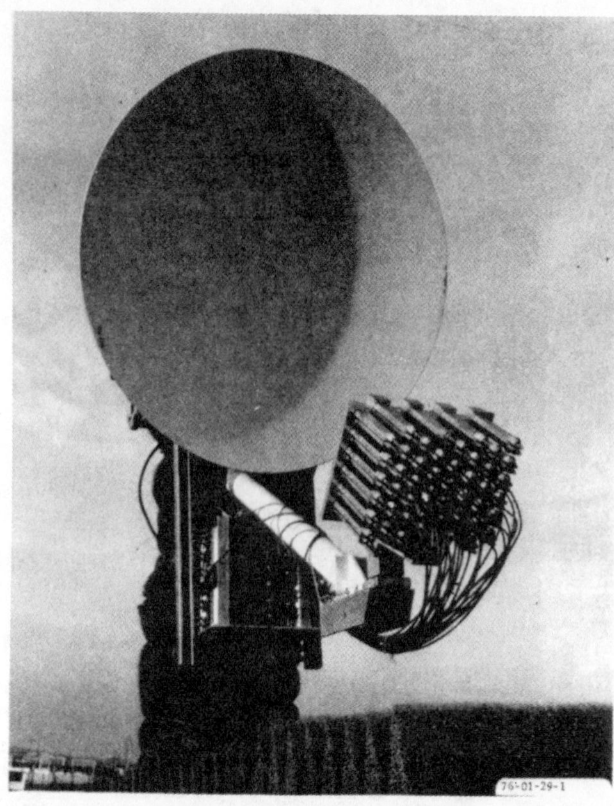

Fig. 35 Antenna Assembly on Test Range

determined by near-field testing. Calculated and measured hemispherical beam contour patterns by TRW
show remarkably good agreement, good sidelobes, and
low cross-polarized levels.

IV. REFERENCES

1. "Multibeam Antenna Study, Phase II," Final Report
on Contract NAS5-21711, Lockheed Missiles and
Space Co., 1973.

2. "Final Report for Advanced General Purpose
Forces Satellite Antenna Study," Rept. No.
2265.30/215, Hughes Aircraft Co., Dec 1973,
pp. 25-29.

3. W. D. Fitzgerald, "Limited Electronic Scanning
With an Offset-Fed Near-Field Gregorian System,"
MIT Lincoln Lab Tech. Rept. 486, 24 Sep 1971.

4. "Final Report for Advanced General Purpose
Forces Satellite Antenna Study," SAMSO TR No.
74-71, Hughes Aircraft Co., Dec 1973,
pp. 187-195.

5. W. A. Imbriale, "Adaptive and Phased Arrays,"
lecture notes from UCLA Short Course on Communication Satellite Antenna Technology, Mar 1976.

6. H. S. Lu, et al, "A Constrained Lens Antenna for
Multiple Beam Satellites," presented at AIAA 5th
Communications Satellite Systems Conference,
Los Angeles, Apr 1974.

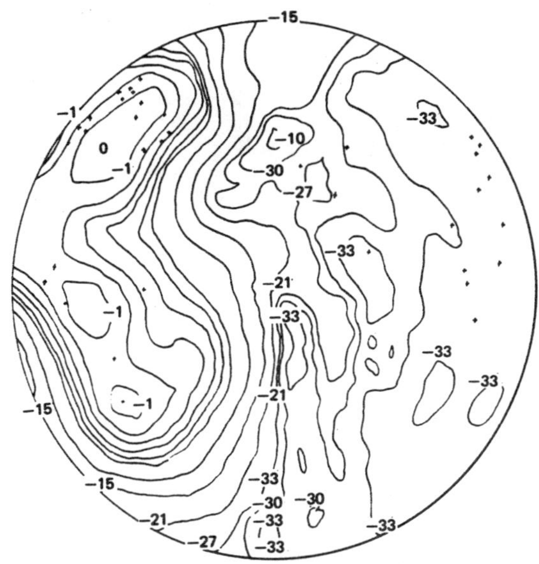

a. Principal Polarization

b. Cross Polarization

Fig. 36. Contour Data (Frequency = 9.54 GHz)

7. C. C. Han, H. W. Bilenko, and A. N. Wickert, "Computer-Aided Array Feed Design for Multiple Beam Lens Antenna," 1975 IEEE AP-S Symposium Digest, pp. 374-377.

8. W. G. Scott, H. S. Luh, and E. W. Matthews, "Design Tradeoffs for Multibeam Antennas in Communication Satellites," Conference Record, Vol. I, presented at 1976 International Conference on Communications, Philadelphia, PA, Jun 1976.

9. E. W. Matthews, "Variable Power Dividers in Satellite Systems," presented at 1976 IEEE-MTT Symposium, Cherry Hill, N. J., Jun 1976.

10. A. R. Dion and L. J. Ricardi, "A Variable Coverage Satellite Antenna System," Proc. IEEE, Feb 1971, pp. 252-262.

11. A. R. Dion, "Optimization of a Communication Satellite Multiple Beam Antenna," MIT Lincoln Lab Tech. Note 1975-39, 27 May 1975.

12. L. J. Ricardi and A. R. Dion, "Beam Scanning With a Multiple Beam Antenna," MIT Lincoln Lab Tech. Memo 61L-0073, 22 Oct 1974.

13. L. J. Ricardi and A. R. Dion, "Earth Coverage Radiation Pattern With a Prescribed Minimum," MIT Lincoln Lab Tech. Memo 61L-0072, 22 Oct 1974.

14. D. W. Prins and D. W. Krejci, "Multibeam Antennas Open a New Era in Satellite Communications," SIGNAL Magazine, Nov/Dec 1975, pp. 6-14.

15. E. F. Binz and D. K. Waineo, "Satellite Multibeam Antenna Concept," AIAA/CASI 6th Communications Satellite Systems Conference, Montreal, Apr 1976.

16. R. H. Turrin, "A Multibeam, Spherical-Reflector Satellite Antenna for the 20- and 30-GHz Bands," Bell System Technical Journal, pp. 1011-1026, Jul-Aug 1975.

17. D. T. Nakatani, et al, "INTELSAT IV-A Communication Antenna - Frequency Reuse Through Spatial Isolation," Conference Record, Vol. I, 1976 International Conference on Communications, Philadelphia, PA, Jun 1976.

18. C. C. Han, et al, "A General Beam Shaping Technique--Multiple-Feed Offset Reflector Antenna System," presented at AIAA/CASI 6th Communications Satellite Systems Conference, Montreal, Apr 1976.

19. J. W. Duncan, S. J. Hamada, and P. G. Ingerson, "Dual Polarization Multiple Beam Antenna for Frequency Reuse Satellites," AIAA/CASI 6th Communications Satellite Systems Conference, Montreal, Apr 1976.

Section 4.3
Communications Satellite Receiver and Transmitter Technology

by R. STRAUSS

INTRODUCTION

Early communication satellites utilized double conversion transponders (Telstar [1], Relay [2], Early Bird/INTELSAT I [3]) based on well-established relatively narrowband double-conversion terrestrial radio-relay concepts [4], [5]. INTELSAT II, in 1967, became the first commercial communications satellite to utilize the single down conversion approach with a single-channel broadband (130 MHz bandwidth) transponder. Furthermore, with INTELSAT II the present "receiver" concepts were also initiated and established [6]. The "receiver" portion of the transponder incorporated the GTD (germanium tunnel diode) low-noise receive amplifiers, broadband matching filters associated with the down converter silicon diode mixer, a crystal temperature-stabilized local oscillator (LO), and the GTD/TWT low-level drive amplifiers preceding the transmitter amplifier(s)—i.e., all the functions which provide broadband multicarrier linear amplification and translation. In INTELSAT III [7] the same basic receiver concept was retained except for a further increase in bandwidth. It was extended to 230 MHz in each of two transponders to cover 90 percent of the allocated 500-MHz band. The basic transponder configurations are shown in Fig. 1.

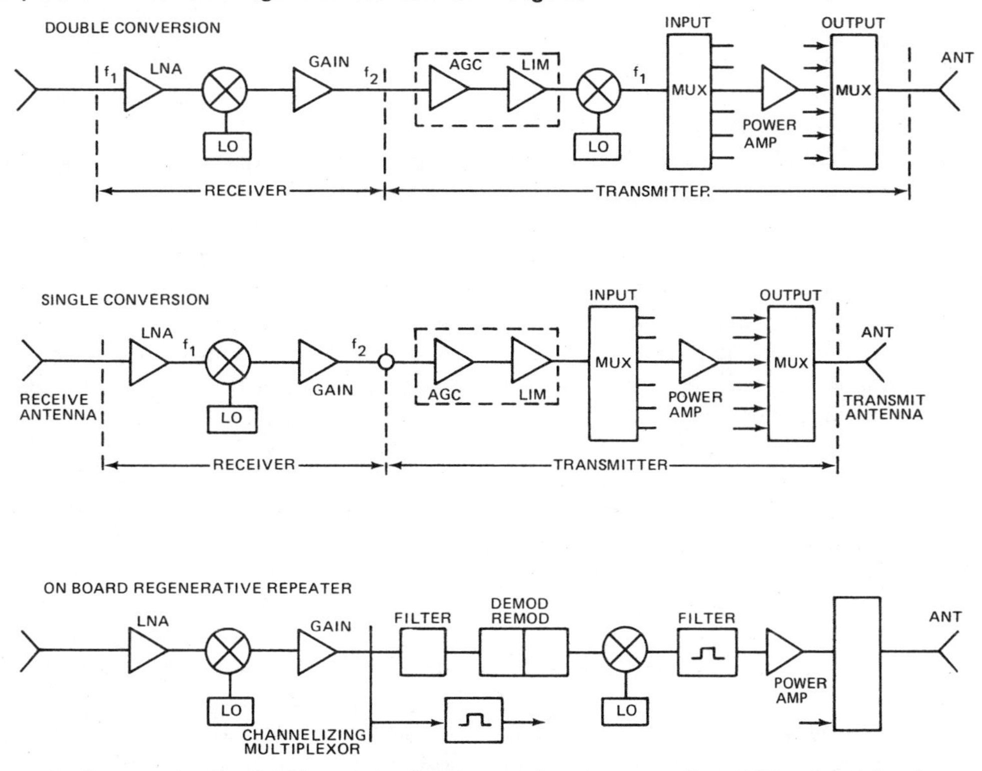

Fig. 1. Generalized communications satellite transponder conceptual arrangements. *Note:* AGC or limiter functions may not be desirable for some applications and may not be required.

Unlike the conceptual changes in the receiver, the transmitter portion of the transponder remained unchanged. The traveling wave tube (TWT) amplifier with its broadband high-gain (50–65 dB) and high-efficiency capability was unchallenged as the sole transmitter device. Like any amplifier, the TWT, when driven into its nonlinear region (near saturation), will generate "noise" components (AM/PM, IM, phase jitter, etc.). The resultant system noise impairments contributed by the transponder are directly related to the TWT drive levels selected and the particular communications signal spectrum chosen (e.g., frequency plan and modulation techniques). This limitation and the coincident development of weight/bandwidth efficient low-loss low-group-delay bandpass filters led, during the late sixties, to the next major transponder evolution—namely, channelization of the transmitter. The INTELSAT IV [8] and TELESAT/WESTAR satellites clearly reflect this concept. A broadband (500 MHz) linear single-conversion receiver drives, for instance, up to twelve 36-MHz bandwidth channelized TWT transmitters. In effect, these satellites carried 12 transponders.

Current Transponder Concepts

During the seventies satellite communications' payload capabilities continued to improve, but the basic transponder concepts remained the same. The number of "channelized" transponders per satellite typically increased—e.g., COMSTAR [9] had 24; INTELSAT IV-A had 20; and INTELSAT V went to 27. Receivers became lighter and more compact. Driver TWTs were initially replaced by silicon bipolar transistor amplifiers at 4 GHz, and these, in turn, are now being replaced by higher gain FET amplifiers both at 4 GHz and 12 GHz. The GTD low-noise "front end" amplifiers [10] are gradually being replaced by low-noise FET transistor amplifiers. The SBS Satellite [11] will, for example, be among the first to utilize low-noise gallium arsenide MESFET amplifiers at 14 GHz with receiver noise figures of about 4 dB.

Furthermore, the progressive application of hybrid circuit miniaturization technologies in the microwave region (e.g., MIC technology—microwave integrated circuits) has resulted in dramatic receiver packaging improvements. For instance, the weight of the SBS satellite receiver package will be 19 percent of the INTELSAT IV receiver weight, and it consumes only 26 percent as much power. Characteristics are more reproducibly achieved and quality control is enhanced by use of photolithographic planar circuit techniques employed in MIC's. This important evolution is discussed in the Reprint Paper 4.3.1.

Net communication transponder gains are in the range of 100 to 120 dB. Approximately half the gain is in the receiver. The receiver is, in turn, followed by low-power-level (e.g., 0.01 to 1.0 MW), linear signal-processing functions such as channelizing filters, delay equalizers, amplitude equalizers, switchable attenuators, and channel interconnect switch matrices before the input of the transponder transmitter amplifier is reached. These passive functions may contribute an additional 10 to 20 dB of loss, which must be compensated for by additional gain. The transmitter amplifier will thus require 50 to 75 dB gain. TWT's can provide 50 dB to 65 dB of stable gain. The TWT is usually driven by a linear $\sim$10 dB gain solid-state amplifier which may have additional signal processing functions included, such as temperature gain control or envelope limiting characteristics. The gain of this driver amplifier is chosen to compensate for the losses in the signal processing section of the transponder.

Transmitter saturation output powers have ranged from 6 to 24 W per transponder. TWTA (tube and power supply) saturation efficiencies with single collector TWT's are about 30 percent depending on the phase shift sought. For multicarrier operation where the typical output power backoff ranges from 3 to 6 dB, the efficiency is correspondingly reduced (i.e., 15 percent and 7.5 percent, respectively). The intermodulation ratio improves with increased backoff [$(C/I)_2$ $\sim$10 dB at saturation $\sim$20 dB at 6 dB output backoff, and 2 dB/dB thereafter]. During the past few years TWT linearity and efficiency improvements have occurred [12], [14]. Multicollector TWT's with current power supply designs attain TWTA saturation efficiencies $\sim$40 percent for output powers of 10 to 20 W. At 3-dB output backoff, TWTA efficiencies now reach 25 percent with $(C/I)_2 \approx 17$ dB. Reprint Paper 4.3.2 reviews the current status of helix-type TWT's.

The various passive elements which constitute transponder input and output multiplex filter and combining networks are discussed in Section 4.4 of this volume. It is a separate and distinct technology.

Transponder Requirements

The primary transponder requirements are set by the system link characteristics [13]. They are the receiver noise figure, the net transponder gain at various illumination levels relative to transmitter saturation and commandable gain settings, the transmitter saturated output power, and transponder channelized bandwidth characteristics. Furthermore, gain flatness, gain slope, and gain stability (without AGC) are specified at large and small signal conditions both for the "end to end" performance and the linear portion of the transponder. Frequency conversion, net translation errors (lifetime, per month, and short-term residual modulation), spurious outputs (harmonics, mixer products, noise, etc.), out-of-band response or rejection (isolation) limits are also specified.

Communications characteristics are limited by several residual signal distortion parameters. Limits on these parameters serve as further requirements. They include deviations from phase linearity at different drive or input flux levels, resultant AM/PM transfer coefficient characteristics, FM intelligible cross talk, AM nonlinearity (usually $[C/I]_2$), channel group delay characteristics, and more recently, for TDMA applications, peak-to-peak pulsed transient responses on amplitude and phase.

Finally, reliability specifications complete the overall requirements. These include, for example, minimum expected redundancy, overdrive protection, channel "patch" switching features, and design life characteristics.

Use of redundancy techniques as applied to the active elements in the basic transponder has been a key factor in attaining excellent operational availability. Both the SBS and Anik C spacecraft will, for example, utilize a form of switch "ring" redundancy developed for enhanced reliability [14].

Not typically specified are compliance of transponder requirements during eclipse periods or at specified temperature extremes. The transponder requirements must be met regardless of the satellite environment. This is an integral part of the system performance approach used in commercial communication satellite procurements.

Transponder performance acceptance testing assures adequate and compliant attainment of the communications requirements. The test plans and automated technology are now well established. Testing is initiated at the unit level (TWTA, receiver, multiplex assembly, etc.), followed by transponder subsystem tests, and complete satellite communications performance tests, both on the ground (thermal/vacuum) and "on station" in space [15]. Temperature cycling and competing temperature coefficient transient effects play a dominant role in the attainment of acceptable overall long-term transponder performance characteristics.

FUTURE TRENDS — EVOLVING TECHNOLOGY

The future of communication satellite repeaters is highly dependent, not only on the evolving technology, but also on the functions and modulation characteristics that system planners expect to impose on the overall satellite communications system. The particular transponder configuration chosen, at some point in time, is a function of many, often conflicting, demands. With the trend towards increased "on-board" processing functions, frequency-band cross-strapping, intersatellite links, etc., it is not clear as to what specific form the transponders of the future will take. It is clear that future transponders will be more complex and that MIC technology will continue to make tremendous inroads into microwave repeater design concepts [16]–[18]. Moreover, MIC technology in conjunction with the FET [19] will make possible the practicality of increased signal processing functions within the repeater. Pertinent examples are the "on-board" MIC format multibeam TDMA switch matrix [20] and "on-board" processing of PSK signals [21], or phase-controlled MIC transistor power amplifier stages associated with multibeam phase array antennas [22], or the predistortion linearization networks preceeding the output amplifiers [23]. Continued emphasis on MIC technology (e.g., active and passive filter or phase networks, lossless dividers or combiners, etc.) should result in marked benefits.

Intimately combined with the MIC technology is the evolution of the microwave transistor [24]. During the past few years the FET in several formats (MESFET, IGFET, and JFET) [25] has become dominant in low-power and low-noise microwave applications between 1 and 30 GHz (Fig. 2). The unique and still developing characteristics of the FET as a usable three [26] or four [27] terminal device

makes their application in MIC microwave repeaters particularly attractive. These remarkable developments are reviewed in two complementary Reprint Papers, 4.3.3 and 4.3.4.

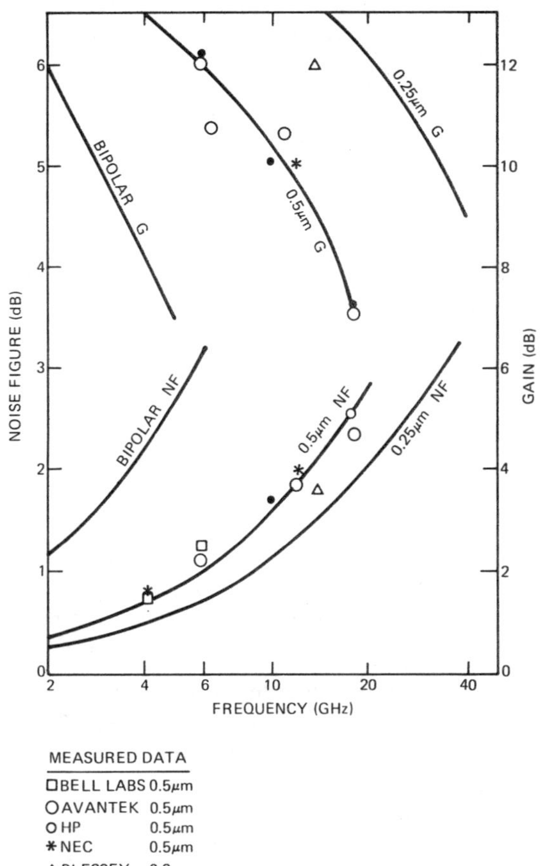

Fig. 2. Measured laboratory noise figures and associated gains are plotted versus frequency for GaAs MESFET's with 0.5-μm and 0.3-μm gate lengths. The solid noise curves are computed by Fukui. They show excellent agreement with measured data for 0.5-μm gates and predict future performance for optimized MESFET's with 0.25-μm gate lengths (e.g., 3.3 dB NF with 0 dB gain at 40 GHz). For comparison, lowest noise data for bipolar transistors are added. (Courtesy C. A. Liechti, Sept. 1978.)

More recently the evolution of power gallium arsenide MESFET's (3 to 16 W RF output per chip) with excellent linearity and efficiency characteristics now also makes possible a solid-state 10-W C-band TWTA replacement [28], [29]. Such an MIC integrated assembly might require seven to ten FET's with several of the output stages operating at 1–2 dB of compression (Fig. 3 indicates per chip power capabilities). Preliminary studies indicate excellent overall characteristics in terms of low intermodulation, phase distortion, and gain slope with reasonable overall efficiency (~20 percent) utilizing class A or AB amplification [30]. Since these studies, although promising, are at an early state of development and power FET reliability factors [31] (characteristic stability at elevated channel temperatures, "hot-spot" thermal resistance factors, excess electric field effects at the gate or drain, etc.) have not yet been thoroughly characterized. It will still take considerable time and testing until solid-state TWTA

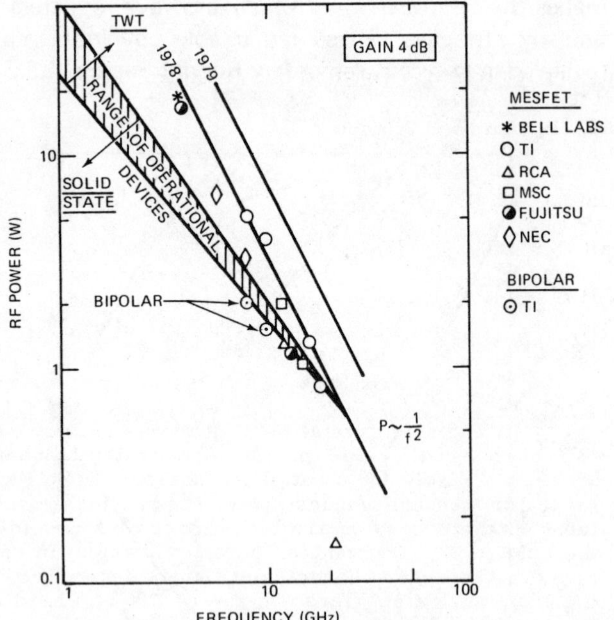

Fig. 3. CW output power of single laboratory MESFET chips are shown versus frequency. The MESFET's are biased for Class A operation, and the output power is measured at 4 dB associated gain. During the next year, it is expected that power output per chip will approach 10W at 10 GHz and 1 W at 26 GHz. In X-band, bipolar transistors are still competing in the power race. (Courtesy of C. A. Liechti, Sept. 1978.)

replacements are qualified for commercial satellite usage.

Other active devices complementary to MIC technology such as Schottky mixer diodes, PIN diodes, step-recovery diodes, and detector diodes, usually in passivated silicon chip or hermetically sealed package format, are available for circuit integration. Furthermore, the continued evolution of submicron technology combined with GaAs is spawning new capabilities, namely, gigabit LSI logic circuitry [32]-[33]. In summary, the continued progress in the ever-widening arsenal of miniaturized microwave circuits and devices should continue to impart enhanced capability to the communication satellite transponders of the future.

REFERENCES

[1] C. G. Davis et al., "Spacecraft communications repeater," Bell Syst. Tech. J., vol. 42, no. 4, Part I, pp. 831–868, July 1963.

[2] J. D. Keisling, "The NASA relay I experimental communications satellite," RCA Review, vol. 25, no. 2, pp. 232–261, June 1964.

[3] S. B. Bennett, "Early Bird communication parameters," Communications Satellite System Technology, vol. 19 of Progress in Astronautics and Aeronautics. New York: Academic, 1966, pp.43–59.

[4] A. A. Roetken et al., "The TD-2 microwave relay system," Bell Syst. Tech. J., vol. 30, no. 4, pp. 1041–1077, Oct. 1951.

[5] A. Harmori et al., "The TH-3 microwave radio system: Microwave transmitter and receiver," Bell System Tech. J., vol. 50, pp. 2117–2135, Sept. 1971.

[6] S. Metzger, "Some aspects of communications satellite systems," Advances in Communication Systems, vol. 3. New York: Academic, 1968, pp. 57–90.

[7] A. J. Grey, "Communication, telemetry, and command subsystem," Electrical Communication, vol. 45, no. 4, pp. 330–342.

[8] Dostis, I., "Communications subsystem—INT IV," COMSAT Tech. Rev., vol. 2, no. 2, pp. 277–294, Fall 1972.

[9] G. E. A. Abutaleb et al., "The COMSTAR satellite system," COMSAT Tech. Rev., vol. 7, no. 1, pp. 52–65, Spring 1977.

[10] A. Revesz et al., "Tunnel diodes in satellite communications," COMSAT Tech. Rev., vol. 8, no. 2, Fall 1978.

[11] M. T. Lyons et al., "Spacecraft design for the SBS system," AIAA Communications Satellite Conf. Proc., San Diego, CA, Apr. 1978.

[12] R. Strauss, "Microwave tube requirements for communications satellite systems," Int. Electron Devices Meeting, Washington, DC, Dec. 1977, Reprint UP-059-CL (COMSAT Laboratories, Clarksburg, MD).

[13] J. C. Fuenzalida et. al., "Summary of the INTELSAT V communications performance specifications," COMSAT Tech. Rev., vol. 7, no. 1, Spring 1977.

[14] F. Assal et al., "Network topologies to enhance the reliability of communication satellites," COMSAT Tech. Rev., vol. 6, no. 2, pp. 309–322, Fall 1977.

[15] I. Dostis et al., "In-orbit testing of communication satellites," COMSAT Tech. Rev., vol. 7, no. 1, pp. 197–226, Spring 1977.

[16] D. A. Cowan et al., Low-noise and linear FET amplifiers for satellite communications," IEEE Trans. MTT, vol. MTT-25, no. 12, pp. 995–1000, Dec. 1977.

[17] A. Hiroyuki et al., "A highly stabilized low-noise GaAs FET integrated oscillator with a dielectric resonator in C-band," IEEE Trans. MTT, vol. MTT-26, no. 3, pp. 156–162, Mar. 1978.

[18] W. H. Childs "Design techniques for bandpass filters using edge-coupled microstrip lines on fused silica," IEEE MTT-S Int. Microwave Symp. Digest, pp. 194–196, June 1976.

[19] C. L. Cuccia "The impact of the FET on the satellite transponder," AIAA Communications Satellite Conf., San Diego, CA, pp. 172–181, Apr. 1978.

[20] X. Rozec, "16 × 16 fast switching matrix in the 3.7–4.2 GHz band for satellite communications," European Int. Microwave Conf. Proc., Sept. 1978.

[21] S. Campanella et al., "Onboard regenerative repeater," ICC '77 Conf. Record, vol. 1 pp. 6.2-121–125.

[22] P. Charas et al., "Multi-beam array, designed for use in a maritime system using an OTS platform," Proc. Int. Astronautics Cong., (IAF No. 28), Sept. 1977, Paper 77-40.

[23] C. Bremenson et al., Predistortion network for traveling wave tube," Thomson-CSF Tech. Rev., vol. 6, no. 2, June 1974.

[24] S. Teszner et al., "Microwave power semiconductor devices, II," Adv. in Electronics and Physics, vol. 44. New York: Academic, 1977, pp. 141–219.

[25] C. A. Liechti "Microwave field effect transistors—1976," IEEE Trans. MTT, vol. MTT-24, no. 6, pp. 279–300, June 1976.

[26] M. Ogawa et al., "Submicron single gate and dual gate GaAs MESFETs with improved low-noise and high gain performance," IEEE Trans. MTT, vol. MTT-24, no. 6, pp. 300–305, June 1976.

[27] T. Furutsuka et al., "GaAs dual-gate MESFET's," IEEE Trans. Electron Devices, vol. ED-25, no. 6, pp. 580–586, June 1978.

[28] M. Fukuta et al., "4-GHz 15-W power GaAs MESFET," IEEE Trans. Electron Devices, vol. ED-25, no. 6, pp. 559–567, June 1978.

[29] C. A. Liechti, "Microwave FET's—What's next?" Microwaves, Oct. 1978.

[30] Y. Arai et al., "A 4-GHz TWT replaceable GaAs MESFET amplifier," Proc. 8th European Microwave Conf., Sept. 1978.

[31] J. C. Irvin et al., "Failure mechanisms and reliability of low-noise GaAs FETs," Bell Syst. Tech. J., Oct. 1978.

[32] R. Zuleeg et al., "Femtojoule high-speed planar GaAs E-JFET logic," IEEE Trans. Electron Devices, vol. ED-25, no. 6, pp. 628–639, June 1978.

[33] P. Fleming et al., "Performance of SAMP devices," Digest IDEM 1978, Dec. 1978.

4.3.1

COMMUNICATIONS SATELLITE RECEIVER DESIGN

Howard Ozaki, Meredith Eick, and Neal Silence

Space and Communications Group

Hughes Aircraft Company

Box 92919

Los Angeles, California 90009

INTRODUCTION

In the 20 years since the introduction of communications satellites, communications technology has been called upon to develop and continually improve component performance and reliability, while lowering component weight and dc power consumption. The early satellite communications receiver was a modification of the dual conversion receiver then used in terrestrial systems. However, as the demand to maximize communications repeater efficiency increased, receivers began to assume a special configuration: single conversion, all microwave.

This chapter describes the evolution of receivers for satellite communications repeaters through the current 6/4 GHz microwave integrated circuit (MIC) receiver.

EVOLUTION OF RECEIVER DESIGN

Dual Conversion

Early satellite communications receivers comprised a significant portion of a limited payload and were dual conversion, with all of the signal amplification occurring at a relatively low IF. This configuration (Figure 1) was dictated primarily by the available hardware.

A common crystal oscillator was usually used to reduce circuitry and improve conversion frequency stability. Since, for a satellite repeater, spectrum inversion is usually undesirable, the two local oscillator (LO) frequencies are both higher or lower than the input/output frequencies (usually lower). In general, the bandpass characteristics are determined by the channel IF amplifiers and limiters. The hard IF limiters provide equal sharing of the downlink traveling wave tube (TWT) power and an excellent intelligible crosstalk ratio between channels despite large amplitude-to-phase conversion numbers in the high level mixer and TWT. Relatively low gain traveling wave tube amplifiers (TWTAs) can be utilized since signal levels of 1 to 10 mW are available from the high level mixer. Operation of the upconverter with the IF input as the larger signal reduces the microwave LO power needed. Since channel filtering is accomplished in the IF, the microwave filtering is minimal, usually accomplished with three section filters for flatness and reduction of LO leakage and image. A separate signal modulated with telemetry data can be conveniently added at IF frequencies at the high level mixer input.

Dual conversion repeaters have design problems where quasilinear operation is desired, that is, where the basic repeater phase and amplitude nonlinearity is determined by the channel filters and output amplifier. The limiter must be removed and separate linear channel upconverters with separate

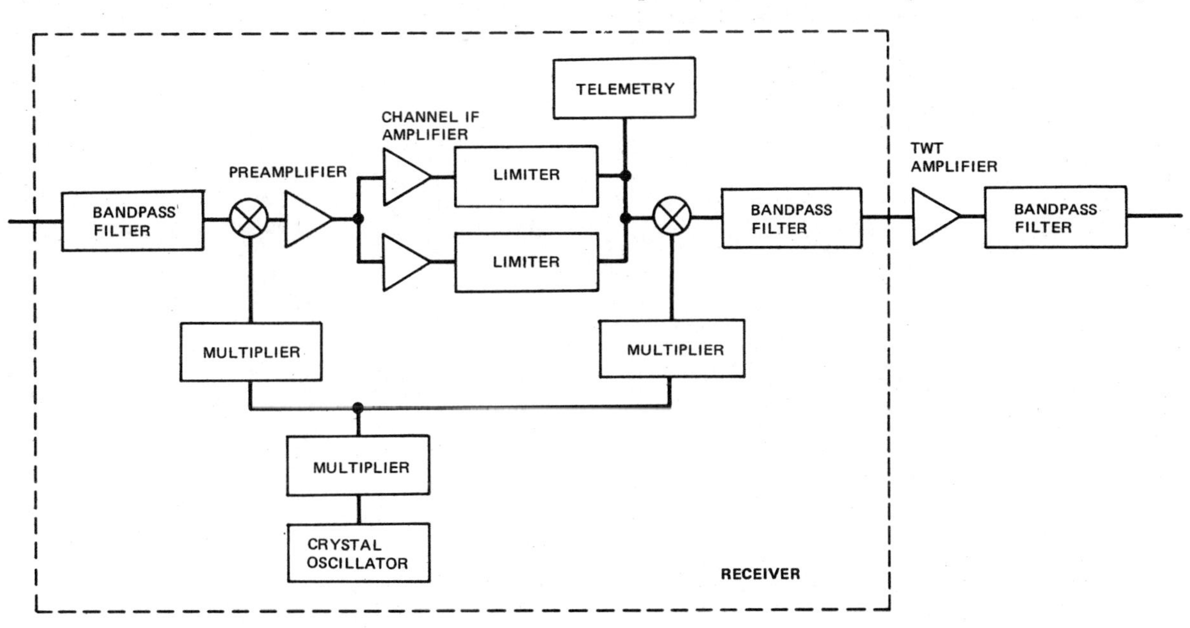

FIGURE 1. DUAL CONVERSION REPEATER

Manuscript received December 1, 1978.

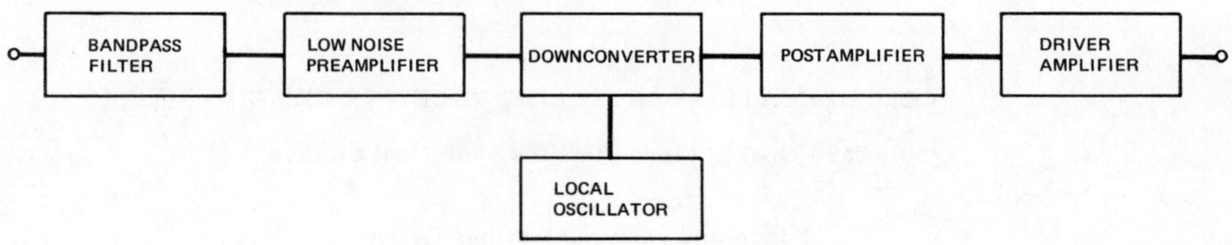

FIGURE 2. SINGLE CONVERSION RECEIVER

output amplifiers provided. More gain is added to the output amplifiers and less gain to the IF amplifier to improve upconverter linearity.

If the total percentage bandwidth is large, then the IF band must be enlarged, thus decreasing the microwave-to-IF frequency ratio, introducing more spurious signal products, and requiring more upconverter linearity. This change also decreases IF amplifier gain. An alternate design requires that separate downconverter mixers be provided for each channel. Preservation of noise figure requires about 15 to 30 dB amplification before mixing. These design constraints tend to increase the total microwave amplifier gain and decrease the required IF gain. Coherent up- and downconverter LO frequencies become difficult on most communications bands without low crystal oscillator frequencies and the resulting large frequency multiplication. Separate LO frequency control means tighter individual oscillator frequency control. If a large number of channels are involved, the hardware for redundant multiple upconverters and downconverters becomes formidable.

Repeaters with small percentage bandwidth channels (less than about 0.1 percent) usually are dual conversion designs, since mechanizing a narrow bandwidth filter with sufficient stability is easier at IF than microwave. Repeater gain is also larger for a given uplink signal-to-noise ratio which can be conviently provided in the IF channel amplifier.

Signal Conversion

Demand for wider bandwidths and quasilinear operation necessitated the use of wideband microwave amplifiers in a single conversion configuration. The introduction of the tunnel diode amplifier (TDA), along with the use of the TWTA, made possible the first wideband receiver design, and still provides the basis for most operational receiver designs, used by commercial communications satellites. The TDA is conceptually a simple, low noise, broadband amplifier that easily provides 10 to 15 dB gain in the 6 and 4 GHz frequency bands. The TWTA is quite linear and provides 30 to 60 dB gain.

In single conversion satellite communications receivers (Figure 2), filtering is provided at the receiver input and as part of the downconverter to minimize unwanted signal inputs, intermodulation products, feedback paths through the satellite repeater, and spurious signal outputs. Level set attenuators are included to provide optimum LO drive and accurately set the overall receiver gain.

The gain distribution through the receiver is adjusted so that the various performance requirements may be primarily established by a single element of the receiver. The objectives are to provide noise figure by the preamplifier, in-band frequency response by the downconverter, overall gain by the postamplifier, and linearity by the driver amplifier.

TABLE 1. RECEIVER PERFORMANCE COMPARISON

Parameter	Intelsat II	Intelsat IV	Intelsat IVA
Bandwidth, MHz	126	500	500
Gain, dB	87	57.5	54.5
Noise figure, dB	5.5	8.5	7.0
Intercept point, dBm	24	31	28
Weight, lb	5.5	9.3	9.1
Power, W	6.0	15.1	7.4

The single conversion configuration provides satellite communications receivers with a relatively simple, straightforward design and the desired communications performance, while offering minimum size, weight, and power consumption.

Typical Receiver Designs

The receiver designs utilized in the Intelsat II, IV, and IVA communications satellites afford typical examples of the evolution of this type of hardware. A comparison of the general parameters for these receivers is shown in Table 1.

The receiver of the Intelsat II spacecraft employed four TDAs at 6 GHz and a driver TWT at 4 GHz. This receiver was the first all-microwave satellite communications receiver to use TDAs. It provided: 1) a significant reduction in the number of required parts over that needed for previous designs, 2) a two and one-half fold increase in bandwidth, 3) single amplification for a single wideband transmitter, and 4) quasilinear operation.

Intelsat IV utilized 12 RF channels, requiring a significant improvement in linearity of the receiver. This improvement, coupled with higher input signal levels, limited the number of permissible preamplifier stages. Also, the linearity of the 4 GHz tunnel diode postamplifier had to be improved against receiver nonlinearity. A X9 diode multiplier was used in the local oscillator, since suitable high frequency transistors were not available. This high multiplication factor necessitated the use of extensive filtering to reduce the spurious frequency response of the LO chain to an acceptable level. A notch filter was provided at 4450 MHz to reduce the level of the second harmonic of the LO introduced into the 4 GHz TDA, thus eliminating unwanted in-band intermodulation products. Active thermal control was used to minimize performance variations caused by spacecraft temperature changes.

The Intelsat IVA receiver (Figure 3) was similar to that used in Intelsat IV. However, the use of microwave transistors in the LO chain , improved performance of the TDAs, and the use of a microwave transistor driver amplifier provided a receiver with exceptional performance characteristics. The receiver is divided into two subunits: an amplifier mixer assembly occupying one compartment of the receiver housing, and a low level driver assembly in

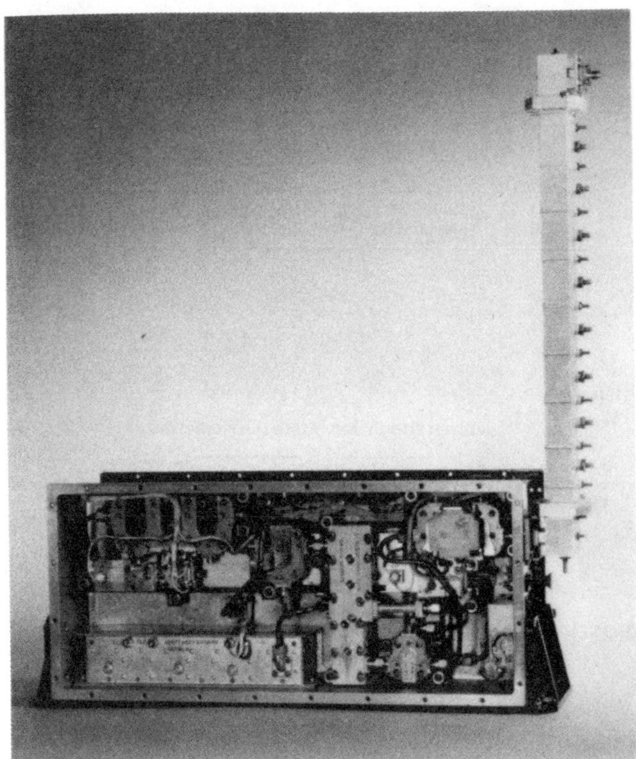

FIGURE 3. INTELSAT IVA COMMUNICATIONS RECEIVER
(PHOTO E043075-2)

a second compartment, providing RF shielding
between the high and low level sections. An incoming
signal passes through the waveguide input bandpass
filter which provides the required out-of-band
attenuation prior to initial amplification. From the
filter, the signal goes through a waveguide-to-coax
transition and enters the 6 GHz TDA. The 6 GHz
TDA provides the first amplification of the incoming
signal and primarily establishes the noise figure of
the receiver. Upon leaving the 6 GHz TDA, the signal undergoes frequency conversion in a single-
ended wideband mixer. Using a solid state LO, the
mixer translates the 6 GHz receive band to the 4 GHz
transmit band. The output of the mixer is amplified
by a high saturation 4 GHz TDA which, in turn,
drives a low level TWTA or microwave transistor
amplifier (MTA). Receiver selection is accomplished
via ground command by energizing the receiver
regulator of the selected receiver. The receiver
regulator supplies the necessary voltages to turn on
the TDA regulator, the two TDAs, the TWTA, and
the mixer LO. Power and command signals enter
the receiver through a filter module, comprised of
RF bypass capacitors, located in the TWTA assembly
compartment. Telemetry information (on/off status)
exits the receiver in the same manner. The overall
gain of this receiver is stabilized over the thermal
environment by controlling the dc bias voltage on the
4 GHz TDA using a temperature sensing circuit in its
bias supply. Passive thermal control is also used to
minimize the temperature changes that occur during
spacecraft eclipse. This technique provides a
typical gain variation of 0.5 dB over an exposure
temperature range of 40 to 110°F. Typical measured
performance parameters are shown in Table 2.

Microwave Integrated Circuitry

During the mid-1960s considerable efforts
were directed toward miniaturizing electronic
circuitry. One approach, an extension of printed
circuit board techniques, was based on the use of

TABLE 2. TYPICAL INTELSAT IVA RECEIVER PERFORMANCE

Gain, dB	54.5 ±0.25
Gain stability, dB/°F	0.008
Noise figure, dB	7.0
Phase shift from no input to P_{in} = -49 dBm, deg	0.8
FM transfer ratio for a total P_{in} = -49 dBm of two carriers, a ratio of modulated-to-unmodulated carrier power of ≤20 dB and baseband frequency of 1 MHz, dB	-85
Carrier to 3 IM ratio for total P_{in} = -49 dBm, dB	50
Gain slope, dB/MHz	0.003

high dielectric constant substrates, bulk and thin film
resistors, chip and small discrete capacitors, and
semiconductors in chip form or in miniature discrete
packages. Variously described as "hybrids" or
MICs, these miniaturized circuits were first used
in digital and VHF electronic circuits. Much of the
circuitry for airborne radar and missile systems,
except for microwave modules, had been changed
over to hybrid circuit designs by the late 1960s.

As MIC technology matured, serious consideration was given to its space application. Space
communications had advanced to the point that the
minimum operating life requirement was 7 years,
with 10 year "birds" as design goals. The advantages
of MICs for earthborne and airborne hardware, low
cost in volume production and extreme miniaturization, would be of secondary importance to light
weight for space hardware. Also, since space hardware production would, for many years, be relatively
low volume (ten spacecraft of one type is considered
an extremely large quantity), designing for mass
production and "learning curve" cost advantages
would be less important than developing designs and
fabrication techniques that permitted easy tuning and
simple rework.

At Hughes' Space and Communications Group,
the first general satellite application of MICs was in
Geostationary Meteorological Satellite 1 (GMS-1) built
for Japan. The program, undertaken after several
years of development at S band, was successfully concluded with launch of the GMS-1 in December 1977,
and its successful operation in orbit since system
turnon. Because the highest frequency in the GMS-1
was 2200 MHz and the hardware consisted mostly of
narrowband circuits, it was clear that the GMS MIC
program, although a valuable introduction, was only a
precursor to the designing of a microwave communications receiver in MIC.

The development of MIC technology in the
lower microwave frequencies did not address many
of the problems that had to be solved for a wideband
communications receiver. These additional considerations arose from the fact that communications
receivers, whether at 6/4 or 14/12 GHz for commercial satellite systems were single conversion types
operating exclusively at high microwave frequencies
except for their LO chains. In addition, their frequency response had to be at least 500 MHz wide and
flat to within 0.2 dB, much more bandwidth than that
needed for typical narrowband telemetry and command (T&C) systems. These receivers would
also have to provide from 50 to 70 dB of linear gain,
not only to eliminate the need for driver TWTs but
to allow the use of output TWTs with reasonably low
gains.

In 1975, after the Intelsat IVA program was
well under way, INTELSAT and COMSAT announced
plans to deploy the Intelsat V, a new satellite system
with much greater communications capability. When

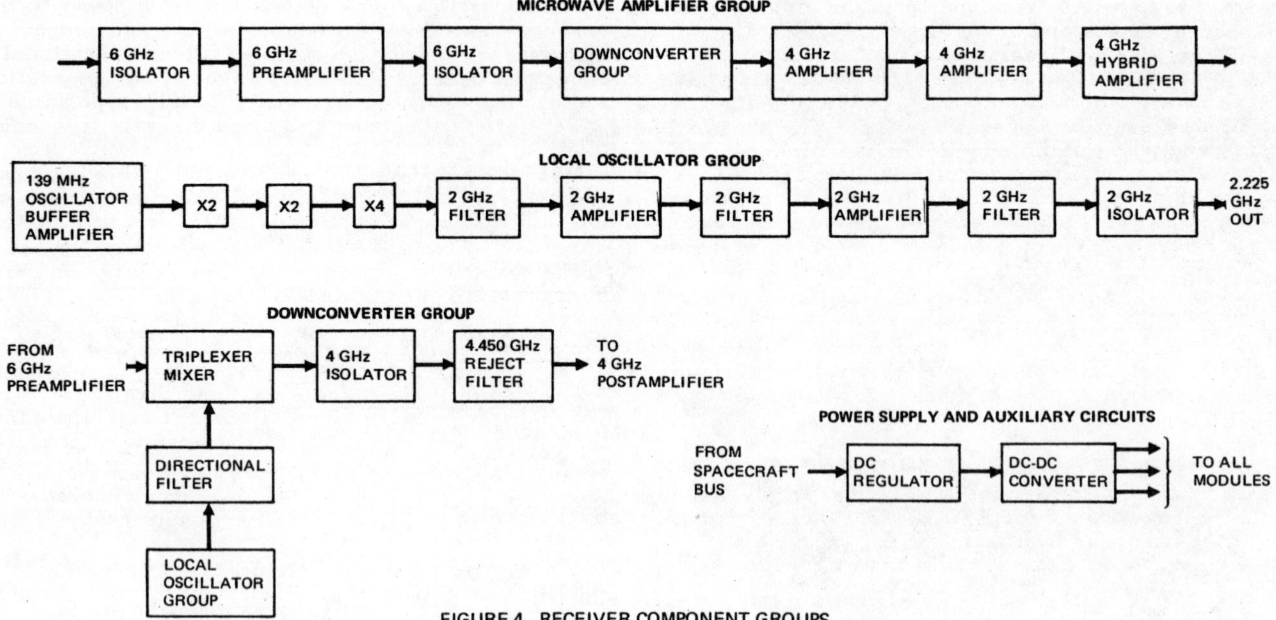

FIGURE 4. RECEIVER COMPONENT GROUPS

existing limitations on satellite boosters and spacecraft weight and power were weighed against the new expanded repeater requirements, it became clear that the discrete receiver designs used in previous INTELSAT spacecraft were much too large and heavy. However, an MIC receiver reasonably miniaturized would be ideal for the Intelsat V. The decision then was made to design an MIC 6/4 GHz communications receiver baseline for the Intelsat V, and this design and associated MIC technology would form the basis for all future single conversion microwave receivers.

6/4 GHZ MIC Communications Receiver

The 6/4 GHz communications receiver integrated design concept evolved from a desire to minimize the integration and test difficulties encountered with previous MIC circuitry. Key receiver elements were separated into four groups: the microwave amplifier, local oscillator, downconverter, and power and auxiliary circuits groups (Figure 4). The receiver requirement was reasonable ease in tuning, testing, and reworking. MIC technology requirements were high reliability and performance, mechanical and thermal integrity, and low weight and cost, in that order. Table 3 summarizes the 6/4 GHz MIC receiver's performance, and the Receiver Design section which follows develops the achievement of that performance.

RECEIVER DESIGN

Receiver Requirements

The receiver performance requirements derived from spacecraft system specifications are detailed in the following paragraphs. (The receiver also has thermal and vibration environmental requirements and packaging constraints.)

Gain. Repeater gain is fixed by the amount of flux density to saturate the output amplifier, antenna gain, and any commandable step attenuation. Net gain is usually about 100 to 120 dB, and since internal losses are about 10 to 15 dB, total active gain is 110 to 135 dB. As a balance between noise figure conservation and mixer linearity, about 15 to 30 dB of gain is assigned to the uplink frequency band, with

TABLE 3. RECEIVER PERFORMANCE PARAMETERS

Gain, dB	73.5
Gain stability, dB/°F	0.008
Noise figure, dB	7.2
Gain slope, dB/MHz (max)	0.004
Phase shift from no input to P_{in} = -68 dBm, deg	0.7
FM transfer ratio for a total P_{in} = -68 dBm of two carriers and a ratio of modulated-to-unmodulated carrier power of $\leq$20 dB and baseband frequency of 1 MHz, dB	-83
Carrier to 3 IM ratio for total P_{in} = -68 dBm, dB	51.5
Weight, lb	2.1
Power, W	5.6

the remainder at the downlink frequency band or channels. Net receiver gain will usually be 50 to 70 dB, so with about 10 dB of internal loss, an active gain of 60 to 80 dB is required. With 15 to 30 dB of gain before the mixer, 30 to 65 dB is required after the mixer. Gain stability without AGC becomes an important consideration with repeater specifications of 1 dB per day for thermal causes and ±3 dB or less per 7 year life (long term drift without gain adjustment). That components and circuits having exceptional thermal and life parameter stability must be selected is a major repeater design problem.

Noise Figure. The required repeater noise figure and receive antenna gain is determined by the specified G/T. Basically, the receiver input amplifier noise figure establishes the repeater noise figure. Reasonably high gain (5 to 15 dB) and low noise figure for the first and second stages are necessary to preserve the total noise figure. In the subsequent repeater circuitry the signal level should be kept large enough to prevent degradation of noise figure by a high noise figure device (such as 15 to 30 dB for a TWT).

Linearity. Receiver linearity is determined by the difference in amplitude or phase between the input and output signals (other than path length time delay) as a function of microwave level and frequency. Quasi-linear repeater operation is defined as channel nonlinearity primarily determined by the channel filters and output power amplifier. The receiver is normally designed to have less than one-tenth the distortion of the channel filters and the output amplifier. The major receiver nonlinearities are input and mixer filter delay distortion near the band edges, and phase change as a function of amplitude, since the effect of amplitude distortion is usually minor. The phase frequency and phase amplitude distortions produce channel-to-channel intermodulation, intelligible crosstalk, and increased error rates at the receiver output. The crosstalk between output channels with single saturated signals is determined in the receiver.

Bandpass. Uplink out-of-band rejection will determine the number of sections in the receiver input filter. The mixer is usually designed with more bandwidth than the receiver input filter, its characteristics determined by the image, LO harmonic, and signal intermodulation rejection required. The receiver amplifiers before and after the mixer are wideband devices, usually with bandwidths much greater than the link bandwidth. Any slope or ripple in the passband is a combination of all the filter and amplifier bandpass characteristics. Limited fine adjustment can be made at the receiver level to improve in-band slope.

Local Oscillator Frequency Stability. Receiver LO frequency short term stability is determined by the quartz crystal loaded Q, the oscillator feedback signal level, oscillator transistor effective noise figure, buffer signal-to-total-noise ratio, multiplication number, and any AM to PM conversion in the LO/mixer circuitry. By selecting oscillator and buffer amplifier transistors for low noise figure and 1/f noise, a high loaded Q overtone crystal, a high signal-to-noise ratio in the oscillator and buffer amplifier, and a properly designed low level limiter and multipliers, the short term instability can be very small. Long term stability is determined by the crystal quartz quality, holder, oscillator, and buffer circuit component stability, including temperature and voltage control. Long term stability can be improved substantially by aging and selecting crystals in actual oscillator circuits at somewhat elevated temperatures. Long term stability has been one of the most difficult parameters to determine prior to spacecraft launch.

Spurious Outputs. The most prevalent spurious outputs are LO harmonics of the crystal oscillator. For fifth overtone crystal oscillator frequencies of about 100 to 150 MHz, several harmonic lines will fall within broadband repeater bandwidths. These lines are suppressed by sufficient filtering in the LO multiplier chain. Also, harmonics of the microwave LO frequency will be present in the mixer output in addition to the image frequency. A multisection mixer output filter and a notch filter are usually required, and more filtering is necessary as the gain after mixing is increased. An additional source of mixer spurious outputs is microwave LO harmonic to signal intermodulation due to mixer nonlinearity, which increases as the signal level in the mixer increases, and usually determines the required mixer linearity. Feedback from the transmitter output to any receiver circuitry can generate spurious outputs and anomalous repeater behavior. For example, a very large number of in-band beat frequencies and greater than normal phase shift and multicarrier intermodulation will be generated if substantial transmitter output is coupled into the receiver input amplifiers. Any coherent feedback mechanism producing a feedback signal greater than 60 dB less than the actual signal at that point will be detectable when the repeater delay distortion is measured, and can be a major single conversion design problem because of the large gain at the output frequency.

DC Bus Interface. The dc bus supplying power to the receiver can exhibit a variable voltage and source impedance as a function of time. Other operational equipment on the spacecraft and the power generation and control equipment can generate a 1/f type noise power density spectrum and large discrete frequency lines on the dc bus. The receiver microwave circuits must be isolated by passive and active low pass filters to prevent any significant phase modulation of the amplified repeater signals. This can be an important factor since the dc bus to microwave AM to PM conversion sensitivity is usually large on most microwave amplifiers and LO circuits.

Component Design

Preamplifiers. Candidates for the 6 GHz preamplifier included tunnel diode, silicon bipolar, and gallium arsenide field effect transistor (GaAs FET) amplifiers. Table 4 compares the performance characteristics of several candidate preamplifiers. The bipolar transistor amplifier was selected as baseline for the preamplifier. Transistors, being three terminal devices, were easiest to tune and most stable in amplifier circuits. The bipolar transistor amplifier, as compared to the TDA, provided superior intermodulation performance and a nearly equal noise figure. It had a history of proven reliability and mature material technology.

Microwave amplifiers in future space systems will utilize GaAs FETs, but the circumstances at that time dictated a conservative choice.

TABLE 4. 6 GHz PREAMPLIFIER CANDIDATE
PERFORMANCE COMPARISON

Parameter	Si Bipolar MIA	GaAs FET Amplifier	TDA
Gain, dB	26 four stages at 6.5 dB/stage	20; two stages at 10 dB/stage	14; one stage
Noise figure, dB	5.5 to 6	3 to 3.5	5 to 5.5
Third order intermodulation intercept point, dBm	+15 min	+17 min	-5 (Ge) +3 (GaAs)
DC power	0.42 W at 15 V	0.3 W at 15 V	0.3 to 0.4 W at 15 V

6 GHz Bipolar Transistor Preamplifier. The baseline design was a four-stage bipolar transistor amplifier with a 6.0 dB noise figure, 26 dB gain, and a third order intercept of +15 dBm. The four stages, including bias circuitry, were packaged on a single 1 x 1-1/2 inch alumina substrate with chip resistor and capacitors. The substrate metallization of the breadboard design was chrome-gold, with a plan to change to chrome-copper-gold metallization to permit soldering when flight amplifiers were produced.

Downconverter. The downconverter included a 6 GHz isolator followed by a filter/mixer module, a 4 GHz isolator, and a notch filter to reject the second harmonic of the LO. A directional filter was selected to couple LO power into the filter/mixer module. The baseline mixer diode was the pill-packaged silicon Schottky barrier diode. The filters in the module were planar MIC realizations of earlier interdigitated designs and were connected to a "triplexer configuration" at whose common junction the mixer diode was located.

TABLE 5. TYPICAL DOWNCONVERTER PERFORMANCE

Parameter	Measured Data
Conversion loss, dB	7.6
Required LO power, dBm	6.0
Conversion loss variation for 1 dB LO power change	±0.1
Passband ripple, dB	0.25
Second harmonic output, dBm	-75
Spurious outputs, dBm	<-100
C/3 IM, two-tone, dB	76
Weight, lb	0.35
Size, in.	3.6 x 3.6 x 0.5

TABLE 6. LOCAL OSCILLATOR PERFORMANCE

Output frequency (fo)	2225 MHz
Crystal frequency (fx)	139.0625 MHz
Frequency stability	
Long term	±0.1 ppm (temperature)
	±1 ppm aging, 7 yr
Short term	Δf = 2 Hz rms (100 to 12 kHz band)
Output power	+6 dBm (at mixer input)

Spurious frequencies relative to 2225 MHz output (at mixer input):

Spurious Frequencies	Predicted Performance, dB
fo ± fx	- 80
fo ± 2 fx	-115
fo ± 4 fx	-124
2 fo - 4 fx	-125

Table 5 shows the typical performance of the completed experimental downconverter. Once the baseline downconverter was completed and integrated into the receiver, alternate designs using doubly balanced diodes and beam lead diodes were evaluated. Extensive history and proven service in space of the pill packaged diode, and the ability to easily burn-in and screen it before integration into a downconverter provide overwhelming arguments for its use in the MIC receiver.

Local Oscillator. The LO configuration was an MIC realization of the most effective of the various early 2.225 GHz LO designs. It consisted of a transistor crystal oscillator at 139.0625 MHz, a pair of X2 multipliers, a X4 multiplier, and two 2.225 GHz amplifiers with interstage filtering to reject spurious sidebands. The oscillator incorporated a thermistor/resistor heater network to maintain frequency stability over temperature. All multipliers were transistor multipliers and included internal filtering. The 2.225 GHz amplifier modules were designed to provide a minimum output of 10 dBm and the interstage filters were designed to assure overall filtering of all spurious signals. All active modules of the LO were packaged on one side of the receiver to minimize interference with the sensitive microwave modules on the other side. The LO signal was passed through an isolator, a directional filter, and an attenuator to be selected during initial tests with the downconverter. Table 6 summarizes typical LO performance.

4 GHz Bipolar Transistor Amplifier. Intelsat IVA and the COMSTAR receivers had a bipolar transistor driver amplifier as backup to the driver TWTA. In the MIC receiver a bipolar amplifier chain satisfies the functions of the 4 GHz TDA postmixer amplifier and the driver TWTA or the driver transistor amplifier. The amplifier chain consisted of a low level module and a driver module. The low level module was required to have 14 dB of gain using two single ended stages. The driver amplifier module was required to produce 38 dB of gain and consisted of three single ended stages followed by two hybrid coupled stages.

Power Supply. The power supply, although the simplest group of all, presented its own problems in MIC implementation. Since its transistors and capacitors were large relative to similar components in the RF and microwave circuits, a planar MIC design was not practical. The decision was made to continue with existing miniaturized point-to-point designs with MIC circuitry used in the low level portion of the dc regulator. In addition, the power supply modules would conform to the outline and mounting dimensions applied to the MIC modules. One departure from previous discrete power supply designs was to avoid the previous practice of foaming the entire module after assembly for additional

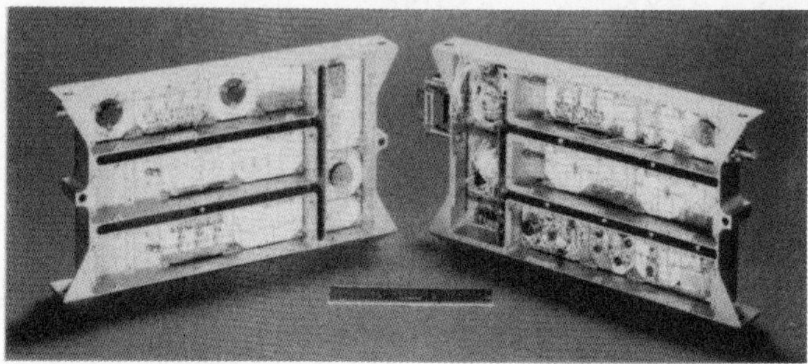

FIGURE 5. DUAL BOX RECEIVER UNIT (PHOTO 76-44844)

mechanical integrity. The final MIC/discrete configuration met all requirements as to regulation, temperature stability, ripple, and vibration.

Receiver Design

Layout. The MIC receiver was packaged in two compartmented boxes as shown in Figure 5. The boxes were bolted together back to back and interconnected through the adjoining walls to form a complete, integrated unit. Individual modules were mounted on raised mounting rails in the pockets in the housing by screws readily accessible to facilitate assembly and removal of individual modules. The low level RF modules were assigned to one of the half-unit boxes and high level RF and dc circuits to the other half-unit box. The layout of the various modules in both boxes is shown in Figure 6.

Interconnections. Since the individual modules were soldered to kovar carriers there is relative motion between the carrier and the aluminum housing as temperature changes. This differential expansion and contraction required that space be allocated between adjacent modules to provide relief against thermal effects. The distance between adjacent modules and the variation under temperature required that all interconnections from substrate to substrate be designed to absorb mechanical motion and to preserve the desired electrical characteristics. Consequently, a series of special microwave interconnects were designed to provide thermal stress relief and low VSWR connections. One of the more successful designs consisted of a shaped ribbon forming a stress relief loop over the gap between substrates and was adopted as baseline. When tested over extreme temperature ranges and vibration, the ribbon interconnects showed such good, stable microwave characteristics that they were also employed as low frequency interconnects. At microwave frequencies, however, an additional degree of freedom

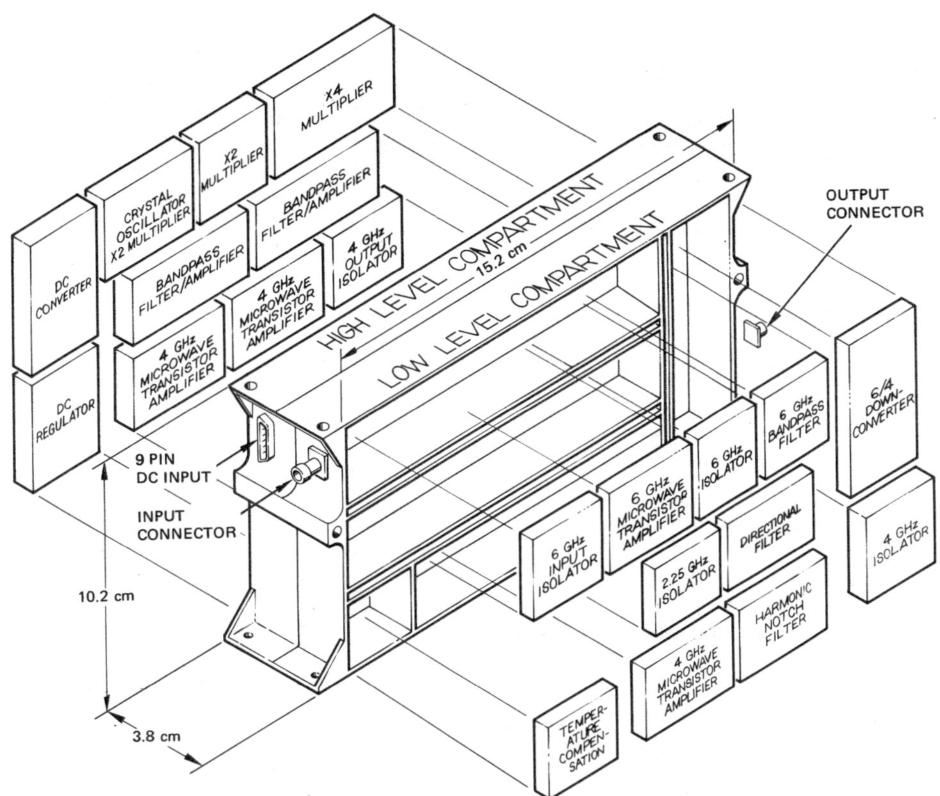

FIGURE 6. COMMUNICATIONS RECEIVER LAYOUT

was provided to facilitate electrical integration. Each substrate was required to have at its input and output a number of tuning pads printed in a cluster on either side of the MIC transmission line to serve as the MIC equivalents of broad wall and narrow wall screw tuners used in waveguide elements.

A more sophisticated interconnect problem was the passing of the LO signal from the high level side to the low level side and the 4 GHz signal in the opposite direction. Various implementations were designed and tested including an all microstrip shielded design. The simplest and most effective feedthrough interconnect was a miniature copper shielded coaxial line with a sufficiently large loop for mechanical and thermal stress relief. Although the losses of these feedthroughs were not small, at 2.225 GHz, the LO frequency, and at 4 GHz, the losses were acceptable.

Mechanical and Thermal Design. Due to the critical nature of mechanical and thermal considerations in an MIC receiver, initial studies for the MIC receiver concentrated on mechanical and thermal design before a single electronic circuit was built. Attention was initially directed to testing the thermal and mechanical characteristics of the substrate to carrier attachment. A test was devised in which all substrate to carrier attachments would be tested in lots of 20 for each size, separated into groups of 10. The first group was subjected to ten cycles of liquid N_2 to boiling water temperature shock and then subjected to 100 g sinusoidal vibration along three axes. The second group was vibration tested first and then thermal shock tested. After initial problems with the design of a proper holding fixture the three standard module sizes of 1 x 1 inch, 1 x 1-1/2 inch, and 1 x 2 inches successfully passed these tests for two different solder systems. Since the final holding fixtures for these test modeled exactly the way each

substrate would be mounted in the receiver, the tests indicated that little trouble would be encountered with thermal and mechanical problems at the receiver level.

As each module design neared completion, time was taken to repeat the thermal/vibration tests with completed electronic modules. Although every module would not be tested in this manner, all LO modules, the downconverter module, and samples of preamplifier and postamplifier modules were successfully tested indicating high thermal and mechanical integrity at the receiver level.

After a third breadboard receiver was integrated and passed all electrical requirements, a series of vibration tests were conducted. These tests were used to isolate mechanical inadequacies and, after appropriate design modifications and repair, were continued until destruction of the unit occurred or the limits of the vibration machine was reached. A failure at 10 g and another at 20 g showed the need for better staking of the power supply transformer. At a higher level, a break in the coaxial line between boxes indicated that a longer service loop and a more secure anchoring near the point of the coax to MIC transition was required. Once these minor flaws were corrected the unit was vibrated to machine limits, a baseplate level of 60 g. In the particular fixture used in vibration, g level amplification factors of up to 5 were present, thus stressing some portions of the MIC receiver to almost 300 g. After several minutes the mounting feet of the receiver housing failed and the test was stopped. Comparison of before and after traces of the receiver frequency response showed no difference, and the receiver was proven mechanically much stronger than anyone had expected. A careful study of the failures at the mounting feet indicated a simple redesign would also prevent such a failure even at such high levels.

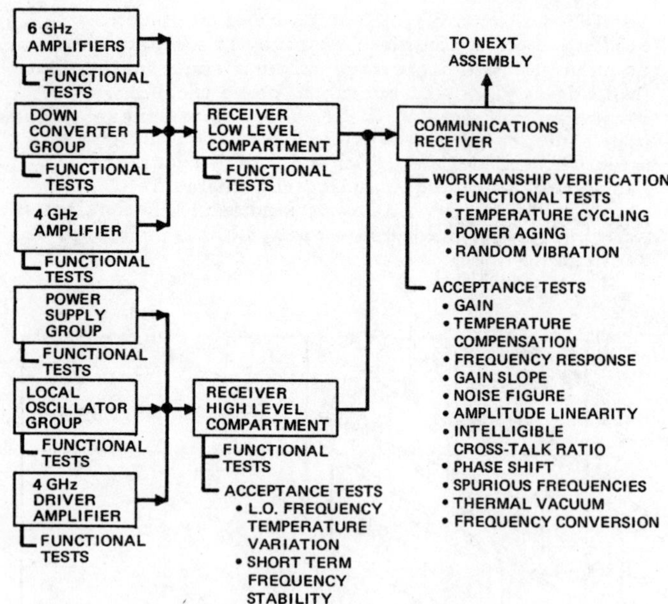

FIGURE 7. RECEIVER INTEGRATION AND TEST

The thermal shock/vibration series conducted at the substrate level demonstrated that the receiver would be excellent from the standpoint of thermal fractures. However, other considerations were involved in the receiver design for temperature rise, temperature control and compensation, and heat sinking to assure low junction temperatures in high current transistors. Of these considerations, minimizing junction temperatures was given top priority. The housing was designed to have additional thermal mass at locations where high level modules such as the 4 GHz hybrid amplifier and the dc regulator were located. In addition, molybdenum carriers were tested as substitutes for Kovar carriers, taking advantage of molybdenum's eight-fold superiority in thermal conductivity. Although molybdenum carriers ultimately were not used they were put through the same shock/vibration series as the Kovar carriers.

At the receiver level, thermal control would or would not be required depending on the overall thermal characteristics of the spacecraft. Additional spaces were reserved in the receiver housing to permit the addition of proportional temperature control circuitry to control receiver temperatures to an acceptable range. A heater blanket located between two half-boxes and controlled by this circuitry was the baseline choice. The receiver mounting technique design thermally isolated the receiver from the shelf, minimizing maximum power demand.

Integration and Test

The assembly and test of a satellite communications receiver must be conducted by a well trained team of engineers and technicians to obtain the performance and reliability required of this unit. Comprehensive tests are conducted throughout the assembly sequence (Figure 7). The functional tests at the various levels are configured according to the device tested and, in addition to the usual electrical testing, may include temperature cycling and random vibration.

After the receiver has been assembled and tuned, it is subjected to RF performance measurements such as input and output voltage standing wave ratio (VSWR), noise figure, passband gain and gain slope, spurious signal generation, and amplitude linearity.

During the temperature cycling and power aging sequence, the gain and passband response is periodically monitored to demonstrate circuit stability and workmanship. In addition, during the final thermal cycle noise figure and gain variation is monitored to show compliance with performance requirements.

Before and after exposure to random vibration and thermal vacuum, the receiver is subjected to RF performance tests. Since the receiver is opened for visual inspection after vibration, RF susceptibility tests are not usually made until the unit is closed for the last time. Pre- and postenvironmental tests normally consist of noise figure, passband gain, spurious output, VSWR, amplitude linearity, phase shift, and FM transfer tests.

For the amplitude linearity tests, the receiver is driven with two equal amplitude carriers and the level of the third-order intermodulation product is measured by comparison with a signal of known amplitude. Carrier frequencies are chosen such that both intermodulation products generally fall within the passband of the receiver. For each set of frequencies, the input power range is varied over approximately a 15 dB range with the maximum total drive level corresponding to at least the maximum total carrier loading of the receiver. The amplitude linearity test, in addition to passband gain and noise figure, is conducted during thermal vacuum exposure.

Total phase shift is measured by using a two carrier AM to PM measurement technique. This method circumvents the problems encountered when single carrier network analyzer measurement methods are attempted on devices that have an input to output frequency translation. With the two carrier technique, one carrier is amplitude modulated while the other is not. The unmodulated carrier is placed in at least three different locations in the passband. The modulated carrier is swept across the passband for each location, and the resulting phase modulation appearing on the unmodulated carrier is detected: a sufficient number of measurements are made at various input levels to permit the plotting of phase shift versus drive.

FM transfer is measured in a manner similar to the two carrier phase shift method mentioned above. For this method both carriers are fixed in frequency and the modulated carrier is frequency modulated rather than amplitude modulated. The measurements are made as a function of input levels at several different sets of carrier frequencies. The selection of the carrier frequencies is based on an analysis of gain slope and phase shift data to ensure the measurement of the worst case crosstalk for the receiver.

The integration and measurement techniques described above have been successfully employed on numerous spacecraft receivers. The performance of the MIC receiver, as measured by these procedures, is shown in Table 3 and Figures 8, 9, and 10. The performance is equal to or better than previous designs.

CONCLUSIONS AND SUMMARY

This chapter has attempted to trace the evolution of the modern satellite communications receiver and has described receiver design, particularizing with the 6/4 GHz MIC receiver.

An indication of the rapid changes in microwave semiconductor technology and the maturation of MICs is that in early 1976 the use of MICs was a reliability concern for spacecraft electronics. Detailed processing of microwave integrated circuitry was a performance problem. Also, concern was expressed for bipolar transistor amplifiers and even greater concern for microwave GaAs FETs for 6 GHz amplifiers.

Today, less than 3 years later, microwave bipolar transistors have become completely acceptable for space electronics applications. However, GaAs FETs have overtaken the bipolars and are replacing them in new microwave amplifiers. Only a few satellite repeaters will be launched with 6 GHz bipolar transistor amplifiers. The detailed processing of microwave integrated circuitry has now matured with a number of well established and tested bonding, attaching, and integration techniques. These techniques allow the receiver designer much more latitude in designing high performance MIC modules at microwave frequencies.

The extremely rapid growth of FET technology has brought with it the promise of even more completely integrated communications receivers. Work at laboratories throughout the world promises multistage microwave amplifiers, complete down-converters, and entire LO chains on one or more semiconductor chips. Designs are also under consideration for a complete receiver in a single chip, an example of a microwave medium scale integration (MSI) device. There appear to be no technology limitations to prevent the development of these higher order integrated receivers. The only criteria for their use will be whether these future receivers can perform to the demanding requirements of a communications repeater and whether such receivers are cost effective solutions to the unique problems of satellite communications systems.

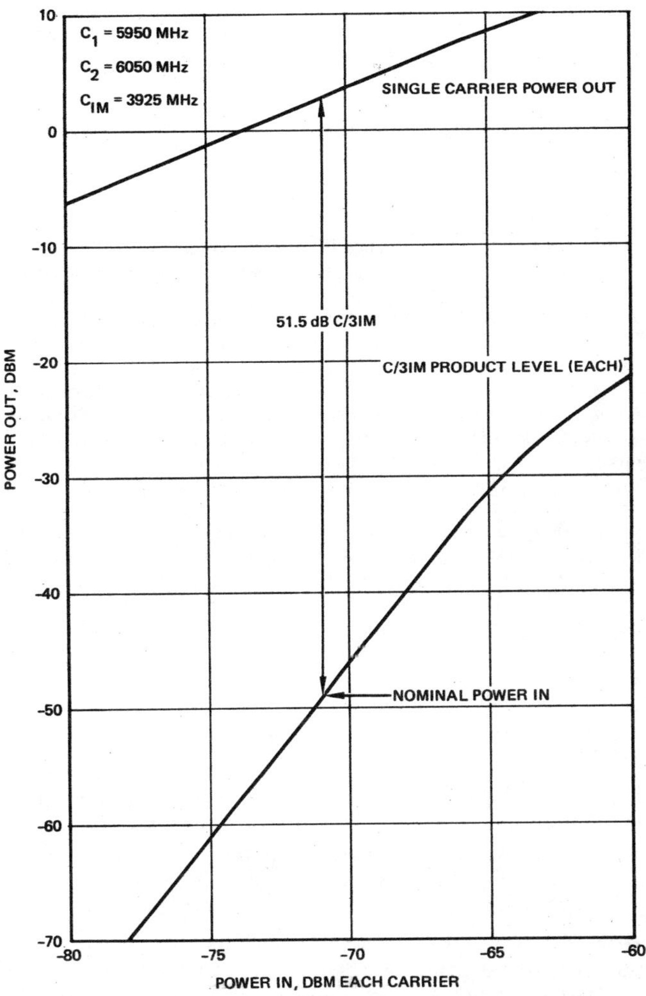

FIGURE 9. AMPLITUDE LINEARITY FOR C/3IM (TWO CARRIER)

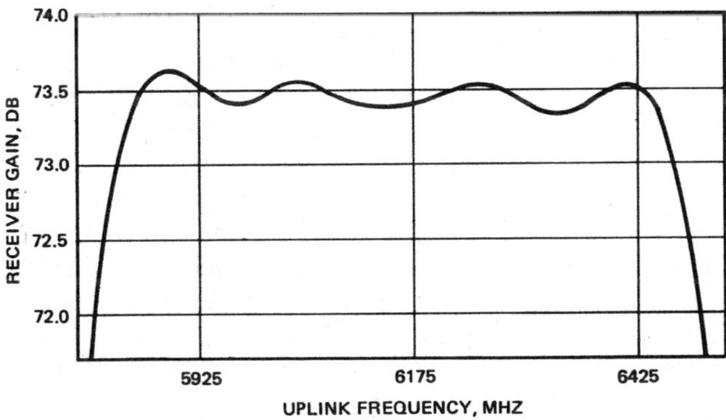

FIGURE 8. 6/4 GHz MIC RECEIVER FREQUENCY RESPONSE

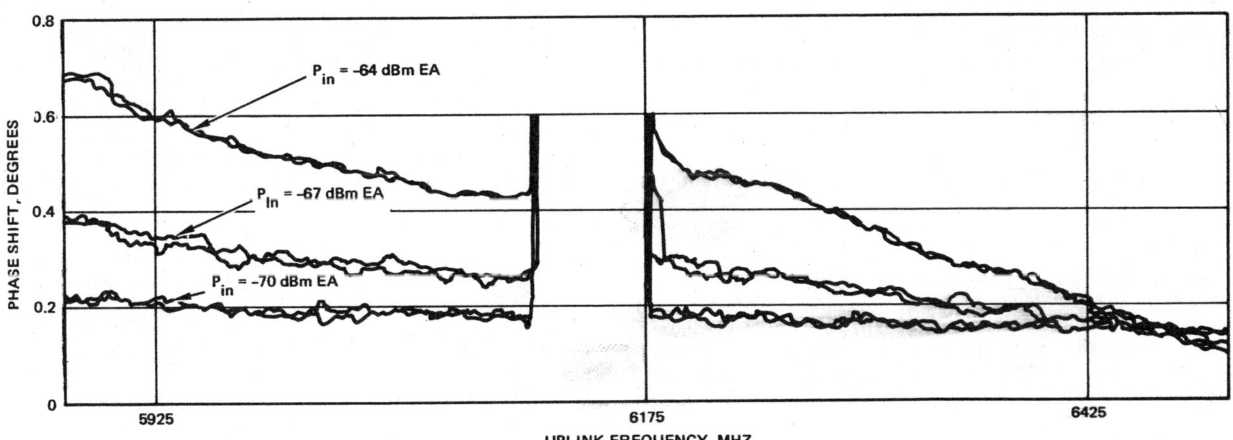

FIGURE 10. 6/4 GHz MIC RECEIVER PHASE SHIFT DATA

Traveling Wave Tubes for Communication Satellites

ROBERT STRAUSS, SENIOR MEMBER, IEEE, JORK BRETTING, AND ROBERT METIVIER

Abstract—Traveling wave tubes (TWT's) have contributed markedly to the development of communications satellites. As the prime-power consuming and transmitting device, the major transponder gain element, and the largest contributor to transmission nonlinearities, the TWT has been the focal point for continuous but carefully measured evolutionary improvements. Efficiency improvements continue to be made without compromising desired communications characteristics or tube lifetimes. These improvements have been made primarily in the RF circuit through loss reduction and phase-velocity tapering techniques, and in the spent-beam region through better multielement collector designs. Traveling wave tubes developed for satellites at 4 and 12 GHz are used as examples.

Since TWT's are life-limited devices, emphasis has been placed on techniques ensuring long life in satellite applications. Both oxide- and dispenser-type cathodes are discussed and data on life characteristics are presented. During the past decade, while generally demonstrating excellent space lifetimes, operating TWT's continue to approach their potential cathode wear-out life, which is theoretically of the order of 10^5 h.

I. INTRODUCTION

SINCE the inception of active communications satellites, traveling wave tubes (TWT's) have played a fundamental role in their development. Typically, the TWT constitutes the major satellite power-consuming element as well as the sole telecommunications microwave power transmitter. It is also the transponder's major amplification element, and it is the largest contributor to transmission nonlinearities. Along with radiation-degradable solar cells, batteries, and station keeping fuel, TWT's remain among the satellite's basic life-limiting items. These features, in addition to the TWT's presently untaxed broad band capability, will continue to focus attention on this device.

The paper, after a brief review, discusses the most important design aspects of current space-type TWT's. First is the maximization of dc-to-RF tube conversion efficiency at a given level of RF output power. The output-power operating point relative to saturation is drive selected, and is based on an acceptable level of intermodulation "noise" contribution to the total noise budget in the transmission path. Therefore, improvements in the tube's linearity which also preserve efficiency are of interest. The second major issue is the TWT lifetime and the techniques used to obtain acceptable life performance. The output power required (typically in the range of 4 to 20 W) determines, in part, the selection of the TWT's long-lived design. To date all *C*-band satellite TWT's have, for example, used oxide-coated cathodes with electron emission densities ranging from 100 to 300 mA/cm^2.

Although other TWT characteristics, such as gain (typically 40 to 70 dB), gain flatness, group delay, VSWR match, noise figure, overdrive response, dissipation, and weight, are important, each in its own way, they do not constitute sources of significant potential improvement at the present high level of device maturity. The major competitive incursion to TWT's comes from the longer lifetime potential and reduced weight of solid-state amplifiers. An indication of this trend is the use of solid-state driver amplifiers (3.7 to 4.2 GHz, 40-dB gain, $\sim$1-W saturation power) to "back up" or replace driver TWT's. The recently launched SATCOM and COMSTAR satellites, for example, carried bipolar, transistor driver amplifiers. Future satellites are gradually expected to have ever greater complements of solid-state microwave devices as advances in power handling, combining, and efficiency occur.

II. BACKGROUND

In the early 1960's, when the first experimental active communications satellites (TELSTAR, RELAY, and SYNCOM) were launched, TWT's provided the RF output amplification. By that time, the TWT (invented 20 years earlier by Kompfner [1]) was already an established device. It had been mathematically characterized by Pierce in 1950 [2], and by the mid-1950's "line-of-sight" radio relay stations began using TWT's as transmitter output amplifiers [3], [4]. These early TWT's typically housed the electron gun, the helix slow-wave structure, and the beam collector in a glass envelope surrounded by a beam-focusing solenoid. This format, although usable in maintainable ground applications, was not attractive for weight-constrained space applications.

Manuscript received July 20, 1976; revised October 14, 1976. This paper is based upon work done in COMSAT Laboratories, Clarksburg, MD; AEG-Telefunken, Ulm, Federal Republic of Germany; and Thomson-CSF, Velizy, France. It was supported in part by COMSAT Corporation and (INTELSAT) Organization. Views expressed in this paper are not necessarily those of INTELSAT or COMSAT.

R. Strauss is with COSMAT Laboratories, Clarksburg, MD 20734.

J. Bretting is with AEG-Telefunken, Fachbereich Rohren, Soeflinger Strasse 100, 79 Ulm (Donau), Federal Republic of Germany.

Robert Metivier is with Thomson-CSF, Tubes Electronics, 2, Rue Latecoere, 78140 Velizy-Villacoublay, France.

Reprinted from *Proc. IEEE,* vol. 65, pp. 387–400, Mar. 1977.

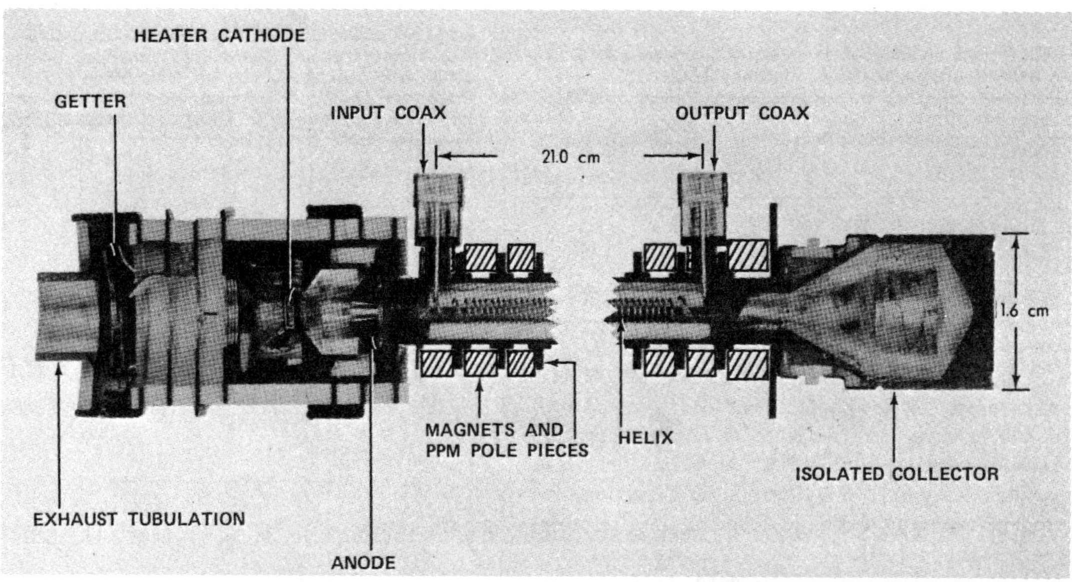

Fig. 1. Section model of 261H INTELSAT IV transmitter TWT single-isolated-collector design (courtesy of the Hughes Aircraft Electron Dynamics Division, Torrance, CA).

TABLE I
SUMMARY OF *C*-BAND TWT SPACE EXPERIENCE (THROUGH MARCH 1976)

Spacecraft and Launch Period	Tube Type[a]	Band-width (MHz)	Saturated Output. Power (W)	Gain (dB)	Nominal Overall Efficiency[b] (%)	Cathode Loading[c] (mA/cm^2)	Space Operation No.[d]	Total (~hr)
Early Bird (1965)	215H	35	6.0	41	36	180	2	56,000
INTELSAT II (1966–67)	226H	130	0.06[e]	40	3	110	8	200,000
	215H	130	6.0	41	36	180	16[f]	200,000
INTELSAT III (1968–69)	233H	225.	0.15[e]	47	10	130	10	416,000
	235	225	12.0	42	33	140	10	416,000
INTELSAT IV (1971–75)	262H	500	1.5[e]	36	15	280	26	180,000
	272H	500	1.5[e]	36	15	200	2	8,000
	261H	36	6.0	58	30	190	168	2,300,000
TELESAT/WESTAR (1972–75)	276H	500	0.4[e]	30	12	150	10	95,000
	275H	36	5.0	55	36	180	60	1,100,000
ATS-6 (1974)	233HC	30	0.2[e]	52	10	140	2	15,000
	235HD	700/ 400	12.0	42	33	140	2[f]	15,000
SYMPHONY	TL 4003	90	13.0	46	34	120	4	15,000
INTELSAT IV-A (1975–)	276H	500	0.4[e] [g]	30	12	150	12	7,000
	271H	36	6.0	58	30	190	20	32,000
	275HA	36	5.0	55	36	180	44	128,000
SATCOM (1975–)	296H	36	5.0	55	36	180	48	60,000
MARISAT (1976 –)	275H	1	5.0	55	36	180	2	1,000
TOTAL:							446	5,240,000

[a]H = Hughes Aircraft Co., Electron Dynamics Div., Torrance, CA; TL = AEG/Telefunken, Ulm, West Germany.
[b]Single-collector TWTs.
[c]Oxide cathode.
[d]Most spacecraft use 2-for-1 redundancy.
[e]Driver TWTs.
[f]Phase combined.
[g]Solid state amplifier back up

Significant technology improvements continued to be made. The major developments included weight and power saving electron beam-focusing techniques such as periodic permanent-magnet (PPM) structures [5] and the bundling of precision helix slow-wave structures with ceramic support rods selec-tively coated with a virtually reflectionless, high-attenuation, and stable pyrolytically deposited carbon attenuator. High-temperature processing, made possible by the adaption of metal/ceramic vacuum seals, contributed to a "cleaner" and better controlled vacuum within the tube envelope. This pro-

TABLE II
CHARACTERISTICS OF THE 291H HELIX TWT

Characteristics at 1.5 GHz	Mode (anode controlled)		
	1	2	3
Saturated Power Output: P_2 (W)	9	30	65
Saturated Gain: G (dB)	25	37	54
Nominal Overall Efficiency: η_{ov} (%)	35	52	48
Cathode Loading: (mA/cm^2)	70	130	250

vided a more conducive environment for elevating the oxide-cathode current-density levels by a factor of 20 (from approximately 10 to 200 mA/cm^2) relative to those of the necessarily ultraconservative submarine cable tubes [6] having projected lives of more than 25 years. TWT's utilizing these features were fabricated in the early sixties in time for the first communications satellites [7]–[10].

The basic aspect and design of TWT's have, by intent, remained virtually unchanged during the past decade. For example, at *C*-band over 500 tubes are now in space and several hundred more are scheduled for launch during the next few years. Total communications TWT space operating hours are well in excess of 6 million with at least an additional 3 million hours of "burn-in" screening and ground life testing. The basic tube parameters are listed in Table I. A representative design is the 261H (Fig. 1), which, as the 6.0-W INTELSAT IV output amplifier [11], has now logged in excess of 40 000 h each on 11 of 12 channels in the F2 spacecraft launched in January 1971.

In addition, during the past decade considerable tube development has occurred at frequencies other than *C*-band. A noteworthy development is a narrowband helix TWT at 1.5 GHz for the MARISAT program; this tube (291H) is a 3-mode device whose characteristics are given in Table II. Its high efficiency is made possible by a high-interaction-impedance helix design, helix resynchronization, and a 2-depressed-electrode, 3-stage collector [12].

At 12 GHz unusual improvements have also occurred. Two tube types, the Th 3535 and the TL 12022, have been developed in Europe for use on the OTS satellite. Both tubes achieve efficiencies in excess of 40 percent with double depressed-collector designs at power-output levels of approximately 20 W. Variations of these tubes will be described later in this paper to illustrate developments which represent advanced approaches.

III. TWT EFFICIENCY

Efficiency is the ratio of the RF power at the fundamental frequency (or frequencies) into a matched load and the electrical power absorbed by the TWT. The following will refer to TWT's utilizing an electron gun with a thermionic cathode that requires filament power ($U_f I_f$). The cathode current (I_k) is controlled by a voltage (U_{g2}) applied to the nonintercepting electrode grid-2 in the gun. As shown in Fig. 2, the grid is followed by a helix delay line for medium-power TWT's operated at voltage U_h, a collector assembly using one to three different stages operated at $U_{c1} \cdots {}_3$, and a periodic permanent magnet beam-focusing system. There are six different voltages which may be varied independently. Each set of voltages together with the RF drive (P_1) applied to the TWT input determines the distribution of current I_k on the delay line and the collector stages.

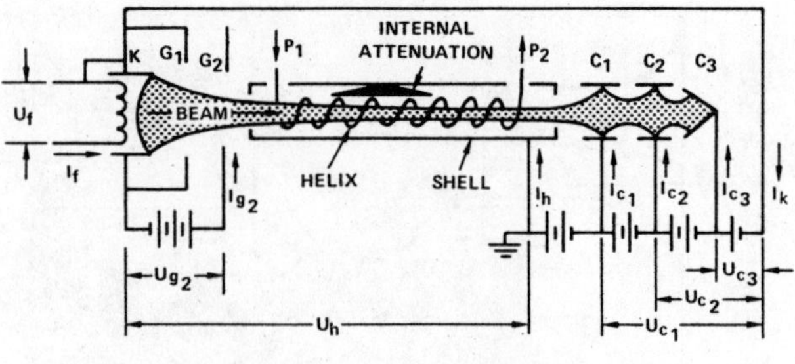

Fig. 2. Three-electrode collector TWT schematic.

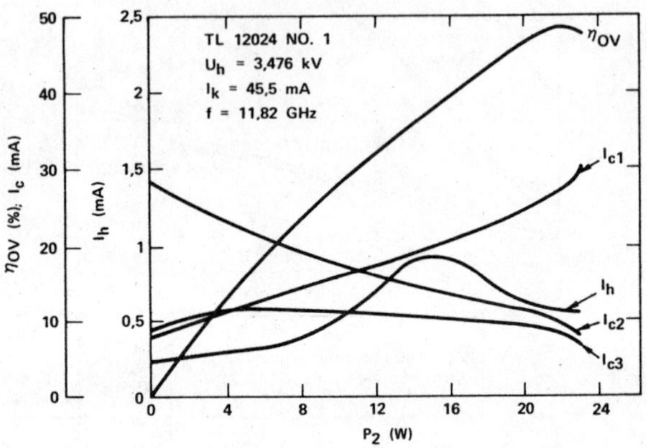

Fig. 3. Overall efficiency of the 3-electrode collector satellite TWT TL 12024 versus output power (uniform helix) showing level-dependent distribution of the beam current on the three collectors and the helix.

As an example, Fig. 3 shows for a 3-electrode collector, the output power and current distribution versus RF input power P_1 at fixed helix and collector voltages. The voltages applied to the TWT are generated by a power supply; by design, each terminal of this supply is characterized by a specific voltage and internal resistance. As a result of varying currents, TWT voltages also depend on RF drive (P_1). The overall efficiency η_{ov} for any value P_1 at a frequency (f) is

$$\eta_{ov} = \frac{GP_1(f)}{U_f I_f + U_h I_h + \sum_1^3 U_{cv} I_{cv}}. \tag{1}$$

[$U_{g2} I_{g2}$ does not appear in the denominator since it is usually negligible relative to the remaining terms, and G is the amplifier power gain at $P_1(f)$.] The largest value of η_{ov} is obtained at saturation drive. Of all possible values of U_{cv} there exists a maximum efficiency, η_{max}, for a fixed helix voltage. Further, as a function of helix voltage there is a maximum value of η_{max}, i.e., the optimum efficiency (η_{opt}) of the TWT.

The TWT efficiency depends on two separate processes.

a) During the interaction processes, part of the kinetic electron energy is converted to RF energy at the fundamental frequency (or frequencies), at its harmonics, and at frequencies corresponding to intermodulation products in the case of multicarrier operation. Part of the RF energy is transformed into heat as a result of RF losses on the delay line, in the internal attenuator, and in the RF matching section. The electronic efficiency (η_{el}) is the ratio of the total generated RF power to the beam power $U_h I_k$, whereas the beam efficiency

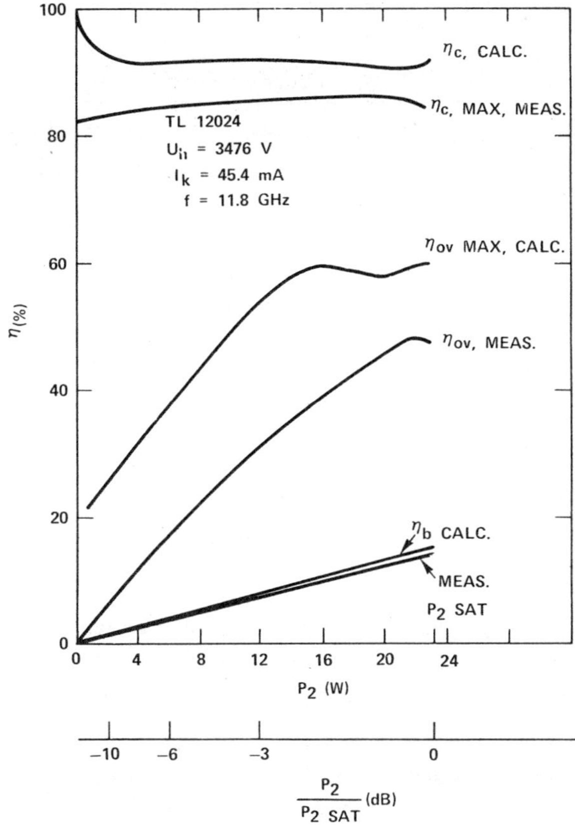

Fig. 4. Measured and calculated values of beam efficiency, overall efficiency, and collector efficiency versus output power for the 3-collector TWT TL 12024.

(η_b) is the ratio of the output power P_2 at the fundamental frequency (or frequencies) to the beam power.

b) Efficiency is improved by means of a multistage collector configuration. In this process, electrons are decelerated to reduce the heat generated by the electrons landing on the collector, sometimes referred to as "soft landing" electrons. The extent to which this deceleration is achieved is measured by the collector efficiency (η_c)

$$\eta_c = \frac{P_{\text{spent beam}} - P_{\text{dissipated}}}{P_{\text{spent beam}}}.$$

It can be shown that

$$P_{\text{spent beam}} = U_h I_k (1 - \eta_{el})$$

$$P_{\text{dissipated}} = \sum U_{cv} I_{cv} - [U_h I_k \eta_{el}]$$

and

$$\eta_c = \frac{1 - (1/U_h I_k) \left[\sum U_{cv} I_{cv} \right]}{1 - \eta_{el}}. \tag{2}$$

It is possible to calculate η_c from the electron-velocity spectrum of the spent beam and the applicable collector configuration. Difficulties arise when η_c must be evaluated from measurements. In this case, direct thermal measurement of collector loss and spent-beam power or a reasonable estimate of η_{el} from the measurement of η_b is required. Calculated values from a computer program derived by Bobisch [13] are shown in Fig. 4 as an example of the variation of these efficiencies for a specific TWT design.

A. Nonlinearities

Since TWT's are used as power amplifiers, it is desirable that they be operated at or close to saturation. In the transmission of broadband information-carrying signals at or near saturation, limitations for acceptable distortion are imposed. These limitations result from systems considerations such as the kind of modulation utilized, e.g., single-carrier FM, multicarrier FM, FDMA, or TDMA. Hence, system requirements are transformed into tube requirements for phase nonlinearities and carrier intermodulation product ratios.

Phase nonlinearities are measured by the total nonlinear phase shift between zero drive and TWT saturation or its derivatives, such as AM/PM conversion and AM/PM transfer coefficients [14]. The actual efficiency, limited by nonlinearities, is defined as the operating efficiency (η_{op})

$$\eta_{op} = \frac{GP_1 \, (f, \text{nonlinearities} \leqq \text{limit})}{U_f I_f + U_h I_h + \sum U_{cv} I_{cv}}. \tag{3}$$

This operating efficiency is obtained by varying the operating parameters (U_h, P_1, U_{cv}) of the TWT while simultaneously requiring that nonlinearities do not exceed the established limits. It is obtained by plotting efficiency and the required nonlinearity bound as two sets of curves versus power (output or input) and eliminating power to obtain one set of curves relating efficiency to nonlinearity. The envelope of this set of curves relates the operating efficiency (η_{op}) to the desired values of the nonlinearities.

Fig. 5 shows the envelope curve for a typical sample of a single-collector TWT developed in 1968 and the equivalent curve for a triple-collector TWT recently developed. Reasonable values of AM/PM conversion coefficient (Kp) are indicated. The comparative benefit of multicollector tubes is significant. Similar curves can be obtained for η_{op} versus total phase shift or 2-carrier intermodulation.

IV. TWT Design

The question now arises as to how to design a TWT for the best operating efficiency. As indicated, this problem may be separated into two parts, namely, the beam/slow-wave interaction process (beam efficiency) and the energy-recovery process (collector efficiency). Depending on the operating frequency, the bandwidth required, and the output-power generating point, a choice of electron-gun design, helix/barrel, PPM structure, and collector(s) is made.

For space TWT's, the helix has been by far the most suitable slow-wave interaction structure. In practice, the helix radial propagation constant

$$\gamma a = \frac{2\pi f}{v_p} a \tag{4}$$

where a is helix radius (in m), f is midband operating frequency (in Hz), v_p is circuit phase velocity (in m/s) = 6×10^5 $(U_h)^{1/2}$ (in V), is typically selected between 1.0 and 1.7 rad at midband with γa of $\sim$1.35 nearly optimum [15]. For fixed-beam characteristics, a lower γa choice will result in a higher interaction impedance [12] and beam efficiency with a more nearly constant RF field across the enclosed beam. A higher γa choice is more suitable for broadband uniform operation with less harmonic content and a lower beam efficiency. Beam efficiencies may range from circa 10 to 25 percent. Computer calculations [16] show that there is generally a

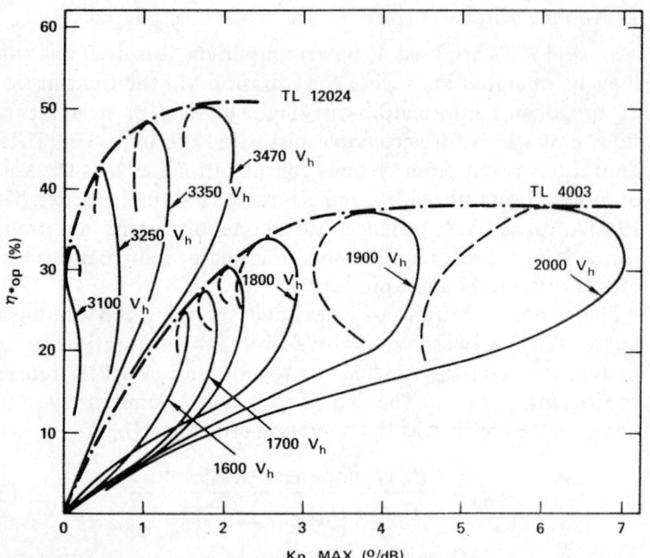

Fig. 5. Operating efficiency versus maximum tolerable AM/PM Conversion coefficient (Kp) for TL 4003 (single collector) and TL 12024 (three collectors) with helix voltage as a parameter [(η_{op}^*)] excludes filament power for comparative purposes. The 4-GHz TL 4003 uses an oxide cathode which requires slightly less power than the dispenser cathode used in the 12-GHz TL 12024).

TABLE III

BEAM EFFICIENCY, OVERALL EFFICIENCY, AND COLLECTOR EFFICIENCY FOR THREE COLLECTOR TWT'S

Efficiency	TL 12023[a]	TL 4012[b]	TL 12023[c]
η_b (at saturation)	13	15	21
η_{ov} (at saturation)	45	45	44
η_c	85	82	71

[a]11 GHz, 20 W, three collectors, uniform helix.
[b]4 GHz, 12 W, three collectors, uniform helix.
[c]11GHz, 20W, three collectors, tapered helix.

slight reduction of theoretical operating efficiency with increasing beam efficiency. Higher beam efficiencies result in a greater electron-velocity spectrum in the spent beam and make the attainment of higher collector efficiencies more difficult. In other words, additional multicollector stages are necessary to capitalize on high beam efficiency so that optimum overall efficiency, as defined in (3), can be attained. This trend is indicated in Table III for 12- to 20-W helix TWT's.

V. HELIX/GUN DESIGN TECHNIQUES

The choice of helix dimensions is fundamental and, as will be shown, represents a compromise between conflicting requirements. It is generally accepted that the interaction efficiency of a uniform helix TWT is proportional to Pierce's parameter C [2] as follows:

$$\eta_b \sim C = \left[\frac{I K}{4 U_h} \right]^{1/3} \qquad (5)$$

where I is the beam current, U_h the helix voltage, and K the beam-to-helix coupling impedance. Hence, the tube should be operated at high beam current and low voltage for greater efficiency and high gain per unit length. Also, the coupling impedance must obviously be as high as possible.

A fundamental consideration in the choice of the operating voltage and in turn the current is the ability to adequately focus the electron beam. The current intercepted by the helix, disregarding thermal and reliability considerations for the moment, heavily penalizes the overall efficiency. For example, an increase of 1 mA in a TWT delivering 20 W at 11 GHz, with a cathode current of 40 mA and an efficiency of 40 percent, will raise the power use by 2.2 W and drop the overall efficiency by 1.7 percentage points.

There is an approximate limit to the maximum value of the perveance of a PPM focused electron beam

$$P_{\mu max} \sim 0.15 \left(\frac{B}{1.8} \right)^2 (\gamma b)^2 \lambda^2 \qquad (6)$$

where P_μ is beam perveance $= (I/U_h^{3/2})$, I is cathode (beam) current (in A), U_h is helix voltage (in V), B is peak PPM axial magnetic field (in T), b is mean beam radius (in m), λ is wavelength in vacuum (m), and

$$\gamma b \sim \frac{\gamma a}{2}.$$

This limiting equation is derived from the Brillouin field beam focusing condition, modified for PPM focusing and combined with (4).

Efficiency considerations lead to a value of γb of the order of 0.5 to 0.6 rad. For $B = 0.2$ T, corresponding to the maximum field obtainable from Alnico 8 (permanent magnet) material, the following result is obtained at 11 GHz:

$$0.3 \times 10^{-6} \lesssim P_{\mu max} \lesssim 0.5 \times 10^{-6}.$$

With an interaction efficiency of 15 percent, a 20-W tube, for example, would operate at 2900 V and 46 mA of beam current. The winding diameter of the helix would be about 0.9 mm (0.035 in). This limiting perveance value can now be extended by using cobalt/rare-earth magnets. Peak magnetic field values of 0.3 T are attainable. It should be noted that in the case chosen, the magnetic focusing field does not depend on the power, but only on the frequency and the perveance at a given γb.

The choice of the helix diameter, and thus (as will be shown later) the TWT helix voltage, is mainly governed by the quality of the electron gun, as measured by the cathode current, the emission homogeneity, the beam convergence, and the laminarity. For a potential operating life of over 100 000 h, the cathode current density should probably not exceed ~ 200 mA/cm^2 with an oxide cathode, or ~ 0.8 A/cm^2 with an impregnated cathode ("dispenser" type). A high-interaction-efficiency 20-W 11-GHz TWT requires a beam of about 40 mA, which would in turn require a beam convergence greater than 100 for an oxide cathode. Such a gun would be difficult to produce in a controlled and reproducible manner. The dispenser cathode therefore becomes the logical choice at this frequency and power level. The necessary convergence is reduced to about 25 for a cathode load of 800 mA/cm^2 and a beam diameter (b) of 0.5 mm (0.020 in).

This combination makes it possible to design a gun having good laminarity at the "crossover" (transition from electrostatic to magnetic focusing) and a constant emission density (to within 10 percent) over the cathode surface. The electron gun is designed with the aid of a computer program that solves Poisson's equation in the presence of a magnetic field. An important point is the effect of the radial velocities of the elec-

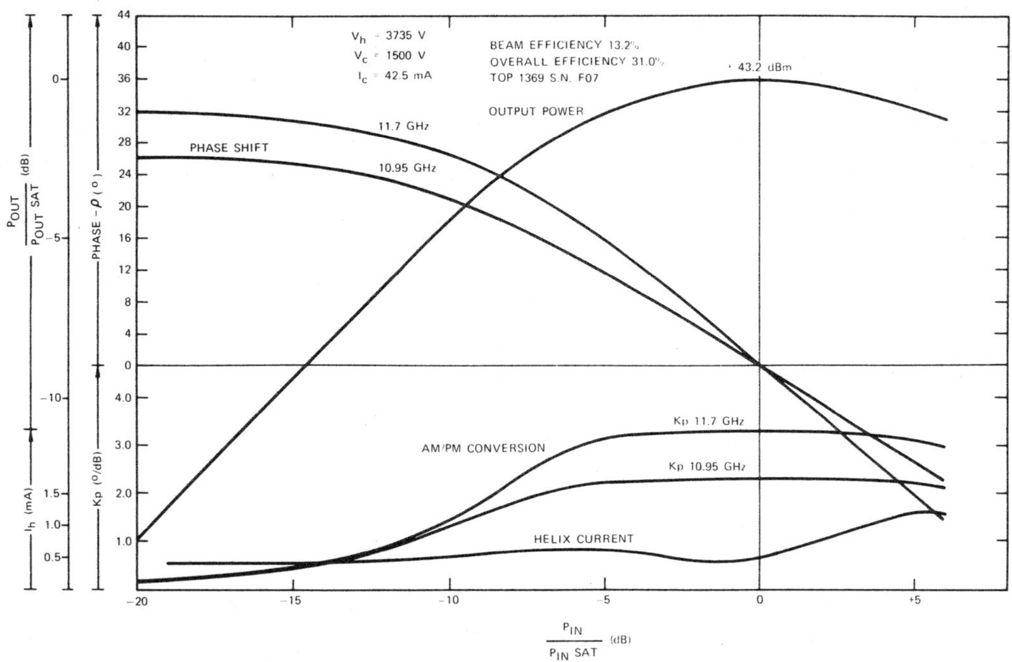

Fig. 6. 20-W, 11-GHz constant pitch helix, single-collector TWT (TOP 1369).

trons emitted by the "hot" cathode [17]. The electrons are emitted from the ~1300-K temperature cathode surface with velocity components in all directions, following Lambert's law and with a statistical averaged energy given by

$$U = kT/e = 0.113 \text{ eV}$$

where T is the absolute cathode temperature, e the electron charge (1.60×10^{-19} C), k is Boltzmann's constant (1.38×10^{-23} J/K), and U is the accelerating potential expressed in volts.

Statistical analysis shows that several percent of the electrons can be emitted with a radial velocity equal to or greater than $\frac{1}{2}$ V. An idea of the effect of such a velocity component on the beam diameter at crossover (where the beam comes under the control of the magnetic field) can be obtained. Assume that, during the time when the electrons travel from the cathode to the crossover point, they are moving radially with an energy of 0.5 eV. Computation gives a deflection of 0.25 mm for a cathode-to-crossover distance of 1 cm and an accelerating potential of 3700 V. Hence, to obtain better than 99-percent beam transmission, the helix diameter must be at least 0.5 mm more than the computed beam diameter. Nevertheless, the homogenous theoretical beam diameter, taken into consideration when computing the interaction with the helix, remains near that originally computed; i.e., it is increased by only about 10 percent, due to electron spreading and PPM beam ripple.

Pierce's theory holds that flat TWT gain is obtained when the radial propagation constant γa (as indicated previously) has a value of approximately 1.5 rad. For tubes with relatively narrow instantaneous bandwidths (about 10 percent or less), as used in communications satellites, this value is too high for optimum bean efficiency. Experience, confirming the computations, has shown that it is possible to obtain a flat gain in a 1-GHz band centered on 11 GHz, with a slow-wave structure consisting of a uniform pitch helix and $\gamma a = 1.2$. The increase in the coupling impedance (K) from this decreased diameter with respect to Pierce's optimum value is of the order of 50

percent, for a constant ratio of beam diameter to helix diameter. The result is a substantial increase in the interaction efficiency.

The specific operating (helix) voltage remains to be chosen. The helix voltage of a TWT delivering, for example, 20W at 11 GHz should be between 3 and 4 kV, which corresponds to a 15-percent helix/diameter variation, whereas the perveance varies by 50 percent, from 0.27 to 0.13×10^{-6}. The variation in Pierce's parameter C and in turn beam efficiency corresponding to this large change in perveance is only 20 percent. However, the final operating voltage within this range will be conditioned by the quality of the electron gun that can be fabricated so that the excellent beam transmission through the smallest possible helix diameter is reliably achievable.

Fig. 6 shows an example of a 20-W TWT at 11 GHz. This tube uses a constant-pitch helix (pitch approximately 450 μm). The inner-helix diameter is 1.1 mm (0.043 in) for an operating voltage of 3.75 kV ($\gamma a \simeq 1.2$). The helix is a tungsten tape of 0.1- $\times$ 0.2-mm (0.004-in $\times$ 0.009-in) cross section. The overall efficiency is typically 31 percent with a single-stage depressed collector, and the beam efficiency is near 13 percent with a maximum small signal to saturation phase shift of less than $35°$.

Improvements in coupling impedance (K) were obtained by confining more of the electrical field within the cylindrical space inside the helix. For example, with the same inner helix diameter, the use of a "flat" wire (tape) instead of a "round" wire increases the coupling impedance by 15 to 20 percent, and at the same time improves the thermal contact with the dielectric helix supports. Winding a tape tungsten helix with good accuracy and reproducibility is mechanically more difficult than with round wire.

The helix is supported by three ceramic rods, usually in a triangulated thin-wall metallic shell. The effect of this bundling is to "load" the helix, causing a reduction in phase velocity and interaction impedance due to the effective dielectric constant plus circuit losses from the dielectric losses. In addition, high rod thermal conductivity and contact pressure are necessary to provide a suitable heat path for the heat generated on the helix. The choice of rod materials which have suitable mechanical, thermal, and dielectric properties is limited. They are as follows.

a) *Alumina* (Al_2O_3) has a high dielectric constant, about 9.3, and a fair thermal conductivity, about 0.3 W/cm/°C.

b) *Beryllia* (BeO) has a lower dielectric constant, about 6.3, but excellent thermal conductivity, 2 W/cm/°C.

c) *Boron nitride* (BN), in the anisotropic form, has a plane dielectric constant of 5.3 and good thermal conductivity, 0.7 W/cm/°C.

d) *Isotropic boron nitride* (IBN) has the lowest dielectric constant (3.0) and a fair thermal conductivity of 0.21 W/cm/°C.

As an example, the tube shown in Fig. 6 utilizes IBN support rods and a relatively high midband impedance (70 ohms) is retained. Simultaneously, a good thermal transfer characteristic is confirmed by the temperature gradient between the helix and the envelope, which is less than 30°C/dissipated W/cm of helix length. Under these conditions, the tungsten helix temperature is controlled during operation, and the RF losses are minimized.

It would, of course, be possible to reduce these losses even more by making the helix from a metal having better electrical conductivity than tungsten (copper, for example), or by coating the output section tape with tungsten and ground shell with copper (using sputtering or sintered electrolytic deposition). These two techniques are usable, but the resulting efficiency gain of one to two percentage points must be carefully evaluated against the chance of vaporizing copper due to local transient beam interception.

The influence of the gain of the last helix section on the efficiency was studied. It was found that the efficiency is nearly independent of this gain as long as it exceeds 30 dB. Below that gain figure, the beam efficiency falls off very sharply at the same time that the phase characteristics become degraded. Two percentage points are lost from the beam efficiency at a gain of 25 dB and four points at 23 dB. Simultaneously, the phase shift increases by 6° and 12°, respectively.

VI. The "Tapered" Helix

The energy conservation law indicates that, in a TWT delivering 20 W with a 40-mA electron beam current, the electrons have lost an average of 500 eV of their energy. This means that near the RF output, where the electrons have begun to give up a considerable fraction of their energy and velocity, the RF wave and the beam are no longer well synchronized.

A recognized method for increasing the efficiency consists of progressively changing (tapering) the helix pitch toward the RF output so that the best possible synchronism is maintained between the RF wave and the electron beam. The reduction in the helix pitch must be concentrated in a short section of helix because the beam decelerates over a distance corresponding to only the last few wavelengths of the slow wave. As an example of this technique, a tube with a 1-cm-long "taper" (linear pitch reduction) helix in the output section was developed. The interaction efficiency was increased to 15 percent, for an increase of about two percentage points with a maximum pitch reduction of only 8 percent. The small signal to saturation phase shift increases with frequency, from 29° at 10.9 GHz to 36° at 11.8 GHz. The AM/PM conversion coefficient did not exceed 3.2°/dB.

TABLE IV
PERFORMANCE CHARACTERISTICS OF SINGLE-COLLECTOR Th 3512
[S/N 29, f = 11.35 GHz, Ik = 37 mA (CONSTANT)]

U_h (V)	Saturation Phase Shift Delay (degrees)	Beam Efficiency, η_b (percent)	Overall Efficiency, η_{ov} (percent)
3930	46	16.4	38
3830	39	15.7	36.5
3730	32	14.8	34.5
3630	25	13.7	32.6
3530	19	12.3	31.5

An important TWT characteristic is the "overvoltage effect" which relates efficiency to phase delay. If the helix voltage is varied while the beam current is held constant, the output power increases with the beam power (and thus with the helix voltage), but the efficiency also increases. Unfortunately, the small signal to saturation phase shift, and thus the AM/PM conversion coefficient, increases also. Under these conditions, the helix voltage operating point must be chosen as a function of the phase shift specification, as shown in Table IV.

A corresponding effect is observed when the helix pitch is reduced. This occurs because a helix pitch reduction amounts to an effective increase in the difference between the beam velocity and the RF wave phase velocity. This increase is the equivalent of an increase in the helix voltage in the "large signal" region of the tube. Therefore, the effects of increased nonlinearity with increased beam efficiency due to "overvoltage" and "phase helix tapering" are analogous [18].

VII. The Double-Taper Helix

A new procedure has been developed to at least partially overcome the problem of increased nonlinearity with increased beam efficiency. The initial idea came from observing the operation of a TWT with low helix voltage (undervoltage). For example, the TH 3512 normally operates with 3730 V on the helix, but if that value is lowered to 3330 V, an interesting result is obtained. The interaction efficiency is obviously reduced somewhat (to about 8 percent for an output power of 10 W at saturation), but the overall phase shift from small signal operation to saturation is zero, and actually takes on positive values below saturation. This is explained by the fact that under these conditions the slow-wave structure is coupled to the fast space charge wave of the beam, whereas in normal operation coupling is only to the slow space-charge wave [19].

Thus a helix section of increased pitch would have the same effect on the phase as a reduction of the helix voltage and the result could be made to partially compensate for the phase shift caused by the deceleration of the beam as it surrenders energy to the RF field. The "increased pitch" section would have to be removed from the output to avoid detracting from the efficiency, but it would also have to be placed where the signal is already large, since the phase shift results from nonlinearity effects. The section of increased phase velocity would, in turn, be followed by a reduced velocity section, nearer the output, to retain the efficiency. Of course, for acceptance this "double-taper" concept would have to produce an overall increase in the efficiency and linearity.

The concept of the double-taper, as shown in Fig. 7, divides the helix into five distinct sections.

a) In the first section, the fixed helix pitch corresponds to

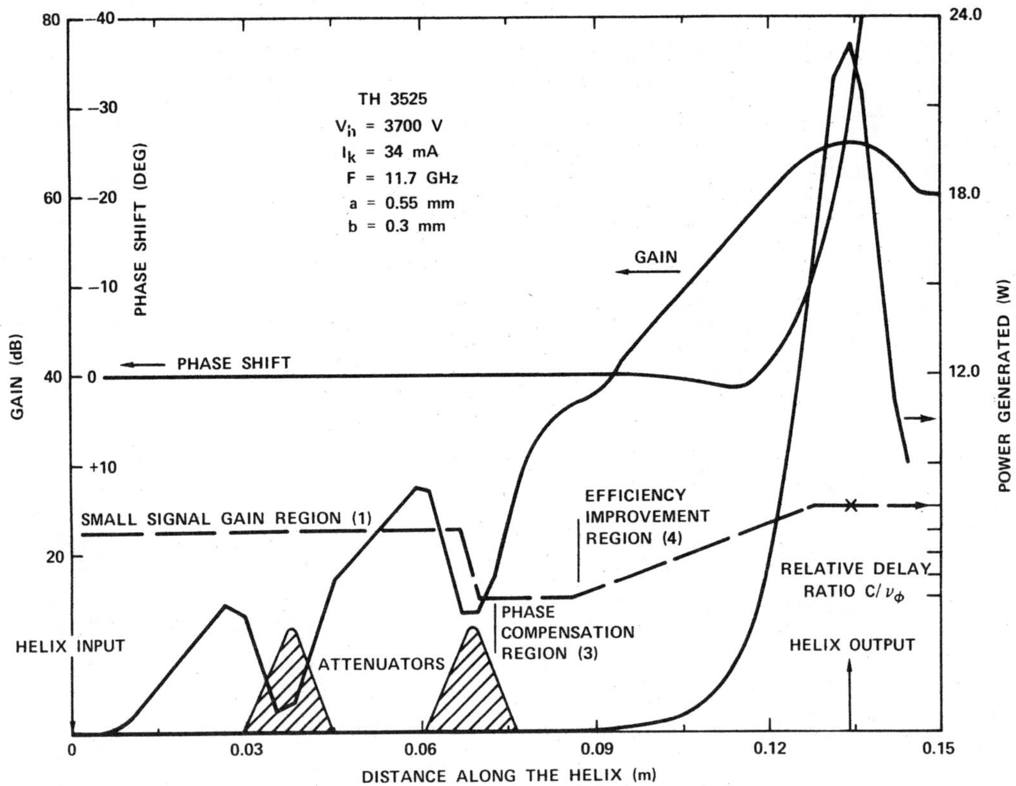

Fig. 7. Double-taper helix profile and computation sample, large signal analysis.

the maximum small signal gain for the selected operating voltage.

b) The next section is the transition zone (under the attenuator) with a fast change in the delay ratio (C/V_p), where C is the speed of light and v_p is the circuit phase velocity.

c) The third section is the zone of positive phase compensation whose depth and length ("hole") define the greater or lesser coupling to the fast space-charge wave, and thus the amount of phase compensation introduced.

d) In the fourth section, the delay-ratio slope is constant. In other words, the helix pitch decreases linearly versus distance from the zone of phase compensation to nearly the end of the helix. At the beginning of this section the wave carried by the helix and the slow space-charge wave carried by the electron beam are relatively far from synchronism, resulting in a weaker interaction. Interaction becomes stronger when moving farther along the helix. As a result, the beam bunching (RF component of the beam current) is progressively greater for a smaller velocity spread within the electron beam. Consequently, both interaction and depressed collector efficiencies are considerably enhanced. The bunching or "phase focusing" is optimized when the slope of the taper chosen provides a constant phase difference between RF beam current and helix RF voltage all along the variable-pitch helix region.

e) Finally, the very short fifth section of constant delay ratio makes it possible to attain maximum taper effectiveness (2-3 wavelengths). The choice of the pitch for this last section is conditioned by the usual efficiency versus phase shift compromise.

The main performance characteristics of the TWT using this helix are given in Fig. 8. Comparison of Figs. 6 and 8 indicates that the beam efficiency of the double-taper TWT is 17.2 percent versus 13.2 percent for the constant-pitch helix tube, de-

spite nearly the same helix voltage and band edge phase shifts. The overall efficiency has been raised from 31 to 43 percent. The two results are not directly comparable because the double-taper TWT has been operated with a 2-stage depressed collector. Nevertheless for comparison if the second collector stage is operated at the same potential as the first, the efficiency still exceeds 36 percent.

In addition, since the helix pitch up to the second attenuator is chosen for constant maximum gain, the gain variation in the 1-GHz band is extremely small (less than 1 dB for a saturated gain of 55 dB). The peak-to-peak group-delay distortion introduced by the double-taper tube is less than 0.1 ns in 100 MHz.

What are limitations of the double-taper? First, determination of the helix profile becomes a costly operation, necessitating the use of a proven computer program, followed by numerous trial and error steps to properly mate the helix to the attenuators and output. Winding a complex helix with great precision is a difficult mechanical problem. In addition, certain operating characteristics, such as the output hot VSWR, become more critical. The double-taper technique is also not applicable to tubes with large instantaneous bandwidth. The weak interaction mechanism, with a constant phase difference between the RF wave and the beam current, provides optimum efficiency for a relatively critical helix length.

The above results do not constitute the ultimate performance. To further improve the performance characteristics, it became necessary to have a longer helix to compensate for the gain lost in the output section. For instance, the TH 3535 delivers 20 W with 60 dB gain at an efficiency of better than 44 percent and a maximum phase shift of only 28°, in the same 10.95- to 11.7-GHz band utilizing a longer helix. A third col-

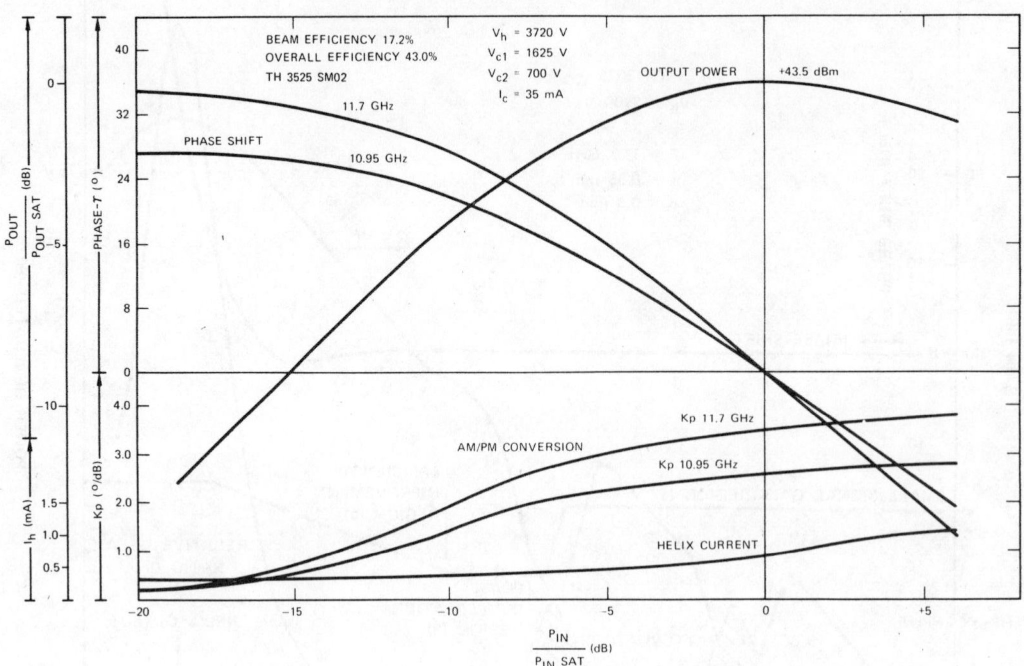

Fig. 8. 20-W, 11-GHz, double-taper, double-collector TWT (Th 3525).

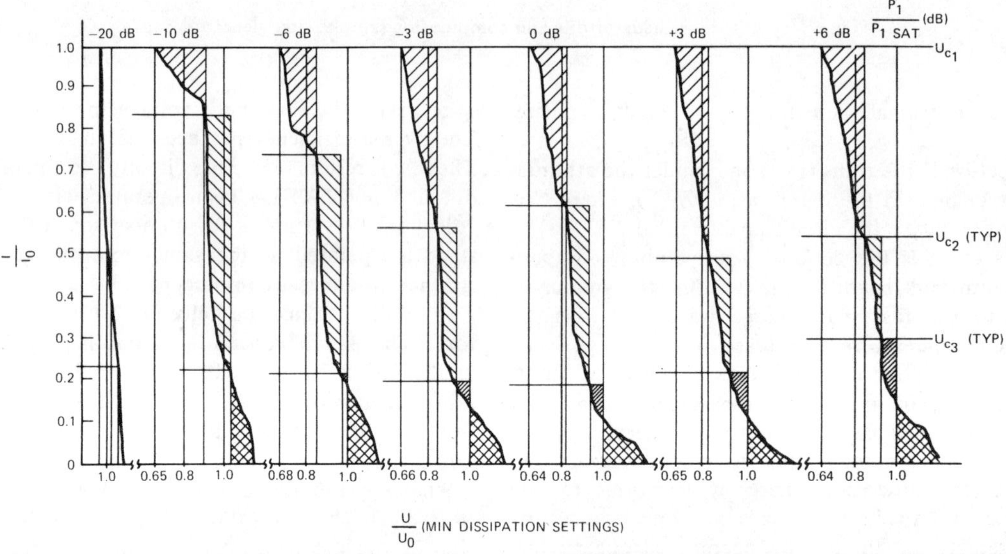

Fig. 9. Calculated velocity spectra of the TWT TL 12024 at different drive levels (U_0 = accelerating voltage of undisturbed beam, I_0 = cathode current of electron beam).

lector added to this tube did not, as expected, provide a significant efficiency improvement at saturation (0.5–1.0 percent) but made overall efficiency at backoff more effective.

VIII. NATURE OF SPENT BEAM, MULTICOLLECTOR PRINCIPLES

Due to the interaction process in a TWT, electrons are subjected to retarding or accelerating forces from the traveling RF fields. Electron bunches are formed which change their position relative to the RF wave as a result of phase slippage and interacting forces. After the RF output of the TWT has been passed by the electrons, the spent or decoupled beam contains a wide and differing velocity spectrum as shown in Fig. 9 for different levels of drive (P_1).

Theoretically, the collector dissipation for a 3-stage collector is

$$P_{\text{dissipation}} = U_{c1}I_{c1} + U_{c2}I_{c2} + U_{c3}I_{c3} - P_{RF}. \qquad (7)$$

This loss is represented by the shaded area of Fig. 9. U_{c1}, U_{c2}, and U_{c3} must be chosen so that $P_{\text{dissipation}}$ becomes a minimum. Analog considerations apply for different numbers of collectors. The design of a multistage collector requires an arrangement for velocity sorting which approaches as nearly as possible the above mentioned conditions. The electron optical device must be designed so that fast electrons are not collected at a collector stage with a higher voltage (with respect to the cathode) than necessary. There are a variety of concepts for multistage collectors.

The tilted electric-field collector (TEFC) investigated by Okoshi [20] uses an axial magnetic field for focusing the

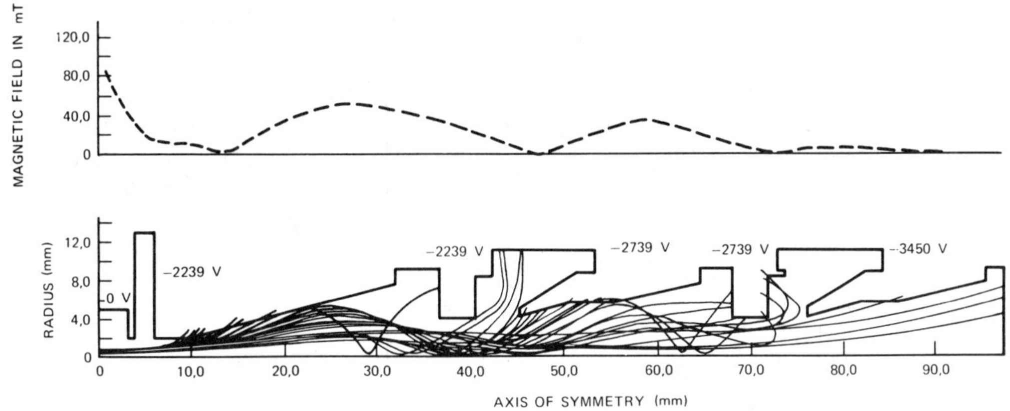

Fig. 10. Calculated electron trajectories in a magnetic lens collector (MLC) at saturation drive used in the TL 12024 together with rotationally symmetric electrode configurations and axial magnetic field (T = Tesla, $P_2 = P_2\,\mathrm{SAT}, I_{C1}$ = 18.0 mA, I_{C2} = 22.5 mA, I_{C3} = 4.5 mA, and η_C = 90.3 percent).

beam and an electrical transverse component for velocity-dependent deflection of the electrons, so that fastest electrons are least deflected. Other concepts use rotationally symmetric devices with purely electrostatic fields [electrostatic collectors (ECs)]. In this case, velocity sorting is achieved by space charge forces deflecting the slowest electrons away from the axis first. Finally, a magnetic lens collector (MLC) is used with superimposed field RF magnetic lenses and a rotationally symmetric electrostatic field.

IX. Collector Designs and Collector Efficiency

The TEFC configuration results in considerable technological complications, while the more frequently used ECs require a large volume for high efficiency. The MLC configuration combines small size with high collector efficiency. Therefore, it is useful for medium-power TWT's. The design follows an iterative process starting with an electron-velocity spectrum determined from large signal computations and a reasonable geometrical collector configuration. The results of these computations are used to change the shape of collector electrodes in an empirical manner for improved collector performance. Fig. 10 shows computed electron trajectories for an MLC with three stages and the electron-beam spectra from Fig. 9. Collector efficiencies and experimental results are shown in Fig. 3 and 4. The results fall between the theoretical optimum and the measured data. Measured collector efficiencies are lower than calculated values because of several factors.

a) The parameters used in the calculations are simplified. The interaction process neglects, for instance, radial variations of space charge density and the RF electrical field and radial components of the RF field. In addition, it does not include the influence of rotational energy resulting from PPM beam focusing.

b) Due to transverse magnetic field components, the currents intercepted by the helix and first collector stage without RF drive are not zero (see Fig. 3). These unwanted currents contribute to increased dc loss.

c) Computations do not allow for reflected or secondary electrons. Although care has generally been exercised to "catch" electrons only on metallic surfaces which provide a decelerating electric field, the disturbance to the proper current distribution on the collector stages caused by secondary or isolated reflected electrons cannot be avoided.

Variation of RF drive changes the composition of the velocity spectrum as shown in Fig. 9. Maintaining constant collector voltages when varying the RF input does not provide the

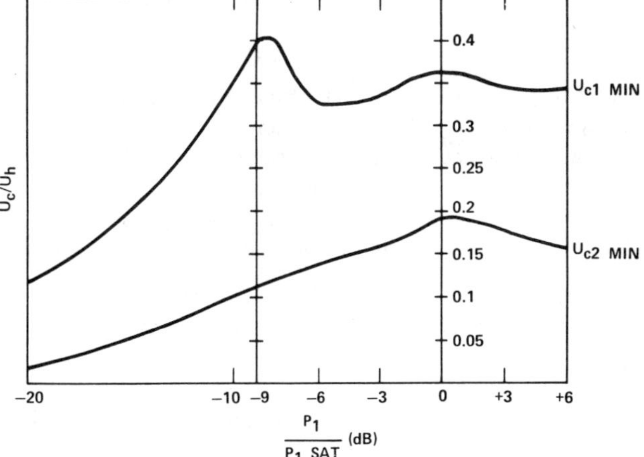

Fig. 11. Calculated values for U_{c1} and U_{c2} resulting in maximum efficiency at various drive levels (see Fig. 12 for MLC conditions).

highest possible efficiency for a specific drive level. For example collector 1 voltage does not require very much change with RF drive down to −10 dB below saturation drive. However, the required U_{c1} is largest at ∼ − 9-dB drive. Collector 2 voltage for optimum efficiency increases with increasing drive up to saturation (Fig. 11).

Selecting collector voltages for variable drive requires adjustment of U_{c1} to the value required for maximum efficiency at −9-dB drive and U_{c2} for saturation drive. Efficiency may be improved for any RF power level by designing the power supply so that U_{c1} and U_{c2} sources have a properly designed internal resistance. In this case, U_{c1} increases with decreasing RF drive because I_{c1} is also decreasing; U_{c2}, however, decreases while I_{c2} increases with decreasing drive.

X. Operating Efficiency

For combined beam and collector efficiencies the overall optimum efficiency as a function of collectors is shown in Fig. 12. These values are for 11-GHz 20-W helix space TWT's with a saturated phase shift delay of ∼30° (max K_p ∼ 3°/dB) [21]. As can be seen, additional collectors beyond three can improve the overall tube saturation efficiency but at a diminishing slope. Thus the change from the single-collector space TWT of the past decade to the present multiple-collector TWT represents a significant increase in communications capacity with all other factors being equal.

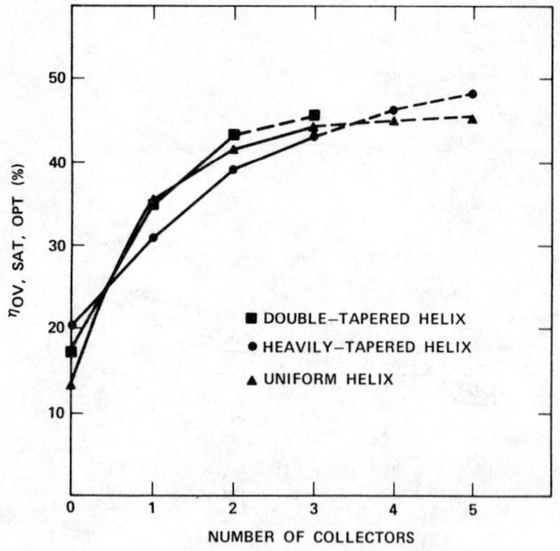

Fig. 12. Typical overall efficiency versus number of collectors.

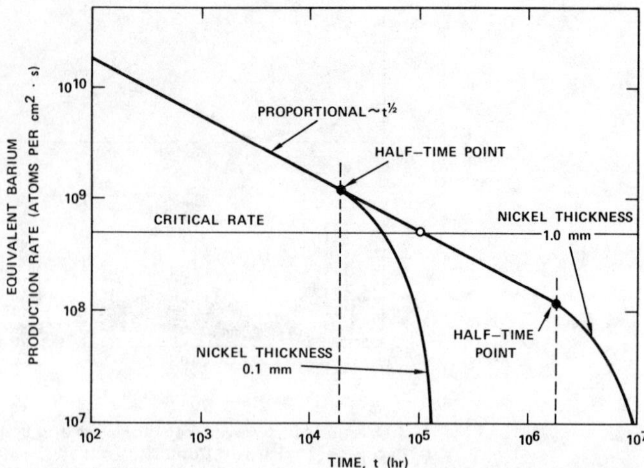

Fig. 13. Calculation of cathode life from the production rate of free Ba (j_k = 120 mA/cm^2, T_K = 650°C).

XI. CATHODE DESIGN

Cathode life essentially determines the actual TWT lifetime. It depends on the cathode itself and the cathode environment during tube processing and operation. The cathode must deliver the necessary current for the TWT. The two following types of thermionic cathodes are presently used in long-life satellite TWT's, depending on the required current density:

a) oxide cathodes (0.1 to 0.2 A/cm^2);
b) dispenser-type cathodes (0.4 to 0.8 A/cm^2).

A. Oxide Cathodes

Oxide cathodes consist of a heater, a nickel base, and an alkaline oxide coating (BaO/SrO or BaO/SrO/CaO). The base nickel is usually doped with a small fraction of Zr, W/Zr, or Mg. These activators diffuse to the Ni surface and undergo a chemical reaction with the alkaline oxide [22]. Hence, for instance, BaO is partially reduced and excess free Ba exists within the oxide coating. This combination lowers the work function to provide "space-charge-limited" electron emission above 520°C (T_{Kn}, the cathode transition or "knee" temperature) for a well-activated cathode. The cathode emits uniformly as long as the work function is consistently maintained. The actual cathode operating temperature is set higher (650°C to 750°C) for a given level of emission-current density to ensure a uniform work function plus sufficient protection against poisoning, but not so high that excessive depletion of the Ba or other cathode constituents results. Theoretically, a decrease of 25°C will generally double the life of a cathode [8].

Free Ba on the cathode evaporates due to cathode temperature and reacts primarily with residual oxygen in the environment. Therefore, a continuing production of free Ba is required. Early in life the activation content of the base nickel is high enough to produce a large amount of excess Ba. Due to diffusion and reaction of activators, their content in the base material decreases. Simultaneously the alkaline oxide is consumed. A sufficient amount of activators and oxide coating must be available. This requires an adequate thickness of the base material ($\gtrsim$1 mm) and the coating ($\gtrsim$50 μm). For a given "doped" nickel base material, Ba production is calculated as a function of time [23], [24]. The end of the theoretical cathode life is approached when Ba production falls below the required value. Fig. 13 shows this calculation for the cathode used in the TL 4003 at 10^5 h [j_k = 120 mA/cm^2, T_k = 650°C > 1-mm base Ni (A) thickness, 0.03-percent Mg content].

The actual life of a cathode is much more difficult to evaluate. The required Ba production is in a state of dynamic equilibrium determined by the residual gas pressure, temperature, and cathode current density. The tube's residual gas pressure (essentially oxygen partial pressure) depends on several factors: leaks, gas sources in the material used within the vacuum envelope, cathode gas release, and getter action. During operation the cathode both absorbs gas molecules and delivers gas due to the outgassing of the base material. Therefore, gas cycling occurs in the tube with an equilibrium pressure dependent on and increasing with cathode temperature. Part of the gas molecules are ionized in the electron beam and, by means of the operating voltages of the TWT, accelerated toward the cathode along the beam axis. Ion bombardment occurs in the center of the cathode, where the ions impinge. Bombardment may damage the active surface and in extreme cases evaporate cathode Ni after localized wear-out of the active surface. This sputtering occurs only in a relatively poor vacuum. An essential contribution to TWT life occurs when the residual gas pressure during processing and operation is minimized. Improvements obtained during recent years have resulted in an internal tube working pressure of 10^{-11} to 10^{-10} torr, which is a crucial factor for long-life oxide cathodes.

Thus the gas cycling process limits the life of the cathode. When the cathode is operated at a sufficiently low temperature ($\sim$650°C), the gas cycling process exerts little influence on life and may be much less than that determined by the activators in the nickel base. At more elevated temperatures ($\sim$750°C), a more active dynamic equilibrium exists and this process becomes more significant. The effects on cathode life are greater with higher gun electrode voltages due to greater ion impact energy. Finally, a qualitative limit (combining cathode operating temperature, residual gas, and beam voltage settings) is reached where a lifetime consistent with satellite requirements ($\sim$5–10 years) is no longer achievable. In this case, use of a dispenser-type cathode becomes necessary.

B. Dispenser-Type Cathodes

Dispenser-type cathodes consist of a matrix of porous tungsten impregnated with alkaline-earth (Ba, Ca) aluminates [25]. Free Ba, in this case, is obtained by chemical reaction and subsequent migration through the pores to the surface. Such cathodes, however, had never been used in the space TWT's and it was necessary to determine their optimum operating

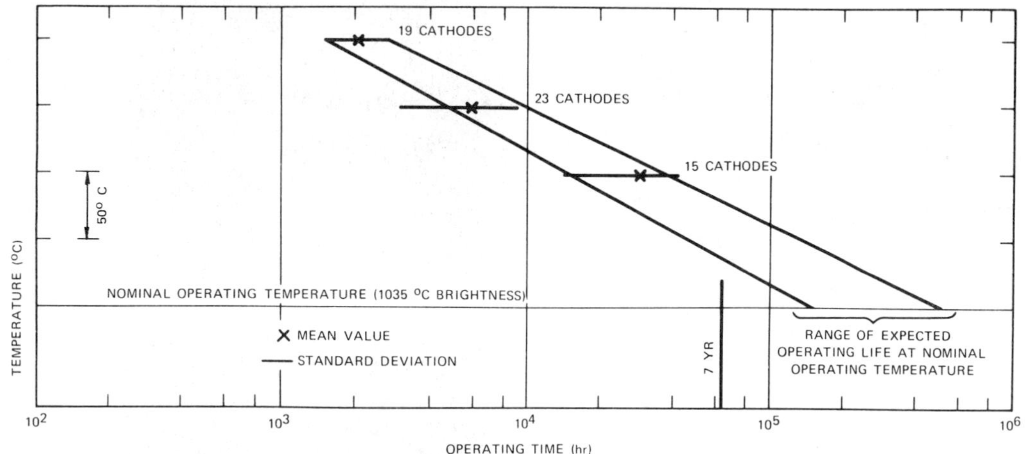

Fig. 14. Results of accelerated life test of space tube cathodes in diode mountings (nominal operating temperature = 1035°C brightness).

temperature and probable useful emission life although virtually all high-power ground station TWT's use these cathodes.

It is well-known that, if the cathode's temperature is too low, the cathode is susceptible to poisoning by the residual gases in the tube, while if the temperature is too high, the active cathode emitter (Ba) will evaporate too rapidly. During operation, the active barium rich (BaO) layer on the cathode surface is continually replenished by Ba + BaO migrating from the interior of the impregnated tungsten cathode to replace evaporated barium. Also, the combination of easily replenishable barium on a tungsten matrix surface makes dispenser-type cathodes much more resistant to high-energy ion bombardment than oxide cathodes (i.e., higher helix voltages, as well as the effects of the residual gas cycling process).

There are, of course, numerous parameters in the design of a reliable impregnated cathode, such as the size of the tungsten particles, the porous tungsten density, and the mole ratio of the impregnant's constituents. Typically the porous matrix is made from tungsten particles whose average size is six microns, sintered to a theoretical tungsten density of 82 percent with an impregnated mole ratio of $BaO:4$, $CaO:1$, and $Al_2O_3:1$.

Dispenser-type cathodes are usually operated at temperatures higher than $1000°C$. Due to the higher current density the cathodes are smaller, but the power per unit square of heater wire surface is considerably increased relative to that of the oxide cathode. It has been shown that heat transfer by radiation from the heater to the cathode results in insufficient heater life. Therefore, the heater is potted with Al_2O_3 into a cavity on the bottom of the cathode. With this kind of heater construction a total of 3.2×10^6 operating switching hours have been accumulated without one failure.

To determine the optimum cathode operating temperature, a large number of diodes with identical cathode/filament structures were fabricated and placed in life test at elevated temperatures. A plot of the experimentally determined end of life for the different test temperatures was then made (Fig. 14). Extrapolation of the results to lower temperatures yielded the optimum operating temperature for a cathode life of at least 150 000 h, and the range of expected operating life at that temperature.

C. Cathode Testing

Cathodes and their heater exert a stronger influence on life and reliability than any other subassembly in the TWT. There-fore, careful tests must be performed to select cathodes to be used in a TWT and to select flight model TWT's from production. Since the investigations of satellite TWT's by Bodmer et al. [8], "dip" testing is widely used. Variation of cathode current versus cathode temperature is observed when the cathode cools down and heats up. The transition or "knee" temperature (T_{Kn}) from saturated to space-charge-limited temperature and the shape of the transition curve are used as selection criteria.

A hybrid "dip" test and a dual-pulse activity test has been investigated for TWT's (INTELSAT Contract 676). Two pulse programs are involved. The first superimposes short rectangular anode voltage pulses (10-μs, 0.1 percent duty cycle) on the normal operating voltage, leading to cathode current pulses four to five times the normal operating current. At this elevated current, transition between space-charge-limited and saturated current occurs at higher temperatures while the rms helix interception is kept sufficiently low. This provides greater potential selection discrimination.

Fig. 15 shows two examples of the first pulse program: dc dip test, pulse dip test, and helix interception are shown versus time after heater power switch off for two samples of the TL 4003. Fig. 15 (a) shows the result of a tube selected as a flight model operated at a cathode temperature of $620°C$. The transition point for the pulse current at $560°C$, while the dc transition occurs at $515°C$. Fig. 15 (b) shows the result of a tube whose cathode has been poisoned by increasing the residual gas pressure to 10^{-7} torr. This cathode must be operated at $750°C$ to give full emission and has a transition temperature of $735°C$ where little difference between dc and pulse transition is observed.

Results of the dual-pulse second program after Tischer [26] are shown in Fig. 16. Fig. 16 (a) shows a cathode operating at $650°C$ in an environmental pressure of 3.5×10^{-10} torr. During the long impulse (50 ms, 1-percent duty cycle) a slight increase of I_K occurs first, most probably due to heating in the interface layer resistance, followed by a slight decrease due to cathode poisoning. When subsequent short pulses (10 μs every 10 ms) are applied, the initial current value of the cathode is immediately reached. Increasing the residual gas pressure to 10^{-8} torr [Fig. 16 (b)] results in a much stronger effect of cathode poisoning followed by a more extended recreation period. Further evaluation of these two pulse test programs is necessary, but it has been demonstrated that they provide

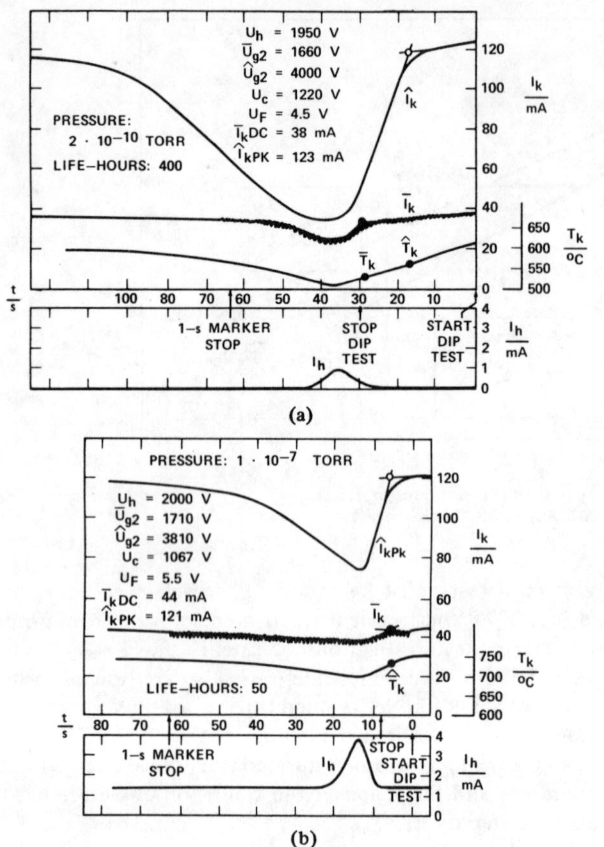

Fig. 15. Dc ($\overline{I}_k$) and pulsed ($\hat{I}_k$) on different cathodes showing cathode current, temperature, and helix interception current versus time ($\circ$ = $\hat{T}_{kn}$, pulse transition temperature, $\bullet$ = $\overline{T}_{kn}$, dc transition temperature. (a) Properly activated cathode for a flight model TWT. (b) Partially poisoned cathode by residual gas pressure of 10^{-7} torr.

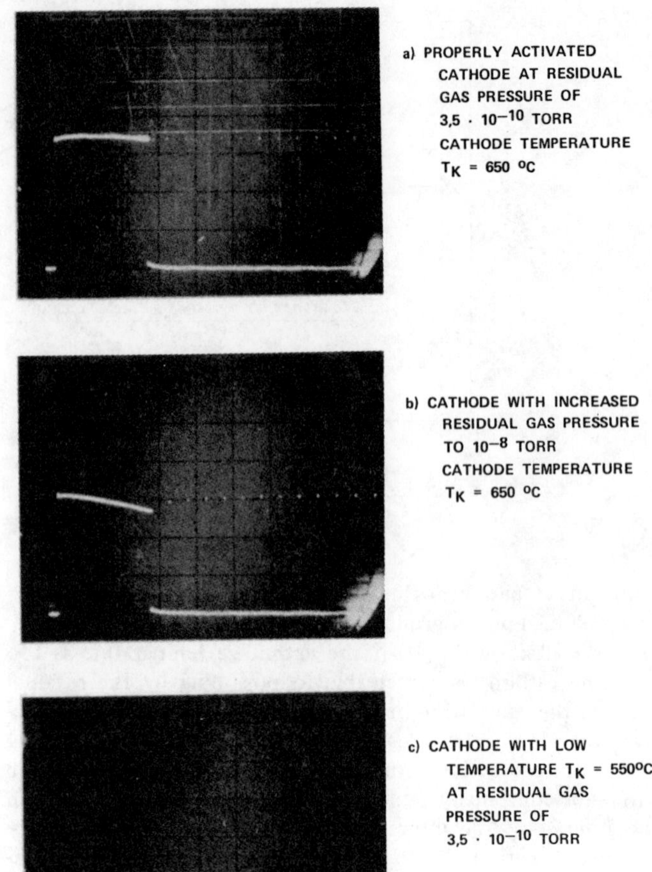

Fig. 16. TWT (TL 4003) testing of oxide cathodes with long/short cathode current pulses.

better insight into cathode performance while preserving cathode operating characteristics during the test.

XII. TWT Life Test Results (Ground Tests)

As an example, fifty oxide-cathode TWT's (TL 4003) out of a production lot have been operated in a life test program. During a 2000 h burn-in period, two tubes were eliminated due to the drift of intercepted helix current. In the subsequent life test period, 50 tubes accumulated approximately 0.5×10^6 h. The test program was continued with 20 tubes having the longest individual life (presently approximately 30 000 h). Cathode activity was tested by dc dip test. A total of 960 000 h have now been accumulated without any failure or variation of characteristics outside the specified range. The transition temperature (T_{Kn}) has been slowly increasing in a quasi-linear manner from ~525°C to ~570°C in more than three years of operation. An average margin of above 20°C is projected for a seven year mission.

Results of dispenser cathode activity tests for 46 20-W 11-GHz, TWT's on continuous life test (Fig. 17) show a regular increase of the cathode knee temperature (T_{Kn}) during the tube's operating life. Cooling down to T_{Kn} during filament switch off dip test is a linear function of time with a slope of 20°C per second. The temperature margin at beginning of life is about 100°C and falls to 50°C after 20 000 hours of operation, following a constant slope on a semilogarithmic scale. Extrapolating to the end of the seven year mission projects a margin of more than 20°C.

XIII. TWT Life Test Results (Space)

For the most part, life-test results in space have been extremely satisfactory. Some TWT's have now exceeded 100 000 h of operation. The INTELSAT I through INTELSAT IV spacecraft series have not been life limited by transmitter TWT's and there are indications of average lives in excess of 40 000 h for tube types such as the 261H and 275H (both utilizing oxide cathodes). At this writing only one 261H out of a population of 168 (approximately three to six years in space) has deactivated after two years of operation. Telemetry information is inconclusive regarding the failure source (i.e., tube or supply).

Unfortunately, in spite of the extensive processing and selection precautions applied, one tube type did not live up to projected expectations. This tube, the 262H, is used as a driver TWT in the INTELSAT IV spacecraft. Present deactivation lifetime estimates for this device are of the order of two years. The spacecraft objective lifetime (7 years) was, however, not compromised, since quadruple redundancy had been specified. The major factors pertinent to the limited life of this device are a higher than usual cathode current density (280 mA/cm^2), an associated elevated cathode temperature (740°C), and an as yet unexplained possible correlation factor, a high ratio of anode aperture diameter to cathode diameter (1.05). This latter characteristic originates with the shallow-angle gun re-

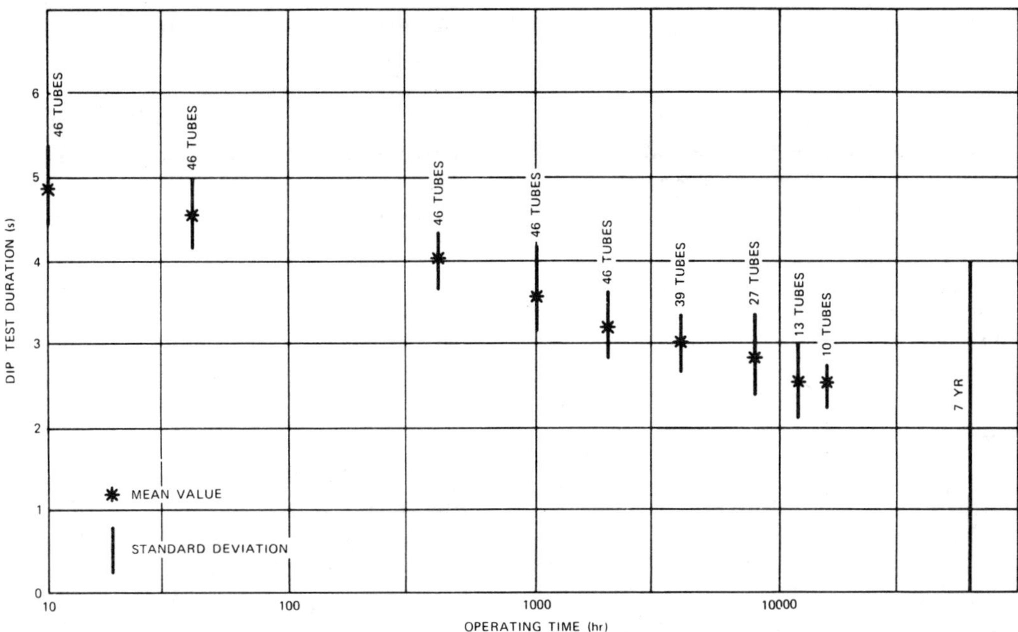

Fig. 17. Th 3525 ($\eta = 4$) and TOP 1369 ($\eta = 42$) life tests (as of January 1976).

quired for a low-noise driver TWT (approximately 18-dB NF). The implication is that a larger than usual percentage of the oxide cathode area is exposed to "ion" bombardment from the ions generated in the cathode/anode region with a possible resultant gradual "poisoning" as discussed earlier. Examination of failed life test devices has not as yet revealed any obvious ion-sputtering damage. Testing is continuing.

XIV. Conclusion

Methodical advances in efficiency and linearization of the characteristics continue to be made. These advances have been reviewed in terms of the double-tapered helix and triple-collector designs utilizing magnetic lenses for electron-velocity sorting. Overall efficiencies in the 45- to 48-percent region for a 10- to 20-W device are now achievable with phase delays of less than 30°.

Current TWT's are representative of a mature device technology yielding exceptional microwave and efficiency characteristics. This makes them unique for communications satellite transmitters. Not withstanding certain limiting causes, actual lifetime and reliability have increased and the acceptance of TWT's in satellites is a matter of record. Hundreds of TWT's are now in space and hundreds more are expected in the next few years.

References

[1] R. Kompfner, *The Invention of the Traveling Wave Tube*. San Francisco, CA: The San Francisco Press, 1964.

[2] J. R. Pierce, *Traveling Wave Tubes*. Princeton, NJ: Van Nostrand, 1950.

[3] N. Sawazaki *et al.*, "New microwave repeater system using traveling wave tubes," *Proc. IRE*, vol. 44, pp. 19–24, 1956.

[4] J. P. Laico *et al.*, "Medium power traveling wave tube for 6000 Mc radio relay," *Bell Syst. Tech. J.*, vol. 35, pp. 1285–1346, Nov. 1956.

[5] J. T. Mendel, "Magnetic focusing of electron beams," *Proc. IRE*, vol, 43, pp. 327–331, 1955.

[6] V. L. Holdaway *et al.*, "Electron tubes for the SD submarine cable system," *Bell Syst. Tech. J.*, vol. 43, pp. 1311–1339, July 1964.

[7] J. Bretting, "13 W traveling-wave tube for communication satellite application," presented at AIAA 2nd Communications Satellite Systs. Conf., San Francisco, CA, 1968.

[8] M. G. Bodmer *et al.*, "The satellite traveling wave tube," *Bell Syst. Tech. J.*, vol. 42, pp. 1703–1749, July 1963.

[9] J. Benton *et al.*, "Latest advances in space traveling wave tubes," presented at AIAA 3rd Communications Satellite Systs. Conf., Los Angeles, CA, Apr. 1970.

[10] J. T. Mendel, "Helix and coupled cavity traveling wave tubes," *Proc. IEEE*, vol. 61, pp. 280–298, Mar. 1973.

[11] E. T. Jilg, Ed., "The INTELSAT IV Spacecraft," *COMSAT Tech. Rev.*, vol. 2, no. 2, pp. 287–289, Fall 1972.

[12] M. K. Scherba, "High efficiency triple mode space traveling wave tube," presented at IEDM Proceedings, Washington, DC, 1975.

[13] P. Bobisch, Internal Report, AEG-Telefunken.

[14] P. Treytl and J. Bretting, "Grössen Zur Charakterisierung Der Nichtlinearitaeten Von Mikrowellen-Sendeverstaerkern Und Deren Zusammenhang," *Nachrichtentech. Z.*, Mar. 1972.

[15] M. J. Schindler, "Advances in traveling wave tubes for spacecraft communications systems," in *Communication Satellite System Technology*. New York: Academic Press, 1966.

[16] J. E. Rowe, *Nonlinear Electron-Wave Interaction Phenomena*. New York: Academic Press, 1965.

[17] G. Hermann, "Optical theory of thermal velocity effects in cylindrical electron beams," *J. Appl. Phys.*, vol. 29, no. 2, pp. 127–136, Feb. 1958.

[18] O. Nilsson, "Nonlinear distortions in TWT," Research Lab of Electronics, Chalmers University of Technology, Gothenburg, Sweden, Res. Rep. 67, 1966.

[19] E. Ezura *et al.*, "On the mechanism of nonlinear distortion in TWT," presented at the Proceedings of MOGA, Amsterdam, 1970.

[20] T. Okoshi *et al.*, "A high efficiency *X*-band TWT equipped with a tilted electronic field (TEFC) collector," in *Proc. Inter. Electron Devices Meeting*, Washington, DC, p. 498, 1974.

[21] H. Thienel, "Wirkungsgradverbesserung Bei Der Wanderfeldroehre TL 12022," *Die Telefunken Roehre*, no. 48, pp. 18–22, Apr. 1976.

[22] E. S. Rittner, "A theoretical study of the chemistry of the oxide cathode," *Philips Res. Rep.*, vol. 3, pp. 184–238, June 1953.

[23] H. E. Kern *et al.*, "Thermionic emission and diffusion studies on zirconium doped nickel cathodes," in *Advances in Electron Tube Tech.*, 6th Nat. Conf., Sept. 1962, pp. 235–239.

[24] F. Dlouhy *et al.*, "Magnesium im kathodennickel" *Die Telefunken Roehre*, no. 46, pp. 39–66, Nov. 1966,

[25] V. L. Stout, "Dispenser cathodes," in *Proc. 4th Nat. Conf. on Tube Tech.*, Sept. 1958, pp. 178–190.

[26] V. M. Tischer, "Die Beurteilung Der Kathodenguete Durch Impulsmessung," *Int. Elektron, Rundschau*, no. 11, 1963.

Microwave Field-Effect Transistors—1976

CHARLES A. LIECHTI, SENIOR MEMBER, IEEE

Abstract—A review of recent and current work on microwave FET's and amplifiers is presented, and an extensive bibliography of recent articles is appended (250 references). First, the various FET structures (MESFET's, JFET's, and IGFET's) and their performances are reviewed. Second, the principle of operation is outlined for Si- and GaAs-MESFET's; the basic device physics, equivalent circuit, high-frequency limitations, and noise behavior are treated. Third, the design principles and performance of microwave MESFET amplifiers are summarized.

I. INTRODUCTION

IT TOOK seventeen years from the initial introduction of the bipolar transistor in 1948 until microwave transistors with practical gain and noise figure became available. In 1965 germanium transistors invaded L band with noise figures under 6 dB. A milestone was reached in 1968 when the AT&T System adopted the balanced transistor amplifier, developed by Engelbrecht and Kurokawa [A1], for use in its S- and C-band microwave communication links. Since 1968 significant progress has been made in obtaining low-noise performance and high-power capability from bipolar transistors for frequencies reaching up to X band. At 8 GHz a silicon n-p-n transistor with 3.9-dB noise figure and 3.8-dB associated gain [A2], and a power transistor with 1-W CW output power[1] and 6-dB power gain [A3] have been reported. At 2 GHz a single (silicon) transistor chip delivers up to 30-W CW output power with 7-dB power gain and 32-percent power-added efficiency [A7]. Bipolar transistors fabricated in GaAs are still in an early stage of development [A8]–[A10].

By 1971, however, breakthroughs had been made in the development of field-effect transistors. Today, GaAs *me*tal *s*emiconductor *f*ield-*e*ffect *t*ransistors (MESFET's) have higher gain, higher power-amplification efficiency, and lower noise figure than bipolar transistors above 4 GHz. More significant is the fact that FET's promise a great deal of potential for further advances. Substantial progress can be expected in the near future because of the following.

1) A large variety of FET structures (MESFET, JFET, IGFET) are suitable for microwave amplification and power generation, and some promising candidates are only in an early stage of development.

2) A variety of semiconductor materials (GaAs, InP, $In_xGa_{1-x}As$, $InAs_xP_{1-x}$) with majority-carrier transport properties[2] superior to silicon are competing for application in FET's [F5].

3) Further miniaturization to submicron dimensions can be realized in most FET structures.

4) Monolithic integration of circuits on semi-insulating substrates enables device isolation with low parasitic capacitances, low-loss interconnections, and high packing density.

This paper reviews recent and current developments in high-frequency FET's and FET amplifiers.[3] In Section II the various microwave FET structures are discussed, and their performance characteristics are summarized. Section III gives an introduction to the device physics, the small-signal characteristics, and the noise behavior of MESFET's. In Section IV the design principles and the performance of microwave amplifiers are reviewed. An extensive bibliography of recent articles is appended. Because of limited space, topics such as fabrication, packaging, reliability, oscillators, mixers, modulators, and digital applications are not treated here.

II. MICROWAVE FET STRUCTURES AND THEIR PERFORMANCE

Today MESFET's, p-n junction FET's (JFET's), and insulated-gate FET's (IGFET's) are used at microwave frequencies [B2]. The cross sections of the various FET structures are illustrated in Fig. 1. Their performance characteristics are tabulated in Table I. Fig. 1 serves as a guide for the following discussion; i.e., first the MESFET's, then the JFET's, and finally the IGFET's are described.

A. MESFET

In 1969 Middelhoek realized a silicon MESFET with 1-μm gate length by projection masking [X1], [X2]. This FET had a 12-GHz maximum frequency of oscillation [C1] which was considerably higher than for previously known FET's and comparable to f_{max} of the best bipolar transistors at that time. The next significant step was the fabrication of 1-μm MESFET's on GaAs. As a result, FET's with f_{max} of 50 GHz and useful gain up to 18 GHz became available in 1971 [C2]–[C6]. This substantial improvement in device performance is due to the following reasons. 1) in GaAs the conduction electrons have a six times larger mobility and a two times larger peak drift velocity than in silicon [F1]. Therefore, parasitic resistances are smaller, the transconductance is larger, and the transit time of electrons in the high-field region is shorter. 2) The active layer is grown on a semi-insulating GaAs substrate with resistivity larger than $10^7 \, \Omega \cdot cm$. The large parasitic capacitance of the gate bonding pad can thus be eliminated by positioning the pad on the substrate.

In 1972 it became apparent that GaAs-MESFET's are capable of very low-noise amplification [C4]–[C6]. Liechti

Manuscript received October 1, 1975; revised January 15, 1976.

The author is with the Solid State Laboratory, Hewlett-Packard Company, Palo Alto, CA 94304.

[1] Combined power from two chips.

[2] Minority-carrier lifetime and mobility are of no concern in unipolar transistors. Therefore, materials with optimized transport properties for electrons but poor performance for holes can be chosen.

[3] The reader is also referred to a review paper on the same topic by Turner [B1].

Reprinted from *IEEE Trans. Microwave Theory Techn.*, vol. MTT-24, pp. 279–300, June 1976.

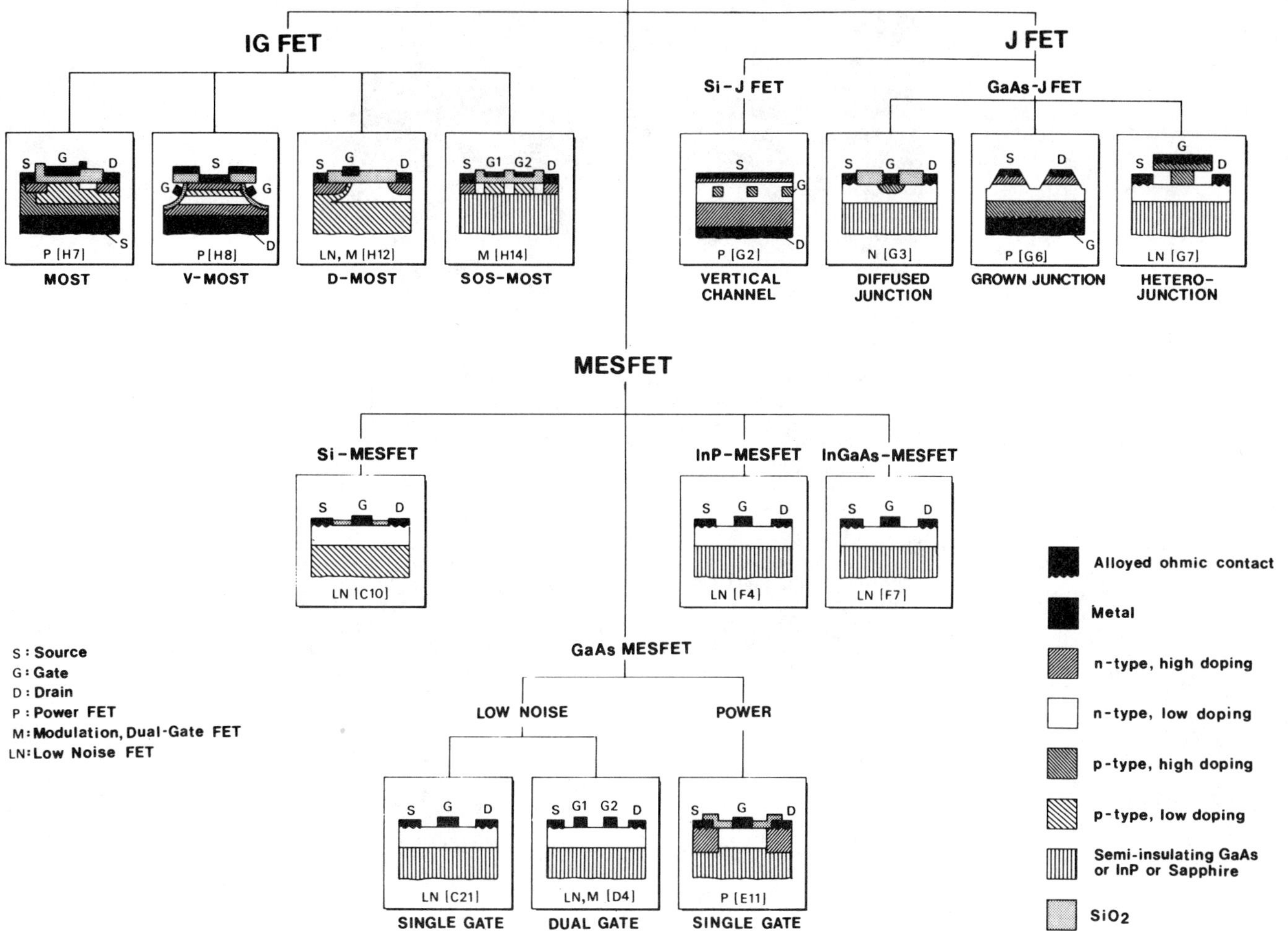

Fig. 1. This "family tree" of microwave FET's shows the cross sections of the various FET structures. Their performance characteristics are listed in Table I.

TABLE I
PERFORMANCE DATA OF THE MICROWAVE FET STRUCTURE SHOWN IN FIG. 1

Type*	Semi-conductor	Single/ Dual Gate	Channel			Appli-cation*	Fre-quency (GHz)	Output Power CW (W)	Assoc. Power Gain (dB)	Small Signal Gain (dB)	Power** Added Effic. (%)	Noise Figure (dB)	Assoc. Gain (dB)	Casc. Noise Figure† (dB)	Max. Avail. Gain (dB)	Reference
			Type	Length (µm)	Width (mm)											
MESFET																
Silicon	Si	SG	n	0.5		LN	10					5.8			5.9	[C10]
LN SG GaAs	GaAs	SG	n	0.5		LN	10					2.7	10.5	3.1	13	[C21]
LN SG GaAs	GaAs	SG	n	1		LN	10					3.2	8.0	3.6	10	[D4]
LN DG GaAs	GaAs	DG	n	1		LN,M	10					4.0	12	4.2	18	[D4]
P SG GaAs	GaAs	SG	n	1.5	5.2	P	8	2.2	3.2	4.2	22					[E11]
P SG GaAs	GaAs	SG	n	1.2	0.6	P	22	0.14	4.8	5.6	9					[E17]
LN SG InP	InP	SG	n	1		LN	10					4.7	6.6	5.4	7.8	[F4]
LN SG InGaAs	InGaAs	SG	n	1		LN	7					5.7	5.0	7.0	18	[F7]
JFET																
Silicon	Si	SG	n	1	1.8	P	2.7	0.2	6.0		19					[G2]
Diff. Junction	GaAs	SG	n	2		LN	4					2.5	10	2.7	10	[G3]
Grown Junction	GaAs	SG	n	1.5	6.1	P	6	1.0	6.0	7.0	26					[G6]
Heterojunction	GaAs	SG	n	2		LN										[G7]
IGFET																
MOST	Si	SG	n	5	20	P	0.7	16	6	10	26					[H7]
V-MOST	Si	SG	n	1	18	P	2	4.0	5	6	32					[H9]
SG D-MOST	Si	SG	n	1		LN	1					3.0	9.0	3.3	–	[H13]
DG D-MOST	Si	DG	n	1		LN,M	1					4.5	14	4.6	15	[H13]
DG SOS-MOST	Si	DG	n	4		M	0.5								25	[H14]

* LN - low noise amplification; P - power amplification; M - modulation, switching or amplification with controlled gain; SG - single-gate FET; DG - dual-gate FET.

** The power added efficiency is defined as the RF output power minus the RF input power divided by the dissipated dc power.

† The cascaded noise figure is defined by Eq. (15).

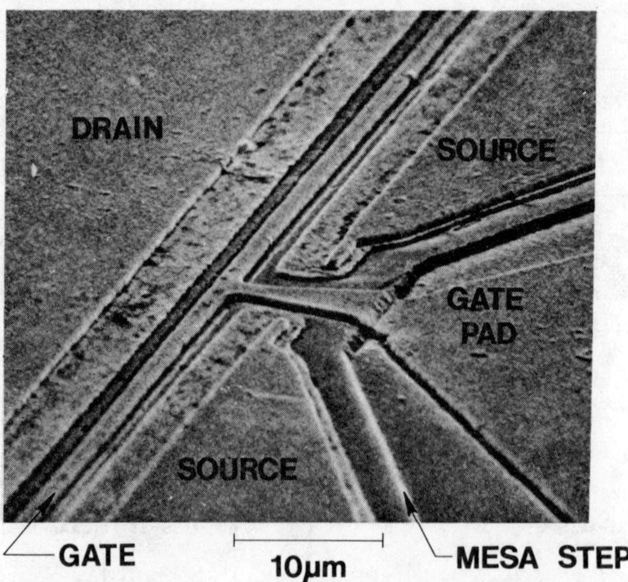

Fig. 2. This scanning electron micrograph shows the center section of a low-noise MESFET. The source and drain are alloyed ohmic contacts to the underlying conductive layer. The gate is the narrow metal stripe forming a Schottky contact. The width of the depletion layer under the gate controls the current flowing from drain to source. To the right, the gate metal runs over a mesa step (edge of the conductive layer) and widens on the semi-insulating substrate into a large bonding pad.

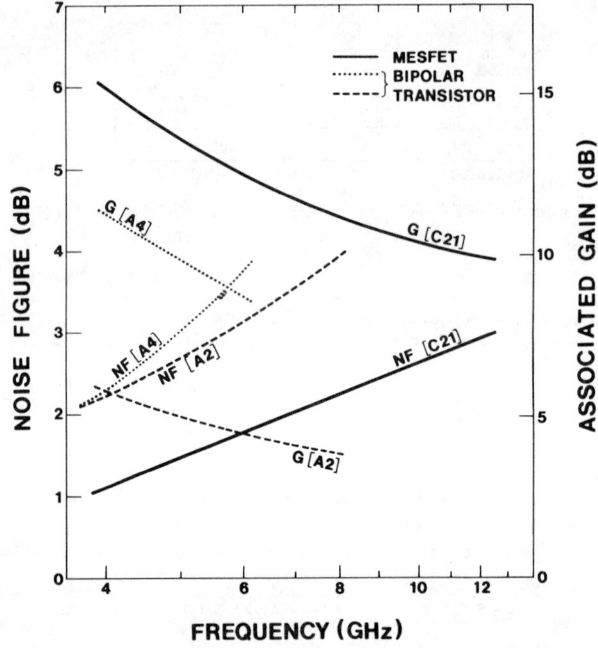

Fig. 3. Lowest reported noise figures and associated gains of microwave transistors are plotted versus frequency. The GaAs-MESFET reported by Ogawa et al. [C21] has 0.5-μm gate length. The dashed line represents the Si bipolar transistor with the lowest published noise figure [A2]. The bipolar transistor with the dotted line [A4] has a considerably higher gain.

et al. [C5] reported a noise figure of 3.5 dB with 6.6-dB associated gain at 10 GHz. Baechtold [M7], [M8] proposed a noise model that agrees well with measurements taking carrier velocity saturation and intervalley scattering into account. In the past few years, small-signal GaAs-MESFET's have been fabricated and characterized by various laboratories [C7]–[C28]. A scanning-electron micrograph of a typical low-noise MESFET is shown in Fig. 2. Lowest reported noise figures and associated gains are plotted in Fig. 3 versus frequency. Best noise performance is achieved with 1) a high-purity buffer layer between the substrate and the active layer [C26]; 2) a high doping level in the active n-layer (2.5×10^{17} cm^{-3} [C21]); 3) smallest possible source and gate-metal resistance [C21], [C22]; and 4) short gate length (0.5 μm [C10], [C21]).

A MESFET structure with two gates is shown schematically

in Fig. 1. This FET has a higher gain and a lower feedback capacitance than the single-gate counterpart [C21], [D1]–[D7]. In addition, the gain can be controlled over a wide range (44 dB [D4]) by varying the dc bias of the second gate. This feature can be used for automatic gain control in amplifiers [D6]. The gain modulation response is very fast. Pulse-amplitude modulation of an RF carrier with less than 100-ps fall and rise times has been demonstrated [D3], [D4].

The GaAs-MESFET is not limited to small-signal low-noise applications. The first power MESFET's appeared in 1973 and were of planar construction as shown in Fig. 4. Fukuta *et al.* [E1] designed a MESFET with 20 gates, each 1 μm long and 400 μm wide, operated in parallel and interconnected with a second metallization layer. At 2 GHz this MESFET exhibited 1.6-W output power[4] with 5-dB power gain and 21-percent power-added efficiency. At the same time, Napoli *et al.* [E2] presented a power transistor with self-aligned[5] gates [Fig. 4(c)]. Multiple gate, source, and drain pads had to be interconnected with bonding wires. The planar power FET has been further developed by various laboratories [E3]–[E19]. The techniques yielding high-power capability per unit gate width are:

1) the use of a high-resistivity epitaxial buffer layer to isolate the active layer from the bulk-grown substrate [E16];
2) the addition of inlaid n^+-regions under the source and drain electrodes, shown in Fig. 4(b), to increase the drain-source breakdown voltage and decrease the parasitic contact resistances [E11];
3) the design of short gate branches to prevent current crowding and to lower the gate-metal resistance [E11];
4) the flip-chip mounting of the transistor to decrease the thermal resistance and the source-to-ground lead inductance [E7], [E15].

A photomicrograph of a power MESFET from Fujitsu Laboratories [E9] with 104 gate branches, each 1.5 μm long and 50 μm wide, is shown in Fig. 5. Recently reported performance data of power MESFET's are summarized in Table II. The output powers from single chips, operated CW Class A, range from 4 W at 4 GHz to 0.14 W at 22 GHz and power-added efficiencies vary between 44 and 9 percent. Highest efficiency has been obtained by Huang *et al.* [E18] in Class B operation; 68 percent at 4 GHz and 41 percent at 8 GHz have been measured. These are the highest reported efficiencies of all microwave solid-state devices in this frequency range. The performance figure of merit M_F proposed by Drukier *et al.* [E15] is also listed in Table II. M_F measures the added RF power per unit gate width times the square of the operating frequency. The MESFET

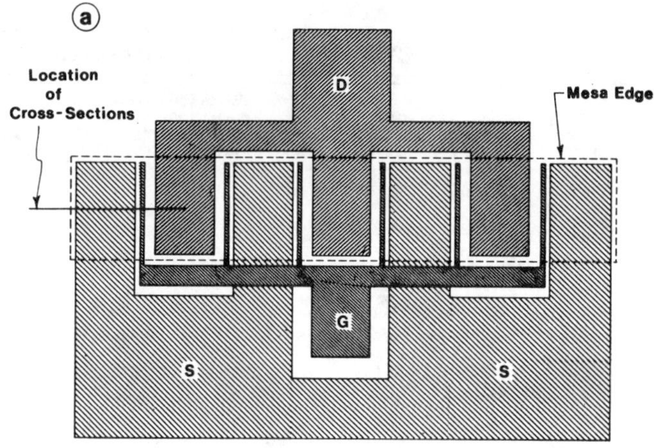

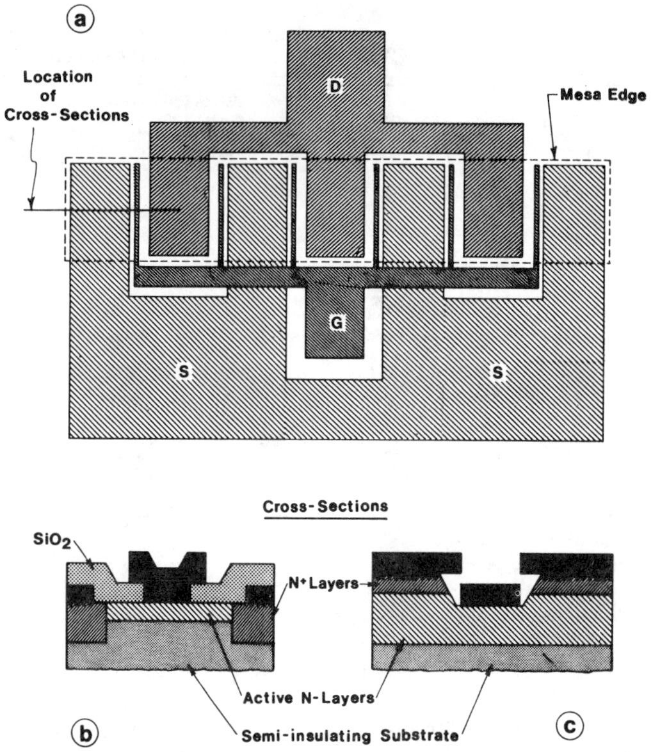

Fig. 4. (a) Illustrates schematically the metallization layout of the planar power MESFET shown in Fig. 5. The gate branches are interconnected with a metal line that crosses over the source. Multigate MESFET's with cross sections (b) [E11] and (c) [E1], [E12]–[E18] have been realized.

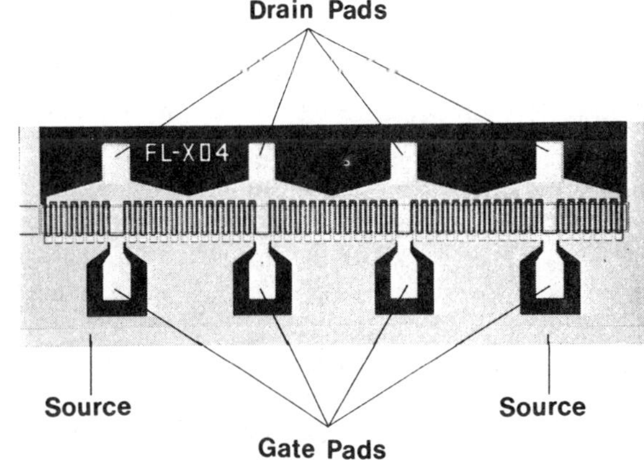

Fig. 5. This photomicrograph shows a power MESFET with 104 gate branches, each 1.5 μm long and 50 μm wide, connected in parallel. The chip delivers 0.9-W output power with 6-dB power gain and 30-percent power-added efficiency at 6 GHz [E9]. (Courtesy of M. Fukuta, Fujitsu Laboratories.)

with the largest M_F delivers 140-mW output power with 4.8-dB power gain and 9-percent power-added efficiency at 22 GHz [E17].

A new device with better heat-sink properties has been proposed by Blocker *et al.* [E5] for power MESFET's.[6] The metallization layout and the cross section of this MESFET are

---

[4] Measured at 1-dB gain compression.
[5] The location of the gates between the source and drain edges is defined by a fabrication method which does not require a critical realignment step.

[6] A similar structure has been realized by Vergnolle *et al.* [G6] in the form of a GaAs p-n junction FET (see below).

TABLE II
PERFORMANCE OF EXPERIMENTAL MICROWAVE POWER TRANSISTORS
CW DATA OF SINGLE-CHIP GaAs-MESFET's,[1] GaAs-JFET's,[2] AND Si BIPOLAR TRANSISTORS[3]

Frequency (GHz)	Transistor Type	Output Power (W)	Power Gain[4] (dB)	Small-Signal Gain[5] (dB)	Power Added Efficiency[6] (%)	Figure of Merit[7] ($WGHz^2$/mm)	Gate Length[8] (μm)	Total Gate Width[9] (mm)	Number of Cells on Chip	Operating Conditions (Class A,B,C)	Company [Reference]
2.0	Bipolar	30	7.0	–	32	2.2	1.5	43	10	C	MSC [A7]
4.0	Bipolar	8.0	7.0	–	25	4.3	1.0	24	8	C	TRW [A6]
	MESFET	4.0	6.0	7.0	44	9.2	1.5	5.2	2	A	Fujitsu [E11]
6.0	Bipolar	1.5	4.4	–	23	11	1.5	3.1	4	C	Hewlett-Packard [A5]
	JFET	1.0	6.0	7.0	26	4.4	1.5	6.1	2	A	Thomson-CSF [G6]
	MESFET	2.7	5.0	6.0	31	13	1.5	5.2	2	A	Fujitsu [E11]
8.0	Bipolar	0.5	6.0	–	22	22	0.5	1.1	2	B	Texas Instruments [A3]
	MESFET	0.6	6.0	7.5	34	21	1.5	1.4	3	A	Plessey [E19]
	MESFET	2.2	3.2	4.2	22	14	1.5	5.2	2	A	Fujitsu [E11]
15.0	MESFET	0.45	5.2	6.7	13	59	1.5	1.2	2	A	RCA [E17]
22.0	MESFET	0.14	4.8	5.6	9	76	1.5	0.6	1	A	RCA [E17]

[1] Operated with common source.

[2] Operated with common gate.

[3] Operated with common base.

[4] Associated with the stated output power.

[5] Only applicable for Class A operation.

[6] Defined as RF output power minus RF input power divided by the dissipated dc power.

[7] Defined as added RF power x $(frequency)^2$ divided by the total gate width or emitter periphery.

[8] Or emitter finger width.

[9] Or total emitter periphery.

shown in Fig. 6. Interdigitated source and drain fingers are located on the top side of the chip, and the gate with plated heat sink is located on the bottom side. The channel is confined to an area within the constricted cross section in the n-layer. This structure has the following advantages: 1) it places the active region in intimate thermal contact with the heat sink; 2) it enables an interdigitated structure without overcrossing, since the ohmic contacts and the gate are located on different sides of the chip; 3) it reduces the parasitic source-to-gate resistance; and 4) it makes a self-aligned process possible. Disadvantages are the higher gate-to-drain and gate-to-source capacitances.

Besides Si and GaAs, InP has been investigated for application in MESFET's. InP has a 50-percent higher maximum drift velocity than GaAs[7] [F2], [F3]. Therefore, one expects a higher current-gain bandwidth f_T for InP-MESFET's. Barrera and Archer [F4] have measured an f_T that is 1.6 times larger than in analogous GaAs devices. However, the maximum frequency of oscillation, f_{max}, is 20 percent lower. Degenerate feedback resulting from a large gate-to-drain capacitance and a small output resistance degrades the gain performance. Noise figures of InP-MESFET's are slightly higher.

The ideal semiconductor for FET's has simultaneously a large mobility, large maximum drift velocity, and large avalanche breakdown field. Consequently, a small electron effective mass, a large intervalley separation, and a large energy gap are required. The first and last requirements are conflicting, since a large energy gap implies a large effective mass; but it is possible to design a better compromise than is found in the binary compounds GaAs and InP.

[7] The low field mobility of InP is 25 percent lower, however.

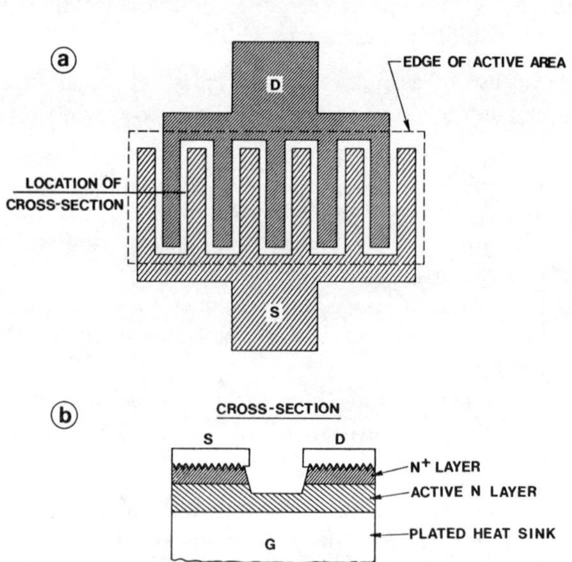

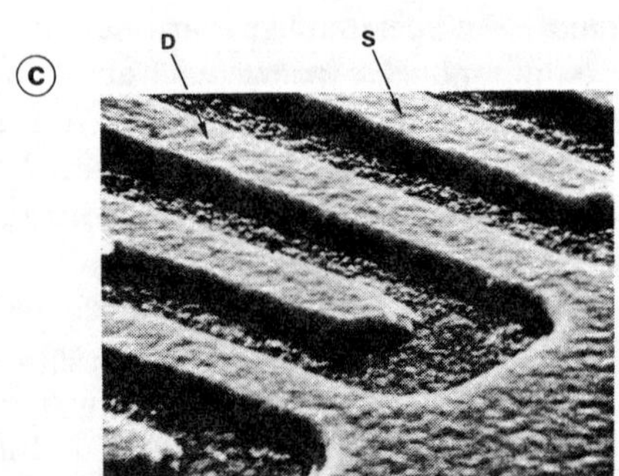

Fig. 6. Power MESFET structure with plated heat sink on the gate as proposed in [E5]. (a) Shows the metallization on the top side of the chip with interdigital source and drain fingers. Outside the active area, the GaAs is converted into semi-insulating material by proton bombardment. (b) Shows the cross section of the device with the source and drain on top and the gate with plated heat sink on the bottom of the chip. (c) Shows the top view of the FET chip [G6]. (SEM courtesy of C. Vergnolle, Thomson-CSF.)

Two systems which merit consideration are the mixed crystals InAs–InP and InAs–GaAs [F5], [F6]. Decker *et al.* [F7] have grown thin films of $In_{0.04}Ga_{0.96}As$ directly on GaAs. MESFET's fabricated on this material are very similar to their GaAs counterpart with the exception of a larger output resistance. The step in the bandgap at the InGaAs–GaAs interface constrains the electrons to the InGaAs layer and prevents penetration of hot electrons into the substrate at the high-field channel region.

B. JFET

A great deal of effort has been spent on developing p-n junction FET's for microwave applications. In the past, JFET's have not moved to higher frequencies as rapidly as MESFET's primarily because of the difficulty in realizing closely defined p-conducting regions by diffusion or implantation.[8] In more recent developments, such as the power JFET with ion-milled channel [G6] or the heterojunction JFET [G7], these limitations do not apply and rapid advances can be expected.

In 1972 Teszner [G1] described an Si-JFET with a vertical channel for power amplification. This multielement transistor has buried p^+-gate fingers, and the ohmic contacts for source and drain are located on opposite sides of the chip (Fig. 1). The structure was later realized by high-energy ion implantation of the gate [G2]. In this vertical geometry, no implantation-induced defects are generated in the channel. Consequently, low annealing temperatures were used and patterns with 1-μm-long channels were realized. Test devices, with 1.8-mm total gate width, delivered 200-mW output power with 6-dB power gain at 2.7 GHz.

In GaAs, Zuleeg *et al.* [G3] have realized a low-noise JFET with a diffused p-n junction of 2-μm length. The transistor exhibits 2.5-dB noise figure and 10-dB associated gain at 4 GHz. This JFET has a high tolerance to fast neutron radiation. A 1-MeV neutron fluence of 5×10^{16} neutrons/cm^2 is required to degrade the transconductance by 10 percent [G4], [G5].

A power GaAs-JFET has been built by Vergnolle *et al.* [G6] [Fig. 6(c)]. The construction is similar to the one shown in Fig. 6(b), except that the Schottky barrier is replaced by a p-n junction grown on a p^+-substrate. This structure has the same advantages as the MESFET of Fig. 6(a) and (b), except that the thermal resistance is limited by the spreading resistance in the GaAs substrate. This JFET with 1.5-μm gate length and 6-mm total gate width delivers 1-W output power[9] with 6-dB power gain and 26-percent power-added efficiency at 6 GHz.

A low-noise FET with a heterojunction gate is now under development at Matsushita Electric [G7]. The junction consists of a p-type $Ga_{0.5}Al_{0.5}As$ layer grown on top of an n-type GaAs layer. The p-layer can be selectively etched to the shape of a 1-μm gate strip. This feature combined with a self-aligned process enables simple fabrication of GaAs-JFET's.

C. IGFET

Field-effect transistors with insulated gates are of interest for power amplification. They offer the following advantages over MESFET's or JFET's. 1) In the active region of an enhancement-mode MOSFET, the input capacitance and the transconductance are almost independent of gate voltage, and the output capacitance is independent of the drain voltage.[10] This leads to very linear (Class A) power amplification with low amplitude and phase distortion. 2) The active gate-voltage range can be larger because n-channel depletion-type IGFET's can be operated from the depletion-mode region $(-V_{GS})$ to the enhancement-mode region $(+V_{GS})$. Unfortunately, practical IGFET's have only been made on Si. Attempts to realize a usable device in GaAs have had only limited success [H1], [H2]. It is difficult to fabricate an insulating film which produces an interface to n-GaAs with a low density of electron states [H2]–[H4]. Recent advances with anodic native oxides of GaAs show promising interface and dielectric properties [H5]. A density of fast interface states of 1 to 2×10^{11} cm^{-2}eV^{-1} has been observed. This is still 10 times higher than the best results in Si [H6], but sufficiently low for MOSFET operation.

Morita *et al.* [H7] have developed an n-channel depletion-type Si-MOSFET which delivers 16-W output power[11] with 6-dB associated power gain and 26-percent power-added efficiency at 700 MHz. The device has a conventional planar structure with diffused n^+ drain and source regions and a 5-μm-long and 20-mm-wide channel. A very different approach is the vertical-gate MOS transistor (VMOST) designed for power amplification [H8]. This device consists of mesa strips with control gates on both sides of each strip (Fig. 1). The source contact is made to the top surface of the mesa, and the n^+-substrate acts as a common drain. The length of the vertical n-channel is controlled by epitaxial growth and the metal-gate length by angular metal deposition. For a 1-μm-long and 18-mm-wide channel, a CW output power of 4 W[12] with 5-dB power gain has been measured at 2 GHz by Heng [H9]. Also, a double-diffused MOSFET (D-MOST) can be fabricated with channel lengths of less than 1 μm using standard photolithography [H10]. The short channel results from subsequent diffusions of the channel and source impurities under the same oxide layer. This technique gives accurate control of the channel length comparable to the control over the base width in diffused bipolar transistors. The D-MOST has been developed for low-noise UHF amplification by Sigg *et al.*

---

[8] Lateral diffusion or lateral spreading of the ions during implantation and during the subsequent high-temperature annealing step widen the p-regions with respect to mask dimensions.

[9] Measured at 1-dB gain compression. For a performance comparison with MESFET's, see Table II.

[10] This applies to MOSFET's with n^- drift region.

[11] Measured at 4-dB gain compression.

[12] Measured at 1-dB gain compression.

[H11], [H12]. Single-gate FET's exhibit 3.0-dB noise figure and 9.0-dB associated gain at 1 GHz [H13]. Dual-gate FET's suitable for mixers and for amplifiers with automatic gain control have 4.5-dB noise figure with 14-dB gain [H13]. Silicon-on-sapphire MOS transistors (SOS-MOST) have also been developed for UHF applications with the advantage of very low drain-to-channel and drain-to-source capacitances. A small-signal dual-gate SOS-MOST has been built by Ronen and Strauss [H14] with a 4-μm channel length yielding 25-dB gain at 0.5 GHz.

D. Conclusion

The MESFET's have been the most successful among the microwave FET's in low noise and in power amplification above 2 GHz. The reasons for this are easy realization on GaAs and the fact that the two critical dimensions, the gate length and channel thickness, can be accurately controlled. With advanced fabrication processes, such as E-beam lithography [C3], [C17], ion implantation [C13], [C16], [C27], [W4], [X4], and molecular-beam epitaxy [C28], further improvement in dimensional control is obtained. In comparison with silicon bipolar transistors, MESFET's have higher maximum frequency of oscillation, lower noise figures, and larger associated gain at microwave frequencies (Fig. 3). They also have higher reverse isolation and lower third-order intermodulation distortion. Above 4 GHz, GaAs MESFET's have better efficiency as power amplifiers than bipolar transistors (Table II). In addition, FET's do not exhibit secondary breakdown, are self-ballasting,[13] and have inherently higher input impedance. Also, the gate-to-source and drain-to-source impedances are fairly insensitive to temperature variations [S6]. As majority-carrier devices, FET's are more immune to the effects of neutron and gamma radiation than bipolar transistors [G4], [G5], [Q13]. Currently, much effort is spent in determining the reliability, failure modes, and stability of GaAs MESFET's [I1]–[I9]. Preliminary results indicate a meantime to failure in excess of 10^7 h at 70°C channel temperature as reported by Irie *et al.* [I3] and Abbott and Turner [I5]. Ch'en *et al.* [I7], [I8] observed an improvement in the long-term stability of MESFET parameters after passivating the GaAs surface with a thin coating of polycrystalline GaAs.

III. MESFET PRINCIPLE OF OPERATION

In this section, the physical principles in the operation of a silicon MESFET are explained. Then the differences between silicon and GaAs-MESFET's are outlined and effects occurring in FET's with very short gate length are discussed. Next, the equivalent circuit is presented and high-frequency limitations are described. Finally, the principles of the noise behavior are presented.

---

[13] With rising temperature, the channel and source resistances increase, preventing "thermal runaway."

A. Principles of Silicon MESFET Operation

The current-voltage characteristic of a thin n-type silicon layer in which electrons are carrying the current is plotted in Fig. 7(a). This layer is supported by an insulating silicon substrate. At the surface of the conducting layer, two ohmic contacts are made, called the source and drain. A cross section of this device is shown in Fig. 7(a).[14] If a positive voltage V_{DS} is applied to the drain, electrons will flow from source to drain. Hence the source acts as the origin of carriers and the drain as a sink. For small voltages, the silicon layer behaves like a linear resistor. For larger voltages, the electron drift velocity does not increase at the same rate as the electric field E (Fig. 8). As a result, the current-voltage characteristic falls below the initial resistor line. As V_{DS} is further increased, E reaches a critical field, E_c,[15] for which the electrons reach a maximum velocity, v_s (Fig. 8). At this drain voltage, the current starts to saturate.

In Fig. 7(b) a metal-to-semiconductor contact, called the gate, has been added between source and drain. This contact creates a layer in the semiconductor that is completely depleted of free-carrier electrons. This depletion layer acts like an insulating region and constricts the cross section available for current flow in the n-layer. The width of the depletion region depends on the voltage applied between the semiconductor and the gate. In Fig. 7(b) the gate is shorted to the source and a small drain voltage is applied. Under these conditions, the depletion layer has a finite width and the conductive channel beneath has a smaller cross section d than in Fig. 7(a). Consequently, the resistance between source and drain is larger, as shown in Fig. 7(b). The current I_{DS} flowing from drain to source is given by

$$I_{DS} = wqn(x)v(x)d(x) \tag{1}$$

where w is the gate width (see Fig. 11), q the charge of an electron, n the density of conduction electrons, v their drift velocity, d the conductive layer thickness, and x the co-ordinate in the direction of the electron drift. The electron density n is equal to the constant donor density N_D as long as the field does not exceed the critical value E_c. The voltage along the channel increases from zero at the source to V_{DS} at the drain. Thus the metal-to-semiconductor junction becomes increasingly reverse biased, and the depletion layer becomes wider as we proceed from source to drain. The resulting decrease in conductive cross section d must be compensated by an increase of electric field and electron velocity v to maintain a constant current through the channel. As the drain voltage is increased further, the electrons reach the maximum limiting velocity v_s under the drain end of the gate. This is illustrated in Fig. 7(c). The channel is constricted to the smallest cross section d_0 under the gate

---

[14] For simplicity, the bending of the bands at the free surface of the n-layer and the depleted region at the substrate interface are neglected. Also the electric field is assumed to be uniform in the n-region between the contacts.

[15] In silicon, the value of E_c cannot be accurately defined.

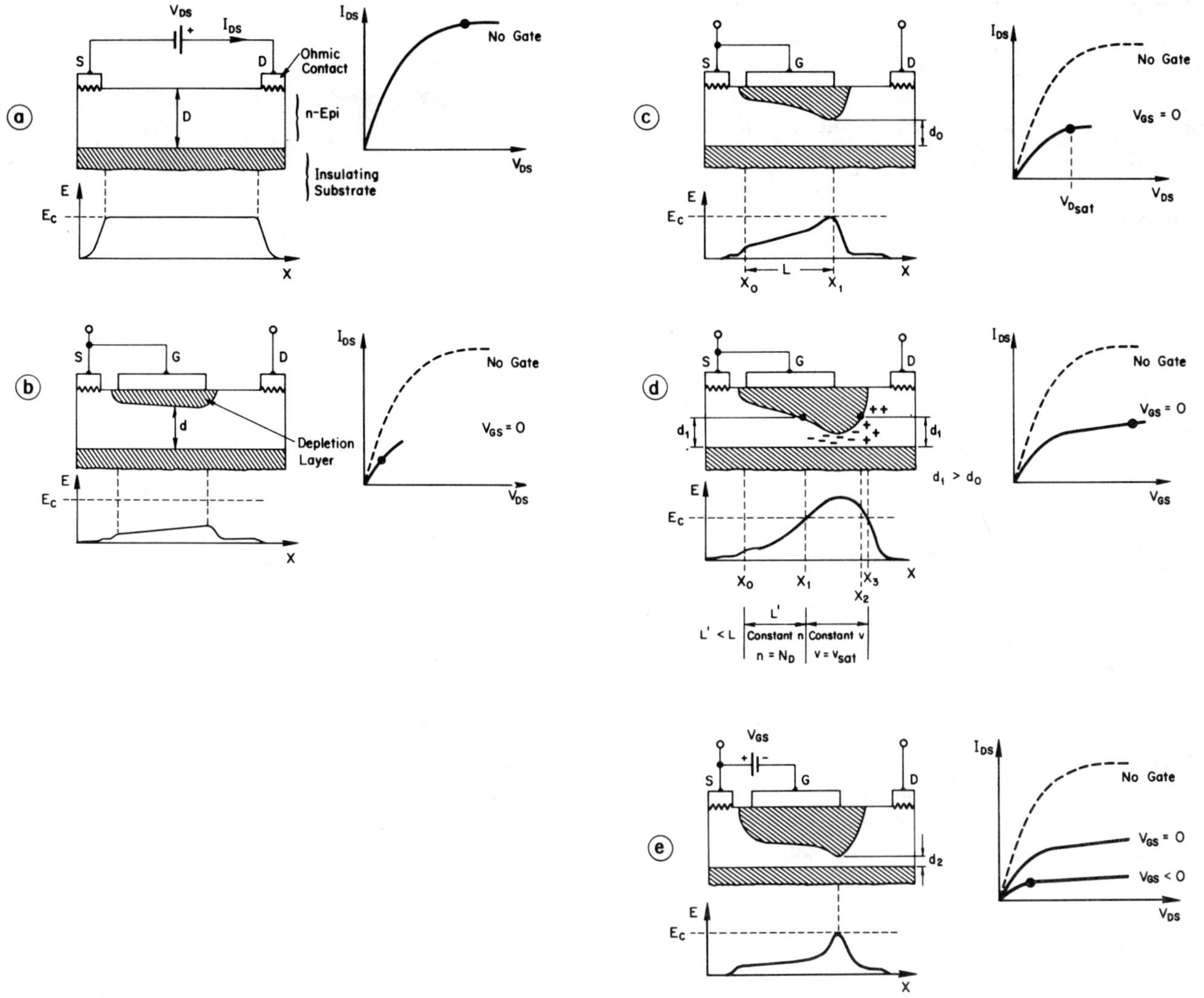

Fig. 7.   (a) Shows the $I-V$ characteristic of an n-type silicon layer with two ohmic contacts. The current saturates because the electrons reach a maximum drift velocity at the critical field E_c. In (b)–(d), the current is controlled by the depletion layer under a Schottky gate, shorted to the source. In (c) the current starts to saturate at $V_{D_{sat}}$, and (d) shows the formation of a stationary dipole layer in the channel for $V_{DS} > V_{D_{sat}}$ [J12], [K1]. (e) Illustrates the condition for a negative gate bias. The depletion layer is wider, it constricts the conductive cross section further, and causes the current to saturate at a lower level.

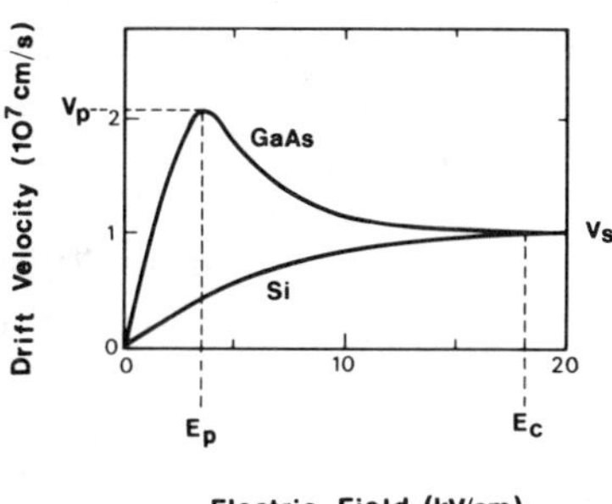

Fig. 8.   Equilibrium electron drift velocity versus electric field in GaAs and silicon (after Ruch [K11]).

edge, the electric field reaches the critical value at this point, and the current starts to saturate.

If the drain voltage is increased beyond $V_{D_{sat}}$, the depletion region widens toward the drain. The point x_1, where the electrons reach the limiting velocity, moves slightly toward the source [Fig. 7(d)]. As x_1 moves closer to the source, the voltage at x_1 decreases.[16] Consequently, the conductive cross section d_1 widens and more current is injected into the velocity-limited region. This results in a positive slope of the I_{DS} curve and a finite drain-to-source resistance beyond current saturation [J8], [J12]. The effect is particularly pronounced in microwave MESFET's with short gate lengths.

Proceeding from x_1 toward the drain, the channel potential increases, the depletion layer widens, and the channel cross section d becomes narrower than d_1. Since the electron velocity is saturated, the change in channel width must be compensated for by a change in carrier concentration to maintain constant current. According to (1), an electron accumulation layer forms between x_1 and x_2, where d is smaller than d_1. At x_2 the channel cross section is again d_1 and the negative space charge changes to a positive space charge to preserve constant current. The

---

[16] The average field between x_0 and x_1 remains nearly unchanged while the distance between x_0 and x_1 decreases.

positive space charge is caused by partial electron depletion. The electron velocity remains saturated between x_2 and x_3 due to the field added by the negative space charge. In short, the drain voltage applied in excess of $V_{D_{sat}}$ forms a dipole layer in a channel that extends beyond the drain end of the gate [J12], [K1].

When a negative voltage is applied to the gate [Fig. 7(e)], the gate-to-channel junction is reverse biased, and the depletion region grows wider. For small values of V_{DS}, the channel will act as a linear resistor, but its resistance will be larger due to a narrower cross section available for current flow. As V_{DS} is increased, the critical field is reached at a lower drain current than in the $V_{GS} = 0$ case, due to the larger channel resistance. For a further increase in V_{DS}, the current remains saturated. In essence, the MESFET consists of a semiconducting channel whose thickness can be varied by widening the depletion region under the metal-to-semiconductor junction. The depletion region widening is the effect of a field or voltage applied between gate and channel of the transistor.

Various analytical solutions for the voltage-current characteristics of short-gate MESFET's with field-dependent electron velocity have been developed. The majority [J2]–[J7] follow a one-dimensional analysis based on the gradual-channel approximation proposed by Shockley [J1]. They compute the drain current at the onset of current saturation [Fig. 7(c)]. Two-dimensional approximations of the field distribution for large drain voltages have also been derived [J8]–[J12]. These analytical solutions make allowance for space charges in the channel and enable calculations of the small-signal drain-to-source resistance in the saturated current region. Much effort has also been concentrated on accurate two-dimensional numerical solutions for Si [K1]–[K7], for GaAs and InP [K8]–[K14].

B. Principles of Gallium Arsenide MESFET Operation

In GaAs, the analysis in the high-field region is considerably more complicated than in Si because 1) the equilibrium electron velocity versus electric field reaches a peak value at about 3 kV/cm, then decreases and levels off at a saturated velocity that is about equal to the limiting velocity in silicon (Fig. 8) [F1], [K11]; 2) for gate lengths shorter than 3 μm, a nonequilibrium velocity-field characteristic has to be considered [K13].

A rigorous treatment of the electron transport in GaAs-MESFET's, based on the equilibrium velocity-field characteristic, has been carried out by Himsworth [K9]. Fig. 9 summarizes the key features of a transistor with 3-μm gate length operated far in the saturated current region. The narrowest channel cross section is located under the drain end of the gate. The drift velocity rises to a peak at x_1, close to the center of the channel, and falls to the low saturated value under the gate edge. To preserve current continuity according to (1), heavy electron accumulation has to form in this region because the channel cross section is narrowing and, in addition, the electrons are moving progressively slower with increasing x. Exactly the opposite occurs between x_2 and x_3. The channel widens and the

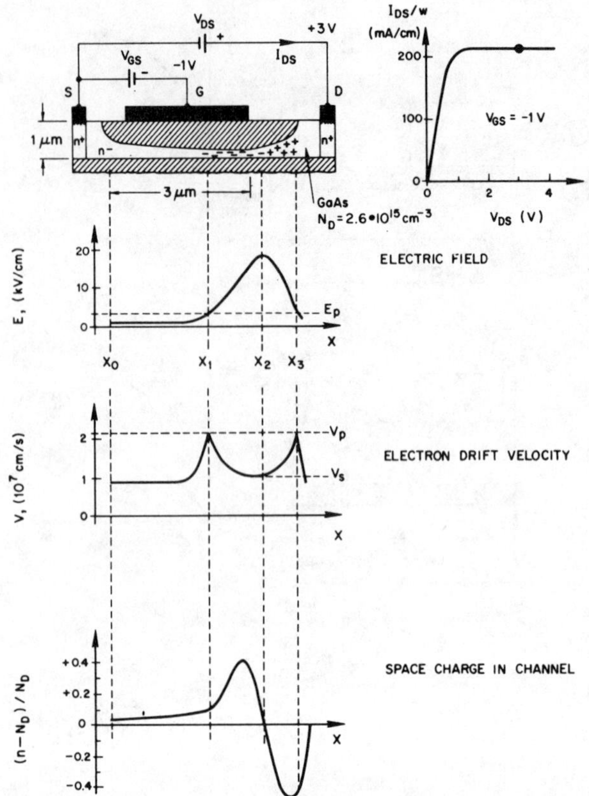

Fig. 9. The channel cross section, electric field, electron drift velocity, and space-charge distribution in the channel are illustrated for a GaAs-MESFET operated in the current-saturated region (data from Himsworth [K9]). Proceeding from x_1 to x_2, the channel cross section becomes narrower and, in addition, the electrons "slow down." To preserve current continuity, a heavy electron accumulation has to form. The opposite occurs between x_2 and x_3.

electrons move faster causing a strong depletion layer.[17] The charges in the accumulation and depletion layers are nearly equal and most of the drain voltage drops in this stationary dipole-layer.

In microwave FET's with very short gate length, the electrons do not reach equilibrium transport conditions in the high-field region of the channel. Nonequilibrium velocity-field characteristics in GaAs have been studied by computer simulations using Monte Carlo methods [K11]–[K14]. In a simplified approach, slow electrons are injected into a constant-field region and their drift velocity is monitored [K11]–[K13]. The situation is schematically illustrated in Fig. 10. As long as E is below the threshold field E_p, the electrons remain in equilibrium conditions. If the electrons enter a high-field region ($E > E_p$), they are accelerated to a higher velocity before relaxing to the equilibrium velocity.[18] This overshoot to more than twice the

---

[17] This region is not fully depleted of free electrons in contrast to the cross-hatched depletion layers.

[18] For $E < E_p$, electrons remain in the "lower valley" where they have a high mobility. For $E \gg E_p$, almost all electrons are transferred to a "satellite valley," a state in which they have a low mobility; i.e., low velocity at a given field. If the field changes suddenly from a value below to above E_p, a time period of approximately 1 ps passes before the carriers are transferred from the lower to the upper valley. During this time, the electrons remain in the high-mobility state in which they can acquire a high velocity in the high field.

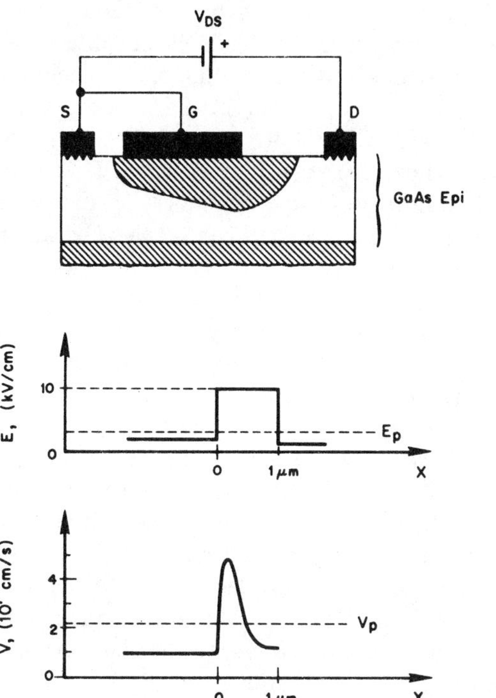

Fig. 10. This figure shows schematically the velocity overshoot of electrons as they enter the high-field region ($E > E_p$) under the gate (data from Ruch [K11]). Under equilibrium conditions, the maximum drift velocity is v_p (Fig. 8).

peak equilibrium velocity v_p, and the relaxation to the equilibrium condition, after traveling over a 0.6-μm path length [K11], is shown in Fig. 10. The effect is only noticeable in MESFET's with less than 3-μm gate length [K12], [K13]. The overshoot shortens the electron transit time through the high-field region and shifts the accumulation layer into the gap between gate and drain [K14].

C. Small-Signal Equivalent Circuit

An RF equivalent circuit of the MESFET should model the channel as a distributed RC network. However, a simple lumped-element circuit is capable of describing the FET's s-parameters accurately up to 12 GHz [C1], [C5], [L1]–[L4]. The equivalent circuit for operation in the saturated current region in common-source configuration is shown in Fig. 11(a). The location of the elements in the FET structure is illustrated in Fig. 11(b). In the intrinsic FET model, the elements $(C_{dg} + C_{gs})$ represent the total gate-to-channel capacitance; C_{dc} models the capacitance of the dipole layer; R_i and R_{ds} show the effects of the channel resistance; and i_{ds} defines the voltage-controlled current source. The transadmittance y_m relates i_{ds} to the voltage across C_{gs}. Up to 12 GHz, y_m is characterized by a frequency-independent magnitude, the transconductance g_m, and by a phase delay τ_0, reflecting the carrier transit time in the channel section where $E > E_p$. The extrinsic (parasitic) elements are: R_s the source resistance, R_d the drain resistance, R_g the gate-metal resistance, and C_{ds} the substrate capacitance. Typical element values for a GaAs-MESFET with 1-μm gate length and 500-μm gate width are listed in Table III.

The analysis of the equivalent circuit yields a critical frequency f_k, above which the MESFET is unconditionally

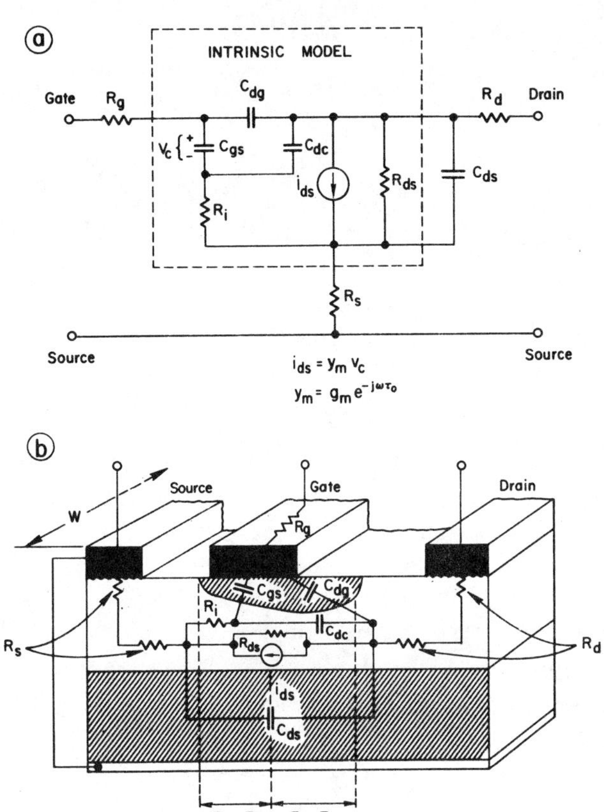

Fig. 11. (a) Is the equivalent circuit of a MESFET. Typical element values are listed in Table III. (b) Shows the physical origin of the circuit elements.

TABLE III
EQUIVALENT-CIRCUIT PARAMETERS OF A LOW-NOISE GaAs-MESFET WITH A 1-μm × 500-μm GATE (HP EXPERIMENTAL, $N_D = 1 \times 10^{17}$ cm^{-3})

Intrinsic Elements	Extrinsic Elements
g_m = 53 mmho	C_{ds} = 0.12 pF
τ_0 = 5.0 ps	R_g = 2.9 Ω
C_{gs} = 0.62 pF	R_d = 3 Ω
C_{dg} = 0.014 pF	R_s = 2.0 Ω
C_{dc} = 0.02 pF	L_g = 0.05 nH*
R_i = 2.6 Ω	L_d = 0.05 nH*
R_{ds} = 400 Ω	L_s = 0.04 nH*

dc Bias

V_{DS} = 5 V

V_{GS} = 0

I_{DS} = 70 mA

* Contacting inductances of the test fixture in series with R_g, R_d and R_s, respectively.

stable. f_k can be approximated by[19]

$$f_k \approx \frac{1}{2\pi(\tau_0 + \tau_1 + \tau_2)} \qquad (2)$$

[19] Equations (2)–(4) were derived as outlined in [B4].

where τ_0 is defined in Fig. 11 and

$$\tau_1 = \frac{C_{dg}(2R_g + R_i + R_s)}{\dfrac{C_{dg}}{C_{gs}} + \dfrac{R_s}{R_{ds}}} \tag{3}$$

$$\tau_2 = \frac{2}{\dfrac{g_m}{C_{gs}}\left[\dfrac{C_{dg}}{C_{gs}} + \dfrac{R_s}{R_{ds}}\right]} \frac{R_g + R_i + R_s}{R_{ds}}. \tag{4}$$

f_k is 6.1 GHz for the MESFET with the parameters listed in Table III. The MESFET with a complex-conjugate-matched input port becomes unstable with decreasing frequency because a larger fraction of the output voltage is fed back to the input over the $C_{dg} - R_{in}$[20] voltage divider. With decreasing frequency, R_{in} rises as $1/\omega^2$ while the reactance of C_{dg} increases only as $1/\omega$.

Mason's unilateral gain [Q1] is approximately

$$G_u \approx \left(\frac{f_u}{f}\right)^2 \tag{5}$$

where f_u is the maximum frequency of oscillation [B4], [C1]

$$f_u \approx \frac{f_T}{2\sqrt{r_1 + f_T\tau_3}} \tag{6}$$

f_T the frequency at unity current gain

$$f_T \approx \frac{1}{2\pi}\frac{g_m}{C_{gs}} \tag{7}$$

r_1 the input-to-output resistance ratio

$$r_1 = \frac{R_g + R_i + R_s}{R_{ds}} \tag{8}$$

and τ_3 the time constant

$$\tau_3 = 2\pi R_g C_{dg}. \tag{9}$$

Equation (5) shows a gain decreasing with 6 dB/octave as the frequency increases. At f_u, unity gain is reached.[21] To maximize f_u, the frequency f_T and the resistance ratio R_{ds}/R_i must be optimized in the intrinsic MESFET. In addition, the extrinsic resistances R_g and R_s and the feedback capacitance C_{dg} have to be minimized.

D. High-Frequency Limitations

The high-frequency limitations of MESFET's are dependent on device geometry and material parameters. In silicon and GaAs, electrons have a higher mobility than holes. Therefore, only n-channel FET's are used in microwave applications (see Table I). Electrons have six times higher low-field mobility[22] and two times higher maximum drift velocity in GaAs as opposed to silicon. The saturated velocities are about equal in both materials. As a consequence, the realized current-gain bandwidths f_T are about two times higher and the maximum frequencies of oscillation f_u three times higher in GaAs- as opposed to Si-MESFET's [C4], [C9], [C10]. In the device geometry, the most critical parameter is the gate length L. Decreasing the gate length decreases the capacitance C_{gs} and increases the transconductance g_m; consequently, there is an improved current-gain bandwidth f_T. For the short-gate-length microwave MESFET's, f_T is proportional to $1/L$ [J5]. High-speed operation is achieved by shrinking the gate length to the minimum size that can be realized with a given technology. Conventional photolithographic contact or projection-masking limits the smallest features to approximately a 1-μm size. An order of magnitude smaller gate length can be realized with X-ray or electron-beam lithography [X3]. A computer study of submicron Si-MESFET's by Reiser and Wolf [K6] reveals that f_T increases while the output resistance R_{ds} decreases with shrinking gate length. The limit for useful gate reduction is reached when the gate length is about equal to the channel thickness D. To keep $L/D > 1$, the channel thickness has to be decreased together with the gate length. This implies a higher doping level. In practical devices, the highest doping level is about 4×10^{17} cm^{-3} because of breakdown phenomena. The conclusion is that the gate length for Si-MESFET's should be larger than 0.1 μm. This geometry limits the current-gain bandwidth to about 70 GHz [K6]. In GaAs, quantitative high-frequency limitations need to be established. In very short gate devices ($L < 0.2$ μm), the field is above the threshold value E_p over the entire gate length [K9] and electrons are expected to remain in their high-mobility state for the entire flight through the channel [K12], [K13].

E. Principles of Noise Behavior

The noise properties of any linear two-port can be represented by a noiseless two-port with noise-current generators connected across the input and output ports [M1]. This is a physically meaningful way of describing the noise behavior of the intrinsic MESFET (Fig. 12).

The noise-current generator at the output represents the short-circuit channel noise generated in the drain-source path. The mean square of i_{nd} can be expressed by [M2]

$$\overline{i_{nd}^2} = 4kT_0\Delta f g_m P \tag{10}$$

with k the Boltzmann constant, T_0 the lattice temperature, Δf the bandwidth, g_m the transconductance, and P a factor depending on the device geometry and the dc bias. For zero drain voltage, i_{nd} characterizes the thermal noise generated by the drain conductance G_{ds}; i.e., $P = G_{ds}/g_m$. For positive drain voltages, the noise generated in the channel is larger than the thermal noise generated by G_{ds} for the following reasons. First, a thermal noise voltage generated locally in the channel modulates the conductive cross section of the channel and results in an amplified noise voltage at the drain [M2]. Second, the electrons are accelerated in the electric field, then scattered in all directions due to interactions with lattice phonons. Their random drift-velocities and the attributed free-carrier temperature

---

[20] R_{in} is the effective resistance between gate and source after the conjugate-impedance-matched generator has been connected; i.e.,

$$R_{in} \approx \frac{1}{2\omega^2 C_{gs}^2(R_g + R_i + R_s)}.$$

[21] f_u is 46 GHz for the MESFET of Table III.

[22] The comparison is made for a doping density of 1×10^{17} cm^{-3} [F1], [F8].

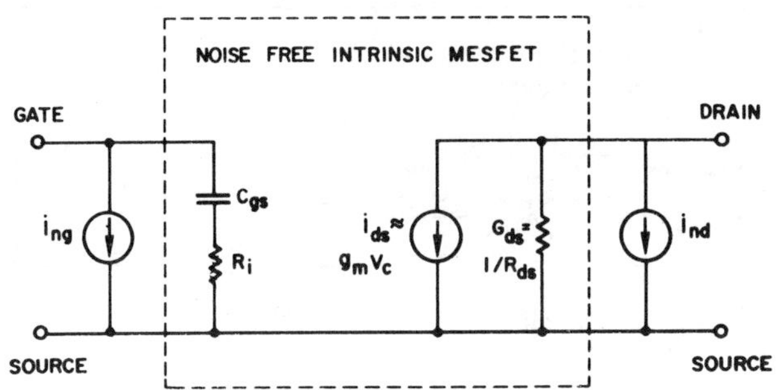

Fig. 12.   Equivalent circuit of the (simplified) intrinsic MESFET with noise-current sources at the input and output port.

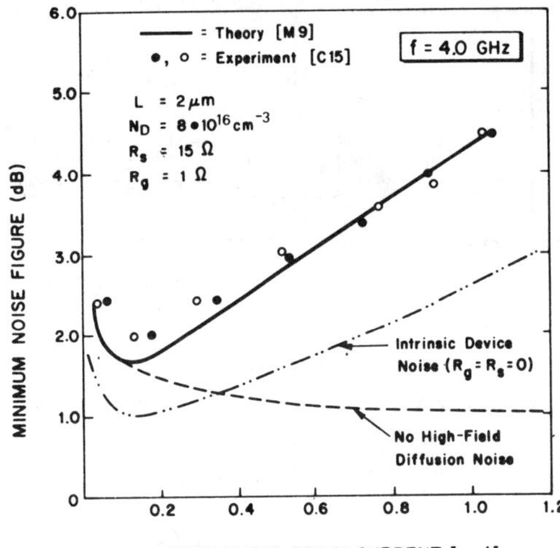

Fig. 13.   Theoretical and measured minimum noise figures as a function of normalized drain current for a GaAs-FET with a 2-μm gate. (Courtesy of R. Pucel [C22].)

increase with the applied field to values considerably higher than the lattice temperature (hot-electron noise [M6], [M7]). Third, in GaAs carriers undergo field-dependent transitions from the central valley in the conduction band to satellite valleys and vice versa. A transferred electron experiences an abrupt velocity change. These transitions cause statistical drift-velocity fluctuations and thus generate field-dependent "intervalley-scattering noise" [M8]. Fourth, for large drain voltages, the electrons reach their limiting velocity on the drain side of the channel. In this region, the field has no influence on the carrier drift velocity. Therefore, this channel section cannot be treated as an ohmic conductor. Here, the noise is formulated as high-field diffusion noise[23] [M5], [M9], [M10], and the mean square of the noise current is porportional to the high-field diffusion coefficient in the semiconductor.

A noise voltage, generated locally in the channel, causes a fluctuation in the depletion-layer width. The resulting charge fluctuation in the depletion layer in turn induces a compensating charge variation on the gate electrode. The total induced-gate charge fluctuation is described in Fig. 12 by a noise-current generator i_{ng} at the gate terminal [M3] where

$$\overline{i_{ng}^2} = 4kT_0\Delta f \frac{\omega^2 C_{gs}^2}{g_m} R. \tag{11}$$

C_{gs} is the gate-source capacitance and R a factor depending on the FET geometry and the bias conditions.[24] The two noise currents, i_{nd} and i_{ng}, are caused by the same noise voltages in the channel. Therefore, partial correlation has to be expected.[25] A correlation factor C is defined as [M3]

$$jC = \frac{\overline{i_{ng}^* \cdot i_{nd}}}{\sqrt{\overline{i_{ng}^2} \cdot \overline{i_{nd}^2}}} \tag{12}$$

where j is the imaginary unit and the asterisk defines the complex conjugate. The correlation coefficient is purely

imaginary because i_{ng} is caused by capacitive coupling of the gate circuit to the noise sources in the drain circuit. The factors P, R, and C in (10)–(12) have been computed versus normalized gate voltage by Baechtold [M8] for GaAs-MESFET's with various channel length-to-height ratios operated at the onset of current saturation. Statz et al. [M9] have extended the computation of P, R, and C to large drain voltages taking diffusion noise in the velocity-saturated channel region into account.

Using the model of Fig. 12, the minimum noise figure of the intrinsic MESFET can be expressed by [M7], [M10]

$$F_{\min} = 1 + 2\sqrt{PR(1 - C^2)}\frac{f}{f_T} + 2g_m R_i P \left(1 - C\sqrt{\frac{P}{R}}\right)\left(\frac{f}{f_T}\right)^2. \tag{13}$$

Low-noise MESFET's are normally operated at frequencies below f_T in order to yield sufficient gain. In this case, the linear frequency term in (13) is dominant. Short-gate MESFET's can exhibit very low noise figures for the following reasons. For an optimized drain current ($I_{DS}/I_{DSS} \approx 0.15$), the diffusion-noise contribution is small (Fig. 13), f_T is close to its maximum value [W3], and the correlation coefficient approaches unity ($C \approx 0.9$ [M9]). Substantial noise cancellation occurs at the drain which is expressed by the factor $(1 - C^2)$ in (13). The amplified input-noise current (αi_{ng}) destructively interferes with the correlated i_{nd} if the MESFET's gain and transmission phase are properly adjusted with an optimized input termination

$$\overline{|\alpha i_{ng} + i_{nd}|^2} \ll \overline{|\alpha i_{ng}|^2} + \overline{|i_{nd}|^2}. \tag{14}$$

In a practical MESFET, the parasitic resistances R_g and R_s, shown in Fig. 11, decrease the effectiveness of this noise cancellation [C22], [M9] and, in addition, they generate thermal noise themselves. A comparison between theoretical and experimental noise figures versus drain current is shown in Fig. 13. The increase of $F_{\min}$ with drain current is caused by the diffusion noise in the velocity-saturated

---

[23] The diffusion-noise theory is also valid at low fields and is a more general formulation than Johnson's formula for the thermal noise of a conductance [M5].

[24] For zero drain voltage, i_{ng} is the thermal noise of the input conductance $g_{11}(g_{11} \approx \omega^2 C_{gs}^2 R_i)$, and R is equal to the product $g_m R_i$.

[25] Complete correlation results for a channel with uniform conductive cross section (e.g., for zero drain voltage).

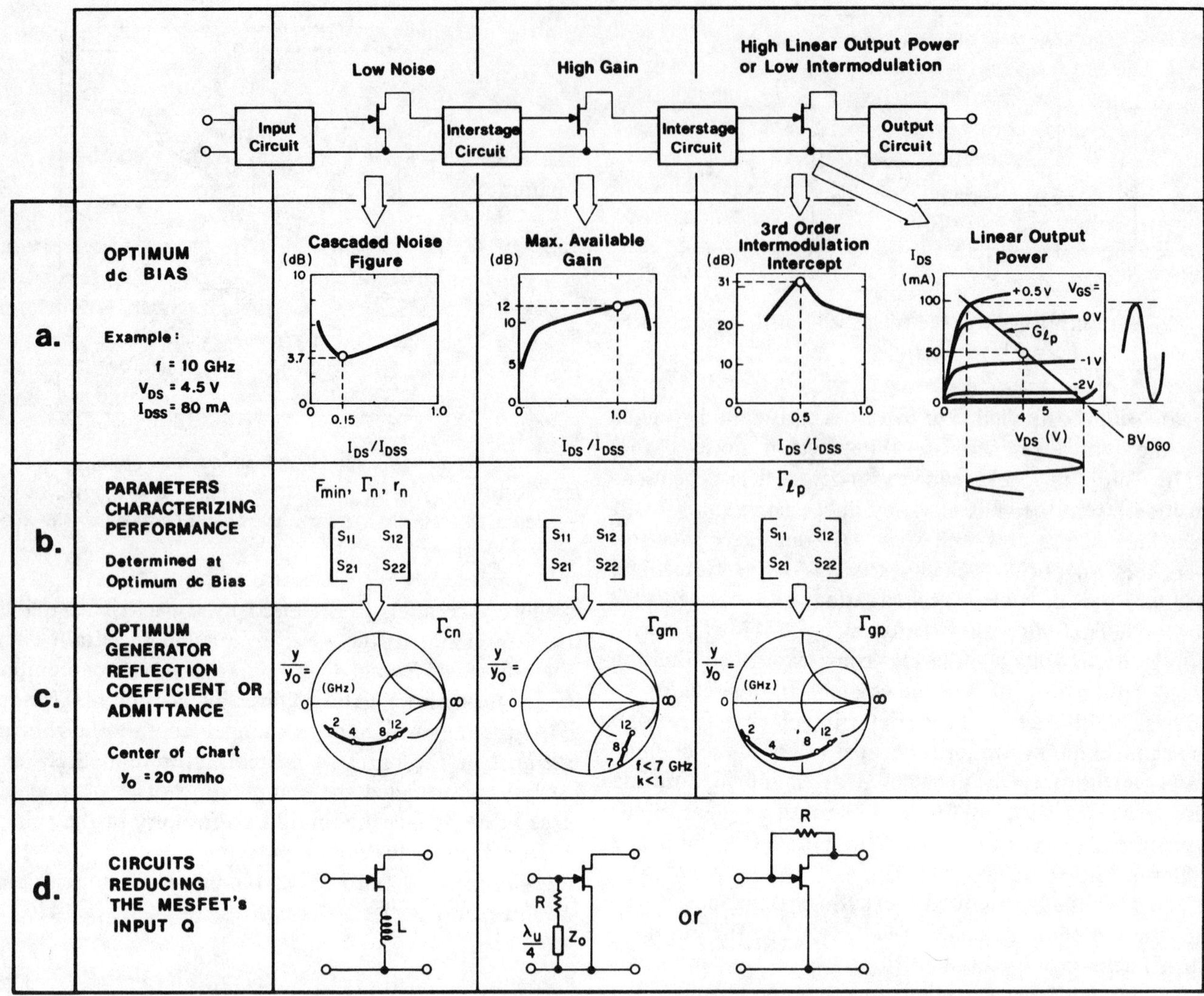

Fig. 14. The figure shows the key parameters in operating a MESFET in either a low-noise front stage, a high-gain stage, or a linear-power stage. (a) Illustrates the FET characteristics leading to the optimum dc bias. (b) Lists the parameters characterizing the performance (e.g., as computer input for CAD). (c) Shows optimum generator reflection coefficient to be synthesized by the circuits. The example is based on an HP MESFET with 1-μm gate length and 500-μm gate width. (d) Shows circuits to reduce the MESFET's input Q or to stabilize the transistor at low frequencies.

region [M9]. The noise-figure rise for small drain currents is caused by the rapid decrease of g_m and consequently of f_T. Also shown is the computed noise figure of the intrinsic MESFET.

IV. MESFET Amplifiers

In this section, basic concepts in the design of low-noise and linear medium-power MESFET amplifiers are reviewed.[26] The optimum dc bias, the parameters for device characterization, and the optimum generator admittance are discussed. Principles for network synthesis are described. Finally, performance characteristics of low-noise MESFET amplifiers are presented. As an example, an amplifier with a low-noise front stage, a high-gain stage, and a power-output stage is considered (Fig. 14). The amplifier shall be designed for low noise figure, flat gain, and high linear output-power capability across a specified band.

A. Optimized Operating Conditions for the MESFET's

The MESFET in the first-amplifier stage has to be operated at the dc bias yielding the lowest cascaded noise figure[27]

$$F_c = \frac{FG - 1}{G - 1} \tag{15}$$

where F_c is the noise figure of an infinite number of cascaded stages each having a noise figure F and gain G. F_c is nearly independent of the drain voltage as long as $V_{DS} > V_{D_{sat}}$; however, strongly dependent on the drain current [C15], [M7], [M9]. Typically, F_c reaches a minimum at $0.1–0.2 I_{DSS}$ [C15], [D4], [M9] where I_{DSS} is the

---

[26] The reader is also referred to recent review papers on solid-state microwave amplifiers by Cuccia [O1], Magarshack [O2], and Osbrink *et al.* [O3].

[27] The term "noise measure," M, proposed by Haus and Adler [N1], is omitted here. Instead, the "cascaded noise figure," F_c is adopted. The relationship between F_c and M is $F_c = 1 + M$.

saturated drain current at zero gate voltage [Figs. 13 and 14(a)]. The drain current is adjusted to this optimum value with the gate bias. In this operating point, the MESFET's noise behavior is characterized by 1) the minimum noise figure F_{min}; 2) the reflection coefficient of the generator Γ_n, which produces F_{min}; and 3) a dimensionless coefficient r_n [N2]–[N9]. For an arbitrary generator reflection coefficient Γ_g, the noise figure is then determined by

$$F = F_{min} + 4r_n \frac{|\Gamma_g - \Gamma_n|^2}{(1 - |\Gamma_g|^2)|1 + \Gamma_n|^2}. \quad (16)$$

If the s-parameters are known, all parameters of interest can be computed. These are, e.g., the optimum generator reflection coefficient for lowest cascaded noise figure Γ_{cn} [N3], [N9], the optimum load reflection coefficient, the associated transistor gain (17), etc. Γ_{cn} is in general different from the reflection coefficient for maximum gain Γ_{gm} [Fig. 14(c)]. A combination of lossless parallel and series feedback is capable of making Γ_{cn} and Γ_{gm} identical without changing F_c.[28] [N10]–[N14].

The MESFET in the second stage is operated with maximum small-signal gain [Fig. 14(a)]. This condition is obtained at approximately zero gate voltage ($I_{DS} \approx I_{DSS}$), where f_T reaches a maximum, and at a drain voltage that maximizes the output resistance and the resistance ratio $1/r_1$ in (8). The small-signal behavior is fully characterized with the four s-parameters [Q3]. The transducer gain [29] for any generator and load reflection coefficients, Γ_g and Γ_l, respectively, is [Q4]

$$G = \frac{|s_{21}|^2(1 - |\Gamma_g|^2)(1 - |\Gamma_l|^2)}{|(1 - s_{11}\Gamma_g)(1 - s_{22}\Gamma_l) - s_{12}s_{21}\Gamma_g\Gamma_l|^2}. \quad (17)$$

The s-parameters determine also Rollett's [Q2] stability factor k

$$k = \frac{1 + |s_{11}s_{22} - s_{12}s_{21}|^2 - |s_{11}|^2 - |s_{22}|^2}{2|s_{12}s_{21}|}. \quad (18)$$

If k is larger than unity,[30] an optimum combination of Γ_{gm} and Γ_{lm} simultaneously image-matches the two MESFET ports and maximizes the gain [Q4]. If k is smaller than unity, the MESFET is only conditionally stable. In this case, the terminations Γ_g and Γ_l must be carefully chosen to operate the transistor in a stable range [Q4], [Q5] or the resistive stabilization networks described below must be applied. Frequently, $k \gg 1$ and $|s_{12}|$ are small enough for the MESFET to be treated as a unilateral two-port ($s_{12} = 0$; $k = \infty$). In this case, the optimum generator and load terminations are

$$\Gamma_{gm} = s_{11}{}^* \qquad \Gamma_{lm} = s_{22}{}^* \quad (19)$$

and the maximum available gain obtained from (17) is

$$G_{max} = \frac{|s_{21}|^2}{(1 - |s_{11}|^2)(1 - |s_{22}|^2)}. \quad (20)$$

The asterisk in (19) defines the complex conjugate of the s-parameters.

The MESFET in the output stage is intended to operate as a linear (Class A) amplifier. The design objectives can be either lowest intermodulation distortion [P1]–[P3], largest added RF power [Q6], or largest linear output power [T1]. In the last case, the dc bias is graphically determined from the static drain-current versus drain-voltage characteristic. The bias and load conductance line are chosen to maximize the product of linear voltage and current swing [Fig. 14(a)]. The limitations are determined by the maximum dc power dissipation, the drain-to-gate breakdown voltage BV_{DGO}, and the positive gate bias ($V_{GS} \approx 0.5$ V) above which appreciable gate current flows. The optimum load conductance is typically much larger than the MESFET's output conductance. Consequently, the MESFET is not matched to the load and does not deliver maximum gain. However, the large load conductance shunting the MESFET's nonlinear output admittance reduces intermodulation distortion.[31] The optimum load conductance and susceptance determine the optimum reflection coefficient of the load Γ_{lp}. The generator reflection coefficient that provides a complex-conjugate match at the input is then

$$\Gamma_{gp} = \left[s_{11} + \frac{s_{12}s_{21}\Gamma_{lp}}{1 - s_{22}\Gamma_{lp}} \right]^* \quad (21)$$

and the associated gain is determined from (17).

Optimum generator reflection coefficients for low noise, high gain, and high linear-power operation are shown in Fig. 14(c). The plotted data are typical for an unpackaged small-signal GaAs-MESFET with 1-μm gate length and 500-μm gate width. At low frequencies, the FET has a high-Q input admittance and wide-band matching is difficult. Simple circuits that lower the Q value are illustrated in Fig. 14(d). A series-feedback inductance between source and ground increases the input series resistance, decreases the input reactance, and leaves F_c unchanged. A resistor in-series with a short-circuited shunt stub, connected between gate and source, lowers the Q and stabilizes the FET at the low-frequency end of the band [T1]. If the shunt stub is a quarter-wavelength long at the upper band edge, the circuit does not load the FET input and does not decrease the gain at this frequency. Also, a parallel feedback resistance has been proposed for stabilization and input Q lowering [Q7].

B. Amplifier Network Synthesis

The input matching network transforms the 50-Ω generator impedance to the optimum impedance with

---

[28] The MESFET's noise figure F and gain G change, but the cascaded noise figure F_c remains invariant.
[29] The transducer gain is defined as the power delivered to the load divided by the available power from the generator.
[30] i.e., the MESFET is operated above the critical frequency f_k discussed in Section III.

[31] If low third-order intermodulation is desired [Fig. 14(a)], a drain current is chosen that minimizes the distortion from the nonlinear transconductance and input capacitance [P3].

reflection coefficient Γ_{cn}. The transformed impedance versus frequency must have a negative-reactance slope, $dX/df < 0$. For narrow-band amplifiers, the circuit is derived from a simple graphical design using the Smith chart [L4], [N3], [Q9]–[Q12]. For moderate-bandwidth MESFET amplifiers, impedance-matching bandpass networks with quarter-wave resonators have been applied successfully [Q13]. These networks are derived from low-pass filter prototypes; the circuit topology is well defined, and the element values are optimized by simple computations. The design procedures treat the MESFET as a unilateral device ($s_{12} = 0$). For more accurate optimizations and for large bandwidths, computer-aided design procedures are used in which the MESFET is characterized by all noise- and s-parameters [Q14]–[Q19]. In general, a particular circuit topology is chosen by the designer. The values of the circuit elements are then optimized by the computer. The optimization routine searches in a systematic way for the global minimum of an error function E defined as

$$E = \sum_{f_1}^{f_n} \left[\text{calculated } H(f) - \text{required } H(f)\right]^2 \quad (22)$$

where f is the frequency and H is a performance function defined as a weighted sum of gain, noise figure, reflection coefficients, etc. The matching networks are normally built as microstrip circuits. A monolithic amplifier stage, combining lumped matching elements with the MESFET on a single GaAs chip, has also been reported [B1], [R13]. The amplifier modules exhibit very broad-band performance (6–12.4 GHz).

In the interstage networks, the insertion loss versus frequency response has to compensate the MESFET's gain slope [Fig. 15(a)] to achieve a flat amplifier gain. Generally, this is done by matching the output of the preceding FET to the input of the following transistor at the upper band edge and providing an increasing mismatch with decreasing frequency [Fig. 15(b)]. Analytical methods have been developed for the synthesis of reactive networks yielding the desired insertion loss versus frequency characteristic [Q20]–[Q24]. Since the gain slope is provided by reactive mismatch, high standing waves result between the stages at the lower band end. The high voltages generated in the standing waves enhance feedback in the FET's and the large reactance versus frequency slopes of the networks cause high group-delay variations. These problems can be avoided with dissipative coupling networks [Q13], [Q25], [Q26] which provide a lossless impedance match at the highest frequency in the band and introduce increasing resistive loss (attenuation) with decreasing frequency [Fig. 15(c)]. The amount of gain compensation has to be individually chosen for each interstage network to achieve lowest amplifier noise figure across the entire band [Q25] and to prevent premature power saturation in the driver stage [Q26].

MESFET amplifiers are built with balanced and unbalanced circuits. Balanced amplifiers consist of a pair of amplifiers or single-tuned stages whose inputs are connected to the

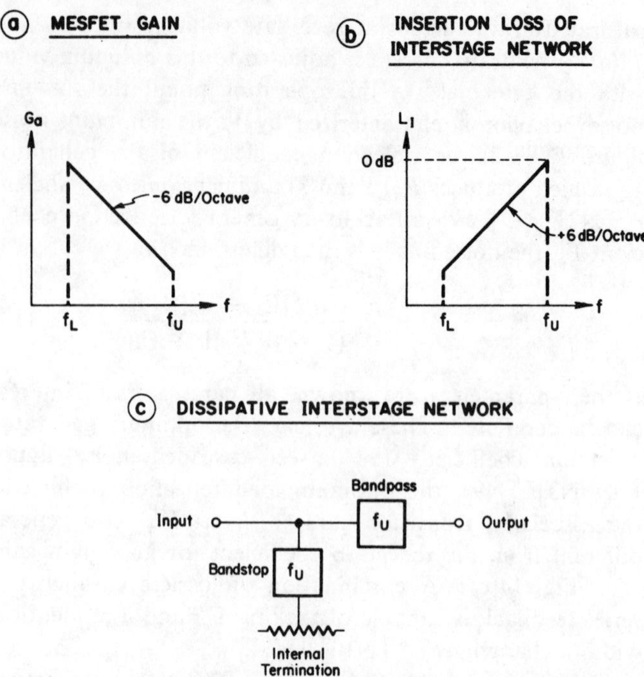

Fig. 15. For a flat amplifier gain, the MESFET's gain versus frequency slope (a) has to be compensated by an inverse slope in the interstage network's insertion loss (b). The insertion-loss characteristic can be obtained with reactive mismatch in an L–C network or with frequency-dependent attenuation in a diplexer resulting in dissipation of the excess power in an internal termination (c).

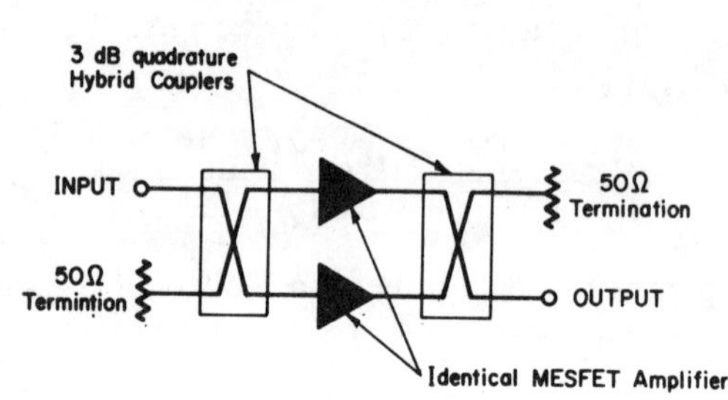

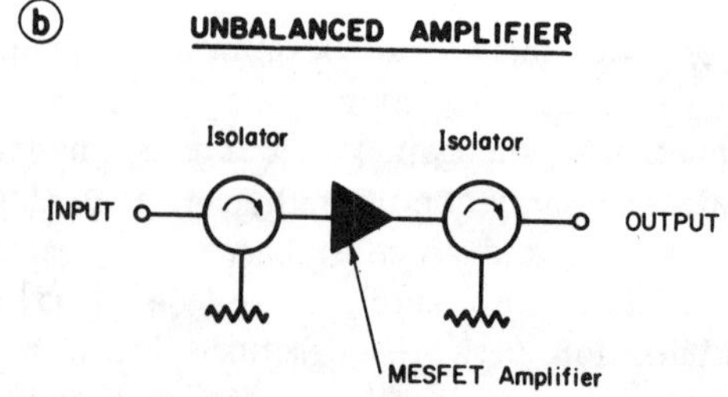

Fig. 16. MESFET amplifiers are built in balanced form with 3-dB hybrid couplers at the input and output of each stage or as an unbalanced chain of stages with isolators on both amplifier ports.

conjugate ports of a 3-dB hybrid coupler and whose outputs are similarly connected to another 3-dB coupler [Fig. 16(a)] [Q27]. The signal applied at the input port of the first coupler splits into two equal parts and is fed to the two amplifiers. The amplified signals from the two amplifier outputs recombine in the second coupler and emerge at one of the coupler's output ports. The advantages of the balanced amplifier over an unbalanced amplifier are improvement in: 1) input and output impedance matching in an amplifier optimized for noise figure or output power, 2) short- and open-circuit stability, 3) phase linearity, 4) gain compression, 5) intermodulation characteristics, and 6) reduced sensitivity to transistor impedance variations, provided the MESFET's are selected in similar pairs. Unbalanced amplifiers, on the other hand, need only half as many MESFET's, matching networks, and dc power. In general, isolators are required at the input and output of broad-band unbalanced MESFET amplifiers to meet low VSWR specifications and to make the noise figure independent of the source admittance [Fig. 16(b)]. 3-dB hybrid couplers are also used for power combining [Q28] as shown in Fig. 17.

C. MESFET Amplifier Performance

Various small-signal FET amplifiers have been described in the literature [C14], [D6], [G3], [P3], [Q7], [Q13], [Q22], [R1]–[R15]. Most are low-noise designs. Noise figures of laboratory prototypes are plotted versus frequency in Fig. 18. The solid lines represent narrow-band amplifiers. Lowest noise figures are 2.2 dB at 4 GHz, 3.6 dB at 8 GHz, and 5.0 dB at 12 GHz. The single data points (circles, squares, etc.) show noise figures of the 1-μm MESFET's used in the first amplifier stages. The data points lie about 1.0 dB below the amplifier noise figures. The noise-figure difference is caused by the insertion loss of the input circuit and by the noise contribution of the following stages. The noise figures of wide-band amplifiers, plotted with dashed lines, are typically 1.0–1.5 dB higher than the narrow-band circuits. At room temperature, thermal-noise sources dominate the noise performance of GaAs MESFET's in microwave amplifiers [M8], [M9]. By cooling the amplifier, a significant noise reduction can be obtained [S1]–[S6]. 1.6-dB noise figure (i.e., 130 K input noise temperature) was measured at 60 K for a 12-GHz amplifier [S5], [S6].

Octave-band MESFET amplifiers covering the 4.0–8.0-GHz frequency range have been built in balanced form [C14], [R5], [R6], [T3]. Typically, a three-stage small-signal amplifier has 24-dB gain, ±0.7-dB gain variation, 1.8:1 maximum VSWR at the input and output port, $+13$-dBm output power for 1-dB gain compression, and $+20$-dBm third-order intermodulation intercept [R16], [R17]. In the 8.0–12.0-GHz band, a three-stage unbalanced amplifier without isolators exhibits 20 ± 1.3-dB gain, 2.5:1 maximum VSWR, $+13$-dBm output power, and $+26$-dBm intermodulation intercept [Q13]. Less gain variation, e.g., 28.5 ± 0.5 dB, can be obtained with a balanced design [R15].

A few medium-power amplifiers have been reported [T1]–[T6]. At 6 GHz, a four-stage amplifier with 26-dB gain and 1-W output power,[32] using a single MESFET chip in the output stage, has been built [T2]. More common are balanced output stages which combine the output power of two transistors [T1], [T3], [T6]. Also, wide-band medium-power amplifiers have been designed [T1], [T3],

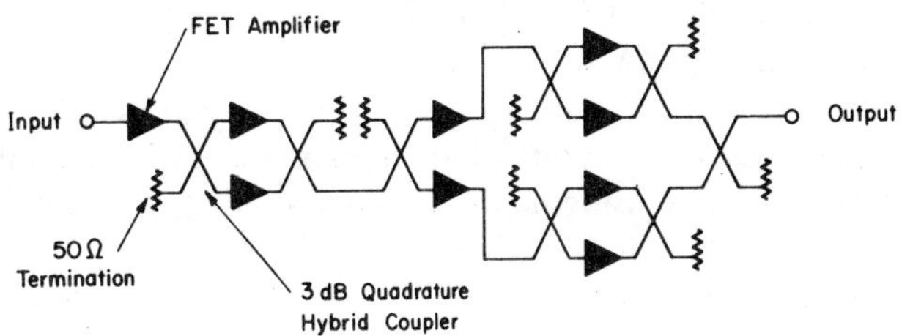

Fig. 17.   In transistor power amplifiers, 3-dB hybrid couplers are also used as power combiners (after S. Lazar [Q28]).

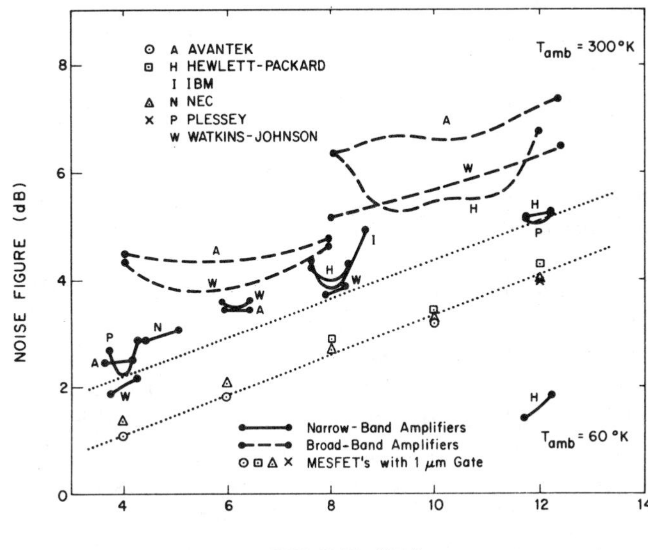

Fig. 18.   This noise figure versus frequency graph summarizes performance of experimental MESFET amplifiers. The solid lines represent narrow-band amplifiers, and the single data points show noise figures of the 1-μm MESFET's used in the first stages. The broken lines illustrate the noise performance of wide-band amplifiers.

and one covers the 2–6-GHz band [T1]. In this frequency range, MESFET's with 1-μm gate length have a high-Q input impedance, and matching circuits with resistive components, illustrated in Fig. 14(d), must be used.

V. OTHER APPLICATIONS

So far, comparatively little effort has been spent on the development of GaAs-MESFET oscillators [U1]–[U8]. This application should prove to be of interest since MESFET's combine the advantages of low bias voltage (< 10 V), relatively low noise measure (< 23 dB),[33] and high efficiency (> 15 percent at 10 GHz [U2]). Also GaAs-MESFET mixers [V1]–[V5] are expected to receive more attention in the near future. Good noise performance ($F = 7.4$ dB)[34] and large dynamic range (third-order intermodulation intercept $= +18$ dBm)[34] can be achieved with conversion gain ($G = 6$-dB)[34] [V3]. Another application is the use

[32] Measured at 1-dB gain compression.

[33] This oscillator noise measure is measured at 1-MHz separation from the X-band carrier [U3]. This noise measure is high in comparison to amplifier noise figures because of upconverted $1/f$ noise. Reduction of traps at the active-layer surface and at the substrate interface is expected to improve the low-frequency noise performance of GaAs-MESFET's in the near future.

[34] Measured at 8 GHz with a balanced MESFET mixer.

of Si-MESFET's [W1], [W2] and GaAs-MESFET's [W3]–[W7] or GaAs-JFET's [W8], [W9] in high-speed digital circuits. Monolithic integrated-logic gates with 100-ps signal-propagation delay (fanout 2) and 4-pJ speed-power product have been built using 1-μm GaAs-MESFET's [W3], [W4]. This is less than half the propagation delay measured on highest speed bipolar logic [W10], [W11]. In addition, the feasibility for medium-scale integration of these GaAs circuits has been demonstrated [W4].

VI. Conclusions and Outlook

GaAs-FET's are capable of low-noise amplification, high-efficiency power amplification and generation, high-speed modulation, and logic. Since these are areas of major microwave-system needs, substantial efforts in device and application developments are anticipated. In the near future, rapid advances are expected in the areas of 1) FET reliability, 2) device fabrication with high yield, uniform and reproducible unit-to-unit parameters by means of ion implantation, and electron-beam lithography, and 3) higher power capability and efficiency due to improved device structures with better heat sinking, thermally stable ohmic contacts, and more burn-out-resistant gates. Looking further ahead, a strong trend toward monolithic integration for digital, analog, and hybrid applications is now apparent. The monolithic approach is attractive because MESFET's fabricated on the same active layer can be used as switches, logic gates with active loads, impedance transformers, amplifiers, oscillators, and mixers; and the devices can be supported, isolated, and interconnected with low parasitic capacitances on the semi-insulating substrate. Monolithic integration is required to handle the complexity of tasks and to serve high-volume low-cast markets. Integrated high-speed logic will be needed in digital communications with gigabit-per-second data rates [W12]–[W14], in multi-phase-shift-keyed modulation and demodulation, time multiplexing, frequency division, counting, frequency synthesis, and waveform synthesis. A need is foreseen for microwave analog and hybrid circuits, on a single chip, such as 1) combinations of a preamplifier, mixer, local oscillator, and IF amplifier, 2) wide-band signal and operational amplifiers, 3) sample and hold circuits, and 4) high-speed D/A and A/D converters. Advances in III–V materials preparation and in submicron device processing will make microwave monolithic circuits with FET's a reality.

Acknowledgment

The author wishes to thank Dr. D. Ch'en, G. Gilbert, Dr. T. Heng, Dr. C. Snapp, Dr. J. Turner, and M. Walker for supplying new, unpublished data, and Dr. R. Archer, Dr. G. Bechtel, Dr. R. Engelmann, Dr. U. Gysel, Dr. R. Lee, Dr. J. Magarshack, Dr. C. Stolte, B. Lizenby, A. Podell, and R. Van Tuyl for many helpful discussions and for their assistance in reviewing the manuscript. The author is also grateful to Mrs. S. Ybarra for drafting the figures and to Ms. E. Miller for typing the manuscript at times of heavy work load.

References

A. Bipolar Transistors

[A1] R. Engelbrecht and K. Kurokawa, "A wide-band low-noise L-band balanced transistor amplifier," *Proc. IEEE*, vol. 53, pp. 237–247, March 1965.

[A2] J. Archer, "Low-noise implanted-base microwave transistors," *Solid-State Electron.*, vol. 17, pp. 387–393, April 1974.

[A3] H. Yuan, J. Kruger, and Y. Wu, "X-band silicon power transistor," in *1975 Int. Microwave Symp., Dig. Tech. Papers*, pp. 73–75.

[A4] C. Snapp, Hewlett-Packard Associates, private communication (HP-515).

[A5] J. Chen and K. Verma, "A 6 GHz silicon bipolar power transistor," in *1974 Int. Electron Devices Meeting, Dig. Tech. Papers*, pp. 299–301.

[A6] J. Steenbergen, A. Harrington, and G. Schreyer, "Broadband power transistor 4.4–5.0 GHz," U.S. Army Electronics Command, Final Technical Report, Contract No. DAAB07-73-C-0283, Fort Monmouth, February 1976.

[A7] G. Gilbert, Microwave Semiconductor Corporation, private communication.

[A8] C. Nuese, J. Gannon, R. Dean, H. Gossenberger, and R. Enstrom, "GaAs vapor-grown bipolar transistors," *Solid-State Electron.*, vol. 15, pp. 81–91, Jan. 1972.

[A9] W. Dumke, J. Woodall, and V. Rideout, "GaAs-GaAlAs heterojunction transistor for high frequency operation," *Solid-State Electron.*, vol. 15, pp. 1339–1343, Dec. 1972.

[A10] M. Konagai and K. Takahashi, "(GaAl)As-GaAs heterojunction transistors with high injection efficiency," *J. Appl. Phys.*, vol. 46, pp. 2120–2124, May 1975.

B. Review Papers on Microwave FET's

[B1] J. Turner, "Microwave FET's and their applications," in *Proc. 1975 Cornell Conference on Active Semiconductor Devices for Microwave and Integrated Optics*, pp. 13–22.

[B2] C. Liechti, "Recent advances in high-frequency field-effect transistors," in *1975 Int. Electron Devices Meeting, Dig. Tech. Papers*, pp. 6–10.

[B3] R. Pringle, "Microwave transistor and monolithic integrated circuit technology," *Microelectron. J.*, vol. 6, pp. 33–41, June 1975.

[B4] S. Ohkawa, K. Suyama, and H. Ishikawa, "Low noise GaAs field-effect transistors," *Fujitsu Sci. Tech. J.*, vol. 11, pp. 151–173, March 1975.

C. Low-Noise Si and GaAs MESFET's (Single-Gate)

[C1] P. Wolf, "Microwave properties of Schottky-barrier field-effect transistors," *IBM J. Res. Develop.*, vol. 14, pp. 125–141, March 1970.

[C2] K. Drangeid, R. Sommerhalder, and W. Walter, "High-speed gallium-arsenide Schottky-barrier field-effect transistors," *Electron. Lett.*, vol. 6, pp. 228–229, April 1970.

[C3] J. Turner, A. Waller, R. Bennett, and D. Parker, "An electron beam fabricated GaAs microwave field-effect transistor," *1970 Symp. GaAs and Related Compounds* (Inst. Phys., Conf. Series No. 9, London, 1971), pp. 234–239.

[C4] W. Baechtold, W. Walter, and P. Wolf, "X and Ku band GaAs MESFET," *Electron. Lett.*, vol. 8, pp. 35–37, Jan. 1972.

[C5] C. Liechti, E. Gowen, and J. Cohen, "GaAs microwave Schottky-gate FET," in *1972 Int. Solid-State Circuits Conf., Dig. Tech. Papers*, pp. 158–159.

[C6] G. Bechtel, W. Hooper, and D. Mock, "X-band GaAs FET," *Microwave J.*, vol. 15, pp. 15–19, Nov. 1972.

[C7] F. Doerbeck, "A planar GaAs Schottky-barrier field-effect transistor with a self-aligned gate," *1970 Symp. GaAs and Related Compounds* (Inst. Phys., Conf. Series No. 9, London, 1971), pp. 251–258.

[C8] M. Driver, H. Kim, and D. Barrett, "Gallium arsenide self-aligned gate field-effect transistors," *Proc. IEEE*, vol. 59, pp. 1244–1245, Aug. 1971.

[C9] W. Baechtold and P. Wolf, "An improved microwave silicon MESFET," *Solid-State Electron.*, vol. 14, pp. 783–790, Sep. 1971.

[C10] W. Baechtold *et al.*, "Si and GaAs 0.5 μm-gate Schottky-barrier field-effect transistors," *Electron. Lett.*, vol. 9, pp. 232–234, May 1973.

[C11] J. Jahncke, "Hoechstfrequenzeigenschaften eines GaAs-MESFET's in Streifenleitungstechnik," *Nachr. Tech. Z.*, vol. 26, pp. 193–199, May 1973.

[C12] R. Hunsperger and N. Hirsch, "GaAs field-effect transistors with ion-implanted channels," *Electron. Lett.*, vol. 9, pp. 577–578, Dec. 1973.

[C13] ——, "Ion-implanted microwave field-effect transistors in GaAs," *Solid-State Electron.*, vol. 18, pp. 349–353, April 1975.

[C14] D. Ch'en and A. Woo, "A practical 4 to 8 GHz GaAs FET amplifier," *Microwave J.*, vol. 17, pp. 26, 72, Feb. 1974.

[C15] G. Brehm and G. Vendelin, "Biasing FET's for optimum performance," *Microwaves*, vol. 13, pp. 38–44, Feb. 1974.

[C16] B. Welch, F. Eisen, and J. Higgins, "Gallium arsenide field-effect transistors by ion implantation," *J. Appl. Phys.*, vol. 45, pp. 3685–3687, Aug. 1974.

[C17] F. Ozdemir *et al.*, "Electron beam fabricated 0.5 μm gate GaAs Schottky-barrier field-effect transistor," *1974 Int. Electron Devices Meeting (late news paper), Suppl. Dig. Tech. Papers*, p. 5.

[C18] E. Kohn, "V-shaped-gate GaAs MESFET for improved high-frequency performance," *Electron. Lett.*, vol. 11, p. 160, April 1975.

[C19] E. Kohn, R. Wueller, R. Stahlmann, and H. Beneking, "High-speed 1 μm GaAs MESFET," *Electron. Lett.*, vol. 11, pp. 171–172, April 1975.

[C20] Data sheets: Fairchild, FMT 940/941, FMT 980/981; Plessey, GAT 3/4; Nippon Electric Company, NE 244/388; Hitachi, HCRL 81–84.

[C21] M. Ogawa, K. Ohata, T. Furutsuka, and N. Kawamura, "Submicron single-gate and dual-gate GaAs MESFET's with improved low noise and high gain performance," *IEEE Trans. Microwave Theory and Techniques*, this issue, pp. 300–305.

[C22] R. Pucel, D. Masse, and C. Krumm, "Noise performance of gallium-arsenide field-effect transistors." in *Proc. 1957 Cornell Conference on Active Semiconductor Devices for Microwaves and Integrated Optics*, pp. 265–276.

[C23] J. Barrera, "The importance of substrate properties on GaAs FET performance," in *Proc. 1975 Cornell Conference on Active Semiconductor Devices for Microwaves and Integrated Optics*, pp. 135–144.

[C24] W. Kellner, H. Kniepkamp, D. Ristow, and H. Boroffka, "Microwave field-effect transistors from sulphur-implanted GaAs," in *1975 Int. Electron Devices Meeting, Dig. Tech. Papers*, pp. 238–242.

[C25] P. Baudet, M. Binet, and D. Boccon-Gibod, "Submicrometer self-aligned GaAs MESFET," *IEEE Trans. Microwave Theory and Techniques*, this issue, pp. 372–376.

[C26] T. Nozaki, M. Ogawa, H. Terao, and H. Watanabe, "Multi-layer epitaxial technology for the Schottky-barrier GaAs field-effect transistor," *1974 Conf. GaAs and Related Compounds* (Inst. Phys., Conf. Ser. No. 24, London, 1975), pp. 46–54.

[C27] J. Higgins, B. Welch, F. Eisen, and G. Robinson, "Performance of ion-implanted GaAs MESFET's," in *1975 Int. Electron Devices Meeting (late news paper), Suppl. Dig. Tech. Papers*, p. 5.

[C28] A. Cho and D. Ch'en, "GaAs MESFET prepared by molecular beam epitaxy (MBE)," *Appl. Phys. Lett.*, vol. 28, pp. 30–31, Jan. 1976.

D. Dual-Gate GaAs MESFET's

[D1] J. Turner, A. Waller, E. Kelley, and D. Parker, "Dual-gate GaAs microwave FET," *Electron. Lett.*, vol. 7, pp. 661–662, Nov. 1971.

[D2] S. Asai, F. Murai, and H. Kodera, "The GaAs dual-gate FET with low noise and wide dynamic range," in *1973 IEEE Int. Electron Devices Conf., Dig. Tech. Papers*, pp. 64–67.

[D3] C. Liechti, "Characteristics of dual-gate GaAs MESFETs," in *Proc. 1974 European Microwave Conf.*, pp. 87–91.

[D4] ——, "Performance of dual-gate GaAs MESFETs as gain-controlled, low-noise amplifiers and high-speed modulators," *IEEE Trans. Microwave Theory and Techniques*, vol. MTT-23, pp. 461–469, June 1975.

[D5] R. Dean and R. Matarese, "Submicrometer self-aligned dual-gate GaAs FET," *IEEE Trans. Electron Devices*, vol. ED-22, pp. 358–360, June 1975.

[D6] M. Maeda and Y. Minai, "Application of dual-gate GaAs FET to microwave variable-gain amplifiers," in *1974 IEEE Int. Microwave Symp., Dig. Tech. Papers*, pp. 351–353.

[D7] S. Asai, F. Murai, and H. Kodera, "GaAs dual-gate Schottky-barrier FET's for microwave frequencies," *IEEE Trans. Electron Devices*, vol. ED-22, pp. 897–904, Oct. 1975.

E. Power GaAs MESFET's

[E1] M. Fukuta, T. Mimura, I. Tujimura, and A. Furumoto, "Mesh source type microwave power FET," in *1973 Int. Solid-State Circuits Conf., Dig. Tech. Papers*, pp. 84–85.

[E2] L. Napoli *et al.*, "High power GaAs FET amplifier—A multi-gate structure," in *1973 Int. Solid-State Circuits Conf., Dig. Tech. Papers*, pp. 82–83.

[E3] M. Driver, M. Geisler, D. Barrett, and H. Kim, "S-band microwave power FET," in *1973 IEEE Int. Electron Devices Meeting, Dig. Tech. Papers*, pp. 393–395.

[E4] H. Yamasaki *et al.*, "S-band power GaAs field-effect transistors," in *Proc. 1975 Cornell Conf. on Active Semiconductor Devices for Microwaves and Integrated Optics*, pp. 287–296.

[E5] T. Blocker, H. Macksey, and R. Adams, "X-band RF power performance of GaAs FET's," in *1974 IEEE Int. Electron Devices Meeting, Dig. Tech. Papers*, pp. 288–291.

[E6] H. Macksey and R. Adams, "Fabrication processes for GaAs power FET's," in *Proc. 1975 Cornell Conf. on Active Semiconductor Devices for Microwave and Integrated Optics*, pp. 255–264.

[E7] J. Angus, R. Butlin, D. Parker, R. Bennett, and J. Turner, "The design and evaluation of GaAs power MESFET's," in *Proc. 1975 European Microwave Conf.*, pp. 291–295.

[E8] M. Fukuta, H. Ishikawa, K. Suyama, and M. Maeda, "GaAs 8 GHz-band high power FET," in *1974 IEEE Int. Electron Devices Meeting., Dig. Tech. Papers*, pp. 285–287.

[E9] M. Fukuta, K. Suyama, H. Suzuki, and H. Ishikawa, "GaAs microwave power FET," *IEEE Trans. Electron Devices*, vol. ED-23, pp. 388–394, Apr. 1976.

[E10] M. Fukuta, K. Suyama, H. Suzuki, Y. Nakayama, and H. Ishikawa, "X-band GaAs Schottky-barrier power FET with a high drain-source breakdown voltage," in *1976 Int. Solid-State Circuits Conf., Dig. Tech. Papers*, pp. 166–167.

[E11] ——, "Power GaAs MESFET with a high drain-source breakdown voltage," *IEEE Trans. Microwave Theory and Techniques*, this issue, pp. 312–317.

[E12] L. Napoli and R. DeBrecht, "Performance and limitations of FETs as microwave power amplifiers," in *1973 IEEE Int. Microwave Symp., Dig. Tech. Papers*, pp. 230–232.

[E13] L. Napoli, J. Hughes, W. Reichert, and S. Jolly, "GaAs FET for high power amplifiers at microwave frequencies," *RCA Rev.*, vol. 34, pp. 608–615, Dec. 1973.

[E14] R. Camisa, I. Drukier, H. Huang, J. Goel, and S. Narayan, "GaAs MESFET linear amplifiers," in *1975 IEEE Int. Solid-State Circuits Conf., Dig. Tech. Papers*, pp. 70–71.

[E15] I. Drukier, R. Camisa, S. Jolly, H. Huang, and S. Narayan, "Medium power J-band MESFET's," in *Proc. 1975 Cornell Conf. on Active Semiconductor Devices for Microwaves and Integrated Optics*, pp. 297–304.

[E16] ——, "Medium power GaAs field-effect transistors," *Electron. Lett.*, vol. 11, pp. 104–105, March 1975.

[E17] H. Huang *et al.*, "GaAs MESFET performance," in *1975 Int. Electron Devices Meeting, Dig. Tech. Papers*, pp. 235–237.

[E18] H. Huang, I. Drukier, R. Camisa, S. Narayan, and S. Jolly, "High-efficiency GaAs MESFET amplifiers," *Electron. Lett.*, vol. 11, pp. 508–509, Oct. 1975.

[E19] J. Turner, The Plessey Company, private communication.

F. Electron Transport Properties in GaAs, InP and InGaAs; InP and InGaAs MESFET's

[F1] J. Ruch and W. Fawcett, "Temperature dependence of the transport properties of gallium arsenide determined by a Monte Carlo method," *J. Appl. Phys.*, vol. 41, pp. 3843–3849, Aug. 1970.

[F2] H. Lam and G. Acket, "Comparison of the microwave velocity field characteristics of n-type InP and n-type GaAs," *Electron. Lett.*, vol. 7, pp. 722–723, Dec. 1971.

[F3] W. Fawcett and D. Herbert, "High-field transport in gallium arsenide and indium phosphide," *J. Phys. C: Solid State Phys.*, vol. 7, pp. 1641–1654, May 1974.

[F4] J. Barrera and R. Archer, "InP Schottky-gate field-effect transistors," *IEEE Trans. Electron Devices*, vol. ED-22, pp. 1023–1030, Nov. 1975.

[F5] W. Fawcett, C. Hilsum, and H. Rees, "Optimum semiconductor for microwave devices," *Electron. Lett.*, vol. 5, pp. 313–314, July 1969.

[F6] M. Glicksman, R. Enstrom, S. Mittleman, and J. Appert, "Electron mobility in $In_xGa_{1-x}As$ alloys," *Phys. Rev. B*, vol. 9, pp. 1621–1626, Feb. 1974.

[F7] D. Decker, R. Fairman, and C. Nishimoto, "Microwave InGaAs Schottky-barrier-gate field-effect transistors—Preliminary results," in *Proc. 1975 Cornell Conf. on Active Semiconductor Devices for Microwaves and Integrated Optics*, pp. 305–314.

[F8] S. Sze, *Physics of Semiconductor Devices.* New York: Wiley, 1969, pp. 40, 59.

G. Si and GaAs JFET's

[G1] S. Tesziner, "Gridistor development for the microwave power region," *IEEE Trans. Electron Devices*, vol. ED-19, pp. 355–364, March 1972.

[G2] D. Lecrosnier and G. Pelous, "Ion-implanted FET for power applications," *IEEE Trans. Electron Devices*, vol. ED-21, pp. 113–118, Jan. 1974.

[G3] R. Zuleeg, E. Bledl, and A. Behle, "Broadband GaAs field-effect transistor amplifier," Air Force Avionics Lab., Technical Report AFAL-TR-73-109, Wright-Patterson AFB, March 1973.

[G4] A. Behle and R. Zuleeg, "Fast neutron tolerance of GaAs JFET's operating in the hot electron range," *IEEE Trans. Electron Devices*, vol. ED-19, pp. 993–995, Aug. 1972.

[G5] R. Zuleeg and K. Lehovec, "Radiation effect on GaAs interface," Air Force Cambridge Research Labs., Scientific Report AFCRL-TR-74-0495, Sep. 1974.

[G6] C. Vergnolle, R. Funck, and M. Laviron, "An adequate structure for power microwave FETs," in *1975 Int. Solid-State*

Circuits Conf., Dig. Tech. Papers, pp. 66–67.

[G7] S. Umebachi, K. Asahi, M. Inoue, and G. Kano, "A new heterojunction gate GaAs FET," *IEEE Trans. Electron Devices*, vol. ED-22, pp. 613–614, Aug. 1975.

H. Si and GaAs IGFET's

[H1] H. Becke and J. White, "Gallium arsenide insulated-gate field-effect transistors," *1966 Symp. GaAs* (Inst. Phys., Conf. Series No. 3, London, 1967), pp. 219–227.

[H2] T. Miyazaki, N. Nakamura, A. Doi, and T. Tokuyama, "N-channel gallium arsenide MISFET," in *1973 Int. Electron Devices Meeting, Dig. Tech. Papers*, pp. 164–166.

[H3] A. Adams and B. Pruniaux, "GaAs surface film evaluation by ellipsometry and its effect on Schottky barriers," *J. Electrochem. Soc.*, vol. 120, pp. 408–414, March 1973.

[H4] R. Singh and H. Hartnagel, "New method of passivating GaAs with Al_2O_3," in *1974 Int. Electron Devices Meeting, Dig. Tech. Papers*, pp. 576–578.

[H5] H. Hasegawa, K. Forward, and H. Hartnagel, "New anodic native oxide of GaAs with improved dielectric and interface properties," *Appl. Phys. Lett.*, vol. 26, pp. 567–569, May 1975.

[H6] G. Declerck, T. Hattori, G. May, J. Beaudouin, and J. Meindl, "Some effects of trichloroethylene oxidation on the characteristics of MOS devices," *J. Electrochem. Soc.*, vol. 122, pp. 436–439, March 1975.

[H7] Y. Morita, H. Takahashi, H. Matayoshi, and M. Fukuta, "Si UHF MOS high-power FET," *IEEE Trans. Electron Devices*, vol. ED-21, pp. 733–734, Nov. 1974.

[H8] J. G. Oakes, R. A. Wickstrom, D. A. Tremere, and T. M. S. Heng, "A power silicon microwave MOS transistor," *IEEE Trans. Microwave Theory and Techniques*, this issue, pp. 305–311.

[H9] T. Heng, Westinghouse Research Laboratories, private communication.

[H10] T. Cauge, J. Kocsis, H. Sigg, and G. Vendelin, "Double-diffused MOS transistor achieves microwave gain," *Electronics*, vol. 44, pp. 99–104, Feb. 1971.

[H11] H. Sigg, G. Vendelin, T. Cauge, and J. Kocsis, "D-MOS transistor for microwave applications," *IEEE Trans. Electron Devices*, vol. ED-19, pp. 45–53, Jan. 1972.

[H12] H. Sigg, D. Pitzer, and T. Cauge, "D-MOS for UHF linear and nanosecond switching applications," in *1974 IEEE INTERCON, Dig. Tech. Papers*, Session 32/4, pp. 1–7, March 1974.

[H13] H. Sigg, Signetics Corp., private communication (see also Signetics Data Sheet, SD-203/SD-308).

[H14] R. Ronen and L. Strauss, "The silicon-on-sapphire MOS tetrode, some small-signal features LF to UHF," *IEEE Trans. Electron Devices*, vol. ED-21, pp. 100–109, Jan. 1974.

I. GaAs MESFET Reliability

[I1] K. Ohta and M. Ogawa, "Degradation of gold-germanium ohmic contact to n-GaAs," in *1974 IEEE Reliability Physics Symp., Dig. Tech. Papers*, pp. 278–283.

[I2] H. Kohzu, I. Nagasako, M. Ogawa, and N. Kawamura, "Reliability studies of one-micron Schottky-gate GaAs FET," in *1975 Int. Electron Devices Meeting, Dig. Tech. Papers*, pp. 247–250.

[I3] T. Irie, I. Nagasako, H. Kohzu, and K. Sekido, "Reliability study of GaAs MESFET's," *IEEE Trans. Microwave Theory and Techniques*, this issue, pp. 321–328.

[I4] D. Abbott and J. Turner, "Some aspects of GaAs FET reliability," in *1975 Int. Electron Devices Meeting, Dig. Tech. Papers*, pp. 243–246.

[I5] ——, "Some aspects of GaAs MESFET reliability," *IEEE Trans. Microwave Theory and Techniques*, this issue, pp. 317–321.

[I6] S. Bellier, R. Haythornthwaite, J. May, and P. Woods, "Reliability of microwave GaAs field-effect transistors," in *Proc. 1975 Reliability Physics Symp.*, pp. 193–199.

[I7] D. Ch'en, H. Cooke, and J. Wholey, "Long-term stabilization of microwave FET's," *Microwave J.*, vol. 18, pp. 60–61, Nov. 1975.

[I8] ——, "Microwave FET's with improved amplifier stability," in *1976 Int. Solid-State Circuits Conf., Dig. Tech. Papers*, pp. 160–161.

[I9] S. Bearse, "GaAs FET's: Device designers solving reliability problems," *Microwaves*, vol. 15, pp. 32–52, Feb. 1976.

J. JFET and MESFET Theory of Operation (Analytic Solutions)

[J1] W. Shockley, "A unipolar field-effect transistor," *Proc. IRE*, vol. 40, pp. 1365–1376, Nov. 1952.

[J2] J. Turner and B. Wilson, "Implications of carrier velocity saturation in a gallium arsenide field-effect transistor," in *Proc. 1968 Symp. GaAs* (Inst. Phys., Conf. Series No. 7, London, 1969), pp. 195–204.

[J3] K. Drangeid and R. Sommerhalder, "Dynamic performance of Schottky-barrier field-effect transistors," *IBM J. Res. Develop.*, vol. 14, pp. 82–94, March 1970.

[J4] K. Lehovec and R. Zuleeg, "Voltage-current characteristics of GaAs J-FET's in the hot electron range," *Solid-State Electron.*, vol. 13, pp. 1415–1426, Oct. 1970.

[J5] P. Hower and G. Bechtel, "Current saturation and small-signal characteristics of GaAs field-effect transistors," *IEEE Trans. Electron Devices*, vol. ED-20, pp. 213–220, March 1973.

[J6] K. Lehovec and W. Seeley, "On the validity of the gradual channel approximation for junction field-effect transistors with drift velocity saturation," *Solid-State Electron.*, vol. 16, pp. 1047–1054, Sep. 1973.

[J7] R. Fair, "Graphical design and iterative analysis of the dc parameters of GaAs FET's," *IEEE Trans. Electron Devices*, vol. ED-21, pp. 357–362, June 1974.

[J8] A. Grebene and S. Ghandi, "General theory for pinched operation of the junction-gate FET," *Solid-State Electron.*, vol. 12, pp. 573–589, July 1969.

[J9] P. Rossel and J. Cabot, "Output-resistance properties of the GaAs Schottky-gate field-effect transistor in saturation," *Electron. Lett.*, vol. 11, pp. 150–152, April 1975.

[J10] D. Mo and H. Yanai, "Current-voltage characteristics of the junction-gate field-effect transistor with field-dependent mobility," *IEEE Trans. Electron Devices*, vol. ED-17, pp. 577–586, Aug. 1970.

[J11] G. Alley and H. Talley, "A theoretical study of the high frequency performance of a Schottky-barrier field-effect transistor fabricated on a high-resistivity substrate," *IEEE Trans. Microwave Theory and Techniques*, vol. MTT-22, pp. 183–189, March 1974.

[J12] K. Lehovec and R. Miller, "Field distribution in junction field-effect transistors at large drain voltages," *IEEE Trans. Electron Devices*, vol. ED-22, pp. 273–281, May 1975.

K. JFET and MESFET Theory of Operation (Numerical Solutions)

[K1] D. Kennedy and R. O'Brien, "Computer-aided two dimensional analysis of the junction field-effect transistor," *IBM J. Res. Develop.*, vol. 14, pp. 95–116, March 1970.

[K2] ——, "Two-dimensional analysis of J-FET structures containing a low-conductivity substrate," *Electron. Lett.*, vol. 7, pp. 714–716, Dec. 1971.

[K3] C. Kim and E. Yang, "An analysis of current saturation mechanism of junction field-effect transistors," *IEEE Trans. Electron Devices*, vol. ED-17, pp. 120–127, Feb. 1970.

[K4] C. Kim, "Differential drain resistance of field-effect transistors beyond pinchoff: A comparison between theory and experiment," *IEEE Trans. Electron Devices*, vol. ED-17, pp. 1088–1089, Dec. 1970.

[K5] M. Reiser, "Two-dimensional analysis of substrate effects in junction FET's," *Electron. Lett.*, vol. 6, pp. 493–494, Aug. 1970.

[K6] M. Reiser and P. Wolf, "Computer study of submicrometre FET's," *Electron. Lett.*, vol. 8, pp. 254–256, May 1972.

[K7] M. Reiser, "A two-dimensional numerical FET model for dc, ac, and large-signal analysis," *IEEE Trans. Electron Devices*, vol. ED-20, pp. 35–45, Jan. 1973.

[K8] B. Himsworth, "A two-dimensional analysis of indium-phosphide junction field-effect transistors with long and short channels," *Solid-State Electron.*, vol. 16, pp. 931–939, Aug. 1973.

[K9] ——, "A two-dimensional analysis of gallium-arsenide junction field-effect transistors with long and short channels," *Solid-State Electron.*, vol. 15, pp. 1353–1361, Dec. 1972.

[K10] J. Barnes and R. Lomax, "Two-dimensional finite element simulation of semiconductor devices," *Electron. Lett.*, vol. 10, pp. 341–343, Aug. 1974.

[K11] J. Ruch, "Electron dynamics in short channel field-effect transistors," *IEEE Trans. Electron Devices*, vol. ED-19, pp. 652–654, May 1972.

[K12] T. Maloney and J. Frey, "Effects of nonequilibrium velocity-field characteristics on the performance of GaAs and InP field-effect transistors," in *1974 Int. Electron Devices Meeting, Dig. Tech. Papers*, pp. 296–298.

[K13] ——, "Frequency limits of GaAs and InP field-effect transistors," *IEEE Trans. Electron Devices*, vol. ED-22, pp. 357–358, June 1975; also corrections in *IEEE Trans. Electron Devices*, vol. ED-22, p. 620, Aug. 1975.

[K14] R. Hockney, R. Warriner, and M. Reiser, "Two-dimensional particle models in semiconductor device analysis," *Electron. Lett.*, vol. 10, pp. 484–486, Nov. 1974.

L. Equivalent Circuits for MESFET's

[L1] R. Dawson, "Equivalent circuit of the Schottky-barrier field-effect transistor at microwave frequencies," *IEEE Trans. Microwave Theory and Techniques*, vol. MTT-23, pp. 499–501, June 1975.

[L2] G. Vendelin and M. Omori, "Circuit model for the GaAs MESFET valid to 12 GHz," *Electron. Lett.*, vol. 11, pp. 60–61, Feb. 1975.

[L3] ——, "Try CAD for accurate GaAs MESFET models," *Microwaves*, vol. 14, pp. 58–70, June 1975.

[L4] G. D. Vendelin, "Feedback effects in the GaAs MESFET

model," *IEEE Trans. Microwave Theory and Techniques*, this issue, pp. 383–385.

M. MESFET Noise Theory

[M1] H. Rothe and W. Dahlke, "Theory of noisy fourpoles," *Proc. IRE*, vol. 44, pp. 811–818, June 1956.

[M2] A. Van der Ziel, "Thermal noise in field-effect transistors," *Proc. IRE*, vol. 50, pp. 1808–1812, Aug. 1962.

[M3] ——, "Gate noise in field-effect transistors at moderately high frequencies," *Proc. IEEE*, vol. 51, pp. 461–467, March 1963.

[M4] A. Van der Ziel and J. Ero, "Small-signal, high-frequency theory of field-effect transistors," *IEEE Trans. Electron Devices*, vol. ED-11, pp. 128–135, April 1964.

[M5] A. Van der Ziel, "Thermal noise in the hot electron regime in FET's," *IEEE Trans. Electron Devices*, vol. ED-18, p. 977, Oct. 1971.

[M6] F. Klaassen, "On the influence of hot carrier effects on the thermal noise of field-effect transistors," *IEEE Trans. Electron Devices*, vol. ED-17, pp. 858–862, Oct. 1970.

[M7] W. Baechtold, "Noise behavior of Schottky barrier gate field-effect transistors at microwave frequencies," *IEEE Trans. Electron Devices*, vol. ED-18, pp. 97–104, Feb. 1971.

[M8] ——, "Noise behavior of GaAs field-effect transistors with short gate lengths," *IEEE Trans. Electron Devices*, vol. ED-19, pp. 674–680, May 1972.

[M9] H. Statz, H. Haus, and R. Pucel, "Noise characteristics of gallium arsenide field-effect transistors," *IEEE Trans. Electron Devices*, vol. ED-21, pp. 549–562, Sep. 1974.

[M10] R. Pucel, H. Haus, and H. Statz, "Signal and noise properties of gallium arsenide microwave field-effect transistors," in *Advances in Electronics and Electron Physics*, vol. 38. New York: Academic Press, 1975, pp. 195–265.

N. Characterization of FET Noise Performance

[N1] H. Haus and R. Adler, *Circuit Theory of Linear Noisy Networks*. New York: Wiley, 1959.

[N2] H. Haus, "IRE standards on methods of measuring noise in linear two ports, 1959," *Proc. IRE*, vol. 48, pp. 60–68, Jan. 1960.

[N3] H. Fukui, "Available power gain, noise figure and noise measure of two-ports and their graphical representations," *IEEE Trans. Circuit Theory*, vol. CT-13, pp. 137–142, June 1966.

[N4] A. Leupp and M. Strutt, "High-frequency FET noise parameters and approximation of the optimum source admittance," *IEEE Trans. Electron Devices*, vol. ED-16, pp. 428–431, May 1969.

[N5] R. Lane, "The determination of device noise parameters," *Proc. IEEE*, vol. 57, pp. 1461–1462, Aug. 1969.

[N6] R. Kaesser, "Noise factor contours for field-effect transistors at moderately high frequencies," *IEEE Trans. Electron Devices*, vol. ED-19, pp. 164–171, Feb. 1972.

[N7] A. Anastassiou and M. Strutt, "Experimental gain and noise parameters of microwave GaAs FET's in the *L* and *S* bands," *IEEE Trans. Microwave Theory and Techniques*, vol. MTT-21, pp. 419–422, June 1973.

[N8] ——, "Experimental and computed four scattering and four noise parameters of GaAs FET's up to 4 GHz," *IEEE Trans. Microwave Theory and Techniques*, vol. MTT-22, pp. 138–140, Feb. 1974.

[N9] J. Eisenberg, "Designing amplifiers for optimum noise figure," *Microwaves*, vol. 13, pp. 36–44, April 1974.

[N10] J. Engberg, "Simultaneous input power match and noise optimization using feedback," in *Proc. 1974 European Microwave Conf.*, pp. 385–389.

[N11] G. Vendelin, "Feedback effects on the noise performance of GaAs MESFET's," in *1975 Int. Microwave Symp., Dig. Tech. Papers*, pp. 324–326.

[N12] S. Iversen, "The effect of feedback on noise figure," *Proc. IEEE*, vol. 63, pp. 540–542, March 1975.

[N13] L. Besser, "Stability considerations of low-noise transistor amplifiers with simultaneous noise and power match," in *1975 Int. Microwave Symp., Dig. Tech. Papers*, pp. 327–329.

[N14] A. Anastassiou and M. Strutt, "Effect of source lead inductance on the noise figure of a GaAs FET," *Proc. IEEE*, vol. 62, pp. 406–408, March 1974; also, for corrections, see S. Iversen, *Proc. IEEE*, vol. 63, pp. 983–984, June 1975.

O. Review Papers on Solid-State Microwave Amplifiers

[O1] L. Cuccia, "Status report: Modern low noise amplifiers in communication systems," *Microwave System News (MSN)*, vol. 4, pp. 120–132, Aug/Sep 1974 and pp. 79–90, Oct/Nov 1974.

[O2] J. Magarshack, "Design and applications of solid-state microwave amplifiers," in *Proc. 1975 European Microwave Conf.*, pp. 153–167.

[O3] N. Osbrink *et al.*, "Review of microwave amplifiers," *Microwave J.*, vol. 18, pp. 27–34, Nov. 1975.

P. Signal Distortion in Transistor Amplifiers

[P1] R. Meyer, M. Shensa, and R. Eschenbach, "Cross modulation and intermodulation in amplifiers at high frequencies," *IEEE J. Solid-State Circuits*, vol. SC-7, pp. 16–23, Feb. 1972.

[P2] R. Fair, "Harmonic distortion in the junction field-effect transistor with field-dependent mobility," *IEEE Trans. Electron Devices*, vol. ED-19, pp. 9–13, Jan. 1972.

[P3] C. Liechti and R. Tillman, "Application of GaAs Schottky-gate FET's in microwave amplifiers," in *1973 IEEE Int. Solid-State Circuits Conf., Dig. Tech. Papers*, pp. 74–75.

Q. Amplifier Network Analysis and Synthesis

[Q1] S. Mason, "Power gain in feedback amplifier," *IRE Trans. Circuit Theory*, vol. CT-1, pp. 20–25, June 1954.

[Q2] J. Rollett, "Stability and power-gain invariants of linear two ports," *IRE Trans. Circuit Theory*, vol. CT-9, pp. 29–32, March 1962.

[Q3] K. Kurokawa, "Power waves and the scattering matrix," *IEEE Trans. Microwave Theory and Techniques*, vol. MTT-13, pp. 194–202, March 1965.

[Q4] G. Bodway, "Two port power flow analysis using generalized scattering parameters," *Microwave J.*, vol. 10, pp. 61–69, May 1967.

[Q5] C. Gledhill and M. Abulela, "Scattering parameter approach to the design of narrow-band amplifiers employing conditionally stable active elements," *IEEE Trans. Microwave Theory and Techniques*, vol. MTT-22, pp. 43–48, Jan. 1974.

[Q6] K. Kotzebue, "Microwave-transistor power-amplifier design by large-signal *y* parameters," *Electron. Lett.*, vol. 11, pp. 240–241, May 1975.

[Q7] L. Besser, "Design considerations of a 3.1–3.5 GHz GaAs FET feedback amplifier," in *1972 IEEE Int. Microwave Symp., Dig. Tech. Papers*, pp. 230–232.

[Q8] W. Leighton, R. Chaffin, and J. Webb, "RF amplifier design with large-signal *s*-parameters," *IEEE Trans. Microwave Theory*, vol. MTT-21, pp. 809–814, Dec. 1973.

[Q9] F. Weinert, "Scattering parameters speed design of high-frequency transistor circuits," *Electron.*, vol. 39, pp. 78–88, Sep. 1966.

[Q10] R. Anderson, "S-parameter techniques for faster, more accurate network design," *Hewlett-Packard J.*, vol. 18, pp. 13–24, Feb. 1967.

[Q11] W. Froehner, "Quick amplifier design with scattering parameters," *Electron.*, vol. 40, pp. 100–109, Oct. 1967.

[Q12] R. Tucker, "Low-noise design of microwave transistor amplifiers," *IEEE Trans. Microwave Theory and Techniques*, vol. MTT-23, pp. 697–700, Aug. 1975.

[Q13] C. Liechti and R. Tillman, "Design and performance of microwave amplifiers with GaAs Schottky-gate field-effect transistors," *IEEE Trans. Microwave Theory and Techniques*, vol. MTT-22, pp. 510–517, May 1974.

[Q14] M. Mokari-Bolhassan and T. Trick, "Computer-aided design of distributed-lumped-active networks," *IEEE Trans. Circuit Theory*, vol. CT-18, pp. 187–190, Jan. 1971.

[Q15] J. Bandler, "Computer aided circuit optimization," in *Modern Filter Theory and Design*, G. Temes and S. Mitra, Ed. New York: John Wiley, 1973.

[Q16] J. Bandler, J. Popovic, and V. Jha, "Cascaded network optimization program," *IEEE Trans. Microwave Theory and Techniques*, vol. MTT-22, pp. 300–308, March 1974.

[Q17] E. Sanchez-Sinencio and T. Trick, "CADMIC-computer aided design of microwave integrated circuits," *IEEE Trans. Microwave Theory and Techniques*, vol. MTT-22, pp. 309–316, March 1974.

[Q18] C. Charalambous, "A unified review of optimization," *IEEE Trans. Microwave Theory and Techniques*, vol. MTT-22, pp. 289–300, March 1974.

[Q19] N. Kuhn, "CAD with graphics make circuit design a science," *Microwaves*, vol. 13, pp. 42–50, June 1974.

[Q20] W. Ku, W. Petersen, and A. Podell, "New results on the design of broadband microwave bipolar and FET amplifiers," in *1974 IEEE Int. Symp. Microwave Theory and Techniques, Dig. Tech. Papers*, pp. 357–359.

[Q21] W. Ku *et al.*, "Microwave octave-band GaAs FET amplifiers," in *1975 IEEE Int. Symp. Microwave Theory and Techniques, Dig. Tech. Papers*, pp. 69–72.

[Q22] D. Mellor, "Insertion loss synthesis of matching networks for microwave amplifiers," in *1975 IEEE Int. Solid-State Circuits Conf., Dig. Tech. Papers*, pp. 68–69, 214.

[Q23] D. Mellor and J. Linvill, "A complete computer program for the synthesis of matching networks for microwave amplifiers," in *1975 IEEE Int. Symp. Microwave Theory and Techniques, Dig. Tech. Papers*, pp. 191–193.

[Q24] ——, "Synthesis of interstage networks of prescribed gain versus frequency slopes," *IEEE Trans. Microwave Theory and Techniques*, vol. MTT-23, pp. 1013–1020, Dec. 1975.

[Q25] N. Marshall, "Optimizing multi-stage amplifiers for low noise," *Microwaves*, vol. 13, pp. 62–64, April 1974.

[Q26] ——, "Optimizing multi-stage amplifiers for linearity," *Microwaves*, vol. 13, pp. 60–64, May 1974.

[Q27] K. Kurokawa, "Design theory of balanced transistor amplifiers," *Bell System Tech. J.*, vol. 44, pp. 1675–1698, Oct. 1965.

[Q28] S. Lazar, "Solid-state power amplifiers for *S*-band phased array radar," *Microwave System News*, vol. 5, pp. 77–84, Feb./Mar. 1975.

R. Performance of Small-Signal MESFET Amplifiers

[R1] P. Clouser and V. Risser, "*C*-band FET amplifiers," in *1970 IEEE Int. Solid-State Circuits Conf., Dig. Tech. Papers*, pp. 52–53.

[R2] W. Baechtold, "*Ku*-band GaAs FET amplifier and oscillator," *Electron. Lett.*, vol. 7, pp. 275–276, May 1971.

[R3] ——, "*X*- and *Ku*-band amplifiers with GaAs Schottky-barrier field-effect transistors," *IEEE J. Solid-State Circuits*, vol. SC-8, pp. 54–58, Feb. 1973.

[R4] S. Arnold, "Single and dual-gate GaAs FET integrated amplifiers in *C*-band," in *1972 IEEE Int. Microwave Symp., Dig. Tech. Papers*, pp. 233–234.

[R5] J. Eisenberg and R. Disman, "Design a 4 to 8 GHz FET amplifier with a 7 dB noise figure," *Microwaves*, vol. 12, pp. 52–56, Feb. 1973.

[R6] C. Ch'en and A. Woo, "A low-noise *C*-band GaAs FET amplifier," 1974 IEEE Int. Microwave Symposium, Session 20, Late News Paper.

[R7] G. Vendelin, J. Archer, and G. Bechtel, "A low-noise integrated *S*-band amplifier," in *1974 IEEE Int. Solid-State Circuits Conf., Dig. Tech. Papers*, pp. 176–177.

[R8] D. James, R. Douville, R. Breithaupt, and A. Van Koughnett, "A 12 GHz field-effect transistor amplifier for communications satellite applications," in *Proc. 1974 European Microwave Conf.*, pp. 97–101.

[R9] N. Slaymaker and J. Turner, "Microwave FET amplifiers with centre frequencies between 1 and 11 GHz," in *Proc. 1973 European Microwave Conf.*, vol. 1, Paper A.5.1.

[R10] N. A. Slaymaker, R. A. Soares, and J. A. Turner, "GaAs MESFET small-signal *X*-band amplifiers," *IEEE Trans. Microwave Theory and Techniques*, this issue, pp. 329–337.

[R11] H. Luxton, "Gallium-arsenide field-effect transistors—Their performance and application up to *X*-band frequencies," in *Proc. 1974 European Microwave Conf.*, pp. 92–96.

[R12] R. Pengelly, "Broadband lumped-element *X*-band GaAs FET amplifiers," *Electron. Lett.*, vol. 11, pp. 58–60, Feb. 1975.

[R13] ——, "Broadband lumped-element *X*-band GaAs FET amplifiers," in *Proc. 1975 European Microwave Conf.*, pp. 301–305.

[R14] R. Soares and J. Turner, "Tunable *X*-band GaAs FET amplifier," *Electron. Lett.*, vol. 11, pp. 474–475, Sep. 1975.

[R15] M. Walker, F. Mauch, and T. Williams, "Cover *X*-band with an FET amplifier," *Microwaves*, vol. 14, pp. 36–45, Oct. 1975.

[R16] D. Ch'en, Avantek, private communication.

[R17] M. Walker, Watkins-Johnson Company, private communication.

S. Performance of MESFET Amplifiers at Low Temperatures

[S1] B. Loriou, M. Bellec, and M. LeRouzic, "Performances à basse température d'un transistor hyperfréquences faible bruit à effet de champ," *Electron. Lett.*, vol. 6, pp. 819–820, Dec. 1970.

[S2] J. Leost and B. Loriou, "Propriétés à basse température des transistors GaAs à effet de champ hyperfréquence," *Hyperfréquences*, vol. 54, pp. 514–522, December 1974.

[S3] J. Jimenez, J. Oliva, and A. Septier, "Very low noise cryogenic MESFET amplifier," in *Proc. 1973 European Microwave Conf.*, vol. 1, Paper A.5.2.

[S4] P. Bura, "Operation of 6 GHz FET amplifier at reduced ambient temperature," *Electron. Lett.*, vol. 10, pp. 181–182, May 1974.

[S5] C. Liechti, R. Larrick, and D. Mellor, "A cooled GaAs MESFET amplifier operating at 12 GHz with 1.6 dB noise figure," in *Proc. 1975 European Microwave Conf.*, pp. 306–309.

[S6] C. A. Liechti and R. B. Larrick, "Performance of GaAs MESFET's at low temperatures," *IEEE Trans. Microwave Theory and Techniques*, this issue, pp. 376–381.

T. Performance of Medium-Power MESFET Amplifiers

[T1] D. Hornbuckle and L. Kuhlman, "Broad-band medium-power amplification in the 2–12.4-GHz range with GaAs MESFET's," *IEEE Trans. Microwave Theory and Techniques*, this issue, pp. 338–342.

[T2] Y. Arai, T. Kouno, T. Horimatsu, and H. Komizo, "A 6-GHz four-stage GaAs MESFET power amplifier," *IEEE Trans.*

Microwave Theory and Techniques, this issue, pp. 381–383.

[T3] R. E. Neidert and H. A. Willing, "Wide-band gallium arsenide powder MESFET amplifier," *IEEE Trans. Microwave Theory and Techniques*, this issue, pp. 342–350.

[T4] P. Bura and D. Cowan, "Highly linear medium power 11 GHz FET amplifier," in *1976 Int. Solid-State Circuits Conf., Dig. Tech. Papers*, pp. 158–159.

[T5] F. Sechi, "High-gain 1 W FET amplifier operating in *G* band," in *1976 Int. Solid-State Circuits Conf., Dig. Tech. Papers*, pp. 162–163.

[T6] R. Camisa, J. Goel, and I. Drukier, "GaAs MESFET linear power amplifier stage giving 1 W," *Electron. Lett.*, vol. 11, pp. 572–573, Nov. 1975.

U. MESFET Oscillators

[U1] M. Maeda, S. Takahashi, and H. Kodera, "CW oscillation characteristics of GaAs Schottky-barrier gate field-effect transistors," *Proc. IEEE*, vol. 63, pp. 320–321, Feb. 1975.

[U2] M. Maeda, K. Kimura, and H. Kodera, "Design and performance of *X*-band oscillators with GaAs Schottky-gate field-effect transistors," *IEEE Trans. Microwave Theory and Techniques*, vol. MTT-23, pp. 661–667, Aug. 1975.

[U3] P. Pucel, R. Bera, and D. Masse, "Experiments on integrated gallium-arsenide FET oscillators at *X*-band," *Electron. Lett.*, vol. 11, pp. 219–220, May 1975.

[U4] N. Slaymaker and J. Turner, "Alumina microstrip GaAs FET 11 GHz oscillator," *Electron. Lett.*, vol. 11, pp. 300–301, July 1975.

[U5] M. Omori and C. Nishimoto, "Common-gate FET oscillator," *Electron. Lett.*, vol. 11, pp. 369–371, Aug. 1975.

[U6] D. James, G. Painchaud, E. Minkus, and W. Hoefer, "Stabilized 12 GHz MIC oscillator using GaAs FET's," in *Proc. 1975 European Microwave Conf.*, pp. 296–300.

[U7] H. Abe, Y. Takayama, A. Higashisaka, R. Yamamoto, and M. Takeuchi, "A high-power microwave GaAs FET oscillator," in *1976 Int. Solid-State Circuits Conf., Dig. Tech. Papers*, pp. 164–165.

[U8] R. Pucel, R. Bera, and D. Masse, "An evaluation of GaAs FET oscillators and mixers for integrated front-end applications," in *1975 IEEE Int. Solid-State Circuits Conf., Dig. Tech. Papers*, pp. 62–63.

V. MESFET Mixers

[V1] J. Sitch and P. Robson, "The performance of GaAs field-effect transistors as microwave mixers," *Proc. IEEE*, vol. 61, pp. 399–400, March 1973.

[V2] R. Pucel, D. Masse, and R. Bera, "Integrated GaAs FET mixer performance at *X*-band," *Electron. Lett.*, vol. 11, pp. 199–200, May 1975.

[V3] ——, "Performance of GaAs MESFET mixers at *X*-band," *IEEE Trans. Microwave Theory and Techniques*, this issue, pp. 351–360.

[V4] O. Kurita and K. Morita, "Microwave MESFET mixer," *IEEE Trans. Microwave Theory and Techniques*, this issue, pp. 361–366.

[V5] S. Komaki, O. Kurita, and T. Memita, "GaAs MESFET regenerator for phase-shift keying signals at the carrier frequency," *IEEE Trans. Microwave Theory and Techniques*, this issue, pp. 367–372.

W. High-Speed Logic

[W1] K. Drangeid *et al.*, "A memory-cell array with normally off-type Schottky-barrier FET's," *IEEE J. Solid-State Circuits*, vol. SC-7, pp. 277–282, Aug. 1972.

[W2] O. Cahen, G. Cachier, and J. Puron, "A subnanosecond switching circuit," in *1974 IEEE Int. Solid-State Circuits Conf., Dig. Tech. Papers*, pp. 110–111.

[W3] R. Van Tuyl and C. Liechti, "High speed integrated logic with GaAs MESFET's," *IEEE J. Solid-State Circuits*, vol. SC-9, pp. 269–276, Oct. 1974.

[W4] ——, "High speed GaAs MSI," in *1976 IEEE Int. Solid-State Circuits Conf., Dig. Tech. Papers*.

[W5] E. Kohn, "Normally-off MESFET with fast switching behavior," *Electron. Lett.*, vol. 10, p. 505, Nov. 1974.

[W6] H. Beneking and E. Kohn, "High-speed GaAs MESFET differential amplifier stage with integrated current source," in *1974 Int. Electron Devices Meeting, Dig. Tech. Papers*, pp. 292–295.

[W7] H. Beneking and W. Filensky, "The GaAs MESFET as a pulse regenerator in the gigabit per second range," *IEEE Trans. Microwave Theory and Techniques*, this issue, pp. 385–386.

[W8] J. Notthoff and R. Zuleeg, "High speed, low power GaAs JFET integrated circuits," in *1975 Int. Electron Devices Meeting, Dig. Tech. Papers*, p. 624.

[W9] V. Vodicka and R. Zuleeg, "Ion implanted GaAs enhancement mode JFET's," in *1975 Int. Electron Devices Meeting, Dig. Tech. Papers*, pp. 625–628.

[W10] T. Sudo *et al.*, "A monolithic 8 pJ/2 GHz logic family," *IEEE J. Solid State Circuits*, vol. SC-10, pp. 524–529, Dec. 1975.

[W11] D. DiPietro, "A 5 GHz f_T monolithic IC process for high-speed digital circuits," in *1975 Int. Solid-State Circuits Conf., Dig. Tech. Papers*, pp. 118–119.

[W12] L. Cuccia, J. Spilker, and D. Magill, "Digital communication at gigahertz data rates," Part I, *Microwave J.*, vol. 13, pp. 80–93, Jan. 1970; Part II, *Microwave J.*, vol. 13, pp. 87–92, Feb. 1970; Part III, *Microwave J.*, vol. 13, pp. 75–80, April 1970.

[W13] L. Cuccia, "A technology status report on high-speed MPSK digital modulation systems," in *Proc. 1974 European Microwave Conf.*, p. 505.

[W14] C. Ryan, "Bipolar IC's for microwave signal processing," in *1975 IEEE Int. Microwave Symposium, Dig. Tech. Papers*, pp. 37–39.

X. Fabrication Technologies

[X1] S. Middelhoek, "Projection masking, thin photoresist layers and interface effects," *IBM J. Res. Develop.*, vol. 14, pp. 117–124, March 1970.

[X2] ——, "Metallization processes in fabrication of Schottky-barrier FET's," *IBM J. Res. Develop.*, vol. 14, pp. 148–151, March 1970.

[X3] H. Smith, "Fabrication techniques for surface-acoustic-wave and thin-film optical devices," *Proc. IEEE*, vol. 62, pp. 1361–1387, Oct. 1974.

[X4] C. Stolte, "Device quality n-type layers produced by ion implantation of Te and S into GaAs," in *1975 Int. Electron Devices Meeting, Dig. Tech. Papers*, pp. 585–587.

GaAs FET Development– Low Noise and High Power

The rapid pace of GaAs FET development shows little sign of abating. This paper describes work done at Bell Laboratories on material structures and fabrication techniques for both low noise and high power devices. Performance data representing the best results of various laboratories are also presented along with projections of device noise and power limitations.

J. V. DiLORENZO
Bell Laboratories
Murray Hill, NJ

The GaAs MESFET has been the most active area of research in the microwave device field for the last five years. The evidence for this is clear in that over 250 papers have been published on the subject of the GaAs FET since 1970.[1] In addition the GaAs FET has been the subject of panel discussions at various conferences over the last few years.[2,3,4] The reason for all this activity is twofold. The first is that as a low noise three-terminal device over the frequency range between 4-20 GHz, GaAs FETs demonstrate higher gain and lower noise figures than any other devices. Secondly, as a power device, the GaAs FET is similar in power but with higher gain when compared to silicon bipolars at frequencies of 4 to 6 GHz and exceeds silicon bipolar performance at the higher frequencies.

This paper deals mainly with the work conducted in our laboratory on GaAs FET materials, and low noise and high power devices. Where suitable, comparisons with data from other laboratories will be made.

GaAs FET EPITAXIAL MATERIAL

Chemical Vapor Deposition (CVD)

The epitaxial material structure which has been used for both low noise and high power devices consists of semi-insulating (SI) substrate/high-resistivity buffer/active/N^+ contact structures on substrates misoriented off the $\langle 100 \rangle$ by $6°$. The material was grown by use of the $AsCl_3/Ga/H_2$ system first described by Knight, et al.[5] but modified to include a second $AsCl_3$ bubbler as described by Nozaki, et al.[6] and Cox and DiLorenzo.[7] A schematic of the doping profile used for both high power and low noise devices is shown in **Figure 1**. The three layer structure is grown without interruption by adjusting the level of H_2S present in the reactor as previously described.[7]

CVD Active Layer Characteristics

There are two crucial parameters relevant to the active layer; the mobility of the active layer as a function of depth as well as the uniformity of the doping and thickness across the wafer. We have achieved uniformity of doping and thickness of approximately ±3% across a 9 cm^2 wafer. In **Figure 2** the Hall mobility variation as a function of depth for a CVD epi layer on a high

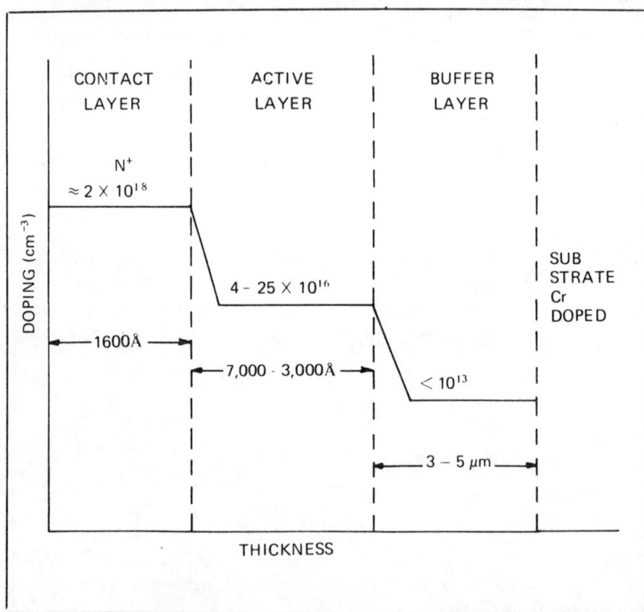

Fig. 1 Schematic doping profile of epitaxial material used for LN & HP GaAs FETs.

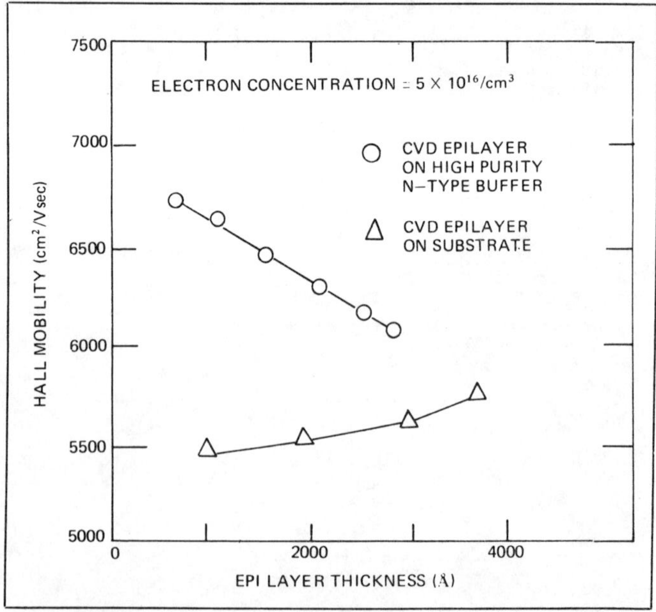

Fig. 2 Variation of Hall mobility with depth for CVD EPI layers.

Reprinted with permission from *Microwave J.*, vol. 21, pp. 39–42, 44, Feb. 1978.

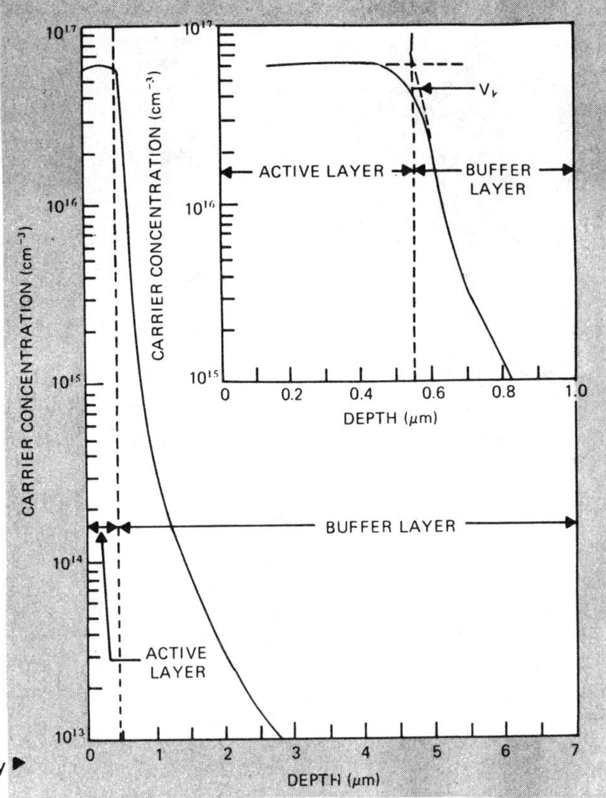

Fig. 3 Doping profile of FET structure with high resistivity buffer layer.

purity buffer layer and a substrate is shown. In this figure zero thickness represents the buffer-active layer interface. It can be seen that the mobility actually rises as the buffer layer is approached. The results for growth directly on a substrate shows decreasing mobility at the active layer — substrate interface.

Our results are consistent with those of Nozaki, et al.[6] who first observed the difference in active layer mobility as a function of depth. This increase of mobility is believed to be partly responsible for the better performance of low noise FETs on buffer versus non-buffer material.

CVD Buffer Layer Characteristics

In **Figure 3**, doping profiles are shown for a buffer layer of approximately 7 μm thickness. A complete description of the properties of growth of undoped buffers has been given previously including the discussion of outdiffusion of acceptors from the substrate.[7] The point to re-emphasize in this paper is that a high resistivity undoped buffer is obtained by making use of acceptors outdiffusing from the substrate to compensate a background doping level of 5–9x10^{14} cm^{-3}. If growth is allowed to proceed too long, a "buffer knee" will result due to the fact that the acceptor concentration from outdiffusion falls below the background level in the system.

All Cr doped and undoped substrates that we have looked at show outdiffu-

sion of acceptors from the substrate. This is consistent with recent observations of both Nozaki, et al.[7] who concluded that buffer layers grown by CVD are compensated and those of Bozler et al.[8] who observed outdiffusion of compensating impurities or defects from the substrate.

LOW NOISE AND HIGH POWER GaAs FETs

Fabrication

For both the low noise (LN) and high power (HP) GaAs FETs, the processing used is quite similar. The major difference between the two is that for the low noise device 10:1 reduction projection photolithography is used to

obtain short gate lengths of $\approx$ 0.5 μm and a high degree of alignment accuracy of ±0.1 μm in short source-drain spacings of 3 μm as previously described.[9] In the case of the power device, contact lithography is used for the more relaxed design rules of gate lengths of 2 μm in source drain spacings of 6 μm.[10]

The processing sequence is listed in **Table I**. All steps are based on a positive photoresist lift-off process for metal definition with only one critical alignment of gate positioning within the source drain space. We have made two types of HP devices, one employing a crossover structure (source over gate high power integrated chip) —

TABLE I PROCESSING SEQUENCE FOR HIGH POWER AND LOW NOISE GaAs FETs		
LN	HP	HPIC
1. MESA FORMATION Ion Milling	Same	Same
2. SOURCE & DRAIN AuGe formed by lift-off.	Same	Same
3. NO	N$^+$ Epi Removal	Same
4. GATE METAL Al formed by lift-off.	Same	Same
5. NO	NO	CROSSOVER DIELECTRIC (open windows to source & drain)
6. SOURCE & DRAIN OVERLAY & CROSSOVER METALLIZATION Ti-Pt-Au	Same	Same

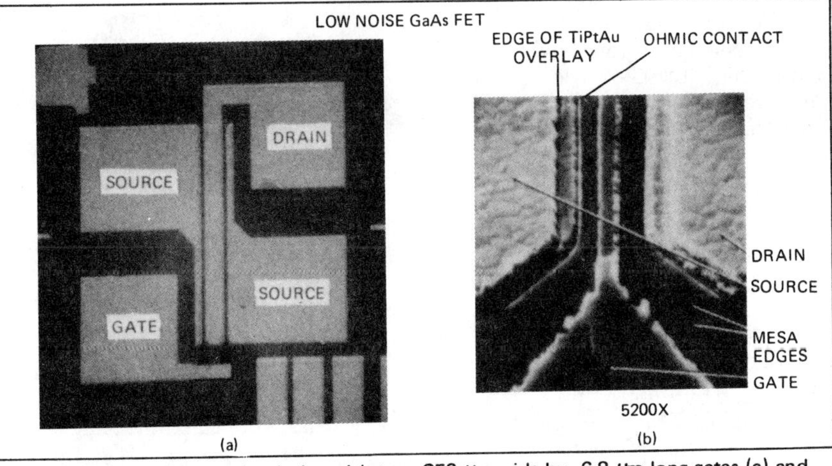

Fig. 4 Layout of low noise device with two, 250 μm wide by .6-8 μm long gates (a) and details of device at high magnification (b).

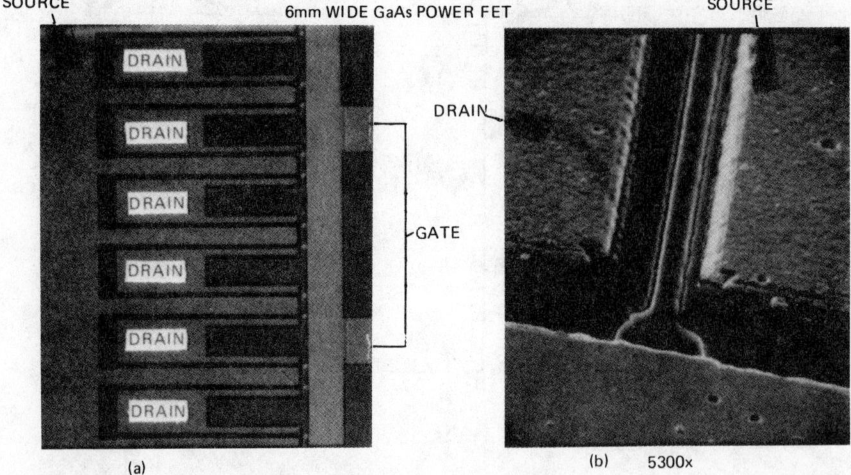

Fig. 5 Overview of 6mm wide power FET (a) and details of channel geometry (b).

TABLE II

GEOMETRICAL PARAMETERS FOR LOW NOISE AND HIGH POWER GaAs FETs

Device Type	Gate Length (μm)	Unit Gate Width (μm)	Total Gate Width (mm)	S-D Spacing (μm)
LN	0.5 - 0.8	250	0.5	3
MP	1	125 or 250	1	5
HP	2	250	3	6 - 9
HP	2	500	6	6 - 9
HPIC	2	250	6	6 - 9
	2	250	14	6 - 9

HPIC) and one which uses wire bonds to interconnect sources. **Table I** points out the main differences in processing between the three types of devices.

Device Geometry

Table II lists the geometrical parameters for LN and HP GaAs FETs. The device layout is similar to that presented[9,10] previously. In **Figure 4**, the layout for our low noise FET is shown in overview (a) and at high magnification (b). The device consists of two, $250\,\mu$m wide by 0.6-$0.8\,\mu$m long Al gates. The details of the device at high magnification are shown in **Figure 4b**. In **Figure 5**, the layout for our present high power FET is shown in overview (a) with the details of the channel region shown in the magnified view (b). The device consists of twelve, $500\,\mu$m wide by $2\,\mu$m long gates, in source-drain spacings which have been varied from 6-9 μm.

DEVICE RESULTS

Low Noise Data and Comparisons with Theory

In **Figure 6**,[13] the noise figure as a function of frequency for some selected experimental points as well as theory is presented. The theoretical lines A and B are calculated by Fukui[11] for two prototype devices using a model which has been previously presented.[9] Line A is representative of what might be termed a totally optimized device, with a gate length of $0.5\,\mu$m, and two gates each having a unit gate width of $100\,\mu$m. The notation used in **Figure 8** for this device (0.5 μm, 100x2) represents gate length, unit gate width and the number of gates in the device. In addition, device parasitics are minimized by the use of N^+ contacts. Line B is representative of our standard device with a gate length of $\approx 1\,\mu$m, and two gates having unit widths of $250\,\mu$m. Using a device with a $0.8\,\mu$m G_L (see **Figure 6**) and N^+ contacts we have obtained the following results with CVD material.

4 GHz		6 GHz	
NF(dB)	G(dB)	NF(dB)	G(dB)
0.73	10	1.25	9

The NF's of 0.73 dB at 4 GHz and 1.25 dB at 6 GHz are in close agreement to that calculated by theory for an $0.8\,\mu$m G_L.

It is interesting to note that our results on Molecular Beam Epitaxy (MBE) material and those of RSRE (see **Figure 6**) suggest that at the present time MBE is inferior to CVD material for low noise devices. In **Figure 6**, data is also presented for devices fabricated on ion implanted channels (Hughes, Rockwell, Avantek). The data reported by Hughes of 2.3 dB at 14 GHz and 2.0 dB at 10 GHz, while not up to the best NF reported at 10 GHz (HP, 1.6 dB), and the Avantek point at 18 GHz suggests that ion implanted channels have arrived as a technique for producing good GaAs FETs. Finally, it should be noted that a noise figure of 0.4 dB at 4 GHz has been obtained at Bell Laboratories from a device cooled to 77°K.

High Power Data and Comparisons

In **Table III**[14], a summary of the best results we have obtained are listed together with data from outside laboratories. The results obtained from a 1 mm wide device at 4 GHz were 0.7

TABLE III

SUMMARY OF BEST HIGH POWER GaAs FET RESULTS

Freq.	Company	Device Size (mm)	Power (W)	Gain at Power (dB)	Figure of Merit (W/mm)
4 GHz	BTL	1.0	.70	17	.58
	BTL	1.0	1.15	10.0	1.05
	NEC	5.4	1.4	8.0	.26
	BTL	8.0	6.2	8.6	.78
	BTL	8.0	7.2	3.0	.90
	FUJITSU	26.0	15.0	5.0	.58
	BTL	16.0	14.4	3.0	.90
	BTL	16.0	10.7	8.1	.67
	RCA	9.6	3.0	4.7	.31
8-9 GHz	BTL	1.0	1.1	4.0	1.1
	BTL	1.0	.63	7.7	.77
	TI	2.4	2.5	4.0	1.0
	TI	6.4	5.1	5.0	.80
	FUJITSU	5.2	2.3	4.2	.44
16 GHz	TI	1.2	1.0	5	.83
	TI	1.2	1.1	4	.92

watt power output with 17 dB associated gain. The same device when tuned differently produced 1.05W and 10.0 dB gain at 4 GHz. We have measured similar 1 mm wide devices at 8 GHz and have obtained 1.1W at 4.0 dB gain. Note one very interesting point. If we define a figure of merit for a device, in terms of W/mm at 3-4 dB gain, it is possible to obtain greater than 1 W/mm up to 8-9 GHz, suggesting very little frequency dependence of power at the 3-4 dB gain level.

This is also supported by results from TI, listed in **Table III**, which shows up to 1 W/mm can be obtained from a 2.4 mm wide device at 8 GHz. However, the 6.4 mm wide TI device at 8 GHz gives only .80 W/mm. Of great significance is a recent result from TI which shows up to .92 W/mm at 4 dB gain can be obtained at 16 GHz.

In **Figure 7**, the behavior of output power as a function of source-drain bias voltage for a 6 mm wide device is shown. The present burnout voltages of our devices are in the 35-40V range. The output power at 3 dB gain, continues to increase with source-drain voltage, and a continued increase in gain at fixed input power is also observed.

Projections for the Future

In **Figure 8**, the best power/width in W/mm as a function of frequency is plotted for selected data points as well as outside laboratory results. It can be seen that as much as 1.2 W/mm at 4 GHz and 1.1 W/mm at 8 GHz at 3-4 dB gain is obtained, with a slow fall off in power above 10 GHz to a value equal to .92 W/mm at 16 GHz. It is significant that approximately 1 W/mm has been obtained up to 16 GHz suggesting that a $P \approx f^{-2}$ relationship is not yet dominating GaAs FETs power at 16 GHz with gains of 3-4 dB.

In **Figure 9**, the maximum width for scaling Wd_{max} as a function of frequency is compiled. The data points were obtained from our laboratory and from published results by asking up to what device width does the power scale to within 30% of a very small device of similar characteristics from the same laboratory. For example, in the case of the Bell Laboratories data at 4 GHz, we have obtained up to .90 W/mm at 16 mm total width, and conclude, therefore, that scaling up to 16 mm in width is possible for a GaAs FET at 4 GHz.

In **Figure 10**, we have plotted the projected power as a function of frequency, obtained by $(W/mm)_{max}$ x Wd_{max} of **Figures 8** and 9. In addition, some selected data points are listed. The projection suggests that up to 20W of power are possible at 4 GHz from a GaAs FET and 6-7W possible in

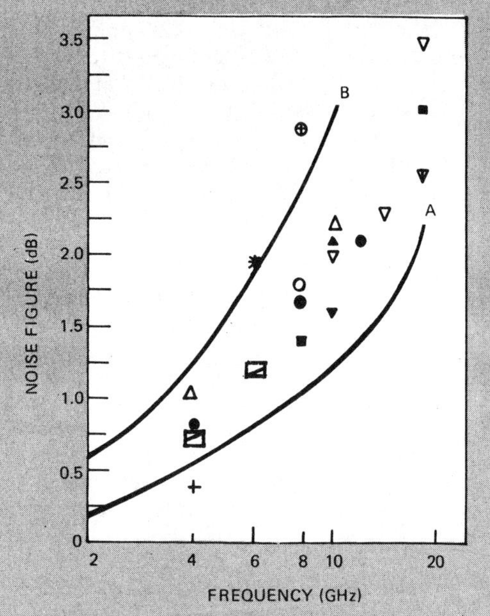

Fig. 6 Theoretical NF as a function of frequency (Lines A & B) with selected experimental data.

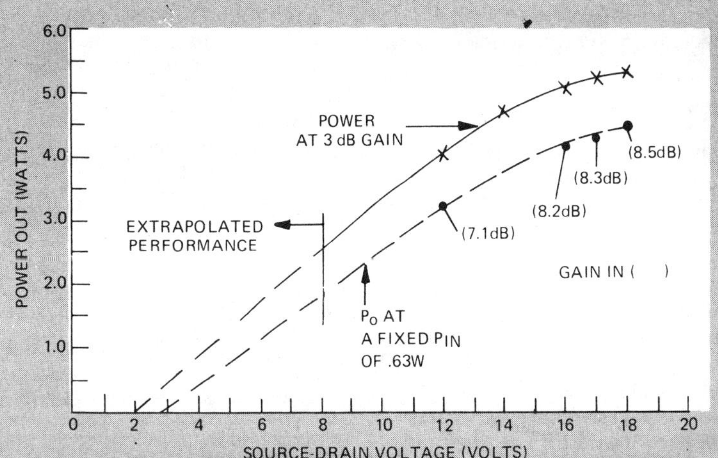

Fig. 7 Power out as a function of source-drain voltage for a 6mm wide device.

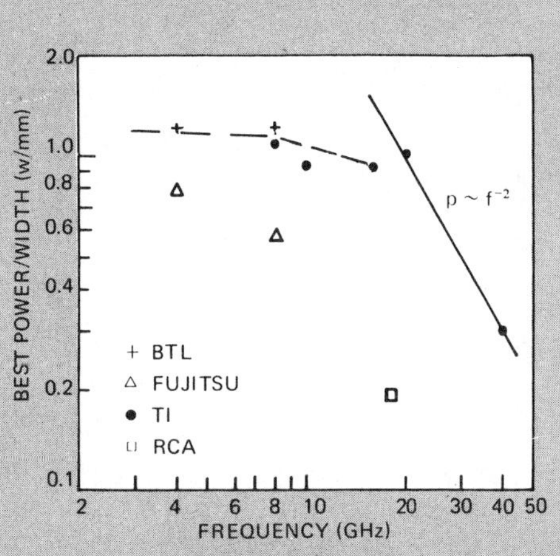

Fig. 8 Best power/width (W/mm) at 3-4 dB gain as a function of frequency.

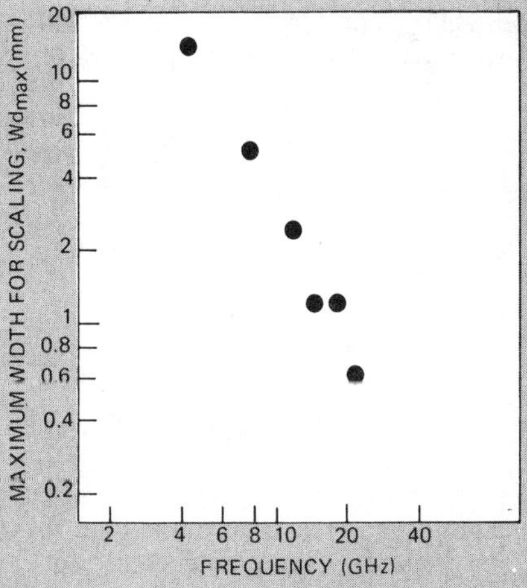

Fig. 9 Maximum width for scaling Wd_{max} (mm) as a function of frequency.

537

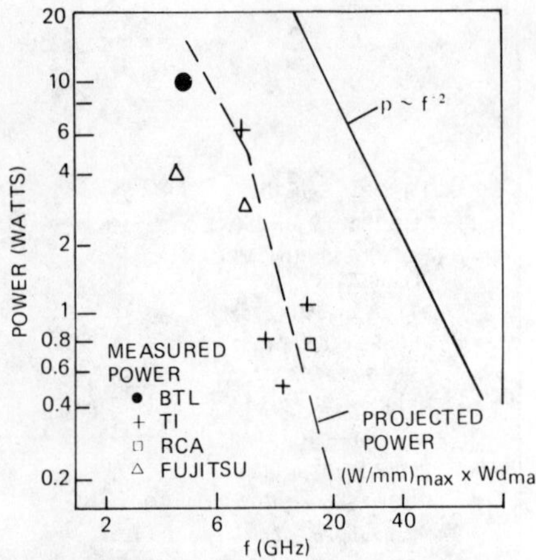

Fig. 10 Projected and measured power as a function of frequency.

the 8-10 GHz range. The fact that approximately 15W at 4 GHz (Fujitsu, Bell Laboratories, Table III) and 5.1 at 8 GHz (TI, Table III) have been obtained suggests that the projections are reasonable and perhaps pessimistic. In any event, it will be exciting to observe the continuing improvements in devices, especially in the frequency range above 10 GHz.

SUMMARY AND CONCLUSIONS

This paper has dealt with the material and device parameters important for obtaining good low noise and high power GaAs FETs. It has been shown that buffer layers are essential for good low noise performance for CVD technology. Growth of undoped buffer layers on semi-insulating substrates is complicated by the diffusion of acceptors from the substrate, which if not recognized, could lead to a conductivity region under the active FET channel.

Low noise devices have been fabricated at Bell Laboratories which show a NF of .73 dB at 4 GHz and 1.25 dB at 6 GHz with associated gains of 9-11 dB. High power devices have also been fabricated at Bell Laboratories which

show 0.7W power with 17 dB gain from a 1 mm wide device, 6.2W at 8.6 dB gain and 7.2W at 3 dB gain from a 8 mm wide device, and 10.7W at 8.1 dB gain and 14.4W at 3 dB gain from a 16 mm wide device. Analysis of all published data suggests that up to 20W at 4 GHz and 6-7W at 10 GHz are possible from GaAs FETs.

ACKNOWLEDGEMENTS

The results reported in this paper represent the contributions of many colleagues at Bell Laboratories; namely, H. M. Cox and A. Y. Cho on on epi materials, L. A. D'Asaro on chip fabrication, H. Fukui on modeling and fabrication of low noise devices, D. E. Iglesias on low noise RF characterization, L. C. Luther on dielectric deposition, W. C. Niehaus on HP FET fabrication and optimization, F. M. Magalhaes on high power RF device characterization, and S. H. Wemple and W. O. Schlosser on numerous contributions, to the design, characterization and understanding of HP and LN GaAs FETs.

In addition to those mentioned above, I wish to acknowledge the processing expertise of J. R. Velebir, P. F. Sciortino and W. R. Robertson on LN and HP GaAs FETs, J. A. Seman for device and mask design and J. P. Beccone for his packaging expertise. In addition, thanks are extended to L. J. Varnerin and J. E. Kunzler for their continuing support of this work.

REFERENCES

1. Hawkins, D. T., Bell Laboratories, private communication.
2. Microwave GaAs FETs: Reliability and Reproducibility, Session THE 10, *1976 IEEE Int. Solid-State Circuits Conference Digest.*
3. Microwave GaAs Power FETs, Session THE 11, *1977 IEEE Int. Solid-State Circuits Conf. Digest.*
4. GaAs Applications Workshop, 1977 IEEE-MTT-S, San Diego CA, June, 1977.
5. Knight, J. R., D. Effer and P. R. Evans, *Solid-State Electron*, Vol. 8, 1965, pp. 178.
6. Nozaki, T., M. Ogawa and H. Watanabke, Gallium Arsenide and Related Compounds 1974, *Inst. Phys. Conf. Soc.* Vol. 24, 1975, pp. 46.
7. Cox, H. M., and J. V. DiLorenzo, Gallium Arsenide and Related Compounds (St. Louis) 1976, *Inst. Phys. Conf Soc.* Vol. 33b, 1977, pp. 11.
8. Bozler, C. O., J. P. Donnelly and W. T. Lindley, *Appl. Phys. Lett.*, Vol. 29, 1976, pp. 698.
9. Hewitt, B. S., H. M. Cox, H. Fukui, J. V. DiLorenzo, W. O. Schlosser and D. E. Iglesias, Gallium Arsenide and Related Compounds (Edinburgh) 1976, *Inst. Phys. Conf. Soc.* Vol. 33a, 1977.
10. Niehaus, W. C., H. M. Cox, B. S. Hewitt, S. H. Wemple, J. V. DiLorenzo, W. O. Schlosser and F. M. Magalhaes, Gallium Arsenide and Related Compounds (St. Louis) 1976, *Inst. Phys. Conf. Soc.* Vol. 33b, 1977, pp. 271.
11. Fukui, H., Bell Laboratories, private communication.
12. Fukuta, M., T. Mimura, I. Tsujimura, A. Furumoto and K. Dasai, *IEEE Int. Solid-State Circuits Conf. Digest.*, pp. 84-85.
13. References for low noise data listed in Fig. 6:
    Plessey - R. S. Butlen, et al., Gallium Arsenide and Related Compounds (Edinburgh) 1976, *Inst. Phy. Conf. Soc.* Vol. 33a, 1977, pp. 237.
    Plessey - J. A. Turner, private communication.
    NEC - T. Nozaki and K. Ohata, private communication.
    Rockwell, J. Higgens et al., WOCSEM-MAD 1977.
    Hughes - G. Ladd, private communication.
    H. P. - J. Barrera, private communication.
    RSRE - J. C. Vokes et al., *1977 IEEE MTT-S. Int. Microwave Symp Digest* 1977, pp. 185.
    Hitachi - H. Kodera et al., ibid pp. 277.
    Avantek - W. W. Hooper, et al., *1977 Int. Electron. Devices Meeting, Digest* pp. 601.
14. References for high power data listed in Table III.
    RCA - H. Hwang, private communication.
    TI - W. R. Wisseman, *1977 Int. Electron Devices Meeting Digest*, pp. 326-328.
    NEC - F. Hasegawa, private communication.
    Fujitsu - Reference 12 and W. R. Wisseman, *1977 Int. Electron Devices Meeting, Digest*, pp. 326-328.

Section 4.4
Filters and Multiplexers

The importance of filters and multiplexers in the communication system was discussed in Reprint Paper 3.2.2. In this section, techniques for realizing these elements are discussed. Reprint Paper 4.4.1 discusses the current state of the art of filter and multiplexer technology in the 12-GHz frequency band. Reprint Paper 4.4.2 discusses the design of dual-mode filters using square and circular geometries. These filters have resulted in significant weight savings in the INTELSAT IV-A and other satellites. References [1] and [2] provide general discussions of filters. References [3]–[9] discuss various aspects of the design of filters and multiplexers.

REFERENCES

[1] C. M. Kudsia and V. O'Donovan, "Microwave filters for communications systems," Dedham, MA: Artech House Inc., 1974.

[2] G. C. Terms and S. K. Mitra, "Modern filter theory and design," New York: Wiley, 1973.

[3] A. E. Williams and A. E. Atia, "Dual mode canonical waveguide filters," *IEEE Trans. Microwave Theory Tech.*, vol. MTT-25, pp. 1021-1026, Dec. 1977.

[4] J. D. Rhodes, "The design and synthesis of a class of microwave band pass linear phase filters," *IEEE Trans. Microwave Theory Techniques,* vol. MTT-17, Apr. 1969.

[5] A. E. Atia, A. E. Williams, and R. W. Newcomb, "Narrow-band multiple-coupled cavity synthesis," *IEEE Trans. Circuits Syst.,* vol. CAS-21, pp. 649-654, Sept. 1974.

[6] A. E. Atia, "Computer-aided design of waveguide multiplexers," *IEEE Trans. on Microwave Theory and Techniques,* vol. MTT-22, pp. 332-336, Mar. 1976.

[7] R. M. Kurzrok, "General four-resonator filters in waveguide," *IEEE Trans. MTT,* vol. MTT-16, pp. 46-47, 1966.

[8] G. L. Matthaei, L. Young, and E. M. T. Jones, *Microwave Filters, Impedance Matching Networks and Coupling Structures.* New York: McGraw-Hill, 1964.

[9] G. L. Ragan, *Microwave Transmission Circuits,* (MIT Radiation Laboratory Series, vol. 9). New York: McGraw-Hill, 1948.

STATUS OF FILTER AND MULTIPLEXER TECHNOLOGY IN THE
12 GHZ FREQUENCY BAND FOR SPACE APPLICATION

C.M. Kudsia
S. Kallianteris
K.R. Ainsworth
M.V. O'Donovan
Com Dev Ltd.
Dorval, Quebec, Canada

Abstract

This paper discusses the latest techniques of filtering and multiplexing in the 12 GHz frequency band. Tradeoffs include mode selection, realizable unloaded Q's and efficiencies, classes of filter response, filter structure and the generation and control of spurious response. Material selection, processes and amenability to construction for space application are emphasized. The ancillary units that are necessary to ensure filter or multiplexer performance as part of the overall transponder subsystem are also described.

Introduction

This paper discusses the progress and status of filter and multiplexer technology in the 12 GHz frequency band. The requirements of 12 GHz input and output multiplexers for Intelsat V and Canadian ANIK-B satellites presently under construction have established a bench mark. The filtering requirements of the next generation of communications satellites dedicated to the 14/12 GHz frequency bands, viz., SBS and ANIK-C, represent an advance in the state-of-the-art. These are highlighted in the paper. Tradeoffs are provided for the selection and the ancillary components required to ensure multiplexer performance as part of the overall communications subsystem. The future of 12 GHz filter technology is briefly discussed.

Mode Selection

The unloaded Q that can be realized in a filter network depends upon the propagating mode and the resistive losses associated with the filter structure. The latter depend upon the surface roughness and conductivity of waveguide walls, tuning screws and imperfect contact areas. The ratio of practically realizable unloaded Q to its theoretical maximum value for a given mode of propagation is referred to as the efficiency of the filter network. In addition, the efficiency is found to depend upon the filter configuration as well, that is, whether the filter is configured using a single mode or two or higher orthogonal modes per waveguide cavity. The efficiency of dual-mode structures is typically 15 to 20% lower than that of comparable single-mode structures but they provide significant advantages in terms of size, weight and realizable functions. Based on these considerations, Table 1 provides a summary of the tradeoffs involved for a variety of propagating modes. There are modes with

higher Q's not included here for reasons of practical realizability at the present time. References 1 to 4 provide additional information in this area.

Filter Structure Tradeoffs

Selection of filter structure is often one area which is given less emphasis by electrical designers. For space application, the mechanical design of the filters and their amenability to construction as flight units singly or as part of a multiplexer channel is as critical as the electrical performance. In the 12 GHz frequency band, the machining tolerances have a significant impact on electrical response. Deviations of half-a-mil in slot dimensions of irises can cause major changes in response. The mechanical design of filters and multiplexers at 12 GHz therefore deserve greater attention. Table 2 summarizes the various tradeoffs involved in the selection of filter structures.

Mechanical Design Considerations and Manufacturing Techniques for Flight Hardware

The mechanical design of flight hardware is frequently subject to conflicting requirements. On the one hand, the unit must be mechanically rigid and stable to withstand severe environmental conditions without compromising the electrical performance, and on the other hand, a strict weight budget must be met.

Traditional methods of fabricating filters using copper clad invar waveguide with soldered flanges and bushings can cause problems in production of flight components in the 12 GHz frequency band. Solder joints must be carefully designed and even more carefully monitored during production to ensure maximum strength and reliability. Square section invar waveguide used for 12 GHz dual-mode cavities has a typical tolerance of ±.002 and such variations can cause noticeable variations in filter performance in a production run. Depending upon the specification requirements, this could mean either increased tuning time or even rejection of imperfect waveguide cavities. In addition, there are the stringent tolerances required of coupling irises. It has been found that half-a-mil difference in slot dimensions of these irises can cause a significant change in the filter response. For this reason, methods of construction which allow accessibility to irises up to the final stages are considered as the most practical way to build flight hardware.

Reprinted with permission from the *Record of Am. Inst. Aeronautics & Astronautics, AIAA 7th Commun. Satellite Syst. Conf.,* Apr. 24-27, 1978, pp. 194-202.

Table 1 Mode selection tradeoffs

Mode of Propagation	Filter Configuration	Realizable Unloaded Q	Unloaded Q* Theoretical	Efficiency	Design Maturity
TE_{111} or TE_{101}	Dual-mode	5,000	9,000	55%	Standard design. Similar designs qualifed at 6 and 14 GHz.
TE_{103}	Dual-mode	9,000	16,300	55%	Proven design. Triplexers with 4-pole elliptic filters and 80 MHz bandwidth have been space qualified on ANIK-B and Intelsat V programs.
TE_{105}	Dual-mode	11,000	19,500	55%	Similar to TE_{103} mode except for longer length. Breadboard units have demonstrated an unloaded Q of 11,000 and close correlation with predicted response.
TE_{113}	Dual-mode	12,000	20,500	58%	Design similar to TE_{103} except for circular cavities. Requires development to achieve spurious free windows in the bands 11.7 to 12.2 GHz and 14 to 14.5 GHz.
TE_{115}	Dual-mode	13,200	23,000	58%	Similar to TE_{113} except for longer lengths. Same comments apply as for TE_{113}. Spurious problem is more severe than TE_{113} filter.
TE_{011}	Single-mode	16,000	24,600	65%	Cumbersome construction. Single-mode configuration. Breadboard work indicates sensitivity to bandwidth and tuning
TE_{211} or TE_{311}	Single-mode	11,000	20,200	60%	Same comments apply as for TE_{011} mode filters.

* Computed Q's assume cross sections which provide or likely to provide spurious free response over 500 MHz bandwidth in the 12 and 14 GHz frequency band.

Another method of fabricating square section cavities has been used successfully on the Intelsat V triplexer. The use of broaching and N/C milling techniques has allowed the construction of cavities where solder joints are eliminated except for one bushing per cavity. The resulting cavities are lighter and stronger and more reliable than conventional soldered units. These techniques have also resulted in closer tolerances being maintained without greatly increased production costs.

It seems probable that this method of construction could be improved by using electron beam welding to join cavities and coupling irises together, thus eliminating heavy flanges and fixing hardware. This welding procedure has the significant advantages that different metals may be joined together and the depth of weld can be very closely controlled; allowing parts to be welded after silver plating. It is possible that this method may also offer a slight increase in Qo due to the better contact at the joints but this has not as yet been fully evaluated.

The combining manifold used in these multiplexers must, because of weight restrictions, be fabricated from aluminum. The construction method employed makes extensive use of precision milling and dip brazing techniques to achieve a rigid lightweight structure. The large difference in thermal expansion coefficients between the aluminum manifold and the invar filters caused a considerable effort to be expended in the design of the filter/manifold interface joint. The adopted design maintains sufficient contact pressure over the operating temperature range so that the electrical performance is maintained; but still allows each filter to be easily installed and removed for tuning purposes.

These construction techniques allowed

Table 2 Filter structure tradeoffs

	Filter Configuration		
Parameters	Single-mode structures (one mode per cavity).	Dual-mode structures (two orthogonal modes per waveguide cavity).	Folded structures with single or dual-mode propagation per waveguide cavity.
Propagation Mode	Wide variety of propagation modes are feasible. Modes of interest listed in Table 1 viz., TE011, TE211, TE10n or TE11n can all be realized.	Variety of propagation modes is restricted. TE11n or TE10n modes can be realized.	Offer the same variety of propagation modes as the single or dual-mode structures respectively.
Realizable Class of Filter Responses	Chebyshev or Butterworth or other minimum-phase functions with monotonic increase in amplitude response.	Chebyshev, Butterworth, quasi-elliptic (equiripple or non equiripple classes), elliptic, or linear-phase response with monotonically increasing amplitude.	Offers the maximum variety of response functions. Includes all response functions included in other columns plus linear phase functions with or without transmission zeros in stopband.
Realizable Unloaded Q	Realized unloaded Q's are 65 to 80% of theoretical Q's, depending upon mode of propagation and geometry of structure.	Realized unloaded Q's range from 50% to 65% of theoretical Q's, depending upon the mode of propagation and geometry of structure.	Realized unloaded Q's are typically 5% lower than those achieved by single or dual-mode structures respectively.
Size, Weight and Amenability to Construction as Flight Hardware	Realatively simple structures but bulky. Good amenability for construction as flight units by itself or part of a mux channel.	Compact structure compared to single-mode with advantage of size and weight. Marginally more complex compared to single-mode. Good amenability to construction as flight unit by itself or part of a mux channel.	Structure is complex and therefore can be expensive. Not easily amenable to construction as flight units, especially if part of a mux channel. Requires further hardware oriented tradeoffs.

the Intelsat V triplexer to withstand and operate without any degradation in performance throughout qualification vibration testing at levels exceeding 40g RMS random and 30g peak sinusoidal.

12 GHz Multiplexing
- Present State-of-the-Art

References 5 to 10 describe the recent developments in the 12 GHz filter and multiplexer technology for space applications.

Typical measured data on qualified 3-channel output multiplexer units for ANIK-B and Intelsat V satellites is described in Tables 3 and 4. Channel filters utilize TE_{103} as the mode of propagation with unloaded Q's of 9,000 or greater. The 241 MHz wideband channel of Intelsat V mux utilizes TE_{111} mode for an unloaded Q of 5,500.

Figures 1 and 2 show the flight quality ANIK-B and Intelsat V output triplexers respectively. The basic material of all filters is invar to achieve the desired temperature stability. The material used for the manifolds is aluminum to conserve weight.

Results of breadboard 4-channel multiplexers for the ANIK-C satellite are described in Table 5. The mode of propagation of these breadboard units is TE_{105} exhibiting an average unloaded Q of 11,000.

Results of breadboard input mux channel for ANIK-C are described in Table 6. This assembly consists of an 8-pole dual-mode filter followed by a 2-pole dual-mode all-pass equalizer, both units utilizing TE_{103} as the mode of propagation. Filter response exhibits eight equi-ripple peaks in passband and a single pair of transmission zeros in the stopband. Results of a 10-pole dual-mode self-equalized filter operating in TE_{103} mode are also shown for comparison.

Figure 3 shows the breadboard ANIK-C quadruplexer. The measured insertion loss response of the quadruplexer over the frequency band 11.7 to 12.2 GHz is shown in Figure 4. The 10-pole linear phase and the externally equalized filters are shown

Table 3 Performance summary of typical ANIK-B output triplexer

Filter Configuration	:	Dual-mode filters in optimized square invar waveguide using TE103 as the mode of propagation.
Average Unloaded Q of Waveguide Cavities:		>9000
Frequency Band	:	11.7 to 12.2 GHz Downlink 14.0 to 14.5 GHz Uplink
Usable Bandwidth of Each Channel	:	72 MHz
Typical Isolation in the Receive Band (14-14.5 GHz)	:	>25 dB

CHANNEL NUMBER	ISOLATION - dB		INSERTION LOSS (dB)		GROUP DELAY NANOSECONDS	
	Spec at fo ±MHz	Measured (Ambient)	Budgeted Loss at fo ±MHz	Measured (Ambient)	Spec at fo ±MHz	Measured (Ambient)
1	20 at ±80	22	<.3 at fo	.3	3 at ±28	1.9
	40 at ±124	>40	<.45 at ±36	.42	4.5 at ±36	4.3
3	20 at ±80	21	<.3 at fo	.3	3 at ±28	1.3
	40 at ±124	>40	<.45 at ±36	.43	4.5 at ±36	3.0
5	20 at ±80	22	<.3 at fo	.3	3 at ±28	1.6
	40 at ±124	>40	<.45 at ±36	.41	4.5 at ±36	2.9

Table 4 Performance summary of typical Intelsat V output triplexer

Frequency Band: 10.95 to 11.2 GHz (Downlink
11.45 to 11.7 GHz (

14 to 14.5 GHz   Uplink

Channel Number	Realized Average Unloaded Q	Usable Band-width MHz	Isolation-dB		Insertion Loss dB		Group Delay nsec	
			dB at ±MHz Spec	Meas. Amb.	Budgeted Loss	Meas. Amb.	Spec	Meas. Amb.
1-2			15 at ±90	22	0.38 dB at 10966-11018 MHz	0.35	1.0 nsec at 10967-11018 MHz	1.0
TE103 4-pole Elliptic Filter	9000	77	30 at ±120	42	0.46 dB at 10954-11031 MHz	0.38	4.5 nsec at 10954-11031 MHz	3.5
			40 at ±130	>45				
5-6			15 at ±90	18	0.38 dB at 11131-11179 MHz	0.32	1.0 nsec at 11131-11179 MHz	0.9
TE103 4-pole Elliptic Filter	9000	72	30 at ±120	33	0.46 dB at 11119-11191 MHz	0.38	4.5 nsec at 11119-11191 MHz	2.5
			40 at ±130	42				
7-12			-390 30 at +420	32	0.22 dB at 11497-11658 MHz	0.18	.5 nsec at 11497-11658 MHz	0.4
TE111 4-pole Elliptic Filter	5500	241	40 at ±600	>45	0.34 dB at 11457-11698 MHz	0.22	1.2 nsec at 11457-11698 MHz	0.9

Fig. 1   Flight Quality ANIK-B Triplexer

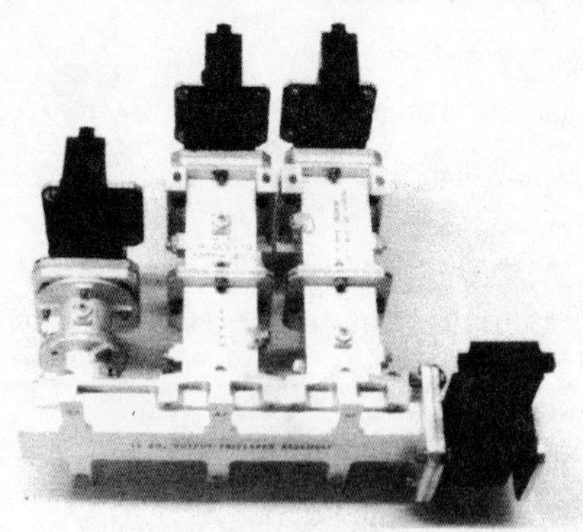

Fig. 2   Flight Quality Intelsat V
Output Triplexer

Table 5   Performance summary - 12 GHz output quadruplexer
evaluation model for ANIK-C spacecraft

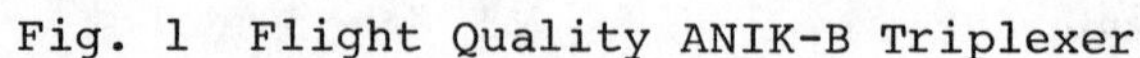

Filter Configuration :   Dual-mode filters in optimized square invar waveguide
using TE105 as the mode of propagation.

Average Unloaded Q
of Waveguide Cavities: >11,000

Frequency Band   : 11.7 to 12.2 GHz Downlink
14.0 to 14.5 GHz Uplink

Usable Bandwidth
of Each Channel   : 54 MHz

Typical Isolation in
the Receive Band   : (14-14.5 GHz) >30 dB

Channel Number	Isolation – dB		Insertion Loss – dB		Group Delay – nsec	
	ANIK-C Spec at ±MHz	Measured (Ambient)	ANIK-C Spec at ±MHz	Measured (Ambient)	ANIK-C Spec at ±MHz	Measured (Ambient)
1	20 at ±37	29.5	None at fo	0.58	6.0 at ±15	2.8
	28 at ±60	35	-0.5 at ±23	-0.20	12.0 at ±20	5.8
			-1.0 at ±27	-0.52	44.0 at ±27	20.0
2	20 at ±37	29.5	None at fo	0.54	6 at ±15	2.6
	28 at ±60	32.0	-0.5 at ±23	-0.26	12 at ±20	5.8
			-1.0 at ±27	-0.46	44 at ±27	21.3
5	20 at ±37	30.0	None at fo	0.54	6 at ±15	2.9
	28 at ±60	32.0	-0.5 at ±23	-0.24	12 at ±20	6.0
			-1.0 at ±27	-0.46	44 at ±27	20.0
7	20 at ±37	33.0	None at fo	0.55	6 at ±15	2.9
	28 at ±60	31.5	-0.5 at ±23	0.27	12 at ±20	6.2
			-1.0 at ±27	-0.47	44 at ±27	20.5

Table 6   Performance summary - 12 GHz input mux channel for ANIK-C spacecraft

Mux Channel Configuration

<u>Alternate 1</u>
8-pole dual-mode TE103 filter with eight equi-ripple peaks in passband and a single pair of transmission zeros in stopband followed by a 2-pole dual-mode TE103 allpass equalizer.

<u>Alternate 2</u>
10-pole dual-mode TE103 linear phase filter with ten equi-ripple peaks in passband and two pairs of real-axis zeros.

Average Unloaded Q of W/G Cavities   >9000

Usable Bandwidth of Each Channel   54 MHz

Parameter	Specification	Measured Performance (Ambient)	
		Externally Equalized Filter Structure	Linear Phase Filter Structure
Insertion Loss at Centre Frequency (CF) in dB	–	2.1	1.9
Loss Variation, dB CF ±27 MHz	1.4	0.6	0.50
Gain Slope at CF ±13 MHz in dB/MHz	.01	<.005	.006
Rejection; dB			
CF ±36 MHz	25	28.5	18.5
CF ±50 MHz	45	49.5	50.0
Group Delay, nsec			
CF ±18	2	.2	1.2
CF ±21	5	2.0	3.2
CF ±24	14	7.5	7.5
CF ±27	36	23.5	15.0
Group Delay Ripple nsec	1.5	.8	.5

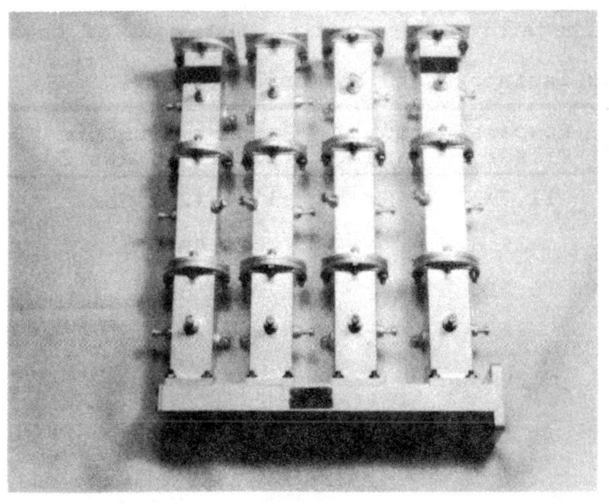

Fig. 3 Breadboard Quadruplexer for ANIK-C Satellite

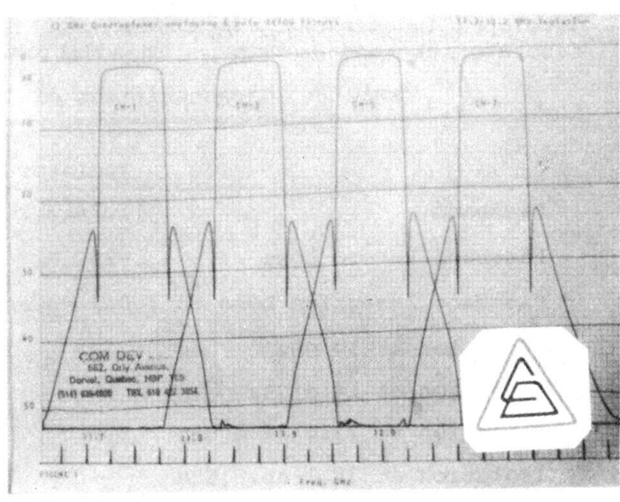

Fig. 4 Frequency Response of Breadboard ANIK-C Quadruplexer

in Figures 5 and 6 respectively.

Fig. 5 10-pole Linear Phase Dual-Mode
Filter Operating in TE_{103} Mode

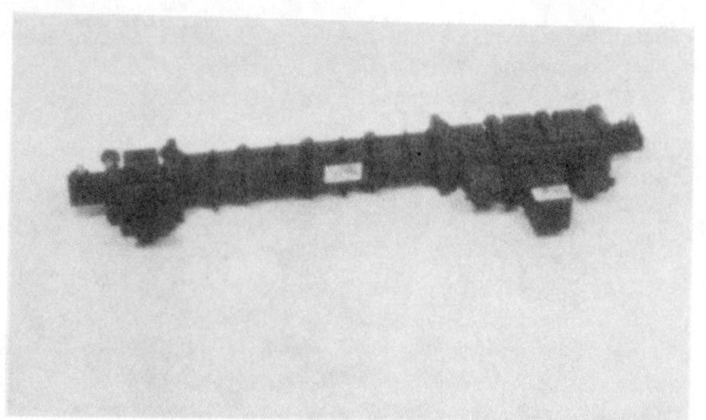

Fig. 6 Externally Equalized
Prototype Filter

Filtering requirements stemming from the SBS satellite are difficult to achieve. Most likely, it will require the newly developed dual-mode filter exhibiting true elliptic response by Hughes Aircraft Company. Similar filters were also developed independently by COMSAT Laboratories[5] and Siemens[6]. In addition, the filter and multiplexer assemblies will require near perfect tuning to achieve the specified performance.

Material Selection

The three basic materials used for waveguide structures for space application are invar, aluminum and graphite fiber epoxy composite (GFEC). The properties of invar and aluminum filter structures are well established and have a history of flight experience. The advent of GFEC is relatively recent and most of the work that has been done with this material is at 4 GHz.

Hard data on GFEC filters and multiplexers at 12 GHz, in terms of achievable unloaded Q, mechanical tolerances and short and long term creep effect, if any, is required and it may then establish it as a candidate design. Use of aluminum or invar (or GFEC) is dictated by the thermal environment as well as the electrical specifications - especially the required bandwidth. Narrowband channel filters of input and output muxes invariably require use of invar whereas broadband filters are generally made of aluminum.

<u>Spurious Control and Ancillary Components</u>

The performance of multiplexers is sensitive to input and output terminations. It is genrally a good practice to use isolators at the input and output of multiplexer channels. The penalty of additional loss and weight is typically more than offset by the lack of performance degradation of channels when the units are integrated as part of the overall communications subsystem.

Channel combining filters in the output circuitry are typically optimized for lowest possible insertion loss in their passband and consequently provide little or no isolation in the receive band of 14-14.5 GHz and the second or third harmonics of the transmit band. A lowpass filter is therefore necessary to take care of these

Table 7 Characteristics of 12 GHz isolators and harmonic filters

	Isolator	Lowpass Harmonic Filter - Class 1	Lowpass Harmonic Filter - Class 2
Passband	11.7–12.2 GHz	11.7–12.2 GHz	11.7–12.2 GHz
Passband Return Loss	–29 dB	–29 dB	–29 dB
Passband Insertion Loss	0.1 dB max.	0.22 dB max.	0.15 dB
Isolation at 14 GHz	–	55 dB	45 dB
Isolation at 14.5 GHz	–	70 dB	65 dB
Isolation at 23.4–24.4 GHz	–	70 dB	60 dB
Isolation at 35.1–36.6 GHz	–	60 dB	10 dB
Operating Temperature Range	–5°C to +45°C for all units		
Weight	3 ounces	2 ounces	1.6 ounces

isolation requirements. Typical characteristics of isolators and lowpass filters
qualified at 12 GHz on ANIK-B and Intelsat V
programs by Com Dev Ltd. are given in
Table 7.

Figure 7 shows a typical flight quality
low to medium power isolator. Figure 8
shows the high power 12 GHz isolator. This
unit was qualified by Com Dev Ltd., at
250 watts CW power handling capability and
is currently operating, satisfactorily, in
the communication technology satellite
(CTS).

Fig. 7 Low Power 12 GHz Isolator
for Space Application

Fig. 8 High Power 250 Watt 12 GHz Isolator
Unit currently operating in
Communications Technology Satellite.

Conclusions and Discussion

Intelsat V at 11 GHz and the Canadian
domestic satellite, ANIK-B, in the 12 GHz
frequency band use, a) conventional dominant mode Chebyshev filters in invar with
allpass equalizers with unloaded Q's in the
range 5,000 to 6,000 for the channel
separating filters of the input multiplexers
and, b) dual-mode filters in invar with
TE_{103} as the mode of propagation in optimized square waveguides with unloaded Q of
9,000 or greater for the output multiplexers.
This establishes a sort of bench mark. The
performance requirements of the new SBS and
ANIK-C satellite systems predicate unloaded
Q's of 9,000 or greater for the input mux
filters and 11,000 or greater for output
multiplexer filters. These unloaded Q's
can be achieved by using TE_{103}/TE_{113} and
TE_{105}/TE_{115} as the mode of propagation
respectively, in a dual-mode configuration.

To realize unloaded Q's greater than
11,000, the contest will likely be between
the single mode TE_{011} folded structure and
TE_{105}/TE_{115} or TE_{107} dual-mode structure
with optimized cross-sections. The ultimate
tradeoff would involve the weight and
complexity of output multiplexers on the
one hand versus the weight per unit of rf
power of the TWTAs.

For input multiplexing networks, the two
competing candidates are the externally
equalized minimum phase filters using allpass equalizers and the self equalized
linear phase filter, both designs configured
in a dual-mode structure. Externally
equalized filter gives about 10 dB higher
isolation than a linear phase filter of
equivalent order and group delay response.
On the other hand, linear phase structure
is elegant and requires no coupling circulator used for dual-mode allpass equalizers
with the associated penalty of 0.15 lbs.
per channel. The tradeoffs involved in
selecting the optimum design for a typical
satellite application are described in a
paper to be presented shortly[11].

Looking ahead to the mid 1980's, the
trend towards satellites with output powers
of up to 1 kW linking central communications
systems with many thousands of inexpensive
earth stations[12] will place ever more
pressure on the multiplexer designers for
more efficient filter networks. The magnitude of the problem can be appreciated by
considering that with 1 kW of rf output
power every 0.5 dB loss costs more than
100 watts of rf energy. Heat dissipation
also becomes an important design constraint
and the susceptibility of narrow band
filters to multipactor at high power levels
is another factor which will have an impact
on the design approach.

The types of filter networks described
herein show promise of being likely candidates for such applications. The output
multiplexer network for Intelsat V combines
three 10 watt TWTAs. While this is not a

high power level, it is worth noting that during qualification testing it was operated in a vacuum with a 20 W swept signal through each channel without evidence of multipactor and was operated through critical pressure at the full rated power of 10 W per channel without evidence of corona breakdown.

More advanced versions of this family of filters are currently being qualified at 14 GHz as multiplexing filter for the high power segment of earth terminals to be used with the ANIK-C and SBS systems. Transmitter power levels of up to 2 kW are being used on this program.

While this work does not involve vacuum testing, it will, nevertheless, provide valuable guidelines on the limits of this particular technology approach and may establish the necessary way station from which a successful program can be launched.

Acknowledgements

Much of the work described herein was funded as part of the Industrial Research Assistance Program of the National Research Council of Canada to whom the authors are indebted.

References

1. D.A. Taggart and R.D. Wanselow, "Mixed Mode Filters," IEEE Trans. MTT, Vol. 22, pp. 898-902, October 1974.

2. A.E. Atia and A.E. Williams, "General TE011-Mode Waveguide Bandpass Filter," IEEE Trans. MTT, Vol. 24, October 1976.

3. G.L. Matthaei, L. Young and E.M.T. Jones, "Microwave Filters, Impedance-Matching Networks and Coupling Structures," McGraw-Hill, pp. 921-934.

4. G.L. Matthaei and D.B. Weller, "Circular TE011 Trapped Mode Bandpass Filters," IEEE Trans. MTT, Vol. 31, pp. 581-589.

5. A. Atia and A. Williams, "Dual-Mode Canonical Waveguide Filters," International Microwave Symposium, June 1977.

6. G. Pfitzenmaier, "An Exact Solution for a Six-Cavity Dual-Mode Elliptic Bandpass Filter," International Microwave Symposium, June 1977.

7. S. Kallianteris and M.V. O'Donovan, "Technology Advances in the Realization of Filter Networks for Communications Satellites Operating at Frequencies Above 10 GHz," AIAA Paper No. 76-292.

8. S. Kallianteris, "Low Loss Linear Phase Filters," International Microwave Symposium, June 1977.

9. S. Kallianteris, C.M. Kudsia and M.N.S. Swamy, "A New Class of Dual-Mode Microwave Filters for Space Application," European Microwave Conference, September 1977.

10. S. Kallianteris, "Design of Low Loss 12 GHz Output Multiplexer for the Second Generation Telesat Canada Satellite," 6th European Microwave Conference, September 1976.

11. C.M. Kudsia, S. Kallianteris and M.N.S. Swamy, "Linear Phase Versus Externally Equalized Longitudinal Dual-Mode Filters for Space Application," Paper submitted for International Microwave Symposium, June 1978.

12. Outlook for Space - Report to the NASA Administrator by the Outlook for Space Study Group published by Scientific and Technical Information Office, Washington, D.C., 1976.

Waveguide Filters for Space — Smaller, Lighter . . . and Fewer Cavities*

Frequency selectivity requirements of satcom programs like INTELSAT IV have inspired the design of some sophisticated filters in square, rectangular and circular geometries. Dual-mode and other types of narrowband multiple coupled-cavity configurations are discussed.

A. E. Atia &
A. E. Williams
COMSAT Laboratories
Clarksburg, Maryland 20734

High-capacity communications satellite transponders usually require many transmit channelizing filters to distribute power over the communications band. To utilize the allocated frequency spectrum as efficiently as possible, guard bands should be very narrow and, hence, sharp cutoff filters are desirable. Furthermore, the filters must have flat in-band group delay and small gain slope to minimize distortion and crosstalk.

As an example, the INTELSAT IV specifications required 12 36-MHz-bandwidth channels, separated by 40 MHz center-to-center over the 3.7–4.2 GHz band. During the design phase of the INTELSAT IV program in 1968, it became apparent that these specifications could be met only by a 10-cavity waveguide Chebychev filter followed by a five-pole time delay equalizer. In addition, to achieve thermal stability, both filter and equalizer had to be constructed from thin wall Invar, resulting in a per-channel weight of approximately eight pounds. Since there were 12 channels on INTELSAT IV—with the possibility of more channels required on future spacecraft—the need for improved filters was urgent.

The possibility of achieving this objective was first mentioned in the literature by R. M. Kursrok,[1] who published design information on 3- and 4-coupled-cavity resonators having real zeros of transmission. In general, filters that have equi-ripple in both the passband and the

stopband can be considered to be optimum. Consequently, they require a smaller number of resonators to achieve a desired selectivity, thereby reducing the in-band gain slope, together with the weight and volume of the filter.

Dual-Mode Filters. Although the authors were initially unaware of Kursrok's papers, in 1969, they began an active search at COMSAT Laboratories for a waveguide structure which would enable elliptic responses to be generated. This was finally achieved by using the following physical line of reasoning:

Consider a cascaded set of four cavities and allow a small amount of energy to couple directly from cavity one to four. If the phase of the direct coupled field is opposite that coupled via cavities 1, 2, 3, and 4, then cancellation at frequencies in the stopband is possible. The geometry that realizes this objective was first introduced in 1970[2] and generalized in 1972.[3]

Figure 1 shows the dual-mode filter, which utilizes either square or circular cavities operating in the orthogonal TE_{101} or TE_{111} mode, respectively. It can readily be seen that, for the four-pole filter, electrical cavities 1 and 4 can be directly coupled and, in general, extra cross couplings 3 and 6, 5 and 8, . . ., may be realized. Coupling between electrical cavities occupying the same physical cavity is achieved by 45° coupling screws, and the sign of the extra cross couplings can be made positive or negative, depending on the relative orientation of adjacent 45° coupling screws.

To realize the frequency selectivity of the INTELSAT IV channelizing filter, the dual-mode filter requires only eight poles or four physical cavities. Consequently, weight and volume savings are achieved together with improved in-band loss and group delay. A measured response is compared with theory in Figure 2. One of the latest communications satellites, INTELSAT

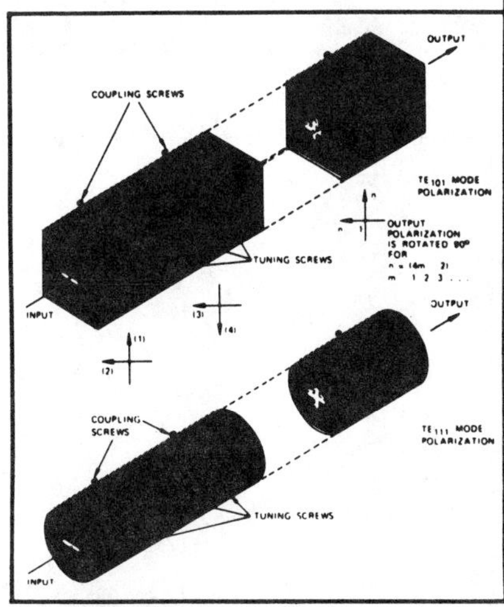

1. Dual-Mode square and circular geometries.

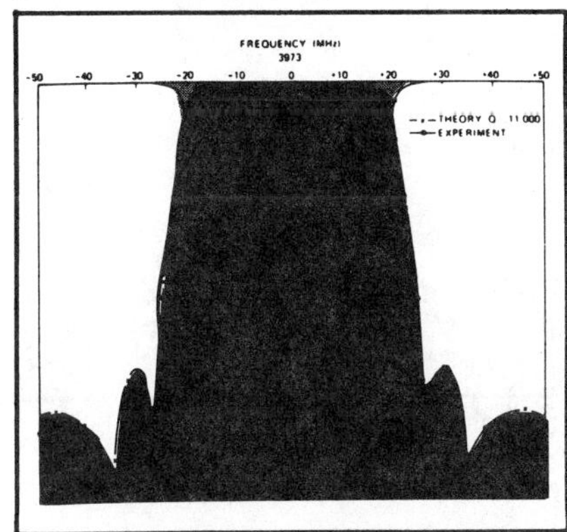

2. Transmission and return loss of 8-pole dual-mode bandpass filter.

IV-A, employs this new type of filter.

Canonical Filter. While the dual-mode filter represents a significant improvement over the Chebychev filter, it is not optimum since a true elliptic response cannot be achieved for orders >4. For example, the true eighth order elliptic response has three real zeros of transmission, while the dual-mode filter has only two. An optimum response can be achieved only by a structure for which extra couplings 1 to n, 2 to n −2, . . ., are possible.

---

*This article is based upon work performed at COMSAT Laboratories under the sponsorship of the International Telecommunications Satellite Organization (INTELSAT). Views expressed herein are not necessarily those of INTELSAT.

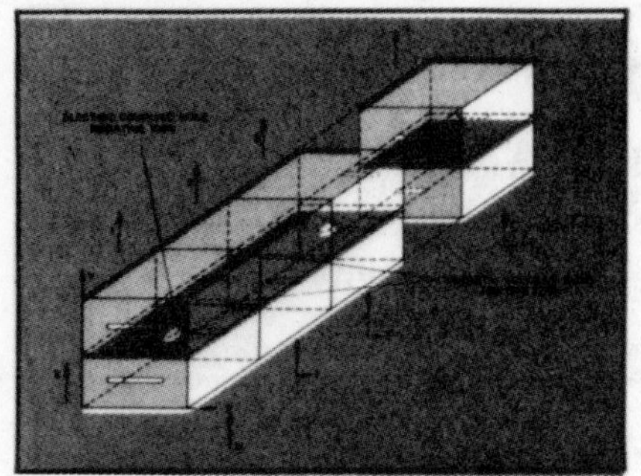

3. Rectangular canonical bandpass filter.

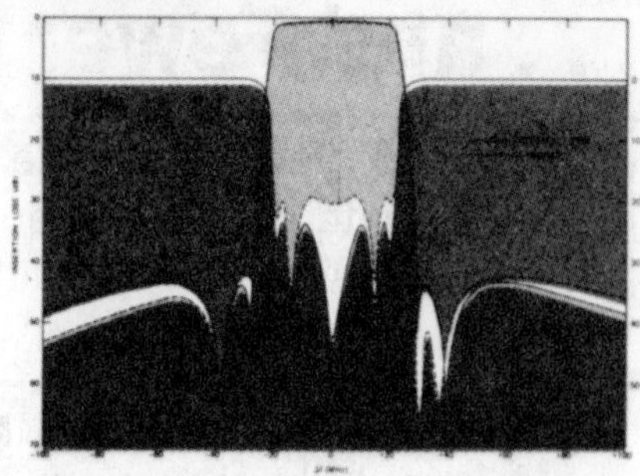

4. Sixth order bandpass filter response.

This arrangement of couplings can be realized by a folded rectangular (or square) geometry such as that of the generalized rectangular structure shown in Figure 3, which was first introduced in 1974.[4] A measured response of a sixth order elliptic filter is shown in Figure 4. Opposite shunt coupling signs are achieved by either electric or magnetic couplings through the broad or common waveguide cavity wall.

Generalized Square Waveguide Filter. The general square dual-mode waveguide shown in Figure 5 was first described in 1973.[5] This geometry simultaneously allows the couplings achieved in both the longitudinal dual-mode filter and the canonical cross-coupling set. Although some of the couplings are redundant, their sensitivity is reduced; therefore, more complex responses may be easily realized. As an example, this structure has been used to generate a 12-pole bandpass filter with an amplitude response having two real zeros of transmission together with flattened time delay (see Figures 6a and 6b). This filter exceeds the INTELSAT IV channel specification with an estimated weight of 1.5 lbs. using an Invar construction as opposed to 8 lbs. for the 10-cavity Chebychev filter and five-pole equalizer. In Figure 7, which compares these two types of filter-equalizers, the volume and size reduction is clearly apparent.

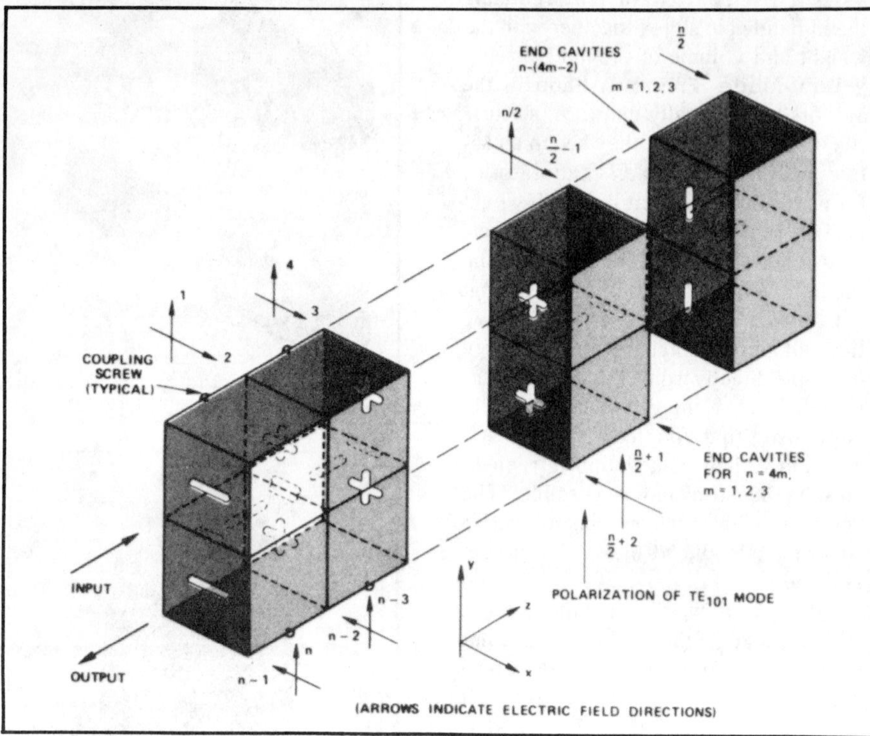

5. Generalized square bandpass filter.

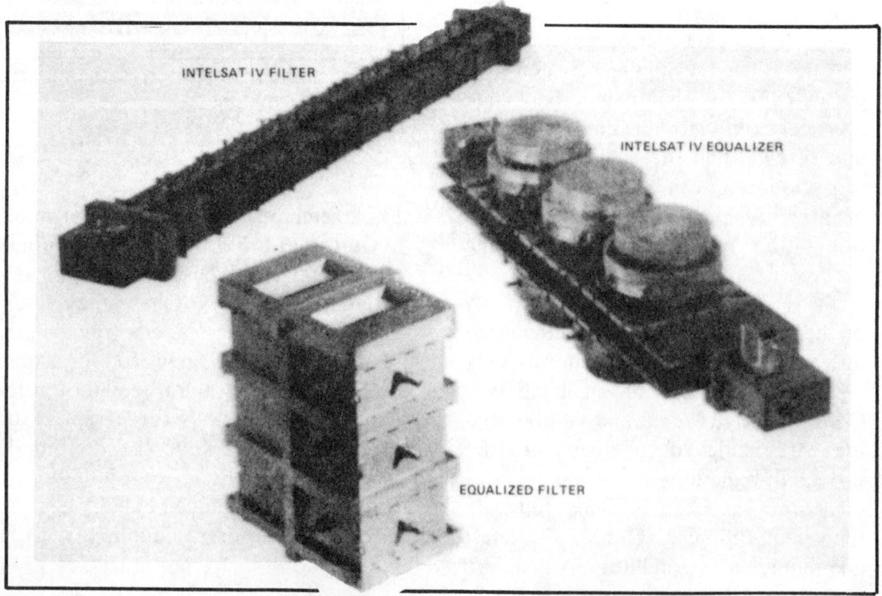

7. INTELSAT IV channelizing 10-pole filter and 5-pole equalizer compared to 12-pole equalized filter.

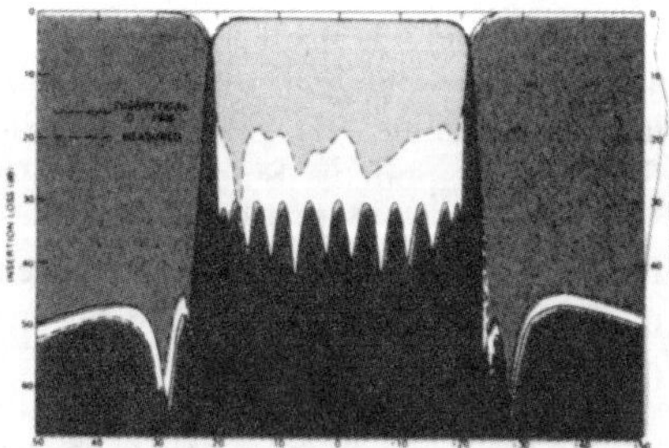

6a. Transmission and return loss of 12-pole filter.

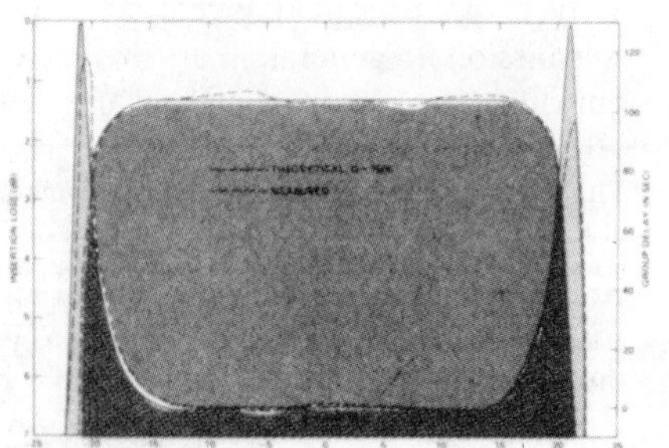

6b. Time delay of 12-pole filter.

General High-Q Filter. All the preceding filters employ the square (rectangular) or circular fundamental TE_{101} or TE_{111} mode. Unloaded Q's of 10,000 are obtained at 4 GHz, but due to the $1/(\text{frequency})^{1/2}$ dependence, Q's of only 5,500 and 4,000 are realized at 12/14 and 20/30 GHz, respectively. The need for a higher Q is therefore of prime importance at these higher frequencies.

One obvious way to achieve this objective is to utilize the higher order TE_{011} circular mode, which enables Q's of at least 16,000 to be realized at 12/14 GHz.[6] A canonical realization of the TE_{011} mode filter is shown in Figure 8. Opposite coupling signs are achieved by radially coupling cavities that are either directly opposite or displaced by one-half a diameter. Measured results of an eighth order 40-MHz bandpass filter centered at 12 GHz are shown in Figure 9. Figure 10 is a photograph of the filter. This filter's bandwidth is 0.33 percent. However, bandwidths up to 1 percent—i.e., 120-MHz wide bandpass filters at 12 GHz—have been successfully tuned with excellent agreement between theory and measurement. A significant advantage of this type of filter is that temperature compensation can be achieved with a very-high-expansion silicone resin at the back of the non-contacting tuning plunger. This approach allows the use of aluminum construction to yield an Invar thermal equivalent and minimizes the weight of the filter.

Conclusion. The new types of bandpass filters discussed herein possess electrical characteristics that are superior to those of the Chebychev design, and enable the use of the minimum number of cavities to achieve a given selectivity. First choice in terms of minimizing weight for spacecraft applications is, therefore, the most efficient electrical design. Second choice is the use of a lightweight, thermally stable material. Conventional spacecraft designs have employed Invar, but in recent years considerable effort has been directed toward the lightweight, thermally stable, graphite fiber epoxy materials. Initial problems such as long-term stability and plating peel strength have recently been overcome, resulting in very lightweight filters. Figure 11 pictures a graphite fiber, eight-pole, dual-mode filter, which has a Q of 10,000 and weighs only 6 oz. The weight and volume reduction apparent in this filter as compared to the INTELSAT IV 10-cavity Chebychev filter, which weighed 3 lbs., clearly indicates the advances in waveguide filter design during the past six years. ∎

References

1. R. M. Kursrok, "General Four-Resonator Filters at Microwave Frequencies," *IEEE Transactions on Microwave Theory and Techniques*, Vol. MTT-14, June 1966, pp 295–296.
2. A. E. Williams, "A Four-Cavity Elliptic Waveguide Filter," *IEEE Transactions on Microwave Theory and Techniques*, Vol. MTT-18, No. 12, December 1970, pp. 1109–1114.
3. A. E. Atia and A. E. Williams, "Narrow Bandpass Waveguide Filters," *IEEE Transactions on Microwave Theory and Techniques*, Vol. MTT-20, 1972, pp. 258–265.
4. A. E. Atia and A. E. Williams, "Non-Minimum Phase-Optimum Amplitude Bandpass Waveguide Filters," *IEEE Transactions on Microwave Theory and Techniques*, Vol. MTT-22, 1974, pp. 425–432.
5. A. E. Atia and A. E. Williams, "Non-Minimum Phase-Optimum Amplitude Bandpass Waveguide Filters," International Microwave Symposium, Boulder, Colorado, June 1973.
6. A. E. Williams and A. E. Atia, "Generalized TE_{011} Mode Waveguide Bandpass Filters," International Microwave Symposium, Palo Alto, California, May 1975.

Information Retrieval No. 76

10. Eight-Pole TE_{011} mode bandpass filter.

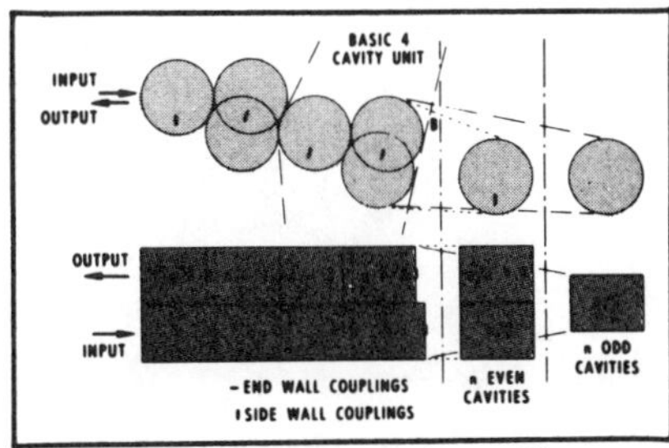

8. Generalized TE_{011} bandpass filter.

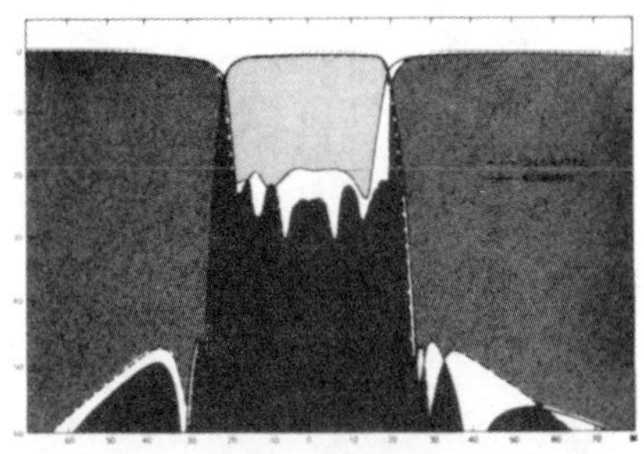

9. Transmission and return loss of TE_{011} mode 8-pole bandpass filter.

11. Eight-Pole dual-mode graphite fiber epoxy filter. *(Photo courtesy of Marconi Space and Defence Systems Ltd., Stanmore, U.K.)*

551

Part V
Propagation

This part deals with various propagation effects that influence the performance of a satellite communications system. For lower frequencies (e.g., 100 MHz–4 GHz), ionospheric scintillation is a problem. References [1] and [2] provide good discussions of scintillation. Reference [3] discusses scintillation at 4 and 6 GHz in the INTELSAT systems. Tropospheric phenomena such as molecular absorption and rain attenuation and scattering must be considered above 4 GHz. References [4]–[6] discuss various aspects of this problem. The dominant effects above 4 GHz are due to rain. Reprint Paper 5.1 discusses methods of measuring rain attenuation at various frequencies and shows how site diversity (receiving the signal at two earth stations separated enough to give uncorrelated rain statistics on the satellite-ground paths) can improve the availability. The other degradations caused by rain, such as depolarization, interference between systems due to scattering, increase in earth-station noise, and degradation of earth-station antenna performance are also discussed.

In recent years, a significant amount of effort has been devoted to measuring rain attenuation and depolarization statistics by the use of beacons on ATS-6, CTS, COMSTAR, ETS-2, and SIRIO. References [7]–[13] discuss these experiments and other aspects of the rain problem.

There are several methods that can be used to combat propagation problems. These include site diversity (e.g., [14]), polarization correction devices (e.g., [15]), and error correcting codes with interleaving to combat scintillations.

As satellite communications systems move to higher frequencies and utilize more sophisticated signal processing, an adequate theoretical and experimental understanding of propagation will become increasingly important.

REFERENCES

[1] J. Aarons, H. E. Whitney, and R. S. Allen, "Global morphology of ionospheric scintillations," *Proc. IEEE,* vol. 59, pp. 159-172, 1971.

[2] R. K. Crane, "Ionospheric Scintillation," *Proc. IEEE,* pp. 180-199, 1977.

[3] R. R. Taur, "Ionospheric scintillation at 4 and 6 GHz," *COMSAT Tech. Rev.,* vol. 3, pp. 145-163, 1973.

[4] R. K. Crane, "Propagation phenomena affecting satellite communications systems operating in the centimeter and millimeter wavelength bands," *Proc. IEEE,* vol. 59, pp. 173-188, 1971.

[5] R. K. Crane, "Prediction of the effects of rain on satellite communication systems," *Proc. IEEE,* vol. 65, pp. 456-474, Mar. 1977.

[6] T. Oguchi, "Rain depolarization studies at centimeter and millimeter wavelengths: Theory and measurement," *J. Radio Res. Lab.,* (Japan), vol. 22, no. 107, pp. 165-211, 1975.

[7] L. J. Ippolito, Ed., "20- and 30-GHz experiments with the ATS-6 satellite," *NASA TN D-8197,* Apr. 1976.

[8] H. W. Arnold and D. C. Cox, "Some results from 19 and 28 GHz COMSTAR beacon propagation experiments," AIAA Paper No. 78-619, AIAA 7th Communications Satellite Conf., San Diego, CA., Apr. 23-27, 1978.

[9] H. Kaneda, K. Tsukamoto, and H. Fuketa, "Experiments in the Japanese CS program," AIAA Paper No. 78-616, AIAA 7th Communications Satellite Conf., San Diego, CA, Apr. 23-27, 1978.

[10] J. L. King and G. Hyde, "The COMSAT 13 and 18 GHz propagation experiment," *Proc. IEEE Int. Conf. Communications,* pp. 18-18-18-21, June 16-18, 1975.

[11] G. Perrotta, "The Italian SIRIO 12-18 GHz experiment: The forerunner of 20-30 GHz preoperational satellites," AIAA Paper No. 78-631, AIAA 7th Communications Satellite Conf., San Diego, CA, Apr. 23-27, 1978.

[12] URSI, "*Proc. 1st Open Symp. URSI Commission F.,* La Baule, France, Apr. 28-May 6, 1977.

[13] P. J. Bartholome, "Propagation experiments at 11/14 GHz as part of the European communications satellite program, "*Proc. IEEE,* vol. 65, no. 3, pp. 475-477, Mar. 1977.

[14] E. E. Reber and D. V. McDonald, "On rainfall and space diversity for millimeter-wave earth-satellite communications systems," TR-0073(3230-01)-1, The Aerospace Corporation, El Segundo, CA, Mar. 15, 1973.

[15] D. DiFonzo, A. E. Williams, and W. S. Trachtman, "Adaptive polarization control for satellite frequency reuse systems," *COMSAT Tech. Rev.,* vol. 6, no. 2, pp. 253-284, Fall 1976.

The Role of Rain in Satellite Communications

DAVID C. HOGG, FELLOW, IEEE, AND TA-SHING CHU, MEMBER, IEEE

Invited Paper

Abstract—The most fundamental obstacle encountered in design of satellite communication systems at frequencies above 10 GHz is attenuation by rain. The microwave power radiated toward an earth station, being limited by factors such as available primary power and size of antenna on the satellite, is insufficient, with present technology, to overcome the large attenuation produced by intense rain cells on the earth–space path. The resultant loss of signal makes for unreliable transmission. In what follows, methods of measurement of this attenuation at various frequencies and a technique called path diversity that substantially improves the reliability are presented. Other degradations produced by rain, such as depolarization, interference, increase in earth-station noise, and deterioration of earth-station antenna performance, are also discussed.

I. Introduction

IN EARLY EXPERIMENTS using microwaves for broadband transmission via satellite, it was quickly recognized that rain influenced performance of the system. For example, in the Telstar experiment [1], in which 4-GHz

Manuscript received February 28, 1975; revised April 12, 1975.
The authors are with Bell Laboratories, Crawford Hill, Holmdel, N.J. 07733.

signals from the satellite were received with sensitive maser amplifiers [2], it was found that the level of noise increased significantly when it was raining in the vicinity of the receiving station. This increase stemmed primarily from two sources: blackbody radiation from the raindrops in the sky [3], and emission and reflection from water layers that formed on the radomes used to protect the earth-station antennas [4]. It was also observed that interfering signals could enter such systems by way of scattering from the raindrops [5]. We now know that all of these effects can be explained by theories of electromagnetic wave interaction with liquid water in its various forms. All of these theories rely upon knowledge of the basic microwave properties of liquid water, first studied in depth by Saxton [6]. Best estimates [6], [7] of the real (refractive) and imaginary (dissipative) components of the refractive index of water are shown in Fig. 1 for the wavelength range 1 mm to 10 cm. The corresponding frequency scale, 300 to 3 GHz, is shown on the upper abscissa. The curves are a typical loss-dispersion pair representing a resonance in the liquid water at a wavelength of about 1 cm (30

Reprinted from *Proc. IEEE,* vol. 63, pp. 1308–1331, Sept. 1975.

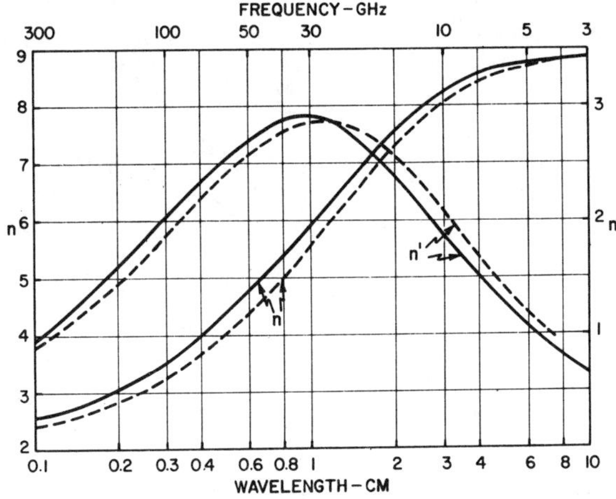

Fig. 1. Refractive index of water at 20°C; n and n' are the real and imaginary components, representing the refractivity and heat loss respectively; —— [6] and – – – [7] are best estimates based on measurements.

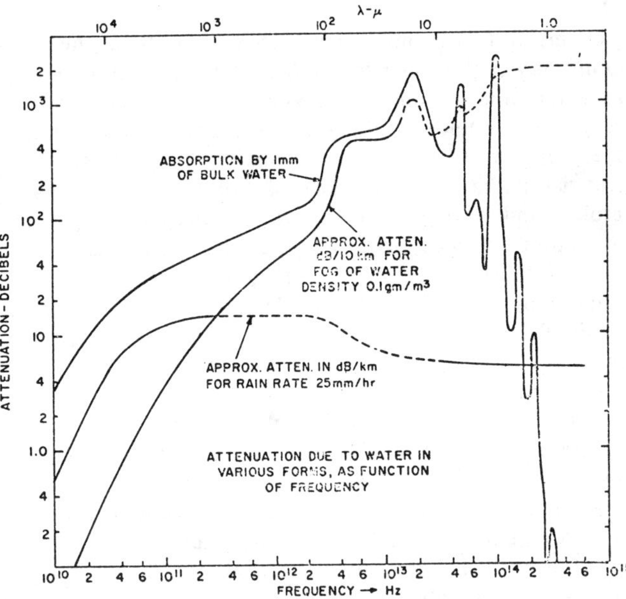

Fig. 2. Comparison of attenuation by water in various forms; the integrated thickness of water on the transmission path is 1 mm in all cases.

GHz). This absorption by the liquid should not be confused with absorption by water in gaseous form (water vapor) which will be mentioned later on.

Satellite communication systems utilizing the 4 and 6 GHz common carrier bands are now fairly commonplace. Provided suitable care is taken with the quality of earth station antennas and selection of the sites to avoid interference with terrestrial systems sharing the same frequency bands (discussed in Section VII), further expansion may take place. However, saturation at these frequencies will eventually occur. Moreover, the bandwidth allocated to these systems is 500 MHz which may soon be exhausted by increasing demands for telephone, television and data traffic. To overcome this saturation, the use of higher frequency bands was proposed [8] and new common carrier bands, more than 2 GHz wide, were subsequently allocated at 19 and 29 GHz. Unfortunately, it turns out that rain becomes a much more serious factor at these higher frequencies than at 4 and 6 GHz and recently there have been several international meetings on the subject [9]–[11]. The thrust of this paper is to present a background of the research in propagation through rain that has pointed the way toward practical design of higher frequency systems [12].

Before getting into the details of the various effects of rain on propagation, there are four statements that should be made regarding snow, fog, the troposphere, and the ionosphere. When water is frozen, as in the case of particles in many clouds, Fig. 1 no longer applies and the corresponding loss resonance appears at much longer wavelengths [7]. The net result is that ice and dry snow exhibit very low loss in the microwave band, and, therefore, we will not consider them further in this discussion. On the other hand, fog is, of course, comprised of small drops of liquid; but the density of liquid water in a heavy fog [13] is less than one-twentieth that of a heavy rain so the attenuations encountered are small and will be neglected here. Another propagation effect, very familiar to those concerned with terrestrial radio-relay systems, is signal fading caused by layers and other aberrations of the refractivity profile of the troposphere. These attenuations of the signal occur on microwave beams that are essentially horizontal, thereby interacting at near-grazing angles with the layers. But for typical earth–satellite paths at elevation angles of more than a few degrees, this type of fading [14] is not troublesome and will not be discussed in detail, nor will the relatively small scintillations that are also created by nonuni-

form refractivity. Earth–satellite paths operating at the lower frequencies in the microwave band, e.g., 4 GHz, are subject to some rotation of polarization by the ionosphere via the Faraday effect, and some signal variations [15]. The latter are most prominent at earth stations located near the equator; but it is believed that the ionosphere will have negligible effect on the higher frequency systems that are our main concern here.

It is both interesting and instructive to compare our familiar visual experiences during rain and fog with the microwave properties. For example, the visibility in a heavy rain is not nearly as restricted as in fog [13], whereas, as mentioned above, the liquid-water content is considerably lower in the fog and it attenuates microwaves much less than does rain. A direct comparison can be made by plotting the attenuation as a function of frequency for a given amount of integrated liquid water on a path. Fig. 2[1] shows the attenuation properties of water in three forms, where the total amount of water involved in each case is 1 mm. On the abscissa, the frequency runs from 10^{10} Hz in the microwave band to beyond the optical band at 6×10^{14} Hz. The curve with the multiple resonances is absorption by liquid water in the form of a slab 1 mm thick. The absorption is about 4 dB at 10 GHz and increases rapidly with frequency in the infrared region. On the other hand, at optical frequencies the loss in the slab of water is less than 0.1 dB. This plot does not take into account reflection at the surfaces of the layer, but is useful as a reference. The other two plots are attenuation of a 25 mm/h rain (of content 1 g/m³) over a 1-km path and of 0.1-g/m³ fog over a 10-km path; in both cases the liquid water integrated along the path is equal to the thickness of the slab, 1 mm. Over the microwave band, the attenuation is considerably greater for the rain than for the fog, but in the visible band, the reverse is true. This behavior arises because the water is absorptive in the microwave band whereas at visible wavelengths the absorption is negligible and the attenuation is caused by scattering; the latter is inversely proportional to the size of the drops. It is of interest that water in the form of a slab pro-

---

[1] The rain attenuation decreases rapidly below 10 GHz. The curves for rain and fog in Fig. 2 will almost coalesce at frequencies below 1 GHz.

duces much higher microwave attenuation than the same amount of water in other forms. This behavior is an important consideration in antenna design, as discussed in Section VIII.

Much of our understanding of the fundamentals of microwave propagation through rain stems from measurements on terrestrial paths. The segment of this information that is useful in understanding the earth–space transmission problem is therefore reviewed first. Of prime importance is the reliability of the transmission, as determined by the attenuation produced by rain. Preservation of polarization is also important since it relates to frequency reuse in an orthogonal mode; designs such as the ATT-Comsat domestic system are based on utilization of both polarizations. Methods of measurement of transmission on earth–space paths and typical measured data are discussed in Section III. In geographical areas that experience heavy rain rather often, it is found that reliability of transmission on a single path is insufficient to meet the objectives of highly reliable communication systems; experiments using path diversity, a technique which improves reliability considerably, are discussed in some detail in Section IV. The questions of extrapolation of measured rain attenuation to other frequencies and of the influence of scattering by rain on interference and frequency sharing between systems are also addressed. Since rain water on antenna structures degrades performance, we give a brief resume of those effects.

II. THEORY AND EXPERIMENT FOR TERRESTRIAL PATHS

A. Calculated Attenuation and Comparison with Observations

Rain increases the noise temperature of a low-noise receiving system, as mentioned earlier, but at frequencies above 10 GHz the rain attenuation *per se* becomes a major factor in the design of a radio system. Calculation of attenuation by rain will be briefly summarized.

The decrease in the magnitude S of the Poynting vector in passing through a precipitation layer of thickness Δl is

$$-\Delta S = S\Delta l \int_0^\infty n(a)Q(a,\lambda)\,da \qquad (1)$$

where $Q(a,\lambda)$ is the extinction cross section (square centimeter) of a spherical raindrop with radius a (centimeter) and $n(a)\,da$ is the number of drops per unit volume (cubic meter) in range da. Integrating (1) yields

$$S = S_0 \exp\left(-\int \alpha\,dl\right) \qquad (2)$$

where

$$\alpha = \int_0^\infty n(a)Q(a,\lambda)\,da. \qquad (3)$$

The attenuation in decibels per kilometer is simply $0.434\,\alpha$ with the parameters given in the above indicated units.

The extinction cross sections Q can be calculated using Mie's scattering solution for spheres [16] and Saxton's measured liquid-water refractive indices (Fig. 1). In the attenuation calculation, measured drop terminal velocities [17] and measured drop-size distributions, often the Laws and Parsons distribution [17], are used. During World War II Ryde and Ryde [18] carried out calculations of microwave rain attenuation; these have since been extended by others [19] with the

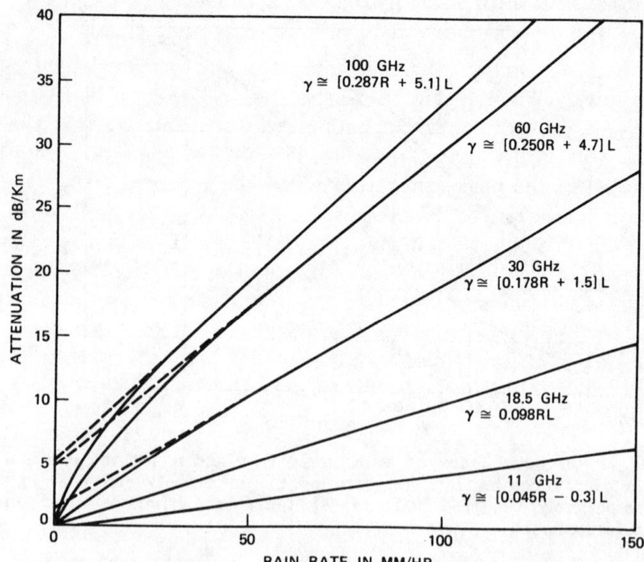

Fig. 3. The solid curves are computed attenuation as a function of rain rate and frequency; approximate linear equations are given; dotted lines are linear extrapolations.

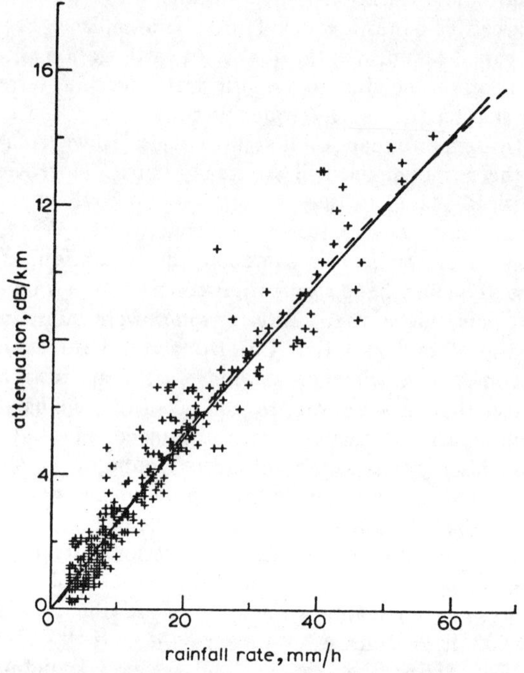

Fig. 4. Attenuation as a function of rainfall rate measured for ten showers during the summer of 1971 at Slough, England; – – – best-fit curve — theoretical curve. (Courtesy of Her Majesty's Stationery Office, U.K.)

aid of modern computers. Recently both attenuation and phase delay through rain have been calculated for centimeter and millimeter wavelengths [20], [21]; these data are presented in Fig. 3 for several frequencies of interest in common carrier radio systems. The attenuation increases significantly with both frequency and rain rate.

Using rapid-response rain gauges, theoretical predictions based upon summing the Mie solution over the Laws and Parsons drop-size distributions have shown agreement with attenuation over short terrestrial paths. An example of this agreement at 35 GHz measured [22] over a short path, 448 m, is given in Fig. 4. However, some comparisons show that the measured attenuation can be considerably higher than the calculated value; such discrepancies can be explained as follows. When a rain shower misses rain gauges distributed over a path, the apparent path-average rain rate is lower than the true value. Also, the slow response of some rain gauges may miss the peak rain rate whereas the recorded peak attenuation is taken as the measured point.[2] Furthermore, owing to the oblate shape of many of the large raindrops, horizontally polarized waves suffer greater attenuation than that calculated for spherical raindrops (see Section II-C). Sometimes the updraft phenomenon [23] in a rain storm can explain abnormally high attenuation for relatively low rain rates on the ground. Inhomogeneity of the rain over a long path causes considerable scatter in the data and leads to uncertainty when measured and calculated values are compared.

Neglect of multiple scattering effects along the path is often suspected [19], [24] as a possible source of error in prediction of rain attenuation. However, scattering by a raindrop at microwave frequencies is essentially isotropic, and near forward scattering contributes very little to a receiving antenna with a narrow beam. It is implicit in (2) that the incident wave at each layer of precipitation takes into account attenuation by preceding layers, therefore, the bulk of the multiple scattering effects are included in (2). One also notes that the liquid water density for a heavy rain of 100 mm/h is only about 4 g/m^3, which corresponds to an average spacing of about 10 cm between drops of the order 1-mm radius. The rain medium is simply not as dense as intuitive vision might suggest. Therefore, the simple summation in (3) of the lost power, extracted in absorption and scattering by each isolated drop, over a typical drop-size distribution, will predict the average rain attenuation for a given rain rate.

The rain-fading data needed for design of a radio communication system is best determined by measuring the attenuation statistics for the given frequency, geographical location and path length. Since acquisition of such data is costly and time-consuming, it is obviously desirable to extrapolate any available attenuation statistics to other frequencies, locations and path lengths. In addition to the capability for calculating the attenuation, a systematic prediction procedure needs the measured point rain-rate statistics of various locations. The measured data for several locations are shown in Fig. 5 [21]. The dependence of attenuation upon path length will be discussed in the next section. Let us now demonstrate the feasibility of predicting attenuation statistics.

An example [25] of measured attenuation data is given in Fig. 6 where percent-of-time distributions are plotted (as dashed curves) for 30.9 GHz over a path length of 1.9 km at Holmdel, N.J., during 1968 and 1969. Since most of the heavy showers usually occur in the summer months, the probability of a given attenuation in the worst month may be about five times that of the yearly distribution. For comparison with measured attenuation statistics, we have also plotted in Fig. 6 a solid curve predicted by calculating attenuation from the measured point rain rate distribution. The agreement in this comparison shows that the attenuation statistics for short transmission hops may be estimated from the measured point rain-rate distributions. The discrepancy between the measured and calculated curves often becomes larger at higher attenuation values or over longer paths where rain cell sizes may be smaller than the path length.

---

[2] Rain on antenna weather covers or radomes (see Section VIII) can also bias measured attenuations on the high side.

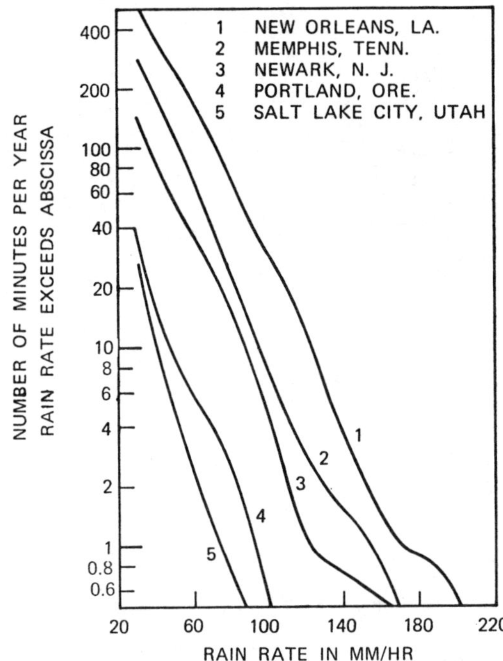

Fig. 5. One-minute rain rate distributions for a five year period (1966–1970) measured at several locations in the U.S.

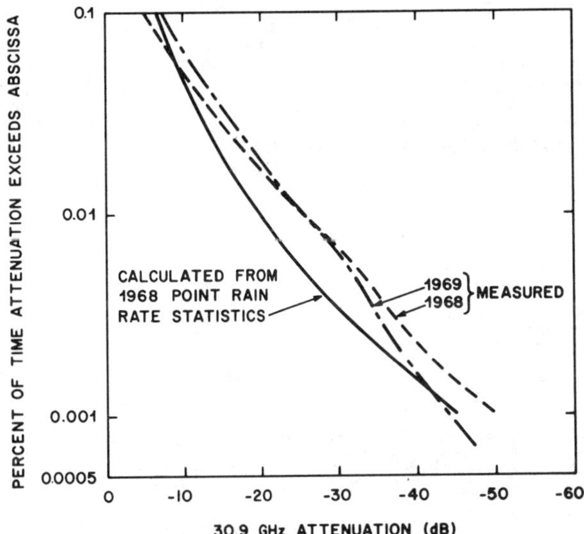

Fig. 6. Percentage-of-time distributions of 30.9-GHz attenuation measured over a 1.9-km path for 1968 and 1969; —— calculated attenuation from point rain rate distribution of 1968.

Since rain attenuation statistics at lower microwave frequencies for an earth-satellite path can be obtained by radiometer measurements as shown later, it is especially interesting to examine extrapolation of lower frequency data to higher frequencies where a saturation problem curtails the usefulness of radiometers (see Section III-B). The frequency extrapolation technique was first checked on a short terrestrial path (2.6 km) where 18.5- and 30.9-GHz attenuations were measured simultaneously [26]. In Fig. 7(a), the ratios of 30.9 to 18.5-GHz measured attenuations are plotted versus the total path attenuation at 30.9 GHz; the dashed line is the theoretical ratio obtained using the Laws and Parsons drop-size distribution. The theoretical ratios represent the body of the data rather well, especially for attenuations exceeding 10 dB at 30.9 GHz. At lower attenuation values, much of the scatter in the measured instantaneous ratios was found

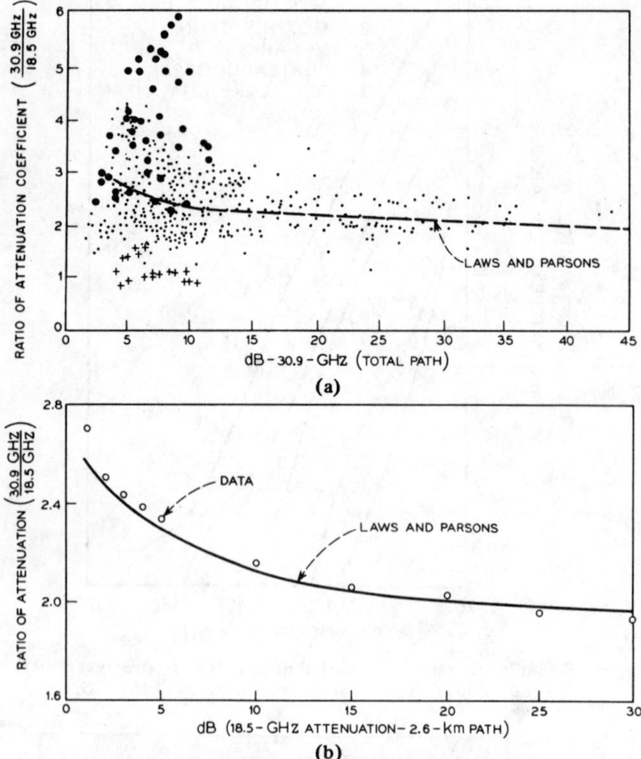

Fig. 7. (a) Ratio of 30.9 to 18.5-GHz attenuation versus the total path attenuation at 30.9 GHz; the dashed curve is the theoretical ratio using Laws and Parsons raindrop-size distributions. The large dots and the crosses are data from two particular storms believed to have special drop-size characteristics. (b) The ratio of measured attenuations taken at the same percent of time from the 30.9- and 18.5-GHz cumulative distributions, plotted as circles, versus the 18.5-GHz total path attenuation; the solid curve is the theoretical ratio using Laws and Parsons drop-size distributions.

to be caused by two particular storms indicated by dots and crosses, with anomalous drop-size distributions. When the percent of time distributions were obtained for 18.5 and 30.9 GHz attenuations, the ratios of 30.9 to 18.5 GHz attenuation taken at given percent of time levels show almost perfect agreement with the theoretical ratios as indicated in Fig. 7(b). A method for extrapolating data obtained on earth–space paths is discussed in Section VI.

B. Dependence of Attenuation on Path Length

Where short path lengths of the order 1 km are involved, the path attenuation is directly proportional to the path length, even for very high rain rates. The attenuation behaves that way because the diameter of the cells is generally larger than the length of the path, and in such cases the percent-of-time distribution of attenuation can be determined from the cumulative distribution of rainrate measured at a point, as discussed in the previous section. A further example of that behavior is given by measurements of 60-GHz attenuation [27], made for one year over a one km path at Holmdel, N.J.; these are compared with attenuation calculated from point rainrates measured for the same period with a rapid response rain gauge at the same location [21]. As shown in Fig. 8, the comparison is favorable to at least the 35-dB attenuation level. Note also that the distribution in Fig. 8 is approximately exponential.

However, the portion of an earth–satellite path that may encounter rain is much longer than 1 km. For example, rain can originate at altitudes of more than 10 km in the tropos-

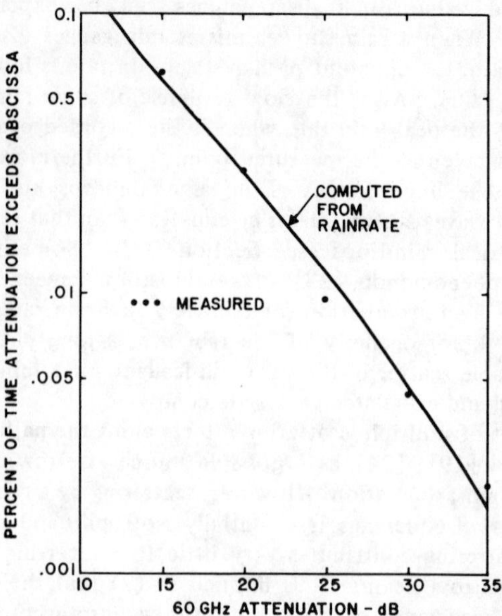

Fig. 8. Cumulative distribution of 60-GHz attenuation by rain measured for one year on a 1-km path compared with calculation using measured point rain rate.

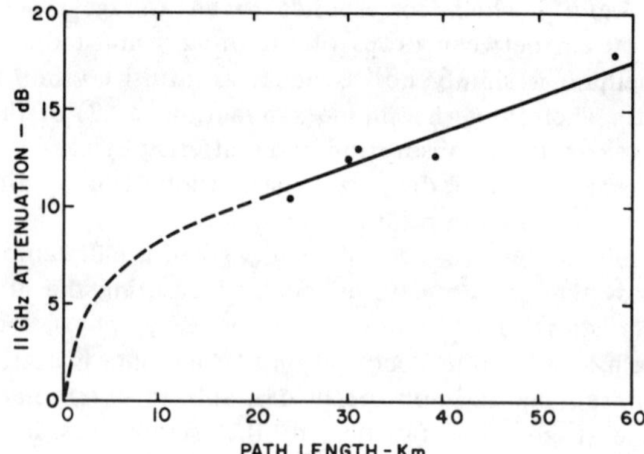

Fig. 9. · · · · measured 11-GHz attenuations exceeded 0.01 percent of the time on five paths in England during 1964; see text for description of curve.

phere[3] as has been determined by weather-radar studies. Therefore, if we assume that the elevation angle of an antenna beam looking from some earth station toward a satellite is 30°, the path length upon which a rain cell may appear is about 20 km, and intense showers, being of limited extent [29], occupy only a fraction of that length. It is because of this limited cell size that path diversity, a technique discussed in Section IV, is viable in satellite systems.

Some understanding of the dependence of rain attenuation on path length can be derived from a set of 11-GHz measurements [30] made in England during 1964 on five terrestrial paths ranging from 24 to 58 km. From the cumulative distributions of attenuation obtained from each path, the attenuation level exceeded for some given amount of time, say 50 min during the year, is determined. The result of that exercise is plotted in Fig. 9. Drawn through the points obtained from the experiment is a line which is extended by a dashed curve to zero attenuation at zero path length; for short path lengths, the dashed curve is essentially linear in accordance with the discussion in Section II-A. On the other hand,

---

[3] The altitude at which rain originates varies considerably over the U.S., being quite high in the central region [28].

considerable curvature is required at path lengths of about 5 km in order to link the two "linear" portions into a well-behaved curve. We believe that this curvature takes place at path lengths of the order of the diameter of the rain cell. Thus the relationship does not continue linearly with the initial slope to large path lengths because the cell diameter becomes smaller than the path length; in this case the attenuation is less than would have been anticipated from assuming point rainrates to be uniform over the entire path. Similar nonlinear behavior has also been observed in attenuations derived from rain-gauge network measurements [31]. That the cell sizes are relatively small for high rainrates has been substantiated in a recent publication from the U.K. [32]. Over a two-year period, 77 rain showers with peak rates greater than 20 mm/h exhibited a mean $1/e$ width of only 3 km, but this is not necessarily valid at other geographic locations. More is said about effective path lengths on earth-satellite paths in Section VI.

C. Polarization-Dependent Attenuation and Depolarization

The assumption of spherical raindrop shape used in the Mie calculation discussed in Section II-A is only a first-order approximation. Close examination by photography reveals that many of the larger drops are better represented by oblate spheroids. The ratio of minor to major axes of the oblate spheriodal raindrop, as determined from the experimental data [33], is approximately $a/b = 1 - \overline{a}$ where $\overline{a}$ is the radius (in centimeter) of an equivolumic spherical drop.[4] Oguchi [34] first investigated the effect of oblate raindrops on microwave propagation using perturbation calculations. Point-matching procedures [35]–[37] and improved perturbation techniques [37] now provide extensive numerical results for the scattering of plane electromagnetic waves by oblate spheroidal raindrops. The rain-induced attenuation and phase shift obtained from the forward scattering functions [38] $S(0)$ are

$$A_{\mathrm{I,II}} = 0.434 \frac{\lambda^2}{\pi} \sum \mathrm{Re}\, S_{\mathrm{I,II}}(0)\, n(\overline{a}), \quad \mathrm{dB/km} \qquad (4)$$

$$\Phi_{\mathrm{I,II}} = -36 \frac{\lambda^2}{\pi} \sum \mathrm{Im}\, S_{\mathrm{I,II}}(0)\, n(\overline{a}), \quad \mathrm{deg/km} \qquad (5)$$

where $n(\overline{a})$ is the number of drops with equivolume spherical radius $\overline{a}$, per cubic meter; the subscripts I and II designate electric fields parallel and perpendicular to the plane containing the axis of symmetry of the raindrop and the direction of propagation of the incident wave, and the summation is taken over all drop sizes. The derivation of (4) and (5) is based upon the effective index of refraction of a scattering medium proposed by Van de Hulst [38]. One notes that (3) of Section II-A and (4) are equivalent to each other and imply the following relation between the extinction cross-section and the forward-scattering function

$$Q = \frac{\lambda^2}{\pi} \mathrm{Re}\, \{S(0)\}.$$

The calculated [39] differential attenuation (II-I) and differential phase shift between polarizations II and I for various rain rates and frequencies from 4 to 100 GHz are given in Figs. 10 and 11; they agree with those obtained by Oguchi and

[4] The large drops, having a lesser component of tension normal to the surface, are most nonspherical.

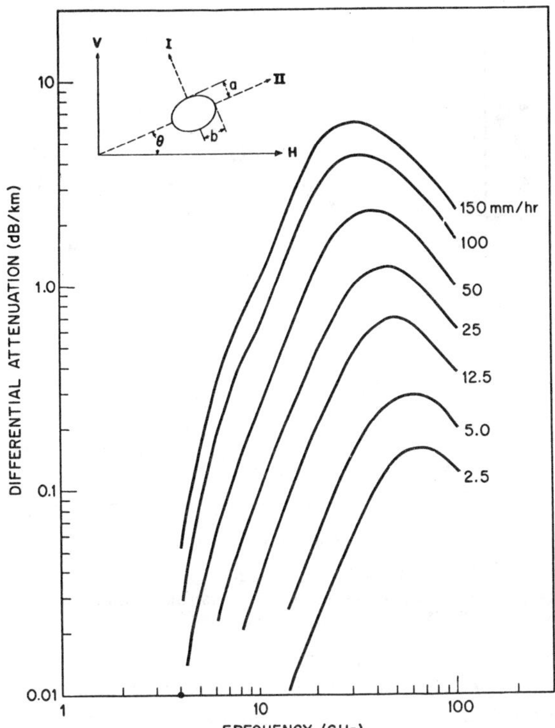

Fig. 10. Rain-induced differential attenuation between polarizations I and II for various rainrates.

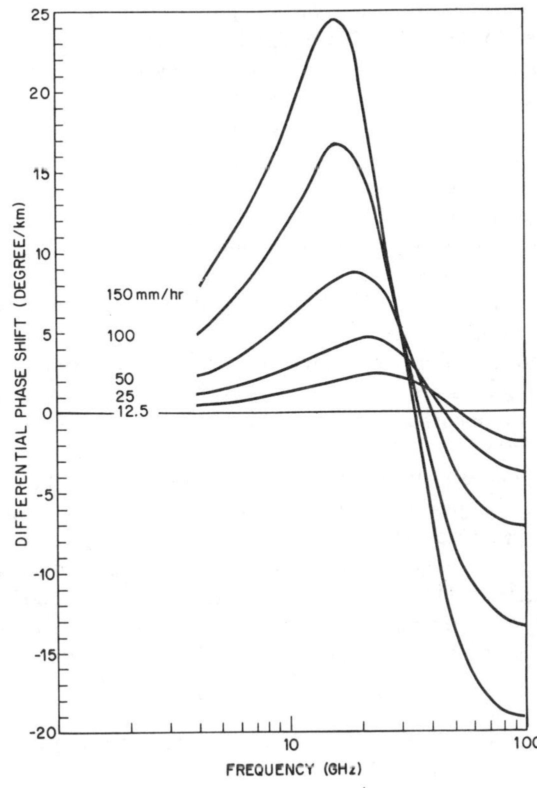

Fig. 11. Rain-induced differential phase shift between polarizations I and II for various rainrates.

Hosoya [40]. It is assumed in these calculated data that the direction of propagation is perpendicular to the axis of symmetry of the oblate raindrop. The differential attenuation between horizontal and vertical polarizations have been measured at 30 GHz [41] as shown in Fig. 12, and at 18 GHz [42] in Fig. 13. Also shown in the figures are calculated

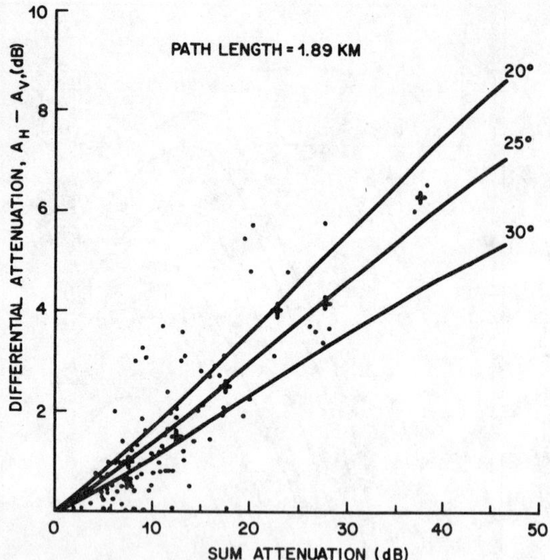

Fig. 12. 30-GHz differential attenuation in New Jersey; curves are calculated assuming the raindrops to be canted at the indicated angles; crosses are medians in 5-dB intervals.

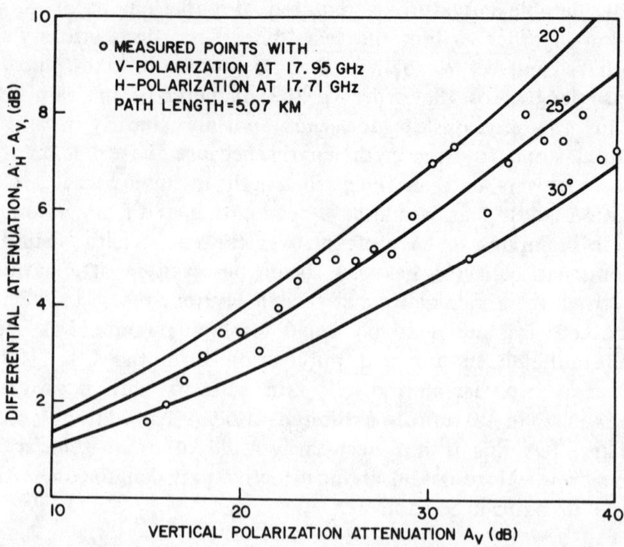

Fig. 13. 18-GHz differential attenuation measured in Georgia; curves are calculated assuming the raindrops to be canted at indicated angles.

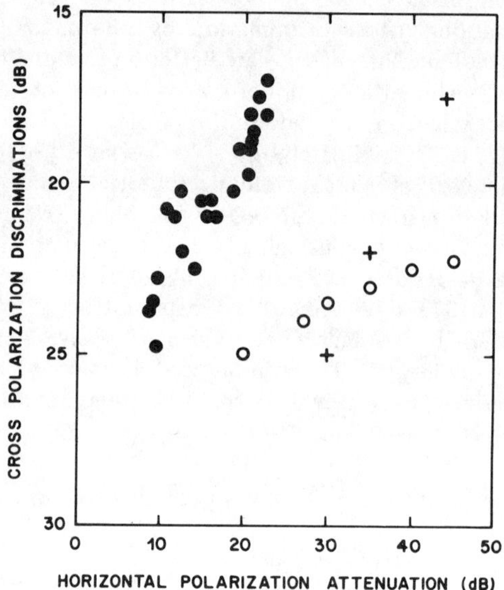

Fig. 14. Measured rain-induced cross polarization for horizontal polarization transmitted; · · · 11 GHz, +++ 17.71 GHz, ooo 60 GHz.

curves [43] with the raindrops assumed to be canted at the angles indicated.[5] The agreement between the measured data and the calculated curve with a canting angle of 25° is consistent with calculation using a photographically measured canting angle distribution [44]. Measured differential attenuation between horizontal and vertical polarizations is found to be smaller at 60 GHz [27] and 11 GHz [45] than at 30 and 18 GHz as predicted by the theory given in Fig. 10. It is especially interesting that the measured differential attenuation at 60 GHz remains below 2 dB for rain fades up to 40 dB [27]; this occurs because small raindrops of lesser ellipticity contribute heavily to the attenuation at shorter wavelengths; surface tension maintains sphericity in the small drops. Differential attenuation at lower microwave frequencies is less significant because the total rain attenuation is small. Recent measurements[6] of differential phase shift through rain has also given similar confirmation of the calculated data.

Depolarization by canted raindrops is a result of both differential attenuation and differential phase shift between the components along polarizations I and II. Numerous measurements of rain-induced cross polariation at centimeter and millimeter wavelengths have been reported [27], [45]–[54] during the past few years, the surge of activity in this area being prompted by efforts to realize frequency reuse of the orthogonal polarization. Measured cross-polarization discrimination[7] for transmitted horizontal polarization at 11, 17.71, and 60 GHz is plotted in Fig. 14. The 11-GHz data are samples measured in England [46], the 17.71-GHz data being measured medians of 16 rainstorms in Georgia, [42], [43]; the latter agree with 18.35-GHz measurements in New Jersey [47], and 20-GHz measurements in Japan [48], [49] (both not shown). The 60-GHz data are taken from the measured statistics [27] of rain-induced fading and cross polarization for the same percent of time level; they are believed to show

somewhat lower discrimination than expected because of a low cross-polarization discrimination level ($\simeq 30$ db) in the measuring system. Because the rain attenuation for vertical polarization is less than that of horizontal, the cross polarization discrimination at a given attenuation for transmitted vertical polarization is expected to be better than for transmitted horizontal polarization as has been experimentally confirmed [46]. The measured depolarization of a linearly polarized wave oriented at 45° with respect to the vertical direction [50] and of a circularly polarized wave [51] have been found to be very severe, as shown in Fig. 15. In all cases, measured cross-polarization discrimination shows a considerable variance about the median value at a given attenuation.

Theoretical prediction of the rain-induced cross polarization is hampered by uncertainty in the raindrop canting angle distribution. It was found from photographs [44] of falling raindrops that the canting angle is not far from an even distribution with respect to the vertical (gravity) direction. In the case of a horizontally or vertically polarized wave, the cross-polarization produced by positive and negative canting angles

<hr>

[5] The canting angle is the angle between the major axis of the image of an oblate raindrop (shown in Fig. 10) projected on a plane normal to the direction of propagation, with respect to the horizontal.

[6] Private communication from P. A. Watson and N. J. McEwan, University of Bradford, Bradford, England.

[7] Cross-polarization discrimination (XPD) is the ratio of the power received in the principal polarization to that converted into the orthogonal polarization.

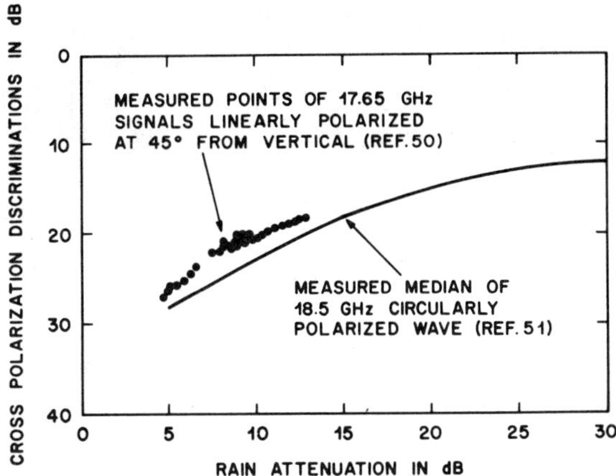

Fig. 15. Measured depolarization versus attenuation for circular polarization and for linear polarization at 45° from vertical transmitted.

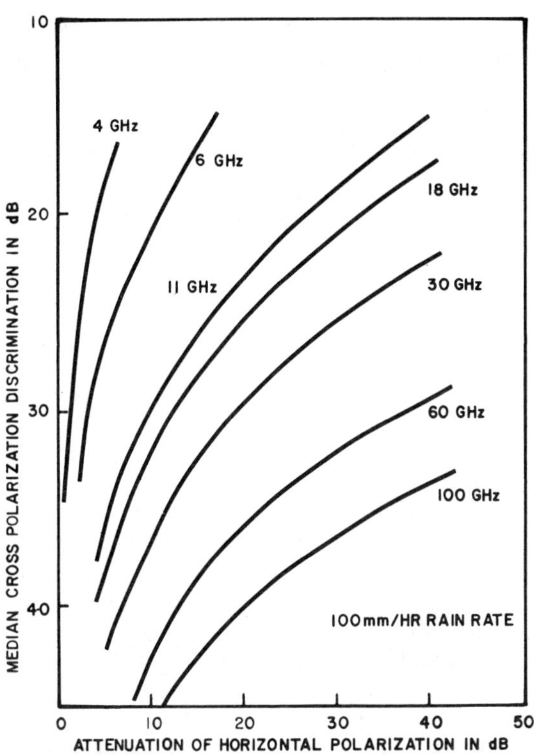

Fig. 16. Semiempirical estimates of median cross polarization discrimination versus attenuation for horizontally polarized transmitted waves at 100-mm/h rain rate.

tends to cancel. But it is found that systematic predictions can be obtained [43], [55] by comparing measured differential attenuation and cross polarization at one frequency with calculated values to determine two empirical parameters: an effective average of the absolute value of the canting angle, and the imbalance in the number of drops with positive and negative canting angles.[8] Such empirical median estimates based upon measured data at 18 GHz [42] are shown in Fig. 16 for the cross polarization of horizontally polarized waves at various frequencies. Comparison between Figs. 14 and 16 shows fair agreement for frequencies below 30 GHz. The lack of precise agreement stems not only from the imperfection of the theoretical model but also from the background levels in the experiments caused in part by ground reflection and antenna depolarization.

A 4-GHz experimental observation [53], [54] of only 9-dB cross-polarization discrimination has been reported for a 40-km path during an extremely heavy rain; however, the significant cause of cross polarization for a long terrestrial path at low microwave frequencies is primarily multipath propagation.

Normalizing the differential attenuation and differential phase shift with respect to the copolar attenuation gives the following physical insight. For a given fade, the differential phase shift increases sharply with decreasing frequency whereas the normalized differential attenuation is relatively insensitive to frequency. The sharp increase of the differential phase shift is caused in part by the increase in refractive index at low microwave frequencies (see Fig. 1) and implies less depolarization at higher microwave frequencies for a given fade. At low microwave frequencies, the very high normalized differential phase shift indicates that significant depolarization is possible even for only a few decibels of attenuation, and that differential phase shift is indeed the dominant cause of depolarization. Differential attenuation plays an increasing role in depolarization as the frequency increases.

The cross-polarization coupling in a dual-polarization communication system produces cochannel interference. Unfortunately, the rain-induced depolarization occurs when rain fades have already depressed the signal-to-noise ratio (SNR). It is desirable, therefore, that a dual-polarization system use a modulation scheme with high resistance to interference. In any case, the cochannel interference increases the SNR re-

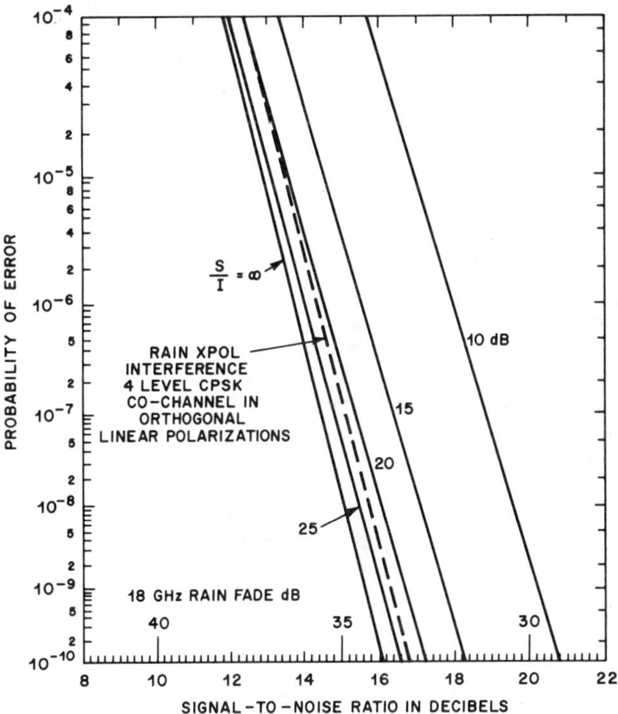

Fig. 17. Calculated bit error rates for 4-phase coherent PSK modulation; versus SNR for constant signal-to-interference ratios from a single interferer, – – – versus 18-GHz rain fade with interference from rain-induced cross polarization (a fade margin of 50 dB is assumed).

quired for maintaining a given probability of error. As a numerical example, Fig. 17 shows calculated bit error rates [56] versus SNR for a 4-phase coherent phase-shift keyed (PSK) system with a single interferer. The solid curves are calculated for a constant level of interfering signal as indicated.

[8] The imbalance is apparently caused by wind component normal to the propagation path [52], [119].

The dashed curve shows the limitation imposed by rain cross polarization at 18 GHz. The depth of the rain fade is also indicated on the abscissa.

III. Methods of Measurement of Rain Attenuation on Earth-Space Paths

In the beginning there were no strong sources in space available for measurement of large rain attenuation at frequencies above 10 GHz, except the sun. Therefore, we will discuss the experimental measuring systems and results obtained in more or less chronological order beginning with suntrackers,[9] followed by the passive radiometer method which is simpler but has less dynamic range, and concluding with receiving systems which can monitor signals from synchronous satellites such as the NASA ATS-5 (15.3 GHz) and ATS-6 (20 and 30 GHz), the Canadian CTS (12 GHz) and the ATT-Comsat (19 and 28 GHz). The clear air background attenuation is comparatively small in the "window" regions and is given in [64].

A. Suntrackers

The diameter of the sun at millimeter wavelengths subtends an angle of about 0.5°, viewed from the earth. If a microwave antenna, perfect in the sense that it receives all of the energy only within angles less than ±0.25° from the beam axis, were pointed at the sun, it would observe a brightness temperature, T_s; Fig. 18, taken largely from [57], shows the effective temperature of the quiet sun for the microwave band.[10] The power from the sun that passes through a receiver of bandwidth B fed by such an antenna is $kT_s B$, k being Boltzmann's constant. In this relationship, the sun is treated like a blackbody emitting noise, at temperature T_s, and it is assumed that the atmosphere of the earth is transparent. But the latter is never true; indeed, if the attenuation constant is α along a path of length l through the atmosphere, the power from the sun in the receiver is reduced to $kT_s Be^{-\alpha l}$

The capability of a system to measure attenuation is usually limited by noise in the system itself. In a suntracker the system noise originates in the receiving equipment and, fortunately or unfortunately as the case may be, in the material on the path, the attenuation of which one wants to measure. In the troposphere, the attenuating material, constituted primarily of oxygen, water vapor and at times liquid water, is itself at some apparent absorber temperature T_a. If the material is absorbing, as opposed to scattering, it produces an effective noise power

$$kT_b B = kB \int_0^l \alpha T_a e^{-\alpha r}\, dr = kT_a B(1 - e^{-\alpha l}). \quad (6)$$

Therefore, even with a noiseless receiver, the power observed when looking at the sun is

$$kT_s Be^{-\alpha l} + kT_b B = kB[T_s e^{-\alpha l} + T_a(1 - e^{-\alpha l})] \quad (7)$$

where T_b is the brightness temperature of the troposphere.

---

[9] Suggested to us in the present application by R. Kompfner, now of Stanford University, Stanford, Calif., and Oxford University, Oxford, England.

[10] Solar flares which make the sun such an erratic source of radiation at meter wavelengths have relatively little effect at wavelengths of order one centimeter where the sun is essentially a constant source. There is, however, an 11 year cycle in the microwave intensity [58]; at sunspot maximum, for a wavelength of 10 cm, this can be a factor of four times the value given in Fig. 18.

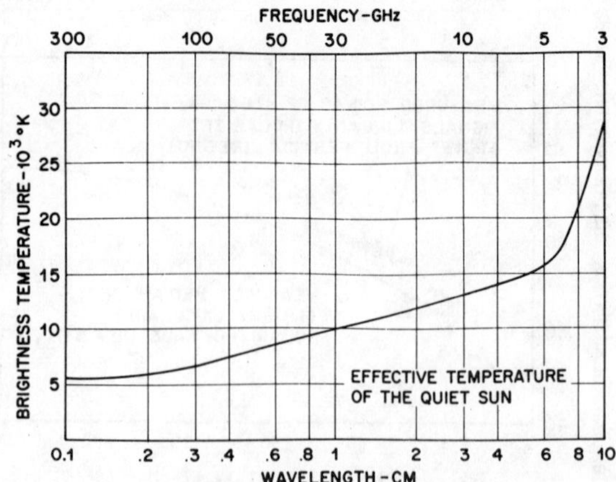

Fig. 18. Microwave brightness temperature of the quiet sun; the optical value of about 6000 K obtains for frequencies above 200 GHz.

Consider for example, an ideal receiver operating at a wavelength of 1 cm (frequency 30 GHz) for which the sun temperature is about 10^4 degrees (see Fig. 18). If a very heavy rainstorm with absorber temperature of, say $T_a = 285°$, then enters the antenna beam and completely occludes the sun, αl is very large resulting in $T_b = T_a$. Thus the ratio of noise powers in the absence and presence of the rain is $10^4/285 = 35$ which is a SNR of 15 dB. Therefore, with basic tracking of the sun, this ideal 30-GHz receiving system is capable of measuring attenuations no larger than 15 dB. In practice, a nonideal antenna and noise in the receiver *per se* limit the measuring range to even lesser values.

A good example of the results obtainable by basic suntracking is the three-year sample of 19-GHz attenuation measured in England [59], a cumulative distribution of which is shown in Fig. 19. The data were taken from 1968 to 1971 for periods when the sun was at least 5° above the horizon; this represents about 10500 hours of recording. The attenuation, under the constraints mentioned in the previous paragraph, is measured up to 10 dB. That value is exceeded about 0.01 percent of the observing time which, if converted to a yearly basis, is almost 1 h. Therefore, if an outage objective of 1 h/y were suitable to a radio engineer designing a 19-GHz satellite communication system in England, he would need to provide a fading margin of about 10 dB. But that is an average value; departures of ±3 dB appear to occur depending on whether the storms during a particular year produce intense or light rainfall.

There is a mode of suntracker operation, other than the basic suntracking just discussed, that provides a large dynamic measuring range. It involves spatial switching, or lobing, of the antenna beam on and off the sun, a technique suggested by Conway [60] that is now commonly used in radio astronomy. To understand this method, assume, as above, that a rainstorm has entered an ideal narrow microwave antenna beam pointed toward the sun, in which case the effective antenna temperature is, from (7)

$$T_e = T_s e^{-\alpha l} + T_a(1 - e^{-\alpha l}) \quad (8)$$

where $e^{-\alpha l}$ is the transmission factor for the particular storm. But when the narrow beam is switched just off the sun, it sees only the rain and the antenna temperature is

$$T_e = T_b = T_a(1 - e^{-\alpha l}). \quad (9)$$

If the switching is rapid enough, (9) can be subtracted from

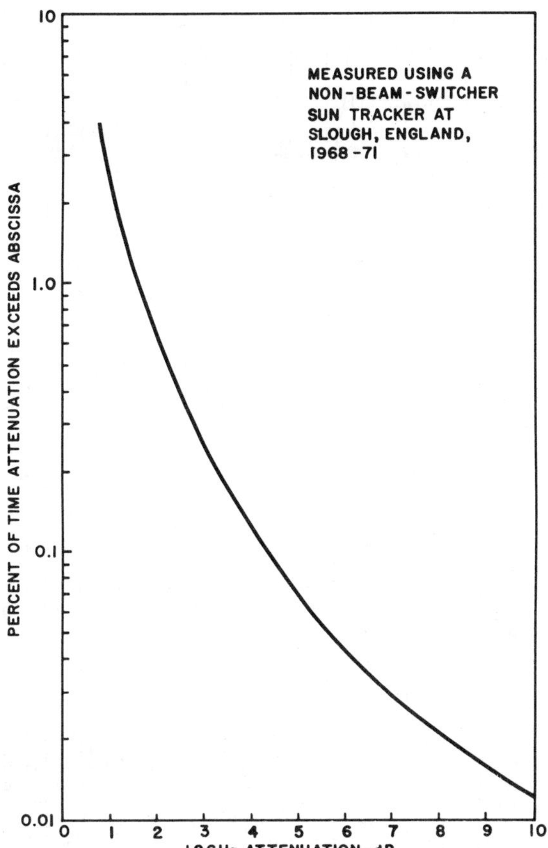

Fig. 19. Cumulative distribution of 19-GHz attenuation measured for three years in England using the basic suntracking (non-beam-switched) method.

Fig. 20. A microwave suntracker using a polar-mounted flat reflector that nods about the declination axis for beam switching.

(8) electronically in the receiver resulting in

$$T_e = e^{-\alpha l} T_s$$

allowing the attenuation αl to be deduced from the measured value of T_e.

There are at least two ways in which an antenna beam can be switched on and off the sun efficiently: by using two feeds in the focal plane of a reflector, or by mechanically nodding the beam. One way of achieving the latter is by using a horn-reflector antenna in conjunction with a nodding flat reflector

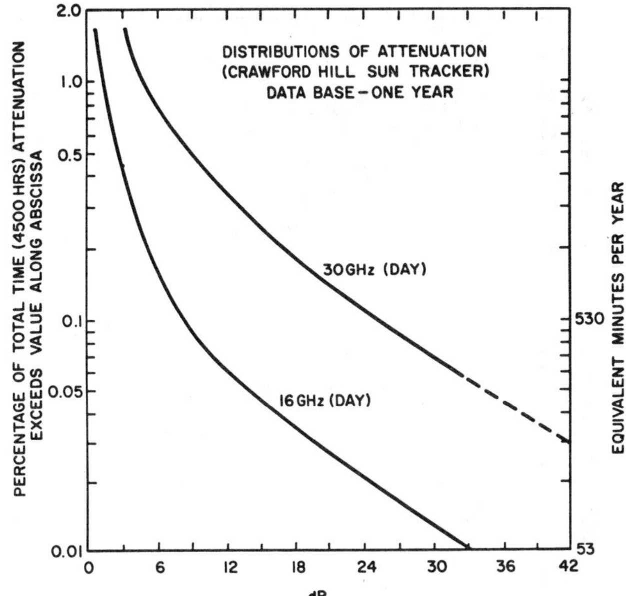

Fig. 21. Cumulative distributions of 16- and 30-GHz attenuation measured during 1968 in New Jersey using the suntracker of Fig. 20.

[61] as shown in Fig. 20. An advantage of this method is that the rain attenuation can be measured simultaneously at more than one frequency on the same antenna. In the arrangement of Fig. 20, the flat reflector moves about the polar axis with time of day, reflecting the solar radiation into the fixed antenna; but it is also mounted on bearings on the orthogonal axis about which it oscillates with an angular amplitude of 2.5° at a rate of 1 Hz. In this way, simultaneous measurements of rain attenuation, reliable to at least 30 dB, were made at 16 and 30 GHz for more than a year. Part of these results, presented as cumulative distributions of attenuation, is shown in Fig. 21 for the measuring period 1967–1968. As anticipated, a much larger attenuation occurs at 30 GHz, for a given percentage of the measuring period, than at 16 GHz.

In the design of broad-band satellite systems for the 18- and 30-GHz bands, it turns out [8] that reasonable choices for power on the satellite, bandwidth, type of modulation, antenna size, satellite stability etc., result in a clear-day fading margin of about 10 dB on a "down" link at 18 GHz. That margin is the amount of fading that the down link can withstand. But we observe from Fig. 21 that the rain attenuation exceeds 10 dB for 400 min, or about 7 h/y which is not a mark of high reliability. For that reason, systems designed to operate in locations which experience heavy rain would need to employ path diversity to improve reliability of transmission, as discussed in Section IV.

It is also important that the system designer employing frequency extrapolation have a measure of confidence for distributions of attenuation such as those in Fig. 21. In other words, for a given 16-GHz attenuation, over what range of values does the 30-GHz attenuation extend? This question was examined by obtaining conditional probabilities for a nine month sample of attenuation data measured by the suntracker discussed above; the ratio of simultaneous values of 30- and 16-GHz attenuation were obtained on an instantaneous basis for given values of the 16-GHz attenuation. The medians and 10- to 90-percent values obtained at 30 GHz from the conditional distributions were then plotted as shown in Fig. 22; the measured cumulative distribution, obtained in the usual

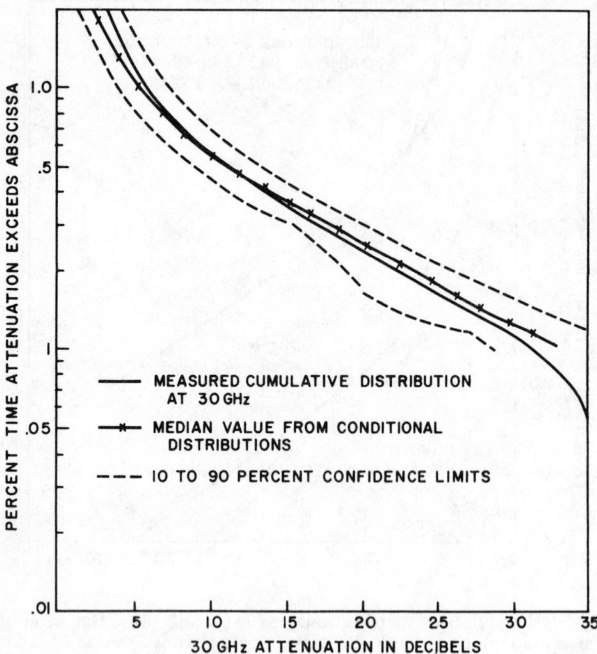

Fig. 22. Median values of instantaneous 30-GHz attenuation, conditional upon given 16-GHz attenuations for fixed percent-of-time levels, are shown as crosses, and the 10- to 90-percent confidence range by dashed lines.

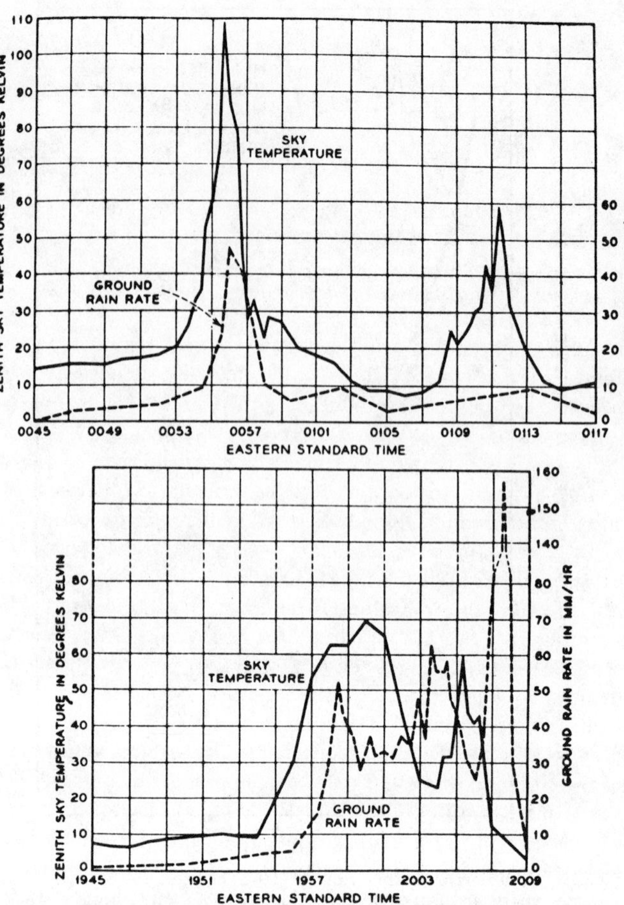

Fig. 23. 6-GHz sky temperature produced by rain cells passing through a zenith-oriented beam in New Jersey; point rain rates measured at the antenna site are shown dashed.

way, is also given to serve as a reference. Clearly, the medians which resulted from the conditional distributions (analysis of instantaneous values), agree well with the reference 30-GHz distribution. The 10- to 90-percent confidence limits are shown as dashed lines. The range defined by the limits expands with increasing attenuation attaining a spread of about 10 dB when the median attenuation at 30 GHz is 25 dB (corresponding to 8 dB at 16 GHz, see Fig. 21). Whether this rather large range will obtain on a fixed path to a synchronous satellite (rather than on the ever-changing path to the sun) is not yet known. This matter is discussed further in Section VI.

The moon could also be used as a source, but its brightness temperature [62] is only of the order 220 K over the 3- to 300-GHz band, about 17 dB below that of the sun at 30 GHz.

B. Passive Radiometers

As mentioned in the previous section, the rain medium emits noise in the fashion of a blackbody[11] at some temperature T_a. It turns out that this characteristic can be utilized to measure the attenuation to levels of about 10 dB. It is because there is no source, such as the sun or a satellite beacon transmitting through the rainstorm, that we refer to this method of measurement as passive radiometry.

Basically there are two types of radiometer receivers, so called dc and switched. A dc radiometer system is comprised of very low noise receiving electronics and a high quality antenna. In that case, the overall background noise of the system achieves a very small value which can, however, be predetermined by calibration. Any additional noise that appears in the receiver, for example by way of a rainstorm entering the antenna beam, is simply measured by observing the increase in

the direct current of the final detector in the receiver. Examples of such measurements [3] at 6 GHz using a zenith-pointing horn-reflector antenna with 1.8-m aperture and a traveling-wave maser input stage are illustrated in Fig. 23. The noise temperature of the receiver *per se* is about 14 K. However, Fig. 23 only shows the sky noise measured by the antenna. About 3 K of the sky noise is contributed by the troposphere [64], and 3 K by the cosmic background radiation [65]. The large changes in noise temperature are produced by rain in the zenith-pointed beam; in the first example, a maximum value of nearly 110 K is attained. The rain-rate measured at the site, Holmdel, N.J., is also plotted in Fig. 23, as a dashed line. It is interesting that in both examples, the rainrate lags the noise temperature; this occurs in most observations using a zenith-oriented beam because a typical rain cell, moving with the wind, slopes forward under the influence of a wind profile that increases with height, thereby causing the rain to enter the antenna beam before reaching the rain gauge at the site, i.e., the rain is detected before it reaches ground.

The other type of receiver, introduced by Dicke [66], switches between the antenna and a reference waveguide load. In this way, synchronous detection can be used and receivers of relatively high noise figure can resolve small changes in antenna temperature.[12] A measurement of the 15.5-GHz

---

[11] The question as to whether collisions of charged raindrops produce significant microwave noise has often been raised. Apparently not much, since measurements at 1400 MHz [63] during heavy showers show that the intensities are very small, typically 1 to 2 K.

[12] An interesting variation of the Dicke system is the noise-adding radiometer [67] in which a small known amount of noise is added in an alternating fashion to the "signal". Frequency switched radiometers are also used in special applications.

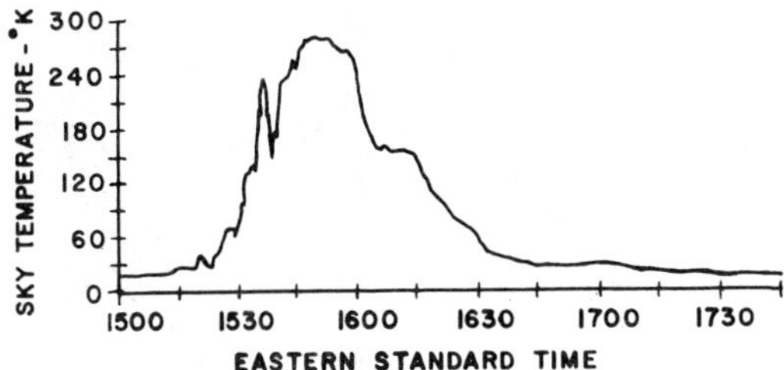

Fig. 24. 15.5-GHz antenna temperature produced by a New Jersey shower passing through a beam at an elevation angle of 30°.

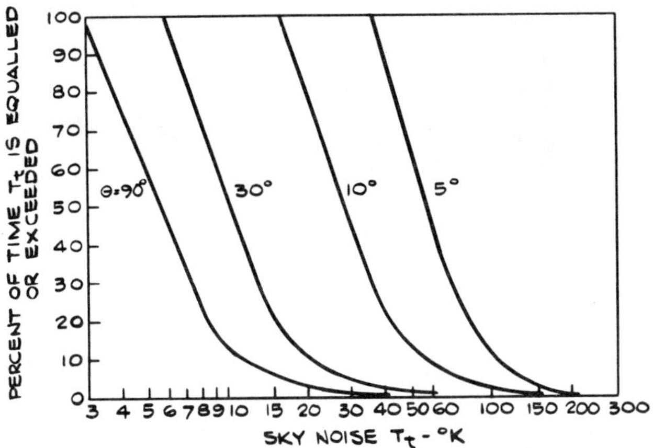

Fig. 25. Cumulative distributions of sky noise at 15-GHz measured under various weather conditions for several elevation angles of an antenna beam in Massachussetts; the tails occur during very humid weather and rain.

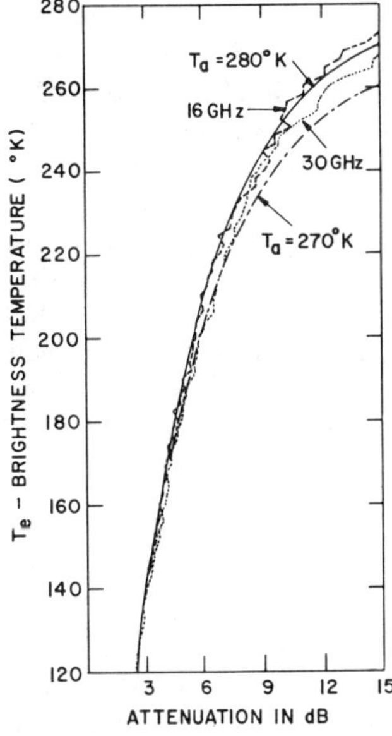

Fig. 26. Calibration curves for 16- and 30-GHz radiometers measured by a suntracker in New Jersey; the smooth curves show the calculated relationship between brightness temperature and attenuation assuming apparent absorber temperatures of 270 and 280 K for the rain.

antenna temperature produced by a shower, using a switched receiver, is shown in Fig. 24. In this case the beam of the antenna, a horn-reflector with 1.2-m aperture, is at an elevation angle of 30°. With the beam at such a relatively low angle, there is little point in measuring the rainrate at the site of the antenna to obtain detailed correlation between the antenna noise and individual storms since, more often than not, the rain shower in the beam will be at some distance from the site. Because of the high 15-GHz attenuation produced by this shower, the radiometer becomes saturated for 5 min beginning at 1550 EST in Fig. 24. The temperature at which saturation occurs, in this case 275 K, is somewhat lower than the kinetic temperature of the raindrops, an effect discussed later in this section.

In Fig. 24, a background noise of about 10 K is generated by the antenna and the sky when the rain is not present. This background temperature is quite dependent on the amount of water vapor in the air and, of course, on the elevation angle of the antenna beam. The variability in sky noise introduced by the day to day changes in water vapor content is well demonstrated by a set of 15-GHz measurements [68] taken over a six-month period at Waltham, Mass., shown as a percentage-time distributions in Fig. 25. Portions of the curves, between the 100 and 20 percent-time values, are quite linear; at an elevation angle of 30° for example, weather changes from a very dry to a humid condition are accompanied by sky temperature changes from 6 to 15 K. The measured dry-air value of 6 K agrees with the calculated sky temperature based solely on absorption by the oxygen [64] in the atmosphere. In radiometric measuring systems the background level therefore varies somewhat, but it represents only a small attenuation which is not our prime concern here.

The examples of radiometric data discussed so far have been expressed in terms of the sky brightness temperature produced by rain, but to make these useful for communication system design, the radiometer must be calibrated in units of attenuation. The only way of doing that properly is to measure directly the attenuation of numerous rainstorms by means of the sun or a satellite signal, simultaneously observe the effective brightness temperature produced by the rain medium (T_b) in a radiometer at about the same microwave frequency, and plot one against the other. Such a calibration [69], using attenuation measured by a sun tracker as reference, is shown by the dotted irregular plots in Fig. 26 for the frequencies 16 and 30 GHz. These plots are average values of the brightness temperatures evidenced by all of the heavy storms during 1970 at Holmdel, N.J. The smooth curves in the figure are calculated from (9)

$$T_b = T_a (1 - e^{-\alpha l})$$

for two values of the apparent absorber temperature $T_a = 270$ and 280 K. Conversely, the attenuation in decibels, obtained

from this equation, is

$$A = -10 \log_{10} (1 - T_b/T_a). \tag{10}$$

It is in this way that one determines the value of T_a most appropriate for conversion of radiometric measurements to attenuation at a given frequency and geographic location. In the present case, Fig. 26 shows that $T_a \simeq 270$ K is appropriate at 30 GHz for attenuations up to 8 dB, and about 275 K for attenuations above that value whereas $T_a = 280$ K pretty well satisfies the data over the whole range at 16 GHz. The reason for the lower value of T_a at the higher frequency is discussed later. Data taken in Hawaii [70] using a suntracker operating at both 15 and 35 GHz showed good agreement between directly measured attenuation and those computed from radiometric measurements using $T_a = 284$ K.

The attenuation of signals from satellites, particularly ATS-5 [71], has also been used to determine apparent absorber temperature. But, because of the intermittent mode of operation of the beacon, the data are given on a storm-by-storm basis rather than statistically. An example [72] of such

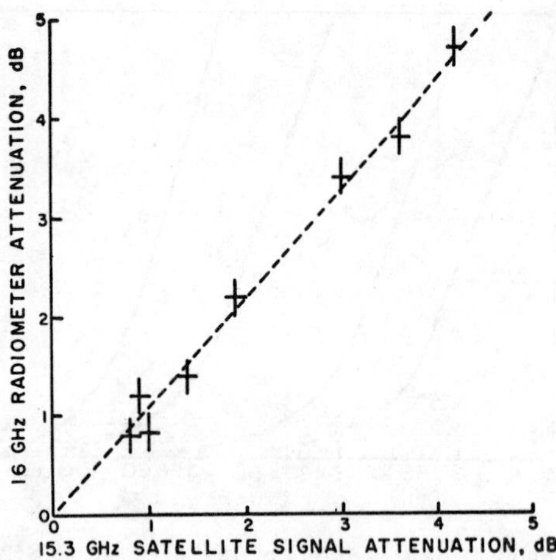

Fig. 27. Calibration of a 16-GHz radiometer using a 15.3-GHz satellite beacon signal for measuring attenuation by a rain shower in New Jersey.

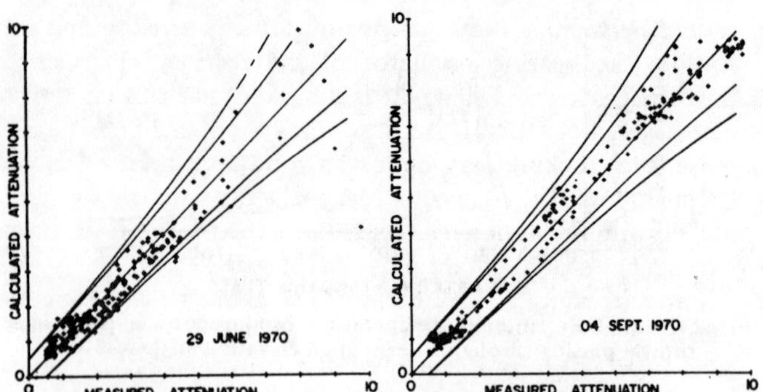

Fig. 28. Calibration of a 15.3-GHz radiometer using a 15.3-GHz satellite signal for measuring attenuation by rain showers at Ottawa, Canada; see text for discussion of the curves.

an event is given in Fig. 27 where the 15.3-GHz attenuation experienced by the ATS-5 satellite signal is compared with attenuation measured simultaneously using a 16-GHz passive radiometer, both receivers being mounted on the same antenna at Holmdel, N.J. An apparent absorber temperature of 275 K was used to compute the radiometer attenuation; a dashed line which takes into account a factor 1.1 representing the expected larger attenuation at 16 GHz (above 15.3 GHz) is also shown. Two events measured [73] at Ottawa, Canada, are shown in Fig. 28. Again, the narrow band 15.3 GHz ATS-5 signal was used for measuring the attenuation directly, and a broadband radiometer at the same frequency was also mounted on the receiving antenna. In the initial comparisons of the directly measured attenuation with those computed from the antenna temperatures, an apparent absorber temperature of 278 K was assumed; this value obtains for the data in Fig. 28. The two sets of bounding curves, computed from (10), show how the relationship changes when T_a is assumed ±10 and ±20 K with respect to 278 K; a system measuring uncertainty of ±0.5 dB is also included in these curves. However, detailed analysis of 65 events measured at Ottawa led to a value $T_a = 272$ K as most appropriate. Individual storms measured at Rosman, N.C. [74], at 15.3 and 16 GHz, produced data compatible with an apparent absorber temperature 270 to 272 K; similar measurements have been made at the Universities of Texas and Ohio, Virginia Polytechnic Institute, and in Maryland.

Comparison of radar return [75] with antenna noise temperatures measured during rain has also been successfully used to calibrate a radiometer.

It is fascinating that the apparent absorber temperature is near or below freezing for observations in North America whereas the rain is in fact usually fairly warm, say 285 K, especially for summer showers near the surface of the earth. This discrepancy [76] is due to scattering[13] by the raindrops; it is a function of operating frequency, drop size, the geometry of the storm, the aspect of the antenna beam axis with respect to the strom cell, the temperature profile of the rain, and the kinetic temperature and microwave properties of the ground surrounding the site of the antenna. Because the scattering coefficient of raindrops increases with frequency, the apparent absorber temperature decreases [77] as suggested by the

measurements in Fig. 26. We forego analysis of this effect but will make a few descriptive statements.

A raindrop in a vertical antenna beam scatters radiation incident upon it more or less isotropically [78]. The radiation incident on the drop originates over a 4π solid angle. If only one raindrop were present, the radiation incident from the half space below the drop would be primarily emission from an absorbing earth, at, say 290 K. But the half space above the drop is very cold, less than 10 K. Therefore, in this highly idealized single-drop model, the radiation, incident over 4π, and available for scattering, is equivalent to an average temperature of only 150 K, which, depending on the magnitude of the scattering coefficient, manifests itself as a sizable decrease in apparent absorber temperature. In reality, the radiation incident on the drop is influenced by the emission and attenuation of numerous neighboring drops and, therefore, by the geometry of the storm cell. For example, if a heavy rain cell encompasses a vertical beam, the radiometer will essentially measure the kinetic temperature of the first kilometer or so of rain and scattering is of little consequence. However, if the beam is at a low elevation angle and a cell enters the beam at some distance from the antenna, the drops nearest the antenna have a most significant scattering effect, resulting in a decreased apparent absorber temperature.

We conclude from the measurements discussed above that an apparent absorber temperature $T_a \simeq 273$ K is suitable for conversion of antenna temperature to rain attenuation (equation (10)) at frequencies of order 20 GHz in many parts of North America. This conclusion applies to statistical data since for individual storms, the most appropriate value of T_a may depart from 273 K. Rather high attenuations (6 to 9 dB) at 6.5 GHz have been observed [79] in New Jersey using the radiometric method, but these are rare events.

C. Beacons on Synchronous Satellites

The usefulness of signals from ATS-5 for calibration of radiometers is evident from the above discussion, but it is appropriate to give more detail on techniques for receiving satellite beacons and some of the results obtained. We will also discuss some plans for future experiments.

A beacon signal is derived from a well stabilized crystal oscillator. Therefore, if the signal is unmodulated, all of the power can be received within a very narrow band, exhibiting potential for the large SNR necessary for measuring the deep fading due to rain attenuation. But synchronous satellites are not perfectly stationary with respect to an earth station, therefore the signal frequency is doppler shifted over a diurnal period. For that reason, and also to overcome frequency drifts

---

[13] Suggested to us in 1968 by J. R. Pierce, now of California Institute of Technology, Pasadena, Calif.

caused by temperature changes in both the satellite and earth station, the beating oscillator of the receiver must employ AFC to track the signal. It is desirable that this control circuit have a hold feature because when the signal fades near the noise level, locking may be lost; a good decision is to hold the frequency until the signal is regained. Even so, the signal may not be within the narrow band after a long fade, in which case a frequency-search circuit must be activated.

Statistics of rain attenuation [80] have been obtained at Clarksburg, Maryland using the 15.3-GHz signals from the ATS-5 satellite, supplemented by sun tracker and passive radiometer measurements. Data for the year 1970–1971 have been analyzed to the 10-dB attenuation level. Comparison with New Jersey data (Fig. 21) shows a slightly higher probability of occurrence of these attenuations in Maryland.

It is planned that three ATT-Comsat satellites, each with beacons [81] at 19.04 and 28.56 GHz be launched beginning in 1976. The beacons will operate continuously and illuminate the continental U.S., thereby affording compilation of attenuation statistics in a variety of rain environments. It is expected [82] that a dynamic range of at least 45 dB will be possible at both frequencies using an earth-station antenna of 7-m diameter. Such dual-frequency measurements, other than of great usefulness in their own right, should provide a sound basis for determining rain attenuation at other frequencies as discussed in Section VI.

Intelsat has been used to measure 4- and 6-GHz rain attenuation [83] at a low elevation angle (8°) in Japan. During 1972–1973, attenuations of about 1 dB at 4 GHz and 3 dB at 6 GHz were observed for 0.01 percent of the time. Values considerably higher than that have been observed for short periods of time in the eastern U.S. as discussed in Section III-B.

An aircraft has been equipped with a transmitter and flown beyond heavy rain clouds to measure attenuation at Ottawa, Canada [84]. Observed attenuations at 15 GHz were compared with values calculated from backscatter measured by a 2.9-GHz radar; these were found in good agreement during stratiform rain, but for a heavy cell, the attenuation derived from the radar method resulted in attenuation six times the actual measured value. Melting hail is believed to have caused this large discrepancy. We believe, therefore, that conversion of most radar backscatter measurements to attenuation is not, at present, an accurate enough procedure for design of satellite communication systems.

IV. Path Diversity in Earth-Space Transmission

The high attenuations observed on earth–space paths are caused by cells of intense rain, and relatively poor reliability of transmission results, as discussed in Section III-A. But by taking advantage of the finite cell size [85], considerable improvement in reliability can be obtained by switching between two earth stations separated by a distance larger than the cell size. This arrangement, called path diversity [86], [87], requires a broadband terrestrial link between the two earth stations and appropriate switching circuitry. The inset of Fig. 29 is a plan view showing antennas separated on a baseline normal to the direction of the path to the satellite. If a rain cell in a given antenna beam produces attenuation exceeding the fading margin[14] of that station, the transmission is switched through another station whose antenna beam does not simultaneously encounter the cell.

[14] Or degrades significantly the signal-to-crosstalk ratio, as governed by cross polarization.

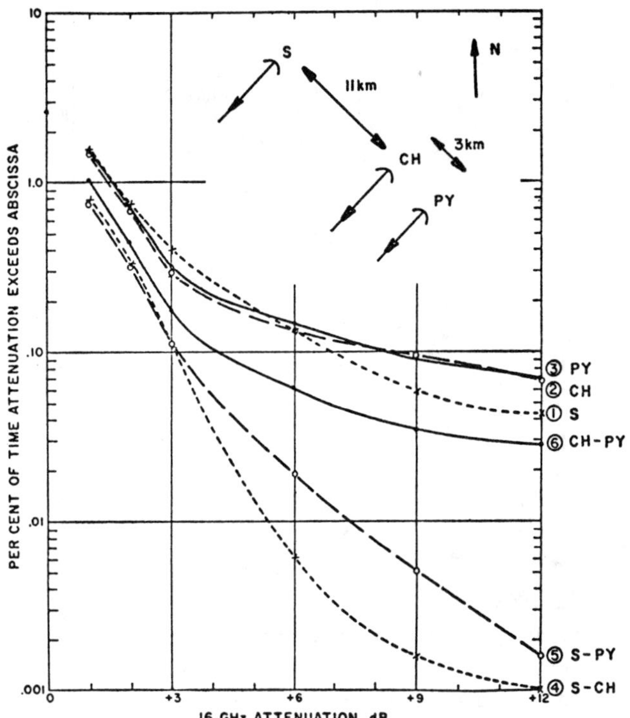

Fig. 29. Statistics of path diversity for beams with elevation angle of 30° separated by 3, 11, and 14 km, measured at 16 GHz in New Jersey.

Initial measurements on the diversity advantage obtainable in this way were made with passive radiometers at 16 GHz [88], no synchronous satellite beacons being then available. There were at least three reasons for this choice: a fixed-path orientation (rather than a variable one as with suntrackers) is desirable to simulate the geometry of a synchronous satellite system; radiometers measure attenuations to only 10 dB, but that is a value of interest in system design as discussed in Section III-A; radiometers are simple, with no moving parts, so they can remain unattended for long periods of time provided the electronics is reliable, and they can be calibrated to good accuracy as discussed in Section III-B. The relative positions of three (horn-reflector) antennas are indicated in the inset in Fig. 29, separations of 3, 11, and 14 km applying to the various pairs. The elevation angle of the beams is about 30°, in the direction of a fictitious satellite located in synchronous orbit south of the central U.S. The figure shows several cumulative distributions of data taken during April through August 1969 at Holmdel, N.J.. The three upper plots, all quite similar, are simply the distributions for the three individual sites; they show, for example, that an attenuation of 10 dB is exceeded about 0.08 percent of the time on the individual paths. The switched diversity pair with separation 3 km (curve 6), at the 10-dB level, shows an improvement of only about a factor of two in reliability, but the pairs with 11- and 14-km separation evidence a diversity improvement of about twenty; the greater improvement for 11-km rather than 14-km separation is believed to be a consequence of insufficient sample size.

More recently, measurements using the same radiometers spaced 11, 19 and 30 km were taken [69] for the full year 1970. The attenuation distributions for the individual paths, the various combinations of pairs, and the diversity triplet are shown in Fig. 30. In the case of the pairs, the diversity improvement clearly increases with separation between

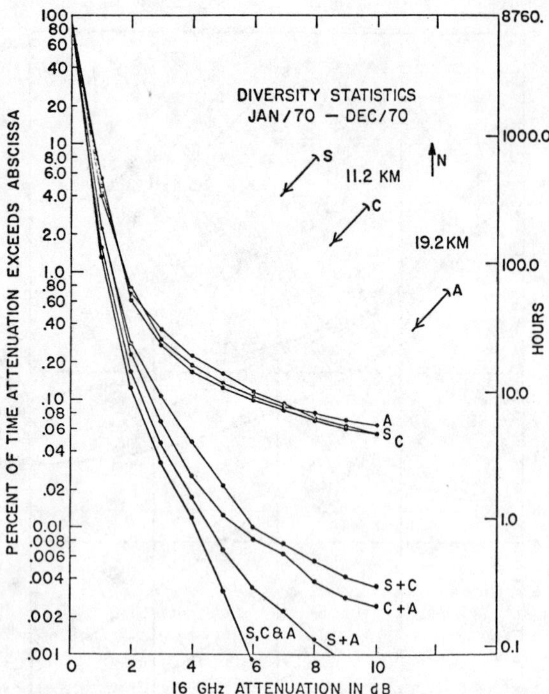

Fig. 30. Statistics of path diversity for beams with elevation angle of 30° separated by 11, 19, and 30 km, measured at 16 GHz in New Jersey.

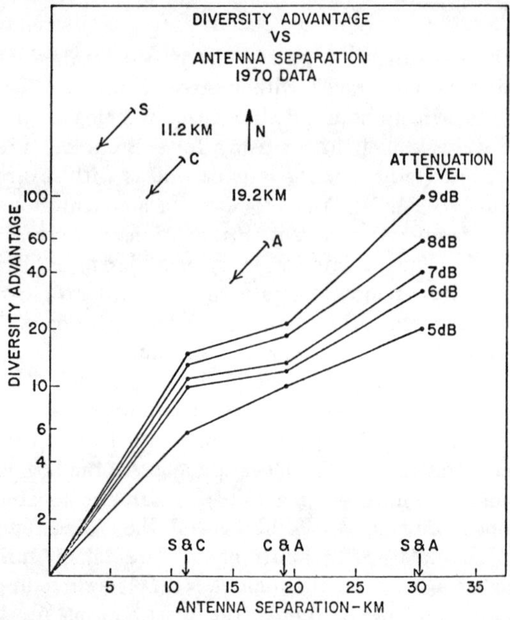

Fig. 31. Diversity advantage (improvement factor relative to a single path) for dual-path diversity at various 15.5-GHz attenuation levels versus path separation.

antennas. The diversity advantage, i.e., the ratio of the percent of time level for a diversity combination to that of a single path, is plotted versus separation, with attenuation as a parameter, in Fig. 31. For the pair separated by 30 km, and attenuation level of 9 dB, the diversity improvement factor is about one hundred which, on a yearly basis, means that the outage of system with a 9-dB fade margin is reduced from eight hours to less than 5 min. Measurements with these radiometers also encompassed a year which included hurricane Doria, in which case the improvement factor at the 9-dB level was 25 for the pair with 30-km spacing. A count of the number of fades of various lengths which occured at the 5-dB

level during 1970 was also made; for the 30-km diversity pair, there were twelve of 5-min and one of 10-min duration.

Path diversity studies have also been made using the ATS-5 satellite beacon on a few storms which occured during the summer months of 1970 and 1971 at Columbus, Ohio [89]. In that case the 15.3-GHz receivers were separated by 4 and 8.3 km along a baseline oriented 50° with respect to the vertical plane containing the path to the satellite. Greater diversity advantage was observed for these small separations in Ohio than for the larger ones in New Jersey. For example, with the 4-km separation, the improvement factor was 200 at the 9-dB attenuation level. Although the total annual rainfall is about the same (1000 mm) in New Jersey and Ohio, thunderstorm-type showers occur more often in Ohio than in New Jersey; this may give rise to the greater diversity improvement; on the other hand, the lengths of the two measuring periods are quite different.

At Waltham, Mass., two 35-GHz suntrackers [90] separated by 11 km on a NNW–SSE baseline were used to determine diversity advantage during 1971. In this case attenuations up to 27 dB could be resolved, but as mentioned above, the beam orientation constantly changes and, of course, measurements can only be taken during the day. Nevertheless, by proper normalization of the data, it was found that 27-dB attenuation was exceeded for 2.5 h on the single paths, but for only about 10 min using the diversity pair; this constitutes an improvement factor of 15 at that attenuation level. An experiment using two suntrackers in Japan [117] observed substantial diversity advantage for a pair separated 15 km but saw very little diversity improvement for a pair separated 2.6 km.

A standing question with regard to path diversity improvement relates to orientation of the baseline with respect to the earth–satellite path vis-a-vis the direction of persistent weather fronts at the particular location. The answer to this is of interest in system design because real estate suitable for sites may not be available on a baseline normal to the earth–space path. To examine the effect of baseline orientation, three radiometers were arranged in a right-angle geometry [91], one edge of the angle, 27 km long, being below the path and the other, 19 km normal to the path, as shown by the inset in Fig. 32. 15.5-GHz attenuation was measured continuously for ten months at Holmdel, N.J.. Interestingly, even though the separation of the stations below the path (P-CH) is 8 km more than for the pair normal to the path (A-P), the probability of occurrence of 8-dB attenuation is more than ten times larger for the former pair. The higher correlation of attenuation between the in-line stations P and CH can be partly explained by the fact that intense showers originating at great altitudes above P will also intersect the antenna beam from CH. However, it is also true that most convective storm fronts approach central New Jersey from the northwest (see inset of Fig. 32) thus any ellipticity in the shape of the cells, with major axis parallel to the front, would result in stronger correlation on the in-line pair. It is also worth noting in Fig. 32 that diversity on the pair A-P, in spite of the relatively large separation of 33 km, is no better than on pair A-CH separated only 19 km. A recent U.K. experiment [118] also showed considerable greater diversity improvement for the baseline perpendicular rather than parallel to the propagation path. Thus on the basis of experiments to date,[15] it is believed that baselines normal to the path and, secondarily, to the direction of convective

[15] A program is also under way in Virginia [92] utilizing radar measurements to determine the effect of eccentricity and orientation of rain cells on path-diversity statistics.

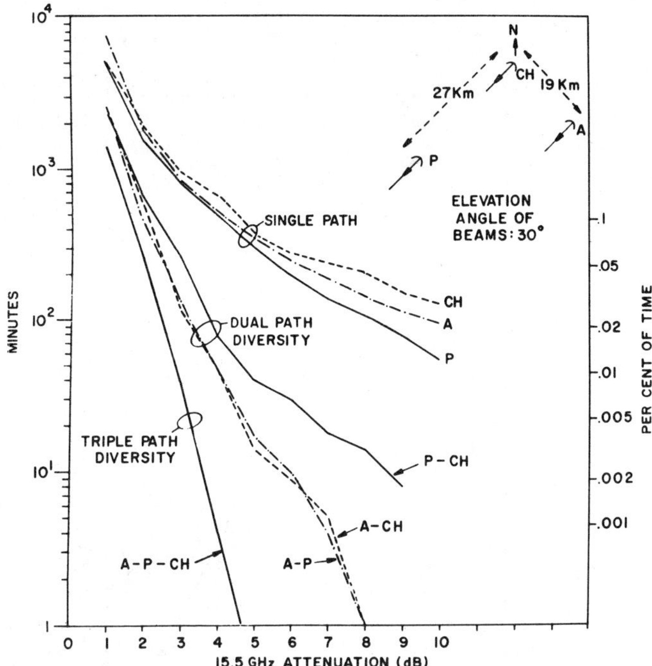

Fig. 32. Statistics of path diversity for antennas located in a right-angle configuration; location, New Jersey.

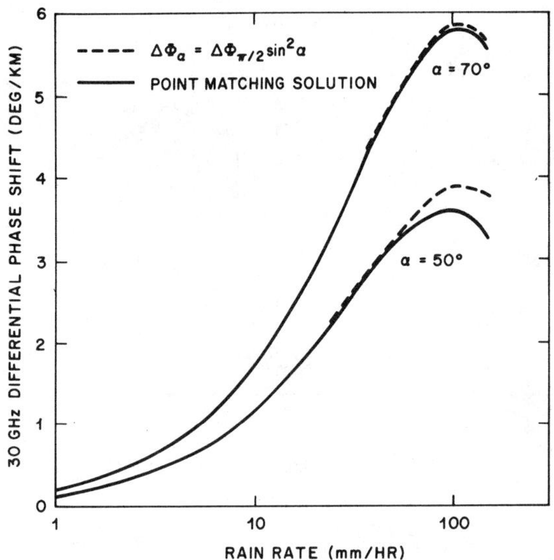

Fig. 33. Comparison between approximate and exact calculations of 30-GHz differential phase shift; α is the angle between the direction of propagation and the axis of symmetry of oblate raindrops.

weather fronts, are most efficient for path diversity. In system design, that is an important matter because minimization of the separation between stations reduces the cost of the broadband terrestrial route that links them.

The synchronous satellite ATS-6 [93] was successfully positioned in orbit, with 20- and 30-GHz beacons on board, on May 30, 1974. Although these beacons are not continuous but operate in an "on-demand" mode, several receivers in the U.S. are observing, and further measurements on path diversity may be forthcoming. There are also 13- and 18-GHz receivers [94] on the satellite which are used to monitor attenuation on paths from several earth-based transmitters arranged in diversity configurations. The signal levels received at the satellite are then frequency translated to 4 GHz and transmitted to earth for analysis.

V. Depolarization by Rain in Earth-Space Transmission

The calculated differential attenuation and differential phase shift discussed in Section II-C were confined to the case of $\alpha = \pi/2$, where α is the angle between the direction of propagation and the axis of symmetry of oblate raindrops, a situation appropriate to terrestrial paths. However, in earth–space transmission, other values of α apply and there exist the following approximate relations [43], [95] for the differential attenuation and phase between orthogonal linear polarizations:

$$(A_{II} - A_I)_\alpha = (A_{II} - A_I)_{\pi/2} \sin^2\alpha$$

$$(\phi_{II} - \phi_I)_\alpha = (\phi_{II} - \phi_I)_{\pi/2} \sin^2\alpha.$$

These equations are derived by assuming the uniform rain as an anisotropic medium and they have been verified by point-matching solutions for $\alpha = 50°$ and $70°$ at 30 GHz. Fig. 33 shows a comparison between the above approximate relations and the point-matching solution for the differential phase shift, the corresponding comparison for the differential attenuation not being shown because the curves coincide. The above equations indicate that depolarization on the earth-space path should be less than on a terrestrial path for the

same attenuation level provided the incident linear polarization from the satellite is parallel or perpendicular to the plane containing the direction of propagation and the local gravity direction, the latter being the axis of symmetry of horizontally disposed oblate spheroidal raindrops, at the ground station. This condition can be met in a multiple-beam satellite system [8] by properly orienting the polarization of each feed on a multiple beam satellite antenna[16] for the corresponding ground station.

In the case of an area-coverage satellite antenna (for example a CONUS coverage),[17] the polarization vectors of the radiation from the satellite can not be parallel and perpendicular to the plane defined above simultaneously for all ground stations. The angle between the unit polarization vector $\hat{p}$ and this preferred plane is given by

$$\xi = \text{arc } \sin\left(\frac{\hat{p} \cdot (\hat{k} \times \hat{r})}{|\hat{k} \times \hat{r}|}\right) \tag{11}$$

where $\hat{k}$ is the direction of propagation and $\hat{r}$ is the local gravity direction at the ground station. If the aperture field of the satellite antenna is linearly polarized in the direction $\hat{e}$, $\hat{p}$ is determined by

$$\hat{p} = (\hat{k} \times \hat{e}) \times \hat{k}/|(\hat{k} \times \hat{e}) \times \hat{k}|.$$

As a numerical example, the contours of constant ξ are plotted in Fig. 34 for a synchronous communication satellite located at $105°$ west longitude with $\xi = 0$ at Denver, Colo. If the satellite antenna beam covers the continental United States, ξ exceeds $30°$ in the fringe area. Therefore, the rain-induced depolarization of linearly polarized radiation from an area-coverage satellite in the fringe area is expected to be worse than that of Fig. 16 but better than that of Fig. 15.

It is tempting to use circular polarization to avoid polarization tracking and also to produce equal attenuations in the orthogonal polarizations (zero differential attenuation). However, both experimental data [51] (see Fig. 15) and theoretical considerations [43] show that rain-induced depolarization is much worse for circularly polarized waves. Measurements [96] of a 4-GHz circularly polarized signal from Intelsat IV

[16] See, for example, [115].
[17] See, for example, [116].

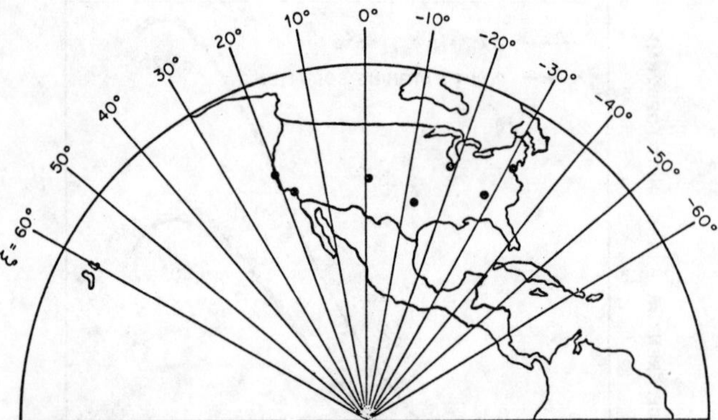

Fig. 34. Contours of constant polarization orientation ξ with respect to the plane containing the local vertical; satellite at 105° west longitude radiating linear polarization in the meridian plane.

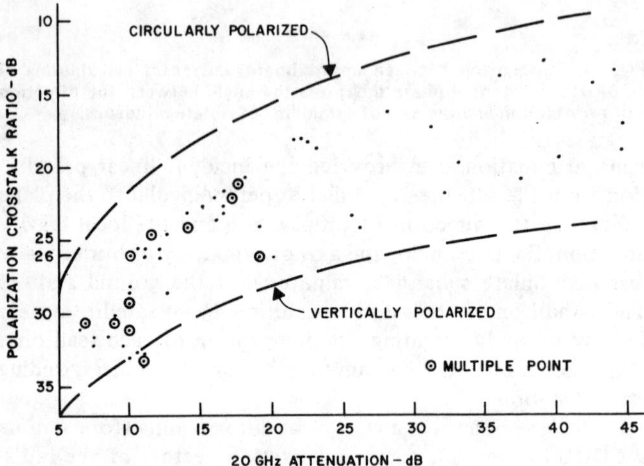

Fig. 35. Measured cross polarization versus attenuation of a 20-GHz signal from ATS-6; the incident linear polarization is oriented 20° from the plane containing the local vertical, – – – calculated curves from [43].

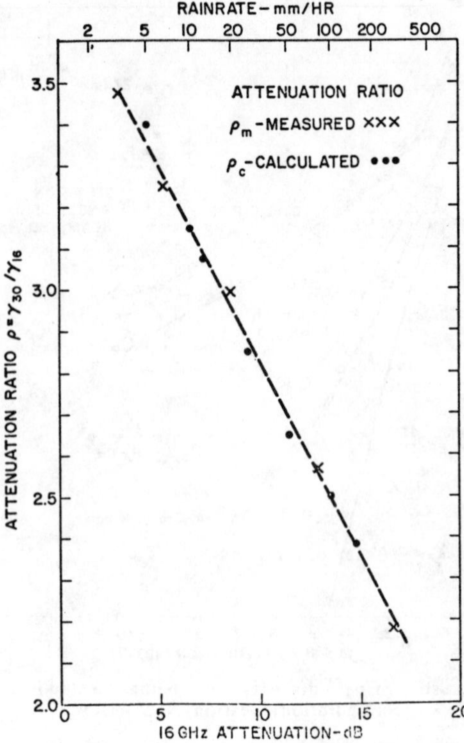

Fig. 36. Ratios of 30- to 16-GHz attenuation derived from the cumulative distributions of Fig. 21 along with calculations using the Laws and Parsons raindrop-size distribution.

showed that rain-induced cross polarization as high as −11 dB relative to the copolarization can occur.

Recent measurements of the 20-GHz signal from ATS-6 have produced experimental data on the rain-induced depolarization of a linearly polarized wave radiated from a satellite. Fig. 35 shows the data obtained [97] during a rain-storm at Holmdel, N.J.;[18] in this case, the incident polarization is oriented 20° from the plane containing the propagation direction and the local vertical. The polarization crosstalk ratio in clear weather, about 26 dB, is caused mainly by the off-axis cross-polarization characteristic of the satellite antenna. The two dashed curves in Fig. 35 are the expected rain-induced crosstalk ratios at 18 GHz for circularly and vertically polarized waves respectively on a terrestrial path; the experimental points are mostly confined within the two curves, as predicted in the earlier discussion. These preliminary data indicate that when significant rain-induced cross-polarization begins to appear (e.g., 20-dB crosstalk ratio), the rain fading probably already exceeds the fading margin (e.g., 10 dB) of a single-path 20-GHz satellite communication system; path-diversity, discussed in Section IV, could be operative under those conditions.

In the propagation experiment using the ATT-Comsat domestic satellites with beacons [81] at 19 and 28 GHz, the

19-GHz beacon signal will be switched at a 1-kHz rate between two orthogonal linear polarizations, horizontal and "vertical" with respect to a receiving site at Holmdel, N.J., and the satellite–earth path. It is believed that this forthcoming experiment will provide a comprehensive set of statistics on cross-polarization produced by rain.

VI. Frequency Dependence of Attenuation by Rain on Earth–Space Paths

The statistics of attenuation by rain measured at a given location are unique for several reasons: the total amount of precipitation varies from one place to another, as does the duration of the storms, their geometry, and the height at which the rain originates. It is also true that attenuation measurements must be made continuously for several years to obtain data sufficient for an engineer to design a system with confidence. Just as discussed for terrestrial paths in Section II-A, when long-term statistics have been measured at a given frequency, it is meaningful to examine methods of extrapolating the attenuation to other frequencies.

The relative attenuation at 16 and 30 GHz, measured simultaneously on a single antenna, was discussed briefly in connection with Fig. 21 of Section III-A. Those data can be interpreted in another way [98] by taking the ratio of the attenuations at various percent-of-time levels and plotting them versus attenuation[19] as shown by the crosses in Fig. 36. But a given ratio of attenuation for two specific frequencies only occurs at a particular rainrate according to the theory, using the Laws and Parsons drop-size distribution, discussed in Section II-A. Therefore, the theoretical ratio of 30- to 16-GHz attenuation can be plotted versus rainrate as shown by the solid dots in Fig. 36. Thus, for this particular

[18] Elevation angle of the antenna beam is about 40°.

[19] As was done for the 30.9- to 18.5-GHz attenuation ratio on a short terrestrial path, Fig. 7 of Section II-A.

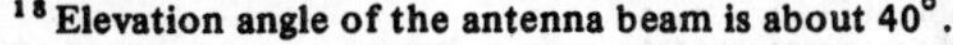

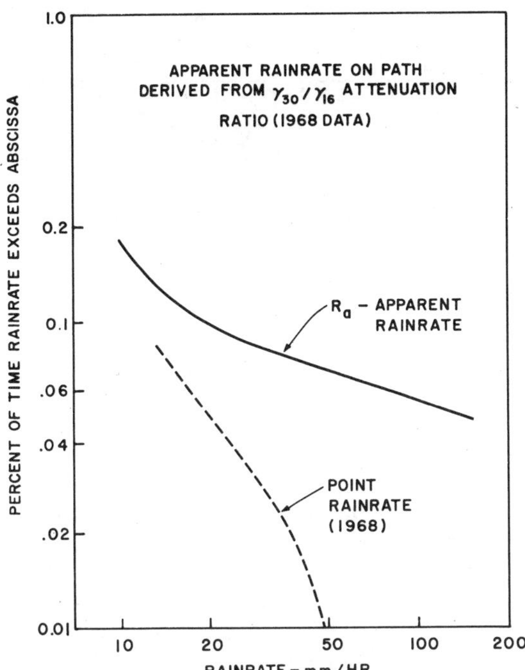

Fig. 37. Apparent rainrate on the earth–space path derived from the cumulative distributions of Fig. 21; point rain rate measured at the New Jersey location is shown dashed.

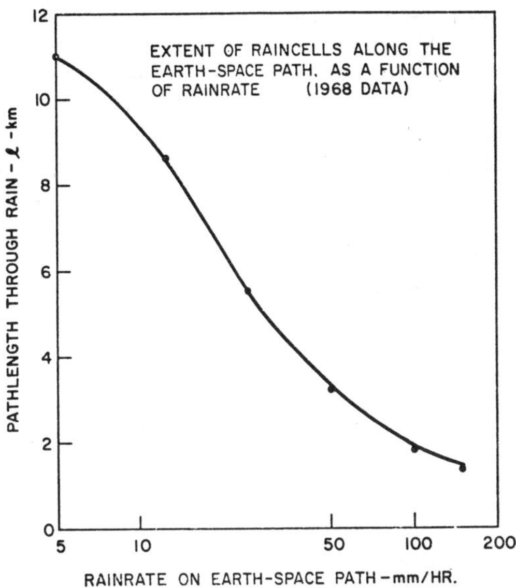

Fig. 38. Rain extent versus apparent rain rate derived from the cumulative distributions of attenuation in Fig. 21.

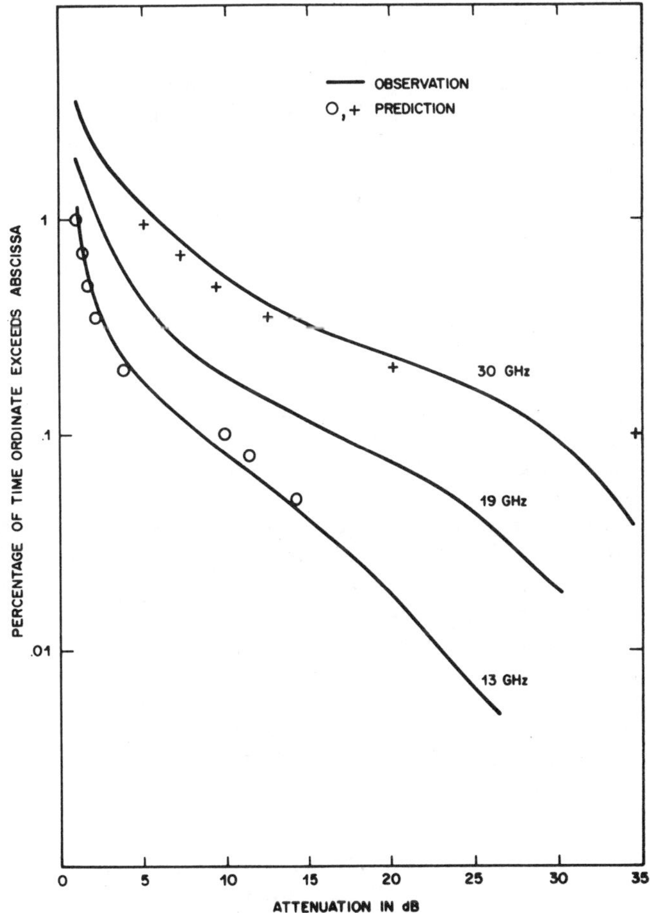

Fig. 39. Cumulative distributions of attenuation measured by a suntracker in New Jersey during 1972–1973; predictions of attenuation at a third frequency from measurements at two other frequencies are shown.

set of data, when the measured 16-GHz attenuation is, for example, 8 dB, the 30- to 16-GHz ratio is 3.0, therefore the apparent rainrate in the cell on the earth–space path is (on the average) 20 mm/h. In this way one constructs a cumulative distribution of apparent rainrate corresponding to the data of Fig. 21; the result is shown in Fig. 37. Plotted as a dashed line on that same figure is the point rainrate distribution measured at the site. Clearly the apparent rainrate on the earth-space path has a much greater probability of occurence than the point rainrate, especially for the higher values. But that is not a surprising result since the former applies to occurence on a path of order 20 km whereas the latter applies only at a point. In other words, a shower or cell is much more likely to intercept a long path than to include a point. This matter was discussed briefly with regard to short terrestrial paths in Section II-B; but simultaneous long-term dual-frequency measurements over *long* terrestrial paths have, to the authors' knowledge, not yet been made.

The apparent rainrate of Fig. 37 is not to be confused with the path-average rainrate. Indeed, so far, nothing has been said about the effective length of the rain medium on the path, i.e., the extent of the rain cell. But this effective length can be obtained from the absolute value of the attenuation; in the above example, from the 16-GHz attenuation of 8 dB and apparent rainrate of 20 mm/h, we calculate a rain extent, or equivalent cell size, of 6.5 km. When this procedure is carried out for all of the measured ratios in Fig. 36, a relationship between the extent and the intensity of the rain is obtained, as shown in Fig. 38. This result is intuitively satisfying because it broadly tells us that light rains usually occur over large areas in New Jersey whereas heavy rains are in the form of relatively small cells. Because a given point in Fig. 38 relates to a certain percent-of-time value, one can, using the standard theory of Section II, derive a cumulative distribution of attenuation for any other desired frequency. This hypothesis has been checked by a three-frequency experiment discussed in the following paragraphs.

Simultaneous measurements [99] of attenuation at 13, 19 and 30 GHz were made using a suntracker (Fig. 20) at Holmdel, N.J., for the year 1972–1973. The cumulative distributions are given in Fig. 39; the 30 GHz distribution does not differ much from its 1967–1968 counterpart in Fig. 21. But the main purpose of the experiment was to utilize attenuations

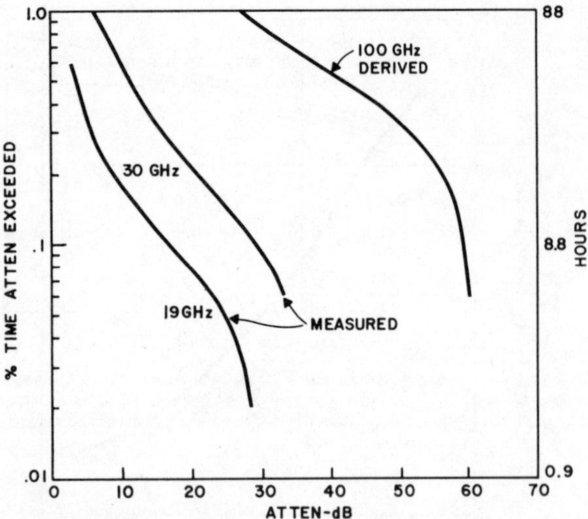

Fig. 40. Prediction of attenuation at 100 GHz from suntracker measurements of 19- and 30-GHz attenuation in New Jersey.

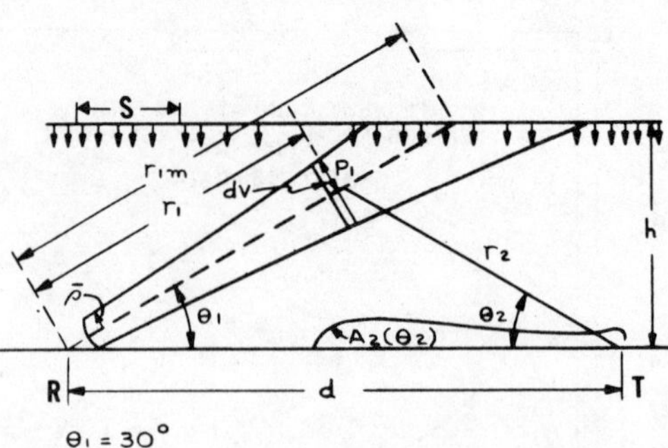

Fig. 41. Geometry for calculation of rain-scattering interference between an earth station R and a terrestrial radio-relay antenna T.

measured at two frequencies for predicting the attenuation at a third frequency and to examine agreement of that value with the actual measurement at the third frequency. This exercise was carried out, as discussed above, by determining the apparent rainrate distribution from two sets of data and applying standard theory to predict the third set. Fig. 39 shows the result of predictions to both lower and higher frequencies, i.e., the open dots are predictions of 13-GHz attenuation, given 19 and 30 GHz, and the crosses predictions of 30 GHz given the 13- and 19-GHz measured data. In both cases, the predicted values are within about 1 dB of the measured values except for the point of highest attenuation, i.e., 35 dB at 30 GHz, in which case the curvature of the measured distribution is open to some question [99] because of dynamic-range limitations.

An example of the result obtained when the apparent rainrate distribution is used to predict attenuation at a much higher frequency, 100 GHz, is shown in Fig. 40. The measured 19 and 30 GHz attenuations used to determine the apparent rainrate are those discussed above. The 100-GHz distribution shows a distinct and rapid downward trend at the higher values of attenuation; this is a consequence of a somewhat slower increase of 100-GHz attenuation at high rainrate. However, in spite of that behaviour, the 100-GHz attenuation reaches values of order 60 dB which, in the light of present technology, appear to be insurmountable in design of a communication system, even if path diversity were used. It should be pointed out that Figs. 39 and 40 are based on only one year of measurements in New Jersey.

VII. Sharing of Frequency Bands between Terrestrial and Satellite Systems (Interference)

In order that terrestrial and satellite communication systems successfully share the same frequency band, the interference between them must be kept under control. In clear weather conditions, one must contend with transmission beyond the horizon [100], [101] due to scattering from atmospheric turbulence, superrefraction, and layer reflection. On the other hand, when the signal is depressed by heavy rain fading, communication systems may be more vulnerable to energy propagated beyond the horizon and also be susceptible to interference caused by scattering from the rain.

If two antenna beams have a common volume,[20] the main beam to main beam coupling via rain scattering is expected to produce strong interference. But in the absence of a common volume, coupling from the main beam of one system into the sidelobes of another is still a potential source of interference. Both cases have been estimated by approximate numerical calculations [102], [103] which will be discussed later. Measured data on the sidelobe to main-beam coupling are very scarce [5], whereas several experiments have been carried out in the common-volume case [104]–[106]. The comparison between the latter experimental results and calculated predictions is not only interesting in its own right but also serves as a bench mark for estimating the sidelobe to main-beam coupling.

Measured ratios [104] of received to transmitted power at 4.5 and 7.7 GHz have been found to be in good agreement with values calculated using the bistatic radar equation, measured radar data, and a rain-scattering coefficient based upon the Rayleigh approximation. The measured statistics [105], [106] of the coupling loss at 3.6, 6, and 7.8 GHz also agree with predictions based upon rain statistics measured at a location on the earth beneath the common volume. The coupling caused by common volume rain scattering can be of the order −120 dB for 0.01-percent probability in the eastern US [105]; intersecting main beams should, therefore, be avoided in the planning of radio communication systems.

Starting from the bistatic radar equation and assuming a well behaved Fresnel zone beam cross section for the earth station antenna, as shown in Fig. 41, the power coupling between a terrestrial radio-relay antenna and the earth-station antenna can be predicted by

$$\frac{P_R}{P_T} = \int \frac{\sigma_v}{4\pi r_2^2} A_2(\theta_2) \exp\left[-\alpha(r_1 + r_2)\right] dr_1 \qquad (12)$$

where σ_v is the rain scattering coefficient (the scattering cross section per unit volume), α the rain attenuation coefficient, and $A_2(\theta_2)$ the pattern function for the effective aperture area of the terrestrial repeater antenna. In the frequency range of interest ($\lesssim 30$ GHz) the scattering by the raindrops can be taken to be isotropic [78], [107]. Previous calculations have been based upon the following relation [108] which involves the Rayleigh approximation:

$$\sigma_v = 0.18 \frac{\pi^5}{\lambda^4} 10^{-15} R^{1.6} \ (\text{m}^{-1}) \qquad (13)$$

where R is the rain rate in millimeters per hour. Since the

<hr>

[20] A common-volume geometry is one in which the main beams of the two antennas intersect.

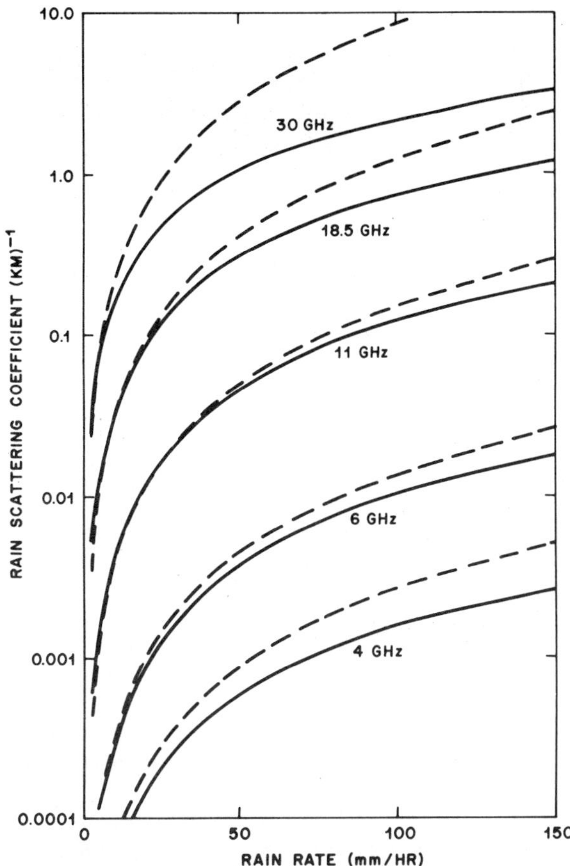

Fig. 42. Calculated scattering coefficients of rain; – – – empirical relation using the Rayleigh wavelength dependence, —— Mie's scattering solution using the Laws and Parsons drop size distribution

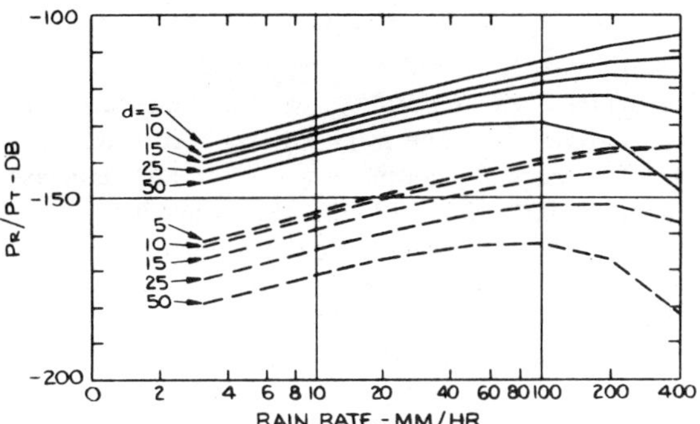

Fig. 43. Calculated coupling at 4 GHz for a uniform rain of 4 km height, —— radio relay antenna beam directed toward earth station; – – – isotropic level sidelobes directed toward the earth-station.

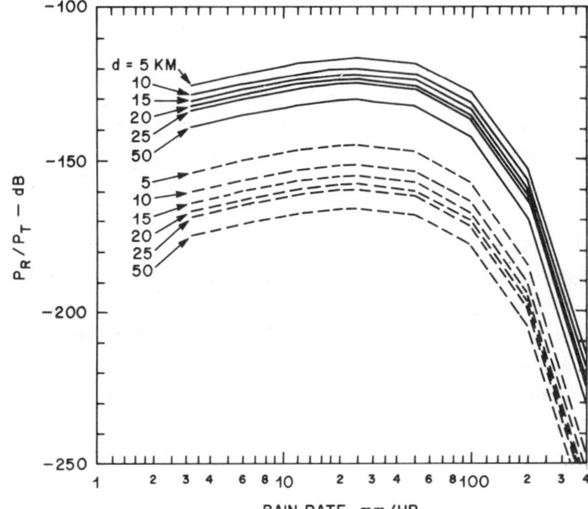

Fig. 44. Calculated coupling at 18.5 GHz for a localized shower extending 1 km from the earth station; —— radio-relay antenna beam directed toward earth station; – – – – isotropic level sidelobes directed toward the earth station.

Rayleigh approximation is known to be invalid at higher microwave frequencies, (13) is compared in Fig. 42 with recent calculations [20] using Mie's scattering solution and Laws and Parsons drop-size distribution. For rainrates up to 150 mm/h, Fig. 42 shows that the dashed lines for the approximate relation (13) are within a factor of two of the exact calculation for the frequencies 4 through 18 GHz. At 30 GHz, the exact result is a factor of four lower at 100 mm/h. Using (12) and (13), numerical calculations of the power coupling loss have been made for various geometries and frequencies. Fig. 43 shows an example at 4 GHz for uniform rain originating at a height of 4 km, and Fig. 44 an example at 18.5 GHz for a localized shower extending $S = 1$ km from the earth station. The upper set of curves in Figs. 43 and 44 apply to the situation where a terrestrial radio relay antenna beam is directed toward an earth station (see Fig. 41), the gain of the antenna in terrestrial system being 43 dB at 4 GHz and 36 dB at 18.5 GHz which are typical of existing designs. The lower set applies where isotropic-level sidelobes of the terrestrial antenna intercept the beam of the earth-station antenna. The coupling is relatively insensitive to frequency because the effects of increased scattering and increased attenuation tend to cancel one another. For the lower frequencies, the maximum coupling occurs at high rain rates, and the interference caused by large-area uniform rain is generally of the same order of magnitude as that of a localized shower. At higher frequencies the rain attenuation reduces the interference for large-area uniform rain, and the maximum coupling occurs at even lower rainrates than for the localized rain condition shown in Fig. 44.

The significance of the cochannel interference resulting from these couplings (of the order −140 dB) is judged by comparison with the thermal noise in the receiver of a typical earth station. If the system noise temperature of a 4 GHz earth station is 70 K within a 40-MHz band, the noise power referred to the input will be −134 dBw. The interference in the earth-station receiver from a terrestrial radio repeater of 10-W transmitting power, with −140 dB coupling, is therefore −130 dBw, 4 dB above the thermal noise. Thus the combined interference of several radio repeaters with sidelobes near the isotropic level could become a serious threat to the performance of a 4-GHz earth-station receiver, therefore, earth stations of 4- and 6-GHz satellite systems have to be located far from metropolitan centers which already accomodate terrestrial systems at these frequencies. Unfortunately, the connecting link between the earth station and the urban traffic center is a significant added cost to the overall system.

It is desirable to locate the earth station of a satellite system in an urban area with terrestrial systems occupying the same frequency bands, thereby reducing the cost of the connecting end link; this is much easier to accomplish at 18 and 30 GHz than at 4 and 6 GHz. Because an 18-GHz earth-station receiver can have both higher noise temperature (150 K) and greater noise bandwidth (say 300 MHz) than at 4 GHz, the total thermal noise of an 18-GHz ground station receiver is expected to be more than 10 dB above that of 4 GHz. The relatively strong satellite signal provided by high gain of the earth station antenna and a spot beam antenna on the satellite will provide more immunity to the rain-scatter coupling (which is about the same as at 4 GHz). Also, one notes that beyond the horizon coupling (in clear air) decreases with wavelength [100]. Furthermore, the higher frequency system will likely utilize digital modulation and hence be more tolerant of interference. For example, analysis shows [56] that the bit error rate for a 4-level CPSK system is less than 10^{-6} for a signal to interference ratio of 15 dB, and a signal to thermal noise ratio of 16 dB (see Fig. 17).

VIII. Degradation of Antenna Performance by Rain

In the introduction it was mentioned that one of the first rain effects observed at 4 GHz was caused by a wet radome. Subsequently, analysis of rain falling on a hemispherical radome showed that a water layer of constant thickness [109] would be formed over such a surface. An electromagnetic wave incident on a layer of water experiences both absorption and reflection loss [110]; the resulting overall degradation in transmission is shown in Fig. 45 as a function of layer thickness for several frequencies of interest. At 18 GHz for example, an attenuation of 10 dB is introduced by a layer of water one quarter of a millimeter in thickness. Therefore, if layers that thick are formed by rain falling on a radome, the system designer is faced with another source of attenuation of the same order as the attenuation produced by rain on the propagation path. Unfortunately, we know little about the thickness of the layer formed at a given rain rate since it depends on the geometry, frictional and wetting properties of the radome surface.

In the past, attenuation by rain on radomes and weather covers of antennas has not been studied experimentally in its own right because it occurs simultaneously with attenuation by rain on a propagation path; separation of the two effects is a difficult process. For that reason, a near-field experiment was implemented [111] in which 20 GHz was transmitted through a portion of a segment of a radome over a path of only nine meters. Over such a short distance, the attenuation on the path itself is negligible even for the heaviest of rains, therefore the observed attenuations are only caused by water on the radome. Fig. 46 shows data measured during rains which occurred after the radome surface had been exposed to the sun and weather for about eight months. The median values of attenuation are plotted (as crosses) versus rainrate measured by a gauge near the radome, along with dashed lines which include the maximum and minimum observed attenuations. The measured data have a behavior similar to the theoretical relationship shown by the solid curve, but are of lower absolute value. Clearly, attenuations of 6 dB or more can occur for rainrates greater than 10 mm/h. Referring back to Fig. 45, the thickness of the layer would be about 0.1 mm to produce that attenuation at 20 GHz.[21]

It is desirable therefore that an earth-station antenna be operable without a radome, assuming that problems introduced by wind forces and ice formation can be overcome in other ways. However, without a radome, rain produces a water layer on the reflecting surfaces which constitute the

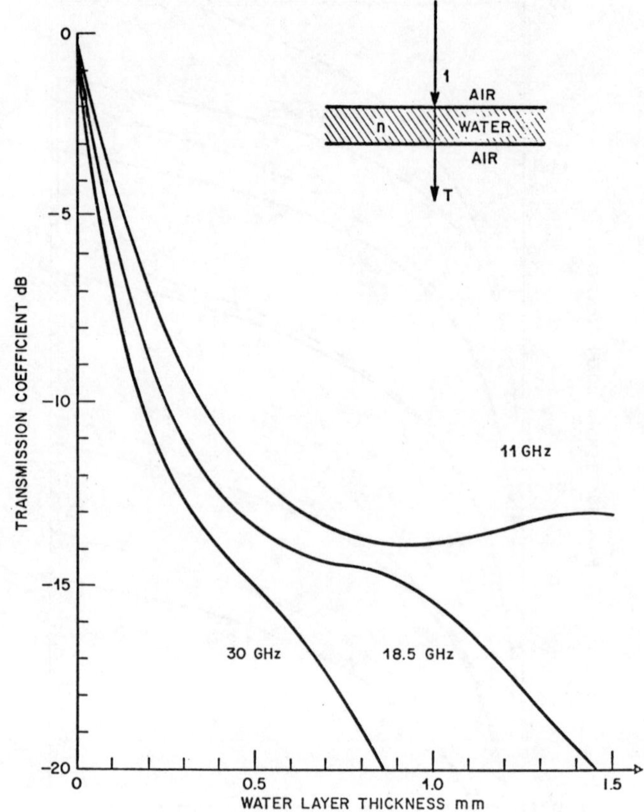

Fig. 45. Transmission through a layer of water at various common-carrier frequencies.

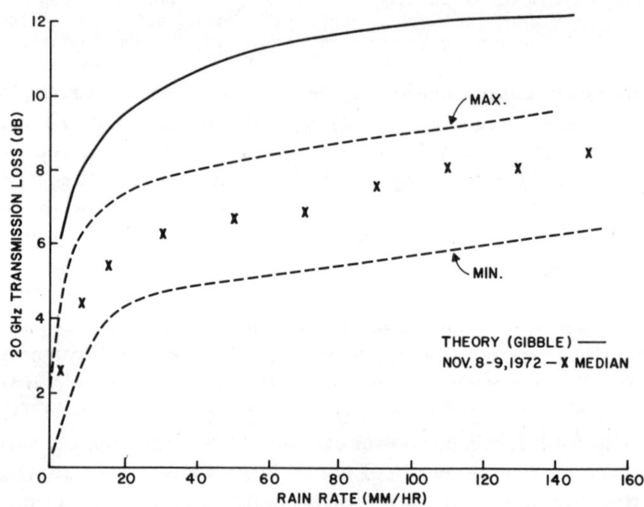

Fig. 46. 20-GHz attenuation through a section of a large spherical radome versus measured rainrate.

antenna and one must examine how much attenuation is introduced. We had occasion to calculate the reflection coefficient of an 18.5-GHz plane wave incident at 45° on water layers of various thickness covering a flat conducting surface. As shown in Fig. 47, the attenuation in the reflection process is a function of the thickness of the water and the wave polarization, a wave polarized normal to the plane of incidence exhibiting about half the degradation of polarization in the plane. However, the significant point is that if we assume the same thickness of water on the reflector (0.1 mm) as on the radome discussed above, the attenuation is, from Fig. 47, less than 0.05 dB for both polarizations. Comparing that value with the 6-dB loss involved in transmission *through* a 0.1-mm

---

[21] Radome surfaces of special design may prevent formation of a layer by causing the water to flow in rivulets; this will decrease the attenuation but presumably new problems such as aperture phase errors and cross polarization may be introduced.

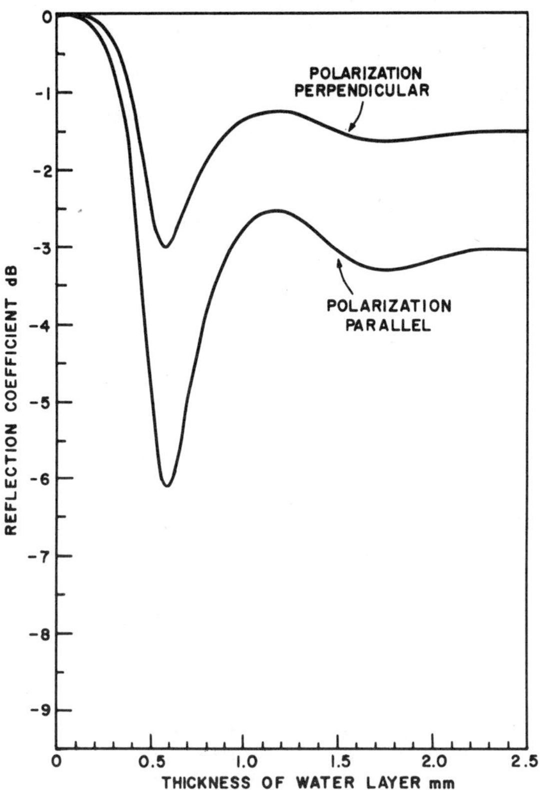

Fig. 47. 18.5-GHz attenuation computed for a wave incident at 45° on a reflecting plane covered by a layer of water.

layer, it is concluded that water layers formed by rain on reflecting surfaces introduce relatively little attenuation at frequencies of order 20 GHz; however, water and snow on antenna surfaces may generate some cross polarization [112].

It is for the above reasons that horn-reflector antennas [113] without weather covers prove very useful in radiometer measuring systems (which resolve rather small attenuations). But regardless of whether the application is in an experiment or a system, the antenna design should be such that the waves do not propagate through a wetted surface, especially where the flux density is high.

It would be inappropriate to finish this writing without making some general comments about the quality of antennas used in both satellite and terrestrial radio relay systems. Because of economics, and to some extent because of tradition, it has been the habit of system designers to use antennas (usually axially symmetric) with only high gain in mind. However, because of the high edge illumination and aperture blockage in many designs, there is significant radiation (or response to radiation) at angles well removed from the intended direction of transmission. That kind of antenna performance compounds the interference problems which arise when frequency bands are shared by the various systems as discussed in Section VII. Suggestions for designs of offset antennas [114] which have potential for obviating these problems, have been made.

ACKNOWLEDGMENT

Much of the content of this paper encompasses the industry and ingenuity of many of our colleagues at Bell Laboratories, past and present, who deserve considerable credit for stimulating studies in the use of high-frequency microwaves for satellite communication systems.

REFERENCES

[1] "The Telstar experiment," *Bell Syst. Tech. J.*, vol. 42, pp. 739–1908, July 1963.
[2] R. W. DeGrasse, E. O. Shultz-DuBois, and H. E. D. Scovil, "Three-level solid-state traveling-wave maser," *Bell Syst. Tech. J.*, vol. 38, pp. 305–334, Mar. 1959.
[3] D. C. Hogg and R. A. Semplak, "The effect of rain and water vapor on sky noise at centimeter wavelengths," *Bell Syst. Tech. J.*, vol. 40, pp. 1331–1348, Sept. 1961.
[4] A. J. Giger, "4Gc transmission degradation due to rain at the Andover, Maine, satellite station," *Bell Syst. Tech. J.*, vol. 44, pp. 1528–1533, Sept. 1965.
[5] D. C. Hogg, R. A. Semplak, and D. A. Gray, "Measurement of microwave interference at 4Gcs due to scatter by rain," *Proc. IEEE* (Corresp.), vol. 51, p. 500, Mar. 1963.
[6] J. A. Saxton, "Dielectric dispersion in pure liquids at very high radio frequencies," *Proc. Roy. Soc. A*, vol. 213, pp. 400–408 and pp. 473–492, Mar. 1952.
[7] P. S. Ray, "Broadband complex refractive indices of ice and water," *Appl. Opt.*, vol. 11, pp. 1836–1844, Aug. 1972.
[8] L. C. Tillotson, "A model of a domestic satellite communication system," *Bell Syst. Tech. J.*, vol. 47, pp. 2111–2138, Dec. 1968.
[9] "Telecommunications aspects on frequencies between 10 and 100 GHz," in *NATO-AGARD-Conf. Proc.*, no. 107, Sept. 1972.
[10] "Propagation of radio waves at frequencies above 10 GHz," in *IEE Conf. Publ. 98*, Apr. 1973.
[11] "The fine structure of precipitation and EM propagation," in *I.U.C.R.M. Colloquium Proc.*, Oct. 1973.
[12] H. W. Evans, "Attenuation on earth–space paths at frequencies up to 30 GHz," in *Proc. IEEE Int. Conf. Communications*, pp. 27-1–27-5, June 1971.
[13] T. S. Chu and D. C. Hogg, "Effects of precipitation on propagation at 0.63, 3.5 and 10.6 microns," *Bell Syst. Tech. J.*, vol. 47, pp. 723–759, May–June 1968.
[14] K. S. McCormick, R. L. Olsen, and L. A. Maynard, "Amplitude fading of satellite communications signals at SHF," in *NATO-AGARD Conf. Proc. 107*, pp. 18-1–18-8, Sept. 1972.
[15] R. R. Taur, "Ionospheric scintillation at 4 and 6 GHz," *COMSAT Tech. Rev.*, vol. 3, pp. 145–163, Spring 1973.
[16] J. A. Stratton, *Electromagnetic Theory*. New York: McGraw-Hill, 1941, pp. 563–573.
[17] D. E. Kerr, *Propagation of Short Radio Waves*. Lexington, Mass.: Boston Tech., 1964, pp. 680–681.
[18] *Ibid*, pp. 674–683.
[19] R. G. Medhurst, "Rainfall attenuation of centimeter waves: Comparison of theory and experiment," *IEEE Trans. Antennas Propagat.*, vol. AP-13, pp. 550–563, July 1965.
[20] D. E. Setzer, "Computed transmission through rain at microwave and visible frequencies," *Bell Syst. Tech. J.*, vol. 49, pp. 1873–1892, Oct. 1970.
[21] W. F. Bodtmann and C. L. Ruthroff, "Rain attenuation on short radio paths: Theory, experiment, and design," *Bell Syst. Tech. J.*, vol. 53, pp. 1329–1350, Sept. 1974.
[22] J. R. Norbury and W. J. K. White, "Microwave attenuation at 35.8 GHz due to rainfall," *Electron. Lett.*, vol. 8, pp. 91–92, Feb. 1972.
[23] R. A. Semplak and R. H. Turrin, "Some measurements of attenuation by rainfall at 18.5 GHz," *Bell. Syst. Tech. J.*, vol. 48, pp. 1767–1788, July–Aug. 1969.
[24] R. K. Crane, "Propagation phenomena affecting satellite communication systems operating in the centimeter and millimeter wavelength bands," *Proc. IEEE*, vol. 59, pp. 173–205, Feb. 1971.
[25] R. A. Semplak, "The influence of heavy rainfall on attenuation at 18.5 and 30.9 GHz," *IEEE Trans. Antennas Propagat.*, vol. AP-18, pp. 507–511, July 1970.
[26] ——, "Dual frequency measurements of rain-induced microwave attenuation on a 2.6-kilometer propagation path," *Bell Syst. Tech. J.*, vol. 50, pp. 2599–2606, Oct. 1971.
[27] O. E. DeLange, A. F. Dietrich, and D. C. Hogg, "An experiment on propagation of 60 GHz waves through rain," *Bell Syst. Tech. J.*, vol. 54, pp. 169–180, Jan. 1975.
[28] L. J. Battan, *Radar observation of the atmosphere*. Chicago, Ill.: Univ. Chicago Press, 1973, pp. 176–178.
[29] J. S. Marshall and C. D. Holtz, "Pattern analysis of one summer's multilevel maps of Montreal rain," *Month. Weather Rev.*, vol. 98, pp. 335–345, May 1970.
[30] B. J. Easterbrook and D. Turner, "Prediction of attenuation by rainfall in the 10.7–11.7 GHz communication band," *Proc. Inst. Elec. Eng.*, vol. 114, pp. 557–565, May 1967.
[31] D. C. Hogg, "Statistics on attenuation of microwaves by intense rain," *Bell Syst. Tech. J.*, vol. 48, pp. 2949–2962, Nov. 1969.
[32] D. N. Harden, J. R. Norbury, and W. J. K. White, "Model of intense convective rain cells for estimating attenuation on terrestrial millimetric radio link," *Electron. Lett.*, vol. 10, pp.

483–484, Nov. 14, 1974.

[33] H. R. Pruppacher and R. L. Pitter, "A semi-empirical determination of the shape of cloud and rain drops," *J. Atmos. Sci.*, vol. 28, pp. 86–94, Jan. 1971.

[34] T. Oguchi, "Attenuation of electromagnetic waves due to rain with distorted raindrops," *J. Radio Res. Labs* (*Tokyo*), part I in vol. 7, pp. 467–485, Sept. 1960; part 2 in vol. 11, pp. 18–44, Jan. 1964.

[35] T. Oguchi, "Attenuation and phase rotation of radio waves due to rain: Calculations at 19.3 and 34.8 GHz," *Radio Sci.*, vol. 8, pp. 31–38, Jan. 1973.

[36] J. A. Morrison, M. J. Cross, and T. S. Chu, "Rain-induced differential attenuation and differential phase shift at microwave frequencies," *Bell Syst. Tech. J.*, vol. 52, pp. 599–604, Apr. 1973.

[37] J. A. Morrison and M. J. Cross, "Scattering of a plane electromagnetic wave by axisymmetric raindrops," *Bell Syst. Tech. J.*, vol. 43, pp. 955–1019, July–Aug. 1974.

[38] H. C. Van deHulst, *Light Scattering by Small Particles.* New York: Wiley, pp. 28–36, 1957.

[39] J. A. Morrison and T. S. Chu, "Perturbation calculations of rain-induced differential attenuation and differential phase shift at microwave frequencies," *Bell Syst. Tech. J.*, vol. 52, pp. 1907–1913, Dec. 1973.

[40] T. Oguchi and Y. Hosoya, "Differential attenuation and differential phase shift of radio waves due to rain: Calculations at microwave and millimeter wave regions," in *I.U.C.R.M. Colloquium Proc.*, pp. I.1–I.1-6, Oct. 1973.

[41] R. A. Semplak, "Effect of oblate raindrops on attenuation at 30.9 GHz," *Radio Sci.*, vol. 5, pp. 559–564, Mar. 1970.

[42] W. T. Barnett, "Some experimental results on 18 GHz propagation," in *Conf. Rec. 1972 Nat. Telecommunications Conf.*, (IEEE Publication 72 CHO 601-5-NTC), pp. 10E-1-10E-4.

[43] T. S. Chu, "Rain induced cross-polarization at centimeter and millimeter wavelengths," *Bell Syst. Tech. J.*, vol. 53, pp. 1557–1579, Oct. 1974.

[44] M. J. Saunders, "Cross polarization at 18 and 30 GHz due to rain," *IEEE Trans. Antennas Propagat.*, vol. AP-19, pp. 273–277, Mar. 1971.

[45] P. A. Watson and M. Arbabi, "Rainfall cross polarization at microwave frequencies," *Proc. Inst. Elec. Eng.*, vol. 120, pp. 413–418, Apr. 1973.

[46] ——, "Rainfall cross polarization, comparison of theory and measurement," in *I.U.C.R.M. Colloquium Proc.*, pp. I.3-1–I.3-6, Oct. 1973.

[47] R. A. Semplak, "Simultaneous measurements of depolarization by rain using linear and circular polarization at 18 GHz," *Bell Syst. Tech. J.*, vol. 53, pp. 400–404, Feb. 1974.

[48] H. Yamamoto, K. Morita, and Y. Nakamura, "Experimental considerations on 20 GHz high-speed digital radio-relay system," in *Conf. Rec. 1973 IEEE Int. Conf. Communications* (IEEE Publication 73 CHO 744-3-CSCB), pp. 28-37-28-42.

[49] M. Shimba, K. Morita, and A. Akeyama, "Radio propagation characteristics due to rain at 20 GHz band," *IEEE Trans. Antennas Propagat.*, vol. AP-22, pp. 507–509, May, 1974.

[50] C. W. Bostian, W. L. Stutzman, P. H. Wiley, and R. E. Marshall, "Millimeter wave rain depolarization: Some recent 17.65 GHz measurements," in *1973 IEEE/AP Int. Symp. Dig.*, pp. 289–292.

[51] R. A. Semplak, "The effect of rain on circular polarization at 18 GHz," *Bell Syst. Tech. J.*, vol. 52, pp. 1029–1030, July–Aug. 1973.

[52] ——, "Measurements of rain-induced polarization rotation at 30 GHz," *Radio Sci.*, vol. 9, pp. 425–429, Apr. 1974.

[53] S. H. Lin, "An occurrence of very heavy rain on a 42-km path," *IEEE Trans., Commun.*, vol. COM-22, pp. 708–710, May 1974.

[54] W. T. Barnett, "Determination of cross-polarization discrimination during rain and multipath fading at 4 GHz," in *Conf. Rec. 1974 Int. Conf. Communications* (IEEE Publication 74 CHO 859-9-CSCB), pp. 12D-1-12D-4.

[55] D. T. Thomas, "Cross polarization discrimination in microwave radio transmission due to rain," *Radio Sci.*, vol. 6, pp. 833–839, Oct. 1971.

[56] V. K. Prabhu, "Error rate considerations for coherent phase-shift keyed systems with co-channel interference," *Bell Syst. Tech. J.*, vol. 48, pp. 743–767, Mar. 1969.

[57] B. L. Ulich, "Absolute brightness temperature measurements at 2.1 mm wavelength," *Icarus*, vol. 21, pp. 254–261, 1974.

[58] A. E. Covington, "Solar emission at ten centimeter wavelength, 1947–1960," *J. Roy. Astron. Soc. Can.*, vol. 48, pp. 167–172, 1961.

[59] P. G. Davies, "Radiometric measurements of atmospheric attenuation at 19 and 37 GHz along sun-earth paths," *Proc. Inst. Elec. Eng.*, vol. 120, pp. 159–164, Feb. 1973.

[60] R. G. Conway, "Measurement of radio sources at centimeter wavelengths," *Nature*, vol. 199, p. 1177, Sept. 21, 1963.

[61] R. W. Wilson, "Sun tracker measurements of attenuation by rain at 16 and 30 GHz," *Bell Syst. Tech. J.*, vol. 48, pp.

1383–1404, May–June 1969.

[62] J. L. Linsky, "The Moon as a proposed radiometric standard for microwave and infrared observations of extended sources," *Astrophys. J.* (*Suppl. Series no. 216*), vol. 25, pp. 163–204, Feb. 1973.

[63] A. A. Penzias and R. W. Wilson, "Microwave noise from rain storm," Science, vol. 169, pp. 583–584, Aug. 1970.

[64] D. C. Hogg, "Effective antenna temperature due to oxygen and water vapor in the atmosphere," *J. Appl. Phys.*, vol. 30, pp. 1417–1419, Sept. 1959.

[65] A. A. Penzias and R. W. Wilson, "A measurement of excess antenna temperature at 4080 MCs," *Astrophys. J.*, vol. 142, pp. 419–421, Aug. 1, 1965.

[66] R. H. Dicke, "The measurement of thermal radiation at microwave frequencies," *Rev. Sci. Instrum.*, vol. 17, p. 268, July 1946.

[67] E. A. Ohm and W. W. Snell, "A radiometer for a space communications receiver," *Bell Syst. Tech. J.*, vol. 42, pp. 2047–2080, Sept. 1963.

[68] K. N. Wulfsberg, "Sky noise measurements at millimeter wavelengths," *Proc. IEEE*, vol. 52, pp. 321–322, Mar. 1964.

[69] R. W. Wilson and W. L. Mammel, "Results from a three-radiometer path-diversity experiment," in *IEE Conf. Publ. 98*, pp. 23–27 and supplementary notes, Apr. 1973.

[70] K. N. Wulfsberg and E. E. Altshuler, "Rain attenuation at 15 and 35 GHz," *IEEE Trans. Antennas Propagat.*, vol. AP-20, pp. 181–187, Mar. 1972.

[71] L. J. Ippolito, "Millimeter wave propagation measurements from the applications technology satellite (ATS-5)," *IEEE Trans. Antennas Propagat.*, vol. AP-18, pp. 535–552, July 1970.

[72] A. A. Penzias, "First result from 15.3 GHz earth–space propagation study," *Bell Syst. Tech. J.*, vol. 49, pp. 1242–1244, July–Aug. 1970.

[73] J. I. Strickland, "The measurements of slant path attenuation using radar, radiometers and a satellite beacon," in *I.U.C.R.M. Colloquium Proc.*, pp. III.6-1–III.6-9, Oct. 1973.

[74] L. J. Ippolito, "Effects of precipitation on 15.3 and 31.6 GHz earth–space transmissions with the ATS-V satellite," *Proc. IEEE*, vol. 59, pp. 189–205, Feb. 1971.

[75] R. K. Crane, "Simultaneous radar and radiometer measurements of rain shower structure," MIT Lincoln Lab., Lexington, Mass., Tech. Note 1968-33, ASTIA Doc AD-678079, Sept. 1968.

[76] D. A. Gray, "The effect of scatter on radiometric observations of rainfall," in *USNC-URS1 Abstracts*, p. 33, Apr. 1972.

[77] A. M. Zavody, "Effect of scattering by rain on radiometric measurements at millimeter wavelengths," *Proc. Inst. Elec. Eng.*, vol. 121, pp. 257–263, Apr. 1974.

[78] D. E. Setzer, "Anisotropic scattering due to rain at radio-relay frequencies," *Bell Syst. Tech. J.*, vol. 50, pp. 861–868, Mar. 1971.

[79] D. A. Gray, "Unusually high Earth–space path attenuations measured using a 6.4 GHz radiometer," *Proc. IEEE*, vol. 61, pp. 138–139, Jan. 1973.

[80] H. D. Craft, Jr., "Attenuation statistics at 15.3 GHz for Clarksburg, Maryland," *COMSAT Tech. Rev.*, vol. 1, pp. 221–225, Fall 1971.

[81] R. B. Briskman, R. F. Latter, and E. E. Muller, "Call for help," *IEEE Spectrum*, vol. 11, pp. 35–36, Oct. 1974.

[82] D. C. Cox, "Design of the Bell Laboratories 19 and 28 GHz satellite beacon propagation experiment," in *IEEE ICC Rec.*, pp. 27E-1-27E-5, June 1974.

[83] H. Yokoi, M. Yamada, and A. Ogawa, "Measurement of precipitation attenuation for satellite communication at low elevation angles," *I.U.C.R.M. Colloquium Proc.*, pp. III.4-1–III.4-6, Oct. 1973.

[84] K. S. McCormick, "A comparison of precipitation attenuation and radar backscatter along Earth–space paths," *IEEE Trans. Antennas Propagat.*, vol. AP-20, pp. 747–754, Nov. 1972.

[85] D. C. Hogg, "Millimeter-wave communication through the atmosphere," *Science*, vol. 159, pp. 39–46, Jan. 5, 1968.

[86] L. C. Tillotson, "Millimeter waves: Can they provide relief from spectrum crowding," *Microwave J.*, vol. 10, p. 32, Nov. 1967.

[87] D. C. Hogg, "Path diversity in propagation through rain," *IEEE Trans. Antennas Propagat.*, vol. AP-15, pp. 410–415, May 1967.

[88] R. W. Wilson, "A three-radiometer path-diversity experiment," *Bell Syst. Tech. J.*, vol. 49, pp. 1239–1241, July–Aug. 1970.

[89] D. B. Hodge, "A 15.3 GHz satellite-to-ground path-diversity experiment utilizing the ATS-5 satellite," *Radio Sci.*, vol. 9, pp. 1–6, Jan. 1974.

[90] K. N. Wulfsberg, "Path diversity for mm-wave Earth-to-satellite links," *Radio Sci.*, vol. 8, pp. 1–5, Jan. 1973.

[91] D. A. Gray, "The relationship between satellite signal attenuations and sky temperature," in *USNC-URSI Abstracts*, pp. 22–23, Apr. 1972.

[92] T. G. Konrad, "A statistical model of rain showers derived from radar observations," in *USNC-URSI Abstracts*, pp. 166–167, Oct. 1974.

[93] L. J. Ippolito, "Attenuation and wideband coherence measurements at 20 and 30 GHz with the ATS-6 satellite," *USNC-URSI Abstracts*, p. 166, Oct. 1974.

[94] L. H. Westerlund, J. L. Levatich, and A. Buige, "ATS-F COMSAT millimeter wave propagation experiment," *COMSAT Tech. Rev.*, vol. 3, pp. 323–334, Fall 1973.

[95] B. G. Evans and J. Troughton, "Calculation of cross-polarization due to precipitation," in *IEE Conf. Publ. 98*, pp. 162–171, Apr. 1973.

[96] R. R. Taur, "Rain depolarization: Theory and experiment," *COMSAT Tech. Rev.*, vol. 4, pp. 187–190, Spring 1974.

[97] D. A. Gray, "Measurements of depolarization by rain at 20 GHz on the Earth–space path," submitted to *Proc. IEEE* (Lett.).

[98] D. C. Hogg, "Intensity and extent of rain on Earth–space paths," *Nature*, vol. 243, pp. 337–338, June 8, 1973.

[99] P. S. Henry, "Measurement and frequency extrapolation of microwave attenuation statistics on the Earth–space path at 13, 19, and 30 GHz," *IEEE Trans. Antennas Propagat.*, vol. AP-23, pp. 271–274, Mar. 1975.

[100] A. B. Crawford, D. C. Hogg, and W. H. Kummer, "Studies in tropospheric propagation beyond the horizon," *Bell Syst. Tech. J.*, vol. 38, pp. 1067–1178, Sept. 1959.

[101] P. L. Rice, A. G. Longley, K. A. Norton, and A. P. Barsis, "Transmission loss predictions for tropospheric communication circuits," *NBS Tech. Note 101*, vol. 2, revised, May 1966.

[102] L. T. Gusler and D. C. Hogg, "Some calculations on coupling between satellite-communications and terrestrial radio-relay systems due to scattering by rain," *Bell Syst. Tech. J.*, vol. 49, pp. 1491–1512, Sept. 1970.

[103] "Influence of scattering from precipitation on the siting of Earth stations," in *ITU Doc. XIII Plenary Assembly* (CCIR Rep. 339-1 of Study Group No. 5), 1974.

[104] R. K. Crane, "Bistatic scatter from rain," *IEEE Trans. Antennas Propagat.*, vol. AP-22, pp. 312–320, Mar. 1974.

[105] A. Buige and J. L. Levatich, "Measurement of precipitation scatter effects on propagation at 6, 12, and 18 GHz," in *Proc. AIAA 3rd Communications Satellite Systems Conf.*, paper No. 70-499, Apr. 6, 1970.

[106] R. W. Hubbard, "Measurement and prediction of bistatic radio scattering due to precipitation," *I.U.C.R.M. Colloquium Proc.*, pp. III.3-1–III.3-11, Oct. 1973.

[107] L. H. Doherty and S. A. Stone, "Forward scatter from rain," *IEEE Trans. Antennas Propagat.*, vol. AP-8, pp. 414–418, July 1960.

[108] K. L. S. Gunm and T. W. R. East, "The microwave properties of precipitation particles," *Quart. J. Roy. Meteorol. Soc.*, vol. 80, pp. 522–545, Oct. 1954.

[109] D. Gibble, "Effects of rain on transmission performance of a satellite communication system," in *IEEE Int. Conv. Rec.*, pt. 6, p. 52, 1964.

[110] J. Ruze, "More on wet radomes," *IEEE Trans. Antennas Propagat.*, vol. AP-13, p. 823, Sept. 1965.

[111] I. Anderson, "Measurements of 20 GHz transmission through a wet radome," *IEEE Trans. Antennas Propagat.*, vol. AP-23, Sept. 1975.

[112] B. G. Evans, A. J. Fryatt, J. Read, and P. T. Thompson, "Investigation of the effects of precipitation on parabolic antennas employing linear orthogonal polarization at 11 GHz," *Electron. Lett.*, vol. 7, pp. 375–377, July 1971.

[113] A. B. Crawford, D. C. Hogg, and L. E. Hunt, "A horn-reflector antenna for space communication," *Bell Syst. Tech. J.*, vol. 40, pp. 1095–1116, July 1961.

[114] C. Dragone and D. C. Hogg, "The radiation pattern and impedance of offset and symmetrical near-field Cassegrainian and Gregorian antennas," *IEEE Trans. Antennas Propagat.*, vol. AP-22, pp. 472–475, May 1974.

[115] R. H. Turrin, "A multibeam satellite antenna for the 20 and 30 GHz radio bands," *Bell Syst. Tech. J.*, July–Aug. 1975.

[116] C. Dragone, "An improved antenna for microwave radio systems consisting of two cylindrical reflectors and a corrugated horn," *Bell Syst. Tech. J.*, vol. 53, pp. 1351–1379, Sept. 1974.

[117] K. Funakawa and Y. Otsu, "Characteristics of the slant path rain attenuation at 35 GHz obtained by solar radiation and atmospheric emission observations," in *I.U.C.R.M. Colloquium Proc.*, pp. III 9-1–III 9-5, Oct. 1973.

[118] J. E. Allnutt and J. E. Hall, "Site-diversity advantage for satellite communication at 11.6 GHz," *Electron. Lett.*, vol. 10, pp. 527–528, Dec. 12, 1974.

[119] G. Brussard, "Rain-induced cross polarization and raindrop canting," *Electron. Lett.*, vol. 10, pp. 411–412, Oct. 3, 1974.

Part VI
Earth Stations

In early communications satellite systems, the satellite capabilities were modest, and the burden of achieving the necessary communication performance was placed on the earth stations. The INTELSAT system is a prime example of a system which historically placed reliance on high-performance high-capacity earth stations. Over the years both satellite and earth-station technology have progressed so that in present and future systems, meaningful trade-offs can be made between the space segment and the earth segment.

Communications satellites are now used for a large variety of purposes with earth terminals ranging from the 33-m antenna INTELSAT stations operating at 6/4 GHz to the 1-m antenna terminals operating with the Japanese broadcast satellite at 30/20 GHz.

Reprint Paper 6.1 provides a general discussion of earth-station technology with emphasis on large earth stations. The various subsystems of a typical ground station—antenna, antenna feed, tracking subsystem, low-noise receiver, high-power amplifier, ground communications equipment, interface equipment, and monitoring and control equipment—are discussed, and the evolution of ground stations in the INTELSAT system is described. Reference [1] discusses the 14/11 GHz earth stations in the INTELSAT system.

Reprint Paper 6.2 describes the changes in earth station technology required for dual polarization operation with INTELSAT V. Dual polarization will become increasingly important in the future as more systems become band-limited and have to exploit the second polarization for frequency reuse.

A major area of growth in the next decade will be in satellite communications systems utilizing small earth terminals. Reprint Paper 6.3 is a status report on low cost earth terminals. References [2]-[4] discuss various aspects of small terminals.

There are a number of areas of interest that are omitted from this brief collection. These include:

1) higher frequency terminals (e.g., [5]);

2) mobile terminals, including land, maritime, and airborne (e.g., [6], [7]);

3) military terminals (e.g., [8]-[12]);

4) unattended terminals (e.g., [11], [12]).

Other aspects of earth stations are discussed in various papers in the Bibliography at the end of this volume.

REFERENCES

[1] M. P. Brown, Jr., "Intelsat V standard "C" earth station performance objectives applied to a possible eastern United States location (14/11 GHz frequency band)," *EASCON 1977,* pp. 13-1A–13-1H.

[2] "The evolution of cost effective [small aperture] earth terminals for communications," *Session 21 of the 1976 International Conference on Communications,* 1976.

[3] D. J. O'Neill, P. M. Fitzgerald, and L. R. Englebrecht, "Single access user earth terminal system design," *ICC 1975 Conf. Record,* vol. 1, pp. 12-1–12-5, 1975.

[4] N. R. Helm and J. Kaiser, "Small earth terminals for medical/educational appliances," AIAA Paper 75-917, *AIAA Conf. Communication Satellites for Health/Education Applications,* July 1975.

[5] L. Cuccia, W. Quan, and C. Hellman, "Above 10 GHz Satcom bands spur new earth terminal development," *Microwave System News,* pp. 37-56, Mar. 1977.

[6] I. L. Lebow, K. L. Jordan, and P. R. Drouilhet, "Satellite communications to mobile platforms," *Proc. IEEE,* Feb. 1971.

[7] D. W. Lipke, D. W. Swearingen, J. F. Parker, E. E. Steinbecker, T. O. Calvit, and H. Dodd, "MARISAT, A maritime satellite communications system," *Comsat Tech. Rev.,* Fall 1977.

[8] F. M. Knipp and J. A. Buegler, "Military satellite communications terminals," *Signal,* vol. 30, no. 6, Mar. 1976.

[9] P. L. McGivern, "Tacsatcom for the U.S. Army," *Signal,* vol. 28, no. 7, Mar. 1974.

[10] J. L. Moore, "Whiskey-3—UHF radio terminal for naval satellite communication," *Signal,* vol. 31, July 1977.

[11] C. L. Cuccia, C. Hellman, and W. Slivkoff, "The technology of unattended remotely monitored earth terminals," AIAA Paper No. 78-610, *AIAA 7th Communications Satellite Systems Conf.,* San Diego, CA, pp. 562-571, Apr. 24-27, 1978.

[12] L. Pollack and W. Sones, "An unattended earth terminal for satellite communications," *COMSAT Tech. Rev.,* vol. 4, no. 2, pp. 205-230, Fall 1974.

6.1
Earth Station Technology

E. R. WALTHALL
RCA
Princeton, N.J., U.S.A.

INTRODUCTION

Since the advent of the first experimental communications satellites TELSTAR and RELAY launched in 1962, more than a decade of communication satellite system development has passed with the development of the geostationary satellites typified by the Intelsat communications satellites. During this period both the space segments and earth station technology have evolved from a performance, reliability and cost standpoint. This evolution has contributed to the expanding use of satellites for communications on a worldwide basis. The most recent event in this expanding use has been the exploitation of domestic communications satellites for interstate telecommunications and television distribution. The basic economy of providing these functions by satellite systems has been demonstrated especially for areas of the world where there is a requirement to reach small, widely dispersed settlements with few transportation facilities and frequently over terrain whose features impede normal construction techniques (1,2).

Thus, the earth-station technology has developed to encompass not only the high-performance, high-capacity stations of the Intelsat network but smaller, much less expensive ground stations designed for totally unattended operation (3). While the evolution of the Intelsat system has led to the large increase in the number of earth stations in that system (approximately 80) the major expansion in terms of numbers of earth stations for the next decade will be dominated by the regional and domestic systems providing both point-to-point communications and direct broadcast for television distribution (4,5). As a result, a major impetus for the development of earth-station technology will be provided by the need for low-cost earth terminals. Mass-production techniques

of simple designs employing small non-tracking antennas, uncooled parametric amplifiers, transistorized receivers, maximum use of integrated-circuit technology and large scale integration (LSI) techniques, efficient modulation schemes and demand-assignment network switching will typify the earth stations, particularly for domestic and regional systems (6).

Table 1 lists some of the key features that presently characterize earth stations ranging from the large Intelsat stations, to medium-sized stations for use with domestic systems such as the Canadian Telesat system, to minimum-sized stations that might be used for thin-route telecommunications. Table 2 lists the characteristics of a typical direct-broadcast receive-only earth station being developed for the Federation of Rocky Mountain States, Inc. (7). These stations are being used for an educational TV experiment in conjunction with the ATS-6 satellite.

EARTH-STATION SYSTEMS PLANNING

The major requirement for the satellite telecommunication system is that it provides reliable, telephone-quality, circuits to remote areas. Earth stations must be rugged and easy to install in remote areas. Since there is usually a large number of earth stations, unit station cost has a great impact on total system cost, and considerable emphasis is placed on the design of low-cost stations. Thus, in developing a system approach, tradeoffs are made between the ground-station performance and cost for a given satellite capability; i.e., the ground-station figure of merit, station capacity and link quality is optimized for a given satellite EIRP and bandwidth availability (8,9,19). Typically, the satellite antenna gain is restricted by the desired area coverage and

Reprinted from *Commun. Satellite Syst.: An Overview of the Technol., IEEE Press*, 1976, pp. 121-128.

TABLE 1. Earth Station Characteristics

High Performance Stations	Medium Performance Stations	Minimum Cost Stations

Antenna
  • Wheel and track azimuth control
  • Single elevation bull gear
  • No upper equipment room
  • No access elevator
  • All electronics in base of elevated foundation
Tracking System
  • Step-track (or "hill climber") tracking system instead of 3-channel monopulse
Feed System
  • Feed horn and feed microwave circuitry at base of antenna
  • 3-reflector focussed-beam system illuminates sub-reflector through aperture of main reflector
Low Noise Receiver
  • 50 K uncooled paramp mounted at antenna base
High Power Amplifier
  • Tuned klystron or wide-band TWT
GCE System
  • Micro-integrated circuitry
  • FDM/FM, PCM/TDM
Redundancy
  • All critical components employ on-line redundancy
	Antenna
	  • Fixed position
	  • Manual steering adjustment in elevation and azimuth
	  • Az-el or polar mounts
	Feed
	  • Sum mode feed
	Low Noise Receiver
	  • 100 K uncooled paramp
	High Power Amplifier
	  • Wide-band TWT
	GCE
	  • For medium to high capacity trunking-FDM/FM techniques
	  • For Thin-Route operations - FM compandored or digital single-voice-channel per carrier technique
	Power System
	  • Battery bank/static inverter backup to commercial power or diesel-generator
	Redundancy
	  • Redundant LNR and HPA
		Antenna
		  • Single element-spun dish fabrication
		  • Fixed position manually steered
		Feed
		  • Primary focus feed
		Low Noise Receiver
		  • 170 to 300 K uncooled paramp
		High Power Amplifier
		  • TWT
		GCE
		  • FM compandored or digital single-voice channel per carrier technique
		Power System
		  • Battery bank/static inverter backup to commercial power or generator
		Redundancy
		  • No redundancy with minimum spare parts

TABLE 2. Receive-Only Terminal Characteristics (ATS-6 HEW Experiment)

Frequency band:	2.5/2.6 GHz
Antenna:	3-m (10-foot) diameter (segmented assembly)
Antenna tracking*:	Operator controlled electric drive elevation track $\pm 3^\circ$
Antenna feed:	Prime focus feed with built-in preamplifier (circular polarization)
Receiver:	Hybrid microwave integrated circuit tuned-radio frequency receiver with 4.2 dB noise figure
Figure-of-merit:	7.1 dB/K

*Tracking is required since north-south stationkeeping is not employed by the ATS-6 spacecraft.

the maximum transmitter power capability is limited by the saturation level of the RF power amplifier and the availability of power from the satellite power supply. Furthermore, the CCIR-allowable flux densities for the allocated frequency bands limit the maximum satellite EIRP to values that do not produce interference with terrestrial microwave facilities (10).

The minimum antenna diameter (and consequently the minimum antenna gain) are determined by the requirement for spatial isolation of satellites in close proximity on orbit and by the desire for non-tracking antennas for thin-route operations and remote sites. The availability of receiver front-end noise temperatures ranging from 20 K for cooled parametric amplifiers (used in the high-capacity Intelsat stations) to the 50 to 300 K range uncooled parametric amplifiers (used in the smaller and less costly domestic-satellite earth terminals) offers a range of station cost and performance capabilities that varies by more than two orders of magnitude (8,9,11,12,20). In some cases transistorized front ends with a noise figure of 3 to 4 dB have been used to reduce cost (7). Thus, commercially planned satellite earth-station figures of merit range from 18 dB/K to 41 dB/K.

Modulation techniques and access methods such as those given in reference (13) and shown in Table 3 may be selected to optimize channel capacity and system cost based on the satellite and ground-station performance capabilities. Figure 1, taken from reference 14, compares several modulation techniques with various ground-station figures of merit for a thin-route telecommunications service. While the data presented are based on a particular satellite and ground-system configuration, similar tradeoffs can be conducted to optimize any particular set of telecommunications requirements.

Computer programs have been developed which analyze and evaluate the major system parameters of a domestic satellite communications system such as type and quality of service, earth-station figure of merit, satellite EIRP, and component costs (17). These programs combined with other programs that analyze the mix between terrestrial facilities and satellite facilities provide a means for optimizing the overall space-system design. Reference (18) provides a description of the telecommuni-

TABLE 3. Modulation and Access Methods

Access Method	Modulation Technique	Channels Per Transponder B_{IF} = 36 MHz	RF Power Per Transponder	Availability	Comments
FDMA	AM SSB SC SCC	7200	$\sim$ 150 W	10 Years	Requires NPR of 42 dB for tolerable intermodulation
	AM DSB SC SCC	3600	$\sim$ 75 W	10 Years	Same as above
	FM SCC	360	$\sim$ 10 W	2 Years	VOX required to reduce power and intermodulation
	PCM SCC 4 ϕ PSK	800	$\sim$ 5 W	1 Year	VOX required, 7 bit PCM with synchronous 64 kb/s
	VSDM SCC 4 ϕ PSK	1200	$\sim$ 5 W	1 Year	VOX required. VSDM at 40 kb/s
	VSDM SCC DC PSK	600	$\sim$ 5 W	1 Year	VOX required. VSDM at 40 kb/s
TDMA	PCM 4 ϕ PSK	800 / 892	$\sim$10 W	4 Years	Limited to 50 M b/s by logic circuits. 72 M b/s possible with 4 ϕ PSK
	VSDM 4 ϕ PSK	1500 / 1784	$\sim$ 10 W	4 Years	Same as above

General Conditions: U.S. Continental Coverage, 10-m (33-ft) Antenna and Uncooled Paramp in Earth Station Demand Assignment is fully-variable for all access methods

SYMBOLS

FDM	Frequency Division Multiplex	PCM	Pulse Code Modulation
TDM	Time Division Multiplex	DC	Differentially Coherent
AM	Amplitude Modulation	PSK	Phase Shift Keying
SSB	Single Side Band	VSDM	Variable Slope Delta Modulation
SC	Suppressed Carrier	B_{IF}	If Bandwidth
SCC	Single Channel Per Carrier	RF	Radio Frequency
FM	Frequency Modulation	VOX	Voice Activated Carrier
ϕ	Phase		

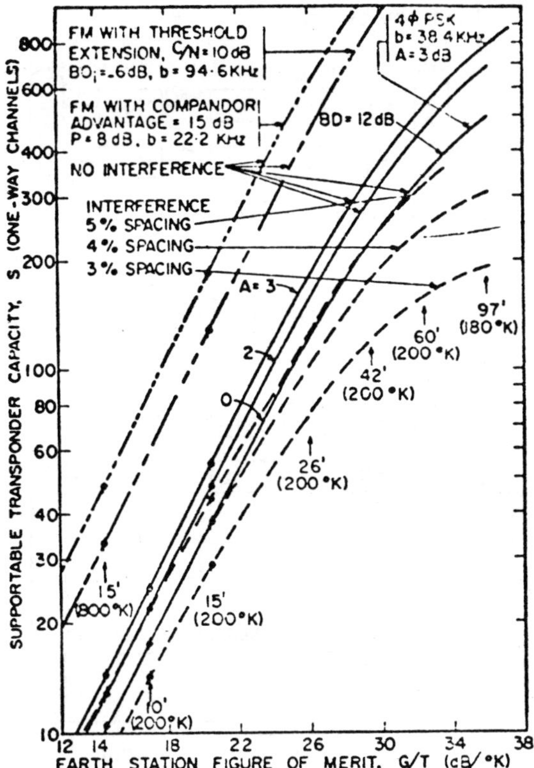

NOTES:
EXCEPT WHERE STATED OTHERWISE ALL CURVES ARE
FOR BOᵢ = INPUT BACK-OFF = -10 dB
b = RF BANDWIDTH = 76·3 KHz
FREQUENCY MODULATION VOICE CHANNEL QUALITY = 37·5 dBrnCo
P = PRE EMPHASIS = 4 dB PEAK VOICE SIGNAL = 3 dBmO
NO COMPANDOR ADVANTAGE
A = INTERMOD ADVANTAGE = 3 dB

SINGLE CARRIER SATURATED VALUES:
UP-PATH C/No = 104·6 dB-Hz
DOWN-PATH C/No = 101·1 dB-Hz (G/T = 36)

Fig. 1. Supportable Transponder Capacity Versus Earth Station G/T

cations requirements for Brazil and the planning for a satellite system that is presently being considered for a major expansion of its existing network.

REVIEW OF EARTH STATION DESIGN AND DEVELOPMENT

Figure 2 is a block diagram of a typical ground station. The major hardware subsystems can be listed as follows:

Antenna
Antenna feed
Low-noise receiver (LNR)
High-power amplifiers (HPA)
Ground communication equipment
Interface equipment
Monitoring and control equipment

A description of a recently designed Intelsat station can be found in reference 15. Descriptions of ground stations related to domestic satellite applications and small earth terminals for direct broadcast reception may be found in references 3, 5, 6, 7, 11, 12, 19 and 20.

Evolution of Commercial Satellite Earth Stations for Operations with the Intelsat Global Network

The complement of personnel required for operation and maintenance of Intelsat-type commercial satellite earth stations has been greatly reduced with the new terminals now available in comparison to the pioneer stations of the early 1960's; for example, the initial installations at Goonhilly Downs, Cornwall, England, Andover, Maine, Pleumeur Bodou, Brittany, France, Raisting (near Munich, FRG) and Mill Village, Nova Scotia, Canada. For these pioneer stations the average complement of personnel for operation and maintenance was about 35. Some new terminals require a staff of about 15; some require a staff of less than 10, and the most recent are unattended, requiring only periodic visits for routine maintenance work.

Over the last 8 years improvements in design of earth stations have permitted over a 50% reduction in the operating and maintenance staff, and have reduced the cost of the earth stations by more than 50%.

More than 80 Intelsat earth stations in the developing nations stimulated the demand for low-cost earth stations that are easy and economical to operate and maintain. This need has been fulfilled by competition from earth-station contractors from Great Britain, Canada, Italy, Japan and the United States.

Earth-station design has evolved in steps to the present level of capital costs and operating and maintenance costs. Four design phases have been selected to describe the rapid evolutionary process in earth-station development. The design phases necessarily overlap to a certain degree.

Phase 1 Earth Stations

Phase 1 was represented by the early Intelsat stations in Great Britain, U.S., France, Germany and Canada, some of which participated in the Telstar and Relay Satellite experiments of 1962 and 1963 that demonstrated the feasibility of a global satellite communication system. These stations started commercial trans-Atlantic satellite service with INTELSAT I (Early Bird), the first Intelsat satellite. All these stations, except Goonhilly, Great Britain, had temperature-controlled radome protection to the antenna to assure tracking accuracy in the presence of wind gusts, icing and wide variations in ambient temperature, all of which would bring a degree of deformation to an exposed reflector. The radomes, made of hyperlon-coated dacron, were air inflated at slightly above atmospheric pressure. These stations also had a control building for the station separated from the antenna complex by about 480 meters (1500 feet). This separation was designed for the possibility of track-

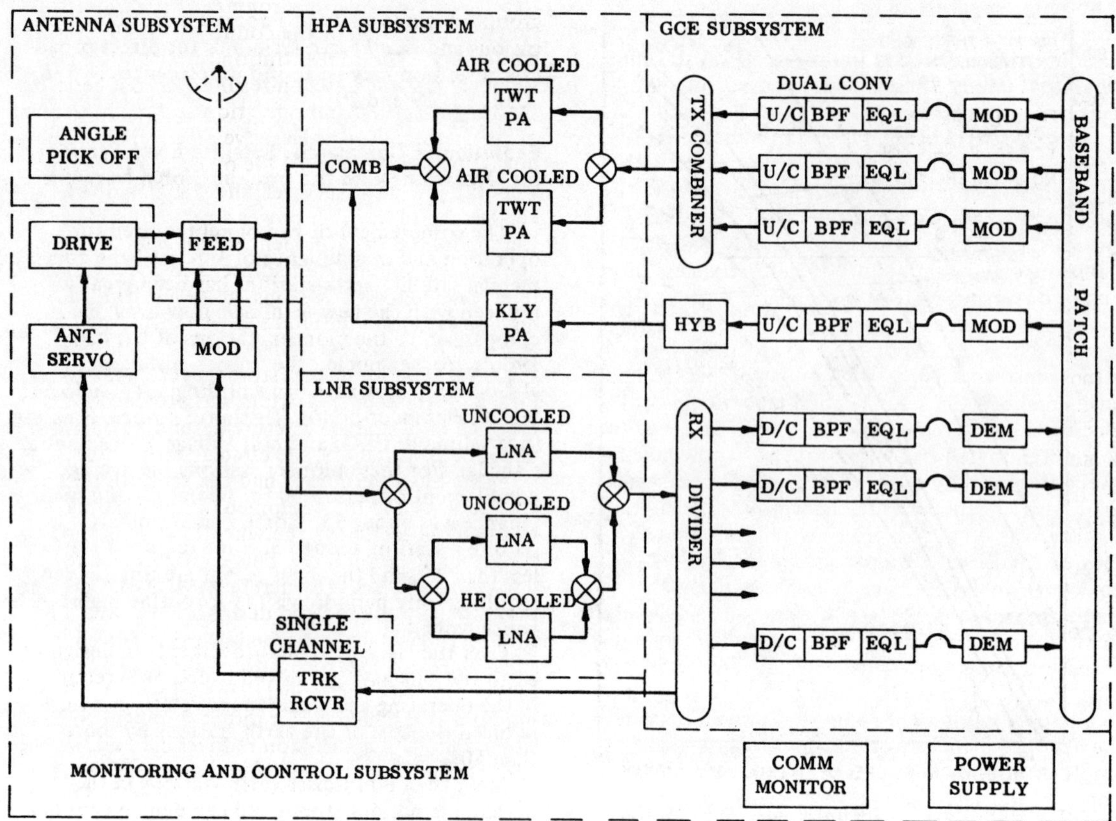

Fig. 2. Typical Ground Station Block Diagram

ing sub-synchronous satellites where two or more antennas, with a centrally located control building, would not shadow another antenna at low radiation angles. Receiving capacity was restricted to only a few RF carriers and the interfacility link was at the baseband frequency level, from RF and IF signal processing equipment at the antenna location to control consoles in the control building. Antenna reflector sizes varied from 26 m (85 feet) to 30 m (100 feet). The high-power amplifier used liquid-cooled 8 kW klystrons—a power level necessary for operation with INTELSAT I. The low-noise receiver subsystems for these early stations were narrow-band masers except for Mill Village No. 1 in Canada which used a liquid-helium parametric amplifier.

Some of the Phase 1 stations had a cone type of upper equipment room for feed, tracking downconverter and low-noise receiver which moved in elevation and azimuth. This made access and maintenance of these subsystems particularly awkward.

Phase 2 Earth Stations

Phase 2 stations dispensed with a radome-protected antenna by improvements in the antenna structure and tracking system, and by the ability to obtain the exact degree of antenna reflector deformation caused by winds, ice and tem-perature extremes. Electric de-icing systems consisting of thermostatically controlled electric blankets behind the reflector panels were used in locations where ice and snow were of concern. During this phase the interfacility link between the antenna complex and the control building became a continuous, flexible, elliptical waveguide system carried on a trough a few feet above ground level—one waveguide for the 6-GHz transmit signals and one for the 4-GHz receive carriers. All signal processing was carried out at the control building, from baseband to RF in the transmit path, and RF to baseband in the receive path. The 4-GHz waveguide could carry over 20 receive carriers, or one from each station to which communications were required. This reconfiguration greatly reduced the operation and maintenance effort in that the RF and baseband signal processing equipment (up and downconverters, modulators, demodulators, etc.) were rack mounted directly adjacent to the control console and in view of the operating personnel, in the control building. Visits to the remote antenna complex were reduced to periodic maintenance checks of the servo-drive subsystem, the low-noise receiver and the high-power amplifier.

Phase 2 stations generally used 3-kW air-cooled klystrons or travelling-wave tubes which represented a significant reduction in maintenance from the liquid-cooled versions. All stations used a helium parametric amplifier with 500-MHz instantaneous bandwidth, 30 dB gain, and 17 to 20

K noise temperature in the low-noise receiver. This revision brought a considerable reduction in station maintenance and improvement in station reliability. Many Phase 2 stations used the double-conversion principle in IF/RF signal processing to provide 500-MHz bandwidth and to enable rapid frequency change by crystal selection alone without replacing or retuning filters. Additionally, the double-conversion principle reduced in-band spurious signal generation to an undetectable level. (Spurious signals can reduce the effectiveness of a single-conversion system, particularly if more than about five carriers are present.) The double-conversion scheme in the up- and down-converters permits instead a 1-for-10 redundancy level in receive chains of the 1-for-1 inherent in single conversion. Logic circuitry permits automatic insertion of the standby chain, with frequency and capacity level automatically selected in place of any faulty chain. Improved access to the low-noise receiver and tracking downconverter mounted behind the apex of the antenna was necessary for Phase 2 stations with an exposed antenna, particularly for northern locations where rain, snow and ice would make staircase access hazardous. Accordingly, most of these stations were fitted with an elevator (in the king post of the azimuth mount) from ground level to an upper equipment room that moved only in the azimuth plane. In this room the parametric amplifier and the feed moved in an elevation arc only and remained accessible to maintenance personnel. Phase 2 stations were represented by second stations at Goonhilly, Raisting, Pleumeur Bodou, Fuccino (Italy), and Mill Village; Comsat stations at Paumalu, Hawaii, Brewster Flat, Washington, Etam, West Virginia, and Jamesburg California. Other Phase 2 stations were built in Chile, Hong Kong, Bahrain, Kenya, Brazil, Argentina, Malaysia, Singapore, Australia, New Zealand, India, Pakistan, Morocco, Greece, Spain, Scandinavia, Kuwait and Israel.

Phase 3 Earth Stations

Phase 3 earth stations appeared in about 1968, primarily in the developing areas of the world where the communications traffic capability was restricted to approximately six receive carriers. The king-post configuration of the antenna pedestal was replaced with a wheel-and-track configuration for all Phase 3 stations as an economy measure, although many Phase 2 stations used the wheel-and-track configuration. The main difference between Phase 3 and Phase 2 was the integration of all equipment and station control functions, as well as administration, within the antenna complex. This was provided by an extension of the antenna foundation from ground level to a height of about 4 meters (12 feet) to provide a circular room of 167 to 232 m^2 (1800 to 2500 ft^2) with the azimuth circular-rail system mounted on the roof of the foundation structure. The basement-type room at the base of the antenna was finished with panelling, walls and partitions to provide a pleasing environment for administration and operation of the country's satellite terminal facility. The concentration of all station personnel at a single location resulted in cost benefits for administration and operation and maintenance of the station. The only active electronic elements not in immediate view of the operating personnel were the low-noise receiver, the tracking downconverter and the high-power amplifier, which were mounted in equipment rooms behind the apex of the antenna and adjacent to the feed system. Some Phase 3 stations, however, had the high-power amplifier mounted at the antenna base and readily accessible by an elevator. Additional space for repairs, offices, stores, conference room, kitchen and sleeping quarters was provided as required by a simple structure extending outward from the periphery of the circular main room.

Some Phase 3 stations had the azimuth rails at ground level and the equipment and operating rooms mounted above the wheels between the legs of the alidade structure. This configuration reduced the amount of concrete work at the site and permitted a prefabricated design which helped to shorten the installation schedule. Furthermore, such a building lends itself to modular design so that equipment can be installed and tested in the modules at the factory for direct shipment to the site. Modular construction reduces cabling, wiring and testing work at the site, and also reduces shipping costs.

The emergency diesel power plant was contained in a separate building adjacent to the antenna base.

Reliability was improved in the Phase 3 stations and spare parts and maintenance costs reduced by replacing the pump and line-amplifier TWT's in the front end. These are replaced with a solid-state pump source for the low-noise receiver and transistorized line amplifiers with 40-dB gain.

Phase 3 type earth stations are typified by those in Iran, Lebanon, Alaska, Guam, Thailand and Philippines.

Phase 4 Earth Stations

The latest generation of earth stations, (Phase 4) concentrated all of the electronic equipment at the antenna base. The Japanese earth-station contractors made this configuration popular, although Raisting No. 2 was the first to use an offset focussed-beam feed system with an elliptical reflector. In Phase 4 stations the radiating source of the feed subsystem is mounted at the antenna base, as is the low-noise receiver, high-power amplifier, and tracking downconverter. The RF connection to the feedhorn is made by either a 4-reflector focussed-beam system contained in a shroud, as used by the Japanese, or by an oversize-waveguide system with transition elements at either end, as used by RCA. This concentration of all electronic/electric equipment at the antenna base eliminates the upper equipment rooms with their associated lighting and air conditioning, the access elevator, remote-control units for the high-power amplifier and low-noise receiver subsystems, and control, power and signal cabling in the antenna tower.

The 4-reflector focussed-beam system has had limited service to date. Problems were experienced in maintaining identical beam axes for the 4- and 6-GHz antennas. With the oversize waveguide approach, only the feed horn is mounted at the apex of the main reflector, the radiating source being mounted at ground level along with the low-noise receiver and high-power amplifier subsystems.

Some configurations of a Phase 4 station use a square building approximately 20 m (64 feet) on a site instead of a circular structure, to enhance the interior appearance and provide increased space for all station functions.

Stations completed by a Japanese contractor in Nicaragua and Ecuador are typical Phase 4 stations. Some new versions of Phase 4 stations utilize an on-off step auto-tracking system instead of the earlier monopulse autotrack system which operated continuously. Step-track locks the antenna to the satellite by maxima seeking of the beacon or communications signal. The step-track system used on the heavy-route Telesat Canada stations and proved in practice at Comsat's Paumalu station, eliminates the tracking downconverter, the 3-channel tracking receiver and the difference-mode circuitry of the feed system. It simplifies the servo-drive system and extends the lifetime of the drive by the on-off mode of operation. Thus, maintenance of the station is simplified considerably through elimination of the units no longer required by the incorporation of step-track. The simplification is of such an extent as to permit operation of the earth station from a remote control center. This type of operation was pioneered by stations in Nicaragua and Ecuador.

With automatic switchover to redundant standby units in the case of the low-noise receiver, high-power amplifier and other units in the transmit and receive path, periodic maintenance is practical. The time and type of failure of the main unit at the time of transfer to the standby unit is shown at the remote-control center. Remote operation of low-capacity earth stations is improved also by simplification of the high-power amplifier. Power levels of only 30 W to 300 W per channel are required for limited message transmission. This enables the use of 500-MHz broadband TWT's without the necessity for retuning when new transmit-frequency assignments are made.

The helium-cooled parametric amplifier remains as the one element in the earth station that is particularly sensitive to maintenance, although with solid-state pump sources and transistorized driver amplifiers this subsystem is achieving high reliability. However, the ground communications equipment, the multiplex and the rearward microwave link are fully solid state and have a high level of reliability. Some new stations are attaining simplification and cost reduction in the low-noise receiver subsystem by having an uncooled standby unit with 50 K noise temperature as back-up to a

cooled main unit with 20 K noise temperature that operates with a reflector about 31 m (95 feet) in diameter. Due to the increased EIRP levels of the latest satellite (INTELSAT IV), systems now have over 3 dB more weather margin than required and the necessity for a high G/T has been diminished. This effect is utilized for stations with satellite look angles of 25° or more, where an uncooled parametric amplifier is adequate and imposes no degradation in receive signal level. The 50 K paramp is achieved with a new order of high pump frequencies and indicates eventual full adoption of uncooled units.

Earth stations that have been installed recently for high-density traffic in the developed nations are designed with particular emphasis on high reliability. Consequently, the newest earth stations of networks such as Intelsat join power stations, audio and TV broadcast stations and microwave relay systems in their ability to operate without interruption for long, if not indefinite, periods.

In the case of stations with low traffic requirements, particularly in remote areas or developing countries where experienced maintenance personnel are scarce, unattended reliable operation is of increasing importance. The Canadian domestic system was planned for unattended operation and has been operating since early 1973. It now has over 50 unattended stations.

Technological advances will affect several areas of commercial satellite earth stations. The antenna diameter will remain in the range of 30 m (100 feet) for the high-capacity Intelsat stations, with some stations using diameters as large as 32 m (105 feet) to permit the use of uncooled low-noise receivers with noise temperatures of about 50 K. The feed systems will require spectrum reuse capability with circular or rotatable linear polarization modes to work with the probable operational mode of the INTELSAT V satellite. Such a feed system needs an axial ratio for linear polarization to achieve the required spectrum reuse isolation in the range of 25 to 35 dB. Two such feed systems for the Telesat Canada heavy-route station at Allan Park, Ontario, and the Canadian West Coast station at Lake Cowichan, B.C., near Vancouver, are presently in use.

Time-division multiple-access (TDMA) and digital-signal modulation will be used in some of the next generation stations, particularly for the heavy route traffic in the Atlantic ocean community of nations. However, emphasis will be placed on the design evolution aspects which will make the station equipment more compact and easier to operate and maintain.

Major design changes will take place in the ground communications equipment (the GCE subsystem). This subsystem, consisting mainly of receiver and exciter chains, control consoles and the TV test and mounting facility, has been full solid-state since the inception of commercial earth stations. It contains a medium traffic-handling station of about ten racks of equipment. The next generation stations will be converted from discrete solid-state components and printed circuits in the baseband, IF and RF circuitry to integrated circuitry. This will bring about an approximate 10-fold reduction in the size of the modules and a 50% reduction in the number of racks. Such units as the modulator/upconverter and the downconverter will each consist of a flat alumina substrate measuring approximately 5.1 × 7.6 cm (2 × 3 in) upon which all components and wiring are printed. Reliability, production and test time will be significantly improved by this application of IC technology. Furthermore, IC's will reduce the size of the subsystem and the number of interconnecting cables, connectors and plugs. The mechanical configuration of the subsystem probably will be changed from the pullout shelves to modules or cards for front-panel insertion.

Another design improvement will be the replacement of the 40-dB TWT driver amplifier in the high-power-amplifier subsystem by a transferred-electron amplifier (TEA) with similar gain. This change, already instituted in some of the Phase 4 stations, will remove all electron tubes from earth stations except the ones used in the high-power amplifiers.

The use of small and large stations makes equalization of the radiated RF power necessary for efficient use of the satellite transponders. Since the network configuration is known and losses predictable, station transmitting power can be preassigned with calibration and adjustment cycles as needed. Other station functions such as antenna pointing and tracking can be performed automatically. Thus, new developments will include the expanded use of computers for monitoring and station control functions, such as switching and supervisory display of equipment status and displaying operational status of unattended stations.

References

1. James Warwick, "Providing Communications to Isolated Areas," *IEEE EASCON '73*, Conv. Record, Washington, D.C., October 1973.
2. P.M.M. Norman and D.E. Weese, "Thin Route Satellite Communications for Northern Canada," *IEEE International Conf. on Comm.*, Montreal, 1971.
3. W.K. Sones and L.E. Gray, "Design Features of an Unattended Earth Terminal for Satellite Communications," *IEEE ICC '73*, Conf. Record.
4. William F. Arnold, "Sales of Earth Terminals Set to Soar," A.D. Little, Inc.
5. Lloyd G. Ludwig, "Some Developments in Semi-Direct Broadcast Satellite and Community Receiving Systems," Proceedings of the 19th Annual Meeting of American Astronautical Society, AAS Paper No. 73-155, June 1973.
6. Andrew M. Werth and Dr. Tadahiro Sekimoto, "Small Earth Terminal Applications for Satellite Communications," Proceedings of the 19th Annual Meeting of American Astronautical Society, AAS Paper No. 73-121, June 1973.
7. J. Janky and J. Potter, "The ATS-F Health-Education Technology Communications System," *IEEE ICC 73*, Conf. Record.
8. A. Dickinson, "The Optimization of System Parameters for Minimum Cost Regional and National Satellite Systems," *IEEE Conference on Earth Station Technology*, London, October 1970.
9. B.G. Evans and R. Walters, "An Economic Satellite Communications System for Small Nations," *IEEE Int. Conf. on Comm.*, Montreal, 1971.
10. "Final Acts of the World Administrative Radio Conference for Space Telecommunications," published by the International Telecommunication Union, Geneva, 1971.
11. C.C. Han, K. Ohkubo, J. Albernaz, J.M. Janky, and B.B. Lusignan, "Optimization in the Design of a 12 GHz Low Cost Ground Receiving System for Broadcast Satellites," *IEEE International Conference on Communications*, Conf. Record, June 1973.
12. A.W. Brook, "An Inexpensive Earth Terminal for TV Reception in Bush Communities," *RCA Engineer*, vol. 19, No. 3, October/November 1973.
13. Bernard J. Mirowsky, "Design of a Cost Effective Satellite/Terrestrial Communication Network," *IEEE EASCON '73*, Washington, D.C., October 1973.
14. Lorne B. Dunn, "Telephony to Remote Communities in Canada, via Satellite (Using Single Channel per R.F. Carrier)," *IEEE International Conf. on Comm.*, Montreal, 1971.
15. H. Morikawa, M. Ogata, and A. Karakami, "A Newly Developed Intelsat Earth Station," *IEEE International Conf. on Comm.*, Conf. Record, Paper No. 8-30, June 1973.
16. J. Almond, "The Telesat Canada Domestic Communications Satellite System," Paper presented at CCITT Latin American Regional Plan Meeting, Brasilia, Brazil, July 1973.
17. Private Communications with Dr. Bruce Lusignan, Associate Professor of Electrical Engineering, the Graduate School of Stanford University.
18. Hygino Caetano Corsetti, "Telecommunications in Brazil," *Telecommunications Journal of Switzerland*, vol. 40, October 1973.
19. Charles C. Sanderson and Lloyd G. Ludwig, "Single Channel Per Carrier Voice Transmission via Communications Satellite," *AIAA 5th Communications Satellite Systems Conference*, Paper No. 74-71, Los Angeles, April 1974.
20. G.P. Petrick and C.M. Abrahamson, "Economic Considerations for Low-Capacity SHF Satellite Communications Earth Terminals" *AIAA 5th Communications Satellite Systems Conference*, Paper No. 74-459, Los Angeles, April 1974.

INTELSAT V EARTH STATION TECHNOLOGY*

R. W. Gruner
COMSAT Laboratories
Clarksburg, Maryland

ABSTRACT

This paper describes the changes in earth station technology required by the application of dual polarization for frequency reuse in the INTELSAT V satellite series. Earth station polarization measurements, rain depolarization effects, and adaptive rainfall polarization correction networks are described, as well as techniques for integrated 11/14-GHz operation on dual polarized 4/6-GHz antenna systems. A simplified analysis for calculating link isolation is also presented.

INTRODUCTION

The INTELSAT V era represents a significant change in earth station technology, particularly in terms of feed design, propagation effects, and polarization measurements. These changes resulted from the application of dual polarization for frequency reuse[1] within the 4- and 6-GHz bands and the introduction of the 11/14-GHz bands. The earth station modifications which are necessary to meet present frequency reuse standards are the major topic of this paper.

The INTELSAT V beam coverage areas[2] are shown in Figure 1. Both hemispheric beams use the same polarization. Zone beams located within each "hemi" beam are cross polarized to the hemi beams. Both the polarization isolation between the cross-polarized hemi and zone beams and the isolation via sidelobe control

between the respective zone and hemi beams must be 27 dB. Two linearly polarized spot beams are formed in the 11/14-GHz bands with a minimum isolation of 33 dB (sidelobe and cross polarization) between beams. A circularly polarized global beam at 4/6 GHz with polarization isolation greater than 32 dB is also required.

SIMPLIFIED POLARIZATION DIAGRAMS

Figure 2 shows the relationship between total link polarization isolation and the earth station isolation assuming a satellite isolation of 27 dB. For minimum isolation the major axes of the satellite and earth station polarization ellipses are perpendicular. Relationships between small axial ratios (<3 dB) may be approximated by forming a vector diagram using the axial ratio as the vector magnitude abd twice the difference in ellipse tilt angles as the angle between the vectors.[3] Figure 3a is a vector diagram for a satellite isolation of 27 dB (0.75-dB axial ratio) and an earth station isolation of 30 dB (0.55-dB axial ratio). Figure 3b is a vector diagram showing relative major axis ellipse orientations of 0°, 90°. and 45°, respectively.

The depolarization caused by conditions along the propagation path such as the differential phase shift of rain may be expressed as an additional axial ratio

Figure 1: ANTENNA PATTERN COVERAGE

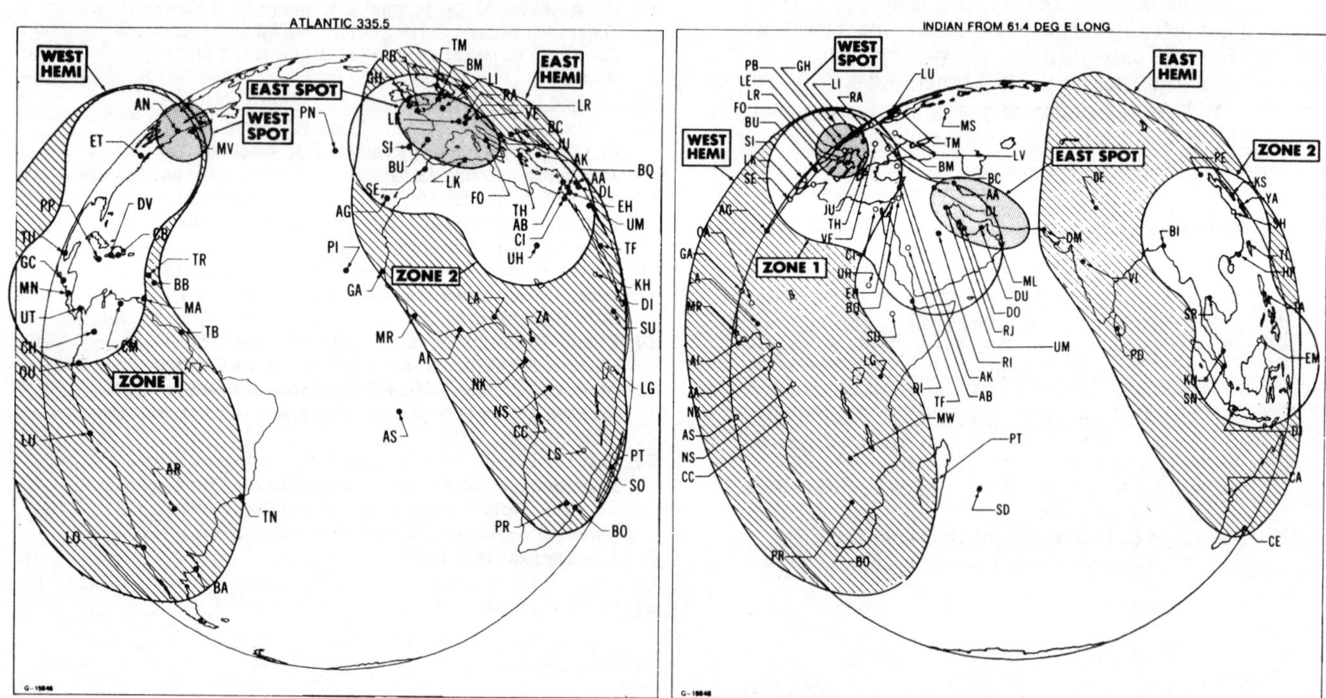

*This paper is based upon work performed in COMSAT Laboratories under the sponsorship of the Communications Satellite Corporation.

Reprinted from the *Record of IEEE Int. Telecommun. Exposition*, vol. I, Oct. 9-15, 1977, pp. 242-248.

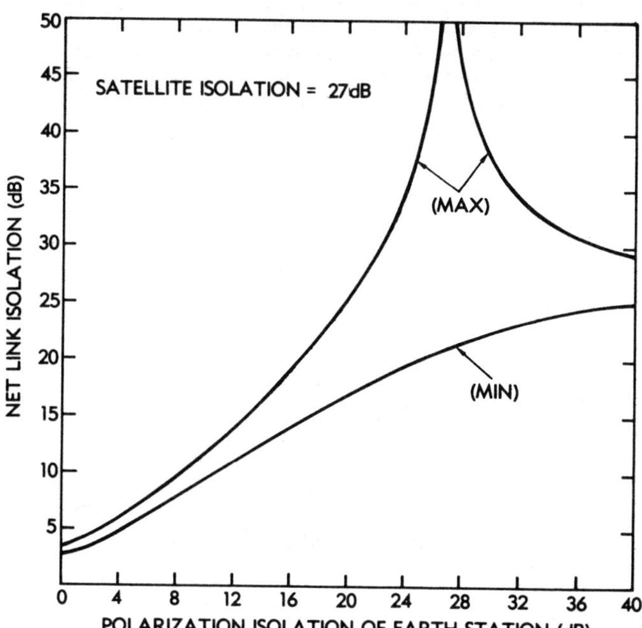

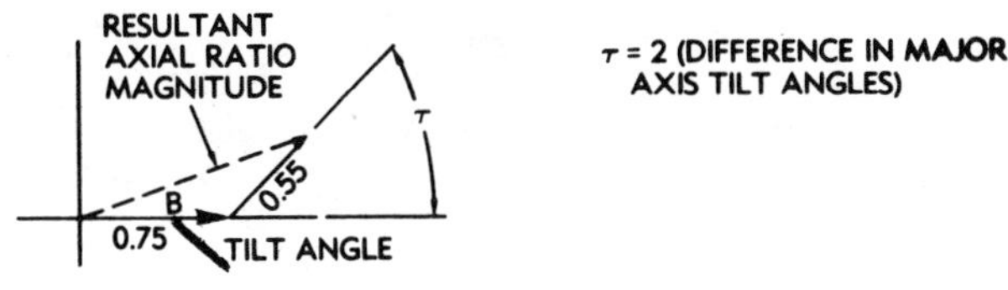

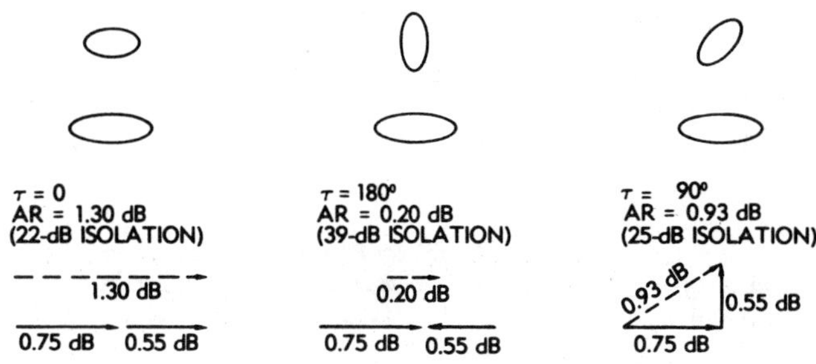

vectorily combining with the satellite and earth station axial ratios. For small phase shifts the relationship between differential phase and axial ratio, AR, is

$$AR \approx 0.151 \, \Delta\phi \ (dB)$$

Various error contributions to overall link isolation may be evaluated by the vector representation. The relationship between isolation and axial ratio is given in Figure 4.

Figure 4:   POLARIZATION ISOLATION VS AXIAL RATIO

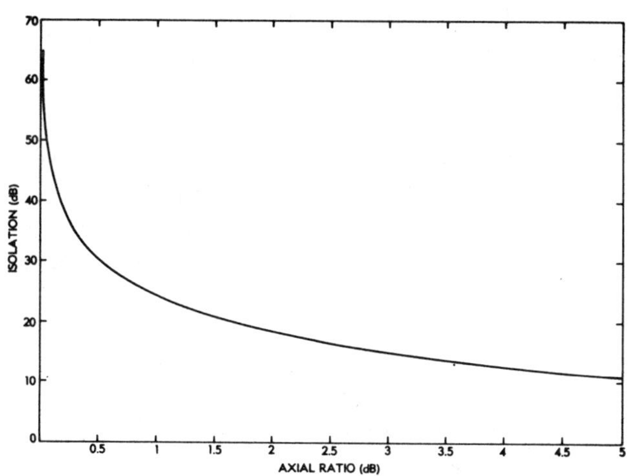

<u>DUAL POLARIZED EARTH STATION PERFORMANCE</u>

Present global system earth stations generally have axial ratios ≤ 3 dB. The revised Standard A earth station specification requires reduction of the axial ratio to 0.75 dB for retrofitted stations and 0.55 dB for newly constructed stations. To comply with the revised specification, an earth station must be retrofitted with a frequency reuse feed system regardless of its location or mode of operation. Stations within the common hemi zone area must be capable of receiving on both senses of polarization; therefore, these stations require two accessible 4-GHz ports. Two ports are also required for simultaneously transmitting both polarizations in the 6-GHz band.

Reflections in a circularly polarized system can depolarize a radiated field unless they are absorbed by a match-terminated port. Without the "extra" match-terminated ports inherent in the frequency reuse feed, the Standard A axial ratio specification would be impossible to meet. However, stations outside the overlapping hemi-zone area do not require an additional receiver and transmitter because the "extra" receive port and transmit port can be match terminated. These reflections may be a more significant factor than the axial ratio of the polarizing device. Depolarization is equal to the sum of the pre-polarizer and post-polarizer return losses. For example, with a 1.30 input voltage standing wave ratio (VSWR) at each side of the polarizer, the polarization isolation is limited to 35.5 dB or an axial ratio contribution of 0.3 dB. In-plant feed data with match-terminated loads predict feed polarization performance which is better than that obtained in operation, since actual receivers and transmitters have poorer VSWR characteristics.

<u>Feed Systems</u>

Numerous polarization network approaches are being pursued for frequency reuse applications. Figure 5 shows a 4/6-GHz network using a broadband (4/6-GHz) polarizer, a simple 6-GHz orthomode transducer (OMT), and a 4-GHz OMT assembly. Four phase equalized arms on the 4-GHz OMT provide a balanced junction that preserves waveguide mode purity within the 6-GHz band.

Figure 5:   FEED SYSTEM WITH BROADBAND 4/6-GHz POLARIZER

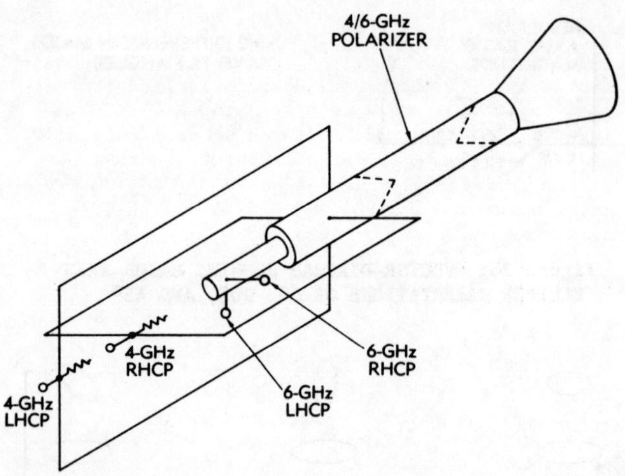

A second approach shown in Figure 6 separates the 4- and 6-GHz signals into two waveguides using two narrowband polarizers. This circuit allows independent control of the 4- and 6-GHz polarizations for rain depolarization networks. It generally has a lower axial ratio because each narrowband polarizer can be independently optimized. The circuit is usually larger than in alternate approaches and requires phase equality in the four 4-GHz interconnecting lines to preserve the axial ratio.

Figure 6:   FEED SYSTEM WITH NARROWBAND POLARIZERS

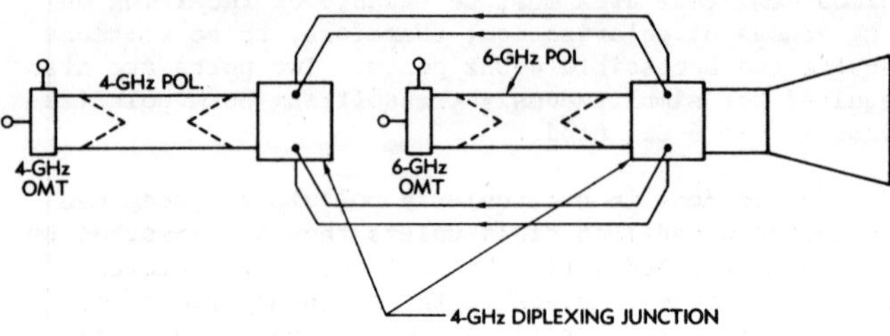

Wideband orthogonal mode junctions (both 4 and 6 GHz) are used in the third approach, shown in Figure 7, to separate the signals by polarization. The signals are then separated by a diplexer and combined. A circularly polarized wave is formed by using separate 4- and 6-GHz quadrature hybrids. Line length balancing preserves the polarization quality. Higher order modes in the 6-GHz bands may be generated by the asymmetric OMT port effecting off-axis polarization quality. Although this configuration is not as flexible in terms of adaptive capability, it yields a moderately compact and lightweight feed system.

Figure 8 shows a feed system using a multi-hole or "unity" coupler for injecting the transmit signal. The 6-GHz signal is coupled into four equal-length radially disposed waveguides. Quadrature hybrids may be used to excite the CP wave. With this concept, the design of the 4-GHz polarizing network using a narrowband polarizer and OMT is straightforward.

An approach using a coaxial waveguide[4] and a corrugated coupling region which eliminates the four balanced arms is shown in Figure 9. Changing the

corrugated surface reactance between the 4- and 6-GHz bands effectively couples the 6-GHz field to the inner 6-GHz circular waveguide and the 4-GHz field to the outer coaxial waveguide. After the transmit and receive signals have been separated, narrowband low-axial-ratio polarizers may be used. This approach requires close mechanical tolerances but results in an extremely compact and lightweight feed assembly.

Figure 7:   FEED SYSTEM WITH WIDEBAND OMT

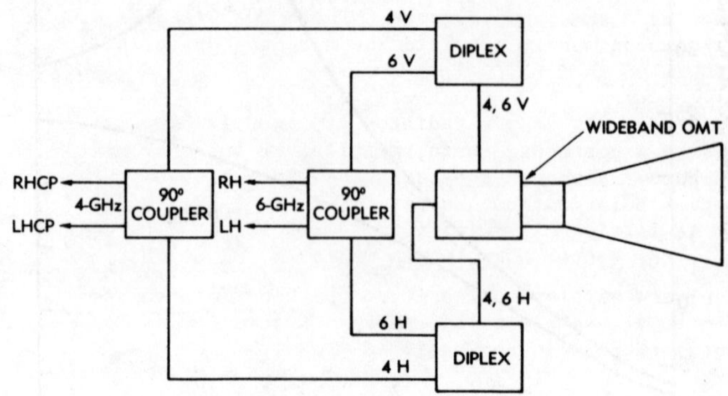

Figure 8:   FEED SYSTEM WITH MULTI-HOLE OR UNITY COUPLER

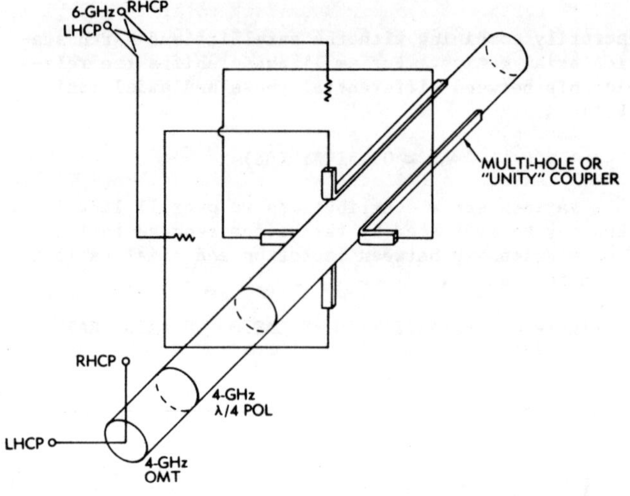

Figure 9:   FEED SYSTEM USING COAXIAL WAVEGUIDE AND A CORRUGATED COUPLING REGION

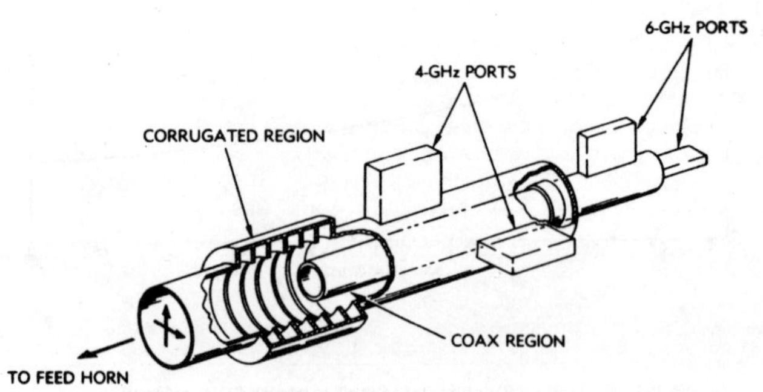

Feed Horns

The radiating characteristics of the feed horn
affect the beam off-axis polarization isolation of the
earth station. A circularly symmetric feed horn with
equal E- and H-plane amplitude and phase patterns which
approximates a Huygen's source radiator is desired.
Properly designed square cross-section horns also approximate Huygen's source radiators. The amplitude of the
off-axis cross polarization is also a function of the
reflector optics. Horn radiation can be viewed as a
difference pattern error, whereas the polarizer can be
viewed as a sum pattern error. These two patterns add
in a complex manner to yield the net cross-polarization
contour.

A Huygen's source radiator is generally approximated by a corrugated horn, multi-flare horn, or multi-
mode horn. Although the corrugated horn probably has
superior polarization quality, it is large and difficult to fabricate. Multi-flare and multi-mode horns
(i.e., the Potter horn, which radiates $TE_{11} + TM_{11}$
modes) are simpler to fabricate but have reduced bandwidth because a prescribed phase difference must be
maintained between the radiated modes.

Reflector Optics

Secondary polarization characteristics are a function of feed polarization and reflector optics. If a
perfect Huygen's source feed is used in a symmetric
parabolic reflector system, secondary polarization will
be perfect regardless of the reflector f/D ratio. The
polarization properties of a Cassegrain antenna system
may be derived from the polarization properties of the
equivalent (equal f/D ratio) parabolic surface.[5] Since
the polarization quality of the feed can generally be
better controlled in the narrower illumination angles
used for Cassegrain reflectors, the presently available
Cassegrain systems are an excellent choice for frequency
reuse applications. Shaping, which is a departure from
the hyperbolic subreflector and parabolic main reflector, is also beneficial. The off-axis cross polarization generated by the feed horn is a difference pattern
distribution that is transformed into the far-field
with reduced efficiency. Depolarization due to the
polarizer is essentially a sum pattern distribution
that is transformed into the far field of the earth
station on a one-for-one basis.

Offset reflector optical systems, including beam
waveguide feed systems with large f/D, should be
acceptable for CP frequency reuse applications. Well
chosen beam waveguide parameters[6] will not degrade the
cross-polarization levels. Offset reflectors with high
f/D ratios operating in CP yield a tolerable polarization-sensitive beam squint that has negligible impact
on the gain or polarization quality of the earth station.

Tracking Couplers

Tracking couplers may generate differential phase
shift and higher order modes that degrade both on-axis
and off-axis polarization isolation. Step tracking
eliminates the need for tracking couplers. Improving the
off-axis polarization characteristics of the feed horn
and thus allowing more tracking error appears more advantageous than designing a tight RF tracking loop with
broadband tracking couplers. Couplers are expensive
and too large for the smaller Standard B stations.
Narrowband tracking couplers would diminish the problem,
but eliminating RF tracking in frequency reuse applications seems preferable, particularly in Standard B
stations.

Rainfall along the earth station to satellite path
depolarizes the signal. Because falling raindrops are
elliptic rather than circularly symmetric, they cause
polarization-sensitive amplitude and phases in the satellite signal. The calculated cross-polarized isolation as a function of rain rate assuming a 5-km rain
cell is given in Figure 10.[7] It can be seen that differential phase is the major contributor to cross polarization. Polarization isolation is a function of
rain rate, rain cell length, and elevation angle of
the earth station.

Figure 10: PHASE AND ATTENUATION COMPONENTS OF
THEORETICAL RAIN DEPOLARIZATION EFFECT

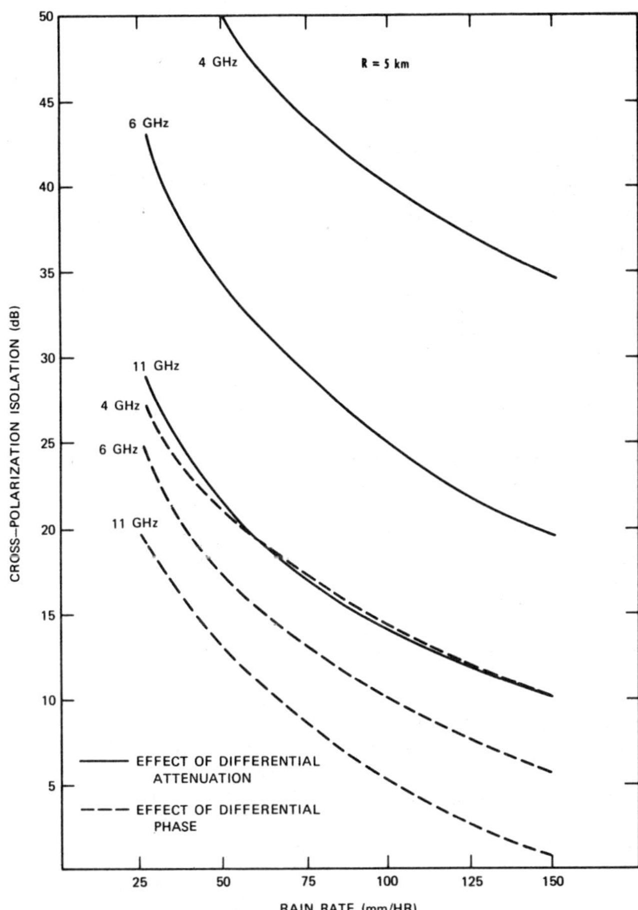

Differential phase shift causes depolarization;
that is, the nominally circular polarization from the
satellite or earth station is changed to an elliptically polarized wave. The right- and left-hand
circularly polarized signals have the same axial ratio
magnitude and the ellipse tilt angles are inclined 90°
apart. When a phase correcting compensating network
corrects one sense of polarization, the other sense is
simultaneously corrected.

The differential amplitude effect causes deorthog-
onalization. The right- and left-hand circular polarization ellipse orientations are no longer 90° apart. A
lossless correction network can provide infinite isolation in only one sense of polarization with a degraded
carrier-to-interference (C/I) ratio in the other channel. A lossy network is needed to reorthogonalize the
signals.[8]

Numerous circuits have been proposed for the correction of differential amplitude and phase due to rainfall.[9] Neglecting the differential attenuation effect of rain, which is a second-order error at 4 and 6 GHz, reduces the complexity of the circuits. Further, eliminating circuit concepts using correction after the low-noise receiver networks (those requiring matched phase and amplitude receivers) resulted in the selection of a particularly simple circuit (shown in Figure 11) with rotating half- and quarter-wave polarizers as the best approach. When this circuit was used in conjunction with a microprocessor, it corrected rainfall polarization during two severe rainstorms.[10] The circuit maintained a 30-dB polarization isolation in the presence of rain which could have reduced the polarization isolation to about 10 dB.

Figure 11: CIRCUIT WITH ROTATING HALF- AND QUARTER-WAVE POLARIZERS

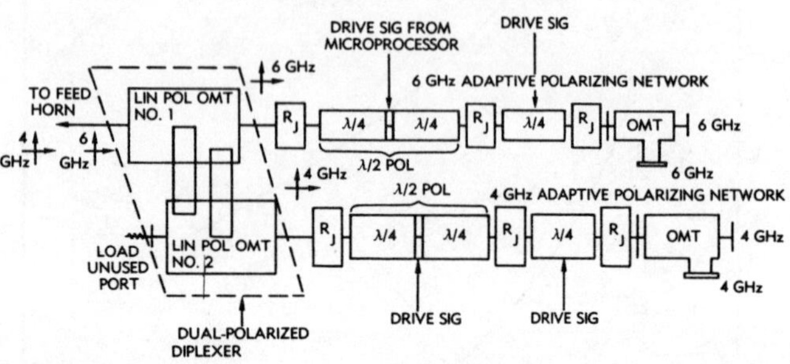

Rotating the quarter-wave polarizer angle away from its nominal clear weather angle of 45° with respect to the OMT ports creates a variable axial ratio. This polarization ellipse is then rotated by the half-wave polarizer to "polarization match" the rain-induced polarization ellipse, thus minimizing the interfering or cross-coupled signal.

The 4-GHz polarizers are positioned with a microprocessor to "null out" rain-induced interference, thus correcting the down-link polarization. Further, if the phase shift of rainfall follows a deterministic frequency relationship, the angular position of the 6-GHz polarizers may be derived strictly from the positional information of the 4-GHz polarizers, thus simultaneously correcting the up-link polarization. Data are presently being gathered at COMSAT Labs and at other earth stations to determine the correlation between up- and down-link cross-polarization.

EARTH STATION POLARIZATION MEASUREMENTS

Earth station polarization measurements use radio star, boresight, or satellite flux sources. The radio star technique has the advantage of being a constantly available, far-field source visible from all earth station locations. However, it has serious disadvantages, including the requirement for a reference antenna for precise polarization calibration of the star, and precise tracking with complex and expensive test instrumentation to accommodate long integration times.

The advantage of boresight range measurements is the simplicity of the measurement instrumentation and techniques. Full frequency capability (including swept frequency) can be implemented over the transmit and receive frequency bands. Antenna tracking is not required during the measurement. The primary disadvantage is that there may not be an adequate boresight range at a particular earth station antenna. Extensive range evaluation is required to ensure that reflected range components arrive off the main beam axis so that they are suppressed by the pattern factor of the earth station antenna. The antenna is typically evaluated at low elevation angles. Generally, it is difficult to put the source in the antenna far field (corresponding to about 22 km at 4-GHz for a 30-m aperture). Measurement data and computational experience reveal that failure to locate the source antenna in the far field is not a serious problem in defining relative polarization isolation levels.[11] Measurement programs using boresight ranges have been successfully conducted.[12,13]

Satellite sources offer a far-field source visible from numerous earth stations, and permit the earth station antennas to be evaluated close to their operating elevation angles. If a high-quality polarization source can be placed on the satellite, earth station down-link polarization isolation can be established by measuring the received power in the two 4-GHz OMT ports. Up-link polarization can be measured by switching the transmitter between the two 6-GHz OMT ports and measuring the down-link looped-back signal. The INTELSAT IC-A F2 and F3 spacecraft presently being used for earth station polarization measurements carry 4/6 GHz global beam radiators with low axial ratios. Without a high-quality satellite source, a reference antenna calibrates the axial ratio and tilt angle of the satellite. Satellite error can then be eliminated from the earth station measurement.

The disadvantage of satellite sources is that a cooperating earth station antenna is required to make the typical copolarized and cross-polarized pattern measurements. However, on-axis and limited angle contour data can be generated on a loop-back basis, requiring only the station under test. A representative measured isolation contour taken at the Mill Village earth station is shown in Figure 12. The contour values represent the ratio of the copolarized response to the cross-polarized response as a function of the off-axis angles.

Figure 12: MILL VILLAGE ANTENNA DOWN-LINK ANTENNA CONTOUR

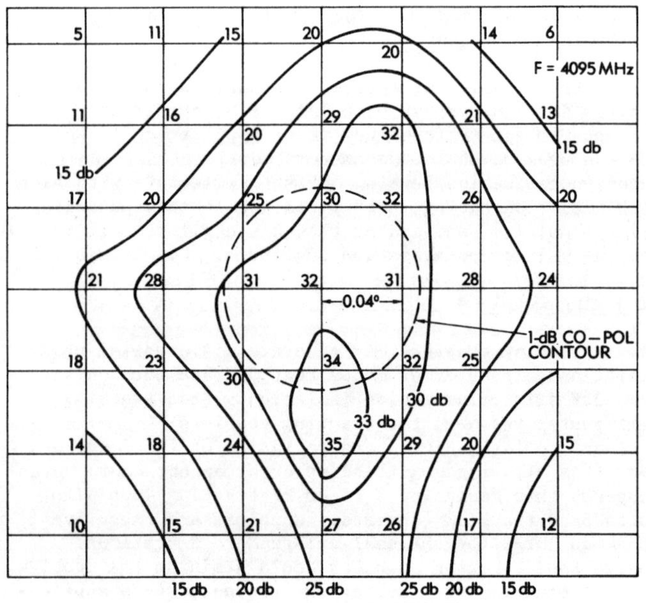

When all aspects of system planning, operation, and maintenance are considered, it may be preferable to build separate 4/6- and 11/14-GHz earth stations. However, the integration of an 11/14-GHz capability which is compatible with the dual polarized 4/6-GHz capability into existing reflector systems is being studied as an alternative.

There are two probable techniques for this integration: the design of a single broadband feed system with full 4/6- and 11/14-GHz capability, and the use of two separate feeds multiplexed together with a frequency selective reflecting (FSR) surface. Figure 13 shows a possible broadband feed solution which uses waveguide cutoff and low-pass filters to route the signals to the correct feed ports. An extremely broadband 4/14-GHz feed horn is also required.

FIGURE 13: BROADBAND FEED SYSTEM

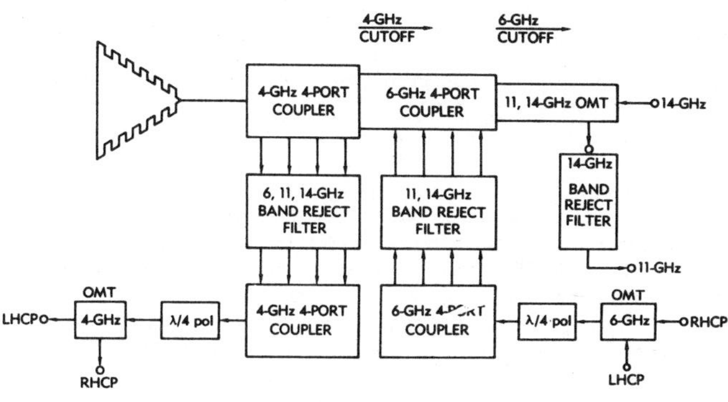

Figure 14 shows the use of an FSR in a beam wave-guide feed system to spatially multiplex the two bands. This concept simplifies feed and feed horn design because separate narrowband 4/6- and 11/14-GHz feed systems can be used. However, the FSR must reflect the 4/6-GHz signal without affecting its polarization quality and have low insertion loss to the 11/14-GHz signal. Numerous feed-FSR combinations are possible, depending upon the specific earth station configuration.

CONCLUSION

The INTELSAT V era represents a major change in earth station technology. Dual polarization at 4 and 6 GHz will require more complex feed systems. For the first time, accurate earth station polarization measurements will be required; these measurements can probably be made most effectively from a satellite source. The use of rainfall corrective networks largely depends on the required C/(N + I) ratios and the specific frequency plan; however, the technology to provide adaptive correction is being developed. Techniques for forming dual circularly polarized 4/6-GHz beams and linearly polarized 11/14-GHz beams on a common reflector surface are also being developed.

FIGURE 14: FREQUENCY SELECTIVE REFLECTING SURFACE IN A BEAM WAVEGUIDE FEED SYSTEM

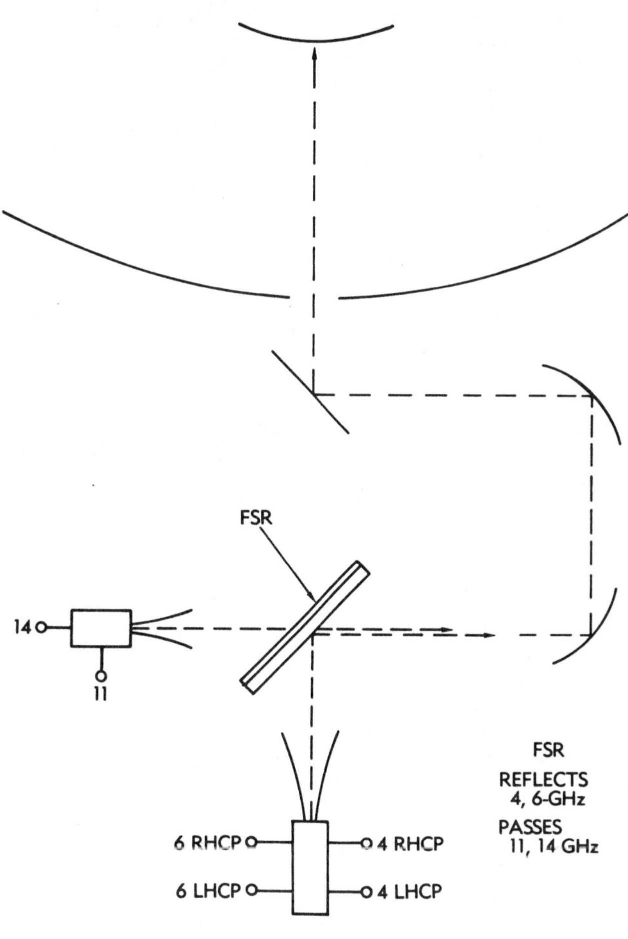

REFERENCES

1. R. W. Kreutel, et al., "Antenna Technology for Frequency Reuse Satellite Communications," Proc. IEEE, Vol. 65, No. 3, March 1977.

2. E. A. Robertson, "Communications Requirements for the INTELSAT V Spacecraft," 1976 IEEE Electronics and Aerospace Systems Convention, September 26-29, 1976.

3. W. J. English, D. F. DiFonzo, and W. S. Trachtman, "Polarization Diagrams for CP Antenna Analysis," 1976 International AP-S Symposium, October 11-14, 1976.

4. R. W. Gruner, "Compact Dual-Polarized Duplexers for 4/6 GHz Earth Station Applications," 1977 International IEEE AP-S Symposium, June 20-24, 1977.

5. Samir I. Ghobrial, "Cross-Polarization in Satellite and Earth Station Antennas," _Proc. IEEE_, Vol. 65, No. 3, March 1977, pp. 378–387.

6. D. J. Sommers, L. I. Parad, and J. G. DiTullio, "Beam Waveguide Feed with Frequency-Reuse Diplexer for Satellite Communication Earth Station." _Microwave Journal_, November 1975.

7. T. Oguchi, "Attenuation of Electromagnetic Waves Due to Rain with Distorted Raindrops," Pt. 1 and 2, _Journal of Radio Research Labs (Japan)_, No. 33, September 1960, pp, 467–485; No. 53, January 1965, pp. 19–44.

8. T. S. Chu, "Restoring the Orthogonality of Two Polarizations in Radio Communications Systems, I," _Bell System Technical Journal_, Vol. 50, No. 9, November 1971, pp. 3063–3069.

9. D. DiFonzo, A. E. Williams, and W. S. Trachtman, "Adaptive Polarization Control for Satellite Frequency Reuse Systems," COMSAT Technical Review, Vol. 6, No. 2, Fall 1976, pp. 253–284.

10. A. E. Williams and F. L. Frey, "Adaptive Polarization Control on a Dual-Polarized 4/6 GHz Satcom Link in the Presence of Rain," _Electronics Letters_, Vol. 12, No. 25, December 29, 1976, pp. 686–687.

11. D. DiFonzo and W. J. English, "Far Field Criteria for Reflector Antennas with Phased Array Feeds," 1974 IEEE AP-S Symposium, June 1974.

12. W. J. English and R. W. Gruner, "Earth Station Antenna System Depolarization Measurements with Boresight Ranges," _1974 International IEEE/AP-S Symposium Digest_, pp. 383–386.

13. R. Briskman et al., "Frequency Reuse Doubles Satcom Link Capacity," _Microwave System News_, Vol. 7, No. 6, June 1, 1977.

Status Report:
The Low-Cost Low-Capacity Earth Terminal

Here's an exhaustive study of a new international business area for RF
and microwave companies. Included are typical specifications for antennas,
HPAs, LNAs, SCPC equipment and TV receivers. Highlight of this report
is a no-nonsense, dollar-and-cents look at earth terminal costs as a
function of such factors as antenna size and EIRP.

Louis Cuccia & Carl Hellman
Aeronutronic-Ford
Palo Alto, California

During its approach toward maturity the communications
satellite earth terminal business has gone through three significant steps. During the 1960's and into the early 1970's, Comsat
and the International Consortium (Intelsat) built large antennas to meet a G/T of 40.7 dB, which provided CCIR quality
utilization of multiplexed voice channel super groups (up to
almost 1000 channels, or a high quality wide-band TV signal
received from any channel of the Intelsat IV satellite.) By mid-
1975 about 100 of these large antenna systems have been constructed in 59 countries.

Growth of the domestic communications satellite system
has provided satellites such as Anik and Westar that can place
spot beams into particular countries or regions for which communication is required. With an increase of approximately 8 dB
in receive signal power from Anik or Westar as compared to
the global beams of say, Intelsat IV, it is possible to utilize a
much lower cost 10-meter antenna with a relatively unsophisti-
cated uncooled paramp to handle almost 800 voice channels
through a standard 36-MHz satellite repeater.

Cost impact of the 10-meter antenna system can be appreciated by the fact that the $5M cost for an Intelsat-type termi-
nal has given way to around $500,000 cost for a 10-meter ter-
minal plus around $1000 per channel for the various multiplex
equipment used—or a cost savings to between 10 to 20% of the
equivalent global beam terminal cost.

The 10-meter antenna business is now well off the ground
and flourishing in various parts of the world. However, still
another business area with great dollar potential is coming into
focus. That is, the small 10–15 ft antenna terminal with G/T's
from 14 to 20 dB. Such terminals can receive and display a
medium quality TV image or can handle a small number of
voice channels (up to 12.) They are primarily intended for thin
root areas where the small communications capacity is sufficient to handle a variety of applications from educational television, teleconferencing, disaster warning and health and welfare communicaions, commerce and industrial communications, and so on. In each of these cases, the minimum capacity
requirements complement the low cost and small terminal that
is being developed to meet these needs.

Small earth terminals have an enormous future; some of the
counties of the world are now predicting requirements from
500 to 10,000 terminals as a means of introducing educational
TV into various remote areas and/or providing at least one
voice channel for local communication where up to now communication is virtually impossible. The impact of numbers of
terminals will drive costs even lower. A small 10–15 ft antenna

system—geared to a TV receive-only or to a variety of uses
from TV receive-only to TV transmit and receive, plus a small
number of voice channels—will cost from $5,000 up to $100,000.
As a result, the small earth terminal era will follow the development and implementation of the medium-sized terminal and
promises to become a major industry during the next decade.

One application of the use of a small earth erminal is illustrated in Figure 1, which depicts a Hughes HS721 remote communications terminal on line with the Canadian ANIK satellite from Richards Island in Northern Canada. This C-band
terminal uses a 10-ft parabolic dish, a 10-watt TWT HPA and a
transistor amplifier with a 4.5 dB noise figure. It is built to provide a single 4-wire telephone channel.

Types of Small Antenna Earth Terminals. The large
capacity of the 97-foot Intelsat earth terminal enables it to
handle 1000 voice channels from each satellite channel and
simultaneously transmit or receive wideband high quality TV
images. On the other hand, the small antenna terminal provides a variety of low capacity services to its users. Types of
small antenna earth terminals include the following:

- *Two-way TV and telephony:* a small terminal that will allow
  both the transmission and reception of TV in addition to the
  transmission and reception of a number of voice channels.

- *Exclusive two-way TV:* a small earth terminal that will handle
  a single television channel on a receive and transmit basis.

- *TV receive-only:* this will probably be one of the more important of the terminals provided by this new business area since
  it serves the educational TV requirements of many parts of
  the world.

- *TV receive only and telephony transmit and receive:* this type
  of terminal will provide both telephony communication and

Figure 1. A small C-band earth Terminal designed by Hughes
for remote locations. *Courtesy of D. Rasmussen.*

Reprinted with permission from *Microwave Syst. News,* vol. 5, pp. 19-20, 22, 26, 30-31, 36-40, 42, June/July 1975.

educational TV for many of the small hamlets of the world.

• *Telephony receive and transmit only:* this type provides much needed communications using very small terminals, well under 10 feet in diameter, which can be built at very low cost.

In the handling of telephone communications, the 10- and 30-meter high-capacity antenna systems can handle super groups of voice channels up to 1200 channels and for the most part interface with the multiplexed FM systems of terrestrial communications and most of the public telephone organizations of the world. The very small antenna earth terminal, on the other hand, will in general utilize, in its thin root varieties, single channel per carrier (SCPC) service, which will make more efficient use of the channel of each domestic communications satellite and which will handle up to 12 voice channels either upon pre-assignment or upon demand assignment.

It is important to note that the commerical 10- and 30-meter high-capacity antenna earth terminals must meet all of the critical CCIR standards of local distribution systems in order to introduce received telephone messages or TV transmission into local regional telephone or broadcasting systems. In contrast, the small antenna system usually interfaces directly with the user and the requrements for voice intelligibility and TV picture quality may be considerably relaxed.

Small Terminal Design. Figures 2 and 3 show, respectively, the system design and capabilities of a small terminal utilizing a 4.5-meter reflector which is being built by Aeronutronic-Ford for system test of single channel per carrier transmission using either ANIK or WESTAR satellites.

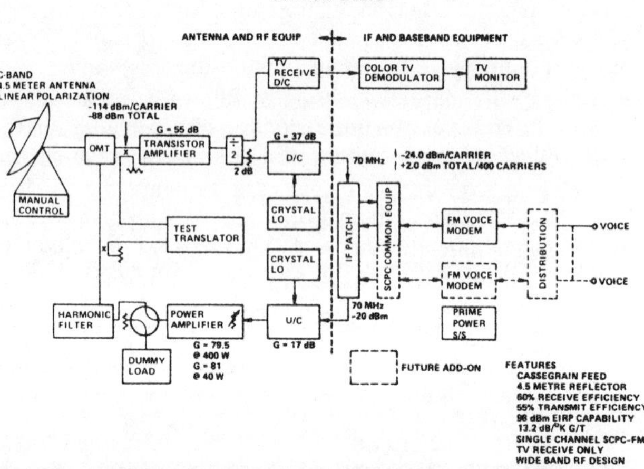

Figure 2. System design of an Aeronutronic-Ford 4.5 meter (15 ft) C-band terminal for TV-receive and SCPC (FM) voice communications.

As shown in Figure 2, a 400-watt power amplifier is utilized to drive the antenna for up-link while a transistor amplifier is utilized as the low noise amplifier. This particular system also includes a separate color TV demodulator and TV monitor and SCPC FM voice modems supplied by California Microwave Inc.

As depicted in Figure 3, the small terminal can handle up to 32 carriers utilizing a 400-watt power amplifier and an uncooled paramp. By utilizing 2400 bits per second PSK carrier with ¾ rate coding, up to 40 carriers can be utilized with a transistor amplifier (transamp) as the low noise amplifier.

Note that the expected TV performance varies from fine to passable and, in general, provides service that is roughly the equivalent of the TV quality found in most U.S. homes without the presence of degradation or multipath effects or interference, which is usually typical of commercial TV transmissions

MSN: JUNE/JULY 1975

SYSTEM CAPABILITIES

| | | TERMINAL CARRIER CAPACITY | | | | | | EXPECTED TV RECEIVE PERFORMANCE | | | | | |
| | | NARROW BAND FM VOICE | | | 2400 b/s PSK DATA (10^{-8} b.e.r., RATE-3/4 CODING) | | | BLACK AND WHITE TV | | | COLOR TV | | |
SATELLITE TYPE	TERMINAL G/T	10-W PA	100-W PA	400-W PA	10-W PA	100-W PA	400-W PA	S/N	GRADE	QUALITY	S/N WTD (C)	TASO GRADE	QUALITY
ANIK/WESTAR	13.2 dB/°K (WITH TA)	1 CARRIER	2 CARRIERS	8 CARRIERS	1 CARRIER	10 CARRIERS	40 CARRIERS	43.8	3 (d)	PASSABLE EQUIVALENT TO FCC CLASS B (RURAL) BROADCAST (D)	40.1	2	"FINE" FOR 50% OF VIEWERS
	20 dB/°K (WITH UNCOOLED PARAMP)	1 CARRIER (b)	8 CARRIERS	32 CARRIERS	8 CARRIERS	85 CARRIERS	280 CARRIERS	50.6	2	FINE EQUIVALENT TO FCC CLASS A (URBAN) BROADCAST	46.9	1	"FINE" FOR 99% OF VIEWERS "EXCELLENT" FOR 50% OF VIEWERS
INTELSAT IV (GLOBAL)	13.2 dB/°K (WITH TA)	(a)	(a)	(a)	(a)	(a)	1 CARRIER						
	20 dB/°K (WITH UNCOOLED PARAMP)	(a)	(a)	(a)	(b) 0 CARRIER (b)	1 CARRIER	4 CARRIERS						

(a) NOT EVALUATED

(b) LIMITED BY TERMINAL TRANSMITTER POWER

(c) PEAK TO PEAK VIDEO SIGNAL TO RMS WEIGHTED NOISE RATIO

(d) S/N WTD = 45 dB CORRESPONDS TO INDICATED GRADE AND QUALITY, HENCE ABOUT 1 dB BELOW INDICATED GRADE AND QUALITY

NOTE TABULATED DATA BASED ON 45% ANTENNA EFFICIENCY EFFICIENCIES NOW PROJECTED TO BE 55% TRANSMIT, 60% RECEIVE

Figure 3. System capabilities for FM voice, 2400 b/s PSK and TV reception for the terminal of Fig. 2 using various available satellites.

from a high antenna site into a metropolitan community.

Some New Domestic Satellite Systems. The domestic satellite communication business has been aided greatly by the success of the Canadian satellite, ANIK, and the U.S. satellite, Western Union's WESTAR, both of which were designed and built by Hughes.

Very shortly a number of new domestic C-band (5.925 to 6.425 GHz up-link, 3.7 to 4.2 GHz down-link) satellites will appear on the scene. These satellites include: 1) a Maritime satellite that will operate with two-way voice communications at UHF, L- and C-bands; 2) the RCA SATCOM system, which will employ three low-weight high-capacity satellites, each having 24 channels 36 MHz wide to beam all 24 channels to Alaska, the contiguous 48 states, and 12 channels to Hawaii; and 4) the Indonesian communication satellite which at C-band will link 5,000 inhabited islands. Figure 4 shows some of the antenna sites involved.

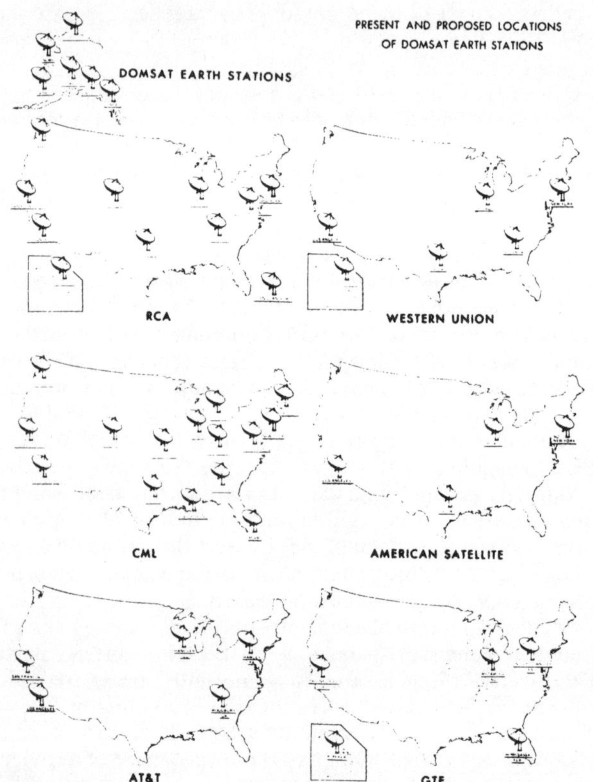

Figure 4. Present and proposed location of various domestic satellite communication terminals, circa 1974. *Courtesy of Sid Topol, Scientific-Atlanta.*

By 1977 Japan will fly the experimental communications satellite now being built by Aeronutronic-Ford to provide a high capacity domestic communications system utilizing the 18- and 30-GHz frequency bands. Within the next few years, the 11- and 14-GHz frequency bands will be implemented by the European orbiting test satellite now destined for 1977 and by a satellite now in the plans of the CML Satellite Corporation, which according to its FCC filing, will transmit high-speed digital data accumulated from a variety of users into single digital carriers that are then transmitted via the satellite.

Some countries, such as Algeria and Norway, will not be putting up their own unique satellites. Algeria, for example, will lease one channel of one of the Intelsat IV-A satellites, and Norway has leased a portion of an Intelsat IV channel to serve the numerous Arctic oil drilling installations now in operation. Some of the U.S.-based antenna systems planned for use with the U.S. domestic communication satellites mentioned above are presented in Figure 4. Significant new earth terminal systems are now being planned with some of these aforesaid new satellites. These include: 10-meter antennas being built for the RCA SATCOM system; 30-meter antennas being built by Aeronutronic-Ford for the AT&T, COMSTAR satellite; and 12-meter antennas being build in Japan for use with the Japanese communications satellite.

Figure 5. This "footprint" of the Indonesian satellite shows how radiated power can be focused on a specific region, thus enabling communication with small terminals.

The latest major effort on the domestic communications satellite scene is the Indonesian system (Figure 5) for which Hughes Aircraft, ITT, and Aeronutronic-Ford are presently supplying a total of 39 ground terminals in the 10-meter variety, which will operate with two satellites (of the ANIK/WESTAR class) being built by Hughes. Following the implementation of these earth terminals and satellites within the next year and a half, it is expected that Indonesia will then go into a procurement of hundreds of small 15-foot antenna terminals to beam educational TV into outlying islands that are not now reachable by even the presently configured domestic satellite net.

In addition, the state of Alaska is presently procuring a number of terminals in C-band utilizing ANIK or RCA GLOBCOM; these terminals—reportedly from 100 to 150 will be purchased—will be both 10- and 15-foot in diameter, will utilize low noise amplifiers with 90–300° noise temperatures and be built to handle SCPC transmission at a variety of sites in Alaska.

Another application of communications satellites soon to be implemented by Comsat General is the MARISAT system,

which involves Atlantic and Pacific L-band satellites being built by Hughes. It will use leased Comsat General shipboard terminals built by Scientific-Atlanta which will feature 4-foot antenna dishes and be capable of handling single channel per carrier FM for voice between the shipboard terminal and a ground terminal in continental United States at Santa Paula, California, or Southbury, Connecticut via the proper satellite. The MARISAT system will also be able to handle low data rate PSK digital communications for teletype on a SCPC mode between ships utilizing the Comsat General marine and shore terminals.

TV Broadcasting from Space. In addition to the telephony and TV utility of a domestic communications satellite, special satellites can be built that provide television broadcasting from space. Specific frequencies are allocated for such TV broadcasting, including 2.54 GHz and the band between 11.7–12.2 GHz.

One advantage of such direct broadcast systems from space is that transmitters of fairly high power can be installed in the spacecraft. Watkins-Johnson has developed a traveling-wave tube at 2.54 GHz that can apply between 50 and 100 watts into the satellite antenna; several companies have developed special high power TWTs that can be utilized at 11.7 GHz for space broadcasting. These companies include Hughes Electronic Dynamics Division, which has developed a 100-watt high efficiency TWTA; Litton Industries, which has developed a special 200-watt TWTA; and Siemens of Munich, Germany, which has developed a tube in the 500–700 watt range.

By developing such a high power tube in the satellite, consistent with a high gain satellite antenna, the effective radiated power (ERP) can be so high that receiving terminals can be built to utilize dishes with diameters in the 1–2-meter range with LNAs of very modest performance.

Typical of this technology has been the so-called Rocky Mountain Experiment, which utilized the ATS-F satellite that beamed down into the Rocky Mountains a telecast at 2.45 GHz which had originated from NASA's Rosman antenna broadcasting up-link at C-band. For this application, Westinghouse developed a special low-cost antenna terminal, including: a 10-foot plastic dish made by the Prodelin Company; a special TRF RF amplifier and video receiver at 2.54 GHz (without down conversion) developed by Hewlett-Packard. Over 150 of these special low-cost antenna terminals were used, involving costs in the 1973 era of approximately $3000–$5000.

A major milestone will soon be achieved when the Canadian Technology Satellite (CTS) will be orbited later this year. This satellite operates at approximately 11 GHz and represents a joint effort between the Canadian government and NASA. CTS will utilize a 200-watt tube, especially developed by Litton Industries, and a large selection of experimentors has been delegated in the U.S. to provide meaningful experiments that can demonstrate the utility of this satellite for direct TV broadcasting and transmission.

U.S. experimentors that will utilize the Canadian satellite, include Comsat Labs, which will study its use for disaster and emergency occasions in cooperation with the American Red Cross. Others on the American team are the Dept. of Health, Education and Welfare (HEW) the Veterans Administration, the Association of Western Hospitals, Stanford University, the Archdiocese of San Francisco in combination with the Pasadena Public School System and the Archdiocese of Salt Lake City, and many other medical and public health facilities.

These participants expect to procure and operate between 50 and 100 small 11-GHz antenna terminals, which will be of the order of 10 feet in diameter, will utilize HPA's from 400 watts to 1.5 kW, and will vary between having a full utility of two-way television receive-transmit to operating as a TV receive-only terminal. Some experimentors in the vicinity of San Francisco expect to utilize the local cable TV as a means to route the information received from the space broadcast.

One interesting aspect of the CTS experiment is that Westinghouse will not only provide a significant experiment in terms of efficient and effective means of utilizing separate conference rooms—in what is called teleconferencing—but will also provide a special low-noise mixer/down-converter operating from 11-GHz to 2.54-GHz, thereby allowing the present ground terminals designed for the ATS-F satellite to be useable with the CTS.

It is expected that the use of the CTS in both Canada and the U.S. will demonstrate the use of such a satellite from the standpoint of making possible public health, medical and health service conferences, which will not only allow doctors in one hospital to see the EKG's and the color slides of cancer segments as made in another hospital, but will also allow lectures given in one place by a public health officer or by a doctor to be received and utilized for education in other comparable facilities.

Teleconferencing is believed to be one of the great outgrowths of TV broadcasting from space, and is seen as one means of reducing the travel now required by many public health officials and also of providing more rapid dissemination of valuable information that is produced by the research into a variety of diseases presently being studied.

Small Earth Terminal Equipment. Four procurements are in the process of crystalizing the hardware availability of the small earth terminal market. These include—1) the precursor HEW procurement of 150 TV receive-only ground terminals for the Rocky Mountain experiment; 2) the three new procurements include the 11-GHz TV receive-only terminals for the Canadian Technology Satellite; 3) roughly 100 small SCPC terminals being procured by the Alaskan state government; and 4) 200 marine terminals for the MARISAT system.

Tables I through V list some of the principal specifications of the five key elements of a small earth terminal: i.e., the antenna; the HPA or high power amplifier; the LNA or low noise

Figure 6. Photograph of the Andrew 4.5-meter antenna, which is capable of limited motion. *Courtesy of Bob Kasik.*

amplifier; the single channel per carrier equipment, which includes the up/down converter; and a TV receive system.

Antennas. Note in Table I the small earth terminal specifications include a 15-foot antenna for C-band and a 10-foot antenna for low K_u-band. In general, the use of the 10-foot antenna in C-band has been seriously questioned by the FCC in terms of the potentiality for interference between co-located antennas and the 15-foot dish for land use would appear to be the principal candidate for minimum capacity applications. At

Table I
Typical Antenna Specifications

	15 feet	10 feet	4 feet
Reflector Size	15 feet	10 feet	4 feet
Frequency (GHz)			
Receive	3.7–4.2	11.7–12.5	1.535 –1.5435
Transmit	5.925–6.425	14.0–14.3	1.6365–1.645
Midband Gain (dB)			
Receive	43.7	49	23.5
Transmit	46.8	50.4	
VSWR Max	1.3:1	1.5:1	1.5:1
Half Power Beamwidth (°)			
Receive	1.2	.57	10
Transmit	0.92	.49	
First Sidelobe	−15 dB	−17 dB	
Average Transmit Power (kW)	2	1.5	G/T = −4 dB
Wind Loading (mi/hr)			
Operate	60	50	Radome
Survival	125	110	
Noise Temperature			
10° El	35°	25°	>100°K
40° El	20°	10°	

Representative Manufacturers

Metal—*Andrew, Datron, Aeronutronic-Ford, Radiation Systems, Inc., Scientific-Atlanta, Commco, NEC*

Plastic—*Prodelin, Ancom*

11- and 14-GHz, the 10-foot antenna gives promise of being the prime candidate, supplying around 50 dB of gain in both receive and transmit modes. Note that the half-power beamwidth for both the receive and transmit modes of the 15-foot C-band antenna is around 1°. Since most satellites of the WESTAR or ANIK type are maintained in a position that varies no greater than ±0.3°, it is evident that the small antenna can be operated without requiring significant auto-tracking systems. simple manual tracking will suffice for most cases. Accordingly, such antenna structures are provided with mechanical tracking features which are, in some cases (CTS Terminals), remotely controlled. The 4-foot antenna operating in L-band is for the Marisat system.

Figures 6 through 10 show some of the representative antennas that are being built to serve the small terminal market. Figure 6 shows an Andrew antenna approximately 15 feet in diameter (4.5 meters). The simplicity of this antenna and the moderate pedestal structure, which is keyed to three concrete blocks, indicates the vast structural simplification that this antenna has achieved over the corresponding large antenna with its huge pedestal, counterweight and base-complex.

Figure 7. The Prodelin 10-ft plastic small antenna used in the Rocky Mountain educational TV experiment. *Courtesy of Mr. Hasselman.*

Figure 7 shows the 10-foot plastic antenna built by Prodelin for Westinghouse in the HEW experiment. This antenna illustrates the new plastic technologies presently being used by both Prodelin and Ancom for manufacturing small antennas. Figure 8 pictures the MARISAT fully-stabilized marine satellite antenna, which is manufactured by Scientific-Atlanta. This antenna, with a 4-foot dish in combination with a 35-watt solid state HPA and a low-noise transistor LNA, operates on a 4-axis slaved pedestal that is controlled by gyroscopes sensitive to the ship's rolling motion. The MARISAT antenna will operate under a radome and 200 of them have been manufactured for lease by Comsat General to marine users who will expect to use the new maritime satellite.

In Figure 9 is shown a special L-band antenna built by G. P. Heckert (now Aeronutronic-Ford) of Automated Marine International in Newport Beach for use on a Coast Guard cutter to determine the problems of satellite reception aboard ships using a signal received from the ATS-6 satellite. This antenna complex utilizes a unique short-backfire antenna in a 16-inch

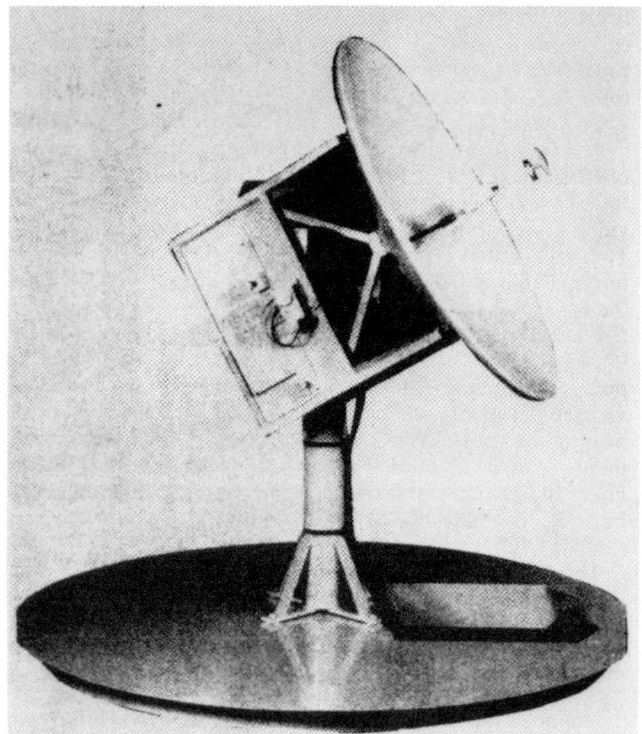

Figure 8. 4-ft diam Marine satellite communications antenna built by Scientific-Atlanta.

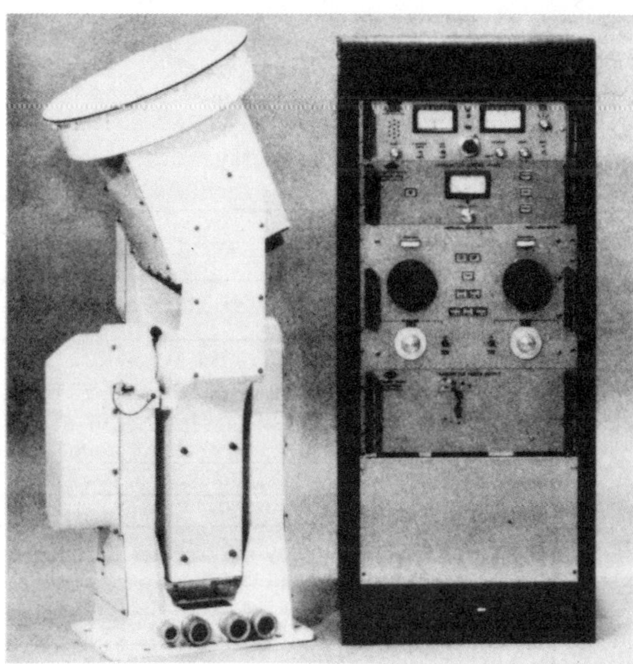

Figure 9. AMI L-band back-fire antenna designed for shipboard experiments involving 1500 MHz reception from ATS-F. *Courtesy of G. Heckert.*

dish. This antenna has between 12 and 14 dB gain at 1540 MHz, an aperture efficiency of 70% and a beamwidth of 35°. It is circularly polarized and represents one of the most simple antennas that can be built for such applications.

Figure 10 illustrates the Comsat Laboratories Unattended Earth Terminal (UET) that has been designed to demonstrate a unique approach to small antenna complexes, which can work simultaneously with a number of satellites and which can

Figure 10. The Comsat Labs unattended Earth Station which uses a single stationary antenna and several remotely-controlable feeds. *Courtesy of P. Bargellini.*

be utilized with a minimum personnel. The hut encloses a number of feeds with independent electronic systems. These feeds can be moved around inside the hut to track different satellites within 10° to 20° of orbital arc. This gives the station operator the ability to move from one satellite to another and provides the unique function of multiple satellite operation with remote control and monitoring.

HPA Systems. Table II shows typical HPA specifications for preset equipments that have been developed for the low power small antenna system.

The MARISAT system at 1500 MHz will utilize power transistor amplifiers at the 35–50 watt level. These amplifiers are now available from a variety of companies. Amplifiers at this power level have been used in the ATSF satellite at this frequency by Aeronutronic-Ford.

Note in Table II the variety of HPA specifications that can be achieved, including HPA's in the 10–40 watt, the 400–600 watt, and in the 1.5–3 KW range. These HPA's have gains typically of 60–70 dB and are designed to conform to gain and group delay specifications derived from the Intelsat and Terrestrial Radio Relay systems presently utilized for handling multiplexed FM.

The 10–40 watt HPA is a fairly small package as evidenced by Figure 11. This shows a Hughes 35-watt TWTA in combination with an up/down converter. As the HPA output rises to 400-watts, a considerable amount of instrumentation must be built in to safeguard the tube. Figures 12 and 13 show HPAs manufactured by Comtech and Varian, respectively. These

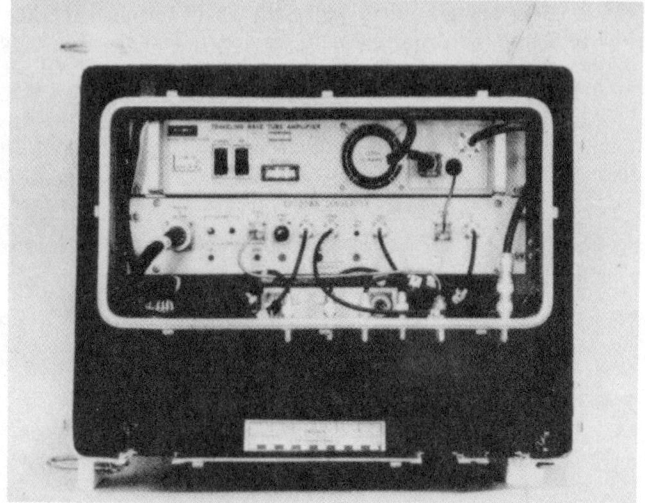

Figure 11. Housing of the Hughes low power (10 to 40 watts) HPA including the up and down converter in Fig. 1. *Courtesy of D. Rasmussen.*

	Table II		
Typical HPA Specifications			
Power (Saturated)	40 watt	400 watt	1.5 kW
Frequency Range (GHz)	5.9–6.425	5.9–6.425	5.9–6.425
1-dB Bandwidth	(Broadband)	(Broadband)	TWT—Broadband
			Klystron—45 MHz Tunable
HPA Gain Plus Driver Amp	60–70 dB	60–70 dB	60–70dB
Input Driver Power (dBm)	0–5	0–5	0–5
RF Level Adjust (dB)	20	20	20
Gain Stability (dB/month)	0.5	0.5	0.25
Gain Slope (dB/MHz)	0.3	0.3	0.65
Spurious Output in any 4-kHz Band in Transmit Band	−65 dBw	−65 dBw	−65 dBw
Load VSWR	1.5:1	1.5:1	1.5:1
Input VSWR	1.2:1	1.3:1	1.3:1
Group Delay			
Linear	±0.5 ns/MHz	±0.5 ns/MHz	±0.5 ns/MHz
Parabolic	0.05 ns/MHz^2	0.05 ns/MHz^2	0.05 ns/MHz^2
Ripple	2 ns p-p	2 ns p-p	2 ns p-p

Representative Manufacturers

10–40 watt TWT–*Hughes, Varian*
400–600 watt TWT–*Litton, Watkins-Johnson, MEC, Varian*
1.5 kW TWT–*Thomson, Varian*
1.5–3 kW Klystron–*Varian, NEC*
HPA's–*Comtech, Varian, ESI, Cober*

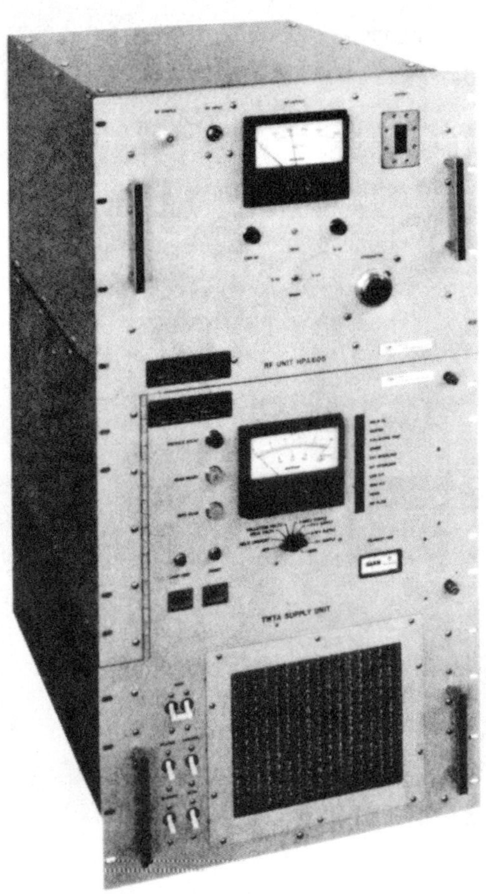

Figure 12. ComTech C-band transmitter (HPA) for 400-watt operation. *Courtesy of J. Greene.*

Figure 13. Varian 400-watt HPA with 75 db gain. *Courtesy of M. Wytyshyn.*

are air-cooled systems; air-cooled tubes can be built up to 3 kW. Beyond this power water cooling must be utilized. Table II includes some of the specifications for both 1.5 kW TWT and klystrons. These spceifications include the use of the IPA (intermediate power amplifier) that brings the gain up to the 60–70 dB range.

As long as powers in the 40-watt to 3-kW range are required, TWTs or klystron-type devices will be utilized. However, below 10 watts, many new devices are appearing on the market and are presently finding their way into existing radio relay

Table III
Typical LNA Specifications

	Thermoelectric Cooled Paramp	Hot Paramp	FET Amplifier
Frequency Range (GHz)	3.7–4.2	3.7–4.2	3.7–4.2
Noise Temperature (°K)	45–55	90–150	200–250 Wideband
	(50)	(100)	150 in 40-MHz BW
Gain (db)	60	60	60
Gain Stability—dB per day	0.5	0.5	0.5
dB per month	1.0	1.0	1.0
Ambient Temperature (°C)	−50 to +50	−50 to +50	−50 to +50
0.5 dB Gain Compression, max (dBm)	−60	−60	−60
Group Delay			
Linear delay in any 40 MHz band	4 ns	4 ns	4 ns
Ripple p-p	0.5 ns	0.5 ns	0.5 ns
Inter-mods			
Two equal input carriers −64 dBm	60 dB down	60 dB down	60 dB down

Representative Manufacturers

50° Paramp—*LNR, AIL, SCI, NEC, Comtech, LCT–Paris*
100° Paramp—*LNR, AIL, NEC, Comtech, Micromega*
FET Amp—*Amplica, Ancom, Avantek, W-J, IMC, Varian*

systems. These devices include Impatt amplifiers and multiplier systems. It is important to realize from the standpoint of system performance and reliability, that the replacement of the TWT, which presently has a lifetime of the order of 20,000 hours or more, will only occur with the availability of devices of significantly operating lives and lower cost. Such replacement has not occurred yet except in special AT&T and Bell-Northern systems.

LNA Systems. Table III shows typical LNA specifications applicable to small antenna terminals. Three devices are used: the thermoelectrically-cooled paramp, the "hot" paramp (which operates at elevated temperature of about 58°C); and the FET amplifier.

Thermoelectrically-cooled paramps operate with a paramp diode and circulator cooled to a temperature range between 0°C and −50°C. Such a paramp also utilizes a Gunn diode pump source, operating at approximately 60 GHz to provide noise temperatures in the 45–55°K range. Such paramps have special circulator developments that not only have extremely low loss but also have significant reverse isolation to prevent deterioration of noise temperature due to the possibility of 1.5:1 antenna VSWR, which, for a 50° paramp, could degrade the noise temperature to almost 70°K. Such paramps are presently in use in most Intelsat earth terminals of the latest vintage and have been installed by Aeronutronic-Ford in the TT&C terminals presently under construction for Comsat General for use with both the AT&T Comstar satellite and the maritime satellite.

The "hot" paramp, utilizing a 60-GHz pump, can provide noise temperatures between 90°K and 150°K, depending upon the system's complexity. Such amplifiers are usually followed by transamps to make up the 60 dB gain requirement. Hot paramps are, in general, very reliable units and are designed for external environments with minimum maintenance.

The Comtech low noise paramp of Figure 14 represents the type of commerical hardware that can be obtained in today's market.

Figure 14. C-band ComTech paramp designed for out-of-doors operation on the rear surface of the antenna dish. *Courtesy of J. Greene.*

The ultra low-noise FET amplifier is the result of pioneering work by Mr. Charles Abronson of Amplica and has become a major factor in the concept and procurement of LNAs for small earth terminals. Mr. Abronson was the first to realize the low noise potential of the field-effect transistor and built FET amplifiers with roughly 2.5 dB noise figure (approximately 220°K noise temperature) across the full 3.7–4.2 GHz band. By utilizing thermoelectric cooling of the FET amplifier module, he achieved a noise temperature of under 2 dB when operated at a device temperature of −50°C.

Such FET amplifiers are presently built using reliable one-micron FET's. With the introduction of the half-micron FET, it is entirely possible that within a year thermoelectrically-cooled FET amplifiers will be available to the 100°K noise temperature range and the prospects of such amplifiers for 12 GHz, where 2 dB noise figures have already been achieved with experimentally cooled FET amps, is starting to be recognized.

One advantage of the FET amplifier, as exemplified by the Amplica unit in Figure 15, is its extremely small size. It is obvious that the FETamp has size, weight, and economic factors that are in its favor.

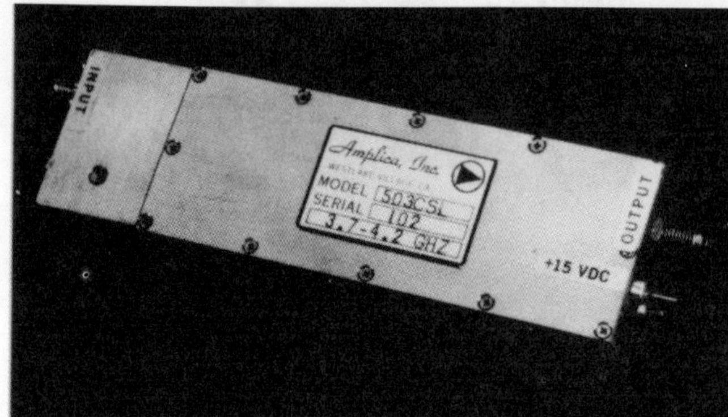

Figure 15. Amplica GaAs FET amplifier. *Courtesy of Charles Abronson.*

At present, the equivalent hot paramp and the FET amplifier have a cost differential of approximately two to one. However, according to M. Lebenbaum of AIL, it will soon be possible to manufacture in quantity 150° hot paramps at a cost of approximately $4500, which would make the hot paramp economically competitive with the FET amplifier in the small earth terminal market.

Ground Control Equipment. In most small earth terminals the up/down converter is integrated into the signal processing equipment. In larger terminals, such as an Intelsat terminal or the military AN/MSC-60 terminal for receiving the DSCS II satellite, the up/down converter must be equipped to tune to any one of a variety of satellite channels. In some cases this tuning must be performed quickly and accurately to within one kHz in order to provide suitable reception for a variety of complex signal modulations, including spread spectrum.

In most small antenna terminals, tunability over a variety of channels simply is not required. In order to receive from a single satellite channel, it is economic to use a filter for that particular satellite channel and then utilize simple single conversion techniques to convert down to an IF frequency of say, 130 MHz or 70 MHz. The same follows for the up-link.

Accordingly, small terminal ground control equipments have combined the up and down converter functions with the signal processing systems to provide equipments such as the Scientific-Atlanta receiver (pictured in Figure 16) which will be used in many terminals in Indonesia and which is the prototype receiver of the MARISAT system. Figure 17 shows the Comtech Laboratories receiver that has been built for Intelsat-type terminals and will be used by Aeronutronic-Ford in the 15

Figure 16. Scientific-Atlanta satellite communication receiver. *Courtesy of Sid Topol.*

Figure 17. ComTech C-band earth terminal receiver which includes down conversion and a threshold demodulator. *Courtesy of J. Greene.*

terminals now being built for the Indonesian communications systems. Both the Scientific-Atlanta and the Comtech receivers are adaptable for FM/FDM communications up to 800 or 900 voice channels in addition to being adaptable to SCPC applications.

Figure 18 shows a Comtech enclosure, which is actually a modular packing container utilized by either a DC-8 or a 747 aircraft. This enclosure is designed to include all of the key small earth RF equipment and even includes space to package a disassembled antenna. With such an enclosure, an entire small earth terminal can be shipped anywhere in the world aboard a commercial jet aircraft.

Table IV shows typical single channel per carrier equipment specifications while Table V shows typical television receiver specifications.

SCPC Hardware. Single channel per carrier is much more efficient than FDM/FM from both the cost and utilization spectrum standpoint for thin root systems and can be expected to become a major communication technique for areas that involve less than 12 telephone users. It is difficult at this time to exactly specify any SCPC equipment, its configuration will be dependent upon the specific system, the G/T of the particular antennas being used and the voice quality required in the terminal user. On the other hand, Table IV includes characteristics specified for the Alaskan system which relate to both the use of FM transmission and also to PSK transmission. More detailed specifications are being established by such individuals as Dr. J. Campanella of Comsat Labs and Dr. W. Pohlman, consultant to the Alaskan System.

Single channel per carrier equipment presently developed by California Microwave for the Algerian domestic communication satellite is shown in Figure 19. Whereas this system is more complex than one would find in most thin root systems, it indicates the type of instrumentation that is typical of SCPC modems.

TV Receivers. Table V specifies performance for a major product area, namely the television receiver that can be utilized to receive TV broadcasts using double sideband FM from a satellite.

Figure 18. Transportable ground terminal in which all electronic receiving and transmitting equipment is permanently housed in a standard FAA air cargo container. *Courtesy of J. Greene, ComTech.*

Table IV
Typical Specifications
Single Channel Per Carrier Equipment (SCPC)

- SCPC equipment includes one up/down converter.
- Exact specifications are difficult to provide at this time.

The Alaskan system has specified the following signal-to-noise objectives:

1) For a nominal carrier-to-noise density ratio (C/N_o) of 55 dB/Hz, the voice circuit performance will be subjectively equivalent to a 10,000 picowatt c-message weighted reference circuit.

2) In an FM system, the test tone-to-noise ratio whall be 33 dB with C-message weighting and pre-emphasis for a C/N_o of 55 dB/Hz. Curve showing test tone-to-noise ratios for 49 dB C/N_o ≤61 dB should be supplied.

3) In a PSK system the objective is that the signal-to-quantizing noise ratio be 30 dB or greater for a 1-kHz test tone at a −10 dBm 0 input level for a BER of 10^{-3}. This performance must be met for a BER of 10^{-4}. A curve showing signal to quantizing noise ratios for 49 dB ≤C/N_o ≤61 dB should be supplied.

Representative Manufacturers
Digital-Delta Mod/PSK—

FM—

Figure 19. FM single channel par carrier (SCPC) system manufactured by California Microwave Inc. *Courtesy of P. Smith.*

Table V
Typical Television Receiver Specifications

Input Frequency	3.7–4.2 GHz
Input Power Range	−85 dBm to −60 dBm
Intermediate Frequency	120 MHz
Video Frequency Response	±0.5 dB from 50 Hz to 4.2 MHz ±1 dB max. at 4.5 MHz
Video De-emphasis	CCIR recommendation 405-1 for 525 line system
Video Output Level	1 volt ±3 dB peak-to-peak
Audio Output Level	8 dBm ±2 dB into 600 ohms for p-p deviation 200 kHz
Audio Frequency Response	±2 dB, 50 Hz to 7500 Hz ±3 dB, 7500 Hz to 15,000 Hz

Representative Manufacturers

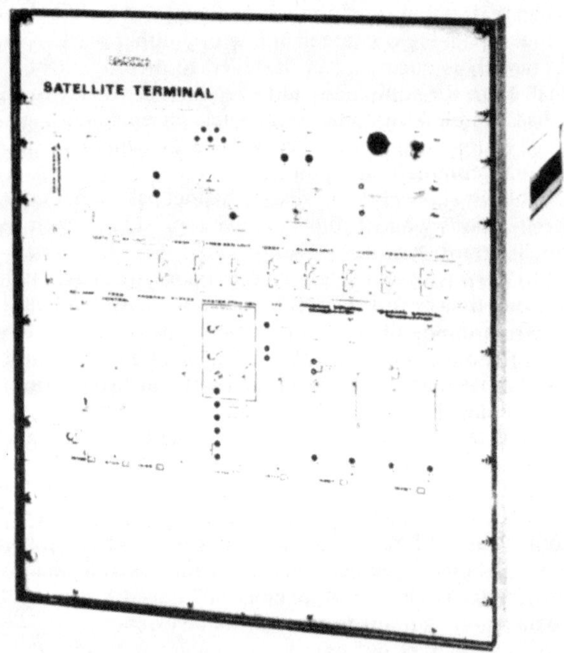

Figure 20. Receiver for TV receive-only applications in C-band satellite earth terminals by California Microwave. *Courtesy of P. Smith.*

TV receivers were first developed by Hewlett-Packard to operate at 2.54 GHz with the Rocky Mountain experiment. Since then, other companies—notably Varian, Microdyne, California Microwave and terrestrial communication companies such as Lenkurt—are offering low cost($4.5–$6.5K) TV receivers that can be utilized to follow the paramp in either a C-band or a K_u-band terminal and provide both video and audio outputs which can be utilized to drive a TV receiver.

Figure 20 shows the satellite terminal television receiver Model SR42 (developed by California Microwave), which is tunable.

Small Antenna Pricing Considerations. An early aspect of the small antenna terminal business that caused many problems was a publication of unrealistic prices against which such terminals could be built. During recent months, particularly under the influence of the Alaskan and CTS systems, many responsible manufacturers have now given cost estimates for a variety of equipments that more clearly indicate the ranges of costs which are representative of such terminals, particularly in 100 terminal quantities.

Consider the C-band single channel per carrier terminals, it is now possible to project a per-terminal cost of around $50,000 (in quantities of 100) which represents an equipment cost of almost $40,000 and an installation cost of approximately $10,000. This is for a terminal with a single voice channel. With the inclusion of additional voice channels and with TV reception, it is evident that the cost of such a terminal can quickly grow to in excess of $100,000.

Some preliminary cost estimates have already been made for terminals of the 10-foot variety for use in the Canadian technology satellite. These cost estimates include approximately $7000–$8000 for TV receive-only terminals in quantities of 100. Costs of approximately $20,000 are typical of TV-receive and telephony-transmit terminals in quantities of 100. Costs of approximately $75K in quantities of ten would be representative of two-way television terminals; and approximately $200K would be required of each of a small quantity to two-way TV with some telephony transmit-receive capability.

These receive-only costs are for a fairly sophisticated terminal. Toshiba of Japan projects a $500 cost for a 1.5-meter TV receive terminal (12-GHz) for reception of signals from the upcoming Japanese TV broadcast satellite.

Figure 21 shows a variety of costs representing quantities of around 100 units for the antenna, HPA, LNA, and single channel per carrier including up and down converter units.

Note in Figure 21(a) that a typical antenna cost for a 15-foot dish is approximately $7.5K. As the antenna diameter increases to 32 feet, the cost increases to almost $50K. The greater size represents a significantly larger overall structure, necessitating additional expense to precisely place the various antenna panels. In addition, shipment and on-site erection are more costly. However, as the antenna size decreases under 10 feet, the cost of the materials and its pedestal becomes less of a factor than the costs of administration, packaging and handling.

HPA costs are depicted in Figure 21(b). A cost of between $5–8K is typical of transmitters in the 10-watt range, regardless of whether they utilize TWTs or solid state devices. However, as the power level rises to the 500-watt range, costs between $25–30K are typical. This cost not only includes the tube but also the increased amount of protective circuitry needed to safeguard the tube's operation. As the HPA power increases to over 1000-watts, its costs increase well over $50K. This is only true of a TWT type of HPA when depressed collector operation is utilized. For klystrons, the simplicity of the HPA is reflected in substantially lower costs—about $45–50K for a 1.5 kW unit. All of the costs shown in Figure 21(b) represent air-cooled systems. As the power level increases above 1.5 kW and, indeed, above 3 kW, liquid cooling must then be introduced and the cost rises significantly.

In Figure 21(c) are presented typical costs of LNA's as related to noise temperature where quantities of approximately

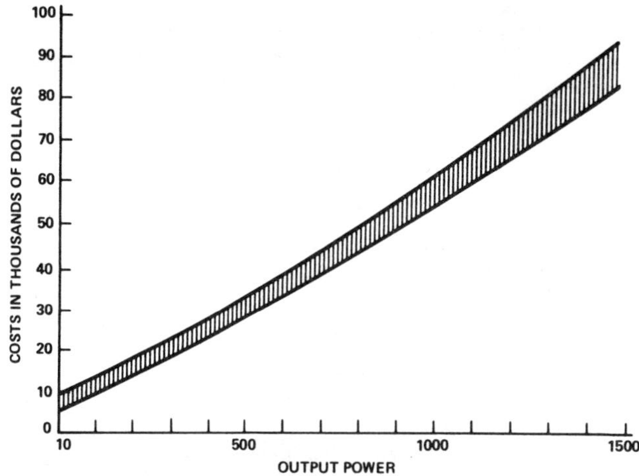
Figure 21b. HPA cost profile

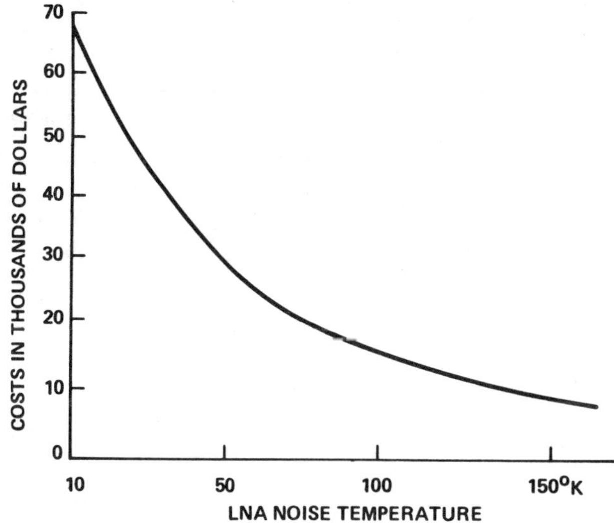
Figure 21c. LNA cost profile

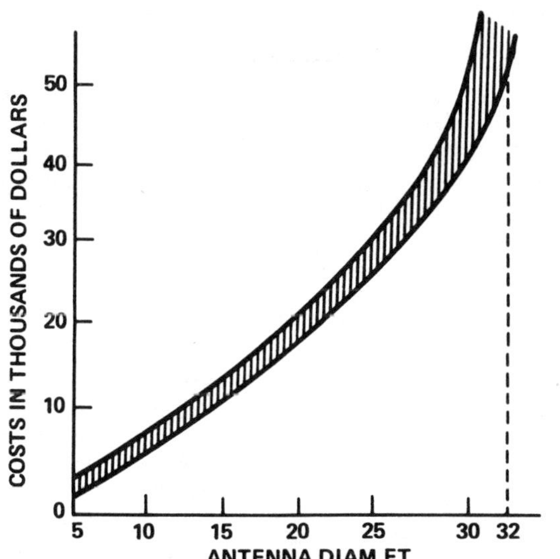
Figure 21a. Antenna Cost profile

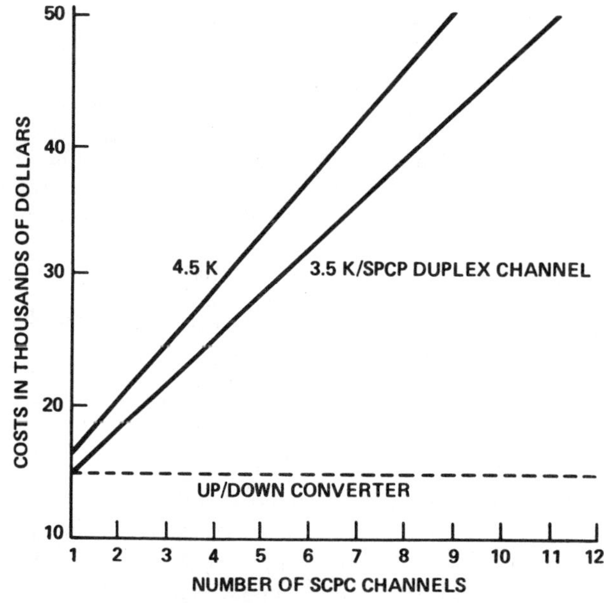

Figure 21d. SCPC cost profile

Table VI
Cost vs. Antenna Diameter for G/T = 20

Antenna Diameter (ft)	5	10	15	20	25	32
Antenna Gain (dB) at 60% Eff and at 4 GHz	34	40	43.5	46	48	50
Nominal Antenna + Feed — Cost (K$)	2	4.5	7.5	16	33	50
$G(dB) - T_s(dB)$ T_s is System Noise Temp in dB	17.2	20	23.5	26	26	30
T_s in °K	53	100	225	400	630	1000
$T_r = T_s - 55°$ Where Antenna Noise Temp is 40°K and Feed Loss is 15°K. T_r = LNA Noise Temp	—	45	170	345 (3.4 dB)	585 (5 dB)	945 (6.5 dB)
Cost of T_r (Cost of LNA)	—	$35K	$8K	$5K	$3K	$2K
Cost of Antenna Plus LNA	—	$39.5K	$15.5K	$21K	$36K	$52K

Table VII
Cost vs. EIRP vs. Antenna Diameter

Antenna Diameter (ft)	5	10	15	20	25	32
Antenna Gain (dB) at 6 GHz	37.2	43.5	47	49.5	51.5	53.5
Nominal Antenna Cost ($K)	2	4.5	7.5	16	33	50
Using 50 Watt TWT & EIRP = 64 dBw						
Required HPA Power in dBw	26.8	20.5	17	14.5	12.5	10.5
HPA Power in Watts	480	112	50	28	17	11
HPA Cost ($K)	28	12	10	8	7	7
Antenna HPA Cost ($K)	30	16.5	17.5	24	40	57
Using 500 Watt TWT & EIRP = 74 dBw						
Required HPA Power in dBw	36.8	30.5	27	24.5	22.5	20.5
HPA Power in Watts	4600	1100	500	300	175	110
HPA Cost ($K)	200	60	28	22	15	12
Antenna + HPA Cost ($K)	202	64.5	35.5	38	48	62

100 are involved. Notice that for a 10°K amplifier, the cryogenics of the amplifier bring the cost to well over $70K. As the temperature rises to 50°K, costs of approximately $30K are realized, largely due to the thermoelectrics, the very high pump frequency, and the technology of the circulator and the special varactors. For LNA temperatures between 100–150°K, costs of between $7–10K are realized due to the easing of some of the circulator, varactor, and pump requirements. Elimination of thermoelectrics and their associated control panel provides a means of significant reduction in LNA costs.

Relative costs of a single channel per carrier channel, including the up/down converter, are shown in Figure 21(d). At present a budgetary cost can be established around $17K for an up/down converter and one receive-transmit single channel per carrier channel. Additional channels, utilizing the single up/down converter complex can be added at approximately $3.5–4.5K.

The line-of-sight communications business, using TV receivers to check LOS transmissions, has resulted in TV receivers in the $4.5–6.5K price range that can down-convert from either 4 or 12 GHz to video and audio baseband with characteristics generally representative of commercial television transmissions.

Table VI shows the cost vs. antenna diameter for G/T = 20 at 4 GHz. Such an antenna would be ideal for use as a TV receiver-only terminal. Note that the cost of the antenna, in combination with the cost of the LNA, is such that a minimum cost will occur at around 15-foot diameter with LNA noise temperature of approximately 170°K. As the antenna diameter is reduced to 10 feet, a significantly lower noise paramp is required, resulting in an increase in terminal cost. On the other hand, any increase in antenna size to utilize lower cost amplifiers will see the less costly LNA overwhelmed by the additional cost of the antenna.

Figure VII compares the cost vs. the effective radiated power at 4 GHz vs. antenna diameter for ERP's of 64 dBW and 74 dBW. These ERP's represent the use of 50-watt TWT and 500-watt TWT, respectively. Observe in Table VII that the cost has a minimum at approximately 15 feet. Any reduction in antenna diameter is met by a significant increase in HPA cost, while an increase in antenna diameter to utilize a lower HPA cost will see the cost of antenna structure far outweigh the savings in HPA cost.

MSN: JUNE/JULY 1975

Part VII
The Future

In this part, the future of satellite communications is discussed. The discussion can be divided into three parts:

1) Advanced Systems,
2) Advanced Concepts,
3) Advanced Technology.

Examples of advanced communications satellite systems include the future INTELSAT System (INTELSAT VI), the next-generation TELESAT system, the Advanced WESTAR System, and the future military system (e.g., [1]-[4]) as well as planned or proposed experimental satellites [5]. Reprint Paper 7.1 discusses the current status of future INTELSAT system planning. The Bibliography contains several papers on future systems.

The Advanced Concepts section includes papers containing ideas and projections for future satellites and systems which are more speculative or further in the future than the advanced systems discussed in the previous section. These papers fall into two categories: 1) large-scale or global system concept studies and 2) specific satellite or system concepts. Typical references in this first category include the NASA "Outlook for Space" report [6], [7] which includes advanced communication satellite concepts as part of a more general study. References in the second category discuss such advanced concepts as sophisticated signal processing in the satellite, spot-beam antennas, and intersatellite links (e.g., [8]).

The papers in the Advanced Technology Section focus on technology for the post-1985 time frame. Although various papers and reports attach different priorities and projections to various technologies, the overall lists are reasonably consistent. This consistency is not surprising given that the authors are extrapolating from a similar current technology base. It will be the unexpected technology breakthroughs that alter these extrapolations. Reprint Paper 1.2 in the first part of the book categorizes the various technology areas. Areas discussed in the various papers (e.g., [9]-[11]) include:

1) modulation and multiple-access techniques,
2) on-board processing,
3) multiple-beam antennas,
4) large deployable antennas,
5) solid-state device technology: higher power, lower noise,
6) higher frequency bands,
7) power generation and energy storage,
8) lower cost earth stations,
9) intersatellite links.

It is perhaps appropriate in a part called "The Future" to reference the fascinating book by Arthur Clarke, *Profiles of the Future—An Inquiry into the Limits of the Possible* [12], which includes a discussion of some of the problems of forecasting. His quotation of testimony in 1945 on intercontinental missiles is particular germane to those who try to predict the future of space communications. Other books and papers discussing various aspects of the future are listed in Section 7.4 of the Bibliography at the end of this volume.

REFERENCES

[1] H. L. Van Trees, "Future Intelsat system (1986–1993) planning," in *Satellite Communications: Future Systems,* D. Jarett, Ed., vol. 54, *Progress in Astronautics and Aeronautics.* New York: AIAA, 1977, pp. 95–142.

[2] D. G., Thorpe, "Evolution of the Telesat Canada system into the mid-1980's," *Ibid.,* pp. 39–57.

[3] J. Ramasastry, W. Heiser, E. Levine, and R. Markham, "Western Union's satellite—Switched TDMA Advanced WESTAR system, *Proc. AIAA 7th Comm. Satellite Systems Conf.,* San Diego, CA, pp. 497–506, Apr. 1978.

[4] I. S. Haas and A. T. Finney, "The DSCS III satellite—A defense communications system for the '80's," *Ibid.,* pp. 351–358.

[5] E. A. Wolff, "Space shuttle communications experiments," *Progress in Astronautics and Aeronautics,* vol. 54, D. Jarett, Ed. pp. 407–422, 1977.

[6] NASA, "Outlook for space," NASA SP-386, Washington, DC, Jan. 1976.

[7] I. Bekey and H. Mayer, "1980–2000: Raising our sights for advanced space systems," *AIAA Astronautics and Aeronautics,* pp. 34–63, July/Aug. 1976.

[8] D. Jarett, "Metting the twin challenges of demand and conservation of spectrum and orbit through technology," *EASCON 1977,* Paper 25-1, Sept. 1977.

[9] Aerospace Corp., "Communication satellite technology: State of the art and development opportunities," *Aerospace Report* No. *ATR-78 (7735) -1,* July 1978.

[10] J. E. Burtt *et al.,* "Technology requirements for communication satellites in 1980's," NASA Report NAS2-7073, Lockheed Missiles and Space Col., Sept. 1973.

[11] NASA, "A forecast of space technology 1980-2000," NASA SP-387, Washington, DC, 1976.

[12] A. C. Clarke, "Profiles of the future—An inquiry into the limits of the possible." New York: Popular Library, Sept. 1977.

PLANNING FOR THE POST-1985
INTELSAT SYSTEM

H. L. Van Trees, J. C. Fuenzalida
M. Mohajeri, J. E. Martin
Communications Satellite Corporation
Washington, D.C.

Abstract

The objectives of the future INTELSAT System Planning Study are to develop and analyze alternative system and spacecraft concepts for the post-1985 period, to identify the key technologies necessary to implement these concepts, and to develop the necessary methodology and analysis and synthesis tools to conduct the required planning.

The traffic in the INTELSAT system is projected to increase to 300,000 channels by year-end 1993. New system concepts and technologies are required to provide the necessary capacity efficiently. Several system concepts and spacecraft designs are described. The effect of intersatellite links, TDMA, small antennas, new frequency allocations, and intersatellite traffic distribution are discussed.

1. Introduction

During the post-1985 period, the global telephony traffic level in the INTELSAT system is expected to grow from its present level of 20,000 channels to 106,000 channels in 1986 and 297,000 channels in 1993. In addition, the demands for leased transponder, television, and digital data transmission will increase. To satisfy these demands, the INTELSAT system must exploit a combination of improved system concepts, advanced spacecraft designs, and new modulation and multiple access techniques.

The objectives of the Future INTELSAT System Planning Study are to develop and analyze alternative system concepts for the post-1985 period, to identify key technologies necessary to implement these concepts, to encourge research and development in the ciritical areas, and to develop the necessary methodology, and analysis and synthesis tools to conduct the required planning. The study in conjunction with the necessary detailed system engineering and spacecraft design will culminate in a system recommendation, defining the INTELSAT system concept for the 1986-1993 period, and the INTELSAT V follow-on spacecraft to be launched in 1985 or 1986 under the current planning assumptions.

Some of the significant study results have been previously presented.[1-4] This paper discusses the current status of work and introduces new results.

2. General System Considerations

The methodology developed in the planning study provides a procedure for the synthesis, analysis, and evaluation of alternative future system designs to satisfy INTELSAT's requirements in the 1986-1993 time frame. These system designs, or composite systems, are composed of three basic elements: future planning scenarios, system concepts, and spacecraft concepts.

2.1 Future Planning Scenarios

A set of INTELSAT planning scenarios has been postulated spanning a broad range of user requirements to account for the inherent uncertainties in the 1986-1993 period. These scenarios include elements of telephony, transponder lease service, occasional use/television service, and digital data service. The earlier studies used three basic user requirements scenarios corresponding to nominal-, high-, and low-growth predictions.[1] The telephony traffic levels for these three scenarios are shown in Table 1. The generation of a new INTELSAT integrated traffic data base and a forecasting study[5] has motivated a slight modification of these scenarios; however their basic growth characteristics are retained.

Table 1. Telephony Traffic Levels of the Nominal-, Low-, and High-Growth Scenarios

a. Nominal Growth

	Atlantic	Indian	Pacific	Total
Year-end 1979	21,266	9,614	3,430	34,310
Year-end 1986	65,984	30,370	9,810	106,164
Year-end 1993	186,634	83,648	26,970	297,252

b. Low Growth

	Atlantic	Indian	Pacific	Total
Year-end 1979	16,180	7,316	2,610	26,106
Year-end 1986	38,058	17,588	5,802	61,448
Year-end 1993	73,908	33,574	11,272	118,754

c. High Growth

	Atlantic	Indian	Pacific	Total
Year-end 1979	21,266	9,614	3,430	34,310
Year-end 1986	71,272	32,514	10,566	114,352
Year-end 1993	234,076	106,672	33,612	374,366

2.2 System Concepts

A system concept is defined by a space segment configuration (i.e., the number of satellite types and their roles, the use of intersatellite links, and whether or not traffic is shared between ocean regions) and by the earth segment equipment allowed (e.g., small antennas and advanced modulation/multiple-access equipment). Table 2 categorizes the system concepts and lists typical examples. Each system concept can be implemented with any of a number of different spacecraft concepts. Parametric system variations, such as the number of small antennas, the number and type of advanced modulation and multiple access techniques, the

This paper is based upon work performed under the sponsorship of the International Telecommunications Organization (INTELSAT). The views expressed herein are strictly those of the authors and do not necessarily reflect the views of INTELSAT.

orbital location of the satellite, and the degree of
connectivity, are then introduced.

The first concept assumes three distinct ocean
regions, a single Primary satellite in each region
to provide the basic connectivity, and an appropriate number of Major Path satellites. The first
example of this concept, in Table 2, uses an
INTELSAT VI for all satellite roles. The second
example uses an INTELSAT VI as a Primary satellite
in the Atlantic Ocean Region (AOR) and Indian Ocean
Region (IOR) and an INTELSAT V in the other roles.
The third example uses an INTELSAT VI as a Primary,
and trunk satellites in the other roles.

Table 2. System Concepts

	System Concept ID	Primary	Examples	Comments
DISTINCT OCEAN REGIONS	I	Single	a. AOR(P, MP1, MP2), IOR(P, MP), POR(P) b. AOR(VI, V , V) c. AOR(VI, TS1, TS2, TS3)	Single satellite type. Two satellite types. Two satellite types.
	II	Distributed	a. AOR(P1, P2, MP1, MP2), IOR(P, MP1) b. AOR(P1, P2, MP), IOR(P1, P2)	Connectivity with ISL. Connectivity with small antennas.
INTERREGIONAL	III	Single	a. AOR(P, MP1, MP2), IOR(P) b. AOR(P, MP1, MP2), IOR(P, MP)	Common region traffic.
	IV	Distributed	a. Dual Primary in AOR and IOR, 10° separation, ISL connectivity.	

Note: Number of Major Path satellites varies with the System
Concept and Ocean Region.

The second concept assumes three distinct ocean
regions, and two or more Primary satellites to provide the connectivity. In the first example shown
in Figure 1, two INTELSAT VIs are used as a dual-
Primary with an intersatellite link to provide the
connectivity. The users in the AOR are divided into
two communities of interest, and the traffic within
these communities is assigned to Primary A and
Primary B. The traffic between the communities is
carried on an intersatellite link. The second ex
ample is similar except that the traffic between the
communities of interest is carried by small antennas
as shown in Figure 2.

The third concept assumes an interregional system with a single Primary in each region and an appropriate number of Major Path satellites. The
first example of this concept entails a (P, MP1,
MP2) configuration in the AOR and a Primary in the
IOR. The common region of visibility between the
IOR Primary and the AOR is shown in Figure 3 as a
function of the location of AOR MP2. Traffic in
this region is allocated to one of the four satellites to utilize the system efficiently. In the
second example, a Major Path satellite is added in
the Indian Ocean; and the allocation of traffic to
the five satellites is considered; and the satellite locations are varied parametrically.

The fourth concept assumes an interregional
system with a dual-Primary in each ocean region
and an appropriate number of Major Path satellites.
The first example of this concept, uses a (P1, P2,
MP) configuration in the AOR and a (P1, P2) configuration in the IOR. The connectivity between the
Primary satellites in each region is provided by
intersatellite links. Figure 4 shows the coverage
regions for representative satellite locations.
This configuration has the advantage of significantly expanding the possible single-hop link
coverages.

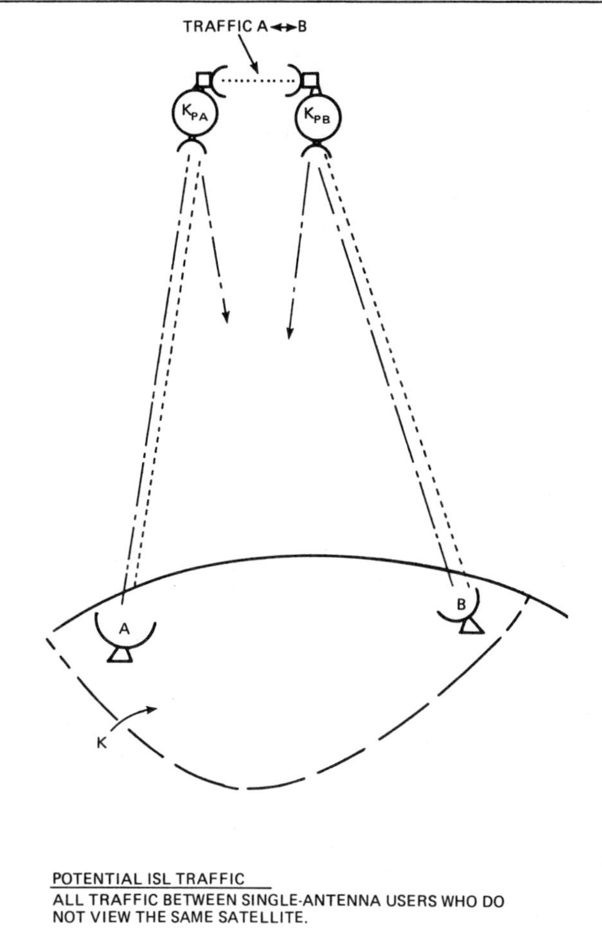

Figure 1. ISL Used in Dual-Primary
Configuration Serving
Region K

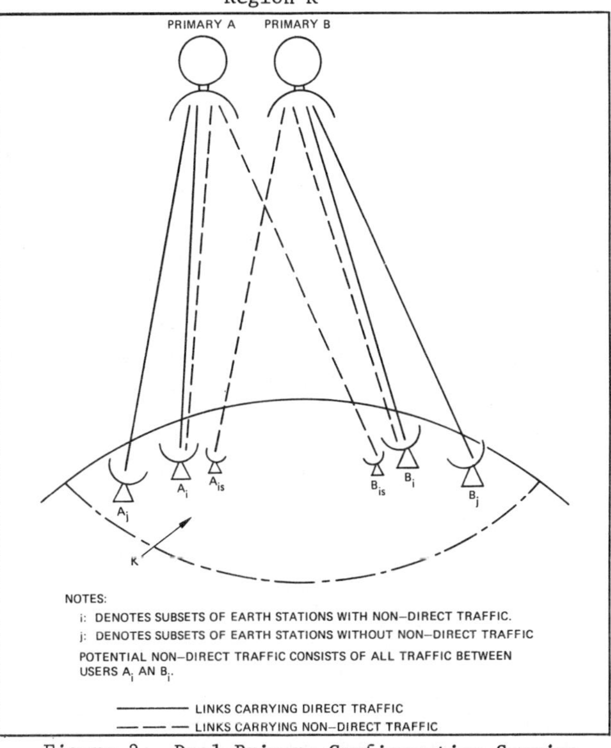

Figure 2. Dual-Primary Configuration Serving
Region K Relying on Small Antennas
for Connectivity

609

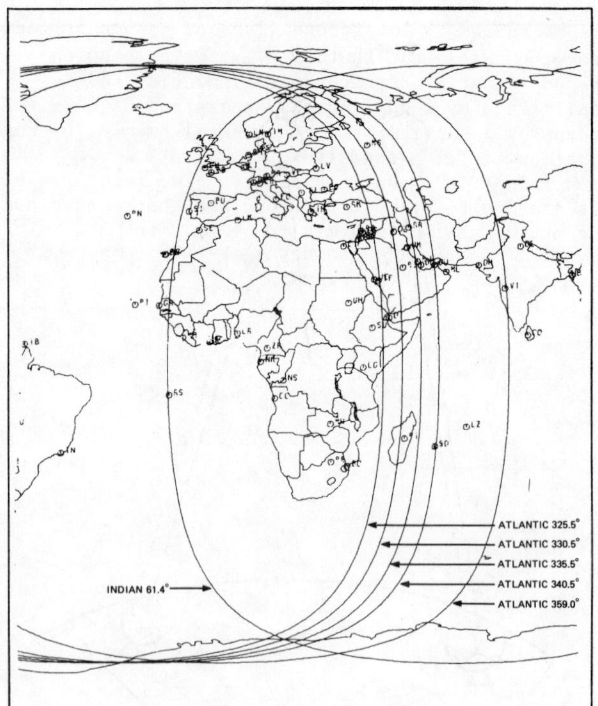

Figure 3. Areas Common to both AOR and IOR within 5 Degree Elevation Contours

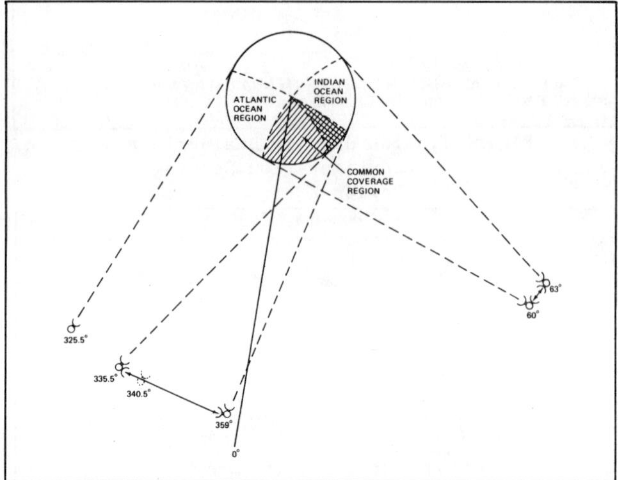

Figure 4. Example of AOR and IOR Satellite Locations and Common Coverage for System Concept IV

2.3 Spacecraft Concepts

Five principal spacecraft concepts for the post-1985 period and a secondary (or auxiliary) spacecraft concept are described in this section. The spacecraft were defined with minimum detail necessary to investigate their suitability in the INTELSAT system context. The spacecraft concepts were defined by coverage or beams; bandwidth availability to each beam; channelization of bandwidth in each beam (transponders); connectivity of transponders (switches); and bandwidth use efficiency by transmission mode, which is a highly simplified parameter representing many combinations of e.i.r.p. levels, beam coupling, and other communications subsystem characteristics.

2.3.1 Principal Spacecraft Concepts. Table 3 summarizes the characteristics of the five principal spacecraft concepts.

Table 3. Characteristics of the Principal Spacecraft Concepts

Spacecraft Concept	Example	Coverages				Transponder Bandwidth (MHz)			
		6/4 GHz			14/11 GHz	Global (6/4 GHz)	Hemi and Zone (6/4 GHz)	Spot (14/11 GHz)	Total
		Global	Hemi	Zone	Spot				
S1	A2	1 DP	2 SP	5 SP	2 DP + 1 SP	160	2800	2400	5360
	B3	1 DP	2 SP	6 SP	2 SP	240	2800	960	4080
S2	A4	1 DP	2 SP	6 SP	2 DP	160	3200	1920	5280
S3	B3 B	1 DP	2 SP	6 SP	2 SP	1200	2880	1920	6000
	B3 E	1 DP	2 SP	6 SP	2 SP	880	4160	1920	6960
	B3 C	1 DP	2 SP	6 SP	2 SP	1200	2880	1920	6000
S4	A3	1 DP	2 SP	9 SP	2 DP	160	4400	1920	6480
S5	INTELSAT V	1 SP	2 SP	2 SP	2 SP	120	1360	800	2280
	V-A*	1 DP	2 SP	2 SP	2 SP	240	1440	960	2560
	V(modified)*	1 DP	2 SP	3 SP	2 SP	240	1800	960	3000

SP: Single Polarized
DP: Dual Polarized

*These have not been fully defined and are presented as examples of possible INTELSAT V-A and a modified INTELSAT V concepts.

Spacecraft concept S1 currently contains two types of satellites, A2* and B3, which are similar in concept. At 6/4 GHz, they utilize five or six high-isolation, high-polarization purity steerable beams in addition to two hemispheric beams. The Atlantic Ocean Region (AOR) coverage patterns are shown in Figures 5 and 6. The B3 design is simpler because it relies on extensive use of SS/TDMA to achieve the required capacity.[6]

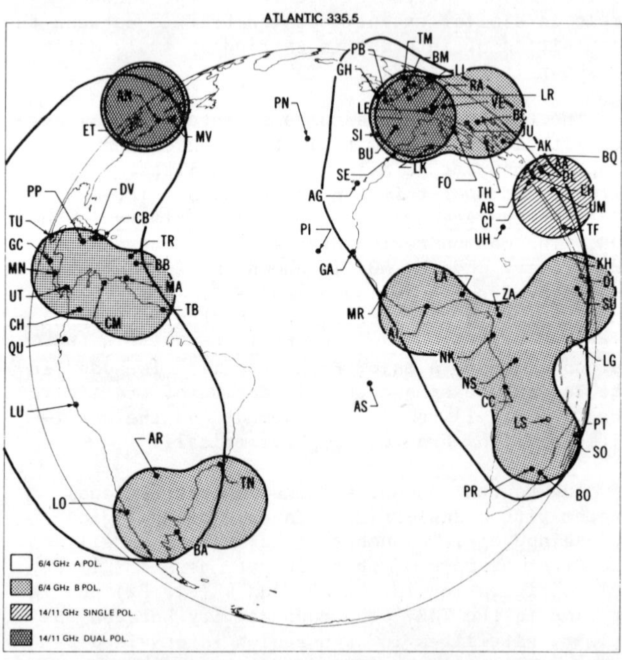

Figure 5. Example of AOR Coverages of Principal Spacecraft Concept S1 (A2)

*These spacecraft identifying letters are used for consistency with the overall study. The basis for the nomenclature is not important for this paper.

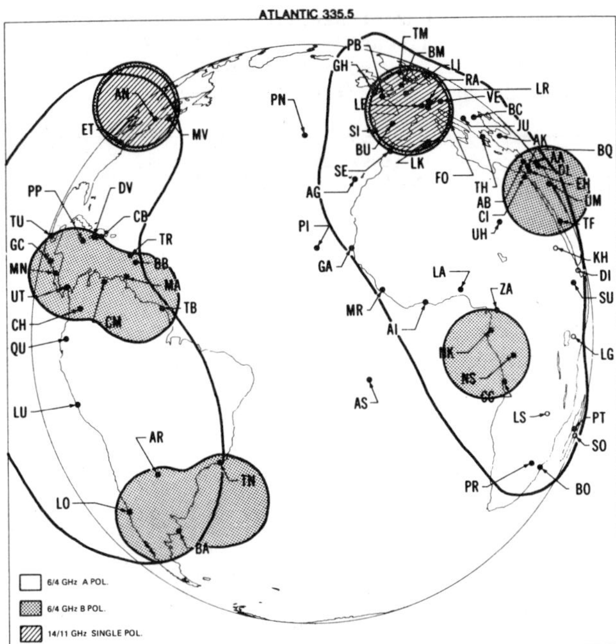

Figure 6. Example of AOR Coverages of
Principal Spacecraft Concept
S1 (B3)

Spacecraft concept S2 is designed with the
ability to reconfigure its coverage pattern in orbit.
Figure 7 shows the AOR coverage patterns for this
spacecraft concept. The zone coverages are generated by an array of 85 feeds arranged in a "close
pack" configuration to provide the required coverage flexibility (shown in Figure 8). The feed array is divided into six segments. Each feed can be
interconnected with any number of feeds from either
the same or the two adjacent segments to form a maximum of six independent beam groups. Beam groups

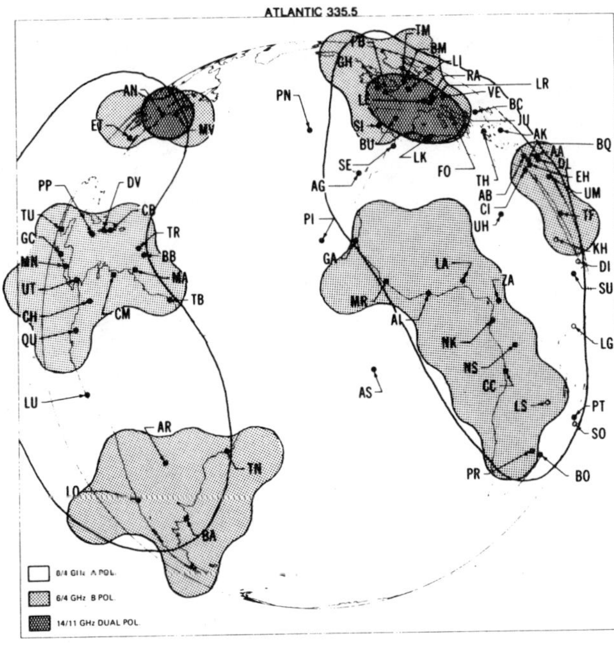

Figure 7. Example of AOR Coverage of
Principal Spacecraft Concept
S2 (A4)

can be formed by interconnecting any number of feeds
from three adjacent sectors. Isolation between
groups is derived from spatial isolation and is
facilitated by requiring the separation of beam
groups by a minimum of one elemental beam. The combination of feeds used to generate each of the AOR
beam groups is shown in Figure 9. Two feed arrays
are assumed to be required, one for the receive and
the other for the transmit function. The feed arrays illuminate separate deployable reflectors to
generate the required beams.

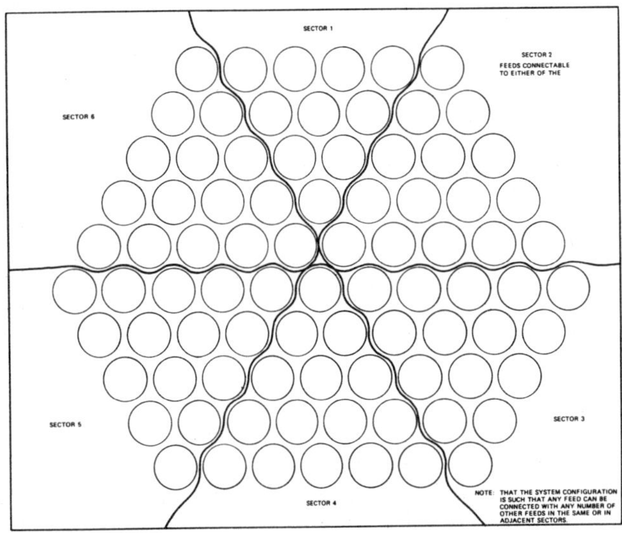

Figure 8. Feed Arrangement for Zone Coverages
of Spacecraft Concept A4

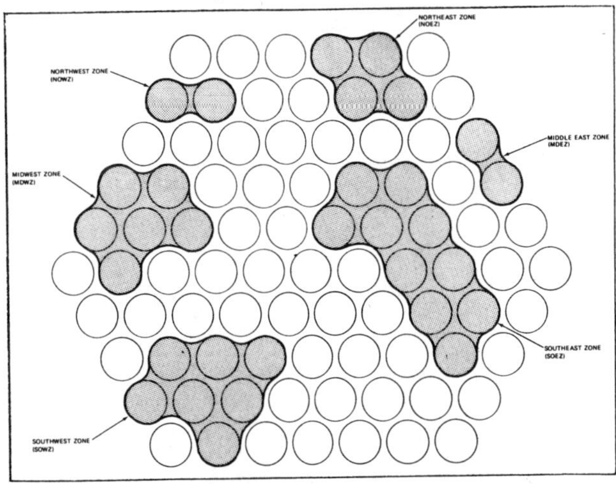

Figure 9. Groupings of Feed Clusters Used in
Achieving the AOR Zone Coverages of
Spacecraft Concept A4

Spacecraft concept S3 is based on the hypothesis
that new frequency bands below 20 GHz will be allocated to the fixed-satellite service at the 1979
WARC. The frequencies of interest, shown in Figures
10 and 11, represent an additional 890 MHz in the
6/4-GHz vicinity and 750 MHz in the 14/11-GHz vicinity.

Utilization schemes to introduce and implement
the new frequency bands in the INTELSAT system were
categorized into three strategies:

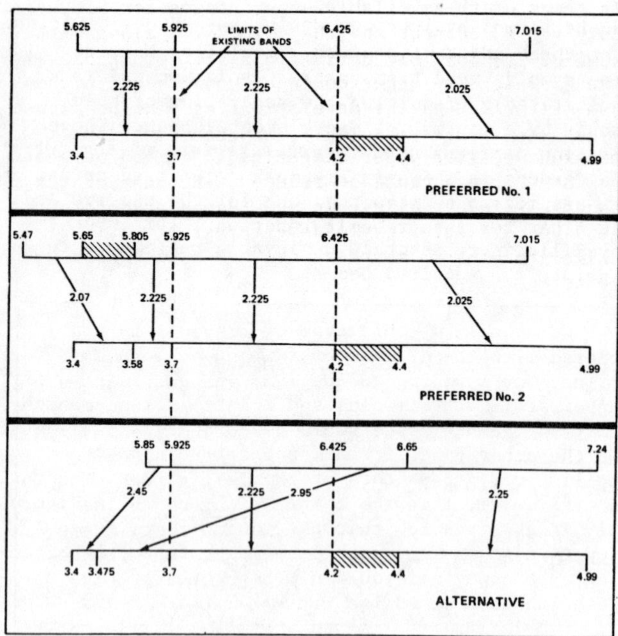

Figure 10. Possible Allocations Below 10 GHz
of Interest to INTELSAT

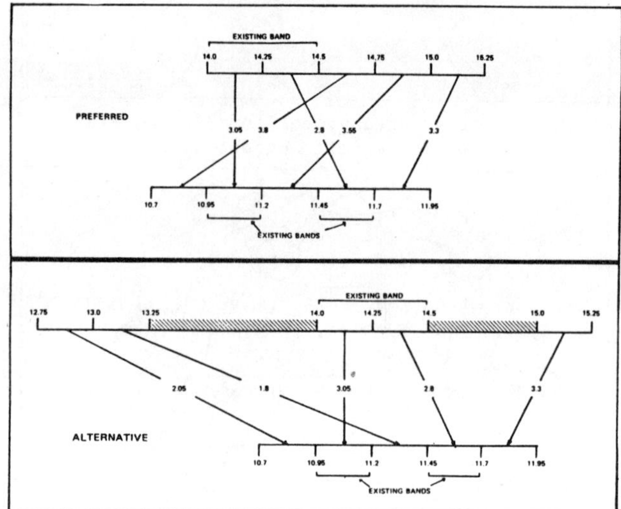

Figure 11. Possible Allocations Above 10 GHz
of Interest to INTELSAT

a. "Bolt-on" strategy. The new bands would be
accessed through a separate communications
package which could be installed on the
spacecraft as an option.

b. "Extension" strategy. The bandwidth of
the existing (or planned) coverage beams
would be increased to incorporate the new
allocations.

c. "Cross-strap" strategy. The new bands
could be interconnected with the presently-
used bands through appropriate frequency
translations on the spacecraft.

Within spacecraft concept S3, three example
spacecraft have been defined, viz. the B3B, B3E and
B3C concepts which would use the B3 spacecraft
concept as a basis and would incorporate the new
bandwidth in the bolt-on, extension, and cross-
strap strategies, respectively. In each case, an

additional 500 MHz of bandwidth was considered
to be the new allocation at both the 6/4- and
14/11-GHz bands. Figure 12 shows a typical
transponder plan, for satellite B3E.

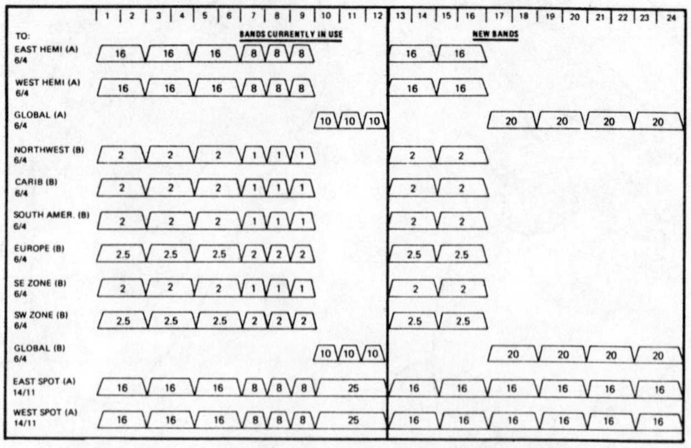

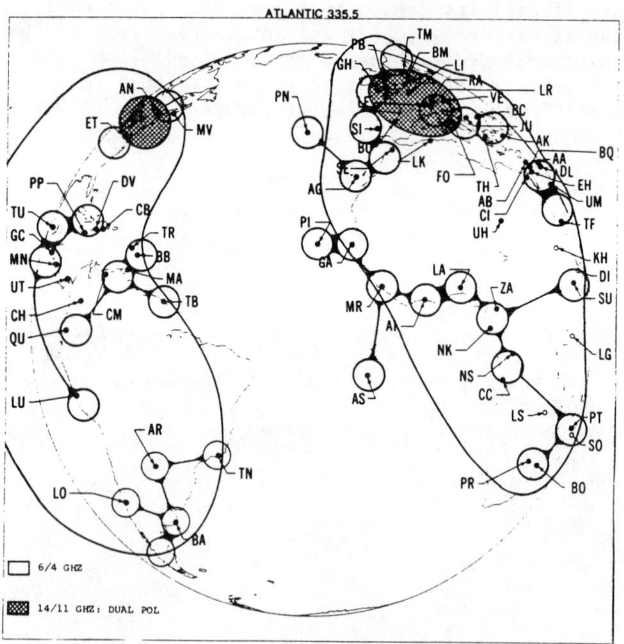

Figure 12. Example of Transponder Plan,
Showing Transmitter Powers,
for Principal Spacecraft
Concept S3(B3E)

Spacecraft concept S4 relies on extensive reuse
of the 6/4-GHz band to achieve the required capacity.
An example spacecraft, A3, utilizes nine reconfigurable zone beam groups with high-polarization purity. Two very large offset-fed reflectors with a
complex feed system are required in addition to the
hemispheric portion of the INTELSAT V antenna complement. The coverage pattern for the AOR is shown
in Figure 13.

Figure 13. Example of AOR Coverages of
Principal Spacecraft Concept
S4 (A3)

Spacecraft concept S5 includes the INTELSAT V
and its various modifications (INTELSAT V-A).
Figure 14 is the coverage pattern of the INTELSAT V.

A possible implementation for a typical spacecraft from these five principal spacecraft concepts
is described in the next section. It should be

noted that all of the spacecraft will include an intersatellite link option for use in System Concepts II and IV.

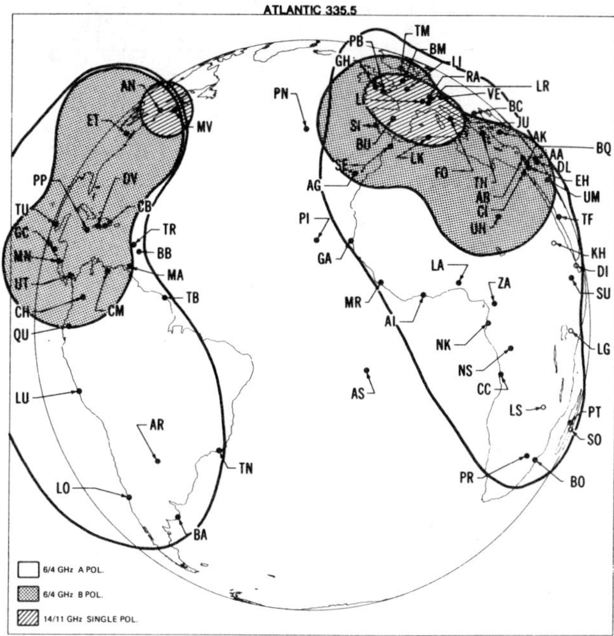

Figure 14.   Example of AOR Coverages of Principal Spacecraft Concept S5 (INTELSAT V)

2.3.2 Implementation of Principal Spacecraft Concepts

The technologies necessary to implement these spacecraft concepts have been identified. Mass and power estimates have also been generated which indicate the approximate "size" of the satellites and their launch vehicle requirements. Cost estimates based on the mass and power have been generated using the SAMSO cost model.

Table 4 shows a summary of the principal implementation characteristics of the various INTELSAT VI spacecraft. The various antenna configurations represent possible realizations. Several configurations for various INTELSAT VI spacecraft that would be suitable for a shuttle launch have been studied.*

Table 4.   Principal Implementation Characteristics of the Various Spacecraft Concepts

Spacecraft Concept	Hemi Reflectors	Zone Reflectors	Spot Reflectors	No. of 6/4-GHz Transponders	No. of 14/11-GHz Transponders	S/C Mass BOL	S/C BOL Power (W)
A2	60-inch Rev 96-inch TX	Uses Hemi Reflectors	3 Reflectors 33-inch each	60	45	1471	3235
A3	60-inch Rev 96-inch TX	14.1-ft Rev 21.3-ft Tx	2 Reflectors 44-inch each	92	36	1709	2776
A4	60-inch Rev 96-inch Tx	80-inch Rev 120-inch Tx	2 Reflectors 44-inch each	60	32	1976	3308
B3	60-inch Rev 96-inch Tx	Uses Hemi Reflectors	2 Reflectors 33-inch each	54	14	1016	2027
INTELSAT V	60-inch Rev 96-inch Tx	Uses Hemi Reflectors	2 Reflectors 44-inch each	24	6	993	1025

---

*This work has been performed by COMSAT Laboratories as part of an INTELSAT sponsored project, and is not yet available in the literature.

In terms of the available upper stages for STS, only the B3 concept and the INTELSAT V along with some of its possible derivatives are compatible with the SSUS-A. The other concepts would require the use of the IUS or the development of mission-peculiar upper stage vehicles. The consideration of mission-peculiar upper stages would depend on whether launch costs could be reduced. Because of the shuttle pricing policy, in some cases tradeoffs leading to lighter payloads are cost effective, while, in others, shorter payloads save launch costs.

An example of this type of tradeoff is associated with the A2 spacecraft concept. When designed for mounting on the IUS, the combination of upper stage vehicle plus spacecraft is long enough that the launch costs would be a function of length. On the other hand, if the spacecraft is designed to utilize only the motors of the IUS, rather than the entire stage, then the combination, shown in Figure 15, is short enough that the launch costs are a function of mass. In this example, further shortening of the payload would not be cost-effective unless the mass were also reduced.

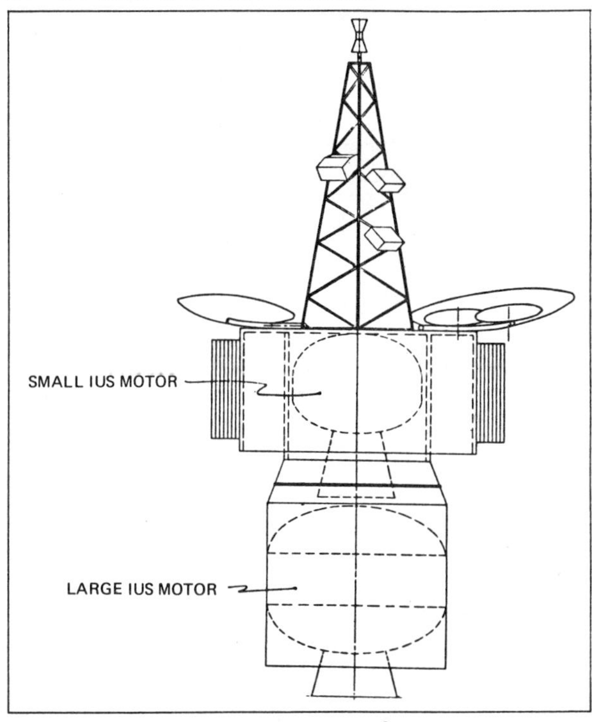

Figure 15.   Spacecraft Concept A2 Using Only the Motors of the IUS in a Shorter Configuration than Is Possible with the IUS

The synchronous orbit payload capacity of the stages to be available within the STS program result in a large gap between the SSUS-A and the IUS. For payloads that lie within the gap, liquid engines may offer a flexible staging solution. This staging concept could also be useful in some applications to minimize payload length since the liquid tanks could be mounted within the spacecraft body (an example of this design is the SYNCOM IV[7]).

2.3.3 Secondary Spacecraft Concepts

The use of a spacecraft concept based on existing non-INTELSAT satellite designs in conjunction with an INTELSAT VI satellite concept is of potential benefit for the future INTELSAT system.

To date, such spacecraft concepts have been considered to carry the large traffic streams (trunks). Some of the potential advantages of small trunk traffic satellites are as follows: a. use of efficient single-carrier-per-transponder transmission is possible in many transponders and use of small earth stations (G/T of 31.7 dB/K) as second or third antennas becomes practical with considerable savings in the earth segment cost; and b. significant potential savings in development costs and lower spacecraft and launch costs can be expected, possibly resulting in lower overall space segment cost.

As an initial approach, three types of "trunk" spacecraft were investigated: global-, hemispheric- and zone-coverage spacecraft, all subject to a spacecraft weight constraint of 907 kg (2,000 lb), the capability of a Delta launch vehicle.

Transponder capacities were estimated under various combinations of antenna coverage, transponder bandwidth, earth station G/T, and number of carriers per transponder. In this paper, a trunk satellite using zone beam coverages will be described.

Figure 16 shows the frequency plan, the transmitter powers and transponder capacity for the zone-type trunk spacecraft. A total capacity of about 36,000 telephone channels can be achieved with half of the transponders operating in the single-carrier mode and the other half in the multiple-carrier mode. This capacity can be achieved with the use of small-diameter non-standard earth stations operating in the FDM/FM/FDMA mode. The antenna coverage for the Atlantic Region in this spacecraft can be arranged into three different coverage schemes as shown in Figures 17, 18, and 19. When zones 3 and 4 are illuminated, the bandwidth available to zones 1 and 2 is reduced.

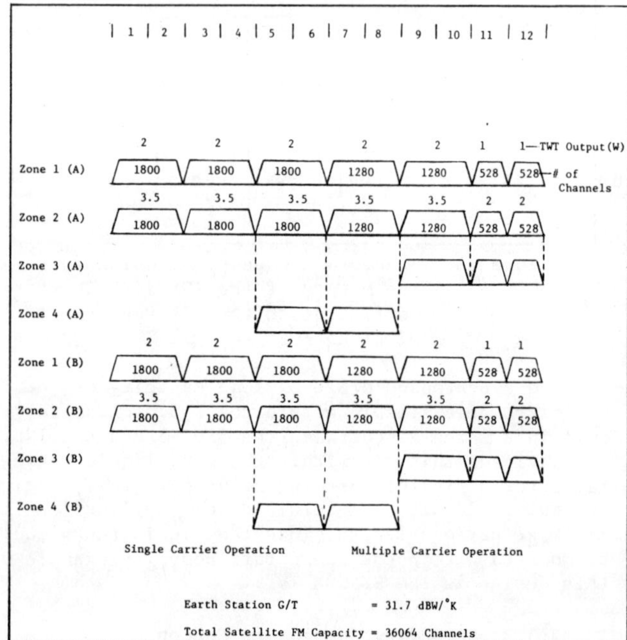

Figure 16. Frequency Plan, Transmitter Outputs, and Transponder Capacities for a Secondary Spacecraft Concept (Zone Trunk)

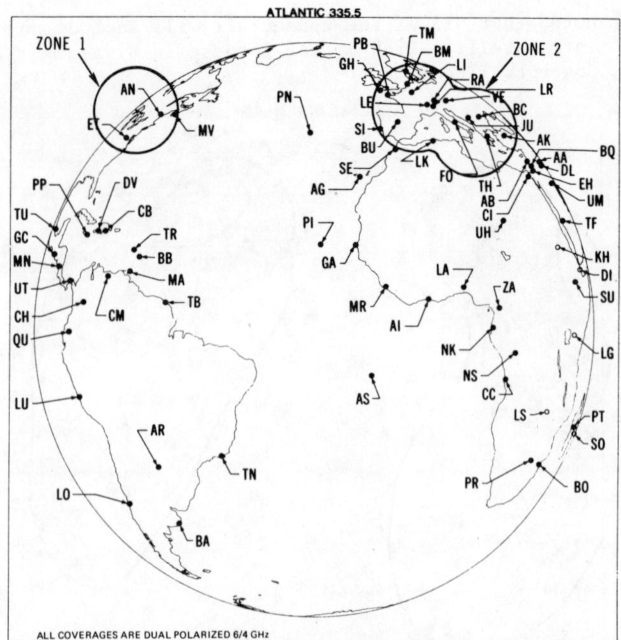

Figure 17. AOR Coverages of a Zone Trunk Spacecraft (Configuration 1)

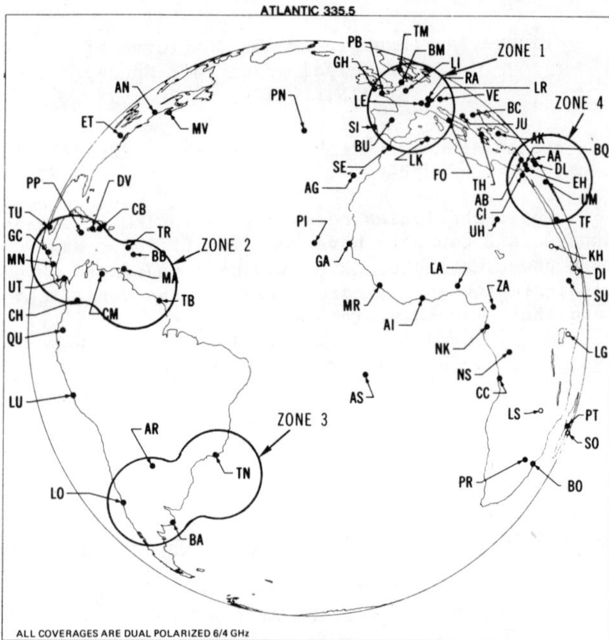

Figure 18. AOR Coverages of a Zone Trunk Spacecraft (Configuration 2)

3. Comparison of Alternative Systems

Using the scenarios, system concepts, and spacecraft concepts described in Section 2, a number of alternative systems have been designed to satisfy the traffic requirements at the end-point of the planning period, 1993. In each case, the satellite on-station plan was generated and the earth station complement was determined. The space segment incremental net present worth (NPW) was estimated based on the satellite on-station plan and the spacecraft cost estimates.

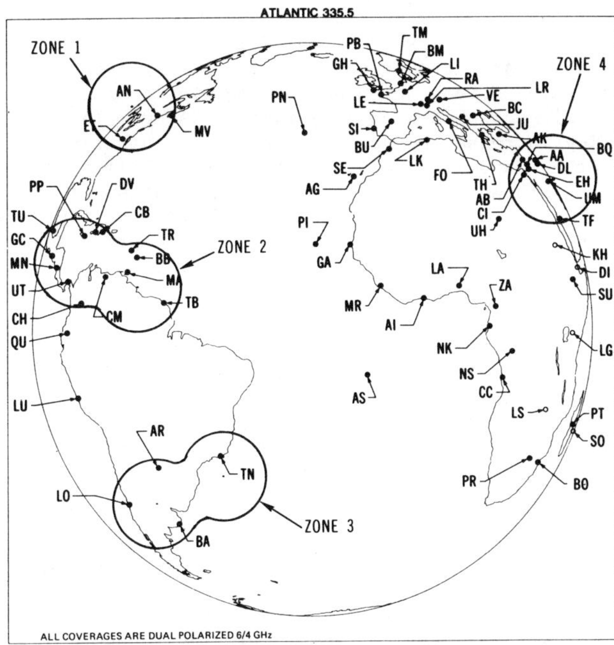

Figure 19. AOR Coverages of a Zone Trunk
Spacecraft (Configuration 3)

3.1 System Concept I

A system configuration with distinct ocean
regions, single Primary in each region, and multiple Major Path satellites was considered, for
the nominal-growth scenario. The same type of
satellite was used throughout the system. First, a
hypothetical "absolute minimum system" was defined
to provide a baseline. This system assumes an all-
FM, 6/4-GHz global satellite with unconstrained
capacity for each ocean region. Hence, each system
user in each ocean region requires one 6/4-GHz antenna, one up-chain, and a minimum number of down-
chains and FM channel equipment. Then, systems
utilizing spacecraft A2, A4, and B3 and designated
as systems 1, 2, and 3, respectively, were designed
and analyzed.

A summary of the earth segment complement and
space segment incremental NPW is shown in Table 5.
These results show that system 2 requires the least
number of antennas. On the other hand, in terms of
space segment NPW, system 3 is the most economical.

Table 5. Summary Results for the
Minimum System and Systems 1,
2, and 3

Item \ System	Minimum System Using Infinite Capacity Spacecraft				System 1 Using A2 Spacecraft				System 2 Using A4 Spacecraft				System 3 Using B3 Spacecraft			
	Atl.	Ind.	Pac.	Total	Atl.	Ind.	Pac.	Total	Atl.	Ind.	Pac.	Total	Atl.	Ind.	Pac.	Total
6/4-GHz antenna	96	58	32	186	149	77	32	258	150	79	32	261	155	79	32	266
14/11-GHz antenna	0	0	0	0	23	19	3	45	12	9	0	21	13	12	0	25
TDMA common equipment	0	0	0	0	31	0	0	31	27	0	0	27	83	18	0	101
Space Segment Inc. NPW ($M)	N/A				334				423				277			

3.2 System Concept I Relying on Two Satellite Types

Systems with two satellite types are of interest because the Atlantic Primary satellite is
usually overdesigned for use in other regions and
for other roles.

Two examples of this type of system have been
considered: an INTELSAT VI used with additional
INTELSAT V satellites; and an INTELSAT VI to be
used with an additional satellite type which
would be a modification of another (non-INTELSAT)
spacecraft. Only the second example will be
discussed in this paper.

In this example, the zone type spacecraft was
analyzed in a composite system with A2 as the
Primary and with four trunk satellites as Major
Paths in the Atlantic Region. The system was
analyzed for year-end 1993 using the nominal-
growth scenario (Scenario I).

The following results were obtained: The
total earth segment requirement for this system
is twelve 14/11-GHz antennas, 96 6/4-GHz standard
A antennas, 63 6/4-GHz Type D* earth station antennas, and 59 TDMA common equipment. TDMA was
introduced on the Primary satellite only.

This type of composite system could provide
a low cost earth segment together with a low cost
space segment with the penalty of the loss of some
path diversity. A higher average channel density
is achievable with this composite system due to the
utilization efficiency of the trunk spacecraft, resulting in a better utilization of the space segment resources than with other systems.

3.3 System Concept III: Inter- regional Systems

A simple example of an interregional system
is based on an extrapolation of the current system configuration. When a fully-connected Primary
is assumed in the Atlantic Region and 2- and 3-path
diversity is given to users with two or three
mutual antennas, the Primary satellite becomes
heavily loaded, and hence the Major Path satellites are not fully utilized. A number of users
in the Indian Ocean Region have access to the
Atlantic and Pacific Region satellites, and it is
possible to off-load some of the Indian Ocean
Region traffic and alleviate the need for the Major
Path satellite in the Indian Ocean Region. A composite system relying on this concept was developed
using satellite concept A4, though other satellite
concepts are also suitable.

Table 6 summarizes the earth segment complement
and the space segment incremental NPW for this
system and for system 2 with fully-connected

---

*A "Type D" antenna has a G/T = 31.7 dB/K, but the
modulation and multiple access scheme is arbitrary
(Note that a Standard B earth station uses SCPC/PCM.).

Primaries* A comparison of the earth segment complement for these two systems shows that the interregional system requires 31 more TDMA common equipment, and offers a savings of 22 antennas. The space segment in this system requires one less spacecraft.

Table 6. Summary Results for System 2 with Fully-connected Primaries and for the Interregional System

Item \ System	System 2 with Fully-connected Primaries				Interregional System			
	Atl.	Ind.	Pac.	Total	Atl.	Ind.	Pac.	Total
6/4-GHz Antenna	167	81	32	280	167	58	32	257
14/11-GHz Antenna	12	9	0	21	15	7	0	22
TDMA Common Equipment	64	0	0	64	64	31	0	95
Space Segment Inc. NPW ($M)	423				392			

3.4 System Concept II: Dual-Primary Systems Relying on Intersatellite Links

A number of systems with distinct Ocean Regions and dual-Primary and Major Path satellites relying on ISLs were designed for the nominal scenario. These systems utilize spacecraft concept A4 with a 25/32-GHz ISL package relying on analog (FM) modulation techniques. Five different systems with ISL capacities of 1,200, 1,800, 2,400, 3,000, and 3,600 voice channels were designed to obtain tradeoffs between the ISL capacity and earth segment requirements.

The space segment configuration for each system consisted of a dual-Primary and a Major Path in the Atlantic Region, a dual-Primary in the Indian Ocean Region and a single Primary in the Pacific Region.

A summary of the space segment incremental NPW, as well as the antenna and TDMA common equipment requirements for these systems, compared to two other cases of interest is given in Table 7. The concept of a fully-connected single Primary system was considered here to provide a baseline for comparison. Note that dual-Primary systems in the Atlantic and Indian Ocean Regions offer full connectivity.

Table 7. Summary Results for System 2 with Fully-connected Primaries, and Dual-Primary System

Item \ System	System 2	System 2 with Fully-connected Primary	Dual-Primary Systems with ISL Capacities in # Voice Channels				
			1200 ch.	1800 ch.	2400 ch.	3000 ch.	3600 ch.
6/4-GHz Antenna	261	280	252	254	240	238	236
14/11-GHz Antenna	21	21	22	23	21	21	20
TDMA Common Equipment	27	64	27	27	26	26	25
Space Segment Inc. NPW ($M)	423	423	432	432	433	434	434

*In the fully-connected Primary system it was assumed that all system users in an ocean region have access to the Primary satellite. In systems 1, 2, and 3, this assumption was not imposed and hence, some communities of interest were moved to Major Path satellites.

It should be noted that the use of an ISL to provide connectivity in a dual-Primary system can provide a large reduction in the number of antennas and TDMA equipment which would otherwise be required in a fully-connected single-Primary system. The resulting increase in space segment incremental NPW,on the other hand,is 2 percent for the dual-Primary system. These figures are based on the assumption of using the A4 spacecraft as the basic satellite concept. For other basic spacecraft concepts these figures will be different but the general conclusions will be valid.

3.5 System Concept II: Use of Small Earth Stations for Connectivity

Use of small earth stations in the INTELSAT System has the potential of significant savings in the earth segment requirements. Small earth stations can be used either to carry thin-route traffic on the Primary and provide connectivity, or to carry diversity traffic on trunk satellites.

An example of the use of small earth stations to access the Primary satellites in the Atlantic and Indian Ocean Regions and to provide connectivity by carrying thin-route traffic is presented here. A system configuration with distinct ocean regions and single Primary and multiple Major Paths utilizing the basic spacecraft concept A4 was designed and evaluated. It was assumed that each system user has at least one large earth station and that most or all of its traffic is carried via its large earth station(s). However, to minimize the earth segment requirements, communities of interest on each of the satellites were formed (as if the Primary and one of the Major Path satellites were connected via ISL), and the residual or non-direct (ISL) traffic, which consists of thin-route traffic totalling approximately 1,100 voice channels in each region, was assumed to be carried by auxiliary small antennas (Type D) accessing the Primary satellites in these two regions.

To transmit and receive the non-direct traffic via small antennas a total of 36 Type D earth stations (18 in each ocean region) in addition to the earth segment requirements of the dual-Primary system, with an ISL, are needed. In addition a few of the spacecraft transponders with large output power, if needed, are dedicated to use by small antennas.

The space segment incremental NPW for this system is $428M which is slightly lower than that for the dual-Primary system.

3.6 System Concept I: Systems Utilizing New Frequency Bands

The composite systems studied for the evaluation of the new bands utilization strategies generally fall in the category of systems which make use of two different types of spacecraft. System concept I was assumed in all cases (three, two, and one operational spacecraft in the Atlantic, Indian Ocean, and Pacific Regions, respectively). Growth Scenario I was used in all the analyses and the level of TDMA implementation was allowed to vary. It was assumed that the new bands would be implemented such that FDM/FM/FDMA would remain the prevalent modulation/ multiple access scheme.

In the Atlantic Ocean Region, three all-FM cases were studied corresponding to the use of the three spacecraft concepts utilizing new bands in the role of the Primary and Major Path 1 spacecraft with the B3 concept for Major Path 2. In the Indian Ocean Region, two all-FM cases were studied. In the first case, the B3 spacecraft concept was assumed for both the Primary and Major Path. In the second case, the B3C spacecraft concept was assumed for the Primary with the B3 concept remaining for the Major Path.

The composite system used for comparison purposes is one wherein spacecraft concept B3 is used throughout the system with the implementation of as much TDMA/DSI as is required to carry the traffic of growth Scenario I in 1993 under system concept I.

Table 8 shows the results for the systems which were studied and indicates that the use of new frequency bands can reduce the earth segment requirements and that the benefits are about the same for each of the three new bands utilization strategies in the Atlantic Ocean Region. The Indian Ocean Region can also benefit in terms of earth segment requirements if new frequency bands are introduced on the Primary spacecraft.

Since the B3B, B3E, and B3C spacecraft concepts can directly replace the B3 (leave the new bands capacity unused), a net present worth analysis was made wherein it was assumed that only one type of spacecraft would be used for all ocean regions and roles rather than using the B3 where new bands are not required. As Table 8 indicates, the lower development costs of such a strategy tend to make it less costly than if two different spacecraft were developed.

In the course of the study, it was determined that certain coverage areas would provide only minimum additional benefit if their bandwidths were extended with new frequency bands in the extension strategy. This is particularly true for the beams in the southern hemisphere. Since the cross-strap strategy provides the most flexibility and since a limited extension strategy has the potential of providing the most capacity, it appears that a combination of these two strategies would be appropriate for introduction of new bands in the system.

4. Summary

This paper has addressed the general system considerations related to the future INTELSAT system planning as well as comparison of several alternative systems. It was shown how the general approach followed consists of three basic elements: future planning scenarios, system concepts and spacecraft concepts.

4.1 Spacecraft Concepts

This paper has stressed the description of the spacecraft concepts under consideration for the INTELSAT V Follow-on system. Five principal spacecraft concepts as well as one secondary (or auxiliary) spacecraft concept have been described, as follows:

a. Spacecraft concept S1, utilizing five or six steerable beams (pol B) overlaid over two hemispheric beams (pol A) at 6/4 GHz in addition to a few 14/11-GHz beams.

b. Spacecraft concept S2, characterized by an extensive ability to reconfigure its 6/4-GHz coverage pattern in orbit.

c. Spacecraft concept S3, based on the hypothesis that new frequency bands below 20 GHz will be allocated to the fixed satellite service at the 1979 WARC.

d. Spacecraft concept S4, relying on extensive reuse of the 6/4-GHz band to achieve the required capacity. A particular example formulated incorporates nine zone beam groups (pol B) overlaid over two hemispheric coverage beams (pol A) at 6/4 GHz.

Table 8. Summary Results for Systems Associated with
New Bands Utilization Evaluation

| SYSTEM | Atl | | | | Ind | | | | Pac | TOTALS | | | | Space Segment Inc. NPW ($M) |
| | Earth Stations | | New Bands Access | | Earth Stations | | New Bands Access | | E.S. | Earth Stations | | New Bands Access | | |
	6/4	14/11	6/4	14/11	6/4	14/11	6/4	14/11	6/4	6/4	14/11	6/4	14/11	
Bolt-on	151	11	47	8	86	16	*	*	32	269	27	47	8	336(326)**
Extension	152	9	42	6	86	16	*	*	32	270	25	42	6	336(325)**
Cross-Strap (A)	151	10	45	7	86	16	*	*	32	269	26	45	7	335(325)**
Cross-Strap (A-I)	151	10	45	7	78	9	10	4	32	261	19	55	11	344(325)**
Reference	155	13	TDMA Common Equipment 83		79	12	TDMA Common Equipment 18		32	266	25	TDMA Common Equipment 101		277

*For these systems, the Indian Ocean Region traffic was assumed to be carried in the presently used bands in an all-FM mode using the B3 spacecraft concept.

**Space segment incremental NPW if only one type of spacecraft is purchased (B3B, B3E, or B3C).

e. Spacecraft concept S5, including the INTELSAT V and various modifications of it.

A summary of the principal physical characteristics of the spacecraft concepts under consideration, including alternative configurations for various INTELSAT VI spacecraft that would be suitable for STS launch were studied. Most spacecraft concepts under consideration exceed the capability of the SSUS-A but fall considerably short of the IUS capability. Considering the large gap between the capability of these two stages it appears that liquid engines may offer a more flexible staging option.

The secondary spacecraft concept investigated is a small (Delta class) trunk traffic satellite using reconfigurable zone beam coverages. Such a satellite can be used to carry as many as 36,000 channels in FM/FDMA with small earth stations as long as half of the transponders operate in the single carrier mode and the other half in the multiple-carrier mode.

4.2 Comparison of Alternative Systems

Using the user requirement scenarios developed in the course of the study, and the system and spacecraft concepts, a number of alternative systems have been designed to satisfy the traffic requirements in 1993. For each case the associated satellite on-station plan was generaged and the necessary earth station complement was determined. Using this information, a number of economic evaluators were calculated. The following systems have been presented:

a. systems relying on distinct ocean regions and single Primary and multiple Major Paths,

b. systems relying on two satellite types,

c. interregional systems,

d. dual-Primary systems relying on intersatellite links,

e. systems relying on the use of small earth stations, and

f. systems utilizing new frequency bands.

A summary of different attributes of these systems is given in Table 9 for comparison purposes.

5. Conclusions

Based on the results of the INTELSAT VI Planning Study, the following conclusions can be made:

a. It is more cost effective to achieve the required capacity by additional reuses of the 6/4-GHz band than by additional reuses of the 14/11-GHz band. This is because the increased power required in the 14/11-GHz band increases the spacecraft mass and cost and because the earth terminals are more expensive.

Table 9. Summary of System Results

System Concept	Spacecraft Concept	Number of Operational Spacecraft			Earth Segment Complement				Space Segment Inc. NPW ($M)	Comments
		Atl.	Ind.	Pac.	6/4-GHz Antenna	14/11-GHz Antenna	TDMA Common Equip.	Type D Antenna		
Distinct Ocean Region with Unlimited Capacity Satellite	6/4-GHz Global	1	1	1	186	0	0	0	N/A	For reference only.
Distinct Ocean Region with Single Primaries	A2	3	2	1	258	45	31	0	334	
	A4	3	2	1	261	21	27	0	423	
	B3	3	2	1	266	25	101	0	277	
Distinct Ocean Region with Fully-connected Single Primaries	A2 and Trunk	1 and 4	N/A	N/A	96*	12*	59*	63*	N/A	Designed and evaluated for Atlantic Region only.
	A4	3	2	1	280	21	64	0	423	
Interregional with Fully-connected Single Primaries	A4	3	1	1	257	22	95	0	392	
Distinct Ocean Region with Dual-Primaries	A4 with ISL	3	2	1	252-236	22	27-25	0	432-434	ISL capacity 1200-3600 voice channels.
Distinct Ocean Region with Single Primaries and Use of Type D Antennas for Connectivity	A4	3	2	1	244	23	27	36	428	With a few high power transponders for Type D antenna reception.
Distinct Ocean Region with Single Primaries	B3B and B3	2 & 1	0 & 2	0 & 1	269	27	0	0	336	Requires a total of 55 new band accesses.
	B3E and B3	2 & 1	0 & 2	0 & 1	270	25	0	0	336	Requires a total of 48 new band access.
	B3C and B3	2 & 1	0 & 2	0 & 1	269	26	0	0	335	Requires a total of 52 new band accesses.
	B3C and B3	2 & 1	1 & 1	0 & 1	261	19	0	0	344	Requires a total of 66 new band accesses.

* Atlantic Ocean Region only.

b. Systems relying on the interregional
concept result in significant savings
in the earth and space segment. This
conclusion is also true for the intermediate time frame (1980-85) with the
INTELSAT V satellite.

c. Systems utilizing intersatellite links
for connectivity purposes offer significant earth segment savings. Systems
with microwave ISL's are currently in
operation.[8] Because of its high impact
on system costs, development of a reliable
ISL system should be emphasized.

d. The cost of the TDMA equipment has a
major impact on the earth segment costs
and therefore changes the relative
attractiveness of various alternative
systems. Recent developments of low-
cost TDMA equipment should be exploited
when the INTELSAT TDMA system specification is finalized.

e. Development and introduction of advanced
source coding, modulation, and multiple
access techniques can significantly improve the system efficiency.[9] Source
coding includes single channel techniques
such as companding, PCM, and delta mod-
ultation; and multiple channel techniques
such as TASI and DSI.

f. Systems utilizing SS-TDMA improve the
efficiency of systems with multiple
frequency reuse satellites. Development and testing of an SS-TDMA system
should be pursued.

g. Systems relying on small antennas either
for connectivity or for path diversity on
trunk satellites are more economical than
those relying on large stations only.

h. Systems relying on the use of new frequency bands offer significant savings
in the earth segment requirements.

Acknowledgment

The results reported here represent a collec-
tive effort. In addition to the authors, contributions were made by H. Weiss, O. Atia, L. Palmer,
C. Pentlicki, P. Rivalan, and W. Schnicke.

References

1. Van Trees, H. L., "Future INTELSAT System
(1986-1993) Planning", paper No. 76-233,
AIAA/CASI 6th Communications Satellite Systems
Conference, April 5-8, 1976, Montreal, Canada.

2. Van Trees, H. L., "Planning for the Future
INTELSAT System", paper No. IAF 76-220, XXVII
IAF Congress, October 1976, Anaheim, California.

3. Van Trees, H. L., "Future INTELSAT System
(1986-1993) Planning", Progress in Astronautics
and Aeronautics, Volume 54, AIAA.

4. Van Trees, H. L., "Future INTELSAT System
(post-1985) Planning; A Status Report", paper
No. 77-09, International Astronautical
Federation (I.A.F.) XXVIIIth Congress,
September 25-October 1, 1977, Prague.

5. Van Trees, H. L., Casey, J. P., Chidambaram,
T. S., Parthsarathy, R. and Chasia, H.,
"Development of Traffic Scenarios for the
Future INTELSAT System", AIAA 7th Communica-
tions Satellite Systems Conference, April 24-27,
1978, San Diego, California.

6. Dill, G., Tsuji, Y. and Muratani, T., "Appli-
cation of SS-TDMA in a Channelized Satellite",
ICC 76, June 14-16, 1976, Philadelphia,
Pennsylvania.

7. Rosen, H. A. and Jones, C. R., "An STS Optimized
Spin Stabilized Concept", EASCON 77, pp. 25-
5A-E.

8. Snider, D. M. and Coomber, D. B., "Satellite-
to-Satellite Data Transfer and Control",
AIAA 7th Communications Satellite Systems
Conference, April 24-27, 1978, San Diego,
California.

9. Welti, G. R. and Kwan, R. K., "Comparison of
Signal Processing Techniques for Satellite
Telephony", NTC 77, December 5-7, 1977,
Los Angeles, California.

Part VIII
Regulatory, Political, Legal, and Economic Issues

Most of the papers up to this point have emphasized the technical aspects of satellite communications. The political, regulatory, and economic issues are of equal, or perhaps greater, importance.

The first area of interest is orbit and spectrum utilization. As Fig. 1 (from [1]) graphically illustrates, the number of satellites in geostationary orbit is increasing rapidly. In the 4- and 6-GHz band, there are already segments of the orbital arc where intersatellite interference is an issue. Reprint Paper 8.1 provides an overview of the problem. Some of the measures identified in the paper which will increase the efficiency of orbit-spectrum utilization include segregation of networks according to their interference potential, improvement of satellite and earth-station antenna sidelobe patterns, and various standardization guidelines. Reference [2] provides a survey of interference problems and their application to the orbit utilization problem. The reader should note the similarity between these interference results and those in the digital modulation section (3.3). References [3]–[6] discuss various aspects of the spectrum and orbit utilization problem.

A second related area is the management of the spectrum and orbit. Decisions made at the World Administrative Radio Conference scheduled for 1979 (WARC-79) could have a significant impact on satellite communications. Reference [7] discusses some of the issues that will be important at the 1979 WARC.

A third area of interest is the availability and costs of

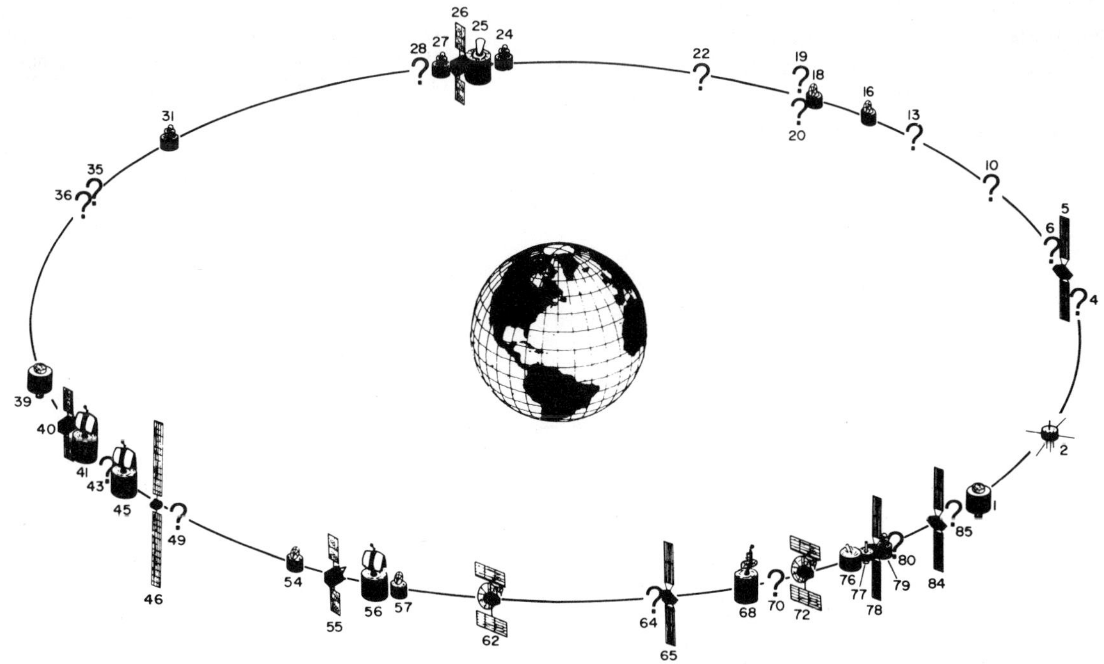

Fig. 1.   Geosynchronous satellites launched or to be launched after January 1, 1976 from [1].

launch vehicles. There are three factors in the launch vehicle area which will be significant in the 1980's. They are 1) the charging policy for the Space Transportation System (Shuttle) and the availability, capability, reliability, and any limitations associated with the shuttle orbit to synchronous orbit booster; 2) the continued availability of U.S. launch services for expendable launch vehicles; and 3) the development of non-U.S. launch capacity. Assuming that the Shuttle costs and availability meet its goals, it will clearly play a dramatic role in the development of communication satellite capability, including being a significant influence on technological priorities and satellite design practices.

Currently, Japan and the U.S.S.R. have their own launch capabilities, and ESA is developing the ARIANE launch vehicle. The U.S.S.R. launch capabilities could potentially impact strongly in the marketplace if they decide to sell launch services.

A fourth area is the growth of regional systems. The questions with respect to regional systems concern the number of such systems to be developed and their impact on and relation to INTELSAT. The Arab Telecommunication Union is moving rapidly to implement a regional system for the Arab countries, and an African system has been proposed. The ECS system, managed by EUTELSAT, is preceeding toward operation.

The 1980's can be expected to be a period of significant sorting out of the relative advantages and economics of the various system options available (e.g., dedicated satellites versus leased transponders, regional versus domestic systems). National goals and priorities, as well as technology and economics, will play a significant role in the decisionmaking process.

A fifth, and perhaps dominant, area is the legal and regulatory environment. There are a number of current actions that will have a major impact on the future of satellite communications, e.g., the cable-satellite question and the appeal of the original authorization of the SBS. However, the dominant long-term factor will be the revision of the Communications Act of 1934, the "Van Deerlin" bill. The outcome of this legislation may affect the entire communications industry in a dramatic manner. References [8]-[10] discuss various aspects of the legal, political, and regulatory environment.

A sixth area can be labeled "government decisions." The question of future NASA involvement in research and development in the communications satellite area is currently being reexamined. The outcome of these discussions will affect both the rate and nature of various technological developments.

Currently, the U.S. Navy is leasing part of the capacity on MARISAT. It has recently made a commitment for a LEASAT which will serve as a follow-up to this service. NASA is proceeding with a lease arrangement on the TDRSS system. Future government policies regarding leasing of services and/or systems will influence the structure and economic viability of various proposed systems.

Various papers in the Bibliography discuss numerous aspects of these problems.

REFERENCES

[1] W. L. Morgan, "Satellite utilization of the geosynchronous orbit," *Comsat Tech. Rev.,* vol. 6, no. 1, Spring 1976.

[2] M. C. Jeruchim, "A survey of interference problems and applications to geostationary satellite networks," *Proc. IEEE,* vol. 65, no. 3, pp. 317–329, Mar. 1977.

[3] W. E. Bradley, "Communications strategy of geostationary orbit," *Astronautics and Aeronautics,* vol. 6, pp. 34–41, Apr. 1968.

[4] J. C. Fuenzalida, "A comparative study of the utilization of the geostationary orbit," *Proc. Int. Conf. Digital Satellite Communication, Intelsat/IEEE,* pp. 213–225, 1969.

[5] J. L. Hult and E. E. Reinhardt, "Satellite spacing and frequency sharing for communication and broadcast services," *Proc. IEEE,* pp. 118–128, Feb. 1971.

[6] J. K. S. Jowett and A. K. Jefferis, "Ultimate capacity of the geostationary-satellite orbit," *Proc. IEEE,* vol. 116, pp. 1304–1310, Aug. 1969.

[7] Federal Communications Commission, "Report and Order in Docket 20271: Preparation for the 1979 General World Administrative Radio Conf. (GWARC), Dec. 6, 1978.

[8] R. G. Gould, "Regulatory aspects of digital communications," *IEEE Trans. Commun.,* vol. 24, no. 1, Jan. 1976.

[9] W. R. Hinchman, "Public policy and the domestic satellite industry," *Int. Conf. Communications,* June 1972.

[10] "Space Broadcasting," *Journal of Space Law,* vol. 3, Spring–Fall 1975.

Effective Utilization of the Geostationary Orbit for Satellite Communication

DAVID J. WITHERS

Abstract—The geostationary satellite orbit has a finite capacity for communication satellites operating in the frequency spectrum available, but saturation could be delayed for a long time if various practices and principles of system design and use were agreed internationally and applied. Three broad areas are identified where action is needed, namely system engineering for an interference-limited environment, intersystem coordination of the use of orbit and the spectrum, and the ordering of the use of orbit and spectrum so that systems with radically different characteristics do not interfere. In each of the three areas, the main factors of concern are identified, the benefits that would be achieved are deduced in broad terms, and the practical and economic feasibility of actually securing the benefit is reviewed for each factor.

I. INTRODUCTION

THE NUMBER of communication satellites in service has increased considerably in the past 2 or 3 years and the upward trend is accelerating [1]. Most of these satellites are geostationary. In some parts of the geostationary orbit it is already necessary to pay careful attention to intersatellite interference problems when choosing the location for a new satellite using the 4- and 6-GHz bands. The time is fast approaching when new geostationary satellites using these bands will be constrained as to their location and technical parameters if they are not to cause unacceptable interference to satellites already in service.

Extensive additional bandwidth is already allocated to satellite services in higher frequency bands but it is little used at present. Radio propagation in these higher bands is more dependent upon meteorological conditions than at 4 and 6 GHz and currently, hardware is in a less advanced state of readiness. However, the higher frequency bands also have advantages when compared with 4 and 6 GHz, making them acceptable or even preferable for some applications. These bands will come into use on their own merits or as a consequence of congestion in the lower frequency bands and eventually the interference problem will appear there also.

Ultimately, it can be foreseen that there will be no room for further general growth of satellite communications services. When that happens, it will be necessary to decide which satellite services should be given priority. We can perhaps foresee that priority will be given to services for ships, aircraft, and isolated communities on land and to various broadcast, quasi-broadcast, and multi-terminal services for which the satellite medium is particularly well-suited. Heavy-traffic point-to-point services may be progressively squeezed out of the more desirable frequency bands and the geostationary orbit, and perhaps ultimately out of space radio altogether into terrestrial transmission media. However, this problem should surely not become acute for several decades if the space medium is used wisely.

A great deal can be done, by both technical and administrative means, to optimise the use made of the geostationary sat-ellite orbit. The limitation of interference between satellite networks is by its nature an international problem and the International Telecommunication Union (ITU) has played an important role, originally at the Extraordinary Administrative Radio Conference - Space Radiocommunication in 1963 and more emphatically at the World Administrative Radio Conference (WARC) in 1971, in dealing with it. Throughout this period CCIR Study Group 4 (Fixed-Satellite Services) has been working on this problem, and more recently Study Groups 8 (Mobile Services), 10 and 11 (Sound and television broadcasting). A WARC with wide terms of reference is planned for 1979 at which it can be foreseen that the ITU provisions for regulating the use of the geostationary orbit will be revised to incorporate the results of recent experience and studies. The approach of this latter conference, which may become a turning-point in the international management of orbit utilisation, makes the present a good time to review the problems and outline some possible solutions.

There are various ways of stretching the satellite communication medium beyond the capacity of the geostationary orbit used in conventional ways.

1) Satellites in phased inclined-circular medium-altitude orbits (typically 20 000 km above the earth's surface)[2], inclined-circular geosynchronous orbits [3], or inclined-elliptical subsynchronous orbits [4] could operate without encroaching upon that unique resource, the geostationary orbit. In general, costs and operational complexity would be greater than for geostationary systems, but some of these orbits also have advantages which offset the disadvantages to some degree and for some applications.

2) Intersatellite links, operating at the higher millimeter-wave frequencies or even in the infrared, could be used to enable the microwave traffic links between earth and space to be used to the greatest advantage by concentrating traffic for an earth station, which might otherwise be distributed between several satellites, into a single high-capacity high-efficiency link. Intersatellite links might also relax the geographical constraints on satellite positioning and increase the operating angle of elevation at earth stations, enabling parts of the orbit and spectrum to be used that might otherwise be unsuitable for serving all the earth stations in any one wide-coverage network.

3) Earth-space frequency bands could be used in both directions of transmission, Satellite A receiving from earth in the frequency band that its near neighbor, Satellite B, is transmitting to the earth and vice versa. Given accurate satellite station-keeping, reasonable satellite antenna sidelobe control and well-separated earth stations, the interference between the networks using Satellites A and B need not be significant and the total potential traffic capacity of the orbit could therefore be doubled.

However, the basic problem is to make the best use of geostationary satellites, used conventionally, and it is this problem that the rest of this paper addresses.

For some services it may be possible to draw up a carefully optimized plan for the utilisation of the orbit and specified

Manuscript received July 12, 1976; revised September 28, 1976.
The author is with the Space Communication Systems Division, Telecommunications Development Department, Post Office, London EC1, England.

Reprinted from *Proc. IEEE,* vol. 65, pp. 308–317, Mar. 1977.

frequency bands. This may be possible, for example, for satellite broadcasting [5], [6]. A similar situation may arise in the maritime and aeronautical mobile services via satellite. However, the wide variety of satellite point-to-point services that can be expected to jostle for the use of the spectrum and the speed and unpredictability of advances in technology will prevent efficient large-scale long-term planning of the fixed-satellite frequency bands. This case demands a continuous process of *ad hoc* adjustment of short-term localized plans.

If the geostationary satellite orbit is to be used effectively for services for "fixed" earth stations, (that is, stations providing permanent communication facilities) systems must be designed and used, not only so that the objectives of individual systems are met but also so that satellite networks using neighbouring positions in the orbit live harmoniously together without waste. Such harmony will have to be bought by careful orbit and spectrum planning, good system engineering and good operational control. The price may include some loss of network capacity and flexibility. Nevertheless, in considering what sacrifices may be required of individual systems for the good of all systems, we must not lose sight of the fact that international consent is unlikely to be secured if the cost to individual systems is too high. It is necessary to seek an acceptable middle way between excessive restriction that would cramp development and self-destructive freedom.

Effective orbit utilization demands three things.

1) Engineering for an interference-limited environment. The basic characteristics of the equipment on the ground and in space and the techniques used for modulation and multiple access should be chosen to minimise the interference power level received from and injected into other networks. That having been done, it is desirable to increase the tolerability of the interference that does occur. These measures would increase the number of networks that can use the orbit, without increasing the number of networks with which each new network must be coordinated.

2) Effective intersystem coordination. Efficient administrative machinery should be used, constructively, to coordinate the use of the orbit and the spectrum by neighboring satellites.

3) The avoidance of extreme inhomogeneity in orbit and spectrum sharing. The satellites of networks that differ greatly in spectral power density and radio channel bandwidth should be segregated from one another into different frequency bands or, where possible, into different parts of the orbit. When this rough sorting process has been completed, a further improvement in orbit loading might be achieved by the standardisation of key system parameters.

The factors arising from these three requirements will now be examined in more detail.

II. ENGINEERING FOR AN INTERFERENCE-LIMITED ENVIRONMENT

Almost all of the frequency bands used so far for satellite communication are shared with terrestrial line-of-sight radio relay systems. These radio relay systems are widely spread and heed has had to be paid to limiting interference between the two media. This has involved endless planning effort, but operational constraints have not been severe so far. The use of carrier energy dispersal and care in earth station siting has been enough to keep interference in check in most situations. The agreed limits on down-path power flux density from satellites

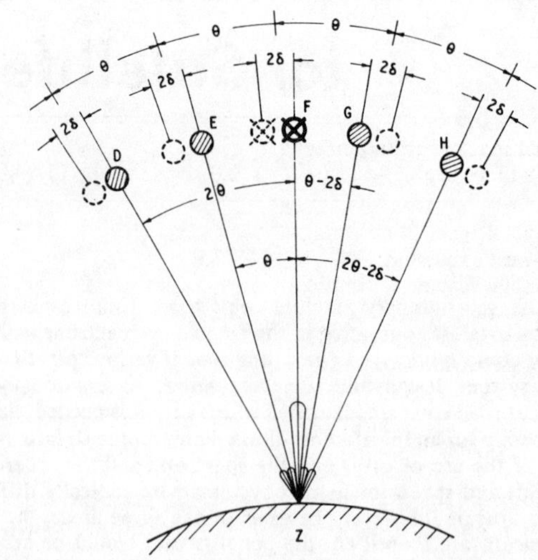

Fig. 1. The worst satellite location situation for earth station Z served by satellite F and receiving interference from neighboring satellites D, E, G, and H.

and of eirp for radio relay stations [7] have provided further protection against interference, but these limits have not set bounds on practical commercial systems until quite recently.

However, satellite system characteristics have been determined largely without regard to interference between satellite networks. Satellite antenna radiation patterns have been optimised, within payload mass limits, to provide maximum gain towards the earth stations served. Earth station antenna radiation characteristics have been chosen, within cost limits, to maximize the G/T ratio. The choice of techniques for modulation and multiple access has been a compromise between a low-cost earth/space interface and high per-transponder traffic capacity. No doubt these objectives will continue to occupy a prominent place in system designers' minds. Nevertheless in the future it will no longer be possible to ignore the need to design also for limitation of interference between satellite networks.

A. Satellite Station-Keeping

Solar radiation pressure and irregularities in the earths' gravitational field, acting upon a satellite which is nominally geostationary, cause the orbit of the satellite to depart to some extent from the ideal. These orbital errors cause the satellite, as seen from the earth, to move to east or west off its nominal longitudinal station. The nominal spacing between satellites must exceed the minimum necessary spacing by a margin which allows for these station-keeping errors. Inclination of the orbital plane also causes a daily east-west motion, but this is negligibly small for moderate angles of inclination.

Under favorable circumstances (i.e., with relatively homogeneous orbit utilization), and assuming perfect station-keeping, the minimum necessary east-west spacing between satellites serving the same region would typically be between $2°$ and $5°$, although spacings as great as $10°$ might be necessary if the earth station antennas were particularly small. The ITU radio regulations require a satellite to be kept within $\mp 1.0°$ of its nominal longitude, but a spacing margin big enough to guard against such large excursions involves a quite significant loss of useful orbit-space.

Fig. 1 shows Satellites D, E, F, G, and H, part of an array of similar satellites all of which illuminate an earth station at Z served by Satellite F. Each satellite is assumed to move $\pm\delta°$

TABLE I

Required spacing, given perfect station-keeping	2.0°	5.0°	10.0°
Required nominal spacing, assuming ±1.0° east–west excursion	2.85°	6.25°	11.25°
Required nominal spacing, assuming ±0.1° east–west excursion	2.13°	5.11°	10.1°
Increase in orbit capacity given by a reduction of excursion from ±1.0° to ±0.1°	25%	22%	10%

about its nominal orbital position and the angular separation between the nominal locations is $\theta°$. (Here and elsewhere in the paper, presentation is simplified by treating angular relationships in orbit as seen from the earth's surface as if they were the same as the corresponding angle seen from the center of the earth. The error involved is fairly small.) In the diagram these five satellites are shown to have drifted into the positions where the interference entering the sidelobes of the earth station antenna at Z will be worst. Then, making the assumption, discussed below, that the gain of the sidelobe envelope of that earth station can be approximated by

$$32 - 25 \log_{10} \phi \text{ dBi}$$

where ϕ is the off-boresight angle, the total interference at Z from satellites $D, E, G,$ and H is a function of

$$\theta^{-2.5}[1.18 + (1 - 2/\theta)^{-2.5} + 0.18(1 - \delta/\theta)^{-2.5}].$$

Interference from more distant similar satellites, that is further from F than D or H, will be small compared with these four main components. This expression can therefore be used to derive the relationship between the required nominal spacing θ and the positional tolerance δ. Equating δ to 0.1, chosen as a good but already attainable operational standard of station-keeping, provides a measure of the waste of orbit allowed by the present international agreement. This calculation has been done for various values of θ and the results, shown in Table I, indicate that orbit loading efficiency could be increased, typically by over 20 percent, if east–west excursions were limited to ±0.1°. (This example is somewhat simplified because it does not take into account the statistics of satellite positions; nevertheless, it contains the substance of the problem.)

In recent years, several operational satellites in the INTELSAT and TELESAT systems have been kept within 0.1° of their nominal longitude. This demands careful and relatively frequent adjustment of the orbit period, perhaps as often as once per week for satellites located at longitudes where the east–west component of the earth's gravitational field is strong. The cost of such activity at the control earth station may not be negligible but it is very small compared with the value of the extra orbital space it would release for use. For these spin-stabilized satellites, INTELSAT III, INTELSAT IV, and ANIK, there is no other significant cost, since little more thruster fuel is required for frequent small adjustments of orbital period than for less frequent large adjustments. For body-stabilized satellites with deployed solar arrays, seasonal correction of orbital ellipticity caused by solar radiation pressure may be necessary if east–west movement from all causes is to be kept within ±0.1° and this will require a small expenditure of fuel.

Clearly, an improvement in required satellite east–west station-keeping to perhaps ±0.1° would be an effective, feasible and economical way of making a substantial improvement in orbit utilization efficiency.

B. Satellite Antenna Radiation Characteristics

Subject to the constraints of payload mass, attitude stability and accuracy of station-keeping, the communications capacity of a satellite will usually be maximised by its designer by making the satellite antenna gain as high as is consistent with enclosing the whole of the service area by the -3 dB or -4-dB contour relative to the peak gain of the main lobe. It may be worthwhile to tailor the cross-section of the beam, departing from a simple circular or elliptical form in order to fit more accurately the projected shape of the service area. The mean gain within the service area may also be increased by modifying the typical $\sin^2 X/X^2$ distribution of power across the main lobe. Limitation of antenna response outside the service area is not a major factor in system design at present, except for satellites using spaced-beam frequency reuse, where the objective is likely to be specifically to achieve an adequate carrier-to-interference ratio in each of the satellite's own service areas, not a general minimization of out-of-beam response.

Effective orbit utilization demands proper attention to satellite antenna response outside the service area and this demand will have to be imposed administratively. Various approaches to the problem are conceivable and the problem is to find one that will be feasible in practice and will substantially increase the practical capacity of the orbit without unduly restricting the development of technology or of new facilities. This will not be an easy combination to achieve and the approach indicated below is offered with some diffidence.

Two typical interference situations arise from satellite antenna characteristics.

1) Two satellites may be well-separated in orbit but their service areas are close together and satellite antenna main-lobe overspill is more than the directivity of the earth-station antennas can control.
2) Two satellites are relatively close together in orbit but their service areas are well-separated and interference arises via the sidelobe response of the satellite antennas.

These two situations call for different treatment.

It would be severely restrictive to require the satellite antenna main-lobe response (case 1) to fall away rapidly in all directions outside the service area. Such a requirement might also limit undesirably the width of the orbital arc within which the satellite could be located and still fulfill its mission (that is, the "service arc"). Furthermore, the maximum gain slope that could reasonably be required in all directions, whether necessary to control interference or not, would be less than could be achieved in selected directions to control specific instances of foreseeable interference. An alternative would be to seek to achieve the best match between the operational needs of the networks serving a region and the available satellite antenna technology by negotiation between the owners of the systems involved, such negotiations forming part of the coordination process. This latter approach has the disadvantage that it would take no account of the requirements of networks that are not in service or planned at the time of coordination; nevertheless it seems more likely to produce good results that the application of rigid main-lobe gain-slope regulations. The guidance of an internationally approved radiation pattern for the main-lobe

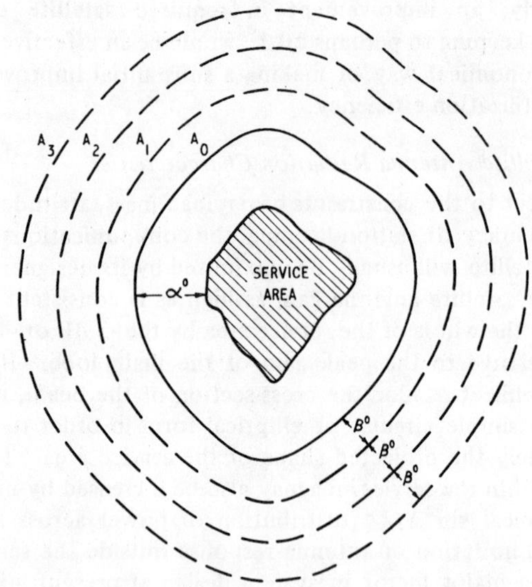

Fig. 2. Limitation of satellite antenna sidelobe radiation; suggested contours relative to the service area.

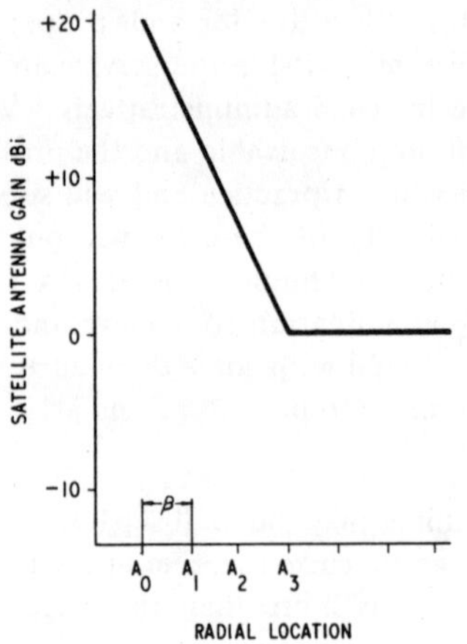

Fig. 3. Limitation of satellite antenna sidelobe radiation; suggested sidelobe gain limits at contours A_0, A_1, etc.

and the first one or two sidelobes might, nevertheless, exert a useful influence on the standards attained.

It is not likely to be feasible to use a coordination process to control interference via satellite antenna sidelobes (case 2); perhaps the most substantial source of difficulty would be the sheer number of parties that would be involved in such discussions when the orbit is approaching saturation. It seems, therefore, that the best solution would be to apply an internationally agreed limit to satellite antenna gain (or related characteristics such as satellite receiving sensitivity and radiated power spectral density) in directions well removed from the service area. The limit might take the following form.

Let a contour A_0 be drawn outside the service area of a beam so that there is an angular clearance of $\mathcal{L}^0$ between the contour and the nearest point in the service area as seen from the nominal location of the satellite. A family of three contours A_1 to A_3 might then be drawn, outside the first contour, each separated from its neighbours by β^0 as seen from the nominal satellite location (Fig. 2). Then the antenna gain might be required to fall below +20 dB relative to isotropic at contour A_0, and so linearly to 0 dBi at contour A_3, the sidelobe peaks remaining below 0 dBi outside A_3 in all directions intercepted

by the earth (Fig. 3). The value of $\mathcal{L}$ might be related to the size of the service area, allowing a generous margin within which the radiation pattern would be determined by system economics, coordination with networks serving nearby areas, and the provision of an adequate service arc. It is, however, important that the response should cut off rapidly outside contour A_0. The limiting gain slope need not be determined by the size of the service area, since large service areas may be covered by overlapping beams generated by multiple feeds serving an antenna of large aperture. The value of β should therefore be related to the frequency and the largest satellite antenna aperture that might reasonably be used; possible values are 1.0^0 at 4 and 6 GHz, 0.4^0 at 11/12 and 14 GHz and 0.2^0 at 20 and 30 GHz.

C. Polarization

There are prospects of improving the efficiency of orbit-spectrum utilization by a factor approaching 2 by dual polarization operation. The technical problems of dual polarization are being explored in a number of experimental programmes involving ATS-6, INTELSAT IVA and the ESA Orbital Test Satellite (OTS) and its use is foreshadowed in the RCA and COMSTAR US domestic systems, INTELSAT V and ESA European Communication Satellite (ECS) systems. The benefit is most evidently realizable for networks, such as those mentioned above, having a traffic capacity requirement that is larger than the bandwidth allocated in the frequency bands used (typically 500 MHz) can readily supply in the single polarization mode.

When the network traffic requirement is less than that which the allocated bandwidth could carry with dual polarization, orbit-spectrum utilization could be improved in the same way, the bandwidth occupied by the system being cut to 100 MHz, 200 MHz, or whatever is sufficient in the dual polarization mode. An apparent alternative, namely for such networks to operate in the single polarization mode but with high purity of polarization, so that interference from another network using a nearby orbital location in the reverse polarization mode could be rejected, seems unlikely to be feasible owing to several difficulties, perhaps the greatest of which is the rapid deterioration in the rejection of cross-polar signals by a typical earth station antenna as the unwanted satellite moves away from the boresight direction. However, while experimental results obtained so far give confidence that dual polarization will be feasible, it is clear that the increase of cost of some elements in systems will not be insignificant, including:

1) dual-polar feeds;
2) antenna systems of good polarization purity;
3) additional earth station high power amplifiers and low noise amplifiers where access is needed to both polarization modes;
4) adaptive interference suppressors if needed to deal with rain depolarization.

Where the network traffic capacity requirement is high enough, these costs would be accepted because dual polarization will enable one satellite and one set of earth terminals to do the work of two; there will therefore be no need for external pressure to urge the adoption of dual polarization in such cases. Where the traffic demand is less, the cost penalty arising from a requirement to use dual polarization would not, however, be negligible. At some time in the future the congestion of the orbit and spectrum may become so severe that general use of dual polarization will be necessary. In anticipation of this, it would be desirable for new earth terminals to be designed so that they could radily be converted to dual polarization operation. For the time being, however, it does not seem to be necessary to make dual polarization a universal requirement; there are more cost-effective ways of improving orbit-spectrum utilization in the short term.

D. Earth-Station Antenna Sidelobe Radiation

INTELSAT requires the antennas of earth stations with unqualified access to the system ("Standard A") to have a figure of merit G/T not less than 40.7 db/K at 4 GHz. Lower values of G/T, typically between 27 and 32 db/K, are specified for other systems. Such antennas have D/λ ratios between 200 and 500 and if realized in ideal form their sidelobes would be negligible a few degrees from boresight. Practical antennas have quite considerable sidelobe skirts, due to their various imperfections, and interference to and from unwanted satellites, coupled through these sidelobes, plays a major part in determining how closely satellites sharing the same frequency band may be located.

The symmetrical Cassegrainian configuration is used almost invariably today for large earth station antennas and Fig. 4 shows the form of the envelope of the sidelobe peaks for a typical antenna. The sidelobe envelope of the great majority of large Cassegrainian antennas would fall within a few decibels of this curve. It is convenient to fit a simple law to the curve, for the purpose of analysis. A convenient form would be

$$E = A - B \log_{10} \phi \text{ dBi}$$

where A and B are numerical coefficients and ϕ is the angle between the direction of interest and the boresight axis. In 1965, the CCIR adopted $E = 32 - 25 \log \phi$ dBi as the reference radiation pattern for use in large-antenna interference calculations when specific antenna data are not available [8], this envelope being assumed to include 90 percent of the sidelobe peaks and to apply from $1°$ off-boresight out to the points where $E = -10$ dBi (that is, to $48°$ off-boresight). Beyond $48°$ it is assumed that the envelope continues constant at -10 dBi. That law was based upon the antenna data then available. It still fits the available information on operational antennas reasonably well, and is widely used in generalized studies of interference.

It is instructive to consider briefly what change would be effected in the calculated maximum load of the orbit if these coefficients A and B were altered. Thus, given homogeneous orbit loading with satellites accessed by earth stations to the INTELSAT standard, using large-capacity FDM/FM carriers, the minimum acceptable spacing would be about $2.9°$ when $E = 32 - 25 \log \theta$. Figs. 5 and 6 show the substantial effects on minimum satellite spacing that would follow if the coefficients were improved relative to the CCIR reference radiation pattern.

However, to see what order of improvement might be obtainable in practice, it is first necessary to consider why the practical radiation patterns of large Cassegrainian antennas are so much worse than the ideal case [9]. Out to about $2°$ off-boresight, high sidelobes are caused mainly by large-scale inaccuracies of the profile of the main reflector. The use of insufficient illumination taper also has a significant effect here. Between $2°$ and about $10°$ off-boresight, small-scale profile errors of the main reflector are the main source, although scat-

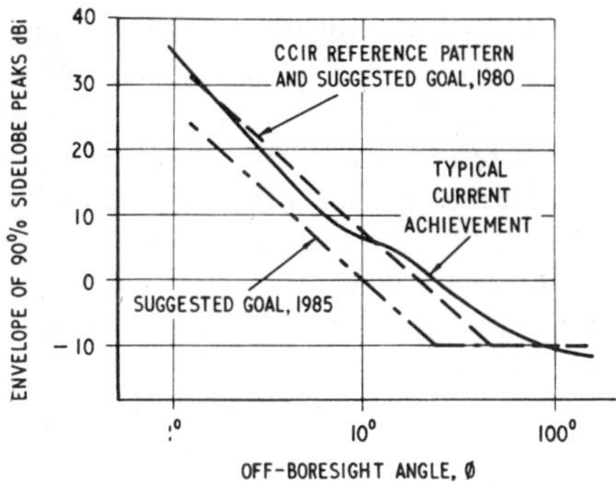

Fig. 4. Sidelobe envelopes for large earth station antennas.

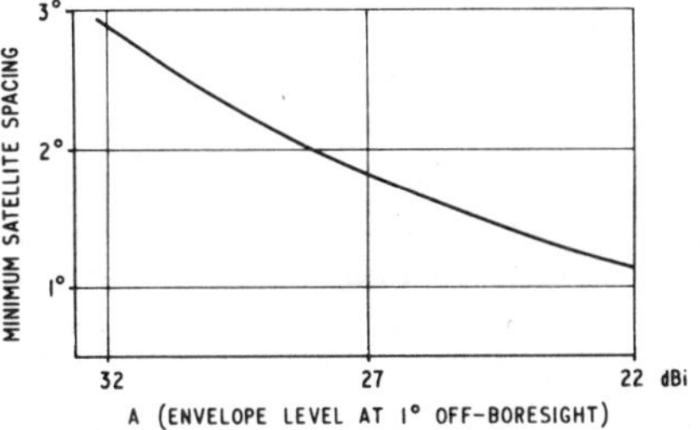

Fig. 5. The effect on minimum required spacing of satellites of homogeneous networks with large earth station antennas if the gain slope of the sidelobe peak envelope remains constant at the value given by $E = A - 25 \log \phi$ but its level varies.

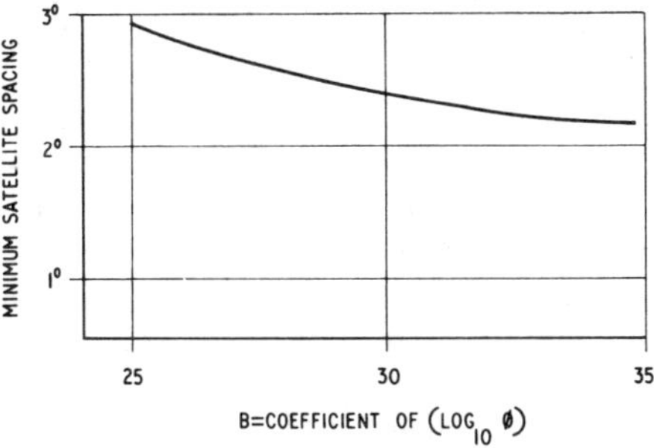

Fig. 6. The effect on minimum required spacing of satellites of homogeneous networks with large earth station antennas if the level of the sidelobe peak envelope at $1°$ off-boresight remains at 32 dBi and B, the coefficient of $\log_{10} \phi$, varies.

tering at the subreflector supports and diffraction at the edge of the subreflector are also significant. Between $10°$ and about $30°$ off-boresight, the main cause is subreflector spillover. Finally, beyond $30°$ the sidelobes are fed from many sources but particularly from scattering at the subreflector and its supports and the sidelobes of the feed radiation pattern.

Thus, two important sidelobe-generating mechanisms emerge for large symmetrical Cassegrainian antennas over the angles where the sidelobes are likely to do most harm, namely small-scale main reflector profile errors and over-illumination of the subreflector. It does not seem to be possible, at present, to provide an exact quantification of the relationship between these causes and their effects, and the following should be considered as approximate estimates only. An improvement in the small-scale profile tolerance from 0.5-mm rms (which is typical of current practice) to 0.25-mm rms (probably corresponding to the best current commercial practice) could be expected to improve the envelope of the sidelobe peaks between 2° and 10° off-boresight from a level somewhat below $32–25 \log \phi$ dBi to about 3 dB lower; the cost would not be negligible but it would be small relative to the cost of an earth station. Further refinement of the profile would be beneficial, but it might be necessary to introduce new techniques of fabrication if the cost is to remain acceptable. An adjustment of subreflector illumination to make sure that the edge-illumination is at least 20 dB below the maximum should bring the sidelobe envelope level between 10° and 30° off-boresight down by several decibels to the CCIR reference level for a trifling cost. A further improvement could be obtained by the use of an offset-fed configuration, but the cost of this would be difficult to estimate, pending the evolution of suitable commercially available designs.

In short, the sidelobe performance of large earth station antennas operating at 4 and 6 GHz could be improved by say 3 dB over the important arcs of the polar diagram for a modest cost, and corresponding improvements could, no doubt be obtained with smaller antennas and at higher frequencies. A general improvement of this order could lead to a reduction in minimum satellite spacings of about 25 percent and a corresponding improvement in orbit-spectrum utilisation.

Improving the sidelobe radiation pattern of the earth station antenna has little effect on boresight gain and so it is not cost-effective in the optimization of equipment design unless some cash value can be attached to increasing the effectiveness of orbit utilization. Thus, improvement is only likely to be achieved by internationally agreed administrative action. Furthermore, benefit in terms of orbit utilization will not be obtained until most earth station antennas in service have better performance. In view of the large number of earth station antennas already in service and the huge investment that they represent, it would be unrealistic to expect a very rapid improvement in orbit-spectrum utilization efficiency to arise from action taken now. Finally, the achievement of an improvement in sidelobe response by administrative action is made more uncertain by the fact that it is far from easy to measure the radiation pattern of a large earth-station antenna or of any antenna, large or small, that has limited steerability.

However, these admitted difficulties are worth overcoming. Just what would be a realistic target to aim for is a matter for further study, but it may not be unreasonable to require all large new antennas taken into service after, perhaps, 1985, to achieve the standard shown on Fig. 3, and for none which fail to achieve it to remain in service after 1995. The $E = 32 – 25 \log \phi$ law might serve as an interim goal, to be achieved by all new antennas from 1980. The same standards might be found suitable for smaller antennas also.

Finally, it may be noted that the minimum separation between two satellites is often determined, not by widespread interference, but by interference on one or two carriers at a few earth stations only. Adaptive interference cancellation has already been demonstrated to be feasible when a terrestrial station interferes with the reception of a satellite signal [10], and this technique should be readily applicable in the much simpler case of interference between satellite networks, especially when the interference arises in the down-path.

E. Modulation Technique and Spectral Energy Distribution

Depending upon the down-path carrier power available and the modulation, multiplexing, and multiple-access techniques used, the traffic capacity of a satellite per unit of bandwidth will vary considerably, for example from 10 telephone channels per megahertz when FDM/FM is used in FDMA to 50 channels in TDMA with 4-phase PSK and digital speech interpolation. The choice of modulation techniques also has a considerable effect upon the level of interference at the receiver input that can be borne without the post-demodulator interference exceeding the permissible level. It would not be practicable to review all the significant aspects of this complicated matter in a general paper[1] nor would it be easy to draw out generalized conclusions as to the techniques which optimise orbit-spectrum utilization. It may not even be helpful to generalise in this way, since the choice may be too strongly influenced by economic factors arising from the use to which the satellite network is to be put and the nature of the terrestrial system with which it interfaces. Nevertheless, it is useful to consider whether there are factors in this area which are conducive to efficient orbit-spectrum utilisation, yet do not enter fundamentally into the economic viability of the system.

The carrier power required for satisfactory reception of an emission is a function of the predemodulator bandwidth, but its potential for causing interference is a function of the interference power falling into the bandwidth of the link suffering interference, and in many cases to the spectral distribution of interference power within that bandwidth. Thus, for FDM/FM wanted emissions in general and for narrow-band wanted emissions regardless of modulation technique, a wide-band unwanted signal may cause interference out of proportion to its mean flux density if the distribution of energy in the interfering spectrum is strongly nonuniform. In the absence of special arrangements, the spectral energy distribution of emissions is usually markedly nonuniform, particularly when traffic loading is light, the energy tending to be concentrated about the carrier frequency for analogue emissions and about a series of discrete spectral lines in digital emissions. Artificial spectral energy dispersal techniques, such as the addition of a low-frequency sawtooth waveform to an FM baseband and the addition of a long-cycle pseudo-random sequence to digital traffic signals, can improve the uniformity of the distribution of the spectral energy of emissions markedly without high cost or significant degradation of the performance of the wanted signal.

Three conclusions can be drawn from this:

1) effective artifical spectral energy dispersal should be used whenever the energy dispersal provided by the traffic signal is inadequate,
2) a measure of uniformity of predemodulator bandwidth is desirable,
3) particular care must be taken to deal with the susceptibility of FDM/FM systems to interference from unwanted signals with insufficiently dispersed spectra.

Of these, the first has the greatest impact on orbit-spectrum utilisation and least impact on system design. At present, most satellite carriers have some degree of artificial energy dispersal applied to prevent the down-path power flux density at the earth's surface from exceeding limits imposed by international agreement to prevent interference to terrestrial radio services. In the future, it would be desirable for all satellite emissions to have as large a degree of artificial energy dispersal as is consistent with efficient reception as a wanted signal.

---

[1] A more extensive discussion of this topic is contained elsewhere in this issue; see "A survey of interference problems and application to geostationary satellites" by M. Jeruchim.

F. The Interference Degradation Budget

At present, designers of analog satellite systems are recommended to allow 10 percent of all channel noise for interference from other satellite networks and a further 10 percent for interference from terrestrial radio systems [11]. The makeup of the remaining 80 percent will include allocations for such components as up-path and down-path thermal noise, earth station and satellite intermodulation noise, cross-polar and interbeam interference within the same network, multipath and group delay distortion, but the allocation for each component will depend on the circumstances of each network and each carrier. Recommendations for the interference budget for digital systems are still under study; when they emerge they are likely to be similar in broad effect to those already existing for analog systems, although differences in detail are bound to be made necessary by differences in the nature of the interference mechanism.

It has been estimated that the total traffic capacity of a busy arc of the orbit could be increased by 75 percent if the proportion of the degradations budget allocated to interference were increased from 10 to 50 percent, circuit performance being restored to the nominal state by making corresponding reductions in the allocations to other degradations. It will sometimes be feasible to do this, although at the cost of a modest loss of satellite capacity, where the thermal noise allocation is initially large, since post-detector thermal noise can be reduced by increasing carrier power and/or by using modulation techniques or parameters more resistant to interference. However, it may not be feasible to do this in networks where the thermal noise allocation has already been made small in order that the allocation to intermodulation noise or to interference from other frequency reuse modes within the same network may be large. In the present, relatively early stage in the technical and operational development of satellite communication it seems premature to increase the internetwork interference allowance beyond 10 percent. In the future, when the orbit is more crowded, and the whole of the 10 percent allowance has already been taken up by interference it may well be useful to allow some additional networks to open service by increasing the internetwork interference budget to perhaps 20 percent of all degradations. However, it might be necessary to apply this increase selectively, excluding its application to networks where the frequency spectrum is reused within the network.

III. Intersystem Coordination

A satellite network needs several years from conception to launch, it is always costly and, once the spacecraft design has been determined in detail, many of the important characteristics cannot readily be changed. Interference between networks which might be severe can often be reduced to a tolerable or even negligible level by minor adjustments to characteristics if the need can be identified early enough. What is needed is a system for bringing together the various entities who are using or planning networks that could interfere, so that the adjustments required can be identified in time and agreed. In practice, this process of coordination will have to be repeated occasionally throughout the lifetime of a satellite as new satellite projects designed to use the same frequencies and the same broad arc of the orbit come into being.

Machinery for the coordination of geostationary satellites was agreed at an ITU conference in 1971 [12]. In brief it is as follows. As soon as an organization planning a new satellite network has determined the technical basis of the system, but not more than five years before the first launch, provisional system characteristics are passed to the ITU and circulated to all national frequency authorities and other interested parties. The system characteristics are given in sufficient detail for any other user or intending user of the orbit to calculate whether interference at a significant level might be suffered by his own network. If the risk is significant (the agreed threshhold condition being when the maximum interference energy at an earth station receiver due to the unwanted network would be equivalent to 2 percent of the total noise arising within the wanted system), the parties concerned collaborate to refine the calculation of interference probability and to identify any necessary and acceptable changes in the characteristics of one system or the other that would reduce interference to a permissible level. The agreed characteristics are eventually recorded by the International Frequency Registration Board, an organ of the ITU.

No doubt these administrative provisions will be improved in detail at the 1979 WARC. For example, it may be possible to devise a method, more selective than the 2 percent-of-thermal-noise criterion, that will identify all the networks for which a proposed new network is a potential source of significant interference without needlessly complicating the coordination process by bringing into it many cases when interference will not be significant. These processes will, however, become extremely complex as the number of systems, operational and planned, grows, and the prospects of success must diminish as complexity increases. Some reduction in the number of parties could be expected from the application of the principles advocated in section II above. Nevertheless, if coordination is to yield solutions which satisfy the parties immediately involved while leaving opportunities for the entry into the orbit of yet further satellites, not then announced, the process must surely be simplified by the adoption of additional technical guidelines to facilitate the achievement of solutions and to eliminate options which are not essential and complicate negotiation. These guidelines might include the following.

A. Frequency Band Pairing

Most present-day fixed-satellite spacecraft use the 3700–4200 MHz (down-path) and the 5925–6425 MHz (up-path) frequency bands. Several projected satellites couple down-path bands at 11 or 12 GHz, with up-path bands at 14 GHz. Soon, no doubt, satellites transmitting at 20 GHz and receiving at 30 GHz will be in service. However, these band pairings arise from technical, economic, or even historical circumstances; the internationally agreed frequency allocations identify some bands as up-path bands and others as down-path bands but do not specify pairings. If for one particular application it were technically preferable to use 14 GHz as the up-band and 4 GHz as the down-band, there is no international impediment to that pairing, even though such an unusual satellite, in the company of others using conventional pairings, would provide the same communications capacity as one conventional satellite while using as much orbit-spectrum as two and while likewise demanding as great a coordination effort as for two conventional satellites. In principle the waste of orbit and spectrum that this would involve could be minimised if another satellite which paired the 6-GHz band (up-path) and the 11 or 12 GHz (down-path) were colocated with the 14-GHz/4-GHz satellite, but such arrangements provide no general solution in practice.

There is an urgent need to determine preferred frequency band pairings and to use them in all normal applications. Complications can be foreseen.

1) Some high capacity satellites will be designed to use two or more pairs of bands. Such satellites may need to be cross-strapped; thus, for example, in a satellite that uses the 4- and 6-GHz bands and the 11- and 14-GHz bands, it may be desirable for some transponders to pair 14 and 4 GHz or 6 and 11 GHz. This practice may be economically advantageous since it may provide the best match to service requirements, and it will not necessarily use the spectrum inefficiently for a network which occupies the full bandwidth of all four bands.

2) National frequency allocations differ in detail in different parts of the world. Thus, for example, some systems operating at 4- and 6-GHz pair 3700–4200 MHz with 5925–6425 MHz, while others pair 3400–3900 MHz with 5725–6225 MHz.

3) There is an imbalance between the up-path and down-path frequency allocations, and this is aggravated by the fact that up-path bands allocated for the fixed-satellite service may be used for the up-path connections to satellites of other services, for example, broadcasting satellites. Thus, in ITU Region 2 (North and South America) it would be natural to use the 14.0–14.5 GHz up-path band for fixed-satellites transmitting in the 10.95–11.2 GHz and 11.45–11.7 GHz bands and also for both fixed-service and broadcasting satellites transmitting in the 11.7–12.2 GHz band.

These various complexities suggest that no simple, inflexible code of band pairing rules is likely to be acceptable. Nevertheless, it should be feasible to agree a set of guidelines that would help to promote efficient orbit-spectrum utilisation.

B. A Standard Frequency Translation Between Up-Path and Down-Path Frequency Bands

Many satellites use the whole width of the allocated frequency band and the mean frequency translation introduced in the satellite is therefore made equal to the difference between the up-path and down-path frequency allocations. Where the bandwidth occupied is much less than the allocated bandwidth, it should be quite feasible for two or more satellites, displaced in frequency, to use the same orbital position without interference. Obviously this economical arrangement will be arrived at more easily if all these satellites also have the same frequency difference between up-path and down-path frequencies. However, the technical mechanism which imposes standardisation automatically on full-bandwidth satellites does not operate for narrow-band satellites and it would be desirable to substitute for it some form of international agreement.

C. An Agreed Scale of Permissible Single-Entry Interference Noise Allocations

By international agreement, intersystem interference may account for up to 10 percent of all the noise degradations that a satellite link suffers. In the process of frequency coordination it is necessary to limit the interference from the particular interfering network under consideration to some part of that 10 percent. The CCIR recommends that the allocation to any one interference source should not exceed 40 percent of the total intersystem interference budget, this being roughly the fraction of the total that would arise from the nearest of a hypothetical homogenous array of equally spaced satellites to either side in orbit of the wanted satellite. This recommendation is insufficient to provide a good basis for coordination. It would be unwise to allocate 40 percent of the total interference budget to a powerful satellite perhaps $20°$ away from the wanted satellite and serving a different region since one or several other satellites may be established later, closer in orbit to the wanted satellite and serving the same region. Clearly the simple "40 percent maximum" figure needs to be elaborated into a scale of allocations, taking account of orbital separation, and perhaps other factors such as the geographical relationship of the service areas so that coordination agreements, once arrived at, can remain in effect for a long time.

D. Maximization of Service Arc

The most elementary and often the most attractive way of reducing internetwork interference is to move the satellites further apart. This solution is available only if one satellite or the other may have it nominal location changed without losing its ability to fulfill its mission. This freedom to change location should preferably be retained throughout a satellite's operating lifetime, but at least the option should be kept open until shortly before launch. The range of orbital location over which a satellite could fulfill its mission is called the "service arc."

A satellite serving earth stations spread on a global scale may have a very small service arc; if the satellite moves far to east and west, it will go below the horizon as seen at some of its earth stations. There are, however, many other circumstances that limit the service arc.

Some of these circumstances are basically economic, operative regardless of whether the spacecraft has been built or launched. For example, it may be necessary to locate a satellite west of the orbital position where eclipse occurs when it is midnight in the service area, so as to provide maximum commercially useful communication capacity without full battery cover. Other limitations are technical; for example a satellite with multiple spot beams may have the shape and relative directions of the beams optimised for a narrow orbital arc, and the coverage pattern might become suboptimum or even totally unacceptable if the satellite location were changed significantly; it may be possible to adjust the coverage pattern up to a few months before launch if the coordination process is concluded in time, but once launched the option may be completely lost. A third kind of problem is operational. Once a satellite is in service, operational continuity may be very important. It may be extremely difficult to maintain service while a satellite is being moved from one location to another, particularly if other satellites working in the same frequency bands must be passed on the way round the orbit or if the wanted satellite is used by numerous unattended earth stations; such difficulties could often be overcome if a spare satellite were available in orbit.

These are reasons why a satellite may have a narrow service arc, and there are others. Some of these difficulties can be overcome at moderate expense, but a rigid requirement to overcome them all might be unacceptably expensive to implement. Nevertheless, a code of guidelines could be drawn up for maximising the service arc; if this were endorsed internationally and applied to the extent that circumstances permit the prospects of successful coordination would be significantly improved.

IV. REDUCTION OF INHOMOGENEITY IN ORBIT-SPECTRUM SHARING

The range of application for satellite communication is very wide, and system parameters cover a correspondingly wide range. It follows that there are marked differences in the ability of systems to inflict interference on one another. The only substantial defence against interference between networks serving the same or neighboring territories and using the same frequency bands is the directivity of the earth station antenna. Where the gain slope of the side-lobe envelope is high, good use can be made of a few degrees of satellite separation in orbit. Thus, for example, an antenna conforming to the sidelobe reference pattern $E = 32 - 25 \log \phi$ dBi has an envelope gain slope of 11 dB per degree at $1°$ off-boresight. and 3.6 dB per degree at $3°$ off-boresight; measurements made on good large antennas show that such performance can be achieved, and perhaps somewhat exceeded. However, at large angles off-boresight, the reference pattern gain slope falls to such low values as 1.1 dB per degree at $10°$ off-boresight, and Cassegrain antennas in use tend to have even worse envelope gain slopes at $10°$ off-boresight owing to the presence of subreflector spillover radiation. In short, if a network causes another network so much interference that the two satellites cannot operate without excessive interference with an orbital separation of a few degrees, then the required spacing tends to escalate to an unmanageably large figure.

This situation can be illustrated by an admittedly extreme hypothetical example involving two satellite networks. Network A provides narrow-band links to sensitive earth terminals with large antennas, as in the SPADE transponder of the INTELSAT Atlantic Primary satellite. Network B provides wide-band links to low-cost earth terminals, for example a TV distribution system using the maximum permitted down-path flux density at 4GHz and serving earth terminals with 5-m antennas. If all earth station antennas are assumed to meet the CCIR sidelobe reference pattern and both systems have effective carrier energy dispersal, it is estimated that an array of satellites like Network A, all serving the same region, could be spaced as closely as $1.5°$ without the intersystem interference exceeding 10 percent of all degradations, while the corresponding figure for an array of satellites like Network B would be about $9°$. However, Network B could cause such bad interference to Network A that satellites would have to be spaced by no less than $33°$ if they served the same region.

Gross inhomogeneity of systems that can interfere is clearly very undesirable. Inhomogeneity cannot be eliminated altogether because of the varied nature of the opportunities for service that the medium provides. Some relief could be obtained if limits were applied to system parameters which primarily determine liability to cause interference and susceptibility to suffer it, but this would probably lead to unsatisfactory compromises which would circumscribe application without greatly reducing the inhomogeneity. A better solution, if it proves to be feasible, would be to classify systems according to their tendency to cause or suffer interference, and to segregate the satellites of the various classes so that interference between networks with very different characteristics does not occur. The details of such a system would need to be worked out. Classification might be based on a small number of parameters, perhaps as follows:

1) the off-beam up-path spectral e.i.r.p. (dBW/Hz);
2) the down-path spectral p.f.d. (dBW/m^2/Hz);
3) the up-path interference power flux level (dBW/m^2) received at the wanted satellite within the bandwidth of a wanted signal, which causes a threshold of interference noise to be exceeded at an earth station;
4) the off-beam down-path interference p.f.d. (dBW/m^2)

received at an earth station within the bandwidth of a wanted signal, which causes a threshold of interference noise to be exceeded.

Segregation could be by orbit-division or spectrum-division. However, having regard to the practical needs of some systems

1) to carry at different frequencies in the same satellite a variety of services that would have widely different ratings according to the parameters listed above;
2) to have access to wide bandwidths for high capacity;
3) to have minimal restrictions on the location of satellites used for any purpose;

it may be supposed that a combination of both orbit and spectrum division will be necessary. Having carried out a rough network sorting process by classification, it may be feasible to bring about a further reduction of inhomogeneity by the standardization of some characteristics within the classes. This might include standardization of up-path eirp spectral density, and earth station G/T and possibly the identification of preferred RF channelling plans.

V. Conclusion

The geostationary orbit is capable of supporting vast traffic flows or a wide variety of services with flexibility. However, it is not likely to achieve both at the same time, and the effective achievement of either depends upon designing and deploying each network having regard to the presence of other networks. The long term problem, therefore, is to find ways of securing action by all users of the orbit that will lead towards effective orbit utilisation. This will cost the users money and constrain their freedom. If such policies are to succeed in the real world, the costs and constraints must not be too great and must be seen to be justifiable in the not-too-distant future.

In this paper a number of measure have been identified that should be acceptable to the users and should be evidently constructive. They are:

1) better satellite station-keeping;
2) limitation of satellite antenna sidelobe radiation;
3) improvement of earth-station antenna sidelobe patterns;
4) carrier energy dispersal;
5) standardized frequency band pairing;
6) standardization of frequency translation in satellites;
7) the adoption of a graduated scale of single-entry interference allowances;
8) the adoption of guidelines for the maximization of service arc;
9) classification and segregation of networks according to their interference potential.

No one of these measures, except the last, can be expected to provide a large individual contribution to more effective orbit-spectrum utilisation, but collectively they could increase it several fold. Hopefully it will be possible to reach international agreement on them within the next few years.

Acknowledgment

Acknowledgement is made to the Senior Director of Development of the British Post Office for permission to make use of the information contained in this paper. The advice of many colleagues is also gratefully acknowledged.

REFERENCES

[1] W. L. Pritchard, "Satellite communication—An overview of the problems and programs," this issue, pp. 294–307.
[2] D. I. Dalgleish and A. K. Jefferis, "Some orbits for communications-satellite systems affording multiple access," *Proc. Inst. Elec. Eng.*, vol. 112, p. 21, Jan. 1965.
[3] H. E. Rowe and A. A. Penzias, "Efficient spacing of synchronous communication-satellites," *Bell Syst. Tech. J.*, vol. 47, no. 10, pp. 2379–2433, Dec. 1968.
[4] I. Petrov, "INTERSPUTNIK International space communication system and organisation," *Telecommun. J.*, vol. 29, p. 679, Nov. 1972.
[5] A. K. Jefferis, D. G. Pope, and P. C. Gilbert, "Satellite television distribution; service from geostationary satellites to community antennas in multiple-coverage areas," *Proc. Inst. Elec. Eng.*, vol 116, no. 9, p. 1501, Sept. 1969.
[6] H. Mertens, "Satellite broadcasting; design and planning of 12 GHz systems," EBU Rep. Tech 3220 E.
[7] ITU Radio Regulations Article 7 (revised 1971).
[8] CCIR Recommendation 465-1. Reference earth station radiation pattern for use in coordination and interference assessment in the frequency range from 2 to about 10 GHz. Documents of the XIIIth Plenary Assembly, Geneva, Switzerland. vol. IV, p. 155, 1974.
[9] J. Dijk and E. J. Maanders, "Some aspects of near and far angle sidelobes in double-reflector antennas," presented at AGARD Conf. Rome, Italy, May 1973.
[10] N. White, D. Brandwood, and G. Raymond, "Some application of interference cancellation to an earth station," in IEE Conf. Publ. 126, *Satellite Communication Systems Technology*, p. 233.
[11] CCIR Recommendations 356-3, 466-1, and 483. Documents of the XIIIth Plenary Assembly, Geneva, Switzerland, vol. IX, p. 360, and vol. IV, pp. 185 and 186, 1974.
[12] ITU Radio Regulations Article 9A (revised 1971).

Bibliography

Bibliography Index

I. General

Balderston, M., "An Historical Survey of Communications Satellite Systems," (in 3 parts), *Telecommunications Journal of Australia,* vol. 25, Nos. 1–3, (1975).

Bargellini, P. L., "Experimental Communications Satellite Programs," *International Conference on Communications,* (June 1973).

Bargellini, P. L., "Advances in Satellite Communications," *Advances in Electronics and Electron Physics,* L. Marton, (Ed.), vol. 31, (1972).

Bekey, I., "Communications Satellites—Issues and Trends," *IEEE Communication Systems and Technology Conference,* (April 1974).

Clarke, A. C., "The World of the Communications Satellite," *Astronautics and Aeronautics,* vol. 2, No. 2, (February 1964).

Defense Communications Agency, "A Digest of Satellite Communications Systems," 4 Volumes, (September 1973).

Defense Communications Agency, "Satellite Communications Reference Data Handbook," (September 1973).

Edelson, B. I., "Global Satellite Communications," *Scientific American,* vol. 236, No. 2, (February 1977).

Edelson, B. I., and Bargellini, P. L., "Technology Development for Global Satellite Communications," AIAA Paper 76-234, *AIAA/CASI 6th Communications Satellite Systems Conference,* (April 1976).

Feldman, N. E., and Kelly, C. M., "The Communication Satellite—A Perspective for the 1970s," *Astronautics and Aeronautics,* vol. 9, No. 9, (September 1971).

Golding, L. S., and Ball, J. E. D., "Satellite Television Covers the World," *Spectrum,* vol. 10, No. 8, (August 1973).

"International Conference on Satellite Communication Systems Technology," *IEEE Conference Publication* 126, (April 1975).

"Introduction to Satellite Communications," CCP-105-5, HQ, U.S. Army Communications Command, (April 1974).

Johnston, J. W., "Status of Military Satellite Communications Research and Development," *IEEE Transactions on Military Electronics,* vol. 9, No. 2, (April 1965).

Kadar, I., "Satellite Communications Systems," *AIAA Selected Reprint Series,* vol. XVIII, (January 1976).

Lagarde, J. B., "Communications Satellites, Dreams for Engineers? Gluttons for Government Funds? Money Makers?", *Journal of the British Interplanetary Society,* vol. 29, No. 5, (May 1976).

Love, D., "Satellites and Cables: Competitive or Complementary," *Flight International,* vol. 104, (November 1973).

McKenzie, A. A., "What's up in satellites," *IEEE Spectrum,* (May 1972).

Miya, K., (Ed.), "Satellite Communications Engineering," Lattice Co., Tokyo, (1975).

Morgan, W. L., "Design Criteria for Communications Satellites," *EASCON '74 Conference Record,* (October 1974).

Proceedings of INTELSAT/IEEE International Conference on Digital Satellite Communications, London, (1969).

Proceedings of the INTELSAT/IEEE Conference on Digital Satellite Communications, Paris, (1972).

Wheelon, A. D., "The Future Outlook for Communication Satellite Applications," World Telecommunication Forum Conference Proceedings, (October 1975), Reprinted in *Telecommunications Journal,* vol. 43, No. 2, (February 1976).

Williams, O. G., "Communication satellite systems," Ch. 5, in *Advanced Communication Systems,* B. J. Halliwell, (Ed.), Newnes-Butterworths, London, pp. 192–242, (1974).

II. Systems

2.1 GENERAL

Aerospace Corp., "Communications Satellites, 1958 to 1980," Report SAMSO-TR-77-76, 1 February 1977, (Distribution limited to U.S. Gov't agencies.)

Clarke, A. C., "Extra-Terrestrial Relays," *Wireless World,* vol. 51, No. 10, (October 1945), Reprinted in Communication Satellite Systems Technology, *Progress in Astronautics and Aeronautics,* vol. 19, R. B. Marsten, (Ed.), (1966).

Davis, M. I., and Krassner, G. N., "SCORE—First Communication Satellite," *Journal of the American Rocket Society,* vol. 4, (May 1959).

Gatland, K. W., "Telecommunication Satellites, Prentice-Hall, New York, (1964).

Gibson, R., "European Communications Satellite Systems," IAF XXVIII Congress, Prague, Czech. (September 25—October 1, 1977).

Gould, R. G., "Commercial Communications Satellites: Operational Experimental and Planned," *National Telecommunications Conference,* (1976).

Jaffe, L., "Project Echo Results," *Astronautics,* vol. 6, No. 5, (May 1961).

LaVean, G. E., and Martin, E. J., "Communication Satellites: The Second Decade," *Astronautics and Aeronautics,* vol. 12, No. 4, (April 1974).

Metzger, S., and Pickard, R. H., "Relay," *Astronautics and Aerospace Engineering,* vol. 1, No. 8, (September 1963).

Moulton, P. D., "Current Status of Commercial Satellite Communications," *Signal,* (April 1975).

"NASA Compendium of Satellite Communications Programs," Document X-751-73-178, Goddard Space Flight Center, Greenbelt, MD, (August 1975).

Norsell, P. E., "Syncom," *Astronautics and Aerospace Engineering,* vol. 1, No. 8, (September 1963).

Pierce, J. R., and Kompfner, R., "Transoceanic Communications by Means of Satellite," *Proceedings of the IRE,* vol. 47, (March 1959).

Pritchard, W. L., and Bargellini, P. L., "Trends in Technology for Communications Satellites," *Astronautics and Aeronautics,* vol. 10, No. 4, (April 1972).

"Space Electronics Issue," *Proceedings of the IRE,* vol. 48, No. 4, (April 1960).

"Special Issue on Project Echo," *Bell System Technical Journal,* vol. 40, No. 4, (July 1961).

"Special Issue on Project West Ford," *Proceedings of the IEEE,* vol. 52, No. 5, (May 1964).

"Special Telstar Issue," *Bell System Technical Journal,* vol. 42, No. 4, (July 1963), Reprinted as Telstar I, NASA SP-32, Vols. 1-3, (July 1963), with vol. 4 (including Telstar II supplement), (December 1965).

2.2 INTERNATIONAL

Astrain, S., "Early Bird to Intelsat-IVA (a decade of growth)," *Telecommunications Journal,* vol. 42, No. 10, (October 1975).

Benedict, R. N., and Kolsrud, J. E., "Digital Service in the Intelsat Network," *EASCON,* (1976).

Bennett, S., and Dostis, I., "Design of the Intelsat IV Transponder," *Communications Satellite Technology, Progress in Astronautics and Aeronautics,* vol. 33, P. L. Bargellini, (Ed.), (1974).

Bentley, R. M., "Early Bird," *Astronautics and Aeronautics,* vol. 3, No. 3, (March 1965).

Blackwell, C. A., and Brown, Jr., M. P., "Communications Satellite System Design," *IEEE Communication Systems and Technology Conference,* (April 1974).

Browne, S., "The INTELSAT Global Satellite Communications System," *COMSAT Technical Review,* vol. 4, No. 2, (Fall 1974).

Dicks, J. L., "Domestic and/or Regional Services Through INTELSAT IV Satellites," *COMSAT Technical Review,* vol. 4, No. 1, pp. 91–117, (Spring 1974).

Dicks, J. L., and Brown, Jr., M. P., "INTELSAT IV—A Satellite Transmission Design," *COMSAT Technical Review,* vol. 5, No. 1, pp. 73–104, (Spring 1975).

Dicks, J. L., and Brown, M. P., "INTELSAT IV—A Satellite Transmission Design," AIAA Paper 74-474, *AIAA 5th Communications Satellite System Conference,* (April 1974).

Edelson, B. I., Strauss, R., and Bargellini, P., "INTELSAT System Reliability," *International Astronautical Federation XXVth Congress Amsterdam,* (October 1974).

Feigen, M., "The INTELSAT III Satellite," *IEEE International Conference on Communications,* (June 1968).

Fortushenko, A. D., "The Soviet Communication Satellite Molniya 1," *Telecommunications Journal,* vol. 32, No. 10, (October 1965).

Gray, L. F., "Experimental Performance of the Early Bird Communication System," *Progress in Astronautics and Aeronautics,* vol. 19, R. B. Marsten, (Ed.), (1966).

"The Global Satellite Communications System," *Interavia,* vol. 26, No. 19, (October 1971).

Hammer, P. R., "The Development of Amateur Communications Satellites," *Proceedings of the IEEE,* vol. 36, No. 5, (May 1975).

"INTELSAT 2 Communications Satellite," *Telecommunications Journal,* vol. 34, No. 2, (February 1967).

Jacobs, G., and Klein, P., "Satellites in the Amateur Radio Service," *Telecommunications Journal,* vol. 38, No. 5, (May 1971).

Johnsen, K., "Soviets Plan Seven-Satellite Global System," *Aviation Week & Space Technology,* (December 15, 1975).

Kantor, L. Ya., Polukhin, V. A., and Talyzin, N. V., "New Relay Stations of the Orbita-2 Satellite Communications System," *Telecommunications and Radio Engineering,* vol. 27, No. 5, (May 1973).

Kasser, J., "OSCAR 7 and Its Capabilities," *Radio Communication,* (November 1973).

Klein, P. I., and King, J. A., "The AMSAT-OSCAR-B Series of Radio Amateur Satellites," AIAA Paper 72-521, *AIAA 4th Communications Satellite Systems Conference,* (April 1972).

Klein, P. I., and King, J. A., "Results of the Amsat-Oscar 6 Communications Satellite Experiment," Paper 6/3, *IEEE Intercon Record,* (March 1974).

Lustiberg, V., "Satellite Radiocommunication in the USSR," *Telecommunications Journal,* vol. 33, No. 12, (December 1966).

Meredith, C. O., "Lessons Learned from the Intelsat III Satellite Program," *Communications Satellite Systems, Progress in Astronautics and Aeronautics,* vol. 32, P. L. Bargellini, (Ed.), (1974).

Metzger, S., "The Commercial Communications Satellite System—1963 to 68," *Astronautics and Aeronautics,* vol. 6, No. 4, (April 1968).

Pilcher, L. S., "Intelsat IV-A as a Communication Capability," AIAA Paper 74-473, *AIAA 5th Communications Satellite Systems Conference,* (April 1974).

Podraczky, E., and Bennett, S. B., "Intelsat Planning for the Next Decade," *WESCON,* (September 1975).

Podraczky, E., and Chitre, N. K. M., "Future Intelsat Services and Operations," AIAA Paper 76-231, *AIAA/CASI 6th Communications Satellite Systems Conference,* (April 1976).

Rees, D. W. E., "Operational Transitions to a New Satellite Series," *EASCON,* 1977.

Robertson, E. A., "Planning for the INTELSAT V System," *National Telecommunications Conference,* (December 1974).

"Special Issue on INTELSAT IV," *COMSAT Technical Review,* vol. 2, No. 2, (Fall 1972).

Taormina, F., McCarty, D. K., Crail, T., and Nakatani, D., "INTELSAT IVA Communications Antenna—Frequency Reuse through Spatial Separation," *International Conference on Communications,* (June 1976).

Votaw, M. J., "The Early Bird Project," *IEEE Transactions on Communications Technology,* vol. 14, No. 4, (August 1966).

Wells, W. C., "Computer Aided Frequency Planning, Transmission Impairment and Performance Analysis in the INTELSAT System," *EASCON,* 1977.

Withers, D. J., "The Problem of Growth in the INTELSAT System," *Journal of the British Interplanetary Society,* vol. 29, No. 1, (January 1976).

2.3 DOMESTIC/REGIONAL

2.3.1 Telesat

Chinnick, R. F., "Canadian Satellite Telecommunications Plans for the 1980s," *IAF XXVIII Congress,* Prague, Czech. (September 25-October 1, 1977).

Chinnick, R. F., "The Canadian Telecommunications Satellite System," *Journal of the British Interplanetary Society,* vol. 26, No. 4, (April 1973).

Greenquist, R. E., "First Generation Domestic Satellite Systems," AIAA Paper 71-842, *AIAA Space Systems Meeting,* (July 1971).

Harrison, L., "Canadian Domestic Satellite (TELESAT), A General Description," *International Conference on Communications,* (June 1971).

Hoedemaker, R. W., and Thorpe, D. G., "Anik B, the new Canadian Domestic Satellite," *RCA Technical Communications,* (1977).

Kowalik, H., "Telesat Satellite Control System," AIAA Paper 74-451, *AIAA 5th Communications Satellite Systems Conference,* (April 1974).

Kwan, R.K., "The TELESAT TDMA System," *International Conference on Communications,* (June 1975).

Miller, A. D. D., "Operational Experience with Small Unattended Television Receive Earth Stations," AIAA Paper 74-454, *AIAA 5th Communications Satellite Systems Conference,* (April 1974).

Thorpe, D. G., "Evolution of the TELESAT Canada System Into the Mid-1980s," No. 76-220, *AIAA/CASI 6th Communications Satellite Systems Conference,* Montreal, (April 5-8, 1976).

Wadham, P. N., "Operational Experience with the Canadian Domestic Satellites," AIAA Paper 74-453, *AIAA 5th Communications Satellite Systems Conference,* (April 1974).

Weese, D. E., "The Canadian Domestic Satellite Communications System—Present and Future," *EASCON 77,* Washington, DC, (September 26-27, 1977).

Weese, D. E., and Smart, F. H., "Measured Communication Performance of the TELESAT Satellite System," AIAA Paper 74-455, *AIAA 5th Communications Satellite System Conference,* (April 1974).

2.3.2 WESTAR

Lee, D. J., "System Performance of America's First Domestic Communications Satellite—Westar," *EASCON '74 Convention Record,* (October 1974).

Nowakowski, D. B., "The Western Union Integrated Satellite/Voice/Data Network," Paper 18B, *National Telecommunications Conference,* (November 1973).

Ramasastry, J., Heiser, W., Levine, E., and Markham, R., "Western Union's Satellite—Switched TDMA Advanced WESTAR System," AIAA 7th Communications Satellite Systems Conference, San Diego, CA, (April 24-27, 1978).

Verma, S. N., "WESTAR Communication Characteristics," *National Telecommunications Conference,* (1974).

2.3.3 Comstar

American Telephone & Telegraph Company (AT&T), "Application for a Domestic Communications Satellite System," New York, (March 29, 1973).

Cox, D. C., "Design of the Bell Laboratories 19 and 28 GHz Satellite Beacon Propagation Experiment," *International Conference on Communications,* (June 1974).

Rusch, R. J., "COMSAT General Domestic Communications Satellite," *National Telecommunications Conference,* (1974).

2.3.4 RCA

Hawkins, H. R., "RCA Satcom System Begins New Communications Era in United States," *Signal,* vol. 28, No. 7, (March 1974).

Napoli, J., and Greenspan, D., "RCA Satcom, The Next Generation Domestic Communications Satellite System," *WESCON Technical Papers,* (September 1975).

Napoli, J., and Christopher, J., "RCA SATCOM System," *National Telecommunications Conference,* (December 1974).

O'Donovan, M. V., "A Lightweight Transponder Design for the U.S. Domestic Communications Satellite," International Conference on Communications, (June 1973).

Schneider, P., "New Approaches Make RCA Satcom Most Cost Effective System for Domestic Satellite Communications for United States," *EASCON '75 Convention Record,* (Sept. 1975).

2.3.5 Foreign

"Arabs New Satcom Network Go-Ahead," *Aviation Week and Space Technology,* pp. 56-59, (July 21, 1975).

Obeid, I., "Arabsat's Role in International Telecommunications," *World Telecommunication Forum Proceedings,* Geneva, (October 1975).

Osborne, W. P., Smith, R. L., and Stopor, H. J., "Sudosat, The National Domestic Satellite Communications Systems for the Government of the Democratic Republic of the Sudan," *EASCON-1977,* Washington, DC, (September 26-27, 1977).

2.3.6 TDRSS

Barritt, P. F., "Tracking and Data Relay Satellite System (TDRSS)," Paper 7B, *National Telecommunications Conference,* (November 1973).

Bernstein, D. M., "Computer Simulation of the Digital K-Band Shuttle/TDRSS Return Link," *National Telecommunications Conference,* (1977).

Clark, G., Fellerman, K., and Schwartz, J., "Current Concepts for a Tracking and Data Relay Satellite System," Paper 28A, *National Telecommunications Conference,* (December 1972).

Deerkoski, L. F., "Tracking and Data Relay Satellite System (TDRSS) Telecommunication Services," *EASCON '75 Convention Record,* (September 1975).

Godfrey, R. D., "The Evolution of the Tracking and Data Relay Satellite System," *EASCON '75 Convention Record,* (September 1975).

Holmes, Jr., W. M., "TDRSS System Design," *National Telecommunications Conference,* (1977).

Melnick, M., "TDRSS Frequency Management," *National Telecommunications Conference,* (1977).

Poza, H. B., "TDRSS Telecommunications Payload: An Overview," *National Telecommunications Conference,* (1977).

2.3.7 SBS

Barnla, J. D., and Zitzmann, F. R., "Digital Communications Satellite System of SBS," *EASCON 1977.*

Crandall, K. H., "The 12 and 14 GHz Bands in Domestic Satellite Communications," Paper 31D, *National Telecommunications Conference,* (November 1973).

2.3.8 Miscellaneous

Ashton, S., and Silverman, D., "The American Satellite Communications System," *AIAA Paper No. 74#482, AIAA 5th Communications Satellite Systems Conference,* Los Angeles, CA, (April 22-24, 1974).

Bairi, A. K., "The Algerian Domestic System," AIAA Paper 74-493,

AIAA 5th Communications Satellite Systems Conference, (April 1974).

Bairi, A., and Leonhard, J., "A Domestic Satellite Communications System for Algeria," *International Conference on Communications,* (June 1975).

Briskman, R. D., "A Domestic Satellite Design," *IEEE Trans. AES,* pp. 263-268, (March 1971).

Brown, J. P., "Public Service Communication Satellite Program," *EASCON 1977,* Washington, DC, (September 26-27, 1977).

Cheadle, G., "Philippine Domestic Satellite System," AIAA Paper 74-491, *AIAA 5th Communications Satellite Systems Conference,* (April 1974).

Fudge, R. E., Ballantine, R. J., and Bayliss, A. J., "Communications Aspects of Broadcast TV Satellites," *Journal of the British Interplanetary Society,* vol. 26, No. 4, (May 1973).

Kale, P. P., Nickelson, R. L., and Sarles, Jr., F. W., "A Design for Insat," AIAA Paper 72-576, *AIAA 4th Communications Satellite Systems Conference,* (April 1972).

Kellock, A., "Domestic Satellite Communications for Australia," *International Conference on Communications,* (June 1970).

Kilty, L., "American Satellite Builds a Major Network for Government Users," *EASCON 1975 Convention Record,* (September 1975).

Kreisel, Jr., G. R., "Control Concept for a Domestic Satellite Network," *AIAA Paper No. 74-478,* Los Angeles, CA, (April 1974).

Menon, M. G. K., "Insat in Perspective," AIAA Paper 72-583, *AIAA 4th Communications Satellite Systems Conference,* (April 1972).

Puente, J. G., "Designing the American Satellite Corporation System," *Proceedings of EASCON,* pp. 75-83, (1973).

Sarkar, S. K., "Policy Planning of the Broadcasting Satellite Service," *World Telecommunication Forum Conference Proceedings,* (October 1975).

Schultze, P. H., Itohara, S., and Dicks, J. L., "Use of the Intelsat Space Segment for Domestic Systems," AIAA Paper 76-305, *AIAA/CASI 6th Communications Satellite Systems Conference,* (April 1976).

2.4 MOBILE

2.4.1 Maritime

Cacciamani, E., and Wilson, R. J., "Shipboard Data Transmission Service," *INTELCOM-1977,* Atlanta, GA, (October 1977).

Calvit, T. O., "Marisat—Prelude to a Global Commercial Maritime Satellite Communications System," *WESCON Technical Papers,* (September 1975).

Conchie, P. J., "The Adoption of OTS to the MAROTS Role," *Journal of the British Interplanetary Society,* vol. 27, No. 10, (October 1974).

Dorian, C., Calvit, T. O., and Lipke, D. W., "Marisat: Design and Operational Aspects," *Journal of the British Interplanetary Society,* vol. 29, No. 5, (May 1976).

Durrani, S. H., "A maritime communications concept using spaceborne phased arrays, *Satellite Systems for Mobile Communications and Surveillance, IEEE Conf. Pub., 95,* Institution of Electrical Engineers, London, pp. 71-74, (1973).

Ellowitz, H. I., "Maritime Satellite Communications," *Microwave Journal,* (June 1975).

ESA, "MAROTS—European Maritime Satellite, Performance Specifications for the MAROTS System," *ESA Document 75-10-02.*

Fenton, R. E., "Operational Marsat—An Evolving Concept," AIAA Paper 75-282, *AIAA 11th Annual Meeting and Technical Display,* (February 1975).

Gleadle, G. H. M., "Maritime Satellites—A Survey," *Journal of the British Interplanetary Society,* vol. 27, No. 10, (October 1974).

Giorgio, F., Knight, I., and Matthew, R., "A Maritime Mobile Terminal for Commercial Communications Satellite Application," *International Conference on Communications,* (June 1974).

Hirata, Y., Kyogoku, M., and Isomura, E., "A Study on Satellite Communications for Mobiles," AIAA Paper No. 72-565, *AIAA Fourth Communications Satellite Systems Conference,* Washington, DC, (April 24-26, 1972).

Howell, T. F., "Marots—The Mission," AIAA Paper 76-258, *AIAA/CASI 6th Communications Satellite Systems Conference,* (April 1976).

LaRosa, R. M., "The Benefits and Applications of Maritime Satellites," *EASCON 1974 Convention Record,* (October 1974).

LaRosa, R. N., "Maritime Satellite Requirements and Utilization," *IEEE Journal of Oceanic Engineering,* (July 1977).

Lipke, D., Swearingen, D., Parker, J., Steinbrecher, E., Calvit, T., and Dodel, H., "MARISAT—A Maritime Satellite Communications System," *COMSAT Technical Review,* vol. 7, No. 2, (1977).

Lipke, D. W., and Swearingen, D. W., "Communication System Planning for Marisat," *International Conference on Communications,* (June 1974).

Lovell, W. M., "The Marots Communications Payload," Proceedings of the Conference on Satellite Communications Systems Technology, *IEEE Conference Publication No. 126,* (April 1975).

McLauchlin, D., and Stainforth, R., "MAROTS Communication Satellite," *Journal of the British Interplanetary Society,* vol. 29, No. 5, (May 1976).

Pope, D. G., "Modulation and speech processing techniques for a maritime-satellite service," *Satellite Systems for Mobile Communications and Surveillance,* IEE Conf. Pub. 95, Institution of Electrical Engineers, London, pp. 56-61, (1973).

Swearingen, D. W., and Lipke, D. W., "Marisat Multiple Access Capabilities," *International Conference on Communications,* pp. 51-10-13, (1976).

Swearingen, D. W., "Marisat Commercial Communications System Status," *EASCON 75, Convention Record,* (September 1975).

Steciw, A. A., "Parametric study of a maritime satellite system," *Satellite Systems for Mobile Communications & Surveillance,* IEEE Conf. Pub. 95, Institution of Electrical Engineers, London, pp. 176-179, (1973).

Vandenkerckhove, J. A., "The ESRO MAROTS Programme," *Journal of the British Interplanetary Society,* vol. 27, No. 10, (October 1974).

Van Hover, A. J. E., and Gribbin, W. J., "Design of a Ground Control System to Operate Domestic and Maritime Satellites," *AIAA Fifth Communications Satellite Systems Conference,* Los Angeles, CA, (April 22-24, 1974).

Wilson, M., "Maritime Satellites," *Flight International,* (April 10, 1976).

Zimmer, T. M., and Calvit, T. O., "A Future Global Satellite System for Commercial Maritime Services," *EASCON,* (1975).

2.4.2 Aeronautical

Braybrook, L. J., "The use of satellites for aircraft communications and air traffic control," *Satellite Systems for Mobile Communications & Surveillance,* IEEE Conf. Pub. 95, Institution of Electrical Engineers, London, pp. 75-83, (1973).

Campanella, S. J., and Sciulli, J. A., "A comparison of voice communication techniques for aeronautical and marine applications," *COMSAT Technical Review 2,* p. 204, (1973).

Carr, F. S., "Aerosat—Current Status and the Test and Evaluation Program," *EASCON 1975,* Washington, DC, (September 29-October 1, 1975).

Cornett, D. E., Haspert, J. K., Diehl, J. H., and Nardone, S. C., "A Study of the communications requirements for a 1985 to 2000 operational aeronautical satellite system," vol. I., Atlantic Ocean area. Report No. FAA-RD-75-80 on Contract No. DOT-FA75WA-3544, ARINC Research Corp., Annapolis, MD.

Cragie, J. H., Garabedian, A., Morrison, D. D., and Zipper, I., "The applications of satellites to communications, navigation and surveillance for aircraft operating over the contiguous United States," Report No. NASA-CR-117760, Contract No. NAS5-21535, TRW Systems Group, Redondo Beach, CA, N71-22339#.

Devieux, C., and Bisaga, J. J., "An efficient multiplexing approach for adaptive aircraft communications via a relay satellite," *IEEE Trans. Comm.,* COM-21, pp. 647-652.

IEEE, *Satellite Systems for Mobile Communications & Surveillance,* IEEE Conference Publication No. 95, Institution of Electrical Engineers, London, (March 1973).

Israel, D. R., "Aerosat—An Aeronautical Communications Satellite for Oceanic Areas," AIAA Paper 73-46, *AIAA 9th Annual Meeting and Technical Display,* (January 1973).

Israel, D. R., "Aerosat—An Aeronautical Satellite Program," Paper 27B, *National Telecommunications Conference,* (November 1973).

Israel, D. R., "AEROSAT Overview," Proceedings, *EASCON 1973,* pp. 135-138.

Kinal, G. V., and Bisaga, J. J., "Multibeam satellite EIRP adaptability for aeronautical communications," *IEEE Trans. Comm.,* COM-21, pp. 653-655.

Lebow, I. L., Jordan, Jr., K. L., and Drouilhet, Jr., P. R., "Satellite communications to mobile platforms," Proc. IEEE 59, No. 2, pp. 139-159.

Lucier, E., "Aerosat System," *National Telecommunications Conference,* (December 1975).

Robinson, J. J., "Aeronautical Satellites—Progress Report on the Joint Aerosat Evaluation Programme," *Journal of the British Interplanetary Society,* vol. 29, No. 5, (May 1976).

Vandenkerckhove, J. A., and Contzen, J. P., "The Aerosat Programme," *Journal of the British Interplanetary Society,* vol. 27, No. 3, (March 1974).

Woodford, J. B., "AEROSAT Performance Specifications," *Proceedings for EASCON 1973,* pp. 139-145.

2.4.3 General

Haviland, R. P., "Aero-Marine Communications by Satellite," *ITU Telecommunication Journal,* vol. 42, pp. 97-104, (February 1975).

Satellite Systems Engineering, Inc., "A Study of Technical and Economic Aspects of Combining Maritime and Aeronautical Communication Satellite Services 1985-2000," Washington, DC, (May 1977).

2.5 MILITARY

2.5.1 General

"The Advent Stationary Communications Relay Satellite," *Interavia,* vol. 17, (June 1962).

Babcock, J. H., "Architecture and Management of DOD Satellite Communications Systems," *Signal,* vol. 30, No. 6, (March 1976).

Bond, F. E., "Overview—DOD Satcom Systems," *AIAA Paper 74-456,* Los Angeles, CA, (April 1974).

Curry, W. H., "The Military Satellite Communications System Architecture," *AIAA/CASI Sixth Communications Satellite Systems Conference,* AIAA Paper 76-268, (April 1976).

Wall, V. W., "Military Communication Satellites," *Astronautics and Aeronautics,* vol. 6, No. 4, (April 1968).

Wall, V. W., "Military Communication Satellites," *International Telecommunications Conference,* (October 1973).

Webb, P., "Military Satellite Communications Using Small Earth Terminals," *IEEE Trans. on Aerospace and Electronic Systems,* vol. AES-10.

2.5.2 DSCS

Drummond, R. L., "Network Control and Coordination with the U.S. Defense Satellite Communications System (DSCS)," *AIAA 5th Communications Satellite Systems Conference,* AIAA Paper 74-477, (April 1974).

Finney, A. T., "A Phase II Satellite for the Defense Satellite Communications System," *Communications Satellites for the 70s: Systems,* Progress in Astronautics and Aeronautics, vol. 26, N. E. Feldman and C. M. Kelly, (Eds.), (1971).

Gould, G. T., "DCA's Role in Satellite Communications," *Signal,* vol. 28, No. 7, (March 1974).

Kucheman, H. B., Pritchard, W. L., and Wall, V. W., "The Initial Defense Communication Satellite Program," AIAA Paper 66-267, *AIAA Communications Satellite Systems Conference,* (May 1966).

LaVean, G. E., "The Defense Satellite Communications System," *AIAA Paper 74-457,* Los Angeles, CA, (April 1974).

LaVean, G. E., and Neikirk, R. O., "Satellite Communications Requirements in the Defense Communications System," *Signal,* vol. 28, No. 7, (March 1974).

Sims, R. J., and Sherwin, R. P., "Communication Technology Trends in the DSCS," *National Telecommunications Conference 1974,* (December 1974).

Witherspoon, J. T., and Sherwin, R. P., "Real-Time Adaptive Control for the Defense Satellite Communications System," *AIAA/CASI 6th Communications Satellite Systems Conference,* AIAA Paper 76-272, (April 1976).

Wynne, H., and Kendall, D. E., "Defense Satellite Communications System in the 1980s," *WESCON Technical Papers,* (September 1975).

Yakabe, H. M., "Formulation of the Follow-on Defense Satellite Communication System Concept," *IEEE 1975 Intercon,* Session 29/4, New York, NY, (April 8-10, 1975).

2.5.3 Tactical

Brandes, R., "The Tactical Communications Satellite," *IEEE Transactions on Aerospace and Electronic Systems,* vol. AES-6, No. 4, (July 1970).

Boyes, J. L., and Harden, T. H., "Navy's Fleet Satellite and Gap-filler Satellite Communications Programs," *Signal,* vol. 28, No. 7, (March 1974).

Boyes, J. L., " A Navy Satellite Communication System," *Signal,* vol. 30, No. 6, (March 1976).

Greiner, R. D., "Air Force Satellite Communications Program," *Signal,* vol. 28, No. 11, (August 1974).

Keane, L. M., and Martin, E. J., "Marisat," *Signal,* vol. 29, No. 3, (November-December 1974).

Keane, L. M., and Martin, E., "The Marisat Spacecraft," *International Conference on Communications,* (June 1974).

Reid, C. E., "A New Era in Worldwide Tactical Communications," Naval Electronics Laboratory Center, "FLTSATCOM System Description, Revision 1," *Technical Document-280,* Naval Electronics Systems Command, San Diego, CA, (July 10, 1974). *International Telecommunications Conference,* (October 1973).

Wardle, N. L., "U. S. Navy Fleet Satellite Communication," AIAA Paper 74-458, *AIAA 5th Communications Satellite Systems Conference,* (April 1974).

Wheeler, J. B., "Fleet Satellite Communications," *Signal,* vol. 28, No. 6, pp. 38-39, (February 1974).

2.5.4 NATO

Bobak, E. T., and Clabaugh, R. G., "NATO Phase III Satellite Design," *EASCON 1977.*

Gray, N., "NATO and the NATO Integrated Communications System (NICS)—An Overview," *EASCON 77,* Washington, DC, (September 26-28, 1977).

Ince, A. N., "Design, Testing and Operation of an X-Band Satellite Communications System," *IEEE Transactions on Communications,* vol. COM-22, No. 9, pp. 1338-1353, (September 1974).

Kissinger, H. A., "NATO Satellite Communications," *Signal,* vol. 30, No. 6, (March 1976).

Latour, C., "Naval Communications," *NATO's Fifteen Nations,* vol. 20, No. 5, (October-November 1975).

Latour, C., "Skynet II," *NATO's Fifteen Nations,* vol. 20, No. 1, (February-March 1975).

Lovell, W. M., "Design of the Skynet II Communications Satellite," *Journal of Science and Technology,* vol. 39, No. 1, (1972).

Lovell, W. M., "The Skynet System, The Satellite Communications Network Built in Britain," *Journal of the British Interplanetary Society,* vol. 29, No. 1, (January 1976).

Simmons, N., "The United Kingdom Programme of Communications Satellites," *Communications Satellite Systems,* Progress in Astronautics and Aeronautics, vol. 32, P. L. Bargellini, (Ed.), (1974).

"Skynet Tries Again," *Spaceflight,* vol. 16, No. 12, (December 1974).

Valentine, D. R., "NATO's Communications Satellite System," *NATO's Fifteen Nations,* vol. 15, No. 5, (October-November 1970).

2.5.5 Lincoln Experimental Satellites (LES)

Berg, R., Chick, R., and Snider, D., "LES-7 Transponder," *AIAA Paper 70-511, AIAA 3rd Communications Satellite Systems Conference,* (April 1970).

Collins, L. J., Jones, L. R., McElroy, D. R., Siegel, D. A. Ward, W. W., and Willim, D. K., "LES-8/9 Communications System Test Results," *AIAA,* (1978).

Dion, A. R., "Variable-Coverage Communications Antenna for LES-7," AIAA Paper 70-423, *AIAA 3rd Communications Satellite Systems Conference,* (April 1970).

MacLellan, D., MacDonald, H., and Waldron, P., "Lincoln Experimental Satellites 5 and 6," *Communication Satellites for the 70s: Systems,* Progress in Astronautics and Aeronautics, vol. 26, N. E. Feldman and C. M. Kelly, (Eds.), (1971).

Sarles, Jr., F. W., Bowles, L. W., Farnsworth, L. P., and Waldron, P., "The Lincoln Experimental Satellites LES 8 and 9," *EASCON 1977,* Washington, DC, (September 26-27, 1977).

Sherman, H., *et al.,* "The Lincoln Experimental Satellite Program (LES-1, -2, -3, -4)," *Journal of Spacecraft and Rockets,* vol. 4, No. 11, (November 1967).

Sherman, H., *et al.,* "The Lincoln Experimental Satellite Program (LES 1, -2, -3, -4)," AIAA Paper 66-271, *AIAA Communications Satellite Systems Conference,* (May 1966).

"USAF Studies Spacecraft Survivability," *Aviation Week and Space Technology,* p. 41, (August 4, 1975).

2.6 EXPERIMENTAL

2.6.1 ATS

"The ATS-F and -G Data Book," Goddard Space Flight Center, (October 1971), Revised, (September 1972).

Berman, A. L., "The ATS-F Comsat Propagation Experiment Transponder," *COMSAT Technical Review,* vol. 3, No. 2, (Fall 1973).

Howard, M., "ATS-6: The First Twelve Months," *Spaceflight,* vol. 17, No. 11, (November 1975).

Ippolito, L. J., "The GSFC 20 and 30 GHz Millimeter Wave Propagation Experiment," *International Conference on Communications,* (June 1975) and *EASCON 75 Convention Record,* (September 1975).

Johnston, W. A., "ATS-6 Experimental Communications Satellite: A Report on Early Orbital Results," *National Telecommunications Conference,* (December 1974).

King, J. L., and Hyde, G., "The Comsat 13 and 18 GHz Propagation Experiment," *International Conference on Communications,* (June 1975).

Redisch, W. N., "ATS-6 Description," *International Conference on Communications,* (June 1975), *EASCON 75 Convention Record,* (September 1975).

Redisch, W. N. and Hall, R. L., "ATS 6 Spacecraft Design/Performance," *EASCON 74, Conference Record,* (October 1974).

Sabelhaus, A. B., "Applications Technology Satellites F and G Communications Subsystem," *Proceedings of the IEEE,* vol. 59, No. 2, (February 1971).

Special Issue on ATS 6, *IEEE Transactions on Aerospace and Electronic Systems,* vol. 11, No. 6, (November 1975).

Westerlund, L. H., "ATS-F Comsat Millimeter Wave Propagation Experiment," *COMSAT Technical Review,* vol. 3, No. 2, (Fall 1973).

2.6.2 CTS

Booth, G. H., "The Canadian/U.S. High Power Communications Technology Satellite," *Satellite Systems for Mobile Communications and Surveillance, IEEE Conference Publication No. 95,* (March 1973).

Donoughe, P. L., "United States Societal Experiments via the Communications Technology Satellite," *International Conference on Communications,* (June 1976).

Franklin, C., and Davison, E., "A High Power Communications Technology Satellite for the 12 and 14 GHz Bands," AIAA Paper 72-580, *AIAA 4th Communications Satellite Systems Conference,* (April 1972), Reprinted in *Communications Satellite Systems, Progress in Astronautics and Aeronautics,* vol. 32, P. L. Bargellini, (Ed.), (1974).

Ippolito, L. J., "Characterization of the CTS 12 and 14 GHz Communication Links—Preliminary Measurements and Evaluation," International Conference on Communications, (June 1976).

Miller, E. F., Fiala, J. L., and Hansen, I. G., "Performance Characteristics of the 12 GHz, 200 Watt Transmitter Experiment Package for CTS," *EASCON 75 Convention Record,* (September 1975).

O'Donovan, "Design of a 14/12 GHz Transponder for the Communications Technology Satellite," AIAA Paper 72-734, *CASI/AIAA Meeting: Space—1972 Assessment,* (July 1972).

Wright, D. L., and Day, J. W. B., "The Communications Technology Satellite and the Associated Ground Terminals for Experiments," AIAA Paper 75-904, *AIAA Conference on Communications Satellites for Health/Education Applications,* (July 1975).

2.6.3 Symphonie

Hierquist, F., "The Earth Segment in the Symphonie Project," *Progress in Astronautics and Aeronautics,* vol. 32, P. L. Bargellini, (Ed.), (1974).

Moesel, G., and Muller, J., "In-Orbit Performance and Experimental Utilization of the Symphonie Satellites," AIAA Paper 76-306, *AIAA/CASI 6th Communications Satellite Systems Conference,* (April 1976).

Pfieffer, B., and Viellard, P., "The Experimental Telecommunication Satellite Project Symphonie," *Journal of the British Interplanetary Society,* vol. 26, No. 2, (February 1973).

Pfeiffer, B., and Viellard, P., "The Franco-German Telecommunication Satellite Symphonie," *Communication Satellites for the 70s: Systems, Progress in Astronautics and Aeronautics,* vol. 26, N. E. Feldman and C. M. Kelly, (Eds.), (1971).

Schroeter, W., "SHF Performances of the Symphonie Satellite," *World Telecommunication Forum Conference Proceedings,* (October 1975).

2.6.4 OTS

Ball, H. K., and Mondre, E., "The Modular Repeater of the European Communication Satellite (ECS) and the Orbital Test Satellite (OTS)," *Journal of the British Interplanetary Society,* vol. 27, No. 5, (May 1974).

Bartholome, P. J., and Dinwiddy, S. E., "Concept and Characteristics of the European Communication Satellite System," *1975 IEEE Intercon Record,* (April 1975).

Bartholome, P., "The European Communications Satellite System—A Review of Current and Planned Activities," AIAA Paper 76-243, *AIAA/CASI 6th Communications Satellite Systems Conference,* Montreal, Canada, (April 1976).

Collette, R. C., "The European Communications Satellite Programme and the Orbital Test Satellite," *Journal of the British Interplanetary Society,* vol. 29, No. 6, (May 1976).

Collette, R. C., and Stockwell, B., "The OTS Project," AIAA Paper 74-495, *AIAA 5th Communications Satellite System Conference,* (April 1974).

European Conference on Electrotechnics: *EUROCON 74,* (April 1974), reprints of Session C-12, European Experimental Satellite System: (a) Bartholome, P., "OTS—A Forerunner of a European Communication Satellite System," (b) Cantarella, G. P., Preukschat, A. W., and Wearmouth, C., "OTS and the Orbital Test Programme," (c) Mondre, E., and Greiner, W., "Repeater Subsystem for OTS," (d) Bayliss, A., "A Guide for OTS Communications Experiments," (e) Mahner, H., "Earth Stations for the OTS System," (f) Hanell, S., Bartholome, P., and Lothaller, W., "Experimental Data Transmission Capability of OTS."

Falk, H., "European Satellites to Fly," *IEEE Spectrum,* vol. 11, No. 8, (August 1974).

Howell, T. F., "Satellite Communications in the 1980s—A European View," AAS Paper 73-148, *AAS 19th Annual Meeting,* (June 1973), Reprinted in *Advances in the Astronautical Sciences,* vol. 30.

Lopriore, M., Ball, H. K., and Mondre, E., "Design of the 14/11 GHz Repeater for the European Orbital Test Satellite," *International Conference on Communications,* (June 1973).

Tirro, S., and Bayliss, A., "The Utilization Programme of the Orbital Test Satellite," AIAA Paper 76-247, *AIAA/CASI 6th Communications Satellite Systems Conference,* (April 1976).

Wearmouth, C., "The Current Status of the Orbital Test Satellite Programme," *World Telecommunication Forum Conference Proceedings,* (October 1975).

Wearmouth C., and McLaurin, D., "The Development of the Orbital Test Satellite," No. 76-246, *1976 AIAA Satellite Communications Conference,* Montreal, Canada, (April 1976).

2.6.5 SIRIO

Carassa, F., "The Italian Satellite Sirio," *Communication Satellites for the 70s: Systems, Progress in Astronautics and Aeronautics,* vol. 26, N. E. Feldman and C. M. Kelly (Eds.), (1971).

Fanti, P., and Tirro, S., "The Italian Sirio Experiments: Satellite Ground Equipment," *Communication Satellites for the 70s: Systems, Progress in Astronautics and Aeronautics,* vol. 26, N. E. Feldman and C. M. Kelly, (Eds.), (1971).

Perrotta, G., "The Italian SIRIO 12–18 Ghz Experiment: The Forerunner of 20–30 Ghz Preoperational Satellites," AIAA 7th Communications Satellite Systems Conference, San Diego, CA, pp. 736–745, (April 1978).

2.6.6. Japanese Programs

Elson, B. M., "U.S. — Built Satellite Readied for Japan," *Aviation Week and Space Technology,* pp. 52–54, (March 1, 1976).

Hyams, H., "Design of the CS Communications Sub-System," No. 76-244, *1976 AIAA Satellite Communications Conference,* Montreal, (April 1976).

Ikegami, F., and Morimoto, S., "Plans for the Japanese Domestic Satellite Communication System," Paper 4G.1, *1972 IEEE Intercon Record,* (March 1972).

Ishida, T., Tsukamoto, K. I., Hirai, M., and Okamoto, H., "Program of Medium-Capacity Communications Satellite for Experimental Purpose," AIAA Paper 76-244, *AIAA/CAST 6th Communications Satellite Systems Conference,* (April 1976).

Ishida, T., Tsukamoto, K. I., Hirai, M., and Ichikawa, Y., "Program of Medium-Scale Broadcasting Satellite for Experimental Purpose," AIAA Paper 76-255, *AIAA/CASI 6th Communications Satellite Systems Conference,* (April 1976).

Johnson, W., and Reichert, H., "Development of a Medium Scale Broadcasting Satellite for Experimental Purposes," *WESCON Technical Papers,* (September 1975).

Kudo, M., Nemoto, Y., Wickert, A. N., Han, C. C., and Ford, D., "The Design of the Communications Antenna for CS," AIAA Paper 76-252,

AIAA/CASI 6th Communications Satellite Systems Conference, (April 1976).

Ohtake, T., Reichert, H., and Seaman, L. T., "Japanese Broadcast Satellite," *Microwave Journal,* (September 1977).

III. The Communication System

3.1 SOURCE CODING

3.1.1 Voice

Abate, J. E., "Linear and Adaptive Delta Modulation," *Proceedings of the IEEE,* vol. 55, pp. 298-308, (March 1967).

Atal, B. S., and Hanauer, S. L., "Speech Analysis and Synthesis by Linear Prediction of the Speech Wave," *The Journal of the Acoustical Society of America,* vol. 50, pp. 637-655, (April 1971).

Atal, B. S., and Schroeder, M. R., "Adaptive Predictive Coding of Speech Signals," *Bell System Technical Journal,* vol. 49, pp. 1973-1986, (October 1970).

Barnwell, T. P., Brown, J. E., Bush, A. M., and Patisaul, C. R., "Pitch and Voicing in Speech Digitization," *Research Report No. E-21-620-74-BU-1,* Georgia Institute of Technology, (August 1974).

Bayless, J. W., Campanella, S. J., and Goldberg, A. J., "Voice Signals: Bit-by-Bit," *IEEE Spectrum,* vol. 10, pp. 28-39, (October 1973).

Bially, T., and Anderson, W. M., "A Digital Channel Vocoder," *IEEE Transactions on Communications Technology,* vol. COM-18, pp. 435-442, (August 1970).

Birch, J. N., "Low Bit Rate Voice Processors," *EASCON 74 Record,* (October 1974).

Birch, J. N., and Getzin, N. R., "Voice Coding and Intelligibility Testing for a Satellite Based Air Traffic Control System," *Final Report, NASA NAS 5-20168,* (April 1971).

Cohn, D. L., and Melsa, J. L., "The Residual Encoder—An Improved ADPCM System for Speech Digitization," *IEEE Transactions on Communications,* vol. COM-23, No. 9, (September 1975).

Crochiere, R. E., Goodman, D. J., Rabiner, L. R., and Sambur, M. R., "Tandem Connections of a 2.4 KBIT/SEC Vocoder and a 16 KBIT/SEC Delta Modulator," *EASCON 1977.*

Cummiskey, P., Jayant, N. S., and Flanagan, J. L., "Adaptive Quantization in Differential PCM Coding of Speech," *Proceedings of the ICC '73,* (June 1973).

Gold, B., and Rader, C. M., "Digital Processing of Signals," McGraw-Hill Book Co., Inc., New York, (1969).

Goldberg, A. J., "Adaptive Predictive Coding at 6400 Bits/Second," *EASCON '74 Record,* (October 1974).

Goodman, D. J., and Gersho, A., "Theory of an Adaptive Quantizer," *IEEE Transactions on Communications,* vol. COM-22, pp. 1037-1045, (August 1974).

deBrouwer, J., "Voice Coding for Digital Communication," *National Telecommunications Conference,* (1976).

Donaldson, R. W., and Chan, D., "Analysis and Subjective Evaluation of Differential Pulse-Code Modulation Voice Communications Systems," *IEEE Transactions on Communications Technology,* pp. 10-19, (Feb. 1969).

Donaldson, R. W., and Douville, R. J., "Analysis, Subjective Evaluation, Optimization, and Comparison of PCM, DPCM, DM, AM, PM Voice Communication Systems," *IEEE Transactions on Communications Technology,* pp. 421-431, (August 1965).

Dooley, D. J., "LSI Implementation of PCM Codecs," *National Tele-Communications Conference,* (1976).

Dudley, H., "The Carrier Nature of Speech," *The Bell System Technical Journal,* vol. 19, pp. 495-515, (October 1940).

Dunn, J. G., "An Experimental 9600-bit/s Voice Digitizer Employing Adaptive Prediction," *IEEE Trans. Comm. Tech.,* vol. COM-19, pp. 1021-1032, (December 1971).

Flanagan, J. L., "Focal Points in Speech Communication Research," *IEEE Trans. Commun. Tech.,* vol. COM-19, pp. 1006-1015, (December 1971).

Flanagan, J. L., "Speech Analysis, Synthesis and Perception," 2nd Edition, Springer-Verlag, New York, (1972).

Flanagan, J. L., Coker, C. H., Rabiner, L. R., Schafer, R. W., and Umeda, N., "Synthetic Voices for Computers," *IEEE Spectrum,* vol. 7., pp. 22-45, (October 1970).

Flanagan, J. L., and Golden, R. M., "Phase Vocoder," *The Bell System Technical Journal,* vol. 45, pp. 1493-1509, (November 1966).

Fulghum, D. P., "An All-Digital Vocoder," *EASCON 74 Record,* (October 1974).

Gersho, A., and Goodman, D. J., "A Training Mode Adaptive Quantizer," *IEEE Trans. on Information Theory,* (November 1974).

Gibson, J. D., Janes, S. K., and Melsa, J. L., "Sequentially Adaptive Prediction and Coding of Speech Signals," *IEEE Trans. on Communications, (November 1974).*

Gold, B., "Computer Program for Pitch Extraction," *The Journal of the Acoustical Society of America,* (July 1972).

Gold, B., and Rader, C. M., "The Channel Vocoder," *IEEE Transactions on Audio and Electroacoustics,* vol. AU-15, pp. 148-161, (December 1967).

Goldberg, A. J., "Adaptive Predictive Coding at 6400 Bits/Second," *EASCON 74 Record,* (October 1974).

Goodman, D. J., and Gersho, A., "Theory of an Adaptive Quantizer," *IEEE Trans. Commun.,* vol. COM-22, pp. 1037-1045, (August 1974).

Goodman, D. J., "The Application of Delta Modulation to Analog to PCM Encoding," *Bell System Technical Journal,* vol. 48, No. 2, (February 1969).

Greefkes, J. A., and Riemens, K., "Code Modulation with Digitally Controlled Companding for Speech Transmission," *Phillips Tech. Rev.,* vol. 31, No. 11/12, pp. 335-353, (1970).

Hammett, J. C., "An Adaptive Spectrum Analysis Vocoder," Ph.D. Thesis School of Electrical Engineering, Georgia Institute of Technology.

Hammett, J. C., and Bush, A. M., "Adaptive Framing Strategies in Speech Analysis and Synthesis," *Proceedings, NTC, 1972,* pp. 27B-1-27B-6, (December 1972).

Haskew, J. R., Kelly, J. M., and McKinney, T. H., "Results of a Study of the Linear Prediction Vocoder," *IEEE Transactions on Communications,* vol. COM-21, No. 9, pp. 1008-1014, (September 1973).

Itakura, F., et al., "An Audio Response Unit Based on Partial Autocorrelation," *IEEE Transactions on Communications,* vol. COM-20, No. 4, pp. 792-797, (August 1972).

Itakura, F., and Saito, S., "On the Optimum Quantization of Feature Parameters in the PARCOR Speech Synthesizer," *IEEE 1972 Conference on Speech Communication and Processing Paper L4,* New York, (1972).

Jayant, N. S., "Adaptive Delta Modulation with a One-Bit Memory," *Bell System Technical Journal,* vol. 49, No. 3, (March 1970).

Jayant, N. S., "Adaptive Quantization With a One-Word Memory," *Bell System Technical Journal,* (September 1973).

Jayant, N. S., "Delta Modulation of Pitch, Formant, and Amplitude Signals for the Synthesis of Voiced Speech," *IEEE Transactions on Audio and Electroacoustics,* vol. AU-21, No. 3, (June 1973).

Jayant, N. S., "Characteristics of a Delta Modulator," *Proceedings of the IEEE,* (March 1971).

Jayant, N. S., "Digital Coding of Speech Waveforms: PCM, DPCM and DM Quantizers," *Proceedings of the IEEE,* (May 1974).

Jayant, N. S., "Pitch-Adaptive Coding of Speech with Two-Bit Quantization and Fixed Spectrum Prediction," *Bell System Technical Journal,* vol. 56, No. 3, (March 1977).

Jayant, N. S., and Rosenberg, A. E., "The Preference of Slope-Overload to Granularity in the Delta Modulation of Speech," *Bell System Technical Journal,* pp. 3117-3125, (December 1971).

Kang, G. S., "Application of Linear Prediction Encoding to a Narrowband Voice Digitizer," *NRL Report 7774,* Naval Research Laboratory, Washington, DC, (October 31, 1974).

Kang, G. S., "Linear Predictive Voice Digitizer," *EASCON 74 Record,* (October 1974).

LoCicero, J. L., and Schilling, D. L., "An All-Digital Technique for ADM to PCM Conversion," *National Telecommunications Conference,* (1976).

Magill, D., "Adaptive Speech Compression for Packet Communications Systems," *Proceedings NTC 73,* (November 1973).

Makhoul, J. I., and Sutherland, W. R., "Natural Communication with Computers, Final Report, Vol. II: Speech Compression Research at BBN," *BBN Report No. 2976,* Bolt Beranet and Newman, Inc., Cambridge, MA, (December 1974).

Makhoul, J., "Linear Prediction: A Tutorial Review," *Proceedings of the IEEE,* vol. 63, No. 4, (April 1975).

Makhoul, J. I., and Wolf, J. J., "Linear Prediction and the Spectral Analysis of Speech," *BBN Report No. 2304,* Bolt, Beranek and Newman, Inc., Cambridge, MA, (August 1972).

Markel, J. D., "Basic Formant and F_0 Parameter Extraction from a Digital Inverse Filter Formulation," *IEEE Transactions on Audio and Electroacoustics,* vol. AU-21, No. 3, pp. 154-160, (June 1973).

Markel, J. D., "Digital Inverse Filtering, A New Tool for Formant Trajectory Estimation," *IEEE Transactions on Audio and Electroacoustics,* vol. AU-20, No. 2, pp. 129-137, (June 1972).

Markel, J. D., "The SIFT Algorithm for Fundamental Frequency Estimation," *IEEE Transactions on Audio and Electroacoustics,* vol. AU-20, No. 5, pp. 367-377, (December 1972).

Markel, J. D., and Gray, Jr., A. H., "A Linear Prediction Vocoder Simulation based upon the Autocorrelation Method," *IEEE Trans. Acoustics, Speech and Signal Processing,* vol. ASSP-22, pp. 124-134, (April 1974).

Markel, J. D., and Gray, Jr., A. H., "On Autocorrelation Equations with Application to Speech Analysis," *IEEE Transactions on Audio and Electroacoustics,* vol. AU-21, No. 2, pp. 69-79, (April 1973).

Markel, J. D., Gray, Jr., A. H., and Wakita, W., "Linear Prediction of Speech-Theory and Practice," *SCRL Monograph No. 10, Speech Communications Research Laboratory, Inc., (September 1973).*

Max, J., "Quantizing for Minimum Distortion," *IRE Trans. Inform. Theory,* vol. IT-6, pp. 7-12, (March 1960).

McDonald, R. A., "Signal-to-Noise and Idle Channel Performance of Differential Pulse Code Modulation Systems—Particular Applications to Voice Signals," *Bell System Technical Journal,* (September 1966).

McNair, Jr., I. M., "Subscriber Loop Multiplexer—A High Pair Gain System for Upgrading and Growth in Rural Areas," *IEEE Trans. Commun. Technol.,* vol. COM-19, pp. 523-527, (August 1967).

Melnick, M., "Intelligibility Performance of Variable Slope Delta Modulation," *Proceedings ICC-73,* (June 1973).

Messerschmitt, D. G., "Speech Encoding for Digital Transmission," *EASCON 1976.*

Miller, J. H., "Digital-to-Digital Conversion between an Adaptive Delta Modulation Format and Companded PCM," *International Conference on Communications,* (1976).

Noll, A. M., "Cepstrum Pitch Determination," *Journal of the Acoustical Society of America,* vol. 41., pp. 293-309, (February 1967).

Nunes, P. R., and Albuquerque, J. P. A., "Statistical Delta Modulation," *International Conference on Communications,* (1976).

Oliver, B. M., Pierce, J. R., and Shannon, C. E., "The Philosophy of PCM," *Proc. IRE,* vol. 36, pp. 1324-1331, (October 1948).

O'Neal, Jr., J. B., "Delta Modulation of Data Signals," *IEEE Trans. on Communications,* (March 1974).

O'Neal, Jr., J. B., "Predictive Quantizing Systems (Differential Pulse Code Modulation) for the Transmission of Television Signals," *Bell System Technical Journal,* (May–June, 1966).

Oppenheim, A. V., and Schafer, R. W., "Homomorphic Analysis of Speech," *IEEE Transactions on Audio and Electroacoustics,* vol. AU-16, pp. 221-226, (June 1968).

Oppenheim, A. V., Schafer, R. W., and Stockham, T. G., "The Non-linear Filtering of Multiplied and Convolved Signals," *Proceedings of the IEEE,* vol. 56, pp. 1264-1291, (August 1968).

Oppenheim, A. V., "Speech Analysis-Synthesis System Based on Homomorphic Filtering," *Journal of the Acoustical Society of America,* vol. 45, pp. 458-465, (February 1969).

Paez, M. D., and Glisson, T. H., "Minimum Mean Squared-Error Quantization in Speech, PCM and DPCM Systems," *IEEE Trans. Commun.,* vol. COM-20, pp. 225-230, (April 1972).

Rabiner, L. R., Schmidt, C. E., and Atal, B. S., "Evaluation of a Statistical Approach to Voiced-Unvoiced-Silence Analysis for Telephone-Quality Speech," *Bell System Technical Journal,* vol. 56, No. 3, (March 1977).

Schafer, R. W., and Rabiner, L. R., "System for Automatic Formant Analysis of Voiced Speech," *Journal of the Acoustical Society of America,* vol. 47, pp. 634-648, (February 1970).

Schroeder, M. R., "Vocoders: Analysis and Synthesis of Speech," *Proceedings of the IEEE,* vol. 54, No. 5, pp. 720-734, (May 1966).

Slepian, D., "On Delta Modulation," *Bell System Technical Journal,* vol. 51, No. 10, (December 1972).

Smith, B., "Instantaneous Companding of Quantized Signals," *Bell System Technical Journal,* pp. 653-709, (May 1957).

Stein, S., and Jones, J. J., "Modern Communication Principles," McGraw-Hill, Inc., New York, (1967).

Viswanathan, R., Blackman, E., and Makhoul, J., "Variable Frame Rate Narrowband Speech Transmission Over Fixed Rate Noisy Channels," *EASCON 1977.*

Wolf, J. K., "Effects of Channel Errors on Delta Modulation," *IEEE Transactions on Communication Technology,* vol. COM-14, No. 1, (February 1966).

Yan, J., and Donaldson, R. W., "Subjective Effects of Channel Transmission Errors on PCM and DPCM Voice Communications Systems," *IEEE Trans. Comm. Systems,* pp. 2-1-290, (June 1972).

Zelinski, R., and Noll, P., "Adaptive Transform Coding of Speech Signals," *IEEE Transactions on Acoustics, Speech, and Signal Processing,* vol. ASSP-25, No. 4, (August 1977).

3.1.2 Image Compression

Abbot, R. P., "DPCM Codec for Video Telephony Using 4-Bits Per Sample," *IEEE Trans. on Comm. Tech.,* vol. COM-19, pp. 907-913, (1971).

Ahmed, N., Natarjan, T., and Rao, K. R., "Discrete Cosine Transform," *IEEE Trans. on Computer,* pp. 90-93, (January 1974).

Anderson, G. B., and Huang, T. S., "Piecewise Fourier Transformation for Picture Bandwidth Compression," *IEEE Transactions on Communication Technology,* vol. COM-19, pp. 133-140, (April 1971).

Candy, J. C., Franke, M. A., Haskell, B. G., and Mounts, F. W., "Transmitting Television as Clusters of Frame-to-Frame Differences," *Bell System Technical Journal,* vol. 50, pp. 1889-1917, (July–August 1971).

C. C. I. R., "Comparison of Different Methods for the Digital Coding of Television Signals," *CCIR Study Groups,* (Period 1974-1978), DOC. CMTT/39-E.

C. C. I. R., "Digital Transmission of Television Signals," *13th Plenary Assembly,* Doc. 646, vol. 12, pp. 208-209, (1974).

C. C. I. R., "Method for the Subjective Assessment of the Quality of Television Pictures," *13th Plenary Assembly,* Rec. 500, vol. 11, pp. 65-68, (1974).

C. C. I. R., "The Specification of Performance for Transmission Circuits which Employ Digital Methods," *13th Plenary Assembly,* Rep. 645, vol. 12, pp. 206-208, (1974).

Chen, W., and Smith, C. H., "Adaptive Coding of Monochrome and Color Images," *IEEE Transactions on Communications,* vol. COM-25, No. 11, (November 1977).

Connor, D. J., Brainard, R. C., and Limb, J. O., "Intraframe Coding for Picture Transmission," *Proc. IEEE,* vol. 60, pp. 779-791, (1972).

Connor, D. J., Haskell, B. G., and Mounts, F. W., "A Frame-to-Frame Picturephone Coder for Signals Containing Differential Quantizing Noise," *Bell System Technical Journal,* vol. 52, No. 1, (January 1973).

Davisson, L. D., "Rate-Distortion Theory and Application," *Proceedings of IEEE,* vol. 60, pp. 800-828, (July 1972).

Davisson, L. D., "Universal Noiseless Coding," *IEEE Trans. on Information Theory,* vol. 19, pp. 783-795, (November 1973).

Frei, W., and Baxter, B., "Rate-Distortion Coding Simulation for Color Images," *IEEE Transactions on Communications,* vol. COM-25, No. 11, (November 1977).

Goblick, T. J., and Holzinger, J. L., "Analog Source Digization: A Comparison of Theory and Practice," *IEEE Trans. Inform. Theory,* vol. IT-13, pp. 323-326, (April 1967).

Habibi, A., "Comparison of n-th Order DPCM Encoder with Linear Transformations and Block Quantization Techniques," *IEEE Transactions on Communication Technology,* vol. COM-19, No. 6, pp. 948-956, (December 1971).

Habibi, A., "Delta Modulation and DPCM Coding of Color Signals," *Proceedings of the International Telemetry Conference,* vol. 8, Los Angeles, pp. 333-343, (October 10-12, 1972).

Habibi, A., "Survey of Adaptive Image Coding Techniques," *IEEE Transactions on Communications,* vol. COM-25, No. 11, (November 1977).

Habibi, A., and Hershel, R. S., "A Unified Representation of Differential Pulse-Code Modulation (DPCM) and Transform Coding Systems," *IEEE Transactions on Communications,* vol. COM-22, No. 5, pp. 692-696, (May 1974).

Habibi, A., and Robinson, G. S., "A Survey of Digital Picture Coding," *IEEE Computer,* vol. 7, No. 5, pp. 22-34, (May 1974).

Habibi, A., and Samulon, A. S., "Bandwidth Compression of Multispectral Data," *Proceedings of SPIE,* vol. 66, pp. 23-35, (August 1975).

Habibi, A., and Wintz, P. A., "Image Coding by Linear Transformation and Block Quantization," *IEEE Transactions on Communication Technology,* vol. COM-19, No. 1, pp. 50-63, (February 1971).

Haskell, B. G., "A Low Bit-Rate Interframe Coder for Videotelephone," *Bell System Technical Journal,* vol. 54, No. 8, pp. 1475-1495, (October 1975).

Haskell, B., Gordon, P. L., Schmidt, R. L., and Scattaglia, J. V., "Interframe Coding of 525-Line, Monochrome Television at 1.5 Mbit/s," *IEEE Transactions on Communications,* vol. COM-25, No. 11, (November 1977).

Haskell, B. G., Mounts, F. W., and Candy, J. C., "Interframe Coding of Video-Telephone Pictures," *Proc. IEEE,* vol. 60, pp. 792-800, (1972). *National Telecommunications Conference,* Dallas, (December 1976).

Iinuma, K., Iijima, Y., Ishiguro, T., and Kaneko, H., "Interframe

Coding for 4 MHz Color Television Signals," *Conference Record, ICC 1975,* pp. 23-26–23-30.

Iinuma, K., et al., "Interframe Coding for 4-MHz Color Television Signals," *IEEE Transactions on Communications,* COM-23, No. 12, pp. 1461-1466, (December 1975).

Ishiguro, T., Iinuma, K., Iijima, Y., Koga, T., Azami, S., and Mune, T., "Composite Interframe Coding of NTSC Color Television Signals,"

Limb, J. O., and Mounts, F. W., "Digital Differential Quantizer for Television," *Bell System Technical Journal,* vol. 48, pp. 2583-2599, (1969).

Limb, J. O., and Pease, R. F. W., "A Simple Interframe Coder for Video Telephony," *Bell System Technical Journal,* vol. 50, No. 6, pp. 1877-1888, (August 1971).

Limb, J. O., Rubinstein, C. B., and Thompson, J. E., "Digital Coding of Color Video Signals—A Review," *IEEE Transactions on Communications,* vol. COM-25, No. 11, (November 1977).

Limb, J. O., Pease, R. F. W., and Walsh, K. A., "Combining intraframe and frame to frame coding for television," *Bell System Technical Journal,* vol. 53, pp. 1137-1173, (July 1974).

Limb, J. O., Rubinstein, C. B., and Walsh, K. A., "Digital Coding of Color PICTUREPHONE Signals by Element-Differential Quantization," *IEEE Trans. on Commun. Technol.,* vol. COM-19, pp. 992-1006, (December 1971).

Mannos, J. L., and Sakrison, D. J., "The Effects of a Visual Fidelity Criterion on the Encoding of Images," *IEEE Trans. Inform. Theory,* vol. IT-20, pp. 525-536, (July 1974).

Millard, J. B., and Maunsell, H. I., "Digital Encoding of the Video Signal," *Bell System Technical Journal,* vol. 50, pp. 459-480, (1971).

Natarajan, T. R., and Ahmed, N., "On Interframe Transform Coding," *IEEE Transactions on Communications,* vol. COM-25, No. 11, (November 1977).

Netravali, A. N., "On Quantizers for DPCM Coding of Picture Signals," *IEEE Trans. on Information Theory,* vol. IT-23, pp. 360-370, (May 1977).

Oliver, B. M., "Efficient Coding," *Bell System Technical Journal,* vol. 31, pp. 724-750, (1952).

O'Neal, Jr., J. B., "Predictive quantizing system (differential pulse code modulation) for the transmission of television signals," *Bell System Technical Journal,* vol. 45, pp. 689-721, (May 1966).

Rice, R. F., and Plaunt, J. R., "Adaptive Variable-Length Coding for Efficient Compression of Space Craft Television Data," *IEEE Trans. on Comm. Tech.,* vol. COM-19, pp. 889-897, (December 1971).

Pratt, W. K., "Digital Image Processing," Wiley, New York, (1977).

Reader, C., "Intraframe and Interframe Adaptive Transform Coding," *Proceedings of SPIE,* vol. 66, pp. 108-118, (August 1975).

Roese, J. A., Pratt, W. K., and Robinson, G. S., Interframe Cosine Transform Image Coding," *IEEE Transactions on Communications,* vol. COM-25, No. 11, (November 1977).

Roese, J. A., and Pratt, W. K., "Theoretical Performance Models for Interframe Transform and Hybrid Transform/DPCM Coders," *Proceedings SPIE,* vol. 87, pp. 172-179, (August 1976).

Roese, J. A., and Robinson, G. S., "Combined Spatial and Temporal Coding of Digital Image Sequences," *Proceedings SPIE,* vol. 66, pp. 172-180, (August 1975).

Roese, J. A., et. al., "Interframe Transform Coding and Predictive Coding Methods," *Proceedings 1975 International Conference on Communications,* San Francisco, vol. 2, pp. 17-21, (June 1975).

Sakrison, D. J., "On the Role of the Observer and a Distortion Measure in Image Transmission," *IEEE Transactions on Communications,* vol. COM-25, No. 11, (November 1977).

Scheinberg, N., and Schilling, D. L., "Techniques for correcting transmission errors in video adaptive delta modulation channels, *IEEE Trans. Commun.,* vol. COM-24, pp. 1064-1069, (September 1976).

Sharma, D. K., and Netravali, A. N., "Design of Quantizers for DPCM Coding of Picture Signals," *IEEE Transactions on Communications,* vol. COM-25, No. 11, (November 1978).

"Special Issue on Digital Picture Processing," *Proceedings IEEE,* vol. 6, No. 7, (July 1972).

"Special Issue on Redundancy Reduction," *Proceedings IEEE,* vol. 55, No. 3, (March 1967).

Tasto, M., and Wintz, P. A., "Image Coding by Adaptive Block Quantization," *IEEE Transactions on Communication Technology,* vol. COM-19, No. 6, pp. 957-971, (December 1971).

Tescher, A. G., Andrews, H. C., and Habibi, A., "Adaptive Phase Coding in Two and Three-Dimensional Fourier and Walsh Image Compression," *1974 Picture Coding Symposium,* Goslar, Germany, (August 26-28, 1974).

Wintz, P. A., "Transform Picture Coding," *Proc. IEEE,* vol. 60, pp. 809-820, (July 1972).

Zschunke, W., "DPCM Picture Coding with Adaptive Prediction," *IEEE Transactions on Communications,* vol. COM-25, No. 11, (November 1977).

Zschunke, W., "Digital Transmission of TV by Satellite," Third International Conference Digital Satellite Communications, Kyoto, Japan, pp. 322-329, (November 11-13, 1975).

3.1.3 Facsimile

Ascher, R. N., and Nagy, G., "A Means for Achieving a High Degree Compaction on Scan-Digitized Printed Text," *IEEE Trans. on Computers,* vol. C-23, No. 11, pp. 1174-1179, (November 1974).

Capon, J., "Probabilistic Model for Runlength Coding of Pictures," *IRE Trans. on Info. Theory,* (December 1959).

deCoulon, F., and Kunt, M., "An Alternative to Runlength Coding for Black-and-White Facsimile," *Proc. 1974 International Zurich Seminar on Digital Communications,* Zurich, Switzerland, (March 1974), See also Monk, U.S. Patent #3,588,329 and Lavallee, U. S. Patent #3,726,993.

Federal Republic of Germany, "Relations between transmission time, transmission bit rate, maximum line feed speed, and buffer memory size in run length coded facsimile systems," *CCITT Study Group XIV,* temp. docum., No. 11, (September 1976).

Huang, T. S. "Coding of Two-Tone Images," *IEEE Transactions on Communications,* vol. COM-25, No. 11, (November 1977).

Huang, T. S., and Koeller, H. U., "Coding of Multilevel Graphics," *IEEE Trans. on Comm.,* vol. COM-23, No. 6, pp. 598-606, (June 1975).

Huffman, D. A., "A Method for the Construction of Minimum Redundancy Codes," *Proc. IEEE,* vol. 40, pp. 1098-1011, (September 1952).

Kalle Infotec U. K. Ltd., "Redundancy reduction technique for fast black-and-white facsimile apparatus," *CCITT Study Group XIV,* temp. docum., No. 15, (April 1975).

Kobayashi, H., and Bahl, L. R., "Image Data Compression by Predictive Coding," *IBM Journal Research and Development,* pp. 164-179, (1974).

Kokusai Denshin Denwa Ltd., "Proposals for redundancy reduction technique of digital facsimile signals," *CCIT Study Group XIV,* temp. docum., No. 18, (April 1975).

Meyr, H., Rosdolsky, H. G., and Huang, T. S., "Optimum Runlength Codes," *IEEE Trans. on Comm.,* (June 1974).

Musmann, H. G., and Preuss, D., "Comparison of Redundancy Reducing Codes for Facsimile Transmission of Documents," *IEEE Transactions on Communications,* vol. COM-25, No. 11, (November 1978).

Phillips Telecommunicatie Industrie B. V., "A redundancy reducing run length code for digital facsimile signals with line correcting capabilities," *CCITT Study Group XIV,* temp. docum., No. 4, (October 1976).

Preuss, D., "Comparison of two-dimensional facsimile coding schemes," *Int. Conf. on Communications,* San Francisco, pp. 7/12-7/16, (1975).

Preuss, D., "Comparison of Two-Dimensional Facsimile Coding Schemes," *Proc. ICC 1975,* San Francisco.

Schreiber, W. F., Huang, T. S., and Tretiak, O. J., "Contour Coding of Images" in Picture Bandwidth Compression, (Ed.) by T. S. Huang and O. J. Tretiak, Gordon and Breach, (1972).

Stern, P. A., and Heinlein, W. E., "Facsimile Coding Using Two-Dimensional Runlength Prediction," *Proc. Zurich Seminar on Digital Communications, (1974).*

3.1.4 Multiple Channel Processing

Adoul, J. P., and Dasboul, F., "Digital Tasi Generalization with Voiced/Unvoiced Discrimination for Tripling T1 Carrier Capacity," *International Conference on Communications,* (1977).

Fraser, J. M., Bullock, D. B., and Long, N. G., "Over-All Characteristics of a TASI System," *Bell System Technical Journal,* (July 1962).

Campanella, S. J., "Digital Speech Interpolation," *COMSAT Technical Review,* vol. 6, No. 1, (Spring 1976).

Campanella, S. J., and Suyderhoud, H. G., "Digital Speech Interpolation for Telephone Communications," *EASCON-1975 Convention Record,* (October 1974).

Campanella, S. J., and Suyderhoud, H. G., "Performance of Digital Speech Interpolation Systems for Satellite Telecommunications," *Na-*

tional Telecommunications Conference-1975, (December 1975).

Lombard, D., and Payet, G., "Digital Speech Interpolation in Satellite Systems," *International Conference on Communications-1975,* (June 1975).

Meacham, L. A., and Peterson, E., "An Experimental Multichannel Pulse Code Modulation System of Toll Quality," *BSTJ,* (January 1948).

Sciulli, J. A., and Campanella, S. J., "A Speech Predictive Encoding Communication System for Multichannel Telephony," *IEEE Transactions on Communications,* vol. COM-21, No. 7, (July 1973).

3.2 ANALOG MODULATION

Aein, J. M., "Power balancing in systems employing frequency reuse," *Comsat Technical Review,* vol. 3, No. 2, (Fall 1973).

Babcock, W. C., "Intermodulation Interference in Radio Systems," *Bell System Technical Journal,* (January 1953).

Bakken, P. M., "Feedforward Linearized Traveling Wave Tube Satellite Transponder," *EASCON 1974,* pp. 405–411.

Bennet, W. R., Curtis, H. E., and Rice, S. O., "Interchannel interference in FM and PM systems under noisy loading conditions," *Bell System Technical Journal,* vol. 34, pp. 601–636, (May 1965).

Berman, A., and Podraczky, E., "Experimental determination of intermodulation distortion produced in a wide-band communication repeater," *IEEE Int. Conv. Rec.,* pt. 2, vol. 15, pp. 69–88, (1967).

Berman, A. L., Mahle, C., and Wachs, M. R., "INTELSAT IV Transmission Modeling," *COMSAT Technical Review,* vol. 2, No. 2, (Fall 1972).

Berman, A. L., and Mahle, C. E., "Nonlinear phase-shift in traveling wave tubes as applied to multiple access communications satellites," *IEEE Trans. Commun. Technol.,* vol. COM-18, pp. 37–48, (February 1970).

Blachman, N.M., "Detectors, bandpass nonlinearities, and their optimization: Inversion of the Chebyshev transform," *IEEE Trans. Inform. Theory,* vol. IT-17, pp. 398–404, (July 1971).

Brown, Jr., M. P., "Review of INTELSAT Earth Stations RF Out-of-Band Emission Criteria," *COMSAT Technical Review,* vol. 4, No. 1, (Spring 1974).

Cahn, C. R., "Crosstalk due to finite limiting of frequency multiplexed signals," *Proc. IRE,* vol. 48, pp. 53–59, (January 1960).

Campanella, S. J., Suyderhoud, H. G., and Wachs, M., "Frequency Modulation and Variable-Slope Delta Modulation in SCPC Satellite Transmission," *Proceedings of the IEEE,* vol. 65, No. 3, (March 1977).

Chakraborty, D., "INTELSAT IV Satellite System (Voice) Channel Capacity Versus Earth-Station Performance," *IEEE Transactions on Communication Technology,* (June 1971).

Chapman, R. C., and Millard, J. B., "Intelligible Crosstalk Between Frequency Modulate Carriers Through AM/PM Conversion," *IEEE Trans. on Comm. Tech.,* vol. CS-12, (1964).

Dicks, J. L., and Brown, Jr., M. P., "INTELSAT IV-A Satellite Transmission Design," *A.I.A.A. Paper No. 74-474,* (April 1974).

Dicks, J. L., Schultze, P. H., and Schmitt, C. H., "INTELSAT IV System Planning," *COMSAT Technical Review,* vol. 2, No. 2, (Fall 1972).

Fuenzalida, J. C., and Podraczky, E., "Reuse of the Frequency Spectrum at the Satellite," *AIAA 3rd Communications Satellite Systems Conference,* Los Angeles, CA, (April 6-8, 1970).

Fuenzalida, J. C., Shimbo, O., and Cook, W. L., "Time-Domain Analysis of Intermodulation Effects Caused by Nonlinear Amplifiers," *COMSAT Technical Review,* vol. 3, No. 1, pp. 89–143, (Spring 1973).

Furuys, T., Fujii, A., and Tezuka, K., "Impulse Noise in FM Receivers in the Presence of Adjacent Channel Interference and Thermal Noise," *International Conference on Communications,* (June 1972).

Garrison, G. J., "Intermodulation Distortion in Frequency Division Multiplex FM Systems—A Tutorial Summary," *IEEE Trans. on Communications Tech.,* vol. COM-16, No. 2, (1968).

Hamer, R., "Radio-Frequency Interference in Multichannel Telephony FM Radio Systems," *Proceedings of the IEE* (UK), vol. 108B, No. 37, (January 1961).

Lillemack, A. H., "Intelligible Cross-Talk Between Angle Modulated Carriers," *ICC Conference Record,* (1971).

Lyons, R. G., "Effects of Non-Linear Gain and AM/PM Conversion in FDMA Communications Through a Satellite Repeater," *ICC Conference Record,* (1971).

McClure, R. B., "Analysis of Intermodulation Distortion in a FDMA Satellite Communication System with a Bandwidth Constraint," *IEEE Int. Conf. on Comm.,* (1970).

Moreno, L., "Spectral characteristics and channel spacing criteria in FDM/FM radio links," *CSELT Technical Report,* vol. V, No. 3, (1977).

Ohwaku, S., and Kuwagaki, D., "Common amplification of many signals by TWT amplifier," *J. Inst. Electron. Commun. Eng. Jap.,* vol. 51-B, No. 5, pp. 78–86, (1968).

Pawula, R. F., "The effects of quadratic AM-PM conversion in frequency-division multiplexed multiple-access communication satellite systems," *IEEE Trans. Commun. Technol.,* vol. COM-19, pp. 345–349, (July 1971).

Pawula, R. F., Fong, T. S., and Sullivan, M. R., "Intermodulation distortion in frequency multiplexed satellite repeaters," *Proc. IEEE,* vol. 59, pp. 213–218, (February 1971).

Prabhu, V. K., and Enloe, L. H., "Interchannel Interference Considerations in Angle-Modulated Systems," *Bell System Technical Journal,* (September 1969).

Shaft, P. D., "Intermodulation Interference in Frequency-Division Multiple Access Systems," *IEEE ICC Conf. Record,* (June 1967).

Shaft, P. D., "Limiting of Several Signals and Its Effects on Communication System Performance," *IEEE Trans. on Comm. Tech.,* vol. COM-13, No. 4, pp. 504–512, (December 1965).

Shimbo, O., "Effects of Intermodulation, AM-PM Conversion and Additive Noise in Multicarrier TWT Systems," *Proceedings of the IEEE,* vol. 59, No. 2, pp. 230–238, (February 1971).

Shimbo, D., and Pontano, B. A., "A General Theory for Intelligible Crosstalk Between Frequency-Division Multiplexed Angle-Modulated Carriers," *IEEE Transactions on Communication,* vol. COM-24, No. 9, pp. 999–1007, (September 1976).

Shukla, V., and Keyes, L. A., "Intelligible Crosstalk Due to Nonlinear TWTA Amplifiers in a Satellite Transponder," *National Telecommunications Conference,* (1976).

Sunde, E. D., "Intermodulation Distortion in Multi-Carrier FM Systems," *1965 IEEE Int. Conv. Record,* Part 2, pp. 130–146.

Walsh, D. W., "Theoretical Solutions to Problems Introduced by the TWA in Multiple Access Communication Systems," *PH.D. Thesis, Univ. Microfilms, No. 74-22553,* George Washington University, Washington, DC.

Westcott, R. J., "Investigation of multiple f.m./f.d.m. carriers through a satellite t.w.t. operating near to saturation," *Proceedings of the IEEE,* (UK), vol. 114, No. 6, pp. 726–750, (June 1976).

3.3 DIGITAL MODULATION

3.3.1 General

Lindsey, W. C., and Simon, M. K., "Telecommunication System Engineering," Prentice-Hall, Englewood Cliffs, NJ, (1973).

Lucky, R. W., Salz, J., and Weldon, E. J., "Principles of Data Communication," McGraw-Hill, New York, p. 249, eq. 9.10, (1968).

Palmer, L., "Economic Trends in Digital Transmission by Satellite," *Proceedings of the Fifth Data Communications Symposium,* Snowbird, UT, pp. 2-30-2-36, (September 1977).

Stein, S., and Jones, J. J., "Modern Communication Principles With Application to Digital Signaling," McGraw-Hill, New York, (1967).

Viterbi, A. J., "Principles of Coherent Demodulation," McGraw-Hill, New York, Chs. 7, 10, (1966).

Welti, G. R., and Kwan, R. K., "Companison of Signal Processing Techniques for Satellite Telephony," *National Telecommunications Conference,* pp. 05:1-1-05:1-6, (1977).

Wozencraft, J. M., and Jacobs, I. M., "Principles of Communications Engineering," Wiley, New York, (1965).

3.3.2 Linear Channel: Additive Noise

Bedrosian, E., "Spectrum Conservation by Efficient Channel Utilization," *IEEE Communications Society Magazine,* pp. 20–27, (March 1977).

Benedetto, S., and Biglieri, E., "Digital PM-AM transmission over real channels," *Second Int. Symp. on Digital Satellite Communications,* Paris, pp. 193–198, (November 1972).

Bussgang, J. J., and Leiter, M., "Error rate approximations for differential phase shift keying," *IEEE Trans. on Comm. Syst.,* vol. CS-12, pp. 18–27, (March 1964).

Cahn, C. R., "Performance of Digital Phase-Modulation Communication Systems," *IRE Transactions on Communications Systems,* pp. 219–222, (May 1959).

Cahn, C. R., "Combined digital phase and amplitude modulation communication systems," *IRE Trans. Commun. Syst.,* vol. CS-8, pp. 150–155, (September 1960).

Campopiano, C. N., Glazer, B. G., "A coherent digital amplitude and phase modulation scheme," *IRE Trans. Commun. Syst.,* vol. CS-10, pp. 90–95, (March 1962).

Chan, C. R., May, C. L., and Robatino, V., "Bandwidth-Efficient QPSK For Multiple-Access Satellite Communication," *National Telecommunications Conference*, pp. 42:3-1-42:3-4, (1977).

Dunbridge, B., "Asymmetric signal design for the coherent Gaussian channel," *IEEE Trans. Inform. Theory*, vol. IT-13, pp. 422-431, (July 1967).

Essman, J. E., and Wintz, P. A., "The Effects of Channel Errors in DPCM Systems and Comparison with PCM Systems," *IEEE Trans. Comm. Tech.*, pp. 867-877, (August 1973).

Foschini, G. J., Gitlin, R. D., and Weinstein, S. B., "Optimization of two-dimensional signal constellations in the presence of gaussian noise," *IEEE Trans. Comm.*, vol. COM-22, pp. 28-38, (January 1974).

Glance, B., "Power Spectra of Multilevel Digital Phase-Modulated Signals," *Bell System Technical Journal*, pp. 2857-2878, (September 1971).

Glenn, A. B., "Comparison of PSK vs FSK and PSK-AM vs FSK-AM binary-coded transmission systems," *IRE Trans. Comm. Systems*, vol. CS-8, pp. 87-100, (June 1960).

Hancock, J. C., and Lucky, R. W., "Performance of combined amplitude and phase-modulated communications systems," *IRE Trans. Comm. Systems*, vol. CS-8, pp. 232-237, (December 1960).

Kawai, K., et al., "Optimum combinations of amplitude and phase modulation scheme and its application to data transmission modem," *Proc. Int. Communication Conf.*, pp. 29-6-29-11, (1972).

Lindsey, W. C., "On the Detection of Differentially Encoded Polyphase Signals," *IEEE Trans. Comm. Tech.*, pp. 1121-1127, (December 1972).

Lucky, R. W., and Hancock, J. C., "On the optimum performance of N-ary Systems having two degrees of freedom," *IRE Trans. Commun. Syst.*, vol. CS-10, pp. 185-192, (June 1962).

Nuttall, A. H., "Error probabilities for equicorrelated M-ary signals under phase-coherent and phase-incoherent reception," *IRE Trans. on Inform. Th.*, vol. IT-18, p. 305, (July 1962).

Salz, J., et al., "Data transmission by combined AM and PM," *Bell System Technical Journal*, vol. 50, pp. 2399-2419, (September 1971).

Simon, M. K., and Smith, J. G., "Hexagonal multiple phase-and-amplitude-shift-keyed signal sets," *IEEE Trans. Comm.*, vol. COM-21, pp. 1108-1115, (October 1973).

Smith, J. G., "A Bandwidth-Compressive Modulation System," *Progress in Astronautics and Aeronautiocs*, vol. 54, AIAA, pp. 513-539.

Smith, J. G., "Odd-Bit Quadrature Amplitude Shift Keying," *IEEE Trans. Comm. Tech.*, pp. 385-389, (March 1975).

Smith, J. G., "Spectrally Efficient Modulation," *International Conference on Communications*, pp. 3.1-37-3.1-41, (1977).

Stein, S., "Unified analysis of certain coherent and non-coherent binary communications systems," *IEEE Trans. on Inform. Th.*, vol. IT-10, pp. 43-51, (January 1964).

Thomas, C. M., Weidner, M. Y., and Durrani, S. H., "Digital amplitude-phase keying with M-ary alphabets," *IEEE Trans. Comm.*, vol. COM-22, pp. 168-180.

Weber, C., "New solutions to the signal design problem for coherent channels," *IEEE Trans. Inform. Theory*, vol. IT-12, pp. 161-167, (April 1966).

Weber, W. J., III, Stanton, P. H., and Sumida, J. T., "A Bandwidth Compressive Modulation System Using Multi-Amplitude Minimum Shift Keying (MAMSK)," *National Telecommunications Conference*, pp. 05:5-1-05:5-7, (1977).

Welti, G. R., "PCM/FDMA satellite telephony with 4-dimensionally coded quadrature amplitude modulation," *Comsat Technical Review*, vol. 6, No. 2, pp. 323-338, (Fall 1976).

Wolejsza, C., and Chakraborty, D., "Performance of 120-Channel Hybrid Modems for Satellite Communications," *International Conference on Communications*, pp. 24.2-145-24.2-148, (1977).

Wood, W. A. H., "Modulation and Filtering Techniques for 3 Bits/Hertz Operation in the 6 GHz Frequency Band," *International Conference on Communications*, pp. 5.3-97-5.3-101, (1977).

3.3.3 Intersymbol Interference

Aein, J. M., and Hancock, J. C., "Reducing the Effects of Intersymbol Interference with Correlation Receivers," *IEEE Transactions on Information Theory*, pp. 167-174, (July 1963).

Assal, F., "Approach to a near-optimum transmitter-receiver filter design for data transmission pulse-shaping networks," *Comsat Tech. Rev.*, vol. 3, pp. 301-322, (Fall 1973).

Benedetto, S., DeVincentiis, G., and Luvison, A., "Error probability in the presence of intersymbol interference and additive noise for multilevel digital signals," *IEEE Trans. on Comm.*, vol. COM-21, pp. 181-190, (March 1973).

Benedetto, S., and Biglieri, E., "Performance of M-ary PSK systems in the presence of intersymbol interference and additive noise," *Alta Frequenza*, vol. XLI, pp. 225-239, (1977).

Cantoni, A., and Butler, P., "Linear Minimum Mean Square Error Estimators Applied to Channel Equalization," *IEEE Transactions on Communications*, pp. 441-446, (April 1977).

Celebiler, M. I., and Shimbo, O., "The Probability of Error Due to Intersymbol Interference and Gaussian Noise in Digital Communication Systems," *IEEE Transactions on Communication Technology*, vol. COM-19, No. 2, (April 1971).

Chakraborty, D., "Consideration of 8-phase CPSK-TDMA Signal Transmission Through Band-Limited Satellite Channels," *IEEE Transactions on Communications*, vol. COM-25, pp. 1233-1237, (October 1977).

Gitlin, R. D., and Magee, Jr., F. R., "Self-Orthogonalizing Adaptive Equalization Algorithms," *IEEE Transactions on Communications*, vol. COM-25, No. 7, pp. 666-672, (July 1977).

Glave, F. E., "An Upper Bound on the Probability of Error Due to Intersymbol Interference for Correlated Digital Signals," *IEEE Transactions on Information Theory*, vol. IT-18, No. 3, pp. 356-363, (May 1972).

Ho, E. Y., and Yeh, Y. S., "Error Probability of Multilevel Digital System with Intersymbol Interference and Gaussian Noise," *Bell System Technical Journal*, vol. 50, No. 3, pp. 1017-1023, (March 1971).

Ho, E. Y., and Yeh, Y. S., "A New Approach for Evaluating the Error Probability in the Presence of Intersymbol Interference and Additive Gaussian Noise," *Bell System Technical Journal*, pp. 2249-2265, (November 1970).

Hubbard, W. M., "The effect of intersymbol interference on error rate in binary differentially coherent phase-shift-keyed systems," *Bell System Technical Journal*, vol. 46, pp. 1149-1172, (July-August 1967).

Hubbard, W. M., Mandeville, G. D., and Goell, J. E., "Multilevel modulation techniques for millimeter guided waves," *Bell System Technical Journal*, vol. 49, pp. 33-54, (January 1970).

Jones, J. J., "Filter Distortion and Intersymbol Interference Effects on PSK Signals," *IEEE Trans. on Comm.*, vol. COM-19, pp. 120-132, (April 1971).

Jenq, Y., Thomas, J. B., and Liu, B., "A Minimum-Error Probability Tapped Delay Line Equalizer," *IEEE Transactions on Communications*, vol. COM-25, No. 10, pp. 1120-1127, (October 1977).

Korn, I., "Probability of Error in Binary Communications with Casual Band-Limiting Filters, Part 1, Non Return-to-Zero Signal," *IEEE Trans. Comm. Tech.*, pp. 878-890, (August 1973).

Koubanitsas, T. S., and Turner, L. F., "Simple bound to the error probability in coherent multi-phase phase-shift-keyed systems with intersymbol interference and Gaussian noise," *Proc. IEEE*, vol. 120, pp. 725-732, (July 1973).

Lugannani, R., "Intersymbol interference and probability of error in digital systems," *IEEE Trans. on Inform. Theory*, vol. IT-15, pp. 682-688, (November 1969).

Magee, Jr., F. R., and Proakis, J. G., "Adaptive Maximum Likelihood Sequence Estimation for Digital Signaling in the Presence of Intersymbol Interference," *IEEE Transactions on Information Theory*, pp. 120-124, (January 1973).

McGee, W. F., "A modified intersymbol interference error bound," *IEEE Trans. Comm.*, vol. COM-21, p. 862, (July 1973).

Matthews, J. W., "Sharp error bounds for intersymbol interference," *IEEE Trans. on Inform. Theory*, vol. IT-19, pp. 440-447, (July 1973).

McLane, P. J., "Lower Bounds for Finite Intersymbol Interference Error Rates," *IEEE Trans. on Comm.*, vol. COM-22, pp. 853-857, (June 1974).

Milewski, A., "New Simple and Efficient Bounds on the Probability of Error in the Presence of Intersymbol Interference and Gaussian Noise," *IEEE Transactions on Communications*, vol. COM-25, No. 10, pp. 1218-1222, (October 1977).

Murphy, J. V., "Binary error rate caused by intersymbol interference and noise," *IEEE Trans. on Commun.*, vol. COM-21, pp. 1039-1046, (September 1973).

Park, J. H., "Effect of Band Limiting on the Detection of Binary Signals," *IEEE Trans. AES,* pp. 867–870, (September 1969).

Prabhu, V. K., "Error Probability Performance of M-ary CPSK Systems with Intersymbol Interference," *IEEE Trans. Commun.,* vol. COM-21, pp. 97–109, (February 1973).

Prabhu, V. K., "Error-rate considerations for digital phase-modulation systems," *IEEE Trans. on Comm. Tech.,* vol. COM-17, pp. 33–42, (February 1969).

Prabhu, V. K., "Intersymbol interference performance of system with correlated digital signals," *IEEE Trans. on Comm.,* vol. COM-21, pp. 1147–1152, (October 1973).

Prabhu, V. K., "Some considerations of error bounds in digital systems," *Bell System Technical Journal,* vol. 50, pp. 3127–3151, (December 1971).

Saltzberg, B. R., "Intersymbol interference error bounds with application to the ideal bandlimited signalling," *IEEE Trans. on Inform. Th.,* vol. IT-14, pp. 563–568, (July 1968).

Shimbo, O., Celebiler, M. I., and Fang, R. J., "Peformance Analysis of DPSK Systems in Both Thermal Noise and Intersymbol Interference," *IEEE Transactions on Communication Technology,* vol. COM-19, No. 6, pp. 1170–1188, (December 1971).

Shimbo, O., Fang, R. J., and Celebiler, M., "Performance of M-ary PSK systems in Gaussian noise and intersymbol interference," *IEEE Trans. on Information Theory,* vol. IT-19, (January 1973).

Shimbo, O., and Celebiler, M. I., "The probability of error due to inter-symbol interference and Gaussian noise in digital communication systems," *IEEE Trans. on Comm. Tech.,* vol. COM-19, pp. 113–119, (April 1971).

Yao, K., "On minimum average probability of error expression for a binary pulse-communication system with intersymbol interference," *IEEE Trans. on Inform. Theory,* vol. IT-18, pp. 528–531, (July 1972).

Yao, K., and Biglieri, E., "Moment Inequalities for Error Probabilities in Digital Communications," *Proceedings of the National Telecommunications Conference,* (December 1977).

Yao, K., and Milstein, L. B., "On ML Bit Detection of Binary Signals with Intersymbol Interference in Gaussian Noise," *IEEE Transactions on Communications,* pp. 971–976, (September 1975).

Yao, K., and Tobin, R. M., "Moment Space Upper and Lower Error Bounds for Digital Systems with Intersymbol Interference," *IEEE Transactions on Information Theory,* vol. IT-22, pp. 65–74, (January 1976).

Yeh, Y. S., and Ho, E. Y., "Improved intersymbol interference error bounds in digital systems," *Bell System Technical Journal,* vol. 50, pp. 2585–2598, (October 1971).

Weidner, M. Y., "Degraded error probability for multiphase digital signalling with linear distortion and Gaussian noise," *Proc. IN-TELSAT/Inst. Elec. Eng. Int. Conf., Digital Satellite Communication,* pp. 202–212, London, (November 1969).

3.3.4 Co-Channel Interference

Aein, J. M., and Turner, R., "Effect of Co-Channel Interference on CPSK Carriers," *IEEE Trans. Commun.,* vol. COM 21, pp. 783–790, (July 1973).

Akima, H., and Nesenbergs, M., "Performance of a Coherent Communication System with Constant Amplitude Interference," *Telecommunications Tech. Memo.,* OTM 73-123, ITS, Boulder, CO (January 1973).

Benedetto, S., Biglieri, E., and Castellani, V., "Combined Effects of Intersymbol, Interchannel and Co-Channel Interferences in M-ary CPSK Systems," *IEEE Trans. Commun.,* vol. COM-21, pp. 997–1008, (September 1973).

Calandrino, L., Corazza, G., Krippa, G., and Immovilli, G., "Intersymbol, Interchannel and Co-Channel Interferences in Binary and Quaternary PSK Systems," *Alta Frequenza,* vol. 40, pp. 407–420, (May 1971).

Castellani, V., and Sant'Agostino, M., "Intersymbol and Interchannel Interference Effects in PCM Signal Transmission on Radio Links: Binary Coherent Phase Modulation," *Alta Frequenza,* vol. 40, pp. 389–397, (May 1971).

Celebiler, M., and Coupe, G., "Probability of Error for Coherent PSK in the Presence of Thermal Noise and Intersymbol, Interchannel and Co-Channel Interferences," *IEEE International Conference on Satellite Communication Systems Technology,* London, (April 7-10, 1975).

"Co-Channel Interference Effects on Multiple Phase Shift Keyed (MPSK) System Performance," *Report 528,* pp. 259–264, (1974).

Colavito, C., and Sant'Agostino, M., "Binary and Quaternary PSK Radio Systems in a Multiple-Interference Environment," *IEEE Trans. Commun.,* vol. COM 21, pp. 1056–1066, (September 1973).

Fang, R., and Shimbo, O., "Unified Analysis of Digital Systems in Additive Noise and Interference," *IEEE Trans. Commun.,* vol. COM 21, pp. 1075–1091, (October 1973).

Glave, F., and Rosenbaum, A., "An Upper Bound Analysis for Coherent Phase-Shift-Keying with Co-Channel Adjacent-Channel and Intersymbol Interference," *IEEE Trans. Commun.,* vol. COM 23, pp. 586–597, (June 1975).

Goldman, J., "Multiple Error Performance of PSK Systems with Co-Channel Interference and Noise," *IEEE Trans. Commun.,* vol. COM 19, pp. 420–430, (August 1971).

Jeruchim, M. C., and Lilley, F. E., "Spacing Limitations of Geostationary Satellites Using Multilevel Coherent PSK Signals," *IEEE Trans. on Communications,* vol. COM 20, pp. 1021–1026, (October 1972).

Lucky, R. W., Salz, J., and Weldon, Jr., E. J., "Principles of Data Communication," McGraw-Hill, New York, (1968).

Rosenbaum, A., "Binary PSK Error Probabilities with Multiple Co-Channel Interferences," *IEEE Transactions on Communication Technology,* vol. COM 18, No. 3, pp. 241–253, (June 1970).

Rosenbaum, A. S., "An Upper Bound Analysis for Coherent Phase Shift Keying with Interference," *International Conference on Communications Conference Record,* Part I, Seattle, WA, pp. 18-10-18-16, (June 1973).

Palmer, L., and Lebowitz, S., "Adjacent Channel Interference Between Unfiltered and Filtered QPSK Signals," *IEEE International Conference on Communications Record,* Seattle, WA, pp. 31-25-31-29, (June 1973).

Prabhu, V. K., "Bandwidth Occupancy in PSK Systems," *IEEE Trans. Commun.,* vol. Com 24, No. 4, pp. 456–462, (April 1976).

Prabhu, V. K., "Error Rate Considerations for Coherent Phase-Shift-Keyed Systems with Co-Channel Interference," *Bell System Technology Journal,* vol. 48, pp. 743–767, (March 1969).

Rosenbaum, A. S., PSK Error Performance with Gaussian Noise and Interference," *Bell System Technology Journal,* vol. 48, pp. 413–442, (February 1969).

Shimbo, O., and Fang, R., "Effects of Co-Channel Interference and Gaussian Noise in M-ary PSK systems," *COMSAT Technical Review,* vol. 3, pp. 183–207, (Spring 1973).

Simon, M. K., and Smith, J. G., "Offset Quadrature Communications with Decision-Feedback Carrier Synchronization," *IEEE Transactions on Communications,* vol. COM 22, No. 10, pp. 1576–1584, (October 1974).

Tobin, R. M., and Yao, K., "Upper and Lower Error Bounds for Coherent Phase-Shift Keyed (CPSK) Systems with Co-Channel Interference," *IEEE Transactions on Communications,* pp. 281–287, (February 1977).

3.3.5 Multiple Interferences

Benedetto, S., Biglieri, E., and Castellani, V., "Combined effects of intersymbol, interchannel and co-channel interferences in M-ary CPSK systems," *IEEE Trans. on Comm.,* vol. COM-21, pp. 997–1008, (September 1973).

Celebiler, M. I., and Coupe, G. M., "Effects of Thermal Noise Filtering and Co-channel Interference on the Probability of Error in Binary Coherent PSK Systems," *IEEE Transactions on Communications,* vol. COM-26, No. 2, pp. 257–267, (February 1978).

Fang, R., and Shimbo, O., "Unified analysis of a class of digital systems in additive noise and interference," *IEEE Trans. on Comm.,* vol. COM-21, pp. 1075–1091, (October 1973).

Glave, F. E., and Rosenbaum, A. S., "An Upper Bound Analysis for Coherent Phase-Shift Keying with Cochannel, Adjacent-Channel, and Intersymbol Interference," *IEEE Transactions on Communications,* vol. COM-23, No. 6, pp. 586–597, (June 1975).

Horstein, M., "Satellite adjacent-channel interference due to multi-carrier transponder operation," *IEEE 8th Annual Int. Conf. Communications Rec.*, Philadelphia, PA, pp. 34.1-34.6, (June 19-21, 1972).

Yao, K., and Biglieri, E., "Moment Inequalities for Error Probabilities in Digital Communication Systems," National Telecommunications Conference, pp. 26:4-1-26:4-4, (1977).

3.3.6 MSK, OQPSK

Amoroso, F., and Kivett, J. A., "Simplified MSK Signaling Technique," *IEEE Transactions on Communications*, pp. 433-441, (April 1977).

Anderson, J. B., and Taylor, D. P., "Trellis Phase Modulation Coding: Minimum Distance and Spectral Results," *EASCON 1977*, pp. 29-1A-19-1G.

Brayer, K., "Transition-coded quadrature modulation for synchronous and asychronous data communication," *Int. Conf. on Comm.*, Philadelphia, PA, pp. 29-24-29-29, (June 19-21, 1972).

DeBuda, R., and Anto, H., "About FSK with low modulation index," *Canadian General Electric Report No. RQ69EE11*, (December 1969).

DeBuda, R., "The fast FSK—a new modulation system," *Canadian General Electric Report No. RQ71EE2*, (February 1971).

DeBuda, R., "Coherent Demodulation of Frequency-Shift Keying with Low Deviation Ratio," *IEEE Trans. Comm. Tech.*, pp. 429-435, (June 1972).

DeBuda, R., "The Fast FSK Modulation System," *Inter. Conf. on Comm.*, pp. 41-25-41-27.

Doelz, M. L., and Heald, E. H., "Minimum-shift data communication system," *U. S. Patent No. 2,977,417*, (March 28, 1961) (assigned to Collins Radio Company).

Greenstein, L. J., "Spectra of PSK Signals with Overlapping Baseband Pulses, *IEEE Transactions on Communications*, pp. 523-530, (May 1977).

Kabal, P., and Pasupathy, S., "Partial-Response Signaling," *IEEE Transactions on Communications*, vol. COM-23, No. 9, pp. 921-934, (September 1975).

Kalet, I., and White, B. E., "Suboptimal Continuous Shift Keyed (CSK) Demodulation for the Efficient Implementation of Low Crosstalk Data Communication," *IEEE Transactions on Communications*, vol. COM-25, No. 9, pp. 1037-1041, (September 1977).

Kwan, R. K., "The effects of filtering and limiting a double-binary PSK signal," *IEEE Trans. on Aerospace and Elect. Syst.*, vol. AES-5, pp. 589-594, (July 1969).

Mathwich, H. R., Balcewicz, J. F., and Hecht, M., "The effect of tandem band and amplitude limiting on the E_b/N_0 performance of minimum (frequency) shift keying (MSK)," *IEEE Trans. Commun.*, vol. COM-22, pp. 1525-1540, (October 1974).

Osborne, W. P., and Lutz, M. B., "Coherent and Noncoherent Detection of CPFSK," *IEEE Trans. Comm.*, pp. 1023-1036, (August 1974).

Prabhu, V. K., "PSK-Type Modulation with Overlapping Baseband Pulses," *IEEE Transactions on Communications*, vol. COM-25, No. 9, pp. 980-990, (September 1977).

Rhodes, S. A., "Effects of hardlimiting on bandlimited transmissions with conventional and offset QPSK modulation," *Proc. Nat. Telecommunications Conf.*, pp. 20F/1-7, (1972).

Rhodes, S. A., "Performance of Offset-QPSK Communications with Partially Coherent Detection," *National Telecomm. Conf.*, (1973).

Rhodes, S. A., "Effect of noisy phase reference on coherent detection of offset QPSK signals," *IEEE Trans. Commun.*, vol. COM 22, pp. 1046-1055, (August 1974).

Simon, M. K., "A Generalization of Minimum-Shift Keying (MSK)—Type Signaling Base Upon Input Data Symbol Pulse Shaping," *IEEE Transactions on Communications*, vol. COM-24, pp. 845-856, (August 1976).

Simon, M. K., "An MSK Approach to Offset QASK," *IEEE Transactions on Communications*, pp. 921-923, (August 1976).

Simon, M. K., and Smith, J. G., "Offset quadrature communications with decision feedback synchronization," *IEEE Trans. Commun.*, vol. COM-22, pp. 1576-1584, (October 1974).

Steel, J., and Smith, B. M., "Carrier and Clock Recovery from Transversal Equalizer Tap Settings for a Partial Response System," *IEEE Transactions on Communications*, pp. 976-979, (September 1975).

Viskanta, V. Z., "Effect of filtering on MSK signals," *Proc. Nat. Telecommunications Conf.*, pp. 30/1-5, (1975).

White, B. E., Kalet, I., and Heggestad, H. M., "Offset Quadrature Phase Comparison Modulation Schemes for Low Crosstalk Communications," *International Conference on Communications*, pp. 6.5-133-6.5-137, (1977).

3.3.7 Synchronization, Timing

Boyko, D. F., and McLane, P. J., "Error Rate Bounds for Digital PAM Transmission with Phase Jitter and Intersymbol or Co-channel Interference," *IEEE Transactions on Communications*, pp. 536-541, (May 1977).

Chiu, H. C., and Simpson, R. S., "Effect of Quadrature and Demodulation Phase Errors in a Quadrature PSK System," *IEEE Trans. on Comm. Tech.*, pp. 945-948, (August 1973).

Falconer, D. D., and Salz, J., "Optimal Reception of Digital Data over the Gaussian Channel with Unknown Delay and Phase Jitter," *IEEE Transactions on Information Theory*, vol. IT-23, No. 1, (January 1977).

Goldman, S. L., "Second Order Phase-Lock Loop Acquisition Time in the Presence of Narrow Band Gaussian Noise," *IEEE Trans. on Comm. Tech.*, pp. 297-300, (April 1973).

Goldstein, A. J., and Byrne, C. J., "Pull-In Frequency of the Phase-Controlled Oscillator," *Proc. IRE Letters*, vol. 49, p. 1209, (July 1961).

Henry, J. C., "DPSK versus FSK with Frequency Uncertainty," *IEEE Trans. Comm. Systems*, pp. 814-816, (December 1970).

Jaffe, R., and Rechtin, E., "Design and performance of phase-lock circuits capable of near-optimum performance over a wide range of input signal and noise levels," *IRE Trans. Inform. Theory*, (March 1955).

Lindsey, W. C., and Simon, M. K., "Carrier Synchronization and Detection of Polyphase Signals," *IEEE Trans. Comm. Tech.*, pp. 441-454, (June 1972).

Lindsey, W. C., "Data-Aided Carrier-Tracking Loops," *IEEE Trans. Comm. Tech.*, pp. 152-169, (April 1971).

Lindsey, W. C., and Simon, M. K., "Detection of Digital FSK and PSK Using a First-Order Phase-Locked Loop," *IEEE Transactions on Communications*, vol. COM-25, No. 2, pp. 200-213, (February 1977).

Lindsey, W. C., "Phase-Shift-Keyed Signal Detection with Noisy Reference Signals," *IEEE Trans. AES*, pp. 393-401, (July 1966).

Lindsey, W. C., "Synchronization Systems in Communications," Englewood Cliffs, NJ, Prentice-Hall, Inc. (1972).

Mallony, P., "Maximum Likelihood Bit Synchronizer," *Proc. 1968 Int. Telemetry Conference*, pp. 1-16.

Nielsen, P. T., "On the Acquisition Behavior of Binary Delay-Lock Loops," *IEEE Trans. Aerospace and Elec. Syst.*, pp. 415-418, (May 1975).

Oberst, J. F., and Schilling, D. L., "Performance of Self-Synchronized Phase-Shift Keyed Systems," *IEEE Trans. Comm. Tech.*, pp. 664-669, (December 1970).

O'Sullivan, M. R., "Tracking Systems Employing the Delay-Lock Discriminator," *IRE Trans. SET*, pp. 1-7, (March 1972).

Palmer, L. C., "Characterization of the Effects of Phase and Symbol Timing Errors for QPSK Transmission," *International Conference on Communications*, pp. 16.3-340-16.3-343, (1977).

Preston, G. W., and Tellier, J. C., "The lock-in performance of an AFC circuit," *Proc. IRE*, (February 1953).

Rhodes, S. A., "Carrier Synchronization Techniques for Offset-QPSK Signals," *National Telecommunications Conference*, pp. 937-945, (1974).

Sanneman, R. W., and Rowbotham, J. R., "Unlock Characteristics of the optimum type II phase-locked loop," *IEEE Trans. Aerospace Naval Electron.*, (March 1964).

Sherman, R. J., "Quadraphase Shift-Keyed Detection with a Noisy Reference Signal," *IEEE EASCON Convention Record*, pp. 42-52, (1969).

Simon, M. K., and Smith, J. G., "Carrier synchronization and detection of QASK signal sets," *IEEE Trans. Comm.,* vol. COM-22, pp. 98-106, (February 1974).

Taylor, D. P., and Cheung, D., "The Effect of Carrier Phase Error on the Performance of a Duobinary Shaped QPSK Signal," *IEEE Transactions on Communications,* pp. 738-744, (July 1977).

Viterbi, A. J., "Phase-locked loop dynamics in the presence of noise by Fokker-Planck techniques," *Proc. IEEE,* (December 1963).

Yamamoto, H., Hirade, K., and Watanabe, Y., "Carrier Synchronization for Coherent Detection of High-Speed Four-Phase-Shift-Keyed Signals," *IEEE Trans. Comm. Tech.,* pp. 803-807, (August 1972).

3.3.8 Modem Implementation

Heegard, C., Heller, J. A., and Viterbi, A. J., "A Microprocessor-based PSK Modem for Packet Transmission over Satellite Channels," *IEEE Transactions on Communications,* vol. COM-26, No. 5, pp. 552-564, (May 1978).

Nosaka, K., Ogawa, A., and Muratani, T., "PSK demodulator for the PCM-TDMA system," *IEEE Trans. on Comm. Tech.,* vol. COM-18, pp. 427-434, (August 1970).

Ungerboeck, G., "Fractional Tap-Spacing Equalizer and Consequences for Clock Recovery in Data Modems," *IEEE Transactions on Communications,* vol. COM-24, No. 8, pp. 856-864, (August 1976).

Walker, A. M., 'High Data Rate PSK Modems for Satellite Communications," *Telecommunications,* (July 1976).

Yokoyama, S., Kato, K., Noguchi, T., Kusama, N., and Otani, S., "The Design of a PSK Modem for the Telecast TDMA System," *International Conference on Communications Conference Record,* vol. 3, pp. 44-11-44-15, (1975).

3.3.9 Nonlinear Systems and Analyses

Acampora, A. S., "Maximum Likelihood Decoding of Binary Convolutional Codes on Band-Limited Satellite Channels," *Conference Record, National Telecommunications Conference,* Dallas, TX, (1976).

Aein, J. M., "Error Rate for Peak-Limited Coherent Binary Channels," *IEEE Transactions on Communication Technology,* vol. COM -16, No. 1, pp. 35-44, (February 1968).

Aein, J. M., "SNR and Error Rate Calculations for Coherent Nonlinear Satellite Channels," *Proc. INTELSAT/IEEE Conf. Digital Satellite Communications,* pp. 189-201, (November 1969).

Blachman, N. M., "Detectors, Bandpass Nonlinearities, and their Optimization: Inversion of the Chebyshev Transform," *IEEE Trans. on Inform. Th.,* vol. IT-17, No. 4., pp. 398-404, (July 1971).

Berman, A. L., and Mahle, C. E., "Nonlinear Phase Shift in Traveling-Wave Tubes as Applied to Multiple Access Communications Satellites," *IEEE Transactions on Communication Technology,* vol. COM-18, No. 1, pp. 37-48, (February 1970).

Biederman, L., and Omura, J. K., "The Computational Cut-Off Rate for Nonlinear ISI Channels," *National Telecommunications Conference,* pp. 36:5-1-36:5-7, (1977).

Biederman, L., and Omura, J. K., "Performance of Viterbi Demodulators for Nonlinear Satellite Channels with Intersymbol Interference," *National Telemetry Conference,* (1977).

Bond, F. E., and Meyer, H. F., "Intermodulation Effects in Limiter Amplifier Repeaters," *IEEE Trans. Comm. Tech.,* pp. 127-135, (April 1970).

Davisson, L. D., and Milstein, L. B., "On the Performance of Digital Communication Systems with Bandpass Limiters—Part I: One-link Systems and—Part II: Two-link Systems," *IEEE Trans. Commun.,* vol. COM-20, pp. 972-980, (October 1972).

Degrave, J. C., and Okkes, R. W., "Error Rate in Noisy Multicarrier PSK Systems with Nonlinear Amplification and AM to PM Conversion," *2nd Int. Symp. on Digital Satellite Comm.,* Paris, pp. 113-125, (November 28-30, 1972).

Desblache, "Optimal Short Desired Impulse Response for Maximum Likelihood Sequence Estimation," *IEEE Transactions on Communications,* pp. 735-738, (July 1977).

Forney, Jr., G. D., "Maximum-Likelihood Sequence Estimation of Digital Sequences in the Presence of Intersymbol Interference," *IEEE Transactions on Information Theory,* vol. IT-18, No. 3, pp. 363-378, (May 1972).

Fredricson, S. A., "Optimum Receiver Filters in Digital Quadrature Phase-Shift-Keyed Systems with a Nonlinear Repeater," *IEEE Trans.*

on Comm., vol. COM-23, pp. 1389-1400, (December 1975).

Fritchman, B. D., Hafer, C. D., Mixsell, J. C., and Sattar, M. A., "Approximations to a Joint Detection/Decoding Algorithm," *IEEE Transactions on Communications,* pp. 271-278, (February 1977).

Fuenzalida, J. C., Shimbo, O., and Cook, W. L., "Time-domain Analysis of Intermodulation Effects Caused by Nonlinear Amplifiers," *COMSAT Tech. Review,* vol. 3, pp. 89-143, (Spring 1973).

Hetrakul, P., and Taylor, D. P., "The Effects of Transponder Nonlinearity on Binary CPSK Signal Transmission," *IEEE Trans. Commun.,* vol. COM-24, No. 5, pp. 546-553, (May 1976).

Jacobs, I., "Effects of Video Clipping on Performance of an Active Satellite PSK communication System," *IEEE Trans. Commun. Technol.,* vol. COM-13, pp. 195-201, (June 1965).

Jain, P., and Blachman, N., "Detection of a PSK Signal Transmitted Through a Hard-Limited Channel," *IEEE Transactions on Information Theory,* vol. IT-19, No. 5, pp. 623-630, (September 1973).

Jain, P. C., "Error Probabilities in Binary Angle Modulation," *IEEE Transactions on Information Theory,* vol. IT-20, No. 1, pp. 36-42, (January 1974).

Jain, P. C., "Limiting of Signals in Random Noise," *IEEE Trans. Information Theory,* vol. IT-18, No. 3, pp. 332-340, (May 1972).

Kalet, I., "A Look at Crosstalk in Quadrature-Carrier Modulation Systems," *IEEE Trans. on Communications,* vol. COM-25, No. 9, pp. 884-892, (September 1977).

Kaye, A. R., George, D. A., and Eric, M. J., "Analysis and Compensation of Bandpass Nonlinearities for Communications," *IEEE Trans. Commun. Technol.,* vol. COM-20, pp. 965-972.

Kobayashi, H., "Correlative Level Coding and Maximum-Likelihood Decoding," *IEEE Trans. on Information Theory,* vol. IT-17, pp. 586-594, (September 1971).

Lee, W. U., and Hill, Jr., F. S., "A Maximum-Likelihood Sequence Estimator with Decision-Feedback Equalization," *IEEE Transactions on Communications,* vol. COM-25, No. 9, pp. 971-979, (September 1977).

Lesh, J. R., "Signal to Noise Ratios in Coherent Soft Limiters," *IEEE Transactions on Communications Technology,* vol. COM-22, pp. 803-811, (June 1974).

Lindsey, W. C., et al., "Investigation of Modulation/Coding Trade-Off for Military Satellite Communications," *TR-7808-1276,* LinCom Corp., (July 19, 1977).

Lindsey, W. C., "Performance of Phase-Coherent Receivers Preceded by Bandpass Limiters," *IEEE Transactions on Communication Technology,* vol. COM-16, pp. 245-251, (April 1968).

Lyons, R. G., "Combined Effects of Up- and Down-Link Fading through a Power-Limiting Satellite Repeater," *IEEE Transactions on Communication,* vol. COM-22, pp. 350-352, (March 1974).

Lyons, R. G., "The Effect of a Bandpass Nonlinearity on Signal Detectability," *IEEE Transaction on Communications,* vol. COM-21, pp. 51-60, (January 1973).

Magee, Jr., E. R., "A Comparison of Compromise Viterbi Algorithm and Standard Equalization Techniques over Bandlimited Channels," *IEEE Transactions on Communication Systems,* pp. 361-366, (March 1975).

Manasse, R., Price, R., and Lerner, R. M., "Loss of Signal Detectability in Bandpass Limiters," *IRE Trans. Information Theory,* vol. IT-4, pp. 34-38, (March 1958).

McCreath, D. R., and McLane, P. J., "Filter Parameter Optimization for Linear and Nonlinear BPSK Transmission," *National Telecommunications Conference,* pp. 26:3-1-26:3-7, (1977).

Mesiya, M. F., McLane, P. J., and Campbell, L. L., "Maximum Likelihood Sequence Estimation of Binary Sequences Transmitted Over Bandlimited Nonlinear Channels," *IEEE Transaction on Communications,* vol. COM-25, No. 7, (July 1977).

Mesiya, M. F., McLane, P. J., and Campbell, L. L., "Optimal Receiver Filters for BPSK Transmission Over a Bandlimited Nonlinear Channel," *National Telecommunications Conference,* pp. 13:5-1-13:5-5, (1976).

Messerschmidt, D. B., "Design of a Finite Impulse Response for the Viterbi Algorithm and Decision-Feedback Equalizer," *International Conference on Communications,* pp. 370-1-370-5, (1974).

Mizuno, T., Morinaga, N., and Namekawa, T., "Transmission Characteristics of an M-ary Coherent PSK Signal via a Cascade of N Bandpass Hard Limiters," *IEEE Transactions on Communications,* vol. COM-24, No. 5, pp. 540-545, (May 1976).

Omura, J. K., "Optimal Receiver Design for Convolutional Codes and Channels with Memory via Control Theoretical Concepts," *Information Science,* vol. 3, pp. 243-266, (July 1971).

Proakis, J. G., and Miller, J. H., "An Adaptive Receiver for Digital Signalling through Channels with Intersymbol Interference," *IEEE Trans. Information Theory*, pp. 484-497, (July 1969).

Raghavan, S. H. R., and Ziemer, R. E., "Improving QPSK Transmission in Band-Limited Channels with Interchannel Interference through Equalization," *IEEE Transactions on Communications*, vol. COM-25, No. 10, pp. 1222-1226, (October 1977).

Robinson, G., et al., "PSK Signal Power Spectrum Spread Produced by Memoryless Nonlinear TWTs," *COMSAT Technical Review*, pp. 227-256, (Fall 1973).

Taylor, D. P., and Hetrakul, P., "Receiver Structure for Saturating Channels," *IEEE Transactions on Communications*, vol. COM-26, No. 2, pp. 270-278, (February 1978).

Thomas, C. M., Weidner, M. Y., and Durrani, S. H., "Digital Amplitude-Phase Keying with M-ary Alphabets," *IEEE Transactions on Communications*, vol. COM-22, No. 2, pp. 168-180, (February 1974).

Ungerboeck, G., "Adaptive Maximum-Likelihood Receiver for Carrier-Modulated Data-Transmission Systems," *IEEE Transactions on Communications*, vol. COM-22, No. 5, pp. 624-636, (May 1974).

Viterbi, A. J., and Omura, J. K., *Digital Communication and Coding*, McGraw-Hill, New York, 1979.

Waylan, C. J., "Intermodulation Degradation to PSK Signals in a Communication Satellite," *NTC '76*, 46.5-1-53.5-5.

Weinberg, A., "Effects of a Hard Limiting Repeater on the Performance of a DPSK Data Transmission System," *IEEE Transactions on Communications*, vol. COM-25, No. 10, pp. 1128-1133, (October 1977).

White, B. E., "A Worst-Case Crosstalk Comparison Among Several Modulation Schemes," *IEEE Trans. on Communications*, vol. COM-25, No. 9, pp. 1032-1037, (September 1977).

3.3.10 Simulation

Lebowitz, S., and Palmer, L., "Simulation results for intersymbol interference loss; QPSK transmission through several filter types," *Proc. Nat. Telecommunications Conf.*, pp. 32D/1-7, (1973).

Muratani, T., Matsushita, I., Tsuji, Y., and Hura, T., "Simulation of a PSK transmission system including the nonlinear satellite channel," *Proc. 2nd Int. Conf. Numerical Telecommunications by Satellite*, Paris, France, (November 28-30, 1972).

Taylor, D. P., Chan, H. C., and Haykin, S. S., "A Simulation Study of Digital Modulation Methods for Wide-Band Satellite Communications," *IEEE Transactions on Communications*, pp. 1351-1361, (December 1976).

3.4 ERROR CONTROL

3.4.1 General

Burton, H. O., and Sullivan, D. D., "Errors and Error Control," *Proceedings of the IEEE*, vol. 60, No. 11, (November 1972).

Jacobs, I. M., "Practical Applications of Coding," *IEEE Transactions on Information Theory*, vol. IT-20, No. 3, pp. 305-310, (May 1974).

Odenwalder, J. P., and Viterbi, A. J., "Overview of Existing and Projected Uses of Coding in Military Satellite Communications," *National Telecommunications Conference*, (1977).

Odenwalder, J. P., Viterbi, A. J., Jacobs, I. M., and Heller, J. A., "Study of Information Transfer Optimization for Communication Satellites," *Final Report for NASA Contract NAS2-6810*, prepared by Linkabit Corp., NASA CR #114561, N73-20889.

Wolf, J. K., "A Survey of Coding Theory," *IEEE Trans. Info. Theory*, pp. 381-388, (July 1973).

Elias, P., "Coding for Noisy Channels," *IRE Convention Record*, (March 1955).

3.4.2 Convolutional Codes

Acampora, A. S., "Maximum Likelihood Decoding of Binary Convolutional Codes on Band-Limited Satellite Channels, *National Telecommunications Conference*, pp. 24.5-1-24.5-5, Dallas, TX, (November 30-December 1, 1976).

Batson, B. H., and Huth, G. K., "Convolutional Coding at 50 Mbps for the Shuttle KU-Band Return Link," *International Telemetering Conference Proceedings*, XII, pp. 175-183, (1976).

Bucher, E. A., and Heller, J. A., "Error Probability Bounds for Systematic Convolutional Codes," *IEEE Transactions on Information Theory*, vol. IT-16, pp. 219-224, (March 1970).

Forney, G. D., and Bower, E. K., "A High-Speed Sequential Decoder: Prototype Design and Test," *IEEE Transactions on Communication Technology*, vol. COM-19, pp. 821-835, (October 1971).

Forney, G. D., and Langelier, R. M., "A high-speed sequential decoder for satellite communications," *Proc. IEEE Int. Conf. Communications*, pp. 39.9-39.17, (1969).

Forney, Jr., G. D., "Coding and its application in space communications," *IEEE Spectrum*, vol. 7, pp. 47-58, (June 1970).

Forney, Jr., G. D., "The Viterbi Algorithm," *Proceedings of the IEEE*, vol. 61, pp. 268-278, (March 1973).

Heller, J. A., "Performance and Implementation of the Viterbi Decoding Algorithm for Satellite and Space Communication," *Conference Record of the International Conference on Communication*, pp. 37A-1-37A-5, (June 1974).

Heller, J. A., and Jacobs, I. M., "Viterbi Decoding for Satellite and Space Communications," *IEEE Transactions on Communication Technology*, vol. COM-19, pp. 835-847, (October 1971).

Hurd, W. J., "Spacecraft Demonstration of Sequential Decoding Using Lunar Orbiter V," *IEEE Transactions on Information Theory*, vol. IT-14, No. 5, pp. 761-765, (September 1968).

Jacobs, I. M., "Robustness of Convolutional Coding—Viterbi Decoding," *Conference Record of the International Conference on Communications*, pp. 40E-1-40E-2, (June 1974).

Kohlenberg, A., and Forney, Jr., G. D., "Convolutional Coding for Channels with Memory," *IEEE Transactions on Information Theory*, vol. IT-14, pp. 618-620, (September 1968).

Layland, J. W., and Lushbaugh, W. A., "A Flexible High-Speed Sequential Decoder for Deep Space Channels," *IEEE Transactions on Communication Technology*, vol. COM-19, pp. 813-820, (October 1971).

Omura, J. K., "On the Viterbi Decoding Algorithm," *IEEE Trans. Info. Theory*, pp. 177-179, (January 1969).

Russell, S. P., and Shohara, A., "Feasibility Study of Reduced-State Viterbi Detection," *EASCON*, (1976).

Viterbi, A. J., "Convolutional Codes and Their Performance in Communication Systems," *IEEE Transactions on Communications Technology*, vol. COM-19, No. 5, (October 1971).

Viterbi, A. J., "Error Bounds for Convolutional Codes and an Asymptotically Optimum Decoding Algorithms," *IEEE Trans. Info. Theory*, pp. 260-269, (April 1967).

Viterbi, A. J., and Odenwalder, J. P., "Further Results on Optimal Decoding of Convolutional Codes," *IEEE Trans. Info. Theory*, pp. 732-734, (November 1969).

Wu, W. W., "New Convolutional Codes—Part I," *IEEE Transactions on Communications*, vol. COM-23, No. 9, (September 1975).

3.4.3 Block Codes

Berlekamp, E. R., "A Survey of coding theory for algebraists and combinatorialists," *Inter. Centre for Mech. Sci.*, Udine, Italy, (1970).

Bose, R. C., and Ray-Chaudhuri, D. K., "Further Results on Error Correcting Binary Group Codes," *Information and Control*, (September 1960).

Chen, C. L., "Computer results on the minimum distance of some binary cyclic codes," *PGIT 16*, pp. 359-360, (1970).

Chien, R. T., "Cyclic Decoding Procedures for Bose-Chaduri-Hocquenghem Codes," *IEEE Transactions on Information Theory*, (October 1964).

Forney, Jr., G. D., "Generalized Minimum Distance Decoding," *IEEE Transactions on Information Theory*, (April 1966).

Gallager, R. G., "Low-Density Parity-Check Codes," *IRE Transactions on Information Theory*, (January 1962).

Goppa, V. D., "A new class of linear error-correcting codes," *Prob. Peredaci Inform. 6*, (in Russian), pp. 24-30, (1970).

Helgert, H. J., "Srivastava codes," *PGIT 18*, pp. 292-297, (1972).

Massey, J. L., "Shift Register Synthesis and BCH Decoding," *IEEE Transactions on Information Theory*, vol. IT-15, pp. 122-127, (January 1969).

McEliece, R. J., "On the symmetry of good nonlinear codes," *PGIT 16*, pp. 609-611, (1970).

McEliece, R. J., and Ramsey, Jr., H., "Euler products, cyclotomy, and coding," Space programs summary, *Jet Propulsion Lab., Calif. Inst. Technol.*, vol. 37-65-III, pp. 22-27, (1970).

Muller, D. E., "Application of boolean algebra to switching circuit design and error detection," *IRE Trans. Electronic Computers*, EC3, pp. 6-12, (1954).

Nordstrom A. W., and Robinson, J. P., "An Optimum Nonlinear Code," *Information and Control*, (November-December 1967).

Peterson, W. W., "Encoding and Error-Correction Procedures for the Bose-Chaudhuri Codes," *IRE Trans. Inform. Theory*, IT-6, pp. 459-470, (1960).

Reed, I. S., "A Class of Multiple-Error-Correcting Codes and the Decoding Scheme," *IRE Trans. Inform. Theory,* PGIT-4, pp. 38–49, (1954).

Tong, S. Y., "Burst-trapping techniques for a compound channel," *PGIT 15,* pp. 710–715, (1969).

Vasil'yev, Y. L., "On Nongroup Close-Packed Codes," *Prob. Cybernte.,* 8, pp. 337–339, (1962).

3.4.4 ARQ

Benice, R. J., and Frey, Jr., A. H., "An Analysis of Retransmission Systems," *IEEE Transactions on Communication Technology,* vol. COM-12, pp. 135–145, (December 1964).

Benice, R. J., and Frey, Jr., A. H., "Comparisons of Error Control Techniques," *IEEE Transactions on Communication Technology,* vol. COM-12, pp. 146–154, (December 1964).

Brayer, K., "ARQ and Hybrid FEC-ARQ System Design to Meet Tight Performance Constraints," *National Telecommunications Conference,* (1976).

Gatfield, A., "Error control on satellite channels using ARQ techniques," *Comsat Technical Review,* vol. 6, No. 1, (Spring 1976).

Metzner, J. J., "Sequential signaling and decision techniques for non-binary block codes over feedback channel limited to block retransmission requests," *T-COM 77,* pp. 561–563, (May 1977).

Rocher, E. Y., and Pickholtz, R. L., "An Analysis of the Effectiveness of Hybrid Transmission Schemes," *IBM J. Res. Develop.,* (July 1970).

Sastry, A. R. K., "Hybrid Error Control Using Retransmission and Generalized Burst-Trapping Codes," *IEEE Transactions on Communications,* vol. COM-24, No. 4, (April 1976).

3.5 TERRESTRIAL INTERFACE

3.5.1 Echo Control

Araseki, T., Ochiai, K., and Tajima, M., "Digital Echo Suppressors and the CCITT," *International Conference on Communications,* (1976).

Campanella, S. J., Suyderhoud, H. G., and Onufry, M., "Analysis of an Adaptive Impulse Response Echo Canceller," *COMSAT Technical Review,* vol. 2, No. 1, (Spring 1972).

Campanella, S. J., Suyderhoud, H. G., and Onufry, M., "Subjective Evaluation of Dedicated Multiple-Hop Satellite Communications for Government and Military Users," *IEEE Transactions on Communication Technology,* vol. COM-18, No. 5, (October 5, 1970).

Demytko, N., English, K. S., "Echo Cancellation on Time-Variant Circuits," *Proceedings of the IEEE,* vol. 65, No. 3, (March 1977).

Drechsler, R. C., "Echo Suppressor Terminal for No. 4 ESS," *International Conference on Communications,* (1976).

Emling, J. W., and Mitchell, D., "The Effects of Time Delay and Echoes on Telephone Conversations," *Bell System Technical Journal,* (November 1963).

Falconer, D. D., Mueller, K. H., and Weinstein, S. B., "Echo Cancellation Techniques for Full-Duplex Data Transmission on Two Wire Lines," *National Telecommunications Conference,* (1976).

Gitlin, R. D., and Thompson, J. S., "A New Structure for Adaptive Digital Echo Cancellation," pp. 8.2-1–8.2-7.

Gitlin, R. D., and Thompson, J. S., "A Technique for Adaptive Phase Compensation in Echo Cancellation," *National Telecommunications Conference,* (1977).

Helder, G. K., "Digital Echo Suppressors and the CCITT," *International Conference on Communications,* (1976).

Helder, G. K., and Lopiparo, P. C., "Improving Transmission on Domestic Satellite Circuits," *Bell Laboratories Record,* vol. 55, No. 8, (September 1977).

Horna, O. A., "Echo Canceller Utilizing Pseudo-Logarithmic Coding," *National Telecommunications Conference,* (1977).

Kato, Y., Chiba, S., Ishiguro, T., Sato, Y., Tajima, M., and Ogihara, T., "A Digital Adaptive Echo Canceller," *NEC Research and Development,* No. 31, (October 1973).

3.5.2 TDM-FDM Translators

Andrews, H. W., Johannes, V. I., and Woodzel, W. E., "A 44.736 Mbits/s Codec for U600 Mastergroups," *IEEE Transactions on Communications,* vol. COM-25, No. 2, (February 1977).

Bellanger, M. G., and Daguet, J. L., "TDM-FDM Transmultiplexer: Digital Polyphase and FFT," *IEEE Transactions on Communications,* vol. COM-22, No. 9, (September 1974).

Freeney, S. L., Kieburtz, R. B., Mina, K. V., and Tewksbury, S. K., "Systems Analysis of a TDM-FDM Translator/Digital A-Type Channel Bank," *IEEE Transactions on Communications Technology,* vol. COM-19, No. 6, (December 1971).

Kaneko, H., Katagiri, Y., and Okada, T., "A Design of a PCM Master-group Codec for TELESAT TDMA System," *NEC Research and Development,* No. 44, pp. 23–33, (January 1977).

Kieburtz, R. B., "Introduction to Papers on TDM/FDM Transmultiplexers," *IEEE Transactions on Communications,* vol. COM-26, No. 5, pp. 697–698, (May 1978).

Maruta, R., and Tomozawa, A., "An Improved Method for Digital SSB-FDM Modulation and Demodulation", *IEEE Transactions on Communications,* vol. COM-26, No. 5, (May 1978).

Peled, A., "TDM-FDM Conversion Requiring Reduced Computation Complexity," *IEEE Transactions on Communications,* vol. COM-26, No. 5, (May 1978).

Takahata, F., Hirata, Y., Ogawa, A., and Inagaki, K., "Development of a TDM/FDM Transmultiplexer," *IEEE Transactions on Communications,* vol. COM-26, pp. 726–733, (May 1978).

Tomlinson, M., and Wong, K. M., "Techniques for the digital interfacing of t.d.m.-f.d.m. systems," *Proceedings of IEEE,* vol. 123, No. 12, (December 1976).

3.5.3 Multiplexing and Synchronization

Buchner, Jr., M. M., "An Asymmetric Encoding Scheme for Word Stuffing," *BSTJ,* pp.379–398, (March 1970).

Butman, S., "Synchronization of PCM Channels by the Method of Word Stuffing," *IEEE Trans. Comm. Tech.,* COM-16, No. 2, pp. 252–254, (April 1968).

Dammann, C. L., McDaniel, L. D., and Maddox, C. L., "D2 Channel Bank—Multiplexing and Coding," *BSTJ,* pp. 1675–1700, (October 1972).

Doll, D. R., "Multiplexing and Concentration," *Proceedings of the IEEE,* (November 1972).

Gallager, R., "An Information Theoretic Approach to Concentrators," *International Symposium on Information Theory,* (1974).

Hayes, J. F., and Sherman, D. N., "A Study of Data Multiplexing Techniques and Delay Performance," *Bell System Technical Journal,* pp. 1983–2007, (November 1972).

Johannes, V. I., and McCullough, R. H., "Multiplexing of Asynchronous Digital Signals Using Pulse Stuffing with Added-Bit Signaling," *IEEE Transactions on Communication Technology,* vol. COM-14, No. 5, pp. 562–568, (October 1966).

Nash, S., "Multiplexer Survey—Comments on Frequency and Time Division Multiplexing," *Telecommunications,* pp. 36–39, (January 1970).

Pierce, J. R., "Synchronizing Digital Networks, *BSTJ,* pp. 615–636, (March 1969).

Rudin, H., "Performance of Simple Multiplexer-Concentrators for Data Communication," *IEEE Transactions on Communication Technology,* vol. COM-19, No. 2, pp. 178–187, (April 1971).

Sommer, R. C., "Asynchronously Multiplexed Binary Channel Capacity," *Proceedings of the IEEE,* pp. 80–81, (January 1966).

Sommer, R. C., "The Noisy, Asynchronously Multiplexed, Binary Channel," *IEEE Transactions on Information Theory,* pp. 140–142, (January 1966).

White, W. D., "Theoretical Aspects of Asynchronous Multiplexing," *Proceedings of the IRE,* pp. 270–275, (March 1950).

3.6 MULTIPLE ACCESS

3.6.1 General

Assadurian, F., and Jacoby, D. L., "Multiple-Access Considerations for Communications Satellites," *IEEE, Military Electronics Conference Record (MIL-E-CON9),* Washington, (September 1965).

Bargellini, P. L., et al., "The INTELSAT IV Communications System," *COMSAT Technical Review,* vol. 2, No. 2, (Fall 1972).

Bennett, S., Puente, J., and Stamminger, R., "Multiple-Access for Satellites: Methods and Performance," *Tech. Memo,* SCED-e-64, COMSAT, Clarksburg, MD, (July 9, 1964).

Berman, A., and Mahle, C. E., "Non-Linear Phase Shift in Traveling Wave Tubes as Applied to Multiple Access Communications Satellites," *IEEE Transactions on Communications Technology,* COM-18, No. 1, (February 1970).

Bernstein, S. L., and Drouilhet, Jr., P. R., "TATS-A Bandspread Modulation-Demodulation System for Multiple Access Tatical Satellite Communication," *EASCON-1969 Conv. Rec.,* Washington, pp. 126–132, (October 27–29, 1969).

Bisaga, J. J., and Blank, H. A., "Access Control Techniques for Satellite Mobile Communications Systems," *AIAA Paper 74-441,* Los Angeles, (April 22–24, 1974).

Blank, H. A., Kinal, G. V., and Klein, L., "Aerosat Access Control Summary," *Final Report,* (July 1975–May 1976).

Blasbalg, H., Najjar, H. F., D'Antonio, R. A., and Haddard, R. A., "Air-Ground Ground-Air Communications using Pseudo-Noise through a Satellite," *IEEE Trans. Aerosp. Electronic System,* vol. AES-4, pp. 774–791, (September 1968).

Blasbalg, H., "A Comparison of Pseudo-Noise and Conventional Modulation for Multiple-Access Satellite Communications," *IBM Journal of Research and Development,* vol. 9, No. 4., (July 1965).

Cameron, A. G., "A Method for Determining the Performance of a Multiple-Access Satellite Communication System," *Telecommunications,* pp. 33–36, (October 1968).

Carleial, A. B., "On the Capacity of Multiple-Terminal Communication Networks," *Thesis, Univ. Microfilms Order No. 76-5700,* Stanford University.

Chen, C. C., and Burnett, J. W., "TDRS Multiple Access Channel Design," *National Telecommunications Conference,* pp. 19:2-1–19:2-7, (1977).

Cohen, A. R., Heller, J. A., and Viterbi, A. J., "A New Coding Technique for Asynchronous Multiple Access Communication," *IEEE Transactions on Communication Technology,* vol. COM-19, No. 5, (October 1971).

Cover, T. M., "Broadcast Channels," *IEEE Transactions on Information Theory,* vol. IT-18., No. 1, (January 1972).

Dalgleish, D. I., and Reed, A. G., "Some Comparisons of the Traffic-Carrying Capacity of Communications Satellites using Digital Techniques with the Capacity of Satellites using Frequency Modulation," *Proc. Int. Conf. Digital Satellite Communication,* pp. 226–240, London, England, (November 25–27, 1969).

Haberle, H., and Hein, P., "Combinations of Frequency, Time and Space Division Multiple Access in Multitransponder Satellite Communications," *Telecommunications Numeriques Par Satellite,* Paris, France, (November 1972).

Horwood, D., and Gagliardi, R., "Signal Design for Digital Multiple Access Communications," *IEEE Transactions on Communications,* vol. COM-23, pp. 378–383, (March 1975).

Hughes-Hartogs, D., "The Capacity of the Degraded Spectral Gaussian Broadcast Channel," *National Telecommunications Conference,* (1976).

Kasami, T., and Lin, S., "Coding for a Multiple-Access Channel," *IEEE Transactions on Information Theory,* vol. IT-22, pp. 129–137, (March 1976).

Kashara, M., *et al.,* "A New Encoding and Decoding Algorithm for a Multiple Access Channel," *Proceedings of the International Symposium on Information Theory,* (1974).

Killen, N., "Digital Communications Performance Using Both Temporal and Spatial Modulation," *International Conference on Communications,* vol. 3, San Francisco, CA, pp. 4-18-4-22, (June 1975).

Liao, J., "Multiple Access Channels," *Technical Report A72-2,* University of Hawaii, (1972).

Lovell, C. C., "Signal Detection in a Multi-User Random Access Channel," *Rand Report No. R-771-PR, AD729768,* Prepared for United States Air Force Project Rand, (July 1971).

Lu, S. C., "A Comparison of Multiple-Access and Orthogonal Multiplexing Techniques," *Technical Report B74-3, The Aloha System,* University of Hawaii, (June 1974).

Magill, D. J., "Multiple-Access Modulation Techniques," *Communications Satellite Systems Technology,* New York, Academic Press, pp. 667–680, (1966).

McEliece, R. J., and Rubin, A. L., "Timesharing without Synchronization," *International Telecommunications Conference,* pp. 16–20, (1976).

McGregor, D. N., Jones, M. E., and O'Neill, Jr., P. E., "Comparison of Several Demand Assignment Multiple Access/Modulation Techniques for Satellite Communications," *International Conference on Communications Conference Record,* (1971).

Murakami, H., Reed, I. S., and Welch, L. R., "A Transfer Decoder for Reed-Solomon Codes in Multiple-User Communication Systems," *IEEE Transactions on Information Theory,* vol. IT-23, No. 6, (November 1977).

Odenwalder, J. P., et al., "Study of Information Transfer Optimization for Communication Satellites," *Report No. NASA CR No. 114561,* LINKABIT Corporation, (January 3, 1973).

Puente, J. G., Schmidt, W. G., and Werth, A. M., "Multiple-Access Techniques for Commercial Satellites," *Proceedings of the IEEE,* vol. 59, No. 2, pp. 218–229, (February 1971).

Reinhart, E. E., "Multiple Access Techniques for Communications Satellites: Analog Modulation, FDM and Related Signal Processing Methods," *The Rand Corporation,* RM-5117-NASA, (November 1966).

Savage, J. E., "Signal Detection in the Presence of Multiple Access Noise," *IEEE Transactions on Information Theory,* (January 1974).

Schwartz, J. W., Aein, J.M., and Kaiser, J. "Modulation Techniques for Multiple-Access to a Hard-Limiting Satellite Repeater," *Proceedings of the IEEE,* vol. 54, No. 5, (May 1966).

Slabinksi, V. J., "Variation in Range, Range-Rate Propagation Time Delay, and Doppler Shift for a Nearly Geostationary Satellite," in *Communications Satellite Technology,* P. L. Bargellini, Ed., Cambridge, MA: MIT Press, pp. 3–28, (1974).

Sommer, R. C., "High Efficiency Multiple Access Communications through a Signal Processing Repeater," *IEEE Transactions on Communication Technology,* vol. COM-16, No. 2, pp. 222–232, (April 1968).

Spira, P. M., "A Class of Codes for Multiple-Access Satellite Communications Systems," *Rand Report #RMI-5131.*

Stette, G. R., "A Multiple Access and Access Control System for MAROTS," *EUROCON Conference Proceedings on Communications,* (1977).

Stiglitz, I. G., "Multiple-Access Considerations—A Satellite Example," *IEEE Trans. Commun.,* vol. COM-21, pp. 577–582, (May 1973).

Wittman, J. H., "Categorization of Multiple-Access/Random Access Modulation Techniques," *IEEE Transactions on Communications Technology,* vol. 15, No. 5, pp. 724–725, (October 1967).

Wittman, J. H., "A Comparison of Satellite Multiple Access Techniques," *IEEE EASCON Conference Record,* pp. 165–172, (1967).

Wolverton, C.T., Moeller, T., and Nirenberg, L. M., "Satellite Multiple Access Study," *TOR-0078(3417)-1,* The Aerospace Corporation, El Segundo, CA, (June 15, 1978).

Wu, W. W., "Random Multiple Access Codes," *Ph.D Thesis,* The Johns Hopkins University, (1976).

Wu, W. W., "Multiple-Access Channel Modeling," *5th Annual Pittsburgh Conference on Modeling and Simulation, Modeling and Simulation Proceeding,* vol. 5, Part 2, pp. 887–892, (April 24–26, 1974).

Wyner, A. D., "Recent Results in the Shannon Theory," *IEEE Transactions on Information Theory,* vol. IT-20, No. 1, (January 1974).

3.6.2 FDMA

Bondrean, P. M., and Davies, N. G., "Modulation and Speech Processing for Single Channel per Carrier Satellite Communications," *ICC Conf. 1971.*

Dicks, J. L., "Domestic and/or Regional Services Through INTELSAT IV Satellites," *COMSAT Technical Review,* vol. 4, No. 1, pp. 118–138, (Spring 1974).

Dicks, J. L., and Brown, Jr., M. P., "Frequency Division Multiple Access (FDMA) for Satellite Communications Systems," *EASCON 1974,* Washington, DC, pp. 167–178.

Ferguson, M. E., "Design of FM Single-Channel-per-Carrier Systems," *International Conference on Communications 1975,* (June 1975).

Gagliardi, R. M., "Optimal Channelization in FDMA Communications— Frequency Division Multiplexing Access in Satellite System," *IEEE Transactions on Aerospace and Electronic Systems,* vol. AES-10, pp. 867–870, (November 1974).

Golding, L. S., "Single Channel per Carrier Transmission for Satellite Communications," *National Telecommunications Conference, NTC 1975,* (December 1975).

Holbrook, B. D., and Dixon, J. T., "Load rating theory for multichannel amplifiers," *Bell System Technical Journal,* pp. 624–644, (1939).

Jain, P. C., Hatfield, V. E., Leung, J. K., and Magill, D. T., "Investigation of the Use of Frequency-Division Multiple Access for Application with a Mix of User Terminals," *Multiple-Access Techniques,* vols. 1–3.

Magill, D. T., "Difficulties of Accurately Modeling the FDMA Satellite Communication Channel," *NTC 1975 Conference Record,* vol. 2, pp. 25-1-25-6.

McClure, R. B., "Analysis of intermodulation distortion in an FDMA satellite communications systems with a bandwidth constraint," *IEEE International Conference on Communications,* San Francisco, CA,

(June 8-10, 1970).

Sanderson, C. C., and Ludwig, L. G., "Single channel per carrier voice transmission via communications satellite," *Proc. 5th AIAA Communications Satellite Systems Conf.,* Los Angeles, CA, (April 22-24, 1974).

Shaft, P. D., "Intermodulation Interference in Frequency Division Multiple Access," *Int. Conf. on Communications,* (June 1967).

Takahashi, T., Nogushi, T., and Oshima, G., "Transmission subsystem for space terminal—Single channel per carrier PCM multiple access demand assignment," in *Proc. 2nd Int. Conf. Numerical Telecommun.* by Satellite, Paris, France (November 28-30, 1972) (June 1967).

3.6.3 TDMA

Agarwal, V., "An Efficient TDMA System," *NAECON 1968,* Dayton, pp. 6-8.

Abramson, N., "The ALOHA system—Another Alternative for Computer Communications," *1970 Fall Joint Computer Conference, AFIPS Conf. Proc.,* vol. 37, Montvale, NJ, pp. 281-285, (1970).

Appel, R. L., and Pardoe, C. T., "A Time-Division Multiplexing System for Asynchronous High-Rate Telemetry," *APL Technical Digest,* pp. 2-12, (October 1971).

Asahara, M., Tsuji, Y., Fukui, M., Sohma, Y., and Harra, T., "Synchronization and Acquisition in SDMA Satellite Communications System," *ICC 1974,* Minneapolis, pp. 43E-1-43E-3, (June 1974).

Assal, F., "Approach to a Near-Optimum Transmitter-Receiver Filter Design for Data Transmission Pulse-Shaping Networks," *COMSAT Technical Review,* vol. 3, No. 2, (Fall 1973).

Ayoub, J. N., "Storage and Spotbeam Switching at the Satellite," *IEEE Proc.,* vol. 59, No. 9, pp. 1366-1368, (September 1971).

Bachmann, E., Howe, K. G., and Petry, T. J., "Introduction of Digital Transmission Including TDMA/DSI as Viewed by a Communications Organization," *ICDSC-3,* Kyoto, pp. 394-402, (November 1975).

Baker, D. W., and Robinson, P. F., "A Fast Acquisition Coherent PSK Demodulator Incorporating Passive Narrowband Filters Operable in a Quenching Mode," *IEE Conference on Satellite Communications System Technology,* pp. 201-206, (1975).

Bargellini, P., and Schaffner, M., "Switching in Orbit," *AIAA/CASI 6th Communications Satellite Systems Conference,* Montreal, (April 1976).

Barrett, J. A., Friedheim, H. O., and Hardwick, D. W., "Terrestrial Interfaces for Digital Satellite Transmission System," *ICDSC-2,* Paris, pp. 70-83, (November 1972).

Baxter, C., "Design of an S-Band Transponder Providing Command, Telemetry, Ranging and Range-Rate Functions for Scientific Satellites," *AIAA/CASI, 6th Communications Satellite Systems Conference,* Montreal, (April 1976).

Carter, C. R., "Efficient Synchronisation Signal for the Switched Satellite," *Electronics Letters,* vol. 12, pp. 646-647, (October 13, 1977).

Cuccia, C. L., and Davies, R. S., "Optimum Adjacent Channel Design Considerations for A Communications Satellite Carrying Both FDMA and TDMA," *International Conference on Communications,* (1976).

Deal, J., "TDMA Open-Loop Acquisition and Synchronization," *Joint Automatic Control Conference,* San Francisco, CA, pp. 1163-1169, (June 1977).

Dill, G., Deal, J., and Maillet, W., "The Intelsat Prototype TDMA System," *1975 IEEE Intercon Record,* (April 1975).

Dill, G., and Tomozawa, A., "Time-Domain Multiple-Access Frame and Burst Synchronization for Spot-Beam Satellites," *IEEE EASCON 1974.*

Ekstrom, B., "Optimized UW Detection for TDMA Communications—Unique-Word Reception in Satellite Systems," *International Conference on Digital Satellite Communications,* 3rd, Kyoto, Japan, pp. 265-272, (November 1975).

Fang, R. J. F., "Modulation Transfer from TDMA to FM Carriers in Memoryless Nonlinear Devices," *International Conference on Communications 1977.*

Gabbard, O. G., and Kaul, P., "A Cost Effective TDMA System for Domestic Applications," *Third International Conference on Digital Satellite Communications,* Kyoto, Japan, pp. 1-6, (November 1975).

Gabbard, O. G., "Design of a Satellite Time Division Multiple Access Burst Synchronizer," *IEEE Transactions on Communication Technology,* vol. COM-16, No. 4, pp. 589-596, (August 1968).

Gabbard, O. G., and Kaul, P., "Time Division Multiple Access," *EASCON 1974 Convention Record,* pp. 179-184.

Ganssmantel, H., and Ekstrom, B., "TDMA Synchronization for Future Multitransponder Satellite Communication," *Conference Record of the International Conference on Communications,* pp. 43 C-1-43 C-5, (June 1974).

Hultbert, R. M., Jean, F. H., and Jones, M. E., "Time Division Access for Military Communications Satellites," *IEEE Transactions on Aerospace and Electronic Systems,* vol. AES-1, No. 3, (December 1965).

Inagki, K., Hirata, Y., and Ogawa, A., "A New Technique for Data Transmission Via TDMA Satellite Link," *National Telecommunications Conference, 1977.*

Jacobs, I. M., "Integration of Demodulation, Decoding, Buffering and Control for TDMA Demand Assignment Systems," *Telecommunications Numeriques Par Satellite,* Paris, France, pp. 287-296, (November 28-30, 1972).

Jefferies, A. K., and Hodson, K., "New Synchronisation Scheme for Communication Satellite Time-Division Multiple-Access Systems," *Electronics Letters,* vol. 9, No. 24, (November 29, 1973).

Kwan, R. K., "A TDMA Application in the Telesat Satellite System," *Proceedings of the National Telecommunications Conference,* pp. 31E-1-31E-6, (November 1973).

Kwan, R. K., "The Telesat TDMA System—Time Division Multiple Access," *ICC 1975 Conference Record,* vol. 3, pp. 44-1-44-5.

Lipke, D. W., and Swearingen, D. W., "Marisat Multiple Access Capabilities," *International Conference on Communications 1976.*

Louie, M., and Rubin, I., "On Access Control Disciplines for a TDMA System With Multiple-Rate Real-Time Sources," *International Telemetering Conference,* Los Angeles, CA, pp. 326-336, (Sept. 28-30, 1976).

Lum, Y. F., "A Proposed Time Division Multiple Access (TDMA) Satellite System for ANIK I," *Proceedings of the International Conference on Communications,* pp. 8-1-8-5, (1973).

Nuspl, P. P., Brown, K. E., Steenaart, W., and Ghicopoulos, B., "Synchronization Methods for TDMA," *Proceedings of the IEEE,* vol. 65, No. 3, pp. 434-444, (March 1977).

Puente, J. G., and Sekimoto, T., "A Satellite Time-Division Multiple-Access Experiment," *IEEE Transactions on Communication Technology,* vol. COM-16, No. 4, pp. 581-588.

Reidel, B., Rathgeber, M., and Zilg, J., "The Initial Acquisition Unit of the French/German TDMA System for INTELSAT Application," *Third International Conference on Digital Satellite Communications,* Kyoto, Japan, pp. 301-307, (November 1975).

Reiner, U., and Zilg, J., "Reflections and Proposals Concerning the Solution of TDMA-Typical Measuring and Supervision Problems," *Telecommunications Numeriques Par Satellite,* Paris, France, pp. 306-313, (November 28-30, 1972).

Rhodes, S. A., "Preamble Requirements for Burst-Type QPSK Satellite Communications under Low E_S/N_O Conditions, *National Telecommunications Conference 1977.*

Rydbeck, N., and Sundberg, C. E., "Techniques for Introducing Error-correcting Codes into TDMA Satellite Communication Systems," *Ericsson Technics.,* vol. 31, No. 4, pp. 217-246, (1975).

Schmidt, W. G., "The Application of TDMA to the INTELSAT IV Satellite Series," *COMSAT Technical Review,* vol. 3, No. 2, pp. 257-276, (Fall 1973).

Schmidt, W., et al., "MAT-1: INTELSAT's Experimental 700-Channel TDMA/DA System," *International Conference on Digital Satellite Communications,* (November 1969).

Schmidt, W. G., and Cooperman, R., "Satellite-Switched SDMA and TDMA System for Wide Band Multibeam Satellite," *International Conference on Communications,* pp. 12-7-12-12, (June 1973).

Schmidt, W. G., "Satellite Time-Division Multiple Access Systems, Past, Present and Future," Comsat Labs, *Telecommunications,* vol. 7, No. 8, pp. 21-24, (August 1973).

Schrempp, W., and Sekimoto, T., "Unique Word Detection in Digital Burst Communications," *IEEE Transactions on Communication Technology,* vol. COM-16, No. 4, (August 1968).

Sekimoto, T., Kaul, P., and Poretti, I., "Design Benefits of Software and Processing for TDMA Terminals," *International Conference on Communications,* pp. 51-6-51-9, (June 1973).

Sinha, A., "A Model for TDMA Burst Assignment and Scheduling," *COMSAT Technical Review,* vol. 6, No. 2, (Fall 1976).

Suguri, Y., Doi, H., and Metzger, E. E., "A TDMA/PCM experiment on applications technology satellites," *IEEE Trans. Commun. Tech.,* vol. COM-19, pp.196-205, (April 1971).

Tomozawa, A., "Study of Spectral Energy Dispersion for Use with TDMA," *Telecommunications Numeriques Par Satellite,* Paris, pp. 297-305, (November 28-30, 1972).

3.6.4 CDMA—Spread Spectrum

Aein, J.M., "Multiple Access to Hard-Limiting Communications Satellite Repeater," *IEEE Trans. Space Electronics and Telemetry*, pp. 159-167, (December 1964).

Aein, J. M., and Schwartz, J., (Eds.), "Proceedings of the IDA Multiple-Access Summer Study," in "Multiple access to a communication satellite with a Hard-Limiting Repeater," IDA Rep. R-108, Vol. 2, (April 1965).

Anderson, D. R., and Wintz, P. A., "Analysis of a Spread-Spectrum Multiple-Access System with a Hard Limiter," *IEEE Transactions on Communications Technology*, vol. COM-17, No. 2, pp. 285-290, (April 1969).

Blasbalg, H., "A Comparison Study of Pseudo Noise and Conventional Modulation for Multiple Access Satellite Communications," *ICC* (1965).

Chesler, D., "Performance of a Multiple Address RADA System," *IEEE Transactions on Communication Technology*, vol. COM-14, No. 4, pp. 369-372, (August 1966).

Cuccia, C. L., "Spread Spectrum Techniques are Revolutionizing Communications," *Microwave Systems News*, (September 1977).

Dixon, R. C., *Spread Spectrum Systems*, Wiley, New York, (1976).

Gold, R., "Study of Correlation Properties of Binary Sequences," *Technical Report TR-66-234*, A. F. Avionics Lab., Wright-Patterson Air Force Base, OH.

Huth, G. K., "Optimization of Coded Spread Spectrum System Performance," *IEEE Transactions on Communications*, vol. COM-25, No. 8, (August 1977).

Kaiser, J., (Ed.), "Modulation Techniques and Their Applications," in "Multiple Access to communication satellite with a Hard-Limiting Repeater," IDA Rep. R-108, vol. 1., (January 1965).

Keller, R. D., "A Wideband, Multimode, Spread Spectrum Modem," *EASCON 1977*.

Kochevar, H. J., "Spread Spectrum Multiple Access Communications Experiment Through a Satellite," *IEEE Transactions on Communications*, vol. COM-25, pp. 853-856, (August 1977).

Kratzer, D. I., "Code Division Multiple Access," *EASCON 1974 Convention Record*, pp. 185-188.

Lebow, I. L., Jordan, K. L., and Drouilhet, P. R., "Satellite Communications to Mobile Platforms," *Proceedings of IEEE*, vol. 59, No. 2, pp. 130-159, (February 1971).

McClamont, A. M., "Multiple-Access Discrete-Address Communication Systems," *IEEE Spectrum*, pp. 87-94, (August 1967).

Mersereau, R. M., and Seay, T. S., "The Design of Multiple Access Frequency Hopping Patterns," NTIS AD785381, (September 1974).

Mohanty, N. C., "Spread Spectrum and Time Division Multiple Access Satellite Communications," *IEEE Transactions on Communications*, vol. COM-25, pp. 810-815, (August 1977).

NATO, "Spread Spectrum Communications," National Technical Information Service AD-766-914, (July 1973).

Pursley, M. B., "Recent Advances in Coding for Multiple Access Communication Systems," *International Telemetering Conference*, Los Angeles, pp. 24-33, (September 28-30, 1976).

Reiffen, B., and Sherman, H., "Parametric Analysis of Jammed Active Satellite Links," *IEEE Trans. on Comm.*, pp. 102-103, (1964).

Ristenblatt, M. P., and Davis, J. L., "Performance Criterion for Spread Spectrum Communication," *IEEE Trans. on Communication*, vol. COM-25, No. 8, (August 1977).

Sampson, C. E., "An Adaptive Multiple-Access System," *IEEE Transactions on Aerospace and Electronic Systems*, ACS-3, No. 6, (November 1967).

Schmidt, H., and McAdam, P. L., "Modulation Techniques for Communications, Part 2: Anti-Jam Performance of Spread Spectrum Coded Systems," *NAECON 1975 Record*.

Scholtz, R. A., "The Spread Spectrum Concept," *IEEE Transactions on Communications*, vol. COM-25, No. 8, (August 1977).

Schuleter, G. A., "Effects of Interference for Time-Freqency Coded Communication Systems," *Ph.D. Dissertation*, Department of Electrical Engineering, Purdue University, (August 1973).

Schwartz, J. W., et al., "Modulation Techniques for Multiple Access to a Hard Limited Satellite Repeater," *Proc. IEEE*, vol. 56, (May 1966).

Shaft, P. D., and Spilker, J. J., "Investigation of a Random Access Communication System with Adaptive Protection Against Jamming," *Philco-Ford, Report #TP-1463-3*, (January 1964).

Seyl, J. W., and Kapell, M. H., "Application of Spread Spectrum to the Shuttle Orbiter Communication System," *National Telecommunications Conference*, 1976.

Stette, G. R., "Code Division Message Multiplexing in a Satellite System," AIAA Paper No. 74-472, Los Angeles, *AIAA 5th Communications Satellite Systems Conference*, (April 22-24, 1974).

3.6.5 Demand Assignment

3.6.5.1 Voice

Aein, J. M., and Kosovych, O. S., "Satellite Capacity Allocation," *Proc. of IEEE*, vol. 65, No. 3, pp. 332-342, (March 1977).

Cacciamani, E. R., "The SPADE Concept Applied to a Network of Large and Small Earth Stations," 1970 *AIAA 3rd Communications Satellite Systems Conference*, Los Angeles, CA, Paper 70-420.

Cacciamani, Jr., E. R., "The SPADE System As Applied to Data Communication and Small Earth Station Operation," *COMSAT Technical Review*, vol. 1, No. 1, (Fall 1971).

Collins Report, "Demand Assignment Techniques Study for Military Satellite Communications Applications," Defense Communications Agency, (June 25, 1977).

Deal, J. H., and Buegler, J., "A Demand-Assignment Time-Division Multiple-Access System for Military Tactical Application," AIAA Paper, 76-270, *AIAA/CASI 6th Communications Satellite Systems Conference*, (April 1976).

Dill, G. D., "Comparison of Circuit Call Capacity of Demand-Assignment and Preassignment Operation," *COMSAT Technical Review*, vol. 2, No. 1, pp. 243-256, (Spring 1972).

Edelson, B. I., and Werth, A. M., "SPADE System Progress and Application," *COMSAT Technical Review*, vol. 2, No. 1, pp. 221-242, (Spring 1972).

Puente, J. G., and Werth, A. M., "Demand-Assigned Service for the INTELSAT Global Network," *IEEE Spectrum*, vol. 8, No. 1, pp. 59-69, (January 1971).

Smalley, A. R., "Communications Capacity Upon Demand—A New Dimension in Versatility via Satellites," *EASCON 1974 Convention Record*, (October 1974).

Werth, A. M., "SPACE: A PCM FDMA Demand Assignment System for Satellite Communications," *IEEE 1970 Int. Conf. Communications*, Paper 70-CP-423-COM, pp. 46-22-46-32.

3.6.5.2 Data

Abramson, N., "The Aloha System—Another Alternative for Computer Communications," *Fall Joint Computer Conference, AFIPS Conference Proceedings*, vol. 37, pp. 281-285, (1970).

Abramson, N., "Packet Switching With Satellites," *AFIFS Conference Proceedings*, vol. 42, *1973 National Computer Conference*, pp. 695-702, (June 1973).

Abramson, N., and Cacciamani, E., "Satellites: not just a big cable in the sky," *IEEE Spectrum*, pp. 36-40, (September 1975).

Abramson, N., "The Throughput of Packet Broadcast Channels," *IEEE Trans. on Comm.*, vol. COM-25, No. 1, pp. 117-128, (January 1977).

Baran, P., Boehm, S., and Smith, P., "On Distributed Communications," Series of 11 reports by Rand Corp., Santa Monica, (1964).

Batra, V. K., "Two Approaches to Demand Assignment of Telephone Channels in Satellite Communications," *ICC*, (1977).

Binder, R., Abramson, N., Kuo, F., Okinaka, A., and Wax, D., "Aloha packet broadcasting—A retrospect," *NCC 1975*.

Binder, R., Rettberg, R., and Walden, D., "The Atlantic Satellite Packet Broadcast and Gateway Experiments," *BBN Report No. 3056*, (April 1975).

Binder, R., "A Dynamic Packet Switching System for Satellite Broadcast Channels," *Proceedings of the International Conference on Communications*, San Francisco, CA, pp. 41-1-41-5, (June 1975).

Binder, R., McQuillan, J., and Rettberg, R., "The Impact of Multi-Access Satellites on Packet Switching Networks," *EASCON 1975 Record IEEE Electronics and Aerospace Systems Conference*, New York, pp. 63-A-63-F, (October 1975).

Binder, R., "A Round-Robin Satellite Reservation System," *ARPA Satellite System Note 59*, (July 16, 1974).

Binder, R., "A Simple Mixed-Traffic Technique for Ground-Based Aloha Channel," Technical Report B75-6, University of Hawaii, Honolulu, HI, (January 1975).

Butterfield, S. C., Rettberg, R. D., and Walden, D. C., "The Satellite IMP for the ARPA Network," *Proceedings of the Seventh Annual Hawaii International Conference on System Sciences*, Honolulu, HI, pp. 70-73, (January 1974).

Bridwell, J., "Packet Reservations with Constrained Response Time for Satellite Channels," *Proc. of 1975 National Telecommunication Conference,* (December 1975).

Cacciamani, E. R., "The SPADE System as Applied to Data Communications and Small Earth Station Operation," *COMSAT Technical Review,* vol. 1, No. 1, pp. 171–182, (Fall 1971).

Carleial, A. B., and Hellman, M. E., "Bistable Behavior of ALOHA-type systems," *IEEE Trans. Commun.,* vol. COM-23, No. 4, pp. 401–410, (April 1975).

Chu, S. W., "A Study of Asynchronous Time Division Multiplexing for Time-Sharing Computer Systems," *Spring Joint Computer Conf. AFIPS, Conf. Proc.,* vol. 35, pp. 669–678, (1969).

Collins Report, "Demand Assignment Techniques Study (DATS) for Military Satellite Communication Applications," Collins Gov't. Telecommunications Group, Final Report to MILSATCOM Systems Office, DCA, Collins Report No. 523-0767775-00221M, (June 25, 1977).

Crowther, W., Rettberg, R., Walden, D., Ornstein, S., and Heart, F., "A System for Broadcast Communication: Reservation ALOHA," *Proceedings of the Sixth Hawaii International Conference on System Sciences,* pp. 371–374, (January 1973).

Hsu, N. T., and Lee, L. N., "Synchronization Techniques for CPODA," *Proceedings of the International Communications Conference,* Toronto, (June 4–7, 1978).

Huynh, D., Kobayashi, H., and Kuo, F. F., "Design Issues for Mixed Media Packet Switching Networks," *AFIPS Conference Proceedings,* vol. 45, (June 1976).

Huynh, D., Koyabyashi, H., and Kuo, F., "Optimal Design of Mixed-Media Packet-Switching Networks: Routing and Capacity Assignment," *IEEE Transactions on Communications,* vol. COM-25, No. 1, (January 1977).

"INTELSAT/IEE International Conference on Digital Satellite Communications," London, pp. 25–27, (November 1969).

Jackson, P. E., and Stubbs, C. D., "A study of multiaccess computer communications," *1969 Spring Joint Comput. Conf., AFIPS Conf. Proc.,* vol. 34, AFIPS Press, Montvale, NJ, pp. 491–504, (1969).

Jacobs, I., et al., "CPODA—A Demand Assignment Protocol for SATNET," *Proceedings, Fifth Data Communications Symposium,* (September 1977).

Kaul, A. K., "A Dynamic Model for Satellite Nonsynchronous ALOHA Systems," *Comsat Technical Review,* vol. 7, pp. 263–390, (Spring 1977).

Kaye, A. R., and Richardson, T. G., "A Performance Criterion and Traffic Analysis for Polling Systems," *INFOR,* (June 1973).

Konheim, A. G., and Meister, B., "Waiting Line and Times in a System with Polling," *J. of the ACM,* (July 1974).

Kaiser, J., "Multiple-Access Problems in Data-Relay Satellite Systems," *Proceedings of the Third Hawaii International Conference on System Sciences,* pp. 609–617, (January 1970).

Kleinrock, L., "Performance of Distributed Multi-Access Computer Communication Systems," *Proceedings of the IFIP Congress,* Toronto, (1977).

Kleinrock, L., and Lam, S. S., "Packet Switching in a Multiaccess Broadcast Channel: Performance Evaluation," *IEEE Transactions on Communications,* vol. COM-23, No. 4, pp. 410–423, (April 1975).

Kleinrock, L., and Scholl, M., "Packet Switching in Radio Channels: New Conflict-Free Multiple Access Schemes for a Small Number of Data Users," *International Conference on Communications,* (1977).

Kleinrock, L., and Lam, S. S., "Packet-Switching in a Slotted Satellite Channel," *AFIPS Conference Proceedings,* vol. 42, (June 1973).

Kleinrock, L., and Lam, S. S., "Packet-Switching in a Slotted Satellite Channel," *Proceedings of the 1973 National Computer Conference,* pp. 703–710.

Kleinrock, L., "Queueing Systems," vol. 2, *Computer Applications,* Wiley: New York, pp. 360–407, (1976).

Kosovych, O. S., and Aein, J. M., "Efficient Transponder Utilization," *EASCON 1976.*

Kummerle, K., and Rudin, H., "Packet and Circuit Switching: A Comparison of Cost and Performance," *National Telecommunications Conference,* (1976).

Lam, S. S., "Delay Analysis of a Packet-Switched TDMA System," *NTC 1976,* 16.3-1–16.3-6.

Lam, S. S., and Kleinrock, L., "Dynamic Control Schemes for a Packet Switched Multi-Access Broadcast Channel," *AFIPS Conf. Proc.,* vol. 44, AFIPS Press, Montvale, NJ, pp. 143-153, (1975).

Lam, S. S., "A New Measure for Characterizing Data Traffic," IBM Thomas J. Watson Research Center, Yorktown Heights, NY, Research Report RC6242, (October 1976).

Lam, S., and Kleinrock, L., "Packet Switching in a Multi-access Broadcast Channel: Dynamic Control Procedures," *IEEE Transactions on Communications,* vol. COM-23, No. 9, (September 1975).

Lee, L., and Jacobs, I. M., "A Priority-Oriented Demand Assignment (PODA) Protocol and an Error Recovery Algorithm for Distributively Controlled Packet Satellite Communication Networks," *EASCON 1977.*

Metzner, J. J., "On improving utilization in ALOHA networks," *IEEE Trans. Commun.,* vol. COM-24, pp. 447–448, (April 1976).

Ng, S. F. W., and Mark, J. W., "A Multiaccess Model for Packet Switching with a Satellite Having Some Processing Capability," *IEEE Trans. on Commun.,* vol. COM-25, pp. 128–135, (January 1977).

Pickholtz, R. L., and Vogelzang, W. F., "Dynamic Satellite Slot Allocations for Data Packets via Adaptive TDMA," *National Telecommunications Conference,* (1977).

Rappaport, S. S., "Traffic Capacity of DAMA Systems Using Collision Type Request Channels," *National Telecommunications Conference,* (1977).

Raymond, H. G., and Kliger, J., "Demand Assignment Technique for Mixed User Services," *Progress in Astronautics and Aeronautics,* vol. 54, AIAA.

Roberts, L., "ALOHA Packet System with and without Slots and Capture," *ACM SIGCOM Computer Communications Review,* vol. 5, No. 2, (April 1975).

Roberts, L., "Data by the Packet," *IEEE Spectrum,* (February 1974).

Roberts, L. G., "Dynamic Allocation of Satellite Capacity Through Packet Reservation," *AFIPS Conference Proceedings,* vol. 42, (June 1973).

Rubin, I., "A Group Random Access Procedure for Multi-Access Communication Channels," *Proceedings of the National Telecommunications Conference,* (December 1977).

Rubin, I., "Integrated Random Access Reservation Schemes for Multi-Access Communication Chanels," School of Engineering and Applied Science, *UCLA Report No. UCLA-ENG-7752,* (July 1977).

Rubin, I., "On Message Delays in FDMA and TDMA Communication Channels," Paper presented at *Conference on Information Sciences and Systems,* Johns Hopkins U., (March 1978).

Rubin, I., "Reservation Schemes for Dynamic Packet Access-Control of Multi-Access Communication Channels," School of Engineering and Applied Science, *UCLA, Report No. UCLA-ENG-7712,* (January 1977).

Sastry, A. R. K., "A Simple Reservation Scheme To Improve Excess Capacity of Slotted Aloha Channels," *International Conference on Communications,* (1977).

Scholl, M. O., "Multiplexing Techniques for Data Transmission Over Packet Switched Radio Systems," *UCLA-ENG-76123,* School of Engineering and Applied Science, UCLA, (December 1976).

Schuchman, L., "Broadcast-Multi-Point-Satellite-Packet Communications," *International Conference on Communications,* (1977).

Smalley, A. R., "Communications Capacity Upon Demand—A New Dimension in Versatility via Satellites," *EASCON 1974.*

Tobagi, F. A., and Kleinrock, L., "Packet Switching in Radio Channels: Part IV—Stability Considerations and Dynamic Control in Carrier Sense Multiple Access," *IEEE Transactions on Communications,* vol. COM-25, pp. 1103–1119, (October 1977).

Wu, W. W., "A Coding Algorithm for Random Access Satellite Systems," *International Telemetering Conference Proceedings,* vol. 10, pp. 504–513, (October 1974).

Werth, A. M., "Trends in Digital Communications via Satellite," *Microwave Systems News,* (April/May 1974).

Wolfe, W. H., and Osborne, W. P., "A Dama System for Use in National Satellite Communications Systems," *Intl. 1977.*

3.6.6 On-Board Processing

3.6.6.1 SS-TDMA

Asahara, M., et al., "Synchronization and Acquisition in SDMA Satellite Communications System," *International Conference on Communications,* pp. 43E-1–43E-6, (1974).

Asahara, M., Tsuji, U., Sohma, Y., Hara, T., Sakamoto, Y., and Fukui M., "Synchronisation and Acquisition Problems for SDMA/SS-TDMA-Results of Hardware Studies—Space Division Multiple Access/Spacecraft Switched Time Division Multiple Access," *International Conference on Satellite Communications Systems Technology,* London, England, pp. 220-226, (April 1975).

Asahara, M., Tsuji, Y., Fukui, M., Sohma, Y., Hara, T., and Sakamoto,

Y., "Synchronization and Acquisition in SDMA Satellite Communication System," *Electronics and Communications in Japan,* vol. 57, pp. 119-127, (July 1974).

Carter, C., and Haykin, S., "Precision Synchronization to a Switched Satellite Using PSK Signals," *Conference Record of the International Conference on Communications,* pp. 43D-1-43D-5, (1974).

Cuccia, C. L., Davies, R. S., and Matthews, E. W., "Baseline Considerations of Beam Switched SS-TDMA Satellites Using Baseband Matrix Switching," *International Conference on Communications,* (1977).

Dill, G. D., Tsuji, Y., and Muratani, T., "Application of SS-TDMA in a Channelized Satellite," *International Conference on Communications,* (1976).

Dobson, D. P., and Ring, A. E., "An LSI Controller for Satellite Switched TDMA," *International Telemetering Conference,* Los Angeles, pp. 56-70, (September 28-30, 1976).

Funck, R., "A Comparison of Redundancy Configurations for an SS-TDMA Satellite Microwave Switch Matrix," *International Astronautical Congress,* 28th, Prague, (Sept./Oct. 1977).

Huang, R. Y., and Hooten, P., "Communications Satellite Processing Repeaters," *Proceedings of the IEEE,* vol. 59, No. 2, pp. 238-252, (February 1971).

Masson, G. M., "On Rearrangeable and Nonblocking Switching Networks," *International Conference on Communications,* (1976).

Muratani, T., Ito, Y., and Koga, K., "Future Outlook for Satellite-Switched TDMA System," *International Astronautical Federation, International Astronautical Congress,* 23th, Prague, Czechoslovakia, (September 25-October 1, 1977).

Rapuano, R., and Shimasaki, N., "Synchronization of Earth Stations to Satellite-Switched Sequences," *AIAA Paper No. 72-545,* (1972).

Schmidt, W. G., "An On-Board Switched Multiple-Access System for Millimeter-Wave Satellites," *International Conference on Digital Satellite Communications,* London, England, (November 1969).

Schmidt, W. G., "Satellite Switched TDMA: Transponder-Switched or Beam-Switched?" *AIAA Paper No. 74-460, AIAA 5th Communications Satellite Systems Conference,* Los Angeles, CA, (April 1974).

Schmidt, W. G., and Cooperman, R., "Satellite-Switched SDMA and TDMA System for Wide Band Multibeam Satellite," *Conference Record of the International Conference on Communications,* pp. 12-7-12-12, (June 1973).

Shimasaki, N., and Rapuano, R., "Synchronization for a Communications Distribution Center On-Board a Satellite," *Conference Record of the International Conference on Communications,* pp. 42-20-42-25, (1971).

3.6.6.2 Regenerative Repeaters

Carter, C. R., "Packet Switching Using a Regenerative Switching Satellite," *Midwestern Symposium on Circuits and Systems,* Montreal, pp. 42-46, (August 1975).

Celebiler, M., and Stette, G., "On Increasing the Down-Link Capacity of a Regenerative Satellite Repeater in Point-to-Point Communications," *Proceedings of the IEEE,* vol. 66, No. 1, (January 1978).

Chethik, F., and Davies, R. S., "Application of Regenerative Repeaters for Communication Stellites," *ICC 77.*

Davies, R. S., "Use of a Processing Satellite for Digital Data Transmission Between Low-Cost Earth Terminals," *AIAA 1978.*

DeRosa, J. K., "A Digital Processing Satellite for Widely Disseminated Telecommunications," *Intelcom 1977.*

Huang, R. Y., and Hooten, P., "Communication Satellite Processing Repeaters," *Proceedings of the IEEE,* vol. 59, No. 2, (February 1971).

Lee, Y. S., "Simulation Analysis for Differentially Coherent Quaternary PSK Regenerative Repeater," *COMSAT Technical Review,* 7 (2), pp. 447-474, (Fall 1977).

Lee, Y. S., "Simulation Analysis for Differentially Coherent Quaternay PSK Regenerative Repeater," *Comsat Technical Review,* Vol. 7, pp. 447-474, (Fall 1977).

Nosaka, K., and Muratani, T., "New Satellite Transponder with Regenerative Repeater and Higher Order DPSK for TDMA/SDMA System," *2nd International Conf. on Digital Satellite Communications,* Paris, (November 28-30, 1972).

Schneider, K. S., and Katz, J. L., "A Comparison of Satellite Processing Techniques for the Coherent Additive White Gaussian Noise Channel," *National Telecommunications Conference,* (1977).

Werth, A.M., "Digital Single Channel and Multichannel Per Carrier Transmission for Satellite Service," *28th International Astronautical Congress,* Prague, Czechoslovakia, (September/October 1977).

3.6.7 *Intersatellite Links*

Barry, J. D., Darien, A., Dawkins, B. T., Freedman, P. Heitman, J. M., Kennedy, C. J., Lyon, J. K., Mattassov, G., Matulka, D. C., Whited, C. E., and Zack, D. M., "1000-Mbits/s Intersatellite Laser Communications System Technology," *IEEE Transactions on Communications,* pp. 470-478, (April 1976).

Barry, J. D., "1-GBPS Space Laser Communications System," *EASCON 1976.*

Cuccia, C. L., and Lee, W. E., "Satellite-to-satellite and satellite-to-ground links-Biphase and quadriphase phase shift keying," in *Communications Satellite Technology,* Cambridge, MA, M.I.T. Press, pp. 479-501, (1974).

Forster, D. C., Goodwin, F. E., and Bridges, W. B., "Wide-Band Laser Communications in Space," *IEEE J. Quant. Elec. QE-8,* pp. 263-272, (February 1972).

O'Hern, Jr., W. L., "Applications of Optical Communications in Space," *International Conference on Communications,* San Francisco, CA, vol. 3, pp. 4-28-4-32, (June 1975).

Ohern, W. L., and Rodenberger, C. A., "Laser data links for communications satellites," *Proc. 5th AIAA Communications Satellite Systems Conf.,* Los Angeles, CA, (April 22-24, 1974).

Peyton, B. J., and Nardo, A. J. D., "Systems Measurements of a CO_2 Laser Communications Link," *International Conference on Communications,* San Francisco, CA, vol. 3, pp. 4-13-4-17, (June 1975).

Pratt, W. K., "Laser Communications Systems," Wiley, New York, (1969).

3.7 SYSTEM ENGINEERING

Aasterud, J. P., "Aids for Domsat Communication System Performance Calculations," *National Telecommunications Conference,* (December 1974).

Hadfield, B. M., "Satellite-Systems Cost Estimation," *IEEE Transactions on Communications,* vol. 22, No. 10, (October 1974).

Kiesling, J. D., Elbert, R. R., Garner, W. B., and Morgan, W. L., "A Technique for Modeling Communications Satellites," *COMSAT Tech. Rev.,* pp. 73-101, (Spring 1972).

IV. Spacecraft Design

4.1 GENERAL

Abutaleb, G. E. A., Kim, M. C., Manning, K. F., Phiel, Jr., J. F., and Westerlund, L. H., "The COMSTAR Satellite System," *COMSAT Technical Review,* vol. 7, No. 1, Spring, pp. 35-83, (1977).

Allison, J. F., Arndt, R., and Meulenberg, A., "A Comparison of the COMSAT Violet and Non-reflective Solar Cells," *COMSAT Technical Review,* vol. 5, No. 2, pp. 211-224, (1975).

"Analysis and Design of Space Vehicle Flight Control Systems," Vol. XVI-Abort, NASA Report #CR-835, (May 1969).

Bakeman, D. C., Jenkins, K. W., Rodden, J. J., Iorillo, A. J., and Garber, T. B., "The Relative Merits of Three-Axis and Dual-Spin Stabilization Systems for Future Synchronous Communication Satellites," *AIAA Progress in Astronautics and Aeronautics: Communication Satellites for the 70's—Systems,* vol. 26, N. E. Feldman and C. M. Kelly (Eds.), MIT Press, Cambridge, MA, Chap IX, pp. 605-653, (1971).

Bate, R. R., Mueller, D. D., and White, J. E., "Fundamentals of Astrodynamics," Dover Pubs., NY, (1971).

Bauer, P., "Batteries for Space Power Systems," NASA SP-172, (1968).

Bean, E. E., and Bloomquist, C. E., "Reliability Data From In-Flight Spacecraft," *Proc. Ann. Rel. & Maint. Symposium,* p. 271, (Jan. 1978).

Bleviss, Z. O., "Expendable Launch Vehicles for Synchronous Communications Satellites," *Satellite Communications; Future Systems,* D. Jarrett (Ed.), AIAA, NY, pp. 319-346, (1977).

Brill, Y., et al., "Propulsion for Geostationary Spacecraft," 1971 JANNAF Combined Propulsion Meeting, Las Vegas, NV, (Nov. 1-5, 1971).

Cannon, R. H., "Some Basic Response Relations for Reaction-Wheel Attitude Control," *ARS Journal,* vol. 32, No. 1, pp. 61-74, (January 1962).

Chidester, L. G., "Advanced Lightweight Solar Array Technology," *Proceedings of the AIAA 7th Communications Satellite Systems Conference,* San Diego, CA, pp. 55-60, (April 1978).

Clopton, E. J., and Crum, W. M., "TDRSS Electrical Power Subsystem," *Proceedings of 13th Intersociety Energy Conversion En-

gineering Conference, SAE P-75, vol. 1, (August 20–25, 1978).

"INTELSAT IV Spacecraft," *COMSAT Technical Review,* vol. 2, No. 2, pp. 271–572, (1972).

Dougherty, H. J., Lebsock, K. L., and Rodden, J. J., "Attitude Stabilization of Synchronous Communications Satellites Employing Narrow-Beam Antennas," *Journal of Spacecraft and Rockets,* vol. 8, No. 8, pp. 834–841, (August 1971).

"Electrical Power, Direct Current, Space Vehicle Design Requirements," MIL-STD-1539 (USAF), (1 August 1973).

Free, B., "Chemical and Electric Propulsion Tradeoffs for Communications Satellites," *COMSAT Technical Review,* vol. 2, No. 1, Spring, pp. 123-145, (1972).

Free, B., and Huson, G., "Selected Comparisons Among Propulsion Systems for Communications Satellites," *AIAA Paper 72-517,* Washington, DC, (1972).

Garber, T. B., "Three-Axis or Dual-Spin Stabilization for Future Synchronous Communications Satellites—A Summary," *Communications Satellites for the 70's: Systems,* vol. 26 AIAA Series, N. E. Feldman and C. M. Kelly (Eds.), MIT Press, Cambridge, MA, pp. 639–653, (1971).

Gopin, A. J., and Friedman, A. S., "Thermal Analysis and Control of the Air Force Tactical Communications Satellite," *Communications Satellites for the 70's: Technology* (AIAA Progress in Astronautics and Aeronautics, vol. 25), N. E. Feldman and C. M. Kelly (Eds.), Cambridge, MA, MIT Press, 1971.

"Government—Industry Workshop on Payload Loads Technology," NASA Conf. Publication #NASA CP 2075, Marshall SFC, (Nov. 14–16, 1978).

Granger, G. R., and Leventhal, A., "Communication Satellite System Reliability," *Proc. Annual Rel. & Maint. Sympos.,* (January 1972).

Hall, D. F., and Lyon, W. C., "Low Thrust Propulsion System Effects on Communication Satellites," *Progress in Astronautics and Aeronautics—Communications Satellite Technology,* vol. 33, pp. 279-306.

Hanson, C. W., and Brown, J. V., "Attitude Determination for the Skynet and NATO Synchronous Communications Satellites," Ibid., pp. 127-40.

Harris, L. A., "Advanced Composites, An Assessment of the Future," *Astronautics and Aeronautics,* pp. 22–33, (March 1976).

Hayden, J., "An Electrical Power System for the Defense Satellite Communications System—Phase III," *Proceedings of 13th Intersociety Energy Converstion Engineering Conference,* SAE P-75, vol. I, (August 20–25, 1978).

Herron, B. G., Garth, D. R., Finke, R. C., and Shumaker, H. A., "Power Processing Systems for Ion Thrusters," *Progress in Astronautics and Aeronautics—Communications Satellite Technology,* vol. 33, pp. 261-277.

Holcomb, L. B., "Survey of Auxiliary Electric Propulsion Systems," Addendum to "Satellite Auxiliary-Propulsion Selection Techniques," *Tech. Rept. No. 32-1505,* July 15, 1971, Jet Propulsion Lab., Pasadena, CA.

Huber, W. G., and Huffaker, C. F., "IUS Capability for Earth Orbital Spacecraft," *Satellite Communications; Future Systems,* D. Jarett (Ed.), AIAA, NY, pp. 357–376, (1977).

Iorillo, A. J., "Stabilization of Future Synchronous Communication Satellites," *AIAA Progress in Astronautics and Aeronautics—Communication Satellites for the 70's—Systems,* vol. 26, pp. 633–638.

Isley, W. C., and Duck, K. I., "Propulsion Requirements for Communication Satellites," *Progress in Astronautics and Aeronautics-Communications Satellite Technology,* vol. 33, pp. 165–190.

Jenkins, K. W., "Relative Merits of Three-Axis and Dual-Spin Stabilization Systems for Future Communication Satellites," *AIAA Progress in Astronautics and Aeronautics: Communication Satellites for the 70's—Systems,* vol. 26, pp. 619–624.

Johnson, C. R., "TACSAT I Nutation Dynamics," *AIAA Progress in Astronautics and Aeronautics: Communication Satellites for the 70's—Technology,* vol. 25, N. E. Feldman and C. M. Kelly (Eds.), MIT Press, Cambridge, MA, pp. 65–96, (1971).

Jones, C. R., "Cost-competitive STS for Geostationary Payloads," *Satellite Communications: Future Systems,* D. Jarett (Ed.), AIAA, NY, pp. 389–404, (1977).

Kalley, Jr., J. J., "Control System Design for Double Gimballed Wheels," AIAA Paper 74-895, Anaheim, CA, (Aug. 5–9, 1974).

Kalley, Jr., J. J., and Mork, H. L., "Development and Air Bearing Test of a Double-Gimballed Momentum Wheel Attitude-Control System" *Progress in Astronautics and Aeronautics, Communication Satellite Developments: Technology,* vol. 42, pp. 49–82.

Kaplan, M. H., "Active Attitude and Orbit Control of Body-Oriented Geostationary Communications Satellites," *Progress in Astronautics and Aeronautics—Communications Satellite Technology,* vol. 33, pp. 29-56.

Kaplan, M. H., and Edwards, T. L., "Double-Gimbaled Momentum Wheel Control System for a Body-Stabilized Spacecraft Configuration" TM CL-45-72, COMSAT Laboratories, Clarksburg, MD, (August 31, 1972).

Kaplan, M. H., "Modern Spacecraft Dynamics and Control," John Wiley, (1976).

Kaplan, M. H., "Estimation and Correction of Electric Thruster Misalignment Effects on an Oriented Geostationary Satellite," *COMSAT Technical Review,* vol. 3, No. 1, Spring, (1973).

Keigler, J. E., Lindorfer, W. J., and Muhlfelder, L., "Momentum Wheel Three-Axis Attitude Control for Synchronous Communications Satellites," *Communications Satellite Technology,* vol. 33, AIAA, P. L. Bargellini (Ed.), MIT Press, Cambridge, MA, pp. 141-161, (1974).

Kloves, T. R., "Space Shuttle Program: An Overview," *Satellite Communications: Future Systems,* D. Jarett (Ed.) AIAA, NY, pp. 347-356, (1977).

Kreith, F., "Radiation Heat Transfer for Spacecraft and Solar Power Plan Design," Internat. Text Book Co., (1962).

Lebeau, A. F., and Orye, R. M., "Ariane Launch Vehicle: A European Program," Op.Cit. (Keaves), pp. 377–388.

Lebsock, K. L., "High Pointing Accuracy with a Momentum Bias Attitude Control System," *Proceedings of the AIAA 7th Communications Satellite Systems Conference,* San Diego, CA, pp. 269–277, (April 1978).

Lebock, K. L., Flanders, H. R., and Rodden, J. J., "Simulation and Testing of a Double-Gimballed Momentum Wheel Stabilization System for Communication Satellites," *Journal of Spacecraft and Rockets,* vol. 11, No. 6, pp. 395-400, (June 1974).

Levine, S. E., "A Review of the Mission Success of Communication Satellites and Related Spacecraft," *SAMSO Report TR-77-180, (October 1977).*

Likins, P. W., "Attitude Stability Criteria for Dual Spin Spacecraft," *Journal of Spacecraft and Rockets,* vol. 4, No. 12, pp. 1638-1643, (Dec. 1967).

Lindmayer, J., and Allison, J. F., "The Violet Cell: An Improved Silicon Solar Cell," *COMSAT Technical Review,* vol. 3, No. 1, pp. 1–22, Spring, (1973).

Lucas, J. W., (Ed.), "Fundamentals of Spacecraft Thermal Design" Vol. 29, *AIAA Progress in Astronautics and Aeronautics,* MIT Press, (1972).

Lyons, M. G., Lebsock, K. L., and Scott, E. D., "Double Gimballed Reaction Wheel Attitude Control System for High Altitude Communications Satellites," AIAA Paper No. 71-949, Hempstead, NY, (August 1971).

Martin, E. R., "Experimental Investigations of the Fuel Slosh of Dual Spin Spacecraft," *COMSAT Technical Review,* vol. 1, No. 1, Fall, pp. 1-20, (1971).

McIntyre, J. E., and Gianelli, M. J., "Effect of an Autotracking Antenna on the Nutational Stability of a Dual-Spin Spacecraft," *Progress in Astronautics and Aeronautics—Communications Satellite Technology,* vol. 33, pp. 85-125.

McKinney, H. N., and Briggs, D. C., "Electrical Power Subsystem for the Intelsat V Satellite," *Proceedings of 13th Intersociety Energy Converstion Engineering Conference,* SAE P-75, vol. I, (August 20–25, 1978).

Neer, J. T., "INTELSAT IV Nutation Dynamics," *Communications Satellite Technology,* vol. 33, AIAA, P. L. Bargellini (Ed.), MIT Press, Cambridge, MA, pp. 57–84, (1974).

Perkel, H., "Stabilite—A Three Axis Attitude Control System Using a Single Reaction Wheel," *AIAA Progress in Astronautics and Aeronautics: Communication Satellite Systems Technology,* vol. 19, Academic Press, NY, pp. 375–400, (1966).

"Proceedings of Aerospace Testing Seminar," Institute for the Environmental Sciences and Aerospace Corporation, Los Angeles, CA, (Ibid, 1975 and 1976), (1973).

Proceedings of the Symposium for Long Life Hardware for Space, NASA Marshall SFC, 2 volumes, (March 1969).

Pugmire, T. K., et al., "Electrothermal Hydrazine Engine Performance," AIAA Paper 71-760, Salt Lake City, UT, (1971).

Raab, B., and Karlin, J. J., "Solar vs Nuclear Power: Is there a choice?" *Progress in Astronautics and Aeronautics: Communication Satellite Developments: Technology,* vol. 42, pp. 83–111.

Robinson, J. A., "Thermal Control, the INTELSAT IV Spacecraft,"

COMSAT Technical Review, vol. 2, No. 2, Fall, pp. 376–388, (1972).

Rodden, J. J., "Relative Merits of Three-Axis and Dual-Spin Stabilization Systems for Future Synchronous Communications Satellites," *Progress in Astronautics and Aeronautics—Communication Satellites for the 70's: Systems,* vol. 26, pp. 625–632.

Rusch, R. J., Johnson, J. T., and Baer, W., "INTELSAT V Spacecraft Design Summary," *Proceedings of the AIAA 7th Communications Satellite Systems Conference,* San Diego, CA, pp. 8–20, (April 1978).

Russi, M. J., "A Survey of Monopropellant Hydrazine Thruster Technology," AIAA Paper 73-1293, (1973).

Salvatore, J. O., and Porter, W. W., "ATS-V-Heat Pipe Tests and Dimensional Analysis," Rept. SGC 10319R, Hughes Aircraft Co., Culver City, CA, (June 1970).

"Satellite Auxiliary Propulsion Selection Techniques," JPL Report TR 32-1505, (1970).

Schmidt, G. E., "Magnetic Attitude Control for Geosynchronous Spacecraft," Proceedings of the AIAA 7th Communications Satellite Systems Conference, San Diego, CA, pp. 278–284, (April 1978).

Slabinski, V. J., "Variation in Range, Range-Rate, Propagation Time Delay, and Doppler Shift for a Nearly Geostationary Satellite," *Progress in Aeronautics and Astronautics: Communications Satellite Technology,* vol. 33, pp. 3–28.

"Solar Cell Array Design Handbook," Jet Propulsion Laboratory Report JPL SP 43-38, (October 1976).

"Space Vehicle Electrical Power Processing Distribution and Control Study," TRW Systems, Contract No. NAS8-26270, (June 1972).

"Spacecraft Thermal Control," NASA Report SP-8105, (May 1973).

Sorsenson, A. A., Elbert, J. D., and Wheeler, P. C., "Spacecraft Digital Attitude Control," 1972 WESCON, (Sept. 19–22, 1972).

Sparks, R. H., "Nickel-Cadmium Battery Technology Advancements for Geosynchronous Orbit Spacecraft," *Proceedings of the AIAA 7th Communications Satellite Systems Conference,* San Diego, CA, pp. 61–65, (April 1978).

Stiltz, H. C., "Aerospace Telemetry, Prentice Hall, (1961).

Stiltz, H. C., "Aerospace Telemetry, vol. II, Prentice Hall, (1961).

Stockel, J. F., Dunlop, J. D., and Betz, F., "NTS-2 Nickel Hydrogen Battery Performance," *Proceedings of the AIAA 7th Communications Satellite Systems Conference,* San Diego, CA, pp. 66–71, (April 1978).

"Study and Analysis of Satellite Power Systems Configurations for Maximum Utilization of Power," TRW Systems Report 04898-6001-R000, (30 December 1966).

Strauss, R., and Owens, J., "Satellite Reliability," IAF Congress, Prague, Czech., (1977).

Thompson, W. T., *Introduction to Space Dynamics,* New York, Wiley, (1961).

Timmins, A. R., and Norris, H. P., "Failure Rate Analysis of Goddard SFC Spacecraft Performance During Orbital Life," *Proc. Annual Reliability & Maintainability Symposium,* (Jan. 1976).

"Trends in Reaction Control Propulsion Systems for Satellites and Spacecraft," Hughes Aircraft Co., Space Systems Div. Report 60405R, (1966).

Vigneron, F. R., "Stability of a Dual-Spin Satellite With Two Dampers," *Journal of Spacecraft and Rockets,* vol. 8, No. 4, pp. 386–389, (April 1971).

Weber, F. W., "Telemetry and Command Subsystems, the INTELSAT IV Spacecraft," *COMSAT Technical Review,* vol. 2, No. 2, Fall, pp. 341–358, (1972).

4.2 ANTENNAS

Archer, D., "Lens-Fed Multiple-Beam Arrays," *Microwave Journal,* (October 1975).

Assaly, R. N., et al., "LES-6 Antenna System," Lincoln Laboratory, MIT, Lexington, Mass., *Technical Report 465,* (March 1969).

Bilenko, H. W., Wickert, A. N., and Han, C. C., "Computer-Aided Array Feed Design for Multiple-Beam Lens Antenna," *1975 AP-S International Symposium Digest,* University of Illinois, (1975).

Chuang, C. A., and Han, C. C., "Active Element Pattern of a Rectangular Waveguide in a Periodic Planar Array," *1976 International IEEE/AP-S Symposium and USNC/URSI Meeting,* University of Massachusetts, (October 1976).

Cuccia, C. L., et al., "Transponder and Antenna-Design Problems at Millimeter Wavelengths for 20-30 GHz Communications Satellites," *Procedures of Symposium on Advanced Satellite Communications Systems,* Genoa, Italy, (December 1977).

Das, S., "Multi-Beam Antennas Improve Satellite Communication," *Microwave Systems News,* (December 1977).

Dietrich, F. J., Ford, D. F., and Hillesland, H. L., "A Millimeter-Wave Intersatellite Communication Antenna," *1970 International Conference on Communications,* San Francisco, CA, (June 1970).

Dion, A. R., "Minimum Directivity of Multiple-Beam Antennas," *IEEE Transactions on Antennas and Propagation,* (March 1975).

Dion, A. R., "Variable-Coverage Communications Antenna for LES-7," Lincoln Laboratory, MIT, *Technical Report 471,* (October 1969).

Dion, A. R., and Ricardi, L. J., "A Variable Coverage Satellite Antenna System," *Proc. IEEE,* vol. 59, No. 2, (February 1971).

Dion, A. R., and Ricardi, L. J., "A Variable-Coverage Satellite Antenna System," *Proceedings of the IEEE,* 59, (February 1971).

Duncan, J. W., Hamada, S. J., and Ingerson, P. G., "Dual Polarization Multiple-Beam Antenna for Frequency Reuse Satellites," AIAA Paper 76-248, *AIAA/CASI 6th Communications Satellite Systems Conference,* Montreal, Canada, (April 1976).

Duncan, J. W., Hamada, S. J., and Wong, W. C., "Polarization Isolation Characteristics of a Dual-Beam Reflector Antenna," *AIAA 4th Comm. Sat. Systems Conference,* Paper No. 72-531, (1972).

Foldes, P., "Multiple Beam Antennas for Regional Communication Satellites," *EASCON 1977,* pp. 275–282.

Ghobrial, S. I., "Cross-Polarization in Satellite and Earth Station Antennas," *Proceedings of the IEEE,* vol. 65, No. 3, pp. 378–387, (March 1977).

Gregorwich, W. S., "An Electronically Despun Array Flush-Mounted on a Cylindrical Spacecraft," *IEEE Transactions on Antennas and Propagation Conformal Arrays,* (January 1974).

Gregorwich, W. S., "A Multipolarization Dual-Band Array," *IEEE/AP-s 1975 International Symposium,* University of Illinois, (June 1975).

Gregorwich, W. S., and Kawata, P. K., "A Broadband Conformal Array for Communications Satellites," *1977 Antenna Applications Symposium,* University of Illinois, (April 1977).

Gregorwich, W. S., and Westerman, C. W., "A Near-Isotropic Microwave Antenna for Communications Satellites," *Communications Satellite Technology,* P. L. Bargellini (Ed.), MIT Press, pp. 533–540, (1974).

Han, C. C., and Ford, D. F., et al., "A General Beam Shaping Technique—Multiple-Feed Offset Reflector Antenna System," *AIAA/CASI 6th Communications Satellite Systems Conference,* Montreal, Canada, (April 1976).

Han, C. C., "A New Multimode Rectangular Horn Antenna Generating A Circularly Polarized Elliptical Beam," *IEEE Transactions on Antennas and Propagation,* pp. 746–750, (November 1974).

Hughes Aircraft Co., "Comparison of Design Approaches for Multi-Beam 7-9 GHz Space Antennas," *Air Force Report No. SAMSO-TR-72-284,* Hughes Aircraft Company, Culver City, CA, (December 1972).

Jamieson, A. R., "Offset Fed Zoned Reflectors for Multi-Beam Antenna Systems," *Technical Report 77-20,* Electromagnetics Laboratory, University of Illinois.

Kreutel, R., "A Multiple-Beam Torus Reflector Antenna for 20-30 GHz Satellite Communications Systems," *AIAA/CASI 6th Communications Satellite Systems Conference,* Paper 76-302, (April 1976).

Luh, H. S., "The Effect of Tolerance on the Isolation of Frequency Reuse Antennas," *IEEE Transactions on Antennas and Propagation,* (January 1975).

Luh, H. S., et al., "A Constrained Lens Antenna for Multiple-Beam Satellites," AIAA Paper 74-465, *AIAA/CASI 6th Communications Satellite Systems Conference,* Montreal, Canada, (April 1976).

Matthews, E. W., "Variable Power Dividers in Satellite Systems," MTT-S, Cherry Hill, NJ, (1976).

Matthews, E. W., Scott, W. G., and Han, C. C., "Advances in Multi-Beam Satellite Antenna Technology," *EASCON,* Washington, DC, (October 1976).

Pelchat, M. G., and Baird, C. A., "Simplified Techniques For Adaptive Polarization Crosstalk Reduction," *EASCON 1977,* pp. 31-3A–31-3G.

Powell, R. V., and Hibbs, A. R., "An Entree for Large Space Antennas," *AIAA Journal,* (December 1977).

Reudink, D. O., Acampora, A. S., and Yeh, Y. S., "Spectral Reuse in 12 GHz Satellite Communication Systems," *International Conference on Communications,* pp. 37.5-32–37.5-35, (1977).

Reudink, D. O., and Yeh, Y. S., "A Scanning Spot-Beam Satellite System," *The Bell System Technical Journal,* 56(8), (October 1977).

Schennum, G. H., "Design of Frequency Selective Surfaces Using A Waveguide Simulator Technique," *IEEE International Antenna and Propagation Symposium,* University of Illinois, (June 1975).

Schroeder, K. G., "Characteristics and Applications of Multibeam Satellite Antennas," *Communications Satellite Technology,* P. L. Bargellini, Ed., Cambridge, MA, MIT Press, pp. 503–532, (1974).

Schroeder, K. G., "Characteristics and Applications of Multi-Beam Spacecraft Antennas," *AIAA Paper No. 72-530, AIAA 4th Communications Satellite Systems Conference,* Washington.

Schultz, J. L., and Kadar, I., "A Frequency Reuse Direct-to-Home TV Broadcast Satellite System," *7th AIAA Communications Satellite Systems Conference,* San Diego, CA, pp. 78-577, (April 1978).

Scott, W. G., "Antenna Pattern Topology," *1969 IEEE/G-AP International Symposium,* University of Texas, Austin, (December 1969).

Scott, W. G., Luh, H. S., and Matthews, E. W., "Design Tradeoffs for Multiple-Beam Antennas in Communication Satellites," *IEEE International Conference on Communications,* Philadelphia, PA, (June 1976), IEEE Communications Society, (March 1977).

Scott, W. G., et al., "Wide-Angle Circularly Polarized Antenna Techniques for Spacecraft," *IRE 9th East Coast Conference on Aerospace and Navigational Electronics,* (October 1962).

Sielman, P. F., Schwartz, L. and Noji, T.T., "Multiple-Beam Communicating Satellites with Remote Beam Steering and Beam Shaping," Paper presented 72-566 at *AIAA 4th Communications Satellite Systems Conference,* Washington DC, (April, 1972).

Smoll, A. E., and Han, C. C., "A Split-Mode Horn/Reflector Antenna—A New Concept in Antenna Design," *International Symposium on Antennas and Propagation,* University of Illinois, (June 1975).

Tankersley, B. C., and Bartlett, H. E., "Tracking and Data Relay Satellite Single Access Deployable Antenna ," *National Telecommunications Conference,* pp. 19:4-1-19:4-6, (1977)

Wickert, A. N., et al., "The Design of the Communications Antenna for CS," *1976 AIAA 6th Communications Satellite Systems Conference,* Montreal, Canada (April 1976).

4.3 RECEIVER AND TRANSMITTER TECHNOLOGY

Abronson, C., "FET Amplifiers: The Viable Alternative," *Microwave Systems News,* 8, p. 29, (February 1978).

Adain, J. E., "Millimeter Wave Radiometry Developments," *Microwave Journal 20,* p. 32, (November 1977).

Akaike, M., and Ohnishi, K., "A Non-Linear Analysis of Schottky-Barrier Diode Up-Converters," *IEEE Trans. MIT, MTT-25,* p. 1059, (1977).

Allen, C. M., and Lombardo, P., "Spaceflight-Qualified Miniature S-Band Parametric Amplifier Assembly," *National Aerospace and Electronics Conference,* Dayton, OH, (May 1977).

Arams, F. R., and Lombardo, P., "Recent Developments in C-Band and K_u-Band SatCom Receivers," *INTELCOM 1978,* Atlanta, GA, pp. 217-221.

Arams, F. R., and Okean, H. C., "State-of-the-Art Low Noise Receivers for Microwave and Millimeter-Wave Communications," *Proc. Nat. Telecommunications Conf.,* Dallas, TX, p. 48.5-1, (1978).

Barrera, J. L., "GaAs FET Explosion Keeps Gaining Momentum," *Microwave System News,* 7, p. 73, (December 1977). 1977).

Bearse, S. V., "Impatts, GaAs FETs and TWTs Deemed Most Likely to Succeed," *MicroWaves,* 15, p. 14, (July 1976).

Calviello, J. A., and Wallace, J. L., "Performance and Reliability of an Improved High Temperature GaAs Schottky Junction and Native-Oxide Passivation," *IEEE Trans. on Elec. Dev.,* ED-24, p. 698, (1977).

Cooke, H. F., "Microwave FETs—A Status Report," *IEEE Solid-State Circuits Conference,* THAM 10-1, p. 116, (February 1978).

Cooper, J. F., and Gupta, M. S., "Microwave Characterization of GaAs MESFET and the Verification of Device Model," *IEEE Trans. Solid-State Circuits,* SC-12, p. 325, (1977).

Cowan, D. A., Mercer, P., and Bell, A., "Low-Noise and Linear FET Amplifiers for Satellite Communications," *IEEE Trans. on MTT, MTT-25,* p. 995, (1977).

Cuccia, C. L., "Status Report: Microwave Communications Update—1976," *Microwave System News,* 6, p. 43, (February/March 1976).

Cuccia, C. L., "Status Report: Solid-State Baseline of Telecommunications into the 1980's," *Microwave Systems News,* 6, p. 43, (April/May 1976).

Cuccia, C. L., and Ho, P. T., "The Impact of the FET on the Satellite Transponder—A Technology Perspective," *7th AIAA Communications Satellite Systems Conference,* (April 1977).

Davis, R. T., and Jordan, J., "FET Scoreboard," *Microwave System News 7,* p. 70, (February 1977).

Defense SatCom Office, "TWTA Background and Requirements For Military Communications Spacecraft," *Space and Missile Systems Organization, Report No. TOR-0074(4624-01)-2,* Los Angeles, CA, pp. 1-19, (February 1, 1974).

Dickens, L. E., and Maki, D. W., "An Integrated-Circuit Balanced Mixer, Image and Sum Enhanced," *IEEE MIT, MTT-23,* (1975).

Douville, R. J., "A 12-GHz Low-Cost Earth Terminal for Direct TV Reception from Broadcast Satellites," *IEEE Trans. on MTT, MTT-25,* (1977).

Edrich, J., "A Cryogenically Cooled Two Channel Paramp Radiometer for 47 GHz," *IEEE MTT, MTT-25,* (April 1977).

Emery, F. E., "Low Noise GaAs FET Amplifiers," *Microwave Journal 20,* (June 1977).

Frey, J., "Effects of Intervalley Scattering on Noise in GaAs and InP FETs," *IEEE Trans. Elec. Dev., ED-23,* (December 1976).

Fukuta, M., Mimura, T., Suzuki, H., and Suyama, K., "4-GHz 15-W Power GaAs MESFET," *IEEE Transactions on Electron Devices,* vol. ED-25, No. 6, (June 1978).

Gibbons, G., "InP Devices Promising but Challenging," *Microwave System News,* 8, (February 1978).

Glaze, G. E., and Hobbs, II, G. H., "FET Amplifier Spans 4-12 GHz," *Microwave Systems News,* 8, (February 1978).

Goel, J., and Wolkstein, H. J., "A 4-8 GHz Dual-Gate Amplifier," *IEEE Solid State Circuits Conf.,* THAM 10-6, (February 1978).

Hasegawa, F., et al., "GaAs Power MESFETs with a Simplified Recess Structure," *IEEE International Solid-State Circuits Conference,* (February 1978).

Hennies, S. R., "A 7-18 GHz Amplifier Using GaAs FETs," *IEEE Solid State Circuits Conf.,* THAM 10-5, (February 1978).

Henry, P. S., Glance, B. S., and Schneider, M. W., "Local Oscillator Noise Cancellation in the Subharmonically Pumped Down-Converter," *IEEE Trans. MTT, MTT-24,* (1976).

Hislop, A., and Kihm, R. T., "A Broad-Band 40-60 GHz Balanced Mixer," *IEEE Trans. MTT, MTT-24,* (1976).

Kelly, A. J., and Okean, H. C., "A Low Noise Millimeter MIC Mixer," *IEEE Trans. on MTT, MTT-24,* (1976).

Kennedy, W. K., "FET Technology: An Overview," *Microwave System News 8,* (February 1978).

Kerr, A. R., "Low-Noise Room Temperature and Cryogenic Mixers for 80-120 GHz," *IEEE Trans. on MTT, MTT-23,* (1975).

Knamuri, N., and Kawasaki, R., "Design and Performance of a 60-90 GHz Broadband Mixer," *IEEE Trans. on MTT, MTT-24,* (1976).

Kosmahl, G. H., McNary, B. D., and Sauseng, O., "High-Efficiency, 200-Watt, 12-Gigahertz Traveling Wave Tube," *NASA Technical Note NASA TN D-7709,* pp. 1-17, (June 1974).

Liechti, C. A., and Larrick, R. B., "Performance of GaAs MESFETs at Low Temperatures," *IEEE MTT, MTT-24,* p. 376, (1976).

Madden, D. R., and Smidt, J. J., "European Capabilities in the Field of Microwave Tubes for Space Use," *AIAA 7th Communications Satellite Systems Conference,* (April 1978).

Magarshack, J., "FETSs: Improving Performance and Reliability," *Microwave System News,* 8, (February 1978).

Mazur, D. G., et al., "40 and 80 GHz Technology Assessment and Forecast,"*NASA-CR-135028,* (May 1976).

McColl, M., "Conversion Loss Limitations on Schottky-Barrier Mixers," *IEEE Trans. on MTT, MTT-25,* p. 54, (1977).

NASA, "Microwave Space Receivers," *A Forecast of Space Technology 1980-2000, NASA SP-387,* pp. 3-45, (1976).

Niehenke, E. C., "An Environmentalized Low-Noise Parametric Amplifier," *IEEE Trans. MTT-25,* (1977).

Ohata, K., Nozaki, T., and Kawamura, N., "Improved Noise Performance of GaAs MESFETs with Selectively Ion-Implanted n^+ Source Regions," *IEEE Trans. Elect. Devices,* ED-24, (1977).

Okean, H. C., and Kelly, A. J., "Low Noise Receiver Design Trends Using State-of-the-Art Building Blocks," *IEEE-MTT, MTT-25,* (1977).

Pierro, J., "Cryogenically Cooled GaAs FET Amplifier With a Noise Temperature Under 70 K at 5.0 GHz," *IEEE MTT, MTT-24,* (1976).

Pucel, R. A., Masse, D. J., and Krumm, C. F., "Noise Performance of Gallium Arsenide Field-Effect Transistors," *IEEE Solid State Circuits, SC-11,* (1976).

Raue, J. E., "High Performance Millimeter Wave Receiver and Transmitter Component Development," *AIAA 7th Communications Satellite Systems Conference,* (April 1978).

Rees, H. D., and Gray, K. W., "Indium Phosphide: A Semi-Conductor for Microwave Devices," *IEEE Solid State and Elect. Devices, 1,* (September 1976).

Schellenberg, J. M., and Cohn, M., "A Wideband Radial Power Combiner for FET Amplifiers," *1978 IEEE International Solid-State Circuits Conference Record,* pp. 164-165, (February 1978).

Schneider, M. V., and Snell, W. W., "Harmonically Pumped Stripline Down-Converter," *IEEE Trans. on MTT, MTT-23,* p. 271, (1975).

Sechi, F. N., and Paglione, R. W., "Design of a High-Gain FET Amplifier Operating at 4.4-5.0 GHz," *IEEE Trans. Solid State Circuits,* SC-12, (1977).

Silzars, A., Bates, D., and Ballonoff, A., "Electron Bombarded Semiconductor Devices," *Proceedings of the IEEE,* (August 1976).

Snell, W. W., and Schneider, M. W., "Millimeter-Wave Thin Film Down-Converter," *IEEE Trans. on MTT,* MTT-24, (1976).

Sokolov, V., and Tserng, H. Q., "Characteristics of Solid State Microwave Power Amplifiers for Telecommunication Applications," *National Telecommunications Conference,* pp. 48.2-1-48.2-5, (1976).

Teledyne, "Traveling Wave Tubes—Theory and Operation," Technical Memorandum 96, pp. 1-32, (June 1971).

Vernon, Jr., F. L., et al., "The Super-Schottky Diode," *IEEE Trans. on MTT, MTT-25,* (1977).

Wade, P. C., "A 1-W, 2-6 GHz Amplifier Utilizing a Discretely Packaged GaAs•FET," *IEEE International Solid-State Circuits Conference,* (February 1978).

Walker, M. G., Marki, F. A., and Abrumowitz, H. M., "MESFET Amplifiers go to 18 GHz," *Microwave System News, 6,* (April/May 1976).

Whelehan, J., "Low-Noise Millimeter-Wave Receiver," *IEEE-MTT, MTT-25,* (1977).

Wilson, W. J., "1975 State-of-the-Art of 5 GHz to 250 GHz Receivers," The Aerospace Corporation, *Aerospace Technical Memorandum, No. ATM-75(5030)-1,* (May 1975).

Wisseman, W. R., "Recent Advances in Solid State Devices For Microwave Power Generation," *National Telecommunications Conference,* pp. 48.1-1-48.1-5, (1976).

Yuan, L. T., "Design and Performance Analysis of an Octave Bandwidth Waveguide Mixer," *IEEE Trans. on MTT, MTT-25,* p. 1048, (1977).

4.4 FILTERS AND MULTIPLEXERS

Atia, A. E., and Williams, A. E., "New Types of Waveguide Band Pass Filters for Satellite Transponders," *COMSAT Tech. Review,* vol. 1, No. 1, pp. 21-43, (Fall 1971).

Atia, A. E., and Williams, A. E., "Narrow-Band Pass Waveguide Filters," *IEEE Trans. Microwave Theory and Techniques,* vol. MTT-20, pp. 258-265.

Atia, A. E., and Williams, A. E., "Non-Minimum-Phase Optimum-Amplitude Band Pass Waveguide Filters, *IEEE Trans. Microwave Theory and Techniques,* vol. MTT-22, pp. 425-431, (April 1974).

Atia, A. E., Williams, A. E., and Newcomb, R. W., "Narrow-Band Multiple-Coupled Cavity Synthesis," *IEEE Trans. on Circuits and Systems,* vol. CAS-21, pp. 649-654, (September 1974).

Atia, A. E., "Computer-Aided Design of Waveguide Multiplexers," *IEEE Trans. on Microwave Theory and Techniques,* vol. MTT-22, pp. 332-336, (March 1976).

Atia, A. E., and Williams, A. E., "General TE_{011}-Mode Waveguide Band Pass Filters," *IEEE Trans. on Microwave Theory and Techniques,"* vol. MTT-24, pp. 640-647, (October 1976).

Atia, A. E., and Williams, A. E., "Temperature Compensation of TE_{011}-Mode Circular Cavities," pp. 668-669, Ibid.

Cohn, S. B., "Direct-Coupled-Resonator Filters," *IRE Proc.,* vol. 43, pp. 187-196, (February 1957).

Cohn, S. B., "Dissipation Loss in Multiple-Coupled-Resonator Filters," *Proc. IRE,* No. 8, pp. 1342-1348, (August 1959).

Easter, B., and Powell, K. J., "Direct-Coupled Resonator Filters Employing Additional Couplings," *Proceeding of European Microwave Conference,* London, (September 1969).

Kudsia, C. M., and O'Donovan, V., "Microwave Filters for Communications Systems," Artech House Inc., Dedham, MA, (1974).

Kurzrok, R. M., "General Four-Resonator Filters in Waveguide," *IEEE Trans. MTT,* vol. MTT-16, pp. 46-47, (1966).

Kurzrok, R. M., "General Four-Resonator Filters at Microwave Frequencies," *IEEE Trans. Microwave Theory and Tech.,* vol. MTT-16, pp. 295-296, (June 1966).

Levy, R., "Filters with Single Transmission Zeros at Real or Imaginary Frequencies," *IEEE Trans. on MTT,* vol. 26, (April 1976).

Matthaei, G. L., Young, L., and Jones, E. M. T., "Microwave Filters, Impedance Matching Networks and Coupling Structures," McGraw-Hill Book Company, New York, (1964).

Ragan, G. L., "Microwave Transmission Circuits," (MIT Radiation Laboratory Series, vol. 9), McGraw-Hill Book Company, New York, (1948).

Rhodes, J. D., "The Theory of Generalized Interdigital Networks," *IEEE Trans. Circuit Theory,* vol. CT-16, pp. 280-288, (August 1969).

Rhodes, J. D., "A Low Pass Prototype Network for Microwave Linear Phase Filters," *IEEE Trans. Microwave Theory and Tech.,* vol. MTT-18, pp. 290-301, (June 1970).

Rhodes, J. D., "The Generalized Direct-Coupled Cavity Linear Phase Filter," *IEEE Trans. Microwave Theory and Tech.,* vol. MTT-18, pp. 301-308, (June 1976).

Rhodes, J. D., "The Design and Synthesis of a Class of Microwave Band Pass Linear Phase Filters," *IEEE Trans. on Microwave Theory and Techniques,* MTT-17, (April 1969).

Terms, G. C., and Mitra, S. K., "Modern Filter Theory and Design," John Wiley and Sons, New York, (1973).

Williams, A. E., "A Four-Cavity Elliptic Waveguide Filter," *IEEE Trans. Microwave Theory and Tech.,* vol. MTT-18, pp. 1109-1116, (Dec. 1970).

Williams, A. E., and Atia, A. E., "Dual Mode Canonical Waveguide Filters," *IEEE Trans. on Microwave Theory and Tech.,* vol. MTT-25, pp. 1021-1026, (December 1977).

V. Propagation

5.1 GENERAL

Battan, L. J., "Radar Observations of the Atmosphere," Univ. of Chicago Press, (1973).

Benoit, A., "Signal Attenuation Due to Neutral Oxygen and Water Vapour, Rain and Clouds," *Microwave Journal,* pp. 73-80, (November 1968).

Crane, R. K., "Low Elevation Angle Measurement Limitations Imposed by the Troposphere: An Analysis of Scintillation Observations Made at Haystack and Millstone," *MIT Lincoln Laboratory,* (May 18, 1976).

Crane, R. K., "Propagation Phenomena Affecting Satellite Communication Systems," *Proc. IEEE,* (February 1971).

Crane, R. K., "Propagation Phenomena Affecting Satellite Communications Systems Operating in the Centimeter and Millimeter Wavelength Bands," *Proc. IEEE,* vol. 59, pp. 173-188, (1971).

Fang, D. J., and Jih, J., "A Model for Microwave Propagation Along an Earth-Satellite Path," *COMSAT Technical Review 6,* p. 379, (1976).

Feldman, N. E., and Dudzinsky, Jr., S. J., "A New Approach to Millimeter-Wave Communications," *R-1936-RC, RAND Corporation,* Santa Monica, (April 1977).

Ippolito, L. J., "The GSFC 20 and 30 GHz Millimeter Wave Propagation Experiment," *Proceedings of the IEEE International Conference on Communications,* pp. 18-12-18-17, (June 16-18, 1975).

Lee, L., "A Polarization Control System for Satellite Communications with Multiple Uplinks," *IEEE Transactions on Communications,* vol. COM-26, No. 8, pp. 1201-1211, (August 1978).

Mallinchrodt, A. J., "Ground multipath in Satellite aircraft propagation," *Symp. Application of Atmospheric Studies to Satellite Transmission,* Boston, MA, (September 1969).

Sutton, R. W., Schroeder, E. H., Thompson, A. D., and Wilson, S. G., "Satellite-aircraft multipath and ranging experiment results at L band," *IEEE Trans. Commun.,* vol. COM-21, pp. 639-647, (May 1973).

5.2 RAIN

Allnutt, J. E., and Shutie, P. F., "Space Diversity Results at 30 GHz," *Proceedings of the First Open Symposium of URSI Commission F,* La Baule, France, pp. 371-374, (April 28-May 6, 1977).

Arnold, H. W, and Cox, D. C., "Some Results from 19 and 28 GHz COMSTAR Beacon Propagation Experiments," AIAA Paper No. 78-619, *AIAA 7th Communications Satellite Conference,* San Diego, CA, (April 23-27, 1978).

Brussard, G., "A meterological model for rain-induced cross-polarisation," *IEEE Trans. Antennas Propagat.,* vol. AP-24, No. 1, pp. 5-11, (January 1976).

Chu, T. S., "Rain-Induced Cross-Polarization at Centimeter and Millimeter Wavelengths," *Bell System Technical Journal 53,* p. 1557, (1974).

Craig, E. R., and Jenkinson, G. F., "The study of tropical rain attenuation at 11 GHz using a solar radiometer," *Aust. Telecommun., Res.,* vol. 7, No. 1, pp. 3-10, (1973).

Crane, R. K., "Bistatic scatter from rain," *IEEE Trans. Antennas Prop.,* vol. AP-22, pp. 312-320, (1974).

Crane, R. K., "Prediction of the Effects of Rain on Satellite Communication Systems," *Proceedings of the IEEE,* 65, pp. 456-474, (1977).

deBettencourt, J. T., "Statistics of millimeter-wave rainfall attenuation," *J. Rech. Atmos.,* vol. 8, pp. 89-119, (1974).

Drufuca, G., "Rain attenuation statistics for frequencies above 10 GHz from radar observations," *J. Rech. Atmos.*, vol. 8, pp. 413–420, (1974).

Drufuca, G., "Rain attenuation statistics for frequencies above 10 GHz from rain gauge observations," *J. Rech. Atmos.*, vol. 8, pp. 399–411, (1974).

Fang, D. J., and Harris, J. M., "A new method of estimating microwave attenuation over a slant propagation path based on rain gauge data," *IEEE Trans. Antennas Prop.*, vol. AP-24, pp. 381–384, (1976).

Fugono, N., and Hayashi, R., "Propagation Experiment in 1.7, 11.5, and 34.5 GHz with Engineering Test Satellite Type II (ETS-II), AIAA Paper No. 78-623, *AIAA 7th Communication Satellite Conference*, San Diego, CA, (April 23–27, 1978).

Goldhirsh, J., "Prediction Methods for Rain Attenuation Statistics at Variable Path Angles and Carrier Frequencies Between 13 and 100 GHz," *IEEE Transactions on Antennas and Propagation*, AP-23, pp. 786–791, (1975).

Harris, J., and Steinhorn, J., "Effects of Rain on Communications Satellite Systems Tradeoffs at 14/11 and 29/19 GHz," *International Conference on Communications*, (1976).

Harris, J. M., and Hyde, G., "Preliminary Results of COMSTAR 19/29-GHz Beacon Measurements at Clarksburg, Maryland," *COMSAT Technical Review*, (1977).

Henry, P. S., "Measurement and frequency extrapolation of microwave attenuation statistics on the Earth-Space path at 13, 19, and 30 GHz," *IEEE Trans. Antennas Propagat.*, vol. AP-23, pp. 271–274, (1975).

Ippolito, L. J., "Effects of Precipitation on 15.3, 31.65 GHz Earth Space Transmissions with the ATS V Satellite," *Proc. IEEE*, pp. 189–205, (February 1971).

Ippolito, L. J., (Ed.), "20- and 30-GHz Experiments with the ATS-6 Satellite," *NASA TN D-8197.*

Ippolito, L. J., "Rain Attenuation and Communications Link Characterization Measurements with Hermes (The Communications Technology Satellite, *Royal Society of Canada Symposium*, Ottawa, (November 1977).

Ishida, T., Fugono, N., Tabata, J., Ohara, M., and Ishizawa, Y., "Propagation Experiment with Japanese Satellite ETS-II," *IAF XXVIII Congress*, Prague, (September 25–October 1, 1977).

Jarett, D., and Spilman, L. D., "The Impact of Rain Attenuation on 18/30 GHz Satellite Systems: An Introduction to Propagation Measurements," *AIAA Paper No. 74-496*, Los Angeles, CA, (April 1974).

Kaneda, H., Tsukamoto, K., and Fuketa, H., "Experiments in the Japanese CS Program," AIAA Paper No. 78-616, *AIAA 7th Communications Satellite Conference*, San Diego, CA, (April 23–27, 1978).

King, J. L., and Hyde, G., "The COMSAT 13 and 18 GHz Propagation Experiment," *Proceedings of the IEEE International Conference on Communications*, pp. 18-18–18-21, (June 16–18, 1975).

Lin, S. H., "Error-Correcting Codes," Prentice, Hall, Inc., Englewood Cliffs, NJ, (1970).

Lin, S. H., "Statistical Behavior of Rain Attenuation," *Bell System Technical Journal*, pp. 557–582, (April 1973).

Medhurst, R. G., "Rainfall attenuation of centimeter waves: Comparison of theory and measurement," *IEEE Trans. Antennas Prop.*, vol. AP-13, pp. 550–564, (1965).

Oguchi, T., "Rain depolarization studies at centimeter and millimeter wavelengths: theory and measurement," *J. Radio Res. Lab*, (Japan), vol. 22, No. 107, pp. 165–211, (1975).

Perrotta, G., "The Italian SIRIO 12-18 GHz Experiment: The Forerunner of 20-30 GHz Preoperational Satellites," AIAA Paper No. 78-631, *AIAA 7th Communications Satellite Conference*, San Diego, CA, (April 23–27, 1978).

Reber, E. E., and McDonald, D. V., "On Rainfall and Space Diversity for Millimeter-Wave Earth-Satellite Communications Systems," *TR-0073(3230-01)-1*, The Aerospace Corporation, El Segundo, CA, (March 15, 1973).

Rice, P. L., and Holmbert, N. R., "Cumulative time statistics of surface-point rainfall rates," *IEEE Trans. Commun.*, vol. COM-21, pp. 1131–1136, (October 1973).

Rogers, R. R., "Statistical rainstrom models—their theoretical and physical foundations," *IEEE Trans. Antennas Prop.*, vol. AP-31, pp. 547–566, (1976).

Semplak, R. A., "The Effect of Rain on Circular Polarization at 18 GHz," *Bell System Technical Journal, 52*, (1973).

Semplak, R. A., "The influence of heavy rainfall on attenuation at 18.5 and 30.9 GHz," *IEEE Trans. Antennas Propagat*, vol. AP-18, pp. 507–511, (1970).

Semplak, R. A., "Simultaneous Measurements of Depolarization by Rain Using Linear and Circular Polarizations at 18 GHz," *Bell System Technical Journal, 53*, pp. 400–474, (1974).

Strickland, J. I., "The measurement of slant path attenuation using radar, radiometers, and a satellite beacon," *J. Rech. Atmos.*, vol. 8, pp. 347–358, (1974).

URSI, "Proceedings of the First Open Symposium of URSI Commission F," La Baule, France, (April 28–May 6, 1977).

Vogel, W. J., et al., "ATS-6 Attenuation and Diversity Measurements at 20 and 30 GHz," *NASA TN D-8197*, 20- and 30-GHz Experiments with the ATS-6 Satellite," L. J. Ippolito, (Ed.), (April 1976).

Watson, P. A., and Soutter, C. J., "Dual polarisation frequency reuse in satellite communications systems at 11 GHz," *International Conference Satellite Communication Syst. Tech.*, London, *IEEE Conf. Pub. No. 126*, pp. 294–301, (April 7–10, 1975).

5.3 IONOSPHERIC SCINTILLATIONS

Aarons, J., Whitney, H. E., and Allen, R. S., "Global morphology of ionospheric scintillations," *Proc. IEEE*, vol. 59, pp. 159–172, (1971).

Aarons, J., "A survey of scintillation data and its relationship to satellite communications," Air Force Cambridge Res. Labs, L. G. Hanscom Field, MA, Rep. AFCRL 70-0053, (January 1970).

Craft, Jr., H. D., and Westerlund, L. H., "Scintillation at 4 and 6 GHz Caused by the Ionosphere," *Proc. AIAA 10th Aerospace Sciences Meeting*, San Diego, CA, (January 1972).

Crane, R. K., "Ionospheric Scintillation," *Proceedings of the IEEE*, pp. 180–199, (1977).

Fremouw, E. J., et al., "Design Considerations for a Microwave Scintillation Experiment," *DNA 4092F, SRI International*, Menlo Park, CA, (March 15, 1977).

Paulson, M. R., and Hopkins, R., "Effects of equatorial scintillation fading on Satcom signals," Naval Electron Lab Cent., San Diego, CA, Rep. NELS-TR-1875, (May 1973).

Taur, R. R., "Ionospheric Scintillation at 4 and 6 GHz," *COMSAT Technical Review 3*, pp. 145–165, (1973).

VI. Earth Stations

Barthle, R. C., and Briskman, R. D., "Trends in Design of Communications Satellite Earth Stations," *Microwave Journal*, (October 1967).

Briskman, R. D., and Smith, G. E., "Television Transmission Performance of an Experimental Small Aperture Earth Station," *IEEE Transactions on Communications*, vol. 23, No. 5, (May 1975).

Brown, Jr., M. P., "Intelsat V Standard "C" Earth Station Performance Objectives Applied to a Possible Eastern United States Location (14/11 GHz Frequency Band)," *EASCON 1977*, pp. 13-1A–13-1H.

Chakraborty, D., "INTELSAT IV Satellite System (Voice) Channel Capacity Versus Earth Station Performance," *EASCON Conf. Record*, (October 1970).

Crispin, H.L., "Earth Station Dot the Globe," *Microwave Systems News*, (March 1977).

"Conference on Earth Station Technology Institution of Electrical Engineers," Savoy Place, London *WC2R OBL, Conf. Pub. No. 72*, pp. 14–16, (October 1970).

Cuccia, C. L., Hellman, C., and Slivkoff, W., "The Technology of Unattended Remotely Monitored Earth Terminals," AIAA Paper No. 78-610, *AIAA 7th Communications Satellite Systems Conference*, San Diego, CA, pp. 562–571, (April 24–27, 1978).

Cuccia, C. L., "RF Design of Communication Satellite Earth Stations," (in 3 parts), *Microwaves*, vol. 6, Nos. 5–7, (May, June, July 1967).

Edelson, B. I., "Small Earth Terminals for Satellite Communications," *Astronautics and Aeronautics*, vol. 11, No. 6, (June 1973).

"The Evolution of Cost Effective [Small Aperture] Earth Terminals for Communications," *Session 21 of the 1976 International Conference on Communications*, (1976).

Gruner, R. W., and Williams, E. A., "A Low-loss Multiplexer for Satellite Earth Terminals," *Comsat Technical Review*, vol. 5, No. 1, pp. 157–177, (Spring 1975).

Han, C. C., "Optimization in the Design of a 12 GHz Low Cost Ground Receiving System for Broadcast Satellites," *ICC Conference Record*, Seattle, pp. 36–38, (June 1973).

Helm, N. R., and Kaiser, J., "Small Earth Terminals for Medical/Educational Applications," AIAA Paper 75-917, *AIAA Conference on Communication Satellites for Health/Education Applications*, (July 1975).

Hogg, D. C., "Ground Station Antennas for Space Communications,"

L. Young, Ed., *Advance in Microwaves,* New York, Academic Press, (1968).

Horton, E. D., "An Adaptive Co-channel Interference Suppression System to Suppress High Level Interference in Satellite Communication Earth Terminals," *National Telecommunication Conference,* (1976).

Kelly, R. L., "Low Cost Ku-Band Earth Terminals for Voice/Data/Facsimile," *EASCON,* (1977).

Knipp, F. M., and Buegler, J. A., "Military Satellite Communications Terminals," *Signal,* vol. 30, No. 6, (March 1976).

Kumagai, D., and Ikegami, F., "An Experimental Earth Station," *Electrical Communication Laboratories Review,* Japan, vol. 22, (May-June 1974).

Lebow, I. L., Jordan, K. L., and Drouilhet, P. R., "Satellite Communications to Mobile Platforms," *Proc. IEEE,* (February 1971).

Linker, D. M., and Lazarus, L., "Specifying 4.5 Meter Satellite Earth Station Hardware for Domsat and CATV Application," LNR Communications, Inc., also, *Communications News,* (April 1977).

Lipke, D. W., Swearingen, D. W., Parker, J. F., Steinbecker, E. E., Calost, T. O., and Dodd, H., "MARISAT, A Maritime Satellite Communications System," *COMSAT Technical Review,* (Fall 1977).

Lubell, P. D., and Rebhun, F. D., "Suppression of Co-Channel Interference with Adaptive Cancellation Devices at Communications Satellite Earth Stations," *International Conference on Communications,* (1977).

McGivern, P. L., "Tacsatcom for the U.S. Army," *Signal,* vol. 28, No. 7, (March 1974).

Moore, J. L., "Whiskey-3—UHF Radio Terminal for Naval Satellite Communication," *Signal,* vol. 31, (July 1977).

Morikami, H., et al., "A Newly Developed Intelsat Earth Station," *Int. Conf. on Comm.,* pp. 8-30-8-35, (1973).

Okean, H. C., and Kelly, A. J., "Low-Noise Receiver Design Trends Using State-of-the-Art Building Blocks," *IEEE Trans. Microwave Theory and Techniques,* MTT-25, (4), (April 1977).

O'Neill, D. J., Fitzgerald, P. M., and Englebrecht, L. R., "Single Access User Earth Terminal System Design," *ICC 1975 Conference Record,* vol. 1, pp. 12-1-12-5, (1975).

Pollack, L., and Sones, W., "An Unattended Earth Terminal for Satellite Communications," *COMSAT Technical Review,* vol. 4, No. 2, pp. 205-230, (Fall 1974).

Pope, D. L., "Parametric Representation of Ground Antennas for Communications Systems Studies," *Bell System Technical Journal,* (December 1968).

Price, R., "A Fixed Reflector, Steerable Beam Earth Station Antenna," *Proc. National Telecomm. Conf.,* p. 25B-1, (1972).

Sciambi, Jr., A.F., "Large Earth Station Antennas for Communications and Radar," *International Conference on Communications,* (1976).

Seither, H., "Ground Segment for the [NATO] Satcom Phase II Communications Project," *Electrical Communications,* vol. 49, No. 3, (March 1974).

Shakil, S. M., and Sinha, V., "Ground Terminals in Communication Satellite Systems," *Institution of Electronics and Telecommunication Engineers Journal,* vol. 22, pp. 439-445, (July 1976).

Tyree, B. E., Lawton, R. S., Chewey, V., and Maresca, P., "An Overview of the Small SHF Satellite Ground Terminal Development Program," *RCA Engineer,* vol. 22, pp. 82-87, (June-July 1976).

Wilkinson, E., "New Earth-Station Antenna Feature Low Sidelobes," *Communications News,* pp. 50-51, (February 1978).

VII. The Future

7.1 ADVANCED SYSTEMS

Bond, D. S., and Keigler, J. E., "Diversified, Cost-Effective Satellite Service for the 80's," AIAA Paper 76-221, *AIAA/CASI 6th Communications Satellite Systems Conference,* (April 1976).

Briskman, R. D., "System Considerations for Advanced North American Domestic Satellites," AIAA Paper 76-222, *AIAA/CASI 6th Communications Satellite Systems Conference,* (April 1976).

Garner, W. B., and Stamminger, R., "Future communications satellites: Definition of three basic concepts," *Conf. Rec. Int. Conf. Commun.,* pp. 3-8-3-11, (1973).

Huffman, G. E., "High Power Payloads for Telecommunication," *EASCON 1975.*

Hockenberry, J. H., "Efficient High Capacity Communications Satellites," AIAA Paper 74-462, *AIAA 5th Communications Satellite Systems Conference,* (April 1974).

Jarett, D., "An Advanced 30/42 GHz Point-to-Multipoint Satellite Communication System," *ESA Satellite Broadcasting,* pp. 85-89.

Jarett, D., "Meeting the Twin Challenges of Demand and Conservation of Spectrum of Orbit through Technology," *EASCON 1977,* Paper 25-1, (September 1977).

Knopow, J. J., "Next Generation Communications Satellites," *AIAA Progress in Astronautics and Aeronautics: Communications Satellite Technology,* vol. 32.

Knowpow, J. J., and Klockseim, J. P., "Concepts of High-Capacity Communications Satellites," *International Conference on Communications,* (June 1973).

Thorpe, D. G., "Evolution of the Telesat Canada System into the Mid-1980's," AIAA Paper 76-220, *AIAA/CASI 6th Communications Satellite System Conference,* (April 1976).

Van Trees, H. L., "Future Intelsat System (1986-1993) Planning," AIAA Paper 76-233, *AIAA/CASI 6th Communications Satellite Systems Conference,* (April 1976).

7.2 ADVANCED CONCEPTS

Bekey, I., and Mayer, H., "1980-2000: Raising Our Sights for Advanced Space Systems," *AIAA Astronautics and Aeronautics,* pp. 34-63, (July/August 1976).

Cuccia, C. L., "Communication Satellite Technologies in the Early Twenty-First Century—A Projection into the Post Intelsat-V Era," *28th International Astronautical Congress,* Prague, Czechoslovakia, (September 25-October 1, 1977).

Edelson, B., and Morgan, W., "Orbital Antenna Farms," *Astronautics and Aeronautics,* (September 1977).

Elias, D., "Future World Telecommunication Networks," *Technical Symposium of the World Telecommunication Forum,* Geneva, (1975).

Future Systems, Inc., "A 25-year Forecast for Commercial Communication Satellites and the Congestion of the Geostationary Arc," *Report No. 103,* (November 1977).

Martin de Bustamante, J. L., "Telecommunications 2000," *Telecommunications Journal,* vol. 43, pp. 361-365.

NASA, "Outlook for Space," *NASA SP-386,* Washington, DC, (January 1976).

Satellite Systems Engineering, Inc., "A Study of Technical and Economic Aspects of Combining Maritime and Aeronautical Communication Satellite Services—1985-2000," Washington, (1977).

Wheelon, A. D., "The Future Outlook for Communication Satellite Applications," *World Telecommunication Forum,* Geneva, (1975).

7.3 ADVANCED TECHNOLOGY

Aerospace Corp, "Communication Satellite Technology: State of the Art and Development Opportunities," *Aerospace Report No. ATR-78 (7735) - 1,* (July 1978).

Aukstikalnis, A. J., "Spacecraft Computers," *Astronautics and Aeronautics,* (July/August 1974).

Burtt, J. E., et al., "Technology Requirements for Communication Satellites in 1980's," *NASA Report NAS2-7073,* Lockheed Missiles and Space Co., (September 1973).

Campbell, G., and Damonte, J., "Large Deployable Antenna Forecast," LMSC-D384788, (December 5, 1974).

Davis, R. C., Esch, F. H., Palmer, L., and Pollack, L., "Future Trends in Communications Satellite Systems," *Acta Astronautica,* (March/April 1978).

Edelson, B. I., and Bargellini, P. L., "Technology Development for Global Satellite Communications," AIAA Paper 76-234, *AIAA/CASI 6th Communications Satellite Systems Conference,* Montreal, Canada, (April 1976).

NASA, "Communications Systems Technology Assessment Study," Report NASA Lewis Research Center, (June 1977).

NASA, "A Forecast of Space Technology 1980-2000," *NASA SP-387,* Washington, DC (1976).

NASA, "Spaceborne Digital Computer Systems, NASA Space Vehicle Design Criteria (Guidance and Control)," *NASA SP-8070,* Washington, DC, (March 1971).

Schaefer, D. H., and Strong, J. P., "Tse Computers," Report No. X-943-75-14, Goddard Space Flight Center, (January 1975).

7.4 MISCELLANEOUS

"AT&T Forecasting Methodology," *FCC Docket 18875,* Question B, (December 30, 1975).

Bebee, E. L., and Gilling, E. J. W., "Telecommunications and Economic Development: A Model for Planning and Policy Making," *Telecom-

munications Journal, vol. 43, pp. 537–543.

Christiansen, D., (Ed.), "Technology '75," *IEEE Spectrum,* pp. 37–94, (January 1975).

Golomb, S., "The Future of Communications to the Year 2000," *Minutes of Technology Seminar,* JPL, (November 20, 1974).

Joseph, E. C., "Computers: Trends Toward the Future," *Information Proceedings 68,* North-Holland Publishing Col, Amsterdam, (1969).

Kahn, H., and Wiener, A., "The Year 2000: A Framework for Speculation on the Next Thirty-Three Years," New York, (1967).

Martin, J., "Future Developments in Telecommunications," Prentice-Hall, (1971).

Rappoport, P. N., "Some Alternative Time Series Approaches to Forecasting International Message Telephone Service," *Public Utilities Forecasting Conference,* Bowness-on-the-Widermere, United Kingdom, (1977).

Spekke, A., Ed., "The Next 25 Years: Crisis & Opportunity," Washington, (1975).

Turn, R., "Computers in the 1980's," Columbia University Press, New York and London, (1974).

"An Analysis of Domestic and Foreign Small Earth Station Markets," U. S. Department of Commerce, Office of Telecommunications, (May 1976).

World Future Society, "The Future: A Guide to Information Sources," Washington, (1977).

World Future Society, "An Introduction to the Study of the Future," Washington, (1977).

VIII. Regulatory, Political, Legal, and Economic Issues

8.1 ORBIT-SPECTRUM UTILIZATION

Bradley, W. E., "Communications Strategy of Geostationary Orbit," *Astronautics and Aeronautics,* vol. 6, pp. 34–41, (April 1968).

CCIR, "The Effect of Modulation Characteristics on the Efficiency of use of the Geostationary-Satellite Orbit," Study Programme 2J-1/4, Report 559, (1974).

CCIR, "XIIth Plenary Assembly (Geneva, 1974) of the International Radio Consultative Committee, 13 vols., especially: vol. 4, "Fixed Service Using Communication Satellites," vol. 8, "Mobile Services," (Section 8E: "Satellite Applications," vol. 9, "Coordination and Frequency Sharing Between Systems in the Fixed Satellite Service and Terrestrial Ratio-Relay Systems," vol. 11, "Broadcasting Service," (Section 11F: "Using Satellites").

CCIR, "Technical Factors Influencing the Efficiency of the Geostationary-Satellite Orbit by Radiocommunication Satellites sharing the same frequency bands," Study Programme 2J-1/4, Report 453-1, (1970–1974).

Final Acts of the World Administrative Radio Conference for Space Telecommunications, published by the *International Telecommunications Union,* (1971).

Freibaum, J., "Effects of Propagation Phenomena and Frequency Allocation on the Growth of Satellite Communications," *International Conference on Communications,* (1976).

Fuenzalida, J. C., "A Comparative Study of the Utilization of the Geostationary Orbit," *Proc. Int. Conf. Digital Satellite Communication, Intelsat/IEEE,* pp. 213–225, (1969).

Hult, J. L., and Reinhardt, E. E., "Satellite Spacing and Frequency Sharing for Communication and Broadcast Services," *Proc. IEEE,* pp. 118–128, (February 1971).

Janky, J. M., Sites, J., and Lusignan, B. B., "Technical and Interference Aspects of Satellite Networks with Small-Aperture Earth Stations," *International Conference on Communications,* (1976).

Jeruchim, M. C., "A Survey of Interference Problems and Applications to Geostationary Satellite Networks," *Proceedings of the IEEE,* vol. 65, No. 3, pp. 317–329, (March 1977).

Jowett, J. K. S., "Effective Use for Satellite Communications of the Radio Frequency Spectrum and the Geostationary Satellite Orbit," *World Telecommunications Forum Conference Proceedings,* (October 1975).

Jowett, J. K. S., and Jefferis, A. K., "Ultimate Capacity of the Geostationary-Satellite Orbit," *Proc. IEEE,* vol. 116, pp. 1304–1310, (August 1969).

Long, Jr., W. G., "The Effects of Multiple Band/Service Satellites On-Orbit/Spectrum Utilization," *AIAA 5th Comm. Sat. Systems Conf.,* Paper No. 74-433, Los Angeles, (April 1974).

Morgan, W. L., "Satellite Utilization of the Geosynchronous Orbit," *COMSAT Technical Review,* vol. 6, No. 1, (Spring 1976).

Ramjo, S., and Savitz, P., "Orbital Design Strategy for Domestic Communication Satellite Systems," *Proc. Int. Conf. on Comm.,* pp. 36-1-36-6, (1973).

Reinhart, E. E., "Orbit-Spectrum Sharing Between the Fixed-Satellite and Broadcasting-Satellite Services at 12 GHz," *National Telecommunications Conference,* (December 1974).

Sawitz, P., "Spectrum-Orbit Utilization—An Overview," *National Telecommunications Conference,* (December 1975).

"Table of Frequency Allocations 10 kc/s to 40 Gc/s, modified by the Extraordinary Administrative Radio Conference to Allocate Frequency Bands for Space Radio Communication Purposes," (Geneva 1963), published by the *International Telecommunications Union,* (1966).

8.2 POLITICAL, REGULATORY

Galloway, J. F., "The Politics of Technology of Satellite Communications," Lexington Books, MA, (1972).

Golden, D. A., "Social Impact on Advanced Communications," *International Conference on Communications,* (June 1976).

Goldstein, I., "Intelsat and the Developing World," *IEEE Transactions on Communications,* vol. 24, No. 7, (July 1976).

Gould, R. G., "Regulatory Aspects of Digital Communications," *IEEE Transactions on Communications,* (January 1976).

Hinchman, W. R., "Public Policy and the Domestic Satellite Industry," *International Conference on Communications,* (June 1972).

"Issue on Space Broadcasting," *Journal of Space Law,* vol. 3, (Spring–Fall 1975).

Pelton, J. N., "Global Communications Satellite Policy," Lomond Books, (1974).

Perret, R., "International Law in Space—The Regulation of Satellite Telecommunications," *Interavia,* vol. 30, No. 12, (December 1975).

Editor's Biography

Harry Van Trees received his Doctor of Science degree from the Massachusetts Institute of Technology in 1961. Previously he had received a B.Sc. from the United States Military Academy in 1952.

From 1961 through 1972, Dr. Van Trees was a faculty member in the Department of Electrical Engineering at the Massachusetts Institute of Technology, becoming a Professor in 1970. He was active in graduate course development and was the leader of a research group doing work in digital communications, radar/sonar, array processing, estimation theory, and state variable areas. He authored a three volume set of books in "Detection, Estimation, and Modulation Theory." The first volume is the standard text for this subject in graduate schools throughout the world. He developed two comprehensive self-study courses in applied probability and random processes and made 50 hours of videotaped lecturers for these courses.

In June 1972, Dr. Van Trees became the Chief Scientist and Associate Director, Technology of the Defense Communications Agency. As senior civilian advisor to the Director, he supervised the Defense Communications Agency's research and development programs, was instrumental in the establishment of a centralized engineering center at the Defense Communications Agency, the implementation of Military Satellite Office and was a major contributor to the Worldwide Military Command and Control Systems program.

Dr. Van Trees became the Assistant Vice President, Advanced Systems, at the Communications Satellite Corporation in March 1975. His duties included responsibility for planning the future International Telecommunications Satellite Organization's satellite system, including definition and analysis of various system alternatives, specifying required research and development objectives, and specifying a transition strategy to achieve future system configuration. In the non-international satellite area, his division generated advanced concepts for post-1980 systems.

In November 1978, Dr. Van Trees became the Chief Scientist of the Air Force. In this position he was involved in the Air Force research and exploratory development programs and various aspects of Air Force Command, Control, and Communications.

In June 1979, Dr. Van Trees became the Principal Deputy Assistant Secretary of Defense for Communications, Command, Control & Intelligence.

He is married to the former Diane Enright of Arlington, VA, who is active as a docent in the History and Technology Museum and the Air and Space Museum of the Smithsonian Institute. They have seven children: Stephen, Mark, Kathleen, Patricia, Eileen, Harry, and Julia.